International Table of Atomic Weights (1977)

Based on relative atomic mass of $^{12}C = 12$.

The following values apply to elements as they exist in materials of terrestrial origin and to certain artificial elements. When used with due regard to footnotes, they are reliable to ± 1 in the last digit, or ± 3 when followed by an asterisk (*). Value in parentheses is the mass number of the isotope of longest half-life.

	Symbol	Atomic number	Atomic weight		Symbol	Atomic number	Atomic weight		Symbol	Atomic number	Atomic weight
Actinium	Ac	89	227.0278	Hafnium	Hf	72	178.49*	Promethium	Pm	61	(145)
Aluminum	Al	13	26.98154a	Helium	He	2	4.00260b,c,g	Protactinium	Pa	91	231.0359f
Americium	Am	95	(243)	Holmium	Ho	67	164.9304a	Radium	Ra	88	226.0254f,g
Antimony	Sb	51	121.75*	Hydrogen	H	1	1.0079b,d	Radon	Rn	86	(222)
Argon	Ar	18	39.948b,c,d,g*	Indium	In	49	114.82g	Rhenium	Re	75	186.207c
Arsenic	As	33	74.9216a	Iodine	I	53	126.9045a	Rhodium	Rh	45	102.9055a
Astatine	At	85	(210)	Iridium	Ir	77	192.22*	Rubidium	Rb	37	85.4678*c,g
Barium	Ba	56	137.33g	Iron	Fe	26	55.847*	Ruthenium	Ru	44	101.07*g
Berkelium	Bk	97	(247)	Krypton	Kr	36	83.80e,d	Samarium	Sm	62	150.4g
Beryllium	Be	4	9.01218a	Lanthanum	La	57	138.9055*b,g	Scandium	Sc	21	44.9559a
Bismuth	Bi	83	208.9804a	Lawrencium	Lr	103	(260)	Selenium	Se	34	78.96*
Boron	B	5	10.81c,d,e	Lead	Pb	82	207.2d,g	Silicon	Si	14	28.0855*
Bromine	Br	35	79.904c	Lithium	Li	3	6.941*c,d,e,g	Silver	Ag	47	107.868c,g
Cadmium	Cd	48	112.41g	Lutetium	Lu	71	174.967*	Sodium	Na	11	22.98977
Calcium	Ca	20	40.08g	Magnesium	Mg	12	24.305c,g	Strontium	Sr	38	87.62g
Californium	Cf	98	(251)	Manganese	Mn	25	54.9380a	Sulfur	S	16	32.06d
Carbon	C	6	12.011b,d	Mendelevium	Md	101	(258)	Tantalum	Ta	73	180.9479*b
Cerium	Ce	58	140.12	Mercury	Hg	80	200.59*	Technetium	Tc	43	(98)
Cesium	Cs	55	132.9054a	Molybdenum	Mo	42	95.94	Tellurium	Te	52	127.60*g
Chlorine	Cl	17	35.453c	Neodymium	Nd	60	144.24*g	Terbium	Tb	65	158.9254a
Chromium	Cr	24	51.996c	Neon	Ne	10	20.179*c,e	Thallium	Tl	81	204.37*
Cobalt	Co	27	58.9332a	Neptunium	Np	93	237.0482f	Thorium	Th	90	232.0381f,g
Copper	Cu	29	63.546*c,d	Nickel	Ni	28	58.70	Thulium	Tm	69	168.9342a
Curium	Cm	96	(247)	Niobium	Nb	41	92.9064a	Tin	Sn	50	118.69*
Dysprosium	Dy	66	162.50	Nitrogen	N	7	14.0067b,c	Titanium	Ti	22	47.90*
Einsteinium	Es	99	(252)	Nobelium	No	102	(259)	Tungsten	W	74	183.85*
Erbium	Er	68	167.26*	Osmium	Os	76	190.2g	Unnilpentium	Unp	105	(262)
Europium	Eu	63	151.96g	Oxygen	O	8	15.9994b,c,d	Unnilhexium	Unh	106	(263)
Fermium	Fm	100	(257)	Palladium	Pd	46	106.4g	Unnilquadium	Unq	104	(261)
Fluorine	F	9	18.998403a	Phosphorus	P	15	30.97376a	Uranium	U	92	238.029b,c,e,g
Francium	Fr	87	(223)	Platinum	Pt	78	195.09*	Vanadium	V	23	50.9415*b,c
Gadolinium	Gd	64	157.25*g	Plutonium	Pu	94	(244)	Xenon	Xe	54	131.30a,e,g
Gallium	Ga	31	69.72	Polonium	Po	84	(209)	Ytterbium	Yb	70	173.04*
Germanium	Ge	32	72.59*	Potassium	K	19	39.0983*	Yttrium	Y	39	88.9059a
Gold	Au	79	196.9665g	Praseodymium	Pr	59	140.9077a	Zinc	Zn	30	65.38
								Zirconium	Zr	40	91.22g

[a] Elements with only one stable nuclide.

[b] Element with one predominant isotope (about 99 to 100% abundance).

[c] Element for which the atomic weight is based on calibrated measurements.

[d] Element for which known variation in isotopic abundance in terrestrial samples limits the precision of the atomic weight given.

[e] Element for which users are cautioned against the possibility of large variations in atomic weight due to inadvertent or undisclosed artificial isotopic separation in commercially available materials.

[f] Most commonly available long-lived isotope.

[g] In some geological specimens this element has an anomalous isotopic composition, corresponding to an atomic weight significantly different from that given.

Principles
of Chemistry

Saunders College Publishing
Complete Package for Teaching with
Principles of Chemistry
by Davis, Gailey, and Whitten

Davis Study Guide to Accompany Principles of Chemistry

Lippincott, Meek, Gailey, & Whitten Experimental General Chemistry

Ragsdale Lecture Outline to Accompany Principles of Chemistry

DeKorte Solutions Manual to Accompany Principles of Chemistry

Whitten & Gailey Problem Solving in General Chemistry, 2nd edition

Davis Instructor's Manual to Accompany Principles of Chemistry

Davis, Gailey, & Whitten Overhead Transparencies

Davis, Gailey, & Whitten Test Bank

Wilkie Computer Tutorial for General Chemistry

Shakhashiri, Schreiner, & Meyer General Chemistry Audio-Tape Lessons, 2nd edition

Shakhashiri, Schreiner, & Meyer Workbook for General Chemistry Audio-Tape Lessons, 2nd edition

Principles of Chemistry

Raymond E. Davis
University of Texas, Austin

Kenneth D. Gailey
University of Georgia, Athens

Kenneth W. Whitten
University of Georgia, Athens

SAUNDERS COLLEGE PUBLISHING
Philadelphia   New York   Chicago
San Francisco   Montreal   Toronto
London   Sydney   Tokyo   Mexico City
Rio de Janeiro   Madrid

Address orders to:
383 Madison Avenue
New York, NY 10017

Address editorial correspondence to:
West Washington Square
Philadelphia, PA 19105

Text Typeface: Palatino
Compositor: Progressive Typographers
Acquisitions Editor: John Vondeling
Developmental Editor: Jay Freedman
Project Editors: Lynne Gery and Patrice L. Smith
Managing Editor & Art Director: Richard L. Moore
Art/Design Assistant: Virginia A. Bollard
Text Design: Caliber Design Planning, Inc.
Cover Design: Lawrence R. Didona
New Text Artwork: Philatek
Production Manager: Tim Frelick
Assistant Production Manager: Maureen Iannuzzi

Cover credit: Photomicrograph showing focal conic texture of a smectic A liquid crystal, by Alfred Saupe, Liquid Crystal Institute, Kent State University. Used with permission.

Library of Congress Cataloging in Publication Data

Davis, Raymond E., 1938–
   Principles of chemistry.

   Includes index.

   1. Chemistry. I. Gailey, Kenneth D. II. Whitten, Kenneth W. III. Title.
QD31.2.D377 1984   540   83-19271
ISBN 0-03-060458-3

PRINCIPLES OF CHEMISTRY   ISBN 0-03-060458-3

4567   032   987654321

CBS COLLEGE PUBLISHING
Saunders College Publishing
Holt, Rinehart and Winston
The Dryden Press

To: Sharon, Angela, Laura, and Brian Davis
     Kathy, Kristen, and Karen Gailey
     Betty, Andy, and Kathryn Whitten

To The Professor

As we surveyed the available principles textbooks in general chemistry, we concluded that there were major deficiencies in them. Some texts that claim to emphasize principles simply present theories "out of the blue." Students are thus deprived not only of the link to physical reality, which makes learning and using theories easier and more meaningful, but also of the basis for the intellectual integrity of science. None of these texts appeared to have been written for students, the people who use them most, and so we decided to write a principles textbook *for students*.

We emphasize that chemistry is an experimental science by always presenting first the observational bases upon which the theories depend, followed by descriptions of the classic experiments and their great importance to the evolution of modern chemical theories. We then emphasize the important role, and justify the validity, of modern theories by interpreting and explaining the significance of important observations as we develop each subsequent topic. To accomplish our goals, we have provided

1. The experimental basis for modern chemical theories.

2. Accurate statements of current theories.

3. Clear, concise definitions of important terms.

4. Simple, yet familiar, analogies that clarify fundamental ideas.

5. Carefully graded, detailed explanations of current theories and important concepts.

6. Numerous substantial illustrative examples that are solved and explained in detail.

7. Carefully graded, comprehensive sets of end-of-chapter exercises that progress from routine manipulations to a reasonable level of sophistication.

8. Plenty of descriptive chemistry to illustrate applications of modern theories.

9. Comprehensive appendices.

10. A great deal of flexibility for professors who teach general chemistry.

These important characteristics of *Principles of Chemistry* require amplification.

Because a clear understanding of modern chemical theories is vital to the study of chemistry, we have tried to state the essence of each theory as clearly and accurately as possible at this level. The significance of each theory is then emphasized and illustrated.

Many students experience difficulty in their study of general chemistry because they do not understand the vocabulary used to describe basic concepts. We have been careful to give clear, concise definitions of important terms *as they are introduced.* In the few cases in which this is not practical, we have included appropriate marginal notes.

Some students have difficulty in piecing together declarative statements in order to understand basic concepts. We have provided many simple, familiar analogies to improve students' comprehension. For example, the arbitrary nature of zero potential energy and the idea of negative potential energy in a chemical system are major obstacles for many students. We solve these problems by presenting a simple analogy in Figure 3–2. Many other examples are used throughout the textbook.

Many students have difficulty in gaining an appreciation of the significance of important theories and concepts. We have presented detailed *explanations* of current theories and concepts so that students can understand and appreciate their significance. Throughout we have provided *substantial* explanations.

Numerous carefully graded illustrative examples are invaluable to students, and so we have provided an abundance of them. We have included some simple examples, some of intermediate difficulty, as well as considerable numbers of *difficult* illustrative examples. For many topics, such as bonding and molecular structure, the illustrative examples are woven into the narrative and are not numbered.

Chemical reactions, and associated periodicity, are introduced early in Chapter 9 from a descriptive point of view. Chapter 10, "Chemical Analysis in Aqueous Solution," provides more descriptive chemistry as well as background for quantitative laboratory work.

Comprehensive appendices are included so that students have the data they need as they study and work problems. Professors have ready access to numerical data as they prepare lectures and construct tests and comprehensive examinations.

Flexibility is important because classes are so heterogeneous — some students have had no previous training in chemistry and others are well prepared. In Chapter 1, we start at a very basic level for students with no background in chemistry. Chapter 1 can be used as assigned reading for students with strong backgrounds.

Throughout the text each topic and its vocabulary are introduced at a very basic level, and through a series of carefully graded steps we progress to a reasonably sophisticated treatment of each topic. Some sections in each chapter (and, in fact, some chapters) may be used as assigned reading.

Basic stoichiometry is presented in Chapter 2, together with an introduction to concentrations of solutions and dilution calculations, to provide a firm foundation for meaningful laboratory experiences. We start at a very basic level, and our treatment of basic stoichiometry is the most comprehensive available. The mole method is used throughout.

In keeping with our philosophy of providing a sound background for laboratory work, we present thermochemistry, through Hess' Law, in Chapter 3. We have taken care to describe what energy is and how it is measured. The arbitrary nature of zero potential energy and negative potential energies in chemical systems are illustrated well.

Chapter 4, "Atoms and Subatomic Particles," provides descriptions of the classic experiments that led to our present ideas about the structures of atoms. Nuclear binding energy is included to avoid leaving a significant gap in background information. Students learn why so many positively charged particles can occupy such a small volume in a stable atomic nucleus and still not "blow apart."

Electronic structures of atoms are presented in Chapter 5. Together, Chapters 4 and 5 are the most comprehensive treatments of these important topics on the market. For those who prefer to spend less time on electronic structures of atoms, several sections can be "slipped over" easily.

Chapter 6 is devoted to a detailed discussion of chemical periodicity and an introduction to bonding. It is beautifully illustrated. Many illustrative examples, some unnumbered, are provided. The treatment of Lewis formulas is quite comprehensive. Inorganic nomenclature, a possibility for assigned reading, completes Chapter 6.

Chapters 7 and 8 are devoted to molecular structure and covalent bonding. VSEPR theory, polarity and dipole moments, and bond energies are included in Chapter 7. The order of presentation is logical and pedagogically sound, starting with the simplest possible cases and progressing to the more complex. This chapter is unique.

Chapter 8 includes discussions of the Valence Bond Theory and Molecular Orbital theory for those who prefer to include these topics.

Chapters 9 and 10 include a great deal of descriptive chemistry. The emphasis in Chapter 9 is on chemical periodicity associated with chemical reactions; Chapter 10 emphasizes the quantitative aspects of acid-base reactions and redox reactions in aqueous solutions. Major emphasis is placed on the mole method and molarity; separate *optional* sections on equivalent weights and normality are included. Chapters 9 and 10 may be postponed until after gases and liquids and solids with no loss of continuity. They may also be covered only in part for those who wish to do so.

Chapters 11, 12, and 13 describe the states of matter and the physical properties of solutions (colloids are included). The treatment of the states of matter, the dissolving process, and colligative properties of solutions are detailed and illustrated well.

Chapter 14, "Chemical Thermodynamics: The Driving Force for Change," emphasizes the role of thermodynamics in assessing and predicting the *spontaneity* of chemical and physical changes, and its relation to *equilibrium.* The chapter opens with a descriptive discussion of spontaneity, with many illustrations that students find helpful. Building on the thermochemistry of Chapter 3, the chapter then progresses to the use of changes in entropy and Gibbs free energy as criteria of spontaneity. Rather than depend on abstract definitions or traditional heat-engine efficiency discussions, the chapter develops the idea of entropy so that students are able to predict system entropy changes for common processes. Finally, a qualitative discussion lays the groundwork for the relation between thermodynamics and equilibrium. In Chapter 15, the subject of "Chemical Kinetics" is presented from the same standpoint—experimental observations first, theoretical interpretations second. These two key chapters develop the ideas needed for a strong introduction to "Chemical Equilibrium," Chapter 16.

A comprehensive discussion of the concepts of acid-base behavior and ionic equilibria is presented in Chapters 17 and 18.

Chapter 19 is a widely acclaimed chapter on electrochemistry that completes the main principles section of the text. It enables professors to "tie together" most of the material presented earlier.

Chapters 20 through 23 provide a block of basically descriptive chapters on the metals and metallurgy (Chapter 20), the nonmetals (Chapters 21 and 22), and coordination compounds (Chapter 23). The treatment of coordination chemistry is the most comprehensive available in a "principles" textbook.

Chapter 24, "Nuclear Chemistry," is entirely self-contained and can be covered at any point in the course after Chapter 4.

Chapter 25, "Organic Chemistry," is a strong introduction to organic chemistry that includes alkanes, alkenes, alkynes, aromatic hydrocarbons, organic nomenclature, and functional groups.

Because there is no consensus on the number of answers to the end-of-chapter exercises that should be available to students, we have tried to provide maximum flexibility. Answers to even-numbered *numerical* problems are included after the Appendices. Detailed solutions and answers to *all even-numbered exercises* are available in the *Solutions Manual* prepared by Professor DeKorte. Answers and solutions to all odd-numbered exercises are provided in the *Instructor's Manual* by Professor Davis. The *Instructor's Manual* may be made available to students, if professors wish to do so.

We welcome suggestions for improvements in future editions.

Acknowledgments

The list of individuals who contributed to the evolution of this book is long indeed. First, we would like to express our appreciation to the professors who contributed so greatly to our scientific education: Miss Dorothy Vaughn, Professors Ralph N. Adams, F. S. Rowland, Calvin Vanderwerf, A. Tulinsky, Wm. von E. Doering, and David Harker (RED); Professors R. D. Dunlap, C. R. Russ, H. H. Patterson, R. L. Wells, A. L. Crumbliss, P. Smith, D. B. Chesnut, R. A. Palmer, and B. E. Douglas (KDG); and Professors Arnold Gilbert, M. L. Bryant, W. N. Pirkle, the late Alta Sproull, C. N. Jones, S. F. Clark, and R. S. Drago (KWW).

Our reviewers were very helpful — they have made major contributions to the development of this text with positive suggestions and constructive criticism:

John M. DeKorte, Northern Arizona University
L. O. Gold, The Pennsylvania State University
W. R. Hall, Community College of Allegheny County
Forrest A. Hentz, Jr., North Carolina State University
Lloyd N. Jones, The United States Naval Academy
James Long, The University of Oregon
D. W. Meek, Ohio State University
L. G. Pederson, The University of North Carolina
R. O. Ragsdale, The University of Utah
S. L. Sincoff, United States Air Force Academy
Calvin A. Vanderwerf, The University of Florida
C. A. von Frankenberg, The University of Delaware

We received partial reviews from a number of professors whom we were unable to identify. We express our appreciation to these individuals.

We are especially indebted to the tens of thousands of students with whom we have interacted in our 51 years (cumulative) of teaching introductory chemistry classes. Their concerns, questions, discussions, and suggestions have made us better teachers and, hopefully, better scientists. We extend our special thanks to our students, who provide inspiration to us.

The staff at Saunders College Publishing has contributed immeasurably to the evolution of this book. Our development and copy editor, Jay Freedman, has done a superb job. We are convinced that he has no peer. Rick Moore and Tom Mallon have given us high quality design and artwork, respectively, that contribute to the appearance and the substance of the book. Additionally, we have drawn freely from the excellent artwork in other Saunders College Publishing texts. Our project editors, Lynne Gery and Patrice Smith, have handled innumerable details with skill and aplomb. Tricia Manning and Michelle Glazer, Assistants to John Vondeling, have facilitated communications and the flow of paper cheerfully and efficiently.

We express our deepest appreciation to our editor and friend, John Vondeling, the best editor in the business. John has guided us at every stage in the development of this book, and our respect and admiration for him have grown with each passing day.

Our secretary, Martha Dove, has been patient and skillful through the many revisions. We are indeed grateful for her patience, her skill, and her dedication.

Finally, our appreciation to our families — Sharon, Angela, Laura and Brian Davis; Kathy, Kristen, and Karen Gailey; and Betty, Andy, and Kathryn Whitten.

Raymond E. Davis
Kenneth D. Gailey
Kenneth W. Whitten

Supplements

Excellent ancillary materials have been prepared to assist students in their study and to aid the professor in teaching the courses.

1. *Lecture Outline for Principles of Chemistry*, Professor Ronald O. Ragsdale, The University of Utah. A comprehensive lecture outline that allows professors to use valuable classroom time more effectively. It provides great flexibility for the professor and makes available more time for special topics, increased drill, or whatever the professor chooses to do.

2. *Solutions Manual for Principles of Chemistry*, Professor John M. DeKorte, Northern Arizona University. A pace-setter! It includes detailed answers, solutions, and *explanations* for *all even-numbered* end-of-chapter exercises. In-depth answers are given for discussion questions, and helpful comments that reinforce basic concepts are included, as well as references to illustrative examples and appropriate sections of chapters in the text.

3. *Study Guide for Principles of Chemistry*, Professor R. E. Davis, The University of Texas at Austin. It includes brief summaries of important ideas in each chapter, study goals with references to text sections and exercises, and simple preliminary tests (averaging more than 80 questions per chapter, all with answers) that reinforce basic skills and vocabulary and encourage students to think about important ideas.

4. *Instructor's Manual to Accompany Principles of Chemistry*, Professor R. E. Davis, The University of Texas at Austin. Also includes solutions to *odd-numbered* end-of-chapter exercises and may be made available to students, if the professor chooses.

5. *Experimental General Chemistry*, W. T. Lippincott (The University of Arizona), D. W. Meek (Ohio State University), K. D. Gailey, and K. W. Whitten. A modern laboratory manual with excellent variety that includes descriptive, quantitative, and instrumental experiments. Designed for mainstream courses for science majors.

6. *Problem-Solving in General Chemistry*, 2nd Ed., K. W. Whitten and K. D. Gailey. Covers the common core of general chemistry courses for science majors.

7. *Computer Tutorial for General Chemistry*, Professor Charles A. Wilkie, Marquette University. Comprehensive review and drill in core topics. On diskettes for Apple II+ and Apple IIe computers.

8. *Test Bank*, Davis, Gailey, and Whitten

9. *Overhead Transparencies*, Davis, Gailey, and Whitten. One hundred figures from the text.

10. *Workbook for General Chemistry Audio-Tape Lessons*, 2nd Ed., B. Shakhashiri, R. Schreiner, and P. A. Meyer (all of The University of Wisconsin, Madison).

11. *General Chemistry Audio-Tape Lessons*, Shakhashiri, Schreiner, and Meyer. Adopters of the workbook or of the text will receive up to three free copies of these tapes along with unlimited duplication rights for student use; transcripts of the tapes are also available.

12. *Modern Descriptive Chemistry*, Eugene G. Rochow, Harvard University. A 250-page paperback for those who desire more descriptive chemistry.

To The Student

We have written this text to assist you as you study chemistry, a fundamental science — some call it the central science. As you pursue your career goals, you will find the vocabulary and ideas of chemistry useful in more ways than you may imagine now.

We begin with the most basic vocabulary and ideas. Then we carefully develop increasingly sophisticated ideas that are necessary and useful in all the other physical sciences, the biological sciences, and areas such as medicine, dentistry, engineering, agriculture, and home economics.

We have tried to make many of the early chapters as nearly self-contained as possible, so that the material can be presented in the order considered most appropriate by your professor.

Early in each section we have provided the experimental basis for the ideas we present. By *experimental basis* we mean the observations and experiments on the phenomena that have been most important in developing concepts. We then present explanations of these experimental observations.

Chemistry is an experimental science. We know what we know because we (literally thousands of scientists) have observed it to be true. Successful theories have evolved to explain many experimental observations (facts) fully and accurately; often they enable us to predict the results of experiments that have not yet been performed. Experiment and theory go hand-in-hand; they are intimately related parts of our attempt to understand and explain natural phenomena.

"What is the best way to study chemistry?" is a question we are asked often by our students. Although there is no single answer, the following suggestions may be helpful.

Read the assigned material *before* it is covered in class so that you become generally familiar with important ideas. Take careful class notes. At the first opportunity, recopy your notes, and try to work the illustrative examples without looking at the solutions in your notes. Read the assigned material again. Reading should be more informative the second time.

Review the "key terms" at the end of the chapter so that you know the exact meaning of each. Work the illustrative examples in the text (cover the solutions). If you find it necessary to look at the solutions, look at only one line at a time and try to figure out the next step. Work the assigned exercises at the end of the chapter. Become familiar with the Appendices and their contents so that you may use them whenever necessary. Answers to all even-numbered numerical problems are given at the end of the text so that you may check your work. The *Solutions Manual,* by Professor J. M. DeKorte, includes detailed solutions and explanations for all even-numbered end-of-chapter exercises.

Study Guide for Principles of Chemistry, by Professor Raymond E. Davis, provides an overview of each chapter and emphasizes the threads of continuity that run through chemistry. It lists study goals, tells why concepts are important, and provides references to the text. The *Study Guide* contains many easy-to-moderately-difficult preliminary test questions so you can gauge your progress. They provide excellent practice in preparing for examinations. Answers are provided, and many have explanations.

If you have suggestions for improving this text, please write to us and tell us about them.

RED, KDG, and KWW

Contents Overview

Contents

9 Periodicity and Chemical Reactions   276

10 Chemical Analysis in Aqueous Solution   311

11 Gases and the Kinetic-Molecular Theory   339

12　Liquids and Solids　372

13　Physical Properties of Mixtures: Solutions and Colloids　412

14   Chemical Thermodynamics: The Driving Force for Changes   447

15   Chemical Kinetics   477

16 Chemical Equilibrium 518

17 Ionic Equilibria — I: Acids and Bases 547

18 Ionic Equilibria — II: Buffers and Acid-Base Titrations; Slightly Soluble Compounds 585

19   Electrochemistry   622

20   The Metals and Metallurgy   660

24 Nuclear Chemistry 786

25 Organic Chemistry: The Chemistry of Carbon Compounds 811

Appendices

Fundamental Ideas of Chemistry

Chemistry is *the science that describes matter, its chemical and physical properties, the chemical and physical changes matter undergoes, and the energy changes that accompany these processes.* Consider for a moment the extraordinary breadth of this subject. Matter, after all, includes everything that is tangible: from our bodies and the stuff of our everyday lives to the grandest objects in the universe. Chemistry thus touches almost every aspect of our lives, our culture, and our environment. Its scope encompasses the air we breathe, the food we eat, the fluids we drink, our clothing, our transportation and fuel supplies, our dwellings, our fellow creatures.

Some call chemistry the central science. It rests on the foundation of mathematics and physics, and in turn underlies the sciences of life — biology and medicine. Before we fully understand living systems, we must first understand the chemical reactions that operate within them. In fact, the chemicals of our bodies profoundly affect even the personal world of our thoughts and emotions.

We understand simple chemical systems well; they lie near chemistry's fuzzy boundary with physics and can often be described exactly by the physicist's laws. We fare less well with more complicated systems. Even where our understanding is fairly complete, we must make approximations. And often our knowledge is far from complete. Each year researchers provide new insights into the nature of matter and its interactions, and as chemists find answers to old questions they learn to ask many new ones. Someone once described our scientific knowledge as an expanding sphere, which, as it grows, encounters an ever-enlarging frontier.

Despite its great complexity and its universal importance, chemistry, like all other sciences, begins simply — with its own vocabulary and set of fundamental concepts. This chapter lays the necessary groundwork. We shall be concerned above all with characterizing matter, which is of course the main subject of chemistry. We shall define it, see how it is related to energy, consider what it is made of and what forms it may take, and describe some of the changes it may undergo. We shall then take a close look at measurements — the heart of any exact science.

Our understanding of chemical phenomena has led to the development of atomic theory, which enables us to make predictions about elements and compounds that have not yet been discovered. It is particularly gratifying to chemists when they discover in nature, or make in the laboratory, a new compound whose existence has been predicted by theory. Theories serve a dual role. They help to bring order to experimental observations and they enable us to predict (often quite accurately) the existence of new compounds.

Macroscopic Observations (What We Have Seen)

1–1 Matter and Energy

Matter is anything that has mass and occupies space. **Mass** is a measure of the quantity of matter in a sample of any material. The more massive an object is, the more force is required to put it in motion. Since all bodies in the universe conform to the definition of matter, they all consist of matter. Our senses of sight and touch usually tell us that an object occupies space, but in the case of colorless, odorless, tasteless gases (such as air) our senses sometimes fail us. We might say that we can "touch" air when it blows in our faces, but we depend on other evidence to show that a still body of air fits our definition of matter.

Energy is commonly defined to be *the capacity to do work or to transfer heat.* We are all familiar with many forms of energy in everyday life, including mechanical energy, electrical energy, heat energy, and light energy. We know that light energy from the sun is used by plants as they grow and produce food, electrical energy allows us to light a room by flicking a switch, and heat energy cooks our food and warms our homes. As a matter of convenience, energy can be classified into two principal types: *potential energy* and *kinetic energy.*

Potential energy is the energy an object possesses because of its position or composition. Coal, for example, possesses chemical energy, a form of potential energy, because of its composition. Many electrical generating plants burn coal, producing heat and subsequently electrical energy. A boulder located atop a mountain possesses potential energy because of its height. It can roll down the mountainside and convert its potential energy into kinetic energy, though perhaps not often a useful sort.

A body in motion, such as the rolling boulder, possesses energy because of its motion. Such energy is called **kinetic energy.** Kinetic energy represents the capacity for doing work directly, and is easily transferred between objects.

We discuss different forms of energy because *all chemical processes are accompanied by energy changes.* Many chemical processes are **exothermic,** meaning that as they occur, *energy is released to the surroundings,* usually as heat energy. However, some chemical processes are **endothermic,** i.e., *they absorb energy from their surroundings.*

Nuclear energy is an important kind of potential energy. Nuclear energy is widely used in the production of electricity.

The term comes from the Greek word *kinein* meaning "to move," from which the word *cinema* also is derived.

1   The Law of Conservation of Matter

If we burn a sample of metallic magnesium in the air, the magnesium will combine with oxygen from the air to form magnesium oxide, a white powder. This chemical reaction is accompanied by the release of large amounts of heat and light. If we then weigh the product of the reaction, magnesium oxide, we inevitably find that it is heavier than the original piece of magnesium. The increase in mass is due to the combination of the oxygen with magnesium. Numerous experiments have shown that the mass of the product of the reaction is exactly the sum of the masses of the magnesium and the oxygen that combined. Similar statements can be made for all chemical reactions. These observations are summarized in the **Law of Conservation of Matter:** *There is no observable change in the quantity of matter during an ordinary chemical reaction.* This statement is an example of a **scientific (natural) law,** a general statement based on the observed behavior of matter, to which no exceptions are known. Scientific laws cannot be rigorously proved.

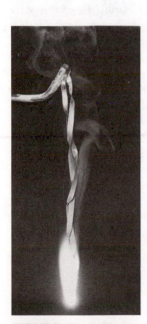

FIGURE 1–1 When magnesium burns, it combines with oxygen from the air to produce magnesium oxide, in a reaction releasing large amounts of light and heat energy. This reaction is used in photographic flashbulbs.

2   The Law of Conservation of Energy

In exothermic chemical reactions, *chemical energy* usually is converted into *heat energy,* but some exothermic processes involve other kinds of energy changes. For example, some liberate light energy without heat, and others may produce electrical energy without heat or light. In endothermic reactions, heat energy, light energy, or electrical energy is converted into chemical energy. While chemical changes always involve energy changes, some energy transformations do not involve chemical changes at all. For example, heat energy may be converted into electrical energy or into mechanical energy without any simultaneous chemical changes. Electricity is produced in hydroelectric plants by converting mechanical energy (from flowing water) into electrical energy. Whatever energy changes we may choose to consider, numerous experiments have demonstrated that all of the energy involved in any change appears in some form after the change. These observations are summarized in the **Law of Conservation of Energy:** *Energy cannot be created or destroyed; it may only be converted from one form to another.*

3   The Law of Conservation of Matter and Energy

The world became aware of the fact that matter can be converted into energy in 1945, when two atomic bombs were exploded over Japan. In nuclear reactions (Chapter 24) matter is transformed into energy. The relationship between matter and energy is given by Albert Einstein's now famous equation:

$$E = mc^2 \qquad\qquad \text{[Eq. 1-1]}$$

Einstein formulated this equation in 1905. Its validity was demonstrated in 1939.

This equation, postulated as a part of the theory of relativity, tells us that *the amount of energy released when matter is transformed into energy is the product of the mass of matter converted and the velocity of light squared.* At the present time, man has not (knowingly) observed the transformation of *energy into matter* on a significant scale. It does, however, happen every day on an extremely small scale in "atom smashers" or particle accelerators used to induce nuclear reactions. Now that the equivalence of matter and energy has been demonstrated, the **Laws of Conservation of Matter and Energy** are combined into a single statement: *The total amount of matter and energy available in the universe is fixed.*

1-2   States of Matter

Matter is conveniently classified in three physical states, although everyone can think of examples that do not fit neatly into any of the three categories. In the **solid state** substances are rigid and have definite shapes, and their volumes do not vary much with changes in temperature and pressure. In some solids, called crystalline solids, the individual particles that make up the solid occupy definite positions in the crystal structure, and the strengths of interaction between the individual particles determine how hard and how strong the solid is. In the **liquid state** the individual particles are confined to a given volume, but may otherwise flow and assume the shapes of their containers up to the volume of the liquid. Although the volumes of most liquids vary more with pressure changes than do volumes of most solids, liquids are still only very slightly compressible. Liquid volumes also generally vary more with temperature changes than do those of solids. **Gases** fill completely any vessel in which they are confined, and assume the shapes of their containers. Gases are capable of infinite expansion and are compressed easily. Because gases are easily compressed, we conclude that they consist primarily of empty space; i.e., the individual particles are quite far apart. The behavior of gases, liquids, and solids will be considered more fully in Chapters 11 and 12.

1-3   Chemical and Physical Properties

In order to distinguish samples of different kinds of matter, we determine and compare their properties. The properties of a person include height, weight, sex, skin and hair coloration, and the many subtle features that constitute that person's general appearance. We recognize different kinds of matter by their properties, which are broadly classified into chemical properties and physical properties.

   Chemical properties *are properties that matter exhibits as it undergoes changes in composition.* These properties of substances are related to the kinds of chemical changes that the substances undergo. For instance, we have already described the

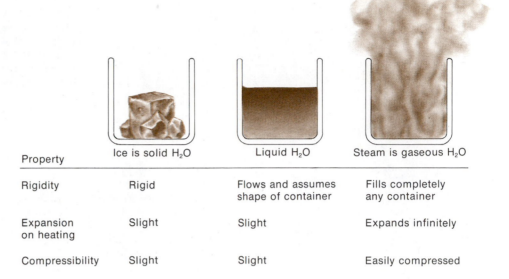

FIGURE 1–2 A comparison of some physical properties of the three states of matter (for water).

Property	Ice is solid H_2O	Liquid H_2O	Steam is gaseous H_2O
Rigidity	Rigid	Flows and assumes shape of container	Fills completely any container
Expansion on heating	Slight	Slight	Expands infinitely
Compressibility	Slight	Slight	Easily compressed

combination of metallic magnesium with gaseous oxygen to form magnesium oxide, a white powder. A chemical property of magnesium is that it can combine with oxygen, releasing energy in the process. Conversely, a chemical property of oxygen is that it can combine with magnesium. An attempt to make the magnesium oxide combine with either magnesium or oxygen will show that magnesium oxide has neither of these chemical properties.

All substances exhibit **physical properties** as well, which can be observed in the *absence* of any change in composition. Color, density, hardness, melting point, boiling point, and electrical and thermal conductivities are physical properties. Some physical properties of a single sample depend on the conditions, such as temperature and pressure, under which they are measured. For instance, water is a solid (ice) at low temperatures, but is a liquid at higher temperatures. At still higher temperatures, it is a gas (steam). As the sample of water is converted from one state to another, its composition is constant; therefore, its chemical properties do not change. On the other hand, the physical properties of ice, liquid water, and steam are quite dissimilar (Figure 1–2).

The physical properties of matter can be further classified as *extensive properties* or *intensive properties.* **Extensive properties** *depend on the amount of material examined.* The volume and the mass of a sample are extensive properties because they depend on, and are directly proportional to, the amount of matter in the sample examined. **Intensive properties** *do not depend on the amount of material examined.* The color and the melting point of a substance, for example, are the same for a small sample as for a large one. All chemical properties are intensive properties.

Since no two substances have identical sets of chemical and physical properties under the same conditions, we are able to identify and distinguish among different substances. For instance, water is the only clear, colorless substance that freezes at 0°C, boils at 100°C at one atmosphere of pressure, dissolves relatively large amounts of ordinary salt, and reacts violently with sodium (Figure 1–3). Table 1–1 compares several intensive properties of a few substances; a sample of any one of them can be distinguished from the others by measurements of its properties.

One atmosphere of pressure is the average atmospheric pressure at sea level.

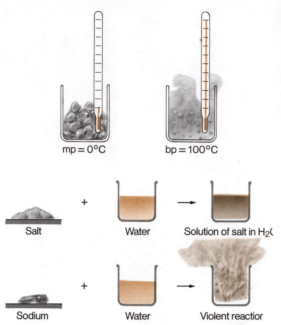

FIGURE 1–3 Some physical and chemical properties of water. Physical: (a) It melts at 0°C; (b) it boils at 100°C (at normal atmospheric pressure); (c) it dissolves salt. Chemical: (d) It reacts with sodium to form hydrogen gas and a solution of sodium hydroxide.

mp = 0°C   bp = 100°C

Salt   +   Water   →   Solution of salt in H₂O

Sodium   +   Water   →   Violent reaction

TABLE 1–1   Physical Properties of a Few Common Substances (at atmospheric pressure)

Substance	Melting Pt. (°C)	Boiling Pt. (°C)	Solubility at 25°C (g/100 g)		Density (g/cm³)
			Water	Ethyl Alcohol	
acetic acid	16.6	118.1	infinite	infinite	1.05
benzene	5.5	80.1	0.07	infinite	0.879
bromine	−7.1	58.8	3.51	infinite	3.12
iron	1530	3000	insoluble	insoluble	7.86
methane	−182.5	−161.5	0.0022	0.033	6.67×10^{-4}
oxygen	−218.8	−183.0	0.0040	0.037	1.33×10^{-3}
sodium chloride	801	1473	36.5	0.065	2.16
water	0	100	—	infinite	1.00

1–4   Chemical and Physical Changes

In any **chemical change,** (1) one or more substances are used up (at least partially), (2) one or more new substances are formed, and (3) energy is absorbed or released. As substances undergo chemical changes they exhibit chemical properties. A **physical change,** on the other hand, occurs with *no change in chemical composition.* Physical properties are altered as matter undergoes physical changes.

Both chemical and physical changes are always accompanied by either the absorption or the liberation of energy. Energy is required to melt ice, and energy is required to boil water. Conversely, the condensation of steam to form liquid water always liberates energy, as does the freezing of liquid water to form ice. The changes in energy accompanying these physical changes are shown in Figure 1–4. All

A process that liberates heat energy is called an exothermic process, while one that absorbs heat energy is an endothermic process.

FIGURE 1-4
Changes in energy that accompany some physical changes (for water).

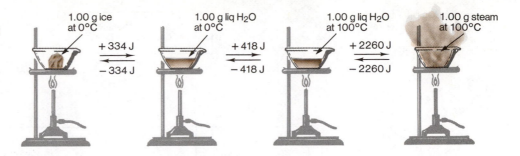

changes are at a pressure of one atmosphere. At this pressure, ice always melts at the same temperature (0°C) and pure water always boils at the same temperature (100°C). The units of energy used in the figure, joules (J), will be defined in Section 3-2. The positive signs preceding joules above the arrows indicate that the changes in the indicated direction are endothermic. Negative signs below the arrows signify exothermic processes.

1-5 Substances, Compounds, Elements, and Mixtures

All samples of pure ethyl alcohol contain exactly the same proportion of carbon, 52.14%, hydrogen, 13.13%, and oxygen, 34.73%, by mass, regardless of the source of the samples. All samples have the same melting point, −117.3°C, boiling point, 78.3°C, and other physical properties. Comparable statements can be made, with different values, of course, for any pure substance. A **pure substance** *is any kind of matter all samples of which have identical composition and, under identical conditions, identical properties.*

A more formal statement of this idea is the Law of Constant Composition, which will be discussed more fully in Section 1-12.

Compounds are pure substances consisting of two or more different elements in a fixed ratio. Several million different compounds are known, and their properties have been determined. Each chemical compound contains its constituent elements in fixed proportions by mass. Every compound possesses definite physical and chemical properties, and these properties can be used to distinguish each compound from all other compounds. Let us describe the decomposition of the compound calcium carbonate, the major component of limestone and marble.

When a sample of pure calcium carbonate is heated to high temperatures, it decomposes, always in the same mass ratio, into calcium oxide, a white solid, and carbon dioxide, a colorless gas. Both calcium oxide and carbon dioxide are compounds.

Calcium oxide can be decomposed, always in the same mass ratio, into two simpler pure substances. One is calcium, a silvery-white metal, and the other is oxygen, a colorless gas. Carbon dioxide can also be decomposed into two pure substances—carbon, a black solid, and oxygen; again we observe that the ratio of these components is the same, regardless of the source of the carbon dioxide. None of the pure substances obtained from the decomposition of calcium oxide and carbon dioxide can be decomposed into simpler substances. Therefore, we conclude that calcium, carbon, and oxygen represent the simplest kinds of pure substances, which we call *elements.*

Elements *are pure substances that cannot be decomposed into simpler substances by chemical changes.* Nitrogen, silver, aluminum, copper, gold, and sulfur are other examples of elements.

The percentages refer to the percent of total mass of the compounds decomposed.

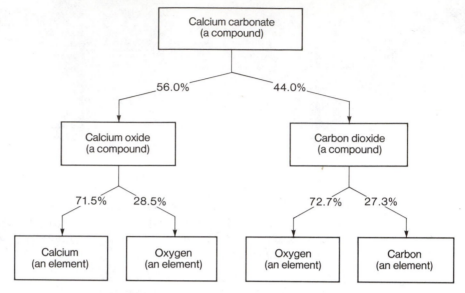

A set of **symbols** represents the known elements. Symbols are used as a matter of convenience because they can be written more quickly than names and they occupy less space. The symbols for the first 103 elements consist of either a capital letter *or* a capital letter followed by a lower case letter, such as C (carbon) or Ca (calcium). Symbols for elements beyond number 103 consist of three letters. A list of the known elements and their symbols is included in the table inside the front cover.

A short list of symbols of common elements is given in Table 1–2. Learning this list will be helpful. Many symbols consist of the first one or two letters of the element's English name. Some are derived from the element's Latin name (indicated in parentheses), and one, W for tungsten, is from the German *Wolfram.* Names and symbols for additional elements should be learned as they are encountered.

Most of the earth's crust is made up of a relatively small number of elements. For example, only ten of the 88 naturally occurring elements make up 99+% by mass of the earth's crust, oceans, and atmosphere (Table 1–3). Oxygen accounts for roughly half, and silicon approximately one fourth, of the whole.

The other known elements have been made artificially in laboratories, as described in Chapter 24. One atom of element number 109 has been reported. Element 108 has not been reported.

TABLE 1–2   Common Elements and Their Symbols

Symbol	Element	Symbol	Element
Ag	silver (argentum)	K	potassium (kalium)
Al	aluminum	Li	lithium
Au	gold (aurum)	Mg	magnesium
Br	bromine	N	nitrogen
C	carbon	Na	sodium (natrium)
Ca	calcium	Ne	neon
Cl	chlorine	Ni	nickel
Cu	copper (cuprum)	O	oxygen
F	fluorine	P	phosphorus
Fe	iron (ferrum)	Pb	lead (plumbum)
H	hydrogen	S	sulfur
He	helium	Si	silicon
Hg	mercury (hydrargyrum)	W	tungsten (wolfram)
I	iodine	Zn	zinc

TABLE 1-3   Abundance of Elements (Earth's Crust, Oceans, and the Atmosphere)

oxygen	O	49.5%		chlorine	Cl	0.19%	
silicon	Si	25.7		phosphorus	P	0.12	
aluminum	Al	7.5		manganese	Mn	0.09	
iron	Fe	4.7		carbon	C	0.08	
calcium	Ca	3.4	99.2%	sulfur	S	0.06	0.7%
sodium	Na	2.6		barium	Ba	0.04	
potassium	K	2.4		chromium	Cr	0.033	
magnesium	Mg	1.9		nitrogen	N	0.030	
hydrogen	H	0.87		fluorine	F	0.027	
titanium	Ti	0.58		zirconium	Zr	0.023	
				All others		<0.1%	

Relatively few elements, approximately one fourth of the naturally occurring ones, occur in nature as free elements. The rest are always found chemically combined with other elements.

Mixtures are combinations of two or more substances in which each substance retains its own composition and properties. The compositions of mixtures can be varied widely. We can make an infinite number of different mixtures of salt and sugar by varying the relative amounts of the two pure substances used. Solutions of salt dissolved in water are mixtures whose composition may vary over a wide range. Air is a mixture of gases that consists primarily of nitrogen, oxygen, argon, carbon dioxide, and water vapor. There are only trace amounts of other substances in the atmosphere.

Mixtures can be classified as either homogeneous or heterogeneous. A **homogeneous mixture,** also called a **solution,** *has uniform composition and properties throughout.* Examples include air (free of particulate matter or mists), salt water, and some alloys, which are homogeneous mixtures of metals in the solid state. Mixtures

By "composition of a mixture" we mean both the identities of the substances present and their relative amounts in the mixture.

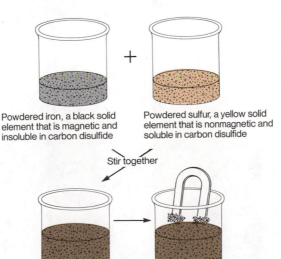

Powdered iron, a black solid element that is magnetic and insoluble in carbon disulfide

Powdered sulfur, a yellow solid element that is nonmagnetic and soluble in carbon disulfide

Stir together

Heterogeneous mixture of iron and sulfur

Separating the mixture by removing the magnetic iron

FIGURE 1-5 A mixture of iron and sulfur is a *heterogeneous* mixture. Like all mixtures, it can be separated by physical means, such as removing the iron with a magnet.

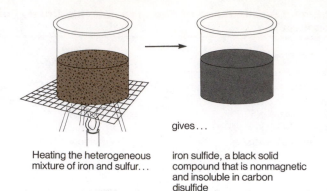

Heating the heterogeneous mixture of iron and sulfur...

gives...

iron sulfide, a black solid compound that is nonmagnetic and insoluble in carbon disulfide

FIGURE 1–6 A chemical reaction occurs when the heterogeneous mixture of iron and sulfur is heated. The result is a compound.

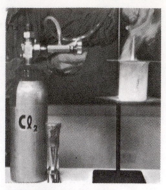

FIGURE 1–7 The reaction of sodium and chlorine to produce table salt, sodium chloride.

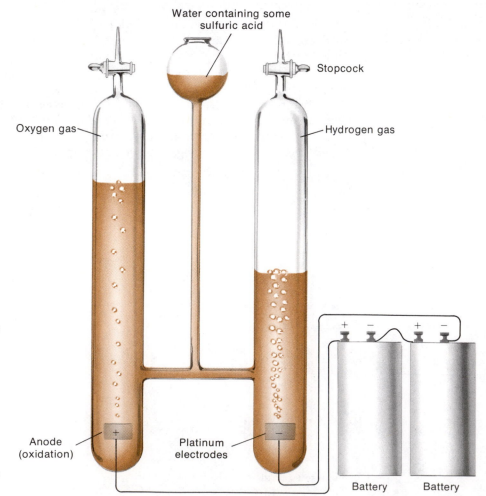

Water containing some sulfuric acid

Stopcock

Oxygen gas

Hydrogen gas

Anode (oxidation)

Platinum electrodes

Battery   Battery

FIGURE 1–8
Apparatus for small-scale electrolysis of water. Electrolysis is the decomposition of substances by electrical energy. Note that the volume of hydrogen obtained is twice that of oxygen. Some dilute sulfuric acid is used to increase the water's conductivity so that electrolysis occurs more rapidly.

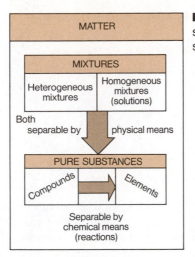

FIGURE 1–9 One classification of matter and some means by which the classes of matter can be separated.

that are not uniform throughout are said to be **heterogeneous.** Examples include mixtures of salt and charcoal (in which one component can be distinguished readily from the other by sight), foggy air (which includes a suspended mist of water droplets), and vegetable soup.

Mixtures can be separated by physical means because each component retains its properties (see Figures 1–5 and 1–9). For example, a mixture of salt and water can be separated by evaporating the water and leaving the solid salt behind. A mixture of sand and salt can be separated by dissolving the salt in water, filtering out the sand, and then evaporating the water to obtain the solid salt. Very fine iron powder can be mixed with powdered sulfur to give what appears to the naked eye to be a homogeneous mixture of the two. However, separation of the components of this mixture is easy. The iron may be removed by a magnet, or the sulfur may be dissolved in carbon disulfide, which does not dissolve iron (Figure 1–5).

The important characteristics of mixtures are (1) that their compositions can be varied, and (2) that each component (compound or uncombined element) of the mixture retains its own properties.

Figure 1–9 summarizes the classification of matter and the methods by which separations can be achieved.

Before we examine the structure of matter in some detail, let us describe how measurements are made and interpreted.

Tools of Scientists (How We Measure What We See)

1–6 Measurements in Chemistry

Measurements in the scientific world are usually expressed in the units of the International System of Units (SI), which was adopted by the National Bureau of Standards in 1964. The SI is based on the seven fundamental units listed in Table 1–4. All other units of measurement are derived from them.

TABLE 1–4   The Seven Fundamental SI Units of Measurement

Physical Property	Name of SI Unit	Abbreviation
length	meter	m
mass	kilogram	kg
time	second	s
electric current	ampere	A
temperature	kelvin	K
luminous intensity	candela	cd
quantity of substance	mole	mol

Just as the general public of the United States has been reluctant to move from such traditional units of distance as the mile and the yard to the "less familiar" meter and kilometer of the metric system, the scientific community has been reluctant to dispense with its own traditional and comfortable metric units and switch completely to SI units.

In many applied sciences such as the health sciences, there has been little if any motion in the direction of using SI units. Although the metric system was endorsed by Congress in 1866, such things as tire gauges and gauges on cylinders of compressed gases are still calibrated in pounds per square inch rather than the metric units, atmospheres or millimeters of mercury of pressure. Only recently has the weather bureau begun to report atmospheric pressure in millimeters of mercury as well as inches of mercury. Thus, we must become familiar with several units of measurement for the same quantity.

In this text we shall use both metric units and SI units. As we shall see later in this chapter, conversions between non-SI and SI units are usually quite straightforward.

Appendix C lists some important units of measurement and their relationship to each other. We'll refer to it often. Appendix D lists several important physical constants for your convenience.

The metric and SI systems are *decimal systems, in which prefixes are used to indicate fractions and multiples of ten.* The same prefixes are used with all units of measurement. The distances and masses in Table 1–5 illustrate the use of some common prefixes and the relationships among them.

In the next three sections, we shall introduce the primary standards chosen for basic units of measurement. These standards were selected because they allow us to make precise measurements, and they are reproducible and unchanging. The values

TABLE 1–5   Common Prefixes Used in the SI and Metric Systems

The prefixes used in the SI and metric systems may be thought of as *multipliers;* e.g., the prefix *kilo* indicates multiplication by 1000, while *milli* indicates multiplication by 0.001 or by 10^{-3}

Prefix	Abbreviation	Meaning	Example
mega-	M	10^6	1 megameter (Mm) = 1×10^6 m
kilo-	k	10^3	1 kilometer (km) = 1×10^3 m
deci-	d	10^{-1}	1 decimeter (dm) = 0.1 m
centi-	c	10^{-2}	1 centimeter (cm) = 0.01 m
milli-	m	10^{-3}	1 milligram (mg) = 0.001 g
micro-	μ^*	10^{-6}	1 microgram (μg) = 1×10^{-6} g
nano-	n	10^{-9}	1 nanogram (ng) = 1×10^{-9} g
pico-	p	10^{-12}	1 picogram (pg) = 1×10^{-12} g

* This is the Greek letter μ (pronounced "mew").

assigned to basic units are arbitrary.* In the United States, all units of measurement are set, by law, by the National Bureau of Standards. Like many other government agencies around the world, the Bureau of Standards accepts the internationally agreed-upon values for basic units.

1–7   Units of Measurement

1   Mass and Weight

We must distinguish between mass and weight. You may recall from Section 1–1 that **mass** *is a measure of the quantity of matter a body contains.* The force required to give a sample of matter a certain acceleration is a measure of its mass. The mass of a body does not vary as its position changes. On the other hand, the **weight** *of a body is a measure of the gravitational attraction of the earth for the body,* and this varies with distance from the center of the earth. An object weighs ever so slightly less high up in an airplane than at the bottom of a deep valley. Nearly everyone has seen pictures of astronauts as they experienced weightlessness during space travel. In stable orbits, they no longer feel the attraction of earth's gravity. Because the mass of a body does not vary with its position, whereas its weight does, the mass of a body is a more fundamental property than its weight. However, we have become accustomed to using the term weight when we mean mass, just because weight is one way of measuring mass. Since we usually discuss chemical reactions at constant gravity, weight relationships are just as valid as mass relationships, but we must keep in mind that *the two are not identical.*

FIGURE 1–10 A duplicate of the meter bar, now replaced as the SI standard of length by a wavelength emitted by the element krypton; and a duplicate of the SI kilogram standard of mass.

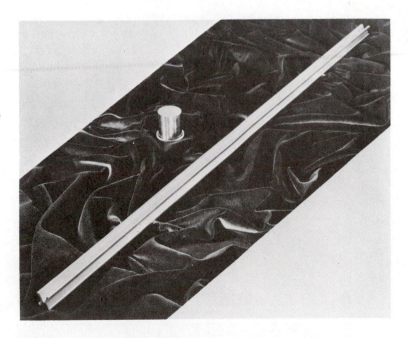

* Prior to the establishment of the U.S. Bureau of Standards in 1906, at least 50 different distances had been used as "1 foot" in measuring land within New York City. Thus, the size of a 100 ft by 200 ft lot in New York depended on the generosity of the seller and did not necessarily represent the expected dimensions.

(a)   (b)   (c)

FIGURE 1–11 Three types of laboratory balances. (a) A triple beam balance used for determining mass to about ±0.01 g. (b) A modern electronic top-loading balance that gives a direct readout of mass to ±0.01 g. (c) A modern analytical balance that can be used to determine mass to ±0.0001 g. Analytical balances are used when masses must be determined as accurately as possible. When less accuracy is required, other kinds of balances are used.

**TABLE 1–6
Some SI Units
of Mass**

*kilo*gram, kg	base unit
gram, g	1000 g = 1 kg
*milli*gram, mg	1000 mg = 1 g
*micro*gram, μg	1,000,000 μg = 1 g

The basic unit of mass in the SI system is the **kilogram** (Table 1–6). The kilogram is arbitrarily defined as the mass of a platinum-iridium cylinder stored in Sèvres, near Paris, France (see Figure 1–10). One kilogram weighs 2.205 pounds, and a one-pound object has a mass of 0.4536 kilogram. The metric system also uses the kilogram and all of its multiples, but the basic mass unit in the metric system is the gram.

We have said that the force required to give a sample of matter a certain acceleration is a measure of its mass. However, it would be very difficult to make such measurements to find the masses of the samples used in the chemistry laboratory. Instead, an instrument called a *balance* is used to compare the force of gravity on the sample to the force of gravity on a standard object of known mass, or to the force exerted by a spring or a twisted fiber (Figure 1–11). In a balance that uses standard "weights" (whose masses are determined by comparisons with those kept by the Bureau of Standards), variations in gravity caused by moving the balance to a different altitude are canceled by the comparisons; spring or fiber (torsion) balances must be *calibrated* by noting the readings obtained from measuring standard masses.

2   Length

The **meter** is the standard of length (distance) in both SI and metric systems. It is approximately 39.37 inches, or a bit more than 3 feet. In situations where the

FIGURE 1–12 The relationship between centimeters and inches: 2.54 cm are exactly equal to 1 inch.

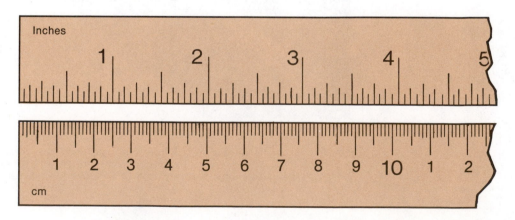

FIGURE 1–13 A comparison of common measures of volume.

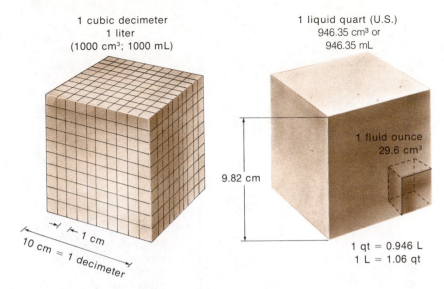

1 cubic decimeter
1 liter
(1000 cm³; 1000 mL)

1 liquid quart (U.S.)
946.35 cm³ or
946.35 mL

1 fluid ounce
29.6 cm³

9.82 cm

1 cm

10 cm = 1 decimeter

1 qt = 0.946 L
1 L = 1.06 qt

English system would use inches, the metric centimeter ($\frac{1}{100}$ meter) is convenient. The relationship between inches and centimeters is shown in Figure 1–12. The meter was originally defined in 1790 as one ten-millionth of the distance from the North Pole to the equator. In 1875 it was redefined as the distance between two lines on a platinum-iridium bar stored in Sèvres (Figure 1–10). It is now defined as 1,650,763.73 times the wavelength of the orange-red light emitted by the element krypton under specified conditions. This is an unvarying standard that is reproducible anywhere with an instrument called a spectrointerferometer. For usual measurements in the laboratory, we still rely on the ordinary meter stick.

3 Volume

Volumes are measured in liters or milliliters in the metric system. One liter (1 L) is defined as one cubic decimeter (1 dm³), or 1000 cubic centimeters (1000 cm³). One milliliter (1 mL) is 1 cm³. In the SI system the cubic meter is the basic volume unit, and the cubic decimeter replaces the metric unit, the liter (Figure 1–13).

Many different kinds of glassware are used to measure the volumes of liquids. The one we choose depends on the accuracy we desire. For example, the volume of a liquid dispensed can be measured more accurately with a buret than with a small graduated cylinder (Figure 1–14).

Equivalences between common English units and metric units are summarized in Table 1–7.

1–8 Significant Figures

There are two kinds of numbers. **Exact numbers** *are numbers that are known to be absolutely accurate.* For example, the exact number of people seated in an orderly arrangement in a closed room can be counted, and there is no doubt about the number of people in the room. A dozen eggs is exactly 12 eggs, no more, no fewer. By definition, the subdivisions of SI and metric base units are also exact; for instance, one meter contains exactly 100 cm.

FIGURE 1–14
Laboratory apparatus used to measure volumes of liquids.

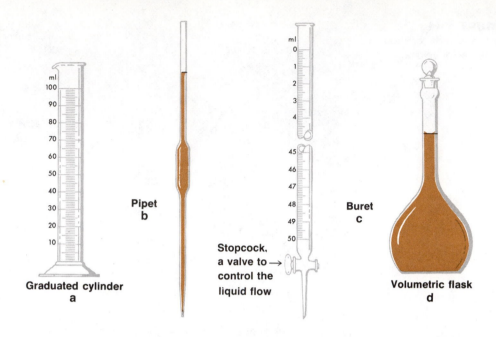

Graduated cylinder
a

Pipet
b

Stopcock, a valve to → control the liquid flow

Buret
c

Volumetric flask
d

Numbers obtained from measurements are not exact. When measurements are made, some estimation is involved. For example, suppose you are asked to measure the length of this page to the nearest 0.1 mm. How do you do it? The smallest divisions (calibration lines) on an ordinary meter stick are 1 mm apart (Figure 1–12). An attempt to measure to 0.1 mm involves estimation. If you measure the length of the page three different times, will you get the same answer each time? Probably not. We deal with this problem by using significant figures.

There is some uncertainty in all measurements.

Significant figures are digits believed to be correct by the person who makes a measurement. We assume that the person is competent to use the measuring device. Suppose one measures a distance with a meter stick and reports the distance as 343.5 mm. What does this number mean? In this person's judgment, the distance is

Significant figures indicate the *uncertainty* in measurements.

TABLE 1–7   Conversion Factors Relating Length, Volume, and Mass Units

Metric		English		Metric-English Equivalents	
Length					
1 km	$= 10^3$ m	1 ft	$= 12$ in	1 in	$= 2.54$ cm
1 cm	$= 10^{-2}$ m	1 yd	$= 3$ ft	1 m	$= 39.37$ in*
1 mm	$= 10^{-3}$ m	1 mile	$= 5280$ ft	1 mile	$= 1.609$ km*
1 nm	$= 10^{-9}$ m				
1 Å	$= 10^{-8}$ cm				
Volume					
1 mL	$= 1$ cm$^3 = 10^{-3}$ liter	1 gal	$= 4$ qt $= 8$ pt	1 liter	$= 1.057$ qt*
1 m^3	$= 10^6$ cm$^3 = 10^3$ liter	1 qt	$= 57.75$ in^3*	1 ft^3	$= 28.32$ liter*
Mass					
1 kg	$= 10^3$ g			1 lb	$= 453.6$ g*
1 mg	$= 10^{-3}$ g	1 lb	$= 16$ oz	1 g	$= 0.03527$ oz*
1 metric tonne	$= 10^3$ kg	1 short ton	$= 2000$ lb	1 metric tonne	$= 1.102$ short ton

* These conversion factors, unlike the others listed, are inexact. They are quoted to four significant figures, which will ordinarily be more than sufficient.

FIGURE 1-15 Two pieces of laboratory equipment for measuring volumes of liquids, a 50 mL graduated cylinder and a 50 mL buret. Because of its smaller diameter, the buret has its markings spaced farther apart, allowing greater precision of measurement than is possible with a graduated cylinder.

greater than 343.4 mm but less than 343.6 mm, and the best estimate is 343.5 mm. The number 343.5 mm contains four significant figures—the last digit, 5, represents a *best estimate* and is therefore doubtful, but it is considered to be a significant figure. In reporting numbers obtained from measurements, *we report one estimated digit, and no more.* Since the person making the measurement is not certain that the 5 is correct, clearly it would be silly, meaningless, and wrong to report the distance as 343.53 mm.

To see more clearly the part significant figures play in reporting the results of measurements, consider Figure 1-15, which shows two pieces of equipment used to measure small volumes of liquid. Graduated cylinders are used to measure volumes of liquids when a high degree of accuracy is not necessary. The calibration lines on a 50 mL graduated cylinder represent 1 mL increments. Estimation of the volume of liquid in a 50 mL cylinder to within 0.2 mL ($\frac{1}{5}$ of the calibration increments) with reasonable certainty is possible. We might measure a volume of liquid in such a cylinder and report the volume as 39.4 mL, i.e., to three significant figures. Clearly, we should not report the volume as 39.42 mL, because it is impossible to read the calibration lines to hundredths of a milliliter, and there is no basis for reporting four significant figures.

Burets are used to measure volumes of liquids when high accuracy is required. The calibration lines on a 50 mL buret represent 0.1 mL increments, allowing us to make estimates to within 0.02 mL ($\frac{1}{5}$ of the calibration increments) with reasonable certainty. Experienced individuals estimate volumes in 50 mL burets to 0.01 mL with considerable success. For example, using a 50 mL buret, we can measure out 36.95 mL (four significant figures) of liquid with reasonable accuracy. A measured volume of 36.955 mL cannot be dispensed from such a buret; there is no basis for reporting five significant figures, since this would imply accuracy of 0.002 mL ($\frac{1}{50}$ of the marked increments).

Accuracy refers to how closely a measured value agrees with the correct value. **Precision** refers to how closely individual measurements agree with each other. Ideally, all measurements should be both accurate and precise. Measurements may be quite precise, yet quite inaccurate, because of some *systematic error*, which is an error repeated in each measurement. (A faulty balance, for example, might produce a systematic error.) Very accurate measurements are seldom imprecise.

Measurements are frequently repeated in an attempt to improve accuracy and precision. Average values obtained from several measurements are usually considered more reliable than individual measurements. Significant figures indicate how accurately measurements have been made (assuming the person who made the measurements was competent).

Some simple rules or conventions govern the use of significant figures in calculations.

Zeroes used just to position the decimal point are not significant figures.

For example, the number 0.0234 g contains only three significant figures, because the two zeroes are used to place the decimal point. The number could be

One of the authors (RED) had an experience with systematic errors obtained on a faulty balance when he was a student taking quantitative analysis. A chain had been improperly replaced on the balance and all the masses he determined were in error by a proportional amount. Near the end of the semester he recognized the problem, recalculated all of his results, and received an ''A'' in the course.

reported equally well as 2.34×10^{-2} g in scientific notation (Appendix A). When zeroes precede the decimal point, but come after other digits, we may have some difficulty in deciding whether the zeroes are significant figures or not. How many significant figures does the number 23,000 contain? We are given insufficient information to answer the question. If all three of the zeroes are used simply to place the decimal point, the number should be written as 2.3×10^4 (two significant figures). If only two of the zeroes are being used to place the decimal point, it is 2.30×10^4 (three significant figures), while if only one zero is being used to place the decimal point, the number should be written 2.300×10^4 (four significant figures). In the event that the number is actually known to be $23,000 \pm 1$, it should be written as 2.3000×10^4 (five significant figures) so that the number of significant figures is obvious.

Life would be much simpler if every person who used numbers did so correctly. As everyone knows, such is not the case. In this text we shall exert a great effort to use numbers correctly so that you will know exactly what every number means.

> In multiplication and division, an answer contains no more significant figures than the least number of significant figures used in the operation.

We must, therefore, distinguish between numbers generated by electronic calculators and the result of an arithmetic operation. Electronic calculators "assume" that all numbers entered are exact numbers. Clearly, such a ridiculous assumption must produce some equally ridiculous answers.

Example 1–1

What is the area of a rectangle 1.23 cm wide and 12.34 cm long? The area of a rectangle is its length times its width.

Solution

$$A = l \times w = (12.34 \text{ cm})(1.23 \text{ cm}) = \underline{15.2} \text{ cm}^2$$

$$\text{calculator result} = (15.1782)$$

The answer should contain only three significant figures, the smallest number of significant figures used in the information given. The number generated by an electronic calculator (15.1782 shown in parentheses) is wrong; the result cannot be more accurate than the information that led to it. Since the calculator has no judgment, you must exercise yours.

$$
\begin{array}{r}
12.3\underline{4} \text{ cm} \\
\times \quad 1.2\underline{3} \text{ cm} \\
\hline
37\ 0\underline{2} \\
2\ 4\underline{6}\ \underline{8} \\
12\ 3\underline{4} \\
\hline
15.1\underline{7}\ \underline{82} \text{ cm}^2 \\
= 15.2 \text{ cm}^2
\end{array}
$$

The first doubtful digit is underlined in the next few calculations.

The step-by-step calculation in the margin demonstrates why the area of the rectangle should be reported as 15.2 cm², rather than as 15.1782 cm². The length, 12.34 cm, contains four significant figures while the width, 1.23 cm, contains only three. Here we underline each uncertain figure, as well as each figure obtained from an uncertain figure. Thus we see that there are only two certain figures (15) in the result. We may report the first doubtful figure (.2), but no more.

Division is just the reverse of multiplication, and the same rules apply. Stated succinctly: in multiplication and division the answer contains no more significant figures than the least number of significant figures given in the information.

> In addition and subtraction, the last digit retained in the sum or difference is determined by the first doubtful digit.

Example 1–2

a. Add 37.24 mL and 10.3 mL.

```
  37.24 mL
+10.3   mL
  47.54 mL is reported as 47.5 mL
          (calculator gives 47.54)
```

b. Subtract 21.2342 g from 27.87 g

```
  27.87   g
−21.2342 g
   6.6358 g is reported as 6.64 g
          (calculator gives 6.6358)
```

Note that in the three simple arithmetic operations we have performed, the number combination generated by an electronic calculator is not the "answer" in a single case! However, the correct result of each calculation can be obtained by "rounding off," and the rules of significant figures tell us where to round off.

In rounding off, certain conventions have been adopted. When the number to be dropped is less than 5, it is just dropped (i.e., 7.34 rounds off to 7.3). When it is more than 5, the preceding number is increased by 1 (i.e., 7.37 rounds off to 7.4). When the number to be dropped is 5, the preceding number is not changed when it (the preceding number) is even (i.e., 7.45 rounds off to 7.4). When the preceding number is odd, it is increased by 1 (i.e., 7.35 rounds off to 7.4). In fairness, we must note that the even-odd rules for rounding terminal 5's are sometimes ignored; instead, the preceding number is increased by 1 when a 5 is dropped.

> The even-odd rules are intended to reduce the accumulation of errors in chains of calculations.

1–9 Dimensional Analysis (The Unit Factor Method)

Many chemical and physical processes can be described by numerical relationships. In fact, many of the most useful ideas in science may be treated mathematically. Let us, then, devote a little time to reviewing problem-solving skills. First, multiplication by unity (by one) does not change the value of an expression. However, when the units of the number "one" are chosen appropriately, many calculations can be done by just "multiplying by one." This method of performing calculations is known as **dimensional analysis,** or the **factor-label method,** or the **unit factor method.** Regardless of the name chosen, it is a very powerful mathematical tool, and is almost foolproof, as we shall demonstrate. We'll illustrate the method with some simple conversions.

Unit factors *may be constructed from any two terms that describe the same or an equivalent "amount" of whatever we may consider.* For example, 1 foot is equal to exactly 12 inches, by definition, and we may write an equation to describe this equality:

$$1 \text{ ft} = 12 \text{ in}$$

Dividing of both sides of the equation by 1 ft gives

$$\frac{1 \text{ ft}}{1 \text{ ft}} = \frac{12 \text{ in}}{1 \text{ ft}} \qquad \text{or} \qquad 1 = \frac{12 \text{ in}}{1 \text{ ft}}$$

The factor (fraction) 12 in/1 ft is a unit factor because the numerator and denominator describe the same distance. We could have divided both sides of the original equation by 12 in, and obtained 1 = 1 ft/12 in, a second unit factor that is the reciprocal of the first. It should be obvious that the reciprocal of any unit factor is also a unit factor. Stated differently, division of an amount by the same amount always yields one!

From our knowledge of the English system of measurements we can write numerous unit factors, such as

Unless otherwise indicated, a "ton" refers to a "short ton," 2000 lb. There are also the "long ton," which is 2240 lb, and the metric tonne, which is 1000 kg.

$$\frac{1 \text{ yd}}{3 \text{ ft}}, \quad \frac{1 \text{ yd}}{36 \text{ in}}, \quad \frac{1 \text{ mile}}{5280 \text{ ft}}, \quad \frac{4 \text{ qt}}{1 \text{ gal}}, \quad \frac{2000 \text{ lb}}{1 \text{ ton}}$$

The reciprocal of each of these is also a unit factor. Items in retail stores are frequently priced with unit factors such as 39¢/lb and $3.98/gal.

Nearly all numbers have units. What does 12 mean? Usually we must supply appropriate units, such as 12 eggs or 12 people. In the unit factor method, the units guide us through calculations in a step-by-step process, because all units except those in the desired result cancel.

We shall express answers to some problems in exponential form to indicate the appropriate number of significant figures. In scientific (exponential) notation, we place one nonzero digit to the left of the decimal as in the following examples.

$$4{,}300{,}000. = 4.3 \times 10^6$$

6 places to the left, ∴ exponent of 10 is 6

$$0.000348 = 3.48 \times 10^{-4}$$

4 places to the right, ∴ exponent of 10 is -4

The reverse process converts numbers from exponential to decimal form. See Appendix A for more detail, if necessary.

Example 1–3

Express 1.47 miles in inches.

Solution

First we write down the units of what we wish to know, preceded by a question mark, and then equate it to whatever we are given.

$\underline{?}$ in = 1.4$\underline{7}$ miles

Then we choose unit factors to convert the given units (miles) to the desired units (inches).

$$\underline{?} \text{ in} = 1.4\underline{7} \text{ miles} \times \frac{5280 \text{ ft}}{1 \text{ mile}} \times \frac{12 \text{ in}}{1 \text{ ft}}$$

$$= 9.3\underline{1} \times 10^4 \text{ in} \quad \text{(calculator gives 93139.2)}$$

Note that both miles and feet cancel, leaving only inches, the desired unit. If either of the unit factors had been written in inverted form, such as 1 mile/5280 ft, the units would not have canceled properly. Thus there is no ambiguity as to how they should be written. The answer contains three significant figures because there are three significant figures in 1.47 miles. The factors 5280 ft/mile and 12 in/ft contain only exact numbers, which are considered to contain an infinite number of significant figures.

Example 1-4

Express 3.934 cubic yards in cubic inches.

Solution

We write down

$? \text{ in}^3 = 3.934 \text{ yd}^3$

and then multiply by the unit factors that convert yd^3 to in^3.

$? \text{ in}^3 = 3.934 \text{ yd}^3 \times \left(\dfrac{3 \text{ ft}}{1 \text{ yd}}\right)^3 \times \left(\dfrac{12 \text{ in}}{1 \text{ ft}}\right)^3$

$= 3.934 \text{ yd}^3 \times \left(\dfrac{27 \text{ ft}^3}{1 \text{ yd}^3}\right) \times \left(\dfrac{1728 \text{ in}^3}{1 \text{ ft}^3}\right)$

$= \underline{1.835 \times 10^5 \text{ in}^3}$

(calculator gives 183544.7)

The answer contains four significant figures because 3.934 yd^3 has four. The other members we used are exact. We could also have used

$\left(\dfrac{36 \text{ in}}{1 \text{ yd}}\right)^3 = \dfrac{46656 \text{ in}^3}{1 \text{ yd}^3}$

instead of the terms with a brace beneath them.

Cancellation of units gives solved problems a cluttered appearance. Therefore, we have not cancelled units in the remainder of the illustrative examples in this text. You may find it useful to cancel units with a colored pencil or pen as you work through the examples.

Conversions within the SI and the metric system are easy, because measurements of a particular kind are related to each other by powers of ten.

Example 1-5

The Ångstrom (Å) is a unit of length, 1×10^{-10} meter, that was defined to provide a convenient scale on which to express the radii of atoms (which are extremely small). The radius of a phosphorus atom is 1.10 Å. What is this distance expressed in centimeters and nanometers?

Solution

$? \text{ cm} = 1.10 \text{ Å} \times \dfrac{1 \times 10^{-10} \text{ m}}{1 \text{ Å}} \times \dfrac{1 \text{ cm}}{1 \times 10^{-2} \text{ m}}$

$= \underline{1.10 \times 10^{-8} \text{ cm}}$

$? \text{ nm} = 1.10 \text{ Å} \times \dfrac{1 \times 10^{-10} \text{ m}}{1 \text{ Å}} \times \dfrac{1 \text{ nm}}{1 \times 10^{-9} \text{ m}}$

$= \underline{0.110 \text{ nm}}$

All the unit factors used in this example contain only exact numbers.

Example 1–6

Assuming a phosphorus atom is spherical, calculate its volume in Å^3, cm^3, and nm^3. The volume of a sphere is $V = (\frac{4}{3})\pi\, r^3$. Refer to Example 1–5.

Solution

$$\underline{?}\ \text{Å}^3 = (\tfrac{4}{3})\pi(1.10\ \text{Å})^3 = \underline{5.58\ \text{Å}^3}$$

$$\underline{?}\ \text{cm}^3 = (\tfrac{4}{3})\pi(1.10 \times 10^{-8}\ \text{cm})^3$$
$$= \underline{5.58 \times 10^{-24}\ \text{cm}^3}$$

$$\underline{?}\ \text{nm}^3 = (\tfrac{4}{3})\pi(1.10 \times 10^{-1}\ \text{nm})^3$$
$$= \underline{5.58 \times 10^{-3}\ \text{nm}^3}$$

Example 1–7

How many millimeters are there in 1.39×10^4 meters?

Solution

$$\underline{?}\ \text{mm} = 1.39 \times 10^4\ \text{m} \times \frac{1000\ \text{mm}}{1\ \text{m}}$$

$$= \underline{1.39 \times 10^7\ \text{mm}}$$

Note that the unit factor 1000 mm/1 m contains two exact numbers.

Example 1–8

A sample of gold has a mass of 0.234 mg. What is its mass in g? in cg?

Solution

$$\underline{?}\ \text{g} = 0.234\ \text{mg} \times \frac{1\ \text{g}}{1000\ \text{mg}} = \underline{2.34 \times 10^{-4}\ \text{g}}$$

$$\underline{?}\ \text{cg} = 0.234\ \text{mg} \times \frac{1\ \text{cg}}{10\ \text{mg}}$$

$$= 0.0234\ \text{cg} \qquad \text{or} \qquad \underline{2.34 \times 10^{-2}\ \text{cg}}$$

Again, we have used unit factors that contain only exact numbers.

Example 1–9

How many square decimeters are there in 215 square centimeters?

Solution

$$\underline{?}\ \text{dm}^2 = 215\ \text{cm}^2 \times \left(\frac{1\ \text{dm}}{10\ \text{cm}}\right)^2$$

$$= 215\ \text{cm}^2 \times \left(\frac{1\ \text{dm}^2}{100\ \text{cm}^2}\right) = \underline{2.15\ \text{dm}^2}$$

Example 1–9 shows that a unit factor *squared* is still a unit factor, i.e.,

$$\left(\frac{1\ \text{dm}}{10\ \text{cm}}\right)^2 = \frac{1\ \text{dm}^2}{100\ \text{cm}^2} = 1.$$

Example 1–10

How many cubic centimeters are there in 8.34×10^5 cubic decimeters?

Solution

$$\underline{?}\ \text{cm}^3 = 8.34 \times 10^5\ \text{dm}^3 \times \left(\frac{10\ \text{cm}}{1\ \text{dm}}\right)^3$$

$$= 8.34 \times 10^5\ \text{dm}^3 \times \left(\frac{1000\ \text{cm}^3}{1\ \text{dm}^3}\right)$$

$$= \underline{8.34 \times 10^8\ \text{cm}^3}$$

Example 1–10 shows that a unit factor *cubed* is still a unit factor, i.e.,

$$\left(\frac{10 \text{ cm}}{1 \text{ dm}}\right)^3 = \frac{1000 \text{ cm}^3}{1 \text{ dm}^3} = 1.$$ We may generalize and say that a unit factor raised to

any power is still a unit factor.

Example 1–11

A common unit of energy is the erg. Convert 3.74×10^{-2} erg to the SI units of energy, joules and kilojoules. One erg is exactly 1×10^{-7} joule.

Solution

$$\underline{?} \text{ J} = 3.74 \times 10^{-2} \text{ erg} \times \frac{1 \times 10^{-7} \text{ J}}{1 \text{ erg}}$$

$$= \underline{3.74 \times 10^{-9} \text{ J}}$$

$$\underline{?} \text{ kJ} = 3.74 \times 10^{-9} \text{ J} \times \frac{1 \times 10^{-3} \text{ kJ}}{\text{J}}$$

$$= \underline{3.74 \times 10^{-12} \text{ kJ}}$$

Conversions between the English and metric or SI systems are conveniently done by the unit factor method. Several conversion factors were listed in Table 1–7. It may be helpful to remember one each for:

length	1 in = 2.54 cm (exact)
mass and weight	1 lb = 454 g (near sea level)
volume	1 qt = 0.946 L
or	1 L = 1.06 qt

Example 1–12

Express 1.0 mL in gallons.

Solution

We write

$$\underline{?} \text{ gal} = 1.0 \text{ mL}$$

and multiply by the appropriate factors.

$$\underline{?} \text{ gal} = 1.0 \text{ mL} \times \frac{1 \text{ L}}{1000 \text{ mL}} \times \frac{1.06 \text{ qt}}{1 \text{ L}} \times \frac{1 \text{ gal}}{4 \text{ qt}}$$

$$= \underline{2.6 \times 10^{-4} \text{ gal}}$$
(calculator gives 0.000265)

The fact that all other units cancel to give the desired unit, gallons, shows that we have used the correct unit factors. The factors, 1 L/1000 mL and 1 gal/4 qt, contain only exact numbers. The factor 1.06 qt/L contains three significant figures, while 1.0 mL contains only two; therefore the answer can contain only two significant figures.

Examples 1–1 through 1–12 show that multiplication by one or more unit factors changes the units and the number of units, but not the amount of whatever we are concerned with.

1–10   Density and Specific Gravity

The **density** of a substance is defined as the mass per unit volume,

$$\text{density} = \frac{\text{mass}}{\text{volume}} \qquad \text{or} \qquad D = \frac{M}{V} \qquad\qquad \text{[Eq. 1–2]}$$

Any unit or number can be transferred from the numerator of an expression to the denominator, or vice versa, merely by reversing the sign of its exponent (that is, $1/L = L^{-1}$).

Ethanol is often called ethyl alcohol or grain alcohol.

Densities may be used to distinguish between two substances or to assist in identifying a particular substance. They are usually expressed as g/cm^3 or g/mL for liquids and solids, and as g/L for gases. These units can also be expressed as $g\ cm^{-3}$, $g\ mL^{-1}$, and $g\ L^{-1}$, respectively.

Density is an *intensive property;* i.e., it does not depend upon the size of the sample. This is because density is constant for a given substance at a given temperature and pressure. The volume of two kilograms of ethanol at a given temperature and pressure is twice the volume of one kilogram of ethanol at the same temperature and pressure. Both samples of ethanol have the same density. Densities of a few substances are listed in Table 1–8.

TABLE 1–8   Densities of Common Substances

Densities are given at room temperature and *one atmosphere* pressure, average atmospheric pressure at sea level. Densities of solids and liquids change only slightly, but densities of gases change greatly, with changes in temperature and pressure.

Substance	Density, g/cm³	Substance	Density, g/cm³
hydrogen (gas)	0.000089	sand*	2.32
carbon dioxide (gas)	0.0019	aluminum	2.70
cork*	0.21	iron	7.86
oak wood*	0.71	copper	8.92
ethyl alcohol	0.79	lead	11.34
water	1.00	mercury	13.59
magnesium	1.74	gold	19.3
table salt	2.16		

* Cork, oak wood, and sand are common materials that have been included to provide familiar reference points. They are *not* pure elements or compounds as are the other substances listed here. High-grade white sand is mostly silicon dioxide, a compound.

Example 1–13

A 47.3 mL sample of liquid has a mass of 53.74 g. What is its density?

Solution

$$D = \frac{M}{V} = \frac{53.74 \text{ g}}{47.3 \text{ mL}} = \underline{1.14 \text{ g/mL}}$$

Example 1–14

If 100 g of the liquid described in Example 1–13 is needed for a chemical reaction, what volume of liquid would you use?

Solution

Recall that the density of the liquid is 1.14 g/mL.

$$D = \frac{M}{V} \qquad \text{so} \qquad V = \frac{M}{D} = \frac{100 \text{ g}}{1.14 \text{ g/mL}} = \underline{87.7 \text{ mL}}$$

Alternatively, we can use the unit factor method to solve the problem.

$$\underline{?} \text{ mL} = 100 \text{ g} \times \frac{1 \text{ mL}}{1.14 \text{ g}} = \underline{87.7 \text{ mL}}$$

The first method for solving the problem requires that one remember a formula, $D = M/V$. The second method doesn't.

Because specific gravity is defined as the *ratio* of two densities, it has no units, as illustrated in Example 1–15.

The **specific gravity** of a substance is the ratio of its density to the density of water.

$$\text{Sp. Gr.} = \frac{D_{\text{substance}}}{D_{\text{water}}}$$

[Eq. 1–3]

The density of water is 1.000 g/mL at 3.98°C, the temperature at which the density of water is greatest. However, variations in the density of water with changes in temperature are small enough that we may use 1.00 g/mL up to 25°C without introducing significant errors into our calculations.

Example 1–15

The density of table salt is 2.16 g/mL at 20°C. What is its specific gravity?

Solution

$$\text{Sp. Gr.} = \frac{D_{\text{salt}}}{D_{\text{water}}} = \frac{2.16 \text{ g/mL}}{1.00 \text{ g/mL}} = \underline{2.16}$$

The density and specific gravity of a substance are numerically equal near room temperature if density is expressed in g/mL (g/cm³), as Example 1–15 demonstrated.

At this point you need not be concerned if you do not know the chemical characteristics of an acid or a base. They will be described in Chapter 9.

Labels on commercial solutions of acids and bases give specific gravities and the percentage by mass of the acid or base present in the solution. From this information, the amount of acid or base present in a given volume of the solution can be calculated.

Example 1–16

Battery acid is 40.0% sulfuric acid, H_2SO_4, and 60.0% water by mass. Its specific gravity is 1.31. Calculate the mass of pure sulfuric acid, H_2SO_4, in 100.0 mL of battery acid.

Solution

We have demonstrated that density and specific gravity are numerically equal at 20°C because the density of water is 1.00 g/mL. Therefore, we may write

density = 1.31 g/mL

The solution is 40.0% H_2SO_4 and 60.0% H_2O by mass. From this information we may construct the desired unit factor.

$$\frac{40.0 \text{ g } H_2SO_4}{100.0 \text{ g soln}} \longleftarrow \begin{bmatrix} \text{because 100.0 g of solution} \\ \text{contains 40.0 g of } H_2SO_4 \end{bmatrix}$$

We can now solve the problem:

$$? \text{ g } H_2SO_4 = 100.0 \text{ mL soln} \times \frac{1.31 \text{ g soln}}{1 \text{ mL soln}}$$

$$\times \frac{40.0 \text{ g } H_2SO_4}{100.0 \text{ g soln}} = \underline{52.4 \text{ g } H_2SO_4}$$

We first used the density as a unit factor to convert the given volume of solution to mass of solution, and then used the percentage by mass to convert the mass of solution to mass of acid.

1–11 Temperature

In Section 1–1 you learned that heat is simply one form of energy. You also learned that the many different forms of energy can be interconverted, and that in chemical processes, chemical energy is converted to heat energy or vice versa. The amount of heat a chemical reaction or a physical change uses (endothermic) or gives off

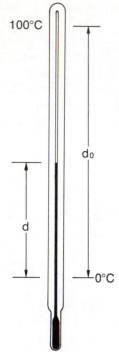

100°C

d_0

d

0°C

FIGURE 1–16 At 45°C, as read on a mercury-in-glass thermometer, d equals 0.45 d_0, where d_0 is the distance from the mercury level at 0°C to the level at 100°C.

x°C and xK refer to temperatures on the Celsius and Kelvin temperature scales, respectively.

(exothermic) can tell us a great deal about that reaction. (This will be discussed in Chapters 3 and 14.) It is important for us to be able to measure both the quantity and the intensity of heat.

Temperature *measures the intensity of heat,* the "hotness" or "coldness" of a body, and it is the property that determines the direction of heat flow. A piece of metal at 100°C feels hot to the touch, while an ice cube feels cold. Why? Because the temperature of the metal is higher, and that of the ice cube lower, than body temperature. As touching a hot surface and a cold surface demonstrate, *heat always flows spontaneously from a hotter body to a colder body*—never in the reverse direction.

Temperatures are commonly measured with mercury-in-glass thermometers. Mercury, more than most other substances, expands as its temperature rises. A mercury thermometer consists of a relatively large reservoir of mercury at the base of a glass tube, open to a very thin (capillary) column extending upward. As mercury expands in the reservoir, its movement up into the thin column is clearly visible.

Anders Celsius, a Swedish astronomer, developed the Celsius temperature scale, formerly called the centigrade temperature scale. When a well-made Celsius thermometer is placed in a beaker of crushed ice stirred with water, the mercury level stands at exactly 0°C, the lower reference point. In a beaker of water boiling at one atmosphere pressure, the mercury level stands at 100°C, the higher reference point. There are 100 evenly spaced calibration marks between these two mercury levels, and they correspond to an interval of 100 degrees between the melting point of ice and the boiling point of water at one atmosphere. Figure 1–16 shows how temperature marks between the reference points are measured.

Temperatures are frequently measured in the United States on the temperature scale devised by Gabriel Fahrenheit, a German instrument maker. On this scale the freezing and boiling points of water are defined as 32°F and 212°F, respectively.

In scientific work, temperatures are often expressed on the Kelvin (absolute) temperature scale. As we shall see in Section 11–5, the zero point of the Kelvin temperature scale is derived from the observed behavior of all matter. The Celsius and Fahrenheit temperature scales have no such rigorous basis.

Relationships among the three temperature scales are illustrated in Figure 1–17. The degree is the same "size" on the Celsius and Kelvin scales because there are 100 degrees between the freezing and boiling points of water in each case. Every Kelvin temperature is 273.15° above the corresponding Celsius temperature. Thus the relationship between the two scales is simply

$$\underline{?}\,K = (x°C + 273.15°C)\,\frac{1.0\,K}{1.0°C} \qquad \text{or} \qquad \underline{?}°C = (xK - 273.15\,K)\,\frac{1.0°C}{1.0\,K} \qquad \text{[Eq. 1–4]}$$

In the SI system, degrees Kelvin are abbreviated simply as K rather than °K, and are called kelvins. Comparing the Celsius and Fahrenheit scales, we find that the interval between the same reference points is 100°C and 180°F, so the Celsius degree is larger than the Fahrenheit degree. That is, it takes 1.8 Fahrenheit degrees to cover the same temperature interval as 1.0 Celsius degree. From this information, we can construct the unit factors

$$\frac{1.8°F}{1.0°C} \qquad \text{and} \qquad \frac{1.0°C}{1.8°F}$$

Additionally, the starting points of the two scales are different, so we *cannot* convert a temperature on one scale to a temperature on the other by just multiplying by the

FIGURE 1–17 The relationships among the Kelvin, Celsius (centigrade) and Fahrenheit temperature scales.

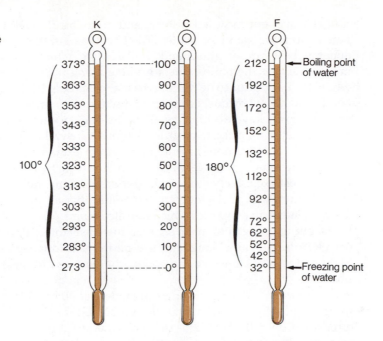

unit factor. To compensate for this, 32 Fahrenheit degrees must be added or subtracted to reach the zero point on the Celsius scale:

$$?°F = \left(x°C \times \frac{1.8°F}{1.0°C}\right) + 32°F \quad \text{and} \quad ?°C = \frac{1.0°C}{1.8°F}(x°F - 32°F) \qquad \text{[Eq. 1–5]}$$

Example 1–17

When the temperature reaches "100°F in the shade," it's hot. What is this temperature on the Celsius scale?

Solution

$$?°C = \frac{1.0°C}{1.8°F}(100°F - 32°F)$$

$$= \frac{1.0°C}{1.8°F}(68°F) = \underline{38°C}$$

Example 1–18

When the absolute temperature is 400 K, what is the Fahrenheit temperature?

Solution

First we convert K to °C, then °C to °F.

$$?°C = (400\ K - 273\ K)\frac{1.0°C}{1.0\ K} = 127°C$$

$$?°F = \left(127°C \times \frac{1.8°F}{1.0°C}\right) + 32°F = \underline{261°F}$$

Atoms, Molecules, and Ions — An Introduction (How We Interpret What We've Seen)

Since chemistry is concerned primarily with the study of the nature of matter, let us now turn our attention to developing a picture of the smallest particles of elements and compounds — atoms, molecules, and ions.

Throughout recorded history, and undoubtedly well before, man has wondered about the nature of matter. The Greek philosophers were among the first to document their observations and "theories" on natural phenomena. We can easily imagine one of the philosophers walking along the seashore and observing that while the sea appears to be continuous matter, as does the beach from a distance, the beach underfoot obviously consists of discrete (discontinuous) particles of sand. He might logically have wondered if all matter that appears to be continuous might not actually consist of discrete particles. Unfortunately, the Greeks did not have available the devices we now have to allow closer examination of the composition of matter. Their minds were very active, but they were forced to restrict testing of their ideas almost exclusively to mental exercise and logic, rather than experiment.

Around 400 BC, Democritus suggested that all matter is composed of tiny, discrete, indivisible particles which he called atoms. His ideas were rejected for 2000 years, but they began to make sense in the late eighteenth century. The term **atom** comes from the Greek language and means "not divided" or "indivisible." Both Plato and Aristotle rejected the notion of atoms, as did most other Greek philosophers.

Before we examine the origin of modern atomic theory, let us consider some of the evidence, i.e., experimental observations, that pointed scientists in the right direction. Robert Boyle (1627–1691) was a very strong promoter of the experimental approach. He performed experiments carefully and kept accurate records so that his experimental results could be verified by others. John Dalton (1766–1844) was scornful of those who proposed theories without subjecting them to experimental verification. The careful experimental approach advocated by Boyle and Dalton is now accepted throughout the scientific world.

1–12   The Laws of Chemical Combination

Several fundamental laws of nature, recognized by about 1800, provided the basis for the first scientifically based atomic theory. Remember that these chemical laws, like all scientific laws, are just statements that summarize our observations.

1. The **Law of Conservation of Matter** states that *there is no observable change in the mass of matter during an ordinary chemical reaction.*

2. As we noted in Section 1–5, some pure substances, called *compounds,* are composed of two or more different elements in a fixed ratio. This generalization, based on many such observations, was recognized by Joseph Proust in 1799, and is known as the **Law of Constant Composition** (or as the *Law of Definite Proportions*). It may be stated as follows: *Different samples of any pure compound contain the same elements in the same proportions by mass.*

A convenient way to express the composition of a sample is to state the percentage of the mass of the sample that is due to each of the elements present; this information is referred to as the "percent composition" of the sample. Example 1–19 illustrates how data on which this law is based could be obtained.

Percent composition is always given on a mass basis.

Example 1–19

One of the earliest chemical reactions to be carefully studied was the decomposition of "red calx of mercury," a reddish powder. (The study of the formation of this compound, when mercury was heated in air, led the great French chemist Antoine Lavoisier in 1789 to propose the Law of Conservation of Matter.) When 50.0 grams of pure red calx of mercury is heated, it decomposes to give 3.7 grams of oxygen gas and 46.3 grams of the metal mercury.

What is the percent composition of red calx of mercury?

Solution

Since the observed masses of the mercury and oxygen add up to give the total mass of the sample, we can see that there can be no other elements present in this compound. That is,

3.7 g O + 46.3 g Hg = 50.0 g total

By definition, the percentage by mass of each component is the mass of that component divided by the total mass, multiplied by 100%. So

$$\% \text{ O} = \frac{3.7 \text{ g O}}{50.0 \text{ g total}} \times 100\% = \underline{7.4\% \text{ O}}$$

and

$$\% \text{ Hg} = \frac{46.3 \text{ g Hg}}{50.0 \text{ g total}} \times 100\% = \underline{92.6\% \text{ Hg}}$$

Analysis of many different pure samples of this compound always leads to the same results (within experimental error).

An alternative method of determining percent composition of a compound involves measuring experimentally the masses of the two (or more) elements that combine to form the compound. Any sample of a pure compound, when analyzed for the amounts of the elements present, is observed to have the same composition as any other sample of that same pure compound. Some further illustrations of the elemental compositions of compounds are

carbon dioxide: 27.3% C, 72.7% O

water: 11.1% H, 88.9% O

sulfuric acid: 2.1% H, 32.7% S, 65.2% O

Observations such as these, with no exceptions (so long as we are careful to deal with pure substances), led to the formulation of the Law of Constant Composition. Once we have *determined* the percent composition of any substance, we can use it as shown in Example 1–20, since we believe that the Law of Constant Composition will always apply.

Example 1–20

By chemical analysis, we determine the percent composition of the compound carbon dioxide to be 27.3% carbon and 72.7% oxygen. (a) What mass of oxygen would combine exactly with 35.0 g carbon to form this compound? (b) What mass of carbon dioxide would be formed in part (a)?

Solution

The known percent composition, together with the Law of Constant Composition, tells us that in *any sample* of carbon dioxide, 27.3/100 of the mass is due to carbon, while the other 72.7/100 of the mass is due to oxygen. For instance, a 100.0 g sample would contain 27.3 g C and 72.7 g O. Another way to describe this is as the ratio of the mass of oxygen to the mass of carbon.

$$\frac{\text{mass O}}{\text{mass C}} = \frac{72.7 \text{ g O}}{27.3 \text{ g C}}$$

Note that this ratio (or its reciprocal) can be considered as a *unit factor* for the compound carbon dioxide. Using this unit factor, we can answer part (a) of the question:

(a) $$? \text{ g O} = 35.0 \text{ g C} \times \frac{72.7 \text{ g O}}{27.3 \text{ g C}} = \underline{93.2 \text{ g O}}$$

(b) According to the Law of Conservation of Matter, the total mass of compound formed must be equal to the sum of the masses of the elements in the compound. (Can you see that this is an equivalent statement of the law?) Thus,

Total mass of compound = mass C + mass O
= 35.0 g C + 93.2 g O
= <u>128.2 g carbon dioxide</u>

Notice that the solution to this problem was based *entirely* on the observed composition of the compound and a belief that the Laws of Constant Composition and Conservation of Matter apply.

3. A third type of observation deals with the fairly common existence of more than one compound containing the same set of elements, but in different proportions. For instance, there are two different compounds consisting (only) of the elements carbon and oxygen. They have the following percent compositions:

carbon monoxide: 42.9% C, 57.1% O

  carbon dioxide: 27.3% C, 72.7% O

Let us express the composition of each of these compounds as the mass of O present per gram of C. Carbon *monoxide* contains

$$\frac{57.1 \text{ g O}}{42.9 \text{ g C}} \quad \text{or} \quad \frac{1.33 \text{ g O}}{1.00 \text{ g C}}$$

In carbon *dioxide* there are

$$\frac{72.7 \text{ g O}}{27.3 \text{ g C}} \quad \text{or} \quad \frac{2.66 \text{ g O}}{1.00 \text{ g C}}$$

Now we see that the amount of oxygen present for the *same amount* of carbon, one gram, is in the ratio 1.33 g : 2.66 g, or the same ratio as the integers 1 : 2.

As a second illustration, let us consider two of the compounds formed between sulfur and oxygen. These have the composition

sulfur dioxide: 50.1% S, 49.9% O

sulfur trioxide: 40.1% S, 59.9% O

As before, let us calculate the mass of O per gram S for each compound. For sulfur dioxide, this is

$$\frac{49.9 \text{ g O}}{50.1 \text{ g S}} \quad \text{or} \quad \frac{0.996 \text{ g O}}{1.00 \text{ g S}}$$

For sulfur trioxide, we obtain

$$\frac{59.9 \text{ g O}}{40.1 \text{ g S}} \quad \text{or} \quad \frac{1.49 \text{ g O}}{1.00 \text{ g S}}$$

Comparing the amount of oxygen present in these compounds for the *same amount* of sulfur, one gram, we obtain the ratio 0.996 : 1.49 or 1.0 : 1.5. Once again, this can be expressed as the ratio of two small integers, 2 : 3.

A study of many such sets of compounds consisting of the same elements always gives similar results. The statement of this general result is the **Law of Multiple Proportions:** *The masses of one element that can combine chemically with a fixed mass of another element are in a small whole-number ratio.*

1–13   Dalton's Atomic Theory

The development of modern ideas about the existence and the nature of atoms was begun in 1803 by an English schoolteacher named John Dalton. In an attempt to explain why matter behaves in such simple and systematic ways as those expressed in the laws of chemical combination (Section 1–12), he summarized and supplemented the nebulous concepts of early philosophers and scientists. Taken together, these ideas form the core of **Dalton's Atomic Theory,** one of the highlights of scientific thought. They were published in 1808. In condensed form, Dalton's postulates may be stated this way:

1. An element is composed of extremely small *indivisible* particles called atoms.
2. All atoms of a given element have identical properties, which differ from those of other elements.
3. Atoms cannot be created, destroyed, or transformed into atoms of another element.
4. Compounds are formed when atoms of different elements combine with each other in simple numerical ratios.
5. The relative numbers and kinds of atoms are constant in a given compound.

Dalton believed that atoms were solid indivisible spheres, an idea we now reject, but he showed remarkable insight into the nature of matter and its interactions. Some of his postulates could not be verified (or refuted) experimentally at the time, but all were in accord with the experimental observations of his day. As we shall see, even with their shortcomings, Dalton's postulates provided a framework that could be modified and expanded, and Dalton is thus considered the father of modern atomic theory.

1–14   Atoms, Molecules, and Formulas

Recall that in Section 1–5 we introduced the symbols that represent the elements. The symbol for a given element may represent not only the name of that element, but also one atom of the element or even a fixed number of atoms of the element. Let us now introduce some ideas and terminology that will be important in our discussion of compounds and elements.

An **atom** is the smallest particle of an element. Historically, an atom has been defined as the smallest particle of an element that can enter into a chemical combination. As we shall see, this historical definition leaves something to be desired, because no chemical reactions of the lighter noble gases (helium, neon, and argon) are known. A **molecule** is the smallest particle of an element or compound that can have a stable independent existence. You may now be wondering "What is the difference between an atom and a molecule?" Actually, an atom and a molecule are sometimes the same. For example, certain elements (especially the noble gases) are capable of existence as single atoms, and therefore their atoms can be regarded as molecules. See Figure 1–18. A molecule that consists of one atom is called a *monatomic* molecule.

The radius of a calcium atom is only 1.97×10^{-8} cm and its mass is 6.66×10^{-23} g.

Statement 3 is true for ordinary chemical reactions. However, it is not true for *nuclear* reactions (Chapter 24).

He

Ne

Ar

Kr

Xe

Rn

FIGURE 1–18
Relative sizes of atoms of the noble gases.

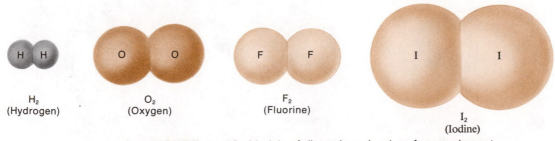

H₂
(Hydrogen)

O₂
(Oxygen)

F₂
(Fluorine)

I₂
(Iodine)

FIGURE 1–19   Models of diatomic molecules of some elements.

FIGURE 1–20 (a) A model of the P_4 molecule of white phosphorus. The radius of the P atom is 0.110 nm. (b) A model of the S_8 molecule of rhombic sulfur. The radius of the S atom is 0.104 nm.

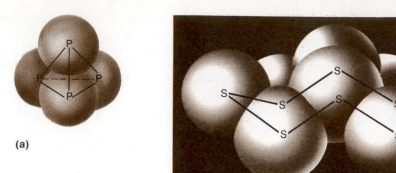

(a)

(b)

An atom of oxygen, on the other hand, cannot exist alone at room temperature and atmospheric pressure. An atom of oxygen can have a stable existence only if it combines with another atom or molecule. The molecule of oxygen we are all familiar with is made up of two atoms of oxygen; it is a *diatomic* molecule. Hydrogen, nitrogen, fluorine, chlorine, bromine, and iodine are also examples of diatomic molecules. See Figure 1–19.

Other elements exist as more complex molecules. Phosphorus exists as molecules consisting of four atoms, while sulfur exists as eight-atom molecules at ordinary temperatures and pressures. Molecules that are composed of more than two atoms are called *polyatomic* molecules. See Figure 1–20.

Molecules of *compounds* are composed of more than one kind of atom. A water molecule, for example, consists of two atoms of hydrogen and one atom of oxygen. A molecule of methane consists of one carbon atom and four hydrogen atoms. The shapes of these and a few other molecules are shown in Figure 1–21.

Methane is the principal component of natural gas.

You should now be aware that atoms are the components of molecules, and that molecules are the components of elements or compounds (the compound water is made up of water molecules, which are made up of hydrogen and oxygen atoms). We are able to see samples of compounds and elements consisting of large numbers of atoms and molecules, but individual atoms and molecules are too small to be seen even with the most powerful optical microscope. To give you an idea of the size of a water molecule, it would take 100 million of them to make a row one inch long.

The **formula** *for a substance shows its chemical composition.* By composition we mean the elements present as well as the ratio in which atoms occur in the substance.

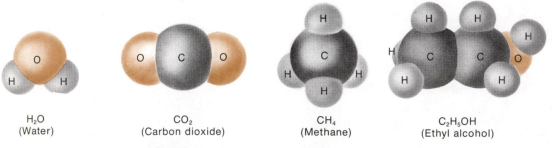

H_2O
(Water)

CO_2
(Carbon dioxide)

CH_4
(Methane)

C_2H_5OH
(Ethyl alcohol)

FIGURE 1–21 Models of molecules of some compounds.

The formula for a single atom is the same as the symbol for the element. Thus, Na represents a single sodium atom. It is unusual to find such isolated atoms in nature, with the exception of the noble gases (He, Ne, Ar, Kr, Xe, and Rn). A subscript following the symbol of an element indicates the number of atoms of that element in a molecule; for instance, F_2 indicates a molecule containing two fluorine atoms, and P_4 represents four phosphorus atoms in a molecule.

Some elements exist in more than one elemental form. Familiar examples include (1) oxygen, found as O_2 molecules, and ozone, found as O_3 molecules, and (2) two different crystalline forms of carbon, diamond and graphite. *Different forms of the same element in the same physical state* are called **allotropic modifications** or **allotropes.**

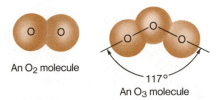

An O_2 molecule   117°
An O_3 molecule

Most compounds, we have seen, contain two or more elements in chemical combination in fixed proportions (although some do not, as we shall explain in Section 2–4). Hence, each molecule of hydrogen chloride, HCl, contains one atom of hydrogen and one atom of chlorine; each molecule of carbon tetrachloride, CCl_4, contains one carbon atom and four chlorine atoms; and a molecule of propane, C_3H_8, contains three carbon atoms and eight hydrogen atoms. An aspirin molecule, $C_9H_8O_4$, contains nine carbon atoms, eight hydrogen atoms, and four oxygen atoms.

Some groups of atoms behave chemically as single entities. For instance, one nitrogen atom and two oxygen atoms may combine into a *nitro* group that can form a part of a molecule. In formulas of compounds containing two or more of the same group, the group is enclosed in parentheses to show its presence. Thus, 2,4,6-trinitrotoluene (often abbreviated TNT) contains three *nitro* groups, and its formula is $C_7H_5(NO_2)_3$. When you count up the number of atoms in this molecule, you must multiply the numbers of nitrogen and oxygen atoms by 3. There are *seven* carbon atoms, *five* hydrogen atoms, *three* nitrogen atoms, and *six* oxygen atoms in a molecule of TNT. Figure 1–22 shows a space-filling model of one TNT molecule.

FIGURE 1–22 A space-filling model of a $C_7H_5(NO_2)_3$ molecule (TNT).

1–15   The Atomic Explanation of the Laws of Chemical Combination

Let us now illustrate how the assumptions of Dalton's atomic theory are consistent with the laws of chemical combination. You should verify that you can apply this reasoning to other compounds discussed in this chapter.

1.   The Law of Conservation of Matter   According to Dalton's theory, the observed combination of elements in chemical reactions is the result of the combination of the atoms of those elements. However, since atoms are constant in their identity, and are not created, transformed, or destroyed, any chemical reaction is just the rearrangement of the same number of "pieces" (i.e., atoms), so that the total

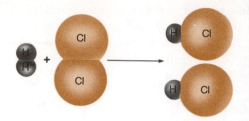

FIGURE 1–23 Illustration of the Law of Conservation of Matter. This is a molecular representation of a reaction in which the atoms present in one H_2 molecule and one Cl_2 molecule are rearranged to form two HCl molecules. Since the same number of atoms of each kind are present before and after the rearrangement (reaction), the *total* mass remains the same. Note that this would still be true even if there were extra H_2 molecules (or Cl_2 molecules) present in the mixture that did not take part in the reaction.

mass remains the same. This is illustrated in Figure 1–23 for a reaction in which hydrogen, H_2, and chlorine, Cl_2, combine to form hydrogen chloride, HCl.

2. The Law of Constant Composition Let us consider the compound carbon monoxide. According to Dalton's theory, this consists of atoms of the two elements, carbon and oxygen, combined in a simple numerical ratio. Suppose that this ratio is 1 : 1 (it is!); that is, there is *one* oxygen atom for every *one* carbon atom, as expressed by the formula CO. Thus, the smallest amount of this compound that could exist is one *molecule,* composed of one carbon atom and one oxygen atom (Figure 1–24). Dalton's theory says that all carbon atoms have the same mass, and that all oxygen atoms have the same mass which is different from that of carbon atoms. We shall see in Chapter 2 that, on an arbitrary scale, a carbon atom has a mass of 12 units while an oxygen atom has a mass of very nearly 16. Thus, some percentage of the mass of the single CO molecule is due to carbon, and the rest is due to oxygen. Numerically, these values would be

$$\frac{12}{12 + 16} \times 100\% = 42.9\% \text{ carbon} \quad \text{and} \quad \frac{16}{12 + 16} \times 100\% = 57.1\% \text{ oxygen}$$

Expressed another way, this molecule contains 16 mass units of oxygen/12 mass units of carbon.

Now think of a collection of many CO molecules, each exactly like the one we have just described. The total mass of carbon in the collection is equal to the number of molecules multiplied by the mass of a single carbon atom. Similarly, the total

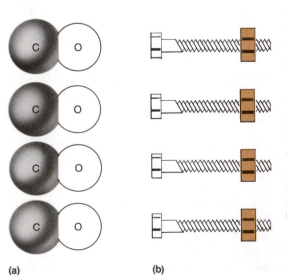

(a)

(b)

FIGURE 1–24 Illustration of the Law of Constant Composition. (a) A representation of several molecules of carbon monoxide (formula CO), all identical. Each molecule has atoms present in the ratio 1 O atom/1 C atom. Therefore, any sample of this substance has equal numbers of C and O atoms. Each C atom weighs 12 units and each O atom weighs 16 units. Thus the *mass ratio* is 16 mass units O to 12 mass units C, no matter how many molecules are present in the sample. (b) A collection of ''molecules'' with the formula $(bolt)_1(nut)_1$, all identical. Each ''molecule'' has components present in the ratio 1 bolt/1 nut. Any collection of these combinations has equal numbers of bolts and nuts. Each bolt weighs 4 grams and each nut weighs 1 gram. Thus the mass ratio is 4 mass units of bolts/1 mass unit of nuts, no matter how many $(bolt)_1(nut)_1$ combinations we have.

mass of oxygen in the collection is equal to the number of molecules multiplied by the mass of a single oxygen atom. The ratio of total mass of oxygen/total mass of carbon is thus

$$\frac{(\text{number of molecules})(\text{mass of oxygen atom})}{(\text{number of molecules})(\text{mass of carbon atom})} = \frac{\text{mass of oxygen atom}}{\text{mass of carbon atom}}$$

$$= \frac{16 \text{ mass units oxygen}}{12 \text{ mass units carbon}}$$

which is the *same* as the ratio in a single carbon monoxide molecule, regardless of how many molecules are present. This is true for any pure sample of any compound, and is the atomic explanation of the Law of Constant Composition.

As an analogy, suppose we have some number of identical bolts, each with a mass of 4 grams, and the same number of identical nuts, each with a mass of 1 gram. Suppose we thread one nut onto each bolt. A single $(\text{bolt})_1(\text{nut})_1$ combination has a total mass of 5 grams; of this, 4 grams (or 80% of the total mass of the combination) is due to the bolt, and the other 20% is due to the nut. This could be expressed as 4 mass units bolt/1 mass unit nut. If we take any number of identical $(\text{bolt})_1(\text{nut})_1$ combinations and "decompose" them, we would find that the collection has the same percent composition as does one $(\text{bolt})_1(\text{nut})_1$ combination. For instance, three $(\text{bolt})_1(\text{nut})_1$ combinations contain $3 \times (4 \text{ g}) = 12$ grams of bolts and $3 \times (1 \text{ g}) = 3$ grams of nuts, or 15 grams total. Of this, $\frac{12 \text{ g}}{15 \text{ g}} \times 100\% = 80\%$ is due to the bolts, and $\frac{3 \text{ g}}{15 \text{ g}} \times 100\% = 20\%$ is due to the nuts—exactly the same percent composition as in the single combination.

---

3. The Law of Multiple Proportions   Let us now consider a second carbon–oxygen compound, carbon dioxide, which has the formula CO_2, i.e., each molecule contains *two* oxygen atoms and *one* carbon atom (Figure 1–25). In CO_2 there are twice as many oxygen atoms per carbon atom as in the previous compound, CO. Numerically, there are now (2×16) mass units of oxygen/12 mass units of carbon. Compare this with the ratio for CO, which is 16 mass units of oxygen/12 mass units of carbon. For the same fixed mass of carbon (12 mass units), the masses of oxygen present in these two compounds are in a simple numerical ratio, $2:1$. This is the explanation, at the atomic level, of the kinds of experimental results summarized in the *Law of Multiple Proportions*.

To illustrate this with nuts and bolts, let us consider a second combination that has *two* nuts for every *one* bolt, or $(\text{bolt})_1(\text{nut})_2$. Now there are twice as many nuts per bolt, and hence twice the mass of nuts for the same mass of bolts. Numerically, there would be 4 grams bolts/$(2 \times 1 \text{ gram})$ nuts. This mass ratio is one half of the 4 grams bolts/1 gram nuts observed for the other "compound" $(\text{bolt})_1(\text{nut})_1$.

It is important to realize that Dalton's theory attempts to explain facts that are *observed*, and that are summarized in the laws of chemical combination. These

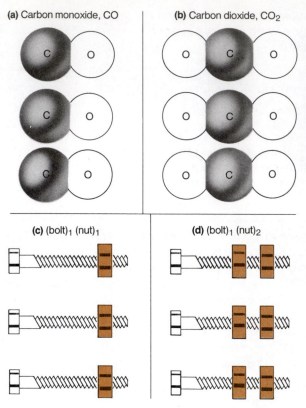

(a) Carbon monoxide, CO

(b) Carbon dioxide, CO_2

(c) $(bolt)_1 (nut)_1$

(d) $(bolt)_1 (nut)_2$

FIGURE 1–25 Illustration of the Law of Multiple Proportions. (a) Representation of several molecules of carbon monoxide, CO, all identical. (b) Representation of several molecules of carbon dioxide, CO_2, all identical. In CO_2 there are twice as many O atoms per C atom as in CO. Thus, in CO_2 the mass of O for a fixed mass of C is twice what it is in CO. (c) Several identical combinations of $(bolt)_1(nut)_1$. (d) Several identical combinations of $(bolt)_1(nut)_2$.

explanations are in terms of entities (atoms) whose existence and properties cannot be proved by direct observation. However, the resulting concepts are so successful in explaining such a broad range of chemical behavior that we accept the fundamental ideas of the atomic theory as correct.

1–16   Ions and Ionic Compounds

It is **absolutely necessary** to include the charge on an ion when writing the formula for the ion. The absence of a number before the sign is understood to mean *one* unit of charge.

So far we have cited only compounds that exist as discrete molecules. Some compounds such as sodium chloride, NaCl, consist of an extended array of ions, as depicted in Figure 1–26. An **ion** *is an atom or group of atoms that carries an electrical charge.* Ions that possess a *positive* charge, such as the sodium ion, Na^+, are called **cations.** Those carrying a *negative* charge, such as the chloride ion, Cl^-, are called **anions.** The superscript sign to the right of the element's symbol indicates whether the charge is positive or negative; if a number appears before the sign, as in Ca^{2+}, it indicates the number of units of charge on the ion (no number means that "1" is understood).

The positive charges are associated with subatomic particles called *protons* (Chapter 4).

As we shall see in Chapter 4, our present picture of the atom is that it consists of a very small, very dense, positively charged *nucleus* surrounded by a diffuse distribution of negatively charged particles called *electrons.* The number of positive charges in the nucleus defines the identity of the element to which the atom corresponds. Electrically neutral atoms contain the same number of electrons

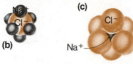

FIGURE 1–26 The arrangement of ions in NaCl. (a) A crystal of sodium chloride consists of an extended array that contains equal numbers of sodium ions, represented by small spheres, and chloride ions, represented by large spheres. (b) Within the crystal, each chloride ion is surrounded by six sodium ions. (c) Within the crystal, each sodium ion is surrounded by six chloride ions.

outside the nucleus as there are positive charges (protons) within the nucleus. Ions are formed when neutral atoms lose or gain electrons. An Na^+ ion is formed when a sodium atom loses one electron, and a Cl^- ion is formed when a chlorine atom gains one electron. **Polyatomic ions** *are groups of atoms that bear an electrical charge.* Examples include the ammonium ion, NH_4^+, the sulfate ion, SO_4^{2-}, and the nitrate ion, NO_3^-.

All **ionic compounds** are electrically neutral (have a net charge of zero), and this must be reflected in the formulas for the compounds. Thus sodium nitrate, $NaNO_3$, contains Na^+ and NO_3^- ions in a 1:1 ratio. Calcium bromide, $CaBr_2$, contains Ca^{2+} and Br^- ions in a 1:2 ratio, because it takes two Br^- ions to counterbalance the charge on one Ca^{2+} ion. Aluminum sulfate, $Al_2(SO_4)_3$, contains Al^{3+} ions and SO_4^{2-} ions in a 2:3 ratio so that the total positive charge (6) is equal to the total negative charge (6).

Reference to the ionic crystal lattice of NaCl (Figure 1–26) shows that it is impossible to identify a particular pair of Na^+ and Cl^- ions as being a "molecule." In fact, within the crystal (though not on the surface) each Na^+ ion is surrounded at equal distances by six Cl^- ions and each Cl^- ion is likewise surrounded by six Na^+ ions. The Na^+ and Cl^- ions are present in a 1:1 ratio, and this is indicated by the formula NaCl. Since we cannot identify "molecules" of ionic substances, we prefer not to use terms such as "a molecule of NaCl." Instead, we refer to a **formula unit** of NaCl, which implies one Na^+ and one Cl^- ion in combination. It is also acceptable to refer to a molecule of a non-ionic (molecular) compound as a formula unit.

The subscript 3 in the formula $Al_2(SO_4)_3$ indicates three SO_4^{2-} ions per formula unit.

The general term *formula unit* can apply to molecular or ionic compounds, whereas the more specific term *molecule* can apply only to elements and compounds that exist as discrete molecules.

TABLE 1–9   Formulas, Ionic Charges, and Names for Some Common Ions

A. Common Cations (Positive Ions)			B. Common Anions (Negative Ions)		
Formula	*Charge*	*Name*	*Formula*	*Charge*	*Name*
Na^+	1+	sodium ion	F^-	1−	fluoride ion
K^+	1+	potassium ion	Cl^-	1−	chloride ion
NH_4^+	1+	ammonium ion	Br^-	1−	bromide ion
Ag^+	1+	silver ion	I^-	1−	iodide ion
			OH^-	1−	hydroxide ion
Mg^{2+}	2+	magnesium ion	CH_3COO^-	1−	acetate ion
Ca^{2+}	2+	calcium ion	NO_3^-	1−	nitrate ion
Cu^{2+}	2+	cupric ion or copper(II) ion	NO_2^-	1−	nitrite ion
Fe^{2+}	2+	ferrous ion or iron(II) ion	O^{2-}	2−	oxide ion
			S^{2-}	2−	sulfide ion
			SO_4^{2-}	2−	sulfate ion
Fe^{3+}	3+	ferric ion or iron(III) ion	CO_3^{2-}	2−	carbonate ion
Al^{3+}	3+	aluminum ion	PO_4^{3-}	3−	phosphate ion

TABLE 1–10 Formulas and Names of Some Common Molecular Compounds

Formula	Name	Formula	Name
HCl	hydrogen chloride (or hydrochloric acid if dissolved in water)	SO_2	sulfur dioxide
		SO_3	sulfur trioxide
		CO	carbon monoxide
H_2SO_4	sulfuric acid	CO_2	carbon dioxide
HNO_3	nitric acid	CH_4	methane
CH_3COOH	acetic acid	C_2H_6	ethane
H_3PO_4	phosphoric acid	C_3H_8	propane
NH_3	ammonia (or aqueous ammonia if dissolved in water)		

1–17 Names and Formulas of a Few Common Ions and Substances

Throughout your study of chemistry you will have many occasions to refer to a compound or ion by name. We shall present and illustrate a set of rules for naming compounds systematically in Chapter 6, after the appropriate foundation has been laid. For now, we provide in Tables 1–9 and 1–10 the names and formulas of some species commonly encountered in the laboratory.

Chemists use *species* as a general term to refer to atoms, ions, molecules, etc.

The names of ionic compounds consist of the name of the positive ion followed by the name of the negative ion, as illustrated in Example 1–21.

Example 1–21

Write the formula and an acceptable name of the ionic compound consisting of (a) K^+ and I^- ions, (b) Ca^{2+} and CO_3^{2-} ions, (c) Zn^{2+} and NO_3^- ions, (d) Fe^{3+} and O^{2-} ions.

Solution

All compounds must be electrically neutral, and this must be reflected in the relative numbers of cations and anions in a formula unit. The correct formulas and names are (a) KI, potassium iodide, (b) $CaCO_3$, calcium carbonate, (c) $Zn(NO_3)_2$, zinc nitrate, and (d) Fe_2O_3, ferric oxide or iron(III) oxide.

Key Terms

Accuracy   how closely a measured value agrees with the correct value.

Allotropic modifications (allotropes)   different forms of the same element in the same physical state.

Anion   an ion that carries a negative charge.

Atom   the smallest particle of an element.

Calorie   the amount of heat required to raise the temperature of one gram of water from 14.5°C to 15.5°C. A unit of energy equal to 4.184 joules.

Cation   an ion that carries a positive charge.

Chemical change   a change in which one or more different substances are formed.

Chemical property   see *Properties.*

Compound   a pure substance composed of two or

more elements in fixed proportions. Compounds can be decomposed into their constituent elements.

Dalton's Atomic Theory theory based on the idea that all matter is composed of extremely small indivisible particles called atoms.

Density mass per unit volume, $D = M/V$.

Element a pure substance that cannot be decomposed into simpler substances by chemical means.

Endothermic describes processes that absorb heat energy.

Energy the capacity to do work or transfer heat.

Exothermic describes processes that release heat energy.

Extensive property a property that depends on the amount of material in a sample.

Formula combination of symbols that indicates the chemical composition of a substance.

Formula unit the smallest repeating unit of a substance, the molecule for non-ionic substances.

Heat a form of energy that flows between two samples of matter because of their difference in temperature.

Heterogeneous mixture a mixture that is not uniform throughout.

Homogeneous mixture a mixture that has uniform composition and properties throughout.

Intensive property a property that does not depend upon the amount of material in a sample.

Ion an atom or group of atoms that carries an electrical charge.

Ionic compound compound consisting of ions.

Joule a unit of energy in the SI system. One joule is 0.2390 calorie.

Kinetic energy energy that matter possesses by virtue of its motion.

Law of Conservation of Energy energy cannot be created or destroyed; it may be changed from one form to another.

Law of Conservation of Matter there is no detectable change in the quantity of matter during an ordinary chemical reaction.

Law of Conservation of Matter and Energy the total amount of matter and energy available in the universe is fixed.

Law of Constant Composition different samples of a pure compound always contain the same elements in the same proportions by mass; also called Law of Definite Proportions.

Law of Definite Proportions see *Law of Constant Composition.*

Law of Multiple Proportions the masses of one element that can combine chemically with a fixed mass of another element are in a small whole-number ratio.

Mass a measure of the amount of matter in an object. Mass is usually measured in grams or kilograms.

Matter anything that has mass and occupies space.

Mixture a sample of matter composed of two or more substances, each of which retains its identity and properties.

Molecule the smallest particle of an element or compound capable of a stable, independent existence.

Physical change a change in which a substance changes from one physical state to another, but no substances with different chemical compositions are formed.

Physical property see *Properties.*

Polyatomic ion a group of atoms that bears an electrical charge.

Potential energy energy that matter possesses by virtue of its position, condition, or composition.

Precision how closely repeated measurements of the same quantity agree with each other.

Properties characteristics that describe samples of matter. Chemical properties are exhibited as matter undergoes chemical changes. Physical properties are exhibited by matter with no changes in chemical composition.

Significant figures digits that indicate the precision of measurements; digits of a measured number that have uncertainty only in the last digit.

Specific gravity the ratio of the density of a substance to the density of water.

Substance any kind of matter all specimens of which have the same chemical composition and physical properties.

Symbol a letter or group of letters that represents (identifies) an element.

Temperature a measure of the intensity of heat, i.e., the hotness or coldness of a substance.

Unit factor a factor in which the numerator

and denominator are expressed in different units but represent the same or equivalent amounts. Multiplying by a unit factor is the same as multiplying by 1.

Weight a measure of the gravitational attraction of the earth for a body.

Work the application of a force to move an object through a distance.

Exercises

Basic Ideas

1. What is chemistry?
2. Define the following terms clearly and concisely. Illustrate each with a specific example.
    (a) matter   (b) mass
    (c) energy   (d) potential energy
    (e) kinetic energy (f) heat energy
    (g) exothermic   (h) endothermic process
        process
3. State the following laws and provide an illustration of each.
    (a) the Law of Conservation of Matter
    (b) the Law of Conservation of Energy
    (c) the combined Law of Conservation of Matter and Energy
4. (a) Write the Einstein equation.
    (b) State the meaning of the Einstein equation in words.
5. Is the following statement true or false? Matter and energy are different forms of a single entity. Why is it true or false?
6. List the three states of matter and some characteristics of each. How are they alike? different?
7. Distinguish between the following pairs of terms and provide two specific examples to illustrate each term.
    (a) chemical properties and physical properties
    (b) chemical changes and physical changes
8. Are the following chemical properties or physical properties? Why?
    (a) the melting point of lead
    (b) hardness
    (c) color of a solid
    (d) color of a flame
    (e) ability to burn in the air
9. Are the following chemical changes or physical changes? Why?
    (a) melting of lead
    (b) burning of gasoline
    (c) rusting of iron

    (d) production of light by a glowworm
    (e) emission of light by an electric light bulb
    (f) emission of light by a whale-oil lamp
10. Which of the following processes are exothermic? endothermic? Why?
    (a) combustion   (b) freezing water
    (c) melting ice   (d) boiling water
    (e) condensing steam
11. Define the following terms clearly and concisely. Provide two specific illustrations of each.
    (a) substance   (b) mixture
    (c) element   (d) compound
12. Classify each of the following as an element, a compound, or a mixture. Justify your classification.
    (a) gold   (b) bronze
    (c) iron   (d) steel
    (e) popcorn   (f) milk
    (g) sawdust   (h) sour cream
    (i) gasoline   (j) natural gas
    (k) shampoo   (l) table salt
13. Write symbols for the following elements.
    (a) silver   (b) calcium
    (c) iron   (d) iodine
    (e) neon   (f) phosphorus
14. Write symbols for the following elements.
    (a) aluminum   (b) chlorine
    (c) hydrogen   (d) potassium
    (e) oxygen   (f) lead
15. Write the name of each of the following elements.
    (a) Br   (b) Cu   (c) He   (d) Mg
    (e) Na   (f) S
16. Write the name of each of the following elements.
    (a) C   (b) F   (c) Hg   (d) N
    (e) Ni   (f) Zn
17. (a) What is the distinction between mass and weight?
    (b) Give a location in which the weight of an object that has a mass of 453.59 grams

would be greater than 1.0000 pound. Justify your choice.

(c) Give a location in which the weight of an object that has a mass of 453.59 grams would be less than 1.0000 pound. Justify your choice.

Significant Figures and Exponential Notation

18. Round off the following numbers to three significant figures. Follow the rules given in Section 1–8 exactly.
    (a) 4325   (b) 6.873×10^3
    (c) 0.17354   (d) 7.8939
    (e) 9.237×10^{-3} (f) 0.0299817

19. Which of the following are likely to be exact numbers? Why?
    (a) 15 eggs
    (b) 10 pounds of potatoes
    (c) 497 live quail
    (d) 11,743 live bees
    (e) 47,398 people
    (f) $47,342.21
    (g) 3 gnus
    (h) 12 square yards of carpet

How many significant figures does each of the numbers in Exercises 20 and 21 contain?

20. (a) 0.0278 meter
    (b) 1.3 centimeters
    (c) 1.00 foot
    (d) 8.021 yards

21. (a) 7.98×10^{-3} pound
    (b) 0.2003 ton
    (c) 4.69×10^4 tons
    (d) 1×10^{12} atoms
    (e) 1.73×10^{24} atoms

22. Express the following numbers in proper exponential form with the indicated number of significant figures.
    (a) 1000 (2 sig. fig.)
    (b) 43,927 (3 sig. fig.)
    (c) 0.000286 (3 sig. fig.)
    (d) 0.000098765 (5 sig. fig.)
    (e) 10,000 (you decide how many significant figures)

23. Express the following exponentials as ordinary numbers.
    (a) 7.23×10^4 (b) 8.193×10^2
    (c) 1.98×10^{-3} (d) 7.51×10^{-7}
    (e) 5.43×10^0

Perform the indicated operations and round off your answers to the proper number of significant figures in Exercises 24 and 25. Assume that all numbers were obtained from measurements.

24. (a) $18.56 + 1.233 =$
    (b) $1.234 \times 0.247 =$
    (c) $4.3/8.74 =$

25. (a) $8.649 - 2.8964 =$
    (b) $0.06936 \times 0.384 =$
    (c) $4567/2.53 =$

In Exercises 26 and 27, perform the indicated operations and round off your answers to the proper number of significant figures. Assume that all numbers were obtained from measurements. Express your answers in exponential notation.

26. (a) $(1.54 \times 10^3) + (2.11 \times 10^3) =$
    (b) $(1.54 \times 10^3) + (2.11 \times 10^2) =$
    (c) $(1.23 \times 10^2)/(4.56 + 18.7) =$
    (d) $(4.56 + 8.7)/(1.23 \times 10^{-2}) =$

27. (a) $(2.11 \times 10^{-3}) + (1.54 \times 10^{-3}) =$
    (b) $(1.54 \times 10^{-3}) + (2.11 \times 10^{-2}) =$
    (c) $(4.56 + 18.7)/(1.23 \times 10^2) =$
    (d) $(1.23 \times 10^{-2})/(4.56 + 1.87) =$

28. Choose the appropriate word prefixes to indicate the multiplier in each of the following numbers.
    (a) 1×10^3 (b) 1×10^{-3}
    (c) 1×10^1 (d) 1×10^{-1}
    (e) 0.01   (f) 0.1
    (g) 0.001   (h) 1×10^{-6}

29. Indicate the multiple or fraction of 10 by which a quantity is multiplied when it is preceded by the following prefixes.
    (a) M   (b) m
    (c) c   (d) d
    (e) k   (f) μ

In Exercises 30 and 31, choose all the units listed that could be used to express (a) length or distance, (b) mass or weight, (c) area, (d) volume.

30. (a) km   (b) mL
    (c) mm^2 (d) dg
    (e) m^3 (f) cm^2

31. (a) mg   (b) mm
    (c) dL   (d) cm^3
    (e) kg   (f) m^2

Dimensional Analysis

32. Fill in the blanks by making the indicated conversions.
    (a) 7.58 km = _____ m
    (b) 758 m = _____ cm

(c) 478 g = _____ kg
(d) 9.87 kg = _____ g
(e) 1386 mL = _____ L
(f) 3.692 L = _____ mL
(g) 1126 cm³ = _____ L
(h) 0.786 L = _____ cm³

33. Express 1.27 feet in millimeters, centimeters, meters, and kilometers.

34. What is the length of a football field (100 yards) expressed in meters?

35. What is the distance around your waist expressed in (a) centimeters and (b) meters?

36. Express 55 miles per hour in kilometers per hour.

37. The capacity of the gasoline tank in an automobile is 12 gallons. How many liters is this?

38. Express:
   (a) 1.00 gallon in milliliters
   (b) 8.00 cubic inches in milliliters (How long is each edge of a cube of this volume?)
   (c) 4.00 cubic yards in cubic centimeters
   (d) 1.00 liter in cubic inches

39. Express:
   (a) 1.00 pint in milliliters
   (b) 1.00 square inch in square centimeters
   (c) 1.00 square foot in square centimeters
   (d) 1.00 cubic inch in cubic centimeters
   (e) 1.00 cubic foot in liters
   (f) 1.00 cubic meter in cubic feet

40. If the price of gasoline is $0.400 per liter, what is its price per gallon?

41. If the price of gasoline is $1.43 per gallon, what is its price per liter?

42. Express the following masses or weights in grams and kilograms (at sea level).
   (a) 1.00×10^6 milligrams
   (b) 4.25×10^5 centigrams
   (c) 3.0 pounds
   (d) 4.00 ounces

43. Express the following in grams.
   (a) 2.35 pounds
   (b) 1.00 ounce
   (c) 1.00 short ton
   (d) 1.00 metric tonne

44. Express:
   (a) 100 pounds in kilograms
   (b) 100 kilograms in pounds
   (c) 8.00 ounces in centigrams
   (d) 2.23 short tons in milligrams

45. If a successful professional athlete is 216 centi-

meters tall and has a mass of 95 kilograms, would you assume that he plays basketball or football? Why?

46. Choose the SI and traditional metric system units that would be most appropriate to indicate the following dimensions. Provide an *approximate* numerical value with each set of units. Example: A standard door is 6 feet and 8 inches high, which is approximately 2 meters high. Justify your choice of units in each case.
   (a) the length of a U.S. football field between the goal lines
   (b) the volume of the gasoline tank in a "compact" automobile
   (c) the area of the floor in your bedroom
   (d) your mass
   (e) your height

47. At a given point the moon is 240,000 miles from the earth. How long does it take for light from a source on the earth to reach a reflector on the moon and then return to earth? The speed of light is 3.00×10^8 m/s.

48. The radius of an aluminum atom is 0.143 nm. How many aluminum atoms would have to be laid side-by-side to give a row of aluminum atoms 1.00 inch long? Assume that the atoms are spherical.

49. Cesium atoms are the largest naturally occurring atoms. The diameter of a cesium atom is 0.524 nm. If 1.00×10^9 cesium atoms were laid side-by-side to make a row, how long would the row be in inches? Assume that the atoms are spherical.

Extensive and Intensive Properties

50. Distinguish between intensive and extensive properties, and cite two specific examples of each.

51. Which of the following are extensive properties and which are intensive properties? Justify your answers.
   (a) temperature   (b) color of copper
   (c) volume   (d) density
   (e) melting point of tin (f) mass

52. Which of the following are extensive properties and which are intensive properties? Justify your answers.
   (a) color of a liquid   (b) mass
   (c) specific gravity   (d) boiling point of
   (e) physical state   water

Density and Specific Gravity

53. Distinguish between density and specific gravity.
54. A 13.5 cubic centimeter piece of chromium has a mass of 97.2 grams. What is its density?
55. What is the mass of 13.5 cubic centimeters of mercury? Its density is 13.6 g/cm³.
56. Refer to the two previous problems. One cm³ of mercury is how many times heavier than one cm³ of chromium?
57. Refer to Table 1-8 and calculate the mass, at room temperature and one atmosphere pressure, of 200 mL of (a) hydrogen, (b) carbon dioxide, (c) ethyl alcohol, (d) aluminum, and (e) gold.
58. Refer to Table 1-8. A 20.0 mL sample of which of the following has the largest mass? hydrogen, sand, gold
59. Refer to Table 1-8. A 20.0 gram sample of which of the following has the smallest volume? cork, table salt, lead
60. The specific gravity of ethyl alcohol is 0.79. What volume of ethyl alcohol has the same mass as 23 mL of water?
61. What is the specific gravity of a liquid if 300 mL of it has the same mass as 400 cm³ of water?
62. A piece of copper is placed in a graduated cylinder containing some water. The total volume increases by 7.43 mL. What is the mass of the piece of copper (density = 8.92 g/cm³)?
63. What is the volume of 479 grams of aluminum (specific gravity = 2.70)?
64. The specific gravity of gold is 19.3. Which of the following contains the greatest mass of gold? 0.50 pound, 0.25 kilogram, 25 milliliters
65. The specific gravity of table salt is 2.16. Which of the following occupies the largest volume? 0.50 pound, 0.25 kilogram, or 0.025 liter of salt
66. (a) What is the volume of a bar of iron that is 4.72 centimeters long, 3.19 centimeters wide, and 0.52 centimeters thick? Its mass is 61.5 grams.
    (b) Calculate the density of iron using the data obtained in (a). Compare with the value in Table 1-8 and explain any difference.
67. The density of gold is 19.3 g/cm³. Suppose someone offered to give you a one-gallon bucket filled with gold if you would carry the bucket of gold up a flight of stairs. Could you accept the gold?
68. (a) Given that the density of gold is 19.3 g/cm³, calculate the volume of 100 pounds of gold in cm³.
    (b) Assume that this sample of gold is a perfect cube. How long is each edge (in inches)?
69. An aqueous solution that is 70.0% nitric acid by mass has a specific gravity of 1.40. What volume of the solution contains (a) 100 grams of solution, (b) 100 grams of pure nitric acid, (c) 100 grams of water? What masses of (d) water and (e) pure nitric acid are contained in 100 milliliters of the solution?

Temperature

In Exercises 70 and 71 express the given temperatures on the indicated scale.

70. (a) 100.000°F as degrees Celsius
    (b) 200.000°F as degrees Celsius
    (c) 100.0 K as degrees Fahrenheit
    (d) 100.0°F as kelvins
71. (a) 0.00°F as degrees Celsius
    (b) 78.0°F as kelvins
    (c) −40.0°C as degrees Fahrenheit
    (d) 0.00 K as degrees Fahrenheit
72. The following data show the relationship between temperature and distance above the surface of the earth on a "standard day." Fill in the blanks by making the appropriate conversions.

	Feet	Meters	°F	°C
(a)	1,000	____	56°	____
(b)	____	1,500	____	5°
(c)	10,000	____	____	−5°
(d)	____	4,500	5°	____
(e)	20,000	____	____	−26°
(f)	____	9,000	−47°	____
(g)	36,087	____	____	−56°

(h) Plot feet versus °F on a sheet of graph paper. Estimate the temperature at altitudes of 7500 feet and 40,000 feet.
(i) Now plot meters versus °C on another piece of graph paper. Compare the two graphs. Are they similar? Why?

The Laws of Chemical Combination and Dalton's Atomic Theory

73. List the basic postulates of Dalton's Atomic Theory and explain the meaning of each.
74. State the Law of Conservation of Matter. How is it explained by Dalton's Atomic Theory?

75. State the Law of Constant Composition. How is it explained by Dalton's Atomic Theory?
76. What parts of Dalton's Atomic Theory were in error?
77. In each of three experiments a sample of pure carbon was burned in oxygen to produce carbon dioxide. In the first experiment, 5.00 grams of carbon produced 18.33 grams of carbon dioxide. In the second experiment, 6.50 grams of carbon produced 23.83 grams of carbon dioxide. In the third experiment, 9.25 grams of carbon produced 33.92 grams of carbon dioxide. Show how these experimental results are consistent with the Law of Constant Composition.
78. How many grams of oxygen reacted in each of the three experiments of Exercise 77?
79. When a sample of calcium chloride was decomposed, 21.7 grams of calcium and 38.3 grams of gaseous chlorine were produced. (a) What mass of calcium chloride decomposed? (b) What is the percent composition of calcium chloride?
80. Suppose that 30.0 grams of calcium are converted completely to calcium chloride (whose percent composition you calculated in Exercise 79) when mixed with an excess of chlorine. (a) What mass of calcium chloride is produced? (b) What mass of chlorine reacts?
81. Sulfur dioxide is 50.1% sulfur and 49.9% oxygen by mass. Suppose that 10.0 grams of sulfur are mixed with 30.0 grams of oxygen and heated until all the sulfur is converted to sulfur dioxide.
    (a) What mass of oxygen remains unreacted?
    (b) What mass of sulfur dioxide is produced?
82. Phosphorus burns in a limited amount of oxygen to produce a compound whose composition is 56.4% phosphorus and 43.6% oxygen by mass. Suppose that 10.0 grams of phosphorus are burned in 5.0 grams of oxygen and that all of one of the elements is consumed.
    (a) Which element is consumed completely?
    (b) What mass of the other element remains unreacted?
    (c) What mass of the compound is produced?
83. Phosphorus reacts with an excess of chlorine to produce a compound that is 14.5% P and 85.5% Cl by mass. Suppose that 10.0 grams of phosphorus are mixed with 151 grams of chlo-

rine and that all of one of the elements is consumed.
    (a) Which element is consumed completely?
    (b) What mass of the other element remains unreacted?
    (c) What mass of the compound is produced?
84. Show that the reactions of Exercises 82 and 83 obey the Law of Conservation of Matter.
85. Phosphorus forms two chlorides. A 30.00 gram sample of one chloride decomposes to give 4.35 g of P and 25.65 g of Cl. A 30.00 gram sample of the other chloride decomposes to give 6.61 g of P and 23.39 g of Cl. Show that these compounds obey the Law of Multiple Proportions.
86. Nitrogen forms several compounds with oxygen. Three of them have the elemental compositions below.

Compound	% N	% O
1	46.7	53.3
2	30.4	69.6
3	25.9	74.1

Show that these three compounds obey the Law of Multiple Proportions. (Hint: Deal with 100.0 grams of each compound for simplicity.)

Atoms, Molecules, and Ions

87. Define and give an example of each of the following terms. (a) atom, (b) molecule, (c) ion, (d) formula unit, (e) allotropism
88. Use words to describe the atomic composition of molecules of the following elements and compounds. (a) F_2, (b) HCl, (c) CH_3OH, (d) Ne, (e) SF_6, (f) H_3AsO_4
89. Write formulas for the following ions. (a) calcium ion, (b) bromide ion, (c) ammonium ion, (d) nitrate ion, (e) ferric ion, (f) aluminum ion, (g) carbonate ion
90. Write formulas for the following ions. (a) silver ion, (b) potassium ion, (c) ferrous ion, (d) fluoride ion, (e) oxide ion, (f) phosphate ion, (g) hydroxide ion
91. Write formulas for the following ionic compounds.
    (a) silver fluoride
    (b) ammonium iodide
    (c) copper(II) oxide

(d) aluminum hydroxide
(e) sodium acetate
(f) magnesium sulfate
(g) ferric sulfate
(h) calcium nitrate

92. Write formulas for the following ionic compounds.
    (a) ferrous sulfide
    (b) iron(II) oxide
    (c) ammonium chloride
    (d) silver acetate
    (e) cupric chloride
    (f) potassium carbonate
    (g) sodium sulfate
    (h) sodium hydroxide

93. Name the following ionic compounds.
    (a) CuO (b) $NaBr$ (c) K_2SO_4
    (d) $Mg(NO_3)_2$ (e) AlF_3 (f) CaI_2
    (g) $(NH_4)_2S$ (h) $Ca(OH)_2$

94. Name the following ionic compounds.
    (a) Fe_2O_3 (b) $AgBr$ (c) Al_2O_3
    (d) $NaCH_3COO$ (e) $Cu(NO_3)_2$ (f) Na_3PO_4
    (g) $FeCO_3$ (h) Ag_2SO_4

95. Name the following molecular compounds.
    (a) NH_3 (b) HCl in water   (c) CO
    (d) CO_2 (e) C_2H_6 (f) H_2SO_4
    (g) SO_3

96. Write formulas for the following molecular compounds.
    (a) carbon dioxide   (b) propane
    (c) acetic acid   (d) nitric acid
    (e) methane   (f) sulfur dioxide

Stoichiometry

2

Mass Relationships Involving Pure Substances — Composition Stoichiometry

Mass Relationships and Chemical Reactions — Reaction Stoichiometry

Concentrations of Solutions — An Introduction

When Dalton postulated the atomic theory in 1803, he provided the framework for our modern picture of the structure of matter, which we shall describe in Chapters 4 and 5. Dalton also provided the basis for a unified view of two previously known scientific laws, (1) the Law of Conservation of Matter and (2) the Law of Definite Proportions, and a third that he advanced, (3) the Law of Multiple Proportions.

The word "stoichiometry" is derived from the Greek *stoicheion*, which means "first principle or element," and *metron*, which means "measure."

The unified view of these laws allows us to understand **stoichiometry,** the quantitative mass relationships among elements in compounds (composition stoichiometry) and among substances as they undergo chemical reactions (reaction stoichiometry). Stoichiometry is the topic of this chapter.

The ideas presented in this chapter are used throughout the scientific world. For example, chemical symbols, formulas, and equations are used in such diverse areas as agriculture, home economics, engineering, geology, physics, the biological sciences, medicine, and dentistry. Many chemical terms are found in your daily newspaper or favorite magazine. It is important to learn this material well so that you can use it precisely and effectively.

Mass Relationships Involving Pure Substances — Composition Stoichiometry

2–1 Atomic Weights

As the chemists of the eighteenth and nineteenth centuries painstakingly (and sometimes painfully!) sought information about the composition of compounds and tried to systematize their knowledge, it became apparent that each element has a characteristic mass relative to every other element. Although these early scientists did not have the experimental means to establish a mass for each kind of atom, they succeeded in defining a *relative* scale of atomic masses.

An early observation, derived from the same experiments that established the Law of Multiple Proportions, was that carbon and oxygen have relative atomic masses, also traditionally called **atomic weights,** of approximately 12 and 16, respectively.

Thousands of experiments on the compositions of compounds have resulted in the establishment of a scale of relative atomic weights based on the **atomic mass unit (amu),** which is defined as *exactly 1/12 of the mass of an atom of a particular isotope of carbon, called carbon-12.*

The term "atomic weight" is widely accepted because of its traditional use, although it is properly a mass rather than a weight. "Atomic mass" is sometimes used, but is not as popular.

On this scale, the atomic weight of hydrogen (H) is 1.0079 amu, that of sodium (Na) is 22.98977 amu, and that of magnesium (Mg) is 24.305 amu. This tells us that Na atoms have nearly 23 times the mass of H atoms, while Mg atoms are about 24 times heavier than H atoms.

2–2 The Mole

Attempts to measure the masses of individual atoms showed that these are very small numbers indeed. Atoms are far too small to be seen or weighed individually; the smallest bit of matter that can be reliably measured contains an enormous number of atoms. Therefore, we must deal with large numbers of atoms in any real situation, and some unit for describing conveniently a large number of atoms is desirable. The idea of a unit to describe a particular number of objects has been around for a long time. We are familiar with the dozen (12 items) and the gross (144 items); some items are grouped in tens or hundreds (think of decades, centuries, and millennia).

Mole is derived from the Latin word *moles*, which means "a mass." Molecule is the diminutive form of this word, and means "a small mass."

The unit chosen for the SI, and used universally by chemists and other scientists, is the **mole.** It is *defined* as the amount of substance that contains as many elementary entities (atoms, molecules, or other particles) as there are atoms in 0.012 kg of the pure carbon-12 isotope. As we noted in Table 1–4, the mole is a fundamental unit of the SI, and is abbreviated "mol". Many experiments, using various techniques, have refined the number described in the definition, and the currently accepted value is

$$1 \text{ mole} = 6.022045 \times 10^{23} \text{ particles}$$

This number, often rounded off to 6.022×10^{23}, is called **Avogadro's number** in honor of Amedeo Avogadro (1776–1856), whose contributions to chemistry will be discussed in Section 11–9.

According to its definition, the mole unit refers to a fixed number of "elementary entities," whose identities must be specified. Just as we may speak of a dozen eggs or a dozen automobiles, we refer to a mole of atoms or a mole of molecules (or a mole of ions, electrons, or other particles). We could even think about a mole of eggs, although the size of the required carton staggers the imagination! Helium exists as discrete He atoms, so one mole of helium consists of 6.022×10^{23} He *atoms.* Hydrogen commonly exists as diatomic (two-atom) molecules, so one mole of hydrogen contains 6.022×10^{23} H_2 *molecules.*

Always be careful to determine the context in which the mole is used.

Every kind of atom, molecule, and ion has a definite characteristic mass. It follows that one mole of a given substance also has a definite mass, regardless of the source of the sample. This idea is of central importance in many calculations throughout the study of chemistry and the related sciences.

Because the mole is defined as the number of atoms in 0.012 kg (or 12 grams) of carbon-12, and the atomic mass unit is defined as 1/12 of the mass of a carbon-12 atom, the following convenient relationship is true:

The mass, in grams, of one mole of atoms of a pure element is numerically equal to the atomic weight of that element in amu.

The atomic weights of the elements are listed in the table inside the front cover.

For instance, if you obtain a pure sample of the metallic element titanium (Ti), whose atomic weight is 47.90 amu, and measure out 47.90 grams of it, you will have one mole or 6.022×10^{23} atoms of titanium.

The symbol for an element can (1) identify the element, (2) represent one atom of the element, or, as we shall see, (3) represent one mole of atoms of the element.

The atomic weight of Fe is 55.847 amu.

A quantity of a substance may be expressed in a variety of ways. For example, consider a barrel containing a dozen ten-kilogram cannon balls and a beaker containing 55.847 grams of iron filings, or one mole of iron. We can express the amount of cannon balls or iron filings present in any of several different units, and we can then construct unit factors to relate an amount of the substance expressed in one kind of unit to the same amount expressed in another unit. This is illustrated in Figure 2–1.

FIGURE 2–1
Representation of amounts of substances in three different ways.

12 balls

or

1 doz balls

or

120 kg balls

6.022×10^{23} Fe atoms

1 mol Fe atoms

55.85 g Fe

Barrel containing twelve 10-kilogram cannon balls

Beaker containing 55.85 g of iron filings

Unit factors:

$$\frac{12 \text{ balls}}{1 \text{ doz balls}}$$

$$\frac{12 \text{ balls}}{120 \text{ kg}}$$

etc.

Unit Factors:

$$\frac{6.022 \times 10^{23} \text{ Fe atoms}}{1 \text{ mol Fe atoms}}$$

$$\frac{6.022 \times 10^{23} \text{ Fe atoms}}{55.85 \text{ g Fe}}$$

etc.

TABLE 2-1   Mass of One Mole of Atoms of Some Common Elements

Element	A sample with a mass of	Contains
carbon	12.01 g C	6.022×10^{23} C atoms or 1 mole of C atoms
calcium	40.08 g Ca	6.022×10^{23} Ca atoms or 1 mole of Ca atoms
titanium	47.90 g Ti	6.022×10^{23} Ti atoms or 1 mole of Ti atoms
gold	196.97 g Au	6.022×10^{23} Au atoms or 1 mole of Au atoms
hydrogen	1.008 g H_2	6.022×10^{23} H atoms or 1 mole of H atoms
		(3.011×10^{23} H_2 molecules or 0.5 mole of H_2 molecules)
nitrogen	14.01 g N_2	6.022×10^{23} N atoms or 1 mole of N atoms
		(3.011×10^{23} N_2 molecules or 0.5 mole of N_2 molecules)
sulfur	32.06 g S_8	6.022×10^{23} S atoms or 1 mole of S atoms
		(0.7528×10^{23} S_8 molecules or 0.125 mole of S_8 molecules)

As Table 2-1 suggests, the concept of a mole as applied to atoms is especially useful because it provides a convenient basis for comparing equal numbers of atoms of different elements.

Figure 2-2 shows what one mole of atoms of each of some common elements looks like. Each of the examples in Figure 2-2 represents 6.022×10^{23} atoms of the element.

FIGURE 2-2

Representations of one mole of atoms of some common elements.

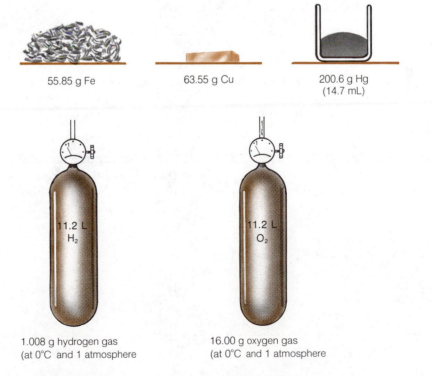

55.85 g Fe

63.55 g Cu

200.6 g Hg
(14.7 mL)

1.008 g hydrogen gas
(at 0°C and 1 atmosphere

16.00 g oxygen gas
(at 0°C and 1 atmosphere

The relationship between the mass of a sample of an element and the number of moles of atoms in the sample is illustrated in Example 2-1.

Example 2–1

How many moles of atoms does 245.2 g of zinc metal contain?

Solution

Since the atomic weight of zinc is 65.38 amu, we know that one mole of zinc atoms is 65.38 grams of zinc.

$$\underline{?}\ \text{Zn atoms} = 0.125\ \text{mol Zn atoms}$$
$$\times \frac{6.022 \times 10^{23}\ \text{Zn atoms}}{1\ \text{mol Zn atoms}}$$
$$= \underline{7.53 \times 10^{22}\ \text{Zn atoms}}$$

Once the number of moles of atoms of an element is known, the actual number of atoms in the sample can be calculated, as Example 2–2 illustrates.

Example 2–2

How many atoms are contained in 0.125 mol of zinc atoms?

Solution

One mole of zinc atoms contains 6.022×10^{23} atoms.

$$\underline{?}\ \text{mol Zn atoms} = 245.2\ \text{g Zn} \times \frac{1\ \text{mol Zn atoms}}{65.38\ \text{g Zn}}$$
$$= \underline{3.750\ \text{mol Zn atoms}}$$

If we know the atomic weight of an element on the carbon-12 scale, we can use the mole concept and Avogadro's number to calculate the average mass of one atom of that element in grams (or any other mass unit we choose).

Example 2–3

Calculate the mass of one aluminum atom in grams.

Solution

We know that one mole of Al atoms has a mass of 26.98 g and contains 6.022×10^{23} Al atoms. Hence

$$\frac{\underline{?}\ \text{g Al}}{\text{Al atom}} = \frac{26.98\ \text{g Al}}{1\ \text{mol Al}} \times \frac{1\ \text{mol Al}}{6.022 \times 10^{23}\ \text{Al atoms}}$$
$$= \underline{4.480 \times 10^{-23}\ \text{g Al/Al atom}}$$

Thus, we see that the mass of one Al atom is only 4.480×10^{-23} gram. To gain some appreciation of how little this is, write 4.480×10^{-23} gram as a decimal fraction and try to name the fraction.

To illustrate just how small atoms are, let's calculate the number of nickel atoms in a micrometeorite that weighs only one billionth of a gram. The radius of a spherical micrometeorite of this mass is 3.0×10^{-4} cm. Micrometeorites of this size constantly shower down on the earth from space.

Example 2-4

Calculate the number of atoms in one billionth of a gram of nickel metal to two significant figures.

Solution

$$\underline{?}\text{ Ni atoms} = 1.0 \times 10^{-9}\text{ g Ni} \times \frac{1\text{ mol Ni}}{58.70\text{ g Ni}}$$

$$\times \frac{6.022 \times 10^{23}\text{ Ni atoms}}{1\text{ mol Ni}}$$

$$= \underline{1.0 \times 10^{13}\text{ Ni atoms}}$$

One billionth of a gram of nickel metal contains approximately 10,000,000,000,000 Ni atoms.

These two examples demonstrate how small atoms are and why we find it necessary to use large numbers of atoms in practical work.

2-3 Formula Weights, Molecular Weights, and Moles

The **formula weight** of a substance is the sum of the atomic weights of the elements in the formula, each taken the number of times the element occurs. That is, it is the mass in amu of one formula unit. Formula weights, like the atomic weights on which they are based, are relative masses, expressed in amu. The formula weight (*rounded to the nearest amu*) for sodium hydroxide, NaOH, is found as follows.

	No. of atoms of stated kind	Mass of one atom
$1 \times \text{Na} =$	1	$\times$ 23 amu
$1 \times \text{H} =$	1	$\times$ 1 amu
$1 \times \text{O} =$	1	$\times$ 16 amu
	Formula weight $=$	40 amu

The approximate formula weight for phosphoric acid, H_3PO_4, is:

	No. of atoms of stated kind	Mass of one atom	
$3 \times \text{H} =$	3	$\times$ 1 amu $=$	3 amu
$1 \times \text{P} =$	1	$\times$ 31 amu $=$	31 amu
$4 \times \text{O} =$	4	$\times$ 16 amu $=$	64 amu
		Formula weight $=$	98 amu

Strictly speaking, the terms *formula mass* and *molecular mass* are correct, but formula weight and molecular weight are widely accepted and used.

The term **molecular weight** is used interchangeably with formula weight when reference is made to molecular (non-ionic) substances, i.e., substances that exist as discrete molecules.

To illustrate how full precision arithmetic can be used to obtain a formula weight, let's calculate the formula weight for ammonium carbonate, $(NH_4)_2CO_3$. The most precisely known values for atomic weights are used and the rules for significant figures (Section 1-8) are applied.

	No. of atoms of stated kind	Mass of one atom
$2 \times N =$	2	$\times 14.0067$ amu $= 28.0134$ amu
$8 \times H =$	8	$\times\; 1.0079$ amu $=\; 8.0632$ amu
$1 \times C =$	1	$\times 12.011\;\;$ amu $= 12.011\;\;$ amu
$3 \times O =$	3	$\times 15.9994$ amu $= 47.9982$ amu
		Formula weight $= 96.086\;\;$ amu

Numerous experiments have shown that the amount of a substance that contains the mass in grams numerically equal to its formula weight in amu contains 6.022×10^{23} formula units. One mole of sodium hydroxide is 40 g of NaOH, one mole of phosphoric acid is 98 g of H_3PO_4, and one mole of ammonium carbonate is 96 g of $(NH_4)_2CO_3$.

These masses are rounded to the nearest whole number of grams here.

One mole of any molecular substance contains 6.022×10^{23} molecules of the substance, as Table 2–2 illustrates. Compare the entries for hydrogen and nitrogen with those in Table 2–1.

TABLE 2–2   One Mole of Some Common Molecular Substances

Substance	Mass of 1 Mole	Contains
hydrogen	2.016 g H_2	6.022×10^{23} H_2 molecules
nitrogen	28.01 g N_2	6.022×10^{23} N_2 molecules
oxygen	32.00 g O_2	6.022×10^{23} O_2 molecules
methane	16.04 g CH_4	6.022×10^{23} CH_4 molecules
oxalic acid	90.04 g $(COOH)_2$	6.022×10^{23} $(COOH)_2$ molecules
propane	44.10 g C_3H_8	6.022×10^{23} C_3H_8 molecules
nitroglycerine	227.1 g $C_3H_5N_3O_9$	6.022×10^{23} $C_3H_5N_3O_9$ molecules

The physical appearance of one mole of each of some molecular substances is illustrated in Figure 2–3. Also demonstrated in this figure are two facts of general interest. First, the volumes of one mole of oxygen and one mole of nitrogen — and, in general, of one mole of any gas — under identical conditions are approximately the same. At 0°C and one atmosphere pressure, one mole of a gas occupies about 22.4 liters, as will be explained in Chapter 11. Second, Figure 2–3 shows two different forms of oxalic acid. The formula unit (molecule) of oxalic acid is $(COOH)_2$, with a formula weight of 90 amu. When oxalic acid is crystallized from an aqueous solution, however, two molecules of water are bound to every molecule of oxalic acid and remain trapped in the crystal when it appears dry. The formula of this **hydrate** if $(COOH)_2 \cdot 2H_2O$, where the dot serves to show that the water is "attached" to the molecule. The formula weight of the hydrate is 126 amu. The water can be driven out of the crystal by heating, to leave the **anhydrous** oxalic acid. (Anhydrous means "without water.") This behavior occurs in many compounds.

Ionic compounds such as sodium chloride (Na^+Cl^-) exist as neutral combinations of ions, rather than as discrete molecules. There are no simple NaCl molecules at ordinary temperatures and pressures and, thus, it is inappropriate to refer to the "molecular weight" of NaCl. One mole of an ionic compound contains 6.022×10^{23} **formula units** of the substance. Recall that one formula unit of sodium chloride consists of one sodium ion, Na^+, and one chloride ion, Cl^-. One mole of NaCl, or 58.44 grams, contains 6.022×10^{23} Na^+ ions and 6.022×10^{23} Cl^- ions. See Table 2–3.

The masses of ions are very nearly identical with the masses of the atoms from which they are derived.

FIGURE 2–3
Representations of one mole of some molecular substances.

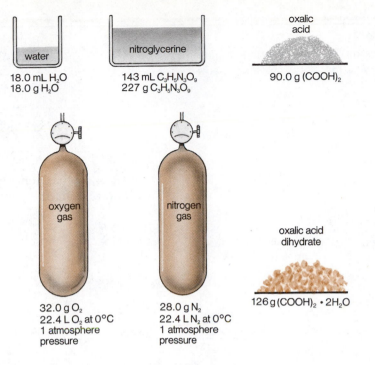

water
18.0 mL H_2O
18.0 g H_2O

nitroglycerine
143 mL $C_3H_5N_3O_9$
227 g $C_3H_5N_3O_9$

oxalic acid
90.0 g $(COOH)_2$

oxygen gas
32.0 g O_2
22.4 L O_2 at 0°C
1 atmosphere pressure

nitrogen gas
28.0 g N_2
22.4 L N_2 at 0°C
1 atmosphere pressure

oxalic acid dihydrate
126 g $(COOH)_2 \cdot 2H_2O$

TABLE 2–3   One Mole of Some Common Ionic Compounds

Compound	Mass of 1 Mole	Contains
sodium chloride	58.44 g NaCl	6.022×10^{23} Na^+ ions 6.022×10^{23} Cl^- ions
calcium chloride	111.0 g $CaCl_2$	6.022×10^{23} Ca^{2+} ions $2(6.022 \times 10^{23})$ Cl^- ions
aluminum sulfate	342.1 g $Al_2(SO_4)_3$	$2(6.022 \times 10^{23})$ Al^{3+} ions $3(6.022 \times 10^{23})$ $SO_4{}^{2-}$ ions

The following examples show the relations between numbers of molecules, atoms, or formula units and their masses.

Example 2–5

What is the mass in grams of twenty billion NH_3 molecules, to two significant figures?

Solution

The approximate formula weight of NH_3 is 17 amu, and therefore one mole, or 6.02×10^{23} molecules of NH_3, has a mass of 17 g. (When fewer than four significant figures are used in calculations, Avogadro's number can be rounded off to 6.02×10^{23}.)

$$\underline{?} \text{ g } NH_3 = 2.0 \times 10^{10} \text{ } NH_3 \text{ molecules}$$
$$\times \frac{1 \text{ mol } NH_3}{6.02 \times 10^{23} \text{ } NH_3 \text{ molecules}}$$
$$\times \frac{17 \text{ g } NH_3}{1 \text{ mol } NH_3}$$
$$= 5.6 \times 10^{-13} \text{ g } NH_3$$

It may be of interest to note that the most commonly used analytical balances are capable of determining masses to within 0.0001 gram. Twenty billion NH_3 molecules have a mass of only 0.00000000000056 gram!

Example 2–6

How many (a) moles of O_2, (b) O_2 molecules, and (c) O atoms are contained in 40.0 grams of gaseous oxygen at 25°C?

Solution

One mole of O_2 contains 6.02×10^{23} O_2 molecules and has a mass of 32.0 g.

(a) $\quad \underline{?} \text{ mol } O_2 = 40.0 \text{ g } O_2 \times \dfrac{1 \text{ mol } O_2}{32.0 \text{ g } O_2}$

$$= 1.25 \text{ mol } O_2$$

(b) $\quad \underline{?} \; O_2 \text{ molecules} = 40.0 \text{ g } O_2 \times \dfrac{1 \text{ mol } O_2}{32.0 \text{ g } O_2}$

$$\times \dfrac{6.02 \times 10^{23} \; O_2 \text{ molecules}}{1 \text{ mol } O_2}$$

$$= \underline{7.52 \times 10^{23} \; O_2 \text{ molecules}}$$

(c) $\quad \underline{?} \text{ O atoms} = 40.0 \text{ g } O_2 \times \dfrac{1 \text{ mol } O_2}{32.0 \text{ g } O_2}$

$$\times \dfrac{6.02 \times 10^{23} \; O_2 \text{ molecules}}{1 \text{ mol } O_2}$$

$$\times \dfrac{2 \text{ O atoms}}{1 \; O_2 \text{ molecule}}$$

$$= \underline{1.50 \times 10^{24} \text{ O atoms}}$$

Or, we can use the number of moles of O_2 calculated in (a) to find the number of O_2 molecules in part (b).

$? \; O_2 \text{ molecules} = 1.25 \text{ mol } O_2$

$$\times \dfrac{6.02 \times 10^{23} \; O_2 \text{ molecules}}{1 \text{ mol } O_2}$$

$$= \underline{7.52 \times 10^{23} \; O_2 \text{ molecules}}$$

Example 2–7

Calculate the number of hydrogen atoms in 39.6 grams of ammonium sulfate, $(NH_4)_2SO_4$.

Solution

One mole of $(NH_4)_2SO_4$ is 6.02×10^{23} formula units and has a mass of 132 g.

$$\underline{?} \text{ H atoms} = 39.6 \text{ g } (NH_4)_2SO_4 \times \dfrac{1 \text{ mol } (NH_4)_2SO_4}{132 \text{ g } (NH_4)_2SO_4} \times \dfrac{6.02 \times 10^{23} \text{ formula units } (NH_4)_2SO_4}{1 \text{ mol } (NH_4)_2SO_4}$$

$$\times \dfrac{8 \text{ H atoms}}{1 \text{ formula unit } (NH_4)_2SO_4} = \underline{1.44 \times 10^{24} \text{ H atoms}}$$

The term millimole (mmol) is useful in laboratory work. As the prefix indicates, one **mmol** is $\frac{1}{1000}$ of a mole. Also, small masses are frequently expressed in milligrams (mg) rather than grams. The relation between millimoles and milligrams is the same as that between moles and grams, as illustrated in Table 2–4.

TABLE 2–4 Comparison of Moles and Millimoles

Compound	1 Mole	1 Millimole
NaOH	40 g	40 mg or 0.040 g
H_3PO_4	98 g	98 mg or 0.098 g
SO_2	64 g	64 mg or 0.064 g
C_3H_8	44 g	44 mg or 0.044 g

Example 2-8

Calculate the number of millimoles of sulfuric acid in 0.147 gram of H_2SO_4.

Solution

One mole of H_2SO_4 has a mass of 98.1 g; thus, 1 mmol of H_2SO_4 is 98.1 mg, or 0.0981 g.

Method 1

? mmol H_2SO_4

$$= 0.147 \text{ g } H_2SO_4$$
$$\times \frac{1 \text{ mmol } H_2SO_4}{0.0981 \text{ g } H_2SO_4}$$
$$= \underline{1.50 \text{ mmol } H_2SO_4}$$

Method 2 Or, using 0.147 g H_2SO_4 = 147 mg H_2SO_4, we have

? mmol H_2SO_4

$$= 147 \text{ mg } H_2SO_4 \times \frac{1 \text{ mmol } H_2SO_4}{98.1 \text{ mg } H_2SO_4}$$
$$= \underline{1.50 \text{ mmol } H_2SO_4}$$

2-4   Percent Composition and Formulas of Compounds

If the formula of a compound is known, its chemical composition can be expressed as the percent by mass of each element in the compound. (As we shall see in the next section, this procedure can also be reversed.) For example, the formula for methane, CH_4, tells us that all samples of pure methane contain four H atoms for every (one) C atom. Therefore, one mole of CH_4 (which has a mass of 16 g) contains one mole of carbon atoms and four moles of hydrogen atoms. Percentage is the part divided by the whole times 100 percent (or simply parts per 100). We can represent the percent composition of methane as:

When chemists use % notation, they mean % by mass, unless otherwise specified.

mass of one mol of C atoms

$$\% \text{ C} = \frac{C}{CH_4} \times 100\% = \frac{12 \text{ g}}{16 \text{ g}} \times 100\% = \underline{75\% \text{ C}}$$

mass of one mol of CH_4 molecules

mass of four mol of H atoms

$$\% \text{ H} = \frac{4 \times H}{CH_4} \times 100\% = \frac{4 \times 1.0 \text{ g}}{16 \text{ g}} \times 100\% = \underline{25\% \text{ H}}$$

mass of one mol of CH_4 molecules

We have calculated that methane is 75% carbon and 25% hydrogen by mass. Strictly speaking, percentages must always add to 100%, although on occasion round-off errors may not cancel and totals such as 99.9% or 100.1% may be obtained.

Example 2–9

Calculate the percent composition of HNO_3.

Solution

The mass of one mole of HNO_3 is calculated first.

	No. of mol of stated kind	Mass of one mol	
$1 \times H =$	1	$\times\ 1.0$ g	$=\ \ 1.0$ g
$1 \times N =$	1	$\times 14.0$ g	$= 14.0$ g
$3 \times O =$	3	$\times 16.0$ g	$= 48.0$ g

Mass of 1 mole of $HNO_3 = 63.0$ g

Now, its percent composition is

$$\% \text{ H} = \frac{H}{HNO_3} \times 100\% = \frac{1.0 \text{ g}}{63.0 \text{ g}} \times 100\%$$
$$= 1.6\% \text{ H}$$

$$\% \text{ N} = \frac{N}{HNO_3} \times 100\% = \frac{14.0 \text{ g}}{63.0 \text{ g}} \times 100\%$$
$$= 22.2\% \text{ N}$$

$$\% \text{ O} = \frac{3 \times O}{HNO_3} \times 100\% = \frac{48.0 \text{ g}}{63.0 \text{ g}} \times 100\%$$
$$= 76.2\% \text{ O}$$

Total $= 100.0\%$

We have calculated that nitric acid is 1.6% H, 22.2% N, and 76.2% O by mass. All samples of pure HNO_3 have this composition, according to the Law of Constant Composition.

Nonstoichiometric Compounds

The Law of Constant Composition (Definite Proportions) is one of the basic tenets of chemistry, and its validity has been demonstrated for many thousands of compounds. However, there are many *solid* compounds that do not obey the "law." They are called *nonstoichiometric compounds* or "Berthollides" in honor of a French chemist, Claude Berthollet, who argued (prior to 1808) that the elemental composition of any compound could vary over wide limits depending on how it was prepared. Berthollet was refuted by Joseph Proust, who showed that nearly all compounds *do* have definite composition, and that Berthollet's "compounds" were actually mixtures. Many other compounds, however, including metal oxides, sulfides, and hydrides, are indeed nonstoichiometric compounds. Typical examples include $Ni_{0.97}O$, $TiO_{1.7-1.8}$, $FeO_{1.055}$, $Cu_{1.7}S$, $CeH_{2.69}$, and $VH_{0.56}$. Many minerals contain nonstoichiometric compounds. A number of these compounds are *superconductors*, substances that conduct electricity very efficiently at low temperatures.

2–5   Derivation of Formulas from Elemental Composition

Each year thousands of new compounds are made in laboratories or discovered in nature. One of the first steps in characterizing a new compound is the determination of its percent composition. A *qualitative* analysis is performed to determine which elements are present in the compound. Then a *quantitative* analysis is performed to determine the percent by mass of each element.

Carbon, hydrogen, and oxygen are found in millions of compounds, and C and H analyses are conveniently performed in a C-H combustion system (Figure 2–4).

A sample of a compound whose mass is carefully determined is burned in a furnace in a stream of oxygen so that all the carbon and hydrogen in the sample are converted to carbon dioxide and water. The resulting increases in masses in the CO_2

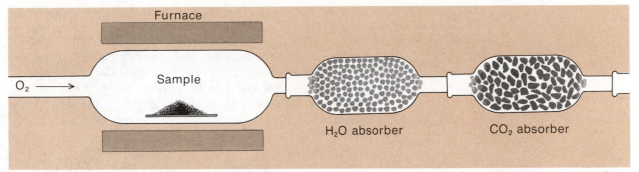

FIGURE 2–4   Combustion train used for carbon-hydrogen analysis. The absorbent for water is magnesium perchlorate, $Mg(ClO_4)_2$. Carbon dioxide is absorbed by finely divided sodium hydroxide supported on asbestos. Only a few milligrams of sample are needed for an analysis.

and H_2O absorbers can then be related to the masses and percentages of carbon and hydrogen in the original sample. Example 2–10 illustrates the procedure.

Example 2–10
A 0.1014 gram sample of purified glucose was burned in a C-H combustion train to produce 0.1486 gram of CO_2 and 0.0609 gram of H_2O. An elemental analysis showed that glucose contains only carbon, hydrogen, and oxygen. Determine the masses of C, H, and O in the sample and the percentages of these elements in glucose. (Glucose, a simple sugar, is one of the products of carbohydrate metabolism and also the main component of intravenous feeding liquids.)

Solution
We first calculate the mass of carbon that was converted to 0.1486 gram of CO_2. There is one mole of carbon atoms, 12.01 grams, in every mole of CO_2, 44.01 grams.

$$\underline{?}\text{ g C} = 0.1486 \text{ g } CO_2 \times \frac{1 \text{ mol } CO_2}{44.01 \text{ g } CO_2}$$

$$\times \frac{12.01 \text{ g C}}{1 \text{ mol } CO_2} = \underline{0.0406 \text{ g C}}$$

Likewise, we can calculate the amount of hydrogen in the original sample from the fact that there are two moles of hydrogen atoms, 2.016 grams, per mole of H_2O, 18.02 grams.

$$\underline{?}\text{ g H} = 0.0609 \text{ g } H_2O \times \frac{1 \text{ mol } H_2O}{18.02 \text{ g } H_2O}$$

$$\times \frac{2.016 \text{ g H}}{1 \text{ mol } H_2O} = \underline{0.00681 \text{ g H}}$$

The mass of oxygen in the sample is calculated by difference, since glucose has been shown to contain only C, H, and O.

$$\underline{?}\text{ g O} = 0.1014 \text{ g sample}$$
$$- [0.0406 \text{ g C} + 0.00681 \text{ g H}]$$
$$= \underline{0.0540 \text{ g O}}$$

The percentages by mass are calculated using the relationship

$$\% \text{ element} = \frac{\text{g element}}{\text{g sample}} \times 100\%$$

for each element in turn:

$$\% \text{ C} = \frac{0.0406 \text{ g C}}{0.1014 \text{ g}} \times 100\% = \underline{40.0\% \text{ C}}$$

$$\% \text{ H} = \frac{0.00681 \text{ g H}}{0.1014 \text{ g}} \times 100\% = \underline{6.72\% \text{ H}}$$

$$\% \text{ O} = \frac{0.0540 \text{ g O}}{0.1014 \text{ g}} \times 100\% = \underline{53.3\% \text{ O}}$$

Total = 100.0%

Once the percentage composition of a compound (or its elemental composition by mass) is known, the simplest formula can be determined. The **simplest** or **empirical formula** for a compound is the smallest whole-number ratio of atoms present in the compound. For ionic compounds the simplest formula is used to represent the compound. For molecular compounds the **molecular formula** indicates the *actual* numbers of atoms present in a molecule of the compound. It may be the same as the simplest formula or else some multiple of it. For example, the simplest and molecular formulas for water are both H_2O. However, for hydrogen peroxide they are HO and H_2O_2, respectively.

Example 2–11

Analysis of a sample of a pure compound reveals that it contains 50.1% sulfur and 49.9% oxygen by mass. What is the simplest formula?

Solution

The ratio of moles of atoms of elements in a compound is the same as the ratio of atoms in that compound, because one mole of atoms of any element is 6.02×10^{23} atoms of the element.* If we calculate the number of moles of atoms of each element in a sample of a compound, we can obtain the ratio that gives the relative number of atoms in the compound. Let's consider 100.0 grams of the compound, which must contain 50.1 grams of S and 49.9 grams of O. We shall calculate the number of moles of atoms of each (as we did in Section 2–3).

$$\underline{?} \text{ mol S atoms} = 50.1 \text{ g S} \times \frac{1 \text{ mol S atoms}}{32.1 \text{ g S}}$$

$$= 1.56 \text{ mol S atoms}$$

$$\underline{?} \text{ mol O atoms} = 49.9 \text{ g O} \times \frac{1 \text{ mol O atoms}}{16.0 \text{ g O}}$$

$$= 3.12 \text{ mol O atoms}$$

Now that we know that 100.0 grams of compound contains 1.56 moles of S atoms and 3.12 moles of O atoms, let's obtain a whole-number ratio between these numbers that gives the ratio of atoms in the simplest formula. A simple and useful way to obtain a whole-number ratio between two or more numbers is (1) divide each number by the smallest, and then (2) multiply the resulting numbers by the smallest whole number that will eliminate fractions, if necessary.

$$\frac{1.56}{1.56} = 1 \text{ S}$$

$$SO_2$$

$$\frac{3.12}{1.56} = 2 \text{ O}$$

This ratio 1:2 tells us that the simplest formula for the compound is SO_2. The solution of Example 2–11 is outlined at the top of the opposite page.

---

* This statement is analogous to the following statement. The ratio of dozens of brown eggs to dozens of white eggs in a box is the same as the ratio of brown eggs to white eggs. Suppose you have a box of eggs containing 50 dozen brown eggs and 25 dozen white eggs. The ratio of brown eggs to white eggs is

$$\frac{50 \text{ doz. brown eggs}}{25 \text{ doz. white eggs}} = \frac{2 \text{ brown eggs}}{1 \text{ white egg}}$$

or a simple 2:1 ratio. Similarly, if a compound contains two moles of atoms of element A and one mole of atoms of element B, the ratio of A atoms to B atoms is

$$\frac{2 \text{ moles A atoms}}{1 \text{ mole B atoms}} = \frac{2(6.02 \times 10^{23} \text{ A}) \text{ atoms}}{1(6.02 \times 10^{23} \text{ B}) \text{ atoms}} = \frac{2 \text{ A atoms}}{1 \text{ B atom}}$$

Element	Percentage by Mass	Relative Number of Moles of Atoms	Divided by Smaller Number	Smallest Whole-number Ratio of Atoms
sulfur	50.1%	$\dfrac{50.1}{32.1} = 1.56$	$\dfrac{1.56}{1.56} = 1$	1
oxygen	49.9%	$\dfrac{49.9}{16.0} = 3.12$	$\dfrac{3.12}{1.56} = 2$	2

SO_2

Example 2-12

A 20.000 gram sample of an ionic compound is found to contain 6.072 grams of Na, 8.474 grams of S, and 6.336 grams of O. What is its simplest formula?

Solution

As before, let's calculate the number of moles of atoms of each element.

$$\underset{?}{\underline{?}} \text{ mol Na atoms} = 6.072 \text{ g Na} \times \frac{1 \text{ mol Na atoms}}{23.0 \text{ g Na}}$$

$$= 0.264 \text{ mol Na atoms}$$

$$\underset{?}{\underline{?}} \text{ mol S atoms} = 8.474 \text{ g S} \times \frac{1 \text{ mol S atoms}}{32.1 \text{ g S}}$$

$$= 0.264 \text{ mol S atoms}$$

$$\underset{?}{\underline{?}} \text{ mol O atoms} = 6.336 \text{ g O} \times \frac{1 \text{ mol O atoms}}{16.0 \text{ g O}}$$

$$= 0.396 \text{ mol O atoms}$$

We now divide each number by the smallest number to obtain the ratio of atoms in the simplest formula.

$$\frac{0.264}{0.264} = 1.00 \text{ Na}; \quad \frac{0.264}{0.264} = 1.00 \text{ S}; \quad \frac{0.396}{0.264} = 1.50 \text{ O}$$

The ratio of atoms in the simplest formula *must be a whole number ratio* (by definition). To convert this ratio, $1:1:1.5$, to a whole number ratio, each number in the ratio must be multiplied by 2, which gives the simplest formula.

$$1.00 \text{ Na} \times 2 = 2 \text{ Na}$$
$$1.00 \quad \text{S} \times 2 = 2 \text{ S} \qquad \longrightarrow \quad Na_2S_2O_3$$
$$1.50 \quad \text{O} \times 2 = 3 \text{ O}$$

In this procedure we often obtain numbers like 0.99 and 1.01 or 1.49 and 1.52. Since there is always some error in results obtained by analysis of samples (as well as round-off errors), we *assume* that 0.99 and 1.01 are equal to 1.0 and that 1.49 and 1.52 are equal to 1.5.

The general procedure for obtaining a *whole-number ratio* among numbers that contain one number that is a fraction is: (1) Express any decimal fraction as a common fraction (for instance, $0.5 = \frac{1}{2}$) and then (2) multiply each number in the ratio by the *denominator of the common fraction*.

Recall that for many compounds the molecular formula is a multiple of the simplest formula. Consider butane, C_4H_{10}. The simplest formula for butane is C_2H_5, but the molecular formula contains twice as many atoms, i.e., $(C_2H_5)_2 = C_4H_{10}$. Benzene, C_6H_6, is another example. The simplest formula for benzene is CH, but the molecular formula contains six times as many atoms, i.e., $(CH)_6 = C_6H_6$.

Percent composition data yield only simplest formulas. In order to determine the molecular formula for a molecular compound, *both* its simplest formula and its molecular weight must be known. Methods for experimental determination of molecular weights will be introduced in Chapter 11.

Example 2–13

In Example 2–10 we found that glucose is 40.0% C, 6.72% H, and 53.3% O. Other experiments show that its molecular weight is approximately 180. Determine the simplest formula and the molecular formula of glucose.

Solution

For simplicity we can deal with 100.0 grams of glucose, which must contain 40.0 grams C, 6.72 grams H, and 53.3 grams O. We first calculate the number of moles of atoms that these masses represent.

$$? \text{ mol C atoms} = 40.0 \text{ g C} \times \frac{1 \text{ mol C atoms}}{12.01 \text{ g C}}$$

$$= 3.33 \text{ mol C atoms}$$

$$? \text{ mol H atoms} = 6.72 \text{ g H} \times \frac{1 \text{ mol H atoms}}{1.008 \text{ g H}}$$

$$= 6.66 \text{ mol H atoms}$$

$$? \text{ mol O atoms} = 53.3 \text{ g O} \times \frac{1 \text{ mol O atoms}}{16.00 \text{ g O}}$$

$$= 3.33 \text{ mol O atoms}$$

We now obtain ratios among these numbers of moles of atoms by dividing each by the smallest number.

$$\frac{3.33}{3.33} = 1.00 \text{ C}, \quad \frac{6.66}{3.33} = 2.00 \text{ H}, \quad \frac{3.33}{3.33} = 1.00 \text{ O}$$

The simplest formula is CH_2O, which has a formula weight of 30.02 amu. Since the molecular weight of glucose is approximately 180 amu, we can determine the molecular formula by (1) dividing the molecular weight by the simplest formula weight, which gives an integer, and (2) multiplying the simplest formula by this integer.

$$\frac{180 \text{ amu}}{30.02 \text{ amu}} = 6$$

The molecular weight is six times the simplest formula weight, so the molecular formula of glucose is $C_6H_{12}O_6$.

Mass Relationships and Chemical Reactions—Reaction Stoichiometry

We have now completed our study of composition stoichiometry, the mass relationships that describe individual substances. Now we shall extend our coverage to mass relationships that describe *chemical changes*. This will allow us to calculate such things as the mass of one substance that will react with a specified mass of another, or the mass of a particular product of a chemical reaction that can be produced from specified masses of reacting substances. Let us begin by considering the use of chemical equations to describe chemical reactions. We shall observe that chemical equations are a very precise as well as a very versatile language. Learning to balance and *to interpret* chemical equations will be useful because chemical equations are used in all areas of science.

2–6 Chemical Equations

Whether we are planning to manufacture a new plastic, trying to discover a new synthetic fuel, or studying the biochemical processes of living things, we must have a way to describe the chemical reactions involved. Chemical reactions always involve changing one or more substances into one or more different substances. That is, they involve regrouping atoms or ions to form other substances.

Chemical equations *are used to describe chemical reactions,* and they show (1) *the substances that react,* called **reactants,** (2) *the substances formed,* called **products,** and (3) *the relative amounts of substances involved.* As a typical example, let's consider the combustion (burning) of natural gas, a reaction used to heat buildings and cook foods. Natural gas is a mixture of several substances, but the principal component is methane, CH_4. The equation that describes the reaction of methane with oxygen is

Reactants are shown on the left and products on the right side of a chemical equation. The arrow may be read as "yields" or "gives."

$$\underbrace{CH_4 + 2O_2}_{\text{reactants}} \longrightarrow \underbrace{CO_2 + 2H_2O}_{\text{products}}$$

What does this equation tell us? In the simplest terms, it tells us that methane reacts with oxygen to produce carbon dioxide, CO_2, and water. More specifically, it says that *one* molecule of methane reacts with *two* molecules of oxygen to produce *one* molecule of carbon dioxide and *two* molecules of water. That is,

A balanced chemical equation may be interpreted on a molecular (or formula unit) basis.

CH_4	+	$2O_2$	$\longrightarrow$	CO_2	+	$2H_2O$
1 molecule		2 molecules		1 molecule		2 molecules

Figure 2–5 shows the rearrangement of atoms described by this equation.

The Law of Conservation of Matter provides the basis for "balancing" chemical equations and for calculations based on those equations. Since matter is neither created nor destroyed during a chemical reaction, and since atoms of one kind are not changed into atoms of another kind in such a change (remember Dalton?), a balanced chemical equation must always include the same number of each kind of atom on both sides of the equation. Since we are not dealing with a mathematical identity, an equal sign is inappropriate; instead we use an arrow (read "yields") to indicate the conversion of reactants into products.

For example, hydrogen and oxygen react to form water. This is a simple statement of fact based on numerous observations. It can be represented by the *unbalanced* "equation" that shows the formulas of the substances involved:

$$H_2 + O_2 \longrightarrow H_2O$$

Notice that substances are represented by formulas that describe them *as they exist.* We write H_2 to represent diatomic hydrogen molecules, not H which represents hydrogen atoms. But as it now stands, the "equation" does not satisfy the Law

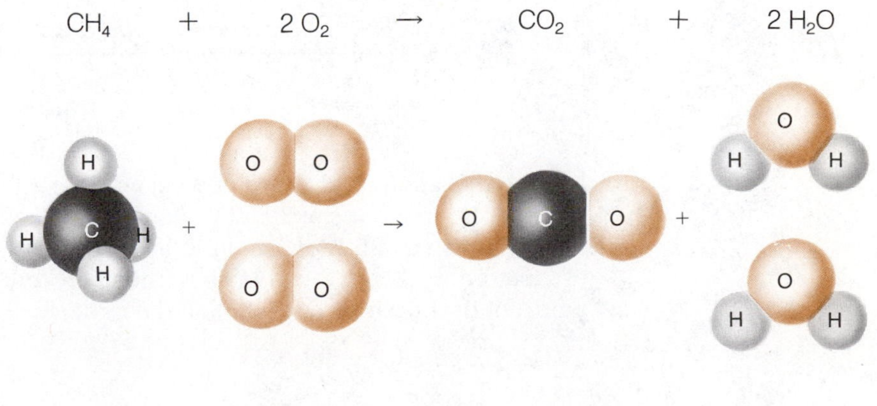

FIGURE 2–5 The reaction of methane with oxygen to form carbon dioxide and water.

$$CH_4 \quad + \quad 2\,O_2 \quad \rightarrow \quad CO_2 \quad + \quad 2\,H_2O$$

of Conservation of Matter because there are two oxygen atoms in the O_2 molecule (left side) and only one oxygen atom in the water molecule (right side).

$$H_2 + O_2 \longrightarrow H_2O$$

Tally:

atoms of	in reactants	in products
H	2	2
O	2	1

In order to account for both oxygen atoms, *two* water molecules must be formed. *Two* H_2O molecules require two H_2 molecules to provide four hydrogen atoms. Thus, the equation is balanced by placing a 2 before the formulas of hydrogen and water. (Numbers placed before formulas are called **coefficients,** and they indicate the number of formula units of that substance involved in the reaction.) The properly balanced equation then becomes:

$$2H_2 + O_2 \xrightarrow{\Delta} 2H_2O$$

Tally:

atoms of	in reactants	in products
H	4	4
O	2	2

The equation is now balanced; that is, it conforms to the Law of Conservation of Matter because there are *equal numbers* of hydrogen and oxygen atoms on both sides of the equation.

Special conditions required for some reactions are indicated by notation placed over the arrow. In the example above, the capital Greek letter delta (Δ) placed over the arrow tells us that heat is necessary to start this reaction.

Formulas must be written correctly for all reactants and products before attempting to balance an equation. Once this is done, the subscripts in the formulas cannot be changed, because doing so would violate the Law of Constant Composition. Formulas represent experimentally determined ratios of masses of elements, and therefore ratios of numbers of atoms, and cannot be changed. Another way to look at it is that different subscripts imply different compounds, so that the equation would no longer describe the same reaction.

Let us generate the balanced equation for the reaction of aluminum metal with hydrochloric acid (hydrogen chloride dissolved in water) to produce aluminum chloride and hydrogen. The procedure is similar to that used for the hydrogen-oxygen reaction, but is slightly more complicated. The unbalanced "equation" is

$$Al + HCl \longrightarrow AlCl_3 + H_2$$

Tally:

atoms of	in reactants	in products
Al	1	1
H	1	2
Cl	1	3

The equation is not balanced.

Note that H atoms and Cl atoms must occur in a 1 : 1 ratio on the left side, while Cl atoms occur in groups of 3 and H atoms occur in groups of 2 on the right side. The least common multiple of 3 and 2 is 6. To balance the equation, we place coefficients of 2 before $AlCl_3$ and 3 before H_2 so that there will be six atoms of chlorine and six atoms of hydrogen (a 1 : 1 ratio) on the right side.

$$Al + HCl \longrightarrow 2AlCl_3 + 3H_2$$

Tally:

atoms of	in reactants	in products
Al	1	2
H	1	6
Cl	1	6

The equation is still not balanced.

Further inspection shows that the coefficient of Al should be 2 and that of HCl should be 6 in the balanced equation:

$$2Al \quad + \quad 6HCl \quad \longrightarrow \quad 2AlCl_3 \quad + \quad 3H_2$$

aluminum   hydrochloric   aluminum chloride   hydrogen
           acid

The equation is now balanced.

Tally:

atoms of	in reactants	in products
Al	2	2
H	6	6
Cl	6	6

2-7 Calculations Based on Chemical Equations

As we indicated, chemical equations represent a very precise and versatile language. We are now ready to use them to calculate the amounts of substances involved in chemical reactions. Let us consider as an example the high-temperature reaction of iron(III) oxide, or ferric oxide, Fe_2O_3, with carbon monoxide, CO, to produce elemental iron, Fe, and carbon dioxide. The balanced equation is

This is an important industrial reaction in which Fe_2O_3 from ores of iron is converted to iron in blast furnaces (Section 20-6).

$$Fe_2O_3 + 3CO \xrightarrow{\Delta} 2Fe + 3CO_2$$

On a quantitative basis, *at the molecular level,* the equation tells us

Fe_2O_3 is an ionic substance ($2Fe^{3+}$, $3O^{2-}$), so we refer to formula units rather than molecules of Fe_2O_3.

$$Fe_2O_3 \quad + \quad 3CO \quad \longrightarrow \quad 2Fe \quad + \quad 3CO_2$$

1 formula unit   3 molecules   2 atoms   3 molecules
of iron(III) oxide   of carbon monoxide   of iron   of carbon dioxide

---

Example 2-14

How many molecules of CO are required to react with 50 formula units of Fe_2O_3 according to the above equation?

Solution

The balanced equation tells us that *one* formula unit of Fe_2O_3 reacts with *three* molecules of CO. We can construct two unit factors from this fact,

$$\frac{1 \text{ formula unit } Fe_2O_3}{3 \text{ molecules CO}} \quad \text{and} \quad \frac{3 \text{ molecules CO}}{1 \text{ formula unit } Fe_2O_3}$$

These are unit factors for *this* reaction because the numerator and denominator are *chemically equivalent,* i.e., they react exactly with each other. We use the second unit factor here.

$$\underline{?} \text{ molecules CO} = 50 \text{ formula units } Fe_2O_3$$
$$\times \frac{3 \text{ molecules CO}}{1 \text{ formula unit } Fe_2O_3}$$
$$= \underline{150 \text{ molecules CO}}$$

The interpretation of chemical equations in terms of molecules is of little practical use, since we cannot handle individual molecules. This interpretation may, however, be multiplied by a convenient factor. Suppose we multiply the interpretation above by, say, 12, which gives us

$$Fe_2O_3 \quad + \quad 3CO \quad \longrightarrow \quad 2Fe \quad + \quad 3CO_2$$

12 formula units + 36 molecules of ⟶ 24 atoms of + 36 molecules of
of iron(III) oxide   carbon monoxide   iron   carbon dioxide

or

1 dozen   3 dozen   2 dozen   3 dozen
formula units + molecules of ⟶ atoms of + molecules of
of iron(III) oxide   carbon monoxide   iron   carbon dioxide

This interpretation is still not very useful, though. Suppose that we multiply the original molecular interpretation by Avogadro's number, 6.022×10^{23}, which gives us

$$Fe_2O_3 \quad + \quad 3CO \quad \longrightarrow \quad 2Fe \quad + \quad 3CO_2$$

$1(6.022 \times 10^{23})$ $3(6.022 \times 10^{23})$ $2(6.022 \times 10^{23})$ $3(6.022 \times 10^{23})$
formula units   molecules of   atoms of iron   molecules of
of iron(III) oxide   carbon monoxide   carbon dioxide

Since Avogadro's number of any kind of chemical entities constitutes *one mole* of that substance, this interpretation can be written in what is certainly its *most useful* form,

The equation is interpreted on a mole basis.

$$Fe_2O_3 + 3\,CO \longrightarrow 2\,Fe + 3CO_2$$

1 mol   3 mol   2 mol   3 mol

which tells us that *one* mole of iron(III) oxide reacts with *three* moles of carbon monoxide to produce *two* moles of iron and *three* moles of carbon dioxide. Since we know the mass of one mole of each of these substances, we can also write

$$Fe_2O_3 + 3CO \longrightarrow 2Fe + 3CO_2$$

1 mol   3 mol   2 mol   3 mol
160 g   3(28 g)   2(56 g)   3(44 g)
       84 g   112 g   132 g

⎵ 244 g reactants   ⎵ 244 g products

We have rounded atomic and formula (molecular) weights to whole numbers for convenience.

The equation now tells us that 160 grams of Fe_2O_3 reacts with 84 grams of CO to produce 112 grams of Fe and 132 grams of CO_2, and that the Law of Conservation of Matter is satisfied. Thus, we have seen how our original interpretation of a balanced chemical equation in terms of relative numbers of formula units, molecules, and atoms of reactants and products (which we cannot actually count or handle) can be reinterpreted in terms of *relative masses* of reactants and products, which are conveniently determined in a laboratory or factory. Chemical equations describe *reaction ratios* (i.e., *the mole ratios of reactants and products*) as well as the *relative masses* of reactants and products, as Example 2–15 illustrates.

Example 2–15

What mass of carbon monoxide is required to react with 279 grams of Fe_2O_3?

Solution

Recall the balanced equation

$$Fe_2O_3 + 3CO \longrightarrow 2Fe + 3CO_2$$

1 mol	3 mol	2 mol	3 mol
160 g	3(28 g)	2(56 g)	3(44 g)

According to the mole method, we first convert 279 g Fe_2O_3 to mol Fe_2O_3 by using the unit factor 1 mol Fe_2O_3/160 g Fe_2O_3.

$$\text{? mol } Fe_2O_3 = 279 \text{ g } Fe_2O_3 \times \frac{1 \text{ mol } Fe_2O_3}{160 \text{ g } Fe_2O_3}$$

$$= 1.74 \text{ mol } Fe_2O_3$$

Then a unit factor relating mol CO consumed to mol Fe_2O_3 consumed, obtained from the balanced equation, is used to convert 1.74 mol Fe_2O_3 to mol CO.

$$\text{? mol CO} = 1.74 \text{ mol } Fe_2O_3 \times \frac{3 \text{ mol CO}}{1 \text{ mol } Fe_2O_3}$$

$$= 5.22 \text{ mol CO}$$

Finally, a third unit factor, 28.0 g CO/1 mol CO, is used to convert 5.22 mol CO to g CO.

$$\text{? g CO} = 5.22 \text{ mol CO} \times \frac{28.0 \text{ g CO}}{1 \text{ mol CO}}$$

$$= 146 \text{ g CO}$$

Alternatively, we could have performed the calculations in one step:

$$\text{? g CO} = 279 \text{ g } Fe_2O_3 \times \frac{1 \text{ mol } Fe_2O_3}{160 \text{ g } Fe_2O_3}$$

$$\times \frac{3 \text{ mol CO}}{1 \text{ mol } Fe_2O_3} \times \frac{28.0 \text{ g CO}}{1 \text{ mol CO}} = 146 \text{ g CO}$$

A third approach is simply to construct a unit factor involving the masses of chemically equivalent amounts of Fe_2O_3 and CO, which must be obtained from the mole ratio given by the balanced chemical equation.

$$\text{? g CO} = 279 \text{ g } Fe_2O_3 \times \frac{3(28.0 \text{ g}) \text{ CO}}{160 \text{ g } Fe_2O_3}$$

$$= 146 \text{ g CO}$$

The same answer, 146 g CO, is obtained by all three methods.

The question in Example 2–15 may be reversed, as in Example 2–16.

Example 2–16

What mass of Fe_2O_3, in grams, is required to react with 146 grams of CO?

Solution

We use the reciprocals of the unit factors used in Example 2–15.

$$\text{? g } Fe_2O_3 = 146 \text{ g CO} \times \frac{1 \text{ mol CO}}{28.0 \text{ g CO}}$$

$$\times \frac{1 \text{ mol } Fe_2O_3}{3 \text{ mol CO}} \times \frac{160 \text{ g } Fe_2O_3}{1 \text{ mol } Fe_2O_3}$$

$$= 278 \text{ g } Fe_2O_3$$

Note that, within round-off accuracy, this is the same amount of Fe_2O_3 as was used to react with 146 grams of CO in Example 2–15.

Let us consider an additional example that relates an amount of a product formed in a reaction to an amount of a reactant consumed.

Example 2–17

Calculate the mass of iron that can be produced by complete reaction of 5.0 moles of Fe_2O_3 with an excess of CO according to the equation of the preceding examples. (An excess of CO is more than enough CO to allow all the Fe_2O_3 to react.)

Solution

Recall the balanced equation

$$Fe_2O_3 + 3CO \longrightarrow 2Fe + 3CO_2$$

1 mol	3 mol	2 mol	3 mol
160 g	3(28 g)	2(56 g)	3(44 g)
		112 g	

which tells us that one mole of Fe_2O_3 produces two moles of Fe (112 g). Since these quantities are *chemically equivalent*, we can construct unit factors that (1) relate mol Fe_2O_3 to mol Fe and (2) relate mol Fe to g Fe.

$$\underline{?} \text{ g Fe} = 5.0 \text{ mol Fe}_2\text{O}_3 \times \frac{2 \text{ mol Fe}}{1 \text{ mol Fe}_2\text{O}_3}$$

$$\times \frac{56 \text{ g Fe}}{1 \text{ mol Fe}} = \underline{5.6 \times 10^2 \text{ g Fe}}$$

In a general way, the equation described above can be written as

Fe_2O_3	+	3CO	$\longrightarrow$	2Fe	+	3CO$_2$
1 mol		3 mol		2 mol		3 mol
160 g		84 g		112 g		132 g

or

$$160 \text{ mass units} + 84 \text{ mass units} \longrightarrow 112 \text{ mass units} + 132 \text{ mass units}$$

The Law of Conservation of Matter allows us to use any mass or weight units we choose, as long as we are consistent. The equation might read

Fe_2O_3	+	3CO	$\longrightarrow$	2Fe	+	3CO$_2$
160 kg		84 kg		112 kg		132 kg

or

$$160 \text{ tons} + 84 \text{ tons} \longrightarrow 112 \text{ tons} + 132 \text{ tons}$$

There are 2000 lb in one short ton of *anything*. Thus the 160:84:112:132 ratio holds for this set, and for any other consistent set, of mass or weight units.

Example 2–18

What weight of CO, in tons, is required to react completely with 12 tons of Fe_2O_3?

Solution

The balanced equation enables us to construct two unit factors relating tons of CO and tons of Fe_2O_3:

$$\frac{160 \text{ tons Fe}_2\text{O}_3}{84 \text{ tons CO}} \quad \text{and} \quad \frac{84 \text{ tons CO}}{160 \text{ tons Fe}_2\text{O}_3}$$

$$\underline{?} \text{ tons CO} = 12 \text{ tons Fe}_2\text{O}_3 \times \frac{84 \text{ tons CO}}{160 \text{ tons Fe}_2\text{O}_3}$$

$$= \underline{6.3 \text{ tons CO}}$$

An alternate solution using the mole method follows:

$$\underline{?} \text{ tons CO} = 12 \text{ tons Fe}_2\text{O}_3$$

$$\times \frac{2000 \text{ lb Fe}_2\text{O}_3}{1 \text{ ton Fe}_2\text{O}_3} \times \frac{454 \text{ g Fe}_2\text{O}_3}{1 \text{ lb Fe}_2\text{O}_3}$$

$$\times \frac{1 \text{ mol Fe}_2\text{O}_3}{160 \text{ g Fe}_2\text{O}_3}$$

$$\times \frac{3 \text{ mol CO}}{1 \text{ mol Fe}_2\text{O}_3} \times \frac{28.0 \text{ g CO}}{1 \text{ mol CO}}$$

$$\times \frac{1 \text{ lb CO}}{454 \text{ g CO}} \times \frac{1 \text{ ton CO}}{2000 \text{ lb CO}}$$

$$= \underline{6.3 \text{ tons CO}}$$

2-8   Percent Purity

Let us now expand our horizons by recognizing that most substances are not 100 percent pure. When impure substances are used, as they frequently are, account must be taken of any and all impurities. Purities (and impurities) are usually indicated by "percent purity" or some such label. For example, a sample of sugar that is 95.00% pure contains 95.00 mass units of pure sugar and 5.00 mass units of impurities. To illustrate the concept, suppose we have a sample that is 95.00 percent sugar and 5.00 percent salt by mass. From the percent composition we can write down several unit factors,

$$\frac{95.00 \text{ g sugar}}{100.00 \text{ g mixture}}, \qquad \frac{5.00 \text{ g salt}}{100.00 \text{ g mixture}}, \qquad \frac{95.00 \text{ g sugar}}{5.00 \text{ g salt}}$$

and, of course, the inverse of each of these. The six unit factors may be used just as we have used other unit factors, as Examples 2-19 and 2-20 illustrate.

Example 2-19

What mass of a mixture that contains 95.00 percent sugar and 5.00 percent salt by mass will provide 230 g of sugar?

Solution

$$? \text{ g mixture} = 230 \text{ g sugar} \times \frac{100.00 \text{ g mixture}}{95.00 \text{ g sugar}}$$

$$= \underline{242 \text{ g mixture}}$$

Example 2-20

Calculate the masses of salt and sugar in 760 g of the mixture of Example 2-19.

Solution

$$? \text{ of sugar} = 760 \text{ g mixture} \times \frac{95.00 \text{ g sugar}}{100.00 \text{ g mixture}}$$

$$= \underline{722 \text{ sugar}}$$

$$? \text{ g salt} = 760 \text{ g mixture} \times \frac{5.00 \text{ g salt}}{100.00 \text{ g mixture}}$$

$$= \underline{38 \text{ g salt}}$$

Observe the beauty of the unit-factor approach to problem-solving! When unit factors are constructed, such questions as "Do we multiply by 0.9500 or do we divide by 0.9500?" never arise. Unit factors always point toward the correct answer because we construct, and use, unit factors so that units *always* cancel until we arrive at the desired unit.

2-9   Percent Yields from Chemical Reactions

Many chemical reactions do not go to completion, i.e., the reactants are not completely converted to products. In some cases, a particular set of reactants undergoes two or more reactions simultaneously, forming undesired products as well as desired products. (Reactions other than the desired one are called "side reactions.") The term **percent yield** is used to indicate how much of a desired product is obtained from a particular reaction. Thus far, we have assumed that the reactions we have considered do go to completion, that is, until one reactant has been used up to form only the desired products. The yields or amounts of products we have calculated are known as **theoretical** or **stoichiometric yields.** The **theoretical yield** of a chemical reaction *is the yield that would be obtained if the stated reaction went to completion.* The percent yield is a way of expressing how nearly complete the reaction was or how effectively the desired product was isolated. This useful quantity is defined as

$$\text{percent yield} = \frac{\text{actual yield of product}}{\text{theoretical yield of product}} \times 100\%$$

[Eq. 2–1]

Consider the preparation of nitrobenzene, $C_6H_5NO_2$, by the reaction of excess nitric acid, HNO_3, with a limited amount of benzene, C_6H_6. The balanced equation for the reaction may be written

$$C_6H_6 + HNO_3 \longrightarrow C_6H_5NO_2 + H_2O$$

| 1 mol | 1 mol | 1 mol | 1 mol |
| 78.1 g | 63.0 g | 123.1 g | 18.0 g |

It is usually not necessary to calculate formula weights for all substances in a balanced chemical equation. For instance, only the formula weights of C_6H_6 and $C_6H_5NO_2$ are needed in the following example.

> Sometimes it is difficult or impossible to isolate all of a desired product from a reaction mixture.

Example 2–21

A 15.6 gram sample of C_6H_6 reacts with excess HNO_3 to produce 18.0 grams of $C_6H_5NO_2$. What is the percent yield of $C_6H_5NO_2$ in this experiment?

Solution

First, let's calculate the theoretical yield of $C_6H_5NO_2$.

$$\underline{?}\ g\ C_6H_5NO_2 = 15.6\ g\ C_6H_6 \times \frac{1\ mol\ C_6H_6}{78.1\ g\ C_6H_6}$$

$$\times \frac{1\ mol\ C_6H_5NO_2}{1\ mol\ C_6H_6}$$

$$\times \frac{123.1\ g\ C_6H_5NO_2}{1\ mol\ C_6H_5NO_2}$$

$$= 24.6\ g\ C_6H_5NO_2$$

This tells us that if *all* the C_6H_6 were converted to $C_6H_5NO_2$, we should obtain 24.6 grams of $C_6H_5NO_2$ (100 percent yield). However, the reaction produces only 18.0 grams of $C_6H_5NO_2$, in *this* experiment, which is considerably less than a 100 percent yield.

$$\% \text{ yield} = \frac{\text{actual yield of product}}{\text{theoretical yield of product}} \times 100\%$$

$$= \frac{18.0\ g}{24.6\ g} \times 100\%$$

$$= 73.2 \text{ percent yield}$$

The amount of nitrobenzene is 73.2% of the amount expected if the reaction had gone to completion, there were no side reactions, and we could recover all of the $C_6H_5NO_2$ produced by the reaction.

2–10   The Limiting Reagent Concept

Thus far we have worked problems in which the presence of an excess of one reactant was stated or implied. The calculations were based on the substance present in lesser amount, called the **limiting reagent.** Before we study the concept of the limiting reagent in stoichiometry, let's develop the basic idea by considering some simple but analogous nonchemical examples.

Suppose you have 20 slices of ham and 36 slices of bread, and you wish to make as many ham sandwiches as possible using only one slice of ham and two slices of bread per sandwich. Obviously, you can make only 18 sandwiches, at which point you run out of bread. Therefore the bread is the "limiting reagent" and the two extra slices of ham are the "excess reagent."

Consider another example. Suppose you are given a box containing 93 bolts, 102 nuts, and 150 washers. How many sets consisting of one bolt, one nut, and two washers can you put together? Seventy-five sets will use up all the washers, and therefore the washers are the "limiting reagent." To be sure, 18 bolts and 27 nuts are left over. They are the "reagents present in excess."

Now let's apply the same reasoning process to chemical reactions.

Example 2–22

What mass of CO_2 can be produced by the reaction of 8.0 grams of CH_4 with 48 grams of O_2 according to the equation below?

$$CH_4 + 2O_2 \longrightarrow CO_2 + 2H_2O$$

Solution

We obtain the following information from the balanced equation:

$$CH_4 + 2O_2 \longrightarrow CO_2 + 2H_2O$$

| 1 mol | 2 mol | 1 mol | 2 mol |
| 16 g | 2(32 g) | 44 g | 2(18 g) |

This tells us that *one* mole of CH_4 reacts with *two* moles of O_2. Since we are given masses of both CH_4 and O_2, let's calculate the number of moles of each reactant.

$$? \text{ mol } CH_4 = 8.0 \text{ g } CH_4 \times \frac{1 \text{ mol } CH_4}{16 \text{ g } CH_4}$$

$$= 0.50 \text{ mol } CH_4$$

$$? \text{ mol } O_2 = 48 \text{ g } O_2 \times \frac{1 \text{ mol } O_2}{32 \text{ g } O_2} = 1.5 \text{ mol } O_2$$

Now, we return to the balanced equation, which tells us that the ratio of reactants that actually react is

| 1 mole of CH_4 | to | 2 moles of O_2 |

or

| 0.5 mole of CH_4 | to | 1 mole of O_2 |

but we actually have

| 0.5 mole of CH_4 | to | 1.5 moles of O_2 |

Thus, once the 0.5 mole of CH_4 has reacted with 1 mole of the O_2, the reaction must stop for lack of CH_4, but there will be 0.5 mole of O_2 left over. The CH_4 is the limiting reagent, and we must base the calculation on it.

$$? \text{ g } CO_2 = 8.0 \text{ g } CH_4 \times \frac{1 \text{ mol } CH_4}{16 \text{ g } CH_4}$$

$$\times \frac{1 \text{ mol } CO_2}{1 \text{ mol } CH_4} \times \frac{44 \text{ g } CO_2}{1 \text{ mol } CO_2}$$

$$= 22 \text{ g } CO_2$$

Thus, 22 grams of CO_2 is the maximum amount of CO_2 that can be produced from 8.0 grams of CH_4 and 48 grams of O_2. If the calculation had been based on O_2 rather than CH_4, the calculated answer would be too big and *wrong*.

Another approach to problems like Example 2–22 is to calculate the mass of one reactant required to react with the given mass of the other reactant. The calculation may be based on either moles or masses of reactants. Recall that Example 2–22 asked, "What mass of CO_2 can be produced by the reaction of 8.0 grams of CH_4 with 48 grams of O_2?"

$$CH_4 + 2O_2 \longrightarrow CO_2 + 2H_2O$$

1 mol   2 mol   1 mol   2 mol
16 g   2(32 g)   44 g   2(18 g)

First, let us calculate the mass of O_2 that reacts with 8.0 grams of CH_4.

$$? \text{ g } O_2 = 8.0 \text{ g } CH_4 \times \frac{1 \text{ mol } CH_4}{16 \text{ g } CH_4} \times \frac{2 \text{ mol } O_2}{1 \text{ mol } CH_4} \times \frac{32 \text{ g } O_2}{1 \text{ mol } O_2} = \underline{32 \text{ g } O_2}$$

This calculation tells us that 8.0 grams of CH_4 react with 32 grams of O_2. We have 48 grams of O_2, which is more than enough (16 grams of O_2 *in excess*) to react with 8.0 grams of CH_4. If we choose to calculate the amount of CH_4 that reacts with 48 grams of O_2, we obtain

$$? \text{ g } CH_4 = 48 \text{ g } O_2 \times \frac{1 \text{ mol } O_2}{32 \text{ g } O_2} \times \frac{1 \text{ mol } CH_4}{2 \text{ mol } O_2} \times \frac{16 \text{ g } CH_4}{1 \text{ mol } CH_4} = \underline{12 \text{ g } CH_4}$$

This calculation tells us that 48 grams of O_2 requires 12 grams of CH_4, and we only have 8.0 grams of CH_4. Both calculations show that CH_4 is the limiting reagent and the calculation *must* be based on CH_4.

Example 2–23

What is the maximum mass of $Al(OH)_3$ that can be prepared by reaction of 13.4 grams of $AlCl_3$ with 10.0 grams of NaOH according to the following equation?

$$AlCl_3 + 3NaOH \longrightarrow Al(OH)_3 + 3NaCl$$

Solution

Interpreting the balanced equation as usual, we have

$$AlCl_3 + 3NaOH \longrightarrow Al(OH)_3 + 3NaCl$$

1 mol   3 mol   1 mol   3 mol
133.4 g   3(40.0 g)   78.0 g   3(58.4 g)

We determine the number of moles of $AlCl_3$ and NaOH present.

$$? \text{ mol } AlCl_3 = 13.4 \text{ g } AlCl_3 \times \frac{1 \text{ mol } AlCl_3}{133.4 \text{ g } AlCl_3}$$

$$= \underline{0.100 \text{ mol } AlCl_3}$$

$$? \text{ mol NaOH} = 10.0 \text{ g NaOH} \times \frac{1 \text{ mol NaOH}}{40.0 \text{ g NaOH}}$$

$$= \underline{0.250 \text{ mol NaOH}}$$

The balanced equation tells us that the reaction ratio is

1 mole of $AlCl_3$ to   3 moles of NaOH

or

0.100 mole of $AlCl_3$ to   0.300 mole of NaOH

but we have

0.100 mole of $AlCl_3$ to   0.250 mole of NaOH

Thus, NaOH is the limiting reagent because there is insufficient NaOH to react with all the $AlCl_3$. The calculation of yield must be based on NaOH.

$$? \text{ g } Al(OH)_3 = 10.0 \text{ g NaOH} \times \frac{1 \text{ mol NaOH}}{40.0 \text{ g NaOH}}$$

$$\times \frac{1 \text{ mol } Al(OH)_3}{3 \text{ mol NaOH}} \times \frac{78.0 \text{ g } Al(OH)_3}{1 \text{ mol } Al(OH)_3}$$

$$= \underline{6.50 \text{ g } Al(OH)_3}$$

2-11   Two Reactions Occurring Simultaneously

The ideas discussed in this chapter are powerful tools that describe the quantitative relationships among substances as they undergo chemical reactions. To gain more appreciation of the versatility of these concepts, let us consider two chemical reactions occurring simultaneously to form a product common to both.

Example 2-24

A sample that contained 1.250 grams of sodium chloride, NaCl, and 1.750 grams of potassium chloride, KCl, was dissolved in water and mixed with an excess of silver nitrate solution. The precipitated silver chloride, AgCl, was then collected by filtration, washed, and dried. What mass of AgCl was produced? Both NaCl and KCl react with $AgNO_3$ to produce AgCl, an insoluble compound, according to the following equations. [Lower case letters in parentheses indicate the physical states in which substances exist: (s) = solid, (ℓ) = liquid, (g) = gas, (aq) refers to substances dissolved in water ("aqueous solution").]

$$NaCl\ (aq) + AgNO_3\ (aq) \longrightarrow$$
$$AgCl\ (s) + NaNO_3\ (aq)$$

$$KCl\ (aq) + AgNO_3\ (aq) \longrightarrow$$
$$AgCl\ (s) + KNO_3\ (aq)$$

Solution

These equations tell us that one mole of NaCl reacts with one mole of $AgNO_3$ to form one mole of AgCl, and that one mole of KCl reacts with one mole of $AgNO_3$ to produce one mole of AgCl, *a very important quantitative observation*. Therefore, we write the mass of one mole of each compound under its formula:

NaCl	+	$AgNO_3$	$\longrightarrow$	AgCl (s)	+	$NaNO_3$
1 mol		1 mol		1 mol		1 mol
58.44 g		169.87 g		143.32 g		84.99 g

KCl	+	$AgNO_3$	$\longrightarrow$	AgCl (s)	+	KNO_3
1 mol		1 mol		1 mol		1 mol
74.56 g		169.87 g		143.32 g		101.11 g

The equations now tell us that one mol or 58.44 g of NaCl produces one mol or 143.32 g of AgCl, and that one mol or 74.56 g of KCl produces one mol or 143.32 g of AgCl, which is the information we need to solve the problem. We can relate the masses of NaCl and KCl to the masses of AgCl they produce. The total mass of AgCl is the sum of the masses of AgCl produced from NaCl and from KCl.

$$\text{total g AgCl} = \underline{\hspace{1cm}} \text{ g AgCl from NaCl}$$
$$+ \underline{\hspace{1cm}} \text{ g AgCl from KCl}$$

First we calculate the mass of AgCl produced from 1.250 g NaCl.

$$\underline{?}\text{ g AgCl} = 1.250 \text{ g NaCl} \times \frac{1 \text{ mol NaCl}}{58.44 \text{ g NaCl}}$$
$$\times \frac{1 \text{ mol AgCl}}{1 \text{ mol NaCl}} \times \frac{143.32 \text{ g AgCl}}{1 \text{ mol AgCl}}$$
$$= 3.066 \text{ g AgCl (from NaCl)}$$

Then we calculate the mass of AgCl produced from 1.750 g KCl.

$$\underline{?}\text{ g AgCl} = 1.750 \text{ g KCl} \times \frac{1 \text{ mol KCl}}{74.56 \text{ g KCl}}$$
$$\times \frac{1 \text{ mol AgCl}}{1 \text{ mol KCl}} \times \frac{143.32 \text{ g AgCl}}{1 \text{ mol AgCl}}$$
$$= 3.364 \text{ g AgCl (from KCl)}$$

Now we add the two masses to get the total mass of AgCl.

$$\underline{?}\text{ g AgCl (total)} = 3.066 \text{ g AgCl} + 3.364 \text{ g AgCl}$$
$$= 6.430 \text{ g AgCl}$$

Example 2–25

A mixture of sodium chloride, NaCl, and potassium chloride, KCl, having a mass of 3.000 grams was dissolved in water, and an excess of silver nitrate, $AgNO_3$, was added. The precipitated silver chloride, AgCl, was collected by filtration, washed, dried, and found to have a mass of 6.184 grams. Calculate the masses and percentages of NaCl and KCl in the sample. (See Example 2–24.)

Solution

We can solve this problem because a common product, AgCl, is produced by both reactions. As in Example 2–24, we see that one mole of NaCl produces one mole of AgCl, and one mole of KCl produces one mole of AgCl, with the following relationships.

$$NaCl + AgNO_3 \longrightarrow AgCl\ (s) + NaNO_3$$

| 1 mol | 1 mol | 1 mol | 1 mol |
| 58.44 g | 169.87 g | 143.32 g | 84.99 g |

$$KCl + AgNO_3 \longrightarrow AgCl\ (s) + KNO_3$$

| 1 mol | 1 mol | 1 mol | 1 mol |
| 74.56 g | 169.87 g | 143.32 g | 101.11 g |

We wish to know the masses of NaCl and KCl in the original sample, so we use an algebraic representation for each. Let

x = g NaCl in original sample

Therefore,

$(3.000 - x)$ = g KCl in original sample

Since we know the *total* mass of AgCl produced, 6.184 g AgCl, we convert g NaCl to g AgCl and g KCl to g AgCl, and then set their sum equal to the total mass of AgCl.

_____ g AgCl from NaCl + _____ g AgCl from KCl

= _____ g AgCl total

$$\underbrace{(x\ \text{g NaCl}) \left(\frac{1\ \text{mol NaCl}}{58.44\ \text{g NaCl}}\right)\left(\frac{1\ \text{mol AgCl}}{1\ \text{mol NaCl}}\right)\left(\frac{143.32\ \text{g AgCl}}{1\ \text{mol AgCl}}\right)}_{\text{g AgCl from KCl}} + \underbrace{[(3.000 - x)\text{g KCl}]\left(\frac{1\ \text{mol KCl}}{74.56\ \text{g KCl}}\right)\left(\frac{1\ \text{mol AgCl}}{1\ \text{mol KCl}}\right)\left(\frac{143.32\ \text{g AgCl}}{1\ \text{mol AgCl}}\right)}_{\text{g Agcl from NaCl}}$$

$$\underbrace{= 6.184\ \text{g AgCl}}_{\text{g AgCl total}}$$

We now cancel units and simplify the equation.

$$\frac{143.32x}{58.44} + \frac{143.32\ (3.000 - x)}{74.56} = 6.184$$

$$\frac{143.32x}{58.44} + \frac{143.32\ (3.000)}{74.56} - \frac{143.32x}{74.56} = 6.184$$

Evaluating fractions gives

$$2.452x + 5.767 - 1.922x = 6.184$$

$$0.530x = 0.417$$

$$x = \frac{0.417}{0.530} = 0.787$$

so

g NaCl = x = 0.787 g NaCl   and

g KCl = $(3.000 - x)$ = 2.213 g KCl

Now that we know the masses of NaCl and KCl in the sample, the percentage of each can be calculated.

$$\% \text{ NaCl} = \frac{\text{g NaCl}}{\text{g sample}} \times 100\%$$

$$= \frac{0.787\ \text{g}}{3.000\ \text{g}} \times 100\% = \underline{26.2\% \text{ NaCl}}$$

$$\% \text{ KCl} = \frac{\text{g KCl}}{\text{g sample}} \times 100\%$$

$$= \frac{2.213\ \text{g}}{3.000\ \text{g}} \times 100\% = \underline{73.8\% \text{ KCl}}$$

More sophisticated chemical calculations are possible, but they can wait until later in the course or until subsequent courses. We have demonstrated that chemical equations are *precise* and *versatile.* Throughout this chapter we have used two fundamental ideas: the *Law of Constant Composition (Definite Proportions)* and the *Law of Conservation of Matter.* These ideas have been very helpful in understanding interactions among countless different kinds of matter.

Concentrations of Solutions — An Introduction

We have illustrated how the amounts of substances involved in chemical reactions can be expressed in moles *or* mass units. Often it is convenient to use solutions of chemicals in reactions. In order to know how much of a given substance is contained in a specified mass or volume of solution, it is necessary to become familiar with methods for expressing concentrations of solutions, our next topic.

A **solution** *is a homogeneous mixture of two or more substances.* Simple solutions usually consist of one substance (the solute) dissolved in another substance (the solvent). The **solute** *may be thought of as the dissolved substance* and the **solvent** *as the dissolving substance.* The solutions used in the laboratory are usually liquids, and the solvent is often water. For example, solutions of hydrochloric acid are prepared by dissolving pure hydrogen chloride (HCl, a gas at room temperature and atmospheric pressure) in water. Solutions of sodium hydroxide are prepared by dissolving solid NaOH in water.

In some solutions, such as a nearly equal mixture of ethyl alcohol and water, the distinction between solute and solvent is arbitrary.

The **concentration** of a solution *may be expressed as the amount of solute present in a given mass or volume of solution or solvent.*

2-12   Percent by Mass

Concentrations of solutions may be expressed in terms of percent by mass of solute, which gives the mass of solute per 100 mass units of solution. The gram is the usual mass unit.

$$\% \text{ solute} = \frac{\text{g solute}}{\text{g solution}} \times 100\%$$

[Eq. 2-2]

Unless we specify otherwise, the solvent is water in all solutions described in this text.

Thus, a solution that is 10.0% NaCl by mass contains 10.0 grams of NaCl in 100.0 grams of *solution.* Note that 100.0 grams of the solution contains 10.0 grams of NaCl in 90.0 grams of water. The density of a 10.0% solution of NaCl is 1.07 g/mL, so 100 mL of a 10.0% solution of NaCl has a mass of 107 grams. Observe that 100 grams of the solution does *not* occupy 100 mL. Unless otherwise specified, percent means percent *by mass.*

Example 2-26
Calculate the mass of sodium chromate, Na_2CrO_4, required to prepare 200 grams of a 20.0% solution of Na_2CrO_4.

Solution
The percentage information given in the problem tells us that the solution contains 20.0 grams of Na_2CrO_4 in 100 grams of solution. The desired information is the mass of Na_2CrO_4 in 200 grams of solution. A unit factor is constructed by placing 20.0

grams of Na_2CrO_4 over 100 grams of solution. Multiplication of the mass of the solution, 200 grams, by the unit factor gives the desired information.

$$? \text{ g } Na_2CrO_4 = 200 \text{ g soln} \times \frac{20.0 \text{ g } Na_2CrO_4}{100 \text{ g soln}}$$

$$= \underline{40.0 \text{ g } Na_2CrO_4}$$

Placing 100 grams of solution over 20.0 grams of Na_2CrO_4 gives another unit factor, as illustrated in the next example.

Example 2–27
Calculate the mass of a 20.0% solution of Na_2CrO_4 that contains 40.0 grams of Na_2CrO_4.

Solution

$$? \text{ g soln} = 40.0 \text{ g } Na_2CrO_4 \times \frac{100 \text{ g soln}}{20.0 \text{ g } Na_2CrO_4}$$

$$= \underline{200 \text{ g soln}}$$

Example 2–28
Calculate the mass of Na_2CrO_4 contained in 200 mL of a 20.0% solution of Na_2CrO_4. The density of the solution is 1.19 g/mL.

Solution

The volume of the solution multiplied by its density gives the mass of the solution. The mass of solution is then multiplied by the fraction of that mass due to

Na_2CrO_4 (20.0 g Na_2CrO_4/100 g soln) to give the mass of Na_2CrO_4 in 200 mL of solution.

$$? \text{ g } Na_2CrO_4 = \underbrace{200 \text{ mL soln} \times \frac{1.19 \text{ g soln}}{1.00 \text{ mL soln}}}_{238 \text{ g soln}}$$

$$\times \frac{20.0 \text{ g } Na_2CrO_4}{100 \text{ g soln}} = \underline{47.6 \text{ g } Na_2CrO_4}$$

Example 2–29
What volume of a solution containing 15.0% iron(III) nitrate contains 30.0 g of $Fe(NO_3)_3$? The density of the solution is 1.16 g/mL.

Note that the answer is not 200 mL, but considerably less because 1.00 mL of solution has a mass of 1.16 g. However, 172 mL of the solution has a mass of 200 g.

Solution

$$? \text{ mL soln} = \underbrace{30.0 \text{ g } Fe(NO_3)_3 \times \frac{100 \text{ g soln}}{15.0 \text{ g } Fe(NO_3)_3}}_{200 \text{ g soln}} \times \frac{1.00 \text{ mL soln}}{1.16 \text{ g soln}} = \underline{172 \text{ mL}}$$

2–13   Molarity (Molar Concentration)

Molarity (M), or molar concentration, is a common unit for expressing the concentrations of solutions. **Molarity** is defined as *the number of moles of solute per liter of solution*. Note that the definition of molarity specifies the amount of solute *per unit of volume,* whereas percent specifies the amount *per unit of mass.* Symbolically, molarity may be represented as:

$$\text{molarity} = \frac{\text{number of moles of solute}}{\text{number of liters of solution}} \qquad \text{[Eq. 2–3]}$$

To prepare one liter of a one molar solution, one mole of solute is placed in a one-liter volumetric flask, enough water is added to dissolve the solute, and water is then added until the volume of the solution is exactly one liter. (See Figure 2-6.) Students sometimes make the mistake of assuming that one molar solutions contain one mole of solute in a liter of water. This is *not* the case; one liter of water *plus* one mole of solute usually has a total volume of more than one liter.

Frequently it is convenient to express the volume of a solution in milliliters rather than in liters. Likewise, it may be convenient to express the amount of solute in millimoles (mmol) rather than in moles. Since one milliliter is $\frac{1}{1000}$ of a liter and one millimole is $\frac{1}{1000}$ of a mole, molarity may also be expressed as the number of millimoles of solute per milliliter of solution, i.e.,

$$\text{molarity} = \frac{\text{number of millimoles of solute}}{\text{number of milliliters of solution}} \qquad \text{[Eq. 2-4]}$$

Example 2-30

Calculate the molarity (*M*) of a solution that contains 3.65 grams of HCl in 2.00 liters of solution.

The concentration of the HCl solution is 0.0500 molar, and the solution is called 0.0500 *M* hydrochloric acid. One liter of the solution contains 0.0500 mole of HCl.

Solution

$$\underline{?}\ \frac{\text{mol HCl}}{\text{liter}} = \frac{3.65\ \text{g HCl}}{2.00\ \text{L}} \times \frac{1\ \text{mol HCl}}{36.5\ \text{g HCl}}$$

$$= \underline{0.0500\ \text{mol HCl/L}}$$

Observe that problems like the one above are worked by the familiar series of operations. The desired information, the molarity of the solution, is the number of moles of HCl in one liter of solution. Therefore, write down (on the left side of the equal sign) "? mol HCl/L." Next examine the problem to determine what information is given. In this case the mass of HCl, 3.65 grams, and the volume of the solution, 2.00 L, are given. In the desired answer the volume term (liters) will be in the denominator. Therefore, we place the mass of HCl, 3.65 grams, over the volume of solution in which it is contained, 2.00 L. The units are now g HCl/L, and the desired units are moles of HCl/L. Multiplication by the unit factor 1 mole HCl/36.5 grams HCl gives the answer, 0.0500 mol HCl/L.

Example 2-31

Calculate the molarity of a solution that contains 49.04 grams of H_2SO_4 in 250.0 mL of solution.

Solution

First we convert g H_2SO_4 to mol H_2SO_4; then we convert mL soln to L soln.

$$\underline{?}\ \frac{\text{mol } H_2SO_4}{\text{liter}} = \frac{49.04\ \text{g } H_2SO_4}{250.0\ \text{mL}}$$

$$\times \frac{1\ \text{mol } H_2SO_4}{98.08\ \text{g } H_2SO_4} \times \frac{1000\ \text{mL}}{1\ \text{L}}$$

$$= \frac{49.04 \times 1 \times 1000}{250.0 \times 98.08 \times 1}\ \text{mol } H_2SO_4/\text{L}$$

$$= \underline{2.000\ M\ H_2SO_4}$$

The concentration of the H_2SO_4 solution is 2.000 molar, and the solution is referred to as 2.000 *M* sulfuric acid. Note that Example 2-31 is worked exactly like Example 2-30, except that one additional factor is used to convert the volume of the solution from milliliters to liters.

(a) Weigh out 1.00 mol (58.4 g) of NaCl.

(b) Transfer NaCl into a 1.00 liter volumetric flask.

FIGURE 2-6 Preparation of 1.00 molar sodium chloride solution.

(c) Add enough distilled H₂O to dissolve NaCl.

(d) Carefully add distilled H₂O until liquid level stands at calibration mark on the neck of the flask. Stopper and invert several times to mix thoroughly.

Example 2-32

Calculate the mass of Ba(OH)₂ required to prepare 2.500 liters of a 0.06000 molar solution of barium hydroxide.

Solution

$$\underline{?} \text{ g Ba(OH)}_2 = 2.500 \text{ L} \times \frac{0.06000 \text{ mol Ba(OH)}_2}{1 \text{ L}}$$

$$\times \frac{171.4 \text{ g Ba(OH)}_2}{1 \text{ mol Ba(OH)}_2}$$

$$= \underline{25.71 \text{ g Ba(OH)}_2}$$

The volume of the solution, 2.500 L, is multiplied by the concentration, 0.06000 mol Ba(OH)₂/L, to give the number of moles of Ba(OH)₂. The number of moles of Ba(OH)₂ is then multiplied by the mass of Ba(OH)₂ in one mole, 171.4 g Ba(OH)₂/mol Ba(OH)₂, to give the mass of Ba(OH)₂ in the solution.

Concentrated commercial acids and other concentrated (or "stock") solutions are used to prepare dilute solutions for laboratory use. The labels on commercial acids usually give the specific gravity and the percent of acid by mass in the solution. Therefore, it is necessary to calculate the concentrations of the concentrated acids before they are diluted.

Example 2-33

Commercial sulfuric acid is 96.4% H₂SO₄ by mass, and its specific gravity is 1.84. Calculate the molarity of commercial sulfuric acid.

Solution

The specific gravity of a solution is numerically equal to its density, i.e., the density of the solution is 1.84 g/mL. The solution is 96.4% H₂SO₄ by mass, and therefore 100 grams of solution contains 96.4 grams of *pure* H_2SO_4. From this information the molarity of the solution can be calculated.

First, we calculate the mass of one liter of solution.

$$\underset{=}{?}\frac{\text{g soln}}{\text{L}} = \frac{1.84 \text{ g soln}}{\text{mL}} \times \frac{1000 \text{ mL}}{\text{L}}$$

$$= 1840 \text{ g soln/L}$$

The solution is 96.4% H_2SO_4 by mass, so the mass of H_2SO_4 in one liter is

$$\underset{=}{?}\frac{\text{g } H_2SO_4}{\text{L}} = \frac{1840 \text{ g soln}}{\text{L}} \times \frac{96.4 \text{ g } H_2SO_4}{100.0 \text{ g soln}}$$

$$= 1.77 \times 10^3 \text{ g } H_2SO_4/\text{L}$$

The molarity is the number of moles of H_2SO_4 per liter of solution.

$$\underset{=}{?}\frac{\text{mol } H_2SO_4}{\text{L}} = \frac{1.77 \times 10^3 \text{ g } H_2SO_4}{\text{L}}$$

$$\times \frac{1 \text{ mol } H_2SO_4}{98.1 \text{ g } H_2SO_4}$$

$$= 18.0 \text{ mol } H_2SO_4/\text{L}$$

Thus the solution is an 18.0 M H_2SO_4 solution. Note that this problem can be solved by using a series of three unit factors.

$$\underset{=}{?}\frac{\text{mol } H_2SO_4}{\text{L}} = \frac{1.84 \text{ g soln}}{\text{mL}} \times \frac{96.4 \text{ g } H_2SO_4}{100 \text{ g soln}}$$

$$\times \frac{1 \text{ mol } H_2SO_4}{98.1 \text{ g } H_2SO_4} \times \frac{1000 \text{ mL}}{\text{L}}$$

$$= 18.0 \ M \ H_2SO_4$$

2-14 Volumes of Solutions Required for Chemical Reactions and Dilution of Solutions

Whenever you plan to carry out a chemical reaction in an aqueous solution, you must calculate the volumes of solutions of known concentration required. If you know the molarity of a solution, you can calculate the amount of solute contained in a specified volume of that solution.

Recall that the definition of molarity is the number of moles of solute divided by the volume of the solution in liters:

$$\text{molarity} = \frac{\text{number of moles of solute}}{\text{number of liters of solution}}$$

Multiplying both sides of this equation by the volume, we get

volume (L) × molarity = number of moles of solute   [Eq. 2-5]

Multiplication of the volume of a solution by its concentration gives the amount of solute in the solution. We may choose to express the volume in milliliters and the amount of solute in millimoles, giving

volume (mL) × molarity = number of mmol of solute   [Eq. 2-6]

Example 2-34 illustrates how we may calculate the amount of solute in a given amount of a solution of known molarity.

Example 2-34
Calculate (a) the number of moles of H_2SO_4, (b) the number of millimoles of H_2SO_4, and (c) the mass of H_2SO_4 in 500 mL of 0.324 M H_2SO_4 solution.

Solution
(a) The volume of a solution in liters times its molarity gives the number of moles of solute, H_2SO_4 in this case (500 mL is more conveniently expressed as 0.500 L in this problem).

$$? \text{ mol } H_2SO_4 = 0.500 \text{ L} \times \frac{0.324 \text{ mol } H_2SO_4}{L}$$

$$= 0.162 \text{ mol } H_2SO_4$$

(b) The volume of a solution in milliliters times its molarity gives the number of millimoles of solute, H_2SO_4.

$$? \text{ mmol } H_2SO_4 = 500 \text{ mL} \times \frac{0.324 \text{ mmol } H_2SO_4}{mL}$$

$$= 162 \text{ mmol } H_2SO_4$$

(c) We may use the results of either (a) or (b) to calculate the mass of H_2SO_4 in the solution.

$$? \text{ g } H_2SO_4 = 0.162 \text{ mol } H_2SO_4 \times \frac{98.1 \text{ g } H_2SO_4}{1 \text{ mol } H_2SO_4}$$

$$= 15.9 \text{ g } H_2SO_4$$

or

$$? \text{ g } H_2SO_4 = 162 \text{ mmol } H_2SO_4 \times \frac{0.0981 \text{ g } H_2SO_4}{1 \text{ mmol } H_2SO_4}$$

$$= 15.9 \text{ g } H_2SO_4$$

Alternatively, the mass of H_2SO_4 in the solution can be calculated without solving explicitly for the number of moles (or millimoles) of H_2SO_4.

$$? \text{ g } H_2SO_4 = 0.500 \text{ L} \times \frac{0.324 \text{ mol } H_2SO_4}{L}$$

$$\times \frac{98.1 \text{ g } H_2SO_4}{1 \text{ mol } H_2SO_4} = 15.9 \text{ g } H_2SO_4$$

Example 2–35 demonstrates how we can relate the volume of a solution of known concentration to the amount of another reactant, expressed in grams.

Example 2–35

What volume of a 0.324 M solution of sulfuric acid is required to react completely with 2.792 grams of Na_2CO_3 by the following reaction?

$$H_2SO_4 + Na_2CO_3 \longrightarrow Na_2SO_4 + CO_2 + H_2O$$

Solution

The balanced equation tells us that one mole of H_2SO_4 reacts with one mole of Na_2CO_3, and we can write

$$\underset{\substack{1 \text{ mol} \\ }}{H_2SO_4} + \underset{\substack{1 \text{ mol} \\ 106.01 \text{ g}}}{Na_2CO_3} \longrightarrow \underset{\substack{1 \text{ mol}}}{Na_2SO_4} + \underset{\substack{1 \text{ mol}}}{CO_2} + \underset{\substack{1 \text{ mol}}}{H_2O}$$

which enables us to solve the problem.

$$? \text{ L } H_2SO_4 = 2.792 \text{ g } Na_2CO_3 \times \frac{1 \text{ mol } Na_2CO_3}{106.01 \text{ g } Na_2CO_3}$$

$$\times \frac{1 \text{ mol } H_2SO_4}{1 \text{ mol } Na_2CO_3} \times \frac{1 \text{ L } H_2SO_4 \text{ soln}}{0.324 \text{ mol } H_2SO_4}$$

$$= 0.0813 \text{ L or } 81.3 \text{ mL } H_2SO_4 \text{ soln}$$

Note that we converted grams of Na_2CO_3 to moles of Na_2CO_3 (first factor), and then related moles of Na_2CO_3 to moles of H_2SO_4 (second factor), which was converted to liters of H_2SO_4 solution (third factor).

Often we must calculate the volume of a solution of known molarity required to react with a specified volume of another solution. We always examine the balanced chemical equation for the reaction to determine the *reaction ratio*, i.e., *the relative numbers of moles (or millimoles) of reactants*.

Example 2–36

What volume of 0.500 M NaOH solution is required to react exactly with 40.0 mL of 0.500 M H_2SO_4 solution? The balanced equation for the reaction is

$$H_2SO_4 + 2NaOH \longrightarrow Na_2SO_4 + 2H_2O$$

Solution

The balanced equation tells us that *one* mole of H_2SO_4 reacts with *two* moles of NaOH.

$$\underset{\substack{1 \text{ mol}}}{H_2SO_4} + \underset{\substack{2 \text{ mol}}}{2NaOH} \longrightarrow \underset{\substack{1 \text{ mol}}}{Na_2SO_4} + \underset{\substack{2 \text{ mol}}}{2H_2O}$$

Therefore, the *reaction ratio is 1 mole H_2SO_4 to 2*

moles NaOH. We are given the volume and the molarity of the H_2SO_4 solution, so we can calculate the number of moles of H_2SO_4 that react.

$$\underline{?}\text{ mol } H_2SO_4 = 0.0400 \text{ L} \times \frac{0.500 \text{ mol } H_2SO_4}{\text{L}}$$

$$= 0.0200 \text{ mol } H_2SO_4$$

The number of moles of H_2SO_4 is related to the number of moles of NaOH by the reaction ratio, 1 mol H_2SO_4/2 mol NaOH.

$$\underline{?}\text{ mol NaOH} = 0.0200 \text{ mol } H_2SO_4 \times \frac{2 \text{ mol NaOH}}{1 \text{ mol } H_2SO_4}$$

$$= 0.0400 \text{ mol NaOH}$$

Now we can calculate the volume of 0.500 *M* NaOH solution that contains 0.0400 mole of NaOH.

$$\underline{?}\text{ L NaOH soln} = 0.0400 \text{ mol NaOH}$$

$$\times \frac{1.00 \text{ L NaOH soln}}{0.500 \text{ mol NaOH}}$$

$$= \underline{0.0800 \text{ L NaOH soln}}$$

which we usually call *80.0 mL of NaOH solution.*

Now that we have worked through the problem stepwise, let us solve it in a single set-up.

$$\underline{?}\text{ L NaOH solution} = 0.0400 \text{ L } H_2SO_4 \text{ soln}$$

$$\times \frac{0.500 \text{ mol } H_2SO_4}{\text{L } H_2SO_4 \text{ soln}}$$

$$\times \frac{2 \text{ mol NaOH}}{1 \text{ mol } H_2SO_4}$$

$$\times \frac{1.00 \text{ L NaOH soln}}{0.500 \text{ mol NaOH}}$$

$$= \underline{0.0800 \text{ L NaOH soln}}$$

In the single set-up solution, we converted (1) liters of H_2SO_4 solution to moles of H_2SO_4, (2) moles of H_2SO_4 to moles of NaOH, and (3) moles of NaOH to liters of NaOH solution, the same steps we used in the stepwise procedure.

When a solution is **diluted** by mixing more solvent with it, *the number of moles of solute present does not change*. The *volume* of the solution and its *concentration* do change. Since the same number of moles of solute is contained in a larger number of liters of solution, molarity decreases. Using a subscript 1 to represent the original, more concentrated solution and a subscript 2 to represent the dilute solution,

$$\text{volume}_1 \times \text{molarity}_1 = \text{number of moles of solute} = \text{volume}_2 \times \text{molarity}_2$$

or

$$V_1 \times M_1 = V_2 \times M_2 \quad \text{(for dilution only)} \qquad \text{[Eq. 2-7]}$$

This expression can be used to calculate any one of the four quantities when the other three are known. For instance, if a certain volume of dilute solution of a given molarity is required for use in the laboratory, and we know the concentration of the stock solution available, we can calculate how much stock solution must be used to make the dilute solution.

CAUTION: Dilution of a concentrated solution, especially of a strong acid or base, frequently liberates a great deal of heat. This can vaporize drops of water as they hit the concentrated solution, and can cause dangerous spattering. As a safety precaution, concentrated solutions are always poured *into* water, allowing the heat to be absorbed by the larger quantity of water. Although calculations are usually simpler to visualize by *assuming* that water is added to the concentrated solution, this is *never* done in practice.

Example 2–37

Calculate the volume of 18.0 M H_2SO_4 required to prepare 1.00 L of a 0.900 M solution of H_2SO_4.

Solution

The volume (1.00 L) and molarity (0.900 M) of the final solution, as well as the molarity (18.0 M) of the original solution, are given. Therefore, the relation $V_1 \times M_1 = V_2 \times M_2$ can be used, with subscript 1 for the concentrated acid solution and subscript 2 for the dilute solution.

$$V_1 \times M_1 = V_2 \times M_2$$

$$V_1 = \frac{V_2 \times M_2}{M_1} = \frac{1.00 \text{ L} \times 0.900 \text{ } M}{18.0 \text{ } M}$$

$$= 0.0500 \text{ L} = \underline{50.0 \text{ mL}}$$

Note that the dilute solution must contain 1.00 L $\times$ 0.900 M = 0.900 mole of H_2SO_4, so 0.900 mole of H_2SO_4 must be present in the original concentrated solution. Indeed, 0.0500 L $\times$ 18.0 mol H_2SO_4/L = 0.900 mole of H_2SO_4.

Example 2–38

If 200 mL of a 4.00 M solution of NaOH and 300 mL of a 6.00 M solution of NaOH are mixed, what will the molarity of the resulting solution be?

Solution

We must calculate the volume of the final solution and the number of moles of sodium hydroxide it contains.

1st solution:

$$0.200 \text{ L} \times \frac{4.00 \text{ mol NaOH}}{L} = 0.800 \text{ mol NaOH}$$

2nd solution:

$$0.300 \text{ L} \times \frac{6.00 \text{ mol NaOH}}{L} = 1.80 \text{ mol NaOH}$$

final solution:

0.500 L contains   2.60 mol NaOH

Hence,

$$? M = \frac{\text{number of mol NaOH}}{\text{volume soln}} = \frac{2.60 \text{ mol NaOH}}{0.500 \text{ L}}$$

$$= \underline{5.20 \text{ } M \text{ NaOH}}$$

Key Terms

Actual yield amount of a specified pure product actually obtained from a given reaction. Compare with *Theoretical yield.*

Atomic mass unit (amu) one twelfth of the mass of an atom of the carbon-12 isotope; a unit used for stating atomic and formula weights.

Atomic weight weighted average of the masses of the constituent isotopes of an element; the relative masses of atoms of different elements.

Avogadro's number see *Mole.*

Chemical equation description of a chemical reaction by placing the formulas of reactants on the left and the formulas of products on the right of an arrow.

Concentration amount of solute per unit volume or mass of solvent or of solution.

Dilution process of reducing the concentration of

a solute in solution, usually simply by mixing with more solvent.

Empirical formula see *Simplest formula.*

Formula combination of symbols that indicates the chemical composition of a substance.

Formula unit the smallest repeating unit of a substance, the molecule for covalent substances.

Formula weight the mass of one formula unit of a substance, in atomic mass units.

Hydrate a solid compound containing a definite percentage of bound water.

Ion an atom or group of atoms that carries an electrical charge.

Law of Constant Composition statement that different samples of a pure compound always contain the same elements in the same proportions by mass; also called Law of Definite Proportions.

Limiting reagent substance that stoichiometrically limits the amount of product(s) that can be formed.

Molarity (*M*) number of moles of solute per liter of solution.

Mole 6.022×10^{23} (Avogadro's number of) formula units of the species under discussion.

Molecular (true) formula formula that indicates the actual number of atoms present in a molecule of a molecular substance. Compare with *Simplest formula*.

Molecular weight the mass of one molecule of a molecular substance, in atomic mass units.

Percent by mass 100% times the mass of the solute divided by the mass of the solution in which it is contained.

Percent composition the mass percent of each element in a compound.

Percent purity the percent of a specified compound or element in an impure sample.

Percent yield 100% times actual yield divided by theoretical yield.

Products substances produced in a chemical reaction.

Reactants substances consumed in a chemical reaction.

Simplest (empirical) formula the smallest whole-number ratio of atoms present in a compound; also called empirical formula. Compare with *Molecular formula.*

Solute the dispersed (dissolved) phase of a solution.

Solution homogeneous mixture of two or more substances.

Solvent the dispersing medium of a solution.

Stoichiometry description of the quantitative relationships among elements in compounds, and among substances as they undergo chemical changes.

Theoretical yield maximum amount of a specified product that could be obtained from specified amounts of reactants, assuming complete consumption of the limiting reagent according to only one reaction, and complete recovery of product. Compare with *Actual yield.*

Exercises

Atomic Weights

1. (a) What is the atomic weight of an element?
   (b) Why can atomic weights be referred to as relative numbers?

2. (a) What is the atomic mass unit?
   (b) The atomic weight of vanadium is 50.942 amu and the atomic weight of ruthenium is 101.07 amu. What can we say about the relative masses of V and Ru atoms?

The Mole Concept

3. What is a mole? Why is a mole a convenient unit?

4. Complete the following table. You may refer to the periodic table.

	Element	Atomic Weight	Mass of 1 Mole of Atoms
(a)	B		
(b)		32.06 amu	
(c)	Fe		
(d)			107.868 g

5. Complete the following table. You may refer to the periodic table.

	Element	Formula	Mass of 1 Mole of Molecules
(a)	H	H_2	
(b)		F_2	
(c)		P_4	
(d)	O		
(e)	He		
(f)		S_8	

6. Calculate:
   (a) the number of moles of boron atoms in 59.4 g of boron.
   (b) the number of calcium atoms in 20 g of calcium.
   (c) the number of iron atoms in 83.7 g of iron.
   (d) the number of lithium atoms in 1.00 g of lithium.
   (e) the number of uranium atoms in 1.00 g of uranium.
   (f) the ratio of the number of lithium atoms in

1.00 g of lithium to the number of uranium atoms in 1.00 g of uranium.

7. Distinguish among the following terms clearly by statements and by calculating the mass of each in grams.
   (a) one nitrogen atom
   (b) one nitrogen molecule
   (c) one mole of nitrogen molecules

8. The atomic weight of carbon is 12.011 amu. Calculate the number of moles of carbon atoms in: (a) 1.00 g of carbon, (b) 12.0 atomic mass units of carbon, (c) 5.66×10^{20} carbon atoms.

9. Which of the following contain eight atoms? (a) one C_2H_6 molecule, (b) one mole of C_2H_6, (c) 30 g of C_2H_6, (d) 4.99×10^{-23} g of C_2H_6.

10. Define and illustrate the following terms. Which ones are frequently used interchangeably? (a) formula, (b) formula weight, (c) molecular weight, (d) mole.

11. A sample of a pure element that had a mass of 1.0 g was found to contain 1.5×10^{22} atoms. If Avogadro's number is taken as 6.0×10^{23}, what is the atomic weight of the element?

Percent Composition and Simplest Formulas

12. Calculate the percent composition of the following compounds to three significant figures:
    (a) MgO,   (b) Fe_2O_3,   (c) Na_2SO_4,
    (d) $(NH_4)_2CO_3$, (e) $Al_2(SO_4)_3 \cdot 18H_2O$

13. Consider 100 g samples of each of the following compounds:
    Li_2O, CaO, CrO_3, As_4O_{10}, U_3O_8
    Which sample contains:
    (a) the greatest mass of oxygen?
    (b) the smallest mass of oxygen?
    (c) the largest total number of atoms?
    (d) the smallest total number of atoms?

14. A compound contains 59.9% titanium and 40.1% oxygen by mass. What is the simplest formula for the compound?

15. A compound contains 66.6% titanium and 33.4% oxygen by mass. What is its simplest formula?

16. A 1.596 g sample of an oxide of iron was found to contain 1.116 g of iron and 0.480 g of oxygen. What is the simplest formula for this oxide?

17. A 2.317 g sample of an oxide of iron was found to contain 1.677 g of iron and 0.640 g of oxygen. What is the simplest formula for this oxide?

18. What law is illustrated by Exercises 14 to 17? State the law.

19. Four compounds were analyzed and found to have the following percentage compositions. What is the simplest formula for each compound?
    (a) 65.20% As, 34.80% O
    (b) 59.72% Ba, 21.72% As, 18.55% O
    (c) 40.27% K, 26.78% Cr, 32.96% O
    (d) 26.58% K, 35.35% Cr, 38.07% O

20. A sample of a compound contains 4.86 g of magnesium, 6.42 g of sulfur, and 9.60 g of oxygen. What is the simplest formula for the compound?

21. A sample of a compound contains 4.86 g of magnesium, 6.42 g of sulfur, and 12.8 g of oxygen. What is the simplest formula for the compound?

22. A sample of a compound contains 4.86 g of magnesium, 12.85 g of sulfur, and 9.60 g of oxygen. What is the simplest formula for the compound?

23. An unidentified organic compound X, containing only C, H, and O, was subjected to combustion analysis. When 228.4 mg of pure compound X was burned in the C-H combustion train, 627.4 mg of CO_2 and 171.2 mg of H_2O were obtained. (a) Determine the masses of C, H, and O in the sample. (b) Determine the simplest formula of compound X.

24. The substance responsible for the sharp odor of rancid butter is butanoic acid, an organic substance consisting entirely of C, H, and O. Combustion analysis of a 0.3164 g sample of butanoic acid yielded 0.6322 g CO_2 and 0.2588 g H_2O. (a) Determine the simplest formula of butanoic acid. (b) In an independent determination, the molecular weight of butanoic acid was shown to be approximately 88 g/mol. What is the molecular formula of butanoic acid?

Formulas and Formula Weights

25. What information does the formula for ethyl alcohol, C_2H_5OH, contain?

26. How many moles of methyl alcohol, CH_3OH, are there in 80 g of CH_3OH?

27. How many moles of ethyl alcohol, C_2H_5OH, are there in 230 g of ethyl alcohol?

28. Calculate the number of oxygen atoms in 160 grams of SO_2.

29. How many hydrogen atoms are contained in 64 g of CH_4?

30. Calculate the number of hydrogen atoms in 69.8 grams of diamminepalladium(II) hydroxide, $Pd(NH_3)_2(OH)_2$.

31. A compound that is 17.2% sulfur by mass contains one sulfur atom per molecule. What is the molecular weight of the compound?

32. Lysine is an essential amino acid. One experiment showed that each molecule of lysine contains two nitrogen atoms. Another experiment showed that lysine contains 19.2% N, 9.64% H, 49.3% C, and 21.9% O by mass. What is the molecular formula for lysine?

Chemical Equations

33. What is a chemical equation? What information does it contain?

34. Balance the following equations:
    (a) $Sn + Cl_2 \rightarrow SnCl_4$
    (b) $Fe + Cl_2 \rightarrow FeCl_3$
    (c) $Fe + O_2 \rightarrow Fe_2O_3$
    (d) $Al + Cl_2 \rightarrow Al_2Cl_6$
    (e) $CaO + HCl \rightarrow CaCl_2 + H_2O$
    (f) $NaOH + SO_2 \rightarrow Na_2SO_3 + H_2O$
    (g) $Mg + HCl \rightarrow MgCl_2 + H_2$
    (h) $Na + H_2O \rightarrow NaOH + H_2$
    (i) $Al + H_2SO_4 \rightarrow Al_2(SO_4)_3 + H_2$
    (j) $NaOH + H_2SO_4 \rightarrow Na_2SO_4 + H_2O$

35. Balance the following equations:
    (a) $Ca(OH)_2 + HNO_3 \rightarrow Ca(NO_3)_2 + H_2O$
    (b) $KOH + H_3PO_4 \rightarrow K_3PO_4 + H_2O$
    (c) $Ca(OH)_2 + H_3PO_4 \rightarrow Ca_3(PO_4)_2 + H_2O$
    (d) $P_4O_{10} + NaOH \rightarrow Na_3PO_4 + H_2O$
    (e) $CuCl_2 + H_2S \rightarrow CuS + HCl$
    (f) $BiCl_3 + H_2S \rightarrow Bi_2S_3 + HCl$
    (g) $CuCl_2 + NH_3 + H_2O \rightarrow Cu(OH)_2 + NH_4Cl$
    (h) $Fe(NO_3)_3 + NH_3 + H_2O \rightarrow Fe(OH)_3 + NH_4NO_3$
    (i) $NaNO_3 \rightarrow NaNO_2 + O_2$
    (j) $(NH_4)_2Cr_2O_7 \rightarrow N_2 + Cr_2O_3 + H_2O$

36. Balance the following equations:
    (a) $Sn + F_2 \rightarrow SnF_4$
    (b) $As + O_2 \rightarrow As_4O_{10}$
    (c) $Al + O_2 \rightarrow Al_2O_3$
    (d) $CH_4 + O_2 \rightarrow CO_2 + H_2O$
    (e) $CS_2 + O_2 \rightarrow CO_2 + SO_2$
    (f) $C_2H_6 + O_2 \rightarrow CO_2 + H_2O$
    (g) $C_2H_6O + O_2 \rightarrow CO_2 + H_2O$
    (h) $KOH + H_2SO_4 \rightarrow K_2SO_4 + H_2O$

    (i) $NaOH + H_3AsO_4 \rightarrow Na_3AsO_4 + H_2O$
    (j) $Cr(OH)_3 + HClO_4 \rightarrow Cr(ClO_4)_3 + H_2O$

37. Balance the following equations:
    (a) $N_2O_5 + H_2O \rightarrow HNO_3$
    (b) $P_4O_{10} + H_2O \rightarrow H_3PO_4$
    (c) $P_4O_{10} + Mg(OH)_2 \rightarrow Mg_3(PO_4)_2 + H_2O$
    (d) $Cl_2O_7 + H_2O \rightarrow HClO_4$
    (e) $Cl_2O_7 + Ba(OH)_2 \rightarrow Ba(ClO_4)_2 + H_2O$
    (f) $H_2SnCl_6 + H_2S \rightarrow SnS_2 + HCl$
    (g) $HSbCl_4 + H_2S \rightarrow Sb_2S_3 + HCl$
    (h) $FeCl_2 + SnCl_4 \rightarrow FeCl_3 + SnCl_2$
    (i) $BiCl_3 + NH_3 + H_2O \rightarrow Bi(OH)_3 + NH_4Cl$
    (j) $NH_3 + O_2 \rightarrow NO + H_2O$

38. Write balanced chemical equations to represent the reactions described by the following statements.
    (a) Fluorine, F_2, combines with hydrogen, H_2, to form hydrogen fluoride, HF.
    (b) Nitrogen, N_2, combines with hydrogen to form ammonia, NH_3.
    (c) Sulfur dioxide, SO_2, combines with oxygen, O_2, to produce sulfur trioxide, SO_3.
    (d) Ferrous chloride, $FeCl_2$, combines with chlorine, Cl_2, to form ferric chloride, $FeCl_3$.
    (e) Heating calcium carbonate, $CaCO_3$, liberates carbon dioxide, CO_2, as a gas and leaves a residue of solid calcium oxide, CaO.
    (f) When ammonium carbonate, $(NH_4)_2CO_3$, is heated it decomposes to form ammonia, NH_3, carbon dioxide, CO_2, and water.
    (g) Ethane, C_2H_6, burns in oxygen to produce carbon dioxide and water.
    (h) Octane, C_8H_{18}, burns in oxygen to produce carbon dioxide and water.
    (i) Ethyl alcohol, C_2H_5OH, burns in oxygen to produce carbon dioxide and water.
    (j) When solid mercury(II) oxide, HgO, is heated it decomposes to produce liquid mercury and gaseous oxygen.
    (k) When solid potassium chlorate, $KClO_3$, is heated it decomposes to produce solid potassium chloride, KCl, and gaseous oxygen.

39. State in words the meaning of the following chemical equations:
    (a) $Ca + Cl_2 \rightarrow CaCl_2$
    (b) $2Mg + O_2 \xrightarrow{\Delta} 2MgO$
    (c) $2Fe + 3F_2 \rightarrow 2FeF_3$
    (d) $CH_4 + 2O_2 \xrightarrow{\Delta} CO_2 + 2H_2O$
    (e) $2C_4H_{10} + 13O_2 \xrightarrow{\Delta} 8CO_2 + 10H_2O$

Calculations Based on Chemical Equations

40. What mass of CaO could be obtained from the thermal decomposition of 2.00 moles of $CaCO_3$?

$$CaCO_3 \xrightarrow{\Delta} CaO + CO_2$$

41. Aluminum and sulfur react at elevated temperatures to form aluminum sulfide as shown by the following equation.

$$2Al + 3S \xrightarrow{\Delta} Al_2S_3$$

Calculate the number of moles of:
(a) aluminum atoms that react with one mole of sulfur atoms. What mass of aluminum is this?
(b) sulfur atoms that react with one mole of aluminum atoms. What mass of sulfur is this?
(c) aluminum atoms that react with 1.00 g of sulfur. What mass of aluminum is this?
(d) sulfur atoms that react with 1.00 g of aluminum. What mass of sulfur is this?

42. A chemist needed 80 g of anhydrous copper(II) sulfate, $CuSO_4$, to perform a particular experiment, but none was readily available. However, he did have available a large supply of copper(II) sulfate pentahydrate, $CuSO_4 \cdot 5H_2O$, which he could dehydrate by heating. How much of the hydrated compound did he heat to obtain 80 g of $CuSO_4$?

43. What mass of Na_3PO_4 can be prepared by the reaction of 4.9 g of H_3PO_4 with an excess of NaOH?

$$H_3PO_4 + 3NaOH \longrightarrow Na_3PO_4 + 3H_2O$$

44. What mass of K_3AsO_4 can be prepared by the reaction of 7.10 g of H_3AsO_4 with an excess of KOH?

$$H_3AsO_4 + 3KOH \longrightarrow K_3AsO_4 + 3H_2O$$

Percent Purity and Percent Yield

45. A particular ore of lead, galena, is 10% lead sulfide, PbS, and 90% impurities by mass. What mass of lead is contained in 50 grams of this ore?

46. What mass of chromium is present in 150 g of an ore of chromium that is 65.0% chromite, $FeCr_2O_4$, and 35.0% impurities by mass? If 90.0% of the chromium can be recovered from 100 grams of the ore, what mass of pure chromium is obtained?

47. Ethylene glycol, $C_2H_6O_2$, is used as antifreeze in automobile radiators. A method of producing small amounts of ethylene glycol in the laboratory is by reaction of 1,2-dichloroethane with sodium carbonate in a water solution, followed by distillation of the reaction mixture to purify the ethylene glycol.

$$C_2H_4Cl_2 + Na_2CO_3 + H_2O \longrightarrow$$
$$C_2H_6O_2 + 2NaCl + CO_2$$

When 26.8 g of 1,2-dichloroethane is used in this reaction, 10.5 g of ethylene glycol is obtained. (a) Calculate the theoretical yield of ethylene glycol. (b) What is the percent yield of ethylene glycol in this process? (c) If the reaction had gone to completion, what mass of Na_2CO_3 would have been consumed?

48. Mesitylene, C_9H_{12}, is used in the synthesis of other organic compounds. It is prepared in poor yield from acetone, C_3H_6O, in the presence of sulfuric acid.

$$3C_3H_6O \xrightarrow{H_2SO_4} C_9H_{12} + 3H_2O$$
$$\text{acetone} \qquad \text{mesitylene}$$

We obtain 12.4 g of mesitylene from 125 g of acetone, according to this reaction. What is the percent yield of mesitylene in this reaction?

49. Chlorobenzene, C_6H_5Cl, is an organic chemical that is manufactured on a large scale and is then used in industrial processes to produce such useful compounds as aspirin, oil of wintergreen, and many dyes, insecticides, and disinfectants. This very useful chemical is produced by the following reaction, under suitable conditions:

$$C_6H_6 + Cl_2 \longrightarrow C_6H_5Cl + HCl$$
$$\text{benzene} \qquad \text{chlorobenzene}$$

Suppose that a particular industrial reactor is known to produce chlorobenzene by this reaction in 73% yield. What mass of benzene, in kg, would be required to produce 775 kg of chlorobenzene?

50. Ethylene oxide, C_2H_4O, a fumigant sometimes used by exterminators, is synthesized in 89% yield by reaction of ethylene bromohydrin with sodium hydroxide:

$C_2H_5OBr + NaOH \longrightarrow$

$\qquad\qquad C_2H_4O + NaBr + H_2O$

ethylene   ethylene
bromohydrin   oxide

How many grams of ethylene bromohydrin would be consumed in the production of 125 g of ethylene oxide, at 89% yield?

Limiting Reagent

51. What mass of $Ca(NO_3)_2$ can be prepared by the reaction of 18.9 g of HNO_3 with 7.4 g of $Ca(OH)_2$?

$2HNO_3 + Ca(OH)_2 \longrightarrow Ca(NO_3)_2 + 2H_2O$

52. What is the maximum amount of $Ca_3(PO_4)_2$ that can be prepared from 7.4 g of $Ca(OH)_2$ and 9.8 g of H_3PO_4?

$3Ca(OH)_2 + 2H_3PO_4 \longrightarrow Ca_3(PO_4)_2 + 6H_2O$

53. Silver nitrate solution reacts with calcium chloride solution according to the equation

$2AgNO_3 + CaCl_2 \longrightarrow Ca(NO_3)_2 + 2AgCl$

All of the substances involved in this reaction are soluble in water, except for silver chloride, $AgCl$, which forms a solid (precipitate) at the bottom of the flask. Suppose we mix together a solution containing 12.6 g of $AgNO_3$ and 8.4 g of $CaCl_2$. What mass of $AgCl$ would be formed?

Unscramble Them

54. What is the total mass of products formed when 19.0 g of carbon disulfide is burned in air? What mass of carbon disulfide would have to be burned to produce a mixture of carbon dioxide and sulfur dioxide with a mass of 34.4 g?

$CS_2 + 3O_2 \longrightarrow CO_2 + 2SO_2$

55. A mixture of calcium oxide, CaO, and calcium carbonate, $CaCO_3$, that weighed 1.727 g was heated until all the calcium carbonate was decomposed according to the following equation. After heating, the sample weighed 1.551 g. Calculate the masses of CaO and $CaCO_3$ present in the original sample.

$CaCO_3 \xrightarrow{\Delta} CaO + CO_2$

56. Analysis of a sample of a compound shows that only carbon, hydrogen, and nitrogen are present. Each molecule contains one nitrogen atom. A 150 mg sample of the compound produces 27.4 mg of NH_3 in which all of the nitrogen comes from the sample. Which of the following is the formula for the compound?
(a) CH_5N,   (b) C_2H_7N,   (c) $C_4H_{11}N$, (d) $C_5H_{13}N$, (e) C_6H_7N

57. Analysis of a sample of a compound revealed the presence of carbon, hydrogen, nitrogen, and oxygen. In one experiment in which all of the N in the compound was converted into NH_3, a 200 mg sample of the compound produced 44.7 mg of NH_3. In another experiment, C was converted into CO_2 and H was converted into H_2O to produce 405 mg of CO_2 and 94.7 mg of H_2O. Which of the following is the formula for the compound?
(a) $C_7H_8N_2O_2$ (b) C_6H_5NO (c) $C_7H_7NO_2$
(d) $C_6H_4N_2O_4$ (e) $C_3H_5N_3O_6$

Two Reactions Occurring Simultaneously

58. A 1.020 g sample known to contain only magnesium carbonate, $MgCO_3$, and calcium carbonate, $CaCO_3$, was heated until the carbonates were decomposed to oxides as indicated by the following equations. After heating, the residue weighed 0.536 g. What masses of $MgCO_3$ and $CaCO_3$ were present in the original sample?

$CaCO_3 \xrightarrow{\Delta} CaO + CO_2$

$MgCO_3 \xrightarrow{\Delta} MgO + CO_2$

59. A mixture that contained 1.8391 g of NaCl and 2.3432 g of KCl was treated with excess $AgNO_3$. The precipitated AgCl was collected, washed, and dried. What mass of AgCl was produced? See Example 2–24 for the equations.

60. A sample that was known to be 47.39% NaCl and 52.61% KCl by mass was treated with excess $AgNO_3$. The precipitated AgCl was collected, washed, dried, and found to have a mass of 4.2367 g. What was the mass of the sample? See Example 2–24 for the equations.

61. A 0.7749 g sample that contained only sodium sulfate, Na_2SO_4, and potassium sulfate, K_2SO_4, was dissolved in water, and an excess of barium chloride solution was added. The following equations represent the reactions that occurred

in the solution. The solid $BaSO_4$ was collected by filtration, dried, and found to have a mass of 1.167 g. What masses of Na_2SO_4 and K_2SO_4 were present in the original sample?

$$Na_2SO_4 + BaCl_2 \longrightarrow 2NaCl + BaSO_4$$

$$K_2SO_4 + BaCl_2 \longrightarrow 2KCl + BaSO_4$$

62. A sample of methane, CH_4, was burned in a quantity of oxygen that was insufficient for complete combustion. The combustion products were a mixture of carbon monoxide, CO, carbon dioxide, CO_2, and water that weighed 24.8 g. The water was separated from the CO and CO_2 and found to weigh 12.6 g. Calculate (a) the mass of CH_4 in the sample, (b) the mass of CO formed, (c) the mass of CO_2 formed, and (d) the mass of oxygen used in the reactions. Equation (1) represents the complete combustion of CH_4 and equation (2) represents the partial combustion of CH_4.

(1) $CH_4 + 2O_2 \longrightarrow CO_2 + 2H_2O$

(2) $2CH_4 + 3O_2 \longrightarrow 2CO + 4H_2O$

Concentrations of Solutions — Percent by Mass

63. What mass of silver nitrate, $AgNO_3$, is required to prepare 500 g of a 3.50% solution of $AgNO_3$?
64. Calculate the masses of potassium dichromate, $K_2Cr_2O_7$, and water in 400 g of a 6.85% solution of $K_2Cr_2O_7$.
65. What mass of a 6.85% solution of potassium dichromate contains 40.0 g of $K_2Cr_2O_7$? What mass of water does this amount of solution contain?
66. Calculate the mass of an 8.30% solution of ammonium chloride, NH_4Cl, that contains 100 g of water. What mass of NH_4Cl does this amount of solution contain?
67. The density of a 7.50% solution of ammonium chloride, NH_4Cl, solution is 1.02 g/mL. What mass of NH_4Cl does 250 mL of this solution contain?
68. The density of a 7.50% solution of ammonium sulfate, $(NH_4)_2SO_4$, is 1.04 g/mL. What mass of $(NH_4)_2SO_4$ would be required to prepare 500 mL of this solution?
69. What volume of the solution of NH_4Cl described in Problem 67 contains 50.0 g of NH_4Cl?

70. What volume of the solution of $(NH_4)_2SO_4$ described in Problem 68 contains 50.0 g of $(NH_4)_2SO_4$?
71. A reaction requires 37.8 g of $(NH_4)_2SO_4$. What volume of the solution described in Problem 68 would you use if you wished to use a 10.0% excess of $(NH_4)_2SO_4$?

Concentrations of Solutions — Molarity

72. What is the molarity of a solution that contains 490 g of phosphoric acid, H_3PO_4, in 2.00 L of solution?
73. What is the molarity of a solution that contains 1.37 g of sodium chloride in 25.0 mL of solution?
74. What is the molarity of a barium chloride solution prepared by dissolving 2.00 g of $BaCl_2 \cdot 2H_2O$ in enough water to make 200 mL of solution?
75. What mass of potassium sulfate, K_2SO_4, is contained in 750 mL of 2.00 molar solution?
76. What mass of calcium chloride, $CaCl_2$, is required to prepare 5.00 L of 0.450 molar solution?
77. What mass of hydrated copper(II) sulfate, $CuSO_4 \cdot 5H_2O$, is needed to prepare one liter of a 0.675 molar solution of $CuSO_4$?
78. What volume of 0.230 M Na_2CO_3 solution can be prepared from 12.4 g of Na_2CO_3?
79. What volume of 0.850 M $(COOH)_2$ solution can be prepared from 75.0 g of oxalic acid dihydrate, $(COOH)_2 \cdot 2H_2O$?

Volumes of Solution Required for Reactions

80. What volume of 0.250 molar nitric acid, HNO_3, is required to react with 3.70 g of calcium hydroxide, $Ca(OH)_2$, according to the following equation?

$$2HNO_3 + Ca(OH)_2 \longrightarrow Ca(NO_3)_2 + 2H_2O$$

81. What volume of a 2.00 molar solution of sulfuric acid, H_2SO_4, is required to react with 250 g of calcium carbonate, $CaCO_3$, according to the following equation?

$$CaCO_3 + H_2SO_4 \longrightarrow CaSO_4 + CO_2 + H_2O$$

82. What volume of a 0.200 molar solution of silver nitrate, $AgNO_3$, is required to react with 12.2 g of potassium phosphate, K_3PO_4, according to the following equation?

$$3AgNO_3 + K_3PO_4 \longrightarrow Ag_3PO_4 + 3KNO_3$$

83. What volume of $0.850\ M$ $(COOH)_2$ solution would be required to react with 40.0 mL of $0.230\ M$ Na_2CO_3 solution?

$$Na_2CO_3 + (COOH)_2 \longrightarrow$$
$$Na_2(COO)_2 + H_2CO_3$$

84. What volume of $0.250\ M$ $FeSO_4$ solution would be required to react with 10.0 mL of $0.200\ M$ $KMnO_4$ (in sulfuric acid solution) according to the following equation?

$$10FeSO_4 + 2KMnO_4 + 8H_2SO_4 \longrightarrow$$
$$5Fe_2(SO_4)_3 + 2MnSO_4 + K_2SO_4 + 8H_2O$$

Dilution of Solutions

85. Commercially available concentrated sulfuric acid is $18.0\ M$ H_2SO_4. Calculate the volume of concentrated sulfuric acid required to prepare 2.00 L of $0.200\ M$ H_2SO_4 solution.

86. Commercial concentrated hydrochloric acid is $12.0\ M$ HCl. What volume of concentrated hydrochloric acid is required to prepare 1.50 L of $1.20\ M$ HCl solution?

87. Calculate the volume of $4.00\ M$ NaOH solution required to prepare 100 mL of a $0.500\ M$ solution of NaOH.

88. Calculate the volume of $0.0500\ M$ $Ba(OH)_2$ solution that contains the same number of moles of $Ba(OH)_2$ as 120 mL of $0.0800\ M$ $Ba(OH)_2$ solution.

89. Calculate the resulting molarity when 150 mL of $0.450\ M$ NaCl solution is mixed with 100 mL of $2.50\ M$ NaCl solution.

90. Calculate the resulting molarity when 200 mL of $3.00\ M$ H_2SO_4 solution is mixed with 120 mL of $6.00\ M$ H_2SO_4 solution.

Thermochemistry

3

Matter and energy are of fundamental interest and importance in chemistry. We were concerned with *matter* and the changes it can undergo in the last two chapters. In this chapter we shall study the nature of *energy* and the changes in energy that accompany chemical reactions. Since these changes in energy are often measured in the form of heat flow into or out of the reaction mixture, we call this subject matter **thermochemistry.**

In Chapter 2 we learned to interpret chemical equations in terms of quantitative mass relationships. In this chapter we shall learn to interpret **thermochemical equations,** *which are chemical equations that also contain information about the amount of heat gained or lost in chemical reactions.* Since modern societies are so critically dependent upon the availability of energy, it is clear that the heat transfer accompanying chemical changes is of great practical importance. In addition, the study of heat transfer is the basis for *thermodynamics* (Chapter 14), which addresses the fundamental question of whether a change can or cannot occur under specified conditions.

3–1   Forms of Energy

Before we consider the energy changes accompanying chemical processes, let us think about the nature of energy and some of its forms. Energy is more abstract than matter, in that we can neither see nor touch energy. In Chapter 1 we defined energy as *the capacity to do work or to transfer heat* and indicated that energy can take many forms. In some instances **work** *is defined as the product of the force, f, applied to move an object and the distance, d, through which the object moves.*

$$w = f \times d$$

[Eq. 3–1]

Energy is the "stuff" that allows work to be performed. Under other circumstances, energy can appear in the form of heat. **Heat** is a form of energy that can flow between two samples of matter because of their difference in temperature.

The energy that a system possesses can be classified as either *kinetic energy* or *potential energy*. **Kinetic energy** *is energy associated with motion.* It is equal to one half the mass, m, of a moving object times the square of its velocity, v.

$$E_{kinetic} = \frac{1}{2} mv^2 \qquad \text{[Eq. 3-2]}$$

Recalling our description of energy as the capacity to do work, we can see that the moving hammer of Figure 3-1a has the capacity of driving the nail into the board — the hammer has both mass and velocity. Notice that the hammer possesses this energy even if it *does not do the work* of which it is capable, as when it misses the nail. The stationary hammer of Figure 3-1b cannot do any work because it has no kinetic energy ($v = 0$). Likewise, it would take many strokes of the very light feather of Figure 3-1c to drive the nail (even though it may have considerable velocity), because it has very little kinetic energy. Another important idea about energy is that it can be transferred from one body to another. For example, the moving foot of Figure 3-1d can transmit (at least part of) its energy to the football, causing it to move; this moving football has the capacity to do work, such as to break a window.

Potential energy *is the energy that a system possesses by virtue of its position or composition.* Because a clear notion of the nature of potential energy is so important in understanding many chemical concepts, we shall illustrate this form of energy in several ways. First consider the rock at the top of the hill in Figure 3-2a. When it is stationary, it has no kinetic energy because its velocity is zero. However, as it rolls down the hill, it attains some kinetic energy (capacity to do work, such as demolish the house at the bottom of the hill). We believe in the conservation of energy, so we might ask where the kinetic energy of the rolling rock "came from." We say that it had a greater amount of *potential energy*, by virtue of its position at the top of the hill, than it would have at the bottom; it is able to convert energy from one form (potential) into another (kinetic) as it rolls down the hill.

It is helpful to analyze differences in potential energy by imagining the work that must be done to change the system from one situation to another. Suppose the rock is at the bottom of the hill (Figure 3-2b) and we wish to push it up to the top (Figure 3-2c), which requires that work be done. The energy that we expend in pushing the rock up the hill can be thought of as being *stored* in the rock, by virtue of

Can you see how these four illustrations may be interpreted in terms of Equation 3-2?

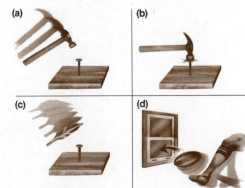

FIGURE 3-1 Some illustrations of kinetic energy. (a) The moving hammer can do work by driving the nail; thus, it has kinetic energy. (b) The stationary hammer ($v = 0$) has no kinetic energy. (c) The moving feather has velocity, but very little mass, so it has little kinetic energy and little capacity to do work. (d) Transfer of kinetic energy from one object to another. The moving foot can transfer part of its kinetic energy to the football. The moving football then has kinetic energy, and could do work such as breaking a window.

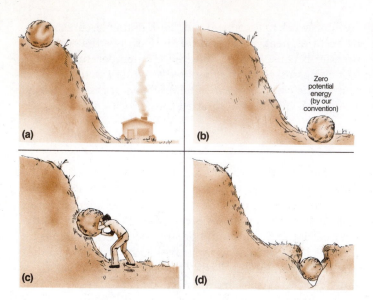

FIGURE 3–2 (a) The rock at the top of the hill has potential energy. This potential energy can be converted into kinetic energy, enabling the rock to do work. (b) The rock on the plain may *arbitrarily* be described as having *zero* potential energy. (c) As we do work to push the rock up the hill, we store energy in it. We increase its potential energy from the arbitrary zero state; thus, its potential energy is positive anywhere above the plain. (d) After an earthquake, the rock falls into the crack. Since we must do work to raise it back up to the condition of zero potential energy, we say that its potential energy is *negative* while it is in the chasm.

its higher elevation. We often describe potential energy as "stored energy." The work that we do in this case is against gravity (which is just another way of saying that the differences in potential energy are due to gravity), so this particular form of potential energy is often referred to as gravitational potential energy. Its magnitude is proportional to the product of the mass of the object and the height of the object above some reference point. For convenience of measuring, suppose we take the position of the rock on the plain (Figure 3–2b) as that in which the rock has *zero* potential energy. Since we increase its potential energy by pushing it up the hill, the numerical value is slightly positive partway up the hill, and more positive at the top. We note that the rock is more stable partway up the hill than at the top (if you let go, which way will it roll?), and still more stable at the bottom of the hill. Thus, lower gravitational potential energy represents a situation of greater stability.

Now let us imagine the rock to be at its position of zero potential energy, at the bottom of the hill, when an earthquake opens a great chasm directly beneath the rock, causing the rock to fall into the crack (Figure 3–2d). How do we describe its potential energy now, considering that we would rather not change our reference point every time some natural disaster comes along? Again, imagine what we must do to restore the rock to some condition. In particular, we must do work (increasing its potential *energy*) to restore the rock to the arbitrarily defined position of *zero* potential energy. Thus, its potential energy down in the chasm must be *less than zero,* or *negative.* Here is a very important idea about potential energy; a zero value of potential energy simply refers to some arbitrarily defined condition — it does not indicate "an absence of potential energy" or "the least potential energy that the object can have." In fact, the only meaningful questions that we can ever ask about potential energy have to do with *differences* in potential energy, and not with its absolute value. In other words, is the object or system of particles more stable (lower potential energy) at one condition than at another?

A situation that will be of more direct significance to our study of chemistry involves systems of charged particles that can either repel one another (e.g., two like charges) or attract one another (e.g., two unlike charges). To establish a scale for measurement of *electrical potential energy,* we adopt the following definition regarding two (or more) charged particles: When the particles are so far apart (infinitely far, actually) that they neither attract nor repel, their potential energy is taken to be *zero*. Let us consider first two particles that repel one another at all distances, such as two idealized point charges of the same sign. In order to bring them closer together than infinite separation, we must do work to overcome their repulsion. Thus, we increase their potential energy as we push them closer together, so it becomes more positive (Figure 3–3a). Again, we see that they are more stable at lower potential energy. If we release them at some close distance, they move to lower potential energy (further apart).

Now consider, instead, two particles that attract one another at all distances, such as two idealized point charges of opposite sign. Imagine that they are some small distance apart. In order to separate them we must overcome their attraction by doing work, which increases their potential energy. If we move them to infinite separation, we increase their potential energy all the way *up to zero*. Thus, their potential energy at distances less than infinite separation must be *less than zero*, or *negative* (Figure 3–3b). It is important to notice the arbitrary nature of the zero value of potential energy in these two examples. In any case, we will ask only "Which way is up, in potential energy?" or "Which situation represents greater or less stability?"

A final situation, which will be of great importance in many subsequent chemical studies, is that in which there are both attractive forces *and* repulsive forces. The latter become more important at smaller distances. For instance, when we deal with *real* oppositely charged particles, which occupy volume, we can see that both of them cannot be in exactly the same place. Another way of saying this is

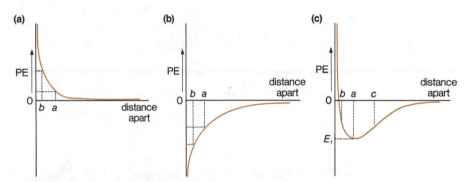

FIGURE 3–3 Potential energy diagram for two charged particles. Potential energy is conventionally taken as zero with the charges at infinite separation. (a) Diagram for two idealized point charges of the same sign, which repel one another at all distances. When the particles are separated by distance *a*, they are more stable (lower potential energy) than at distance *b*. (b) Diagram for two idealized point charges of opposite sign, which attract one another at all distances. When these particles are separated by distance *a*, they are less stable (have higher potential energy) than at distance *b*. (c) Diagram for two charges that attract one another at large distances (note similarity of curvature to part (b)) and that repel one another at small distances (note similarity of curvature to part (a)). The distance *a* represents the most stable distance of separation for these two particles. In order to move them closer together (point *b*), we must do work on them, to increase their potential energy against the net repulsive force; in order to pull them further part (point *c*), we must do work on them to increase their potential energy against the net attractive force. Thus, their potential energy at separation *a*, denoted by E_1, is the lowest value possible for these two particles. In order to completely separate the particles, we must increase their energy by an amount equal to the magnitude of E_1.

that at very small distances, they must repel one another. The potential energy curve that describes this situation is shown in Figure 3–3c. The most important idea here is that for such a system, a *minimum attainable* potential energy exists, representing the *most stable* distance of separation of these particles. We shall see many applications of this idea in our study of chemistry. An analogy, in which the downward "attractive" force of gravity is opposed by an upward "repulsive" force due to the compression of an inner-spring mattress, is illustrated in Figure 3–4.

Let us introduce energy at the molecular level by considering a sample of oxygen at room temperature and atmospheric pressure. The **internal energy** of a collection of gaseous O_2 molecules is the sum of all the kinds of energy it possesses, some kinetic and some potential. The kinetic energy is of three kinds (Figure 3–5): (1) translational kinetic energy due to straight line motion of the molecules, (2) vibrational kinetic energy due to motions that cause separations of the atoms of a molecule that are periodically greater than or smaller than the average separation, and (3) rotational kinetic energy due to tumbling motion.

There are many kinds of potential energy associated with the collection of O_2 molecules. Just a few of these are the energies of electrostatic attraction and repulsion involving the charged particles that comprise the atoms and the energies of interaction among the molecules themselves. Atoms are held together by *chemical bonds* in molecules. The most stable states, or most energetically favorable states, are those of *lowest* potential energy.

The most stable form of oxygen at room temperature and atmospheric pressure is diatomic molecules, O_2. Energy must be absorbed by, and work done on, an O_2 molecule to break the bond between the two O atoms. The two separated O atoms are at higher potential energy than in the O_2 molecule, as a result of having absorbed energy. If the two separated atoms recombine to form an O_2 molecule, the same amount of energy is released, and the system returns to lower potential energy.

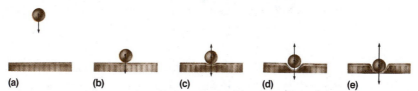

(a)   (b)   (c)   (d)   (e)

FIGURE 3–4 A system analogous to Figure 3–3c. A steel ball suspended above an inner-spring mattress is gradually lowered. The downward "attractive" force due to gravity is represented by a downward arrow; the opposing "repulsive" force, due to the compression of the mattress, is represented by an upward arrow. (a) The ball is suspended above the mattress, so that only the attractive force of gravity acts on it. (b) It has been lowered so that it is about to touch the mattress. (c) It is compressing the mattress slightly, but the downward attractive force due to gravity still exceeds the repulsive force due to the mattress pushing back. (d) If we let go of the ball, it settles into a position in which the downward force of gravity is exactly matched by the upward force of the compressed mattress. This position represents the minimum potential energy of the ball. We would have to do work against the attractive force of gravity to raise the ball back to (c), (b), or (a). Likewise, we would have to do work against the increasing repulsive force of the compressed mattress to push it deeper into the mattress, as in (e).

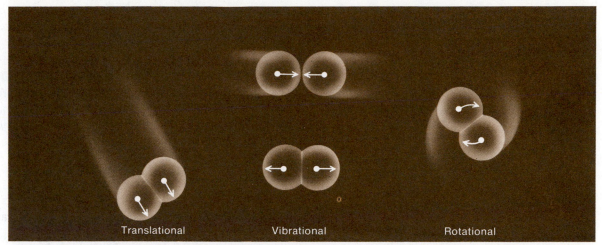

FIGURE 3-5 An oxygen molecule, O_2, has three different kinds of kinetic energy. The energy that it possesses due to its motion through space is called translational energy. The molecule vibrates; i.e., the distance between the two oxygen nuclei changes. Energy associated with vibrations is called vibrational energy. The energy associated with spinning and tumbling motions is called rotational energy.

In exothermic chemical reactions, such as burning gasoline, some of the potential energy of the reactants is converted to heat (thermal) and light (radiant) energy. In such cases the total potential energy of the products is lower than that of the reactants. The difference between the energies of the reactants and products is equal to the amount of thermal energy and radiant energy released.

Work also can be done as chemical reactions occur. For example, gases are produced by the combustion of gasoline in an automobile engine, and these gases do work as they expand to push the pistons in the cylinders. The energy obtained from combustion of gasoline is supplied to the drive train by the moving pistons as *mechanical energy*, which moves the car. Unfortunately, internal combustion engines are only about 10% efficient in converting chemical energy into mechanical energy. The rest of the energy released in combustion is wasted (as heat).

The chemical reactions that occur in automobile batteries result in the conversion of *energy associated with chemical bonds* into *electrical energy*. This electrical energy is used for a variety of purposes, such as starting the engine and the operation of lights, radios, heaters, and other electrical equipment in automobiles.

There are many other forms of potential energy. For example, nuclear energy is converted into heat energy (as well as other forms of energy) in nuclear reactors as unstable atomic nuclei decompose. Thus nuclear energy can be obtained from the potential energy associated with unstable nuclei. Nuclear changes involve the conversion of matter, which is equivalent to energy, into other forms of energy. Some of the heat generated in nuclear reactors can be used to boil water to produce steam (a physical change), which can drive a turbine (mechanical energy) to run a generator that produces electrical energy. The rest of the heat is waste heat that is absorbed by the cooling water, which is discharged into a nearby body of water.

These conversions of energy from one form to another, as well as transfers of energy, obey the Law of Conservation of Energy: *Energy is neither created nor destroyed in ordinary chemical changes or physical changes,* or *the total amount of energy in the universe is constant.*

Carbon dioxide and water (steam) are the principal chemical products of the combustion of gasoline.

Chapter 24 deals with nuclear chemistry and nuclear reactors. Section 4-9 describes nuclear stability and the conversion of matter into other forms of energy.

We call this *thermal pollution.*

In light of this law you may wonder how we can be facing a declining supply of energy. The problem is not really a decline in the *total* supply, but rather a decline in the most readily *useful* forms of energy. As "concentrated" forms of energy (such as fossil fuels or materials that serve as fuels for nuclear reactors) are used, only a portion of the energy released can be put to useful work. A large percentage of the energy released is, in essence, wasted as heat that eventually becomes widely distributed throughout the environment. We could say that this thermal energy, which ultimately heats the atmosphere or natural waters, winds up in very "dilute" form, which is not easily collected or transformed into other forms of energy.

3-2   Heat Transfer and the Measurement of Heat

The so-called "large calorie" or "Calorie," used to indicate energy content of foods in nutrition, is really one kilocalorie, or 1000 calories.

As chemical reactions and physical changes occur, they are accompanied by either the evolution of heat (exothermic processes) or the absorption of heat (endothermic processes). The amount of heat transferred in a process is usually expressed in joules, or in calories. Historically, the calorie was originally defined as the amount of heat necessary to raise the temperature of one gram of water from 14.5°C to 15.5°C. The amount of heat necessary to raise the temperature of one gram of liquid water by one degree Celsius varies slightly with temperature, so it is necessary to specify a particular one degree temperature increment in defining the calorie in this way. For our purposes the variations are sufficiently small that we are justified in ignoring them. **One calorie** *is now defined as exactly* **4.184 joules.**

Recall that $E_{kinetic} = \frac{1}{2}mv^2$.

The SI unit of energy and work is the **joule** (J), which is defined as $1\ kg \cdot m^2/s^2$. One joule is the amount of kinetic energy possessed by a 2 kg object moving at one meter per second. (In English units this corresponds to a 4.4 lb object moving at 197 feet per minute or 2.2 miles per hour.) You may find it more convenient to think in terms of the amount of heat required to raise the temperature of one gram of water from 14.5°C to 15.5°C, which is 4.184 joules.

The **specific heat** *of a substance is the amount of heat required to raise the temperature of one gram of the substance one degree C (which is also one kelvin) with no change in physical state or form.* The specific heat of each substance is an intensive physical property of the substance, and is different for the solid, liquid, and gaseous states of the substance. For example, the specific heat of ice is $2.09\ J/(g \cdot °C)$ near 0°C; for liquid water it is $4.184\ J/(g \cdot °C)$, while the specific heat of steam is 2.03 $J/(g \cdot °C)$ near 100°C. The specific heat of water is quite high compared with that for most common substances. A table of specific heats is provided in Appendix E.

Example 3-1

How much heat is required to raise the temperature of 500 g of water from 40°C to 80°C?

Solution

Since the specific heat of a substance is the amount of heat required to raise the temperature of 1 g of substance 1°C, or

specific heat

$$= \frac{(\text{amount of heat, J})}{(\text{mass of substance, g})(\text{temp. change, °C})}$$

we can rearrange the equation so that

amount of heat = (mass of substance)(sp. ht.)
$$\times \text{(temp. change)}$$
$$= (500\ g)(4.18\ J/g \cdot °C)(40°C)$$
$$= \underline{8.4 \times 10^4\ J}\ \text{or}\ \underline{84\ kJ}$$

Or, we can use the dimensional analysis approach to obtain the same result.

$$?\ J = (500\ g)(4.18\ J/g \cdot °C)(40°C)$$
$$= \underline{8.4 \times 10^4\ J}\ \text{or}\ \underline{84\ kJ}$$

Note that all units except joules cancel, and the answer contains only two significant figures because the temperature difference contains only two.

If 500 g of water were cooled from 80°C to 40°C, it would be necessary to remove exactly the same amount of heat, 84 kJ.

Example 3-2

How much heat, in joules and kilojoules, is required to raise the temperature of 500 g of iron from 40°C to 80°C? The specific heat of iron is 0.444 J/g·°C.

Solution

$? J = (500 \text{ g})(0.444 \text{ J/g} \cdot °C)(40°C)$
$\quad = \underline{8.9 \times 10^3} \text{ J or } \underline{8.9} \text{ kJ}$

Note that the specific heat of iron is much smaller than the specific heat of water. Therefore, much less heat is required to raise the temperature of 500 g of iron than for 500 g of water (Example 3-1).

The purpose of insulation is to prevent the transfer of heat energy from one place to another (for instance, from your home to the outdoor air). When any process takes place within a completely insulated container, the Law of Conservation of Energy implies that all of the energy present in the container before the process is still there afterward. If the process is the transfer of heat energy due to a temperature difference between two objects, then energy conservation means that the heat lost by one object must be gained by the other object. In the absence of any other process, this will result in changes in the temperatures of the two objects, until they come to the same temperature.

Example 3-3

If 200 g of water at 40°C is mixed with 100 g of water at 80°C in an insulated container, what will be the temperature of the mixture?

Solution

The temperature of the warm water will decrease and the temperature of the cool water will increase until the temperatures of the two samples become equal. Call the new temperature t°C. It must be between 80°C and 40°C. The temperature changes are $(t - 80)$°C and $(t - 40)$°C, respectively, or $t_{final} - t_{initial}$. Since the container is insulated, the Law of Conservation of Energy tells us that the amount of heat change experienced by the warm

water plus the amount of heat change experienced by the cool water must be zero.

Heat change of warm water
$\qquad$ + heat change of cool water = 0

$(100 \text{ g})(4.18 \text{ J/g} \cdot °C)(t - 80)°C$
$\qquad + (200 \text{ g})(4.18 \text{ J/g} \cdot °C)(t - 40)°C = 0$

Simplifying and solving for t, we get

$418t - 33440 + 836t - 33440 = 0$
$1254t - 66880 = 0$
$1254t = 66880$
$t = 53$

and the final temperature is $\underline{53°C}$.

Example 3-4

When 200 grams of lead at 170.0°C were dropped into 100 milliliters of water at 40.0°C in an insulated container, both lead and water reached the same temperature of 47.4°C. Calculate the specific heat of lead.

Solution

Since the container is insulated, the amount of heat gained by the water is just equal to the amount of

heat lost by the lead. Since the density of water is 1.00 g/mL, 100 mL of water is 100 grams. The temperature change for lead is $(47.4 - 170.0)°C = -122.6°C$; the temperature change for water is $(47.4 - 40.0)°C = 7.4°C$. We know that *heat gain must equal heat loss,* and we can represent both in an equation with only one unknown, the specific heat of lead.

heat lost by hot Pb (joules)
$\qquad$ + heat gained by cool H_2O (joules) = 0

200 g × sp. ht. × (−122.6°C)

$$+ 100 \text{ g} \times \frac{4.18 \text{ J}}{\text{g} \cdot °\text{C}} \times 7.4°\text{C} = 0$$

$(-2.45 \times 10^4 \text{ g} \cdot °\text{C})(\text{sp. ht.}) + 3.1 \times 10^3 \text{ J} = 0$

$(2.45 \times 10^4 \text{ g} \cdot °\text{C})(\text{sp. ht.}) = 3.1 \times 10^3 \text{ J}$

sp. ht. = 0.13 J/g·°C

We see that only 100 g of water can cool 200 g of lead by 122.6°C, while itself being warmed by only 7.4°C. The unusually high specific heat of water makes it an efficient coolant.

3–3   Some Thermodynamic Terms

Historically, the first studies of energy and the transfer of energy dealt predominantly with heat flow.

The term "thermodynamics" literally means the motion or flow (*dynamics*) of heat (*thermo*). We now use the term in a more general sense. **Thermodynamics** *is the study of all changes in energy or transfers of energy that accompany physical and chemical processes.* We shall introduce the topic here as we first define some of the terms commonly used in thermodynamics. Then we shall extend our coverage of heat flow to chemical reactions.

Several words are given special meanings for use in thermodynamics. The substances involved in a chemical reaction or a physical change are called the **system,** while everything else in the system's environment constitutes its **surroundings.** The **universe** is the system plus its surroundings. The system may be thought of as the part of the universe under investigation. The **First Law of Thermodynamics** tells us that the energy of the universe is constant, i.e., energy is neither created nor destroyed, but is merely transferred between the system and its surroundings. As exothermic processes occur, heat energy is transferred from the system to the surroundings. As endothermic processes occur, the system absorbs heat energy from its surroundings.

Some examples may help to clarify this somewhat formal terminology.

(a) In Example 3–4, we could consider the block of lead as the system, and the water as the surroundings (Figure 3–6a). The process in which the *system* (warm lead) gives off heat to *surroundings* (cool water) is *exothermic*.

(b) Alternatively, the block of lead *and* the water together could be considered as the system, and the rest of the environment (table, room, air, and so on) could be called the surroundings. In this description, the system (lead + water) cannot transfer heat to its surroundings because the system is in an insulated container.

(c) Consider melting ice at 0°C to form liquid water at 0°C (Figure 3–6b). As the ice melts, it absorbs heat from the surrounding air and from the vessel in which it sits. The ice could be considered as the system, and the air and container together would be the surroundings. The melting of ice is an *endothermic* process.

This neutralization reaction will be discussed in more detail in Section 9–6.1.

(d) When an acidic solution is added to a basic solution, a chemical reaction called neutralization occurs, and heat is given off. We could call the solution mixture the system, and the air, container, and so on would be the surroundings (Figure 3–6c). This process is *exothermic*.

The **thermodynamic state of a system** *is a set of conditions that completely specifies all of the properties of the system.* This set includes the temperature, pressure, composition (identity and number of moles of each component), and physical state (gas, liquid, or solid forms) of each part of the system. Once the state has been specified, all other properties—both physical and chemical—are fixed.

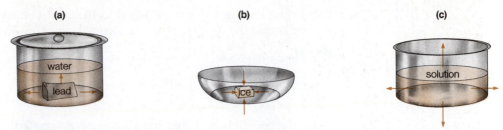

FIGURE 3–6 (a) The direction of heat flow is indicated by arrows. When a warm piece of lead is placed in cool water (in an insulated container), heat flows from the lead to the water. (b) When a piece of ice (the system) is placed in a pan at room temperature, the ice melts as it absorbs heat from the air and the pan (which together make up the surroundings). Heat flows from the surroundings to the ice. (c) When a solution of an acid and a solution of a base are mixed (the system), the resulting chemical reaction is exothermic; it gives off heat to the surroundings (the container and the air).

The conditions that make up the state of a system, and the variables that measure them, are called **state functions.** Their defining characteristic is that the *value* of a state function depends *only* on the state of the system, and not on the manner by which the system came to be in that state (what we might call the history of the system). A *change* in a state function describes the *differences* between two states, and is independent of the process by which the change occurs.

For instance, consider a sample of one mole of pure liquid water at 30°C and 1 atm pressure. If at some later time the temperature of the sample is 22°C at the same pressure and with no change of physical state or composition, then it is in a different thermodynamic state. We can tell that the *net* temperature change is −8°C, and it does not matter whether (1) the cooling took place directly (either slowly or rapidly) from 30°C to 22°C, or (2) the sample was first heated to 36°C, then cooled to 10°C, and finally warmed to 22°C, or (3) any other conceivable path was followed from the initial state to the final state. Also, regardless of the path followed between states, in this case the *net* change in each of the other state functions is *zero*. Of course, it is possible for a change between two states to involve positive or negative changes in two or more (or all) state functions simultaneously.

Any property of a system that depends only on the values of its state functions is also a state function. Thus, for instance, the volume of a sample of matter, which depends only on temperature, pressure, composition, and physical state, is a state function. We shall encounter other state functions as we continue our study of thermodynamics.

It is equally important to recognize variables that are not state functions. One such variable is the work done in sliding an object along a horizontal surface through a fixed distance. If the friction between the object and the surface is low, then the force required to cause the change of position will be relatively small; if the friction is greater, more force will be required. Although the initial and final states of the object are the same for both processes, the work ($w = f \times d$) is greater for the high-friction case.

Most chemical reactions and physical changes that occur in the laboratory (and most natural changes that occur elsewhere, for that matter) do so at constant (atmospheric) pressure. *The change in heat content of a system that accompanies a process at constant pressure is defined as the* **enthalpy change, ΔH,** *of the process.* An enthalpy change is sometimes loosely referred to as a *heat change* or a *heat of* *reaction.* The enthalpy change is equal to the heat content, H, of the substances produced *minus* the heat content of the substances consumed.

$$\Delta H = H_{\text{substances produced}} - H_{\text{substances consumed}} \qquad \text{[Eq. 3-3]}$$

or

$$\Delta H = H_{\text{final state}} - H_{\text{initial state}} \qquad \text{[Eq. 3-4]}$$

It is impossible to know the absolute heat content (enthalpy) of a system. However, *enthalpy is a state function,* and it is the *change in enthalpy* (change in heat content) that is of real interest; this can be measured for many processes. Let us now focus attention on chemical reactions and the enthalpy changes that occur in these processes.

By convention, most ΔH values are reported at one atmosphere, which is called *standard pressure.* Such enthalpy changes are designated with a superscript of zero, ΔH^0. This is called the *standard enthalpy change* at a particular temperature, usually 25°C (298 K), unless otherwise specified by a subscript, as in ΔH^0_{300}, which refers to 300 K.

3-4   Thermochemical Equations

When we include a description of the enthalpy change that accompanies a chemical change, the resulting equation is called a **thermochemical equation.** Consider again the complete combustion of methane at 25°C.

$$CH_4 \text{ (g)} + 2O_2 \text{ (g)} \longrightarrow CO_2 \text{ (g)} + 2H_2O \text{ (ℓ)} + 890.3 \text{ kJ}$$

| CH₄ (g) | 2O₂ (g) | | CO₂ (g) | 2H₂O (ℓ) | |
| 1 mol | 2 mol | | 1 mol | 2 mol | |

We could also convey the same information by writing:

$$CH_4 \text{ (g)} + 2O_2 \text{ (g)} \longrightarrow CO_2 \text{ (g)} + 2H_2O \text{ (ℓ)} \qquad \Delta H^0 = -890.3 \text{ kJ}$$

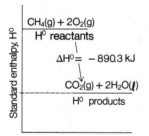

Standard enthalpy, H^0

CH₄(g) + 2O₂(g)
H^0 reactants

$\Delta H^0 = -890.3$ kJ

CO₂(g) + 2H₂O(ℓ)
H^0 products

FIGURE 3-7 The change in heat content for the complete combustion of one mole of methane at 25°C.

This thermochemical equation tells us that when one mole of methane burns in excess oxygen, 890.3 kJ of heat is released from the reacting system (CH_4 and O_2) to the surroundings. When one mole of CH_4 burns in *excess* O_2 it *reacts* with 2 moles O_2. This is because the heat content of one mole of CO_2 and two moles of liquid water is 890.3 kJ less than that of one mole of methane and two moles of oxygen under the same conditions of temperature and pressure (Figure 3-7).

$$\Delta H^0 = H^0_{\text{products}} - H^0_{\text{reactants}} = -890.3 \text{ kJ}$$

The *negative sign of ΔH^0 (or ΔH in general) indicates that this is an exothermic reaction,* as are all reactions in which fuels undergo combustion.

The reverse reaction would require the absorption of 890.3 kJ per mole of CO_2 that reacts; i.e., it is endothermic with $\Delta H^0 = +890.3$ kJ.

$$CO_2 \text{ (g)} + 2H_2O \text{ (ℓ)} + 890.3 \text{ kJ} \longrightarrow CH_4 \text{ (g)} + 2O_2 \text{ (g)}$$

The following conventions apply to thermochemical equations:

1. The coefficients in a balanced equation refer to the *number of moles* of reactants and products involved in the balanced chemical equation to which the specified ΔH value applies. In the thermodynamic interpretation of equations, we *never*

interpret the coefficients as *numbers of molecules*. Thus, coefficients in thermochemical equations are sometimes written as fractions rather than integers.

2. The physical states of all species are important and must be specified.

3. The value of ΔH^0 *does* vary with temperature, though often not very dramatically.

Let us illustrate the significance of these statements. The amount of heat released or absorbed during a reaction depends upon how much reaction occurs, just as the mass of a given product obtained from a reaction depends upon the amounts of reactants consumed. For this reason we can use balanced thermochemical equations to construct unit factors such as -890.3 kJ/mol CH_4, which tells us that 890.3 kJ of heat is released by the reaction that consumes 1.000 mol of CH_4.

Example 3-5

How much heat is liberated by the complete combustion of 24.0 grams of methane at 25°C? The thermochemical equation is

$$CH_4\ (g) + 2O_2\ (g) \longrightarrow CO_2\ (g) + 2H_2O\ (\ell)$$
$$\Delta H^0 = -890.3 \text{ kJ}$$

Solution

We first convert 24.0 g of CH_4 into moles, and then calculate the amount of heat released by this number of moles of CH_4, using the unit factor -890.3 kJ/mol CH_4.

$$\underline{?}\ \text{kJ} = 24.0 \text{ g } CH_4 \times \frac{1 \text{ mol } CH_4}{16.0 \text{ g } CH_4} \times \frac{-890.3 \text{ kJ}}{1 \text{ mol } CH_4}$$

$$\text{g } CH_4 \longrightarrow \text{mol } CH_4 \longrightarrow \text{kJ heat}$$

$$= \underline{-1.34 \times 10^3 \text{ kJ}}$$

This tells us that 1.34×10^3 kJ of heat is *released to the surroundings* during the combustion of 24.0 grams of methane.

Let us now examine Statement 2 by considering the combustion of one mole of methane at 25°C under conditions such that the H_2O is produced as a vapor rather than as liquid water.

$$CH_4\ (g) + 2O_2\ (g) \longrightarrow CO_2\ (g) + 2H_2O\ (g) \qquad \Delta H^0 = -802.3 \text{ kJ}$$

The value of ΔH^0 is *less negative* than in the previous reaction. The reason for this is that the heat content of two moles of water vapor is greater than the heat content of two moles of liquid water; the difference is accounted for by the heat of vaporization of water at 25°C, which is 44.0 kJ/mol. Thus, for the reaction that produces water vapor, $\Delta H^0 = [-890.3 + 2(44.0)]$ kJ $= -802.3$ kJ. You could think of this as a two-stage process: First the chemical reaction occurs, producing liquid water, and then part of the heat released by the reaction is used to vaporize the water. The heat content of the system decreases by 890.3 kJ and then increases again by 88.0 kJ, for a *net* decrease of 802.3 kJ.

Finally, we consider Statement 3. The value of ΔH^0 varies somewhat with temperature. The enthalpy change for an exothermic reaction is the amount of heat *in excess of* the amount necessary to maintain the system (chemical substances) at 25°C, or some other specified temperature. This must be different for the same reaction at two different temperatures. For an endothermic reaction, the enthalpy change is the total amount of heat that must be absorbed to produce the products *and* to maintain constant temperature.

Let us now consider how we actually measure values of ΔH for chemical reactions.

3–5 Calorimetry

Enthalpy changes are measured in **calorimeters.** For now we shall focus our attention on the determination of heats of reaction at *constant pressure.* A "coffee-cup" calorimeter made of Styrofoam is frequently used in teaching laboratories to measure heats of reaction in aqueous solution (Figure 3–8). Reactions in which there are no gaseous reactants or products are usually chosen, so that all reactants and products remain in the cup throughout the experiment. The pressure remains constant because the cup is not sealed. The double walls and top provide enough insulation so that only an insignificant amount of heat escapes.

> Solvents other than water can also be used (if they don't dissolve the Styrofoam).

The amount of heat evolved by the reaction can be calculated from the amount by which it causes the temperature of the system to rise. The heat can be imagined as divided into two parts; one increases the temperature of the known mass of the solution, and the other increases the temperature of the calorimeter. A separate experiment is done to determine the **heat capacity** of the calorimeter, which indicates the amount of heat the calorimeter, itself, absorbs per degree Celsius increase in temperature.

Example 3–6 illustrates the method of determining the heat capacity of a calorimeter.

Example 3–6

When 50.00 g of water at 52.7°C was mixed with 50.00 g of water at 22.3°C in a calorimeter, the final temperature of the system was 36.7°C. (The calorimeter was at 22.3°C initially.) Calculate the heat capacity of the calorimeter in J/°C. The specific heat of water is 4.184 J/g·°C.

Solution

If no heat were lost to the calorimeter, the amount of heat lost by the warm water would equal the amount of heat gained by the cool water. The actual difference represents the heat absorbed by the calorimeter (assuming negligible loss to the surroundings).

Heat change of warm water:

$$\boxed{\begin{array}{l}50.0 \text{ g } H_2O \\ \text{at } 52.7°C\end{array}} \longrightarrow \boxed{\begin{array}{l}50.00 \text{ g } H_2O \\ \text{at } 36.7°C\end{array}}$$

$$? J = 50.0 \text{ g} \times \frac{4.184 \text{ J}}{\text{g·}°C} \times (36.7 - 52.7)°C$$
$$= -3.35 \times 10^3 \text{ J}$$

Heat change of cool water:

$$\boxed{\begin{array}{l}50.00 \text{ g } H_2O \\ \text{at } 22.3°C\end{array}} \longrightarrow \boxed{\begin{array}{l}50.00 \text{ g } H_2O \\ \text{at } 36.7°C\end{array}}$$

$$? J = 50.0 \text{ g} \times \frac{4.184 \text{ J}}{\text{g·}°C} \times (36.7 - 22.3)°C$$
$$= 3.01 \times 10^3 \text{ J}$$

Since 3.35×10^3 J was lost by the warm water and only 3.01×10^3 J was gained by the cool water, the difference is the amount of heat absorbed by the calorimeter.

$$? J = 3.35 \times 10^3 \text{ J} - 3.01 \times 10^3 \text{ J}$$
$$= 0.34 \times 10^3 \text{ J or } 3.4 \times 10^2 \text{ J absorbed by calorimeter}$$

We divide 3.4×10^2 J, the amount of heat absorbed by the calorimeter, by the temperature change for the calorimeter to obtain the heat capacity of the calorimeter.

$$? \frac{J}{°C} = \frac{3.4 \times 10^2 \text{ J}}{(36.7 - 22.3)°C} = \underline{24 \text{ J/}°C}$$

The heat capacity for the calorimeter, 24 J/°C, tells us that the calorimeter absorbs 24 J of heat for each degree Celsius its temperature increases.

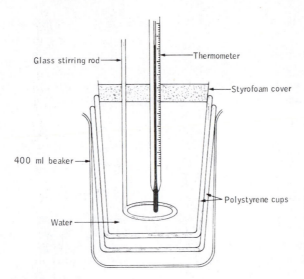

Glass stirring rod

Thermometer

Styrofoam cover

400 ml beaker

Polystyrene cups

Water

FIGURE 3–8 A coffee-cup calorimeter. The stirring rod is moved up and down to insure thorough mixing and uniform heating of the solution during reaction.

After calibration, the calorimeter can be used to measure the amount of heat evolved by the reaction of a known quantity of copper(II) sulfate solution of known concentration, $CuSO_4$ (aq), with a solution containing a stoichiometrically equivalent amount of sodium hydroxide, $NaOH$ (aq), to produce insoluble copper(II) hydroxide, $Cu(OH)_2$ (s), and a solution of sodium sulfate, Na_2SO_4 (aq), as in Example 3–7.

Example 3–7

A 50.0 mL sample of 0.300 M copper(II) sulfate solution at 23.0°C is mixed with 50.0 mL of 0.600 M sodium hydroxide solution, also at 23.0°C, in the calorimeter described in Example 3–6. After the reaction has occurred, the temperature of the resulting mixture is measured to be 26.3°C. Calculate (a) the amount of heat evolved and (b) ΔH for the reaction. The density of the final solution is 1.02 g/mL. Assume that the specific heat of the solution is the same as that of pure water, 4.18 J/g·°C. The equation for the chemical reaction is

$$CuSO_4 \text{ (aq)} + 2NaOH \text{ (aq)} \longrightarrow$$
$$Cu(OH)_2 \text{ (s)} + Na_2SO_4 \text{ (aq)}$$

Solution

(a) The amount of heat released by the reaction is absorbed by the solution *and* the calorimeter (assuming negligible loss to the surroundings). The amount of heat absorbed by the solution is the mass of the solution times its specific heat times its temperature change. We shall assume that the volume of the reaction mixture is just the sum of the vol-

umes of the initial solutions. (When *dilute aqueous solutions* are mixed, their volumes are very nearly additive. This is *not* true for most solutions.) Thus, the mass of the solution is

$$? \text{ g soln} = (50.0 + 50.0) \text{ mL} \times \frac{1.02 \text{ g soln}}{\text{mL}}$$
$$= 102 \text{ g soln}$$

The amount of heat absorbed by the solution plus the calorimeter is

$$? \text{ J} = \underbrace{102 \text{ g} \times \frac{4.18 \text{ J}}{\text{g} \cdot {}^\circ\text{C}} \times (26.3 - 23.0)^\circ\text{C}}_{\substack{\text{amount of heat} \\ \text{absorbed by solution}}}$$

$$+ \underbrace{\frac{24 \text{ J}}{{}^\circ\text{C}} \times (26.3 - 23.0)^\circ\text{C}}_{\substack{\text{amount of heat} \\ \text{absorbed by calorimeter}}}$$

$$= 1.4 \times 10^3 \text{ J} + 79 \text{ J}$$
$$= 1.5 \times 10^3 \text{ J of heat absorbed by solution plus calorimeter}$$

Thus the reaction must have liberated 1.5×10^3 J of heat.

(b) To determine ΔH, we must determine *how much* reaction occurred, that is, how many moles of reactants were consumed. Copper(II) sulfate and sodium hydroxide react in $1:2$ mole ratio.

$$CuSO_4 + 2NaOH \longrightarrow Cu(OH)_2 + Na_2SO_4$$

1 mol   2 mol   1 mol   1 mol

By multiplying the volume of each solution in liters by its concentration in mol/L (molarity), we can determine the number of moles of each reactant mixed.

$$\underset{=}{?} \text{ mol } CuSO_4 = 0.0500 \text{ L} \times \frac{0.300 \text{ mol } CuSO_4}{1.00 \text{ L}}$$

$$= 0.0150 \text{ mol } CuSO_4$$

$$\underset{=}{?} \text{ mol } NaOH = 0.0500 \text{ L} \times \frac{0.600 \text{ mol } NaOH}{1.00 \text{ L}}$$

$$= 0.0300 \text{ mol } NaOH$$

The reactants were mixed in a $1:2$ mole ratio, so all of each reactant is consumed. Thus 1.5×10^3 J of heat was released during the consumption of 0.0150 mol of $CuSO_4$. This corresponds to the release of 1.0×10^2 kJ per mol of $CuSO_4$, as the following calculation shows.

$$\underset{=}{?} \frac{\text{kJ}}{\text{mol } CuSO_4} = \frac{1.5 \times 10^3 \text{ J}}{0.0150 \text{ mol } CuSO_4}$$

$$= 1.0 \times 10^5 \text{ J}$$

$$= 1.0 \times 10^2 \text{ kJ released/mol } CuSO_4$$

The chemical equation written above involves one mole of $CuSO_4$, so we have determined that when the reaction occurs *to the extent indicated by the equation*, 1.0×10^2 kJ of heat is released (reaction is exothermic). Thus, remembering that negative values of ΔH indicate exothermic reactions, we write

$$\Delta H = -1.0 \times 10^2 \text{ kJ}$$

3–6   Standard Molar Enthalpies of Formation, ΔH_f^0

A temperature of 25°C is 77°F (Section 1–11). This is slightly above room temperature, which is about 20°C or 68°F.

The **thermochemical standard state** *of a substance is its most stable state under standard pressure* (one atmosphere) *and at some specific temperature* (usually 25°C *unless otherwise specified*). Examples of elements in their standard states at 25°C are hydrogen, gaseous diatomic molecules, H_2 (g); mercury, a silver-colored liquid metal, Hg (ℓ); sodium, a silvery-white solid metal, Na (s); and carbon, a grayish-black solid called graphite, C (graphite). The designation C (graphite) is used instead of C (s) to distinguish it from another allotropic modification of carbon, C (diamond). Examples of standard states of compounds include ethanol (ethyl alcohol or grain alcohol), a liquid, C_2H_5OH (ℓ); water, a liquid, H_2O (ℓ); calcium carbonate, a solid, $CaCO_3$ (s), and carbon dioxide, a gas, CO_2 (g).

Recall that the superscript zero signifies standard pressure, 1.0 atm. Negative values for ΔH_f^0 describe exothermic reactions, while positive values for ΔH_f^0 describe endothermic reactions.

A special name, **standard molar enthalpy of formation,** ΔH_f^0, *is given to the amount of heat absorbed in a reaction in which one mole of a substance in its standard state is formed from its elements in their standard states.* Standard molar enthalpy of formation is often called **standard molar heat of formation** or, more simply, **heat of formation.** For example, ΔH_f^0 for solid calcium oxide, CaO (s), is -635.5 kJ/mol.

$$Ca \text{ (s)} + \tfrac{1}{2}O_2 \text{ (g)} \longrightarrow CaO \text{ (s)} \qquad \Delta H^0 = \Delta H_{f\,CaO\,(s)}^0 = -635.5 \text{ kJ}$$

The coefficient $\tfrac{1}{2}$ preceding O_2 (g) does not imply half a molecule of oxygen, *because coefficients always refer to the number of moles under consideration in thermodynamic equations.* Because of the definition of ΔH_f^0, the coefficient of the product (CaO (s) in this case) must be 1.

Many stable compounds, such as CaO (s), have large negative values of ΔH_f^0 because they are produced from their constituent elements in their standard states in *exothermic* reactions.

TABLE 3-1 Selected Standard Molar Enthalpies of Formation at 298 K

Substance	ΔH_f^0, kJ/mol	Substance	ΔH_f^0, kJ/mol
Br_2 (ℓ)	0	H_2 (g)	0
Br_2 (g)	30.91	HBr (g)	−36.4
C (diamond)	1.897	H_2O (ℓ)	−285.8
C (graphite)	0	H_2O (g)	−241.8
CH_4 (g)	−74.81	N_2 (g)	0
C_2H_4 (g)	52.26	N_2H_4 (ℓ)	50.63
C_6H_6 (ℓ)	49.03	NO (g)	90.25
C_2H_5OH (ℓ)	−277.7	Na (s)	0
CO (g)	−110.52	NaCl (s)	−411.0
CO_2 (g)	−393.51	O_2 (g)	0
CaO (s)	−635.5	Pb (s)	0
$CaCO_3$ (s)	−1207	PbO (s) yellow	−217.3
Cl (g)	121.7	SO_2 (g)	−296.8
Cl_2 (g)	0	SiH_4 (g)	34
Hg (ℓ)	0	$SiCl_4$ (g)	−657.0
HgS (s) red	−58.2	SiO_2 (s)	−910.9

Some compounds with large positive values of ΔH_f^0 are unstable and may even decompose explosively. An example is hydrazine, N_2H_4 (ℓ), for which $\Delta H_f^0 = 50.63$ kJ/mol. Hydrazine is used as a rocket fuel.

$$N_2 \text{ (g)} + 2H_2 \text{ (g)} \longrightarrow N_2H_4 \text{ (ℓ)} \qquad \Delta H = \Delta H_{f\,N_2H_4(\ell)}^0 = 50.63 \text{ kJ/mol}$$

By definition, ΔH_f^0 values of all elements in their standard states are zero. If we consider this together with the definition of ΔH_f^0 for compounds, we arrive at the following very useful interpretation: ΔH_f^0 for any compound may be viewed as its "heat content per mole," on a scale on which all elements in their standard states are (arbitrarily, but conveniently) taken to have *zero* "heat content." Table 3-1 provides a list of ΔH_f^0 values for a few common elements and compounds. Appendix K contains a longer list.

Since we are concerned only with the *changes* in heat content that accompany a physical or chemical change, any convenient scale of heat content may be used. This scale is appropriate, since one element is never transformed into another in an ordinary chemical reaction.

Example 3-8

The standard molar enthalpy of formation of benzene, C_6H_6 (ℓ), is 49.03 kJ/mol. Present this information in the form of a thermochemical equation.

Solution

The desired reaction involves the formation of one mole of liquid benzene from its constituent elements in their standard states at 25°C and one atmosphere. The standard state for carbon is graphite and that of hydrogen is H_2 (g).

$$6C \text{ (graphite)} + 3H_2 \text{ (g)} \longrightarrow C_6H_6 \text{ (ℓ)}$$
$$\Delta H^0 = 49.03 \text{ kJ (endothermic)}$$

We may also write

$$6C \text{ (graphite)} + 3H_2 \text{ (g)} + 49.03 \text{ kJ} \longrightarrow C_6H_6 \text{ (ℓ)}$$

3-7 Hess' Law

In 1840, G. H. Hess published his **Law of Heat Summation,** which he derived on the basis of numerous thermochemical calculations.

Hess' Law states that the enthalpy change for a reaction is always the same whether it occurs by one step or a series of steps.

This is consistent with the fact that enthalpy is a state function, and, therefore, an enthalpy change is independent of the pathway by which a reaction occurs. We do not need to know whether the reaction does or even can occur by the series of steps used in the calculations. That is, the steps (only formally) must result in the overall reaction. The utility of Hess' Law is that we can often calculate, from tabulated values, enthalpy changes for reactions for which the quantity can be measured only with difficulty, if at all. Consider, for example, the reaction,

The common allotropic forms of carbon are graphite and diamond.

$$C(\text{graphite}) + \tfrac{1}{2}O_2 (g) \longrightarrow CO (g) \qquad \Delta H^0 = ?$$

The enthalpy change for this reaction cannot be measured directly. Even though CO (g) is the predominant product of the reaction of graphite with a limited amount of O_2 (g), some CO_2 (g) is always produced as well. However, the following reactions do go to completion with excess O_2 (g), and therefore ΔH^0 can be measured directly for them.

$$C(\text{graphite}) + O_2 (g) \longrightarrow CO_2 (g) \qquad \Delta H^0 = -393.5 \text{ kJ} \qquad [1]$$

$$CO (g) + \tfrac{1}{2}O_2 (g) \longrightarrow CO_2 (g) \qquad \Delta H^0 = -283.0 \text{ kJ} \qquad [2]$$

The reverse of an exothermic reaction is endothermic.

If we reverse equation [2] (and change the sign of its ΔH^0 value) and then add it to equation [1], cancelling equal numbers of moles of the same species, on each side, we obtain the equation for the reaction of interest. Adding the corresponding enthalpy changes gives us the enthalpy change of interest.

$C(\text{graphite}) + O_2 (g) \longrightarrow CO_2 (g)$	-393.5 kJ	[1]
$CO_2 (g) \longrightarrow CO (g) + \tfrac{1}{2}O_2 (g)$	$-(-283.0 \text{ kJ})$	[-2]
$C(\text{graphite}) + \tfrac{1}{2}O_2 (g) \longrightarrow CO (g)$	$\Delta H^0 = -110.5$ kJ	

In general terms, Hess' Law of Heat Summation may be represented as

$$\Delta H^0 = \Delta H^0_a + \Delta H^0_b + \Delta H^0_c + \ldots \qquad \text{[Eq. 3-5]}$$

where $a, b, c, \ldots$ refer to the equations that can be summed to give the equation for the reaction of interest. ΔH^0 is sometimes represented as ΔH^0_{rxn}.

Example 3-9

Use the equations and ΔH^0 values given below to determine ΔH^0 at 25°C for the reaction
$C(\text{graphite}) + 2H_2 (g) \longrightarrow CH_4 (g)$

	ΔH^0	
$C(\text{graphite}) + O_2 (g) \longrightarrow CO_2 (g)$	-393.5 kJ	[1]
$H_2 (g) + \tfrac{1}{2}O_2 (g) \longrightarrow H_2O (\ell)$	-285.8 kJ	[2]
$CO_2 (g) + 2H_2O (\ell) \longrightarrow CH_4 (g) + 2O_2 (g)$	$+890.3$ kJ	[3]

Solution

If we double (i.e., multiply by 2) equation [2] and add it to equations [1] and [3] we obtain the desired result.

	ΔH^0	
$C(\text{graphite}) + O_2 (g) \longrightarrow CO_2 (g)$	-393.5 kJ	[1]
$2H_2 (g) + O_2 (g) \longrightarrow 2H_2O (\ell)$	$2(-285.8 \text{ kJ})$	$2 \times$ [2]
$CO_2 (g) + 2H_2O (\ell) \longrightarrow CH_4 (g) + 2O_2 (g)$	$+890.3$ kJ	[3]
$C(\text{graphite}) + 2H_2 (g) \longrightarrow CH_4 (g)$	$\Delta H^0_{rxn} = -74.8$ kJ	

Note that ΔH^0_{rxn} is ΔH^0_f for CH_4 (g), as listed in Table 3–3. However, we should not jump to the conclusion that methane can be produced by this series of reactions on a large scale. There are many complicating factors.

Example 3–10

Given the following equations and ΔH^0 values, calculate the heat of reaction at 298 K for the reaction

$$C_2H_4 \text{ (g)} + H_2O \text{ } (\ell) \longrightarrow C_2H_5OH \text{ } (\ell)$$

ethylene   ethanol

	ΔH^0	
$C_2H_5OH \text{ } (\ell) + 3O_2 \text{ (g)} \longrightarrow 3H_2O \text{ } (\ell) + 2CO_2 \text{ (g)}$	-1367 kJ	[1]
$C_2H_4 \text{ (g)} + 3O_2 \text{ (g)} \longrightarrow 2CO_2 \text{ (g)} + 2H_2O \text{ } (\ell)$	-1411 kJ	[2]

Solution

If we reverse equation [1] to give [−1] and then add it to equation [2] as written, the equation for the reaction of interest is obtained.

	ΔH^0	
$3H_2O \text{ } (\ell) + 2CO_2 \text{ (g)} \longrightarrow C_2H_5OH \text{ } (\ell) + 3O_2 \text{ (g)}$	$+1367$ kJ	[−1]
$C_2H_4 \text{ (g)} + 3O_2 \text{ (g)} \longrightarrow 2CO_2 \text{ (g)} + 2H_2O \text{ } (\ell)$	-1411 kJ	[2]
$C_2H_4 \text{ (g)} + H_2O \text{ } (\ell) \longrightarrow C_2H_5OH \text{ } (\ell)$	$\Delta H^0 = -44$ kJ	

Notice that when equation [1] is reversed, the sign of ΔH^0 is also changed because the reverse of an exothermic reaction is an endothermic reaction. The ΔH^0 for the reaction of interest is −44 *kilojoules per mole of* $C_2H_5OH \text{ } (\ell)$ *formed. Note, however, that since this reaction does not involve formation of* $C_2H_5OH \text{ } (\ell)$ *from its constituent elements,* ΔH^0 *is* not *equal to* ΔH^0_f *for* $C_2H_5OH \text{ } (\ell)$.

A more generally useful formulation of Hess' Law enables us to use extensive tables of ΔH^0_f values to calculate the enthalpy change accompanying the desired reaction. Let us illustrate this approach by reconsidering the reaction of Example 3–10.

$$C_2H_4 \text{ (g)} + H_2O \text{ } (\ell) \longrightarrow C_2H_5OH \text{ } (\ell)$$

A table of ΔH^0_f values (Appendix K) gives the values $\Delta H^0_{f \, C_2H_5OH(\ell)} = -277.7$ kJ/mol, $\Delta H^0_{f \, C_2H_4 \text{ (g)}} = 52.3$ kJ/mol, and $\Delta H^0_{f \, H_2O(\ell)} = -285.8$ kJ/mol. We may express this information in the form of the thermochemical equations shown below.

		$\Delta H^0 = \Delta H^0_f$	
for $C_2H_5OH \text{ } (\ell)$:	$2C \text{ (graphite)} + 3H_2 \text{ (g)} + \frac{1}{2}O_2 \text{ (g)} \longrightarrow C_2H_5OH$	-277.7 kJ	[1]
for $C_2H_4 \text{ (g)}$:	$2C \text{ (graphite)} + 2H_2 \text{ (g)} \longrightarrow C_2H_4 \text{ (g)}$	52.3 kJ	[2]
for $H_2O \text{ } (\ell)$:	$H_2 \text{ (g)} + \frac{1}{2}O_2 \text{ (g)} \longrightarrow H_2O \text{ } (\ell)$	-285.8 kJ	[3]

We may generate the equation for the desired net reaction by adding equation [1] to the reverse of equations [2] and [3]. The value of ΔH^0 for the desired reaction is then simply the sum of the corresponding ΔH^0 values:

	ΔH^0	
$2C \text{ (graphite)} + 3H_2 \text{ (g)} + \frac{1}{2}O_2 \text{ (g)} \longrightarrow C_2H_5OH \text{ } (\ell)$	-277.7 kJ	[1]
$C_2H_4 \text{ (g)} \longrightarrow 2C \text{ (graphite)} + 2H_2 \text{ (g)}$	-52.3 kJ	[−2]
$H_2O \text{ } (\ell) \longrightarrow H_2 \text{ (g)} + \frac{1}{2}O_2 \text{ (g)}$	$+285.8$ kJ	[−3]
net rxn: $C_2H_4 \text{ (g)} + H_2O \text{ } (\ell) \longrightarrow C_2H_5OH \text{ } (\ell)$	$\Delta H^0 = -44.2$ kJ	

Note that ΔH^0 for this reaction is given by

$$\Delta H^0_{rxn} = \Delta H^0_{[1]} + \Delta H^0_{[-2]} + \Delta H^0_{[-3]}$$

and *also* by

$$\Delta H^0_{rxn} = \Delta H^0_{f\,C_2H_5OH\,(\ell)} - [\Delta H^0_{f\,C_2H_4\,(g)} + \Delta H^0_{f\,H_2O\,(\ell)}]$$
$$\text{product} \qquad\qquad \text{reactants}$$

In general terms, Hess' Law may be stated

In applying this relationship you must remember that the ΔH^0_f value of each product and reactant must be multiplied by the coefficient of that species, n, in the balanced equation for the reaction.

$$\Delta H^0_{rxn} = \Sigma\, n\Delta H^0_{f\,products} - \Sigma\, n\Delta H^0_{f\,reactants} \qquad\qquad \text{[Eq. 3–6]}$$

The capital Greek letter sigma (Σ) is read "the sum of," and n is the coefficient of each species in the balanced equation.

This general result is a very useful form of Hess' Law: *The standard enthalpy change of a reaction is equal to the sum of the standard molar enthalpies of formation of the products, each multiplied by its coefficient in the reaction of interest, minus the corresponding sum of the standard molar enthalpies of formation of the reactants.*

Let us now work some examples that further illustrate the point.

Example 3–11

Calculate ΔH^0 at 298 K for the decomposition of calcium carbonate, $CaCO_3$ (s), using the ΔH^0_f values in Table 3–1.

$$CaCO_3\ (s) \longrightarrow CaO\ (s) + CO_2\ (g)$$

Solution

$$\Delta H^0 = \Sigma\, n\Delta H^0_{f\,products} - \Sigma\, n\Delta H^0_{f\,reactants}$$

$$= \Delta H^0_{f\,CaO\,(s)} + \Delta H^0_{f\,CO_2\,(g)} - \Delta H^0_{f\,CaCO_3\,(s)}$$

$$= (1\ mol)(-635.5\ kJ/mol)$$
$$+ (1\ mol)(-393.5\ kJ/mol)$$
$$- (1\ mol)(-1207\ kJ/mol)$$

$$\Delta H^0 = \underline{+178\ kJ}\ \text{(an endothermic reaction)}$$

Example 3–12

Calculate ΔH^0 for the following reaction at 298 K.

$$SiH_4\ (g) + 2O_2\ (g) \longrightarrow SiO_2\ (s) + 2H_2O\ (\ell)$$

Solution

$$\Delta H^0 = \Sigma\, n\Delta H^0_{f\,products} - \Sigma\, n\Delta H^0_{f\,reactants}$$

$$\Delta H^0 = \Delta H^0_{f\,SiO_2\,(s)} + 2\Delta H^0_{f\,H_2O\,(\ell)}$$
$$- [\Delta H^0_{f\,SiH_4\,(g)} + 2\Delta H^0_{f\,O_2(g)}]$$

$$= (1\ mol)(-910.9\ kJ/mol)$$
$$+ (2\ mol)(-285.8\ kJ/mol)$$
$$- [(1\ mol)(+34\ kJ/mol)$$
$$+ 2\ mol\ (0\ kJ/mol)]$$

O_2 (g) is standard state of the element

$$\Delta H^0 = \underline{-1516}$$

Notice that there are two moles of water in the balanced equation, so the enthalpy of formation of water, -285.8 kJ/mol, is multiplied by 2 moles.

If we know ΔH^0 at 298 K for a reaction and all but one of the ΔH^0_f values for reactants and products, the unknown ΔH^0_f value can be calculated, as Example 3–13 shows.

Example 3–13

Given the following information:

$$PbO \text{ (s)} + CO \text{ (g)} \longrightarrow Pb \text{ (s)} + CO_2 \text{ (g)}$$
yellow $\Delta H^0_{rxn} = -65.69$ kJ

ΔH^0_f for CO_2 (g) = -393.5 kJ/mol

ΔH^0_f for CO (g) = -110.5 kJ/mol

Determine, without consulting Table 3–1 or Appendix K, the value of ΔH^0_f for yellow PbO (s).

Solution

$$\Delta H^0_{rxn} = \Sigma\, n\Delta H^0_{f\,products} - \Sigma\, n\Delta H^0_{f\,reactants}$$

Since the coefficients of all reactants and products in the balanced equation are ones, each ΔH^0_f is multiplied by 1 mol.

$$\Delta H^0_{rxn} = \Delta H^0_{f\,Pb\,(s)} + \Delta H^0_{f\,CO_2(g)}$$
$$\qquad - [\Delta H^0_{f\,PbO\,(s)} + \Delta H^0_{f\,CO\,(g)}]$$

$$-65.69 \text{ kJ} = (1 \text{ mol})(0 \text{ kJ/mol})$$
$$+ (1 \text{ mol})(-393.5 \text{ kJ/mol})$$
$$- [(1 \text{ mol})(\Delta H^0_{f\,PbO\,(s)})$$
$$+ (1 \text{ mol})(-110.5 \text{ kJ/mol})]$$

Rearranging to solve for the unknown $\Delta H^0_{f\,PbO\,(s)}$:

$$\Delta H^0_{f\,PbO\,(s)} = 65.69 \text{ kJ} - 393.5 \text{ kJ} + 110.5 \text{ kJ}$$
$$= \underline{-217.3 \text{ kJ/mol}}$$

which agrees with the value listed in Table 3–1.

Key Terms

Calorie originally, the amount of heat required to raise the temperature of one gram of water from 14.5°C to 15.5°C. Now a unit of energy defined to be equal to 4.184 joules.

Calorimeter a device used to measure the heat transfer between system and surroundings.

Enthalpy the heat content of a specific amount of a substance.

Enthalpy change change in heat content of a system that accompanies a process at constant pressure.

First Law of Thermodynamics the total amount of energy in the universe is constant; also known as the Law of Conservation of Energy.

Heat a form of energy that can flow between two samples of matter because of their difference in temperature.

Heat capacity the amount of heat required to raise the temperature of a body (of whatever mass) one degree Celsius.

Hess' Law of Heat Summation the enthalpy change of a reaction is the same whether it occurs in one step or a series of steps.

Internal energy the sum of all the kinds of energy that a sample possesses.

Joule a unit of energy in the SI system. One joule is 1 kg·m²/s², which is also 0.2390 calorie.

Kinetic energy energy associated with motion; one half of the mass of a moving object times the square of its velocity.

Potential energy energy that a system possesses by virtue of its position or composition.

Specific heat the amount of heat required to raise the temperature of one gram of a substance one degree Celsius.

Standard molar enthalpy of formation the amount of heat absorbed in the formation of one mole of a substance in its standard state from its elements in their standard states.

State functions conditions that make up the thermodynamic state of a system; their values depend only on the current state of the system, not on its history.

Surroundings everything in the environment of the system.

System the substances of interest in a process; the part of the universe under investigation.

Thermochemical equation chemical equation

that also contains information about the amount of heat gained or lost in a chemical reaction.

Thermochemistry   study of the changes in energy that accompany chemical reactions.

Thermodynamics   study of the changes in energy or energy transfers accompanying physical and chemical processes.

Thermodynamic standard state   the most stable state of a substance under one atmosphere pressure and at some specific temperature (usually 25°C unless otherwise specified).

Thermodynamic state of a system   a set of conditions that completely specifies all of the properties of the system.

Universe   the system plus the surroundings.

Work   can be defined as the product of the force, f, applied to move an object and the distance, d, through which the object moves.

Exercises

Heat Transfer

1. Calculate the amount of heat required to raise the temperature of 25.0 grams of water from 10.0°C to 40.0°C. Express your answer in joules and in calories.

2. How much heat must be removed from 15.0 grams of water at 60.0°C to decrease its temperature to 10.0°C? Express your answer in kilojoules and in calories.

3. How much heat is required to raise the temperature of 25.0 grams of iron from 10.0°C to 40.0°C? Express your answer in joules. The specific heat of iron is 0.444 J/g·°C.

4. If 150 grams of liquid water at 100°C and 250 grams of water at 10.0°C are mixed in an insulated container, what is the final temperature of the mixture?

5. If 150 g of iron at 100.0°C is placed in 250 g of water at 10.0°C in an insulated container, and both are allowed to come to the same temperature, what will that temperature be? The specific heat of iron is 0.444 J/g·°C. Compare this result with the one obtained in Exercise 4.

6. (a) Calculate the amount of heat required to raise the temperature of 100 g of mercury from 10.0° to 75.0°C. The specific heat of mercury is 0.138 J/g·°C. Mercury, commonly called quicksilver, is a liquid metal at room temperature.
(b) Calculate the amount of heat required to raise the temperature of 100 g of water from 10.0°C to 75.0°C.
(c) What is the ratio of the specific heat of liquid water to that of liquid mercury?
(d) What does answer (c) tell us?

7. A 94.4 g piece of metallic iron at 152.6°C is dropped into 140 mL of octane, C_8H_{18}, at 24.6°C. The final temperature is 45.2°C. The density of octane is 0.703 g/mL. The specific heat of iron is 0.444 J/g·°C. (a) What is the specific heat of octane? (b) What is the heat capacity of octane in J/mol?

8. How much water at 80.0°C must be mixed with one liter of water at 15.0°C in an insulated container so that the temperature of the mixture will be 30.0°C?

9. How much iron at 200.0°C must be placed in 200.0 mL of water at 20.0°C so that the final temperature of both will be 30.0°C? Assume no heat is lost to the surroundings.

10. If 75.0 grams of metal at 75.0°C is added to 150 grams of water at 15.0°C, the temperature of the water rises to 18.3°C. Assume no heat is lost to the surroundings. What is the specific heat of the metal?

11. The specific heat of mercury is 0.138 J/g·°C. Express this value in (a) cal/lb·°F and (b) J/g·K.

Fundamental Thermochemical Concepts

12. Distinguish between the terms specific heat and heat capacity.

13. Distinguish among the following kinds of kinetic energy associated with the molecules of a gas: (a) translational, (b) rotational, and (c) vibrational.

14. What is a state function? Would Hess' Law be a law if enthalpy were not a state function?

15. Which of the following are examples of state functions?
(a) your bank balance
(b) your mass
(c) your weight

(d) the heat lost by perspiration during a climb up a mountain along a fixed path.

16. (a) Distinguish between ΔH and ΔH^0 for a reaction.
    (b) Distinguish between ΔH^0 and ΔH_f^0.

17. Can we calculate the total energy content at 25°C of one liter of gasoline of a specified composition? If so, how?

18. Can we calculate the heat of combustion at 25°C of one liter of gasoline of a specified composition? If so, how?

19. Both of the following reactions are exothermic. Which one is more exothermic and why?

$$H_2 \text{ (g)} + Br_2 \text{ (g)} \longrightarrow 2HBr \text{ (g)}$$

$$H_2 \text{ (g)} + Br_2 \text{ (}\ell\text{)} \longrightarrow 2HBr \text{ (g)}$$

20. According to the First Law of Thermodynamics, the total amount of energy in the universe is constant. Why, then, do we say that we are experiencing a declining supply of energy?

21. A student wishes to determine the heat capacity of a coffee-cup calorimeter. After she mixes 100.0 g of water at 61.0°C with 100.0 g of water, already in the calorimeter, at 24.2°C, the final temperature of the mixed water is 41.5°C. (a) Calculate the heat capacity of the calorimeter in J/°C. Use 4.18 J/g·°C as the specific heat of water. (b) Why is it more useful to express this value in J/°C rather than units of J/(g calorimeter·°C)?

22. The heat capacity of a calorimeter is sometimes converted to the *water equivalent* of the calorimeter, which is defined as the mass of (extra) water that would absorb the same amount of heat as the calorimeter, itself, does per °C increase in temperature. What is the water equivalent of the calorimeter of Exercise 21?

23. A coffee-cup calorimeter having a heat capacity of 472 J/°C is used to measure the heat evolved when the following aqueous solutions, both initially at 22.6°C, are mixed: 100 g of solution containing 6.62 g of lead(II) nitrate, $Pb(NO_3)_2$, and 100 g of solution containing 6.00 g of sodium iodide, NaI. The final temperature is 24.2°C. Assume that the specific heat of the mixture is the same as that of water, 4.18 J/g·°C. The reaction that occurs is

$$Pb(NO_3)_2 \text{ (aq)} + 2NaI \text{ (aq)} \longrightarrow$$
$$PbI_2 \text{ (s)} + 2NaNO_3 \text{ (aq)}$$

(a) Calculate the heat evolved in the reaction.
(b) Calculate ΔH for the reaction (per mole of $Pb(NO_3)_2$ (aq) consumed) under the conditions of the experiment.

24. A strip of magnesium metal having a mass of 1.22 g dissolves in 100 mL of 6.02 M HCl, which has a specific gravity of 1.10. The hydrochloric acid is initially at 23.0°C, and the resulting solution reaches a final temperature of 45.5°C. The heat capacity of the calorimeter in which the reaction occurs is 562 J/°C. Calculate ΔH for the reaction (per mole of Mg (s) consumed) under the conditions of the experiment, assuming that the specific heat of the final solution is the same as that for water, 4.18 J/g·°C.

$$Mg \text{ (s)} + 2HCl \text{ (aq)} \longrightarrow MgCl_2 \text{ (aq)} + H_2 \text{ (g)}$$

Standard Molar Enthalpy of Formation, ΔH_f^0

25. Explain the significance of each word in the term "standard molar enthalpy of formation."

26. From the data in Appendix K, determine the form that represents the standard state for each of the following elements.
    (a) fluorine   (b) iron
    (c) carbon   (d) sulfur
    (e) hydrogen   (f) mercury
    (g) bromine

27. Write thermochemical equations that describe the formation of the following compounds from their elements in their standard states.
    (a) PCl_5 (g)   (b) N_2O_4 (g)
    (c) NH_4NO_3 (s)

28. Comment on the statement, "Since ΔH_f^0 is zero for all elements in their standard states, their heat contents are all zero kilojoules per mole."

Thermochemical Equations and Hess' Law

29. Use Hess' Law to calculate ΔH^0 at 25°C for each of the following reactions, using the data in Appendix K.
    (a) SiH_4 (g) $+ 2O_2$ (g) $\rightarrow SiO_2$ (s) $+ 2H_2O$ (g)
    (b) C_2H_6 (g) $\rightarrow C_2H_4$ (g) $+ H_2$ (g)
    (c) Fe_3O_4 (s) $+ CO$ (g) $\rightarrow 3FeO$ (s) $+ CO_2$ (g)

30. Calculate ΔH^0 at 25°C for each of the following reactions.
    (a) SiO_2 (s) $+ Na_2CO_3$ (s) $\rightarrow$
    $$Na_2SiO_3 \text{ (s)} + CO_2 \text{ (g)}$$

(b) H_2SiF_6 (aq) $\rightarrow$ 2HF (aq) + SiF_4 (g)

(c) 2MgS (s) + $3O_2$ (g) $\rightarrow$
$$2MgO (s) + 2SO_2 (g)$$

31. Calculate ΔH^0 at 25°C for the following reactions.

(a) 2NO (g) + H_2 (g) $\rightarrow$ N_2O (g) + H_2O (g)

(b) N_2O (g) + H_2 (g) $\rightarrow$ N_2 (g) + H_2O (g)

32. A common laboratory method for generating O_2 (g) is the thermal decomposition of potassium chlorate, $KClO_3$ (s) (in the presence of a catalyst). Show that this reaction is exothermic.

$$2KClO_3 (s) \longrightarrow 2KCl (s) + 3O_2 (g)$$

33. Calculate ΔH^0 at 25°C, using data in Appendix K, for the preparation of hydrogen fluoride by the reaction of calcium fluoride with a nonvolatile acid, sulfuric acid:

$$CaF_2 (s) + H_2SO_4 (aq) \longrightarrow$$
$$CaSO_4 (s) + 2HF (g)$$

34. Calculate ΔH^0 at 25°C, using data in Appendix K, for the preparation of the World War I poison gas, phosgene, from carbon monoxide and chlorine:

$$CO (g) + Cl_2 (g) \longrightarrow COCl_2 (g)$$

35. Calculate ΔH^0 at 25°C, using data in Appendix K, for the industrial roasting (heating in the presence of oxygen) of zinc sulfide ore prior to smelting (formation of the metal):

$$2ZnS (s) + 3O_2 (g) \longrightarrow 2ZnO (s) + 2SO_2 (g)$$

36. The equation for a reaction used to prepare ultrapure silicon for making semiconductors follows. Calculate ΔH^0 at 25°C for this reaction, using data in Appendix K.

$$SiBr_4 (\ell) + 4Na (s) \longrightarrow 4NaBr (s) + Si (s)$$

37. Using the data in Appendix K, calculate the standard heat of vaporization, ΔH^0_{vap}, at 298 K for tin(IV) chloride, $SnCl_4$, in kilojoules per mole and joules per gram.

$$SnCl_4 (\ell) \longrightarrow SnCl_4 (g)$$

38. Calculate the amount of heat released, in kilojoules, in the reaction of 14.2 g of magnesium with excess chlorine to produce magnesium chloride under standard thermodynamic conditions.

$$Mg (s) + Cl_2 (g) \longrightarrow MgCl_2 (s)$$

39. Calculate the amount of heat released, in kilocalories, in the complete oxidation of 13.2 g of aluminum at 25°C and one atmosphere pressure to form aluminum oxide, the protective coating on aluminum doors and windows.

$$4 Al (s) + 3O_2 (g) \longrightarrow 2Al_2O_3 (s)$$

40. Calculate the amount of heat released in the roasting (heating in the presence of oxygen) of

(a) 1.35 grams of iron pyrite, FeS_2

(b) 1.35 tons of iron pyrite

$$4FeS_2 (s) + 11O_2 (g) \longrightarrow$$
$$2Fe_2O_3 (s) + 8SO_2 (g)$$

41. The air pollutant sulfur dioxide can be partially removed from roaster stack gases (see Exercise 40) and converted to sulfur trioxide, which is used to make commercially important sulfuric acid. Calculate the standard enthalpy change for the reaction of 25.0 grams of SO_2 with O_2 (g).

$$2SO_2 (g) + O_2 (g) \longrightarrow 2SO_3 (g)$$

42. Calculate the standard enthalpy change at 25°C accompanying the reaction of 39.2 grams of sulfur trioxide with a stoichiometric quantity of water.

$$SO_3 (g) + H_2O (\ell) \longrightarrow H_2SO_4 (\ell)$$

43. Solid phosphorus exists in two allotropic forms, red and white. Both react with chlorine to produce phosphorus trichloride, a colorless liquid that fumes in air.

$$P_4 \text{ (white)} + 6Cl_2 (g) \longrightarrow 4PCl_3 (\ell)$$
$$\Delta H^0 = -1.15 \times 10^3 \text{ kJ}$$

$$P_4 \text{ (red)} + 6Cl_2 (g) \longrightarrow 4PCl_3 (\ell)$$
$$\Delta H^0 = -1.23 \times 10^3 \text{ kJ}$$

Calculate the standard enthalpy of reaction for the reaction by which red phosphorus is converted to white phosphorus.

$$P_4 \text{ (red)} \longrightarrow P_4 \text{ (white)}$$

44. From the following equations and ΔH^0 values,

$$H_2 (g) + Cl_2 (g) \longrightarrow 2HCl (g)$$
$$\Delta H^0 = -185 \text{ kJ}$$

$$2H_2 (g) + O_2 (g) \longrightarrow 2H_2O (g)$$
$$\Delta H^0 = -483.7 \text{ kJ}$$

calculate ΔH^0 for the following reaction.

$$4HCl\,(g) + O_2\,(g) \longrightarrow 2Cl_2\,(g) + 2H_2O\,(g)$$

45. Given the following equations and ΔH^0 values,

$$H_3BO_3\,(aq) \longrightarrow HBO_2\,(aq) + H_2O\,(\ell)$$
$$\Delta H^0 = -0.02\ kJ$$

$$H_2B_4O_7\,(s) + H_2O\,(\ell) \longrightarrow 4HBO_2\,(aq)$$
$$\Delta H^0 = -11.3\ kJ$$

$$H_2B_4O_7\,(s) \longrightarrow 2\,B_2O_3\,(s) + H_2O\,(\ell)$$
$$\Delta H^0 = 17.5\ kJ$$

calculate ΔH^0 for the following reaction.

$$2H_3BO_3\,(aq) \longrightarrow B_2O_3\,(s) + 3H_2O\,(\ell)$$

46. The following reaction is one that occurs in a blast furnace when iron is extracted from its ores.

$$Fe_2O_3\,(s) + 3CO\,(g) \longrightarrow 2Fe\,(s) + 3CO_2\,(g)$$

Evaluate ΔH^0 for this reaction at 298 K, given the following enthalpy changes at 298 K.

$$3Fe_2O_3\,(s) + CO\,(g) \longrightarrow$$
$$2Fe_3O_4\,(s) + CO_2\,(g)$$
$$\Delta H^0 = -46.4\ kJ$$

$$FeO\,(s) + CO\,(g) \longrightarrow Fe\,(s) + CO_2\,(g)$$
$$\Delta H^0 = 9.0\ kJ$$

$$Fe_3O_4\,(s) + CO\,(g) \longrightarrow 3FeO\,(s) + CO_2\,(g)$$
$$\Delta H^0 = -41.0\ kJ$$

47. What do the last three equations of Exercise 46 have in common?

48. The reaction that occurs during the discharge of a typical automobile battery, called a lead storage battery, is

$$Pb\,(s) + PbO_2\,(s) + 2H_2SO_4\,(\ell) \longrightarrow$$
$$2PbSO_4\,(s) + 2H_2O\,(\ell)$$

Evaluate ΔH^0 at 25°C for this reaction from the following information.

$$SO_3\,(g) + H_2O\,(\ell) \longrightarrow H_2SO_4\,(\ell)$$
$$\Delta H^0 = -133\ kJ$$

$$Pb\,(s) + PbO_2\,(s) + 2SO_3\,(g) \longrightarrow 2PbSO_4\,(s)$$
$$\Delta H^0 = -775\ kJ$$

49. "Water gas" is an important industrial fuel consisting of a mixture of carbon monoxide and hydrogen. It is produced by the action of superheated steam on red-hot coke.

$$C\,(s) + H_2O\,(g) + 131.3\ kJ \xrightarrow{1000°C}$$
coke
$$CO\,(g) + H_2\,(g)$$
water gas

Given that ΔH_f^0 for $H_2O\,(g)$ is -241.8 kJ/mol, and assuming that coke has the same molar enthalpy of formation as graphite, calculate ΔH_f^0 for CO.

50. Welders use acetylene, C_2H_2, in torches to attain very high temperatures. Calculate ΔH^0 at 25°C for the complete combustion of one mole of acetylene,

$$2C_2H_2\,(g) + 5O_2\,(g) \longrightarrow$$
$$4CO_2\,(g) + 2H_2O\,(\ell)$$

given the following ΔH^0 values at 25°C.

$$H_2\,(g) + \tfrac{1}{2}O_2\,(g) \longrightarrow H_2O\,(\ell)$$
$$\Delta H^0 = -285.8\ kJ$$

$$C\,(graphite) + O_2\,(g) \longrightarrow CO_2\,(g)$$
$$\Delta H^0 = -393.5\ kJ$$

$$2C\,(graphite) + H_2\,(g) \longrightarrow C_2H_2\,(g)$$
$$\Delta H^0 = 226.7\ kJ$$

51. Methanol, CH_3OH, is a possible fuel or fuel supplement of the future (in gasohol, for example). It can be prepared industrially by the reaction of carbon monoxide with hydrogen ("water gas" with excess H_2; see Exercise 49) at high temperatures and pressures in the presence of appropriate catalysts.

$$CO\,(g) + 2H_2\,(g) \xrightarrow[ZnO]{Cr_2O_3} CH_3OH\,(\ell)$$

Evaluate ΔH^0 for this reaction, given the following:

$$2C\,(graphite) + O_2\,(g) \longrightarrow 2CO\,(g)$$
$$\Delta H^0 = -221\ kJ$$

$$C\,(graphite) + O_2\,(g) \longrightarrow CO_2\,(g)$$
$$\Delta H^0 = -393.5\ kJ$$

$$2CH_3OH\,(\ell) + 3O_2\,(g) \longrightarrow$$
$$2CO_2\,(g) + 4H_2O\,(\ell)$$
$$\Delta H^0 = -1453\ kJ$$

$$2H_2\,(g) + O_2\,(g) \longrightarrow 2H_2O\,(\ell)$$
$$\Delta H^0 = -571.6\ kJ$$

52. From the following data, calculate the standard molar heat of formation of gaseous PCl_5.

P_4 (white) $+ 6Cl_2$ (g) $\longrightarrow$ $4PCl_3$ (ℓ)
$\Delta H^0 = -1.15 \times 10^3$ kJ

PCl_3 (ℓ) $+ Cl_2$ (g) $\longrightarrow$ PCl_5 (g)
$\Delta H^0 = -111$ kJ

53. At 25°C, ΔH_f^0 for P_4O_{10} (s) is -2984 kJ/mol, that for HNO_3 (ℓ) is -174.1 kJ/mol, and that for N_2O_5 (g) is 11 kJ/mol. Given the following, also at 25°C,

P_4O_{10} (s) $+ 4HNO_3$ (ℓ) $\longrightarrow$
$4HPO_3$ (s) $+ 2N_2O_5$ (g)
$\Delta H^0 = -102.3$ kJ

calculate $\Delta H_{f\,HPO_3\,(s)}^0$ at 25°C.

54. Silicon carbide or carborundum, SiC, is one of the hardest substances known and is used as an abrasive. It has the structure of diamond with half of the carbons replaced by silicon. It is prepared industrially by reduction of sand (SiO_2) with coke (C) at 3000°C in an electric furnace.

SiO_2 (s) $+ 3C$ (s) $\longrightarrow$ SiC (s) $+ 2CO$ (g)

Given that ΔH^0 for this reaction is 624.6 kJ, and that ΔH_f^0 for SiO_2 (s) and CO (g) are -910.9 kJ/mol and -110.5 kJ/mol, respectively, calculate ΔH_f^0 for silicon carbide.

55. The tungsten used in filaments for light bulbs can be prepared from tungsten(VI) oxide by reduction with hydrogen at 1200°C.

114.9 kJ $+ WO_3$ (s) $+ 3H_2$ (g)
$\longrightarrow$ W (s) $+ 3H_2O$ (g)

The ΔH_f^0 value for H_2O (g) is -241.8 kJ/mol. What is ΔH_f^0 for WO_3 (s)?

Atoms and Subatomic Particles

4

As we have seen in the past two chapters, a variety of types of chemical and physical behavior, both of individual substances and of substances related in chemical reactions, can be summarized and understood in terms of only a few concepts. Dalton's atomic theory proposed a simple description of matter as composed of atoms; with this idea we can describe many macroscopic observations, such as the Law of Constant Composition, the Law of Conservation of Matter, and the amounts of substances related in chemical reactions. It is important to notice, however, that Dalton's theory did not attempt to tell us, for example, *why* atoms combine to form compounds, or *why* they combine in simple numerical ratios, only *that they do* these things. To gain a better understanding of atoms and their properties, we must expand our ideas to include more sophisticated atomic theories.

Much of the development of our modern atomic theory was based on two broad types of research performed during the late 1800s and the early 1900s. The first area dealt with the electrical nature of matter. These studies led to the realization that atoms are composed of still more fundamental particles, and to a description of the approximate arrangements of these particles in atoms. The second broad area of research dealt with the interaction of matter with energy in the form of light, i.e., the colors (wavelengths) of light that substances can give off or absorb. Such observations led to a much more detailed description of the arrangements of particles in atoms, especially electrons, and to an understanding of the relative stabilities of various arrangements.

Chapters 4 and 5 deal with this background and with the development of modern atomic theory. It is in terms of the resulting atomic theory that we shall be able *to organize, understand,* and *predict* a broad range of chemical and physical properties.

TABLE 4–1   Fundamental Particles of Matter

Particle	Mass	Charge
electron (e^-)	0.00055 amu	1−
proton (p or p^+)	1.0073 amu	1+
neutron (n or n^0)	1.0087 amu	none

4–1 Fundamental Particles

Current atomic theory is an especially valuable tool in understanding the forces that hold atoms in chemical combination with each other. It is considerably less than complete, but nonetheless very useful. We shall look first at the basic building blocks of all atoms, which we call *fundamental particles*.

Atoms, and hence all matter, consist principally of three fundamental particles, *protons*, *electrons*, and *neutrons*. Many other smaller particles (such as positrons, neutrinos, pions, and muons) have been discovered, but it is not necessary to study their characteristics to understand the fundamentals of atomic structure that are important in chemical reactions. However, knowledge of the nature and functions of the three fundamental particles is essential to understanding chemical interactions. The important properties, mass and charge, of these three particles are shown in Table 4–1. The mass of an electron is very small compared with the mass of either a proton or a neutron; the latter two have nearly equal masses. The charge on a proton is equal in magnitude, but opposite in sign, to the charge on an electron.

We shall investigate these particles in more detail, but our first consideration will be the experimental evidence that led to their discovery.

4–2 Electrons

In 1859 and the years following, experiments were done along the following lines. Suppose that we evacuate a glass tube such as that shown schematically in Figure 4–1a, and then fill it with a gas. If we then increase the voltage to a very high value, the gas discharges (sparks). If, however, the gas pressure in the tube is lowered, the gas in the tube emits a continuous glow, and a meter shows that an electric current is being passed. This is the glow that we see, for instance, in a neon light. The color of the light depends on the identity of the gas in the tube, so that such emissions can be used to identify elements present in the gas (as we shall see in Section 5–3). Lowering the gas pressure even further causes dark regions called *Crooke's dark spaces* to develop, but the current still flows. When more gas is removed from the tube, the dark regions can be caused to fill the tube, but the current still flows. This flow of current is referred to as **cathode rays.** In a **cathode ray tube,** such as the one in Figure 4–1a, these rays emanate from the cathode (negative electrode) and travel in straight lines toward the anode (positive electrode). If a small object is placed in the path of the cathode rays, it casts a shadow on a fluorescent screen as shown in Figure 4–1b. This proves that the cathode rays travel from the cathode toward the anode, and suggests that they are negatively charged. Further proof of their negative charge may be obtained by observing that the rays are deflected by electric fields (Figure 4–1c) and magnetic fields (Figure 4–1d) in the direction expected for

The dark spaces are named after William Crooke, who carried out many of these early experiments.

FIGURE 4–1 Some experiments with cathode ray tubes that show the nature of cathode rays. (a) A cathode ray (discharge) tube, showing the production of a beam of electrons (cathode rays). This beam is detected by observing the glow of a fluorescent screen. (b) A small object placed in a beam of cathode rays casts a shadow, which shows that cathode rays travel in straight lines. (c) Cathode rays have negative electrical charge, as demonstrated by their deflection in an electric field. (The electrically charged plates produce an electric field.) (d) Interaction of cathode rays with a magnetic field is also consistent with negative charge. The magnetic field goes from one pole to the other. (e) Cathode rays have mass, as shown by their ability to turn a small paddlewheel in their path.

negatively charged particles. Figure 4–1e shows how it can be demonstrated that electrons have mass.

Other kinds of radiation also can be detected.

In 1897, J. J. Thomson carried out a series of experiments designed to characterize cathode rays further. An apparatus that is used for such experiments is shown in Figure 4–2. In this instrument, a crude **mass spectrometer,** charged particles are forced into a very narrow beam and accelerated by a voltage difference into a region containing electric and magnetic fields. The beam of charged particles then strikes a

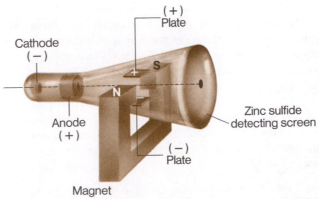

FIGURE 4–2 Apparatus for determining the charge-to-mass ratio of electrons. The apparatus shown here is a cathode ray tube of different design than those shown in Figure 4–1. (A cathode ray tube in electric and magnetic fields becomes a crude mass spectrometer; see Figure 4–10.)

fluorescent screen. Either the electric field alone or the magnetic field alone will deflect the charged particles from their straight-line path. The two fields are arranged so that the deflections they cause are exactly in opposite directions. Then, for some combination of electric and magnetic fields, the effects cancel and the spot on the screen returns to the center position. The *extent* to which the beam of charged particles is deflected depends upon four factors:

1. Magnitude of the accelerating voltage. The range varies from about 500 to about 2000 volts. Higher voltages result in beams of more rapidly moving particles that are deflected less than the beams of slower particles produced by lower voltages.
2. Strengths of the electric and magnetic fields. Stronger fields deflect a given beam more than weaker fields.
3. Masses of the particles. Because of their inertia, heavier particles are deflected less than lighter particles that carry the same charge.
4. Charges on the particles. Highly charged particles interact more strongly with the magnetic and electric fields and are thus deflected more than particles of equal mass with smaller charges.

In the experiments carried out by Thomson, only the strengths of the electric and magnetic fields that cancelled each other could be measured. (The accelerating voltage was fixed.) Thus, the masses and charges of the particles could not both be determined, but only the ratio of charge to mass (e/m).

> As an analogy to Thomson's experiments, consider a pitcher who throws baseballs with great accuracy, always hitting the same target. Suppose that he throws the baseballs (analogous to the acceleration of the charged particles by the voltage difference) from west to east, and a strong but steady wind begins to blow from north to south. This disturbing force, analogous to the electric and magnetic fields, causes the path of the baseballs to be altered. If the pitcher were throwing something the same size as a baseball but with much less mass (say a foam-rubber ball), then the same disturbing force would cause a much larger deflection. Thus, the smaller the mass, the greater the deflection; that is, the deflection is inversely proportional to the mass. Now a ball with the same mass as a baseball but with larger surface would catch the wind better, and would be deflected more. This kind of interaction with the disturbing force is analogous to the interaction of the charge on the particle with a field.

By correlating the amount of deflection observed for selected electric and magnetic field strengths, Thomson was able to calculate the ratio of charge to mass (e/m) of the cathode ray particles. Such experiments showed that the charge-to-mass ratio of cathode rays is the same,

$$e/m = 1.76 \times 10^8 \text{ coulombs*/gram}$$

[Eq. 4–1]

regardless of the metal used to construct the cathode or of the nature of the gas in the tube. This suggests that the negative particles present in the cathode ray tube are *fundamental particles,* present in all matter. Thomson called these particles **electrons.**

---

* The coulomb is the standard unit of *charge* or *quantity of electricity,* and is defined as that amount of electricity that will deposit 0.001118 gram of silver in an apparatus set up for plating silver. It corresponds to a *current of one ampere flowing for one second.*

In the early 1800s, the English chemist Humphrey Davy found that when he passed electrical current through some substances, these substances decomposed (a process called *electrolysis*). This led him to propose that the elements of a chemical compound are held together by electrical forces. In 1832–33, Michael Faraday, Davy's protégé, determined the quantitative relationship between the amount of electricity used in an electrolysis and the amount of chemical reaction that occurs. In 1874, George Stoney carefully studied Faraday's work and suggested that units of electrical charge are associated with atoms. In 1891, he suggested that these electrical charges be named *electrons*, but no determination was made at that time of the properties of these charges, and the correlation with the cathode ray experiments was not recognized.

Once the charge-to-mass ratio for the electron had been experimentally determined, additional experiments were necessary to determine the value of either mass or charge, so that the other could be calculated. This was accomplished in 1909 by the American scientist Robert A. Millikan, using the apparatus illustrated in Figure 4–3. The rates at which tiny, uncharged droplets of oil (sprayed from an atomizer) fall in air are determined by their mass and size, both of which Millikan could measure. When x-rays are used to irradiate the fine oil spray, one or more electrons are dislodged from atoms in the air and can be absorbed by oil droplets. Millikan measured the effect of an electric field on the rates at which charged oil

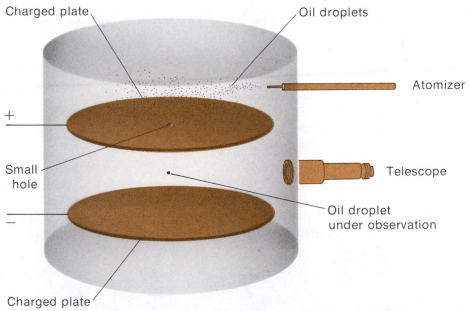

FIGURE 4–3 Millikan oil-drop experiment. Tiny oil droplets are produced by an atomizer and then irradiated with x-rays so that some of the oil droplets become negatively charged. When the voltage between the plates is increased, a negatively charged drop falls more slowly, since it is attracted to the positively charged plate. At one particular voltage, the electrical and gravitational forces on the drop are exactly balanced, and the drop remains stationary. When this voltage and the mass of the drop are known, it is possible to calculate the charge on the drop.

droplets fall under the influence of gravity. By adjusting the strength of the electric field between the charged plates, the force of gravity could be counterbalanced so that a droplet was suspended motionless. From measurements of the electric field, he calculated the charges on the droplets. All charges turned out to be integral multiples of the same charge, 1.60×10^{-19} coulomb (C). He therefore concluded that this smallest observed charge must be the charge on an individual electron.

The charge-to-mass ratio determined earlier by Thomson was then used in inverse form to calculate the mass of the electron:

The capital C here, which represents coulombs, should not be confused with °C for degrees Celsius.

$$e/m = 1.76 \times 10^8 \text{ C/g}$$

$$m = 1.60 \times 10^{-19} \text{ C} \times \frac{1.00 \text{ g}}{1.76 \times 10^8 \text{ C}} = 9.09 \times 10^{-28} \text{ g}$$

It is important to realize that the mass of an electron is only a small fraction (about $\frac{1}{1840}$) of the mass of a hydrogen atom, the lightest known atom. Thus, the electron must be a subatomic particle, and not an atom, molecule, or any such chemical entity.

Millikan's oil-drop experiment stands as one of the most fundamental of all classic scientific experiments. The value of e/m obtained by Thomson and the value of m obtained by Millikan differ slightly from the modern values given in this text. Scientists who work at the forefront of a research area often build their own apparatus, and the values they obtain may not be as accurate as desired. Important determinations are repeated many times by different methods, and values are refined as equipment and techniques are improved.

4–3 Canal Rays

Careful observations of the cathode ray tube indicated that it also generates a stream of positively charged particles that moves from the anode to the cathode; this was first observed by Eugen Goldstein in 1886. The particles were termed **canal rays** because they were observed occasionally to pass through a channel or "canal" drilled in the negative electrode (Figure 4–4). When these positively charged particles were subjected to the same e/m study as has been described for electrons, several important differences were observed.

1. *Different* values for e/m are obtained, depending on the gas in the tube.
2. The lower the atomic weight of the gas in the tube, the higher is the e/m ratio observed for the canal ray particles.
3. The highest e/m ratio is observed when gaseous hydrogen fills the tube, and it is many times less than the e/m ratio observed for electrons.

These observations could be explained in the following way. Atoms are known to be neutral particles. Since they contain electrons (negatively charged particles), they must contain enough positive electrical charge to balance the total negative charge due to their electrons. In the cathode ray experiments, electrons are ripped out of atoms, so that the "leftovers" contain an excess of positive charge. Of course, these are the simplest examples of what we now call *positive ions* or *cations*. For example, when an atom, call it X, loses one electron, it forms an ion, X^+, with positive one charge.

$$X \longrightarrow X^+ + e^-$$

[Eq. 4–2]

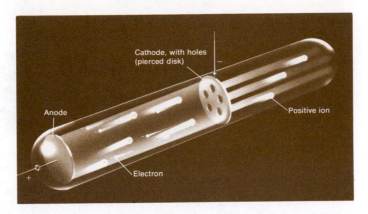

FIGURE 4–4 A cathode ray tube with a perforated cathode was used to demonstrate that *positive rays* are formed in the gas. Like cathode rays, these ''canal rays'' are deflected by magnetic and electric fields, but in opposite directions.

All such ions have the same charge (all having lost the same amount of negative charge); however, those ions formed from atoms of higher atomic weight have greater mass, and thus a *smaller* charge-to-mass ratio. Consideration of the regularity of the e/m values thus obtained for different ions led to the idea that there is a unit of the positive charge, and that it resides in the *proton*. A **proton** is *a fundamental particle with a charge equal in magnitude but opposite in sign to the charge on the electron, and with a much larger mass.*

4–4 Natural Radioactivity

Several experiments that we just described indicate that, contrary to the Daltonian idea of the indivisibility of atoms, we can actually break atoms apart into simpler constituents. Near the end of the nineteenth century, evidence developed that some atoms can spontaneously decompose. In 1896, Henri Becquerel observed that a photographic plate left in a drawer with a sample of uranium was darkened, even though the plate had been protected from light. Further studies of this phenomenon, termed **natural radioactivity,** by Becquerel's student Marie Curie and her husband Pierre Curie established that a number of the heavier elements besides uranium behave in similar fashion. It was shown by Ernest Rutherford that radioactive elements can emit three kinds of radiation, distinguishable by their behavior in electric or magnetic fields (Figure 4–5). These were identified by the first three letters of the Greek alphabet as *alpha* (α), *beta* (β), and *gamma* (γ) rays, having the following characteristics:

1. **Alpha rays (alpha particles)** were found to be identical to He^{2+} ions, i.e., helium atoms that have each lost two electrons.
2. **Beta rays (beta particles)** were shown to be very high energy electrons.
3. **Gamma rays** were found to be very high energy electromagnetic radiation.

This spontaneous disintegration of atoms will be discussed further in Section 4–9, and the general topic of radioactive decay in Chapter 24. However, alpha rays (particles) play a very important role in our further investigation of the nature of atoms, the topic of the next section.

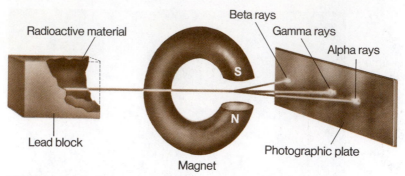

Radioactive material

Beta rays

Gamma rays

Alpha rays

S

N

Lead block

Photographic plate

Magnet

FIGURE 4–5 Behavior of the three kinds of natural radioactivity in a magnetic field. Gamma rays are high energy electromagnetic radiation, and are not deflected by a magnetic field. Beta rays (electrons) have very small mass and are negatively charged, so they are deflected by a large amount. Alpha particles are positively charged, with charge twice as great as that of beta rays; however, they are much more massive, so they are deflected by only a small amount, in the opposite direction from beta rays.

4–5 Rutherford and the Nuclear Atom

By the first decade of this century, it was clear that every atom contains regions of both positive and negative charge. The question was, how are these charges distributed? The dominant view was that summarized in J. J. Thomson's model of the atom. In Thomson's model, the positive charge was assumed to be distributed homogeneously throughout the atom, with the negative charges, electrons, imbedded in the atom like plums in a pudding (hence the name "plum pudding model" of the atom). However, this model was not consistent with the results of further experiments, so it had to be discarded.

In 1910 Ernest Rutherford's research group carried out a series of experiments that had enormous influence on our concept of the nature of the atom. They bombarded a very thin gold foil with alpha particles from a radioactive source. A fluorescent zinc sulfide screen surrounded the foil so that the extent to which the alpha particles were scattered by the gold foil (Figure 4–6) could be observed. Scintillations, or flashes, on the screen caused by the individual alpha particles were counted to determine the relative numbers of alpha particles deflected at various angles.

Gold was used because it can be fabricated into thinner sheets than any other metal. Knowledge of the density of gold and its atomic weight allowed calculation that even the thinnest sheet was many thousands of atoms thick. Alpha particles were known to be extremely dense, much more so than even the heavy metal, gold. Further, they were known to be emitted with very high kinetic energies. In light of the dominant picture of the atom as a more or less continuous distribution of positively charged matter with electrons imbedded in it, the "plum pudding" model, it was expected that any of the alpha particles that passed through the foil would emerge with little or no deflection.

The observations made by Rutherford and his co-workers, Hans Geiger and Ernest Marsden, may be summarized:

Earlier, Rutherford had worked with J. J. Thomson at Cambridge, England, studying the electrical nature of matter.

Geiger is also the inventor of the "Geiger counter," which is used to detect radioactivity.

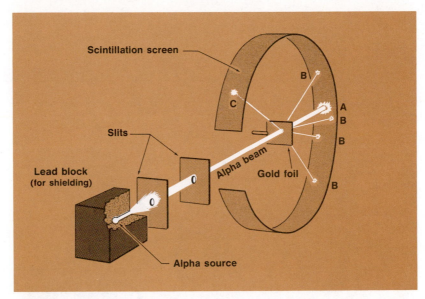

FIGURE 4–6 Rutherford scattering experiment. A narrow beam of alpha particles (helium atoms stripped of their electrons) from a radioactive source was directed at a very thin gold foil. Most of the particles passed right through the foil, striking the screen at point A. Many of the particles were deflected through moderate angles, striking the screen at points such as B. The larger deflections were surprises, but the 0.001% of the total that were reflected at acute angles, C, were totally unexpected. Similar results were observed using foils of other metals.

1. Most of the alpha particles passed straight through the foil undeflected.
2. Somewhat surprisingly, a significant number of the particles were scattered, or deflected, through moderate angles.
3. Astonishingly, a few alpha particles were scattered backward from the gold foil in more or less the direction from which they had come!

> Rutherford was astounded. In his own words,
> "It was quite the most incredible event that has ever happened to me in my life. It was almost as if you fired a 15-inch shell into a piece of tissue paper and it came back and hit you."

Rutherford's interpretation of these results involved an entirely new idea of the arrangement of matter in the atom. The observation that most of the alpha particles passed through the foil undeflected suggested that the gold atom was, in fact, almost entirely empty space. The ease with which electrons could be removed from atoms, for example in the cathode ray experiments, led Rutherford to suggest that the electrons occupied this essentially empty space. However, it would have been impossible for the very high energy alpha particles, which are many hundreds of times heavier than electrons, to be deflected from their path by attraction by the electrons. Rutherford suggested that these deflections were caused by *repulsion* of the positively charged alpha particles by the positively charged centers of the gold atoms (Figure 4–7). On the basis of a mathematical analysis, he concluded that the mass of one of these centers is very nearly equal to that of a gold atom, but that the diameter is no more than $\frac{1}{10,000}$ that of the gold atom.

$\frac{1}{10,000}$ is a *very* small fraction!

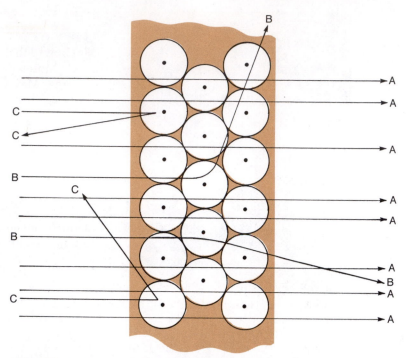

FIGURE 4–7 Interpretation of the Rutherford scattering experiment. The atom is pictured as consisting mostly of open space. At the center is a tiny and extremely dense nucleus that contains all of the positive charge of the atom and nearly all of the mass. The electrons are thinly distributed throughout the open space. Most of the positively charged alpha particles (A) pass through the open space undeflected, not coming near any gold nuclei. Those that pass fairly close to a nucleus (B) are repelled by electrostatic force and thereby deflected. The few particles that are on a "collision course" with gold nuclei are repelled backward at acute angles (C). Calculations based on the results of the experiment indicated that the diameter of the open-space portion of the atom is from 10,000 to 100,000 times greater than the diameter of the nucleus.

After many experiments with foils of different metals yielded similar results, Rutherford concluded that atoms contain tiny, positively charged, massive centers that he called **atomic nuclei.** The few alpha particles that are deflected are the ones whose paths would bring them close to the heavy, highly charged atomic nuclei.

Suppose you are given the atomic weight and the density of gold. Could you calculate the number of gold atoms in a given volume—say 100 nm³?

As a somewhat similar experiment, imagine that we are shooting pellets from a "BB gun" at a chain-link fence. Since the "particles" are much smaller than the "empty spaces" in such a fence, most of the pellets would go through the fence undeflected. However, some would glance off the wires of the fence and be deflected through moderate angles; a few would hit a wire directly enough to bounce back to our side of the fence. If we fired a very large number, say a million such "particles," at the fence, and counted how many went through undeflected and how many were deflected by various angles, we could perhaps calculate the fraction of the area of the fence that is open space, and even something about the size and distribution of the wires. (Of course, the gold foil is many atomic layers thick, so we could suppose that there were several such fences, one behind the other.)

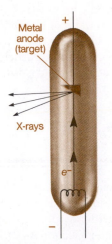

FIGURE 4–8 A simplified representation of the production of x-rays by bombardment of a solid target with a high-energy beam of electrons.

A modern periodic table is shown inside the front cover.

4–6 Atomic Number

Only a few years after Rutherford's experiments showed that atomic nuclei are very heavy, positively charged centers of atoms, H. G. J. Moseley developed a technique to determine the magnitude of the positive charge. Max von Laue had already demonstrated that x-rays could be diffracted by crystals into a spectrum, in much the same way that visible light can be separated into its component colors. (A more detailed discussion of this effect will be presented in Section 5–1.) Moseley generated x-rays by aiming a beam of high-energy electrons at a target made of a block of a single pure element (Figure 4–8).

When the spectra of x-rays produced by targets of different elements were recorded photographically, each was observed to consist of a series of lines representing x-rays at various wavelengths. Comparison of the spectra of different elements revealed that corresponding lines within a given series were displaced regularly toward shorter wavelengths as the atomic weights of the target materials increased (with a few exceptions). On the basis of his mathematical analysis of these x-ray data, Moseley concluded that *each element differs from the preceding element by having one more positive charge in its nucleus.* For the first time, it was possible to arrange all known elements in order of increasing nuclear charge. The positions of Co and Ni in the periodic table, as well as those of Te and I, were established by Moseley's work. Their positions were reversed (and wrong) in earlier periodic tables that were based on atomic weights. A plot summarizing this interpretation of Moseley's x-ray data appears in Figure 4–9.

We now know that every nucleus contains an integral number of protons, which is exactly equal to the number of electrons in a neutral atom of the element. The proton was actually discovered by Rutherford and James Chadwick in 1919 as a particle that is emitted by the bombardment of certain atoms by alpha particles from radium. Every hydrogen atom contains one proton, every helium atom contains two protons, every lithium atom contains three protons, and so on. As we shall see

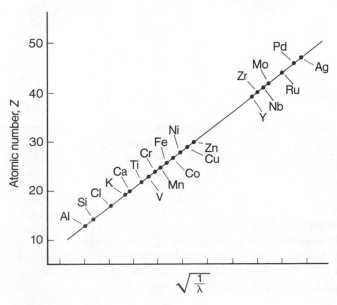

FIGURE 4–9 A plot of some of Moseley's x-ray data. The atomic number of an element is found to be directly proportional to the square root of the reciprocal of the wavelength of a particular x-ray spectral line. Wavelength is represented by λ.

(Section 6–1), evidence already existed, in the form of the periodic behavior of the elements, that the atomic mass is not the fundamental characteristic of elemental identity. However, the interpretation of x-ray wavelengths by Moseley in 1913 was the first direct evidence relating elemental identity to the number of atomic particles. The **atomic number** of an element is defined as *the number of protons or positive charges in its nucleus,* and it determines the identity of the element.

4–7   Isotopes, the Existence of Neutrons, and Mass Number

Nuclear charges and masses of atoms were deduced, for example, from the alpha particle scattering experiments and from Moseley's x-ray spectroscopic results.

From a comparison of known nuclear charges with the known masses of atoms, Rutherford deduced that there was more mass in most atoms than could be accounted for by the existence of just protons and electrons. Further evidence along these lines arose from careful studies, using the mass spectrometer (Section 4–2), of the charge-to-mass ratios of various ions.

When a beam of positive ions produced by excitation of neon gas was passed into a more modern mass spectrometer (Figure 4–10) to determine the e/m ratio of the ions, it was found that the beam consisted of both Ne^+ and Ne^{2+} ions, formed by the loss of one and two electrons, respectively, from neutral Ne atoms. Careful

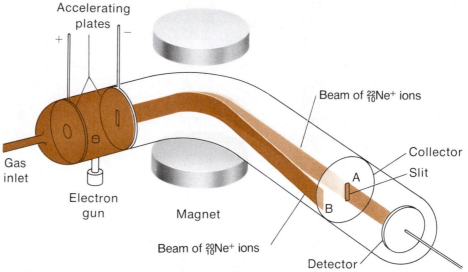

FIGURE 4–10 A simplified representation of a modern mass spectrometer. In this instrument, gas molecules at very low pressure are ionized and accelerated by an electric field. The ion beam is then passed through a magnetic field. In that field the ion beam is separated into components, each containing particles of equal e/m. Since most ions have charge $1+$, the resulting separation is according to mass. Lighter particles are deflected by the magnetic field more strongly than heavy ones. To simplify the illustration, only the beam of $1+$ ions is shown separated into its two components. The spectrometer is adjusted to detect the $^{22}_{10}Ne^+$ ions. By changing the magnitude of the magnetic field or the accelerating electric field, one could move the beam of $^{20}_{10}Ne^+$ ions striking the collector from B to A so that it could be detected. (Although the two beams of Ne^{2+} ions are not shown, they are very similar to the beams of Ne^+ ions, but are deflected much more strongly because of their higher charge.)

examination of the experimental results showed that the beam of Ne^+ ions actually was two beams, consisting of two different types of particles having slightly different e/m ratios. The same result was observed for the beam of Ne^{2+} ions. Since each beam must consist entirely of ions of the same charge, the different e/m ratios within each beam must be due to the presence of neon ions with different masses. Furthermore, this could not be due to the loss of different numbers of electrons, since that would just cause formation of Ne^{2+} ions, for example, as opposed to Ne^+ ions. It must be due, then, to the existence of atoms of different masses in the original pure gaseous sample of neon. (This concept contradicts the Daltonian idea of the atom, in which all atoms of a given element were thought to be identical, and the feature that distinguishes the atoms of one element from those of another was thought to be the atomic mass.)

It is now known that most elements naturally consist of mixtures of atoms having slightly different masses. Atoms of the same element having different masses are called **isotopes** of that element. Rutherford, James Chadwick, and W. D. Harkins explained the existence of isotopes in 1920 by postulating the existence in the nucleus of particles having considerable mass but no charge, called **neutrons.** Notice that this proposed arrangement is still consistent with the idea that most of the mass and all of the positive charge are in the atomic nucleus. Although the existence of neutrons was predicted in 1920, 12 years passed before the prediction was verified by the indirect detection of neutrons by James Chadwick.

> The neutron wasn't fully characterized until 1957 by Richard Hofstadter at Stanford University, although it had been discovered earlier.

Nuclei of all atoms except the common form of hydrogen contain neutrons. However, for most elements, different nuclei of the same element can contain different numbers of neutrons. For example, there are three distinctly different kinds of hydrogen atoms, commonly called hydrogen, deuterium, and tritium. Each hydrogen atom contains one proton in its nucleus; that is, the atomic number of hydrogen is always 1 (recall Section 4–6). The common form of hydrogen contains no neutrons, but deuterium atoms contain one neutron and tritium atoms contain two neutrons in their nuclei.

> The isotopes of most elements do not have special names, as those of hydrogen do.

In terms of Rutherford's theory, which is now well established, **isotopes** *are atoms of the same element containing different numbers of neutrons in their nuclei.* Isotopes of a given element *always* contain the same numbers of protons because they are atoms of the same element; if they are neutral atoms, they also always contain the same number of electrons. As we shall see in subsequent chapters, *the chemical behavior of atoms depends on the number, arrangement, and energies of their electrons.* Isotopes of an element are identical in these respects; therefore, the chemical properties of isotopes of the same element are essentially identical.

The **mass number** of an atom *is the sum of the number of protons and the number of neutrons in its nucleus,* i.e.,

mass number = number of protons + number of neutrons

$\qquad\qquad$ = atomic number + neutron number $\qquad\qquad$ [Eq. 4–3]

Mass numbers are always integers. The mass number for normal hydrogen atoms is 1, for deuterium 2, and for tritium 3. The composition of a nucleus is indicated by its **nuclide symbol.** This consists of the symbol for the element (E), with the atomic number (Z) written as a subscript at the lower left and the mass number (A) as a superscript at the upper left, $^{A}_{Z}E$. By this system, the three isotopes of hydrogen are designated as $^{1}_{1}H$, $^{2}_{1}H$, and $^{3}_{1}H$.

Although some elements, such as fluorine and aluminum, exist in only one isotopic form, most elements occur in nature as isotopic mixtures. Table 4–2 shows

TABLE 4–2   Naturally Occurring Isotopic Abundances for Some Common Elements

Element	Isotope	% Natural Abundance	Mass (amu)
boron	$^{10}_{5}$B	19.6	10.01294
	$^{11}_{5}$B	80.4	11.00931
oxygen	$^{16}_{8}$O	99.759	15.99491
	$^{17}_{8}$O	0.037	16.99914
	$^{18}_{8}$O	0.204	17.99916
magnesium	$^{24}_{12}$Mg	78.70	23.98504
	$^{25}_{12}$Mg	10.13	24.98584
	$^{26}_{12}$Mg	11.17	25.98259
chlorine	$^{35}_{17}$Cl	75.53	34.96885
	$^{37}_{17}$Cl	24.47	36.96590
uranium	$^{234}_{92}$U	0.0057	234.0409
	$^{235}_{92}$U	0.72	235.0439
	$^{238}_{92}$U	99.27	238.0508

The twenty elements that have no other stable naturally occurring isotopes are: $^{9}_{4}$Be, $^{19}_{9}$F, $^{23}_{11}$Na, $^{27}_{13}$Al, $^{31}_{15}$P, $^{45}_{21}$Sc, $^{55}_{25}$Mn, $^{59}_{27}$Co, $^{75}_{33}$As, $^{89}_{39}$Y, $^{93}_{41}$Nb, $^{103}_{45}$Rh, $^{127}_{53}$I, $^{133}_{55}$Cs, $^{141}_{59}$Pr, $^{159}_{65}$Tb, $^{165}_{67}$Ho, $^{169}_{69}$Tm, $^{197}_{79}$Au, $^{209}_{83}$Bi. However, there are artificially produced isotopes of these elements.

the naturally occurring isotopic abundances for a few common elements. The percentages are based on the relative *numbers* of naturally occurring atoms of each isotope, *not on their masses.* These isotopic abundances are determined by comparison of the strengths of the signals caused by their ions in the mass spectrometer.

Neon actually occurs in nature as three isotopes, $^{20}_{10}$Ne, $^{21}_{10}$Ne, and $^{22}_{10}$Ne. This statement is based on the fact that a beam of Ne$^+$ ions is split into *three* segments in a sophisticated and very sensitive mass spectrometer. The mass spectrum of Ne$^+$ ions, a graph of the relative numbers of ions of each mass, is shown in Figure 4–11. Note that the isotope $^{20}_{10}$Ne, of mass 19.99244 amu, is the most abundant isotope (tallest peak) and accounts for 90.9% of the naturally occurring neon atoms, while $^{22}_{10}$Ne accounts for 8.8%, and $^{21}_{10}$Ne only 0.3% of the atoms. (This extremely small abundance explains why the isotope of mass number 21 was not detected in the cruder experiments described earlier in this section.)

4–8   The Atomic Weight Scale and Atomic Weights

As we saw in Section 2–1, it is convenient to establish a special scale for expressing the masses of atoms, since they are so much smaller than the objects with which we usually deal. As a result of action taken by the International Union of Pure and Applied Chemistry (IUPAC) in 1962, the currently accepted atomic weight scale is based on the carbon-12 isotope, $^{12}_{6}$C. Remember that on this scale, one atom of this isotope is taken to have a mass of exactly 12 *atomic mass units* (*amu*), and all other atomic masses are expressed relative to this mass. By definition, then, **one amu** *is exactly $\frac{1}{12}$ the mass of a carbon-12 atom.* You should note that on this scale, the mass of one atom of $^{1}_{1}$H, the lightest isotope of the lightest element, is approximately one amu.

The mass spectrometer is also used to determine the actual masses of individual isotopes. This information, together with isotopic abundances, allows determination of the atomic weights of elements that occur in nature as two or more isotopes, as we shall see next.

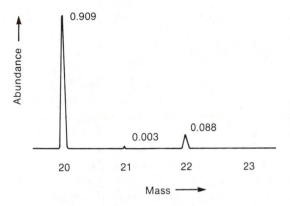

FIGURE 4–11 Mass spectrum of neon (1 + ions only). Neon contains three isotopes, of which neon-20 is by far the most abundant. The mass of that isotope, to five decimal places, is 19.99244 on the carbon-12 scale.

Helium occurs in nature almost exclusively as $^4_2$He. Let's see how its atomic mass is measured. To simplify the picture, we'll assume that only ions with 1 + charge are formed in the experiment illustrated in Figure 4–12.

A carefully measured, fixed accelerating voltage is applied and a sample of vaporized $^{12}_6$C is fed into the mass spectrometer. The magnetic field strength that causes $^{12}_6$C$^+$ ions to arrive at point A is measured. A sample of helium is then fed into the mass spectrometer with exactly the same accelerating voltage, and the lighter $^4_2$He$^+$ ions are deflected more than the heavier $^{12}_6$C$^+$ ions to some point B. The magnetic field strength is then decreased slowly until the beam of lighter ions falls at point A. Now that the field strengths required to focus the two kinds of ions at the same point are known, the mass of the lighter ion can be calculated. The relationship

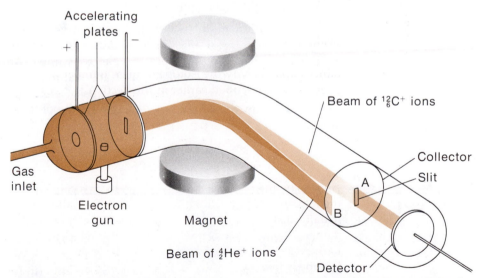

FIGURE 4–12 Determination of atomic weights in a modern mass spectrometer. A sample of vaporized carbon-12, $^{12}_6$C, is fed into the mass spectrometer at a very low pressure. The magnetic field strength required to focus the beam of $^{12}_6$C$^+$ ions at point A is measured carefully. A sample of helium is then fed into the mass spectrometer and the magnetic field strength required to focus the beam of $^4_2$He$^+$ ions at the same point A is measured carefully. From the assigned mass of $^{12}_6$C$^+$ ions and the magnetic field strengths required to focus the two beams of ions at point A, the atomic weight of helium can be calculated.

between masses of particles and magnetic field strengths, $\mathscr{H}$, is:

$$\frac{\text{mass } {}^4_2\text{He}^+}{\text{mass } {}^{12}_6\text{C}^+} = \left(\frac{\mathscr{H} \text{ for } {}^4_2\text{He}^+}{\mathscr{H} \text{ for } {}^{12}_6\text{C}^+}\right)^2$$

The $\mathscr{H}$'s are the magnetic field strengths required to focus the beams of ions at point A. When the experiment is done, the ratio of the two magnetic field strengths is measured as 0.5776, so the mass of ${}^4_2\text{He}^+$ is found to be 0.3336 times the mass of the ${}^{12}_6\text{C}^+$ ion. Since the mass of ${}^{12}_6\text{C}^+$ ions is defined to be *exactly* 12 amu, the mass of ${}^4_2\text{He}^+$ ions is 0.3336×12 amu $= 4.003$ amu.

(0.5776)² = 0.3336

In Section 2–2, we said that one mole of atoms contains 6.022×10^{23} atoms, and that the mass of one mole of atoms of any element, in grams, is numerically equal to the atomic weight of the element. Since the mass of one carbon-12 atom is exactly 12 amu, the mass of one mole of carbon-12 atoms is exactly 12 grams.

Let us now determine the relationship between atomic mass units and grams. To do this we calculate the number of atomic mass units in one gram of any element or isotope, say ${}^{12}_6\text{C}$.

$$\underset{\underbrace{\underbrace{\rule{0pt}{0pt}}_{\text{no. of } {}^{12}_6\text{C atoms in 1.000 g of } {}^{12}_6\text{C}}}_{\text{no. of amu } {}^{12}_6\text{C in 1.000 g of } {}^{12}_6\text{C}}}{\underline{?} \text{ amu } {}^{12}_6\text{C} = 1.000 \text{ g } {}^{12}_6\text{C} \times \frac{6.022 \times 10^{23} \; {}^{12}_6\text{C atoms}}{12 \text{ g } {}^{12}_6\text{C}} \times \frac{12 \text{ amu } {}^{12}_6\text{C}}{1 \; {}^{12}_6\text{C atom}} = 6.022 \times 10^{23} \text{ amu } {}^{12}_6\text{C}}$$

Therefore, $\underline{1.000 \text{ g } ({}^{12}_6\text{C or anything else}) = 6.022 \times 10^{23} \text{ amu } ({}^{12}_6\text{C or anything else})}$ or, dividing both sides by 6.022×10^{23},

You may wish to verify that the same result is obtained regardless of the element or isotope chosen.

$$\underline{1.661 \times 10^{-24} \text{ g} = 1.000 \text{ amu}}$$

The **atomic weight (mass)** that we observe for an element *is the weighted average of the masses of its constituent isotopes.* Please note that the mass of an atom is *not* the sum of the masses of its protons, electrons, and neutrons, as we will see in the next section. Most atomic weights are fractional numbers, not integers. Examples 4–1 and 4–2 illustrate how atomic weights of elements can be calculated from isotopic abundances and masses, data that can be obtained with a mass spectrometer.

Example 4–1

Naturally occurring chlorine consists of two isotopes: 75.53% of the atoms in a sample are ${}^{35}_{17}\text{Cl}$, mass $= 34.96885$ amu, and the other 24.47% are ${}^{37}_{17}\text{Cl}$, mass $= 36.96590$ amu. Calculate the atomic weight of chlorine.

Solution

The contribution that each isotope makes to the atomic weight is its abundance, expressed as a decimal fraction, times its mass.

at. wt. $= 0.7553 \, (34.96885 \text{ amu})$
$\qquad + 0.2447 \, (36.96590 \text{ amu})$
$\qquad = 26.41 \text{ amu} + 9.04 \text{ amu}$
$\qquad = \cancel{35.45 \text{ amu}}$ (to four significant figures)

Note that approximately $\frac{3}{4}$ of naturally occurring chlorine atoms are ${}^{35}\text{Cl}$, while only about $\frac{1}{4}$ are ${}^{37}\text{Cl}$. Thus, the atomic weight of chlorine is much closer to 35 than to 37.

Example 4–2

Three isotopes of magnesium occur in nature. Their abundances and masses are listed below. Use this information to calculate the atomic weight of magnesium.

Isotope	% Abundance	Mass (amu)
^{24}Mg	78.70	23.98504
^{25}Mg	10.13	24.98584
^{26}Mg	11.17	25.98259

Solution

We multiply the fraction of each isotope times its mass and add these numbers to obtain the atomic weight of magnesium.

$$
\begin{aligned}
\text{at. wt.} &= 0.7870\ (23.98504\ \text{amu}) \\
&\quad + 0.1013\ (24.98584\ \text{amu}) \\
&\quad + 0.1117\ (25.98259\ \text{amu}) \\
&= 18.88\ \text{amu} + 2.531\ \text{amu} + 2.902\ \text{amu} \\
&= \underline{24.31\ \text{amu}}\ \text{(to four significant figures)}
\end{aligned}
$$

The two heavier isotopes make small contributions to the atomic weight of magnesium, since most magnesium atoms are the lightest isotope.

Example 4–3 shows how the process can be reversed so that percent abundances can be calculated from isotopic masses and the atomic weight of an element that occurs in nature as a mixture of two isotopes.

Example 4–3

The atomic weight of gallium is 69.72 amu. The masses of the naturally occurring isotopes are 68.9257 amu for ^{69}Ga and 70.9249 amu for ^{71}Ga. Calculate the percent abundance of each isotope.

Solution

We can represent the fraction of each isotope algebraically.

Let: x = fraction of ^{69}Ga
Then: $(1 - x)$ = fraction of ^{71}Ga

Atomic weight is defined as the weighted average of the masses of the constituent isotopes. Therefore, the fraction of each isotope is multiplied by its mass, and the sum of the results is equal to the atomic weight.

$$
x(68.9257\ \text{amu}) + (1 - x)(70.9249\ \text{amu}) = 69.72\ \text{amu}
$$

$$
\begin{aligned}
68.9257x + 70.9249 - 70.9249x &= 69.72 \\
-1.9992x &= -1.20 \\
x &= 0.600
\end{aligned}
$$

$x = 0.600$ = fraction of ^{69}Ga, so $\underline{60.0\%\ ^{69}\text{Ga}}$
$(1 - x) = 0.400$ = fraction of ^{71}Ga, so $\underline{40.0\%\ ^{71}\text{Ga}}$

4–9 Nuclear Stability and Binding Energy

It is an experimentally observed fact that the mass of an atom is always *less* than the sum of the masses of its constituent particles. We now know why this **mass defect** occurs, and we also know that the mass deficiency is in the nucleus of the atom and has nothing to do with the electrons. However, since tables of masses of isotopes include the electrons, we shall include them also.

The mass defect for a nucleus is the difference between the sum of the masses of electrons, protons, and neutrons in the atom, sometimes called calculated mass, and the actual measured mass of the atom, i.e.,

mass defect = (sum of masses of all e^-, p^+, and n^0) − (actual mass of atom)

[Eq. 4–4]

For most naturally occurring isotopes, the mass defect is only about 0.15% or less of the calculated mass of the atom.

Example 4–4

Calculate the mass defect for chlorine-35 atoms. The actual mass of a chlorine-35 atom is 34.9689 amu.

Solution

The mass number of this isotope is 35. Since all chlorine atoms contain 17 protons, there are $35 - 17 = 18$ neutrons. Since there are 17 protons, there must be 17 electrons in a neutral atom. First we sum the masses of the constituent particles,

using the masses listed in Table 4–1:

$$17 \times 1.0073 \text{ amu} = 17.124 \text{ amu of protons}$$
$$17 \times 0.00055 \text{ amu} = 0.0094 \text{ amu of electrons}$$
$$\underline{18 \times 1.0087 \text{ amu} = 18.157 \text{ amu of neutrons}}$$
$$\text{Calculated mass} = 35.290 \text{ amu}$$

Then we subtract the actual mass from the calculated mass to obtain the mass defect.

$$\text{mass defect} = 35.290 \text{ amu} - 34.9689 \text{ amu}$$
$$= \underline{0.321 \text{ amu}}$$

We have calculated the mass defect in amu/atom. Recall from Section 4–8 that one gram is 6.022×10^{23} amu. We can easily show that a number expressed in amu/atom is equal to the same number of g/mol of atoms.

$$? \frac{\text{g}}{\text{mol}} = \frac{0.321 \text{ amu}}{\text{atom}} \times \frac{1 \text{ g}}{6.022 \times 10^{23} \text{ amu}} \times \frac{6.022 \times 10^{23} \text{ atoms}}{1 \text{ mol } ^{35}\text{Cl atoms}}$$

$$= 0.321 \text{ g/mol of } ^{35}\text{Cl atoms}$$

(mass defect, *not* the mass of a mole of Cl atoms)

What has happened to the mass represented by the mass defect? In 1905, Einstein postulated the theory of relativity, in which he stated that matter and energy are equivalent. This was the basis of our combination of the Laws of Conservation of Matter and Energy in Chapter 1. An obvious corollary is that matter can be transformed into energy, and energy into matter. The first transformation occurs in the sun and other stars, and was first accomplished on earth when controlled nuclear fission was achieved in 1939. The reverse transformation, energy into matter, has not yet been accomplished on a large scale.

Einstein stated that the equivalence between matter and energy could be related by a simple mathematical equation, which we first encountered in Chapter 1:

$$E = mc^2$$

where E represents the energy released, m is the mass of matter transformed into energy, and c is the velocity of light in a vacuum, 2.9979249×10^8 m/s, which is usually rounded off to 3.00×10^8 m/s.

A mass defect represents the amount of matter that would be converted into energy and released if a nucleus were formed from initially separated protons and neutrons. This energy is called the **nuclear binding energy.** Specifically, if one mole of ^{35}Cl nuclei were to be formed from 17 moles of protons and 18 moles of neutrons, the resulting mole of nuclei would have 0.321 g less mass than the original protons and neutrons. Stated differently, the nuclear binding energy for ^{35}Cl is the amount of energy (theoretically) required to separate one mole of ^{35}Cl nuclei into 17 moles of protons and 18 moles of neutrons. This has never been done. In Example 4–4 we calculated the mass defect for ^{35}Cl atoms. Let's use this value to calculate their nuclear binding energy.

Example 4–5

The mass defect for ^{35}Cl is 0.321 g/mol of atoms. Calculate the nuclear binding energy of ^{35}Cl in J/mol, where 1 joule = 1 kg·m²/s².

Solution

We substitute into the Einstein equation to obtain the nuclear binding energy. If we express the mass defect as 3.21×10^{-4} kg/mol, we obtain the binding energy in J/mol of ^{35}Cl atoms:

$$E = mc^2 = (3.21 \times 10^{-4} \text{ kg/mol})(3.00 \times 10^8 \text{ m/s})^2$$

$$= 2.89 \times 10^{13} \frac{\text{kg} \cdot \text{m}^2/\text{s}^2}{\text{mol}}$$

$$= \underline{2.89 \times 10^{13} \text{ J/mol}}$$

The nuclear binding energy of a mole of ^{35}Cl nuclei, 2.89×10^{13} J/mol, is an enormous amount of energy, enough in fact to heat 6.9×10^7 kg ($\sim$ 76,000 tons) of water from 0°C to 100°C!

Nuclear binding energies are frequently expressed in J/g of nuclei rather than in J/mol to give a better idea of nuclear stabilities. Figure 4–13 is a plot of average binding energy per gram of nuclei versus mass number. It shows that nuclear binding energies (per gram) increase rapidly with increasing mass number, reach a maximum at about mass number 50, and then decrease slowly. The nuclei with the highest binding energies (mass numbers 40 to 150) are the most stable, because so much energy would be required to separate these nuclei into their component neutrons and protons. Even though these nuclei are the most stable ones, *all* nuclei are stable with respect to complete decomposition into protons and neutrons because all nuclei have mass defects. In other words, all nuclei lost some energy when they were formed from protons and neutrons, and a loss of energy means greater stability.

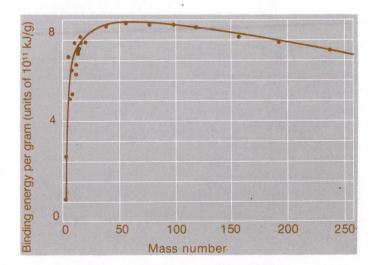

FIGURE 4–13 Plot of binding energy per gram versus mass number. Very light and very heavy nuclei are relatively unstable.

Key Terms

Alpha (α) particle helium ion with 2+ charge; an assembly of two protons and two neutrons.

Anode in cathode ray tube, the positive electrode.

Atomic mass unit (amu) arbitrary mass unit defined to be exactly $\frac{1}{12}$ the mass of the carbon-12 isotope.

Atomic number   number of protons in the nucleus; an integer; defines the identity of an element.

Beta (β) particle   high-energy electron.

Canal ray   stream of positively charged particles (cations) that moves from the positive electrode toward the negative electrode in a cathode ray tube; observed to pass through canals in the negative electrode.

Cathode   in cathode ray tube, the negative electrode.

Cathode ray   beam of electrons going from the negative electrode toward the positive electrode in a cathode ray tube.

Cathode ray tube   closed glass tube containing a gas under low pressure, with electrodes near the ends and a fluorescent screen at the end near the positive electrode; produces cathode rays when high voltage is applied; also called gas discharge tube.

Electron   a subatomic particle having a mass of 0.00055 amu and a charge of 1−.

Gamma (γ) ray   high-energy electromagnetic radiation.

Isotopes   two or more forms of atoms of the same element with different masses; atoms containing the same number of protons but different numbers of neutrons.

Mass defect   amount of matter that would be converted into energy if an atom were formed from its constituent particles.

Mass number   the sum of the numbers of protons and neutrons in an atom; an integer.

Mass spectrometer   an instrument that measures the charge-to-mass ratio of charged particles.

Natural radioactivity   spontaneous decomposition of an atom.

Neutron   a neutral subatomic nuclear particle having a mass of 1.0087 amu.

Nuclear binding energy   energy equivalent of the mass defect; energy released in the formation of an atom from subatomic particles.

Nucleus   the very small, very dense, positively charged center of an atom containing protons and neutrons, as well as other subatomic particles.

Nuclide symbol   symbol for an atom, $_Z^A$E, in which E is the symbol for an element, Z is its atomic number, and A is its mass number.

Proton   a subatomic particle having a mass of 1.0073 amu and a charge of 1+, found in the nuclei of atoms.

Exercises

Subatomic Particles and the Nuclear Atom

1. List the three fundamental particles of matter, and indicate the mass and charge associated with each.

2. A gas in a cathode ray tube sparks when a high enough voltage is applied. Describe what must be done to produce (a) a continuous glow, (b) Crooke's dark spaces, (c) no glow but continued flow of current.

3. What evidence from cathode ray tube experiments shows (a) that electrons are negatively charged and (b) that electrons have mass?

4. What evidence do we have that all atoms contain electrons?

5. Describe how Thomson determined the charge-to-mass ratio for the electron.

6. Describe Millikan's oil drop experiment. What is its significance?

7. How are canal rays produced? Are they due to subatomic particles? How do we know?

8. Describe the "plum-pudding" model of the atom proposed by Thomson.

9. Distinguish among alpha particles, beta particles, and gamma rays. Arrange them in order of increasing degree of deflection by magnetic fields.

10. Outline Rutherford's contribution to our understanding of the nature of atoms.

11. What do we mean when we refer to the nuclear atom?

12. Why were Rutherford and his co-workers so surprised that some of the alpha particles were actually scattered backward in the gold foil experiment?

13. Summarize Moseley's contribution to our knowledge of the structure of atoms.

14. How do the masses of protons and neutrons compare with those of electrons?

15. Compare the properties of cathode rays, protons, neutrons, electrons, and alpha particles.

16. Estimate the percentage of the total mass of a $_{47}^{109}$Ag atom that is due to (a) electrons, (b) protons, and (c) neutrons by *assuming* that the mass of the atom is simply the sum of the

masses of the appropriate numbers of sub-atomic particles. (The mass defect would introduce a small correction.)

17. The radius of a silver atom, Ag, is 0.144 nanometer. Using the results of Exercise 16 for $^{109}_{47}Ag$ and the estimate that the diameter of the nucleus of the $^{109}_{47}Ag$ atom is $\frac{1}{10,000}$ that of the entire atom, estimate the density of the nucleus of a $^{109}_{47}Ag$ atom. Express your answer in g/cm³. Comment on the magnitude of this calculated density.

18. Refer to Exercise 17. Estimate the density of an entire silver atom.

19. The actual density of a macroscopic sample of silver is 10.5 g/cm³. Can the difference between this and the estimated density of Exercise 18 be due only to the assumption of Exercise 16? How can the difference in values be explained? There are two other factors. The *least* of them involves isotopic distributions. [If you can't think of any adequate explanation, you may look ahead to Section 12–11.4, on metallic solids.]

20. Define and illustrate the following terms clearly and concisely.
    (a) atomic number (c) isotope
    (b) mass number   (d) nuclear charge

21. How are isotopic abundances determined experimentally?

22. How do the isotopes of a given element differ?

23. Which of the following pairs of species would have the greatest similarities in chemical properties? Justify your answer.
    (a) ^{39}K and $^{39}K^+$ (c) ^{35}Cl and $^{35}Cl^-$
    (b) ^{15}O and ^{16}O

24. Complete the chart at the bottom of the page for the indicated isotopes.

The Atomic Weight Scale

25. What is the basis for the atomic weight scale?

26. The arbitrary standard for the atomic weight scale is the exact number 12 for the mass of the carbon-12 isotope. Why is the atomic weight of carbon listed as 12.011 amu?

27. Naturally occurring carbon consists of two isotopes, ^{12}C (98.89%) and ^{13}C (1.11%). The mass of the ^{12}C isotope is 12.00000 amu and the mass of the ^{13}C isotope is 13.00335 amu. There are also minute quantities of ^{14}C, but these are too small for us to consider at this point. Using these data, calculate the atomic weight of naturally occurring carbon.

28. Prior to 1962, the atomic weight scale was based on the assignment of an atomic weight of exactly 16 amu to *naturally occurring* oxygen. The atomic weight of bromine is 79.904 amu on the carbon-12 scale. What would it be on the older scale?

29. What is the atomic weight of a hypothetical element that consists of the following isotopes in the indicated relative abundances?

Isotope	Isotopic Mass (amu)	% Natural Abundance
1	39.0	78.8
2	40.0	18.1
3	41.0	3.1

30. Naturally occurring silicon consists of three isotopes with the abundances indicated below. From the masses and relative abundances of these isotopes, calculate the atomic weight of naturally occurring silicon.

Isotope	Isotopic Mass (amu)	% Natural Abundance
^{28}Si	27.97693	92.21
^{29}Si	28.97649	4.70
^{30}Si	29.97376	3.09

Kind of Atom	Atomic Number	Mass Number	Isotope	Number of Protons	Number of Electrons	Number of Neutrons
nitrogen	___	14		___	___	___
fluorine	___	___		___	___	10
	___	___	$^{27}_{13}Al$	___	___	___
sulfur	___	___		___	___	16
___	24	52		___	___	___
	38			___	___	50
___	___	73		32	___	___
___	___			___	50	69
___	___	132		___	54	

31. Calculate the atomic weight of zinc from the following information.

Isotope	Isotopic Mass (amu)	% Natural Abundance
^{64}Zn	63.9291	48.89
^{66}Zn	65.9260	27.81
^{67}Zn	66.9271	4.11
^{68}Zn	67.9249	18.57
^{70}Zn	69.9253	0.62

32. The atomic weight of lithium is 6.941 amu. The two naturally occurring isotopes of lithium have the following masses: ^6Li, 6.01512 amu, and ^7Li, 7.01600 amu. Calculate the percent of ^6Li in naturally occurring lithium.

33. Rubidium consists of two naturally occurring isotopes: ^{85}Rb, which has a mass of 84.9117 amu, and ^{87}Rb, which has a mass of 86.9085 amu. The atomic weight of rubidium is 85.4678 amu. Determine the percent abundance of each isotope in naturally occurring rubidium.

34. Refer to Table 4–2 *only* and estimate the atomic weights of boron and chlorine to one decimal place. Justify your estimates.

35. The following is a mass spectrum of the 1+ charged ions of an element. Calculate the atomic weight of the element. What is the element?

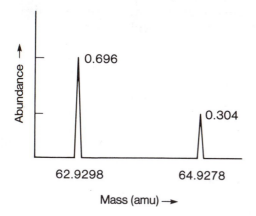

36. Consider the ions $^{14}_7N^+$, $^{15}_7N^+$, $^{14}_7N^{2+}$, and $^{15}_7N^{2+}$ produced in a mass spectrometer. Which ion's path would be deflected (a) the most and (b) the least by a magnetic field? Which one would travel (c) the most rapidly and (d) the least rapidly under the influence of a particular accelerating voltage? Justify your answers.

Mass Defect and Binding Energy

37. What is the equation that relates the equivalence of matter and energy? What does each term in this equation represent?

38. What is mass defect?

39. What is binding energy?

40. The actual mass of a ^{80}Se atom is 79.9165 amu per atom. Calculate the mass defect in amu/atom and in g/mol for this isotope.

41. What is the nuclear binding energy in kJ/mol and in kcal/mol for the isotope described in Exercise 40?

42. The actual mass of a ^{47}Ti atom is 46.9518 amu. (a) Calculate the mass defect in amu/atom and g/mol for this isotope. (b) Calculate the binding energy in kJ/mol.

43. How do nuclear binding energies vary with mass number; i.e., which elements have high and which elements have low nuclear binding energies?

Equation Balancing

44. Balance the following equations by inspection (to keep your skills sharp).
    (a) $Al + O_2 \xrightarrow{\Delta} Al_2O_3$
    (b) $S_8 + O_2 \xrightarrow{\Delta} SO_2$
    (c) $Cl_2O_7 + H_2O \rightarrow HClO_4$
    (d) $PCl_3 + H_2O \rightarrow H_3PO_3 + HCl$
    (e) $NiSO_4 + NH_3 + H_2O \rightarrow$
    $\qquad Ni(NH_3)_6(OH)_2 + (NH_4)_2SO_4$

45. Balance the following equations by inspection.
    (a) $Sb + F_2 \rightarrow SbF_5$
    (b) $PBr_3 + Br_2 \rightarrow PBr_5$
    (c) $SnS_2 + HCl \rightarrow H_2SnCl_6 + H_2S$
    (d) $RbO_2 + H_2O \rightarrow RbOH + O_2$
    (e) $PF_3Br_2 \rightarrow PF_5 + PBr_5$

Electronic Structures of Atoms

5

We have seen in a general way how the three kinds of subatomic particles are arranged in atoms, with the protons and neutrons in the atomic nucleus, which is surrounded by the electrons. However, many questions remain unanswered. Why do different elements have such a broad range of behavior in such aspects as their electrical properties, chemical reactivities, and physical states? Why do certain groups of elements have properties that are so similar, and why are so many of these properties periodic? Why do atoms combine to form chemical compounds? Why do compounds have the particular formulas that they do? Why do different compounds have different physical and chemical properties?

To begin answering such chemically important questions, we must turn our attention to the details of the arrangements of the electrons in atoms, which determine the chemical properties of elements.

Most of our information about the arrangements of electrons in atoms has come from studies of the interaction of matter with light—the light that matter can either absorb or give off (emit). To help us understand the nature of these interactions, and what they tell us about the nature of the atom, let us first describe electromagnetic radiation (light).

5–1 Electromagnetic Radiation— the Wave View of Light

To oscillate means "to change periodically," i.e., to vibrate above and below a mean value.

The phenomenon that we call "light" represents just a small part of a much more general phenomenon known as *radiant energy*. This radiant energy may be thought of as energy that moves by means of electric and magnetic fields, whose strengths oscillate in directions perpendicular to the direction of propagation of the energy.

This interpretation gives rise to another name for radiant energy — **electromagnetic radiation.** As we shall see, this phenomenon includes not only visible light, but also x-rays, microwaves, infrared rays, ultraviolet light, and many other forms of radiant energy which are, in fact, fundamentally the same.

Whenever anything oscillates, we describe it as having *wave character.* All wave phenomena — electromagnetic radiation, sound waves, waves on the surface of water, a vibrating string — can be described in similar mathematical terms. Let us illustrate this description first with waves on a surface of water and with a vibrating string such as a violin or guitar string. Then we shall see how this same description can be applied to light.

One aspect of a wave that we can specify is its **wavelength,** λ, which is the distance between any two consecutive crests of the wave train (Figure 5–1). Wavelength has units of distance such as meters, nanometers, or Ångstroms. At any point of a wave (e.g., a cork floating in the water waves, or a point on a vibrating string) the surface rises and falls regularly and repetitively. The number of cycles of rise-fall-rise that occur in a given unit of time is called the **frequency,** v, of the wave. Frequency has units of time^{-1}, such as min^{-1} or s^{-1}. A third quantity that describes a wave is its intensity, or **amplitude,** which is the maximum displacement of a point on the wave from its center, or equilibrium, position. In the case of a vibrating string, the amplitude determines the loudness of the tone, while the frequency determines the pitch.

For the vibrating string, these three quantities completely describe the wave motion. When the wave is also being propagated, as are the waves on the surface of water, the velocity of the wave must be described. For most waves this velocity is represented by the symbol v, and it has the usual velocity units of distance/time, such as m/s. Comparison of units should convince you of the relationship

$$\lambda v = v$$

[Eq. 5–1]

Thus, for a given constant velocity, wavelength and frequency are inversely proportional to one another; the shorter the wavelength, the higher the frequency. As an illustration of this, consider a floating object in a series of waves traveling at a

> The wavelength can be described equally well as the distance between any two consecutive troughs, or any two consecutive identical points of the set of waves. λ is lambda; v is nu.

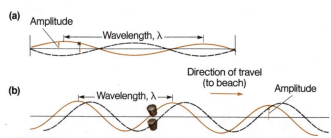

FIGURE 5–1 Two familiar examples of waves. (a) One possible mode of vibration of a violin or guitar string. At a given time, the string is displaced as shown by the solid line, while at a later time, at the other extreme of the vibration, it would be displaced as indicated by the dashed line. The number of times per second that a given point (say X) on the string cycles between its extremes is the frequency, v, of the wave. Since this wave has two fixed points, the ends of the string, the wave does not travel; such a wave is called a standing wave. (b) Waves on the surface of water. A wave moves to the right, at velocity v. The surface of the water at some instant is shown by the solid line; at a slightly later time, it is as shown by the dashed line. A cork bobs up and down repetitively as successive crests and troughs pass. The number of times it bobs up, then down, then up again to its starting point is the frequency of the wave.

constant velocity toward the shore. The closer together the waves are (shorter wavelength), the more rapidly the object bobs up and down (higher frequency).

Example 5–1

A series of waves on the surface of water is traveling toward the shore at a constant velocity of 120 ft/min. The distance between crests of the waves is 15 ft. What is the frequency of the waves?

Solution

The velocity of the waves is $v = 120$ ft/min. The distance between crests of the waves is the wavelength, so $\lambda = 15$ ft. Then, from Eq. 5–1,

$$\lambda v = v \qquad \text{or} \qquad v = v/\lambda$$

$$v = \frac{120 \text{ ft/min}}{15 \text{ ft}} = \underline{8.0 \text{ min}^{-1}}$$

This means that a floating object in this wave train would bob up and down 8.0 cycles every minute.

We have seen that for water waves and for a vibrating string, the regular periodic variations in some quantity (depth of water, displacement of string) can be described in wave terminology. In the case of electromagnetic radiation, the quantity that varies regularly is the strength of the electric and magnetic fields at a point in the beam (Figure 5–2). Thus, the same wave description can be applied to this "wave." Electromagnetic radiation is described in terms of frequency and wavelength (which determines the color of visible light).

The *frequency*, v, of electromagnetic radiation is usually expressed in cycles/s, commonly represented as $1/s$ or s^{-1}. One cycle per second is also called one hertz (Hz), after Rudolf Hertz, who in 1896 discovered electromagnetic radiation outside the visible range and measured its speed and wavelengths. The *amplitude* of the changes in electric and magnetic fields determines the intensity of electromagnetic radiation—this would correspond, for example, to the brightness of visible light. The *velocity* of electromagnetic radiation is usually given the special symbol c. In a vacuum, this velocity is the same for all wavelengths, 2.9979249×10^8 m/s. The relationship between the wavelength and the frequency of electromagnetic radiation, with c rounded to three significant figures, is

$$\lambda v = c = 3.00 \times 10^8 \text{ m/s} \qquad \text{[Eq. 5–2]}$$

Example 5–2

The frequency of violet light is 7.31×10^{14} Hz, and that of red light is 4.57×10^{14} Hz. Calculate the wavelength of each color.

Solution

Since frequency and wavelength are inversely proportional to each other, $\lambda = c/v$, we may substitute the frequencies into the relationship and calculate wavelengths.

$$\text{(violet light) } \lambda = \frac{c}{v} = \frac{3.00 \times 10^8 \text{ m/s}}{7.31 \times 10^{14} \text{ s}^{-1}}$$

$$= \underline{4.10 \times 10^{-7} \text{ m}} \quad (4.10 \times 10^2 \text{ nm})$$

$$\text{(red light) } \lambda = \frac{c}{v} = \frac{3.00 \times 10^8 \text{ m/s}}{4.57 \times 10^{14} \text{ s}^{-1}}$$

$$= \underline{6.56 \times 10^{-7} \text{ m}} \quad (6.56 \times 10^2 \text{ nm})$$

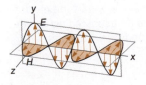

FIGURE 5–2 Electromagnetic radiation has two components: (1) an electric field that is represented by E in this diagram, and (2) a magnetic field that is represented by H. The electric and magnetic fields oscillate (change periodically) at right angles to each other and to the direction in which the electromagnetic radiation travels.

Isaac Newton first separated sunlight, which consists of waves of electromagnetic radiation, into its component colors by passing it through a prism. The resulting display of the component colors (wavelengths) is called a **spectrum.** Since sunlight (white light) contains all wavelengths of visible light, it gives the **continuous spectrum** observed in the rainbow. Of the light visible to the human eye, red light has the longest wavelength (lowest frequency), while blue-violet light has the shortest wavelength (highest frequency). But visible light represents only a tiny segment of the electromagnetic radiation spectrum. In addition to all wavelengths of

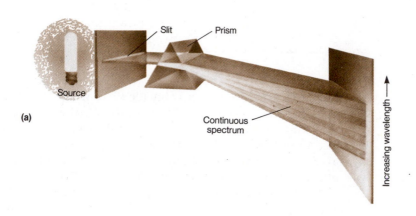

(a)

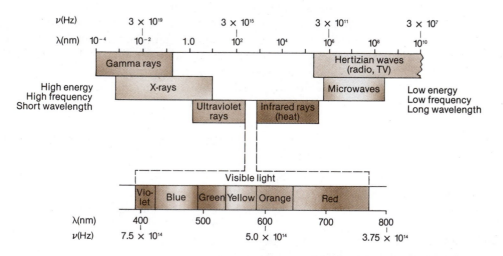

(b)

FIGURE 5–3 (a) Dispersion of visible light by a prism. When light from a source of white light is passed through a slit and then through a prism, it is separated into a continuous spectrum of all wavelengths of visible light, as well as longer and shorter wavelength light that our eyes cannot detect. (b) Visible light is only a very small portion of the electromagnetic spectrum. The upper part shows the approximate ranges on a logarithmic scale. The lower part shows the visible region on an expanded scale. Note that wavelength increases as frequency decreases.

visible light, sunlight also contains shorter wavelength (ultraviolet) radiation as well as longer wavelength (infrared) radiation, which may be detected photographically or by spectrophotometers designed for that purpose. The wavelengths and frequencies associated with the entire electromagnetic spectrum are indicated in Figure 5-3.

To summarize, then, electromagnetic radiation can be described as consisting of regularly repeated changes in electric and magnetic fields. In wave terminology, these field changes can be described in terms of their wavelength and frequency (which determines the color), their amplitude (which determines the intensity or brightness), and the velocity of their propagation. This wave view of light is useful in describing and understanding such properties of light as its reflection by mirrors, its refraction by lenses, its diffraction (reinforcement and/or cancellation of different waves), and its dispersion into various colors by prisms or ruled gratings.

FIGURE 5-4 A sample of molten steel, which is predominantly iron. Experienced steelworkers judge the approximate temperature of the molten metal from its color.

Probably you have seen a piece of iron heated until it is "red-hot," or even "white-hot."

5-2 Electromagnetic Radiation— the Particle View of Light

Some properties of light cannot be properly understood in terms of the wave description. One of these involves the energy emitted when samples of matter are heated. When a solid is heated to a sufficiently high temperature, it begins to glow, at first with a red color. As the temperature is raised, the color changes to yellow, and eventually, at very high temperatures, to bluish-white. Astronomers can tell the temperatures of stars from the color of the light they emit. In fact, for matter at any temperature, the light given off is a mixture of a range of colors (wavelengths), with the maximum intensity corresponding to the color observed.

A basic assumption of all physical theories prior to about 1900 (the so-called *classical* or *Newtonian physics*) was that natural phenomena are continuous. One prediction of this theory was that the intensity of emitted light should increase with increasing frequency of the light. That is, more blue and ultraviolet light should be given off than we actually observe. Thus, this classical theory was unable to explain the observed distribution of colors emitted by a heated body.

In 1900, Max Planck made a suggestion about the nature of light that did allow interpretation of the distribution of wavelengths of a **blackbody radiator.** Planck's suggestion was that radiant energy is not continuous, but consists of individual bundles of energy, called **photons.** The amount of energy associated with each photon depends on the color of the light and is called a **quantum of energy.** A *monochromatic* beam of light (that is, containing only one color) consists entirely of quanta of the same energy. Planck was able to establish a relationship between the energy of a photon and the color of the light. This relationship states that the energy, E, of each photon is proportional to the frequency, v, of the radiation:

An object that absorbs or radiates energy with 100% efficiency is called a *blackbody radiator.*

$$E = hv \qquad \text{[Eq. 5-3]}$$

The proportionality constant, h, is called *Planck's constant* and is the same for radiant energy of all frequencies; it has the units of energy multiplied by time. Expressing energy in joules, $h = 6.6262 \times 10^{-34}$ J·s. Introducing the relation given in Eq. 5-2, we can write

$$E = \frac{hc}{\lambda} \qquad \text{[Eq. 5-4]}$$

This view of radiant energy as composed of individual packets of energy is sometimes called the particle, or corpuscular, theory of light. The importance of the Planck relation (Eqs. 5–3 and 5–4) is that it provides a way to shift between the particle view (energy of the photon) and the wave view (wavelength and frequency of the oscillating fields). This connection is illustrated in Example 5–3.

Example 5–3

In Example 5–2, we calculated the wavelengths of violet light of frequency 7.31×10^{14} s^{-1} and of red light of frequency 4.57×10^{14} s^{-1}. Calculate the energy of an individual photon in each of these two colors of light in joules.

Solution

(violet light) $E = h\nu$
$$= (6.63 \times 10^{-34}\,\text{J}\cdot\text{s})(7.31 \times 10^{14}\,\text{s}^{-1})$$
$$= \underline{4.85 \times 10^{-19}\,\text{J}}$$

(red light) $E = h\nu$
$$= (6.63 \times 10^{-34}\,\text{J}\cdot\text{s})(4.57 \times 10^{14}\,\text{s}^{-1})$$
$$= \underline{3.03 \times 10^{-19}\,\text{J}}$$

You can check these answers by calculating the energies directly from the wavelengths, using Eq. 5–4.

As you can see from the preceding example and from Eqs. 5–3 and 5–4, the greater the frequency or the shorter the wavelength of radiation, the greater is its energy *per photon*. In this particle view, the brightness or intensity of the light depends on the *number* of photons striking a given area in a unit of time.

One application of this effect is in the photoelectric sensors that open some supermarket or elevator doors when the shadow of a person interrupts the light beam.

Another experimental observation that had not been satisfactorily explained in terms of the classical wave model of light was the **photoelectric effect.** An apparatus for the photoelectric effect is shown in Figure 5–5. The negative electrode in the evacuated tube is made of a pure metal such as cesium. When light of a sufficiently high energy is allowed to strike the metal, electrons are knocked off its surface. They then travel to the positive electrode and form a current flowing through the circuit. The current (the number of electrons emitted per second) increases with the intensity of the light. If the light that strikes the metal is too near the red end of the visible spectrum (i.e., of too low energy), no electrons are ejected. Once some threshold frequency (color) of light is passed, however, the current that

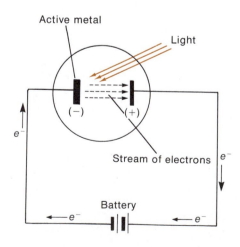

FIGURE 5–5 The photoelectric effect. When electromagnetic radiation of sufficient minimum energy strikes the surface of a metal (negative electrode) inside an evacuated tube, electrons are stripped off the metal to create an electric current. The current increases with increasing radiation intensity.

flows does not depend on the energy of the light. Classical theory said that electrons should accumulate energy and should be released when they have enough energy to get away from the metal atoms. Thus, even lower energy red light would have been expected to be able, eventually, to eject electrons from the metal surface.

The answer to this puzzle was provided by Albert Einstein in 1905, extending Planck's idea that the light behaves as though it were composed of *photons*, each with a particular amount (a quantum) of energy. The picture that Einstein provided is one of a particle of light striking an electron near the surface of the metal and giving up all of its energy to the electron. If that energy is equal to or greater than the amount needed to liberate the electron, then that electron can escape to join the photoelectric current. Of course, the light must have at least some minimum frequency (energy) in order to be absorbed at all. As the intensity of light (number of photons per unit area per second) increases, the number of electrons emitted per second (the current) also increases. For this explanation, Einstein received the Nobel Prize in physics in 1921.

Students often are disturbed by our inability to "decide" whether light is a wave or a particle, and by terminology such as "the dual view of light." Perhaps the following parable will clarify this situation.

Several blind men were asked to describe an elephant. Each tried to determine what the elephant was like by touching it. The first blind man said the elephant was like a tree trunk; he had felt the elephant's massive leg. The second blind man disagreed, saying that the elephant was like a rope, having grasped the elephant's tail. The third blind man had felt the elephant's ear, and likened the elephant to a palm leaf, while the fourth, holding the beast's trunk, contended that the elephant was more like a snake. Of course each blind man was giving a good description of that one aspect of the elephant that he was observing, but none was entirely correct. In much the same way, we use the wave and particle analogies to describe different manifestations of the phenomenon that we call radiant energy, because as yet we have no single description that will explain all of our observations.

5-3 Atomic Spectra and the Bohr Theory of the Atom

When an electric current is passed through a gas in a vacuum tube at very low pressures, the gas emits light that can be separated by a prism into distinct lines. Such a spectrum is called an **emission spectrum** to indicate the *origin* of its light (Figure 5-6a and b). It is also called a **line spectrum** to describe its *appearance*. It is important to note that the appearance of only certain lines in such a spectrum indicates that the sample is emitting only certain wavelengths. This is in contrast to the *continuous spectrum* described in Section 5-2. The lines of an atomic emission spectrum can be recorded photographically, and the wavelength of light that produced each line can be calculated from the position of the line on the photograph. Each element emits a characteristic set of lines (wavelengths), and the observed distribution of lines can serve as a very sensitive "fingerprint" to identify the elements that are present in the sample.

In a similar experiment (Figure 5-6c and d) in which white light is passed through a sample of a gas, almost all the light is transmitted, but some wavelengths

The analytical method that utilizes such observations is called atomic emission spectroscopy.

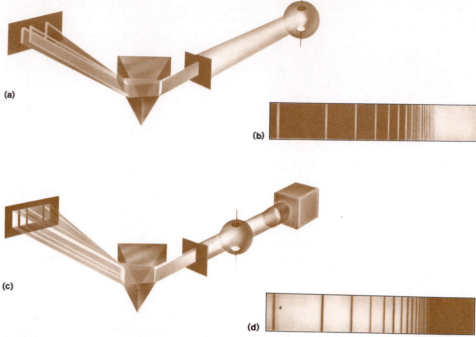

(a)

(b)

(c)

(d)

FIGURE 5–6 (a) Atomic emission. The light emitted by a sample of excited hydrogen atoms (or any other element) can be passed through a prism and separated into certain discrete wavelengths. Thus an emission spectrum (b), which is a photographic recording of the separated wavelengths, is called a line spectrum. Any sample of reasonable size contains an enormous number of atoms. Although a single atom can be in only one excited state at a time, the collection of atoms contains all possible excited states. The light emitted as these atoms fall to lower energy states is responsible for the spectrum. (c) Atomic absorption. When white light is passed through unexcited hydrogen and then through a slit and a prism, the transmitted light is lacking in intensity at the same wavelengths as are emitted in (a). The recorded absorption spectrum (d) is also a line spectrum, and the photographic negative of (b), the emission spectrum.

are absorbed. Analysis by a prism reveals that the wavelengths *absorbed* in this experiment are the *same* as those emitted upon excitation of the same gas. This **absorption spectrum,** like the emission spectrum, is also a *line spectrum.*

Careful studies showed that several series of lines in the spectrum of hydrogen are produced when an electric current is passed through the gas at very low pressures. These lines were studied intensely by many scientists. J. R. Rydberg discovered in the late nineteenth century that the wavelengths of the various lines in the hydrogen spectrum can be related by a mathematical equation, called the **Rydberg equation.**

$$\frac{1}{\lambda} = R \left(\frac{1}{n_1^2} - \frac{1}{n_2^2} \right)$$

[Eq. 5–5]

In this expression $R = 1.097 \times 10^7$ m^{-1} and is known as the *Rydberg constant.* The n's are positive integers, and n_1 is smaller than n_2. The Rydberg equation was derived from numerous observations, not from theory, and is thus an empirical equation.

The origin of this emitted light could be explained as follows. The electric current excites electrons in the atom to higher potential energy. Just as a rock rolls down a hill toward a position of lower potential energy, each excited electron then moves to a state of lower potential energy (closer to the nucleus). In doing so, the energy that it loses is emitted as a single quantum of radiant energy (light), whose frequency and wavelength depend on the amount of energy lost, according to Eqs. 5–3 and 5–4.

Prior to 1913, atomic theories suggested that an electron could be located at any distance from the nucleus in an atom, and thus could have any energy. This interpretation would indicate that an electron could lose any amount of energy, and could emit light of any frequency, to give a continuous spectrum. This model of the atom could not explain the observed existence of line spectra.

In 1913, Neils Bohr, a Danish physicist, provided an explanation for the occurrence of line spectra, such as those described by the Rydberg equation and shown in Figure 5–6b and d. Applying Planck's ideas about quantization of radiant energy, Bohr suggested that the electrons in an atom revolve around the nucleus of an atom *only in certain discrete orbits,* with no other orbits being possible. This means that an electron could have only certain energies (i.e., could occupy only certain *energy levels*), and that it must absorb or emit energy in *discrete amounts* as it moves from one allowed energy level (orbit) to another. When an electron is promoted from a lower energy level to a higher one, it absorbs a definite, or *quantized,* amount of energy. When the electron falls back to the original energy level, it emits exactly the same amount of energy it absorbed in moving from the lower to the higher energy level.

The idea of quantized energy levels is analogous to the possibility of only certain values of gravitational potential energy on a stairway, contrasted with the continuity of energies available on a ramp (Figure 5–7). In terms of energy *changes,* the person on the ramp can, with infinitesimally small energy changes, move by very small amounts up or down the ramp. The person on the steps, however, must exert enough energy to go up one full step (or two, or three, . . .), or she cannot move up at all. Likewise, if she goes downward, she must go by full steps, and her energy decreases by corresponding fixed amounts.

Bohr assumed that electrons revolve around the nucleus of an atom in circular orbits. From mathematical equations describing these orbits for the hydrogen atom,

ENERGY ⟶

FIGURE 5–7 The gravitational potential energy of a person increases with increasing height. (a) A person walking up a ramp has a (gravitational) potential energy that can vary continuously. (b) A person walking up a flight of stairs can have only certain values of gravitational potential energy, since she cannot stand anywhere except on a step.

together with the assumption of quantization of energy, he was able to determine two significant aspects of each allowed orbit:

1. *Where* (with respect to the nucleus) the electron can be—that is, the radius, r, of the circular orbit, as given approximately by

$$r = \frac{n^2 h^2}{4\pi^2 m e^2}$$

[Eq. 5-6]

where h = Planck's constant, e = the charge of the electron, m = the mass of the electron, and n is a positive integer (1,2,3, . . .) that tells us which orbit is being described.

2. *How stable* the electron would be in that orbit—that is, its potential energy, E, as given by

$$E = -\frac{2\pi m e^4}{n^2 h^2}$$

[Eq. 5-7]

where the symbols have the same meaning as before. Note that E is *always* negative.

Results of evaluating these equations for some of the possible values of n (1,2,3, . . .) are shown in Figure 5-8. As you can see, the larger the value of n, the further from the nucleus is the orbit being described, and the radius of this orbit increases as the *square* of n increases. As n increases, n^2 increases, $1/n^2$ decreases, and therefore the electronic energy increases (becomes less negative and smaller in magnitude). For orbits further from the nucleus, the electronic potential energy is higher (less negative—the electron is in a less stable state, or in a *higher* energy level). Going away from the nucleus, the allowable orbits are further apart in distance, but closer together in energy. We should pay close attention to the two possible limits of these equations. One limit is when $n = 1$; this corresponds to the electron at the closest possible distance to the nucleus and at its lowest (most negative) energy. The other limit is for very large values of n, i.e., as n approaches infinity. As this limit is approached, the electron is very far from the nucleus, or effectively removed from the atom, while the energy is as high as possible, approaching zero.

With these equations and the relationship that the Planck equation provides between energy and frequency or wavelength, Bohr was able to predict the wavelengths observed in the hydrogen emission spectrum. Figure 5-9 illustrates the relationship between lines in the emission spectrum and the electronic transitions (changes of energy level) that occur in hydrogen atoms. We now accept the fact that electrons do indeed occupy only certain energy levels in atoms. In most atoms, some

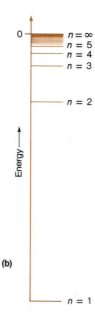

(a)

FIGURE 5-8 (a) The first four Bohr orbits for a hydrogen atom. The dot at the center represents the nucleus. The radius of each orbit is proportional to the square of n. (b) Relative values for the energies associated with the various energy levels in a hydrogen atom. Note that the energies become closer together as n increases. They are so close together for large values of n that they form a continuum. Note also that the energies of electrons in atoms are always negative.

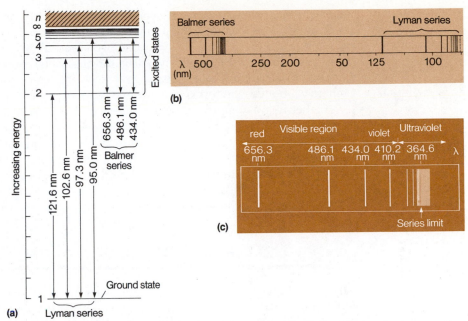

FIGURE 5–9 (a) The energy levels that the electron can occupy in a hydrogen atom and a few of the transitions that cause the emission spectrum of hydrogen. The numbers on the vertical lines show the wavelengths of radiation emitted when the electron falls to a lower energy level. (Radiation of the same wavelength is absorbed when the electron is promoted to the higher energy level.) The difference in energy between two given levels is exactly the same for all hydrogen atoms, so it corresponds to a specific wavelength and to a specific line in the emission spectrum of hydrogen. In a given sample, some hydrogen atoms could have their electron excited to the $n = 2$ level. This electron could then fall to the $n = 1$ energy level, giving off the *difference* in energy in the form of light (the 121.6 nm transition). Other hydrogen atoms might have their electrons excited to the $n = 3$ level, and subsequently fall to the $n = 1$ level (the 102.6 nm transition). Because higher energy levels become closer and closer in energy, *differences* in energy between successive transitions become smaller and smaller. The corresponding lines in the emission spectrum become closer together and result in a continuum, a series of lines so close together that they are indistinguishable. (b) The emission spectrum of hydrogen. The series of lines produced by the electron falling to the $n = 1$ level is known as the *Lyman series;* it is in the ultraviolet region. Transitions in which the electron falls to the $n = 2$ level give rise to a similar set of lines in the visible region of the spectrum, which is known as the *Balmer series.* Not shown are series involving transitions to energy levels with higher values of n. (c) The Balmer series shown on an expanded scale. Note that the line at 656.3 nm (the $n = 3 \rightarrow n = 2$ transition) is much more intense than the line at 486.1 nm (the $n = 4 \rightarrow n = 2$ transition) because the first transition occurs much more frequently than the second. Successive lines in the spectrum become less intense as the series limit is approached because the transitions that correspond to these lines are less probable.

of the energy differences between various levels correspond to the energy of a photon of light in the visible region of the spectrum, and thus colors associated with electronic transitions in such elements can be observed by the human eye. The light produced by the gaseous neon in neon signs and the lightning flashes produced in electrical storms are two familiar examples.

Although the Bohr theory satisfactorily explained the spectrum of hydrogen and those of other (charged) species containing only one electron, it could not explain the observed spectra of more complex species. This model was modified in 1916 by Arnold Sommerfeld, who assumed elliptical rather than circular orbits, but his revised theory still could not explain the observed spectra of any but the simplest atoms. In addition, these theories gave no basis for correlating or understanding chemical bonding in terms of electronic properties.

Let us now approach the big question: How are electrons arranged in more complicated atoms, and how do they behave? The modern picture of the atom is somewhat different from Bohr's picture, but we should not minimize Bohr's contributions, any more than those of Dalton, Rutherford, and the other pioneers in the study of the atom. In particular, we should notice that our major goal in interpreting more modern theories will still be the same — to interpret the electronic energy states to tell something about *where* and *how stable* the electrons are.

5–4   The Wave–Particle View of the Electron

The fact that *light* can exhibit both wave properties and particle properties suggested to Louis de Broglie that *matter* also might be able to exhibit both kinds of properties under the proper circumstances. There are many experiments in which electrons act as particles, such as the cathode ray experiments (Section 4–2). In his doctoral thesis in 1925, de Broglie predicted that such a particle with mass m and velocity v should have some properties interpretable in terms of a wave, and that the numerical value of the wavelength, λ, would be

$$\lambda = \frac{h}{mv}$$

[Eq. 5–8]

where h is Planck's constant (Section 5–2).

Two years after de Broglie's prediction, C. Davisson and L. H. Germer at the Bell Telephone Laboratory demonstrated diffraction of electrons by a crystal of nickel. This sort of behavior is possible only for waves, and shows conclusively that electrons do have wave properties. The wavelength associated with electrons of known energy was found by Davisson and Germer to be just that predicted by de Broglie. Similar diffraction experiments have been successfully performed with other particles, such as neutrons.

5–5   The Quantum Mechanical Picture of the Atom

FIGURE 5–10 The electron microscope is a practical application of the wave behavior of electrons. These crystals are seen 750 times actual size in an image formed by electrons rather than by light.

Through the work of de Broglie, Davisson and Germer, and many others, we now know that electrons in atoms can be treated as waves more effectively than as small compact particles. Large objects such as golf balls and moving automobiles obey the laws of classical mechanics (Isaac Newton's laws). Very small particles, on the other hand, can be described much better by a different kind of mechanics, called **quantum mechanics,** because the main idea of this kind of mechanics is the quantization of energy.

One of the underlying principles of quantum mechanics is that we cannot determine precisely the paths that electrons follow as they move about atomic

Any mechanism for determining either the position or the velocity of an electron necessarily changes both.

nuclei. The **Heisenberg Uncertainty Principle** (proposed in 1927 by Werner Heisenberg) is a theoretical statement that is also consistent with all experimental observations. It states that *it is impossible to determine accurately both the momentum and the position of an electron (or any other small particle) simultaneously.* Momentum is mass times velocity, *mv.* Because electrons are so small, and they move so rapidly, their motion is usually detected by electromagnetic radiation. Photons possess energies of about the same magnitudes as those associated with electrons. Consequently, the interaction of a photon with an electron severely disturbs the motion of the electron. The situation is not unlike detecting the position of a moving automobile by driving another automobile into it.

Since it is not possible to determine accurately and simultaneously both the position and the velocity of an electron, we must use a statistical approach and speak of the *probability* of finding an electron within specified regions in space. With this idea in mind, we can now list the postulates of quantum mechanics.

1. Atoms and molecules can exist only in certain energy states that are characterized by definite energies. When an atom or molecule changes its energy state, it must absorb or emit just enough energy to bring it to the new energy state (the quantum condition).

Atoms and molecules possess various forms of energy, but let us focus on the motions of electrons and the corresponding *electronic potential energies.*

Recall that a delta, Δ, preceding a quantity is read as "change in" that quantity.

2. When atoms or molecules absorb or emit radiation as they change their energies, the frequency of the light is related to the energy change by a simple equation:

$$\Delta E = h\nu \qquad \text{[Eq. 5–9]}$$

which is Eq. 5–3 applied to a *change* in energy. Recall that $\lambda\nu = c$, so Eq. 5–9 can be written as

$$\Delta E = \frac{hc}{\lambda} \qquad \text{[Eq. 5–10]}$$

to give a relationship between the energy change, ΔE, and the wavelength, λ, of radiation absorbed or emitted. The energy gained (or lost) by an atom as it goes from lower to higher (or higher to lower) energy states is equal to the energy of the photon absorbed (or emitted) during the transition.

3. The allowed energy states of atoms and molecules can be described by sets of numbers called *quantum numbers.*

The quantum mechanical treatment of atoms and molecules, formulated in 1926 by Erwin Schrödinger, is highly mathematical because it involves the equation of the electron as a wave. The important point is that the solution of this wave equation gives a set of numbers, called **quantum numbers,** that describe the energies of electrons in atoms. These numbers agree with those deduced from experiment and from empirical equations such as the Rydberg equation. Solutions of the Schrödinger equation also give information about the shapes and orientations of the statistically most probable distributions of the electrons around the nucleus. (Recall that, in light of the Heisenberg Uncertainty Principle, this is how we describe the positions of the electrons.) These *atomic orbitals* (which we shall describe in Section 5–7), deduced from the solutions of the Schrödinger equation, are directly related to the quantum numbers. In 1928, Paul A. M. Dirac reformulated electron quantum mechanics to take into account the effects of relativity. Quantum me-

Schrödinger and Dirac shared the 1933 Nobel Prize in physics for this work.

chanics, as we shall describe its application to atomic systems, is based on the work of these pioneers of modern atomic theory, Schrödinger and Dirac.

Just as we were able, in interpreting the Bohr theory, to find out something about where the electron in the hydrogen atom could be, and how stable it would be, we shall be able to draw similar conclusions about atoms containing many electrons, based on quantum mechanics. We must, therefore, now describe the quantum numbers.

5–6 Quantum Numbers

The solutions of the Schrödinger and Dirac equations for hydrogen atoms give four quantum numbers that describe the various states available to hydrogen's single electron. We can use these quantum numbers to describe the electronic arrangements in all atoms, their **electronic configurations.** The roles these quantum numbers play in describing the energy levels of electrons, as well as the shapes of the regions of space where the probability of an electron's presence is large, will not become completely clear until we present a fuller discussion of atomic orbitals in the following section. Here we shall just define each quantum number, tell a little about its physical significance, and describe the range of values it may take. For now it is sufficient to understand the term **atomic orbital** to mean *a region in space in which the probability of finding an electron is large.*

1. The **principal quantum number, n,** describes the main energy level an electron occupies. It may take any positive integral value:

$$n = 1, 2, 3, 4, \ldots \qquad \text{[Eq. 5–11]}$$

As an analogy, suppose you have purchased a seat in a multilevel arena. Specifying n would tell you whether your seat is in the floor level, the mezzanine, or the upper level; specifying the value of ℓ would tell you the row in which your seat is located.

2. The **subsidiary** (or **azimuthal**) **quantum number, ℓ,** specifies sublevels within the main energy levels. (It is also called the orbital quantum number.)

Each different value of ℓ also designates the geometric shape of the region in space that an electron occupies. The number of different sublevels within each main energy level depends on the value of n; ℓ may take any integral values from 0 up to and including $(n-1)$.

$$\ell = 0, 1, 2, \ldots, (n-1) \qquad \text{[Eq. 5–12]}$$

The maximum value of ℓ is thus $(n-1)$. With each value of ℓ, a letter has become associated, each corresponding to a different *sublevel, or set of atomic orbitals:*

$$\ell = 0, 1, 2, 3, \ldots, (n-1)$$
$$s \quad p \quad d \quad f$$

The s, p, d, f designations arise from the characteristics of spectral emission lines produced by electrons occupying those atomic orbitals: s (sharp), p (principal), d (diffuse), and f (fundamental). Beginning at $\ell = 4$, the designations follow the alphabetical order g, h, i. . . .

In the first energy level ($n = 1$), the maximum value of ℓ is zero, which tells us that there is only an s sublevel, and no p or other sublevels. In the second energy level ($n = 2$), the permissible values of ℓ are 0 and 1, which tells us that there is an s sublevel ($\ell = 0$) and a p sublevel ($\ell = 1$), but no d or other sublevels in this energy level.

3. The **magnetic quantum number, m_ℓ,** designates the spatial orientation of a single atomic orbital. The number of different possible orientations of the orbitals in a sublevel depends on the kind of sublevel. Within each sublevel, m_ℓ takes all integral values from $-\ell$ through zero up to and including $+\ell$:

$$m_\ell = (-\ell), \ldots, 0, \ldots, (+\ell) \qquad \text{[Eq. 5–13]}$$

Clearly, the maximum value of m_ℓ depends on the value of ℓ. For example, when $\ell = 1$, which designates the p sublevel, there are only three permissible values of m_ℓ: -1, 0, and $+1$.

4. The **spin quantum number, m_s,** refers to the spin of an electron and the orientation of the magnetic field produced by the spin of that electron. For every set of n, ℓ, and m_ℓ values, m_s can take the values $+\frac{1}{2}$ and $-\frac{1}{2}$.

5–7 Atomic Orbitals and the Allowed Combinations of Quantum Numbers

The **Pauli Exclusion Principle,** which Wolfgang Pauli derived from relativistic quantum mechanics in 1940, but which had been deduced as an additional postulate by Dirac and by Heisenberg in 1926, states that *no two electrons in an atom may have identical sets of all four quantum numbers.* Each allowable combination of the four quantum numbers describes a different electron in the atom. Let us now consider the allowed combinations of quantum numbers, and develop a clearer definition of what we mean by an *atomic orbital.* Then, in the next two sections, we shall elaborate on these ideas.

The significance of sets of quantum numbers lies in the results that can be calculated from the Schrödinger and Dirac equations. Though the calculations themselves are far beyond the level of this book, we can describe the results as follows. Each different set of the first three quantum numbers, n, ℓ, and m_ℓ (which we can write according to the rules of Section 5–6), corresponds to a different *atomic orbital,* to which we give the following two interpretations.

Note the parallel to our interpretation of the single quantum number of the Bohr theory.

1. Each orbital corresponds to a different probability density distribution in space, i.e., to a different region around the nucleus where the probability of finding the electron is large (Section 5–8).
2. Each orbital also has a particular energy associated with it (Section 5–9), which can be calculated from the quantum mechanical equation. This is the energy that an electron would have if it occupied that orbital.

For an electron in a given orbital, either value of the spin quantum number m_s, $+\frac{1}{2}$ or $-\frac{1}{2}$, would correspond to the same region and the same energy for the electron.

Let us now systematically tabulate a few of the allowed combinations of quantum numbers.

n = 1 Following the rules in Section 5–6, we see that when $n = 1$ the only possible value of ℓ is $\ell = 0$. Recall that this is called an s orbital, or more completely a $1s$ orbital, to indicate the s orbital in the first main energy level. Further, since m_ℓ can range only from $-\ell$ (i.e., 0) to $+\ell$ (i.e., 0), the only allowable value of m_ℓ is 0. Consideration of the rules of Section 5–6 should convince you that we can write no other allowed combination of n, ℓ, and m_ℓ with $n = 1$. Thus, there is only one orbital, the $1s$ orbital, in the first main energy level. We enter this as the first line of Table 5–1.

Any s sublevel consists of only one s orbital.

n = 2 (a) In the second main energy level, one possible value of ℓ is 0. Since ℓ can range up to $(n - 1)$, and n must be a positive integer, the value $\ell = 0$ is allowed in any main energy level. Thus any energy level n contains an ns orbital. As we saw for the $1s$ orbital above, when $\ell = 0$ the only possible value of m_ℓ is 0, so that there can also be only a single $2s$ orbital. This reasoning accounts for line 2 of

TABLE 5–1   Some Permissible Combinations of Values of the Three Quantum Numbers n, ℓ, and m_ℓ

n	ℓ	m_ℓ	Designation		Total Number of Orbitals in Level
1	0	0	$1s$		1
2	0	0	$2s$		
2	1	-1	$2p$ ⎫		
2	1	0	$2p$ ⎬ A set of 3 equivalent orbitals		4
2	1	$+1$	$2p$ ⎭		
3	0	0	$3s$		
3	1	-1	$3p$ ⎫		
3	1	0	$3p$ ⎬ A set of 3 equivalent orbitals		
3	1	$+1$	$3p$ ⎭		
3	2	-2	$3d$ ⎫		9
3	2	-1	$3d$		
3	2	0	$3d$ ⎬ A set of 5 equivalent orbitals		
3	2	$+1$	$3d$		
3	2	$+2$	$3d$ ⎭		
4	0	0	$4s$		
4	1	-1	$4p$ ⎫		
4	1	0	$4p$ ⎬ A set of 3 equivalent orbitals		
4	1	$+1$	$4p$ ⎭		
4	2	-2	$4d$ ⎫		
4	2	-1	$4d$		
4	2	0	$4d$ ⎬ A set of 5 equivalent orbitals		
4	2	$+1$	$4d$		
4	2	$+2$	$4d$ ⎭		16
4	3	-3	$4f$ ⎫		
4	3	-2	$4f$		
4	3	-1	$4f$		
4	3	0	$4f$ ⎬ A set of 7 equivalent orbitals		
4	3	$+1$	$4f$		
4	3	$+2$	$4f$		
4	3	$+3$	$4f$ ⎭		

Table 5–1. (b) Now, since $n = 2$, the additional value of $\ell = 1$ is allowable. Orbitals of this type are called p orbitals, specifically $2p$ orbitals since they are in the second main energy level. Since m_ℓ can range from $-\ell$ to $+\ell$ by integers, there are three allowable values of m_ℓ: -1, 0, and $+1$. Each of these corresponds to a different orbital of the $2p$ type, as shown in Table 5–1. As we shall see in Section 5–9, these three orbitals all have exactly the same energy. We refer to such a collection of orbitals as a set of **equivalent orbitals.** With $n = 2$, no other combinations of the three quantum numbers can be written to conform with the rules of Section 5–6.

A set of (energetically) equivalent orbitals is often described as *degenerate* orbitals, i.e., of the same energy.

$n = 3$ For the third main energy level, there is the single $3s$ orbital ($\ell = 0$, $m_\ell = 0$), by the same reasoning as presented for $n = 1$ and $n = 2$. Likewise, this energy level contains a set of three equivalent $3p$ orbitals ($\ell = 1$, $m_\ell = -1$, 0, and $+1$), as was shown for $n = 2$. In addition, since ℓ can reach a maximum value of $n - 1$, the value $\ell = 2$ is possible (d orbitals). For $\ell = 2$ there are five allowable values of m_ℓ ranging from $-\ell$ to $+\ell$ (-2, -1, 0, $+1$, $+2$), so that there exists a set of *five* equivalent d orbitals. No other combinations are possible for $n = 3$, so that this main energy level consists of nine possible orbitals (Table 5–1).

TABLE 5-2   A Summary of Relationships Between Main Energy Levels and Sublevels

Energy Level, n	Allowed ℓ Values	Sublevel Designations
1	0	$1s$
2	0,1	$2s,2p$
3	0,1,2	$3s,3p,3d$
4	0,1,2,3	$4s,4p,4d,4f$
5	0,1,2,3,4	$5s,5p,5d,5f,5g*$
6	0,1,2,3,4,5	$6s,6p,6d,6f,6g,6h$
7	0,1,2,3,4,5,6	$7s,7p,7d,7f,7g,7h,7i$

TABLE 5-3   Number of Equivalent Orbitals in Sublevels

Type of Subshell	Value of ℓ	Allowed Values of m_ℓ	Number of Equivalent Orbitals in Sublevel
s	0	0	1
p	1	$-,1,0,1$	3
d	2	$-2,-1,0,1,2,$	5
f	3	$-3,-2,-1,0,1,2,3$	7
$g*$	4	$-4,-3,-2,-1,0,1,2,3,4$	9

* None of the known elements, in their ground states, have electrons in g, h, or higher energy orbitals. They are included here and in Table 5-2 only to show the pattern.

Extension of this procedure leads to results which, along with those in Table 5-1, are more compactly summarized in Tables 5-2 and 5-3. You should check your understanding of the quantum number rules by verifying some of the results in those tables. For instance, try to write down all permitted combinations of quantum numbers with $n=4$ or $n=5$ to verify the level-sublevel-equivalent orbital information.

Before we present interpretations of atomic orbitals in terms of locations and energies of electrons, some remarks about the origin of these results might be helpful. The calculations with the quantum mechanical equations have been carried out rigorously only for the simplest atom, hydrogen, which contains only a single electron. Methods of evaluating the solutions of these equations both for electron probability (Section 5-8) and for orbital energies (Section 5-9) are only approximate for more complex, multielectron atoms. With the availability of high-speed computers in recent years, such approximate methods have been applied with increasing accuracy. The details of these calculations would be inappropriate in this text. Qualitative interpretations of them will be presented, however, since they are the basis for our current understanding of the structures of atoms and chemical bonding.

5-8   Shapes of Atomic Orbitals

One kind of information that can be calculated from the quantum mechanical equations for each atomic orbital is the probability of finding an electron at various locations around the nucleus. Note that it is a consequence of the Heisenberg Uncertainty Principle, and of the wave treatment of the electron, that we can never precisely specify *where* an electron of known energy is, but only something about the probability of its being in various places (i.e., in a particular orbital). In a sense, the calculations mathematically describe an electron as a "cloud" of charge, whose density is different at various points about the nucleus.

Let us illustrate the approach first with the 1s orbital. We ask, "What is the probability of finding an electron at a specified distance (actually within a small spherical shell of radius r) from the nucleus?" The answer to this calculation is given in the **radial probability function,** which is plotted versus r in Figure 5–11a. As we can see from this curve, there is a very small probability of finding an electron in this 1s orbital very close to the nucleus. This probability increases to a maximum at a certain distance, denoted by a_0, and then gradually decreases again. For the hydrogen 1s orbital, the distance a_0 is 0.0592 nm.

Thus, the *most probable distance* from the nucleus for the electron is a_0, with slightly lower probabilities at distances that are slightly greater or less than a_0. There is a much lower, but still finite, chance of finding an electron at distances much less than, or much greater than, this preferred distance. Significantly, for the 1s orbital, the radial probability function is the same in *any direction* from the nucleus. Another way of saying this is that the 1s orbital is *spherically symmetrical with respect to the nucleus.*

Suppose we wish to draw a three-dimensional picture of this atomic orbital. We cannot draw any region where the electron *must be,* since it has some probability of being at any finite distance from the nucleus. What we usually do is draw a surface that connects points of equal probability and encloses a volume in which there is a high probability, say 90%, of finding the electron. Such a surface for the 1s orbital is a sphere (Figure 5–11b). In all of our subsequent discussions of "where" an electron in a particular orbital is, we shall rely on this pictorial representation.

Let us now see how the three geometrical aspects of an orbital—size, shape, and directional properties—depend, respectively, on the three quantum numbers n, ℓ, and m_ℓ:

1. The *size* of an orbital depends most on the value of n, the principal quantum number. For a given atom, an orbital with $n = 2$ has its region of maximum probability further from the nucleus than an orbital with $n = 1$, and not so far away as one with $n = 3$ or greater.

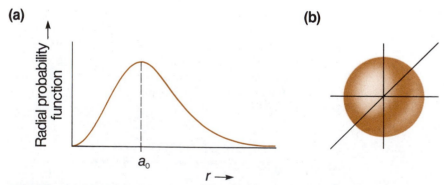

(a)

Radial probability function

a_0

$r \longrightarrow$

(b)

FIGURE 5–11 (a) The radial probability function for the 1s orbital of hydrogen, as a function of distance r from the nucleus. For an s orbital, these plots are the same for any direction from the nucleus. The radial probability function shows the probability of finding the electron in a thin spherical shell as a function of distance from the nucleus. (b) The shape of the region containing 90% probability of finding the 1s electron (i.e., the orbital shape). Note its spherical symmetry.

2. The *shape* of an orbital is determined by the value of ℓ, the subsidiary or orbital quantum number.

3. The *directional properties* of an orbital are described by the value of m_ℓ, the magnetic quantum number.

These latter two features, dependence on ℓ and on m_ℓ, must be discussed together:

a. $\ell = 0$ (*s* orbitals). Recall that when $\ell = 0$, the only possible value for m_ℓ is 0. As we have seen, the 1s orbital is spherically symmetrical with respect to the nucleus. Evaluation of the radial probability function for the combination $n = 2$, $\ell = 0$ (and, of course, $m_\ell = 0$) shows that the 2s orbital is also spherically symmetrical, but now the likelihood of finding an electron is highest at a greater distance from the nucleus than for the 1s orbital. The 3s orbital is larger, but still has the same shape. Compare Figures 5–12a, b, and c. In summary, all *s* orbitals ($\ell = 0$) are spherically symmetrical with respect to the nucleus.

b. $\ell = 1$ (*p* orbitals). Calculation of the probability functions that lead to the orbital shapes for $\ell = 1$ shows that these orbitals are *not* spherically symmetrical. This means that, at a given distance from the nucleus, the electron probability function may be different in different directions. The shape of any 2p orbital is a dumbbell-shaped region, as shown in Figure 5–12d. Notice that the probability of finding an electron in this orbital is highest in two opposite directions from the nucleus, and lowest (zero, in fact) in the plane perpendicular to these directions. A 2p orbital has its highest density at about the same distance from the nucleus as does the 2s orbital.

For the 2p orbitals (i.e., $n = 2$, $\ell = 1$), the probability calculations are carried out for each of the three possible values of m_ℓ (−1, 0, and +1). The resulting three orbitals have exactly the same shapes, but different orientations. For one of them, the two lobes of maximum probability are along the $+x$ and $-x$

A plane in which the probability of finding an electron is zero is called a nodal plane.

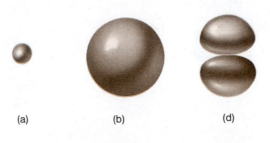

(a)   (b)   (d)

FIGURE 5–12 Shapes and approximate relative sizes of several orbitals in an atom. (a) 1s, (b) 2s, (c) 3s, (d) 2p, (e) 3p.

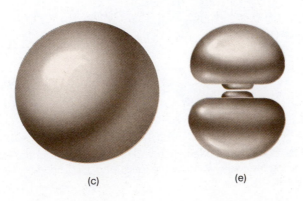

(c)   (e)

(a)

(b)

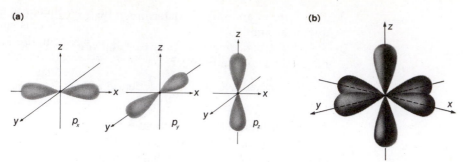

FIGURE 5–13 (a) Relative directional character of a set of atomic *p* orbitals. (b) A model of three *p* orbitals (p_x, p_y, and p_z) on a single set of axes. The nucleus is at the center. (The lobes are actually more diffuse than depicted—see Figures 5–12d and e for comparison.)

Even though these three orbitals are ordinarily equivalent, they become nonequivalent when the atom is in an unsymmetrical environment, such as a magnetic field.

axes; for another, the electron cloud is most dense in regions along the $+y$ and $-y$ axes, while the third orbital in this set has its high probability regions in the $+z$ and $-z$ directions. These three *p* orbitals are sometimes denoted the p_x, p_y, and p_z orbitals (Figure 5–13). They arise from the three possible values of m_ℓ, but do *not necessarily* correspond to the values -1, 0, and $+1$ in this order. The three orbitals are *equivalent* in shape and extent, and, as we shall see in Section 5–9, they are equivalent in energy as well, so which axis is denoted as *x*, which as *y*, and which as *z* is an arbitrary feature of the calculation.

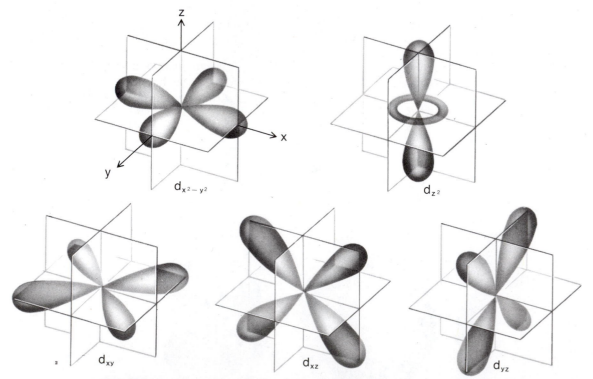

FIGURE 5–14 Spatial orientation of *d* orbitals. Note that the lobes of d_{z^2} and those of $d_{x^2-y^2}$ lie along the axes, whereas the lobes of d_{xy}, d_{xz}, and d_{yz} lie along diagonals between axes.

In the third main energy level ($n = 3$), there are again three equivalent p orbitals. Each $3p$ orbital, shown in Figure 5-12e, is similar in shape to the $2p$ orbitals shown in Figure 5-12d. In a given atom, the $3p$ orbitals are larger than the $2p$ orbitals, with the most dense part of each $3p$ orbital about as far from the nucleus as the maximum density of the $3s$ orbital.

c. $\ell = 2$ (d orbitals). When $\ell = 2$, there are five possible values of m_ℓ ($-2, -1, 0, +1, +2$). The shapes of the resulting five d orbitals in a set are shown in Figure 5-14. Even though the orbital denoted d_{z^2} has a different shape from the others, all five are energetically equivalent. According to the rules in Section 5-6, d orbitals can exist only in main energy levels with $n = 3$ or greater.

d. $\ell = 3$ (f orbitals). Since there are seven possible values of m_ℓ (can you list them?), there are seven equivalent orbitals in an f set. The value of n must be 4 or greater for f orbitals to exist.

Familiarity with this interpretation of the quantum numbers and with their alternative notations, such as $2s$, $2p$, and $3d$, will be necessary to understand such properties as atomic size and ease of removal of electrons (Chapter 6), as well as chemical bonding (Chapters 6 to 8).

5-9 Energies of Atomic Orbitals

The second major feature of atomic orbitals that can be calculated from the quantum mechanical equations is their *relative energies*. In this section, we shall schematically present the results of such calculations. Several features of this approach should be noted:

1. We are interested only in *relative energies* of the various orbitals within a given atom.

2. We shall present here only *approximate* relative energies, which can vary somewhat from atom to atom, depending on such factors as nuclear charge, type and occupancy of the other orbitals that contain electrons in the atom, and the electrical and magnetic environment of the atom.

3. A useful aspect is the approximate *differences* in energies between different kinds of atomic orbitals.

A convenient way to present the results of these calculations is on an energy level diagram such as Figure 5-15. This diagram is similar to the one we saw for the Bohr model of the atom in Figure 5-8b except that we now draw a short dash to represent each orbital. Only the vertical direction (energy) has any meaning in such diagrams. The lower an orbital appears, the more stable an electron would be in that orbital.

Of all the possible combinations of quantum numbers, we find that, for any atom, the combination that gives lowest energy is $n = 1$, $\ell = 0$, $m_\ell = 0$, or the $1s$ orbital. The orbital of next lowest energy for any atom is the $2s$, but please note that the $2s$ orbital is at considerably higher energy than the $1s$ orbital. When we carry out the energy calculations for the three orbitals of the $2p$ type, we find that they all have the same energy. We represent this on the energy level diagram by drawing three dashes, all at equal height. Remember that we earlier called such orbitals *a set of equivalent orbitals*. It is in this important aspect of energy that they are equivalent. For any atom, the $2p$ orbitals are at higher energy than the $2s$ orbital, but this energy gap is never as great as that between $1s$ and $2s$ orbitals.

Recall that the $2p$ orbitals all have $n = 2$, $\ell = 1$, and differ only in their values of m_ℓ (-1, 0, and $+1$).

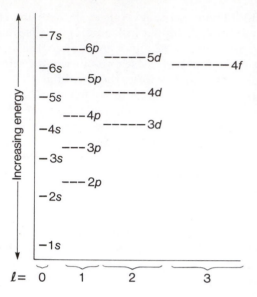

FIGURE 5–15 The usual order of energies of the orbitals of an atom. This is sometimes referred to as the *aufbau* order (Section 5–10). The energy scale varies for different elements, but the following main features should be noted: (1) The energies of orbitals are generally closer together at higher energies. (2) The largest energy gap is between 1s and 2s orbitals. (3) The gap between np and $(n + 1)s$, e.g., between 2p and 3s, is usually fairly large. (4) The gap between nd and $(n + 1)s$, e.g., between 3d and 4s, is quite small. (5) The gap between nf and $(n + 2)s$, e.g., between 4f and 6s, is even smaller.

The next lowest energy orbital is found to be the 3s orbital, at a considerably higher energy than the 2p set. This is followed by the set of three equivalent 3p orbitals. Next comes the 4s orbital, at an energy well above the 3p orbitals. Very close to this in energy, we find the set of five equivalent 3d orbitals. (Can you enumerate the n, ℓ, m_ℓ combinations that they represent?) Note that in this region of the energy level diagram (Figure 5–15), the energies of orbitals are very close together. (As we shall see in Section 5–10, the order of these closely spaced energy levels can vary from one atom to another.) Then, somewhat above these, we find the set of three equivalent 4p orbitals, and so on.

Note that the 4s orbital is usually of slightly lower energy than the 3d orbitals.

In Figure 5–15 we have depicted the approximate order of increasing energies of atomic orbitals. Even though there are several elements for which the order is somewhat different, it is useful to know the usual order of increasing orbital energies:

1s, 2s, 2p, 3s, 3p, 4s, 3d, 4p, 5s, 4d, 5p, 6s, 4f, 5d, 6p, 7s, 5f, 6d, 7p, 8s . . .

A device that can help you remember this order follows. (1) Write down the various orbitals in an array. This array is constructed by *rows,* with each different n value starting a new row, and with the ℓ values increasing left to right within a row. Indent each row by one position. (2) Read this array back by *columns,* as shown by the arrows.

5-10 The Electronic Structures of Atoms

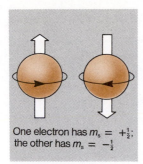

One electron has $m_s = +\frac{1}{2}$; the other has $m_s = -\frac{1}{2}$

FIGURE 5-16
Electron spin. Electrons act as though they spin about an axis through their centers. Because there are two directions in which an electron may spin, the spin quantum number has two possible values, +1/2 and −1/2. This electron spin produces a magnetic field.

Now that we have seen the qualitative interpretations of some calculations with the quantum mechanical equations, we shall use these results to describe the arrangements of electrons in atoms. These electronic arrangements are our key to understanding the physical and chemical properties of the elements and their compounds. Nothing that you have studied so far in chemistry has such far-reaching significance as the material in this section. Your diligent study and development of skill in determining these electronic arrangements will be well repaid in the next several chapters.

For each neutral atom, we must account for a number of electrons equal to the number of protons in the nucleus, i.e., the atomic number, Z, of the atom. Each electron is described by a different set of values for the four quantum numbers, n, ℓ, m_ℓ, and m_s. That is, each electron is in an orbital, which means that (1) it has a particular energy and (2) there is a high probability that it occupies a particular region in space. In addition, each electron has one of the two possible values of spin.

We haven't yet elaborated on the fourth quantum number, the spin quantum number m_s. Since electrons are negatively charged, and they behave as though they were spinning about axes through their centers, they act like tiny magnets. The spins of electrons produce magnetic fields, and these, like all magnetic fields, may interact so that they attract or repel one another. Two electrons may occupy the same region of space, the same orbital, only if they have opposite m_s values, so that their opposing magnetic fields attract one another (Figure 5-16). The magnetic fields of a pair of electrons *in the same orbital* that *have opposite values of m_s interact attractively.* Such electrons are said to be *spin-paired.* This magnetic interaction reduces the force with which the electrons (which are both negatively charged) repel each other, and thus lowers the energy of the atom.

Thus, we see the physical meaning of the *Pauli Exclusion Principle*, which was mentioned in Section 5-7, and which we restate here:

> No two electrons in an atom may have identical sets of four quantum numbers.

The significance of this principle is that each orbital in an atom can hold a maximum of two electrons, *provided* these two electrons have opposite spins. Knowing this fact and using the rules of Section 5-6, we can determine the largest possible number of electrons that can occupy each energy level as follows. The principal quantum number n indicates the energy level. The number of sublevels per energy level is equal to n, the number of atomic orbitals per energy level is n^2, and the maximum number of electrons per energy level is $2n^2$, since each atomic orbital can hold two electrons (Table 5-4).

We shall say much more about the periodic table in the next several chapters.

We shall next consider the elements in order of increasing atomic number, using as our guide the *periodic table* inside the front cover of this text. The approach that we shall use is to imagine "building up" the correct arrangement of electrons in each atom, by "putting in" one new electron at a time and using the idea that the most stable arrangement is the one that has the lowest total energy for the collection of electrons. This arrangement is called the **ground-state electronic configuration** of the element, to distinguish it from other possible arrangements that are not of lowest possible energy and that are called **excited state configurations.**

Usually when the term "electronic configuration" is used without qualification, it refers to the *ground-state configuration.*

TABLE 5–4   The Maximum Numbers of Electrons That Can Occupy Energy Levels (Through $n = 5$)

Energy Level n	Number of Sublevels per Energy Level n	Number of Atomic Orbitals n^2	Maximum Number of Electrons $2n^2$
1	1	1	2
		$\overbrace{(1s)}$	
2	2	4	8
		$\overbrace{(2s, 2p_x, 2p_y, 2p_z)}$	
3	3	9	18
		$\overbrace{(3s, \text{ three } 3p\text{'s, five } 3d\text{'s})}$	
4	4	16	32
5	5	25	50

The German verb "aufbauen" means "to build up or to construct."

The imaginary procedure of building up the electronic configurations one electron at a time is often referred to as the "aufbau" process. The **Aufbau Principle** states that the electron that differentiates an element from the preceding element (i.e., the "last" electron added) enters the available atomic orbital of lowest energy. The usual order of orbitals arranged by increasing energy, which we presented at the end of Section 5–9, is called the **aufbau order of orbitals.**

The first energy level contains only one atomic orbital, the $1s$ orbital, and can hold a maximum of two electrons. Hydrogen, $_1$H, contains just one electron. Helium, $_2$He, is a noble gas; its first energy level is *filled* with two electrons, and the atom is so stable that no chemical reactions of helium are known at the present time. Its electrons can be displaced only by electrical forces, as in excitation by high-voltage discharge.

In the simplified notation we represent with superscripts the number of electrons in each sublevel.

	1s	**Simplified Notation**
$_1$H	↑	$1s^1$
$_2$He	↑↓	$1s^2$

In representing these electronic structures, we have used two different notations. In the first one, each orbital is represented as a dash, with its occupancy shown as __ (no electrons), ↑ (an unpaired electron), or ↑↓ (a pair of spin-paired electrons). By the term "unpaired electron" we mean an electron that occupies an orbital singly. By "unpaired electrons" we mean two or more electrons that occupy different orbitals of a given sublevel singly, and hence electrons whose magnetic fields are not balanced. In a simplified notation, we denote each orbital or set of equivalent orbitals by the notation $1s$, $2p$, and so on, and then indicate with a superscript the number of electrons that occupy that orbital or that set of equivalent orbitals. The simplified notation cannot distinguish between paired and unpaired electrons, although the arrangement can often be deduced (as we will discuss shortly).

Elements of atomic numbers 3 through 10 occupy the second period, or horizontal row, in the periodic table and have the following electronic configurations.

In the simplified notation at the right, [He] is used to indicate two electrons in the 1s orbital, the *helium configuration*.

	1s	2s	2p		Simplified Notation
$_3$Li	↑↓	↑		$1s^2 2s^1$ or	[He] $2s^1$
$_4$Be	↑↓	↑↓		$1s^2 2s^2$	[He] $2s^2$
$_5$B	↑↓	↑↓	↑ __ __	$1s^2 2s^2 2p^1$	[He] $2s^2 2p^1$
$_6$C	↑↓	↑↓	↑ ↑ __	$1s^2 2s^2 2p^2$	[He] $2s^2 2p^2$
$_7$N	↑↓	↑↓	↑ ↑ ↑	$1s^2 2s^2 2p^3$	[He] $2s^2 2p^3$
$_8$O	↑↓	↑↓	↑↓ ↑ ↑	$1s^2 2s^2 2p^4$	[He] $2s^2 2p^4$
$_9$F	↑↓	↑↓	↑↓ ↑↓ ↑	$1s^2 2s^2 2p^5$	[He] $2s^2 2p^5$
$_{10}$Ne	↑↓	↑↓	↑↓ ↑↓ ↑↓	$1s^2 2s^2 2p^6$	[He] $2s^2 2p^6$

Some comments regarding these configurations are required. For lithium, $_3$Li, we find that only two electrons can occupy the 1s orbital without violating the Pauli Exclusion Principle, and the third electron must occupy the orbital of next lowest energy, the 2s orbital. In beryllium, Be, the fourth electron also goes into the 2s orbital, to complete a pair and to fill the 2s orbital.

When we reach boron, $_5$B, the fifth electron can go into any one of the three equivalent 2p orbitals. Since the three are equivalent, it is meaningless to ask whether this electron is in p_x, p_y, or p_z, or which value of m_ℓ corresponds to which dash in the diagram. Usually, though, we systematically fill the orbitals in our diagram "left-to-right," just to make it easier to keep track of them.

Putting the sixth electron into carbon, $_6$C, requires a different kind of decision. Do we put this sixth electron into the 2p orbital already singly occupied, or does it go into an unoccupied p orbital? The first of these two choices would require spin-paired 2p electrons, while the second would give two singly occupied 2p orbitals. Even with pairing of spins, however, two electrons that are in the same orbital repel each other more strongly than do two electrons in different (but equal-energy) orbitals. So, both theory and experimental observation (see the following box) lead to **Hund's Rule:**

Electrons occupy all the orbitals of a given sublevel *singly* before pairing begins.

Thus carbon has two unpaired electrons in its 2p orbitals.

Both paramagnetism and diamagnetism are hundreds to thousands of times weaker than *ferromagnetism,* the effect seen in iron bar magnets.

Substances that contain unpaired electrons are *attracted* weakly into magnetic fields and are said to be **paramagnetic.** By contrast, those in which all electrons are paired are *repelled* (much more weakly) by magnetic fields and are called **diamagnetic.** The paramagnetic effect can be measured by hanging a test tube full of a paramagnetic substance (such as copper sulfate) by a long thread, and bringing it near a strong magnet; the test tube will swing toward the magnet. Alternatively, the test tube can be hung just outside the gap of an electromagnet; when the current is switched on, the test tube will swing toward the magnet. If an electromagnet is placed below a sample suspended from the beam of a balance (Figure 5–17), the paramagnetic attraction per mole of substance can be measured by weighing the sample before and after energizing the magnet. The paramagnetism per mole increases with increasing number of unpaired electrons per formula unit.

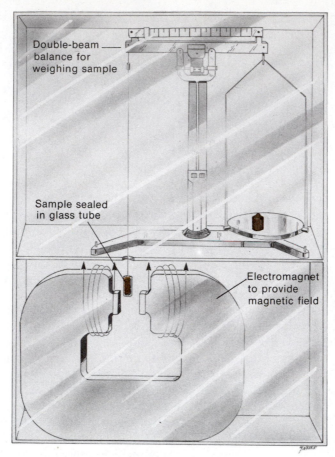

FIGURE 5–17 A magnetic balance, used to measure the magnetic properties of a compound. The sample is weighed first with the electromagnet turned off, and then a current is sent through the electromagnet, providing a magnetic field of known intensity. The balance then measures the force with which the sample is drawn into the magnetic field.

Double-beam balance for weighing sample

Sample sealed in glass tube

Electromagnet to provide magnetic field

Similar reasoning tells us that nitrogen, $_7N$, has three unpaired electrons. The eighth electron of oxygen, $_8O$, then goes into one of the already occupied $2p$ orbitals to make a pair. Continued application of this process lets us deduce the configurations of the other two elements of the second period, fluorine, $_9F$, and neon, $_{10}Ne$. The second energy level is filled completely in neon atoms. Neon is a noble gas, extremely stable, and no reactions of neon are known. However, as with helium, its electrons can be displaced by high-voltage electrical discharge, as is observed in neon signs.

In writing electronic structures of atoms, we frequently use abbreviated notations. In the preceding tabulation, the simplified notation [He] indicates that the $1s$ orbital is completely filled as in helium. In the following tabulation, [Ne] indicates that the $1s$, $2s$, and $2p$ sublevels of an atom are filled with 10 electrons as in neon.

The next element beyond neon is sodium, and we begin to add electrons to the third energy level. Elements 11 through 18 occupy the third period in the periodic table.

Although the third energy level is not yet filled (the d orbitals are still empty), argon is classified as a noble gas (an extremely stable, unreactive element). All noble gases, except helium in which the first energy level is filled with two electrons, have ns^2np^6 electronic configurations (where n indicates the highest occupied energy level), and all noble gases are extremely stable.

The third energy level contains five d atomic orbitals, in addition to the $3s$ and

	3s	3p	Simplified Notation
$_{11}$Na	[Ne] ↑		[Ne] $3s^1$
$_{12}$Mg	[Ne] ↑↓		[Ne] $3s^2$
$_{13}$Al	[Ne] ↑↓	↑ __ __	[Ne] $3s^2 3p^1$
$_{14}$Si	[Ne] ↑↓	↑ ↑ __	[Ne] $3s^2 3p^2$
$_{15}$P	[Ne] ↑↓	↑ ↑ ↑	[Ne] $3s^2 3p^3$
$_{16}$S	[Ne] ↑↓	↑↓ ↑ ↑	[Ne] $3s^2 3p^4$
$_{17}$Cl	[Ne] ↑↓	↑↓ ↑↓ ↑	[Ne] $3s^2 3p^5$
$_{18}$Ar	[Ne] ↑↓	↑↓ ↑↓ ↑↓	[Ne] $3s^2 3p^6$

$3p$ orbitals that are filled in argon and all subsequent elements. These $3d$ orbitals can hold ten electrons. However, in most atoms the $4s$ orbital has slightly lower energy than the $3d$ orbitals, and it is an experimentally observed fact that *an electron enters the available orbital of lowest energy.* Therefore, the $4s$ atomic orbital is filled before electrons enter the $3d$ orbitals.

When the $3d$ sublevel is full, the $4p$ orbitals fill next, followed by the $5s$ orbital and then the five $4d$ orbitals. The $5p$ orbitals fill next to take us to xenon, a noble gas. Remember that we are considering these orbitals to be filled in the order of increasing energy as presented in Section 5–9, the aufbau order.

Let us now examine the electronic structures of the 18 elements in the fourth period in some detail. Note that for the first time we encounter elements that have electrons in d orbitals.

	3d	4s	4p	Simplified Notation
$_{19}$K	[Ar]	↑		[Ar] $4s^1$
$_{20}$Ca	[Ar]	↑↓		[Ar] $4s^2$
$_{21}$Sc	[Ar] ↑ __ __ __ __	↑↓		[Ar] $3d^1 4s^2$
$_{22}$Ti	[Ar] ↑ ↑ __ __ __	↑↓		[Ar] $3d^2 4s^2$
$_{23}$V	[Ar] ↑ ↑ ↑ __ __	↑↓		[Ar] $3d^3 4s^2$
$_{24}$Cr	[Ar] ↑ ↑ ↑ ↑ ↑	↑		[Ar] $3d^5 4s^1$
$_{25}$Mn	[Ar] ↑ ↑ ↑ ↑ ↑	↑↓		[Ar] $3d^5 4s^2$
$_{26}$Fe	[Ar] ↑↓ ↑ ↑ ↑ ↑	↑↓		[Ar] $3d^6 4s^2$
$_{27}$Co	[Ar] ↑↓ ↑↓ ↑ ↑ ↑	↑↓		[Ar] $3d^7 4s^2$
$_{28}$Ni	[Ar] ↑↓ ↑↓ ↑↓ ↑ ↑	↑↓		[Ar] $3d^8 4s^2$
$_{29}$Cu	[Ar] ↑↓ ↑↓ ↑↓ ↑↓ ↑↓	↑		[Ar] $3d^{10} 4s^1$
$_{30}$Zn	[Ar] ↑↓ ↑↓ ↑↓ ↑↓ ↑↓	↑↓		[Ar] $3d^{10} 4s^2$
$_{31}$Ga	[Ar] ↑↓ ↑↓ ↑↓ ↑↓ ↑↓	↑↓	↑ __ __	[Ar] $3d^{10} 4s^2 4p^1$
$_{32}$Ge	[Ar] ↑↓ ↑↓ ↑↓ ↑↓ ↑↓	↑↓	↑ ↑ __	[Ar] $3d^{10} 4s^2 4p^2$
$_{33}$As	[Ar] ↑↓ ↑↓ ↑↓ ↑↓ ↑↓	↑↓	↑ ↑ ↑	[Ar] $3d^{10} 4s^2 4p^3$
$_{34}$Se	[Ar] ↑↓ ↑↓ ↑↓ ↑↓ ↑↓	↑↓	↑↓ ↑ ↑	[Ar] $3d^{10} 4s^2 4p^4$
$_{35}$Br	[Ar] ↑↓ ↑↓ ↑↓ ↑↓ ↑↓	↑↓	↑↓ ↑↓ ↑	[Ar] $3d^{10} 4s^2 4p^5$
$_{36}$Kr	[Ar] ↑↓ ↑↓ ↑↓ ↑↓ ↑↓	↑↓	↑↓ ↑↓ ↑↓	[Ar] $3d^{10} 4s^2 4p^6$

As you study these electronic configurations, you should be able to see how most of them are predicted from the Aufbau Principle, keeping in mind the order of filling atomic orbitals. However, as we proceed to fill the $3d$ set of orbitals, from $_{21}$Sc to $_{30}$Zn, we see that these orbitals are not filled quite regularly. As we mentioned in Section 5–9, some sets of orbitals are quite close in energy (e.g., $4s$ and $3d$), so that minor changes in their relative energies may occasionally change the order of filling.

Chemical and spectroscopic evidence indicate that the configurations of Cr and Cu have only one electron in the 4s orbital, and that their 3d sets are half-filled and filled, respectively, in the ground state. Calculations from the quantum mechanical equations also indicate that *a special stability is associated with half-filled and filled sets of equivalent orbitals.* Apparently this increased stability is sufficient to lower the 3d orbitals below the 4s orbital in energy, so that, in $_{24}$Cr for example, the total energy of [Ar] 3d ↑ ↑ ↑ ↑ ↑ 4s ↑ is lower than that for [Ar] 3d ↑ ↑ ↑ ↑ __ 4s ↑↓. Similar reasoning lets us rationalize the apparent exception of the configuration of $_{29}$Cu from that predicted by the Aufbau Principle. It might be asked why a similar exception does not occur in, for example, $_{32}$Ge or $_{14}$Si, where the possibility exists of an s^1p^3 configuration that would have a half-filled set of equivalent orbitals. The reason this does not occur is the very large energy gap between ns and np orbitals, in contrast to the very small gap between ns and $(n - 1)d$ orbitals (e.g., 4s and 3d orbitals). We shall see evidence in Chapter 6 that does, however, illustrate the enhanced stability of half-filled sets of p orbitals.

> A vertical column in the periodic table is called a group or a family of elements.

The periodic tables in this text are divided into "A" and "B" Groups. (See inside front cover.) The A Groups contain elements in which s and p orbitals are being filled. Elements within any particular A Group have similar electronic configurations and chemical properties, as we shall see in the next chapter. The B Groups are those in which there are one or two electrons in the s orbital of the highest occupied energy level, and d orbitals one energy level lower are being filled.

Observe that hydrogen, lithium, and sodium, elements of Group IA (the leftmost column of the periodic table), have a single electron in their outermost s orbital (ns^1). Beryllium and magnesium, of Group IIA, have two electrons in their highest energy level, ns^2, while boron and aluminum in Group IIIA have three electrons in their highest energy level, ns^2np^1. Similar observations can be made for each A Group.

The electronic configurations of the A Group elements and the noble gases can be predicted accurately from Figure 5–15 and the mnemonic device on page 156. However, there are some irregularities in the B Groups below the fourth period that are not easily predictable. In the heavier B Group elements, the higher energy sublevels in different principal energy levels have energies that are very nearly equal. It is easy for an electron to jump from one orbital to another of nearly the same energy in a different set because the orbitals are *perturbed* (their energies change slightly), as an extra electron is added in going from one element to the next.

The table inside the back cover gives electronic configurations for the known elements. The periodic table is a very useful guide for organization of chemical and physical properties of the elements, which depend, as we shall see, on electronic structure. The periodic table is also a convenient summary of electronic configurations of the elements. Your study of the important material in this section should include gaining an understanding of the relationship between the electronic configuration of an element and its position in the periodic table.

Key Terms

Absorption spectrum spectrum associated with absorption of electromagnetic radiation by atoms (or other species) resulting from transitions from lower to higher energy states.

Amplitude the maximum displacement of a point on a wave from its center, or equilibrium, position.

Atomic orbital region or volume in space in which the probability of finding electrons is highest.

Aufbau ("building up") Principle   describes the order in which electrons fill orbitals in atoms.

Blackbody radiator   an object that radiates energy with 100% efficiency.

Continuous spectrum   spectrum that contains all wavelengths in a specified region of the electromagnetic spectrum.

Degenerate   of the same energy.

Diamagnetism   *weak* repulsion by a magnetic field.

***d* orbitals**   beginning in the third energy level, a set of five degenerate orbitals per energy level, higher in energy than *s* and *p* orbitals of the same energy level.

Electromagnetic radiation   energy that is propagated by means of electric and magnetic fields that oscillate in directions perpendicular to the direction of travel of the energy.

Electron configuration   specific distribution of electrons in atomic orbitals of atoms or ions.

Electronic transition   the transfer of an electron from one energy level to another.

Emission spectrum   spectrum associated with emission of electromagnetic radiation by atoms (or other species) resulting from transitions from higher to lower energy states.

Equivalent orbitals   in an atom, a set of orbitals having the same energy; characterized by the same values of n and ℓ, with different values of m_ℓ.

Excited state   of an atom or molecule, any state other than the ground state.

***f* orbitals**   beginning in the fourth energy level, a set of seven degenerate orbitals per energy level, higher in energy than *s*, *p*, and *d* orbitals of the same energy level.

Frequency   the number of repeating corresponding points on a wave that pass a given observation point per unit time.

Ground state   of an atom or molecule, the lowest energy state or most stable state.

Group   a vertical column in the periodic table; also called a family.

Heisenberg Uncertainty Principle   it is impossible to determine accurately both the momentum and position of an electron simultaneously.

Hund's Rule   all orbitals of a given sublevel are occupied by single electrons before pairing begins. See *Aufbau Principle.*

Line spectrum   a spectrum containing only partic-

ular wavelengths in a specified region of the electromagnetic spectrum.

Magnetic quantum number (m_ℓ)   quantum mechanical solution to a wave equation that designates the particular orbital within a given set (s, p, d, f) in which an electron resides.

Pairing   of electrons, attractive interaction of the spin-induced magnetic fields of two electrons, of opposite spins, in the same orbital.

Paramagnetism   attraction toward a magnetic field, considerably stronger than diamagnetism, but still weak.

Pauli Exclusion Principle   no two electrons in the same atom may have identical sets of four quantum numbers.

Period   a horizontal row in the periodic table.

Photoelectric effect   emission of an electron from the surface of a metal caused by impinging electromagnetic radiation of certain minimum energy; current increases with increasing intensity of radiation.

Photon   a "packet" of light or electromagnetic radiation; also called a quantum of light.

***p* orbitals**   beginning with the second energy level, a set of three (per energy level) mutually perpendicular, equal-arm dumbbell-shaped atomic orbitals; higher in energy than the *s* orbital of the same energy level.

Principal quantum number (n)   quantum mechanical solution to a wave equation that designates the major energy level or shell in which an electron resides.

Quantum   a "packet" of energy.

Quantum mechanics   mathematical method of treating particles on the basis of quantum theory, which assumes that energy (of small particles) is not infinitely divisible.

Quantum numbers   numbers that describe the energies of electrons in atoms; derived from quantum mechanical treatment.

Radial probability function   function that describes variation in the probability of finding an electron at different radii from the nucleus.

Radiant energy   see *Electromagnetic radiation.*

Rydberg equation   empirical equation that relates wavelengths in the hydrogen emission spectrum to integers.

***s* orbital**   a spherically symmetrical atomic orbital; one per energy level.

Spectral line   any of a number of lines corresponding to definite wavelengths in an atomic emission or absorption spectrum; represents the energy difference between two energy levels.

Spectrophotometer   an instrument that measures the absorption or emission of electromagnetic radiation as a function of wavelength.

Spectrum   display of component wavelengths (colors) of electromagnetic radiation.

Spin quantum number (m_s)   quantum mechanical

solution to a wave equation that indicates the relative spins of electrons.

Subsidiary (azimuthal) quantum number (ℓ)   quantum mechanical solution to a wave equation that designates the sublevel or set of orbitals (s, p, d, f) within a given major energy level in which an electron resides.

Wavelength   the distance between two corresponding points of a wave.

Exercises

Electromagnetic Radiation

1. What is electromagnetic radiation?
2. Describe the following terms clearly and concisely in relation to electromagnetic radiation.
   (a) wavelength   (b) frequency
   (c) amplitude   (d) color
3. Distinguish between a traveling wave and a standing wave.
4. A series of waves on the surface of water is traveling toward the shore at a constant velocity of 150 ft/min and with a frequency of 6.5 wave crests per minute. What is the separation between successive wave crests, i.e., the wavelength of the wave?
5. Calculate the wavelengths, in meters, of radiation of the following frequencies.
   (a) 5.00×10^{15} s^{-1} (b) 1.36×10^{14} s^{-1}
   (c) 7.26×10^{12} s^{-1}
6. Calculate the frequencies of radiation of the following wavelengths.
   (a) 10,000 Å   (b) 520 nm
   (c) 5.30×10^{-5} cm   (d) 86.2 cm
   (e) 2.00×10^{-7} cm
7. What is the energy of a photon of each of the radiations in Exercise 6? Express your answer in joules per photon.
8. In which regions of the electromagnetic spectrum do the radiations in Exercise 6 fall?
9. Radio station WSB-AM in Atlanta broadcasts at a frequency of 750 kHz. What is the wavelength of its signal in meters?
10. Alpha Centauri is the star closest to our solar system. It is about 2.5×10^{13} miles away. How many light years is this? A light year is the distance that light travels (in a vacuum) in one year. Assume that space is essentially a vacuum.

The Photoelectric Effect

11. What evidence supports the idea that electromagnetic radiation is (a) wave-like, (b) particle-like?
12. Describe the influence of frequency and intensity of electromagnetic radiation on the current in the photoelectric effect.
13. Cesium is often used in "electric eyes" for self-opening doors in an application of the photoelectric effect. The amount of energy required to ionize (remove an electron from) a cesium atom is 3.89 electron volts. (One electron volt is 1.60×10^{-19} J.) What would be the wavelength, in nanometers, of light with just sufficient energy to ionize a cesium atom?

Atomic Spectra and the Bohr Theory

14. Distinguish between a continuous spectrum and an atomic spectrum.
15. Distinguish between an atomic emission spectrum and an atomic absorption spectrum.
16. What is the Rydberg equation? Why is it called an empirical equation?
17. The Lyman series is the name given to the series of lines in the ultraviolet portion of the emission spectrum of hydrogen. These lines correspond to transitions of electrons from higher energy states to the lowest ($n = 1$) energy state. Use the Rydberg equation to calculate the wavelengths of the three lowest energy lines in the Lyman series.
18. The Rydberg equation also can be written in terms of energies rather than wavelengths,

$$\Delta E = R' \left(\frac{1}{n_1{}^2} - \frac{1}{n_2{}^2} \right)$$

in which R' is a constant. Using the Rydberg

constant, $R = 1.097 \times 10^7$ m^{-1}, determine the value and units of R'. Use the unit of joules for energy.

19. Calculate the energies corresponding to the three lines described in Exercise 17 by using the equation in Exercise 18.

20. Repeat Exercise 19 using the Planck equation (Eq. 5–9) instead of the equation in Exercise 18.

21. The following are prominent lines in the visible region of the emission spectra of the elements listed. The lines can be used to identify the elements. What color is the light responsible for each line?
    (a) sodium, 546 nm   (b) neon, 616 nm
    (c) mercury, 454 nm

22. Outline the Bohr model of the hydrogen atom.

23. Why is the Bohr model for the hydrogen atom referred to as the solar system model?

24. Each of the series of lines (Lyman, Balmer, Paschen, etc.) in the emission spectrum of hydrogen has intense, widely spaced lines at higher wavelengths and less intense, closely spaced lines at lower wavelengths. Explain this appearance.

25. An atom emits a photon of light with wavelength 500 nm. How much energy in joules has the atom lost?

26. If each atom in one mole of atoms emits a photon of wavelength 500 nm, how much energy is lost? Express the answer in J/mol and kJ/mol. As a reference point, burning 1 mole (16 g) of CH_4 produces 819 kJ of heat.

27. The absorption spectrum of atomic hydrogen has a line at 121.6 nm. What is the frequency of the photons absorbed, and what is the energy difference, in joules, between the ground state and this excited state of atomic hydrogen?

The Wave–Particle View of the Electron

28. What evidence supports the idea that electrons are (a) particle-like, (b) wave-like?

29. What was de Broglie's contribution to our understanding of the nature of atoms?

30. Classical Newtonian mechanics cannot be applied to some kinds of systems. What kinds?

31. Using the de Broglie equation, calculate the wavelength of an electron moving at a velocity of 5.0×10^6 m/s. The mass of an electron is 9.11×10^{-31} kg, and 1 J = 1 kg·m^2/s^2. How does this wavelength compare with the diameter of a typical atom?

32. Using the de Broglie equation, show that the wavelength of a 0.13 kg baseball traveling at a velocity of 40 m/s (90 miles/hr) is so short as to be unobservable. How does this wavelength compare with the diameter of a typical atom?

The Quantum Mechanical Picture of Atoms and Atomic Orbitals

33. State the Heisenberg Uncertainty Principle. What does it mean, and what is its significance?

34. State the basic postulates of quantum mechanics and explain the meaning and significance of each.

35. What are atomic orbitals?

36. What is the relationship among the number of electrons in a neutral atom, its atomic number, and the number of protons in its nucleus?

37. Without giving the ranges of possible values of the four quantum numbers, n, ℓ, m_ℓ, and m_s, describe briefly what information each one gives.

38. How are the possible values for the subsidiary quantum number for a given electron restricted by the value of n?

39. How are the values of m_ℓ for a particular electron restricted by the value of ℓ?

40. What is the origin of the designation of atomic orbitals as s, p, d, and f orbitals?

41. What are the values of n and ℓ for the following sublevels? (a) $2s$, (b) $3p$, (c) $4s$, (d) $5d$, (e) $4f$

42. Compare the relative energies of the s, p, d, and f sublevels in a particular major energy level.

43. How many individual orbitals are there in the fourth major energy level? Write out n, ℓ, and m_ℓ quantum numbers for each one and label each set as describing s, p, d, or f orbitals.

44. What is Hund's Rule? What does it mean?

45. State the Pauli Exclusion Principle. What does it mean?

46. Consider the following sets of quantum numbers. Which sets represent impossible combinations? Indicate why they are impossible.

	n	ℓ	m_ℓ	m_s
(a)	1	0	1	+1/2
(b)	2	1	1	−1/2
(c)	3	2	−2	−1/2
(d)	3	3	0	+1/2
(e)	3	1	−2	−1/2
(f)	2	0	0	−1/2

47. What do we mean when we say that $a_0 = 0.0592$ nm for the hydrogen atom?

48. How many nodal planes are there for each of the following orbitals? (a) $2s$, (b) $2p_y$, (c) $3p_z$, (d) $3d_{xy}$, (e) $4d_{x^2-y^2}$, (f) $3d_{z^2}$

49. What do we mean when we say that the highest energy electron of a lithium atom in its ground state is in the $2s$ orbital?

50. What is the meaning of the following statement? The probability of finding an electron at any point is directly proportional to the electron density at that point.

51. Sketch the shapes of $1s$, $2s$, and $3s$ atomic orbitals on the same set of axes.

52. Describe a set of atomic p orbitals.

53. Complete the following table for the first five energy levels. What does each column indicate?

n	n^2	$2n^2$
1	1	2

54. What do we mean when we refer to the "spin of an electron"?

55. What are spin-paired electrons?

Electronic Configurations

56. What is the Aufbau Principle? What evidence do we have that it is valid?

57. Draw representations of ground-state electronic configurations, using ↑↓ notation, for the following elements. (a) He, (b) B, (c) Mg, (d) Ar, (e) Fe, (f) Sb

58. Give the ground-state electronic configurations for the elements of Exercise 57 using shorthand notation, as in $1s^2 2s^2 2p^6$.

59. Draw representations of ground-state electronic configurations using ↑↓ notation, for the following elements. (a) Li, (b) N, (c) Ca, (d) V, (e) Zn, (f) Mo, (g) Pb

60. Give the ground-state electronic configurations for the elements of Exercise 59 using shorthand notation, as in $1s^2 2s^2 2p^6$.

61. What do the electronic structures of elements 24 and 29 illustrate?

62. Why do electrons occupy the $4s$ orbital before they occupy the $3d$ orbitals in elements 19 and 20?

63. Distinguish between the terms diamagnetic and paramagnetic, and provide an example that illustrates the meaning of each.

64. How is paramagnetism measured experimentally?

65. Which of the elements in Exercise 57 are paramagnetic?

66. Which of the elements in Exercise 59 are paramagnetic?

67. Indicate whether each of the following elements is paramagnetic in its atomic ground state. (a) Be, (b) C, (c) Na, (d) Cl, (e) Ar, (f) Mn, (g) Cu, (h) Zn

68. Which one of the elements in Exercise 67 exhibits the highest degree of paramagnetism in its ground state?

69. Construct a table in which you list a possible set of values for the four quantum numbers for each electron in the following atoms in their ground states. (a) Be, (b) Al, (c) Mn

70. Construct a table in which you list a possible set of values for the four quantum numbers for each electron in the following atoms in their ground states. (a) N, (b) Cl, (c) Kr

71. List the n, ℓ, and m_ℓ quantum numbers for the highest energy electron (or one of the highest energy electrons if there are more than one) in the following atoms in their ground states. (a) Li, (b) S, (c) Co, (d) Sr

72. List n, ℓ, and m_ℓ quantum numbers for the highest energy electron (or one of the highest energy electrons if there are more than one) in the following atoms in their ground states. (a) C, (b) Ar, (c) Rh, (d) Ce

73. What is the distinction between the A and B Groups in the periodic table?

74. How do the periodic group numbers relate to the outermost electronic configurations of the elements in the A Groups of the periodic table?

75. Draw general electronic structures for the A Group elements using the ↑↓ notation, where n is the principal quantum number for the highest occupied energy level.

	ns	np		
IA	___			
IIA	___	___	___	___
etc.				

76. Repeat Exercise 75 using $ns^x np^y$ notation.

Periodicity and Chemical Bonding

As we look around us we see that some substances are gases, others are liquids, while still others are solids. Some solids are soft (paraffin), others are quite hard (diamond), and others are quite strong (steel). Clearly, there must be significant differences among the attractive forces that hold atoms together in different substances.

The formation of chemical bonds between atoms of different elements results in the formation of compounds. When the atoms are separated, chemical bonds are broken and the original compounds cease to exist. The attractive forces between atoms are electrical in nature, and chemical reactions between atoms involve changes in their electronic structures.

Chemical bonding theories are inadequate to explain all chemical behavior, but as we shall see, they are useful because they assist in systematizing much information.

In Chapters 4 and 5 we described the structures of atoms in some detail. An element's position in the periodic table is determined by its electronic configuration, which in turn plays a major role in determining the kinds of chemical bonds the element forms. Let us now turn our attention to the evolution of the periodic table, and to *chemical periodicity, the variation in properties of elements with their positions in the periodic table.* Then we shall introduce the two extremes in chemical bonding, ionic bonding and covalent bonding.

6–1   The Periodic Table

Pronounced "men-del-lay'-ev."

In 1869 the Russian chemist Dimitri Mendeleev and Lothar Meyer, a German, independently published arrangements of known elements that closely resemble the periodic table in use today. Mendeleev's classification was based primarily on chemical properties of the elements, while Meyer's classification was based largely on physical properties. The tabulations were surprisingly similar and both emphasized the **periodicity,** or regular repetition, of properties with increasing atomic weight.

Mendeleev arranged the known elements in order of increasing atomic weight in successive sequences so that elements with similar chemical properties fell in the same column. *He noted that both physical and chemical properties of the elements vary in a periodic fashion.* This statement is known as the **periodic law.** His periodic table of 1872, which contained the 62 elements known then, is shown in Figure 6–1.

Consider H, Li, Na, and K, all of which appear in "Gruppe I" of Mendeleev's table. All were known to combine with F, Cl, Br, and I of "Gruppe VII" to produce compounds that have similar formulas such as HCl, LiCl, NaBr, and KI. All these compounds dissolve in water to produce solutions that conduct electricity well. The

FIGURE 6–1 Part of Mendeleev's early periodic table, published in 1872.

TABELLE II

REIHEN	GRUPPE I. — R^2O	GRUPPE II. — RO	GRUPPE III. — R^2O^3	GRUPPE IV. RH^4 RO^2	GRUPPE V. RH^3 R^2O^5	GRUPPE VI. RH^2 RO^3	GRUPPE VII. RH R^2O^7	GRUPPE VIII. — RO^4
1	H=1							
2	Li=7	Be=9,4	B=11	C=12	N=14	O=16	F=19	
3	Na=23	Mg=24	Al=27,3	Si=28	P=31	S=32	Cl=35,5	
4	K=39	Ca=40	—=44	Ti=48	V=51	Cr=52	Mn=55	Fe=56, Co=59, Ni=59, Cu=63
5	(Cu=63)	Zn=65	—=68	—=72	As=75	Se=78	Br=80	
6	Rb=85	Sr=87	?Yt=88	Zr=90	Nb=94	Mo=96	—=100	Ru=104, Rh=104, Pd=106, Ag=108.
7	(Ag=108)	Cd=112	In=113	Sn=118	Sb=122	Te=125	J=127	
8	Cs=133	Ba=137	?Di=138	?Ce=140	—	—	—	— — —
9	(—)	—	—	—	—	—	—	
10	—	—	?Er=178	?La=180	Ta=182	W=184	—	Os=195, Ir=197, Pt=198, Au=199.
11	(Au=199)	Hg=200	Tl=204	Pb=207	Bi=208	—	—	
12	—	—	—	Th=231	—	U=240	—	— — —

TABLE 6–1 Predicted and Observed Properties of Germanium

Property	Eka-Silicon Predicted, 1871	Germanium Reported, 1886	Modern Values
Atomic weight	72	72.32	72.59
Atomic volume	13 cm³	13.22 cm³	13.5 cm³
Specific gravity	5.5	5.47	5.35
Specific heat	0.073 cal/g·°C	0.076 cal/g·°C	0.074 cal/g·°C
Maximum valence*	4	4	4
Color	Dark gray	Grayish white	Grayish white
Reaction with water	Will decompose steam with difficulty	Does not decompose water	Does not decompose water
Reactions with acids and alkalis	Slight with acids; more pronounced with alkalis	Not attacked by HCl or dilute aqueous NaOH; reacts vigorously with molten NaOH	Not dissolved by HCl or H_2SO_4 or dilute NaOH; dissolved by concentrated NaOH
Formula of oxide	EsO_2	GeO_2	GeO_2
Specific gravity of oxide	4.7	4.703	4.228
Specific gravity of tetrachloride	1.9 at 0°C	1.887 at 18°C	1.8443
Boiling point of tetrachloride	100°C	86°C	84°C
Boiling point of tetraethyl derivative	160°C	160°C	186°C

* Valence refers to the number of atoms bonded to a specific atom.

"Gruppe II" elements form compounds such as $BeCl_2$, $MgBr_2$, and $CaCl_2$, as well as compounds with O and S from "Gruppe VI" such as MgO, CaO, MgS, and CaS.

One of the advantages of Mendeleev's periodic table was that it provided for elements that had not yet been discovered. When he encountered "missing" elements, he left blank spaces. Some appreciation of Mendeleev's genius in constructing the table as he did can be gained by examining a comparison of predicted (1871) and observed properties of germanium, which was not discovered until 1886. Mendeleev called the undiscovered element eka-silicon because it fell below silicon in his table. He was familiar with the properties of germanium's neighboring elements, and these served as the basis for his predictions of properties of germanium (Table 6–1). The modern values of properties of germanium differ significantly from those reported in 1886, because many of the data available to Mendeleev were inaccurate.

At this point we might observe that in many areas of human endeavor progress is slow and faltering. However, in most areas there is an occasional individual who is able to develop concepts and techniques that clarify previously confused situations. Mendeleev was such an individual. Unfortunately, although his work was of fundamental importance, it did not clarify entirely the concept of periodicity. Because Mendeleev's arrangement of the elements was based on increasing *atomic weights*, several elements appeared to be out of place in the table. Mendeleev put the controversial elements (Te and I, Co and Ni, and after the noble gases were discovered, Ar and K) in locations consistent with their properties, and attributed the apparent reversal of atomic weights to inaccurate values for those weights. Careful redetermination of the atomic weights showed that these values were correct in the first place. Indeed, resolution of the problem of these out-of-place elements had to await the development and clarification of the concept of *atomic number*, after which the periodic law was evolved in essentially its present form.

As the **periodic law** is now stated, *the properties of the elements are periodic* *functions of their atomic numbers.* This statement tells us that if we arrange the elements in order of increasing atomic number, periodically we encounter elements that have similar chemical and physical properties. The presently used "long form" of the periodic table (Tables 6–2 and 6–3) is such an arrangement. The vertical columns are referred to as **groups** or **families,** and the horizontal rows are referred to as **periods.** Elements in a *group* have similar chemical and physical properties, while those within a *period* have properties that change progressively across the table. Several groups of elements have common names that are used so frequently they should be learned. The Group IA elements (except H) are referred to as **alkali metals,** and the Group IIA elements are called the **alkaline earth metals.** The Group VIIA elements are called **halogens,** which means "salt formers," and the Group 0 elements are called **noble** (or **rare**) **gases.**

Alkaline means basic. The character of basic compounds will be described in Section 9–4.3.

Differences Between A and B Groups

The A and B designations for groups of elements in the periodic table are arbitrary designations, and they are reversed in some periodic tables. The system used in this text is the one commonly used in the United States. Elements with the same group numbers, but with different letter designations, have relatively few similar properties. The origin of this notation is the fact that some compounds of elements with the same group numbers but different letter designations have similar formulas [e.g., $NaCl$ (IA) and $AgCl$ (IB), $MgCl_2$ (IIA) and $ZnCl_2$ (IIB)]. As we shall see, variations in the properties of the B Group elements across a row are not nearly as dramatic as the variations observed across a row of A Group elements because electrons are being added to the $(n-1)$ *d* orbitals, where *n* represents the highest energy level that contains electrons. Since the *outermost* electrons have the greatest influence on the properties of elements, adding an electron to an *inner d* orbital results in less striking changes in properties than adding an electron to an *outer s* or *p* sublevel.

Elements may be classified in a number of ways. One of the most useful classifications follows.

Noble Gases   For many years the Group 0 elements were called *inert* gases because no chemical reactions involving these elements were known, but we now know that the heavier members do form compounds, mostly with fluorine and oxygen. With the exception of helium (in which the first energy level is filled with two electrons), these elements have eight electrons in the highest energy level. Their structures may be represented as . . . ns^2np^6.

Representative Elements   The A Group elements in the periodic table are known as representative elements. They are characterized by unfilled highest occupied energy levels, in which their "last" electron is added to an *s* or *p* orbital. The elements generally show distinct and fairly regular variations in their properties with atomic number.

***d*-Transition Elements**   Elements in the B Groups (except IIB) in the periodic table are known as *d*-transition elements, or more simply as transition elements or transition metals. They were considered as transitions between the alkaline elements (base formers) on the left and the acid formers on the right. All of them are metals and are characterized by electrons being added to *d* orbitals. Stated differ-

TABLE 6–2   Periodic Table, the Long-period Form*

REPRESENTATIVE ELEMENTS

GROUPS

REPRESENTATIVE ELEMENTS

																		VIIA	0

PERIODS

Period	IA	IIA	IIIB	IVB	VB	VIB	VIIB	VIII	VIII	VIII	IB	IIB	IIIA	IVA	VA	VIA	VIIA	0
1	1 H 1766																	2 He 1895
2	3 Li 1817	4 Be 1798											5 B 1808	6 C	7 N 1772	8 O 1772	9 F 1886	10 Ne 1898
3	11 Na 1807	12 Mg 1755											13 Al 1827	14 Si 1823	15 P 1669	16 S	17 Cl 1774	18 Ar 1894
4	19 K 1807	20 Ca 1808	21 Sc 1879	22 Ti 1791	23 V 1830	24 Cr 1797	25 Mn 1774	26 Fe	27 Co 1735	28 Ni 1751	29 Cu	30 Zn 1746	31 Ga 1875	32 Ge 1886	33 As 1817	34 Se 1817	35 Br 1826	36 Kr 1898
5	37 Rb 1861	38 Sr 1790	39 Y 1794	40 Zr 1789	41 Nb 1801	42 Mo 1778	43 Tc 1937	44 Ru 1844	45 Rh 1803	46 Pd 1803	47 Ag	48 Cd 1817	49 In 1863	50 Sn	51 Sb	52 Te 1782	53 I 1811	54 Xe 1898
6	55 Cs 1860	56 Ba 1808	57 La 1839	58→71 Ce→Lu	72 Hf 1923	73 Ta 1802	74 W 1781	75 Re 1925	76 Os 1803	77 Ir 1803	78 Pt 1735	79 Au	80 Hg	81 Tl 1861	82 Pb	83 Bi	84 Po 1898	85 At 1940	86 Rn 1900
7	87 Fr 1939	88 Ra 1898	89 Ac 1899	90→103 Th→Lr	104 Unq 1965	105 Unp 1970	106 Unh 1974	107 Uns 1976	108	109 1982	110	111	112	113	114	115	116	117	118
8			121																

═══ TRANSITION ELEMENTS ═══

			LANTHANIDE SERIES	58 Ce 1803	59 Pr 1885	60 Nd 1843	61 Pm 1947	62 Sm 1879	63 Eu 1896	64 Gd 1880	65 Tb 1843	66 Dy 1886	67 Ho 1879	68 Er 1843	69 Tm 1879	70 Yb 1878	71 Lu 1907

7	ACTINIDE SERIES	90 Th 1828	91 Pa 1917	92 U 1789	93 Np 1940	94 Pu 1940	95 Am 1945	96 Cm 1944	97 Bk 1950	98 Cf 1950	99 Es 1952	100 Fm 1953	101 Md 1955	102 No 1958	103 Lr 1961

8		122	123	124	125	126	127	128	129	130	131	132	133	134	135

* The heavy "stairstep" line approximately separates the metallic elements, to the left, from the nonmetallic elements, to the right. Besides the atomic numbers, the dates of discovery of the elements are given. The elements with no given dates were known to the ancients. Positions of yet undiscovered elements up to 135 are shown.

ently, the *d*-transition elements are building an inner (next to highest occupied) energy level from eight to 18 electrons. They are referred to as

1st Transition Series: $_{21}$Sc through $_{29}$Cu
2nd Transition Series: $_{39}$Y through $_{47}$Ag
3rd Transition Series: $_{57}$La and $_{72}$Hf through $_{79}$Au
4th Transition Series: (not complete) $_{89}$Ac and elements 104 through 111

Strictly speaking, the Group IIB elements, zinc, cadmium, and mercury, are not *d*-transition metals since their "last" electrons go into *s* orbitals; but they are usually discussed with the *d*-transition metals because their chemical properties are similar.

Inner Transition Elements   These elements are sometimes known as *f*-transition elements. They are elements in which electrons are being added to *f* orbitals, and those in which the second from the highest occupied energy level is building from 18 to 32 electrons. All are metals. The inner transition elements are located between Groups IIIB and IVB in the periodic table, and are usually shown in separate rows below the main part of the table.

TABLE 6–3   The s, p, d, and f Blocks of the Periodic Table*

* *n* is the principal quantum number. The d^1s^2, d^2s^2, . . . designations represent known configurations. Designations like d^1s^2 refer to $(n-1)d$ and ns orbitals. Several exceptions to expected configurations are highlighted.

1st Inner Transition Series (**lanthanides**): $_{58}$Ce through $_{71}$Lu
2nd Inner Transition Series (**actinides**): $_{90}$Th through $_{103}$Lr

Table 6–3 shows the divisions of the periodic table according to the kinds of atomic orbitals occupied by the "last added" electron.

6–2   Lewis Dot Representations of Elements

As should become obvious in the next several chapters, the number and arrangement of electrons in the outermost shell determine the chemical and physical properties of the elements as well as the kinds of chemical bonds they form. A convenient bookkeeping method for keeping track of these "chemically important electrons" is the writing of *Lewis dot representations* of the elements, as follows.

We first write the electronic configuration for the element, as studied in Chapter 5. We then determine the "last" noble gas configuration of electrons, and the "last" completely filled *d* and *f* sublevels, *if there are any.* These electrons constitute very stable groups of electrons, which rarely are involved in chemical bonding. These electrons, together with the nucleus, make up the **kernel,** while the remaining, chemically significant electrons in a partially filled outer shell are called the **valence electrons.** In the *Lewis dot representation* of an element, then, *the kernel is represented by the element's symbol, while each valence electron is represented by a dot.* We shall restrict our attention to the representative elements; their valence electrons are simply the ones in the outermost *s* and *p* sublevels. Let us illustrate this procedure for two elements.

Sodium, Na   First we write the electronic configuration, most conveniently in the condensed format.

$$_{11}Na \qquad \underleftarrow{ 1s^2 2s^2 2p^6 } \Big| \underrightarrow{ 3s^1 }$$

$$\qquad\qquad\qquad \text{kernel} \qquad \text{valence electron}$$

The last noble gas configuration in sodium ends with $2p^6$, the neon configuration. Thus, Na has *one valence electron*, its lone $3s$ electron. The Lewis dot formula Na$\cdot$ represents sodium. There is no significance to where (above, below, to the left or right of the symbol) we write the dot — this is a device to aid us in *counting* electrons.

Arsenic, As   Again we begin by writing the electronic configuration:

$$_{33}As \qquad 1s^2 2s^2 2p^6 3s^2 3p^6 4s^2 3d^{10} 4p^3$$

It is important to remember (Section 5 – 10) that a filled set of d orbitals, $3d^{10}$, is quite stable, so that these electrons are unlikely to be involved in bonding. Accordingly, it is convenient to rewrite the electron configuration by collecting together all orbitals having the same value of n (recall that we did this in Chapter 5):

$$_{33}As \qquad \underleftarrow{ 1s^2 2s^2 2p^6 3s^2 3p^6 3d^{10} } \Big| \underrightarrow{ 4s^2 4p^3 }$$

$$\qquad\qquad\qquad \text{kernel electrons} \qquad \text{valence electrons}$$

Now we see that the very stable inner core of electrons includes not only the last noble gas configuration, but also the $3d^{10}$ electrons, so that arsenic has five valence electrons — $4s^2 4p^3$. We write the Lewis dot formula for As as the symbol surrounded by five dots:

paired $4s^2$ electrons $\curvearrowright \rightarrow : \overset{\displaystyle\cdot}{\underset{\displaystyle\cdot}{As}} \cdot \leftarrow$ unpaired $4p^3$ electrons

Notice that paired and unpaired valence electrons also are indicated. All elements in a given A group have the same configuration in their outer (valence) shells, so that they have the same number of dots in their Lewis dot formulas. It will be helpful to remember that the representative elements, except for the noble gases, have the same number of valence electrons (dots in the Lewis formula) as their group number in the periodic table. Thus, since N, P, As, Sb, and Bi all are in Group VA, each has five valence electrons and $: \overset{\cdot}{X} \cdot$ as a dot formula.

Table 6 – 3 shows electronic configurations of the elements in condensed form and Table 6 – 4 gives the Lewis dot representations of the representative elements. Since the purpose of writing such formulas is to provide a convenient way to count electrons, we usually do not write dot formulas for transition or inner transition elements. The large number of dots might become confusing.

6 – 3   Metals, Nonmetals, and Metalloids

In a very general classification scheme, the elements are divided into three classes: *metals, nonmetals,* and *metalloids*. The periodic tables inside the front cover and in Table 6 – 2 show this classification. The elements to the left of those touching the heavy stairstep line (except hydrogen) are **metals,** while those to the right are

About 80% of the elements are metals.

TABLE 6–4   Electron Dot Formulas of the Representative Elements

Group	IA	IIA	IIIA	IVA	VA	VIA	VIIA	0
Number of Electrons in Outer Shell	1	2	3	4	5	6	7	8 (except He)
	H·							He:
	Li·	Be:	B̈·	C̈·	·N̈·	·Ö:	·F̈:	:N̈e:
	Na·	Mg:	Äl·	S̈i·	·P̈·	·S̈:	·C̈l:	:Är:
	K·	Ca:	G̈a·	G̈e·	·Äs·	·S̈e:	·B̈r:	:K̈r:
	Rb·	Sr:	Ïn·	S̈n·	·S̈b·	·T̈e:	·Ï:	:Xe:
	Cs·	Ba:	T̈l·	P̈b·	·B̈i·	·P̈o:	·Ät:	:R̈n:
	Fr·	Ra:						

nonmetals. Such a classification is somewhat arbitrary, and several elements do not fit neatly into either class. The elements adjacent to the heavy line are often called **metalloids** (or *semimetals*), because they show some properties that are characteristic of both metals and nonmetals.

The physical properties that distinguish between metals and nonmetals are summarized in Table 6–5. The most important thing to notice is that the general properties of metals and nonmetals are opposite. Not all metals and nonmetals possess all these properties, but they share most of them to varying degrees. The physical properties of metals can be explained on the basis of metallic bonding within the solids, which will be discussed in Section 12–15. The strength of metallic bonding is influenced by the number of electrons, especially the unpaired electrons, beyond the "last" noble gas core.

The contrasting chemical properties of metals and nonmetals will be summarized later in this chapter, after a discussion of the variations of the properties of elements with position in the periodic table. The most important single factor in determining chemical properties of metals and nonmetals is the number of valence electrons.

As we pointed out earlier, *metalloids* show some properties that are characteristic of both metals and nonmetals. Many of the metalloids, such as silicon, germanium, and antimony, act as *semiconductors*, which are important in solid state electronic circuits. **Semiconductors** *do not conduct electrical current at room temperature but do so at higher temperatures* (Section 12–16).

Aluminum, silicon, germanium, and arsenic are examples of metalloids. All are metallic in appearance.

TABLE 6–5   Physical Properties of Metals and Nonmetals

Metals	Nonmetals
1. High electrical conductivity	1. Poor electrical conductivity
2. High thermal conductivity	2. Good heat insulators
3. Metallic gray or silver luster*	3. No metallic luster
4. Almost all are solids†	4. Solids, liquids, or gases
5. Malleable (can be hammered into sheets)	5. Brittle in solid state
6. Ductile (can be drawn into wires)	6. Nonductile
7. Solid state characterized by metallic bonding	7. Covalently bonded molecules; noble gases are monatomic

* Except copper and gold.

† Except mercury; cesium and gallium melt in protected hand.

Periodic Trends in Properties of Elements

We have learned that physical and chemical properties of the elements recur periodically. Let us now investigate the nature of this *periodicity* in some detail, since some knowledge of periodicity is valuable in understanding bonding in simple compounds. Many physical properties, such as melting points, boiling points, and densities, show periodic variations, but for the moment let us concern ourselves with the variations that are most useful in predicting chemical properties.

As we study trends in the properties of elements, here and in Chapter 9, we shall be concerned with three aspects: (1) *describing* the trends with position in the periodic table, based upon observations, (2) *understanding* (or at least *rationalizing*) the trends by correlating them with atomic properties such as electronic configurations, and (3) *predicting* properties of elements from a knowledge of properties of nearby elements as well as trends within the periodic table.

In the next several sections, we shall compare some particular property for two or more elements or ions, and consider how that property changes with position in the periodic table. Our understanding of these comparisons will usually involve considering the following aspects of the electronic configurations of the elements or ions:

1. How many valence electrons are present?
2. What is the outermost occupied shell—i.e., what is the highest value of n present in each configuration?
3. How many paired and unpaired valence electrons are present in this shell?
4. Are the "new" electrons being added to an inner shell, as in the transition and inner transition elements, or to an outer shell?
5. How do the nuclear charges differ?

6–4 Atomic Radii

It is impossible to describe an atom as having an invariant size because the size of an atom is determined by its interaction with surrounding atoms. Additionally, we cannot isolate an individual atom and measure its diameter the way we can measure the diameter of a golf ball. An indirect approach is required. By analogy, suppose we arrange a specified number of golf balls in a box in an orderly array. If we know how the balls are positioned, the number of balls, and the dimensions of the box, it is possible to calculate the diameter of an individual ball. In determining the "sizes" of atoms, the picture is complicated by the diffuse nature of the electron cloud surrounding an atom and by variations in the size of this cloud with environmental factors. Recognizing, then, that for all practical purposes the size of an individual atom is a quantity that cannot be defined accurately, we can make measurements on samples of pure elements. The data obtained from such measurements indicate the *relative* sizes of individual atoms. Figure 6–2 displays the relative sizes of atoms of the representative elements and the noble gases. It clearly indicates periodicity in atomic radii. [Atomic radii are often stated in the SI unit **nanometers** (1 nm = 10^{-9} m) or in **angstroms** (1 Å = 10^{-10} m).]

As we move from left to right across a period in the periodic table, there is a regular decrease in atomic radii of the representative elements as electrons are added to a

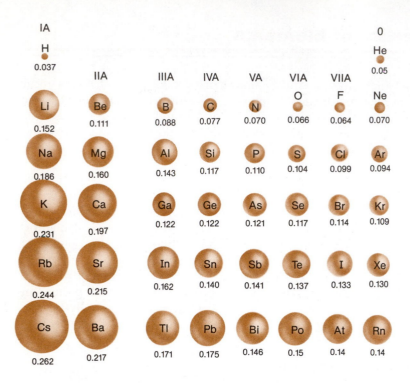

FIGURE 6–2 Atomic radii of the A Group (representative) elements and the noble gases in nanometers. Atomic radii *increase as a group is descended* because electrons are being added to higher energy levels. Atomic radii *decrease from left to right within a given row* due to increasing effective nuclear charge. Hydrogen atoms are the smallest and cesium atoms are the largest.

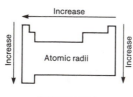

General trends in atomic radii of A Group elements with position in periodic table

particular energy level. As the nuclear charge increases and electrons are added to the *same principal energy level,* the increased nuclear charge draws the electron cloud closer and closer to the nucleus. *As we move down a group, atomic radii increase as electrons are added to larger orbitals in higher energy levels.* For the transition elements, the variations are not so regular because the "last" electrons are being added to an inner shell, but all transition elements have radii smaller than those of the preceding Group IA and Group IIA elements in the same period.

The observed decrease in atomic radii *from left to right across a row of representative elements* is accounted for by **nuclear shielding** or the **shielding effect,** which in turn can be explained in terms of the effective nuclear charge of an atom. The **effective nuclear charge,** Z_{eff}, may be thought of as the difference between the nuclear charge, Z, and the effective number of electrons, S, that shield the outermost electrons from the nuclear charge.

$$Z_{eff} = Z - S$$

[Eq. 6–1]

Electrons in filled sets of s and p orbitals (i.e., in noble gas configurations) between the nucleus and the outer shell electrons fairly effectively shield the outer shell electrons from the effects of an equal number of protons in the nucleus. For instance, a lithium atom contains three electrons, $1s^2 2s^1$. The two electrons in the filled $1s$ orbital fairly effectively screen out the effect on the $2s$ electron of two protons in the nucleus. Thus, the single electron in the $2s$ orbital behaves as if it were under the influence of only slightly more than one proton; the *effective nuclear charge of lithium is somewhat larger than 1+.* As a second example, a beryllium atom contains four electrons, $1s^2 2s^2$. Again the two electrons in the $1s$ orbital nearly shield the valence electrons from the effect of two protons in the nucleus, and the two electrons in the $2s$ orbital behave as if they were under the influence of only about two protons.

Beryllium atoms ($r = 0.111$ nm) are smaller than Li atoms ($r = 0.152$ nm) because the *effective nuclear charge* of Be is larger than that of Li and their valence electrons are in the *same* shell.

As a *group* is descended in the periodic table, the effective nuclear charge increases somewhat, but the outermost electrons are in orbitals of higher principal quantum number, n, and so are further away from the nucleus. The force of attraction of a positively charged nucleus for an electron decreases as the *square* of distance separating them increases: $F \propto 1/d^2$. (The $\propto$ means "is proportional to.") This effect dominates in determining vertical trends in radii. Thus, sodium has the configuration $1s^2 2s^2 2p^6 3s^1$, in which there are ten electrons in filled sets of s and p orbitals in the inner shells, and one outer shell electron. The effective nuclear charge of sodium is somewhat larger than that of lithium, but its radius, 0.186 nm, is larger than that of lithium, 0.152 nm, because the valence electron of sodium is in the third shell while that of lithium is in the second shell.

As we shall see next, the radius of an atom is an especially important factor in determining the ease with which the atom gains or loses electrons to form ions, and therefore in determining its chemical properties.

6–5 Ionization Energy

The **first ionization energy** of an element *is the minimum amount of energy required to remove the most loosely bound electron from an isolated gaseous atom to form an ion with a $1+$ charge.* For calcium, for example, the first ionization energy, IE_1, is 589.8 kJ/mol.

$$\text{Ca (g)} + 589.8 \text{ kJ} \longrightarrow \text{Ca}^+ \text{(g)} + e^- \qquad \text{[Eq. 6–2]}$$

The **second ionization energy** is the amount of energy required to remove the second electron from the *gaseous* ion, which may be represented for calcium as

$$\text{Ca}^+ \text{(g)} + 1145 \text{ kJ} \longrightarrow \text{Ca}^{2+} \text{(g)} + e^- \qquad \text{[Eq. 6–3]}$$

For a given element, IE_2 *is always greater than* IE_1 because it is always more difficult to remove an electron from an ion with a positive charge than from the corresponding neutral atom. Table 6–6 shows successive ionization energies for the first 13 elements.

Ionization energies measure how tightly electrons are bound to atoms. Note that ionization is an *endothermic* process; i.e., it requires energy to remove an electron from the attractive force of the nucleus. Low ionization energies indicate ease of removal of electrons, and hence ease of positive ion (cation) formation. Figure 6–3 shows a plot of first ionization energy versus atomic number for several elements. In general, *first ionization energies increase from left to right and bottom to top of the periodic table, but there are some exceptions.*

Note also that in each periodic segment of Figure 6–3, the noble gases have the highest first ionization energies. This should not be surprising, since the noble gases are known to be very stable elements, and only the heavier members of the family show any tendency to combine with other elements to form compounds. It requires more energy to remove an electron from a helium atom (slightly less than 3.940×10^{-18} J/atom or 2372 kJ/mol) than from an atom of any other element.

$$\text{He (g)} + 2372 \text{ kJ} \longrightarrow \text{He}^+ \text{(g)} + e^- \qquad \text{[Eq. 6–4]}$$

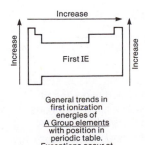

General trends in first ionization energies of A Group elements with position in periodic table. Exceptions occur at Groups IIIA and VIA.

TABLE 6–6   Ionization Energies of Several Elements (kJ/mol)

	First	Second	Third	Fourth	Fifth	Sixth	Seventh	Eighth
H	1,312							
He	2,372	5,250						
Li	520	7,298	11,815					
Be	899	1,757	14,849	21,000				
B	801	2,427	3,660	25,020	32,810			
C	1,086	2,353	4,620	6,223	37,800	47,300		
N	1,402	2,856	4,577	7,473	9,443	53,250	64,340	
O	1,314	3,388	5,301	7,468	10,990	13,320	71,300	84,050
F	1,681	3,374	6,045	8,418	11,020	15,164	17,860	92,000
Ne	2,081	3,952	6,276	9,376	12,190	15,230	—	—
Na	495.8	4,562	6,912	9,540	13,360	16,610	20,110	25,490
Mg	737.7	1,451	7,733	10,550	13,630	18,000	21,700	25,660
Al	577.6	1,817	2,745	11,580	15,030	18,370	23,290	27,460

The Group IA metals (Li, Na, K, Rb, Cs) have very low first ionization energies, as is expected since these elements have only one electron in their highest energy levels (. . . ns^1). The first electron added to a principal energy level is easily removed. Note also that as we move down the group the first ionization energies become smaller. The decrease is due to the fact that filled sets of inner orbitals "shield" the nucleus from the outermost electrons which, therefore, are not very strongly held. Additionally, recall that the force of attraction of the positively charged nucleus for electrons decreases as the square of the separation between

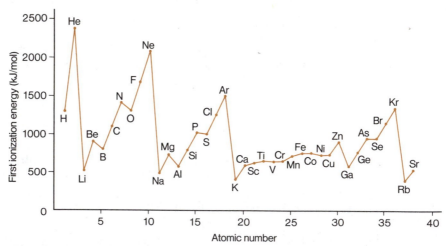

FIGURE 6–3 A plot of first ionization energies for the first 38 elements versus atomic number. Note that the noble gases have very high first ionization energies, whereas the IA metals have low first ionization energies. Note also the similarities between the variations for elements 3 through 10 and those for elements 11 through 18, as well as the later A Group elements. Variations in B Group elements are not nearly so pronounced as those for A Group elements.

them increases. In other words, as atomic radii increase in a given group, first ionization energies decrease because the valence electrons are further from the nucleus, and this effect outweighs the increases in effective nuclear charge.

As we shall see, ionization energies are also important factors in determining crystal lattice energies of ionic solids, *which measure the stabilities of ionic crystals (Section 12–11.3).*

The first ionization energies for the Group IIA elements (Be, Mg, Ca, Sr, Ba) are significantly higher than those of the Group IA elements in the same periods. This is due to the fact that the Group IIA elements have smaller atomic radii and higher effective nuclear charges. Thus, their valence electrons are held more tightly than those of the preceding IA metals of the same periods. Since a pair of electrons in an *s* orbital is more stable than a single electron, considerably more energy is required to remove an electron from the filled outermost *s* orbitals of the Group IIA elements than to remove the single electron from the half-filled outermost *s* orbitals of the Group IA elements.

The first ionization energies for the Group IIIA elements (B, Al, Ga, In, Tl) are *lower* than those of the IIA elements in the same periods because the Group IIIA elements have only a single electron in their outermost *p* orbitals. It requires less energy to remove a *p* electron than an *s* electron from the same principal energy level because an *np* orbital is higher in energy than an *ns* orbital. The major factors that favor low values of ionization energies across a period are low effective nuclear charge and the formation, by removal of an electron, of electronic configurations that have only empty, half-filled, and filled sublevels.

The second peak on each segment of an ionization energy curve occurs at the Group VA elements (N, P, As, Sb, Bi). These elements are characterized by three unpaired electrons in the three outermost *p* orbitals, that is, . . . $ns^2np_x^1np_y^1np_z^1$, a half-filled set of *p* orbitals. The Group VIA elements (O, S, Se, Te, Po) have slightly *lower* first ionization energies than the Group VA elements in the same periods. This tells us that it takes slightly less energy to remove a paired electron from a Group VIA element than to remove an unpaired *p* electron from a Group VA element in the same period. This is due to the greater repulsion between the two paired *p* electrons of the VIA elements, and to the relative stability of the half-filled set of *p* orbitals that results from removal of one electron from the VIA elements.

We have seen other consequences of the special stability of half-filled sets of equivalent orbitals, i.e., "exceptions" to the Aufbau Principle (Section 5–10).

As we shall see, *knowledge of the relative values of ionization energies assists us in predicting which elements are likely to form ionic or molecular (covalent) compounds.* Elements with low ionization energies form ionic compounds by losing electrons to form positively charged ions (cations). Elements with intermediate ionization energies generally form molecular compounds by sharing electrons with other elements.

The "driving force" for an atom of a *representative* element to form a monatomic ion in a compound usually lies in the formation of a stable noble gas electronic configuration. Reference to the ionization energies for the elements in the first two periods (Table 6–6) shows that energy considerations are consistent with this observation. For example, one mole of Li from Group IA forms one mole of Li^+ ions as it absorbs 520 kJ per mole of Li atoms. The second ionization energy is much greater, 7298 kJ/mol, and is prohibitively large for the formation of Li^{2+} ions under ordinary conditions. If Li^{2+} ions were to form, an electron would have to be removed from the filled first energy level, and we recognize that this is unlikely to occur. The same behavior is seen for the other alkali metals, for the same reason.

Noble gas configurations for ions are stable only in compounds. *In fact, Li^+ (g) is less stable than Li (g) by 520 kJ/mol.*

Likewise, the first two ionization energies of Be are 899 and 1757 kJ/mol, but the third ionization energy is much larger, 14,849 kJ/mol, so Be forms Be^{2+} ions, but not Be^{3+} ions. The other alkaline earth metals, Mg, Ca, Sr, Ba, and Ra, behave in a similar way. The stepwise lines in Table 6–6 indicate the amount of energy required to ionize all outer shell electrons of an element. Due to the high energy requirements, simple monatomic ions with positive charges greater than 3+ do not form under ordinary circumstances; only the lower members of Group IIIA, beginning with Al, form 3+ ions (as do Bi and some *d*- and *f*-transition metals). Thus we see that the magnitudes of successive ionization energies support the ideas of electronic configurations discussed in Chapter 5.

6–6   Electron Affinity

The **electron affinity** (EA) of an element is defined as *the amount of energy involved in the process in which an electron is added to an isolated gaseous atom to form an ion with a 1− charge.* The current convention is to assign a negative value when energy is released, as it usually is, and a positive value when energy is absorbed. This is consistent with thermodynamic convention. To provide specific examples, we can represent the electron affinities of beryllium and chlorine as

$$Be\ (g) + e^- + 241\ kJ \longrightarrow Be^-\ (g) \qquad EA = 241\ kJ/mol \qquad \text{[Eq. 6–5]}$$

$$Cl\ (g) + e^- \longrightarrow Cl^-\ (g) + 348\ kJ \qquad EA = -\ 348\ kJ/mol \qquad \text{[Eq. 6–6]}$$

This value of EA for Cl can also be represented as -5.78×10^{-19} J/atom.

Eq. 6–5 tells us that when one mole of gaseous beryllium atoms gain one electron each to form a mole of gaseous Be^- ions, 241 kJ is *absorbed.* Eq. 6–6 tells us that when one mole of gaseous chlorine atoms gain one mole of electrons to form one mole of gaseous chloride ions, 348 kJ of energy is *liberated.* Figure 6–4 shows a plot of electron affinity versus atomic number for several elements.

The processes described by Eqs. 6–5 and 6–6 both involve the *addition* of an electron to a neutral gaseous atom, rather than the *removal* of an electron (as shown in Eq. 6–2, which defines the first ionization energy for a neutral atom). Thus, the electron affinity and the ionization energy for an element are related. However, these two processes are *not* simply the reverse of one another. The reverse of the ionization process for the hypothetical element X,

$$X^+\ (g) + e^- \longrightarrow X\ (g) + IE_1 \qquad \text{[Eq. 6–7]}$$

is not the same as

$$X\ (g) + e^- \longrightarrow X^-\ (g) - EA \qquad \text{[Eq. 6–8]}$$

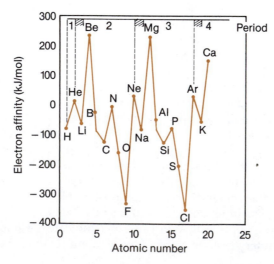

FIGURE 6–4 Plot of electron affinity versus atomic number for the first 20 elements. Electron affinities tend to become more negative (more energy is released as an extra electron is added) from Group IA through Group VIIA for a given period. Exceptions occur at the IIA and VA elements.

TABLE 6-7 Electron Affinity Values (kJ/mol) of Some Elements*

H −72									He (21)
Li −60	Be (241)		B −23	C −122	N 0	O −142	F −322		Ne (29)
Na −53	Mg (231)	Cu −123	Al −44	Si −119	P −74	S −200	Cl −348		Ar (35)
K −48	Ca (156)	Ag −125	Ga (−36)	Ge −116	As −77	Se −194	Br −323		Kr (39)
Rb −47	Sr (119)	Au −222	In (−34)	Sn −120	Sb −101	Te −190	I −295		Xe (40)
Cs −45	Ba (52)		Tl (−48)	Pb −101	Bi −101	Po (−173)	At (−270)		Rn (40)
Fr (−44)									

* Estimated values in parentheses.

because the first one begins with a positive ion, while the second begins with a neutral atom. Thus, IE_1 and EA are *not* simply equal in value with the sign reversed.

Elements with very negative electron affinities gain electrons readily to form negative ions (anions). We see from Figure 6-4 that, generally speaking, electron affinities become more negative (i.e., the elements show a greater attraction for an extra electron) from left to right across a row in the periodic table, excluding the noble gases. Note that the halogens, which have the outer electronic configuration . . . ns^2np^5, have the most negative electron affinities. They form stable anions with noble gas configurations, . . . ns^2np^6, by gaining one electron.

Electron affinity is a precise and quantitative term, like ionization energy, but it is difficult to measure. Table 6-7 shows electron affinities for the representative elements.

For many reasons, the variations in electron affinities are not regular across a period. Exceptions to the general trend that electron affinities of the elements become more negative from left to right in each period are most noticeable for the elements of Group IIA (Be, Mg, Ca, . . .) and Group VA (N, P, As, . . .), which have less negative (more positive) values than the trends suggest (Figure 6-4). The electron affinity of a IIA metal is very positive because it involves the addition of an electron to an atom that already has only completely filled *ns* and empty *np* orbitals. The values for the VA elements are slightly less negative (i.e., less exothermic) than expected because they apply to the addition of an electron to a relatively stable half-filled set of *np* orbitals ($ns^2np^3 \longrightarrow ns^2np^4$).

The addition of a second electron to an ion with a 1− charge is usually endothermic, so electron affinities of anions are usually positive.

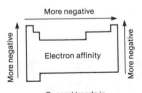

General trends in electron affinities of A Group elements with position in periodic table. Exceptions occur at Groups IIA and VA.

6-7 Ionic Radii

As we have seen in the preceding two sections, many elements (metals) have sufficiently low ionization energies that they form positive ions easily, while the very negative electron affinities of some other elements (nonmetals) indicate that they readily form negative ions.

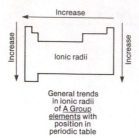

General trends in ionic radii of <u>A Group</u> elements with position in periodic table

Comparison of the experimentally measured sizes of atoms and their common stable ions, as presented in Figure 6–5, leads us to the following generalizations:

(1) Positive ions (cations) are always *smaller* than the the neutral atoms from which they are formed. (2) Negative ions (anions) are always *larger* than the neutral atoms from which they are formed.

Even though Figure 6–5 shows data only for A Group elements, these generalizations are true for all monatomic ions. In this section, we shall explain the observed trends in ionic sizes in terms of general ideas such as the kinds of orbitals that are occupied and the strength of attraction of the nucleus for the outermost electrons.

To convert from nm to Å, move the decimal point right one place; e.g., the atomic radius of Li is 0.152 nm or 1.52 Å. (1 Å = 1×10^{-10} m = 0.1 nm)

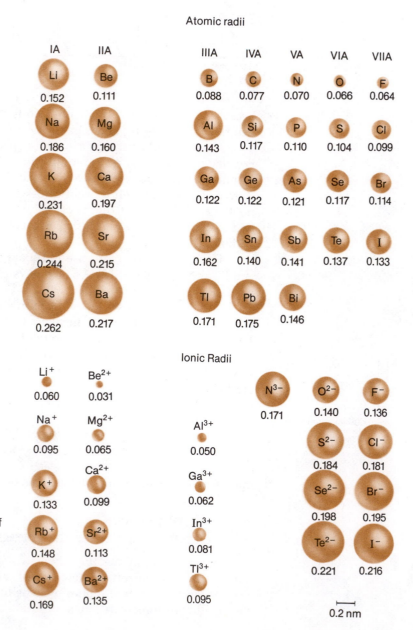

FIGURE 6–5 Sizes of atoms and ions of the A Group elements in nanometers. Positive ions (cations) are always *smaller* than the neutral atoms from which they are formed, whereas negative ions (anions) are always *larger* than the neutral atoms from which they are formed.

Li   Li⁺ ●

$r = 0.152$ nm $r = 0.060$ nm

Na   Na⁺

$r = 0.186$ nm $r = 0.095$ nm

The nuclear charge remains constant when the ion is formed.

Be   Be²⁺ ●

$r = 0.111$ nm $r = 0.031$ nm

Mg   Mg²⁺

$r = 0.160$ nm $r = 0.065$ nm

F   F⁻

$r = 0.064$ nm $r = 0.136$ nm

Cl   Cl⁻

$r = 0.099$ nm $r = 0.181$ nm

Let us first describe the decrease in size of positive ions compared to the neutral atoms from which they are formed. The reasoning will be illustrated by comparing Li⁺ ($r = 0.060$ nm) with Li ($r = 0.152$ nm). The size of the neutral lithium atom is determined by the attraction of the positively charged nucleus (charge 3+, equal to the number of protons) for the outermost electron, which is in the $2s$ orbital. When the ion Li⁺ is formed, the $2s$ electron is lost, so that the outermost electrons now are those that occupy the $1s$ orbital, much closer to the nucleus. The 3+ nuclear charge draws the two remaining electrons much closer to itself, so a lithium ion is much smaller than a neutral lithium atom. All Group IA metals (Li, Na, K, Rb, Cs) have only one electron in their highest energy level (electronic configuration . . . ns^1). They all react with other elements by losing one electron to attain noble gas configurations, and form the ions Li⁺, Na⁺, K⁺, Rb⁺, and Cs⁺, respectively. Applying the same line of reasoning that we used for Li convinces us that for each of the Group IA elements the ion, M⁺, should be smaller than the corresponding atom, M.

We see in Figure 6–5 that the ions formed by the Group IIA elements (Be²⁺, Mg²⁺, Ca²⁺, Sr²⁺, Ba²⁺) are significantly smaller than the *isoelectronic* ions formed by the Group IA elements in the same period. **Isoelectronic** species have the same electronic configuration. For instance, Be²⁺ ($r = 0.031$ nm) is much smaller than Li⁺ ($r = 0.060$ nm). Again, this is just what we might expect. Both ions have the $1s^2$ electronic configuration. A beryllium ion, Be²⁺, is formed when a beryllium atom, Be, loses two electrons while the 4+ nuclear charge remains constant. We expect the 4+ nuclear charge in Be²⁺ to draw the remaining two electrons much closer to itself than the 3+ nuclear charge of Li⁺ can draw two electrons to itself. Comparison of the ionic radii for the IIA elements with the corresponding atomic radii indicates the validity of our reasoning. Similar reasoning leads us to conclude that the ions formed by the Group IIIA metals (Al³⁺, Ga³⁺, In³⁺, Tl³⁺) should be even smaller than isoelectronic ions formed by Group IA and Group IIA elements in the same periods.

Let us now focus our attention on the Group VIIA elements (F, Cl, Br, I), which have the outermost electronic configuration . . . ns^2np^5. These elements can fill their outermost p orbitals completely by gaining one electron to attain noble gas configurations. Thus, when a fluorine atom (atomic number = 9) gains one electron, it becomes a fluoride ion, F⁻, which contains ten electrons. Again, the nuclear charge of the atom remains constant, 9+, but increased electron-electron repulsions cause the effective volume of the (ten) electron cloud to expand greatly. The fluoride ion ($r = 0.136$ nm) is significantly larger than the neutral fluorine atom ($r = 0.064$ nm). Similar reasoning indicates that a chloride ion, Cl⁻, should be larger than a neutral chlorine atom, Cl. Reference to Figure 6–5 verifies our logic.

If we compare the sizes of an oxygen atom and an oxide ion, O²⁻, we find that the negatively charged ion is larger than the neutral atom from which it is formed by gaining two electrons. The oxide ion is also larger than the isoelectronic fluoride ion because the oxide ion contains ten electrons held by a nuclear charge of only 8+, while the fluoride ion has ten electrons held by a nuclear charge of 9+.

As a final illustration about ionic sizes, let us predict the trend in sizes of the isoelectronic species S²⁻, Cl⁻, Ar, K⁺, Ca²⁺. Since these five species all have the configuration $1s^22s^22p^63s^23p^6$, no difference in size would be predicted resulting from differences in the orbitals that are occupied (the same orbitals are occupied in all). The five do differ, however, in the nuclear charge that acts to attract the 18 electrons. Because all five species have the same electronic configurations, there is no difference in shielding due to inner electrons, so that the effective nuclear charge increases just as the actual nuclear charge increases. The nuclear charges are as follows: Ca^{2+}, $20+$; K^+, $19+$; Au, $18+$; Cl^-, $17+$; S^{2-}, $16+$. Thus, the sizes of atoms and ions in this series would be expected to increase as the nuclear charge decreases, leading to the prediction of sizes (smallest to largest) as

$$Ca^{2+} < K^+ < Ar < Cl^- < S^{2-}$$

In summary, *monatomic negatively charged ions (anions) are always larger than the neutral atoms from which they are derived. As the magnitude of the negative charge on isoelectronic ions increases, so do their sizes. Conversely, positively charged ions (cations) are always smaller than the neutral atoms from which they are formed. As the magnitude of the positive charge on isoelectronic ions increases, their sizes decrease.*

As we shall see shortly, chemical bonds can be classified as predominantly ionic (in which electrons are *transferred* from one atom or group of atoms to another) or covalent (in which electrons are *shared* between bonded atoms). In fact, most bonds have some degree of both ionic and covalent character. The ionic radii of the species involved are crucial in determining the degree of ionic or covalent character of a given bond. Additionally, the arrangements of ions in the crystals of ionic solids depend upon the radii of the ions involved. Finally, the ratio of the charge on a cation to its radius, *the charge/radius ratio*, determines such things as the extent to which the cation reacts with water to produce acidic aqueous solutions. Thus we see that there are many ways in which ionic radii are important in understanding chemical and physical properties. These will be discussed in subsequent chapters.

6–8 Electronegativity

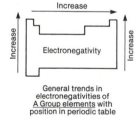

General trends in electronegativities of <u>A Group elements</u> with position in periodic table

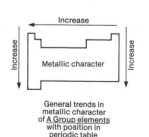

General trends in metallic character of <u>A Group elements</u> with position in periodic table

The **electronegativity** of an element *is a measure of the relative tendency of an atom to attract electrons to itself when it is chemically combined with another atom.* Electronegativities of the elements have been calculated and are expressed in Table 6–8 on a somewhat arbitrary scale, called the *Pauling electronegativity scale,* with a maximum value of 4.0.

Note that the electronegativity of fluorine (4.0) is higher than that of any other element. This tells us that when fluorine is chemically bonded to another element, it has a greater tendency to attract electron density to itself than does any other element. Oxygen is the second most electronegative element.

For the representative elements, *electronegativities usually increase from left to right across periods and from bottom to top within groups.* Variations among the transition elements aren't as regular. Generally both ionization energies and electronegativities are low for elements at the lower left of the periodic table (metals) and high for those at the upper right (nonmetals).

Although the electronegativity scale is somewhat arbitrary, it can be used to make predictions about bonding with a reasonable degree of certainty. Elements with large differences in electronegativity (a metal and a nonmetal) tend to react with each other to form an ionic compound. The less electronegative (more metallic) element gives up one or more electrons to the more electronegative (more nonmetallic) element. Elements with low electronegativity differences (two nonmetals) tend to form covalent bonds with each other, but they share their electrons in such a way that the more electronegative element attains a greater share. This will be

TABLE 6–8 Electronegativity Values of the Elements*

Key:
- ☐ <1.0
- ☐ 1.0–1.4
- ☐ 1.5–1.9
- ☐ 2.0–2.4
- ☐ 2.5–2.9
- ☐ 3.0–4.0

1 H 2.1																	
3 Li 1.0	4 Be 1.5											5 B 2.0	6 C 2.5	7 N 3.0	8 O 3.5	9 F 4.0	
11 Na 1.0	12 Mg 1.2											13 Al 1.5	14 Si 1.8	15 P 2.1	16 S 2.5	17 Cl 3.0	
19 K 0.9	20 Ca 1.0	21 Sc 1.3	22 Ti 1.4	23 V 1.5	24 Cr 1.6	25 Mn 1.6	26 Fe 1.7	27 Co 1.7	28 Ni 1.8	29 Cu 1.8	30 Zn 1.6	31 Ga 1.7	32 Ge 1.9	33 As 2.1	34 Se 2.4	35 Br 2.8	
37 Rb 0.9	38 Sr 1.0	39 Y 1.2	40 Zr 1.3	41 Nb 1.5	42 Mo 1.6	43 Tc 1.7	44 Ru 1.8	45 Rh 1.8	46 Pd 1.8	47 Ag 1.6	48 Cd 1.6	49 In 1.6	50 Sn 1.8	51 Sb 1.9	52 Te 2.1	53 I 2.5	
55 Cs 0.8	56 Ba 1.0	57 La 1.1	72 Hf 1.3	73 Ta 1.4	74 W 1.5	75 Re 1.7	76 Os 1.9	77 Ir 1.9	78 Pt 1.8	79 Au 1.9	80 Hg 1.7	81 Tl 1.6	82 Pb 1.7	83 Bi 1.8	84 Po 1.9	85 At 2.1	
87 Fr 0.8	88 Ra 1.0	89 Ac 1.1															

58 Ce 1.1	59 Pr 1.1	60 Nd 1.1	61 Pm 1.1	62 Sm 1.1	63 Eu 1.1	64 Gd 1.1	65 Tb 1.1	66 Dy 1.1	67 Ho 1.1	68 Er 1.1	69 Tm 1.1	70 Yb 1.0	71 Lu 1.2
90 Th 1.2	91 Pa 1.3	92 U 1.5	93 Np 1.3	94 Pu 1.3	95 Am 1.3	96 Cm 1.3	97 Bk 1.3	98 Cf 1.3	99 Es 1.3	100 Fm 1.3	101 Md 1.3	102 No 1.3	103 Lr 1.5

* The shading of the block indicates the EN value of that element according to the key in the figure.

discussed in detail in the sections that follow and in Chapter 7. Since the noble gases form few compounds, they are usually excluded from discussions of electronegativity.

Now that we have discussed some of the chemical properties of the elements, we can see from Table 6–9 that metals and nonmetals have contrasting chemical properties, just as they have opposite physical properties.

TABLE 6–9 Some Chemical Properties of Metals and Nonmetals

Metals	Nonmetals
1. Outer shells contain few electrons; usually three or fewer	1. Outer shells contain four or more electrons
2. Low ionization energies	2. High ionization energies
3. Slightly negative or positive electron affinities	3. Quite negative electron affinities
4. Low electronegativities	4. High electronegativities*
5. Form cations by losing e^-	5. Form anions by gaining e^-*
6. Form ionic compounds with nonmetals	6. Form ionic compounds with metals* and molecular (covalent) compounds with other nonmetals

* Except the noble gases.

Chemical Bonds

We are now ready to introduce chemical bonding, which addresses the fundamental question of how atoms and ions are held together in compounds. The chemical and physical properties of a compound depend upon the kind of bonding in the compound, as well as upon the strengths of the bonds. Chemical reactions are accompanied by the breaking and making of bonds. As a result, the net energy released or absorbed during a reaction is related, at least in part, to the strengths of the bonds of the reactants and products.

We shall soon see that the character of the elements combining to form a compound, and therefore the positions of these elements in the periodic table, are related to the kinds of bonds they form with each other.

6–9   Kinds of Chemical Bonds

The attractive forces that hold atoms together in compounds are referred to as **chemical bonds.** As a matter of convenience, we usually divide chemical bonds into classes; stated differently, we say that compounds that exhibit certain kinds of properties have one kind of chemical bonding, while compounds with different properties have other kinds of bonding. There are two major classes of bonding: (1) **ionic bonding,** which results from electrostatic interactions among ions, which can be formed by the *transfer* of one or more electrons from one atom or group of atoms to another, and (2) **covalent bonding,** which results from *sharing* one or more electron pairs between two atoms. These represent two extremes, and all bonds have at least some degree of both ionic and covalent character. *Compounds in which the bonding is predominantly ionic are called* **ionic compounds,** *and those in which the bonding is predominantly covalent are called* **covalent compounds.**

Some of the properties associated with many simple ionic and covalent compounds in the extreme cases are summarized in the following table. The differences in properties can be accounted for by the differences in bonding between the atoms or ions.

Ionic Compounds	Covalent Compounds
1. High melting solids	1. Gases, liquids, or low melting solids
2. Many are soluble in water	2. Many are insoluble in water
3. Most are insoluble in hexane, C_6H_{14}, and other hydrocarbons*	3. Most are soluble in hexane, C_6H_{14}, and other hydrocarbons*
4. Molten compounds conduct electricity well because they contain charged particles (ions)	4. Liquid or molten compounds do not conduct electricity
5. Aqueous solutions conduct electricity well because they contain charged particles (ions)	5. Aqueous solutions are *usually* poor conductors of electricity because they do not contain charged particles

As we shall see in Section 9–4, aqueous solutions of *some* covalent compounds do conduct electricity.

* Hydrocarbons are compounds consisting of only C and H.

The strong electrostatic forces of attraction among oppositely charged ions in ionic compounds account for their relatively high melting points. According to Coulomb's Law, the force of attraction, F, between two oppositely charged particles of charge magnitude q is directly proportional to the product of the charges and inversely proportional to the square of the distance separating them, d.

$$F \propto \frac{q^+ q^-}{d^2}$$

[Eq. 6–9]

Thus, the greater the charges on oppositely charged ions, and the closer the ions are to each other, the stronger is the resulting attraction. Of course, like-charged ions repel each other, and a neutral ionic crystal must contain equal numbers of positive

Figure 6–6 has been expanded to show the spatial arrangement of ions. Adjacent ions actually are in contact with each other. The lines do *not* represent formal chemical bonds; they have been drawn to emphasize the spatial arrangement of ions.

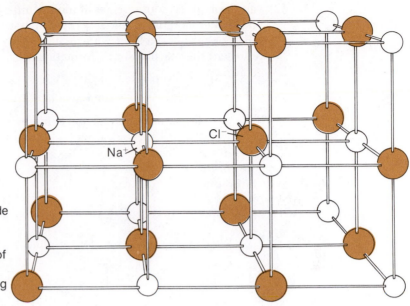

FIGURE 6–6 Crystal structure of NaCl, expanded for clarity. Like the Cl⁻ ion in the center of the cube, each chloride ion is surrounded by six sodium ions. Each Na⁺ ion is similarly surrounded by six Cl⁻ ions. The crystal includes billions of ions in the pattern shown. Compare with Figure 12–18, which shows a space-filling model of the NaCl structure.

Remember that a formula unit shows the simplest whole-number ratio of atoms (or ions) in a compound.

and negative charges, so the distances separating ions in solids are those at which the attractions and the repulsions are exactly balanced. The structure of common table salt, sodium chloride (NaCl), is shown in Figure 6–6. Typical of simple ionic compounds, NaCl exists in a regular, extended array of positive and negative ions, Na^+ and Cl^-.

Distinct molecules of ionic substances do not exist at room temperature and atmospheric pressure, so we must refer to *formula units* instead of molecules. While each individual interaction between two ions in an ionic compound is relatively weaker than a typical interaction between two atoms covalently bonded together, the sum of the forces of all the interactions in an ionic solid is substantial. In predominantly covalent compounds, the bonds between atoms within a molecule (*intra*molecular bonds) are relatively strong but the forces of attraction between molecules (*inter*molecular forces) are relatively weak, accounting for their lower melting and boiling points. It must be emphasized, though, that both covalent and ionic bonding can be strong or weak, depending on the circumstances.

6–10   Ionic Bonding

The first kind of chemical bonding we shall study is *ionic* or *electrovalent* bonding. **Ionic bonding** *is a form of bonding resulting from the transfer of one or more electrons from one atom or group of atoms to another.* As our previous discussions of ionization energy, electronegativity, and electron affinity might lead us to speculate, ionic bonding occurs most easily when elements with low ionization energies (metals) react with elements with high electronegativities and very negative electron affinities (nonmetals). Many metals lose electrons easily, and many nonmetals gain electrons readily.

Consider the reaction of sodium (a Group IA metal) with chlorine (a Group VIIA nonmetal). Sodium is a soft silvery metal (mp 98°C), and chlorine is a yellowish green corrosive gas at room temperature. Both sodium and chlorine react with water, sodium vigorously. By contrast, sodium chloride is a white solid (mp 801°C) that is soluble in water with the absorption of just a little heat. We can represent the reaction for its formation as

$$2Na\ (s) +\ Cl_2\ (g) \longrightarrow\ 2NaCl\ (s)$$
sodium   chlorine   sodium chloride

If we wish to examine this reaction in more detail, we can show electronic configurations for all species. We deal with chlorine as individual atoms rather than molecules, for simplicity.

We can also use Lewis dot formulas (Section 6–2) to represent the reaction.

$$Na\cdot + :\ddot{\underset{\cdot\cdot}{Cl}}\cdot \longrightarrow Na^+ \left[:\ddot{\underset{\cdot\cdot}{Cl}}:\right]^-$$

In this reaction, sodium atoms lose one electron each to form sodium ions, Na^+, which contain only ten electrons, the same number as the *preceding* noble gas, neon. In contrast, chlorine atoms gain one electron each to form chloride ions, Cl^-, which contain 18 electrons and are isoelectronic with the *following* noble gas, argon. Similar observations will be repeated for most ionic compounds formed by reactions between *representative metals and representative nonmetals.*

The formula for sodium chloride, NaCl, indicates that the compound contains Na and Cl in a 1:1 ratio. This is the formula we might predict based on the fact that sodium atoms contain only one electron in their highest energy level while chlorine atoms need only one electron to fill completely their outermost p orbitals.

The chemical formula NaCl does not explicitly indicate the ionic nature of the compound, only the ratio of atoms. So we must learn to recognize, from positions of elements in the periodic table and the known trends in electronegativity, when the difference in electronegativity is large enough to favor ionic bonding. In general, *the further apart across the periodic table two elements are, the more likely they are to form an ionic compound.*

The noble gases are excluded from this generalization.

All the Group IA metals (Li, Na, K, Rb, Cs) react with the Group VIIA elements (F, Cl, Br, I) to form compounds of the same general formula, MX. All the resulting ions have noble gas configurations. We can generalize and write the equation for the reaction of the IA metals with the VIIA elements as

$$2M\ (s) +\ X_2 \longrightarrow 2MX\ (s) \qquad M = Li, Na, K, Rb, Cs$$

$$X = F, Cl, Br, I$$

The Lewis dot representation for the generalized reaction is

$$M\cdot + :\ddot{\underset{\cdot\cdot}{X}}\cdot \longrightarrow M^+ \left[:\ddot{\underset{\cdot\cdot}{X}}:\right]^-$$

Next, consider the reaction of lithium (Group IA) with oxygen (Group VIA) to form lithium oxide, an ionic solid (mp > 1700°C).

$$4Li\ (s) +\ O_2\ (g) \longrightarrow\ 2Li_2O\ (s)$$

lithium   oxygen   lithium oxide

The formula for lithium oxide, Li_2O, indicates that two atoms of lithium combine with one atom of oxygen. Again, if we examine the structures of the atoms before reaction, we predict that two lithium atoms react with one oxygen atom.

Each Li atom has one e^- in its valence shell. Each O atom has 6 e^- in its valence shell and needs 2 e^- to give it a noble gas configuration.

$$_3Li \quad _3Li \quad _8O \longrightarrow Li^+ \ (1\ e^- \text{ lost}) \quad Li^+ \ (1\ e^- \text{ lost}) \quad O^{2-} \ (2\ e^- \text{ gained})$$

The Lewis dot formulas for the atoms and ions are

$$2Li\cdot + :\overset{\cdot\cdot}{\underset{\cdot\cdot}{O}}\cdot \longrightarrow 2Li^+ \left[:\overset{\cdot\cdot}{\underset{\cdot\cdot}{O}}: \right]^{2-}$$

Note that lithium ions, Li^+, are isoelectronic with helium atoms (2 e^-) while oxide ions, O^{2-}, are isoelectronic with neon atoms (10 e^-).

Calcium (Group IIA) reacts with oxygen (Group VIA) to form calcium oxide, a white solid ionic compound with a very high melting point, 2580°C.

$$2Ca\ (s) + O_2\ (g) \longrightarrow 2CaO\ (s)$$

calcium   oxygen   calcium oxide

Again we write out the electronic structures of the atoms and ions, this time representing the inner electrons by the symbol of the preceding noble gas (in brackets).

$$_{20}Ca \quad [Ar]\ 4s \qquad _8O \quad [He]\ 2s\ 2p \longrightarrow Ca^{2+}\ [Ar]\ 4s \ (2\ e^-\text{ lost}) \qquad O^{2-}\ [He]\ 2s\ 2p \ (2\ e^-\text{ gained})$$

In writing equations in which electrons in different energy levels in different atoms are involved, atomic orbitals are more conveniently labeled *under* the lines that represent them.

The Lewis dot notation for the atoms and ions is

$$Ca: + :\overset{\cdot\cdot}{\underset{\cdot}{O}}\cdot \longrightarrow Ca^{2+} \left[:\overset{\cdot\cdot}{\underset{\cdot\cdot}{O}}: \right]^{2-}$$

Calcium ions, Ca^{2+}, are isoelectronic with argon (18 e^-), the preceding noble gas. Oxide ions, O^{2-}, are isoelectronic with neon (10 e^-), the following noble gas.

Calcium oxide melts at a much higher temperature than sodium chloride (2580°C as opposed to 801°C) because it has two charges per ion rather than one, and the ionic bonding is considerably stronger in CaO.

As our final example of ionic bonding, consider the reaction of magnesium (Group IIA) with nitrogen (Group VA) at elevated temperatures to form magnesium nitride, Mg_3N_2, a white ionic solid compound. We can represent the reaction as

$$3Mg\ (s)\ +\ N_2\ (g)\ \overset{\Delta}{\longrightarrow}\ Mg_3N_2\ (s)$$

magnesium   nitrogen   magnesium nitride

The formula, Mg_3N_2, indicates that magnesium and nitrogen combine in a 3 : 2 ratio. We could predict this from the fact that magnesium atoms contain two electrons in their highest energy level, while nitrogen atoms need three electrons to fill theirs completely. Thus, three magnesium atoms lose two electrons each for a total of 6 e^-, while two nitrogen atoms gain three electrons each, also for a total of 6 e^-. As before, examination of the structures makes the picture clearer.

TABLE 6–10   Simple Binary Ionic Compounds

Metal		Nonmetal		General Formula	Ions Present	Example
IA*	+	VIIA	$\longrightarrow$	MX	(M^+, X^-)	$LiBr$
IIA	+	VIIA	$\longrightarrow$	MX_2	$(M^{2+}, 2X^-)$	$MgCl_2$
IIIA	+	VIIA	$\longrightarrow$	MX_3	$(M^{3+}, 3X^-)$	GaF_3
IA*	+	VIA	$\longrightarrow$	M_2X	$(2M^+, X^{2-})$	Li_2O
IIA	+	VIA	$\longrightarrow$	MX	(M^{2+}, X^{2-})	CaO
IIIA	+	VIA	$\longrightarrow$	M_2X_3	$(2M^{3+}, 3X^{2-})$	Al_2O_3
IA*	+	VA	$\longrightarrow$	M_3X	$(3M^+, X^{3-})$	Li_3N
IIA	+	VA	$\longrightarrow$	M_3X_2	$(3M^{2+}, 2X^{3-})$	Ca_3P_2
IIIA	+	VA	$\longrightarrow$	MX	(M^{3+}, X^{3-})	AlP

* Hydrogen is considered a nonmetal, and all binary compounds of hydrogen are covalent except certain metal hydrides such as NaH and CaH_2, in which the hydride ion, H^-, is present.

Using Lewis dot formulas for the atoms and ions, we have

$$3Mg\!:\, + \, 2\cdot\ddot{N}\cdot \longrightarrow 3Mg^{2+}, \, 2\left[:\ddot{N}:\right]^{3-}$$

In this case both magnesium ions, Mg^{2+}, and nitride ions, N^{3-}, are isoelectronic with neon (10 e^-).

Table 6–10 summarizes the general formulas of binary ionic compounds formed by the representative elements. M represents metals and X represents nonmetals from the indicated groups.

In the examples of ionic bonding examined thus far, we note that metal atoms have lost one, two, or three electrons each and nonmetal atoms have gained one, two, or three electrons. Recall that simple (monatomic) ions rarely have charges greater than 3 + or 3 −. Ions with greater charges generally interact so strongly with the electron clouds of other ions in compounds that electron clouds are distorted severely, and considerable covalent character results in the bonds. This point will be developed later.

The d- and f-transition elements form many compounds that are essentially ionic in character. Examination of the electron configurations of the transition metals and their simple ions shows that the ions usually do not have noble gas configurations.

Experimental Determination of Ionic Charges

In many cases the gross properties of a compound, such as melting point, solubility in certain solvents, and physical state, indicate whether the compound is ionic or covalent. For example, a pure liquid that does not conduct electricity and is miscible with hexane, a common but toxic liquid solvent, is likely to be a covalent compound. A high melting crystalline solid that dissolves easily in water, but not in hexane, is likely to be ionic.

Inquisitive students often wonder how the actual charges on ions are determined. There are several ways. In many simple cases they may be deduced by determination of the empirical formula of a compound containing the ions of interest. Another method depends on the fact, presented in Section 6–7, that the sizes of ions vary with their charges. In some cases ionic charges can be inferred from the ionic sizes, which are found during the determination of the structure of an ionic solid by a technique called *x-ray diffraction* (described in the box preceding Section 12–10).

A much simpler and more common method involves determining the **electrical conductivity** of an aqueous solution of known concentration. If it is assumed that ionic compounds dissociate completely when dissolved in water, the extent to which the resulting solutions conduct electrical current is determined by the magnitudes of the charges on the ions and their mobilities. For example, a 0.010 M aqueous solution of chromium(III) chloride, $CrCl_3$, should conduct electrical current better than a 0.010 M solution of calcium chloride, $CaCl_2$, which should, in turn, conduct better than a 0.010 M solution of sodium chloride, NaCl. The cations involved are Cr^{3+}, Ca^{2+}, and Na^+. The fact that the solutions contain Cl^- in a 3:2:1 ratio also contributes to the effect and must be considered. Likewise, a solution that is 0.010 M in sulfate ions, SO_4^{2-}, should conduct better than one that is 0.010 M in chloride ions, Cl^-, if the same cation is present in both cases. As we shall see, no ionic compounds are really 100% dissociated in solution, although many are nearly so. Measurement of comparative conductivities of known concentrations of dissolved compounds can provide useful information as to whether a compound is ionic, and tell something about the charges on its ions. Determination of ionic charge by electrolysis is also quite common.

As we shall see, ions of the same element with different charges, such as Cu^+ and Cu^{2+}, and Fe^{2+} and Fe^{3+}, have very different chemical properties.

6–11 Covalent Bonding—An Introduction

A **covalent bond** *is formed when two atoms share one or more pairs of electrons.* Most covalent bonds involve two, four, or six electrons or one, two, or three *pairs* of electrons. Two atoms form a **single covalent bond** *when they share one pair of electrons,* a **double covalent bond** *when they share two electron pairs,* and a **triple covalent bond** *when they share three electron pairs.* Covalent bonds involving one and three electrons are known, but these are very rare (Chapter 22).

Almost all bonds have both ionic and covalent character. By experimental means, a given bond can usually be identified as being "closer" to one or the other extreme type. We find it convenient to use the labels for the major classes of bonds to describe simple compounds, while keeping in mind the fact that they represent ranges of behavior.

Most compounds resulting from the reaction of a metal with a nonmetal are primarily ionic. However, as we shall see, reactions between two nonmetals result in covalent bonds.

When we refer to single, double, or triple bonds, we mean bonds that are primarily covalent.

1 The Hydrogen Molecule, H_2

The principal attractive forces between atoms in covalent compounds are electrostatic in nature. Quantum mechanical calculations show that the interactions between two atoms vary with the distance between the atoms. Figure 6–7 shows a

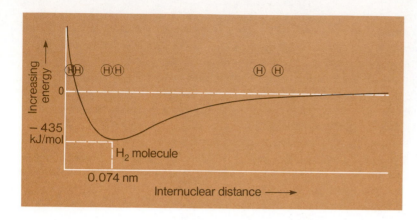

FIGURE 6–7 The potential energy of the H_2 molecule as a function of the distance between two nuclei. The lowest point on the curve, -435 kJ/mol, corresponds to the internuclear distance actually observed in the H_2 molecule, 0.074 nm. (The minimum potential energy, -435 kJ/mol, corresponds to -7.23×10^{-19} joule per H_2 molecule.) Energy is compared to that of two separated hydrogen atoms, which is defined as zero.

plot of potential energy versus internuclear distance for two hydrogen atoms. Note that the minimum energy (-435 kJ/mol), corresponding to the most stable arrangement, occurs at an internuclear distance of 0.074 nm, the actual distance between two hydrogen nuclei in an H_2 molecule. At this internuclear separation, the attractive forces exactly balance the repulsions between the two electrons and between the two nuclei. (Review the discussion accompanying Figure 3–3, if necessary.)

This picture of the covalent bond can be generalized by postulating that bonds form by the **overlap** of two atomic orbitals, each containing a single unpaired electron. This is the essence of the *Valence Bond theory*, which we shall study in detail in Chapter 8. In the case of H_2, a covalent bond forms when two $1s$ atomic orbitals overlap, thus allowing the two electrons to be shared by the two atoms.

We can represent the single bond in the H_2 molecule with a Lewis dot formula, H : H, or its dash formula, in which a shared pair of electrons (a bond) is represented as a dash, H—H. In order to study more complicated molecules we shall need to know how to draw their Lewis formulas. We shall describe the methods for constructing dot formulas in the following subsections.

2   Lewis Formulas for Molecules and Polyatomic Ions

In Sections 6–2 and 6–10 we drew Lewis dot formulas for atoms and monatomic ions. Lewis dot formulas can also be used to represent atoms covalently bonded in a molecule or a polyatomic ion (an ion that contains more than one atom). For instance, a water molecule can be represented by either of the following diagrams.

The oxygen atom has 6 e^- in its valence shell and each hydrogen atom has 1 e^- in its valence shell, so the Lewis formula for H_2O shows a total of 8 e^-.

$$H : \overset{..}{\underset{..}{O}} :$$
$$H$$

$$H — \overset{..}{O} :$$
$$\,\,|$$
$$H$$

two shared electron pairs; two *single* covalent bonds

Lewis dot formula   dash formula

In "dash formulas" a shared pair of electrons is indicated by a dash. There are two *double* bonds in carbon dioxide:

$$: \overset{..}{O} :: C :: \overset{..}{O} : \qquad : \overset{..}{O} = C = \overset{..}{O} :$$

four shared electron pairs; two *double* covalent bonds

The covalent bonds in a polyatomic ion can be represented in the same way. We shall refer to either the dot formula or the dash formula as a "Lewis formula," making a distinction where necessary.

The ammonium ion, NH_4^+, is shown below. Note that the Lewis dot formula shows only eight electrons, even though the N atom has five electrons in its valence shell and each H atom has one, for a total of $5 + 4(1) = 9$. The reason is that the ion, with a charge of 1+, has one less electron than the original atoms.

The (uncharged) molecule NH_3, like the NH_4^+ ion, has eight valence electrons about the nitrogen atom. The Lewis formula is

H:N:H
  H

$$
\begin{array}{cc}
\begin{matrix} & H & \\ & \overset{..}{} & + \\ H: & N & :H \\ & \overset{..}{} & \\ & H & \end{matrix}
&
\begin{matrix} & H & \\ & | & + \\ H- & N & -H \\ & | & \\ & H & \end{matrix}
\end{array}
$$

Lewis dot formula   dash formula

Lewis dot formulas show only the number of valence electrons and the number and kinds of bonds, but are not intended to depict the three-dimensional shapes of molecules and polyatomic ions. As we shall see shortly, construction of the correct Lewis formulas allows us *to predict* the shapes of species. Their shapes, in turn, greatly influence both physical and chemical properties of molecules and polyatomic ions.

3   The Octet Rule

Representative elements usually attain stable noble gas electronic configurations when they share electrons. In the water molecule (p. 192) the oxygen has a share in eight outer shell electrons, giving it the neon configuration, while hydrogen shares two electrons, attaining the helium configuration. Likewise, the carbon and oxygen of CO_2 and the nitrogen of NH_3 and the NH_4^+ ion each have a share in eight electrons in their outer shells, the neon configuration. The hydrogen atoms in NH_3 and NH_4^+ each share two electrons, the helium configuration. We write Lewis formulas based on the fact that *the representative elements achieve a noble gas configuration in most of their compounds.* This rule is usually called the **octet rule,** because these configurations have 8 e^- in their outermost shells (except for H, Li^+, and Be^{2+}, which have 2 e^-).

The rule does not work for some compounds in which the central atom does not achieve a noble gas configuration. Exceptions to the octet rule will be discussed later in the chapter.

For the present, we shall restrict our discussion to compounds of the *representative elements.* The octet rule, alone, does not enable us to write Lewis formulas. We still need to know how to place the electrons around the bonded atoms—that is, how many of the available valence electrons are **bonding** (shared) **electrons** and how many are associated with only a single atom. The latter kind are called **unshared electrons.** A pair of unshared electrons in the same orbital is called a **lone pair.** A simple mathematical relationship is helpful here:

$$S = N - A$$

[Eq. 6-10]

In Eq. 6-10, S is the total number of electrons shared in the molecule or polyatomic ion. In this equation N represents the number of valence shell electrons *needed by all the atoms in the molecule or ion to achieve noble gas configurations* (in other words, $N = 8 \times$ number of atoms *not* including H, plus $2 \times$ number of H atoms). The number A is the number of electrons *available* in the valence shells of all of the (representative) atoms, which is simply the sum of their periodic group numbers.

For example, in CCl_4, A for each chlorine atom is 7 and A for the carbon atom is 4, so A for CCl_4 is $4 + 4(7) = 32$.

The following general steps describe the use of this relationship in constructing dot formulas for molecules and polyatomic ions.*

Writing Lewis Formulas

1. Select a reasonable (symmetrical) "skeleton" for the molecule or polyatomic ion.
   a. The *least electronegative element* is usually the central element, except that hydrogen is never the central element. The least electronegative element is usually the one that needs the most electrons to fill its octet. Example: CS_2 has the skeleton S C S.
   b. Oxygen atoms do not bond to each other except in (1) O_2 and O_3 molecules, (2) the peroxides, which contain the O_2^{2-} group, and (3) the rare superoxides, which contain the O_2^- group. Example: The sulfate ion, SO_4^{2-}, has the skeleton:

   O
   O S O
   O

A ternary acid contains *three* elements—H, O, and another nonmetal.

   c. Hydrogen usually bonds to an O atom, *not* to the central atom *in ternary acids* (*oxyacids*). Example: Nitrous acid, HNO_2, has the skeleton H O N O. However, there are a few exceptions to this rule, such as H_3PO_3 and H_3PO_2.
   d. For polycentered species such as ethene, C_2H_4, and the pyrophosphate ion, $P_2O_7^{4-}$, the most symmetrical skeletons possible are used. Examples: C_2H_4 and $P_2O_7^{4-}$ have the following skeletons:

   H H   O   O
   C C   and   O P O P O
   H H   O   O

2. Calculate N, *the number of outer (valence) shell electrons needed* by all atoms in the molecule or ion to achieve noble gas configurations. For compounds containing only representative elements, N is equal to $8 \times$ number of atoms *not* including H, plus $2 \times$ number of H atoms. Examples: For H_2SO_4,

   $N = 8 \times 1$ (S atom) $+ 8 \times 4$ (O atoms) $+ 2 \times 2$ (H atoms)
   $= 8 + 32 + 4$
   $= 44\ e^-$ needed

   For SO_4^{2-},

   $N = 8 + 32 = 40\ e^-$ needed

3. Calculate A, *the number of electrons available* in the outer (valence) shells of all the atoms. The number of valence shell electrons in an atom is equal to its periodic group number *for the representative elements* (1 for an H atom and 8 for a noble gas atom, except 2 for He). For *negatively charged ions,* add to the total the number of electrons equal to the charge on the anion; for *positively charged ions,* subtract

---

* The authors express their appreciation to Professor Kenneth F. Bean, Northern Arizona University, for suggesting this approach to writing Lewis dot formulas.

the number of electrons equal to the charge on the cation. Examples: For H_2SO_4,

$$A = 2 \times 1 \text{ (H atoms)} + 1 \times 6 \text{ (S atom)} + 4 \times 6 \text{ (O atoms)}$$
$$= 2 + 6 + 24$$
$$= 32 \; e^- \text{ available}$$

For SO_4^{2-},

$$A = 1 \times 6 \text{ (S atom)} + 4 \times 6 \text{ (O atoms)} + 2 \text{ (for } 2^- \text{ charge)}$$
$$= 6 + 24 + 2$$
$$= 32 \; e^- \text{ available}$$

4. Calculate S, *the total number of electrons shared* in the molecule or ion, using the relationship $S = N - A$. Examples: For H_2SO_4,

$$S = N - A = 44 - 32$$
$$= 12 \text{ electrons shared (6 pairs of } e^- \text{ shared)}$$

For SO_4^{2-},

$$S = N - A = 40 - 32$$
$$= 8 \text{ electrons shared (4 pairs of } e^- \text{ shared)}$$

5. Place the *shared pairs of electrons* into the skeleton, using double and triple bonds *only when necessary*. Structures may be represented either by Lewis "dot" formulas or by "dash" formulas, in which a dash represents a shared pair of electrons. Examples:

$$H_2SO_4$$

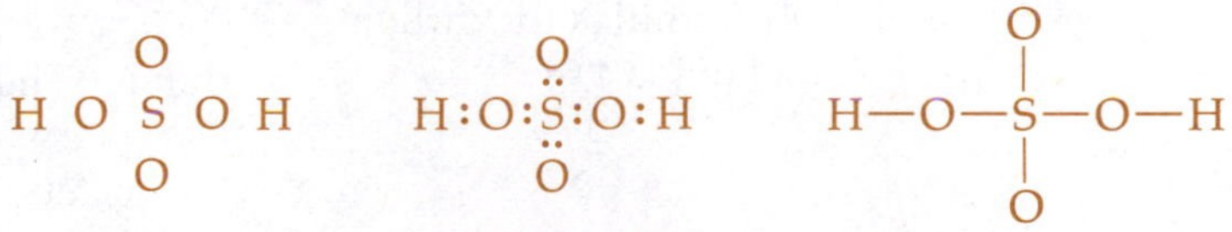

skeleton

dot formula
("bonds" in place,
but incomplete)

dash formula
("bonds" in place,
but incomplete)

$$SO_4^{2-}$$

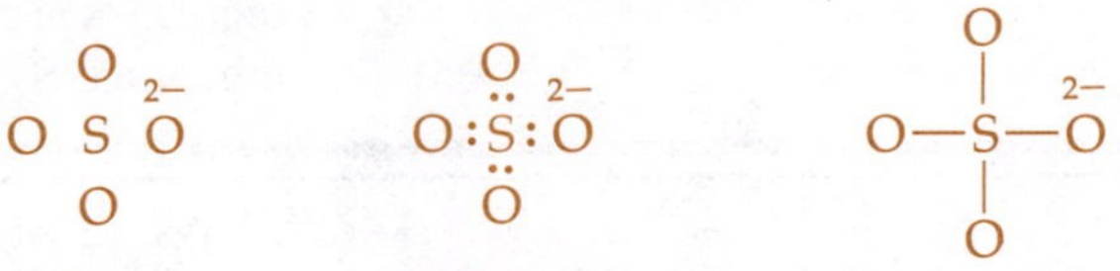

skeleton

dot formula
("bonds" in place,
but incomplete)

dash formula
("bonds" in place,
but incomplete)

Please note that a Lewis dot (dash) formula does *not* indicate the geometry of a molecule or an ion. We shall discuss the actual geometries of molecules and ions in Chapter 7.

6. Place the additional *unshared (lone) pairs of electrons* into the skeleton to fill the octet of every A Group element (except hydrogen, which can share only 2 e^-). Check to determine that the total number of electrons is equal to A, calculated in Step 3. Examples:

For H_2SO_4

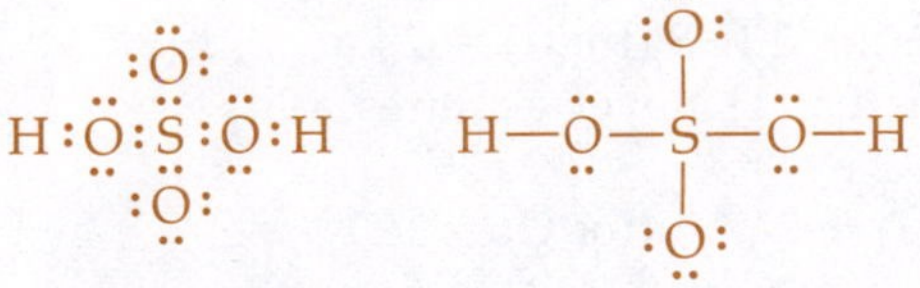

Check: 16 pairs of e^- have been used. $2 \times 16 = 32$ e^- available

For $SO_4{}^{2-}$

$$
\begin{array}{c}
:\overset{..}{\underset{..}{O}}:{}_{2-} \\
:\overset{..}{\underset{..}{O}}:\overset{..}{\underset{..}{S}}:\overset{..}{\underset{..}{O}}: \\
:\overset{..}{\underset{..}{O}}:
\end{array}
\qquad
\begin{array}{c}
:\overset{..}{O}: \\
| \\
:\overset{..}{\underset{..}{O}}{-}\overset{\displaystyle |}{\underset{\displaystyle |}{S}}{-}\overset{..}{\underset{..}{O}}:{}^{2-} \\
| \\
:\overset{..}{\underset{..}{O}}:
\end{array}
$$

Check: 16 pairs of electrons have been used: $2 \times 16 = 32$ e^- available

Let us now illustrate this procedure with more examples.

Example 6–1

Write the Lewis dot formula and dash formula for the nitrogen molecule, N_2.

Solution

Step 1 The skeleton is: N N
Step 2 $N = 2 \times 8 = 16$ e^- needed (total) by both atoms
Step 3 $A = 2 \times 5 = 10$ e^- available (total) for both atoms

Step 4 $S = N - A = 16$ $e^- - 10$ $e^- = 6$ e^- shared
Step 5 N:::N
        6 $e-$ (3 pairs) are shared; a *triple* bond.
Step 6 The additional 4 e^- are accounted for by a lone pair on each N. The complete Lewis diagram is

$:N:::N:$ or $:N{\equiv}N:$

Check: 10 e^- (5 pairs) have been used.

Example 6–2

Write the Lewis dot and dash formulas for carbon disulfide, CS_2, an ill-smelling liquid.

Solution

Step 1 The skeleton is: S C S
Step 2 $N = 1 \times 8$ (for C) $+ 2 \times 8$ (for S) $= 24$ e^- needed by all atoms
Step 3 $A = 1 \times 4$ (for C) $+ 2 \times 6$ (for S) $= 16$ e^- available
Step 4 $S = N - A = 24$ $e^- - 16$ $e^- = 8$ e^- shared

Step 5 S::C::S
        8 $e-$ (4 pairs) are shared; two *double* bonds
Step 6 The remaining 8 e^- are distributed as lone pairs on the sulfur atoms to give each S an octet. (C already has an octet.) The complete Lewis formula is

$:\overset{..}{\underset{..}{S}}::C::\overset{..}{\underset{..}{S}}:$ or $:\overset{..}{S}{=}C{=}\overset{..}{S}:$

Check: 16 e^- (8 pairs) have been used. The bonding picture is similar to that of CO_2; this is not surprising, since S is below O in Group VIA.

Example 6–3

Write the Lewis dot formula and dash formula for ethene (or ethylene), C_2H_4.

Solution

Step 1 The most symmetrical skeleton is (Step 1d):

    H H
    C C
    H H

Step 2 $N = 4 \times 2$ (for H) $+ 2 \times 8$ (for C) $= 24$ e^- needed. Remember that H never acquires a share of more than 2 e^-.
Step 3 $A = 4 \times 1$ (for H) $+ 2 \times 4$ (for C) $= 12$ e^- available

Step 4 $S = N - A = 24$ $e^- - 12$ $e^- = 12$ e^- shared
Step 5
$$
\begin{array}{cc}
H & H \\
\overset{..}{C} & ::\overset{..}{C} \\
H & H
\end{array}
$$
Step 6 All electrons are already accounted for with 12 e^- shared; there are no lone pairs. The Lewis formula is

$$
\begin{array}{cc}
H & H \\
\overset{..}{C} & ::\overset{..}{C} \\
H & H
\end{array}
\quad \text{or} \quad
\begin{array}{c}
H \quad\; H \\
| \quad\;\; | \\
C{=}C \\
| \quad\;\; | \\
H \quad\; H
\end{array}
\quad \text{(one double bond)}
$$

Check: 12 e^- (6 pairs) have been used.

Example 6-4

Write the Lewis dot formula for the carbonate ion, CO_3^{2-}.

Solution

Step 1 The skeleton is: $\quad\overset{O}{\underset{}{}}\; C\; O \quad^{2-}$

Step 2 $N = 1 \times 8$ (for C) $+ 3 \times 8$ (for O)
$\qquad = 8 + 24 = 32\ e^-$ needed by all atoms

Step 3 $A = 1 \times 4$ (for C) $+ 3 \times 6$ (for O) $+ 2$ (for the $2-$ charge)
$\qquad = 4 + 18 + 2 = 24\ e^-$ available

Step 4 $S = N - A = 32\ e^- - 24\ e^- = 8\ e^-$ (4 pairs) shared

Step 5 $O:\overset{O}{\underset{}{C}}::O:\quad^{2-}$
(4 pairs are shared. At this point it doesn't matter which O is doubly bonded.)

Step 6 The Lewis formula is

$$:\overset{..}{\underset{..}{O}}:\ \overset{..}{\underset{..}{O}}:\overset{..}{\underset{..}{C}}::\overset{..}{\underset{..}{O}}: \quad^{2-} \qquad \text{or} \qquad :\overset{..}{\underset{..}{O}}-\overset{\overset{:\overset{..}{O}:}{|}}{C}=\overset{..}{\underset{..}{O}}: \quad^{2-}$$

Check: $24\ e^-$ (12 pairs) have been used.

4 Resonance

In fact, two other Lewis formulas for the CO_3^{2-} ion, in addition to the one shown in Example 6-4, are equally acceptable. In these formulas, $4\ e^-$ could be shared between the carbon atom and either of the other two oxygen atoms.

$$:\overset{\overset{:\overset{..}{O}:}{|}}{\underset{..}{O}}-\overset{..}{\underset{..}{C}}=\overset{..}{\underset{..}{O}}: \quad^{2-} \longleftrightarrow :\overset{\overset{:O:}{\|}}{\underset{..}{O}}-\overset{..}{\underset{..}{C}}-\overset{..}{\underset{..}{O}}: \quad^{2-} \longleftrightarrow :\overset{\overset{:\overset{..}{O}:}{|}}{O}=\overset{..}{\underset{..}{C}}-\overset{..}{\underset{..}{O}}: \quad^{2-}$$

> You must remember that these formulas *do not* imply anything about the shape of the molecule; the carbonate ion is actually triangular.

A molecule or ion for which two or more equally acceptable dot formulas are available to describe the bonding is said to exhibit **resonance.** The three structures above are **resonance structures** of the carbonate ion. The relationship among them is indicated by the double-headed arrows, $\longleftrightarrow$. This symbol should not be taken to imply that the ion flips back and forth among the three illustrated structures, but rather that the true structure is like an average of the three.

The C—O bonds in CO_3^{2-} are really *neither* double nor single bonds, but are intermediate in bond length (and strength). This has been verified experimentally. The typical C—O single bond length (based on measurements in many compounds) is 0.143 nm, and the typical C=O double bond distance is 0.122 nm, while the C—O bond length for each bond in the CO_3^{2-} ion is intermediate at 0.129 nm. Another way to represent this situation is by **delocalization** of bonding electrons,

> When electrons are shared among more than two atoms, the electrons are said to be *delocalized*.

$$O\overset{\overset{O}{\|}}{=\!\!=\!\!=}\overset{}{C}=\!\!=\!\!=O \quad^{2-}$$
(lone pairs on O atoms not shown)

where the dashed lines indicate that some of the electrons shared between carbon and oxygen atoms are *delocalized* among all four atoms; that is, four pairs of electrons are equally distributed among three C—O bonds.

Example 6-5

Draw two resonance structures for the sulfur dioxide molecule, SO_2.

Solution

$$\overset{S}{\searrow} \quad \overset{O}{\searrow}$$

$$
\begin{aligned}
N &= 1(8) + 2(8) = & 24\ e^- \\
A &= 1(6) + 2(6) = & 18\ e^- \\
\hline
S &= \quad N - A \quad = & 6\ e^- \text{ shared}
\end{aligned}
$$

The resonance structures are

$$:\ddot{O}::\ddot{S}:\ddot{O}: \longleftrightarrow :\ddot{O}:\ddot{S}::\ddot{O}:$$

or, using dash formulas,

$$:\ddot{O}=\ddot{S}-\ddot{O}: \longleftrightarrow :\ddot{O}-\ddot{S}=\ddot{O}:$$

We could show delocalization of electrons as follows:

$$O = \ddot{S} = O \qquad \text{(lone pairs on O not shown)}$$

Remember that dot and dash formulas *do not necessarily show shapes.* SO_2 molecules are angular, not linear, as we shall see in Section 7–4.

Formal Charges

An experimental determination of the structure of a molecule or polyatomic ion is necessary to establish unequivocally its correct structure and Lewis formula. However, we often do not have these results at our fingertips. The concept of *formal charges,* which is a system of "electron bookkeeping," allows us to write Lewis formulas correctly in most cases.

Consider the reaction of ammonia, NH_3, with hydrogen ions, H^+, to form ammonium ions, NH_4^+.

$$H:\ddot{N}:H + H^+ \longrightarrow H:\overset{\displaystyle H}{\underset{\displaystyle H}{\ddot{N}}}:H \;\; +$$

The unshared pair of electrons on the N atom in the NH_3 molecule is shared with the H^+ ion to form the NH_4^+ ion, in which the N atom has four covalent bonds. Because N is a Group VA element, we expect it to form three covalent bonds to complete its octet. How can we explain the fact that N has four covalent bonds in species like NH_4^+? The answer is obtained by calculating the *formal charge* on each atom in NH_4^+ by the following rules.

Rules for Assigning Formal Charges to Atoms of A Group Elements

1. a. In a molecule, the sum of the formal charges is zero.
   b. In a polyatomic ion, the sum of the formal charges is equal to the charge on the ion.
2. The formal charge, abbreviated FC, on an atom in a Lewis formula is given by the relationship

   FC = (Group No.) − [(No. of Bonds) + (No. of Unshared e^-)]

   The group number of the noble gases is taken as VIIIA, rather than zero, in calculating formal charges. Formal charges are represented by ⊕ and ⊖ to distinguish between formal charges and real charges on ions.
3. In a Lewis formula, an atom that has the same number of bonds as its periodic group number has no formal charge.
4. Generally, the most energetically favorable structure for a molecule is one in which the formal charge on each atom is zero. (This is not always attainable.)
5. When difficulty arises in deciding which atom should be assigned a negative formal charge, the negative formal charge is assigned to the more electronegative element.
6. Atoms that are bonded to each other should not be assigned formal charges with the same sign (the *adjacent charge rule*), if this can be avoided. Lewis formulas in which atoms bonded to each other are assigned formal charges of the same sign are seldom accurate representations.

Let us apply these rules to the ammonia molecule, NH_3, and to the ammonium ion, NH_4^+. Since N is a Group VA element, its group number is 5.

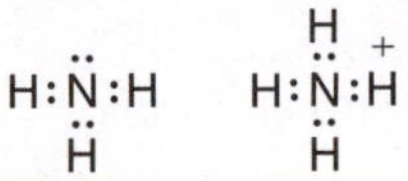

$$\ddot{H:N}:H \qquad H:\overset{\overset{H}{..}}{N}:H \,\,^+$$
$$\overset{\,\,}{H} \qquad \quad \overset{\,\,}{H}$$

In NH_3 the N atom has 3 bonds and 2 unshared e^-, and so for N,

FC = (Group No.) − [(No. of Bonds) + (No. of Unshared e^-)]

FC = 5 − (3 + 2) = 0 (for N)

For H,

FC = (Group No.) − [(No. of Bonds) + (No. of Unshared e^-)]

FC = 1 − (1 + 0) = 0 (for H)

Since the formal charges of N and H are both zero in NH_3, the sum of the formal charges is [0 + 3(0)] = 0, consistent with rule 1a.

In NH_4^+ the N atom has 4 bonds and no unshared e^-, and so for N,

FC = (Group No.) − [(No. of Bonds) + (No. of Unshared e^-)]

FC = 5 − (4 + 0) = 1+ (for N)

Recall that the absence of a sign on a number means that the number is *positive*. Therefore, we attached a "+" sign to the 1.

Calculation of the FC for H atoms gives zero, as above. The sum of the formal charges in NH_4^+ is [(1+) + 4(0)] = 1+, which is consistent with rule 1b.

Recall that FCs are indicated by ⊕ and ⊖ in color, and that the sum of the formal charges in a polyatomic ion is equal to the charge on the ion, 1+ in NH_4^+.

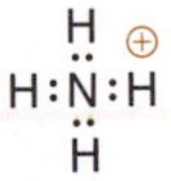

$$H:\overset{\overset{H}{..}}{N}:H \,\,^\oplus$$
$$\overset{\,\,}{H}$$

Thus we see that the octet rule is obeyed in both NH_3 and NH_4^+, and that the sum of the formal charges in each case is that predicted by rule 1, even though nitrogen has four covalent bonds in the NH_4^+ ion.

Let us now write a Lewis formula for, and assign formal charges to, the atoms in thionyl chloride, $SOCl_2$, a compound often used in organic synthesis. Both Cl atoms and the O atom are bonded to the S atom. (If we were not given this information, we could deduce it by application of rule 1a on page 194.) The Lewis formula is

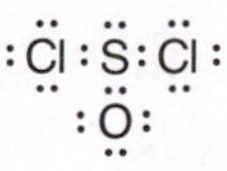

$$:\ddot{\ddot{Cl}}:\ddot{S}:\ddot{\ddot{Cl}}:$$
$$:\ddot{O}:$$

Formal charges on the atoms are calculated by the usual relationship, as shown below.

FC = (Group No.) − [(No. of Bonds) + (No. of Unshared e^-)]

For Cl: FC = 7 − (1 + 6) = 0

For S: FC = 6 − (3 + 2) = 1+

For O: FC = 6 − (1 + 6) = 1−

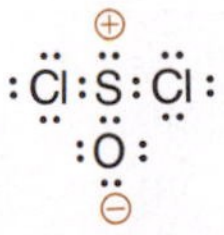

$$:\ddot{\ddot{Cl}}:\overset{\oplus}{\ddot{S}}:\ddot{\ddot{Cl}}:$$
$$:\underset{\ominus}{\ddot{O}}:$$

5   Limitations of the Octet Rule for Lewis Formulas

You may recall that representative elements achieve a noble gas electronic configuration in *most* of their compounds. For cases to which the octet rule is not applicable, of course, the relationship $S = N - A$ is not valid without modification. The following are general cases for which the relationship *must be modified* in constructing Lewis formulas.

Hydrogen always violates the octet rule because it can share only two electrons. (This was considered in Step 2, page 194.)

1. Most covalent compounds of beryllium, Be. Because Be contains only two valence shell electrons, it usually forms only two covalent bonds when it bonds to two other atoms. Therefore, we *use four electrons as the number needed by Be in Step 2, page 194; and in Steps 5 and 6, we use only two pairs of electrons for Be.*
2. Most covalent compounds of the Group IIIA elements, *especially* boron, B. The IIIA elements contain only three valence shell electrons, so they often form three covalent bonds when they bond to three other atoms. Therefore, we *use six electrons as the number needed by the IIIA elements in Step 2; and in Steps 5 and 6, we use only three pairs of electrons for the IIIA elements.*

Recall also that Lewis formulas are not normally written for compounds containing d- and f-transition metals. The d- and f-transition metals utilize d and/or f orbitals in bonding as well as s and p orbitals, and thus can accommodate more than eight valence electrons.

3. Compounds or ions containing an odd number of electrons. Examples are NO with 11 valence shell electrons and NO_2 with 17 valence shell electrons.
4. Compounds or ions in which *the central element must have a share in more than eight valence shell electrons* to accommodate all the available electrons, A. Extra rules are added to Step 4 in the previous list, as well as to Step 6, when this is encountered.

Step 4a.   If S, the number of electrons shared, is less than the number needed to bond all atoms to the central atom, then S is increased by the number of extra electrons needed.

Step 6a.   If S must be increased in Step 4a, then the octets of all the atoms will be satisfied before all A of the electrons have been added. Place the extra electrons on the central element.

Examples 6–6 through 6–9 illustrate some of these limitations and show how Lewis formulas are constructed in such cases.

Example 6–6

Draw the Lewis dot formula and dash formula for gaseous beryllium chloride, $BeCl_2$, a covalent compound.

Solution

This is an example of limitation 1.

Step 1   Skeleton is: Cl Be Cl

Step 2 $N = 2 \times 8$ (for Cl) $+ 1 \times 4$ (for Be) $= 20$ e^- needed

see limitation 1

Step 3 $A = 2 \times 7$ (for Cl) $+ 1 \times 2$ (for Be) $= 16$ e^- available

Step 4 $S = N - A = 20\ e^- - 16\ e^- = 4\ e^-$ shared

Step 5   Cl : Be : Cl

Step 6   :Cl : Be : Cl :   or   :Cl—Be—Cl :

Calculation of formal charges shows:

For Be: FC $= 2 - (2 + 0) = 0$
For Cl: FC $= 7 - (1 + 6) = 0$

Note that the chlorine atoms achieve the argon configuration, [Ar], while the beryllium atom has a share of only four electrons. Compounds such as $BeCl_2$, in which the central atom shares fewer than 8 e^-, are sometimes referred to as **electron deficient** compounds. This "deficiency" refers only to satisfying the octet rule for the central atom. The term does not imply that there are fewer electrons than there are protons in the nuclei, as in the case of a cation; the molecule is neutral.

One might try to predict a similar situation for compounds of the other IIA metals, Mg, Ca, Sr, Ba, and Ra. However, these elements have *lower ionization energies* and *larger radii than Be*, so they usually form ionic compounds by losing two electrons to achieve the configurations of the preceding noble gases.

Example 6–7

Draw the Lewis dot formula and dash formula for boron trifluoride, BF_3, a covalent compound.

Solution

This is an example of limitation 2.

Step 1   Skeleton is: F B F
(see limitation 2)

Step 2 $N = 3 \times 8$ (for F) $+ 1 \times 6$ (for B) $= 30\ e^-$ needed

Step 3 $A = 3 \times 7$ (for F) $+ 1 \times 3$ (for B) $= 24\ e^-$ available

Step 4 $S = N - A = 30\ e^- - 24\ e^- = 6\ e^-$ shared

Step 5   F:B:F

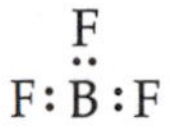

Step 6   :F:B:F: or   :F—B—F:

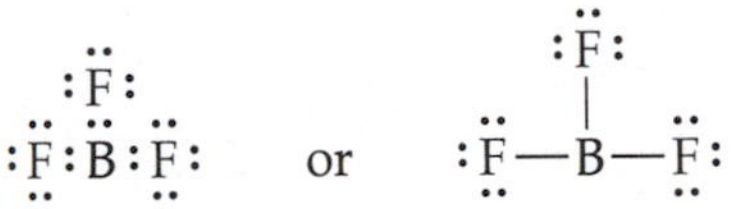

Note that each fluorine atom achieves the Ne configuration, while the boron (central) atom acquires a share of only six valence shell electrons. Calculation of formal charges shows:

For B: FC = 3 − (3 + 0) = 0
For F: FC = 7 − (1 + 6) = 0

Example 6–8

Write the Lewis dot formula and dash formula for the covalent compound phosphorus pentafluoride, PF_5.

Solution

This is an example of limitation 4.

Step 1   Skeleton is:
F F
P
F F
F

Step 2 $N = 5 \times 8$ (for F) $+ 1 \times 8$ (for P) $= 48\ e^-$ needed

Step 3 $A = 5 \times 7$ (for F) $+ 1 \times 5$ (for P) $= 40\ e^-$ available

Step 4 $S = N - A = 8\ e^-$ shared

Step 4a   Five F atoms are bonded to P. This requires the sharing of a minimum of 10 e^-, whereas only 8 e^- have been calculated in Step 4. Therefore, increase S from 8 e^- to 10 e^-. Note that only 40 e^- are available.

Step 5   (Lewis dot structure of PF_5)

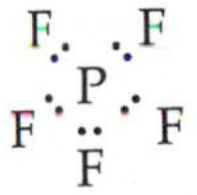

Step 6   (Lewis dot structure)   or   (dash formula)

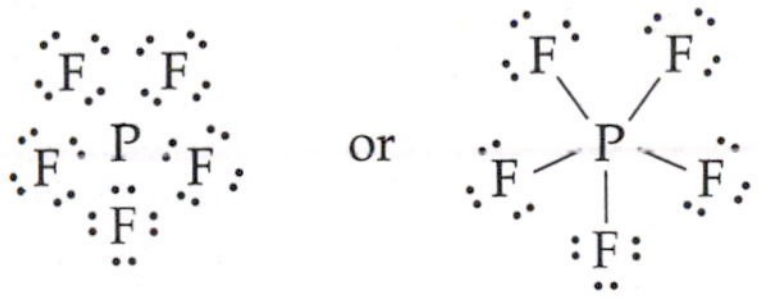

Note that each F attains the Ne configuration, while the phosphorus (central) atom has a share of 10 electrons. Calculation of formal charges shows:

For P: FC = 5 − (5 + 0) = 0
For F: FC = 7 − (1 + 6) = 0

Example 6–9

Even though the noble gases are very unreactive, the heavier ones do form some compounds, especially with fluorine. Write the Lewis dot formula and dash formula for xenon tetrafluoride, XeF_4.

Solution

This is another example of limitation 4.

Step 1   Skeleton is:
F   F
Xe
F   F

Step 2 $N = 1 \times 8$ (for Xe) $+ 4 \times 8$ (for F) $= 40 \; e^-$
needed

Step 3 $A = 1 \times 8$ (for Xe) $+ 4 \times 7$ (for F) $= 36 \; e^-$
available

Step 4 $S = N - A = 4 \; e^-$ shared

Four F atoms are bonded to the Xe atom. This requires the sharing of a minimum of 8 electrons, whereas only $4 \; e^-$ have been calculated in Step 4. Therefore, increase S from 4 to 8 (see Step 4a).

Step 5

Note that all "octets" are filled by using 32 electrons, but 36 electrons are available (see Step 3).

Step 6a   Place the extra four electrons as two lone pairs on the Xe atom.

Calculation of formal charges shows:

For Xe: $FC = 8 - (4 + 4) = 0$
For F: $FC = 7 - (1 + 6) = 0$

As the previous examples demonstrate, atoms attached to the central atom nearly always attain noble gas configurations, even when the central atom does not.

6–12   Oxidation Numbers

The **oxidation number** or oxidation state of an element in a simple binary ionic compound is the number of electrons gained or lost by an atom of that element when it forms the compound. In the cases of single-atom ions it corresponds to the actual charge on the ion. In covalent compounds, oxidation numbers do not have the same physical significance they have in ionic compounds. However, they can be useful as mechanical aids in writing formulas and in balancing equations. In such cases positive and negative oxidation numbers indicate shifts (not transfers) of electron density from one atom toward another. The more electronegative element is assigned a negative oxidation number, while the less electronegative element is assigned a positive oxidation number. We distinguish between actual charges on ions and oxidation numbers by representing the former as $n+$ or $n-$ and the latter as $+n$ and $-n$.

The most common oxidation states of the representative elements are shown in Table 6–11.

The general rules for assigning oxidation numbers are given below. These rules are not comprehensive, but they cover most cases.

A. The oxidation number of any free, uncombined element is zero. This includes polyatomic elements such as H_2, O_3, and S_8.

B. The charge on a simple (monatomic) ion is the oxidation number of the element in that ion. In a polyatomic ion, the sum of the oxidation numbers of the constituent elements must be equal to the charge on the ion.

C. In compounds (whether ionic or covalent), the sum of the oxidation numbers of all elements in the compound is zero.

It follows from these rules and the electronic configurations of the elements that we may expect the following oxidation numbers for the representative elements.

TABLE 6–11   Common Oxidation Numbers for the Representative Elements

Elements in Group	e^- Configuration		Most Common (Nonzero) Oxidation Numbers	
			Ionic Compounds	Covalent Compounds
	ns	np		
IA	↑	— — —	+1	
IIA	↑↓	— — —	+2	+2
IIIA	↑↓	↑ — —	+3, +1	+3
IVA	↑↓	↑ ↑ —	rare	−4, −3, −2, −1, +1, +2, +3, +4
VA	↑↓	↑ ↑ ↑	−3	−3, −1, +1, +3, +5
VIA	↑↓	↑↓ ↑ ↑	−2	−2, +4, +6
VIIA	↑↓	↑↓ ↑↓ ↑	−1	−1, +1, +3, +5, +7

1. The Group IA metals exhibit the +1 oxidation number in *all* their compounds. Hydrogen exhibits the +1 oxidation number in all its compounds except the metal hydrides such as NaH and CaH_2, in which hydrogen is bonded to elements less electronegative than itself; in these it exhibits the −1 oxidation number.

2. The Group IIA metals exhibit the +2 oxidation number in *all* their compounds.

3. The Group IIIA elements exhibit the +3 oxidation number in most of their *common* compounds.

4. The Group VIIA elements exhibit the −1 oxidation number in all their binary compounds with metals (and with NH_4^+). The lower members of the family (Cl, Br, I) also exhibit +1, +3, +5, and +7 oxidation numbers in covalently bonded species that contain more electronegative elements, such as ClO^-, ClO_2^-, ClO_3^-, and ClO_4^-.

5. The Group VIA elements exhibit the −2 oxidation number in binary compounds with metals (and with NH_4^+). Oxygen exhibits the −2 oxidation number in all its compounds except OF_2 (+2), the peroxides such as H_2O_2 and Na_2O_2 (−1), and the rare superoxides such as KO_2 and RbO_2 ($-\frac{1}{2}$). The lower members of the Group VIA elements also commonly exhibit +4 and +6 oxidation numbers in covalently bonded species such as SO_2, SO_3, and SF_6.

6. The Group VA elements form few binary compounds with metals, but they exhibit the −3 oxidation number in these compounds and in species such as NH_3 and NH_4^+. The VA elements exhibit the +3 oxidation number in species such as NO_2^- and PCl_3, and the +5 oxidation number in species such as NO_3^-, PO_4^{3-}, and AsO_4^{3-} as well as in covalent compounds such as PCl_5, PF_5, and P_4O_{10}.

7. The Group IVA elements exhibit the +2 and +4 oxidation states in many of their compounds. They also exhibit a variety of other oxidation states.

Example 6–10

Determine the oxidation numbers of nitrogen in the following species: (a) N_2O_4, (b) NH_3, (c) KNO_3, (d) NO_3^-, (e) N_2.

Solution

(a) Since the oxidation number of oxygen is −2 and the sum of the oxidation numbers of two nitrogens and four oxygens must be zero, the oxidation number of N is +4.

$$\text{ox. no.} \quad \overset{x}{} \quad \overset{-2}{}$$
$$N_2O_4$$

$$\text{sum of ox. no.} = 2x + 4(-2) = 0$$
$$2x - 8 = 0$$
$$x = \underline{+4}$$

(b) The oxidation number of H is $+1$ (Item 1), so that of N must be -3.

ox. no. x $+1$

$$NH_3$$

sum of ox. no. $= x + 3(+1) = 0$
$$x = \underline{-3}$$

(c) The oxidation numbers of K and O are $+1$ and -2, respectively. Therefore, that of N must be $+5$.

ox. no. $+1$ x -2

$$KNO_3$$

sum of ox. no. $= +1 + x + 3(-2) = 0$
$$x = \underline{+5}$$

(d) The charge on the ion is $1-$; therefore, the oxidation number of N is $+5$. This is the same as in KNO_3, which contains the NO_3^- ion.

ox. no. x -2

$$NO_3^-$$

sum of ox. no. $= x + 3(-2) = -1$
$$x = \underline{+5}$$

(Remember that oxidation numbers are represented as $+n$ and $-n$, while ionic charges are represented as $n+$ and $n-$. Both oxidation numbers and ionic charges are treated the same algebraically.)

(e) The oxidation number of any free element is zero.

Naming Inorganic Compounds

The rules for naming inorganic compounds were set down in 1957 by the Committee on Inorganic Nomenclature of the International Union of Pure and Applied Chemistry (IUPAC). To make naming easier we classify simple compounds in two major categories. These are **binary compounds,** those consisting of two elements, and **ternary compounds,** those consisting of three elements.

6-13   Binary Compounds

Binary compounds may be either ionic or covalent. In both cases the general rule is to name the less electronegative element first and the more electronegative element second. The more electronegative element is named by adding an "-ide" suffix to the element's characteristic (unambiguous) stem, which is derived from the name of the element. Stems for the *nonmetals* are given below:

Since millions of compounds are known, it is important to be able to associate names and formulas unambiguously and in a systematic way.

IIIA		IVA		VA		VIA		VIIA	
								H	hydr
B	bor	C	carb	N	nitr	O	ox	F	fluor
		Si	silic	P	phosph	S	sulf	Cl	chlor
				As	arsen	Se	selen	Br	brom
				Sb	antimon	Te	tellur	I	iod

Binary ionic compounds contain metal cations and nonmetal anions. The cation is named first and the anion second according to the rule above. Some examples are:

Formula	Name	Formula	Name
KBr	potassium bromide	Rb_2S	rubidium sulfide
$CaCl_2$	calcium chloride	Al_2Se_3	aluminum selenide
NaH	sodium hydride	SrO	strontium oxide

The above method is sufficient for naming binary ionic compounds containing metals that exhibit only one oxidation number other than zero (Section 6–12). However, most transition elements, and a few of the more electronegative representative metals, exhibit more than one oxidation number. These metals can form two or more binary compounds with the same nonmetal. In order to distinguish among all possibilities, the oxidation number of the metal is indicated by a Roman numeral in parentheses following its name. Some typical examples follow.

Formula	Oxidation Number of Metal	Name	Formula	Oxidation Number of Metal	Name
Cu_2O	+1	copper(I) oxide	$SnCl_4$	+4	tin(IV) chloride
CuF_2	+2	copper(II) fluoride	SnS_2	+4	tin(IV) sulfide
FeS	+2	iron(II) sulfide	PbO	+2	lead(II) oxide
Fe_2O_3	+3	iron(III) oxide	PbO_2	+4	lead(IV) oxide
$SnCl_2$	+2	tin(II) chloride			

Roman numerals are *not* necessary for metals that commonly exhibit only one oxidation number.

An older method, still in use but not recommended by the IUPAC, involves the use of "-ous" and "-ic" suffixes to indicate lower and higher oxidation numbers, respectively. This system can distinguish between only two different oxidation numbers for a metal and therefore is not as useful as the Roman numeral system. However, the older system is still widely used in many scientific, engineering, and medical fields. The advantage of the IUPAC system is that if you know the formula you can write the exact and unambiguous name, and if you are given the name you can write the formula at once.

Formula	Oxidation Number of Metal	Name	Formula	Oxidation Number of Metal	Name
$CuCl$	+1	cuprous chloride	SnF_2	+2	stannous fluoride
$CuCl_2$	+2	cupric chloride	SnF_4	+4	stannic fluoride
FeO	+2	ferrous oxide	Hg_2Cl_2	+1	mercurous chloride
$FeBr_3$	+3	ferric bromide	$HgCl_2$	+2	mercuric chloride

Pseudobinary ionic compounds contain more than two elements. In these compounds one or more of the ions consist of more than one element but behave as if they were simple ions. The most common examples of such anions are the hydroxide ion, OH^-, and the cyanide ion, CN^-. As before, the name of the anion ends in *-ide*. The ammonium ion, NH_4^+, is the most common cation that behaves like a simple metal cation. Some examples follow.

Formula	Name	Formula	Name
NH_4I	ammonium iodide	NH_4CN	ammonium cyanide
$Ca(CN)_2$	calcium cyanide	$Cu(OH)_2$	cupric hydroxide or copper(II) hydroxide
$NaOH$	sodium hydroxide	$Fe(OH)_3$	ferric hydroxide or iron(III) hydroxide

Nearly all **binary covalent compounds** involve two *nonmetals* bonded together. Although many nonmetals can exhibit different oxidation numbers, their oxidation numbers properly are *not* indicated by Roman numerals or suffixes. Instead, elemental proportions in binary covalent compounds are indicated by using a prefix system for both elements. The Greek or Latin prefixes used are mono, di, tri, tetra, penta, hexa, hepta, octa, nona, deca, undeca, and dodeca. The prefix mono is omitted except in the common (trivial) name for CO, carbon monoxide. The minimum number of prefixes required to name a compound unambiguously is used.

Formula	Name	Formula	Name
SO_2	sulfur dioxide	Cl_2O_7	dichlorine heptoxide
SO_3	sulfur trioxide	CS_2	carbon disulfide
N_2O_4	dinitrogen tetroxide	As_4O_6	tetraarsenic hexoxide

Chemists sometimes name binary covalent compounds that contain two nonmetals by the same system used to name compounds of metals that show variable oxidation states; i.e., the oxidation state of the less electronegative element is indicated by a Roman numeral in parentheses. However, we do not recommend this procedure because it is incapable of naming compounds *unambiguously*, which is the principal requirement for a system for naming compounds. For example, both NO_2 and N_2O_4 are called nitrogen(IV) oxide by this system, and the name does not distinguish between the two compounds. The compound P_4O_{10} is tetraphosphorus decoxide, which indicates clearly its composition. Using the Roman numeral system, it would be called phosphorus(V) oxide, which could lead to the incorrect formula, P_2O_5. The simplest formula for P_4O_{10} is P_2O_5, but the name for a covalent compound must indicate clearly the composition of its molecules, not just its simplest formula.

Binary acids are compounds containing hydrogen bonded to one of the more electronegative nonmetals. These compounds act as acids when dissolved in water. The pure compounds are named as typical binary compounds. Their aqueous solutions are named by modifying the characteristic stem of the nonmetal with the prefix "hydro-" and the suffix "-ic" followed by the word "acid." The stem for sulfur in this instance is "sulfur" rather than "sulf." Some typical binary acids are listed below.

Formula	Name of Compound	Name of Aqueous Solution
HCl	hydrogen chloride	hydrochloric acid, HCl(aq)
HF	hydrogen fluoride	hydrofluoric acid, HF(aq)
H_2S	hydrogen sulfide	hydrosulfuric acid, H_2S(aq)
HCN	hydrogen cyanide	hydrocyanic acid, HCN(aq)

6–14 Ternary Compounds

Ternary acids (oxyacids) are compounds of hydrogen, oxygen, and a nonmetal. A nonmetal that can exhibit more than one oxidation state can form more than one ternary acid. These ternary acids differ in the number of oxygen atoms they contain (the higher the oxidation state of the central element, the greater the number of oxygen atoms). As with binary compounds, the suffixes "-ous" and "-ic" indicate lower and higher oxidation states, respectively. These follow the stem name of the central element. One common ternary acid of each nonmetal is (somewhat arbitrarily) designated as the "-*ic acid.*" That is, it is named "stem-ic acid." The common ternary "-ic acids" are:

Periodic Group of Central Elements	IIIA	IVA	VA	VIA	VIIA
	H_3BO_3 boric acid	H_2CO_3 carbonic acid	HNO_3 nitric acid		
		H_4SiO_4 silicic acid	H_3PO_4 phosphoric acid	H_2SO_4 sulfuric acid	$HClO_3$ chloric acid
			H_3AsO_4 arsenic acid	H_2SeO_4 selenic acid	$HBrO_3$ bromic acid
				H_6TeO_6 telluric acid	HIO_3 iodic acid

There are no common ternary "-ic" acids for the omitted nonmetals. It is important to learn the names and formulas of these acids, since the names of all other ternary acids and salts are derived from them.

Acids containing *one fewer oxygen* atom per central atom are named in the same way except that the "-ic" suffix is changed to "-ous" as in the following examples. Notice that the central element has a lower oxidation number in the "-ous" acid than in the "-ic" acid.

Formula	Oxidation Number	Name	Formula	Oxidation Number	Name
H_2SO_3	+4	sulfur*ous* acid	H_2SO_4	+6	sulfur*ic* acid
HNO_2	+3	nitr*ous* acid	HNO_3	+5	nitr*ic* acid
H_2SeO_3	+4	selen*ous* acid	H_2SeO_4	+6	selen*ic* acid
$HBrO_2$	+3	brom*ous* acid	$HBrO_3$	+5	brom*ic* acid

"Hypo" means "lower."

Ternary acids that have one fewer O atom than the "-ous" acids and two fewer O atoms than the "-ic" acids are named by using the prefix "hypo-" *and* the suffix "-ous." These correspond to acids in which the oxidation state of the central nonmetal is even lower than that in the "-ous acids."

Formula	Oxidation Number	Name
$HClO$	+1	*hypo*chlor*ous* acid
H_3PO_2	+1	*hypo*phosphor*ous* acid
HIO	+1	*hypo*iod*ous* acid
$H_2N_2O_2$	+1	*hypo*nitr*ous* acid

Notice that $H_2N_2O_2$ has a 1:1 ratio of nitrogen to oxygen, as would the hypothetical HNO.

Acids containing *one more oxygen atom* per central nonmetal atom than the normal "-ic acid" are named in the same way as the "-ic acid" but with a "per-" prefix.

Formula	Oxidation Number	Name
$HClO_4$	+7	*per*chlor*ic* acid
$HBrO_4$	+7	*per*brom*ic* acid
HIO_4	+7	*per*iod*ic* acid

TABLE 6–12   Formulas, Ionic Charges, and Names for Some Common Ions

A. Common Cations			B. Common Anions		
Formula	*Charge*	*Name*	*Formula*	*Charge*	*Name*
Li^+	$1+$	lithium ion	F^-	$1-$	fluoride ion
Na^+	$1+$	sodium ion	Cl^-	$1-$	chloride ion
K^+	$1+$	potassium ion	Br^-	$1-$	bromide ion
NH_4^+	$1+$	ammonium ion	I^-	$1-$	iodide ion
Ag^+	$1+$	silver ion	OH^-	$1-$	hydroxide ion
			CN^-	$1-$	cyanide ion
Mg^{2+}	$2+$	magnesium ion	ClO^-	$1-$	hypochlorite ion
Ca^{2+}	$2+$	calcium ion	ClO_2^-	$1-$	chlorite ion
Ba^{2+}	$2+$	barium ion	ClO_3^-	$1-$	chlorate ion
Cd^{2+}	$2+$	cadmium ion	ClO_4^-	$1-$	perchlorate ion
Zn^{2+}	$2+$	zinc ion	CH_3COO^-	$1-$	acetate ion
Cu^{2+}	$2+$	copper(II) ion or cupric ion	MnO_4^-	$1-$	permanganate ion
Hg_2^{2+}	$2+$	mercury(I) ion or mercurous ion	NO_2^-	$1-$	nitrite ion
Hg^{2+}	$2+$	mercury(II) ion or mercuric ion	NO_3^-	$1-$	nitrate ion
Mn^{2+}	$2+$	manganese(II) ion or manganous ion	SCN^-	$1-$	thiocyanate ion
Co^{2+}	$2+$	cobalt(II) ion or cobaltous ion			
Ni^{2+}	$2+$	nickel(II) ion or nickelous ion	O^{2-}	$2-$	oxide ion
Pb^{2+}	$2+$	lead(II) ion or plumbous ion	S^{2-}	$2-$	sulfide ion
Sn^{2+}	$2+$	tin(II) ion or stannous ion	HSO_3^-	$1-$	hydrogen sulfite ion or bisulfite ion
Fe^{2+}	$2+$	iron(II) ion or ferrous ion	SO_3^{2-}	$2-$	sulfite ion
			HSO_4^-	$1-$	hydrogen sulfate ion or bisulfate ion
Fe^{3+}	$3+$	iron(III) ion or ferric ion	SO_4^{2-}	$2-$	sulfate ion
Al^{3+}	$3+$	aluminum ion	HCO_3^-	$1-$	hydrogen carbonate ion or bicarbonate ion
Cr^{3+}	$3+$	chromium(III) ion or chromic ion	CO_3^{2-}	$2-$	carbonate ion
			CrO_4^{2-}	$2-$	chromate ion
			$Cr_2O_7^{2-}$	$2-$	dichromate ion
			PO_4^{3-}	$3-$	phosphate ion
			AsO_4^{3-}	$3-$	arsenate ion

Let us recap the system for naming ternary acids, using the oxyacids of chlorine as examples.

Formula	Oxidation Number	Name
$HClO$	$+1$	*hypo*chlor*ous* acid
$HClO_2$	$+3$	chlor*ous* acid
$HClO_3$	$+5$	chlor*ic* acid
$HClO_4$	$+7$	*per*chlor*ic* acid

Ternary salts are compounds that result from replacing the hydrogen in a ternary acid with another ion. They usually contain metal cations or the ammonium ion. As with binary compounds, the cation is named first. The name of the anion is based on the name of the ternary acid from which it is derived. An anion derived from a ternary acid with an "-ic" ending is named by dropping the "-ic acid" and replacing it with "-ate" ion. An anion derived from an "-ous acid" is named by replacing the suffix "-ous acid" with "-ite" ion. The "per-" and "hypo-" prefixes are retained.

Formula	Name		Formula	Name
$(NH_4)_2SO_4$	ammonium sulfate (SO_4^{2-} derived from H_2SO_4)		$NaBrO_2$	sodium bromite (BrO_2^- derived from $HBrO_2$)
KNO_3	potassium nitrate (NO_3^- derived from HNO_3)		$FePO_4$	iron(III) phosphate (PO_4^{3-} derived from H_3PO_4)
$Ca(NO_2)_2$	calcium nitrite (NO_2^- derived from HNO_2)		$NaClO$	sodium hypochlorite (ClO^- derived from $HClO$)
$LiClO_4$	lithium perchlorate (ClO_4^- derived from $HClO_4$)			

Acid salts are salts containing anions derived from ternary acids in which one or more acidic hydrogen atoms remain. These salts are named in the same way as they would be if they were the usual type of ternary salt, except that the word "hydrogen," or "dihydrogen," is inserted after the name of the metal cation to indicate the number of acidic hydrogen atoms:

Formula	Name		Formula	Name
$NaHSO_4$	sodium hydrogen sulfate		K_2HPO_4	potassium hydrogen phosphate
$NaHSO_3$	sodium hydrogen sulfite		$NaHCO_3$	sodium hydrogen carbonate
KH_2PO_4	potassium dihydrogen phosphate			

An older, commonly used method (which is not recommended by the IUPAC) involves the use of the prefix "bi-" attached to the name of the anion to indicate the presence of an acidic hydrogen. According to this system, $NaHSO_4$ is called sodium bisulfate and $NaHCO_3$ is named sodium bicarbonate.

A list of common cations and anions is given in Table 6–12.

Key Terms

Actinides   elements 90 to 103 (after actinium).

Alkali metals   metals of periodic Group IA.

Alkaline earths   elements of periodic Group IIA.

Anion   a negative ion; an atom or group of atoms that has gained one or more electrons.

Atomic radius   radius of an atom.

Binary compound   compound consisting of two elements.

Cation   a positive ion; an atom or group of atoms that has lost one or more electrons.

Chemical bonds   the attractive forces that hold atoms together in compounds; range in character from ionic to covalent, with most bonds having character somewhere between these extremes.

Covalent bond   chemical bond formed by the sharing of one or more electron pairs between two atoms.

Covalent compounds   compounds containing predominantly covalent bonds.

Double bond   covalent bond resulting from the sharing of four electrons (two pairs) between two atoms.

***d*-Transition elements (metals)**   B Group elements other than IIB in the periodic table; sometimes simply transition elements.

Effective nuclear charge   the difference between the nuclear charge and the number of electrons that effectively shield the outermost electrons from the nucleus.

Electrical conductivity   a measure of a substance's ability to conduct electricity.

Electron affinity   the amount of energy involved in the process in which an electron is added to a neutral isolated gaseous atom to form a gaseous ion with a 1– charge; has a negative value if energy is released.

Electron deficient compound   compound in which the central atom achieves a share of fewer than eight electrons in its outermost shell.

Electronegativity   a measure of the relative tendency of an atom to attract electrons to itself when chemically combined with another atom.

Family   see *Group*.

***f*-Transition elements**   see *Inner transition elements*.

Group a vertical column of elements in the periodic table.

Halogens elements of periodic Group VIIA.

Inner transition elements elements 58 to 71 and 90 to 103; also called *f*-transition elements.

Ionic bonding chemical bonding resulting from the transfer of one or more electrons from one atom or group of atoms to another.

Ionic compounds compounds containing predominantly ionic bonding.

Ionic radius radius of an ion.

Ionization energy the minimum amount of energy required to remove the most loosely held electron of an isolated gaseous atom or ion.

Isoelectronic having the same electron configurations.

Kernel nucleus of an atom plus its nonvalence electrons or core electrons.

Lanthanides elements 58 to 71 (after lanthanum).

Lewis dot formula representation of a molecule or formula unit by showing atomic symbols and only outer shell electrons; does not show shape.

Lone pair pair of electrons residing on one atom and not shared by other atoms; unshared pair.

Metal an element below and to the left of the stepwise division (metalloids) in the upper right corner of the periodic table; includes about 80% of all known elements.

Metalloids elements with properties intermediate between metals and nonmetals: B, Al, Si, Ge, As, Sb, Te, Po, and At.

Noble gas configuration the stable electronic configuration of a noble gas.

Noble gases elements of periodic Group 0; also called rare gases; formerly called inert gases.

Nonmetals elements above and to the right of the metalloids in the periodic table.

Nuclear shielding see *Shielding effect.*

Octet rule many representative elements attain a share of eight electrons in their valence shells when they form molecular or ionic compounds; there are limitations.

Overlap of orbitals: the interaction of orbitals of different atoms in the same region of space.

Oxidation numbers arbitrary numbers that can be used as mechanical aids in writing formulas and balancing equations; for single-atom ions they correspond to the charge on the ion; more electronegative atoms are assigned negative oxidation numbers.

Period a horizontal row of elements in the periodic table.

Periodicity regular periodic variations of properties of elements with atomic number (and position in the periodic table).

Periodic law the properties of the elements are periodic functions of their atomic numbers.

Periodic table an arrangement of elements in order of increasing atomic number that also emphasizes periodicity.

Pseudobinary ionic compounds compounds that contain more than two elements but are named like binary compounds.

Rare earths see *Inner transition elements.*

Rare gases see *Noble gases.*

Representative elements A Group elements in the periodic table.

Resonance concept in which two or more equivalent dot formulas (resonance structures) are necessary to describe the bonding in a molecule or ion.

Semiconductor a substance that does not conduct electricity at room temperature but does at higher temperatures.

Shielding effect electrons in filled sets of *s* and *p* orbitals between the nucleus and outer shell electrons fairly effectively shield the outer shell electrons from the attraction of an equal number of protons in the nucleus; also called screening effect.

Single bond covalent bond resulting from the sharing of two electrons (one pair) between two atoms.

Ternary compound compound containing three elements.

Triple bond covalent bond resulting from the sharing of six electrons (three pairs) between two atoms.

Unshared pair see *Lone pair.*

Valence number of electrons that are shared, gained, or lost by an atom of an element.

Valence electrons electrons in the outermost shell of an atom; usually those involved in bonding.

Exercises

The Periodic Table

1. In your own words, what does the term *periodicity* mean?
2. What was Mendeleev's contribution to the construction of the modern periodic table?
3. State the periodic law. What does it mean?
4. Mendeleev's periodic table was based on increasing atomic *weight*, while the modern periodic table is based on increasing atomic *number*. In the modern table argon comes before potassium, yet it has a higher atomic weight. Explain how this can be.
5. Consult a handbook of chemistry and look up melting points of the elements of periods 2 and 3. Show that melting point is a property that varies periodically for these elements.
6. Which of the following is a better example of deductive logic based on experimental observations? Why?
   (a) Noble gas electron configurations must be stable *because* noble gases are quite unreactive.
   (b) Noble gases must be unreactive *because* they have only filled electronic sublevels.
7. Distinguish between the following terms clearly and concisely, and provide specific examples of each: groups or families of elements, and periods of elements. Write symbols for the alkali metals, the Group IIIA elements, and the Group IVB elements.
8. How do properties of elements vary within a group? Cite some specific examples.
9. How do properties of elements vary within a period? Cite some specific examples.
10. Define and illustrate the following terms clearly and concisely: (a) metals, (b) nonmetals, (c) noble gases, (d) representative elements, (e) *d*-transition elements, (f) inner transition elements.
11. Explain why period 1 contains two elements, period 2 contains eight, period 3 contains eight, and period 4 contains eighteen elements.
12. Why does period 3 contain only eight elements? The third major energy level ($n = 3$) has *s*, *p*, and *d* sublevels.
13. How many elements make up period 6? Account for this number of elements.

14. What do Lewis dot representations for atoms show? Draw Lewis dot representations for the following atoms: H, Li, N, Na, S, Ar, Ba, Sn, Se, As.
15. Draw Lewis dot representations for the following atoms: He, C, F, Al, P, Ca, As, Sb, Xe.
16. How do the physical properties of metals differ from those of nonmetals?
17. What are metalloids? List four examples.
18. Compare the metals and nonmetals with respect to (a) number of outer shell electrons, (b) ionization energies, (c) electron affinities, and (d) electronegativities.
19. Of the Group VA elements, one is distinctly metallic and two are distinctly nonmetallic. Identify them and explain why.
20. By referring only to a periodic table, arrange the following in order of increasing metallic character: S, Be, Cl, Ge, and Rb.

Atomic Radii

21. What is the meaning of the following statement? It is impossible to describe an atom as having an invariant radius.
22. Why do atomic radii increase from top to bottom within a group of elements in the periodic table?
23. Why do atomic radii decrease from left to right within a period in the periodic table?
24. Arrange the members of each of the following sets of atoms in order of increasing atomic radii: (a) the Group IA elements, (b) the halogens, (c) the noble gases, (d) the elements in the third period, (e) Be, F, Na, I, and Te.
25. Why are variations in the atomic radii of the transition elements not so pronounced as those of the representative elements?

Ionization Energy

26. Define: (a) first ionization energy, (b) second ionization energy.
27. For a given element, why is the second ionization energy always greater than the first ionization energy?
28. Arrange the members of each of the following sets of elements in order of increasing first ionization energies: (a) the alkaline earth metals, (b) the Group VIA elements, (c) the

elements in the third period, (d) Li, Rb, Be, Sr, F, and I.

29. Explain why there is a general increase in first ionization energy across each period.
30. Explain the trend in first ionization energy upon descending a periodic group.
31. In a plot of first ionization energy versus atomic number for periods 2 and 3, "dips" occur at the IIIA and VIA elements. Account for these dips.
32. What is the general relationship between the sizes of the atoms of period 2 and their first ionization energies? Rationalize the relationship.
33. Why must a zinc atom absorb more energy than a calcium atom in order to ionize a $4s$ electron?
34. On the basis of data in Table 6–6, would you expect a Mg^{3+} ion to be stable? Why or why not? How about Al^{3+}?
35. How much energy, in kJ, must be absorbed by 1.00 mole of gaseous sodium atoms to convert them all to gaseous Na^+ ions?

Electron Affinity

36. What is electron affinity?
37. Arrange the members of each of the following sets of elements in order of increasing electron affinities: (a) the Group IA metals, (b) the Group VIIA elements, (c) the elements in the second period, (d) Li, K, C, F, and I.
38. The electron affinities of the halogen atoms are much more negative than those of the elements of Group VIA. Why is this expected?

Ionic Radii

39. Compare the sizes of cations and the neutral atoms from which they are formed by citing three specific examples.
40. Arrange the members of each of the following sets of cations in order of increasing ionic radii:
    (a) Li^+, Rb^+, Cs^+, Na^+, K^+
    (b) Na^+, Mg^{2+}, Al^{3+}
    (c) Na^+, Cs^+, Be^{2+}, Ga^{3+}
41. Compare the sizes of anions and the neutral atoms from which they are formed by citing three specific examples.
42. Arrange the following sets of anions in order of increasing ionic radii:
    (a) F^-, I^-, Cl^-, Br^-

(b) O^{2-}, N^{3-}, F^-
(c) Se^{2-}, I^-, P^{3-}, O^{2-}

43. Compare and explain the relative sizes of H^+, H, and H^-.
44. Most transition metals can form more than one simple positive ion. For example, iron forms both Fe^{2+} and Fe^{3+} ions, and copper forms both Cu^+ and Cu^{2+} ions. Which is the smaller ion of each pair, and why?

Electronegativity

45. What is electronegativity?
46. Arrange the members of each of the following sets of elements in order of increasing electronegativities:
    (a) Li, F, O, Be   (b) Be, Ba, Ca, Mg
    (c) O, Se, Te, S   (d) Ca, Cs, O, Cl, C
47. Which of the following statements is better? Why?
    (a) The electronegativity of fluorine is high *because* fluorine has a strong attraction for electrons in a chemical bond.
    (b) Fluorine has a strong attraction for electrons in a chemical bond *because* it has a high electronegativity.
48. Comment on the validity of the following statement. Sodium has a low electronegativity because it forms sodium ions, Na^+, readily.

Chemical Bonds — Basic Ideas

49. What are chemical bonds?
50. Distinguish between ionic and covalent bonding. Why are covalent bonds called directional bonds, whereas ionic bonding is called nondirectional?
51. Construct a table in which four properties of ionic and covalent compounds are compared and contrasted.
52. Look up the properties of HCl and $MgCl_2$ in a handbook of chemistry. Why is HCl classified as a covalent compound and $MgCl_2$ classified as an ionic compound?
53. The following properties can be found in a handbook of chemistry:
    potassium bromide, KBr — colorless cubic crystals; specific gravity 2.75 at 20°C; melting point 730°C, boiling point 1435°C; solubility in cold water 53.48 g KCl/100 mL H_2O
    carbon tetrachloride, CCl_4 — colorless liquid; specific gravity 1.5867 at 20°C, melting point

−23°C; boiling point 76.8°C; insoluble in cold water.

Would you classify each of these compounds as ionic or covalent? Why?

54. Based on the positions of the following pairs of elements in the periodic table, predict whether bonding between the two would be primarily ionic or covalent. Justify your answers. (a) O and Cl, (b) K and O, (c) B and O, (d) N and P, (e) Si and Si, (f) Ca and Br

55. Predict whether the bonding between the following pairs of elements would be ionic or covalent. Justify your answers. (a) Ba and Br, (b) C and S, (c) Na and S, (d) P and O, (e) Ca and Se, (f) As and I.

56. Classify the following compounds as ionic or covalent: (a) $SiBr_4$, (b) $Ca(NO_3)_2$, (c) NO_2Cl, (d) SeO_2, (e) Na_2SO_4 (f) H_3PO_4, (g) Br_2, (h) LiF, (i) BF_3

Ionic Bonding

57. Write electronic configurations for the following ions: (a) Ca^{2+}, (b) Cl^-, (c) Al^{3+}, (d) O^{2-}, (e) Bi^{3+}.

58. When a d-transition metal undergoes ionization, it loses its outer s electrons before it loses any d electrons. Using [Noble Gas]$(n-1)d^x$ representations, write the outer electron configurations for the following ions: (a) Fe^{2+}, (b) Cu^+, (c) Cu^{2+}, (d) Mn^{3+}, (e) Cr^{3+}, (f) Ag^+ (g) Cd^{2+}

59. Which of the following do not accurately represent stable binary ionic compounds? Why? RbF, Cs_2S, Mg_2O_3, AlF_2, Li_2O, Ba_3N, $SrCl_2$.

60. Which of the following do not accurately represent stable binary ionic compounds? Why? $CaBr_2$, RbO, $InCl_3$, NaF, Ca_3S_2, MgI_3, BaO.

61. Write chemical equations for reactions between the following pairs of elements. Draw electronic structures of the atoms before reaction as well as electronic structures of the ions formed in the reactions, using both the $\uparrow\downarrow$ and $ns^x np^y$ notations, where n refers to the outermost occupied shell.
    (a) lithium and chlorine
    (b) potassium and sulfur
    (c) calcium and fluorine
    (d) magnesium and bromine
    (e) barium and phosphorus
    (f) lithium and nitrogen

62. Draw Lewis dot formulas for all atoms and ions in Exercise 61.

63. What are isoelectronic species? Cite three examples.

64. All but one of the following species are isoelectronic. Which one is not? O^{2-}, F^-, Ne, Na, Mg^{2+}, Al^{3+}.

65. All but one of the following species are isoelectronic. Which one is not? P^{3+}, S^{2-}, Cl^-, Ar, K^+, Ca^{2+}.

66. Write formulas for two cations and two anions that are isoelectronic with neon.

67. Write formulas for two cations and two anions that are isoelectronic with argon.

68. Write formulas for two cations that have the following electronic configurations in their highest occupied energy level.
    (a) . . . $4s^2 4p^6$ (b) . . . $5s^2 5p^6$

69. Write formulas for two anions that have the electronic configurations listed in Exercise 68.

70. Write chemical equations for the following reactions. Draw electronic structures of the atoms before reaction as well as electronic structures of the ions formed in these reactions, using both the $\uparrow\downarrow$ and $ns^x np^y$ notations.
    (a) reaction of lithium with sulfur
    (b) reaction of magnesium with chlorine
    (c) reaction of aluminum with fluorine

71. Draw Lewis dot formulas for all atoms and ions in Exercise 70.

72. Using M as the general symbol for metals and X as the general symbol for nonmetals, write general equations for the reactions of the following pairs of elements. Don't forget to include charges on ions.
    (a) A Group IA metal with a Group VIIA element.
    (b) A Group IIA metal with a Group VIIA element.
    (c) A Group IIA metal with a Group VA element.

73. Draw Lewis dot formulas for all atoms and ions in Exercise 72.

Covalent Bonding — Basic Concepts

74. What is the nature of the attractive forces between atoms that are covalently bonded together? Describe these forces.

75. What does Figure 6–7 tell us about the attrac-

tive and repulsive forces in a hydrogen molecule?

76. Distinguish between heteronuclear and homonuclear diatomic molecules.

77. Distinguish among single, double, and triple covalent bonds.

Lewis Dot Formulas for Molecules and Polyatomic Ions

78. (a) What are Lewis dot formulas? (b) Write Lewis dot formulas for the following elements and compounds: H_2, N_2, Br_2, HCl, HI.

79. What are polyatomic ions?

80. Write Lewis dot formulas for the following molecules and polyatomic ions: (a) H_2O and H_2S, (b) NH_3 and NH_4^+, (c) PH_3 and PH_4^+

81. What is the octet rule? Is it generally applicable to compounds of the transition metals? Why?

82. (a) What is the simple mathematical relationship that is useful in writing Lewis dot formulas? (b) What does each term in the relationship represent? (c) Show how the rule may be applied to the molecules and ions listed in Exercise 80.

83. What do we mean when we refer to resonance?

84. Draw resonance structures for (a) HCO_3^-, (b) NO_3^-, (c) SO_2, and (d) SO_3. In (a), the H atom is bonded to an O atom and all O atoms are bonded to the carbon atom.

85. (a) What are the limitations of the $S = N - A$ relationship? (b) Cite examples of three species to which the relationship does not apply, and explain why it does not.

86. All of the following are covalently bonded. Which ones violate the octet rule? (a) PF_3, (b) CO_2, (c) HCN, (d) BeI_2

87. All of the following are covalently bonded. Which ones violate the octet rule? (a) SF_4, (b) H_3PO_4

88. Draw the Lewis dot formula for the sulfur molecule, S_8, which is an eight-membered ring.

89. Draw Lewis dot formulas for the following anions: (a) ClO^-, (b) ClO_2^-, (c) ClO_3^-, and (d) ClO_4^-

90. Draw Lewis dot formulas for the following molecules: (a) CCl_4, (b) SiH_4, (c) SO_3, (d) H_2SO_3, (e) NOCl, and (f) XeF_2

91. Draw Lewis dot formulas for the following molecules: (a) Si_2H_6, (b) AsF_5, (c) HCN, (d) $HClO_4$, (e) C_2Cl_4, and (f) OF_2

92. El is the general symbol for a representative element. In each case, state the periodic group in which the element is located. Justify your answers and cite specific examples.

(a) $H - \overset{\cdot\cdot}{\underset{\cdot\cdot}{O}} - \overset{\overset{\textstyle :\overset{\cdot\cdot}{O}:}{|}}{El} - \overset{\cdot\cdot}{\underset{\cdot\cdot}{O}}:$

(b) $:\overset{\cdot\cdot}{\underset{\cdot\cdot}{Br}} - \overset{\cdot\cdot}{\underset{\cdot\cdot}{El}} - \overset{\cdot\cdot}{\underset{\cdot\cdot}{Br}}:$

(c) $:\overset{\cdot\cdot}{\underset{\cdot\cdot}{Cl}} - \overset{\overset{\textstyle \overset{\cdot}{O}\overset{\cdot}{}}{\|}}{El} - \overset{\cdot\cdot}{\underset{\cdot\cdot}{Cl}}:$

93. El is the general symbol for a representative element. In each case, state the periodic group in which the element is located. Justify your answers and cite specific examples.

(a) $:\overset{\cdot\cdot}{\underset{\cdot\cdot}{O}} - \overset{\overset{\textstyle :\overset{\cdot\cdot}{O}:}{|}}{\underset{\underset{\textstyle :\overset{\cdot\cdot}{O}:}{|}}{El}} - \overset{\cdot\cdot}{\underset{\cdot\cdot}{O}}:$ $^{2-}$

(b) $:\overset{\cdot\cdot}{\underset{\cdot\cdot}{O}} - \overset{\overset{\textstyle :\overset{\cdot\cdot}{O}:}{}}{\underset{\underset{\textstyle :\overset{\cdot\cdot}{O}:}{|}}{El}} - \overset{\cdot\cdot}{\underset{\cdot\cdot}{O}}:$ $^{-}$

(c) $:\overset{\cdot\cdot}{\underset{\cdot\cdot}{O}} - \overset{\underset{\textstyle :\overset{\cdot\cdot}{O}:}{|}}{\underset{}{El}}:$ $^{-}$

94. Draw Lewis dot formulas for (a) H_2O_2, (b) ClF_3, (c) N_2H_4, (d) IF_7, (e) $AlCl_4^-$, and (f) CN^-.

95. Many common stains, such as chocolate and those of other fatty foods, can be removed by dry-cleaning solvents such as tetrachloroethylene, C_2Cl_4. Is C_2Cl_4 ionic or covalent? Draw its Lewis dot formula.

96. Draw acceptable dot formulas for the following common air pollutants. (a) SO_2, (b) NO_2, (c) CO, (d) O_3 (ozone), (e) SO_3, and (f) $(NH_4)_2SO_4$. Which one is a solid? Which ones exhibit resonance?

Oxidation Numbers

97. What are oxidation numbers?

98. List the common oxidation numbers for the A Group elements in binary compounds.

99. What oxidation numbers are the following elements expected to exhibit in simple ionic compounds? Li, Mg, O, F, Cs, Al, N.

100. What oxidation numbers are the following elements expected to exhibit in simple ionic compounds? K, Ca, S, Br, Cs, I.

101. Determine the oxidation number of phosphorus in each of the following species.

   (a) PH_3 (b) P_4

   (c) P_4O_6 (d) H_3PO_2

   (e) H_3PO_4 (f) NaH_2PO_4

   (g) $Ca_2P_2O_7$

102. Determine the oxidation number of manganese in each of the following species.

   (a) $MnCl_3$ (b) $KMnO_4$

   (c) Mn_2O_7 (d) K_2MnO_4

   (e) Mn_2O_3 (f) MnO_2

   (g) MnO

Inorganic Nomenclature

103. Write formulas for the compounds that are formed by the following pairs of ions. On a separate sheet, write the name for each compound.

	A	B	C	D	E	F
	OH^-	Cl^-	NO_3^-	SO_4^{2-}	CO_3^{2-}	PO_4^{3-}
1. Na^+						
2. NH_4^+						
3. Ca^{2+}						
4. Fe^{2+}						
5. Cu^{2+}						
6. Zn^{2+}						
7. Cr^{3+}						
8. Fe^{3+}						

104. Write formulas for the following compounds: (a) magnesium sulfide, (b) calcium hydroxide, (c) lithium acetate, (d) hydroiodic acid, (e) aluminum nitrate, (f) copper(II) sulfate, (g) barium phosphate, (h) ferric bromide.

105. Write formulas for the following compounds: (a) copper(I) cyanide, (b) ferrous sulfate, (c) ammonium phosphate, (d) nickel(II) nitrite, (e) magnesium acetate, (f) perbromic acid, (g) chloric acid, (h) mercuric chloride.

106. Write names for the following compounds: (a) FeI_2, (b) $CrCl_3$, (c) $NaNO_3$, (d) $CaSO_4$, (e) BaS, (f) $Al(CN)_3$, (g) $KClO_4$, (h) $Al(OH)_3$, (i) $Sr(MnO_4)_2$, (j) $Na_2Cr_2O_7$.

107. Write names for the following compounds: (a) $Ba(NO_3)_2$, (b) $Cu(SCN)_2$, (c) K_2CrO_4, (d) CrO_3, (e) $Cr(CH_3COO)_3$, (f) Li_2S, (g) $NaClO_3$, (h) $NaClO$.

108. Write names for the following compounds: (a) Hg_2Br_2, (b) $AgNO_3$, (c) P_4O_6, (d) MnO_2, (e) $Zn(ClO_4)_2$, (f) $Ca_3(AsO_4)_2$, (g) K_2CO_3, (h) $Co(HCO_3)_2$.

109. Write names for the following compounds: (a) N_2O_3, (b) $CrAsO_4$, (c) $(NH_4)_2CO_3$, (d) $CoBr_3$, (e) PCl_5, (f) $FeCl_2$, (g) Na_2CrO_4, (h) $CuSO_3$, (i) Li_2SO_4, (j) KNO_2.

110. Write formulas for the following compounds: (a) hydrofluoric acid, (b) chromic hydroxide, (c) silver sulfide, (d) barium hydrogen sulfate, (e) zinc permanganate, (f) calcium iodate, (g) magnesium phosphate, (h) sulfurous acid.

Covalent Bonding I: Molecular and Ionic Geometry

7

Understanding the nature of the covalent bond is one of the central challenges of chemistry. Bonding is the key to molecular structure, and structure is intimately related to the physical and chemical properties of a compound. In Chapter 6 we introduced some of the basic ideas of chemical bonding and showed how Lewis dot and dash formulas can be constructed for molecules and polyatomic ions. We shall now apply the ideas we have developed to predict the *three-dimensional structures* of these species. Then we shall begin to study the ways in which three-dimensional structures affect the properties of substances by studying polarities of molecules. The polarity of a molecule is a measure of its tendency to adopt a preferred orientation in an electrical field. Polarity itself is influenced both by the three-dimensional structure of a molecule and by the distribution of electrons in the molecule.

7–1   Basic Notions of Bonding Theory

The bonding theories that are currently accepted allow us to predict structures and properties that are usually accurate (although they are not always entirely satisfactory). As always, we should keep in mind the truism that whatever we propose *must* be consistent with experimentally determined facts. When there is disagreement between facts and theory, theory must be modified to accommodate *all* known facts. In this chapter, we must account for a large body of knowledge about molecular structure, all of it based on reliable experiments.

We shall discuss three theories. The first is the Valence Shell Electron Pair Repulsion (VSEPR) theory, which assumes that electron pairs are arranged around the central element of a compound in such a way that there is maximum *separation* (and, therefore, minimum repulsion) among electron pairs. Although this statement may appear to be obvious (it is), the idea will prove to be remarkably useful in predicting the geometries of molecules and polyatomic ions. The second theory is the Valence Bond (VB) theory, which will be discussed in the next chapter. The VB theory explains the bonding in terms of overlapping atomic orbitals. These two theories go hand-in-hand in any discussion of covalent molecules. The third theory, also discussed in the next chapter, is called Molecular Orbital theory; it provides an alternate description of bonding.

We shall also introduce one additional concept that will allow us to "mix" the atomic orbitals discussed in Chapter 5 to form new orbitals with different spatial orientations. This process of mixing, called *hybridization,* is often necessary to explain experimental results or to achieve the structures predicted by the VSEPR theory. The VB theory applies to hybrid atomic orbitals just as it does to "pure" atomic orbitals.

Throughout the discussion of covalent bonding we shall focus attention on the electrons in the outer shell, or valence shell, of the atoms because these are the electrons involved in bonding. To make things easier, we shall draw a Lewis dot formula for each molecule we discuss. Recall that dot formulas show only the valence shell electrons, which may be thought of as those that were not present in the preceding noble gas, ignoring filled sets of *d* and *f* orbitals.

7–2   Valence Shell Electron Pair Repulsion (VSEPR) Theory

The qualitative idea that geometries of molecules and ions can be predicted by considering the number of regions of high electron density about the central atom is very useful. This theory is known as the **Valence Shell Electron Pair Repulsion theory,** or **VSEPR theory.** It is based on the following central idea.

The valence shell electron pairs on an atom repel one another, and these repulsions determine the shapes of polyatomic molecules and ions.

From a geometric viewpoint, we are most concerned with the repulsions in the valence shell of the *central atom* of a molecule or polyatomic ion. A **central atom** is *any atom that is bonded to more than one other atom.* The repulsions among the bonding *and* lone (unshared) valence electron pairs on the central atom determine the arrangement of the outer atoms around the central atom. In some molecules, more than one central atom may be present. In such cases, the arrangement around each is determined in turn, to build up a picture of the overall shape of the molecule or ion.

Around the central atom, we determine "electronic geometry" by first counting the number of **stereoactive sets of electrons** in the valence shell, according to the following rules:

1. Each set of electrons that is used to bond another atom to the central atom is counted as *one* stereoactive set of electrons, whether the bonding is single, double, or triple.
2. Each *unshared* pair, or **lone pair,** of valence electrons on the central atom counts as *one* stereoactive set of electrons.

The main idea of the VSEPR theory is that stereoactive sets of electrons repel one another, so that they are arranged as far apart from one another as possible. The resulting *arrangement of stereoactive sets of electrons* is referred to as the **electronic geometry** of the central atom.

Table 7–1 shows the relation between the common numbers of stereoactive sets of electrons and the corresponding electronic geometries. Consideration of the number of stereoactive sets of electrons that connect the central atom to other atoms can then lead to a prediction (or explanation) of the arrangement of the other atoms about the central atom.

TABLE 7–1   Geometry of Stereoactive Sets of Electrons About a Central Atom

Number of Stereoactive Sets of Electrons	Electronic Geometry*	Angles†
2	linear	180°
3	trigonal planar	120°
4	tetrahedral	109°28′
5	trigonal bipyramidal	90°, 120°, 180°
6	octahedral	90°, 180°

* Electronic geometries are illustrated using only single pairs of electrons as stereoactive sets of electrons.
† Angles made by imaginary lines through the central atom's nucleus and the centers of regions of high electron density.

To see how these five electronic geometries are derived, think of a set of sticks hinged at a common point (the central atom) and with each pair of sticks joined by compressed springs. With only two sticks, any bending at the hinge would result in further compressing one spring; the equilibrium position is *linear* or 180° apart. With three sticks, equilibrium occurs when they all lie in the same plane and are 120° apart. If one of the sticks bends toward its neighbor in the plane, or bends up or down out of the plane, one or more springs would be compressed and would push the stick back into position. This is the *trigonal planar* arrangement.

Four sticks assume an equilibrium position in which each stick points toward a corner of a tetrahedron, a four-sided pyramid whose faces are equilateral triangles. Therefore, this is called the *tetrahedral* arrangement. (It is an interesting exercise in solid geometry to show that the angles between the sticks are all 109°28′. Start by recognizing that the corners of a tetrahedron are also four of the eight corners of a cube: for instance, top front left, top back right, bottom front right, and bottom back left.)

With five sticks, the most stable arrangement consists of three in the trigonal planar arrangement and the other two lying at right angles above and below the plane of the triangle. There are two different sets of angles: 120° between sticks in the plane (called "equatorial"), and 90° between any equatorial stick and those at the top and bottom (called "axial"). Of course, there is a 180° angle between the two axial sticks. Together, the five sticks point toward the corners of a trigonal bipyramid, which can be thought of as two distorted tetrahedra joined face-to-face. Thus, this is called *trigonal bipyramidal* geometry.

Finally, six sticks come to equilibrium as three sets of linear pairs at right angles to each other, like a set of three-dimensional axes. Each stick is 90° from its nearest neighbors and 180° from one other stick. This can also be considered as pointing toward the corners of an octahedron, an eight-sided solid whose faces are equilateral triangles, so this is called the *octahedral* geometry.

Any one of the equatorial sticks can be pushed into an axial position; the repulsions will automatically push one of the axial sticks into an equatorial position to compensate.

We are now ready to study the structures of some simple molecules and polyatomic ions, starting with the simplest examples. In Sections 7–3 through 7–7 we shall describe species whose central atoms have 2, 3, 4, 5, and 6 stereoactive sets of electrons, respectively, around them. For each type of molecule and polyatomic ion we shall first give the known (experimentally determined) facts about shape. Then, after drawing the Lewis dot formula, we shall show that the observed shapes are predictable by the VSEPR theory. In Chapter 8 many of the same species also will be treated in terms of Valence Bond theory so that we can develop a more detailed picture of *how* the bonds form in these species.

7–3   Two Stereoactive Sets of Electrons About a Central Atom

BeF_2 is ionic rather than covalent.

There are several molecules and ions that consist of a central atom plus two atoms of another element, in which there are no lone pairs on the central atom. This kind of molecule is abbreviated as AB_2. Typical compounds include CO_2, $BeCl_2$, $BeBr_2$, and BeI_2, as well as CdX_2 and HgX_2 where X = Cl, Br, or I. A polyatomic ion in this category is the nitronium ion, NO_2^+. All of these are known to be **linear** (bond angle = 180°), as shown in the first row of Table 7–1.

1   Beryllium Chloride, BeCl₂

In the solid state, $BeCl_2$ molecules are bonded to each other in a polymeric solid, i.e., a solid in which many small molecules are bonded together into a large molecule. However, $BeCl_2$ exists as discrete molecules in the gaseous state, and we shall focus our attention on this form. We drew the Lewis formula for $BeCl_2$ in Example 6–6. It shows two single covalent bonds, with Be and Cl each contributing one electron to each bond.

> Remember that most covalent compounds of Be violate the octet rule (see Section 6–11.5).

$$:\ddot{C}l:Be:\ddot{C}l: \quad \text{or} \quad :\ddot{C}l{-}Be{-}\ddot{C}l:$$

Beryllium is the central atom. It is bonded to two other atoms and has no lone pairs of valence electrons, so it has *two* stereoactive sets of electrons. Valence Shell Electron Pair Repulsion theory, which assumes that regions of high electron density (electron pairs) will be as far from one another as possible, predicts that the two electron pairs on Be will be 180° apart. Thus VSEPR theory *predicts* a **linear** structure for $BeCl_2$, and all other molecules of this type, which is consistent with experimental observations.

$$\overset{180°}{:\ddot{C}l{-}Be{-}\ddot{C}l:}$$

2   Nitronium Ions, NO₂⁺

Let us now consider the nitronium ion, NO_2^+, as an example of an AB_2 *ion* with two sets of stereoactive electrons and no lone pairs of electrons on the central atom, nitrogen. The electronic configurations of N and O in their ground states are shown below. We first construct the dot formula for NO_2^+, remembering that it contains one fewer electron than the sum of the three atoms.

	1s	**2s**	**2p**		
$_7$N	↑↓	↑↓	↑	↑	↑
$_8$O	↑↓	↑↓	↑↓	↑	↑

$$S = N - A$$
$$= 24\,e^- - 16\,e^-$$
$$= 8\,e^-\text{ shared}$$

$$:\ddot{O}::N::\ddot{O}:\overset{+}{} \quad \text{or} \quad :\ddot{O}{=}N{=}\ddot{O}:\overset{+}{}$$

Again we have a case in which the central atom has two stereoactive sets of electrons. Thus VSEPR theory predicts, and experimental observations confirm, that the ion is **linear,** which gives maximum separation between the two regions of high electron density about the central atom.

> All of the electrons that bond one atom to a central atom count as only *one* stereoactive set, even though the bond is a double bond.

$$\overset{180°}{:\ddot{O}{=}N{=}\ddot{O}:\overset{+}{}}$$

7–4   Three Stereoactive Sets of Electrons About a Central Atom

1   Boron Trifluoride, BF₃

Boron is a Group IIIA element that forms many covalent compounds by bonding to three other atoms. Typical examples include boron trifluoride, BF_3, boron trichloride, BCl_3, boron tribromide, BBr_3, and boron triiodide, BI_3. All are **trigonal planar** (that is, flat molecules in which all three bond angles are 120°).

In Example 6 – 7 we drew the Lewis formula for BF_3. Since F, Cl, Br, and I are members of Group VIIA, the Lewis formulas for BCl_3, BBr_3, and BI_3 are similar.

$$:\ddot{F}:$$
$$\overset{..}{B}$$
$$:\ddot{F}:\quad:\ddot{F}:$$

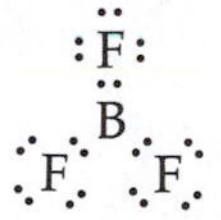

Recall that in BF_3 and other "electron deficient" compounds, the central atoms do *not* attain noble gas configurations by sharing electrons. Boron shares only six electrons in BF_3.

VSEPR theory predicts a *trigonal planar structure* for molecules such as BF_3 because this structure gives maximum separation among the three bonding electron pairs. There are no lone pairs of electrons associated with the boron atom. The structures of BCl_3, BBr_3, and BI_3 are similar.

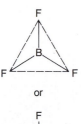

F
B
F   F
or
F
120°   120°
B
F   F
120°

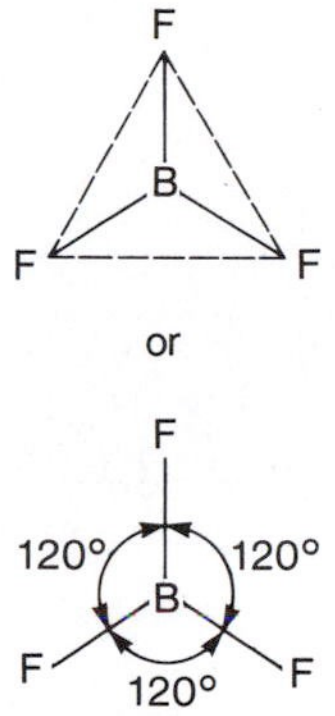

The solid lines represent bonds between B and F atoms. The dashed lines only emphasize the shape of the molecule.

2   Nitrate Ions, NO_3^-

The ground-state electronic configurations of nitrogen and oxygen atoms were shown in Section 7 – 3.2. The Lewis dot and dash formulas for the nitrate ion, NO_3^-, are

$$S = N - A$$
$$= 32\ e^- - 24\ e^- = \underline{8\ e^-\ \text{shared}}$$

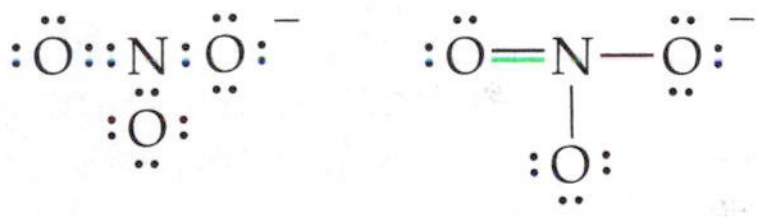

$$:\ddot{O}::N:\ddot{O}:^-$$
$$:\ddot{O}:$$

$$:\ddot{O}=N-\ddot{O}:^-$$
$$|$$
$$:\ddot{O}:$$

The Lewis formula shown above represents only one of three equivalent resonance structures of the NO_3^- ion,

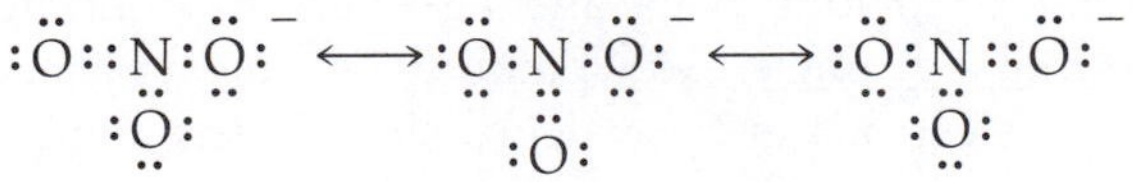

$$:\ddot{O}::N:\ddot{O}:^- \longleftrightarrow :\ddot{O}:N:\ddot{O}:^- \longleftrightarrow :\ddot{O}:N::\ddot{O}:^-$$
$$:\ddot{O}: \qquad\qquad :\ddot{O}: \qquad\qquad :\ddot{O}:$$

which we can represent as follows to emphasize **delocalization** of electrons.

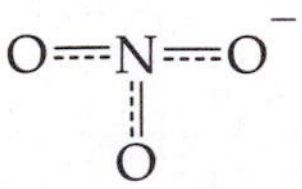

$$O\!=\!=\!=\!N\!=\!=\!=\!O^-$$ (lone e^- pairs on O are not shown)
$$\overset{\|\|}{O}$$

The Lewis formula for the nitrate ion shows three stereoactive sets of electrons about the nitrogen atom (and no lone pairs on N). The VSEPR theory predicts that these three regions will be directed toward the corners of an equilateral triangle. The prediction is in accord with experimental observations.

Experimental evidence, which shows that all three bonds in the NO_3^- ion are equivalent in bond length and bond strength, is consistent with this representation.

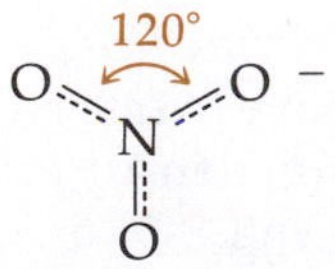

$$O \overset{120°}{\underset{N}{\underset{\|\|}{\ \ }}} O\ ^-$$
$$O$$

trigonal planar
ionic geometry
(lone pairs of e^-
on O atoms not shown)

All molecules and polyatomic ions *with three regions of high electron density and no lone pairs about the central atom* are described as having **trigonal planar** molecular or ionic geometry.

3 Sulfur Dioxide, SO₂

In Example 6–5 we constructed the Lewis formulas for the sulfur dioxide, SO_2, molecule. We found that it has two equivalent resonance structures.

$$:\ddot{O}=\ddot{S}-\ddot{O}: \longleftrightarrow :\ddot{O}-\ddot{S}=\ddot{O}: \quad \text{or} \quad O\!=\!=\!\ddot{S}\!=\!=\!O \qquad \text{(lone pairs of } e^- \text{ on O not shown in last representation)}$$

In SO_2 we have yet another species with three stereoactive sets of electrons about the central atom, two sets for the bonding electrons and one for the lone pair. Again, VSEPR theory predicts that, *to a first approximation*, they will be directed toward the corners of an equilateral triangle.

$$\underset{O\!=\!=\!\overset{\displaystyle\ddot{S}}{}\!=\!=\!O}{}$$

trigonal planar electronic geometry and *angular* molecular geometry

If all three of the regions of electron density about the sulfur atom were *equivalent*, we would expect the O—S—O bond angle to be 120°. Clearly the three regions of electron density are *not* equivalent because two regions are associated with the S—O bonds, while the third region is the lone pair of electrons that resides on the S atom. Experiments show that the bond angle is slightly less, 119.5°, indicating that the lone pair of electrons requires more space than the bonding pairs.

As we have seen, the term **lone pair** refers to a pair of electrons that is associated with (and moves under the influence of) only one nucleus. The known geometries of numerous molecules and polyatomic ions, based on measurements of bond angles, indicate clearly that *lone pairs of electrons occupy more space than bonding pairs.* The theory asserts this is because lone pairs have no other bonded atoms exerting strong attractive forces on them, so they reside closer to the nucleus than do bonding electrons. These observations indicate that the relative magnitudes of the repulsive forces between pairs of electrons on an atom are as follows:

$$\ell p/\ell p \gg \ell p/bp > bp/bp$$

where ℓp refers to lone pairs and bp refers to bonding pairs of valence shell electrons. From a geometric viewpoint, we are most concerned with the repulsions involving the electrons in the valence shell of the central atom of a molecule or polyatomic ion. The angles at which repulsive forces among valence shell electron pairs around the central atom are *exactly balanced* are the angles at which the nuclei of the outer atoms (and bonding pairs and lone pairs) are found in covalently bonded molecules and polyatomic ions.

The compression of the bond angle in SO_2 due to its lone pair is less than might be expected because of the partial double bond character of the S—O bonds. The two electron pairs of a double bond exert greater repulsive effects than the single pair of electrons of a single bond.

At this point we must be very careful to distinguish between *electronic geometry* and *molecular geometry*. **Electronic geometry** refers to *the geometric arrangement of valence shell electrons around the central atom.* **Molecular geometry** refers to *the arrangement of atoms (that is, nuclei), not just pairs of electrons, around the central atom.* For example, both BF_3 and SO_2 have *trigonal planar electronic* geometry because there are three regions of high electron density around the central atom in each case. The molecular geometries of BF_3 and SO_2 are not the same, however. Since all the valence electron pairs of BF_3 are associated with fluorine atoms as well as the central atom, boron, BF_3 molecules have both *trigonal planar electronic geometry and trigonal planar molecular geometry.* In contrast, SO_2 has an *angular*

shape, or molecular geometry, defined by the positions of the atomic nuclei, not the pairs of electrons. Only two atoms other than S are associated with bonding electrons in SO_2. Although the lone pair of electrons influences the molecular geometry of SO_2, it is not part of it our *description* of molecular geometry.

7-5   Four Stereoactive Sets of Electrons About a Central Atom

1   Methane, CH_4, and Carbon Tetrafluoride, CF_4

The Group IVA elements have four electrons in their highest energy levels, and they form numerous covalent compounds by sharing these four electrons with four other atoms. Typical examples include CH_4, CF_4, CCl_4, SiH_4, and SiF_4. All are **tetrahedral** molecules (bond angles = 109°28′). In each of them, the IVA atom is located in the center of a regular tetrahedron. The other four atoms are located at the four corners of the tetrahedron.

The Group IVA elements contribute four electrons in tetrahedral AB_4 molecules, and the other four atoms contribute one electron each to the bonds. The Lewis formulas for methane, CH_4, and carbon tetrafluoride, CF_4, are typical.

Remember that in Lewis dot formulas the emphasis is on pairs of valence electrons and *not* on the geometries of molecules. The fact that these structures have been drawn as planar is a matter of convenience and indicates nothing about their geometries, which are tetrahedral.

$$
\begin{array}{cc}
\text{H} & \ddot{\text{F}} \\
\text{H}:\ddot{\text{C}}:\text{H} & :\ddot{\text{F}}:\ddot{\text{C}}:\ddot{\text{F}}: \\
\ddot{\text{H}} & :\ddot{\text{F}}: \\
CH_4 & CF_4 \\
\text{methane} & \text{carbon tetrafluoride}
\end{array}
$$

The carbon atoms in both molecules have four sets of stereoactive electrons, so VSEPR theory predicts tetrahedral electronic geometry. Since all the valence shell electrons of the carbon atoms are bonding electrons, both molecules also have tetrahedral molecular geometry.

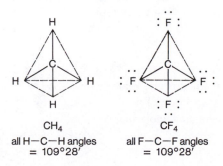

CH_4
all H—C—H angles
= 109°28′

CF_4
all F—C—F angles
= 109°28′

You may wonder whether square planar AB_4 molecules exist. They do, in compounds of the transition metals in which there are unshared pairs of electrons on the metal atom; but there are no *simple* square planar AB_4 molecules or ions with no unshared electron pairs on the central atom. The bond angles in square planar molecules and ions are 90°; nearly all AB_4 molecules and ions are tetrahedral with larger bond angles (109°28′), and therefore greater separation of valence shell electron pairs around A.

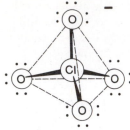

Tetrahedral electronic geometry and tetrahedral ionic geometry. All Cl—O bond lengths = 0.114 nm. The structures of sulfate ions, SO_4^{2-}, and phosphate ions, PO_4^{3-}, are similar to the structure of the ClO_4^- ion.

2 Perchlorate Ions, ClO_4^-

The Lewis formulas of the perchlorate ion, ClO_4^-, are shown below.

$$S = N - A$$
$$= 40\ e^- - 32\ e^-$$
$$= \underline{8\ e^-\text{ shared}}$$

:Ö:
:Ö:Cl:Ö: or   :Ö—Cl—Ö:
:Ö:

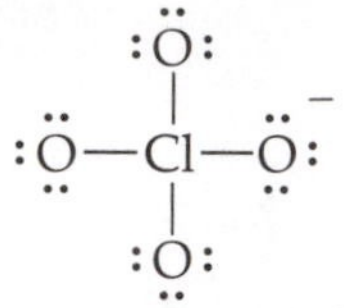

Like the carbon atoms in CH_4 and CF_4 molecules, the chlorine atom in the ClO_4^- ion sits at the center of four stereoactive sets of electrons. Thus the ClO_4^- ion has **tetrahedral electronic geometry.** Since each stereoactive set of electrons is associated with a Cl—O bond, the ion also has **tetrahedral ionic geometry.**

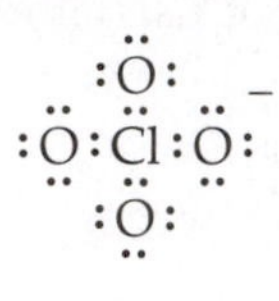

3 Ammonia, NH_3, and Nitrogen Trifluoride, NF_3

The Group VA elements have five electrons in their valence shells, and they form some covalent compounds by sharing three electrons with three other atoms. Two examples are ammonia, NH_3, and nitrogen trifluoride, NF_3. Both are observed to be pyramidal molecules with a lone pair of electrons on the nitrogen atom.

Some Group VA elements also form covalent compounds by sharing all five valence electrons, as we shall see in Section 7–6.

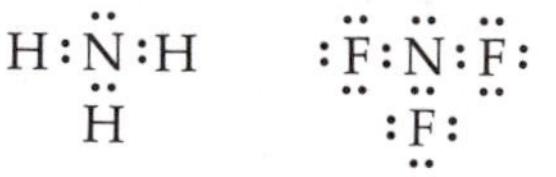

H:N:H   :F:N:F:
H   :F:

NH₃   NF₃

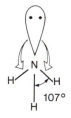

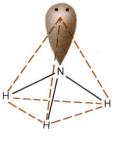

The heavy lines indicate chemical bonds, while the dashed lines emphasize the electronic geometry.

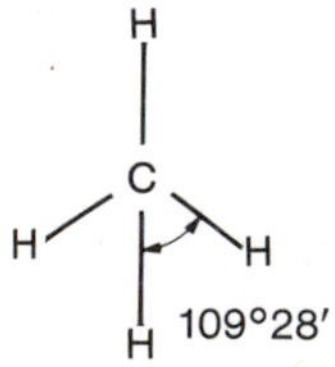

As in the previous examples, VSEPR theory predicts that the four electron pairs will be directed toward the corners of a tetrahedron, since this gives maximum separation. Thus both molecules have **tetrahedral electronic geometry.** The shapes of the molecules are described by specifying the positions of the nuclei (not the lone pairs), and both NH_3 and NF_3 are said to have **pyramidal molecular geometry.** However, the bond angles in NH_3 and NF_3 are not exactly 109°28′ as they are for a regular tetrahedron, because these molecules have a lone pair of electrons in the valence shell of the nitrogen atom. Recall that lone pairs of electrons repel bonding pairs more strongly than bonding pairs repel each other (Section 7–4.3). Thus the bond angles in both NH_3 and NF_3 are *less* than the tetrahedral angle of 109°28′.

tetrahedral electronic geometry

pyramidal molecular geometry

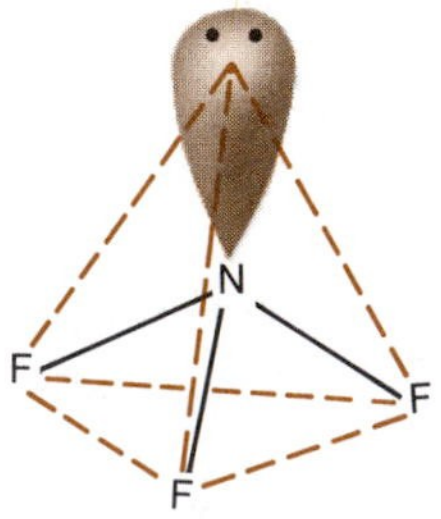

NH₃
H—N—H ∠s = 107°

NF₃
F—N—F ∠s = 102°

4 Water, H_2O

The Group VIA elements have six electrons in their highest energy levels, and they form many covalent compounds by acquiring a share in two additional electrons

H:$\overset{\cdot\cdot}{\underset{H}{O}}$:

Remember that the "shape" of a dot formula does *not* indicate the geometry of the molecule.

from two other atoms. Typical examples are H_2O, H_2S, N_2O, and Cl_2O. All are **angular** molecules. Let's consider the structure of water in detail. The bond angle in water is 104.5°. Oxygen is a Group VIA element, so it contains six electrons in its highest energy level, and its Lewis formula is shown in the margin.

TABLE 7–2   Molecular and Ionic Geometries of Molecules and Ions with Four Stereoactive Sets of Electrons

Molecule or Ion	Lewis Formula	Structure	Molecular or Ionic Geometry
CH_4 methane			tetrahedral (no lone pairs of e^- on C)
CF_4 carbon tetrafluoride			tetrahedral (no lone pairs of e^- on C)
ClO_4^- perchlorate ion			tetrahedral (no lone pairs of e^- on Cl)
SO_4^{2-} sulfate ion			tetrahedral (no lone pairs of e^- on S)
NH_3 ammonia			pyramidal (one lone pair of e^- on N)
NF_3 nitrogen trifluoride			pyramidal (one lone pair of e^- on N)
SO_3^{2-} sulfite ion			pyramidal (one lone pair of e^- on S)
H_2O water			angular (two lone pairs of e^- on O)
HF hydrogen fluoride			linear (three lone pairs of e^- on F)

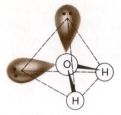

tetrahedral electronic geometry; angular molecular geometry

H—O—H ∠ = 104.5° due to repulsions between unshared e^- pairs and bonding pairs

VSEPR theory predicts that the *four stereoactive sets of electrons* (two bonding pairs and two lone pairs) around the oxygen atom in H_2O should be directed toward the corners of a regular tetrahedron, i.e., tetrahedral electronic geometry with bond angles of 109°28'. However, since two of the sets of electrons are lone pairs, VSEPR theory satisfactorily explains the *angular molecular structure* of water molecules. When we consider the greater repulsions between lone pair and lone pair as well as repulsions between lone pairs and bonding pairs, we can account for the observed bond angle of only 104.5°.

5   Hydrogen Fluoride, HF

The Group VIIA elements have seven electrons in their highest energy levels. They form covalent compounds such as H—F, H—Cl, H—Br, and H—I by sharing one electron with another atom which contributes one electron to the bond. The Lewis dot formula for HF is

Neither VSEPR theory nor Valence Bond theory adds anything to what we already know about the molecular geometry of HF, HCl, HBr, and HI. All diatomic molecules are necessarily **linear.**

Thus far in Section 7–5, we have considered several species that have *four stereoactive sets of electrons about the central atom.* All have tetrahedral electronic geometry, but those with one or more lone pairs do *not* have tetrahedral molecular or ionic geometry, as shown in Table 7–2.

7–6   Five Stereoactive Sets of Electrons About a Central Atom

1   Phosphorus Pentafluoride, PF₅

In Section 7–5.3 we pointed out that the Group VA elements have five electrons in their outermost shells and form some covalent compounds by sharing only three of these electrons with other atoms (e.g., NH_3, NF_3). The heavier Group VA elements, P, As, and Sb, form some covalent compounds by sharing all five of their valence electrons with five other atoms. Phosphorus pentafluoride, PF_5, is such a compound. Each phosphorus atom has five valence electrons to share with five fluorine atoms (Example 6–8). Phosphorus pentafluoride molecules are observed to be **trigonal bipyramidal** molecules.

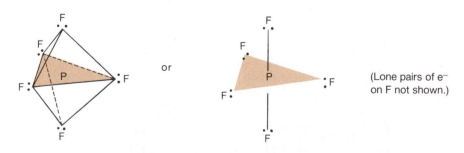

VSEPR theory predicts that the five electron pairs around the phosphorus atom in PF_5 should be as far apart as possible. Maximum separation of five items (the bonding electrons to the five F atoms) around a sixth item (P atom) is achieved when the five items are placed at the corners and the sixth item in the center of a trigonal bipyramid. This is in agreement with experimental observation. The phosphorus atom has trigonal bipyramidal electronic geometry. Since each set of electrons is a bonding pair, PF_5 also has trigonal bipyramidal molecular geometry. The axial P—F bonds are longer than the equatorial ones.

2   Other Examples

A number of simple molecules and ions have five stereoactive sets of electrons about the central atom. All exhibit **trigonal bipyramidal electronic geometry.** Some of them are shown in Table 7–3. As usual, dot formulas are constructed by applying the rules given in Section 6–11.

Note that all the lone pairs of electrons on the central atoms of the species in Table 7–3 are directed toward the corners of the equatorial plane. This is consistent with the fact that repulsions involving lone pairs are stronger than repulsions involving only bonding pairs. Any two equatorial positions are 120° apart, while the angle between an axial and an equatorial position is 90°. When as many lone

TABLE 7–3   Additional Examples of Molecular and Ionic Geometries of Molecules and Ions with Five Stereoactive Sets of Electrons

Molecule or Ion	Lewis Formula	Structure	Molecular or Ionic Geometry
$SnCl_5^-$			trigonal bipyramidal (no lone pairs on Sn)
SF_4			distorted tetrahedral (one lone pair on S)
BrF_4^+			distorted tetrahedral (one lone pair on Br)
ClF_3			T-shaped (two lone pairs on Cl)
XeF_3^+			T-shaped (two lone pairs on Xe)
XeF_2			linear (three lone pairs on Xe)
ICl_2^-			linear (three lone pairs on I)

pairs as possible are in the equatorial plane, their repulsions for other electron pairs are minimized. This accounts for the fact that ClF_3 and XeF_3^+ are T-shaped rather than trigonal planar or distorted pyramidal, and also explains why XeF_2 and ICl_2^- are linear rather than angular. In general, lone pairs are arranged about the central atom so that (1) imaginary lines through the centers of the regions occupied by lone pairs make the largest possible angles, *and* (2) imaginary lines through the centers of the regions occupied by lone pairs and through bonding pairs make as few small (i.e., 90°) angles as possible.

7–7 Six Stereoactive Sets of Electrons About a Central Atom

1 Sulfur Hexafluoride, SF₆

The heavier Group VIA elements form some covalent compounds of the AB_6 type by sharing their six electrons with six other atoms. Sulfur hexafluoride, SF_6, an unreactive gas, is an example. Sulfur hexafluoride molecules are observed to be **octahedral.**

In SF_6 molecules we have six electron pairs and six F atoms surrounding one S atom. Since there are no lone pairs in the valence shell of sulfur, the electronic and molecular geometries are identical. The maximum separation possible for six F atoms surrounding one S atom is achieved when the F atoms are situated at the corners and the S atom in the center of a regular octahedron. Thus, VSEPR theory is consistent with the observed structure.

TABLE 7–4 Additional Examples of Molecular and Ionic Geometries of Molecules and Ions with Six Stereoactive Sets of Electrons

Molecule or Ion	Lewis Formula	Structure	Molecular or Ionic Geometry
SiF_6^{2-}			octahedral (no lone pairs on Si)
IF_5			distorted square pyramidal (one lone pair on I)
SF_5^-			distorted square pyramidal (one lone pair on S)
XeF_4			square planar (two lone pairs on Xe)
ICl_4^-			square planar (two lone pairs on I)

The greater spatial requirement of lone pairs of electrons, compared with bonding pairs, accounts for the geometries of the last four species in this table.

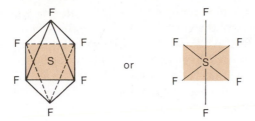

2   Other Examples

The structures of a few other molecules and ions that have *six stereoactive sets of electrons* about the central atom are summarized in Table 7–4. Each has **octahedral electronic geometry,** but most do *not* have octahedral molecular or ionic geometry.

7–8   Polar and Nonpolar Covalent Bonds

Covalent bonds are described as being either *polar* or *nonpolar* bonds. In **nonpolar bonds** such as that in the hydrogen molecule, H_2, the electron pair is *shared equally* between the two nuclei. This means that the shared electrons are equally attracted to both nuclei and therefore spend equal amounts of time near each nucleus (Figure 7–1). Stated differently, in nonpolar covalent bonds the *electron density* is symmetrical about a plane that is perpendicular to a line between the two nuclei. This holds true for all diatomic molecules of identical atoms like H_2, O_2, N_2, F_2, and Cl_2 because the two atoms exert identical attractions for the shared electron pairs. Thus, we can generalize and say that the covalent bonds in all *homonuclear diatomic molecules* must be nonpolar.

When we consider *heteronuclear* diatomic molecules, the situation isn't so simple. As an example, consider the fact that hydrogen fluoride, HF, is a gaseous substance at room temperature. This tells us that it is a covalent compound (remember from Section 6–9 that all ionic compounds are solids at room temperature). We also know that this covalent bond will have some degree of polarity

FIGURE 7–1 Electron density in H_2. The depth of shading is proportional to the probability of finding an electron in a particular region. In chemical bonds there tends to be a concentration of electronic charge between the nuclei.

because H and F are not identical atoms and therefore will not attract the electrons equally. But, how will the electrons in this bond be distributed?

In Section 6–8 we defined electronegativity as the tendency of an atom to attract electrons to itself in a chemical bond. According to Table 6–8, the electronegativity of hydrogen is 2.1 and that of fluorine is 4.0. Clearly, the fluorine atom, with its higher electronegativity, attracts the shared electron pair much more strongly than hydrogen. We represent the structure of HF as shown below. Notice the unsymmetrical distribution of electron density; the electron density is distorted in the direction of the more electronegative fluorine atom.

$$\text{H} \; \overset{\displaystyle ..}{\underset{\displaystyle ..}{:\text{F}}} \; : \quad \textit{or}$$

Covalent bonds, such as the one in HF, in which the *electron pairs are shared unequally* are referred to as **polar covalent bonds.** We use two kinds of notation to indicate polar bonds.

$$\overset{\delta +}{\text{H}} - \overset{\delta -}{\text{F}} \quad \textit{or} \quad \overset{\longmapsto}{\text{H} - \text{F}}$$

The $\delta +$ over the H atom indicates a "partial positive charge." This means that the hydrogen end of the molecule is slightly more positive than the fluorine end of the molecule. The $\delta -$ over the F atom indicates that the fluorine end of the molecule is slightly more negative than the hydrogen end. Please note that we are *not* saying that hydrogen has a charge of 1+ or that fluorine has a charge of 1−! A second way to indicate the polarity of a bond is to draw an arrow so that the head points toward the negative end (F) of the molecule and the crossed tail indicates the positive end (H).

Polar covalent bonds may be thought of as being intermediate between pure (nonpolar) covalent bonds, in which electron pairs are equally shared, and pure ionic bonding, in which there has been complete transfer of electrons from one atom to another. In fact, bond polarity is sometimes described in terms of *partial ionic character*, which usually increases with increasing difference in electronegativity between bonded atoms. Calculations based on the measured dipole moment (see the next section) of *gaseous* HCl indicate approximately 17% "ionic character" of the H—Cl bond.

The separation of charge in a polar covalent bond creates a **dipole.** The word dipole means "two poles" and refers to the positive and negative regions, or poles, that result from the separation of charge within the molecule. If we consider the covalent molecules HF, HCl, HBr, and HI, we should expect their dipoles to be different because F, Cl, Br, and I have different electronegativities and therefore different tendencies to attract electron pairs shared with hydrogen. This is indeed the case. Qualitatively, we can indicate this difference as shown below, where $\Delta(\text{EN})$ is the difference in electronegativity between two atoms that are bonded together.

The longest arrow indicates the largest dipole or largest separation of electron density in the molecule.

	$\overset{\longmapsto}{\text{H}-\text{F}}$	$\overset{\longmapsto}{\text{H}-\text{Cl}}$	$\overset{\longmapsto}{\text{H}-\text{Br}}$	$\overset{\longmapsto}{\text{H}-\text{I}}$
EN:	2.1 4.0	2.1 3.0	2.1 2.8	2.1 2.5
$\Delta(\text{EN})$	1.9	0.9	0.7	0.4

FIGURE 7-2 If polar molecules, such as HF, are subjected to an electric field, they tend to line up in a direction opposite to that of the field. This minimizes the electrostatic energy of the molecules. Nonpolar molecules are not oriented by an electric field.

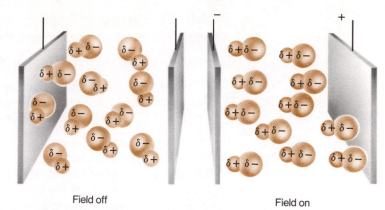

Field off   Field on

7-9   Dipole Moments

As with other concepts, it is convenient to express differences in polarities of bonds on a numerical scale. We indicate the polarity of a molecule by its **dipole moment,** μ, *which measures the net separation of charge within a molecule.* To measure a dipole moment, a sample of the substance is placed between two electrically charged plates. Polar molecules such as HF, HCl, HBr, and HI orient themselves in the electric field, causing the measured voltage between the plates to change. By contrast, nonpolar molecules do not line up, and the voltage between the plates does not change (Figure 7-2).

Generally, as electronegativity differences increase in *diatomic* molecules, the measured dipole moments increase. This can be seen clearly in the data of Table 7-5 for the hydrogen halides and hydrogen, H_2, which is nonpolar.

The **dipole moment, μ,** is defined as *the product of the distance separating charges of equal magnitude and opposite sign, times the magnitude of the charge.* However, the dipole moments of *individual bonds* can be determined only in simple diatomic molecules, because *entire molecules* rather than pairs of atoms must be subjected to measurement. Thus, tabulated values of dipole moments of polyatomic molecules refer to the average effects of all the bond dipoles in the molecules to which they apply, and reflect the *overall* polarities of the molecules. In subsequent sections we shall see that structural features, such as molecular geometry and the presence of lone (unshared) pairs of electrons, affect the polarity of a molecule. Many nonpolar molecules (those with dipole moments of zero) contain polar covalent bonds whose effects cancel one another.

TABLE 7-5   Dipole Moments and $\Delta(EN)$ Values for Hydrogen and Hydrogen Halides

Species*	Dipole Moment† (μ)	$\Delta(EN)$
HF	1.91 D	1.9
HCl	1.03 D	0.9
HBr	0.79 D	0.7
HI	0.38 D	0.4
H_2	0.0 D	0.0

* Hydrogen halides in pure gaseous state.
† Dipole moments are usually expressed in Debye units, D.

7–10   Polarity of Molecules

In order for a molecule to be polar, *both* of the following conditions must be met:

The ozone, O_3, molecule is polar. Can you tell why?

1. There *must* be at least one polar bond or one lone pair on the central atom.
2. The polar bonds, if there are more than one, *must not* be so symmetrically arranged that their bond polarities cancel.

Putting this another way, if no polar bonds or central atom lone pairs are present, the molecule *cannot* be polar. Even if these are present, they may be arranged so that their polarities cancel one another, resulting in a nonpolar molecule.

Consider as examples the H_2O molecule and the CO_2 molecule. The water molecule is angular, with two lone pairs of electrons on the oxygen atom. The carbon dioxide molecule is linear.

Effect of two unshared electron pairs

:O=C=O: no net dipole

	O—H			O—C
EN	3.5 2.1		EN	2.5 3.5
Δ(EN) =	1.4		Δ(EN) =	1.0

polar molecule ($\mu = 1.84$ D)   *nonpolar molecule* ($\mu = 0$ D)

The bond dipoles of H_2O are directed from H to O because O is more electronegative than H. These are reinforced by strong dipoles associated with the two lone pairs on the O atom. The result is that H_2O molecules are very polar. The *bonds* in the CO_2 molecule are also polar, with dipoles directed toward the more electronegative O atoms. However, *the CO_2 molecule is nonpolar* because the equal bond dipoles are opposite in direction, and therefore exactly cancel each other.

Let us now examine a variety of molecules, to account for their observed polar or nonpolar properties. Consider *gaseous* $BeCl_2$ molecules (Section 7–3.1). Examining the bond dipoles in $BeCl_2$, we see that the electronegativity difference (Table 6–8) is large (1.5 units) and the bonds are quite polar.

:Cl—Be—Cl:

	Cl ———— Be ———— Cl
EN	3.0 1.5 3.0
Δ(EN) =	1.5 1.5

However, the bond dipoles are identical in magnitude and opposite in direction, and therefore they cancel to give *nonpolar* molecules.

Two exceptions are BeF_2 and BeO, which are ionic compounds because they contain Be bonded to *very electronegative* elements.

The difference in electronegativity between Be and Cl is large enough that we might expect ionic bonding. However, the radius of Be^{2+} is so small (0.031 nm) and its charge density (ratio of charge to size) is so high that most simple beryllium compounds are covalent rather than ionic. The high charge density of Be^{2+} causes it to attract and distort the electron clouds of monatomic anions of all but the most electronegative elements to such an extent that electrons are *shared* rather than being localized on the negative ions.

Examination of the individual bonds of trigonal planar BF_3 (Section 7–4.1)

In the margin (left):

The B^{3+} ion is so small (radius = 0.020 nm) that boron does not form simple binary ionic compounds. BF_3 is a covalent compound.

shows that the electronegativity difference is very large (2.0 units) and that the bonds are very polar.

$$
\begin{array}{cc}
& B—F \\
EN & 2.0 \quad 4.0 \\
\Delta(EN) = & 2.0
\end{array}
$$

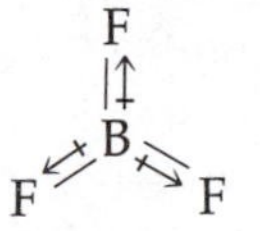

(all F—B—F angles are 120°)

However, because the molecule is symmetrical, the three bond dipoles cancel to give nonpolar molecules.*

In Section 7-5.1 we studied the bonding and geometry of tetrahedral methane, CH_4, and carbon tetrafluoride, CF_4, molecules. All bond angles are 109°28′ in both molecules. In CH_4 the individual C—H bonds are only slightly polar, while in CF_4 the individual C—F bonds are quite polar.

$$
\begin{array}{cc}
& C—H \\
EN & 2.5 \quad 2.1 \\
\Delta(EN) = & 0.4
\end{array}
\qquad\qquad
\begin{array}{cc}
& C—F \\
EN & 2.5 \quad 4.0 \\
\Delta(EN) = & 1.5
\end{array}
$$

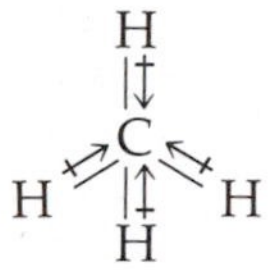

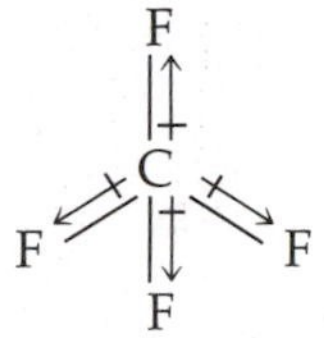

In CH_4 the small bond dipoles are directed toward carbon, while in CF_4 the large bond dipoles are directed away from carbon. *Because both molecules are completely symmetrical, however, the bond dipoles cancel and both molecules are nonpolar.* Similar statements can be made about all tetrahedral AB_4 molecules in which there are no lone pairs of electrons on the central element.

We described *pyramidal* ammonia, NH_3, and nitrogen trifluoride, NF_3, molecules in Section 7-5.3. Their formulas are often written as $:NH_3$ and $:NF_3$ to emphasize the lone pairs of electrons. The presence of these lone pairs of electrons is an extremely important factor, not only in determining much of the reaction chemistry of $:NX_3$ species, but also in accounting for the observed polarity of these substances. The contribution of each lone pair can be depicted as shown in the margin.

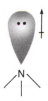

The electronegativity differences in NH_3 and NF_3 are nearly equal, *but* the resulting nearly equal bond polarities are in opposite directions.

$$
\begin{array}{cc}
& N—H \\
EN & 3.0 \quad 2.1 \\
\Delta(EN) = & 0.9
\end{array}
\qquad N—H \qquad\qquad
\begin{array}{cc}
& N—F \\
EN & 3.0 \quad 4.0 \\
\Delta(EN) = & 1.0
\end{array}
\qquad N—F
$$

In the margin (left):

In NH_3 the bond dipoles *reinforce* the effect of the lone pair, so NH_3 is very polar ($\mu = 1.5$ D). In NF_3 the bond dipoles *oppose* the effect of the lone pair, so NF_3 is only slightly polar ($\mu = 0.2$ D).

Thus, we have

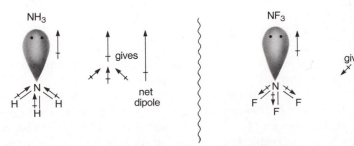

* If you are familiar with vectors from mathematics or physics, you will recognize that the polarity of the entire molecule is the vector sum of the individual bond polarities.

We can now use this information to explain the bond angles observed in NF_3 and NH_3. Because of the direction of the bond dipole in NH_3, the electron-rich end of each N—H bond is at the central atom, N. In NF_3, on the other hand, the fluorine end of each bond is the electron-rich end. As a result, the lone pair can more closely approach the N in NF_3 than in NH_3. Therefore, the lone pair exerts greater repulsion toward the bonded pairs in NF_3 than in NH_3. The net effect is that the lone pair is more successful in reducing the bond angles in NF_3 than in NH_3. We might loosely represent the situation as follows:

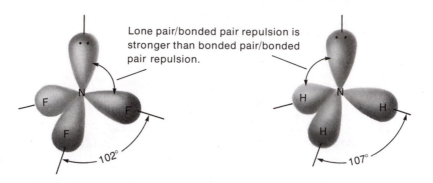

Lone pair/bonded pair repulsion is stronger than bonded pair/bonded pair repulsion.

We might expect the larger fluorine atoms ($r = 0.064$ nm) to repel each other more strongly than hydrogen atoms ($r = 0.030$ nm), leading to larger bond angles in NF_3 than in NH_3. However, this is not the case.

In phosphorus pentafluoride, PF_5, molecules (Section 7–6.1), the large difference in electronegativity, 1.9, indicates that the bonds are very polar.

$$
\begin{array}{c}
P-F \\
\underbrace{2.1 \quad 4.0} \\
\end{array}
$$

EN   2.1   4.0

$\Delta(EN) = \quad 1.9$

Let's consider the bond dipoles in two groups, because there are two different kinds of P—F bonds with different bond lengths in PF_5 molecules.

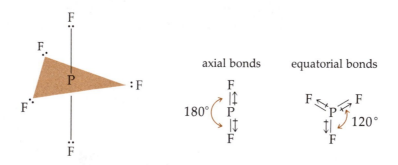

axial bonds   equatorial bonds

The two axial bond dipoles cancel each other, and the three equatorial bond dipoles cancel, so the PF_5 molecules are nonpolar.

In a similar fashion, examination of the octahedral molecule of sulfur hexafluoride, SF_6, reveals that the F—S—F bond angles are 90° and 180°, and the bonds are quite polar. Since the molecule is symmetrical, the bond dipoles cancel, and the molecule is nonpolar. To see this, let's separate the bond dipoles into two groups as we did in the trigonal bipyramid.

Note that the equatorial bonds actually consist of two linear pairs of S—F bonds, whose dipoles cancel individually.

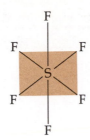

All six bonds in such an octahedral molecule are *equivalent*; if you turn the molecule so that two "equatorial" F's become "axial," the molecule still looks exactly the same.

$$S\text{—}F$$

EN $\quad 2.5 \quad 4.0$

$\Delta(EN) = \quad 1.5$

axial bonds

equatorial bonds

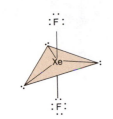

two nonpolar molecules

As we indicated earlier, some molecules with lone pairs of electrons on the central atom also are nonpolar. Consider XeF_4 (Table 7–4) and XeF_2 (Table 7–3) as examples. When each of these molecules is visualized in terms of its axial and equatorial components, we see that all bond dipoles cancel in the same way as they did in PF_5 and SF_6 molecules. The effects of the lone pairs of electrons also exactly cancel.

Molecules with symmetrical structures, but different atoms bonded to the central atom, are usually polar. We know that tetrahedral CH_4 molecules are nonpolar; however, (slightly distorted) tetrahedral CH_3Cl molecules are polar. Not only is the C—Cl bond slightly more polar than the C—H bonds, but the C—Cl bond dipole is directed toward the Cl rather than the C atom, so it reinforces the bond dipoles of the C—H bonds rather than leading to cancellation of effects. Thus the molecule must be polar.

$$C\text{—}H \qquad C\text{—}Cl$$

EN $\quad 2.5 \quad 2.1 \qquad 2.5 \quad 3.0$

$\Delta(EN) = \quad 0.4 \qquad\quad 0.5$

gives $\quad$ net dipole

With these ideas in mind, you should be able to explain why the molecules on the left in Table 7–6 are nonpolar, while those on the right are polar.

7–11 Bond Energies

We have described covalent bonding in several molecules in considerable detail. We have also said, in Chapter 3, that chemical reactions are accompanied by the breakage of bonds or the formation of new bonds, and usually both. The net energy that is released or absorbed in chemical reactions is due (at least in part, and for gas phase reactions completely) to the difference between the energies associated with the chemical bonds of the reactants and the products. A reasonable question to pose at this point is, how strong are the individual covalent bonds in a molecule; i.e., how much energy does it take to break a given covalent bond? This section addresses this fundamental question.

Energy is always required to break a chemical bond. The **bond energy**, which is the same as bond enthalpy for all practical purposes, is the amount of energy necessary to dissociate a bond in a covalent substance in the gaseous state into atoms in the gaseous state. For example, for the reaction

We use the term bond *energy* rather than bond *enthalpy* only because it is common practice to do so. Tabulated values of average bond energies are actually average bond enthalpies.

$$H_2\,(g) \longrightarrow 2H\,(g) \qquad \Delta H^0_{rxn} = \Delta H_{H\text{—}H} = +435\,kJ \quad (\text{or} +104 \text{ kcal})$$

TABLE 7–6 Some Examples of Nonpolar and Polar Molecules

Nonpolar		Polar	

$BeBr_2$ Br—Be—Br

$BeClBr$ $\overset{\delta+}{Br}$—Be—$\overset{\delta-}{Cl}$

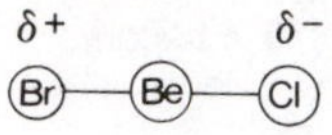

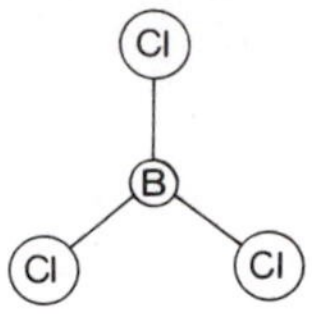

BCl_3

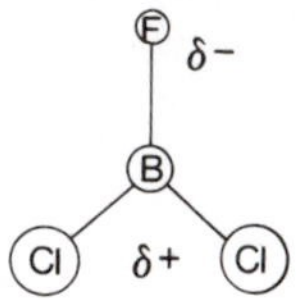

BCl_2F

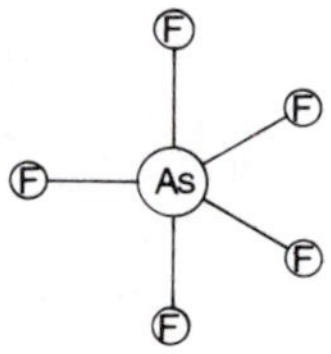

AsF_5

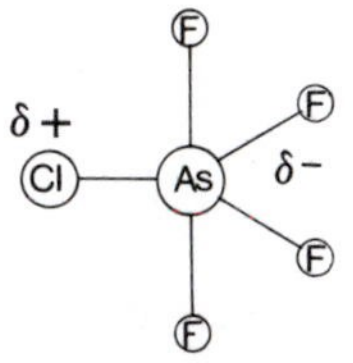

AsF_4Cl

CO_2 O=C=O

COS $\overset{\delta+}{S}$=C=$\overset{\delta-}{O}$

SO_3

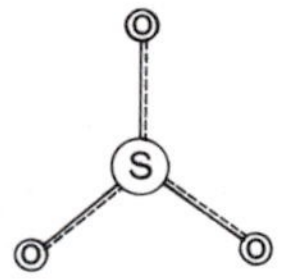

SO_2

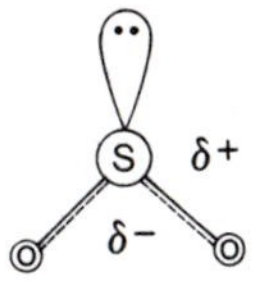

BF_3

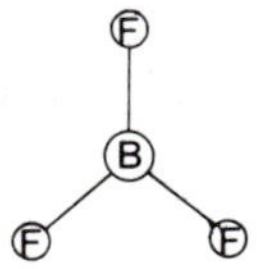

ClF_3

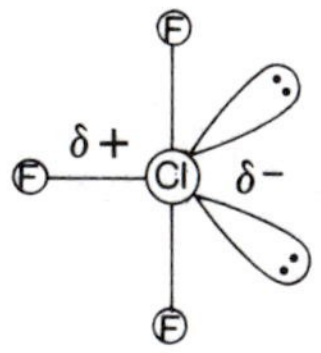

TABLE 7–7 Some Average Single Bond Energies in kJ/mol of Bonds

H	C	N	O	F	Si	P	S	Cl	Br	I	
435	414	389	464	569	293	318	339	431	368	297	H
	347	293	351	439	289	264	259	330	276	238	C
		159	201	272	—	209	—	201	243?	—	N
			138	184	368	351	—	205	—	201	O
				159	540	490	327	255	197?	—	F
					176	213	226	360	289	213	Si
						213	230	331	272	213	P
							213	251	213	—	S
								243	218	209	Cl
									192	180	Br
										151	I

TABLE 7-8 Some Multiple Bond Energies

$N=N$	418 kJ	(100 kcal)	$C=C$	611 kJ	(146 kcal)
$N\equiv N$	946 kJ	(226 kcal)	$C\equiv C$	837 kJ	(200 kcal)
$C=N$	615 kJ	(147 kcal)	$C=O$	741 kJ	(177 kcal)
$C\equiv N$	891 kJ	(213 kcal)	$C\equiv O$	1.07×10^3 kJ	(256 kcal)
			$O=O$	498 kJ	(119 kcal)

the bond energy of the hydrogen-hydrogen bond is 435 kJ per mole of bonds. The reaction is *endothermic* (ΔH^0_{rxn} is positive) and could be written as follows:

$$H_2 \text{ (g)} + 435 \text{ kJ} \longrightarrow 2H \text{ (g)}$$

In Table 7-7, each bond energy refers to the covalent bond between the element at the top of the column and the one at the end of the row. Table 7-8 gives bond energies of some multiple bonds.

Let us consider more complex molecules. For the reaction

$$CH_4 \text{ (g)} \longrightarrow C \text{ (g)} + 4H \text{ (g)}$$

$\Delta H^0_{rxn} = 1.66 \times 10^3$ kJ (398 kcal). Since the four hydrogen atoms are identical, all the C—H bonds are identical in bond length and strength *in methane molecules*. However, the energies required to break the individual C—H bonds differ for successively broken bonds, as shown below.

Since the hydrogen atoms are indistinguishable, it makes no difference which one is removed first.

CH_4 (g)	$\longrightarrow CH_3$ (g)	+ H (g)	$\Delta H^0 = +427$ kJ
CH_3 (g)	$\longrightarrow CH_2$ (g)	+ H (g)	$\Delta H^0 = +439$ kJ
CH_2 (g)	$\longrightarrow CH$ (g)	+ H (g)	$\Delta H^0 = +452$ kJ
CH (g)	$\longrightarrow C$ (g)	+ H (g)	$\Delta H^0 = +347$ kJ
CH_4 (g)	$\longrightarrow C$ (g)	+ 4H (g)	$\Delta H^0 = +1665$ kJ

"average" C—H bond energy $= \Delta H^0/4 = 416$ kJ (99.5 kcal)

We see that the *"average" C—H bond energy* in methane is one fourth of 1665 kJ, or 416 kJ, although no single C—H bond is actually broken by absorption of exactly that amount of energy. Average C—H bond energies differ slightly from compound to compound, as in CH_4, CH_3Cl, CH_3NO_2, and so on, but are nevertheless sufficiently constant to be useful, as illustrated below.

A special case of Hess' Law involves the use of bond energies to *estimate* heats of reaction. Consider the reaction of hydrogen with gaseous bromine to form hydrogen bromide.

The standard state of Br_2 at 25°C is the liquid state, but the liquid exists in equilibrium with its vapor.

$$H_2 \text{ (g)} + Br_2 \text{ (g)} \longrightarrow 2HBr \text{ (g)} \qquad \Delta H^0_{rxn} = ?$$

The bond energies are given by ΔH^0 values for the following reactions:

		ΔH^0	
HBr (g)	$\longrightarrow$ H (g) + Br (g)	368 kJ/mol (ΔH_{H-Br})	[1]
H_2 (g)	$\longrightarrow$ 2H (g)	435 kJ/mol (ΔH_{H-H})	[2]
Br_2 (g)	$\longrightarrow$ 2Br (g)	192 kJ/mol (ΔH_{Br-Br})	[3]

If we reverse equation [1] and multiply it by two, and then add it to equations [2] and [3], we obtain the equation for the reaction of interest and ΔH^0_{rxn}.

FIGURE 7–3 The heat of reaction for the formation of HBr (g) from H_2 (g) and Br_2 (g).

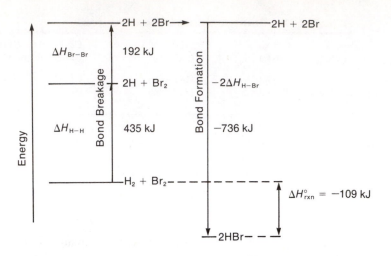

	ΔH^0 (kJ)	
$2H\ (g) + 2Br\ (g) \longrightarrow 2HBr\ (g)$	$2(-368) = 2(-\Delta H_{H-H})$	$2 \times [-1]$
$H_2\ (g) \longrightarrow 2H\ (g)$	$435 = \Delta H_{H-H}$	[2]
$Br_2\ (g) \longrightarrow 2Br\ (g)$	$192 = \Delta H_{Br-Br}$	[3]
$H_2\ (g) + Br_2\ (g) \longrightarrow 2HBr\ (g)$	$\Delta H_{rxn}^0 = -109$ kJ (or -26 kcal)	

From this result we see that

$$\Delta H_{rxn}^0 = \Delta H_{H-H} + \Delta H_{Br-Br} - 2\Delta H_{H-Br}$$

You may wish to verify that if ΔH_{rxn}^0 is calculated from ΔH_f^0 data for H_2 (g), Br_2 (g), and HBr (g) at 298 K, the value obtained is -104 kJ. Since the tabulated bond energies represent average bond energies, the correspondence is quite good. In general terms, ΔH_{rxn}^0 is related to the bond energies of the reactants and products in gas phase reactions by the following version of Hess' Law:

$$\Delta H_{rxn}^0 = \Sigma \text{B.E.}_{\text{reactants}} - \Sigma \text{B.E.}_{\text{products}} \qquad \text{in gas phase reactions only}$$

Note that this involves bond energies of *reactants* *minus* bond energies of *products*, rather than [products] minus [reactants] as when molar enthalpies of formation are used.

The net enthalpy change of a gas phase reaction is the energy required to break all the bonds in reactant molecules *minus* the energy required to break all the bonds in product molecules. Stated in another way, the amount of energy released when a bond is formed is equal to the amount absorbed when the same bond is broken. Therefore, the heat of reaction can also be described as the energy released in forming all the bonds in the product molecules minus the energy released in forming all the bonds in the reactant molecules. The situation is depicted in Figure 7–3.

Example 7–1

Using the bond energies listed in Table 7–7, estimate the heat of reaction at 298 K for the reaction below. All bonds are single bonds.

$$Br_2\ (g) + 3F_2\ (g) \longrightarrow 2BrF_3\ (g)$$

Solution

Each BrF_3 molecule contains three Br—F bonds, so two BrF_3 molecules contain six Br—F bonds. Three F_2 molecules contain a total of three F—F bonds and one Br_2 contains one Br—Br bond. Using the bond energy form of Hess' Law,

$$\Delta H_{rxn}^0 = \Delta H_{Br-Br} + 3\Delta H_{F-F} - [6\Delta H_{Br-F}]$$
$$= 192 \text{ kJ} + 3(159 \text{ kJ}) - 6(197 \text{ kJ})$$
$$\Delta H_{rxn}^0 = \underline{-513 \text{ kJ}} \ (-123 \text{ kcal})$$

Example 7–2

Use the information from Example 7–1 and the following information,

$$Br_2 (\ell) \longrightarrow Br_2 (g) \qquad \Delta H^0_{vaporization} = 31 \text{ kJ}$$

to determine ΔH^0_{rxn} for the following reaction at 298 K.

$$Br_2 (\ell) + 3F_2 (g) \longrightarrow 2BrF_3 (g)$$

Solution

This problem is solved in much the same way as Example 7–1, except that we must also consider the fact that the reaction involves $Br_2 (\ell)$ rather than $Br_2 (g)$.

	ΔH^0
$Br_2 (g) + 3F_2 (g) \longrightarrow 2BrF_3 (g)$	-513 kJ
$Br_2 (\ell) \longrightarrow Br_2 (g)$	$+31$ kJ
$Br_2 (\ell) + 3F_2 (g) \longrightarrow 2BrF_3 (g)$ $\qquad \Delta H^0_{rxn} = -482$ kJ	

The next example shows a variation of the bond energy method: given a reaction for which ΔH^0_{rxn} and all but one of the bond energies are known, we can calculate the missing bond energy.

Example 7–3

Given the following equation and average bond energies:

$$C_3H_8 (g) + 5O_2 (g) \longrightarrow 3CO_2 (g) + 4H_2O (g) \qquad \Delta H^0_{rxn} = -2.05 \times 10^3 \text{ (kJ)}$$

Bond	Energy
C—C	347 kJ/mol of bonds
C—H	414 kJ/mol of bonds
C=O	741 kJ/mol of bonds
O—H	464 kJ/mol of bonds

Without consulting Table 7–8, estimate the energy of the oxygen-oxygen bond in O_2 molecules. Compare this with the value listed in Table 7–8.

Solution

There are 8 moles of C—H bonds and 2 moles of C—C bonds per mole of C_3H_8. As before, we add and subtract the appropriate bond energies:

$$\Delta H^0_{rxn} = 2\Delta H_{C-C} + 8\Delta H_{C-H} + 5\Delta H_{O=O} - [6\Delta H_{C=O} + 8\Delta H_{O-H}]$$
$$-2.05 \times 10^3 \text{ kJ}$$
$$= 2(347 \text{ kJ}) + 8(414 \text{ kJ}) + 5\Delta H_{O=O} - [6(741 \text{ kJ}) + 8(464 \text{ kJ})]$$

Rearranging, we obtain

$$-5\Delta H_{O=O} = 694 \text{ kJ} + 3.31 \times 10^3 \text{ kJ} - 4.45 \times 10^3 \text{ kJ} - 3.71 \times 10^3 \text{ kJ} + 2.05 \times 10^3 \text{ kJ}$$
$$5\Delta H_{O=O} = 2.11 \times 10^3 \text{ kJ}$$
$$\Delta H_{O=O} = 422 \text{ kJ per mole O=O bonds}$$

The value listed in Table 7–8 is 498 kJ. The calculated value is in error by about 15%. While it is a reasonable estimate of the O=O bond energy, it does point out the limitations of using *average* bond energies.

Key Terms

Bond energy   the amount of energy necessary to break one mole of bonds in a substance, dis-   sociating the substance in the gaseous state into atoms of its elements in the gaseous state.

Bonding pair   pair of electrons involved in a covalent bond.

Central atom   an atom in a molecule or polyatomic ion that is bonded to more than one other atom.

Debye   the unit used to express dipole moments.

Delocalization   refers to bonding electrons that are distributed among more than two atoms that are bonded together; occurs in species that exhibit resonance.

Dipole   the separation of charge between two covalently bonded atoms.

Dipole moment (μ)   the product of the distance separating opposite charges of equal magnitude times the magnitude of the charge; a measure of the polarity of a molecule.

Electronic geometry   the geometric arrangement of the stereoactive sets of electrons surrounding the central atom of a molecule or polyatomic ion.

Ionic geometry   the arrangement of atoms, and *not* lone pairs of electrons, about the central atom of a polyatomic ion.

Lone pair   a pair of electrons residing on one atom and not shared by other atoms; also called unshared pair.

Molecular geometry   the arrangement of atoms, and *not* lone pairs, around a central atom of a molecule.

Nonpolar bond   covalent bond in which electron density is symmetrically distributed.

Octahedral   refers to molecules and polyatomic ions that have one atom at the center and six atoms at the corners of an octahedron, which is a polyhedron with eight equal-sized, equilateral triangular faces and six apices (corners).

Polar bond   covalent bond in which there is an unsymmetrical distribution of electron density.

Square planar   refers to molecules and polyatomic ions that have one atom in the center and four atoms at the corners of a square.

Stereoactive set of electrons   a region of high electron density about a central atom; one set may be a lone pair or the electrons of a single, double, or triple bond.

Tetrahedral   refers to molecules and polyatomic ions that have one atom in the center and four atoms at the corners of a tetrahedron, which is a polyhedron with four equal-sized, equilateral triangular faces and four apices.

Trigonal bipyramidal   refers to molecules and polyatomic ions that have one atom in the center and five atoms at the corners of a trigonal bipyramid, which is a six-sided polyhedron with five apices, consisting of two pyramids sharing a triangular base.

Trigonal planar   refers to molecules and polyatomic ions that have one atom in the center and three atoms at the corners of an equilateral triangle.

Valence Shell Electron Pair Repulsion (VSEPR) theory   a theory that assumes that electron pairs are arranged around the central atom of a molecule or polyatomic ion so that there is maximum separation (and minimum repulsion) among electron pairs.

Exercises

Electronic and Molecular (or Ionic) Geometry

1. State in your own words the basic idea of the VSEPR theory.

2. (a) Distinguish between "lone pairs" and "bonding pairs" of electrons. (b) Which has the greater spatial requirement? How do we know this? (c) Indicate the order of increasing repulsions among lone pairs and bonding pairs of electrons.

3. As an exercise in geometry that will be very useful as molecular and electronic geometries are studied, determine how many ways the following number of items ($\times$) may be arranged around a single item ($\bigcirc$) symmetrically. (a) two, (b) three, (c) four, (d) five, (e) six. In those cases in which more than one symmetrical arrangement is possible, which gives maximum separation among $\times$'s?

4. Draw a Lewis dot formula for each of the following molecules and indicate the number of stereoactive sets of electrons, as well as its electronic and molecular geometries. (a) NH_3 (b) CF_4 (c) PF_5 (d) H_2O

5. (a) What would be the ideal bond angles in the molecules in Exercise 4, ignoring lone pair effects? (b) How do these differ, if at all, from the actual values? Why?

6. Draw a Lewis dot formula for each of the fol-

lowing species and indicate the number of stereoactive sets of electrons, as well as its electronic and molecular or ionic geometries. (a) BF_3 (b) SeF_6 (c) ClO_4^- (d) NO_2^+

7. (a) What would be the ideal bond angles in the molecules in Exercise 6, ignoring lone pair effects? (b) How do these differ, if at all, from the actual values? Why?

8. Draw a Lewis dot formula for each of the following molecules and indicate the number of lone pairs about the central atom, the number of stereoactive sets of electrons, and the electronic geometry. (a) $BeBr_2$ (b) H_2Se (c) AsF_3 (d) GaI_3

9. What is the molecular geometry of each molecule in Exercise 8?

10. Draw a Lewis dot formula for each of the following polyatomic ions and indicate the number of lone pairs about the central atom, the number of stereoactive sets of electrons, and the electronic geometry. (a) PH_4^+ (b) AlH_4^- (c) BrO_3^- (d) ClO_2^-

11. What is the ionic geometry of each ion in Exercise 10?

12. Pick the member of each pair that you would expect to have the smaller bond angles and explain why. (a) SF_2 and SO_2 (b) BF_3 and BCl_3 (c) CF_4 and SF_4 (d) NH_3 and H_2O

13. Draw a dot formula, sketch the three-dimensional shape, and name the electronic geometry and molecular geometry for the following molecules. (a) PF_3 (b) AsF_5 (c) $GeCl_2$ (d) IF_3

14. Draw a dot formula, sketch the three-dimensional shape, and name the electronic geometry and molecular geometry for the following polyatomic ions. (a) H_3O^+ (b) $AsCl_4^-$ (c) SiF_6^{2-} (d) ICl_2^-

15. Draw a dot formula, sketch the three-dimensional shape, and name the electronic geometry and molecular geometry for the following molecules. (a) CO_2 (b) H_2CO_3 (c) $COCl_2$ (Cl's bonded only to C) (d) $NOCl$

16. Draw a dot formula, sketch the three-dimensional shape, and name the electronic geometry and ionic geometry for the following polyatomic ions. (a) SO_4^{2-} (b) SO_3^{2-} (c) CN^- (d) $SnCl_3^-$

17. Draw three resonance structures for each of the following species and name the molecular or ionic geometries. (a) SO_3 (b) CO_3^{2-}

18. Draw three-dimensional representations of both of the species of Exercise 17, showing delocalization in each case.

19. Draw two resonance structures for the ozone molecule, O_3, and two for the SO_2 molecule. Then draw a single structure showing delocalization for each, and describe their molecular geometries. Is the similarity surprising? Why or why not?

20. Draw dot formulas and three-dimensional structures for the following molecules that contain the noble gas, xenon. Name the molecular geometry of each. (a) XeF_2 (b) XeF_4 (c) XeO_3 (d) XeO_4

21. As the name implies, the interhalogens are compounds that contain two halogens. Draw dot formulas and three-dimensional structures for the following bromine fluorides. Name the electronic and molecular geometry of each. (a) BrF (b) BrF_3 (c) BrF_5

22. A number of ions derived from the interhalogens (Exercise 21) are known. Draw dot formulas and three-dimensional structures for the following ions of interhalogens. Name the electronic and ionic geometry of each. (a) IF_2^- (b) ICl_4^- (c) BrF_2^+

23. Chlorine forms four oxyanions, ClO^-, ClO_2^-, ClO_3^-, and ClO_4^-. Draw the dot formula for each, and then sketch and compare the electronic and ionic geometries of these oxyanions.

24. Nitrogen dioxide, NO_2, contains one unpaired electron per molecule. Draw a dot formula for the molecule. At low temperatures NO_2 reacts with another NO_2 molecule (dimerizes) to form N_2O_4 molecules. Suggest a reason. Draw a dot formula for N_2O_4 and describe the molecular geometry with respect to the nitrogen atoms, which are bonded to each other.

25. The pyrophosphate ion, $P_2O_7^{4-}$, contains one oxygen atom that is bonded to both phosphorus atoms. Draw a dot formula and sketch the three-dimensional shape of the ion. How would you describe the ionic geometry with respect to the central oxygen atom and with respect to each phosphorus atom?

26. Carbon forms two common oxides, CO and CO_2, both of which are linear (see Exercise 15). It forms a third (very uncommon) oxide, carbon suboxide, C_3O_2, which is also linear. The structure has terminal oxygen atoms on both ends. Draw dot and dash formulas for C_3O_2. How

many stereoactive sets of electrons are there about each of the three carbon atoms?

27. (a) List three important points that were illustrated by the discussion of covalent bonding. (b) What is the significance of each?

Polarities of Bonds and of Molecules

28. Why is an HCl molecule polar while a Cl_2 molecule is nonpolar?

29. Distinguish between polar and nonpolar covalent bonds. Cite three specific examples of each kind.

30. Explain the observed polarities of HF, HCl, HBr, and HI molecules.

31. What do we mean when we refer to the dipole moment of a molecule?

32. Is there a relationship between the polarity of a diatomic molecule and its dipole moment? If so, what is it?

33. Use Table 6–8 to list the following pairs of elements in order of increasing polarities of the *bonds*. S—Cl, Cl—O, Be—I, N—F, As—F

34. Which of the following molecules are polar? Why? (a) PCl_3 (b) BCl_3 (c) H_2O (d) HgI_2 (e) SiF_4 (f) SF_4

35. Which of the following molecules are polar? Why? (a) CH_4 (b) CH_3Br (c) CH_2Br_2 (d) $CHBr_3$ (e) CBr_4

36. Which one of the following statements is a better example of deductive logic? Why?
(a) Because the dot formula for SO_2 shows that there are three stereoactive sets of electrons about the sulfur atom, SO_2 molecules must be angular, and therefore polar.
(b) Because SO_2 molecules are polar, and therefore angular, the dot formula must show three stereoactive sets of electrons about the sulfur atom.

37. Which of the following molecules have dipole moments of zero? Justify your answer. (a) SO_3 (b) BrF_3 (c) CS_2 (d) O_3 (e) S_8 (puckered 8-membered ring) (f) $AsCl_3$

38. The PF_3Cl_2 molecule has a dipole moment of zero. Use this information to sketch its three-dimensional shape. Justify your choice.

39. Why are the bond angles in NF_3 molecules less than those in the more polar NH_3 molecule?

Bond Energies

40. How is the amount of heat released or absorbed in a *gas phase reaction* related to bond energies of products and reactants?

41. Write equations for reactions that illustrate the definitions of (a) bond energy and (b) average bond energy.

42. Suggest a reason for the fact that different amounts of energy are required for the successive removal of the three hydrogen atoms of an ammonia molecule, even though all N—H bonds in ammonia are equivalent.

43. Suggest why the N—H bonds in different compounds such as ammonia, NH_3, methylamine, CH_3NH_2, and ethylamine, $C_2H_5NH_2$, have slightly different bond energies.

44. The structural formula for benzene, C_6H_6, can be represented by two resonance hybrid forms:

How would you expect the carbon-carbon bond energies to compare with those of ethane,

and ethene or ethylene,

45. (a) Calculate the carbon-oxygen bond energies in CO_2 and CO. For CO_2 (g), $\Delta H_f^0 = -393.5$ kJ/mol; for CO (g), $\Delta H_f^0 = -110.5$ kJ/mol. For O (g) and C (g), $\Delta H_f^0 = 249.2$ kJ/mol and 716.7 kJ/mol, respectively.
(b) Compare these bond energies with those listed in Tables 7–7 and 7–8. Would you classify the bonds as single, double, or triple covalent bonds?

46. Use the data in Appendix K to calculate the average bond energy in NH_3.

47. Use the data in Appendix K to calculate the average bond energy in SF_6.

48. Use the bond energies in Table 7–7 to estimate

the heat of reaction at 298 K for

$$ICl_3 (g) + 2H_2 (g) \longrightarrow HI (g) + 3HCl (g)$$

49. Use the bond energies in Tables 7–7 and 7–8 to estimate the heat of reaction at 298 K for

$$2ClF_3 (g) + 2O_2 (g) \longrightarrow Cl_2O (g) + 3OF_2 (g)$$

50. Ethylamine undergoes an endothermic gas phase dissociation to produce ethylene (or ethene) and ammonia.

ethylamine   ethylene   ammonia

The following are average bond energies per

mole of bonds: C—H, 414 kJ; C—C, 347 kJ; C=C, 611 kJ; N—H, 389 kJ. Calculate the C—N bond energy. Compare this with the value in Table 7–7.

51. Use bond energies from Table 7–7 and the following thermochemical equation

$$4HCl (g) + O_2 (g) \longrightarrow$$
$$2Cl_2 (g) + 2H_2O (g) + 114 \text{ kJ}$$

to calculate the value of $\Delta H_{O=O}$. Compare your value with that given in Table 7–8.

Covalent Bonding II: Valence Bond Theory and Molecular Orbital Theory

8

Valence Bond (VB) Theory

In Section 6–11, we explained covalent bonding as the result of *orbital overlap*—the overlap of atomic orbitals from two atoms, and the sharing of electrons in the region of overlap. This is the central tenet of Valence Bond theory, which describes *how* covalent bonding occurs. Valence Bond theory is complementary to VSEPR theory, which does not describe *how* bonding occurs but rather *where* bonding occurs, as well as where lone pairs of valence shell electrons are directed. The one factor that both theories have in common (along with all other theories) is that they must be consistent with, and be able to explain, experimental observations.

In the first part of this chapter we shall use Valence Bond theory to account for the bonding in many of the molecules and ions whose geometries we described in the last chapter, as well as several other examples. (In the last part of the chapter we shall treat covalent bonding by an alternate method, the Molecular Orbital theory.) We shall also assume that each bonding pair and each lone pair occupies a separate orbital. To provide the necessary orbitals, and to be consistent with the geometry (as observed experimentally and as predicted by the VSEPR theory), it usually will be necessary to invoke the concept of *hybridization* , or mixing, of atomic orbitals.

The number of stereoactive sets of electrons about a central atom in a molecule or polyatomic ion indicates which of that atom's valence atomic orbitals are likely to be involved in hybridization, as summarized in Table 8–1. The designations given to sets of hybridized orbitals reflect the number and kinds of atomic orbitals that hybridize to produce the set.

TABLE 8–1   Common Kinds of Hybridization Associated with Particular Electronic Geometries

Number of Stereoactive Sets of Electrons About a Central Atom	Electronic Geometry	Common Hybridization	Atomic Orbitals Mixed
2	linear	sp	one s, one p
3	trigonal planar	sp^2	one s, two p's
4	tetrahedral	sp^3	one s, three p's
5	trigonal bipyramidal	sp^3d	one s, three p's, one d
6	octahedral	sp^3d^2	one s, three p's, two d's

As we did with VSEPR theory, we shall proceed systematically from the simplest examples to the more complex, in order of increasing number of stereoactive sets of electrons. Then, after the appropriate foundation has been laid, we shall apply the VB treatment of bonding to polycentered molecules containing double and triple bonds.

8–1   Two Stereoactive Sets of Electrons (*sp* Hybridization) — BeCl₂

The dot formula and ball and stick model of the beryllium chloride molecule, $BeCl_2$, and many of the properties of gaseous $BeCl_2$ described in Section 7–3 are shown in the margin on page 246. Let us consider the electronic configurations of Be and Cl atoms in their ground states as we begin to describe the bonding.

:C̈l : Be : C̈l :

two stereoactive sets of electrons about Be; linear electronic geometry

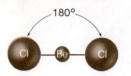

180°

linear molecular geometry; polar bonds, nonpolar molecule

	1s	**2s**	**2p**	**3s**	**3p**
$_4$Be	↑↓	↑↓			
$_{17}$Cl	↑↓	↑↓	↑↓ ↑↓ ↑↓	↑↓	↑↓ ↑↓ ↑

The two electrons in the 1s orbital of Be, the central atom, are nonvalence (inner) electrons and are not involved in bonding. There are two more electrons *paired* in the 2s orbital. How, then, will two Cl atoms bond to Be? The Be atom must somehow make available one orbital for each (bonding) Cl electron (the unpaired p electrons). Recall that the electron configuration shown above is the *ground-state* configuration for an isolated Be atom. Another configuration may be more stable in a bonding environment. Suppose that the Be atom *promotes* one of the paired 2s electrons to one of the 2p orbitals, the next highest energy orbitals:

$$_4\text{Be [He]} \quad \underset{2s}{↑↓} \quad \underset{2p}{\overline{}} \quad \xrightarrow[\text{promote}]{} \quad _4\text{Be [He]} \quad \underset{2s}{↑} \quad \underset{2p}{↑ \; \overline{}}$$

Now there would be two Be orbitals available for bonding, but we find a discrepancy with experimental fact. This "promoted pure atomic" arrangement would predict two *nonequivalent* Be—Cl bonds because the Be 2s and 2p orbitals are not expected to overlap a Cl 3p orbital with equal effectiveness. Yet we observe experimentally that both Be—Cl bonds are *equivalent* in bond length and bond strength. So we reject the idea of simple "promotion" as an explanation.

As we learned in Chapter 5, orbitals are described mathematically as waves. Just as we can combine waves on the surface of water (and their mathematical descriptions) to give new wave patterns (and *their* mathematical descriptions), we can also combine atomic orbitals to give new orbital patterns called *hybridized* orbitals, whose mathematical representations tell us their shapes and orientations. When we do this, combining the mathematical description of one s and one p atomic orbital, we get a new set of two equivalent orbitals called **sp hybrid orbitals** on the Be atom. They are intermediate in energy between the original s and p orbitals. Consistent with Hund's Rule (Section 5–10), the two valence electrons of Be will occupy each of these orbitals individually.

$$_4\text{Be [He]} \quad \underset{2s}{↑↓} \quad \underset{2p}{\overline{}} \quad \xrightarrow[\text{hybridize}]{} \quad _4\text{Be [He]} \quad \underset{sp}{↑ \; ↑} \quad \underset{2p}{\overline{}}$$

The sp hybrid orbitals are mathematically described as *linear orbitals*, and this is consistent with the experimental observation that the molecule is linear. In general, *sp hybridization* occurs at the central atom of a molecule, or a polyatomic ion, whenever there are two regions of high electron density around the central atom.

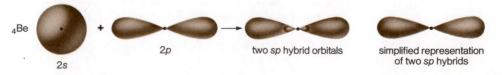

$_4$Be   2s +   2p →   two sp hybrid orbitals   simplified representation of two sp hybrids

The dot in each orbital above represents the Be nucleus. Recall that each Cl atom has a $3p$ orbital that contains only one electron, so that overlap with the sp hybrids of Be is possible. We picture the bonding in $BeCl_2$ in the following diagram, in which only the bonding electrons are represented.

Lone pairs of e^- on Cl
atoms are not shown.

two sp hybrids on Be

$3p$ $3p$

One additional idea about hybridization is worth special emphasis: The number of hybrid orbitals is always the same as the number of atomic orbitals that were combined. Hybrid orbitals are named by indicating the *number and kind* of atomic orbitals hybridized. As we have seen, hybridization of *one 2s* orbital and *one 2p* orbital gives *two sp hybrid orbitals*. We shall see presently that hybridization of *one ns* orbital and *two np* orbitals gives *three sp^2* hybrid orbitals, while hybridization of *one ns* orbital and *three np* orbitals gives *four sp^3 hybrids,* and so on.

Br and I use $4p$ and $5p$
orbitals, respectively, in
bonding in these com-
pounds, whereas Cl uses a
$3p$ orbital.

The structures of beryllium bromide, $BeBr_2$, and beryllium iodide, BeI_2, are similar to that of $BeCl_2$. The chlorides, bromides, and iodides of cadmium, CdX_2, and mercury, HgX_2, are also linear covalent molecules (where X = Cl, Br, or I). Cadmium has two electrons in its $5s$ orbital, and its $5p$ orbitals are vacant. Similarly, mercury has two electrons in its $6s$ orbital, and its $6p$ orbitals are vacant. Thus, the possibility of sp hybridization exists in each metal.

8–2   Three Stereoactive Sets of Electrons (sp^2 Hybridization)

1   Boron Trifluoride, BF_3

In Section 7–4 we found that the BF_3 molecule is trigonal planar with no lone pairs of electrons on B, the central atom. The structure of BF_3 is shown in the margin.

To be consistent with experimental findings, the Valence Bond theory must explain three *equivalent* B—F bonds and 120° bond angles. Again, we invoke hybridization. In this case, the $2s$ orbital and two of the $2p$ orbitals of B must hybridize to form a set of three equivalent sp^2 hybrid orbitals, which can hold the three valence electrons individually.

:F:
|
B
:F: :F:

three stereoactive sets of electrons about B; trigonal planar electronic geometry

F
120° | 120°
F — B — F
120°

trigonal planar molecular geometry; polar bonds, nonpolar molecule

$$_5B\ [He]\ \underset{2s}{\uparrow\downarrow}\ \underset{2p}{\underline{\uparrow}\ \underline{\ }\ \underline{\ }}\ \xrightarrow{\text{hybridize}}\ _5B\ [He]\ \underset{sp^2}{\underline{\uparrow}\ \underline{\uparrow}\ \underline{\uparrow}}\ \underset{2p}{\underline{\ }}$$

The three sp^2 hybrid orbitals are directed toward the corners of an equilateral triangle:

$2s$ $2p$ $2p$ three sp^2 hybrid orbitals

$_5B$

The dots in the orbitals above represent the boron nucleus. Each of the three F atoms has one unpaired electron in a $2p$ orbital, which can overlap one of the three sp^2 hybrid orbitals on boron. Three electron pairs are shared between one boron and three fluorine atoms:

Lone pairs of e^- on F are not shown.

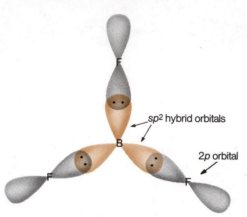

To generalize, *sp^2 hybridization* occurs at the central atom whenever there are three regions of high electron density around the central atom.

Suppose that we had considered *only* the promotion of one of the $2s$ electrons of atomic B to a vacant $2p$ orbital to give three unpaired valence electrons, with no hybridization. In that case it would have been possible to account for all three B—F bonds. However, not only would we predict that two of the bonds would be different from the third (because two $2p$ orbitals on B and one $2s$ orbital on B would be involved), but also we could *not* account for the observed 120° bond angles. Instead, we would expect two F—B—F angles of 90°, because the $2p$ orbitals are perpendicular, and one unknown or at least "undirected" angle involving the $2s$ orbital on B. So the evidence in favor of hybridization is very strong. *Hybridization can account for the number of bonds formed, and for both the relative strengths and the directional characteristics of the bonds.*

2   Nitrate, NO_3^-, and Nitrite, NO_2^-, Ions

In Section 7–4 we found that the nitrate ion, NO_3^-, has some multiple-bond character and has the three resonance structures shown below.

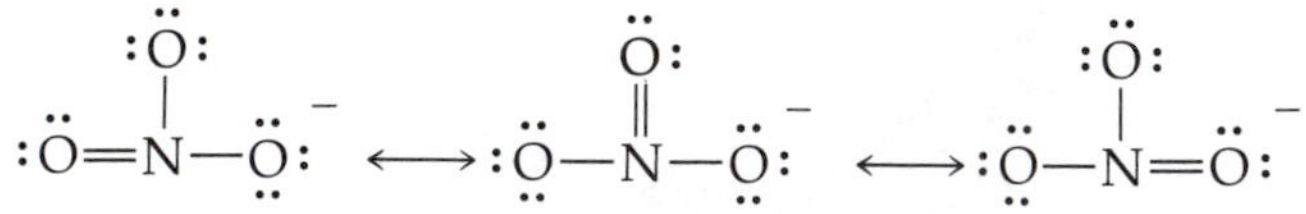

$$
\ddot{O}=N-\ddot{O}:^{-} \longleftrightarrow :\ddot{O}-N-\ddot{O}:^{-} \longleftrightarrow :\ddot{O}-N=\ddot{O}:^{-}
$$

These can be represented as a single trigonal planar structure, showing delocalization of electrons.

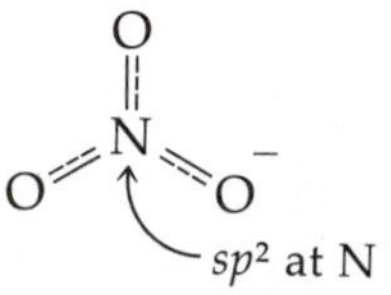

(lone pairs of electrons on O atoms not shown)

sp^2 at N

Like the boron atom in a BF_3 molecule, the nitrogen atom has three stereoactive sets of electrons, and we associate sp^2 hybridization with it.

Another example of an ion with three stereoactive sets of electrons about the central atom, and therefore sp^2 hybridization, is the angular nitrite ion, NO_2^-, which has the following resonance structures.

For NO_2^- ion:
$S = N - A$
$S = 24 - 18$
$S = 6\ e^-$ shared

$$:\!\ddot{O}\!=\!\ddot{N}\!-\!\ddot{O}\!:\ ^- \longleftrightarrow :\!\ddot{O}\!-\!\ddot{N}\!=\!\ddot{O}\!:\ ^-$$ or

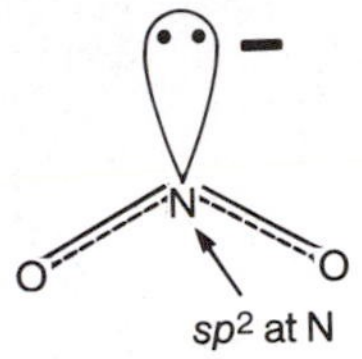

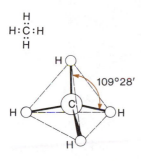

sp^2 at N
(lone pairs on O not shown)

8–3 Four Stereoactive Sets of Electrons (sp^3 Hybridization)

1 Methane, CH_4, and Carbon Tetrafluoride, CF_4

In Section 7–5 we found that both methane, CH_4, and carbon tetrafluoride, CF_4, molecules are tetrahedral, nonpolar molecules with no lone pairs on the carbon atoms.

According to Valence Bond theory, each Group IVA atom (carbon in our examples) must make available for bonding *four equivalent orbitals* and *four unpaired electrons*. To do this, carbon forms four equivalent sp^3 **hybrid orbitals** in this bonding environment by mixing the s and all three p orbitals in its outer shell. This results in four unpaired electrons.

$$_6C\ [He]\ \frac{\uparrow\downarrow}{2s}\ \frac{\uparrow\quad\uparrow\quad\underline{}\ \underline{}}{2p}\ \xrightarrow[\text{hybridize}]{}\ _6C\ [He]\ \boxed{\frac{\uparrow\quad\uparrow\quad\uparrow\quad\uparrow}{sp^3}}$$

These orbitals are directed toward the corners of a regular tetrahedron, which has $109°28'$ angles from (any) corner to center to (any) corner.

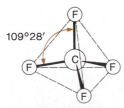

four stereoactive sets of electrons; tetrahedral electronic geometry, tetrahedral molecular geometry

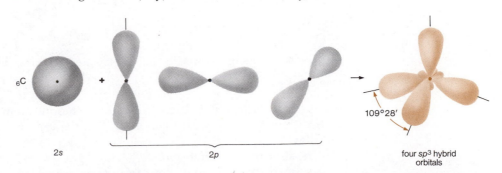

| $2s$ | $2p$ | four sp^3 hybrid orbitals |

$109°28'$

Each of the four atoms that bond to carbon possesses a half-filled atomic orbital that can overlap the half-filled sp^3 hybrids, as illustrated below for methane and carbon tetrafluoride.

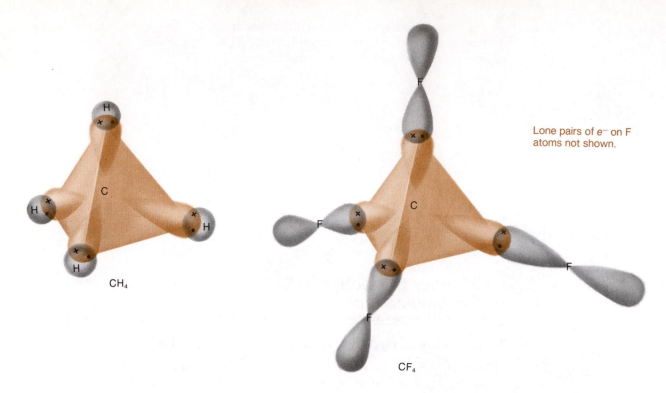

Lone pairs of e^- on F
atoms not shown.

CH_4

CF_4

In general, *sp^3 hybridization* occurs at the central atom of a molecule or ion when there are *four* stereoactive sets of electrons around that atom.

2 Perchlorate Ions, ClO_4^-

The perchlorate ion is also tetrahedral (Section 7–5) with no lone pairs of electrons on the Cl atom, as shown below.

:Ö: ⁻
:Ö:Cl:Ö:
:Ö:

four stereoactive sets of electrons;
tetrahedral electronic geometry

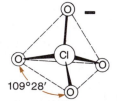

$109°28'$

tetrahedral ionic geometry

Since there are four sets of electrons about the chlorine atom, VB theory postulates that the chlorine atom is sp^3 hybridized, like the carbon atoms in CH_4 and CF_4 molecules.

The structures of sulfate ions, SO_4^{2-}, and phosphate ions, PO_4^{3-}, are similar to the structure of the ClO_4^- ion. They can all be described in terms of sp^3 hybridization at the central atom. So can other ions with four stereoactive sets of electrons, such as ClO_3^- (pyramidal) below, ClO_2^- (angular), and SO_3^{2-} (pyramidal).

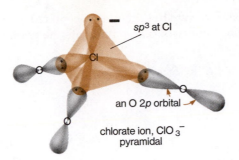

sp³ at Cl

an O 2p orbital →

chlorate ion, ClO_3^-
pyramidal

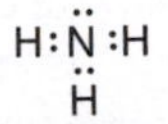

3 Ammonia, NH₃, and Ammonium Ions, NH₄⁺

H:N̈:H
Ḧ

four stereoactive sets of
electrons; tetrahedral
electronic geometry

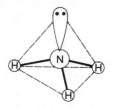

pyramidal molecular
geometry; polar bonds
reinforce effect of lone pair
on N to give very polar
molecules

Let us consider the polar, pyramidal ammonia molecule first. (See Section 7–5 and the margin.) Note there is one lone pair of electrons on the nitrogen atom.

Since experimental results suggest four nearly equivalent orbitals (three involved in bonding, a fourth to accommodate the lone pair), we again need four sp^3 hybrid orbitals.

$$_7N\ [He]\ \underset{2s}{\uparrow\downarrow}\ \underset{2p}{\underline{\uparrow}\ \underline{\uparrow}\ \underline{\uparrow}}\ \xrightarrow[\text{hybridize}]{}\ _7N\ [He]\ \underset{sp^3}{\underline{\uparrow\downarrow}\ \underline{\uparrow}\ \underline{\uparrow}\ \underline{\uparrow}}$$

The reaction of NH₃ with hydrogen chloride, HCl, produces the ionic compound ammonium chloride, NH₄Cl, which contains the ammonium ion, NH₄⁺. This ion is formed as the 1s orbital of a hydrogen ion, H⁺, which has no electrons, overlaps the sp^3 orbital of nitrogen that contains the lone pair. The resulting bond is a **coordinate covalent bond,** which is a covalent bond in which both electrons originated on the same atom. However, since electrons are indistinguishable, we cannot tell one bond from another in the NH₄⁺ ion.

Both ammonia, NH₃, and the ammonium ion, NH₄⁺, are sp^3 hybridized at N. Note that NH₄⁺ is tetrahedral rather than pyramidal, because each of the four stereoactive sets of electrons about the nitrogen atom is a bonding pair. This is consistent with the observed 109°28′ (tetrahedral) bond angles in NH₄⁺.

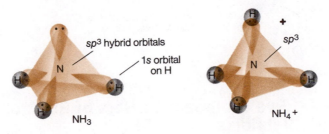

sp³ hybrid orbitals

1s orbital
on H

N

NH₃

sp³

N

NH₄⁺

4 Water, H₂O, and Hydrogen Sulfide, H₂S

The structure and properties of polar, angular water, H₂O, and hydrogen sulfide, H₂S, molecules are shown in the margin on page 252.

In light of the experimental evidence on the bond angle in H₂O, the Valence

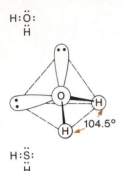

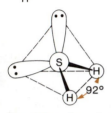

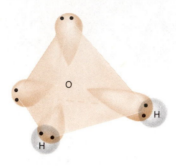

four stereoactive sets of electrons; tetrahedral electronic geometry and angular molecules; polar bonds, polar molecules

Note that sulfur is located directly below oxygen in Group VIA.

Bond theory postulates four sp^3 hybrid orbitals centered on the oxygen atom: two to participate in bonding, and two to accommodate the two lone pairs.

Again, we can explain the observed bond angle of 104.5° easily. The expected bond angle for sp^3 hybridization (tetrahedral electronic geometry) is 109°28′. However, the two lone pairs repel each other and the bonding pairs of electrons strongly. These repulsions force the bonding pairs closer together and result in the decreased bond angle. Since oxygen is very electronegative, the shared electron pairs are shifted toward the O ends of the O—H bonds, and the decrease in the H—O—H bond angle (from 109°28′ to 104.5°) is greater than the corresponding decrease in the H—N—H bond angles in ammonia.

Hydrogen sulfide, H_2S, is also an angular molecule, but the H—S—H bond angle is only 92°. Therefore, we need not postulate hybrid orbitals. The two hydrogen atoms are able to exist at approximately right angles to each other (close to the 90° angles between two unhybridized $3p$ orbitals of sulfur) when they are bonded to the larger sulfur atom.

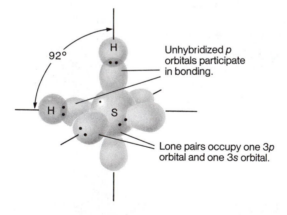

Unhybridized p orbitals participate in bonding.

Lone pairs occupy one $3p$ orbital and one $3s$ orbital.

five stereoactive sets of electrons; trigonal bipyramidal electronic geometry

8–4   Five Stereoactive Sets of Electrons (sp^3d Hybridization) — PF₅

In Section 7–6, we described phosphorus pentafluoride molecules, PF_5, in terms of VSEPR theory. The molecule is trigonal bipyramidal and nonpolar, with no lone pairs of electrons on the phosphorus atom.

Since phosphorus is the central element in PF_5 molecules, it must have available five half-filled orbitals to form bonds with five fluorine atoms. Hybridiza-

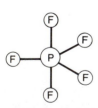

trigonal bipyramidal molecular geometry; polar bonds, nonpolar molecules

tion is again the answer, this time using one *d* orbital from the vacant set of $3d$ orbitals in addition to the $3s$ and $3p$ orbitals of the phosphorus atom.

$$_{15}P\ [Ne]\ \frac{\uparrow\downarrow}{3s}\quad \frac{\uparrow}{}\ \frac{\uparrow}{3p}\ \frac{\uparrow}{}\quad \overline{}\ \overline{}\ \overline{3d}\ \overline{}\ \overline{}$$

$$\xrightarrow{\text{hybridize}}\ _{15}P\ [Ne]\quad \frac{\uparrow}{}\ \frac{\uparrow}{}\ \frac{\uparrow}{sp^3d}\ \frac{\uparrow}{}\ \frac{\uparrow}{}\quad \overline{}\ \overline{}\ \overline{3d}\ \overline{}\ \overline{}$$

The five *sp³d* **hybrid orbitals** are directed toward the corners of a **trigonal bipyramid.** Each is overlapped by the (only) $2p$ orbital of a fluorine atom that contains a single electron. The resulting pairing of P and F electrons forms a total of five covalent bonds.

The *d* orbital involved is the one we designated as the d_{z^2} orbital (see Figure 5–14).

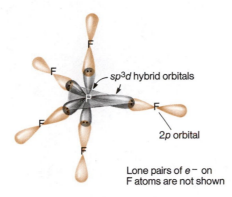

sp³d hybrid orbitals

$2p$ orbital

Lone pairs of *e*⁻ on F atoms are not shown

We should note that *sp³d* hybridization involves utilization of an available *d* orbital in the outermost shell of the central atom, P. We mentioned earlier that the heavier Group VA elements, P, As, and Sb, can form *five-coordinate compounds* utilizing this hybridization. But it is *not* possible for nitrogen (also in Group VA) to form such five-coordinate compounds. Two factors are important in understanding why: (1) nitrogen is too small to accommodate five (even very small) substituents without excessive crowding, which causes instability, and (2) it has no low-energy *d* orbitals. In fact, we can generalize: No elements of the second period can be central elements in five-coordinate molecules (or higher-coordinate ones) because they have no low-energy *d* orbitals available for hybridization, and because they are too small. It is important to realize that the set of *s* and *p* orbitals in a given energy level, and therefore any set of hybrids composed only of them, can accommodate a *maximum* of eight electrons and participate in a *maximum* of four covalent bonds.

All other molecules and ions with five stereoactive sets of electrons about the central atom, such as $SnCl_5^-$, SF_4 (one lone pair), ClF_3 (two lone pairs), and ICl_2^- (three lone pairs), are also *sp³d* hybridized at the central atom. Some were shown in Table 7–3.

octahedral electronic geometry

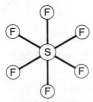

octahedral molecular geometry; polar bonds, nonpolar molecules

8–5 Six Stereoactive Sets of Electrons (*sp³d²* Hybridization)—SF₆

Sulfur hexafluoride molecules, SF_6, are octahedral and nonpolar (Section 7–7), with no lone pairs of electrons on the sulfur atom.

The sulfur atom can form six hybrid orbitals by using all of its $3s$ and $3p$ and two of its empty $3d$ orbitals to accommodate six electron pairs. The Valence Bond theory description can be visualized as:

$$_{16}S\ [Ne]\ \frac{\uparrow\downarrow}{3s}\quad \frac{\uparrow\downarrow}{}\frac{\uparrow}{3p}\frac{\uparrow}{}\quad \overline{\quad}\ \overline{\quad}\ \overline{3d}\ \overline{\quad}\ \overline{\quad}\quad \xrightarrow{\text{hybridize}}\quad _{16}S\ [Ne]\quad \frac{\uparrow}{}\ \frac{\uparrow}{}\ \frac{\uparrow}{}\ \frac{\uparrow}{}\ \frac{\uparrow}{}\ \frac{\uparrow}{}\atop sp^3d^2\quad \overline{\quad}\ \overline{3d}\ \overline{\quad}$$

The six sp^3d^2 **hybrid orbitals** are directed toward the corners of a regular octahedron. Each sp^3d^2 hybrid orbital is overlapped by a half-filled p orbital from fluorine to form a total of six covalent bonds.

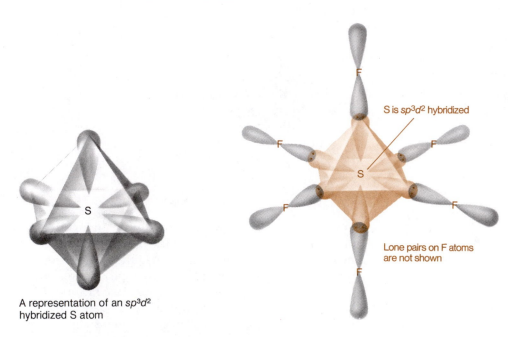

A representation of an sp^3d^2 hybridized S atom

S is sp^3d^2 hybridized

Lone pairs on F atoms are not shown

All the species that were shown in Table 7–4 have six stereoactive sets of electrons about the central atom. These include $SiF_6{}^{2-}$ (no lone pairs), $SF_5{}^-$ (one lone pair), and XeF_4 (two lone pairs). All are sp^3d^2 hybridized at the central atom.

Table 8–2 summarizes the common kinds of hybridization at the central atoms of molecules and ions with different numbers of stereoactive sets of electrons.

Valence Bond Theory — Polycentered Molecules and Multiple Bonding

We describe the geometries of more complex (polycentered) species in terms of *two or more central atoms*. We can usually describe the shapes of such molecules and ions as combinations of shapes with which we are already familiar. Let us consider the hydrocarbon ethane, C_2H_6, as our first example.

TABLE 8–2   Common Kinds of Hybridization Associated with Central Atoms

Number of Stereoactive Sets of e⁻ About Central Atom	Number of Lone Pairs Among the Stereoactive Sets	Example and Shape of Molecule (or Ion)*	Hybridization at Central Atom	Characteristic Angles
2	0	BeCl₂, linear	sp	180°
3	0	BF₃, trigonal planar	sp^2	120°
	1	SO₂, angular		†
4	0	CH₄, tetrahedral	sp^3	109°28′
	1	NH₃, pyramidal		†
	2	H₂O, angular		†
5	0	PF₅, trigonal bipyramidal	sp^3d	90°, 120°, 180°
	1	SF₄, distorted tetrahedral		†
	2	ClF₃, T-shaped		†

TABLE 8–2 (continued)

Number of Stereoactive Sets of e⁻ About Central Atom	Number of Lone Pairs Among the Stereoactive Sets	Example and Shape of Molecule (or Ion)*	Hybridization at Central Atom	Characteristic Angles
	3	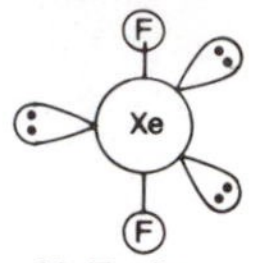 XeF_2, linear		
6	0	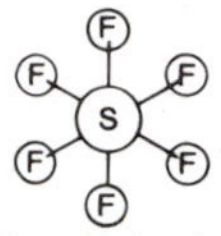 SF_6, octahedral	sp^3d^2	90°, 180°
	1	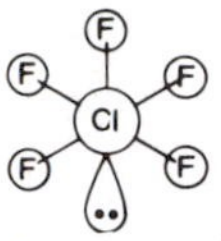 ClF_5, square pyramidal		†
	2	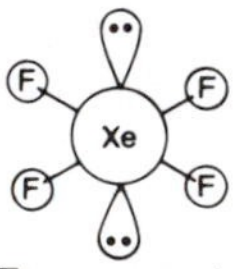 XeF_4, square planar		

* All atom connections are represented as single lines; circles represent atoms.
† The presence of lone pairs decreases bond angles from idealized values in these molecules and ions.

8–6   Ethane, C_2H_6

The Lewis formula for the ethane molecule is shown below.

Remember that we choose the most symmetrical skeleton possible.

$$S = N - A$$
$$= 28\ e^- - 14\ e^-$$
$$= 14\ e^-\ \text{shared}$$

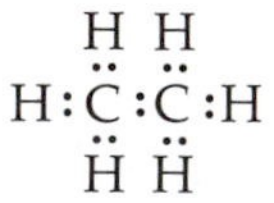

$$
\begin{array}{cc}
\text{H} & \text{H} \\
\text{H}\!:\!\ddot{\text{C}}\!:\!\ddot{\text{C}}\!:\!\text{H} \\
\ddot{\text{H}} & \ddot{\text{H}}
\end{array}
\quad \text{or} \quad
\begin{array}{c}
\ \ \ \text{H}\ \ \ \text{H} \\
\ \ \ | \ \ \ \ | \\
\text{H}\!-\!\text{C}\!-\!\text{C}\!-\!\text{H} \\
\ \ \ | \ \ \ \ | \\
\ \ \ \text{H}\ \ \ \text{H}
\end{array}
$$

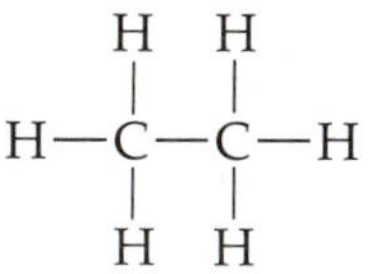

We do this only to see better the relationship of bonding and geometry in C_2H_6 molecules to that in CH_4 molecules.

We consider *each* carbon atom, in turn, as a central atom and deduce the arrangement around it. Each carbon atom has four stereoactive sets of electrons, so VSEPR theory predicts (and numerous experiments have shown) that each carbon atom is at the center of a tetrahedron. We can describe the molecule as two tetrahedra sharing a common corner. The carbon atoms are sp^3 hybridized, as in methane (see Section 8–3.1). Although ethane molecules are *not* formed by the combination of two methane molecules, we might imagine a process in which two methane molecules each lose a hydrogen atom with its electron, leaving two CH_3 fragments, each with one unpaired electron in an sp^3 orbital. The overlapping of these sp^3 orbitals and pairing of electrons would form the C—C bond.

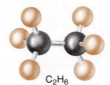

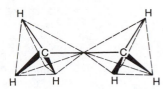

C_2H_6

tetrahedral electronic geometry and a tetrahedral arrangement of atoms about each C atom; sp^3 hybridization at each C atom

The larger hydrocarbons are simply longer chains of linked tetrahedra. For example, propane, C_3H_8, is shown below.

$$\begin{array}{ccc} H & H & H \\ \ddot{} & \ddot{} & \ddot{} \\ H\!:\!\ddot{C}\!:\!\ddot{C}\!:\!\ddot{C}\!:\!H \\ \ddot{} & \ddot{} & \ddot{} \\ H & H & H \end{array}$$

$$\begin{array}{ccc} H & H & H \\ | & | & | \\ H\!-\!C\!-\!C\!-\!C\!-\!H \\ | & | & | \\ H & H & H \end{array}$$

C_3H_8

8-7 Ethene, C₂H₄, and Formaldehyde, CH₂O

Ethene (ethylene), C_2H_4, is the simplest hydrocarbon that contains a *double* bond.

The alternate (common) name for ethene is ethylene.

$S = N - A$
$= 24\ e^- - 12\ e^-$
$= 12\ e^-$ shared

$$\begin{array}{cc} H & H \\ \ddot{} \quad \ddot{} \\ C\!::\!C \\ \ddot{} \quad \ddot{} \\ H & H \end{array}$$

Since there are *three stereoactive sets of electrons around each carbon atom*, VSEPR theory tells us that each carbon is at the center of a *trigonal plane*, as shown. It follows that Valence Bond theory tells us that *each doubly bonded carbon atom is sp^2 hybridized,* with one electron in each sp^2 hybrid orbital and one electron in the unhybridized $2p$ orbital that is *perpendicular* to the plane of the three sp^2 hybrid orbitals.

Relative energies of atomic orbitals and hybridized orbitals are not indicated.

$1s$	$2s$	$2p$		$1s$	sp^2		$2p$

$_6C$ ↑↓ ↑↓ ↑ ↑ ⟶ ↑↓ | ↑ ↑ ↑ | ↑

Recall that sp^2 hybrid orbitals are directed toward the corners of an equilateral triangle. Top and side views of these hybrid orbitals are shown below.

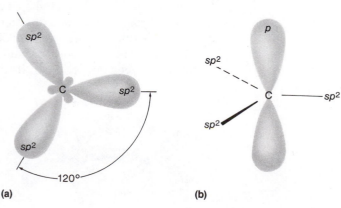

FIGURE 8–1 (a) A top view of three sp^2 hybrid orbitals. (b) A side view of a carbon atom in a trigonal planar environment, showing the remaining p orbital perpendicular to the plane of the three sp^2 hybrid orbitals.

(a)

(b)

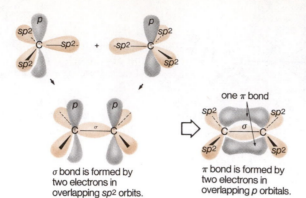

FIGURE 8–2 Schematic representation of the formation of a carbon-carbon double bond. Two sp^2 hybridized carbon atoms form a σ bond by overlap of two sp^2 orbitals, and a pi (π) bond by overlap of properly aligned p orbitals.

σ bond is formed by two electrons in overlapping $sp2$ orbits.

π bond is formed by two electrons in overlapping p orbitals.

The two carbon atoms interact by *end-to-end* (which is called *head-on*) overlap of sp^2 hybrids pointing toward each other to form a *sigma (σ) bond* and by *side-by-side* (which is called *side-on*) overlap of the unhybridized $2p$ orbitals to form a *pi (π) bond*. The sigma and pi bonds together comprise a double bond. The $1s$ orbitals (with 1 e^- each) of four hydrogen atoms overlap the remaining four sp^2 orbitals (with 1 e^- each) on the carbon atoms to form four C—H sigma bonds.

> A **sigma bond** *is a bond resulting from head-on overlap of atomic orbitals, in which the region of electron sharing is along and cylindrically around the imaginary line connecting the bonded atoms.* All single bonds are sigma bonds, and we have seen that many kinds of pure atomic orbitals and hybridized orbitals can be involved in sigma bond formation. A **pi bond** *is a bond resulting from side-on overlap of atomic orbitals, in which the regions of electron sharing are above and below the imaginary line connecting the bonded atoms, and parallel to this line.* A pi bond can form *only* if there is *also* a sigma bond between the same two atoms. A single bond is a sigma bond. A double bond consists of one sigma and one pi bond. A triple bond consists of one sigma and two pi bonds.

As a consequence of the sp^2 hybridization of carbon atoms in carbon-carbon double bonds, each carbon atom is at the center of a trigonal plane. The fact that the p orbitals (perpendicular to the trigonal planes) that overlap to form the π bond must be parallel to each other for effective overlap to occur adds the further restriction that these trigonal planes must also be *coplanar*. Thus, all four atoms attached to the doubly bonded carbon atoms lie in the same plane (Figure 8–3). There are many other important examples of organic compounds containing carbon-carbon double bonds. Several are described in Chapter 25.

Let us now consider formaldehyde, CH_2O, an organic compound containing a carbon-oxygen double bond. Its dot and dash formulas are shown on page 259.

one π bond

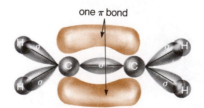

FIGURE 8–3 Four C—H σ bonds, one C—C σ bond, and one C—C π bond in the planar C_2H_4 molecule.

$$S = N - A$$
$$= 20 \ e^- - 12 \ e^-$$
$$= 8 \ e^- \text{ shared}$$

$$H:\overset{..}{C}::\overset{..}{O}:$$

$$\overset{H}{\underset{H}{>}}C = \overset{..}{\underset{..}{O}}$$

The carbon atom sits at the center of three stereoactive sets of electrons with no lone pairs on the carbon atom. Therefore, VSEPR theory predicts (and we observe) trigonal planar electronic and molecular geometries with respect to carbon. Valence Bond theory accounts for the bonding by assuming sp^2 hybridization at the carbon

atom. The bonding is analogous to that in ethene, except that a $=C\overset{H}{\underset{H}{<}}$ unit is

replaced by $=\overset{..}{O}:$ (see Figure 8–4).

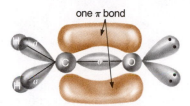

one π bond

FIGURE 8–4 Both C and O are thought to be sp^2 hybridized in the planar formaldehyde molecule, CH₂O.

The oxygen atom also would be pictured as sp^2 hybridized because it is surrounded by three stereoactive sets of electrons. The two lone pairs on the oxygen atom occupy two of the three sp^2 orbitals. The third sp^2 orbital of oxygen (containing one electron) overlaps one sp^2 orbital of the carbon atom (containing one electron) to form a C—O sigma bond. The pi (second) bond between C and O is formed by the overlap of the unhybridized p orbitals of C and O, perpendicular to the plane of the molecule, and consequent pairing of electrons in the region of overlap.

8-8 Ethyne, C₂H₂

The common name for ethyne is acetylene. It is used in oxyacetylene torches for welding metals.

A common compound containing a triple bond is ethyne, C₂H₂, which has the dot formula

$$S = N - A$$
$$= 20 \ e^- - 10 \ e^-$$
$$= 10 \ e^- \text{ shared}$$

$$H:C:::C:H$$

$$H—C≡C—H$$

VSEPR theory predicts that the two stereoactive sets of electrons around each carbon atom will be 180° apart.

The three p orbitals in a set are indistinguishable. We label the one involved in hybridization as "p_x" only to help us visualize the orientations of the two un-hybridized p orbitals on carbon.

Valence Bond theory postulates that each triply bonded carbon atom is sp hybridized (see Section 8–1) because each has two stereoactive sets of electrons. If we designate the p_x orbitals as the ones involved in hybridization, there is one carbon electron in each sp hybrid orbital and one electron in each of the $2p_y$ and $2p_z$ orbitals (before bonding is considered).

$${}_6C \ \underset{1s}{\overset{\uparrow\downarrow}{\rule{0.7em}{0.4pt}}} \ \underset{2s}{\overset{\uparrow\downarrow}{\rule{0.7em}{0.4pt}}} \ \underset{2p}{\overset{\uparrow \ \ \uparrow}{\rule{1.5em}{0.4pt}}} \quad \longrightarrow \quad \underset{1s}{\overset{\uparrow\downarrow}{\rule{0.7em}{0.4pt}}} \ \underset{sp}{\overset{\uparrow \ \ \uparrow}{\rule{1.5em}{0.4pt}}} \ \underset{2p_y}{\overset{\uparrow}{\rule{0.7em}{0.4pt}}} \ \underset{2p_z}{\overset{\uparrow}{\rule{0.7em}{0.4pt}}}$$

Remember that *each p* orbital has *two* lobes of equal size.

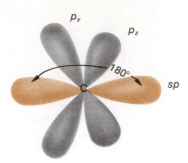

FIGURE 8–5 Diagram of the two linear, hybridized *sp* orbitals of the carbon atom (in color), which lie in a straight line, and the two unhybridized *p* orbitals (gray).

The unhybridized atomic $2p_y$ and $2p_z$ orbitals are perpendicular to each other and to a line through the centers of the two *sp* hybrid orbitals. Each carbon atom forms one sigma bond with the other carbon and a sigma bond with one hydrogen atom.

Additionally, the two half-filled *p* orbitals on different carbon atoms overlap side-to-side, p_y with p_y and p_z with p_z, to form *two* pi bonds (Figure 8–6).

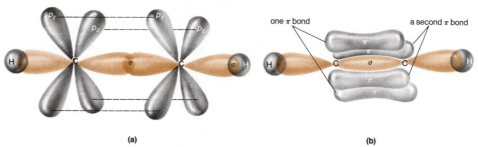

(a)   (b)

FIGURE 8–6 The acetylene molecule. (a) The overlap diagram of two *sp* hybridized carbon atoms and two *s* orbitals from two hydrogen atoms. There are two σ C—H bonds, one σ C—C bond, and two π C—C bonds (making the net carbon-carbon bond a triple bond). The dashed lines, each connecting two lobes, indicate the side-by-side overlap of the four unhybridized *p* orbitals. The hybridized *sp* orbitals are shown in color, and the unhybridized *p* orbitals are shown in gray. (b) The overall outline of the bonding orbitals in acetylene. The π bonding orbitals are positioned with one above and below the line of the σ bonds and the other behind and in front of the line of the σ bonds.

Since the *sp* hybrids are 180° apart and since *sp* hybrids on each carbon atom overlap end-to-end, the entire molecule must be linear.

Molecular Orbital Theory in Chemical Bonding

So far we have considered bonding and molecular geometry in terms of Valence Bond theory and (in Chapter 7) in terms of Valence Shell Electron Pair Repulsion theory. Valence Bond theory postulates that bonds result from the sharing of electrons in overlapping orbitals of different atoms. These orbitals may be *pure atomic orbitals* or *hybridized atomic orbitals* of *individual* atoms. Electrons in overlapping orbitals of different atoms are thought of as being localized in the bonds between the two atoms involved, rather than delocalized over the entire molecule. Hybridization is invoked when necessary to account for either the geometry of a molecule or the number of unpaired electrons that must be available for bonding.

Molecular Orbital theory, on the other hand, postulates the *combination of atomic orbitals of different atoms to form* **molecular orbitals,** *so that electrons in them belong to the molecule as a whole.* The Valence Bond and Molecular Orbital approaches have strengths and weaknesses that are complementary. They represent alternative descriptions of chemical bonding, and, in the limit, produce the same results. Valence Bond theory is descriptively attractive, lending itself well to visualization. Molecular Orbital (MO) theory gives somewhat better descriptions of electron cloud distributions, bond energies, and magnetic properties, but it is harder to visualize.

As an example of a weakness of the Valence Bond theory, its picture of the bonding in the oxygen molecule, O_2, involves a double bond and no unpaired electrons. It predicts sp^2 hybridization at each oxygen because there are three sets of valence electrons on each oxygen atom.

:Ö::Ö: (wrong!)

However, experiments show that O_2 is actually paramagnetic (p. 264) and therefore has unpaired electrons. Molecular Orbital theory *predicts* that O_2 has two unpaired electrons (see below), and it was the ability of MO theory to account for the paramagnetism of O_2 that brought it to the forefront as a major theory in bonding. In the following sections, we shall develop some of the notions of MO theory and then apply them to some relatively simple molecules and ions.

8–9 Molecular Orbitals

The mathematical pictures of hybrid orbitals in Valence Bond theory can be generated by combining the wave functions that describe two or more atomic orbitals on a *single* atom. Similarly, mathematical descriptions of molecular orbitals can be generated by combining wave functions that describe atomic orbitals on *separate* atoms.

The electron waves that describe atomic orbitals have positive and negative phases or amplitudes, just as there are positive (upward) and negative (downward) amplitudes associated with a standing wave. When waves are combined, they may interact either constructively or destructively. If the two identical waves shown at the left are added, they combine constructively to produce the wave at the right. Conversely, if they are subtracted, it is as if the phases of one wave were reversed and the result was added to the first wave. This causes destructive interference, resulting in zero amplitude in the wave at the right.

Likewise, when two atomic orbitals overlap they can be in-phase (added) or out-of-phase (subtracted). When they overlap in-phase, constructive interaction occurs in the region between the nuclei and a **bonding orbital** is produced. The energy of the bonding orbital is always lower (more stable) than the energies of the combining orbitals. When they overlap out-of-phase, destructive interference reduces the probability of finding an electron in the region between the nuclei and an

In some polyatomic molecules, a molecular orbital may extend over only a fragment of the molecule. Molecular orbitals also exist for polyatomic ions such as CO_3^{2-}, SO_4^{2-}, and NH_4^+.

In fact, since O_2 has an even number of valence electrons, at least two must be unpaired.

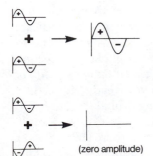
(zero amplitude)

Electrons in bonding orbitals are often called *bonding electrons;* electrons in antibonding orbitals, *antibonding electrons.*

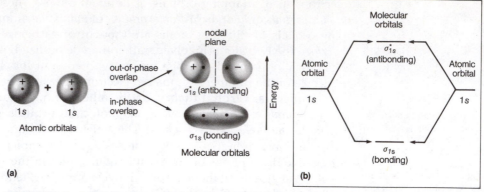

FIGURE 8–7 Molecular orbital (MO) diagram for the combination of two $1s$ atomic orbitals (at the left) to form two molecular orbitals. One is a *bonding* orbital, σ_{1s}, resulting from addition of the wave functions of the $1s$ orbitals. The other is an *antibonding* orbital, σ_{1s}^*, at higher energy resulting from subtraction of the wave functions. In all σ-type molecular orbitals the electron density is symmetrical about an imaginary line connecting the two nuclei. (The terms "subtraction of wave functions," "out-of-phase," and "destructive interference in the region between the nuclei" all refer to the formation of an antibonding molecular orbital.)

antibonding orbital is produced. The energy of an antibonding orbital is higher (less stable) than the energies of the combining orbitals.

We can illustrate this basic principle by considering the combination of two $1s$ atomic orbitals (Figure 8–7). (Don't confuse the $+$ and $-$ designations, which indicate amplitudes of wave functions, with electrical charges.) When these orbitals are occupied by electrons, the shapes of the orbitals are plots of electron density that represent regions of space in which the probabilities of finding electrons in molecules are greatest.

In the bonding orbital, the two $1s$ orbitals have combined their densities in the region between the two nuclei by in-phase overlap, or addition of the wave functions. In the antibonding orbital, they have canceled out one another by out-of-phase overlap or subtraction of the wave functions. We designate both molecular orbitals as sigma (σ) orbitals (which indicates that they are symmetrical about a line drawn through the two nuclei, the internuclear axis), and we indicate with subscripts the atomic orbitals that have been combined. The asterisk denotes an antibonding orbital. All sigma antibonding orbitals have nodal planes bisecting the internuclear axis. A **node** or **nodal plane** *is a region in which the probability of finding electrons is zero.* Thus, two $1s$ orbitals produce a σ_{1s} (read "sigma-1s") bonding orbital and a σ_{1s}^* (read "sigma-1s-star") antibonding orbital. Figure 8–7b shows the relative energy levels of these orbitals.

Placing electrons in bonding molecular orbitals leads to a more stable (lower energy) system than the individual atoms. Placing electrons in antibonding orbitals, which requires adding energy to the molecule, leads to destabilization of the molecule relative to individual atoms.

For any two sets of p orbitals on two different atoms, corresponding orbitals such as p_x orbitals can overlap head-on, giving σ_p and σ_p^* orbitals, as shown in Figure 8–8 for the head-on overlap of $2p_x$ orbitals on two atoms.

FIGURE 8–8
Production of σ_{2p_x} and $\sigma^*_{2p_x}$ molecular orbitals by overlap of $2p_x$ orbitals on two atoms.

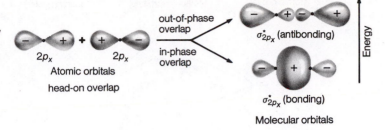

out-of-phase overlap

in-phase overlap

$\sigma^*_{2p_x}$ (antibonding)

$\sigma^*_{2p_x}$ (bonding)

$2p_x$ $2p_x$

Atomic orbitals

head-on overlap

Molecular orbitals

Energy

If we had chosen the z-axis as the axis of head-on overlap of the $2p$ orbitals in Figure 8–8, side-on overlap of the $2p_x$-$2p_x$ and $2p_y$-$2p_y$ orbitals would form the π-type molecular orbitals.

If the remaining p orbitals overlap (p_y with p_y and p_z with p_z), they must do so *side-on*, forming *pi (π) molecular orbitals*. Depending on whether all p orbitals overlap, there can be a total of two π_p and two π^*_p orbitals. Figure 8–9 illustrates the overlap of two corresponding $2p$ orbitals on two atoms to form π_{2p} and π^*_{2p} molecular orbitals. Note that there is a *nodal plane* along the internuclear axis for all pi molecular orbitals.

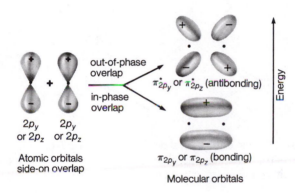

out-of-phase overlap

in-phase overlap

$\pi^*_{2p_y}$ or $\pi^*_{2p_z}$ (antibonding)

π_{2p_y} or π_{2p_z} (bonding)

$2p_y$ $2p_y$
or $2p_z$ or $2p_z$

Atomic orbitals
side-on overlap

Molecular orbitals

Energy

FIGURE 8–9 The π_{2p} and π^*_{2p} molecular orbitals from overlap of one pair of $2p$ atomic orbitals (for instance, $2p_y$ orbitals). There can be an identical pair of molecular orbitals at right angles to these, formed by another pair of p orbitals on the same two atoms (in this case, $2p_z$ orbitals).

8–10   Molecular Orbital Energy-level Diagrams

Homonuclear: containing only atoms of the same element.

Here we are putting *all* of the electrons into MO energy-level diagrams, not just those in the valence shell.

Using what we know about how atomic orbitals combine to produce molecular orbitals, we can now draw molecular orbital energy-level diagrams for simple molecules. The simplest example, shown in Figure 8–10, is for most homonuclear diatomic molecules of elements in the first and second periods of the periodic table. It is simply an extension of the diagram in Figure 8–7b, to which we have added molecular orbitals formed from $2s$ and $2p$ atomic orbitals.

To use the diagram in Figure 8–10, we simply follow the same rules established for atomic energy-level diagrams in Chapter 5. Count the number of electrons in the molecule, and then fill the molecular orbitals, lowest-energy orbitals first, while observing the Aufbau Principle (Section 5–10), Hund's Rule (Section 5–10), and the Pauli Exclusion Principle (Section 5–7).

For the cases shown in Figure 8–10, the two π_{2p} orbitals are lower in energy

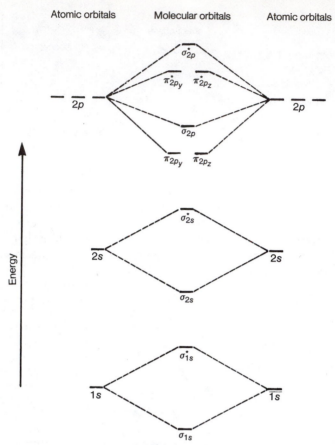

Atomic orbitals   Molecular orbitals   Atomic orbitals

FIGURE 8–10 The energy-level diagram for H_2, He_2, Li_2, Be_2, B_2, C_2 and N_2 molecules and their ions. The solid lines represent the relative energies of the indicated atomic and molecular orbitals; the energies are not shown to scale.

than the σ_{2p} orbital. However, spectroscopic data indicate that for O_2, F_2, and (hypothetical) Ne_2 molecules, the σ_{2p} orbital is lower in energy than the π_{2p} orbitals (Figure 8–11).

8–11   Bond Order and Bond Stability

Now all that is needed is a criterion by which to judge the stability of a molecule, once its energy-level diagram has been filled with the appropriate number of electrons. This criterion is the **bond order,** *which is defined as the number of electrons in bonding orbitals minus the number of electrons in antibonding orbitals, all divided by two.*

$$\text{bond order} = \frac{(\text{number of } e^- \text{ in bonding MO's}) - (\text{number of } e^- \text{ in antibonding MO's})}{2} \qquad \text{[Eq. 8–1]}$$

Generally, the bond order corresponds to the number of bonds established by the Valence Bond theory. Fractional bond orders exist in species that contain an odd number of electrons, such as the nitrogen oxide molecule, NO. The greater the bond

The NO molecule contains a total of 15 electrons.

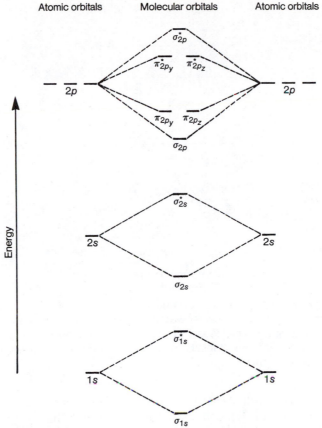

Atomic orbitals   Molecular orbitals   Atomic orbitals

FIGURE 8–11 The energy-level diagram for O_2, F_2, and Ne_2 molecules and their ions. The solid lines represent the relative energies of the indicated atomic and molecular orbitals; the energies are not shown to scale.

order of a diatomic molecule or ion, the more stable we predict it to be. Likewise, for a bond between two given atoms, the greater the bond order, the shorter the bond length, and the greater the bond energy.

8–12 Some Homonuclear Diatomic Molecules

The electron distributions for the homonuclear diatomic molecules of the first and second periods are shown in Figure 8–12, together with the bond order, bond length, and bond energy for each molecule. We shall now look at a few of them.

1 Hydrogen, H_2

The overlap of the $1s$ orbitals of two hydrogen atoms produces σ_{1s} and σ_{1s}^* molecular orbitals. The two electrons of the molecule occupy the lower energy σ_{1s} orbital as shown in Figure 8–13.

Since there is one pair of electrons in a bonding orbital and none in an antibonding orbital of an H_2 molecule, the bond order is one. We conclude that the

INCREASING ENERGY (not to scale)

Orbital	H$_2$	He$_2$††	Li$_2$†	Be$_2$†	B$_2$†	C$_2$†	N$_2$		O$_2$	F$_2$	Ne$_2$††
σ^*_{2p}	—	—	—	—	—	—	—		—	—	↿⇂
$\pi^*_{2p_y}, \pi^*_{2p_z}$	— —	— —	— —	— —	— —	— —	— —		↑ ↑	↿⇂ ↿⇂	↿⇂ ↿⇂
σ_p [right: π_{p_y}, π_{2p_z}]	—	—	—	—	—	—	↿⇂		↿⇂ ↿⇂	↿⇂ ↿⇂	↿⇂ ↿⇂
π_{2p_y}, π_{2p_z} [right: σ_p]	— —	— —	— —	— —	↑ ↑	↿⇂ ↿⇂	↿⇂ ↿⇂		↿⇂	↿⇂	↿⇂
σ^*_{2s}				↿⇂	↿⇂	↿⇂	↿⇂		↿⇂	↿⇂	↿⇂
σ_{2s}			↿⇂	↿⇂	↿⇂	↿⇂	↿⇂		↿⇂	↿⇂	↿⇂
σ^*_{1s}	—	↿⇂	↿⇂	↿⇂	↿⇂	↿⇂	↿⇂		↿⇂	↿⇂	↿⇂
σ_{1s}	↿⇂	↿⇂	↿⇂	↿⇂	↿⇂	↿⇂	↿⇂		↿⇂	↿⇂	↿⇂
Paramagnetic ?	no	no	no	no	yes	no	no		yes	no	no
Bond Order	1	0	1	0	1	2	3		2	1	0
Bond Length (nm)	0.074	—	0.267	—	0.159	0.131	0.109		0.121	0.143	—
Bond Energy (kJ/mol)	435	—	110	—	~270	602	946		498	159	—

† Exists only in vapor state at elevated temperatures.
†† Unstable, i.e., unknown species.

FIGURE 8–12 Electron distribution in molecular orbitals, bond order, bond length, and bond energy of homonuclear diatomic molecules of the first and second row elements. Note that nitrogen molecules, N$_2$, have the highest bond energies listed because they have a bond order of three (triple bonds). The doubly bonded species, C$_2$ and O$_2$, have the next highest bond energies.

H$_2$ molecule is stable, and of course it is. The energy associated with two electrons in the H$_2$ molecule is less than that associated with the same two electrons in 1s atomic orbitals. As we saw in Chapter 3, the lower the energy of a system is, the more stable it is.

2   Helium (Hypothetical), He$_2$

The energy-level diagram for He$_2$ is similar to that of H$_2$ except that it has two more electrons, which occupy the antibonding σ^*_{1s} orbital. (See Figures 8–10, 8–12, and 8–13.) Thus its bond order is zero, and we conclude that the molecule is not stable. In fact, He$_2$ is not known; helium exists only as *monatomic* molecules.

3   Boron, B$_2$

The boron atom has the configuration $1s^2 2s^2 2p^1$ and is the first element with p electrons to participate in bonding. Reference to Figures 8–10 and 8–12 shows that

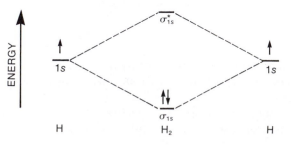

FIGURE 8–13 Molecular orbital diagram for H$_2$.

the π_{2p_y} and π_{2p_z} MO's are lower in energy than σ_{2p} for B_2. Thus its electron configuration is $\sigma_{1s}^2\,\sigma_{1s}^{*2}\,\sigma_{2s}^2\,\sigma_{2s}^{*2}\,\pi_{2p_y}^1\,\pi_{2p_z}^1$. This illustrates the validity of Hund's Rule in dealing with molecular orbital theory. The π_{2p_y} and π_{2p_z} orbitals are equal in energy and contain a total of two electrons. As a result, one electron occupies each orbital. The bond order is one, and experiments verify not only that the molecule exists in the vapor state but that it is paramagnetic with two unpaired electrons.

Orbitals of equal energy are called degenerate *orbitals.*

4 Oxygen, O_2

Among homonuclear diatomic molecules, only N_2 and the very small H_2 have shorter bond lengths than O_2 (0.121 nm). As mentioned earlier, Valence Bond theory predicts that O_2 would be diamagnetic. However, experiments show that it is paramagnetic, with two unpaired electrons. A structure consistent with this observation is predicted by MO theory. Spectroscopic evidence indicates that for O_2, the σ_{2p} orbital is lower in energy than the π_{2p_y} and π_{2p_z} orbitals. Each oxygen atom has eight electrons and the O_2 molecule has 16, distributed as follows:

$$\sigma_{1s}^2 \quad \sigma_{1s}^{*2} \quad \sigma_{2s}^2 \quad \sigma_{2s}^{*2} \quad \sigma_{2p}^2 \quad \pi_{2p_y}^2 \quad \pi_{2p_z}^2 \quad \pi_{2p_y}^{*1} \quad \pi_{2p_z}^{*1}$$

The prediction of unpaired electrons is consistent with the observed paramagnetism of O_2.

The two unpaired electrons reside in the *degenerate* antibonding orbitals, $\pi_{2p_y}^*$ and $\pi_{2p_z}^*$. Since there are four more electrons in bonding orbitals than in antibonding orbitals, the bond order is two (double bond) and we conclude that the molecule should be very stable, as it is (Figures 8-11 and 8-12).

Example 8-1

Predict the stabilities and bond orders of the O_2^+ and O_2^- ions.

Solution

The O_2^+ ion is obtained by removing one electron from the O_2 molecule. Since the electrons that are withdrawn most easily are those in the highest energy orbitals, one of the π_{2p}^* electrons is lost. Thus the configuration of O_2^+ is $\sigma_{1s}^2\,\sigma_{1s}^{*2}\,\sigma_{2s}^2\,\sigma_{2s}^{*2}\,\sigma_{2p}^2\,\pi_{2p_y}^2\,\pi_{2p_z}^2\,\pi_{2p_y}^{*1}$. There are five more electrons in bonding orbitals than in antibonding orbitals, and the bond order is 2.5. We conclude that the ion would be reasonably stable relative to other diatomic ions, and it does exist. In fact, the unusual ionic compound $[O_2^+][PtF_6^-]$ played an important role in the discovery of the first noble gas compound, $[Xe^+][PtF_6^-]$.

The superoxide ion, O_2^-, results from adding an electron to one of the π_{2p}^* orbitals of O_2. The configuration of O_2^- is $\sigma_{1s}^2\,\sigma_{1s}^{*2}\,\sigma_{2s}^2\,\sigma_{2s}^{*2}\,\sigma_{2p}^2\,\pi_{2p_y}^2\,\pi_{2p_z}^2\,\pi_{2p_y}^{*2}\,\pi_{2p_z}^{*1}$. The bond order is 1.5 because there are three more bonding electrons than antibonding electrons. Thus we can conclude that the ion would be stable, but less stable than O_2. It is interesting to note that the superoxides of the heavier Group IA elements, KO_2, RbO_2, and CsO_2, which contain O_2^-, all exist. These compounds are formed by combination of the free metals with oxygen (Section 9-3).

---

5 Heavier Homonuclear Diatomic Molecules

On the surface it would seem reasonable to use the same types of MO diagrams to predict the stabilities or existence of homonuclear *diatomic* molecules of the third and subsequent periods. However, other than the heavier halogens, Cl_2, Br_2, and I_2, containing only single (sigma) bonds, no others exist at room temperature. We would predict from both Molecular Orbital theory and Valence Bond theory that the other (nonhalogen) homonuclear diatomic molecules from below the second period would exhibit multiple bonding, and therefore pi bonding. Some heavier elements exist as diatomic species, such as S_2, in the *vapor* phase at elevated temperatures, but most are not very stable. The instability appears to be directly related to the inability of atoms of the heavier elements to form strong pi bonds with each other. This is because the degree of overlap of atomic *p* orbitals on different atoms, and therefore the strength of pi bonds, decreases quite rapidly with increasing atomic size, and the accompanying increase in sigma bond length. For example, N_2 is much more stable than P_2 because the $3p$ orbitals on one phosphorus atom do not overlap side-by-side in a pi bonding manner with corresponding $3p$ orbitals on another phosphorus atom nearly as effectively as the corresponding $2p$ orbitals on the smaller nitrogen atoms do. Note that multiple bonding is not predicted for Cl_2, Br_2, and I_2, all of which have a bond order of one.

8–13   Hydrogen Fluoride, HF (Heteronuclear)

The corresponding atomic orbitals of two *different* elements, such as hydrogen and fluorine, have different energies because the nuclei of the atoms have different numbers of protons. As a result, molecular orbital diagrams such as Figures 8–10 and 8–11 are inappropriate for *heteronuclear* diatomic molecules. The MO diagram must be skewed so that the combining atomic orbitals of the *more electronegative element* (the $2p$ orbitals of fluorine in this case) are *lower* in energy than the combining orbitals of the less electronegative element (the $1s$ orbital of hydrogen in this case).

The MO diagram of HF is shown in Figure 8–14. The hydrogen fluoride molecule contains a very polar bond because the electronegativity difference between hydrogen (EN = 2.1) and fluorine (EN = 4.0) is very large (ΔEN = 1.9). Figure 8–15 shows the overlap of the $1s$ orbital of hydrogen with a $2p$ orbital of fluorine to form σ_{sp} and σ_{sp}^* molecular orbitals. Since the remaining two fluorine $2p$ orbitals have no net overlap with hydrogen orbitals, they are called **nonbonding orbitals.** The same is true of the fluorine $2s$ and $1s$ orbitals. These nonbonding orbitals retain the characteristics of the fluorine atomic orbitals from which they are formed.

The closer in energy a molecular orbital is to the energy of one of the atomic orbitals of which it is composed, the more of the character of that atomic orbital it shows. Thus, in the case of the HF molecule, the bonding (σ_{sp}) molecular orbital has

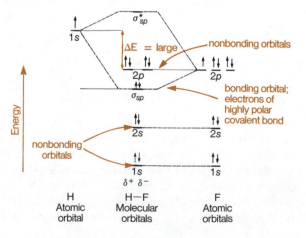

FIGURE 8–14 MO diagram for hydrogen fluoride, HF, a very polar molecule (μ = 1.9 D). Bond order = 1.

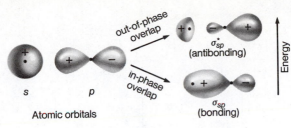

FIGURE 8–15 Formation of σ_{sp} and σ_{sp}^* molecular orbitals in HF by overlap of the 1s orbital of hydrogen with a 2p orbital of fluorine.

more fluorine-like character and the antibonding orbital has more hydrogen-like character. In general, the energy difference, ΔE, between the combining atomic orbitals reflects the difference in electronegativities between the two atoms. The greater this energy difference is, the more polar is the bond joining the atoms and the greater is its ionic character.

8–14 Delocalization and the Shapes of Molecular Orbitals

In Sections 6–11, 7–4, and 8–2 reference was made to resonance in molecules and ions. A term more appropriately used than resonance within the context of molecular orbital descriptions of bonding is "delocalization" of electrons. The shapes of molecular orbitals for species in which delocalization occurs can be determined by "averaging" all the contributing atomic orbitals.

1 Carbonate Ions, CO_3^{2-}

Consider the trigonal planar carbonate ion, CO_3^{2-}, as an example. Valence Bond theory describes the ion in terms of three contributing resonance structures, shown at the top of Figure 8–16. Remember that no one of the three resonance forms adequately describes the bonding, since all the carbon-oxygen bonds in the ion have equal bond length and equal bond energy, intermediate between those of typical C—O and C=O bonds.

According to Valence Bond theory, the carbon atom is described as sp^2 hybridized, and it forms one sigma bond with each of the three oxygen atoms. This leaves one unhybridized 2p atomic orbital, say the $2p_z$ orbital, on the carbon atom, which is capable of overlapping and mixing with the $2p_z$ orbital of any one of the three oxygen atoms. The sharing of two electrons in the resulting localized pi orbital would form a pi bond. Thus three equivalent resonance structures can be drawn.

The MO description of the pi bonding involves the overlap and mixing of the $2p_z$ orbital of the carbon with the $2p_z$ orbitals of all three oxygen atoms *simultaneously*, to form a bonding pi molecular orbital extending above and below the plane of the sigma system, as well as an antibonding pi orbital. Two electrons are said to occupy the entire bonding pi MO, which is depicted at the bottom of Figure 8–16. Note that it has the shape we would obtain by "averaging" the three contributing Valence Bond resonance structures. The bonding in such species as nitrate ion, NO_3^-, nitrite ion, NO_2^-, and sulfur dioxide, SO_2, may be described in comparable terms.

The bond order of each carbon-oxygen bond is $1\frac{1}{3}$ in the CO_3^{2-} ion.

FIGURE 8–16
Resonance structures and the delocalized π MO system for the carbonate ion. A multicenter π system extends over more than two atoms (four atoms in CO_3^{2-}).

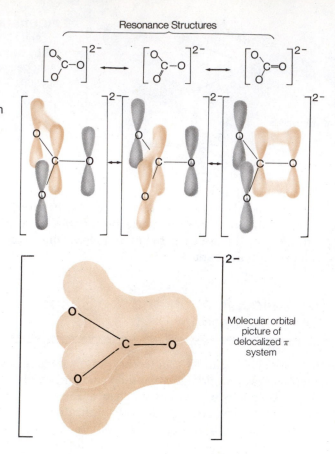

Resonance Structures

Molecular orbital picture of delocalized π system

2   Benzene, C_6H_6

Now let us consider the benzene molecule, C_6H_6, whose two Valence Bond resonance forms are shown at the top of Figure 8–17. The Valence Bond description involves sp^2 hybridization at each carbon atom. Not only is each carbon atom at the center of a trigonal plane, but the entire molecule is known to be planar. There are sigma bonds from each carbon atom to the two adjacent carbon atoms and to one hydrogen atom. This leaves one unhybridized $2p_z$ orbital on each carbon atom and one remaining valence electron for each. According to Valence Bond theory, adjacent pairs of $2p_z$ orbitals overlap side-by-side, and the six remaining electrons occupy the regions of overlap to form a total of three pi bonds in either one of the two ways shown in the middle of Figure 8–17.

The actual structure of the C_6H_6 molecule does *not* contain alternating single and double carbon-carbon bonds. The usual C—C single bond length is 0.154 nm and the usual C=C double bond length is 0.134 nm. Experimentally we find that all six of the carbon-carbon bonds in benzene have the same length, 0.139 nm, intermediate between those of single and double bonds. This is well explained by the MO theory, which produces the prediction that the six $2p_z$ orbitals of the carbon atoms overlap and mix to form three pi bonding and three pi antibonding molecular orbitals. The bottom of Figure 8–17 shows the MO representation of the delocalized pi system obtained from the combination of the three pi bonding MO's. It is this extended system of MOs that contains the six pi electrons, which are distributed

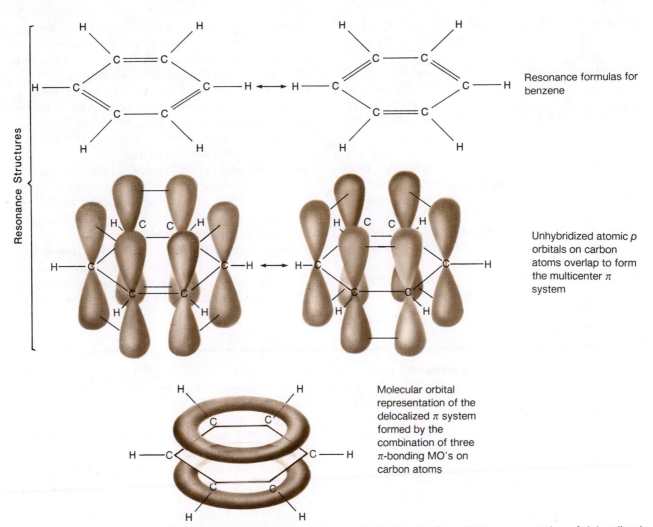

Resonance formulas for benzene

Unhybridized atomic p orbitals on carbon atoms overlap to form the multicenter π system

Molecular orbital representation of the delocalized π system formed by the combination of three π-bonding MO's on carbon atoms

FIGURE 8–17 Resonance structures and molecular orbital representation of delocalized π system of benzene, C_6H_6.

throughout the molecule as a whole, above and below the plane of the sigma-bonded framework. This results in identical character for all carbon-carbon bonds in benzene. Each carbon-carbon bond has a bond order of $1\frac{1}{2}$. Note that, as in the CO_3^{2-} ion, the MO representation of the extended pi system is the same as one would obtain by averaging the two contributing Valence Bond resonance structures.

Key Terms

Antibonding orbital a molecular orbital higher in energy than any of the atomic orbitals from which it is derived; lends instability to a molecule or ion when populated with electrons; denoted with an asterisk (*) on symbol.

Bonding orbital a molecular orbital lower in energy than any of the atomic orbitals from which it is derived; lends stability to a molecule or ion when populated with electrons.

Bond order half the number of electrons in bonding orbitals minus half the number of electrons in antibonding orbitals.

Coordinate covalent bond covalent bond in which both electrons originated on the same atom.

Hybridization mixing of a set of atomic orbitals on a single atom to form a new set of orbitals with the same total electron capacity, and with properties and energies intermediate between those of the original unhybridized orbitals.

Molecular orbital (MO) an orbital resulting from overlap and mixing of atomic orbitals on different atoms, that belongs to the molecule as a whole.

Molecular Orbital theory a theory of chemical bonding based upon the postulated existence of molecular orbitals.

Nodal plane (node) a region in which the probability of finding an electron is zero.

Nonbonding orbital a molecular orbital derived only from an atomic orbital of one atom; lends neither stability nor instability to a molecule or ion when populated with electrons.

Pi (π) bond a bond resulting from the side-on overlap of atomic orbitals, in which the regions of electron sharing are above and below the imaginary line connecting the bonded atoms, and parallel to this line.

Sigma (σ) bond a bond resulting from the head-on overlap of atomic orbitals, in which the region of electron sharing is along and (cyclindrically) symmetrical to the imaginary line connecting the bonded atoms.

Valence Bond (VB) theory a bonding theory that assumes that covalent bonds are formed when atomic orbitals on different atoms overlap and electrons are shared by the two atoms.

Exercises

Valence Bond Theory — Molecules and Ions Having Only Single Bonds

1. (a) Outline the Valence Bond theory. (b) What are hybrid orbitals?

2. What angles are associated with orbitals in the following hybridized sets of orbitals? (a) sp (b) sp^2 (c) sp^3 (d) sp^3d (e) sp^3d^2

3. Describe the electronic geometries of atoms that have undergone the kinds of hybridization listed in Exercise 2.

4. Describe the bonding in each of the following molecules with a Lewis formula and a Valence Bond structure.* Show the orbital orientations and label the orbitals. (a) $BeCl_2$ (b) CF_4 (c) H_2S (d) SF_6

5. Describe the bonding in each of the following species with a Lewis formula and a VB structure. Show the orbital orientations and label the orbitals. (a) BF_3 (b) ClO_4^- (c) NF_3 (d) $(CH_3)_2O$ (e) AsF_5

6. Describe the bonding in each of the following molecules or ions with a Lewis formula and a VB structure. Show the orbital orientations and label the orbitals. (a) CH_3Cl (b) ClO_3^- (c) H_3O^+ (d) $B(OH)_3$ (boric acid)

7. Describe the bonding in each of the following molecules or ions with a Lewis formula and a VB structure. Show the orbital orientations and label the orbitals. (a) $SnCl_4$ (b) $SnCl_6^{2-}$ (c) OF_2 (d) $Te(OH)_6$ (telluric acid)

8. Describe the change in hybridization at the central atom of the reactant at the left that occurs in each of the following reactions.
   (a) $BF_3 + F^- \longrightarrow BF_4^-$
   (b) $AsCl_3 + Cl_2 \longrightarrow AsCl_5$
   (c) $SF_4 + F_2 \longrightarrow SF_6$

9. What changes in hybridization occur in the following reaction?
   $:NH_3 + BF_3 \longrightarrow H_3N:BF_3$

10. The industrial manufacture of sulfuric acid, H_2SO_4, involves the conversion of elemental sulfur, S_8, to SO_2, then to SO_3 and finally to H_2SO_4. The S—S—S bond angles in S_8 are all about 108°. Describe the changes in hybridization that accompany these conversions.

11. Give a reason why phosphorus forms both PF_3 and PF_5 but nitrogen forms only NF_3.

12. Draw Lewis formulas and VB structures of each of the following molecules that contain the noble gas xenon. Label the orbitals and indicate (approximate) bond angles. (a) XeF_2 (b) XeF_4 (c) XeO_3 (d) XeO_4

---

* We use the term Valence Bond (VB) structure to describe a structure that shows hybridized orbitals on the central element and the bond angles of a molecule or polyatomic ion.

13. Draw Lewis formulas and VB structures of each of the following molecules and ions of the interhalogens. Label the orbitals and indicate (approximate) bond angles. (a) ClF_3 (b) BrF_5 (c) ICl_4^+ (d) ICl_4^-

Valence Bond Theory — Polycentered Molecules and Multiple Bonding

14. For each of the molecules shown below, predict the hybridization at each carbon atom.

(a) acetone,

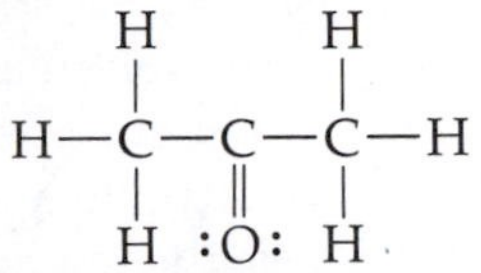

(b) methanol or methyl alcohol (wood alcohol),

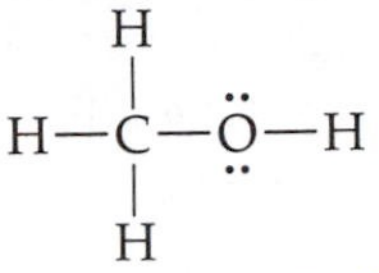

(c) acetic acid (vinegar is about a 5% solution of acetic acid in water),

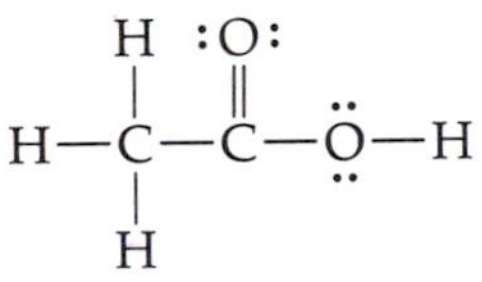

(d) tetrachloroethene (a dry-cleaning solvent),

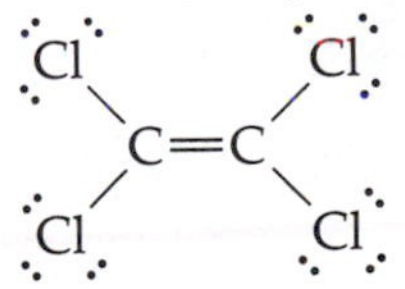

15. For each of the molecules shown below, predict the hybridization at the starred (*) atoms and the approximate bond angles at those atoms.

(a) methylamine,

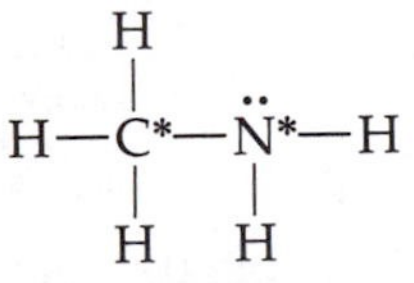

(b) urea (a product of metabolism of proteins),

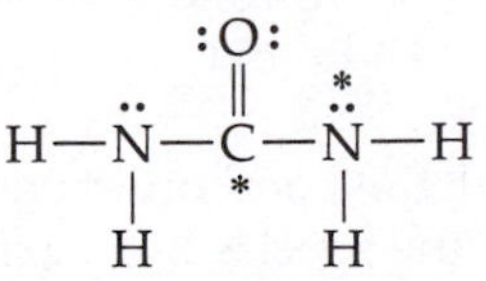

(c) glycine (an amino acid that is a constituent of many proteins),

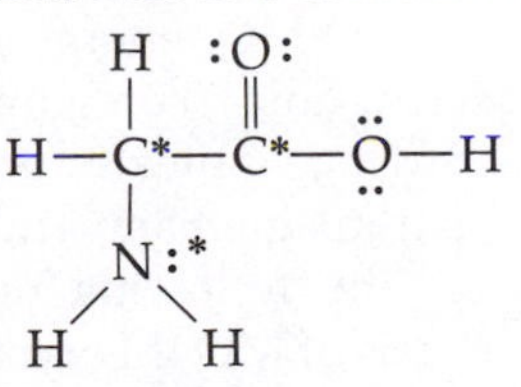

(d) asparagine (another amino acid),

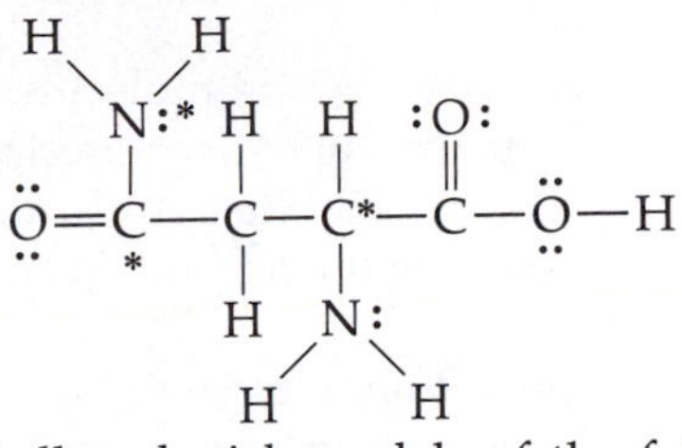

16. Sketch ball-and-stick models of the following ions, and indicate the hybridization and approximate bond angles at the central atoms. (a) nitrate ion, NO_3^- (b) nitrite ion, NO_2^- (c) sulfate ion, SO_4^{2-} (d) sulfite ion, SO_3^{2-}

17. Sodium acetylide, $Na^+C_2H^-$, is an ionic compound. The two C atoms are bonded together. Draw a Lewis formula and a three-dimensional sketch of the acetylide anion, showing hybridization and bond angle.

18. The pyrosulfate ion, $S_2O_7^{2-}$, contains one oxygen atom bonded to both sulfur atoms, called a "bridging oxygen." Draw a dash formula and sketch the shape of the ion, showing the hybridized orbitals at the sulfur atoms. What are the approximate bond angles at these atoms?

19. Describe the bonding in the N_2 molecule with a three-dimensional VB structure. Show the orbital overlap and label the orbitals.

20. Draw a dash formula and a three-dimensional structure for each of the following polycentered molecules. Indicate hybridizations and bond angles at each carbon atom. (a) butane, C_4H_{10} (b) propene, CH_3CHCH_2 (c) propyne, CH_3CCH (d) acetaldehyde, CH_3CHO

21. How many sigma bonds and how many pi bonds are there in each of the compounds of Exercise 20?

Molecular Orbital Theory

22. What differences and similarities exist among (1) atomic orbitals, (2) localized hybridized atomic orbitals according to Valence Bond theory, and (3) molecular orbitals?

23. What is the relationship between the maximum number of electrons that can be accommodated by a set of molecular orbitals and the maximum number that can be accommodated by the atomic orbitals from which the MO's are formed? What is the maximum number of electrons that one MO can hold?

24. Answer Exercise 23 after replacing "molecular

orbitals" with "localized hybridized atomic orbitals."

25. Sketch and describe the molecular orbitals resulting from the following overlap of atomic orbitals: (a) two s orbitals (b) head-on overlap of two p orbitals (c) side-on overlap of two p orbitals

26. State Hund's Rule. Describe the application of Hund's Rule to electron occupancy of molecular orbitals.

27. How are the electron occupancies of bonding, nonbonding, and antibonding orbitals related to the stability of a molecule or ion?

28. From memory, draw the energy-level diagram of molecular orbitals produced by the overlap of orbitals of two identical atoms from the second period. (Show the π_{2p_y} and π_{2p_z} orbitals below the σ_{2p} in energy.)

29. Utilizing sets of MO's such as you drew in Exercise 28, determine the electron configurations of the following molecules and ions: Li_2, Li_2^+, C_2, C_2^-, C_2^{2-}, N_2^-, and B_2^+.

30. (a) What is the bond order of each of the species in Exercise 29? (b) Which are diamagnetic (D) and which are paramagnetic (P)? (c) Apply MO theory to predict relative stabilities of these species. (d) Comment on the validity of the predictions from (c). What else *must* be considered in addition to electron occupancy of MO's?

31. How does the effectiveness of overlap of atomic orbitals on two atoms vary with the relative energies of the overlapping orbitals?

32. How does the polar or nonpolar character of a bond formed by overlap of atomic orbitals on two atoms depend upon the relative energies of the overlapping orbitals? Assume that two electrons occupy the resulting MO.

33. Which homonuclear diatomic molecules or ions of the second period have the following electron distributions in MO's? In other words, identify X in each case below.

(a) X_2 $\sigma_{1s}^2 \sigma_{1s}^{*2} \sigma_{2s}^2 \sigma_{2s}^{*2} \pi_{2p_y}^1 \pi_{2p_z}^1$

(b) X_2^- $\sigma_{1s}^2 \sigma_{1s}^{*2} \sigma_{2s}^2 \sigma_{2s}^{*2} \sigma_{2p}^2 \pi_{2p_y}^2 \pi_{2p_z}^2 \pi_{2p_y}^{*2} \pi_{2p_z}^{*1}$

(c) X_2^+ $\sigma_{1s}^2 \sigma_{1s}^{*2} \sigma_{2s}^2 \sigma_{2s}^{*2} \sigma_{2p}^2 \pi_{2p_y}^2 \pi_{2p_z}^2 \pi_{2p_y}^{*1}$

34. What is the bond order of each of the species of Exercise 33?

35. Assuming that the σ_{2p} MO is lower in energy than the π_{2p_y} and π_{2p_z} MO's for the following species, write out electronic configurations for

all of them. O_2, O_2^-, O_2^{2-}, F_2, F_2^+, F_2^-, Ne_2, and Ne_2^+.

36. (a) What is the bond order of each of the species in Exercise 35? (b) Are they diamagnetic or paramagnetic? (c) What would the application of MO theory predict regarding the stabilities of these species?

37. Is it possible for a molecule or complex ion in its ground state to have a negative bond order? Why?

38. The following is the molecular orbital energy-level diagram for a heteronuclear diatomic molecule, XY, in which both X and Y are from period 2 and Y is the more electronegative element. Use this to fill in an MO diagram for carbon monoxide, CO. What is the bond order of CO? Is it stable? Is it paramagnetic?

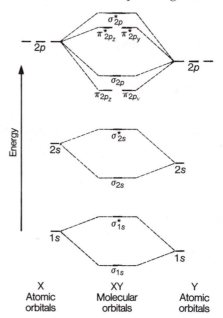

39. Draw and label the complete MO energy-level diagrams for the following species: (a) Li_2 (b) Be_2^+ (c) C_2 (d) F_2 (e) Cl_2 (f) F_2^- (g) HeH (h) HO (i) HCl

40. Determine the bond order and predict the stabilities of the species in Exercise 39 based only on electron occupancy of MO's.

41. Draw MO energy-level diagrams, write out electronic configurations, determine bond orders, indicate predicted diamagnetism or paramagnetism, and predict relative stabilities of the following molecules and ions on the basis of MO theory: OF, OF$^+$, NF, NF$^-$, BC, and

BC⁻. Refer to the diagram in Exercise 38. Assume that the σ_{2p} MO is lower in energy than the π_{2p_y} and π_{2p_z} MO's for all except BC and BC⁻.

42. Rationalize the following observations in terms of the stabilities of sigma and pi bonds. (a) The most common form of nitrogen is N_2, whereas the most common form of phosphorus is P_4 (see structure on p. 32). (b) The most common forms of oxygen are O_2 and (less common) O_3, whereas the most common form of sulfur is S_8 (see structure on p. 32).

43. Considering the shapes of MO energy-level diagrams for nonpolar covalent and polar covalent molecules, what would you predict about MO diagrams, and therefore overlap of atomic orbitals, for ionic compounds?

44. Draw Lewis dot formulas depicting the resonance structures of the following species from the Valence Bond point of view, and then draw MO's for the delocalized π systems: (a) SO_2, (b) HCO_3^-, hydrogen carbonate ion (H is bonded to O), (c) SO_3, sulfur trioxide, (d) O_3, ozone, (e) C_2H_4, ethene, (f) C_2H_2, ethyne.

Periodicity and Chemical Reactions

9–1 Introduction

So far, we have studied the relationship of electronic structure to the position of an element in the periodic table, we have related electronic structure to some of an element's important properties (atomic and ionic radii, ionization energy, electron affinity, and electronegativity), and we have shown how an element's position in the periodic table can be used to predict its tendency to form ionic and covalent compounds.

In this chapter we shall extend our study to include the relationship of periodicity to the kinds of reactions that a few common elements and their compounds undergo. Because it is not possible to discuss, in a single chapter, typical reactions of elements from each group in the periodic table, we shall focus our attention on metals far to the left and nonmetals far to the right in the periodic table. We shall begin with hydrogen, oxygen, and their compounds.

We shall also provide an introduction to reactions in aqueous solution. The Arrhenius theory of acids and bases will be introduced, and we shall study reactions of hydrides, nonmetal oxides and metal oxides with water, and the acidic or basic properties of the resulting solutions. Finally, we introduce oxidation-reduction reactions, which are reactions in which changes in oxidation numbers occur.

In addition to relating periodicity to chemical reactions, the material presented here will serve as a foundation for the next chapter, which deals specifically with acid-base and oxidation-reduction reactions as they are used in the laboratory.

9–2 Hydrogen and the Hydrides

1 Hydrogen

The name means "acid former."

Elemental hydrogen is a colorless, odorless, tasteless, diatomic gas, H_2, with the lowest atomic weight and density of any known substance. Discovery of the element is attributed to the Englishman Henry Cavendish, who prepared it in 1766 by passing steam through a red-hot gun barrel (mostly iron) and by the reaction of various acids with metals. The latter is still a common method for the preparation of small amounts of H_2 in the laboratory.

$$3Fe\ (s) + 4H_2O\ (g) \xrightarrow{\Delta} Fe_3O_4\ (s) + 4H_2\ (g)$$
$$\text{steam}$$

$$Zn\ (s) + 2HCl\ (aq) \longrightarrow ZnCl_2\ (aq) + H_2\ (g)$$

Hydrogen also can be prepared by electrolysis of water (Section 1–5).

$$2H_2O\ (\ell) \xrightarrow{\text{Elect.}} 2H_2\ (g) + O_2\ (g) \qquad \Delta H^0 = +571.6\ kJ$$

Note that the amount of energy liberated by burning 2 moles of H_2 is the same as the amount of energy required to decompose 2 moles of H_2O.

If it becomes economical to convert solar energy into electrical energy that can be used to electrolyze water, H_2 could become an important fuel in the future (although the dangers of storage and transportation would have to be overcome). The combustion of H_2 liberates a great deal of heat. **Combustion** is the highly exothermic reaction of a substance with oxygen, usually with a visible flame (see Section 9–7.3).

Hydrogen is no longer used in blimps and dirigibles; it has been replaced by slightly denser, but much safer, helium.

$$2H_2\ (g) + O_2\ (g) \xrightarrow[\text{or }\Delta]{\text{spark}} 2H_2O\ (\ell) \qquad \Delta H^0 = -571.6\ kJ$$

Hydrogen is very flammable, and was responsible for the *Hindenburg* airship disaster in 1937. A spark is enough to initiate the combustion reaction, which is exothermic enough to provide the heat necessary to sustain the reaction.

A catalyst is a substance that affects the rate at which a reaction occurs, without being consumed itself in the reaction.

Vast quantities of hydrogen are produced commercially each year by the reaction of methane with steam at 850°C in the presence of a nickel catalyst.

$$CH_4\ (g) + H_2O\ (g) \xrightarrow[\text{Ni}]{\Delta} CO\ (g) + 3H_2\ (g)$$

Hydrogen is also prepared by the "water gas reaction" shown below, which results from the passage of steam over white-hot coke (impure carbon, a nonmetal) at 1500°C. "Water gas" is used industrially as a fuel because both components, CO and H_2, undergo combustion.

$$C \text{ (s)} + H_2O \text{ (g)} \xrightarrow{1500°C} CO \text{ (g)} + H_2 \text{ (g)}$$

in coke   steam   water gas

Hydrogen is also prepared from petroleum by the thermal cracking of hydrocarbons. **Thermal cracking** refers to heating a substance with a catalyst in the absence of air. With petroleum this treatment causes the following **decomposition reactions** (among others):

$$CH_4 \text{ (g)} \xrightarrow[\text{catalyst}]{\Delta} C \text{ (s)} + 2H_2 \text{ (g)}$$

methane

$$C_4H_{10} \text{ (g)} \xrightarrow[\text{catalyst}]{600°C} 2C_2H_2 \text{ (g)} + 3H_2 \text{ (g)}$$

butane

2   Reactions of Hydrogen

Hydrogen combines with metals or other nonmetals to form binary compounds called **hydrides.** Atomic hydrogen has the $1s^1$ electronic configuration; it forms (1) *ionic hydrides,* containing hydride ions, H^-, by gaining one electron per atom from a very active metal, or (2) *covalent hydrides* by sharing its electron with an atom of another nonmetal to form a single covalent bond. The H^- ion has the stable He configuration, $1s^2$. In covalent bonds the hydrogen atom attains a *share* of two outer shell electrons, so again it has the He configuration.

The use of the term hydride does not necessarily imply the presence of hydride ion, H^-.

Whether a given binary compound of hydrogen is ionic or covalent depends upon the position of the other element in the periodic table (Figure 9 – 1). The alkali (IA) and alkaline earth (IIA) metals lose electrons much more readily than hydrogen does. As we saw in Chapter 6, they form M^+ and M^{2+} ions, respectively. Thus, the combination reactions of H_2 with the alkali and heavier (more active) alkaline earth metals produce solid *ionic hydrides,* often called *saline (saltlike) hydrides.* A **combination reaction** is a reaction in which two substances, elements or compounds, combine to form one compound. Hydrogen combines with lithium (IA) to form lithium hydride (Li^+, H^-) and with calcium (IIA) to form calcium hydride (Ca^{2+}, $2H^-$).

$$2Li \text{ (molten)} + H_2 \text{ (g)} \longrightarrow 2LiH \text{ (s)} \qquad \text{lithium hydride}$$

$$Ca \text{ (s)} + H_2 \text{ (g)} \longrightarrow CaH_2 \text{ (s)} \qquad \text{calcium hydride}$$

IA	IIA	IIIA	IVA	VA	VIA	VIIA
LiH	BeH$_2$	B$_2$H$_6$	CH$_4$	NH$_3$	H$_2$O	HF
NaH	MgH$_2$	(AlH$_3$)$_x$	SnH$_4$	PH$_3$	H$_2$S	HCl
KH	CaH$_2$	Ga$_2$H$_6$	GeH$_4$	AsH$_3$	H$_2$Se	HBr
RbH	SrH$_2$	InH$_3$	SnH$_4$	SbH$_3$	H$_2$Te	HI
CsH	BaH$_2$	TlH	PbH$_4$	BiH$_3$	H$_2$Po	HAt

FIGURE 9–1 Common hydrides of the representative elements. The ionic hydrides are shaded heavily, covalent hydrides are unshaded, and those of intermediate character are shaded lightly.

Ionic hydrides can serve as sources of H_2 because they react with water to form the corresponding metal hydroxides (compounds that contain metal ions and OH^- ions) and hydrogen. When water is added dropwise to lithium hydride, for example, lithium hydroxide and hydrogen are produced. The reaction of calcium hydride is similar.

Although LiOH and $Ca(OH)_2$ are soluble in water, we write LiOH (s) and $Ca(OH)_2$ (s) here because not enough water is available to act as a solvent.

$$LiH \text{ (s)} + H_2O \text{ (}\ell\text{)} \longrightarrow LiOH \text{ (s)} + H_2 \text{ (g)}$$

$$CaH_2 \text{ (s)} + 2H_2O \text{ (}\ell\text{)} \longrightarrow Ca(OH)_2 \text{ (s)} + 2H_2 \text{ (g)}$$

Hydrogen reacts with nonmetals to form binary *covalent hydrides*. For example, H_2 combines with the halogens to form colorless, gaseous hydrogen halides, which are corrosive and have choking odors.

The hydrogen halides are named by the word hydrogen followed by the stem for the *hal*ogen with an *-ide* ending. Recall that the names of all binary compounds have *-ide* endings (Section 6-13).

$$H_2 \text{ (g)} + X_2 \longrightarrow \qquad 2HX \text{ (g)} \qquad X = F, Cl, Br, I$$
$$\text{hydrogen halides}$$

Specifically, hydrogen reacts with fluorine to form hydrogen fluoride and with chlorine to form hydrogen chloride.

$$H_2 \text{ (g)} + F_2 \text{ (g)} \longrightarrow 2HF \text{ (g)} \qquad \text{hydrogen fluoride}$$

$$H_2 \text{ (g)} + Cl_2 \text{ (g)} \longrightarrow 2HCl \text{ (g)} \qquad \text{hydrogen chloride}$$

Hydrogen burns in an atmosphere of bromine vapor. Hydrogen bromide is formed.

Hydrogen also combines with the Group VIA elements to form binary covalent compounds. You already know that hydrogen and oxygen combine to form water. The hydrides of the heavier members of this group are gases at room temperature, whose formulas resemble that of water.

These compounds are named hydrogen oxide (H_2O), hydrogen sulfide (H_2S), hydrogen selenide (H_2Se), and hydrogen telluride (H_2Te).

The primary industrial use of H_2 is in the synthesis of ammonia, NH_3, a covalent hydride, by the Haber process (Section 16-5). Most of the NH_3 is used directly as a fertilizer or to make fertilizers.

$$N_2 \text{ (g)} + 3H_2 \text{ (g)} \xrightarrow{\Delta} 2NH_3 \text{ (g)} \qquad \text{ammonia}$$

Ammonia may be applied directly to the soil as a fertilizer.

Many of the covalent (nonmetal) hydrides are acidic (Section 9–4.2), especially those of elements in Groups VIIA and VIA; that is, their aqueous solutions produce hydrated hydrogen ions, H^+ (aq). These hydrides include HF, HCl, HBr, HI, H_2S, H_2Se, and H_2Te.

As we shall see in Chapter 17, even H_2O is weakly acidic.

$$HCl \text{ (g)} \xrightarrow{H_2O} H^+ \text{ (aq)} + Cl^- \text{ (aq)} \qquad \text{hydrochloric acid}$$

Most transition metals react with H_2 to form *interstitial hydrides,* substances in which the small H_2 molecules merely fit into spaces (interstices) in the crystal lattices of the metals. It is uncertain whether these should be called compounds, because there is little evidence for chemical reaction and the properties of the interstitial hydrides closely resemble those of the metals themselves. Some transition metals are able to absorb enormous amounts of H_2 under high pressures, which often can be released by reducing the pressure or by heating. For this reason, some transition metals may become very useful for storing hydrogen in a safe, economical way if it ever becomes an important fuel.

9–3 Oxygen and the Oxides

1 Oxygen and Ozone

Oxygen was discovered in 1774 by an English minister and scientist, Joseph Priestley, who observed the thermal decomposition of mercury(II) oxide, a red powder that was called red calx of mercury (Example 1–19).

$$2HgO \text{ (s)} \xrightarrow{\Delta} 2Hg \text{ (ℓ)} + O_2 \text{ (g)}$$

The part of the earth that we see is approximately 50% oxygen by mass. About 2/3 of the mass of the human body is due to oxygen. Elemental oxygen, O_2, is an odorless and nearly colorless gas that makes up about 21% by volume of dry air. In the liquid and solid states it is pale blue. The structure of the O_2 molecule was described in Section 8–12.4. Oxygen also occurs in a second allotropic form, ozone (O_3).

Oxygen is very slightly soluble in water; only about 0.004 gram dissolves in a hundred grams of water at 25°C. Yet this is sufficient to sustain fish and other marine organisms. The greatest single industrial use of O_2 is for oxygen-enrichment in steel blast furnaces (Section 20–11) for the conversion of pig iron to steel.

Oxygen is obtained commercially by the fractional distillation of liquid air. In this process, air is first liquefied under high pressure by cooling it below the boiling points of its various components. The components of liquid air have different boiling points, and they can be vaporized and collected individually by slowly increasing the temperature. Liquid oxygen boils at -183°C under one atmosphere pressure.

Like hydrogen, very pure oxygen is obtained by electrolytic decomposition of water, as described in Section 1–5. Small amounts of very pure oxygen may also be obtained by immersing a platinized nickel foil catalyst in a 30% solution of hydrogen peroxide, H_2O_2, and heating gently.

$$2H_2O_2 \text{ (aq)} \xrightarrow[\text{Pt}]{\Delta} 2H_2O \text{ (}\ell\text{)} + O_2 \text{ (g)}$$

A common laboratory method for the preparation of oxygen is the thermal decomposition of potassium chlorate, $KClO_3$, in the presence of a catalyst, MnO_2.

$$2KClO_3 \text{ (s)} \xrightarrow[\text{MnO}_2]{\Delta} 2KCl \text{ (s)} + 3O_2 \text{ (g)}$$

Ozone: Its Properties and Presence in the Upper Atmosphere

Ozone, O_3, is an unstable, pale blue substance (a gas at room temperature) that is formed by passing an electrical discharge through gaseous oxygen. It has a unique pungent odor that is often noticed during electrical storms and in the vicinity of electrical equipment. With three oxygen atoms per molecule instead of two, its density is about $1\frac{1}{2}$ times that of O_2. At -112°C it condenses to a deep blue liquid. As a concentrated gas or a liquid, ozone can easily decompose explosively.

$$2O_3 \text{ (g)} \longrightarrow 3O_2 \text{ (g)} + 286 \text{ kJ}$$

Oxygen atoms, or *radicals*, appear in the course of this exothermic decomposition. A **radical** is a species containing one or more unpaired electrons; many radicals are very reactive. Substances like ozone that release radicals have such applications as destroying bacteria in water purification.

The ozone molecule is angular and diamagnetic. Both oxygen-oxygen bond lengths (0.128 nm) are identical and are intermediate between typical single and double bond lengths. Ozone is represented by the following resonance formulas:

(lone pairs on end O's are not shown)

Ozone is formed in the upper atmosphere as O_2 molecules absorb high-energy electromagnetic radiation from the sun. Its concentration in the stratosphere (above

Recall that *allotropes* are different forms of the same element in the same physical state.

Liquid O_2 is used as an oxidizer for rocket fuels, and in health fields for O_2-enriched air.

The advantage of using ozone instead of chlorine for water purification is that ozone leaves no residual taste.

about 4 km) is about 10 ppm, whereas it is only about 0.04 ppm just above the surface of the earth. The ozone layer is responsible for absorbing some of the ultraviolet light from the sun which, if it reached the surface of the earth in higher intensity, could cause damage to plants and animals (including humans). It has been predicted that the incidence of skin cancer would increase by 2% for every 1% decrease in the concentration of ozone in the stratosphere. Although it decomposes rapidly in the upper atmosphere, the ozone supply is constantly replenished.

2 Reactions of Oxygen

Oxygen combines directly with almost all other elements — except the noble gases and the noble (unreactive) metals (Au, Pd, Pt) — to form **oxides,** binary compounds that contain oxygen. While such reactions are generally exothermic, many proceed quite slowly and require energy to initiate the reaction. This is due in large part to the high bond energy (498 kJ/mol) of the O_2 molecule. Once a small number of oxygen atoms are liberated, most of these reactions release more than enough energy to be self-sustaining; they sometimes result in incandescence (see Figure 1–1).

The Group IA metals combine with oxygen to form three kinds of solid ionic products called oxides, peroxides, and superoxides. *In general, metallic oxides (and peroxides and superoxides) are ionic solids.* For example, lithium reacts with oxygen to form lithium oxide.

$$4\text{Li (s)} + O_2 \text{ (g)} \longrightarrow 2\text{Li}_2\text{O (s)} \qquad \text{lithium oxide}$$

By contrast, sodium reacts with an excess of oxygen to form sodium peroxide, Na_2O_2, rather than sodium oxide, Na_2O, as the major product.

$$2\text{Na (s)} + O_2 \text{ (g)} \longrightarrow Na_2O_2 \text{ (s)} \qquad \text{sodium peroxide}$$

Peroxides contain the $:\overset{..}{\underset{..}{O}}{-}\overset{..}{\underset{..}{O}}:^{2-}$ group, in which the oxidation number (Section 6–12) of oxygen is −1, whereas **normal oxides** of metals, such as lithium oxide, Li_2O, contain oxide ions, $:\overset{..}{\underset{..}{O}}:^{2-}$. The heavier members of Group IA (K, Rb, Cs) react with an excess of oxygen to form **superoxides,** which contain the superoxide ion, O_2^-, in which the oxidation number of oxygen is −1/2. The reaction with potassium is

$$\text{K (s)} + O_2 \text{ (g)} \longrightarrow KO_2 \text{ (s)} \qquad \text{potassium superoxide}$$

Thus, the tendency of the Group IA metals to form oxygen-rich compounds increases upon descending the group, i.e., with increasing cation radius and decreasing charge density on the metal ion. A similar trend is observed in the reactions of the Group IIA metals with oxygen.

With the exception of Be and Ba, the Group IIA metals react with oxygen at normal temperatures to form normal ionic oxides, MO, and at high pressures of oxygen the heavier ones form ionic peroxides, MO_2 (Table 9–1).

$$2\text{M (s)} + O_2 \text{ (g)} \longrightarrow 2(M^{2+}, O^{2-}) \text{ (s)} \qquad M = \text{Be, Mg, Ca, Sr, Ba}$$

$$\text{M (s)} + O_2 \text{ (g)} \longrightarrow (M^{2+}, O_2^{2-}) \text{ (s)} \qquad M = \text{Ca, Sr, Ba}$$

For example, the equations for the reactions of calcium and oxygen are:

$$2\text{Ca (s)} + O_2 \text{ (g)} \longrightarrow 2\text{CaO (s)} \qquad \text{calcium oxide}$$

$$\text{Ca (s)} + O_2 \text{ (g)} \xrightarrow{\text{high pressure of } O_2} \text{CaO}_2 \text{ (s)} \qquad \text{calcium peroxide}$$

This satellite, launched in 1979, is used to measure the amount of ozone at various latitudes from a circular orbit nearly 400 miles above the earth.

TABLE 9–1   Oxygen Compounds of the IA and IIA Metals*

	IA					IIA				
	Li	Na	K	Rb	Cs	Be	Mg	Ca	Sr	Ba
Oxide	Li$_2$O	Na$_2$O	K$_2$O	Rb$_2$O	Cs$_2$O	BeO	MgO	CaO	SrO	BaO
Peroxide	Li$_2$O$_2$	Na$_2$O$_2$	K$_2$O$_2$	Rb$_2$O$_2$	Cs$_2$O$_2$			CaO$_2$	SrO$_2$	BaO$_2$
Superoxide		NaO$_2$	KO$_2$	RbO$_2$	CsO$_2$					

* The compounds underlined represent the principal product of the direct reaction of the metal with oxygen.

The other metals, with the exceptions of Au, Pd, and Pt, react with oxygen to form solid metal oxides. Since many metals to the right of Group IIA in the periodic table show variable oxidation states, several oxides may be formed. For example, iron reacts with oxygen to form the following oxides in a series of reactions.

> Fe_3O_4 is a mixed oxide containing FeO and Fe_2O_3 in a 1:1 ratio. Ordinary iron rust is primarily Fe_2O_3.

$$2Fe\ (s) + O_2\ (g) \xrightarrow{\Delta} 2FeO\ (s) \qquad \text{iron(II) oxide or ferrous oxide}$$

$$6FeO\ (s) + O_2\ (g) \xrightarrow{\Delta} 2Fe_3O_4\ (s) \qquad \text{magnetic iron oxide (a mixed oxide)}$$

$$4Fe_3O_4\ (s) + O_2\ (g) \xrightarrow{\Delta} 6Fe_2O_3\ (s) \qquad \text{iron(III) oxide or ferric oxide}$$

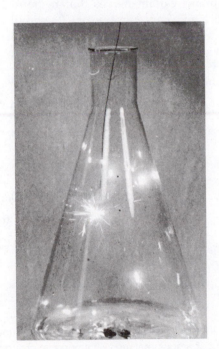

Iron burns brilliantly in pure oxygen to form the iron oxide Fe_3O_4.

Metals that exhibit variable oxidation states (Section 6–12) react with a limited amount of oxygen to give lower oxidation state oxides (such as FeO), while reaction with an excess of oxygen gives higher oxidation state oxides (such as Fe_2O_3).

Oxygen combines with many nonmetals to form covalent oxides. For example, carbon burns in oxygen to form carbon monoxide or carbon dioxide, depending on the relative amounts of carbon and oxygen:

$$2C \text{ (s)} + O_2 \text{ (g)} \xrightarrow{\Delta} 2\overset{\curvearrowright \;+2 \text{ ox. no.}}{C}O \text{ (s)} \qquad \text{(excess C and limited } O_2)$$

carbon monoxide

$$C \text{ (s)} + O_2 \text{ (g)} \xrightarrow{\Delta} \overset{\curvearrowright \;+4 \text{ ox. no.}}{C}O_2 \text{ (g)} \qquad \text{(limited C and excess } O_2)$$

carbon dioxide

Sulfur burns in oxygen to form sulfur dioxide gas.

Carbon monoxide is also produced by the incomplete combustion of carbon-containing compounds, such as those in gasoline and diesel fuel. It is a poisonous substance, and each year there are many reports of carbon monoxide asphyxiation. The Lewis formula for carbon monoxide, :C:::O:, shows one unshared pair of electrons on the carbon atom. This pair of electrons can be shared with the iron atom in hemoglobin in blood. The resulting bond is stronger than the bond that oxygen molecules form with the iron atom in hemoglobin. Attachment of the CO molecule to the iron atom destroys the ability of hemoglobin to pick up oxygen in the lungs and carry it to the brain and muscle tissues.

Carbon dioxide is not toxic. It is one of the products of the respiratory process. In fact, your body contains certain nerves that respond to the amount of CO_2 dissolved in your blood; a high concentration of CO_2 stimulates the breathing reflex. Carbonated beverages are saturated solutions of carbon dioxide in water. A small amount of the carbon dioxide combines with the water to form carbonic acid, H_2CO_3, which is a very weak acid.

Sulfur, another nonmetal, burns in air to form sulfur dioxide.

$$S_8 \text{ (s)} + 8O_2 \text{ (g)} \xrightarrow{\Delta} 8SO_2 \text{ (g)} \qquad \text{sulfur dioxide}$$

A limited amount of oxygen reacts with phosphorus to form tetraphosphorus hexoxide, P_4O_6,

$$P_4 \text{ (s)} + 3O_2 \text{ (g)} \longrightarrow \overset{\curvearrowright \;+3 \text{ ox. no.}}{P}_4O_6 \text{ (s)} \qquad \text{tetraphosphorus hexoxide}$$

while an excess of oxygen reacts with phosphorus to form tetraphosphorus decoxide, P_4O_{10}.

$$P_4 \text{ (s)} + 5O_2 \text{ (g)} \longrightarrow \overset{\curvearrowright \;+5 \text{ ox. no.}}{P}_4O_{10} \text{ (s)} \qquad \text{tetraphosphorus decoxide}$$

Nonmetals exhibit more than one oxidation state in their compounds. In general, the *most* common oxidation states of a nonmetal are (1) its periodic group number, (2) its periodic group number minus two, and (3) its periodic group number minus eight. The reactions of nonmetals with a limited amount of oxygen give products that contain the nonmetals (other than oxygen) in lower oxidation states, usually case (2), while reactions with an excess of oxygen give products in which the nonmetals exhibit higher oxidation states, case (1).

In some instances, the *true* formulas of the oxides are not easily predictable but the *simplest* formulas are. For example, the two most common oxidation states of phosphorus (Group VA) in covalent compounds are +3 and +5. The simplest formulas for the corresponding phosphorus oxides are P_2O_3 and P_2O_5, respectively. The true formulas are twice these, P_4O_6 and P_4O_{10}.

Having described reactions by which the *ionic metal oxides* and *covalent nonmetal oxides* are formed, we shall now study the characteristics of these compounds when they dissolve in water to give aqueous solutions, as well as the classical definitions of acids and bases. This will allow us to describe the acidic and basic properties and reactions of the oxides and related compounds.

9–4 Aqueous Solutions—An Introduction

We introduced solution terminology in Chapter 2.

Many chemical reactions, including most of those of biological and industrial importance, occur in aqueous (water) solutions. Thus it is useful to know the kinds of substances that are soluble in water, and the forms in which they exist in solution, before we continue our study of reactions.

1 Electrolytes and Extent of Ionization

Electrolytes *are substances whose aqueous solutions conduct electricity.* **Strong electrolytes** *are substances that conduct electricity well in dilute aqueous solution,* while **weak electrolytes** *are substances that conduct electricity only poorly in dilute aqueous solution.* Aqueous solutions of **nonelectrolytes** *do not conduct electricity* to any appreciable extent. Since electricity is carried through aqueous solutions by the movement of ions, the strength of an electrolyte depends both upon its tendency to ionize and upon the magnitudes of the charges on the ions (see Figure 9–2).

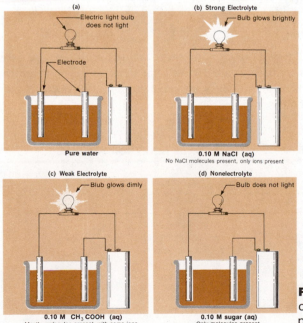

(a) Electric light bulb does not light — Electrode — Pure water

(b) Strong Electrolyte — Bulb glows brightly — 0.10 M NaCl (aq) — No NaCl molecules present, only ions present

(c) Weak Electrolyte — Blub glows dimly — 0.10 M CH_3COOH (aq) — Mostly molecules present, with some ions

(d) Nonelectrolyte — Bulb does not light — 0.10 M sugar (aq) — Only molecules present

FIGURE 9–2 Representation of an experiment to determine the presence of ions in solution.

Acids and bases are described below.

Strong acids, strong soluble bases, and most soluble salts are strong electrolytes because they are completely or nearly completely ionized or dissociated in dilute aqueous solutions.

A **salt** *is a compound that contains a cation other than H+ and an anion other than hydroxide ion, OH−, or oxide ion, O²⁻.* As we shall see shortly, salts are formed when acids react with bases. Solutions of salts containing highly charged ions, like magnesium sulfate, $MgSO_4$, are better conductors of electricity than equally concentrated solutions of salts with less highly charged ions, like potassium bromide, KBr. Aqueous solutions of $MgSO_4$ contain equal concentrations of Mg^{2+} and SO_4^{2-} ions; aqueous solutions of KBr contain equal concentrations of K^+ and Br^- ions.

$$MgSO_4 \text{ (s)} \xrightarrow{H_2O} Mg^{2+} \text{ (aq)} + SO_4^{2-} \text{ (aq)}$$

$$KBr \text{ (s)} \xrightarrow{H_2O} K^+ \text{ (aq)} + Br^- \text{ (aq)}$$

2   The Arrhenius (Classical) Theory of Acids and Bases

In 1680, Robert Boyle noted that acids (1) dissolve many substances, (2) change the colors of some natural dyes (indicators), and (3) lose their characteristic properties when mixed with alkalies (bases). By 1814, Joseph Gay-Lussac concluded that acids *neutralize* bases, and that *the two classes of substances can be defined only in terms of their reactions with each other.* The latter is an extremely important idea!

Svante Arrhenius, a Swedish chemist, presented his theory of acid-base reactions in 1884. According to this view, an **acid** is defined as *a substance that contains hydrogen and produces H+ in aqueous solution.* It is now known that all ions are hydrated, or bonded to water molecules, in aqueous solution. The species that Arrhenius represented as H^+ is now represented as H^+ (aq). In Arrhenius's theory, a **base** is defined as *a substance that contains hydroxyl groups, —OH, and produces hydroxide ions, OH−, in aqueous solution.* **Neutralization** is defined as *the reaction of hydrogen ions with hydroxide ions to form water molecules:*

The hydrated hydrogen ion is sometimes written as H_3O^+ [or $H(H_2O)^+$]. However, it really exists in varying degrees of hydration, such as $H(H_2O)^+$, $H(H_2O)_2^+$, and $H(H_2O)_3^+$.

$$H^+ \text{ (aq)} + OH^- \text{ (aq)} \longrightarrow H_2O \text{ (}\ell\text{)} \qquad \text{neutralization}$$

The following broader and more useful definitions are extensions of the Arrhenius theory: **acids** are substances that increase the concentration of H^+ (aq) ions in aqueous solution; **bases** are substances that increase the concentration of OH^- ions in aqueous solutions.

TABLE 9−2   Strong Acids and Their Anions

Common Strong Acids		Anions Derived from Common Strong Acids	
Formula	Name	Formula	Name
HCl	hydrochloric acid	Cl^-	chloride ion
HBr	hydrobromic acid	Br^-	bromide ion
HI	hydroiodic acid	I^-	iodide ion
HNO_3	nitric acid	NO_3^-	nitrate ion
$HClO_4$	perchloric acid	ClO_4^-	perchlorate ion
$HClO_3$	chloric acid	ClO_3^-	chlorate ion
H_2SO_4	sulfuric acid	HSO_4^-	hydrogen sulfate ion
		SO_4^{2-}	sulfate ion

Many properties of aqueous solutions of acids are due to H^+ (aq) ions. These properties are described in Section 9-4.3.

Classification of Acids   As a matter of convenience we place acids into two classes: strong acids and weak acids. **Strong acids** *ionize completely, or very nearly completely, in dilute aqueous solution.* The seven common strong acids and their anions are listed in Table 9-2. We usually write formulas for inorganic acids with hydrogen first. Organic acids can often be recognized by the presence of the —COOH group.

Dilute aqueous solutions of strong acids contain (predominantly) the ions of the acid rather than acid molecules. The equation for the ionization of hydrochloric acid illustrates the point. Pure hydrogen chloride, HCl, is a polar *covalent* (molecular) compound that is a gas at room temperature and atmospheric pressure. It dissolves in water to produce a solution that contains equal concentrations of *hydrogen ions* and *chloride ions.*

Unless we specify otherwise, solutions are *aqueous* solutions.

$$HCl\ (g) \xrightarrow{H_2O} H^+\ (aq) + Cl^-\ (aq)$$

Similar equations can be written for all strong acids.

Weak acids *ionize only slightly in dilute aqueous solution.* The list of weak acids is very long. Many common ones are listed in Appendix F, and a few of them and their anions are given in Table 9-3.

The equation for the ionization of acetic acid, CH_3COOH, in water is typical of weak acids.

$$CH_3COOH\ (\ell) \xrightleftharpoons{H_2O} H^+\ (aq) + CH_3COO^-\ (aq)$$

The double arrow ($\rightleftharpoons$) generally signifies that the reaction occurs in both directions and that the forward reaction (reading left to right) does not go to completion. All of us are familiar with solutions of acetic acid. Vinegar is 5% acetic acid by mass. Our use of oil and vinegar as a salad dressing tells us that acetic acid is a weak acid. We wouldn't think of using oil and 5% hydrochloric acid (which is "stomach acid") as a salad dressing! To be specific, acetic acid is 1.3% ionized (and 98.7% nonionized) in 0.10 molar solution.

Our stomachs have linings that are much more resistant to attack by acids than our other tissues.

TABLE 9-3   Common Weak Acids and Their Anions

Common Weak Acids		Anions Derived from Common Weak Acids	
Formula	*Name*	*Formula*	*Name*
HF*	hydrofluoric acid	F^-	fluoride ion
CH_3COOH	acetic acid	CH_3COO^-	acetate ion
HCN	hydrocyanic acid	CN^-	cyanide ion
HNO_2†	nitrous acid	NO_2^-	nitrite ion
H_2CO_3†	carbonic acid	HCO_3^-	hydrogen carbonate ion
		CO_3^{2-}	carbonate ion
H_2SO_3†	sulfurous acid	HSO_3^-	hydrogen sulfite ion
		SO_3^{2-}	sulfite ion
H_3PO_4	phosphoric acid	$H_2PO_4^-$	dihydrogen phosphate ion
		HPO_4^{2-}	hydrogen phosphate ion
		PO_4^{3-}	phosphate ion
$(COOH)_2$	oxalic acid	$H(COO)_2^-$	hydrogen oxalate ion
		$(COO)_2^{2-}$	oxalate ion
HClO†	hypochlorous acid	ClO^-	hypochlorite ion

* Note that HF is a weak acid, while HCl, HBr, and HI are strong acids.
† Free acid molecules exist only in dilute aqueous solution, or not at all. However, many salts of these acids are common, stable compounds.

Acetic acid is an example of an *organic* acid, most of which are weak acids. Organic acids contain carbon, specifically the carboxylate grouping of atoms,

$$-\text{COOH} \quad \text{or} \quad -C\begin{smallmatrix}\ddot{\text{O}}: \\ \\ \ddot{\text{O}}-\text{H}\end{smallmatrix}$$

. They can ionize slightly by breakage of the O—H bond.

$$\text{H}_3\text{C}-\text{C}\begin{smallmatrix}\ddot{\text{O}}: \\ \\ \ddot{\text{O}}-\text{H}\end{smallmatrix} \;\underset{}{\overset{\text{H}_2\text{O}}{\rightleftharpoons}}\; \text{H}_3\text{C}-\text{C}\begin{smallmatrix}\ddot{\text{O}}: \\ \\ \overset{..}{\underset{..}{\text{O}}}:^-\end{smallmatrix} + \text{H}^+ \text{(aq)}$$

Organic acids are discussed as a class in Chapter 25. Carbonic acid, H_2CO_3, and hydrocyanic acid, HCN (aq), are the common acids that contain carbon but that are considered *inorganic* acids. Inorganic acids are often called **mineral acids** because they are obtained primarily from mineral (nonliving) sources.

Some acids are capable of ionizing in two stages to give two H^+ ions per molecule of acid. Sulfuric acid provides an example of such behavior.

$$\text{H}_2\text{SO}_4 \text{ (aq)} \longrightarrow \text{H}^+ \text{ (aq)} + \text{HSO}_4^- \text{ (aq)}$$

sulfuric acid   hydrogen sulfate ion

This is followed by

$$\text{HSO}_4^- \text{ (aq)} \longrightarrow \text{H}^+ \text{ (aq)} + \text{SO}_4^{2-} \text{ (aq)}$$

hydrogen sulfate   sulfate ion
ion

Such a two-stage ionization may be represented by the single overall equation

$$\text{H}_2\text{SO}_4 \text{ (aq)} \longrightarrow 2\text{H}^+ \text{ (aq)} + \text{SO}_4^{2-} \text{ (aq)}$$

Acids that ionize in two stages are called **diprotic acids;** they may be strong (e.g., sulfuric acid in dilute solution) or weak (e.g., carbonic acid or oxalic acid). Some acids contain three ionizable hydrogens, and are termed **triprotic acids;** phosphoric acid, H_3PO_4, is a prominent example.

Classification of Bases   Most common bases are metal hydroxides, and most are insoluble in water. The exceptions to this are the **strong soluble bases,** *which are soluble in water and are dissociated into ions completely in dilute aqueous solution.* The strong soluble bases are the hydroxides of the Group IA metals and the heavier members of Group IIA, listed in Table 9–4. The equation for the dissociation of sodium hydroxide in water is typical, and similar equations can be written for all strong soluble bases.

$$\text{NaOH (s)} \xrightarrow{\text{H}_2\text{O}} \text{Na}^+ \text{ (aq)} + \text{OH}^- \text{ (aq)}$$

TABLE 9–4   **Strong Soluble Bases**

LiOH	lithium hydroxide		
NaOH	sodium hydroxide		
KOH	potassium hydroxide	Ca(OH)$_2$	calcium hydroxide
RbOH	rubidium hydroxide	Sr(OH)$_2$	strontium hydroxide
CsOH	cesium hydroxide	Ba(OH)$_2$	barium hydroxide

Inorganic acids may be strong or weak; most organic acids are weak.

Monoprotic acids contain only one ionizable hydrogen per molecule.

Strong soluble bases are ionic compounds in the solid state.

Solutions of bases have a set of common properties due to the OH$^-$ (aq) ion. These properties are described in Section 9–4.3.

Other metals form ionic hydroxides, but these are so sparingly soluble in water that they cannot produce strongly basic solutions. They are called **insoluble bases.** On the other hand, **weak bases** such as ammonia, NH_3, are very soluble in water but ionize only slightly in solution.

$$NH_3 \ (aq) + H_2O \ (\ell) \rightleftharpoons NH_4^+ \ (aq) + OH^- \ (aq)$$

The weak bases are *covalent* compounds that are sometimes called *molecular bases* because they exist in aqueous solution primarily as nonionized molecules.

3 Properties of Acids and Bases

Protonic acids are those that contain a hydrogen atom capable of ionization to form H^+ (aq). Aqueous solutions of most protonic acids exhibit the following properties, which are properties of hydrated hydrogen ions.

1. *Acids have a sour taste.* Pickles are usually preserved in vinegar. Many pickled condiments also contain large amounts of sugar so that the sour taste of the acetic acid isn't so pronounced. Lemons contain citric acid, which is responsible for their characteristic taste.
2. *Acids change the color of many indicators* (highly colored dyes). Litmus changes from blue to red, and bromothymol blue changes from blue to yellow, in acidic solutions.
3. *Acids react with metals that have a greater tendency to form cations than does hydrogen (see Section 9-7.2 on the activity series) to liberate hydrogen, H_2.*
4. *Acids react with (neutralize) metal oxides and metal hydroxides to form salts and water.*
5. *Acids react with salts of weaker or more volatile acids (i.e., acids with lower boiling points) to form the weaker or more volatile acids and a new salt.*
6. *Aqueous solutions of acids conduct an electric current because the acids are wholly or partly ionized.*

Aqueous solutions of soluble metal hydroxides (that is, the strong soluble bases listed in Table 9-4) and weak bases also exhibit certain general properties, which are due to hydrated hydroxide ions.

1. *Bases have a bitter taste.*
2. *They have a slippery feeling.* Soaps are common examples; they are mildly basic. A solution of household bleach also feels very slippery because it is strongly basic.
3. *They change the colors of many indicators.* Litmus changes from red to blue, and bromothymol blue changes from yellow to blue, in basic solutions.
4. *They react with (neutralize) acids to form salts and water.*
5. *Their aqueous solutions conduct an electric current because they are dissociated into separate ions. The weak bases are slightly ionized.*

Acids, bases, and salts are indispensable compounds. Table 9-5 lists the 18 acids, bases, and salts that were included in the top 50 chemicals produced in the United States in 1981. It is of interest to note that the 1981 production of sulfuric acid (81.35 billion pounds) was more than twice the production of ammonia (38.07 billion pounds), the second chemical in the top 50 list. Sixty-five percent of the H_2SO_4 is used in the production of fertilizers, as is most of the NH_3.

Many acids, bases, and salts occur in nature and serve a wide variety of

TABLE 9–5   1981 Production of Acids, Bases, and Salts in the United States

Formula	Name	Billions of Pounds	Major Uses
H_2SO_4	sulfuric acid	81.35	manufacture of fertilizers and other chemicals
NH_3	ammonia	38.07	fertilizer; manufacture of fertilizers and other chemicals
CaO, $Ca(OH)_2$	lime (calcium oxide and calcium hydroxide)	35.99	manufacture of other chemicals; steelmaking; water treatment
$NaOH$	sodium hydroxide	21.30	manufacture of other chemicals, pulp and paper, soap and detergents, aluminum, textiles
H_3PO_4	phosphoric acid	19.83	manufacture of fertilizers and detergents
HNO_3	nitric acid	18.08	manufacture of fertilizers, explosives, plastics and lacquers
NH_4NO_3	ammonium nitrate	17.58	fertilizer and explosive
Na_2CO_3	sodium carbonate (soda ash)	16.56	manufacture of glass, other chemicals, detergents, pulp and paper
$C_6H_4(COOH)_2$*	terephthalic acid	6.35	manufacture of fibers (polyesters), films, and bottles
HCl	hydrochloric acid	4.89	manufacture of other chemicals and rubber; metal cleaning
$(NH_4)_2SO_4$	ammonium sulfate	4.22	fertilizer
CH_3COOH*	acetic acid	2.71	manufacture of acetate esters
$Al_2(SO_4)_3$	aluminum sulfate	2.41	water treatment; dyeing textiles
Na_2SO_4	sodium sulfate	2.33	manufacture of paper, glass, and detergents
$CaCl_2$	calcium chloride	1.83	de-icing roads in winter, controlling dust in summer; concrete additive
Na_2SiO_3	sodium silicate	1.48	manufacture of detergents, cleaning agents, and adhesives
$Na_5P_3O_{10}$	sodium tripolyphosphate	1.37	builder in detergents
$C_4H_8(COOH)_2$*	adipic acid	1.21	manufacture of Nylon 66

* Organic compounds.

purposes. For instance, your "digestive juice" contains approximately 0.01 mole of hydrochloric acid per liter.

4   Reversible Reactions

Reactions that can occur in both directions, and thus do not go to completion, are called **reversible reactions.** Let us illustrate the difference between reactions that go to completion and reactions that are reversible. We have seen that the ionization of acetic acid, CH_3COOH, does not go to completion, whereas the ionization of HCl in water is nearly complete. Suppose we dissolve some table salt, NaCl, in water and add some dilute nitric acid to it. The resulting solution contains hydrogen ions from the nitric acid and chloride ions from the salt, which happen to be the products of the ionization of HCl (the solution also contains sodium and nitrate ions). The H^+ and Cl^- ions do *not* react to form nonionized molecules of HCl, which would be the reverse of the ionization of HCl. The complete and irreversible ionization of a strong electrolyte, like HCl, is indicated by a single arrow ($\rightarrow$).

In contrast, when a sample of sodium acetate, $NaCH_3COO$, is dissolved and mixed with nitric acid, the resulting solution contains hydrogen ions from the nitric acid and acetate ions from the sodium acetate (as well as sodium ions and nitrate ions). Most of the H^+ and CH_3COO^- ions *do* combine to produce nonionized molecules of acetic acid, which is the reverse of the ionization of the acid. Thus, the ionization of acetic acid is reversible, which we represent with a double arrow as before.

$$CH_3COOH \text{ (aq)} \rightleftharpoons H^+ \text{ (aq)} + CH_3COO^- \text{ (aq)}$$

The equation for the ionization of any weak electrolyte may be written this way.

5 Solubility Rules for Compounds in Aqueous Solutions

We shall consider some of the factors that influence solubility in Chapter 13.

Solubility is a complex phenomenon, and it is not possible to state simple rules that cover all cases. Although the following rules for aqueous solution are not comprehensive, they will be of great value for most acids, bases, and salts encountered in general chemistry. Compounds that dissolve in water to the extent of approximately 0.02 mole per liter or more are usually classified as "soluble" compounds, while those that are less soluble are classified as "insoluble" compounds. No gaseous or solid substances are infinitely soluble in water.

1. The common inorganic acids are soluble in water. Low-molecular-weight organic acids are soluble.
2. The common compounds of the Group IA metals (Li, Na, K, etc.) and the ammonium ion, NH_4^+, are soluble in water.
3. The common nitrates, NO_3^-, acetates, CH_3COO^-, chlorates, ClO_3^-, and perchlorates, ClO_4^-, are soluble in water.
4. a. Most common chlorides, Cl^-, are soluble in water; the exceptions are AgCl, Hg_2Cl_2, and TlCl. $PbCl_2$ is sparingly soluble.
   b. The common bromides, Br^-, and iodides, I^-, show approximately the same solubility behavior as chlorides, but there are some exceptions. As the halide ions (Cl^-, Br^-, I^-) increase in size, the solubilities of their slightly soluble compounds decrease. Although $HgCl_2$ is readily soluble in water, $HgBr_2$ is only slightly soluble and HgI_2 is even less soluble.

They are called pseudo-halides because their reactions and properties are somewhat similar to those of halide ions.

   c. The solubilities of compounds containing the pseudo-halide ions, CN^- (cyanide) and SCN^- (thiocyanate), are quite similar to those of the corresponding iodides. Additionally, both CN^- and SCN^- show strong tendencies to form soluble compounds containing complex ions.
5. The common sulfates, SO_4^{2-}, are soluble in water except $PbSO_4$, $BaSO_4$, and $HgSO_4$; $CaSO_4$ and Ag_2SO_4 are sparingly soluble.
6. The common fluorides, F^-, are *insoluble* except those of the Group IA metals, NH_4^+, Ag^+, and Tl^+.
7. The common metal hydroxides are *insoluble* in water except those of the Group IA metals and the heavier members of the Group IIA metals, beginning with $Ca(OH)_2$.
8. The common carbonates, CO_3^{2-}, phosphates, PO_4^{3-}, and arsenates, AsO_4^{3-}, are *insoluble* in water except those of the Group IA metals and NH_4^+. $MgCO_3$ is fairly soluble.
9. The common sulfides, S^{2-}, are *insoluble* in water except those of the Group IA and Group IIA metals and the ammonium ion.

The fact that a substance is insoluble in water does *not* mean that it can't take part in a reaction that occurs in contact with water.

Now that we have distinguished between strong and weak electrolytes and soluble and insoluble compounds, let us continue our study of chemical reactions, with those of the oxides of the metals and nonmetals.

9–5 Reactions of Oxides

1 Reactions of Oxides of Metals with Water

Oxides of metals are called **basic anhydrides** because many of them combine with water to form bases *with no change in oxidation state of the metal. Anhydride* means

Although Na_2O_2 is the common product of the reaction of Na with O_2, it is possible to prepare Na_2O.

"without water"; the metal oxide may be thought of as a hydroxide base with the water "removed." Metal oxides that are soluble in water react to produce the corresponding hydroxides. For example, sodium oxide dissolves in water to produce sodium hydroxide, and barium oxide dissolves in water to produce barium hydroxide. (The numbers above the equations show oxidation numbers.)

$$\overset{+1}{Na_2O} (s) + H_2O (\ell) \longrightarrow 2\overset{+1}{NaOH} (aq) \qquad \text{sodium hydroxide}$$

$$\overset{+2}{BaO} (s) + H_2O (\ell) \longrightarrow \overset{+2}{Ba(OH)_2} (aq) \qquad \text{barium hydroxide}$$

The oxides of the Group IA metals and the heavier Group IIA metals dissolve in water to give solutions of strong soluble bases. Most other metal oxides are insoluble in water.

2 Reactions of Oxides of Nonmetals with Water

Ternary means "three elements."

Nonmetal oxides are called **acid anhydrides** because many of them dissolve in water to form **ternary acids** *with no change in oxidation state of the nonmetal.* Consider the combination of carbon dioxide with water to form carbonic acid:

$$\overset{+4}{CO_2} (g) + H_2O (\ell) \longrightarrow \overset{+4}{H_2CO_3} (aq) \qquad \text{carbonic acid}$$

Similarly, the dissolution of sulfur dioxide in water produces sulfurous acid,

$$\overset{+4}{SO_2} (g) + H_2O (\ell) \longrightarrow \overset{+4}{H_2SO_3} (aq) \qquad \text{sulfurous acid}$$

and the dissolution of sulfur trioxide in water produces sulfuric acid.

$$\overset{+6}{SO_3} (\ell) + H_2O (\ell) \longrightarrow \overset{+6}{H_2SO_4} (aq) \qquad \text{sulfuric acid}$$

Nitric acid may be produced by the reaction of dinitrogen pentoxide with water.

$$\overset{+5}{N_2O_5} (s) + H_2O (\ell) \longrightarrow 2\overset{+5}{HNO_3} (aq) \qquad \text{nitric acid}$$

Phosphoric acid solutions are prepared by dissolving tetraphosphorus decoxide in water.

$$\overset{+5}{P_4O_{10}} (s) + 6H_2O (\ell) \longrightarrow 4\overset{+5}{H_3PO_4} (aq) \qquad \text{phosphoric acid}$$

Except for the oxides of boron and silicon, which are insoluble, *nearly all oxides of nonmetals dissolve in water to give solutions of ternary acids.*

Figure 9–3 shows formulas for the oxides of the representative elements in their maximum oxidation states. It also shows the acidic character of solutions of nonmetal oxides and the basic character of solutions of metal oxides.

9–6 Metathesis Reactions

Metathesis reactions are those in which two compounds react to form two new compounds and no changes in oxidation number occur. They are frequently described as reactions in which the ions of two compounds simply "change partners," although it is not necessary that either or both reactants be ionic. One of the most common kinds of metathesis reactions is the reaction of an acid with a base to form a salt and water.

Increasing acidic character →

	IA	IIA	IIIA	IVA	VA	VIA	VIIA
	Li_2O	BeO	B_2O_3	CO_2	N_2O_5		F_2O
	Na_2O	MgO	Al_2O_3	SiO_2	P_4O_{10}	SO_3	Cl_2O_7
	K_2O	CaO	Ga_2O_3	GeO_2	As_2O_5	SeO_3	Br_2O_7
	Rb_2O	SrO	In_2O_3	SnO_2	Sb_2O_5	TeO_3	I_2O_7
	Cs_2O	BaO	Tl_2O_3	PbO_2	Bi_2O_5	PoO_3	At_2O_7

(Increasing base character ↓)

FIGURE 9–3 The normal oxides of the representative elements in their maximum oxidation states. Acidic oxides are shaded heavily, amphoteric oxides are shaded lightly, and basic oxides are unshaded.

1 Acid-Base Reactions (The Preparation of Salts)

Consider the reaction of sodium hydroxide (sometimes called lye) and hydrochloric acid (sometimes called muriatic acid) to form sodium chloride and water.

$$HCl \, (aq) \quad + \quad NaOH \, (aq) \quad \longrightarrow \quad NaCl \, (aq) \quad + H_2O \, (\ell)$$

hydrochloric acid   sodium hydroxide   sodium chloride

Recall that a salt *contains a cation derived from a base and an anion derived from an acid.*

In this reaction two very corrosive compounds react to form water and a neutral compound, ordinary table salt. The reaction of an acid and a base to form a salt and water is called **neutralization** because the corrosive properties of both acids (H^+) and bases (OH^-) are eliminated as the reaction occurs.

The equation above is called a **molecular equation** because formulas are written as though all species existed as molecules in contact with water. They don't. Sodium hydroxide is a strong soluble base (Table 9–4) and NaCl is a soluble ionic salt (solubility rules, Section 9–4.5). Additionally, HCl, a covalent gaseous compound in the pure form, is a strong acid that is completely ionized in dilute aqueous solution (Table 9–2).

In **total ionic equations,** formulas are written to show the *predominant* form of the substances in, or in contact with, aqueous solution. The ions produced from a given source are enclosed within brackets. The total ionic equation for this reaction is:

$$H^+ \, (aq) + Cl^- \, (aq)] + [Na^+ \, (aq) + OH^- \, (aq)] \longrightarrow [Na^+ \, (aq) + Cl^- \, (aq)] + H_2O \, (\ell)$$

Examination of the total ionic equation shows that two species, sodium ions, Na^+ (aq), and chloride ions, Cl^- (aq), appear on both sides of the equation; neither is involved in the reaction. Elimination of Na^+ (aq) and Cl^- (aq), the so-called **spectator ions,** gives the **net ionic equation** for this reaction.

Brackets are not used in net ionic equations, which show only the species involved in reactions.

$$H^+ \, (aq) + OH^- \, (aq) \longrightarrow H_2O \, (\ell)$$

The only reaction that occurs when hydrochloric acid and sodium hydroxide solutions are mixed is the combination of hydrogen ions from the acid with hydroxide ions from the base to form water. This net ionic equation describes *all reactions of strong acids with strong soluble bases to produce soluble salts and water.*

In general, you must answer two questions about a substance to determine whether or not it should be written in ionic form in an ionic equation.

1. Is it soluble in water?
2. If it is soluble, is it highly ionized or dissociated in water?

If *both* answers are yes, the substance is a soluble strong electrolyte and its formula is written in ionic form. If *either* answer is no, its formula is written as if it exists primarily as molecules. To answer these questions it is necessary to know the lists of strong acids (Table 9–2) and strong soluble bases (Table 9–4). These acids and bases are completely, or nearly completely, ionized in dilute aqueous solutions. Other common acids and bases are either insoluble or only slightly ionized. In addition, the solubility rules allow you to determine which compounds are soluble in water. Most soluble salts are also strong electrolytes. Exceptions such as lead acetate, $Pb(CH_3COO)_2$, which is soluble but predominantly nonionized, will be noted as they are encountered.

The common substances that are written in ionized or dissociated form in ionic equations are (1) strong acids, (2) strong soluble bases, and (3) soluble ionic salts.

Why are we concerned with the distinctions among molecular, total ionic, and net ionic equations? In stoichiometry and thermochemistry, it is usually important to specify the complete formulas of all reactants and products and, therefore, we must use molecular equations. However, the net ionic equation allows us to focus attention on the essence of a reaction, that is, the species that undergo change during the reaction. The total ionic equation merely provides the bridge between molecular and net ionic equations.

When an acid and a base are mixed in the stoichiometric proportions indicated by the balanced molecular equation, the only compound in the resulting solution (other than water) is a salt, which can be obtained by boiling away the water. In reactions that produce insoluble salts, the salts can be separated from the water by filtration.

Let us consider the reaction of acetic acid, CH_3COOH, a weak acid, with sodium hydroxide, NaOH, a strong soluble base, to produce sodium acetate, $NaCH_3COO$, a soluble ionic salt. The molecular, total ionic, and net ionic equations for this reaction are

$$CH_3COOH\ (aq) + NaOH\ (aq) \longrightarrow NaCH_3COO\ (aq) + H_2O\ (\ell)$$

$$CH_3COOH\ (aq) + [Na^+\ (aq) + OH^-\ (aq)] \longrightarrow$$
$$[Na^+\ (aq) + CH_3COO^-\ (aq)] + H_2O\ (\ell)$$

$$CH_3COOH\ (aq) + OH^-\ (aq) \longrightarrow CH_3COO^-\ (aq) + H_2O\ (\ell)$$

Na⁺ ions are the only species common to both sides of the equation.

In general, the reactions of weak *monoprotic* acids, HA (where A represents the anion), with strong soluble bases to form soluble salts may be written as

$$HA\ (aq) + OH^-\ (aq) \longrightarrow A^-\ (aq) + H_2O\ (\ell) \qquad \text{(net ionic equation)}$$

Aqueous ammonia is the most common weak soluble base, and all simple ammonium salts are soluble ionic compounds. Nitric acid reacts with aqueous ammonia to form ammonium nitrate, which is used as a fertilizer. The equations are

$$HNO_3\ (aq) + NH_3\ (aq) \longrightarrow NH_4NO_3\ (aq)$$

$$[H^+\ (aq) + NO_3^-\ (aq)] + NH_3\ (aq) \longrightarrow [NH_4^+\ (aq) + NO_3^-\ (aq)]$$

$$H^+\ (aq) + NH_3\ (aq) \longrightarrow NH_4^+\ (aq)$$

In reactions of strong acids with weak bases to form soluble salts, the anion of the strong acid is the only species common to both sides of the equation.

In general terms, the equation for the reaction of a weak base with a strong acid to form a soluble salt can be represented as (where B represents the weak base)

$$H^+ (aq) + B (aq) \longrightarrow BH^+ (aq)$$

Example 9–1

Write balanced (a) molecular, (b) total ionic, and (c) net ionic equations for the following acid-base reactions in, or in contact with, water. Assume that complete neutralization occurs.

(1) phosphoric acid + calcium hydroxide
(2) nitric acid + copper(II) hydroxide
(3) nickel hydroxide + hydrosulfuric acid

Solution

(1) H_3PO_4 is a weak acid, $Ca(OH)_2$ is a strong soluble base, and the salt produced, calcium phosphate, $Ca_3(PO_4)_2$, is insoluble in water.

(a) $2H_3PO_4 (aq) + 3Ca(OH)_2 (aq) \rightarrow$
$$Ca_3(PO_4)_2 (s) + 6H_2O (\ell)$$

(b) $2H_3PO_4 (aq) + 3[Ca^{2+} (aq) + 2OH^- (aq)] \rightarrow$
$$Ca_3(PO_4)_2 (s) + 6H_2O$$

(c) There are no species common to both sides.

$2H_3PO_4 (aq) + 3Ca^{2+} (aq) + 6OH^- (aq) \rightarrow$
$$Ca_3(PO_4)_2 (s) + 6H_2O$$

(2) HNO_3 is a strong acid, $Cu(OH)_2$ is an insoluble base, and $Cu(NO_3)_2$ is a soluble ionic salt.

(a) $2HNO_3 (aq) + Cu(OH)_2 (s) \rightarrow$
$$Cu(NO_3)_2 (aq) + 2H_2O (\ell)$$

(b) $2[H^+ (aq) + NO_3^- (aq)] + Cu(OH)_2 (s) \rightarrow$
$$[Cu^{2+} (aq) + 2NO_3^- (aq)] + 2H_2O (\ell)$$

(c) $NO_3^- (aq)$ ions are spectator ions.

$$2H^+ (aq) + Cu(OH)_2 (s) \rightarrow Cu^{2+} (aq) + 2 H_2O (\ell)$$

(3) $Ni(OH)_2$ is an insoluble base, H_2S is a weak acid, and NiS is an insoluble salt, so all three equations are identical.

(a) $Ni(OH)_2 (s) + H_2S (aq) \rightarrow NiS (s) + 2H_2O (\ell)$

(b) same as (a), (c) same as (a)

There are other kinds of reactions between acids and bases, but we have provided enough examples to illustrate how chemical equations for them are written.

In Section 9–5.1 we illustrated the reactions of active metal oxides with water to form metal hydroxides that are strongly basic. The active metals have low first ionization energies and low electronegativities. In Section 9–5.2 we illustrated the reactions of nonmetal oxides with water to form ternary acids. We now examine the hydroxides of metals with intermediate electronegativities as well as the hydroxides of the metalloids, both of which exhibit behavior intermediate between the two extremes.

2 Amphoterism

The Greek word *amphos* means "both."

Amphoterism is the ability to react with both acids and bases. Several *insoluble* metal hydroxides are amphoteric. They act as bases when they react with acids to form salts and water, as expected, but they also act as acids when they react with (and dissolve in) solutions of excess strong soluble bases. Generally, *elements of intermediate electronegativity form amphoteric hydroxides;* those of high and low electronegativity form acidic and basic "hydroxides," respectively.

TABLE 9–6 Amphoteric Hydroxides

Metal or Metalloid Ions	Insoluble Amphoteric Hydroxide	Complex Ion Formed in an Excess of a Strong Soluble Base
Be^{2+}	$Be(OH)_2$	$[Be(OH)_4]^{2-}$
Al^{3+}	$Al(OH)_3$	$[Al(OH)_4]^{-}$
Cr^{3+}	$Cr(OH)_3$	$[Cr(OH)_4]^{-}$
Zn^{2+}	$Zn(OH)_2$	$[Zn(OH)_4]^{2-}$
Sn^{2+}	$Sn(OH)_2$	$[Sn(OH)_3]^{-}$
Sn^{4+}	$Sn(OH)_4$	$[Sn(OH)_6]^{2-}$
Pb^{2+}	$Pb(OH)_2$	$[Pb(OH)_4]^{2-}$
As^{3+}*	$As(OH)_3$	$[As(OH)_4]^{-}$
Sb^{3+}*	$Sb(OH)_3$	$[Sb(OH)_4]^{-}$
Si^{4+}*	$Si(OH)_4$	$SiO_4{}^{4-}$ and $SiO_3{}^{2-}$
Co^{2+}†	$Co(OH)_2$	$[Co(OH)_4]^{2-}$
Cu^{2+}†	$Cu(OH)_2$	$[Cu(OH)_4]^{2-}$

* As, Sb, and Si are metalloids.

† $Co(OH)_2$ and $Cu(OH)_2$ are only slightly amphoteric; a very large excess of strong soluble base is required to dissolve small amounts of these insoluble hydroxides.

Aluminum hydroxide is a typical amphoteric metal hydroxide. Its behavior as a base may be illustrated by its metathesis reaction with nitric acid to form a *salt*.

$$Al(OH)_3 \text{ (s)} + 3HNO_3 \text{ (aq)} \longrightarrow Al(NO_3)_3 \text{ (aq)} + 3H_2O \text{ (\ell)}$$

$$Al(OH)_3 \text{ (s)} + 3[H^+ \text{ (aq)} + NO_3^- \text{ (aq)}] \longrightarrow [Al^{3+} \text{ (aq)} + 3NO_3^- \text{ (aq)}] + 3H_2O \text{ (\ell)}$$

$$Al(OH)_3 \text{ (s)} + 3H^+ \text{ (aq)} \longrightarrow Al^{3+} \text{ (aq)} + 3H_2O \text{ (\ell)}$$

As seen from the net ionic equation, this is a typical neutralization reaction, with $Al(OH)_3$ as the base.

When an *excess* of sodium hydroxide solution (or any other strong soluble base) is added to solid aluminum hydroxide, the aluminum hydroxide dissolves. The equation for the reaction is usually written as

$$Al(OH)_3 \text{ (s)} + NaOH \text{ (aq)} \longrightarrow NaAl(OH)_4 \text{ (aq)}$$

    an acid   a strong   sodium aluminate,
               soluble base   a soluble compound

The total ionic and net ionic equations are

$$Al(OH)_3 \text{ (s)} + [Na^+ \text{ (aq)} + OH^- \text{ (aq)}] \longrightarrow [Na^+ \text{ (aq)} + Al(OH)_4^- \text{ (aq)}]$$

$$Al(OH)_3 \text{ (s)} + OH^- \text{ (aq)} \longrightarrow Al(OH)_4^- \text{ (aq)}$$

Table 9–6 contains a list of the common amphoteric hydroxides. Three are hydroxides of the metalloids.

3 Precipitation Reactions

Precipitation reactions are a common kind of *metathesis* reaction in which one of the products is an insoluble solid, called a **precipitate,** which separates from solution. An example is the formation of insoluble lead(II) iodide as a result of mixing solutions of the soluble ionic compounds lead(II) nitrate and potassium iodide.

An ionic reaction. When aqueous potassium iodide is added to an aqueous solution of lead(II) nitrate, lead(II) iodide is precipitated.

$$Pb(NO_3)_2 \text{ (aq)} + 2KI \text{ (aq)} \longrightarrow 2KNO_3 \text{ (aq)} + PbI_2 \text{ (s)}$$

$$Pb(NO_3)_2 \text{ (aq)} + 2KI \text{ (aq)} \longrightarrow PbI_2 \text{ (s)} + 2KNO_3 \text{ (aq)}$$
$$\text{yellow ppt}$$

$$[Pb^{2+} \text{ (aq)} + 2NO_3^- \text{ (aq)}] + 2[K^+ \text{ (aq)} + I^- \text{ (aq)}] \longrightarrow$$
$$PbI_2 \text{ (s)} + 2[K^+ \text{ (aq)} + NO_3^- \text{ (aq)}]$$

$$Pb^{2+} \text{ (aq)} + 2I^- \text{ (aq)} \longrightarrow PbI_2 \text{ (s)}$$

The reaction of barium hydroxide with dilute sulfuric acid to produce insoluble barium sulfate is another common example of a precipitation reaction. The balanced molecular and total ionic equations for this reaction are

$$Ba(OH)_2 \text{ (aq)} + H_2SO_4 \text{ (aq)} \longrightarrow BaSO_4 \text{ (s)} + 2H_2O \text{ (}\ell\text{)}$$

In very dilute solution the second step ionization of H_2SO_4 is nearly complete. In 0.10 M and higher concentrations it is not. *Concentrated* H_2SO_4 is represented as $[H^+ \text{ (aq)} + HSO_4^- \text{ (aq)}]$ in ionic equations.

$$[Ba^{2+} \text{ (aq)} + 2OH^- \text{ (aq)}] + [2H^+ \text{ (aq)} + SO_4^{2-} \text{ (aq)}] \longrightarrow BaSO_4 \text{ (s)} + 2H_2O \text{ (}\ell\text{)}$$

Examination of the total ionic equation shows no species common to both sides of the equation, so the net ionic equation for this reaction is the same, except that brackets are not used. Note that in addition to being a *metathesis reaction*, this is also a *neutralization reaction* and a *precipitation reaction*.

4 Removal of Ions from Aqueous Solution

Many reactions that occur in aqueous solutions result in the removal of ions from solution. This is often accomplished in one of three ways: (1) formation of predominantly nonionized molecules (weak electrolytes or nonelectrolytes) in solution from ions, (2) formation of a precipitate from its ions, and (3) formation of a gas, which bubbles out of the reaction mixture. You have seen several examples of the first two in acid-base neutralization and precipitation reactions.

A gas that bubbles out of a reaction mixture is an element or compound that is quite insoluble in the reaction medium, which is usually water.

Since we have just seen examples of the first two cases, let us illustrate the third case. When an acid (say, hydrochloric acid) is added to aqueous sodium carbonate, a metathesis reaction occurs in which carbonic acid, a weak acid, is produced. The reaction is exothermic.

$$Na_2CO_3 \text{ (aq)} + 2HCl \text{ (aq)} \longrightarrow H_2CO_3 \text{ (aq)} + 2NaCl \text{ (aq)}$$

$$CO_3^{2-} \text{ (aq)} + 2H^+ \text{ (aq)} \longrightarrow H_2CO_3 \text{ (aq)} \qquad \text{(net ionic equation)}$$

The heat generated by the reaction causes thermal decomposition of carbonic acid to carbon dioxide and water.

$$H_2CO_3 \text{ (aq)} \xrightarrow{\Delta} CO_2 \text{ (g)} + H_2O \text{ (}\ell\text{)}$$

Because CO_2 is not very soluble in water, most of the CO_2 bubbles out of the reaction mixture and the reaction goes to completion (with respect to the limiting reagent). The net effect is the conversion of ionic species into nonionized molecules of gaseous CO_2 and water.

Example 9–2

Write balanced (a) molecular, (b) total ionic, and (c) net ionic equations for the reactions that occur when aqueous solutions of the following compounds are mixed, if a reaction occurs. Identify the reactions that do occur as precipitation, acid-base, or both.

(1) ammonium sulfide + iron(II) chloride
(2) sodium perchlorate + lithium sulfate
(3) potassium hydroxide + chromium(III) nitrate (3 : 1 mole ratio)

Solution

(1) The solubility rules (Section 9–4.5) indicate that the reactants are soluble ionic salts and that one product, iron(II) sulfide, is insoluble. This is a precipitation reaction, and the driving force is the formation of a precipitate.

(a) $(NH_4)_2S \text{ (aq)} + FeCl_2 \text{ (aq)} \longrightarrow$
$$FeS \text{ (s)} + 2NH_4Cl \text{ (aq)}$$

(b) $[2NH_4^+ \text{ (aq)} + S^{2-} \text{ (aq)}] + [Fe^{2+} \text{ (aq)} + 2Cl^- \text{ (aq)}] \longrightarrow$
$$FeS \text{ (s)} + 2[NH_4^+ \text{ (aq)} + Cl^- \text{ (aq)}]$$

(c) NH_4^+ (aq) and Cl^- (aq) are spectator ions, so the net ionic equation is

$$S^{2-} \text{ (aq)} + Fe^{2+} \text{ (aq)} \longrightarrow FeS \text{ (s)}$$

(2) The four ions present in this reaction mixture, Na^+ (aq), ClO_4^- (aq), Li^+ (aq), and SO_4^{2-} (aq), do not react to form any insoluble or nonionized substances. Therefore, no reaction occurs when aqueous solutions of $NaClO_4$ and Li_2SO_4 are mixed.

(3) This is a precipitation reaction. Such reactions are often employed to prepare insoluble compounds like chromium(III) hydroxide, $Cr(OH)_3$, which can be separated from the solution by filtration.

(a) $3KOH \text{ (aq)} + Cr(NO_3)_3 \text{ (aq)} \longrightarrow$
$$Cr(OH)_3 \text{ (s)} + 3KNO_3 \text{ (aq)}$$

(b) $3[K^+ \text{ (aq)} + OH^- \text{ (aq)}] + [Cr^{3+} \text{ (aq)} + 3NO_3^- \text{ (aq)}] \longrightarrow$
$$Cr(OH)_3 \text{ (s)} + 3[K^+ \text{ (aq)} + NO_3^- \text{ (aq)}]$$

(c) K^+ (aq) and NO_3^- (aq) are spectator ions, so the net ionic equation is

$$3OH^- \text{ (aq)} + Cr^{3+} \text{ (aq)} \longrightarrow Cr(OH)_3 \text{ (s)}$$

9–7   Oxidation-Reduction Reactions

1   Basic Concepts

Reactions in which substances undergo changes in oxidation number are called **oxidation-reduction reactions** or **redox reactions.** Many involve, or appear to involve, electron transfer. We shall need to use the rules for assigning oxidation numbers, which were treated in Section 6–12. Figure 9–4 contains a list of some common oxidation numbers of elements.

       The term *oxidation* originally referred to the combination of a substance with oxygen, which implied an increase in the oxidation state of an element in that

These rules should be reviewed if necessary.

Groups

IA	IIA	IIIB		IVB	VB	VIB	VIIB	VIIIB			IB	IIB	IIIA	IVA	VA	VIA	VIIA	0
1 H +1 −1																	1 H +1 −1	2 He
3 Li +1	4 Be +2												5 B +3	6 C +4 +2 −4	7 N +5 +4 +3 +2 +1 −3	8 O −1 −2	9 F −1	10 Ne
11 Na +1	12 Mg +2												13 Al +3	14 Si +4 −4	15 P +5 +3 −3	16 S +6 +4 +2 −2	17 Cl +7 +5 +3 +1 −1	18 Ar
19 K +1	20 Ca +2	21 Sc +3		22 Ti +4 +3 +2	23 V +5 +4 +3 +2	24 Cr +6 +3 +2	25 Mn +7 +6 +4 +3 +2	26 Fe +3 +2	27 Co +3 +2	28 Ni +2	29 Cu +2 +1	30 Zn +2	31 Ga +3	32 Ge +4 −4	33 As +5 +3 −3	34 Se +6 +4 −2	35 Br +5 +3 +1 −1	36 Kr +4 +2
37 Rb +1	38 Sr +2	39 Y +3		40 Zr +4	41 Nb +5 +4	42 Mo +6 +4 +3	43 Tc +7 +6 +4	44 Ru +8 +6 +4 +3	45 Rh +4 +3 +2	46 Pd +4 +2	47 Ag +1	48 Cd +2	49 In +3	50 Sn +4 +2	51 Sb +5 +3 −3	52 Te +6 +4 −2	53 I +7 +5 +3 +1 −1	54 Xe +6 +4 +2
55 Cs +1	56 Ba +2	57 La +3	58 Ce → 71 Lu +3	72 Hf +4	73 Ta +5	74 W +6 +4	75 Re +7 +6 +4	76 Os +8 +4	77 Ir +4 +3	78 Pt +4 +2	79 Au +3 +1	80 Hg +2 +1	81 Tl +3 +1	82 Pb +4 +2	83 Bi +5 +3	84 Po +2	85 At −1	86 Rn

FIGURE 9–4 The most common nonzero oxidation states of the elements.

substance. According to the original definition, the following reactions involve oxidation of the element or compound shown on the far left side of each equation.

	oxidation state of Fe
$4Fe\ (s) + 3O_2\ (g) \longrightarrow 2Fe_2O_3\ (s)$ rust	$0 \longrightarrow +3$

	oxidation state of C
$C\ (s) + O_2\ (g) \longrightarrow CO_2\ (g)$	$0 \longrightarrow +4$
$2CO\ (g) + O_2 \longrightarrow 2CO_2\ (g)$	$+2 \longrightarrow +4$
$C_3H_8\ (g) + 5O_2\ (g) \longrightarrow 3CO_2\ (g) + 4H_2O\ (g)$	$-8/3 \longrightarrow +4$

Since oxidation number is a bookkeeping method adopted for our convenience, the numbers are determined by reliance upon rules, which can result in a fractional oxidation number. This does not mean that electronic charges are split, nor do oxidation numbers necessarily represent charges on ions.

Originally the term *reduction* described the removal of oxygen from a compound. Ores (many of which are oxides) are reduced to metals (a very real reduction in mass). For example, tungsten for use in light bulb filaments can be prepared by reduction of tungsten(VI) oxide with hydrogen at 1200°C:

Tungsten is reduced as its oxidation state decreases from +6 to zero.

$$WO_3\ (s) + 3H_2\ (g) \longrightarrow W\ (s) + 3H_2O\ (g) \qquad \text{oxidation state of W: } +6 \longrightarrow 0$$

The terms oxidation and reduction are now applied much more broadly. **Oxidation** is defined as an *algebraic increase in oxidation number,* and often corresponds to the *loss of electrons.* **Reduction** refers to an *algebraic decrease in oxidation number,* and often corresponds to a *gain of electrons.* Since electrons cannot be created or destroyed, oxidation and reduction always occur simultaneously in ordinary chemical reactions, and to the same extent. In the four equations cited as examples of oxidation, observe that not only do the oxidation numbers of iron and carbon atoms increase as they are oxidized, but in each case oxygen is reduced as its oxidation state decreases from zero to −2. In the reduction of WO_3, hydrogen also is oxidized from the zero to the +1 oxidation state.

> **Oxidizing agents** *are the species that (1) contain elements that decrease in oxidation number, (2) are reduced, and (3) oxidize other substances.* **Reducing agents** *are the species that (1) contain elements that increase in oxidation number, (2) are oxidized, and (3) reduce other substances.*

The following equations represent other examples of redox reactions. Oxidation numbers are shown above the formulas in color, and oxidizing and reducing agents are indicated.

$$\underset{\text{reducing agent}}{\overset{0}{Fe\ (s)}} + \underset{\text{oxidizing agent}}{\overset{0}{3Cl_2\ (g)}} \longrightarrow \overset{+3\ -1}{2FeCl_3\ (s)}$$

$$\underset{\text{reducing agent}}{\overset{+3-1}{2FeBr_3\ (aq)}} + \underset{\text{oxidizing agent}}{\overset{0}{3Cl_2\ (g)}} \longrightarrow \overset{+3-1}{2FeCl_3\ (aq)} + \overset{0}{3Br_2\ (\ell)}$$

Equations for redox reactions involving ionic species may be written as total ionic and net ionic equations as well. For example, the last equation may also be written as

$$2[Fe^{3+}\ (aq) + 3Br^-\ (aq)] + 3Cl_2\ (g) \longrightarrow 2[Fe^{3+}\ (aq) + 3Cl^-\ (aq)] + 3Br_2\ (\ell) \qquad \text{(total ionic equation)}$$

$$2Br^-\ (aq) + Cl_2\ (g) \longrightarrow 2Cl^-\ (aq) + Br_2\ (\ell) \qquad \text{(net ionic equation)}$$

Notice that the spectator ions, Fe^{3+} ions in this case, do not participate in electron transfer. Cancellation of these ions allows us to focus more clearly on the oxidizing agent, Cl_2 (g), and the reducing agent, Br^- (aq).

Example 9–3

Write each of the following molecular equations as a net ionic equation if the two differ. Which ones represent redox reactions? In those that are redox reactions, identify the oxidizing agent, the reducing agent, the species oxidized, and the species reduced.

(a) $Cu(NO)_3\ (aq) + Zn\ (s) \longrightarrow Zn(NO_3)_2\ (aq) + Cu\ (s)$

(b) $2KClO_3\ (s) \xrightarrow{\Delta} 2KCl\ (s) + 3O_2\ (g)$

(c) $3AgNO_3\ (aq) + K_3PO_4\ (aq) \longrightarrow Ag_3PO_4\ (s) + 3KNO_3\ (aq)$

Solution

(a) According to the solubility rules (Section 9–4.5), both copper(II) nitrate, $Cu(NO_3)_2$, and zinc nitrate, $Zn(NO_3)_2$, are soluble ionic compounds. The total ionic equation and oxidation states are

$$[\overset{+2}{Cu^{2+}}(aq) + 2\overset{+5-2}{NO_3^-}(aq)] + \overset{0}{Zn}(s) \longrightarrow$$
$$[\overset{+2}{Zn^{2+}}(aq) + 2\overset{+5-2}{NO_3^-}(aq)] + \overset{0}{Cu}(s)$$

Cancelling NO_3^- gives the net ionic equation.

$$\overset{+2}{Cu^{2+}}(aq) + \overset{0}{Zn}(s) \longrightarrow \overset{+2}{Zn^{2+}}(aq) + \overset{0}{Cu}(s)$$

This is a redox reaction. The oxidation number of Cu^{2+} (aq) ion decreases from $+2$ to zero; copper(II) ion is reduced, and is the oxidizing agent. The oxidation number of zinc metal, Zn (s), increases from zero to $+2$; zinc is oxidized, and is the reducing agent.

(b) This reaction involves two solids and a gas, so the molecular and net ionic equations are identical. It is a redox reaction.

$$2\overset{+1+5-2}{KClO_3}(s) \overset{\Delta}{\longrightarrow} 2\overset{+1-1}{KCl}(s) + 3\overset{0}{O_2}(g)$$

The $KClO_3$ is both the oxidizing agent and the reducing agent. It contains the chlorine, whose oxidation number decreases, and the oxygen, whose oxidation number increases. Reactions in which the same compound is both oxidized and reduced are called **disproportionation reactions.**

(c) The solubility rules indicate that all these compounds are soluble and ionic except for silver phosphate, Ag_3PO_4. The total ionic equation is

$$3[Ag^+(aq) + NO_3^-(aq)] + [3K^+(aq) + PO_4^{3-}(aq)] \longrightarrow$$
$$Ag_3PO_4(s) + 3[K^+(aq) + NO_3^-(aq)]$$

Elimination of the spectator ions, NO_3^- (aq) and K^+ (aq), generates the net ionic equation.

$$3\overset{+1}{Ag^+}(aq) + \overset{+5-2}{PO_4^{3-}}(aq) \longrightarrow \overset{+1+5-2}{Ag_3PO_4}(s)$$

This is not a redox reaction because there are no changes in oxidation numbers. It is a precipitation reaction.

2 Displacement Reactions

Displacement reactions are a common kind of oxidation-reduction reaction in which one element displaces another from a compound. *Active metals* displace less active metals or hydrogen from their compounds in aqueous solution, and also in many solids. **Active metals** are those with low ionization energies that readily lose electrons to form cations (see Table 6–6).

When metallic copper is added to a solution of (colorless) silver nitrate, the more active metal, copper, displaces silver ion from the solution. The resulting solution contains blue copper(II) nitrate, and metallic silver is produced as a finely divided solid.

Since we have not studied trends in properties of transition metals, it would be difficult for you to predict that Cu is more active than Ag. The fact that this reaction occurs indicates that this is true.

$Cu(s) + 2AgNO_3(aq) \rightarrow Cu(NO_3)_2(aq) + 2Ag(s)$

A displacement reaction. Copper atoms displace silver ions from an aqueous solution of silver nitrate.

TABLE 9–7
Activity Series
of the Metals

	Displace H from water	Displace H from steam	Displace H from acids
Li			
K			
Ca			
Na			
Mg			
Al			
Mn			
Zn			
Cr			
Fe			
Cd			
Co			
Ni			
Sn			
Pb			
H (a nonmetal)			
Sb (a metalloid)			
Cu			
Hg			
Ag			
Pt			
Au			

$$2AgNO_3 \text{ (aq)} + Cu \text{ (s)} \longrightarrow 2Ag \text{ (s)} + Cu(NO_3)_2 \text{ (aq)}$$

$$2[Ag^+ \text{ (aq)} + NO_3^- \text{ (aq)}] + Cu \text{ (s)} \longrightarrow 2Ag \text{ (s)} + [Cu^{2+} \text{ (aq)} + 2NO_3^- \text{ (aq)}]$$

$$2Ag^+ \text{ (aq)} + Cu \text{ (s)} \longrightarrow 2Ag \text{ (s)} + Cu^{2+} \text{ (aq)}$$

We noted in Section 9–2 that a common method for the preparation of small amounts of hydrogen involves the reaction of active metals with acids, such as hydrochloric acid and sulfuric acid. These are also displacement reactions. For example, when freshly brushed magnesium is placed into sulfuric acid, the reaction produces magnesium sulfate; hydrogen is displaced from the acid, bubbling off as gaseous H_2.

$$\underset{\text{active metal}}{Mg \text{ (s)}} + \underset{\text{strong acid}}{H_2SO_4 \text{ (aq)}} \longrightarrow \underset{\text{soluble salt}}{MgSO_4 \text{ (aq)}} + H_2 \text{ (g)}$$

$$Mg \text{ (s)} + [2H^+ \text{ (aq)} + SO_4^{2-} \text{ (aq)}] \longrightarrow [Mg^{2+} \text{ (aq)} + SO_4^{2-} \text{ (aq)}] + H_2 \text{ (g)}$$

$$Mg \text{ (s)} + 2H^+ \text{ (aq)} \longrightarrow Mg^{2+} \text{ (aq)} + H_2 \text{ (g)}$$

Table 9–7 is a short form of the **activity series.** When any metal listed above hydrogen in this series is added to solutions of acids such as HCl or H_2SO_4, the metal dissolves to produce gaseous hydrogen and a salt.

Very active metals can even displace hydrogen from water. However, the reactions of very active metals of Group IA with water are dangerous because they can generate enough heat to cause explosive ignition of the hydrogen in the presence of air.

$$2K \text{ (s)} + 2H_2O \text{ (ℓ)} \longrightarrow 2KOH \text{ (aq)} + H_2 \text{ (g)}$$

$$2K \text{ (s)} + 2H_2O \text{ (ℓ)} \longrightarrow 2[K^+ \text{ (aq)} + OH^- \text{ (aq)}] + H_2 \text{ (g)}$$

Potassium, like the other Group I metals, reacts vigorously with water.

Active nonmetals form monatomic anions readily.

This is the means by which bromine is obtained commercially from sea water.

Many *nonmetals* displace less active nonmetals from combination with a metal or other cation. For example, when chlorine is bubbled through a solution containing bromide ions (derived from a soluble ionic salt such as sodium bromide, NaBr), chlorine displaces bromide ions to form elemental bromine and chloride ions:

$$\underset{\text{chlorine}}{Cl_2 \text{ (g)}} + \underset{\text{bromide ions}}{2Br^- \text{ (aq)}} \longrightarrow \underset{\text{chloride ions}}{2Cl^- \text{ (aq)}} + \underset{\text{bromine}}{Br_2 \text{ (ℓ)}}$$

Similarly, when bromine is added to a solution containing iodide ions, the iodide ions are displaced by bromine to form iodine and bromide ions.

$$Br_2\ (\ell)\ +\ 2I^-\ (aq)\ \longrightarrow\ 2Br^-\ (aq)\ +\ I_2\ (s)$$

bromine   iodide ions   bromide ions   iodine

Each halogen will displace less electronegative halogens from their binary salts. Conversely, a halogen will *not* displace more electronegative halogens from their salts.

$$I_2\ (s) + 2F^-\ (aq)\ \longrightarrow\ \text{No Reaction}$$

> **Electronegativity of the halogens decreases as the group is descended.**

3 Combustion Reactions

Combustion, or burning, is simply an oxidation-reduction reaction in which oxygen combines rapidly with a substance in a highly exothermic reaction, with a visible flame. The complete combustion of hydrocarbons produces carbon dioxide and water (steam) as the major products.

> **Hydrocarbons are compounds that contain only hydrogen and carbon.**

$$\overset{-4+1}{CH_4}\ (g)\ \ +2\overset{0}{O_2}\ (g)\ \xrightarrow{\Delta}\ \overset{+4-2}{CO_2}\ (g)\ +2\overset{+1\ -2}{H_2O}\ (g) + 802\ kJ$$

excess

$$\overset{-2+1}{C_6H_{12}}\ (\ell)\ +9\overset{0}{O_2}\ (g)\ \xrightarrow{\Delta}\ 6\overset{+4-2}{CO_2}\ (g) + 6\overset{+1\ -2}{H_2O}\ (g) + 4127\ kJ$$

cyclohexane   excess

As we have seen, the origin of the term *oxidation* lies in just such reactions, in which oxygen "oxidizes" another species.

4 Combustion of Fossil Fuels and Air Pollution

Fossil fuels are mixtures of variable composition that consist primarily of hydrocarbons. We burn them because they release energy, rather than to obtain chemical products. The incomplete combustion of hydrocarbons yields undesirable products, carbon monoxide and elemental carbon (soot), which pollute the air. Also, unfortunately, all fossil fuels—natural gas, coal, gasoline, kerosene, oil—contain undesirable nonhydrocarbon impurities, which undergo combustion to produce oxides (primarily of nitrogen and sulfur) that act as additional air pollutants. At this time it is not economically feasible to remove all of these impurities.

> **Carbon or soot is one of many kinds of *particulate matter* in polluted air.**

Fossil fuels result from the decay of animal and vegetable matter, and since all living matter contains some sulfur and nitrogen, fossil fuels also contain sulfur and nitrogen impurities to varying degrees.

Combustion of sulfur produces sulfur dioxide, SO_2, probably the most harmful pollutant.

$$\overset{0}{S_8}\ (s) + 8\overset{}{O_2}\ (g)\ \xrightarrow{\Delta}\ 8\overset{+4}{SO_2}\ (g)$$

Sulfur dioxide is corrosive; it damages structural materials and living things. Sulfur dioxide is slowly oxidized to sulfur trioxide, SO_3, by oxygen in air.

$$2\overset{+4}{SO_2}\ (g) + O_2\ (g)\ \longrightarrow\ 2\overset{+6}{SO_3}\ (\ell)$$

Sulfur trioxide combines with moisture in the air to form the strong, corrosive acid, sulfuric acid, the major contributor to "acid rain."

> **SO_3 is an *acid anhydride*. No changes in oxidation numbers accompany this kind of reaction.**

$$\overset{+6}{SO_3} + H_2O\ (\ell)\ \longrightarrow\ \overset{+6}{H_2SO_4}\ (aq)$$

A marble statue, the victim of acid rain. When fossil fuels are burned, oxides of nitrogen and (usually) sulfur are released into the atmosphere, where they dissolve in moisture to form acidic solutions. These acids react with marble, a form of calcium carbonate, and slowly corrode it.

The oxides of sulfur also result from roasting ores that contain metal sulfides, in the process of extracting the free (elemental) metals. **Roasting** involves heating an ore in the presence of air. For many metal sulfides this produces a metal oxide and SO_2. The metal oxide is subsequently reduced to the free metal. Consider the roasting of lead sulfide, PbS, as an example.

$$2\ PbS\ (s) + 3\ O_2\ (g) \longrightarrow 2\ PbO\ (s) + 2\ SO_2\ (g)$$

Compounds of nitrogen are also impurities in fossil fuels, and they undergo combustion to form nitric oxide, NO. However, most of the nitrogen in the NO in exhaust gases from furnaces and engines comes from the air that is mixed with the fuel.

Remember that "clean air" is about 80% N_2 and 20% O_2 by mass. This reaction does *not* occur at room temperature, but does at the high temperatures of furnaces, internal combustion engines, and jet engines.

$$\overset{0}{N_2}\ (g) + O_2\ (g) \xrightarrow{\Delta} \overset{+2}{2NO}\ (g)$$

The NO can be oxidized further by oxygen to nitrogen dioxide, NO_2; this reaction is enhanced in the presence of ultraviolet light from the sun.

$$\overset{+2}{2NO}\ (g) + O_2\ (g) \xrightarrow[\text{light}]{uv} \overset{+4}{2NO_2}\ (g) \qquad \text{(reddish-brown gas)}$$

The NO_2 is responsible for the reddish-brown haze that hangs over many cities in the afternoons of sunny days, and for many respiratory problems associated with air pollution. It can react to produce other oxides of nitrogen and other secondary pollutants. In addition to being a pollutant itself, nitrogen dioxide reacts with water in the air to form nitric acid, which is another contributor to "acid rain."

$$\overset{+4}{3NO_2}\ (g) + H_2O\ (\ell) \longrightarrow \overset{+5}{2HNO_3}\ (\ell) + \overset{+2}{NO}\ (g)$$

Federal and state regulations regarding permissible limits of pollutants in exhaust gases have been enacted over the last several years, and this has resulted in the lowering, to some extent, of the concentrations of pollutants in the air.

Photochemical smog (a brown haze) enveloping the city of Los Angeles.

Key Terms

Acid a substance that increases the concentration of H^+ (aq) ions in aqueous solution. Strong acids are completely, or nearly completely, ionized in dilute aqueous solution. Weak acids are only slightly ionized.

Acid anhydride a nonmetal oxide that reacts with water to form an acid, with no change in oxidation number.

Active metal metal with low ionization energy that loses electrons readily to form cations.

Activity series a listing of metals (and hydrogen) in order of decreasing activity.

Amphoterism the ability to react with both acids and bases.

Base a substance that increases the concentration of OH^- (aq) ions in aqueous solution. Strong soluble bases are soluble in water and are completely dissociated.

Basic anhydride a metal oxide that reacts with water to form a base, with no change in oxidation number.

Binary acid an acid that contains only hydrogen and another nonmetal.

Catalyst a substance that speeds up a chemical reaction without being consumed itself in the reaction.

Combination reaction reaction in which two substances (elements or compounds) combine to form one compound.

Combustion reaction reaction of a substance with oxygen in a highly exothermic reaction, usually with a visible flame.

Decomposition reaction reaction in which a single compound decomposes into one or more other compounds and/or one or more elements.

Displacement reaction reaction in which one element displaces another from a compound.

Disproportionation reaction   redox reaction in which the oxidizing agent and the reducing agent are the same species.

Electrolyte   a substance whose aqueous solutions conduct electricity.

Hydride   a binary compound of hydrogen.

Insoluble base   hydroxide of a metal other than the Group IA metals or the heavier members of Group IIA.

Metathesis reaction   reaction in which two compounds react to form two new compounds, with no changes in oxidation number.

Mineral acid   an inorganic acid; an acid that can be obtained from mineral sources.

Molecular equation   equation for a chemical reaction in which all formulas are written *as if* all substances existed as molecules; only complete formulas are used.

Net ionic equation   equation that results from cancelling spectator ions and eliminating brackets from a total ionic equation.

Neutralization   the reaction of an acid with a base to form a salt and water; the reaction of hydrogen ions with hydroxide ions to form water molecules.

Nonelectrolyte   a substance whose aqueous solutions do not conduct electricity.

Normal oxide   a compound containing oxide ion, O^{2-}, or oxygen in the -2 oxidation state.

Oxidation   an algebraic increase in oxidation number; may correspond to a loss of electrons.

Oxidation-reduction reaction   reaction in which oxidation and reduction occur; also called redox reaction.

Oxide   a binary compound of oxygen.

Oxidizing agent   a substance that oxidizes another substance and is reduced.

Peroxide   a compound containing an O_2^{2-} ion, or oxygen in the -1 oxidation state.

Polyprotic acid   an acid that contains more than one ionizable hydrogen per molecule; a diprotic acid contains two and a triprotic acid contains three.

Precipitate   an insoluble solid that forms and separates from a solution.

Precipitation reaction   reaction in which a precipitate forms.

Radical   a species containing one or more unpaired electrons; many radicals are very reactive.

Redox reaction   see *Oxidation-reduction reaction.*

Reducing agent   a substance that reduces another substance and is oxidized.

Reduction   an algebraic decrease in oxidation number; may correspond to a gain of electrons.

Reversible reaction   reaction that occurs in both directions; indicated by double arrow ($\rightleftharpoons$).

Roasting   heating an ore of an element in the presence of air.

Salt   a compound that contains a cation derived from a base and an anion derived from an acid.

Spectator ions   ions in solution that do not participate in a chemical reaction.

Strong acid   an acid that is nearly completely ionized in dilute aqueous solution.

Strong electrolyte   a substance that conducts electricity well in dilute aqueous solution.

Strong soluble base   a metal hydroxide of the Group IA metals or the heavier members of Group IIA; these are soluble in water and highly dissociated in dilute aqueous solution.

Superoxide   a compound containing an O_2^- ion, or oxygen in the $-\frac{1}{2}$ oxidation state.

Ternary acid   an acid containing three elements, H, O, and another nonmetal.

Thermal cracking   decomposition by heating a substance in the presence of a catalyst and in the absence of air.

Total ionic equation   equation for a chemical reaction written to show the predominant form of each species in, or in contact with, aqueous solution.

Weak acid   an acid that is only slightly ionized in dilute aqueous solution.

Weak base   a (water-soluble) base that is only slightly ionized in dilute aqueous solution; ammonia and the amines are the most common examples.

Weak electrolyte   a substance that conducts electricity only poorly in dilute aqueous solution.

Exercises

Hydrogen and the Hydrides

1. Write balanced molecular equations for (a) the reaction of iron with steam, (b) the reaction of calcium with hydrochloric acid, (c) the electrolysis of water, (d) the "water gas" reaction, (e) the thermal cracking of a hydrocarbon such as propane, C_3H_8.

2. Write a balanced molecular equation for the preparation of (a) an ionic hydride, (b) a covalent hydride.

3. Classify the following hydrides as covalent, ionic, or interstitial: (a) NaH (b) H_2S (c) BaH_2 (d) $TiH_{1.7}$ (e) NH_3

4. Explain how a formula like $TiH_{1.7}$ can represent a hydride.

5. Write molecular equations for the reactions of (a) NaH and (b) BaH_2 with a limited amount of water.

6. Name the following compounds: (a) H_2S (b) HF (c) KH (d) NH_3 (e) H_2Se (f) MgH_2

Oxygen and the Oxides

7. Draw Lewis dot formulas for O_2 and O_3. Why is that for O_2 inadequate?

8. Briefly compare and contrast the properties of oxygen with those of hydrogen.

9. Write molecular equations to show how oxygen can be prepared from (a) mercury(II) oxide, HgO, (b) hydrogen peroxide, H_2O_2, and (c) potassium chlorate, $KClO_3$.

10. Which of the following elements form normal oxides as the *major* products of reactions with oxygen? (a) Li (b) Na (c) Rb (d) Mg (e) Zn (exhibits only one common oxidation state) (f) Al

11. Write molecular equations for the primary reactions of oxygen with the following elements: (a) Li (b) Na (c) K (d) Ca

12. Write molecular equations for the reactions of the following elements with a *limited* amount of oxygen: (a) Sr (b) Fe (c) Mn (d) Cu

13. Write molecular equations for the reactions of the following elements with an *excess* of oxygen: (a) Sr (b) Fe (c) Mn (d) Cu

14. Write molecular equations for the reactions of the following elements with a *limited* amount of oxygen; (a) C (b) As_4 (c) Ge

15. Write molecular equations for the reactions of the following elements with an *excess* of oxygen: (a) C (b) As_4 (c) Ge

Aqueous Solutions — An Introduction

16. Define acids, bases, and neutralization according to the Arrhenius theory, or classical theory.

17. Give broader definitions of acids and bases than those given in Exercise 16.

18. List several properties of (a) classical acids and (b) classical bases.

19. Define and distinguish among (a) strong electrolytes, (b) weak electrolytes, and (c) nonelectrolytes.

20. Distinguish between strong and weak acids.

21. Classify each of the following as a strong or weak acid and name each one: (a) HNO_3 (b) HNO_2 (c) HCl (d) HF (e) H_2SO_4 (f) CH_3COOH

22. Classify each of the following as a strong or weak acid and name each one: (a) H_3PO_4 (b) HI (c) $HClO_4$ (d) $(COOH)_2$ (e) H_2SO_3

23. Write equations for the ionization of the following acids in water: (a) HBr (b) CH_3COOH (c) dilute H_2SO_4 (d) concentrated H_2SO_4 (e) HClO

24. What group of atoms characterizes an organic acid?

25. Distinguish among strong soluble bases, insoluble bases, and weak bases. Give an example of each.

26. Which of the following are strong soluble bases? (a) $Al(OH)_3$ (b) KOH (c) $Mg(OH)_2$ (d) $Ca(OH)_2$ (e) $Cu(OH)_2$

27. Which of the following are insoluble bases? (a) $Fe(OH)_2$ (b) $Fe(OH)_3$ (c) NaOH (d) $Cu(OH)_2$ (e) $Be(OH)_2$ (f) NH_3

28. Which of the following are classified as water-soluble salts? (a) $NaClO_4$ (b) $CaCl_2$ (c) $(NH_4)_2S$ (d) AgI (e) $Ca_3(PO_4)_2$ (f) K_2CO_3

29. Which of the following are classified as water-soluble salts? (a) $PbSO_4$ (b) $CaCO_3$ (c) NaBr (d) FeS (e) BaS (f) $Mg_3(AsO_4)_2$

30. Which of the following are classified as water-*insoluble* salts? (a) $NiSO_4$ (b) CuS (c) $Fe_2(CO_3)_3$ (d) $BaSO_4$ (e) $Hg(CH_3COO)_2$ (f) Hg_2Cl_2

31. Which of the following are classified as water-*insoluble* salts? (a) $Cr(NO_3)_3$ (b) $Fe(ClO_3)_3$ (c) KF (d) NH_4CN (e) AgBr (f) Li_2SO_3

32. Which of the following substances are strong

electrolytes? (a) HClO (b) LiOH (c) $KClO_3$ (d) HBr (e) $MgCO_3$ (f) $Cr(OH)_3$

33. Which of the following substances are strong electrolytes? (a) $Sr(OH)_2$ (b) HNO_3 (c) $PbCO_3$ (d) HCN (e) NaOH (f) HNO_2

34. Which of the following are strong electrolytes? (a) Na_2SO_4 (b) HF (c) NH_4F (d) $HClO_3$ (e) $NaNO_3$ (f) $CuCl_2$

35. Which of the following can be classified as basic anhydrides? (a) SO_2 (b) Li_2O (c) SeO_3 (d) CaO (e) N_2O_5

36. Which of the following can be classified as acid anhydrides? (a) CuO (b) CO_2 (c) Cl_2O_7 (d) Na_2O_2 (e) As_2O_5

37. Write balanced molecular equations for the following reactions and name the products:
(a) sulfur dioxide, SO_2, with water
(b) sulfur trioxide, SO_3, with water
(c) selenium trioxide, SeO_3, with water
(d) dinitrogen pentoxide, N_2O_5, with water
(e) dichlorine heptoxide, Cl_2O_7, with water

38. Write balanced molecular equations for the following reactions and name the products. (a) sodium oxide, Na_2O, with water (b) calcium oxide, CaO, with water (c) lithium oxide, Li_2O, with water

39. Identify the acid anhydride of each of the following ternary acids: (a) H_2SO_4 (b) H_2CO_3 (c) H_2SO_3 (d) H_3AsO_4 (e) HNO_2

40. Identify the basic anhydride of each of the following metal hydroxides: (a) NaOH (b) $Mg(OH)_2$ (c) $Fe(OH)_2$ (d) $Al(OH)_3$

Identifying Reaction Types

The following reactions apply to Exercises 41–48.

a. H_2SO_4 (aq) $+$ 2KOH (aq) $\rightarrow$ K_2SO_4 (aq) $+$ $2H_2O$ (ℓ)

b. 2Rb (s) $+$ Br_2 (ℓ) $\rightarrow$ 2RbBr (s)

c. 2KI (aq) $+$ F_2 (g) $\rightarrow$ 2KF (aq) $+$ I_2 (s)

d. CaO (s) $+$ SiO_2 (s) $\overset{\Delta}{\rightarrow}$ $CaSiO_3$ (s)

e. S (s) $+$ O_2 (g) $\overset{\Delta}{\rightarrow}$ SO_2 (g)

f. $BaCO_3$ (s) $\overset{\Delta}{\rightarrow}$ BaO (s) $+$ CO_2 (g)

g. HgS (s) $+$ O_2 (g) $\overset{\Delta}{\rightarrow}$ Hg (ℓ) $+$ SO_2 (g)

h. $AgNO_3$ (aq) $+$ HCl (aq) $\rightarrow$ AgCl (s) $+$ HNO_3 (aq)

i. Pb (s) $+$ 2HBr (aq) $\rightarrow$ $PbBr_2$ (s) $+$ H_2 (g)

j. 2HI (aq) $+$ H_2O_2 (aq) $\rightarrow$ I_2 (s) $+$ $2H_2O$ (ℓ)

k. RbOH (aq) $+$ HNO_3 (aq) $\rightarrow$ $RbNO_3$ (aq) $+$ H_2O (ℓ)

l. N_2O_5 (s) $+$ H_2O (ℓ) $\rightarrow$ $2HNO_3$ (aq)

m. H_2O (g) $+$ CO (g) $\overset{\Delta}{\rightarrow}$ H_2 (g) $+$ CO_2 (g)

n. MgO (s) $+$ H_2O (ℓ) $\rightarrow$ $Mg(OH)_2$ (s)

o. $PbSO_4$ (s) $+$ PbS (s) $\overset{\Delta}{\rightarrow}$ 2Pb (s) $+$ $2SO_2$ (g)

p. 2BaO (s) $+$ O_2 (g) $\overset{\Delta}{\rightarrow}$ $2BaO_2$ (s)

q. CO_2 (g) $+$ H_2O (ℓ) $\rightarrow$ H_2CO_3 (aq)

r. 2ZnS (s) $+$ $3O_2$ (g) $\overset{\Delta}{\rightarrow}$ 2ZnO (s) $+$ $2SO_2$ (g)

s. H_2CO_3 (aq) $\overset{\Delta}{\rightarrow}$ CO_2 (g) $+$ H_2O (ℓ)

t. $Fe(OH)_3$ (s) $+$ 3HCl (aq) $\rightarrow$ $FeCl_3$ (aq) $+$ $3H_2O$ (ℓ)

u. CuO (s) $+$ H_2SO_4 (aq) $\rightarrow$ $CuSO_4$ (aq) $+$ H_2O (ℓ)

v. $5Ba(IO_3)_2$ (s) $\overset{\Delta}{\rightarrow}$ $Ba_5(IO_6)_2$ (s) $+$ $4I_2$ (s) $+$ $9O_2$ (g)

w. $TiCl_4$ (ℓ) $+$ 2Mg (s) $\overset{\Delta}{\rightarrow}$ $2MgCl_2$ (s) $+$ Ti (s)

x. Al_2S_3 (s) $+$ $6H_2O$ (ℓ) $\rightarrow$ $2Al(OH)_3$ (s) $+$ $3H_2S$ (aq)

y. $2H_3PO_4$ (aq) $+$ $3Mg(OH)_2$ (s) $\rightarrow$ $Mg_3(PO_4)_2$ (s) $+$ $6H_2O$ (ℓ)

z. N_2 (g) $+$ $3H_2$ (g) $\overset{\Delta}{\rightarrow}$ $2NH_3$ (g)

aa. 4HCl (g) $+$ O_2 (g) $\overset{\Delta}{\rightarrow}$ $2Cl_2$ (g) $+$ $2H_2O$ (g)

bb. $2PbO_2$ (s) $\overset{\Delta}{\rightarrow}$ 2PbO (s) $+$ O_2 (g)

cc. Fe_2O_3 (s) $+$ 2Al (s) $\overset{\Delta}{\rightarrow}$ Al_2O_3 (s) $+$ 2Fe (s)

dd. $2SO_3$ (g) $\overset{\Delta}{\rightarrow}$ $2SO_2$ (g) $+$ O_2 (g)

41. Identify the metathesis reactions.

42. Identify the metathesis reactions that are also precipitation reactions.

43. Identify the metathesis reactions that are also classical acid-base reactions.

44. Identify the oxidation-reduction reactions.

45. Identify the oxidizing agent and reducing agent for each oxidation-reduction reaction.

46. Identify the oxidation-reduction reactions that are also displacement reactions.

47. Identify the decomposition reactions.

48. Identify the combination reactions.

Equations for Metathesis Reactions and Identification of Products

49. Write balanced molecular equations for the following reactions occurring in water, or in contact with water. For acid-base reactions, assume complete neutralization.
(a) nitric acid $+$ calcium hydroxide $\rightarrow$
(b) silver nitrate $+$ ammonium bromide $\rightarrow$
(c) hydrocyanic acid $+$ sodium hydroxide $\rightarrow$
(d) dilute sulfuric acid $+$ barium chloride $\rightarrow$

50. Write balanced total ionic equations for the reactions of Exercise 49.

51. Write balanced net ionic equations for the reactions of Exercise 49.

52. Write balanced molecular equations for the following reactions in water, or in contact with water. For acid-base reactions assume complete neutralization.
    (a) copper(II) hydroxide + hydrochloric acid $\longrightarrow$
    (b) sulfurous acid + potassium hydroxide $\longrightarrow$
    (c) aqueous ammonia + perchloric acid $\longrightarrow$
    (d) lithium phosphate + calcium chloride $\longrightarrow$

53. Write balanced total ionic equations for the reactions of Exercise 52.

54. Write balanced net ionic equations for the reactions of Exercise 52.

55. Write balanced molecular equations for the preparation of the following salts from acids and bases in aqueous solution or in contact with water.
    (a) magnesium sulfate
    (b) lithium carbonate
    (c) calcium nitrate
    (d) copper(II) sulfide
    (e) barium acetate

56. Write balanced total ionic equations for the reactions of Exercise 55.

57. Write balanced net ionic equations for the reactions of Exercise 55.

58. Which of the following metal hydroxides are amphoteric? (a) $Fe(OH)_3$ (b) KOH (c) $Zn(OH)_2$ (d) $Fe(OH)_2$ (e) $Pb(OH)_2$

59. Write balanced molecular, total ionic, and net ionic equations for the reactions of beryllium hydroxide, $Be(OH)_2$, with (a) hydrochloric acid and (b) an excess of aqueous sodium hydroxide.

60. Repeat Exercise 59 for chromium(III) hydroxide, $Cr(OH)_3$, instead of $Be(OH)_2$.

61. For each of the following reactions, indicate whether ions are being removed from solution. If so, how?
    (a) K_2SO_3 (aq) + 2HBr (aq) $\longrightarrow$
        $2KBr$ (aq) + H_2SO_3 (aq)
        $\overset{\Delta}{\longrightarrow}$ SO_2 (g) + H_2O (ℓ)
    (b) $Ba(OH)_2$ (aq) + 2HCl (aq) $\longrightarrow$ $BaCl_2$ (aq) + $2H_2O$ (ℓ)
    (c) $BaCl_2$ (aq) + H_2SO_4 (aq) $\longrightarrow$ $BaSO_4$ (s) + $2HCl$ (aq)
    (d) 2NaOH (aq) + $NiCl_2$ (aq) $\longrightarrow$ $Ni(OH)_2$ (s) + 2NaCl (aq)

Oxidation-Reduction Reactions

62. Define (a) oxidation, (b) reduction, (c) oxidizing agent, (d) reducing agent.

63. Write balanced molecular equations for the following redox reactions.
    (a) aluminum with sulfuric acid, H_2SO_4, to produce aluminum sulfate, $Al_2(SO_4)_3$, and hydrogen
    (b) nitrogen, N_2, with hydrogen, H_2, to form ammonia, NH_3
    (c) zinc sulfide, ZnS, with oxygen, O_2, to form zinc oxide, ZnO, and sulfur dioxide, SO_2
    (d) carbon with nitric acid, HNO_3, to produce nitrogen dioxide, NO_2, carbon dioxide, CO_2, and water
    (e) sulfuric acid with hydrogen iodide, HI, to produce sulfur dioxide, SO_2, iodine, I_2, and water

64. Identify the oxidizing agents and reducing agents in the oxidation-reduction reactions given in Exercise 63.

65. Write total ionic and net ionic equations for the following redox reactions occurring in aqueous solution or in contact with water.
    (a) $Fe + 2HCl \longrightarrow FeCl_2 + H_2$
    (b) $2KMnO_4 + 16HCl \longrightarrow 2MnCl_2 + 2KCl + 5Cl_2 + 8H_2O$
    (c) $4Zn + 10HNO_3 \longrightarrow 4Zn(NO_3)_2 + NH_4NO_3 + 3H_2O$

66. Of the possible displacement reactions below, which one(s) will occur?
    (a) $2Cl^-$ (aq) + Br_2 (ℓ) $\longrightarrow$ $2Br^-$ (aq) + Cl_2 (g)
    (b) $2Br^-$ (aq) + F_2 (g) $\longrightarrow$ $2F^-$ (aq) + Br_2 (ℓ)
    (c) $2I^-$ (aq) + Cl_2 (g) $\longrightarrow$ $2Cl^-$ (aq) + I_2 (s)
    (d) $2Br^-$ (aq) + Cl_2 (g) $\longrightarrow$ $2Cl^-$ (aq) + Br_2 (ℓ)

67. Which of the metals listed below will displace hydrogen from water or steam? (a) K (b) Ca (c) Cr (d) Ni (e) Hg (f) Pt

68. Refer to the metals of Exercise 67. Which one(s) will *not* displace hydrogen from sulfuric acid? (Assume that any oxide coatings on the metals have been brushed off.)

69. Which of the reactions below are disproportionation reactions?
    (a) $KClO_4 + H_2SO_4 \longrightarrow KHSO_4 + HClO_4$
    (b) $CH_4 + 2O_2 \longrightarrow CO_2 + 2H_2O$
    (c) $3NaClO \longrightarrow NaClO_3 + 2NaCl$
    (d) $2NO_2 \longrightarrow N_2O_4$
    (e) $3NaOH + P_4 + 3H_2O \longrightarrow 3NaH_2PO_2 + PH_3$

70. Write equations for the complete combustion of the following compounds: (a) ethane, C_2H_6 (b) propane, C_3H_8 (c) ethanol, C_2H_5OH

71. Write equations for the incomplete combustion of the following compounds to produce carbon monoxide: (a) ethane, C_2H_6 (b) propane, C_3H_8

72. Write equations for the complete combustion of the following compounds. Assume sulfur is converted to SO_2 and nitrogen is converted to NO. (a) C_6H_5N (b) C_2H_5SH (c) $C_7H_{10}NO_2S$

73. Describe the formation of the reddish-brown haze of some cities experiencing this kind of air pollution.

74. Account for the occurrence of "acid rain."

Chemical Analysis in Aqueous Solution

10–1 Introduction

Many chemical reactions in nature, particularly those in living systems, occur in aqueous solutions. We often carry out reactions in aqueous solutions in the laboratory. Solutions are used (1) for the convenience of dispensing volumes rather than masses of reactants and (2) because reactions are generally more rapid in solution than when pure solid reactants are mixed, due to the immediate intimate mixing of reactants in solutions. Distilled water is often used because it is an excellent solvent for many ionic and molecular solutes, and because it is inexpensive and readily available.

In Chapter 9 we studied many kinds of reactions that occur in aqueous solution, as well as some that do not. Among them are acid-base reactions, precipitation reactions, and oxidation-reduction reactions. In this chapter we shall study the application of these classes of reactions to quantitative chemical analysis in the laboratory. A **quantitative analysis** *is one in which the amount or concentration of a particular species in a sample is determined accurately and precisely.* Such analyses are frequently performed in many areas of the physical and biological sciences. We shall deal with two broad areas of analysis—*volumetric analysis* and *gravimetric analysis.*

Volumetric analysis is a quantitative analysis based upon the measurement of the volume of a solution required to react with a definite volume of a second solution of known concentration or a definite mass of another sample. In **gravime-**

In some gravimetric analyses, a gas is evolved and the mass loss due to evolution of the gas is determined.

tric analysis a solid is usually collected by filtration, washed free of impurities, dried, and then weighed. The mass of the solid is then related to the amount of a reactant that must have been consumed to produce it.

In Section 2–13 we introduced molarity as a unit to express concentrations of solutions, and we related it to stoichiometric calculations and to the dilution of concentrated solutions. You may wish to review that material before proceeding. In this chapter we shall expand our treatment of molarity and introduce another unit of concentration, *normality,* which some industrial and biological chemists prefer to use. We shall illustrate both the mole method, using molarity as a concentration unit, and then the equivalent weight method, using normality as a concentration unit.

10–2 Calculations Involving Molarity (A Brief Review)

The most common method of expressing concentrations of solutions is *molarity* (M). Recall that molarity is defined as the number of moles of solute dissolved per liter of solution (not solvent). It is also the number of millimoles of solute per milliliter of solution.

$$\text{Molarity} = \frac{\text{number of moles solute}}{\text{liter of soln}} = \frac{\text{number of millimoles solute}}{\text{milliliter of soln}} \qquad \text{[Eq. 10–1]}$$

Example 10–1

The directions for an experiment call for the addition of 50.0 mL of a 0.0150 M solution of potassium hydroxide, KOH, to another solution. How many millimoles and how many grams of KOH are contained in 50.0 mL of this solution?

Solution

Rearranging Eq. 10–1, the product of the volume (mL) of a solution times its concentration (M) is the amount of solute present (*mmol*).

number of mmol solute = mL $\times$ M

$$\underline{?}\ \text{mmol KOH} = 50.0\ \text{mL} \times \frac{0.0150\ \text{mmol KOH}}{\text{mL}}$$

$$= 0.750\ \text{mmol KOH}$$

The mass of KOH in 0.750 mmol can be calculated easily because we know that 1 mol of KOH is 56.1 g, or 1 mmol of KOH is 0.0561 g.

$$\underline{?}\ \text{g KOH} = 0.750\ \text{mmol KOH} \times \frac{0.0561\ \text{g KOH}}{\text{mmol KOH}}$$

$$= 0.0421\ \text{g KOH}$$

Example 10–2

Commercial nitric acid, HNO_3, is 70.0% HNO_3 by mass, and its specific gravity is 1.42. What is its molarity?

Solution

$$\underline{?}\ \frac{\text{mol } HNO_3}{\text{L soln}} = \underbrace{\frac{1.42\ \text{g soln}}{1\ \text{mL soln}} \times \frac{1000\ \text{mL}}{1\ \text{L}}}_{\text{g soln/L soln}} \times \frac{70.0\ \text{g } HNO_3}{100\ \text{g soln}} \times \frac{1\ \text{mol } HNO_3}{63.0\ \text{g } HNO_3} = \underline{15.8\ M\ HNO_3}$$

g HNO_3/L soln

mol HNO_3/L soln

Example 10–3

Commercial nitric acid, 15.8 M HNO_3, is also known as "concentrated nitric acid." An experiment requires a student to prepare 100 mL of 3.00 M HNO_3 from concentrated HNO_3. How many milliliters of concentrated acid does the student use?

Solution

Since the number of moles (or millimoles) of HNO_3 does not change during dilution, the following relationships apply [where the subscripts "1" and "2" refer to the concentrated and diluted solutions, respectively (Section 2–14)].

$$\text{no. of mmol}_1 \; HNO_3 = \text{no. of mmol}_2 \; HNO_3$$
$$mL_1 \times M_1 = mL_2 \times M_2$$
$$mL_1 = \frac{mL_2 \times M_2}{M_1}$$

Substitution of the given data gives

$$mL_1 = \frac{100 \text{ mL} \times 3.00 \; M}{15.8 \; M} = \underline{19.0 \text{ mL}}$$

The addition of 19.0 mL of concentrated nitric acid to sufficient water to make 100 mL of solution gives a 3.00 M solution of HNO_3.

10–3   Acid-Base (Neutralization) Reactions

In *some cases* one mole of an acid reacts with one mole of a base as indicated in the following examples.

$$HCl + NaOH \longrightarrow NaCl + H_2O$$

$$HNO_3 + KOH \longrightarrow KNO_3 + H_2O$$

Recall that the reaction of an acid with a base is called *neutralization*.

$$HClO_4 + NH_3 \longrightarrow NH_4ClO_4$$

Since one mole of each acid reacts with one mole of each base in these examples, *one liter of a one molar solution of any of these acids* reacts with *one liter of a one molar solution of any of these bases*. Note that the acids used in the examples have only one acidic hydrogen per formula unit, and that the bases have one hydroxide ion per formula unit or (in the case of ammonia) that one formula unit of base reacts with one hydrogen ion.

Example 10–4

If 100 mL of 1.00 M HCl solution and 100 mL of 1.00 M NaOH are mixed, what is the molarity of the salt in the resulting solution? You may assume that the volumes are additive, because experiments have shown that volumes of dilute aqueous solutions are very nearly additive and no significant error is introduced by making this assumption.

Solution

The following tabulation shows that equal numbers of moles (or millimoles) of HCl and NaOH are mixed, and therefore the resulting solution contains only NaCl, the salt formed by the reaction, and water.

	NaOH	+	HCl	$\longrightarrow$	NaCl	+ H_2O
reaction ratio:	1 mol		1 mol		1 mol	1 mol
start:	100 mL $\left(\dfrac{1.00 \text{ mmol}}{\text{mL}}\right)$		100 mL $\left(\dfrac{1.00 \text{ mmol}}{\text{mL}}\right)$			
	100 mmol NaOH		100 mmol HCl			
change:	−100 mmol		−100 mmol		+ 100 mmol	
after reaction:	0 mmol		0 mmol		100 mmol NaCl	

The HCl and NaOH neutralize each other exactly, and the resulting solution contains 100 mmol of NaCl in 200 mL of solution. Its molarity is:

$$? \frac{\text{mmol NaCl}}{\text{mL}} = \frac{100 \text{ mmol}}{200 \text{ mL}} = \underline{0.500 \ M \text{ NaCl}}$$

Note that the reaction in Example 10–4 also produces 100 mmol of H_2O, which has a mass of 1.8 g and a volume of 1.8 mL. However, the volume occupied by this H_2O is not significantly different from the volume occupied by the H^+ (aq) and OH^- (aq) ions that produced it. Therefore, we assume that the final volume of the solution (200 mL) is very nearly equal to the sum of the volumes of HCl and NaOH solutions that were mixed.

Example 10–5

If 100 mL of 1.00 M HCl solution and 100 mL of 0.75 M NaOH are mixed, what is the molarity of the resulting solution?

Solution

The following tabulation shows that some of the HCl remains unreacted and some NaCl is produced, but no NaOH remains.

Since two solutes are present in the solution after reaction, we must calculate the concentrations of both.

$$? \frac{\text{mmol HCl}}{\text{mL}} = \frac{25 \text{ mmol HCl}}{200 \text{ mL}} = \underline{0.12 \ M \text{ HCl}}$$

$$? \frac{\text{mmol NaCl}}{\text{mL}} = \frac{75 \text{ mmol NaCl}}{200 \text{ mL}} = \underline{0.38 \ M \text{ NaCl}}$$

Both HCl and NaCl are strong electrolytes, so the solution is 0.12 M in H^+ (aq), (0.12 + 0.38) M = 0.50 M in Cl^-, and 0.38 M in Na^+ ions.

	HCl	+	NaOH	⟶	NaCl	+	H_2O
reaction ratio:	1 mol		1 mol		1 mol		1 mol
start:	100 mmol		75 mmol				
change:	−75 mmol		− 75 mmol		+ 75 mmol		
after reaction:	25 mmol HCl		0 mmol		75 mmol NaCl		

In many cases one mole of an acid will not neutralize one mole of a base, or one mole of a base will not neutralize one mole of an acid, as shown in the following examples.

$$H_2SO_4 + 2NaOH \longrightarrow Na_2SO_4 + 2H_2O$$

1 mol   2 mol   1 mol

$$2HCl + Ca(OH)_2 \longrightarrow CaCl_2 + 2H_2O$$

2 mol   1 mol   1 mol

The first equation shows that one mole of H_2SO_4 reacts with two moles of NaOH, and therefore *two* liters of 1 molar NaOH solution are required to neutralize one liter of 1 molar H_2SO_4 solution. The second equation shows that two moles of HCl react with one mole of $Ca(OH)_2$, and therefore *two* liters of HCl solution are required to neutralize one liter of $Ca(OH)_2$ solution of equal molarity.

Example 10–6

What volume of 0.00300 M HCl solution will just neutralize 30.0 mL of 0.00100 M $Ca(OH)_2$ solution?

Solution

The balanced equation for the reaction is:

$$2HCl + Ca(OH)_2 \longrightarrow CaCl_2 + 2H_2O$$

2 mol   1 mol   1 mol   2 mol

We can convert (a) mL of $Ca(OH)_2$ solution to mol of $Ca(OH)_2$ using molarity as a unit factor, 0.00100 mol $Ca(OH)_2$/1000 mL $Ca(OH)_2$ solution; (b) mol of $Ca(OH)_2$ to mol of HCl using the unit factor 2 mol HCl/1 mol $Ca(OH)_2$ (from the balanced equation); and (c) mol of HCl to mL of HCl solution using the unit factor 1000 mL HCl/0.00300 mol HCl.

$$? \text{ mL HCl} = 30.0 \text{ mL Ca(OH)}_2 \times \underbrace{\frac{0.00100 \text{ mol Ca(OH)}_2}{1000 \text{ mL Ca(OH)}_2}}_{\text{mol Ca(OH)}_2} \times \frac{2 \text{ mol HCl}}{1 \text{ mol Ca(OH)}_2} \times \frac{1000 \text{ mL HCl}}{0.00300 \text{ mol HCl}}$$

$$\underbrace{}_{\text{mol HCl}}$$

$$\underbrace{}_{\text{mL HCl}}$$

$$= \underline{20.0 \text{ mL Hcl}}$$

Note that in Example 10-6, we used the unit factor 2 mol HCl/1 mol $Ca(OH)_2$ to convert moles of $Ca(OH)_2$ to moles of HCl, because the balanced equation for the reaction shows that two moles of HCl are required to neutralize one mole of $Ca(OH)_2$. We must always write balanced equations for reactions and determine the *reaction ratio*, i.e., *the relative numbers of moles of reactants*, from the balanced equation for the reaction of interest.

Example 10-7

If 100 mL of 1.00 M H_2SO_4 solution is mixed with 200 mL of 1.00 M KOH, what salt is produced, and what is its molarity?

Solution

	H_2SO_4	+	2KOH	$\longrightarrow$	K_2SO_4	+ 2H$_2$O
reaction ratio:	1 mol		2 mol		1 mol	
start:	100 mmol		200 mmol		0 mmol	
change:	−100 mmol		− 200 mmol		+ 100 mmol	
after reaction:	0 mmol H_2SO_4		0 mmol KOH		100 mmol K_2SO_4	

The reaction produces 100 mmoles of potassium sulfate, which are contained in 300 mL of solution, and therefore the concentration is

$$? \frac{\text{mmol K}_2\text{SO}_4}{\text{mL}} = \frac{100 \text{ mmol K}_2\text{SO}_4}{300 \text{ mL}} = \underline{0.333 \; M \; \text{K}_2\text{SO}_4}$$

10-4 Standardization of Solutions and Acid-Base Titrations

Solutions of accurately known concentrations are called **standard solutions.** Standard solutions of some substances, called *primary standards,* can be prepared by dissolving a carefully weighed solid sample in enough water to give an accurately

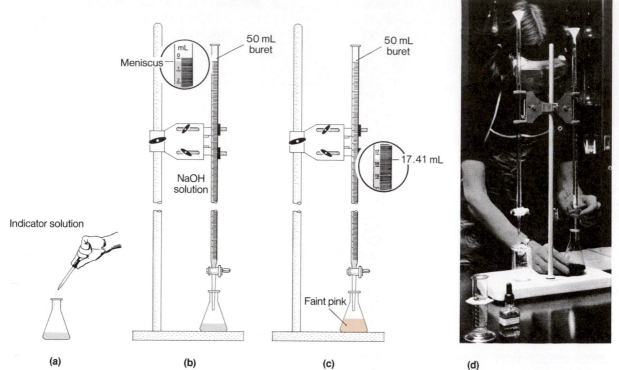

FIGURE 10–1 The titration process. (a) The solution to be titrated is placed in an Erlenmeyer flask, and a few drops of indicator are added. (b) The buret is filled with a standard solution (or the solution to be standardized). The meniscus is the curved surface of the liquid in the buret. The position of the bottom of the meniscus is read and recorded. (c) The solution in the buret is added (dropwise near the end point) to the Erlenmeyer flask until the end point in the titration is reached. The end point is signaled by the appearance (or sometimes disappearance) of color in the solution being titrated. The volume of the solution is read again; the difference between the final and initial buret readings is the volume of solution used in the titration. (d) A typical setup for titration in a teaching laboratory.

known volume of solution. Other substances cannot be weighed out accurately and conveniently because they react with the atmosphere (primarily with CO_2 and/or H_2O). Solutions of such substances are prepared and then their concentrations are determined by titration. **Titration** is the process by which one determines the volume of a solution required to react with an exactly known amount of a primary standard.

Suppose we titrate an acid solution of unknown concentration by adding a standard solution of a base dropwise from a **buret** (Figure 10–1). A common buret is graduated at intervals of 1 mL and at smaller intervals of 0.1 mL, so that it is possible to estimate the volume of a solution dispensed to within ±0.02 mL. Experienced individuals can often read a buret to within ±0.01 mL.

How does one know when to stop a titration, that is, when the chemical reaction is just complete? In most cases a few drops of an *indicator* solution are added to the solution to be titrated. An **indicator** for an acid-base titration is a substance that can exist in solution in two different forms, with different colors that depend upon the concentration of hydrogen ions in the solution. At least one of these forms must be very intensely colored so that very little indicator is necessary for one to detect its presence visually. The analyst tries to choose an indicator that changes color clearly at the point at which stoichiometrically equivalent amounts of acid and base have reacted, the **equivalence point.** The point at which the indicator changes color and the titration is stopped is called the **end point.** Ideally, the end point should coincide with the equivalence point.

Phenolphthalein is a very common indicator for the titration of a solution of a strong acid with a solution of a base. It is colorless in acidic solution and red in basic solution. The end point in a titration in which a base is added to an acid, and in which phenolphthalein is used as the indicator, is signaled by the first appearance of a faint pink coloration that persists for at least 15 seconds (Figure 10–1). Indicators will be described in more detail in Section 18–4.

Now let us consider the properties of *ideal* **primary standards.** They include the following:

1. must not react with or absorb the components of the atmosphere, such as water vapor, oxygen, or carbon dioxide
2. must react according to one invariable reaction
3. known high percentage purity
4. high formula weight to minimize error in weighing
5. soluble in the solvent of interest
6. nontoxic

Each of the first five of these characteristics minimizes the errors involved in analysis. An additional factor of low cost is desirable but not necessary. Because primary standards are often costly and difficult to prepare, secondary standards are usually used in day-to-day work. A **secondary standard** is a solution to be used in analysis of an unknown, the concentration of which is determined by titrating it with a primary standard.

1   The Mole Method and Molarity

Let us now describe a few primary standards for acids and bases and illustrate how they are used. A common primary standard for solutions of acids is sodium carbonate, Na_2CO_3, a solid compound. Consider the following reaction:

$$H_2SO_4 + Na_2CO_3 \longrightarrow Na_2SO_4 + CO_2 + H_2O$$

1 mol	1 mol	1 mol	1 mol	1 mol
98.08 g	106.0 g			

$$1 \text{ mol } Na_2CO_3 = 106.0 \text{ g} \qquad \text{and} \qquad 1 \text{ mmol } Na_2CO_3 = 0.1060 \text{ g}$$

Sodium carbonate is a salt (of carbonic acid and sodium hydroxide). However, since a base can be broadly defined as a substance that reacts with hydrogen ions, Na_2CO_3 can be thought of as a base in *this* reaction. Notice that H_2SO_4 and Na_2CO_3 react in a 1:1 mole ratio.

Example 10–8

Calculate the molarity of a solution of H_2SO_4 if 40.0 mL of the solution reacts with 0.364 g of Na_2CO_3.

Solution

We know from the balanced equation that one mole of H_2SO_4 reacts with one mole of Na_2CO_3, 106.0 g. This provides the unit factors that convert 0.364 g of Na_2CO_3 to the corresponding number of moles of H_2SO_4.

? mol H_2SO_4 = 0.364 g Na_2CO_3

$$\times \frac{1 \text{ mol } Na_2CO_3}{106.0 \text{ g } Na_2CO_3}$$

$$\times \frac{1 \text{ mol } H_2SO_4}{1 \text{ mol } Na_2CO_3}$$

$$= 0.00343 \text{ mol } H_2SO_4$$

Now we calculate the molarity of the H_2SO_4 solution.

$$\underline{?} \frac{\text{mol } H_2SO_4}{L} = \frac{0.00343 \text{ mol } H_2SO_4}{0.0400 \text{ L}}$$

$$= 0.0858 \text{ } M \text{ } H_2SO_4$$

Most inorganic bases are metal hydroxides, all of which are solids. However, most inorganic bases react rapidly with CO_2 (an acid anhydride) from the atmosphere, even in the solid state, and most metal hydroxides also absorb H_2O from the air. In order to obtain solutions of bases of accurately known concentration, they are commonly standardized against an acidic salt, potassium hydrogen phthalate, $KC_6H_4(COO)(COO\underline{H})$, which is produced by neutralization of one of the two ionizable hydrogens of an organic acid, phthalic acid.

phthalic acid
$C_6H_4(COO\underline{H})_2$

potassium hydrogen
phthalate (KHP)
$KC_6H_4(COO)(COO\underline{H})$

The P in KHP stands for the phthalate ion, $C_6H_4(COO)_2{}^{2-}$, not phosphorus.

This acidic salt, abbreviated as KHP, has one acidic hydrogen (underlined), and reacts with bases. It is easily obtained in a high state of purity, is soluble in water, and is used as a primary standard for bases.

Example 10–9

A 20.00 mL sample of a solution of NaOH reacts with 0.3641 gram of KHP. Calculate the molarity of the basic solution.

NaOH + KHP $\longrightarrow$ NaKP + H_2O

1 mol   1 mol   1 mol   1 mol
         204.2 g

Solution

We first calculate the number of moles of NaOH that react with 0.3641 gram of KHP. Note that

NaOH and KHP react in a 1:1 mole ratio.

$$\underline{?} \text{ mol NaOH} = 0.3641 \text{ g KHP} \times \frac{1 \text{ mol KHP}}{204.2 \text{ g KHP}}$$

$$\times \frac{1 \text{ mol NaOH}}{1 \text{ mol KHP}}$$

$$= 0.001783 \text{ mol NaOH}$$

Then we calculate the molarity of the NaOH solution.

$$\underline{?} \frac{\text{mol NaOH}}{L} = \frac{0.001783 \text{ mol NaOH}}{0.02000 \text{ L}}$$

$$= 0.08915 \text{ } M \text{ NaOH}$$

Impure samples of acids can be titrated with standard solutions of bases, and the results can be used to determine percentage purity of the acid, as Example 10–10 illustrates.

Example 10–10

A 0.2000 gram sample of impure oxalic acid, $(COOH)_2$, required 39.82 mL of 0.1023 M NaOH solution for complete neutralization. No acidic impurities were present. Calculate the percentage purity of the $(COOH)_2$. Note that each molecule of $(COOH)_2$ contains two acidic H's,

$$H—O \qquad O—H$$
$$C—C$$
$$O \qquad O$$

Solution

The equation for the reaction of NaOH with $(COOH)_2$,

$$2NaOH + (COOH)_2 \longrightarrow Na_2(COO)_2 + 2H_2O$$

2 mol	1 mol	1 mol	2 mol
	90.04 g		

shows that *two* moles of NaOH are required to neutralize completely one mole of $(COOH)_2$. The number of moles of NaOH that react is the volume times the molarity of the solution.

$$? \text{ mol NaOH} = 0.03982 \text{ L} \times \frac{0.1023 \text{ mol NaOH}}{L}$$

$$= 0.004074 \text{ mol NaOH}$$

Now we calculate the mass of $(COOH)_2$ that reacts with 0.004074 mol NaOH.

$$? \text{ g } (COOH)_2 = 0.004074 \text{ mol NaOH}$$

$$\times \frac{1 \text{ mol } (COOH)_2}{2 \text{ mol NaOH}}$$

$$\times \frac{90.04 \text{ g } (COOH)_2}{1 \text{ mol } (COOH)_2}$$

$$= 0.1834 \text{ g } (COOH)_2$$

The sample contained 0.1834 g of $(COOH)_2$, and its percentage purity is

$$\% \text{ purity} = \frac{0.1834 \text{ g } (COOH)_2}{0.2000 \text{ g sample}} \times 100\% = 91.70\%$$

Some solutions can be standardized by precipitation reactions in *gravimetric analyses*, which are based on the accurate determination of the mass of a pure substance (usually a solid, but sometimes a gas) produced in a reaction. For example, silver nitrate, $AgNO_3$, is sometimes used in the standardization of solutions of hydrochloric, hydrobromic, or hydroiodic acids (or solutions of other compounds containing ionizable chloride, bromide, or iodide ions). In each case a precipitate (AgCl, AgBr, or AgI) is produced, and the reaction goes to completion because an

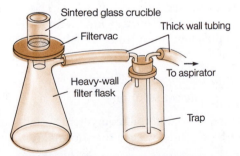

FIGURE 10–2 Filtration apparatus for collecting a precipitate whose mass is to be determined. A sintered glass crucible is one kind of filter crucible used to collect precipitates quantitatively. The mass of the clean, dry sintered glass crucible is determined. The precipitate is collected, and then washed free of impurities. The crucible and precipitate are dried in an oven, and the mass of the crucible with precipitate is determined. The mass of the precipitate is the difference between the mass of the crucible that contains the precipitate and the mass of the empty crucible.

excess of $AgNO_3$ is used. (The analysis is based on the mass of the silver halide precipitated, and we want to be sure that all the halide ions have been precipitated.) The reaction is complete when the addition of $AgNO_3$ solution produces no more cloudiness (precipitation). At this point the solution is filtered through a sintered glass filter crucible (Figure 10–2), and the precipitate is rinsed with distilled water and dried in an oven. The mass of the precipitate is the mass of the crucible with the precipitate minus the mass of the empty crucible. This mass can then be related to the number of moles of acid (or other compound) that must have reacted to produce the precipitate. The molarity of the acidic (or other) solution can be calculated from the known volume of solution used.

Example 10–11

A 25.00 mL sample of an HCl solution was treated with excess $AgNO_3$ solution. The precipitated AgCl was collected by filtration, dried, and found to have a mass of 0.3214 gram. Calculate the molarity of the HCl solution.

$$HCl + AgNO_3 \longrightarrow AgCl\ (s) + HNO_3$$

$$\begin{array}{llll} 1\ mol & 1\ mol & 1\ mol & 1\ mol \\ & & 143.3\ g & \end{array}$$

Solution

We first calculate the number of moles of HCl that react to produce 0.3214 gram of AgCl.

$$\underline{?}\ mol\ HCl = 0.3214\ g\ AgCl \times \frac{1\ mol\ AgCl}{143.3\ g\ AgCl}$$

$$\times \frac{1\ mol\ HCl}{1\ mol\ AgCl}$$

$$= 0.002243\ mol\ HCl$$

Then we calculate the molarity of the HCl solution.

$$\underline{?}\ \frac{mol\ HCl}{L} = \frac{0.002243\ mol\ HCl}{0.02500\ L}$$

$$= 0.08972\ M\ HCl$$

---

2   Equivalent Weights and Normality

Because one mole of an acid does not necessarily neutralize one mole of a base, many chemists choose a method of expressing concentration other than molarity in order to retain a one-to-one relationship. Concentrations of solutions of acids and bases are frequently expressed as *normality* (N). The **normality** of a solution is defined as the *number of equivalent weights* (defined below) *of solute per liter of solution.* (The term equivalent mass is not widely used.) Normality may be represented symbolically as

An equivalent weight is often referred to simply as an equivalent (eq).

$$normality = \frac{number\ of\ equivalent\ weights\ of\ solute}{liter\ of\ solution}$$

[Eq. 10–2]

By definition there are 1000 milliequivalent weights (meq) in one equivalent weight of an acid or base. Thus, we see that normality may also be expressed as

A milliequivalent weight is often referred to simply as a milliequivalent (meq).

$$normality = \frac{number\ of\ milliequivalent\ weights\ of\ solute}{milliliter\ of\ solution}$$

[Eq. 10–3]

In *acid-base* reactions the **equivalent weight,** or **one equivalent (eq),** of an acid is defined as the mass of the acid (expressed in grams) that will furnish 6.022×10^{23} hydrogen ions (1 mole) or that will react with 6.022×10^{23} hydroxide ions (1 mole). One mole of an acid contains 6.022×10^{23} formula units of the acid.

Consider hydrochloric acid as a typical monoprotic acid:

$$HCl \xrightarrow{\text{H}_2\text{O}} H^+ (aq) + Cl^- (aq)$$

1 mol	1 mol	1 mol
36.46 g	1.008 g	35.45 g
6.022×10^{23}	6.022×10^{23}	6.022×10^{23}

This equation shows that one mole of HCl can produce 6.022×10^{23} H^+, and therefore *one mole of HCl is one equivalent*. The same is true for all monoprotic acids.

Sulfuric acid is a diprotic acid because it can furnish two H^+ ions per molecule of H_2SO_4:

$$H_2SO_4 \xrightarrow{\text{H}_2\text{O}} 2H^+ (aq) + SO_4^{2-} (aq)$$

1 mol	2 mol	1 mol
98.08 g	2(1.008 g)	96.06 g
6.022×10^{23}	$2(6.022 \times 10^{23})$	6.022×10^{23}

This equation shows that one mole of H_2SO_4 can produce $2 \times 6.022 \times 10^{23}$ H^+ ions; therefore, one mole of H_2SO_4 is two equivalents in all reactions in which both acidic hydrogen atoms react with a base. Or, one equivalent of H_2SO_4 is one half of a mole.

> The equivalent weight of an acid is obtained by dividing its formula weight in grams by the number of acidic hydrogen atoms furnished by one formula unit of the acid, or by the number of hydroxide ions with which one formula unit of the acid reacts.

The equivalent weight of a base is defined as the mass of the base (expressed in grams) that will furnish 6.022×10^{23} hydroxide ions or the mass of the base that will react with 6.022×10^{23} hydrogen ions.

> The equivalent weight of a base is obtained by dividing its formula weight in grams by the number of hydroxide ions furnished by one formula unit, or by the number of hydrogen ions with which one formula unit of the base reacts. Likewise, we define one equivalent weight of a salt as the mass of salt produced by the reaction of one equivalent of an acid or base.

The equivalent weights of some common acids and bases are listed in Table 10-1.

TABLE 10-1 Equivalent Weights* of Common Acids and Bases

Acids		Bases	
Symbolic Representation	One eq	Symbolic Representation	One eq
$\dfrac{HNO_3}{1}$	$= \dfrac{63.02 \text{ g}}{1} = 63.02$ g HNO_3	$\dfrac{NaOH}{1}$	$= \dfrac{40.00 \text{ g}}{1} = 40.00$ g NaOH
$\dfrac{CH_3COOH}{1}$	$= \dfrac{60.03 \text{ g}}{1} = 60.03$ g CH_3COOH	$\dfrac{NH_3}{1}$	$= \dfrac{17.04 \text{ g}}{1} = 17.04$ g NH_3
$\dfrac{KC_6H_4(COO)(COOH)}{1}$	$= \dfrac{204.2 \text{ g}}{1} = 204.2$ g $KC_6H_4(COO)(COOH)$	$\dfrac{Ca(OH)_2}{2}$	$= \dfrac{74.10 \text{ g}}{2} = 37.05$ g $Ca(OH)_2$
$\dfrac{H_2SO_4}{2}$	$= \dfrac{98.08 \text{ g}}{2} = 49.04$ g H_2SO_4	$\dfrac{Ba(OH)_2}{2}$	$= \dfrac{171.36 \text{ g}}{2} = 85.68$ g $Ba(OH)_2$

* Complete neutralization is assumed.

From the definitions of one equivalent of an acid and a base, we see that *one equivalent of any acid reacts with one equivalent of any base*. It is not true that one mole of any acid reacts with one mole of any base, but as a consequence of the definition of equivalents, 1 eq acid ≎ 1 eq base. In general terms, we may write the following relationship for all acid-base reactions that go to completion:

The symbol ≎ means "is chemically equivalent to."

no. eq of acid ≎ no. eq of base   [Eq. 10–4]

The product of the volume of a solution and its normality is equal to the number of equivalents of solute contained in the solution. For a solution of an acid,

Remember that the product of volume and concentration equals the amount of solute.

$$L_{acid} \times \frac{eq\ acid}{L_{acid}} = eq\ acid,$$

or $L_{acid} \times N_{acid} = eq\ acid,$ *or* $mL_{acid} \times \frac{meq\ acid}{mL_{acid}} = meq\ acid$

Similar relationships can be written for bases. Since 1 eq of acid always reacts with 1 eq of base, we may write

no. eq of acid ≎ no. eq of base,   *or*

$$L_{acid} \times N_{acid} ≎ L_{base} \times N_{base},   or   mL_{acid} \times N_{acid} ≎ mL_{base} \times N_{base},$$

three different forms of a very useful relationship.

Example 10–12

What volume of 0.100 N HNO$_3$ is required to neutralize 50.0 mL of a 0.150 N solution of Ba(OH)$_2$?

Solution

Since no. meq acid ≎ no. meq of base, we can write

$$
\begin{aligned}
?\ mL_{acid} &= \frac{mL_{base} \times N_{base}}{N_{acid}} \\
&= \frac{50.0\ mL \times 0.150\ N}{0.100\ N} \\
&= 75.0\ mL\ of\ HNO_3\ solution
\end{aligned}
$$

Example 10–13

Calculate the normality of a solution that contains 4.202 grams of HNO$_3$ in 600 mL of solution.

Solution

$$N = \frac{number\ of\ eq\ HNO_3}{L}$$

$$?\ \frac{eq\ HNO_3}{L} = \underbrace{\frac{4.202\ g\ HNO_3}{0.600\ L} \times \frac{1\ mol\ HNO_3}{63.02\ g\ HNO_3}}_{M_{HNO_3}} \times \frac{1\ eq\ HNO_3}{mol\ HNO_3} = \underline{0.111\ N\ HNO_3}$$

Notice that *normality* is *molarity* times the number of equivalents per mole of solute; it is always equal to or greater than molarity.

$$M \times \frac{\text{number eq}}{\text{mol}} = N$$

[Eq. 10-5]

Example 10-14

What is the normality of a solution of 9.50 grams of barium hydroxide in 2000 mL of solution? What is the molarity?

Solution

$$\underset{?}{?}\, \frac{\text{eq Ba(OH)}_2}{\text{L}} = \underbrace{\frac{9.50 \text{ g Ba(OH)}_2}{2.00 \text{ L}} \times \frac{1 \text{ mol Ba(OH)}_2}{171.4 \text{ g Ba(OH)}_2}}_{M_{\text{Ba(OH)}_2}} \times \frac{2 \text{ eq}}{\text{mol}} = 0.0554 \text{ N Ba(OH)}_2$$

The $Ba(OH)_2$ solution is 0.0554 N and 0.0277 M.

Let us solve Example 10-8 again in Example 10-15, using normality rather than molarity. The balanced equation for the reaction of H_2SO_4 with Na_2CO_3 interpreted in equivalent weights is

$$H_2SO_4 + Na_2CO_3 \longrightarrow Na_2SO_4 + CO_2 + H_2O$$

1 mol	1 mol	1 mol	1 mol	1 mol
2 eq	2 eq			
98.08 g	106.0 g			

1 eq Na_2CO_3 = 53.0 g   and   1 meq Na_2CO_3 = 0.0530 g

Example 10-15

Calculate the normality of a solution of H_2SO_4 if 40.0 mL of the solution reacts with 0.364 g of Na_2CO_3.

Solution

First we calculate the number of milliequivalents of Na_2CO_3 in the sample:

no. meq Na_2CO_3 = 0.364 g Na_2CO_3

$$\times \frac{1 \text{ meq Na}_2\text{CO}_3}{0.0530 \text{ g Na}_2\text{CO}_3}$$

= 6.87 meq Na_2CO_3

Since no. meq H_2SO_4 = no. meq Na_2CO_3, we may write

$$\text{mL}_{H_2SO_4} \times N_{H_2SO_4} = 6.87 \text{ meq } H_2SO_4$$

$$N_{H_2SO_4} = \frac{6.87 \text{ meq } H_2SO_4}{\text{mL}_{H_2SO_4}} = \frac{6.87 \text{ meq } H_2SO_4}{40.0 \text{ mL}}$$

$$= 0.172 \text{ N } H_2SO_4$$

Note that the value of the normality of this H_2SO_4 solution is twice the value of the molarity obtained in Example 10-8 because 1 mol of H_2SO_4 is two equivalents.

10–5   Oxidation-Reduction Reactions

Rules for assigning
oxidation numbers were
given in Section 6–12.

Many volumetric analyses involve oxidation-reduction reactions, or redox reactions. Recall (Section 9–7) that such reactions involve changes in oxidation number. All balanced equations must satisfy two criteria:

1. There must be mass balance. That is, the same number of atoms of each kind must be shown as reactants and products.
2. There must be charge balance. The sums of actual charges on the left and right sides of the equation must be equal.

Although equations for redox reactions can often be balanced by simple inspection, many require considerable effort. A systematic method is therefore helpful. Before we describe redox titrations we shall learn two general methods for balancing redox equations.

1   Balancing Oxidation-Reduction Equations

Our rules for assigning oxidation numbers are constructed so that *the total increase in oxidation numbers must equal the total decrease in oxidation numbers in all redox reactions.* This equivalence provides the basis for balancing redox reactions. Although there is no single "best method" for balancing all redox equations, two methods are particularly useful, (1) the change-in-oxidation-number method and (2) the ion-electron method.

Most redox equations can
be balanced by both
methods, but in some in-
stances one may be easier
to use than the other.

Change-in-Oxidation-Number Method   Example 10–16 illustrates this method, which is based on *equal total increases and decreases in oxidation numbers.* The general procedure is

1. Write the overall unbalanced equation.
2. Assign oxidation numbers to determine which elements undergo changes in oxidation number.
3. Insert coefficients to make the total increase and decrease in oxidation numbers equal.
4. Balance the remaining atoms by inspection.

Example 10–16

Aluminum dissolves in hydrochloric acid to form aqueous aluminum chloride and gaseous hydrogen. Balance the net ionic and molecular equations and identify the oxidizing and reducing agents.

Solution

The unbalanced molecular equation and oxidation numbers are:

$$\overset{0}{\text{Al (s)}} + \overset{+1\,-1}{\text{HCl (aq)}} \longrightarrow \overset{+3\,-1}{\text{AlCl}_3\text{ (aq)}} + \overset{0}{\text{H}_2\text{ (g)}}$$

The oxidation number of aluminum increases from zero to +3. Therefore, aluminum is the reducing agent and is oxidized. (We represent an increase in

oxidation number by $n\uparrow$ and a decrease in oxidation number by $n\downarrow$, where n is the magnitude of the change.)

$$\overset{0}{\text{Al}} \xrightarrow{3\uparrow} \overset{+3}{\text{Al}}$$

There is a decrease in the oxidation number of hydrogen, which is reduced and is the oxidizing agent. Since *at least* two hydrogen atoms in the zero oxidation state are produced (H_2), we must have at least two hydrogen atoms on the left at this point.

$$\overset{+1}{2\text{H}} \xrightarrow{2\downarrow} \overset{0}{\text{H}_2}$$

The total increase in oxidation number for aluminum must equal the total decrease in oxidation number for hydrogen. The lowest common multiple of two and three is six. Therefore we multiply the aluminum "helping" equation by 2 and the hydrogen "helping" equation by 3, so that there are six units of increase in oxidation number for every six units of decrease.

$$2(\overset{0}{Al} \xrightarrow{3\uparrow} \overset{+3}{Al}) \quad \text{or} \quad 2\overset{0}{Al} \xrightarrow{6\uparrow} 2\overset{+3}{Al}$$

$$3(2\overset{+1}{H} \xrightarrow{2\downarrow} \overset{0}{H_2}) \quad \text{or} \quad 6\overset{+1}{H} \xrightarrow{6\downarrow} 3\overset{0}{H_2}$$

From these numbers we can generate the balanced net ionic equation.

$$2Al \text{ (s)} + 6H^+ \text{ (aq)} \longrightarrow 2Al^{3+} \text{ (aq)} + 3H_2 \text{ (g)}$$

(balanced net ionic equation)

In this example there are 2 Al and 6 H on each side, so the first criterion is satisfied; the second is satisfied by charges of 6+ on each side.

The ionic equation is easily converted to a molecular equation that is also balanced with respect to atoms (2Al, 6H, 6Cl) and charge (zero) by adding $6Cl^-$ to each side.

$$2Al \text{ (s)} + 6HCl \text{ (aq)} \longrightarrow 2AlCl_3 \text{ (aq)} + 3H_2 \text{ (g)}$$

(balanced molecular equation)

Adding H^+, OH^-, or H_2O Frequently it is necessary to complete the mass balance for reactions in aqueous solution by adding hydrogen atoms or oxygen atoms. For reactions in *acidic solution*, H^+ (aq) or H_2O or both may be added as needed, since both are readily available. In *basic solution*, OH^- (aq) or H_2O or both may be added as needed. This is illustrated in the next two examples.

Example 10–17

Potassium dichromate reacts with hydroiodic acid to produce potassium iodide, chromium(III) iodide, and iodine. Write balanced net ionic and molecular equations for the reaction, and identify the oxidizing and reducing agents.

Solution

The incomplete and unbalanced molecular equation and oxidation states are:

$$\overset{+1\ +6\ -2}{K_2Cr_2O_7} + \overset{+1-1}{HI} \longrightarrow \overset{+1-1}{KI} + \overset{+3-1}{CrI_3} + \overset{0}{I_2}$$

Chromium is reduced from the +6 to the +3 oxidation state. *Some* of the iodine is oxidized from the −1 to the zero oxidation state, while some remains in the −1 state in KI and CrI_3.

$$\overset{+6}{Cr} \xrightarrow{3\downarrow} \overset{+3}{Cr}$$

$$2\overset{-1}{I} \xrightarrow{2\uparrow} \overset{0}{I_2}$$

Only the I^- that undergoes a change in oxidation number is converted to I_2.

We now balance the changes in oxidation numbers and add the two resulting "helping" equations.

$$2(\overset{+6}{Cr} \xrightarrow{3\downarrow} \overset{+3}{Cr})$$

$$3(2\overset{-1}{I} \xrightarrow{2\uparrow} \overset{0}{I_2})$$

$$\overline{2\overset{+6}{Cr} + 6\overset{-1}{I} \longrightarrow 2\overset{+3}{Cr} + 3\overset{0}{I_2}}$$

There are two chromium atoms per formula unit of $Cr_2O_7^{2-}$, so the coefficient of $Cr_2O_7^{2-}$ is one:

$$Cr_2O_7^{2-} \text{ (aq)} + 6I^- \text{ (aq)} \longrightarrow$$
$$2Cr^{3+} \text{ (aq)} + 3I_2 \text{ (s)} \quad \text{(incomplete)}$$

The equation is not yet balanced with respect to atoms or charge. Now we must add H_2O as suggested in the beginning of this section. There are seven atoms of oxygen in $Cr_2O_7^{2-}$ on the left, and none on the right; these oxygen atoms must appear in the product as $7H_2O$ (since none of the other products of the reaction contain oxygen). Fourteen H^+ can then be added to the left (since this reaction is occurring in an acidic solution) to achieve atom (and charge) balance in the net ionic equation.

$$Cr_2O_7^{2-} \text{ (aq)} + 14H^+ \text{ (aq)} + 6I^- \text{ (aq)} \longrightarrow$$
$$2Cr^{3+} \text{ (aq)} + 3I_2 \text{ (s)} + 7H_2O \text{ (}\ell\text{)}$$

Two things must now be done to obtain the molecular equation from this net ionic equation. First, note that the ionic equation shows a total charge of 6+ on each side, whereas the molecular equation must be neutral on both sides; therefore, we must add species with a total charge of 6− to both sides. Second, the spectator ions (the potassium ions in potassium dichromate and the iodide ions that counterbalance Cr^{3+} and K^+ ions) must be accounted for. Thus, we add $2K^+$ and $8I^-$ to each side.

$$Cr_2O_7{}^{2-}\text{ (aq)} + 14H^+\text{ (aq)} + 6I^-\text{ (aq)} \longrightarrow 2Cr^{3+}\text{ (aq)} + 3I_2\text{ (s)} + 7H_2O\ (\ell) \qquad \text{(net ionic equation)}$$

$$\underline{+ 2K^+\text{ (aq)} + 8I^-\text{ (aq)} \longrightarrow 6I^-\text{ (aq)} + 2K^+\text{ (aq)} + 2I^-\text{ (aq)}} \qquad \text{(spectator ions)}$$

$$K_2Cr_2O_7\text{ (aq)} + 14HI\text{ (aq)} \longrightarrow 2CrI_3\text{ (aq)} + 2KI\text{ (aq)} + 3I_2\text{ (s)} + 7H_2O\ (\ell) \qquad \text{(molecular equation)}$$

Example 10−18

In basic solution, permanganate ions oxidize iodide ions to iodine and form manganese(IV) oxide. Write and balance the net ionic equation for the reaction.

Solution

We first write the incomplete unbalanced equation and assign oxidation numbers. MnO_2 is insoluble in water and is written in molecular form.

$$\overset{+7}{MnO_4{}^-}\text{ (aq)} + \overset{-1}{I^-}\text{ (aq)} \longrightarrow \overset{0}{I_2}\text{ (s)} + \overset{+4}{MnO_2}\text{ (s)}$$

Manganese is reduced from the +7 to the +4 oxidation state. The oxidation number of iodine increases from −1 to zero as iodide ions are oxidized to iodine molecules, I_2.

$$\overset{+7}{Mn} \xrightarrow{3\downarrow} \overset{+4}{Mn}$$

$$\overset{-1}{2I} \xrightarrow{2\uparrow} \overset{0}{I_2}$$

To balance the changes in oxidation numbers, the first "helping" equation is multiplied by two, and the second by three, and then they are added:

$$2(\overset{+7}{Mn} \xrightarrow{3\downarrow} \overset{+4}{Mn})$$

$$\underline{3(\overset{-1}{2I} \xrightarrow{2\uparrow} \overset{0}{I_2})}$$

$$\overset{+7}{2Mn} + \overset{-1}{6I} \longrightarrow \overset{+4}{2Mn} + \overset{0}{3I_2}$$

The dichromate ion, which contains chromium in the +6 oxidation state, is the oxidizing agent. Iodine in the −1 oxidation state, I^-, is the reducing agent.

We next begin to balance the net ionic equation using these coefficients.

$$2MnO_4{}^-\text{ (aq)} + 6I^-\text{ (aq)} \longrightarrow 3I_2\text{ (s)} + 2MnO_2\text{ (s)}$$
$$\text{(unbalanced)}$$

The changes in oxidation number have been balanced, but we do not yet have mass balance or charge balance. (It is important to realize that at this point the charge *cannot* be balanced by adding or removing electrons.) Both mass and charge balance are accomplished by adding OH^- and/or H_2O since the reaction occurs in basic solution. The net charge on the left side of the equation is 8−. Therefore, we add 8 OH^- to the right side to balance charges. Four H_2O are required on the left to give mass balance.

$$4H_2O\ (\ell) + 2MnO_4{}^-\text{ (aq)} + 6I^-\text{ (aq)} \longrightarrow$$
$$3I_2\text{ (s)} + 2MnO_2\text{ (s)} + 8OH^-\text{ (aq)}$$

The equation now has atom balance (8H, 12O, 2Mn, 6I on each side) and charge balance (8− on each side). Although there must be some spectator ions, we aren't told what they are.

Ion-Electron Method (Half-Reaction Method) The ion-electron method relies upon the separate and complete balancing of equations describing so-called oxidation and reduction **half-reactions.** This is followed by equalizing the numbers of electrons gained and lost in each and, finally, adding the resulting half-reactions to give the overall balanced equation. The general procedure follows.

1. Write as much of the overall unbalanced equation as possible.
2. Construct unbalanced oxidation and reduction half-reactions (these are usually incomplete as well as unbalanced).
3. Balance the atoms in each half-reaction.
4. Balance the charge in each half-reaction by adding electrons.
5. Balance the electron transfer by multiplying the balanced half-reactions by appropriate integers.
6. Add the resulting half-reactions, and eliminate any common terms to obtain the balanced equation.

Example 10-19

Permanganate ions oxidize iodide ions to iodine in acidic solution, and are reduced to manganese(II) ions. Write and balance the net ionic equation for the reaction by the ion-electron method.

Solution

Applying the first step in the procedure to the statement of the problem, we can write:

$$MnO_4^- \, (aq) + I^- \, (aq) \longrightarrow Mn^{2+} \, (aq) + I_2 \, (s)$$
(incomplete and unbalanced)

(Note that the reactions between MnO_4^- and I^- in basic and acidic solutions are *not* the same.) From the information we begin to construct half-reactions for the oxidation and reduction parts of the reaction.

$$2I^- \longrightarrow I_2$$

$$MnO_4^- \longrightarrow Mn^{2+}$$

The first half-reaction, $2I^- \, (aq) \rightarrow I_2 \, (s)$ (unbalanced), shows mass balance but not charge balance. Addition of 2 electrons to the right gives a balanced half-reaction; this must be the oxidation because the iodide ions lose electrons.

$$2I^- \, (aq) \longrightarrow I_2 \, (s) + 2e^-$$
(balanced *oxidation* half-reaction)

The other half-reaction is balanced by:

(1) adding $4H_2O$ to the right because there are 4 oxygen atoms on the left,

$$MnO_4^- \longrightarrow Mn^{2+} + 4H_2O \qquad \text{(unbalanced)}$$

(2) adding $8H^+$ to the left to give mass balance,

$$MnO_4^- + 8H^+ \longrightarrow Mn^{2+} + 4H_2O \qquad \text{(unbalanced)}$$

(3) adding 5 electrons to the left to give charge balance and a *balanced* reduction half-reaction.

$$MnO_4^- + 8H^+ + 5e^- \longrightarrow Mn^{2+} + 4H_2O$$

(balanced *reduction* half-reaction)

Examination of the balanced half-reactions

$$2I^- \, (aq) \longrightarrow I_2 \, (s) + 2e^-$$

$$MnO_4^- \, (aq) + 8H^+ \, (aq) + 5e^- \longrightarrow$$
$$Mn^{2+} \, (aq) + 4H_2O \, (\ell)$$

indicates that the first should be multiplied by 5 and the second by 2 to balance the electron transfer. Addition of the resulting half-reactions term-by-term, and cancellation of $10e^-$ from each side, gives

$$5(2I^- \longrightarrow I_2 + 2e^-)$$

$$\underline{2(MnO_4^- + 8H^+ + 5e^- \longrightarrow Mn^{2+} + 4H_2O)}$$

$$10I^- + 2MnO_4^- + 16H^+ + 10e^- \longrightarrow 5I_2 + 2Mn^{2+} + 8H_2O + 10e^-$$

$$10I^- \, (aq) + 2MnO_4^- \, (aq) + 16H^+ \, (aq) \longrightarrow 5I_2 \, (s) + 2Mn^{2+} \, (aq) + 8H_2O \, (\ell)$$

Example 10-20

The chromate ion, CrO_4^{2-}, oxidizes the stannite ion, $HSnO_2^-$, to the stannate ion, $HSnO_3^-$, and is reduced to chromite ion, CrO_2^-, in basic solution. Balance the net ionic equation for this reaction.

Solution

From the statement of the problem we may write:

$$CrO_4^{2-} \, (aq) + HSnO_2^- \, (aq) \longrightarrow$$
$$HSnO_3^- \, (aq) + CrO_2^- \, (aq) \qquad \text{(unbalanced)}$$

Let us deal with oxidation of the tin-containing species first.

$$HSnO_2^- \longrightarrow HSnO_3^- \quad \text{(unbalanced)}$$

The hydrogen and tin atoms are balanced, but one more oxygen is required on the left. We add two OH^- to the left (since the solution is basic) and one H_2O to the right to maintain the hydrogen balance:

$$2OH^- + HSnO_2^- \longrightarrow HSnO_3^- + H_2O$$
$$\text{(unbalanced)}$$

The net charge on the left is $3-$, and on the right it is $1-$, so 2 electrons are added to the right.

$$2OH^- + HSnO_2^- \longrightarrow HSnO_3^- + H_2O + 2e^-$$
$$\text{(balanced } oxidation \text{ half-reaction)}$$

Reduction involves conversion of chromate ions to chromite ions.

$$CrO_4^{2-} \longrightarrow CrO_2^- \quad \text{(unbalanced)}$$

We add $4OH^-$ to the right and $2H_2O$ to the left to balance oxygen and hydrogen atoms.

$$2H_2O + CrO_4^{2-} \longrightarrow CrO_2^- + 4OH^- \quad \text{(unbalanced)}$$

To balance the charge ($2-$ left, $5-$ right), 3 electrons are added to the left.

$$3e^- + 2H_2O + CrO_4^{2-} \longrightarrow CrO_2^- + 4OH^-$$
$$\text{(balanced } reduction \text{ half-reaction)}$$

We now balance the electron transfer, add the two half-reactions, and eliminate common terms.

$$3(2OH^- + HSnO_2^- \longrightarrow HSnO_3^- + H_2O + 2e^-) \quad \text{(oxidation)}$$
$$2(3e^- + 2H_2O + CrO_4^{2-} \longrightarrow CrO_2^- + 4OH^-) \quad \text{(reduction)}$$

$$6OH^- + 3HSnO_2^- + 6e^- + 4H_2O + 2CrO_4^{2-} \longrightarrow 3HSnO_3^- + 3H_2O + 6e^- + 2CrO_2^- + 8OH^-$$

$$3HSnO_2^- \text{ (aq)} + H_2O \text{ } (\ell) + 2CrO_4^{2-} \text{ (aq)} \longrightarrow 3HSnO_3^- \text{ (aq)} + 2CrO_2^- \text{ (aq)} + 2OH^- \text{ (aq)}$$

2 Oxidation-Reduction Titrations

One method of analyzing samples quantitatively for the presence of oxidizable or reducible substances is by **redox titration.** In such analyses, the concentration of the oxidizable or reducible substance in a solution is determined by allowing the solution to react with an accurately determined amount of an oxidizing or reducing agent. Alternatively, a solution of known concentration (a standard solution) of oxidizing agent or reducing agent can be used to determine the purity of any sample with which it exactly reacts.

Concentrations of solutions involved in redox titrations can be expressed in terms of either molarity or normality, and amounts of solutes can be described in terms of either moles or equivalents. We shall illustrate redox titrations by both the mole/molarity method and the equivalent/normality method.

The Mole Method and Molar Solutions of Oxidizing and Reducing Agents As in other kinds of chemical reactions, we must pay particular attention to the mole ratio, i.e., the reaction ratio, in which oxidizing agents and reducing agents react. The examples that follow will re-emphasize the importance of the reaction ratio.

Potassium permanganate, $KMnO_4$, is a strong oxidizing agent, and it has been the "workhorse" of redox titrations over the years. For example, in acidic solution $KMnO_4$ can be used to react with iron(II) sulfate, $FeSO_4$, according to the following balanced molecular and net ionic equations. (A strong acid, such as H_2SO_4, is used to make the solution acidic.)

$$2KMnO_4 + 10FeSO_4 + 8H_2SO_4 \longrightarrow 2MnSO_4 + 5Fe_2(SO_4)_3 + K_2SO_4 + 8H_2O$$

$$MnO_4^- \text{ (aq)} + 5Fe^{2+} \text{ (aq)} + 8H^+ \text{ (aq)} \longrightarrow Mn^{2+} \text{ (aq)} + 5Fe^{3+} \text{ (aq)} + 4H_2O \text{ } (\ell)$$

We have provided the balanced molecular and net ionic equations to emphasize the fact that the reaction ratio is 1 mol $KMnO_4$: 5 mol $FeSO_4$ or 1 mol MnO_4^- : 5 mol Fe^{2+}. (Examples 10–21 and 10–23 are based on this reaction.)

We often refer to "permanganate solutions" as a sort of shorthand. The source of MnO_4^- ions is the soluble ionic compound $KMnO_4$. Clearly such solutions contain cations, K^+ in this case. Likewise, we often refer to "iron(II) solutions" without specifying that the anion is SO_4^{2-}.

Example 10–21

What volume of 0.0200 M $KMnO_4$ solution is required to oxidize 40.0 mL of 0.100 M $FeSO_4$ in sulfuric acid solution?

Solution

From the above discussion, we know the balanced equation and the reaction ratio, 1 mol MnO_4^- : 5 mol Fe^{2+}.

$$MnO_4^- (aq) + 8H^+ (aq) + 5Fe^{2+} (aq) \longrightarrow$$
$$5Fe^{3+} (aq) + Mn^{2+} (aq) + 4H_2O$$

We first determine the number of moles of Fe^{2+} to be titrated.

$$? \text{ mol } Fe^{2+} = 40.0 \text{ mL} \times \frac{0.100 \text{ mol } Fe^{2+}}{1000 \text{ mL}}$$
$$= 4.00 \times 10^{-3} \text{ mol } Fe^{2+}$$

Now we use the reaction ratio to determine the number of moles of MnO_4^- required.

$$? \text{ mol } MnO_4^- = 4.00 \times 10^{-3} \text{ mol } Fe^{2+}$$
$$\times \frac{1 \text{ mol } MnO_4^-}{5 \text{ mol } Fe^{2+}}$$
$$= 8.00 \times 10^{-4} \text{ mol } MnO_4^-$$

Finally we calculate the volume of 0.0200 M MnO_4^- solution that contains 8.00×10^{-4} mol MnO_4^-.

$$? \text{ mL } MnO_4^- \text{ soln} = 8.00 \times 10^{-4} \text{ mol } MnO_4^-$$
$$\times \frac{1000 \text{ mL } MnO_4^- \text{ soln}}{0.0200 \text{ mol } MnO_4^-}$$
$$= 40.0 \text{ mL } MnO_4^- \text{ soln}$$

Because it has an intense purple color, $KMnO_4$ acts as its own indicator. One drop of 0.020 M $KMnO_4$ solution imparts a pink color to a liter of pure water. When $KMnO_4$ solution is added to a solution of a reducing agent, the end point in the titration is taken as the point at which a pale pink color appears in the solution being titrated and persists for at least 30 seconds.

Potassium dichromate, $K_2Cr_2O_7$, is another frequently used oxidizing agent. However, an indicator must be used when reducing agents are titrated with dichromate solutions because $K_2Cr_2O_7$ is not intensely colored enough to serve as its own indicator, and the reduction product, Cr^{3+}, is intensely colored (green).

Consider the oxidation of sulfite ions, SO_3^{2-}, to sulfate ions, SO_4^{2-}, by $Cr_2O_7^{2-}$ ions in the presence of a strong acid such as sulfuric acid. We shall balance the equation by the ion-electron method, beginning with the reduction of $Cr_2O_7^{2-}$.

$$Cr_2O_7^{2-} \longrightarrow Cr^{3+}$$
$$Cr_2O_7^{2-} \longrightarrow 2Cr^{3+}$$
$$14H^+ + Cr_2O_7^{2-} \longrightarrow 2Cr^{3+} + 7H_2O$$
$$6e^- + 14H^+ + Cr_2O_7^{2-} \longrightarrow 2Cr^{3+} + 7H_2O \qquad \text{(balanced } reduction \text{ half-reaction)}$$

Now we balance the oxidation half-reaction by the usual procedure:

$$SO_3^{2-} \longrightarrow SO_4^{2-}$$
$$SO_3^{2-} + H_2O \longrightarrow SO_4^{2-} + 2H^+$$
$$SO_3^{2-} + H_2O \longrightarrow SO_4^{2-} + 2H^+ + 2e^- \quad \text{(balanced } oxidation \text{ half-reaction)}$$

We now equalize the electron transfer, add the balanced half-reactions, and eliminate common terms:

$$
\begin{array}{ll}
(6e^- + 14H^+ + Cr_2O_7^{2-} \longrightarrow 2Cr^{3+} + 7H_2O) & \text{(reduction)} \\
3(SO_3^{2-} + H_2O \longrightarrow SO_4^{2-} + 2H^+ + 2e^-) & \text{(oxidation)}
\end{array}
$$
$$8H^+ (aq) + Cr_2O_7^{2-} (aq) + 3SO_3^{2-} (aq) \longrightarrow 2Cr^{3+} (aq) + 3SO_4^{2-} (aq) + 4H_2O (\ell)$$

The balanced equation tells us that the reaction ratio is 3 mol SO_3^{2-}/mol $Cr_2O_7^{2-}$ or 1 mol $Cr_2O_7^{2-}$/3 mol SO_3^{2-}. Potassium dichromate is the usual source of $Cr_2O_7^{2-}$ ions and Na_2SO_3 is the usual source of SO_3^{2-} ions. Thus, the reaction ratio could also be expressed as 1 mol $K_2Cr_2O_7$/3 mol Na_2SO_3.

Example 10–22

A 20.00 mL sample of Na_2SO_3 was titrated with 36.30 mL of 0.05130 M $K_2Cr_2O_7$ solution in the presence of sulfuric acid. Calculate the molarity of the Na_2SO_3 solution.

Solution

From the above discussion we know the balanced equation and the reaction ratio:

$$3SO_3^{2-} + Cr_2O_7^{2-} + 8H^+ \longrightarrow$$
$$3SO_4^{2-} + 2Cr^{3+} + 4H_2O$$

The number of moles of $Cr_2O_7^{2-}$ used is

$$\underline{?} \text{ mol } Cr_2O_7^{2-} = 0.03630 \text{ L}$$
$$\times \frac{0.05130 \text{ mol } Cr_2O_7^{2-}}{L}$$
$$= 0.001862 \text{ mol } Cr_2O_7^{2-}$$

The number of moles of SO_3^{2-} that reacted with 0.001862 mole of $Cr_2O_7^{2-}$ is

$$\underline{?} \text{ mol } SO_3^{2-} = 0.001862 \text{ mol } Cr_2O_7^{2-}$$
$$\times \frac{3 \text{ mol } SO_3^{2-}}{1 \text{ mol } Cr_2O_7^{2-}}$$
$$= 0.005586 \text{ mol } SO_3^{2-}$$

The Na_2SO_3 solution contained 0.005586 mole of SO_3^{2-} (and of necessity 0.005586 mole of Na_2SO_3). Its molarity is

$$\underline{?} \frac{\text{mol } Na_2SO_3}{L} = \frac{0.005586 \text{ mol } Na_2SO_3}{0.02000 \text{ L}}$$
$$= \underline{0.2793 \ M \ Na_2SO_3}$$

Equivalent Weights and Normal Solutions of Oxidizing and Reducing Agents When we studied acid-base reactions, we defined the term *equivalent weight* so that one equivalent weight, or one equivalent (eq), of any acid reacts with one equivalent of any base. We now define the term as it applies to redox reactions.

In redox reactions, **one equivalent weight** *of a substance is the mass of the oxidizing or reducing substance that gains or loses* 6.022×10^{23} *electrons.* This definition tells us that one equivalent of any oxidizing agent reacts exactly with one equivalent of any reducing agent. For all redox reactions we may write.

number of eq of oxidizing agent = number of eq of reducing agent   [Eq. 10–6]

Using the definition of normality, Eq. 10–6 may also be expressed as

$$L_O \times N_O = L_R \times N_R$$

or as

number of meq of oxidizing agent = number of meq of reducing agent

$$mL_O \times N_O = mL_R \times N_R$$

Note that from a *stoichiometric* point of view, the only difference between calculations on acid-base reactions and those on redox reactions is the *definition* of the equivalent.

We saw earlier that the balanced net ionic equation for the reaction of $KMnO_4$ with $FeSO_4$ in acidic solution is

$$MnO_4^- (aq) + 8H^+ (aq) + 5Fe^{2+} (aq) \longrightarrow Mn^{2+} (aq) + 5Fe^{3+} (aq) + 4H_2O (\ell)$$

The balanced half-reactions are:

$$Fe^{2+} (aq) \longrightarrow Fe^{3+} (aq) + 1e^- \qquad \text{(oxidation)}$$

$$MnO_4^- (aq) + 8H^+ (aq) + 5e^- \longrightarrow Mn^{2+} (aq) + 4H_2O (\ell) \qquad \text{(reduction)}$$

Each formula unit of $FeSO_4$ loses one electron in this reaction, and one mole of $FeSO_4$ loses 6.022×10^{23} electrons. Thus, one mole of $FeSO_4$ (151.9 grams) is one equivalent *in this reaction.*

Each permanganate ion gains 5 electrons in the above reaction, and each mole of $KMnO_4$ that reacts gains $5(6.022 \times 10^{23}$ electrons). *In this reaction,* one mole of $KMnO_4$ (158.0 grams) provides 5 equivalents.

We may summarize our discussion of this reaction as follows.

> The equivalent weight of an oxidizing agent or reducing agent depends upon the specific reaction it undergoes.

	Compound	e^- Transferred per Formula Unit	One mole	One eq
oxidizing agent:	$KMnO_4$	5	158.0 g	$\dfrac{158.0 \text{ g}}{5} = 31.60$ g
reducing agent:	$FeSO_4$	1	151.9 g	$\dfrac{151.9 \text{ g}}{1} = 151.9$ g

In this reaction 31.60 grams of $KMnO_4$ react with 151.9 grams of $FeSO_4$.

Example 10–23

What volume of 0.1000 N $KMnO_4$ is required to oxidize 40.0 mL of 0.100 N $FeSO_4$ in sulfuric acid solution?

Solution

From the preceding discussion we know that the balanced equation is

$$MnO_4^- (aq) + 8H^+ (aq) + 5Fe^{2+} (aq) \longrightarrow$$
$$5Fe^{3+} (aq) + Mn^{2+} (aq) + 4H_2O (\ell)$$

Since

$$mL_{MnO_4^-} \times N_{MnO_4^-} = mL_{Fe^{2+}} \times N_{Fe^{2+}}$$

we can solve for the number of milliliters of $KMnO_4$ solution:

$$? \, mL_{MnO_4^-} = \frac{mL_{Fe^{2+}} \times N_{Fe^{2+}}}{N_{MnO_4^-}}$$

$$= \frac{40.0 \text{ mL} \times 0.100 \, N}{0.1000 \, N}$$

$$= \underline{40.0 \text{ mL } KMnO_4 \text{ solution}}$$

Notice that this is the same as Example 10–21 except that concentrations are expressed as *normality* rather than as *molarity.*

Key Terms

End point the point at which an indicator changes color and a titration is stopped.

Equivalence point the point at which chemically equivalent amounts of reactants have reacted.

Equivalent weight the mass of an acid or base that furnishes or reacts with 6.022×10^{23} H^+ or OH^- ions; of oxidizing or reducing agent, the mass that gains (oxidizing agents) or loses (reducing agents) 6.022×10^{23} electrons in a redox reaction.

Gravimetric analysis the process in which a solid (or occasionally a gas) is isolated and its mass is determined in order to measure the mass of the reactant that produced it.

Indicators for acid-base titrations, organic compounds that exhibit different colors in solutions of different acidities; used to determine the point at which reaction between two solutes is complete.

Monoprotic acid an acid that contains one ionizable hydrogen atom per formula unit.

Neutralization (commonly) the reaction of H^+ from an acid with OH^- from a base to form H_2O.

Normality the number of equivalents of solute per liter of solution.

Oxidation an algebraic increase in oxidation number; a loss, or apparent loss, of electrons.

Oxidation-reduction reaction a reaction in which substances undergo changes in oxidation number; one undergoes an increase and another a decrease in oxidation number.

Oxidizing agent a species that is reduced in a redox reaction; gains (or appears to gain) electrons.

Polyprotic acid an acid that contains more than one ionizable hydrogen atom per formula unit.

Precipitation reaction a reaction in solution in which a precipitate (insoluble substance) is formed.

Primary standard a substance of a known high degree of purity that undergoes one invariable reaction with the other reactant of interest.

Quantitative analysis the process in which the amount or concentration of a particular species in a sample is determined accurately and precisely.

Redox reaction an oxidation-reduction reaction.

Redox titration the quantitative analysis of the amount or concentration of an oxidizing or reducing agent in a sample by observing its reaction with a known amount of a reducing or oxidizing agent.

Reducing agent a species that is oxidized in a redox reaction; loses (or appears to lose) electrons.

Reduction an algebraic decrease in oxidation number; a gain, or apparent gain, of electrons.

Secondary standard a solution that has been titrated against a primary standard. A standard solution is a secondary standard.

Standardization the process by which the concentration of a solution is accurately determined by titrating it against an accurately known amount of a primary standard.

Standard solution a solution of accurately known concentration.

Titration procedure in which a solution of known concentration, a standard solution, is added to a solution of unknown concentration until the chemical reaction between the two solutes is complete.

Volumetric analysis a process based upon the measurement of the volume of a standard solution required to react with a definite volume of another solution or a definite mass of another sample.

Exercises

General Ideas

1. Define and illustrate the following terms clearly and concisely: (a) standard solution, (b) titration, (c) primary standard, (d) secondary standard.

2. Distinguish between the terms *equivalence point* and *end point* of a titration.
3. Distinguish between a gravimetric and a volumetric analysis.
4. Distinguish between an acid-base titration and a redox titration.
5. Define and illustrate the following terms: (a) oxidation, (b) reduction, (c) oxidizing agent, (d) reducing agent, (e) equivalent weights of oxidizing and reducing agents.
6. (a) What are the desirable characteristics of a primary standard? (b) What is the significance of each?
7. Why can sodium carbonate be used as a primary standard for solutions of acids?
8. (a) What is potassium hydrogen phthalate? (b) How is it prepared? (c) For what is it used?

Molarity

9. Calculate the molarities of solutions that contain the following masses of solute in the indicated volumes.
   (a) 75 g of H_3AsO_4 in 500 mL of solution
   (b) 8.3 g of $(COOH)_2$ in 600 mL of solution
   (c) 13.0 g of $(COOH)_2 \cdot 2H_2O$ in 750 mL of solution
10. Calculate the molarities of solutions that contain the following:
    (a) 17.5 g of $(NH_4)_2SO_4$ in 300 mL of solution
    (b) 143 g of $Cr_2(SO_4)_3$ in 3.00 liters of solution
    (c) 143 g of $Cr_2(SO_4)_3 \cdot 18H_2O$ in 3.00 liters of solution
11. Calculate the mass of NaOH required to prepare 2.5 liters of 1.50 M NaOH solution.
12. What mass of $(NH_4)_2SO_4$ is required to prepare 10.0 liters of 1.5 M $(NH_4)_2SO_4$ solution?
13. Calculate the mass of $Mg(NO_3)_2 \cdot 6H_2O$ required to prepare 1200 mL of 0.375 M $Mg(NO_3)_2$ solution.
14. Calculate the molarity of a solution that is 20.0% H_2SO_4 by mass. The specific gravity of the solution is 1.14.
15. What is the molarity of a solution that is 19.0% HNO_3 by mass? The specific gravity of the solution is 1.11.
16. If 200 mL of 2.00 M HCl solution is added to 400 mL of 1.00 M NaOH solution, the resulting solution will be _____ molar in NaCl.
17. If 400 mL of 0.400 M HCl solution is added to 800 mL of 0.100 M $Ba(OH)_2$ solution, the resulting solution will be _____ molar in $BaCl_2$.

18. If 200 mL of 2.00 M H_3PO_4 solution is added to 600 mL of 2.00 M NaOH solution, the resulting solution will be _____ molar in Na_3PO_4.
19. What volumes of 1.00 M KOH and 0.750 M HNO_3 solutions would be required to produce 4.80 g of KNO_3?
20. What volumes of 1.00 M NaOH and 0.500 M H_3PO_4 solutions would be required to form 1.00 mole of Na_3PO_4?
21. What volumes of 0.0200 M $Ca(OH)_2$ and 0.0300 M H_3AsO_4 solutions would be required to prepare 1.00 g of $Ca_3(AsO_4)_2$?
22. Ammonia reacts with sulfuric acid to produce ammonium sulfate. What volume of sulfuric acid solution (sp. gr. = 1.82; 90.0% H_2SO_4 by mass) would be required to prepare 3.00 pounds of ammonium sulfate?

Dilution

23. Calculate the volume of 6.00 M NaOH required to prepare 500 mL of 0.120 M NaOH solution.
24. Calculate the volume of 0.200 M H_2SO_4 required to prepare 200 mL of 0.100 M H_2SO_4 solution.
25. Calculate the volume of 3.00 M H_3PO_4 required to prepare 900 mL of 0.100 M H_3PO_4.

Standardization and Acid-Base Titrations — Mole Method

26. What volume of 0.100 molar sodium hydroxide solution would be required to react with 20.0 mL of 0.400 molar sulfuric acid solution? Assume complete neutralization.
27. A 20 mL sample of 0.10 molar phosphoric acid solution required 120 mL of barium hydroxide solution for complete neutralization. Calculate the molarity of the barium hydroxide solution.
28. What volume of 0.10 molar barium hydroxide would be required to react completely with 20 mL of 0.10 molar phosphoric acid? Assume complete neutralization.
29. Calculate the molarity of a KOH solution if 20.0 mL of the KOH solution reacted with 0.408 g of KHP.
30. An impure sample of $(COOH)_2 \cdot 2H_2O$ that weighed 2.00 g was dissolved in water and titrated with standard NaOH solution. The titration required 40.0 mL of 0.200 M NaOH solution. Calculate the percentage of $(COOH)_2 \cdot 2H_2O$ in the sample.

31. A 50 mL sample of 0.050 M Ca(OH)$_2$ is added to 10 mL of 0.20 M HNO$_3$. (a) Is the resulting solution acidic or basic? (b) How many moles of excess acid or base are present? (c) How many mL of 0.050 M Ca(OH)$_2$ or 0.050 M HNO$_3$ would be required to neutralize the solution?

32. A solution of sodium hydroxide was dispensed from a buret and standardized against potassium hydrogen phthalate. From the following data, calculate the molarity of the NaOH solution.

mass of KHP	0.8407 g
buret reading before titration	0.06 mL
buret reading after titration	44.68 mL

33. The secondary standard solution of NaOH of Exercise 32 was used to titrate a solution of unknown concentration of HCl. A 30.00 mL sample of the HCl solution required 24.21 mL of the NaOH solution for complete neutralization. What is the molarity of the HCl solution?

34. A 34.53 mL sample of a solution of sulfuric acid, H$_2$SO$_4$, is neutralized by 28.24 mL of the NaOH solution of Exercise 32. Calculate the molarity of the sulfuric acid solution.

35. Benzoic acid, C$_6$H$_5$COOH, is sometimes used as a primary standard for the standardization of solutions of bases. A 1.862 gram sample of the acid is neutralized by 31.62 mL of a NaOH solution. What is the molarity of the base solution?

$$C_6H_5COOH \text{ (s)} + NaOH \text{ (aq)} \longrightarrow$$
$$C_6H_5COONa \text{ (aq)} + H_2O \text{ (}\ell\text{)}$$

36. An antacid tablet containing magnesium hydroxide as an active ingredient requires 22.6 mL of 0.597 M HCl for complete neutralization. What mass of Mg(OH)$_2$ did the tablet contain?

37. Butyric acid, whose empirical formula is C$_2$H$_4$O, is the acid that is responsible for the odor of rancid butter. It has one ionizable hydrogen per molecule. A 1.000 gram sample of butyric acid is neutralized by 54.42 mL of standardized 0.2088 M NaOH solution. What are (a) the molecular weight and (b) the molecular formula of butyric acid?

38. The typical concentration of HCl in stomach acid (gastric juice) is about 8.0 × 10^{-2} M. One experiences "acid stomach" when the stomach contents reach a concentration of about 1.0 × 10^{-1} M HCl. One ROLAIDS® tablet (an antacid) contains 334 mg of active ingredient, NaAl(OH)$_2$CO$_3$. Assume that you have "acid stomach" and that your stomach contains 300 mL of 1.0 × 10^{-1} M HCl. What percentage of a tablet should be just sufficient to return the molarity of HCl to 8.0 × 10^{-2} M? Assume that the neutralization reaction produces NaCl, AlCl$_3$, CO$_2$, and H$_2$O.

Standardization and Acid-Base Titrations — Equivalent Weight Method

39. Calculate the normality of a solution that contains 4.9 g of H$_2$SO$_4$ in 100 mL of solution.

40. What is the normality of a solution that contains 7.08 g of H$_3$AsO$_4$ in 100 mL of solution? Assume that we are referring to a reaction involving complete neutralization of the H$_3$AsO$_4$.

41. Calculate the molarity and the normality of a solution that was prepared by dissolving 34.2 g of barium hydroxide in enough water to make 4000 mL of solution.

42. Calculate the molarity and the normality of a solution that was prepared by dissolving 19.6 g of phosphoric acid in enough water to make 600 mL of solution. Assume that we are referring to a reaction involving complete neutralization of the phosphoric acid.

43. What are the molarity and normality of a sulfuric acid solution that is 19.6% H$_2$SO$_4$ by mass? The density of the solution is 1.14 g/mL.

44. A 20 mL sample of 0.20 normal nitric acid solution required 40 mL of barium hydroxide solution for neutralization. Calculate the molarity of the barium hydroxide solution.

45. Vinegar is an aqueous solution of acetic acid, CH$_3$COOH. Suppose you titrate a 25.00 mL sample of vinegar with 17.62 mL of a standardized 0.1060 N solution of NaOH. (a) What is the normality of acetic acid in the vinegar? (b) What is the mass of acetic acid contained in 1.000 liter of vinegar?

46. A 44.4 mL sample of sodium hydroxide solution was titrated with 40.0 mL of 0.100 N sulfuric acid solution. A 36.0 mL sample of hydrochloric acid required 40.0 mL of the sodium hydroxide solution for titration. What is the normality of the hydrochloric acid solution?

47. Calculate the normality and molarity of an H_2SO_4 solution if 20.0 mL of the solution reacted with 0.212 g of Na_2CO_3.

$$H_2SO_4 + Na_2CO_3 \longrightarrow$$
$$Na_2SO_4 + CO_2 + H_2O$$

48. Calculate the normality and molarity of an HCl solution if 30.0 mL of the solution reacted with 0.318 g of Na_2CO_3.

$$2HCl + Na_2CO_3 \longrightarrow 2NaCl + CO_2 + H_2O$$

49. What are the normality and molarity of a $Ca(OH)_2$ solution of 50.0 mL if the $Ca(OH)_2$ solution reacted with 0.102 g of KHP?

50. An impure sample of KHP that weighed 1.000 g was dissolved in water and titrated with 30.00 mL of 0.1000 N NaOH solution. Calculate the percentage of KHP in the sample.

51. A 20 mL sample of 0.20 M H_2SO_4 is added to 20 mL of 0.20 M NaOH. (a) Is the resulting solution acidic or basic? (b) How many meq of excess acid or base are present? (c) What volume of 0.10 N H_2SO_4 or NaOH would be required to neutralize the solution?

Gravimetric Analysis

52. An HCl solution was standardized by precipitating the chloride ion from a sample of HCl by adding excess $AgNO_3$, and then weighing the precipitated AgCl. A 10.00 mL sample of HCl gave 0.2868 g of AgCl. Calculate the molarity of the HCl solution.

$$HCl + AgNO_3 \longrightarrow AgCl\ (s) + HNO_3$$

53. An H_2SO_4 solution was standardized by precipitating the sulfate ion from a sample of H_2SO_4 by adding excess $BaCl_2$, and then weighing the precipitated $BaSO_4$. A 10.0 mL sample of H_2SO_4 gave 0.7005 g of $BaSO_4$. Calculate the molarity of the H_2SO_4 solution.

$$H_2SO_4 + BaCl_2 \longrightarrow BaSO_4\ (s) + 2HCl$$

54. A 0.6000 gram sample of a solid mixture of $BaCl_2 \cdot 2H_2O$ and KCl of unknown composition was heated until all the water was driven off. The remaining solid had a mass of 0.5468 gram. (a) How many moles of water were liberated? (b) What mass of $BaCl_2 \cdot 2H_2O$ was present in the mixture? (c) What was the percentage of $BaCl_2 \cdot 2H_2O$ in the mixture?

55. An excess of silver nitrate solution is added to 28.00 mL of a solution of hydrobromic acid of unknown concentration. The precipitated silver bromide, AgBr, was collected by filtration, dried, and found to have a mass of 0.7111 gram. What is the molarity of the HBr solution?

56. An excess of a solution of barium chloride, $BaCl_2$, was added to 40.00 mL of a solution of sulfuric acid to produce a precipitate of barium sulfate, $BaSO_4$, having a mass of 0.7921 gram. Calculate the molarity of the sulfuric acid solution.

57. When the sky blue salt copper(II) sulfate is crystallized from aqueous solution, the salt is hydrated. We could represent its formula as $CuSO_4 \cdot xH_2O$. A gravimetric analysis is performed on a 2.600 gram sample of $CuSO_4 \cdot xH_2O$, during which it is heated in a crucible until all the water is driven off. Only 1.662 grams of nearly colorless (greenish) anhydrous $CuSO_4$ remains in the crucible. Determine the percentage of water in the hydrated salt and the value of x.

58. A crystal of calcite, pure $CaCO_3$, having a mass of 1.900 grams is dissolved in 49.28 mL of a solution containing an *excess* of hydrochloric acid, and the solution is boiled to expel carbon dioxide. The unneutralized acid is then titrated with 17.16 mL of a sodium hydroxide solution. In a separate titration 25.00 mL of the sodium hydroxide solution is just neutralized by 37.52 mL of the original HCl solution. Determine (a) the molarity of the NaOH solution, (b) the molarity of the HCl solution.

59. What is the primary standard in Exercise 58? What is the secondary standard?

Assigning Oxidation Numbers

60. Assign oxidation numbers to the element specified in each group of compounds.
   (a) N in NO, N_2O_3, N_2O_4, NH_3, N_2H_4, NH_2OH, HNO_3
   (b) C in CO, CO_2, CH_2O, CH_4O, C_2H_6O, $(COOH)_2$, Na_2CO_3
   (c) S in S_8, H_2S, SO_2, SO_3, Na_2SO_3, H_2SO_4, K_2SO_4

61. Assign oxidation numbers to the element specified in each group of compounds.
   (a) P in PCl_3, P_4O_6, P_4O_{10}, HPO_3, H_3PO_4, $POCl_3$, $H_4P_2O_7$, $Mg_3(PO_4)_2$

(b) Cl in Cl_2, HCl, HClO, $HClO_2$, $KClO_3$, Cl_2O_7, $Ca(ClO_4)_2$

(c) Mn in MnO, MnO_2, $Mn(OH)_2$, K_2MnO_4, $KMnO_4$, Mn_2O_7

(d) O in OF_2, Na_2O, Na_2O_2, KO_2

62. Assign oxidation numbers to the element specified in each group of ions.

(a) S in S^{2-}, SO_3^{2-}, SO_4^{2-}, $S_2O_3^{2-}$, $S_4O_6^{2-}$

(b) Cr in CrO_2^-, $Cr(OH)_4^-$, CrO_4^{2-}, $Cr_2O_7^{2-}$

(c) B in BO_2^-, BO_3^{3-}, $B_4O_7^{2-}$

63. Assign oxidation numbers to the element specified in each group of ions.

(a) N in N^{3-}, NO_2^-, NO_3^-, N_3^-, NH_4^+

(b) Br in Br^-, BrO^-, BrO_3^-, BrO_4^-

Identification of Redox Reactions

64. Determine which of the following are oxidation-reduction reactions. For those that are, identify the oxidizing and reducing agents.

(a) $HgCl_2$ (aq) + 2KI (aq) →
$$HgI_2 \text{ (s)} + 2KCl \text{ (aq)}$$

(b) $4NH_3$ (g) + $3O_2$ (g) →
$$2N_2 \text{ (g)} + 6H_2O \text{ (g)}$$

(c) $CaCO_3$ (s) + $2HNO_3$ (aq) →
$$Ca(NO_3)_2 \text{ (aq)} + CO_2 \text{ (g)} + H_2O \text{ (}\ell\text{)}$$

(d) PCl_3 (ℓ) + $3H_2O$ (ℓ) →
$$3HCl \text{ (aq)} + H_3PO_3 \text{ (aq)}$$

65. Determine which of the following are oxidation-reduction reactions. For those that are, identify the oxidizing and reducing agents.

(a) $2H_2O_2$ (ℓ) → $2H_2O$ (ℓ) + O_2 (g)

(b) ICl (s) + H_2O (ℓ) → HCl (aq) + HOI (aq)

(c) 3HCl (aq) + HNO_3 (aq) →
$$Cl_2 \text{ (g)} + NOCl \text{ (g)} + 2H_2O \text{ (}\ell\text{)}$$

(d) Fe_2O_3 (s) + 3CO (g) $\xrightarrow{\Delta}$
$$2Fe \text{ (s)} + 3CO_2 \text{ (g)}$$

Balancing Redox Equations

66. Balance the following as net ionic equations.

(a) MnO_4^- (aq) + H^+ (aq) + Br^- (aq) →
$$Mn^{2+} \text{ (aq)} + Br_2 \text{ (}\ell\text{)} + H_2O \text{ (}\ell\text{)}$$

(b) $Cr_2O_7^{2-}$ (aq) + H^+ (aq) + I^- (aq) →
$$Cr^{3+} \text{ (aq)} + I_2 \text{ (s)} + H_2O \text{ (}\ell\text{)}$$

(c) MnO_4^- (aq) + SO_3^{2-} (aq) + H^+ (aq) →
$$Mn^{2+} \text{ (aq)} + SO_4^{2-} \text{ (aq)} + H_2O \text{ (}\ell\text{)}$$

(d) $Cr_2O_7^{2-}$ (aq) + Fe^{2+} (aq) + H^+ (aq) →
$$Cr^{3+} \text{ (aq)} + Fe^{3+} \text{ (aq)} + H_2O \text{ (}\ell\text{)}$$

67. Balance the following as net ionic equations.

(a) $Cr(OH)_4^-$ (aq) + OH^- (aq) + H_2O_2 (aq) →
$$CrO_4^{2-} \text{ (aq)} + H_2O \text{ (}\ell\text{)}$$

(b) MnO_2 (s) + H^+ (aq) + NO_2^- (aq) →
$$NO_3^- \text{ (aq)} + Mn^{2+} \text{ (aq)} + H_2O \text{ (}\ell\text{)}$$

(c) $Sn(OH)_3^-$ (aq) + $Bi(OH)_3$ (s) + OH^- →
$$Sn(OH)_6^{2-} \text{ (aq)} + Bi \text{ (s)}$$

68. Balance the following as net ionic equations.

(a) Al (s) + NO_3^- (aq) + OH^- (aq) + H_2O →
$$Al(OH)_4^- \text{ (aq)} + NH_3 \text{ (g)}$$

(b) NO_2 (g) + OH^- (aq) →
$$NO_3^- \text{ (aq)} + NO_2^- \text{ (aq)} + H_2O \text{ (}\ell\text{)}$$

(c) MnO_4^- (aq) + H_2O (ℓ) + NO_2^- (aq) →
$$MnO_2 \text{ (s)} + NO_3^- \text{ (aq)} + OH^- \text{ (aq)}$$

(d) I^- (aq) + H^+ (aq) + NO_2^- (aq) →
$$NO \text{ (g)} + H_2O \text{ (}\ell\text{)} + I_2 \text{ (s)}$$

(e) Hg_2Cl_2 (s) + NH_3 (aq) → Hg (ℓ)
$$+ HgNH_2Cl \text{ (s)} + NH_4^+ \text{ (aq)} + Cl^- \text{ (aq)}$$

69. Balance the following as net ionic equations.

(a) CrO_4^{2-} (aq) + H_2O (ℓ) + $HSnO_2^-$ (aq) →
$$CrO_2^- \text{ (aq)} + OH^- \text{ (aq)} + HSnO_3^- \text{ (aq)}$$

(b) C_2H_4 (g) + MnO_4^- (aq) + H^+ (aq) →
$$CO_2 \text{ (g)} + Mn^{2+} \text{ (aq)} + H_2O \text{ (}\ell\text{)}$$

(c) H_2S (aq) + H^+ (aq) + $Cr_2O_7^{2-}$ (aq) →
$$Cr^{3+} \text{ (aq)} + S \text{ (s)} + H_2O \text{ (}\ell\text{)}$$

(d) ClO_3^- (aq) + H_2O (ℓ) + I_2 (s) →
$$IO_3^- \text{ (aq)} + Cl^- \text{ (aq)} + H^+ \text{ (aq)}$$

(e) Cu (s) + H^+ (aq) + SO_4^{2-} (aq) →
$$Cu^{2+} \text{ (aq)} + H_2O \text{ (}\ell\text{)} + SO_2 \text{ (g)}$$

70. Balance the following as net ionic equations for reactions in acidic solution. H^+ or H_2O (but not OH^-) may be added as necessary.

(a) Fe^{2+} (aq) + MnO_4^- (aq) →
$$Fe^{3+} \text{ (aq)} + Mn^{2+} \text{ (aq)}$$

(b) Br_2 (ℓ) + SO_2 (g) → Br^- (aq) + SO_4^{2-} (aq)

(c) Cu (s) + NO_3^- (aq) →
$$Cu^{2+} \text{ (aq)} + NO_2 \text{ (g)}$$

(d) PbO_2 (s) + Cl^- (aq) → $PbCl_2$ (s) + Cl_2 (g)

(e) Zn (s) + NO_3^- (aq) → Zn^{2+} (aq) + N_2 (g)

71. Balance the following as net ionic equations for reactions in acidic solution. H^+ or H_2O (but not OH^-) may be added, as necessary.

(a) P_4 (s) + NO_3^- (aq) →
$$H_3PO_4 \text{ (aq)} + NO \text{ (g)}$$

(b) H_2O_2 (aq) + MnO_4^- (aq) →
$$Mn^{2+} \text{ (aq)} + O_2 \text{ (g)}$$

(c) HgS (s) + Cl^- (aq) + NO_3^- (aq) →
$$HgCl_4^{2-} \text{ (aq)} + NO_2 \text{ (g)} + S \text{ (s)}$$

(d) HBrO (aq) → Br^- (aq) + O_2 (g)

(e) Cl_2 (g) → ClO_3^- (aq) + Cl^- (aq)

72. Balance the following as net ionic equations in basic solution. OH^- or H_2O (but not H^+) may be added as necessary.

(a) $MnO_4^- (aq) + NO_2^- (aq) \longrightarrow$
$$MnO_2 (s) + NO_3^- (aq)$$
(b) $Zn (s) + NO_3^- (aq) \longrightarrow$
$$NH_3 (aq) + Zn(OH)_4^{2-} (aq)$$
(c) $N_2H_4 (aq) + Cu(OH)_2 (s) \longrightarrow$
$$N_2 (g) + Cu (s)$$
(d) $Mn^{2+} (aq) + MnO_4^- (aq) \longrightarrow MnO_2 (s)$
(e) $Cl_2 (g) \longrightarrow ClO_3^- (aq) + Cl^- (aq)$

73. Balance the following as net ionic equations in basic solution. OH^- or H_2O (but not H^+) may be added as necessary.
    (a) $Mn(OH)_2 (s) + H_2O_2 (aq) \longrightarrow MnO_2 (s)$
    (b) $CN^- (aq) + MnO_4^- (aq) \longrightarrow$
    $$CNO^- (aq) + MnO_2 (s)$$
    (c) $As_2S_3 (s) + H_2O_2 (aq) \longrightarrow$
    $$AsO_4^{3-} (aq) + SO_4^{2-} (aq)$$
    (d) $CrI_3 (aq) + H_2O_2 (aq) \longrightarrow$
    $$CrO_4^{2-} (aq) + IO_4^- (aq)$$

74. Balance the following molecular equations.
    (a) $H_2SO_4 (aq) + C (s) \longrightarrow$
    $$CO_2 (g) + SO_2 (g) + H_2O (\ell)$$
    (b) $MnO_2 (s) + HCl (aq) \longrightarrow$
    $$Cl_2 (g) + MnCl_2 (aq) + H_2O (\ell)$$
    (c) $KMnO_4 (aq) + NaI (aq) + H_2SO_4 (aq) \longrightarrow$
    $$I_2 (s) + MnSO_4 (aq) + Na_2SO_4 (aq)$$
    $$+ K_2SO_4 (aq) + H_2O (\ell)$$
    (d) $K_2Cr_2O_7 (aq) + KBr (aq) + H_2SO_4 (aq) \longrightarrow$
    $$Br_2 (\ell) + K_2SO_4 (aq)$$
    $$+ Cr_2(SO_4)_3 (aq) + H_2O (\ell)$$

75. Balance the following molecular equations.
    (a) $HNO_3 (aq) (dil) + ZnS (s) \longrightarrow$
    $$S (s) + NO (g) + Zn(NO_3)_2 (aq) + H_2O (\ell)$$
    (b) $HNO_3 (aq) (conc) + Cu (s) \longrightarrow$
    $$Cu(NO_3)_2 (aq) + NO_2 (g) + H_2O (\ell)$$
    (c) $KMnO_4 (aq) + H_2O_2 (aq) + H_2SO_4 (aq) \longrightarrow$
    $$O_2 (g) + MnSO_4 (aq) + K_2SO_4 (aq) + H_2O (\ell)$$
    (d) $(NH_4)_2Cr_2O_7 (s) \xrightarrow{\Delta}$
    $$Cr_2O_3 (s) + H_2O (g) + N_2 (g)$$

Redox Titrations — Mole Method

76. What volume of $0.10\ M$ $KMnO_4$ would be required to oxidize 20 mL of $0.10\ M$ $FeSO_4$ in acidic solution? Refer to Exercise 70(a).

77. What volume of $0.10\ M$ $K_2Cr_2O_7$ would be required to oxidize 60 mL of $0.10\ M$ Na_2SO_3 in acidic solution? The products include Cr^{3+} and SO_4^{2-} ions.

78. What volume of $0.10\ M$ $KMnO_4$ would be required to oxidize 50 mL of $0.10\ M$ KI in acidic solution?

79. What volume of $0.10\ M$ $K_2Cr_2O_7$ would be required to oxidize 50 mL of $0.10\ M$ KI in acidic solution?

80. A 5.026 gram sample of an ore of iron was dissolved in an acid solution. The iron was converted to the $+2$ oxidation state. The resulting solution was titrated by 30.68 mL of 0.06402 molar $KMnO_4$. What are the mass and the percentage of iron in the ore?

81. A 0.683 gram sample of an ore of iron is dissolved in acid and converted to the ferrous form. The sample is oxidized by 38.50 mL of $0.161\ M$ ceric sulfate, $Ce(SO_4)_2$, solution; the ceric ion, Ce^{4+}, is reduced to Ce^{3+} ion. (a) Write a balanced equation for the reaction. (b) What is the percentage of iron in the ore?

82. Limonite is an ore of iron that contains $Fe_2O_3 \cdot 1\frac{1}{2}H_2O$ (or $2Fe_2O_3 \cdot 3H_2O$). A 0.5166 gram sample of limonite is dissolved in acid and treated so that all the iron is converted to ferrous ion, Fe^{2+}. This sample requires 42.96 mL of $0.02130\ M$ sodium dichromate solution, $Na_2Cr_2O_7$, for titration. Fe^{2+} is oxidized to Fe^{3+} and $Cr_2O_7^{2-}$ is reduced to Cr^{3+}. What is the percentage of iron in the limonite?

83. Given the following unbalanced equation:

$$KMnO_4 + Na_2(COO)_2 + H_2SO_4 \longrightarrow$$
$$K_2SO_4 + MnSO_4 + Na_2SO_4 + H_2O + CO_2$$

(a) Balance the molecular equation and write the balanced total ionic and net ionic equations for this reaction. All salts in this equation may be assumed to be ionic compounds.

(b) Calculate the molarity of a $KMnO_4$ solution if 20.0 mL of the $KMnO_4$ solution reacts with 0.268 g of solid $Na_2(COO)_2$.

(c) Calculate the mass of $Na_2(COO)_2$ required to prepare 12.0 liters of $0.600\ M$ $Na_2(COO)_2$ solution.

(d) Calculate the mass of solid $KMnO_4$ required to prepare 10.0 liters of $0.0400\ M$ $KMnO_4$ solution.

Redox Titrations — Equivalent Weight Method

84. Calculate the molarity and normality of a solution that contains 15.8 g of $KMnO_4$ in 500 mL of solution to be used in the reaction in Exercise 74(c).

85. Calculate the molarity and normality of a solu-

tion that contains 2.94 g of $K_2Cr_2O_7$ in 100 mL of solution to be used in the reaction in Exercise 74(d).

86. Calculate the molarity and normality of a solution that contains 16.2 g of $FeSO_4$ in 200 mL of solution to be used in the reaction in Exercise 70(a).

87. Calculate the molarity and normality of a solution that contains 12.6 g of Na_2SO_3 in 1.00 liter of solution to be used in the reaction in Exercise 66(c).

88. Arrange in a tabular form (a) formula weight, (b) equivalent weight, and (c) the number of grams per liter of a 0.1500 N solution for the following.
    $KMnO_4$ (in acid solution)
    $K_2Cr_2O_7$

$Na_2S_2O_3 \cdot 5H_2O$ (which will be converted to $S_4O_6^{2-}$ ion)
SO_2 (which will be converted to H_2SO_4)
HNO_2 (which will be converted to HNO_3)

89. What is the percentage of iron in an ore sample with a mass of 0.6835 gram if the sample, when dissolved in acid and converted to iron(II), requires 38.50 mL of 0.1607 N ceric sulfate, $Ce(SO_4)_2$, solution for complete titration? The Ce^{4+} is reduced to Ce^{3+} and Fe^{2+} is oxidized to Fe^{3+}.

90. A sample containing vanadium is to be analyzed by converting the vanadium to V_2O_5, which is dissolved in acid and reduced to VO^{2+}. The VO^{2+} is then oxidized by a standardized $KMnO_4$ solution to VO_2^+ ion. Calculate the equivalent weight of V_2O_5 for this reaction.

Gases and the Kinetic-Molecular Theory

11

Matter is generally regarded as existing in three physical states: solids, liquids, and gases. Many samples do not fit neatly into any of these categories, but this classification enables us to deal with large amounts of information systematically. In the solid state water is known as ice, in the liquid state it is called water, while in the gaseous state it is known as steam or water vapor. Most, but not all, substances can exist in three states. Most solids change to liquids and most liquids change to gases as they are heated.

11–1   Comparison of Solids, Liquids, and Gases

Liquids and gases are known as **fluids** because they flow freely. Solids and liquids are referred to as **condensed states** because they have much higher densities than gases. Table 11–1 displays the densities of a few common substances in different states.

TABLE 11–1 Comparison of Densities (g/mL) of Substances in Different States at Atmospheric Pressure

Substance	Solid	Liquid (20°C)	Gas (100°C)
water (H_2O)	0.917 (0°C)	0.998	0.000588
benzene (C_6H_6)	0.899 (0°C)	0.879	0.00255
carbon tetrachloride (CCl_4)	1.7 (-25°C)	1.59	0.00503

The density of ice is less than the density of liquid water, which is quite unusual. Most substances are more dense in the solid state.

As the data in Table 11–1 indicate, the densities of solids and liquids are many times greater than those of gases. The molecules must be very far apart in gases and much closer together in liquids and solids. For example, the volume of one mole of liquid water is slightly more than 18 milliliters, whereas one mole of steam occupies 30,600 milliliters at 100°C at atmospheric pressure. The possibilities for interaction among gaseous molecules would be minimal (because they are so far apart) were it not for their rapid motion. Because their molecules are so far apart, and because interactions among their molecules are relatively ineffective, gases fill completely any container in which they are placed and are easily compressible.

All substances that are gases at room temperature may be liquefied by cooling and compressing them. Volatile liquids, those of low boiling point, are easily converted to gases at room temperature or slightly above. The term **vapor** describes gases formed by boiling or evaporating liquids.

A comparison of some other properties of the three states of matter will be made in Chapter 12.

11–2 Composition of the Atmosphere and Some Properties of Gases

Many important chemical substances are gases. The atmosphere we breathe is a mixture of gases (Table 11–2). The major components are oxygen, O_2 (bp, -182.98°C), and nitrogen, N_2 (bp, -195.79°C). All gases are miscible; that is, they mix completely in all proportions *unless* they react with each other.

Several scientists, notably Torricelli (1643), Boyle (1660), Charles (1787), and

TABLE 11–2 Volume Percentage Composition of Dry Air

Gas		Percent by Volume
N_2, nitrogen		78.09
O_2, oxygen		20.94
Ar, argon		0.93
CO_2, carbon dioxide		0.03 (variable)
He, Ne, Kr, Xe (noble gases)		0.002
CH_4, methane		0.00015 (variable)
H_2, hydrogen		0.00005
all others*	less than	0.00004

* Atmospheric moisture (expressed as relative humidity) varies.

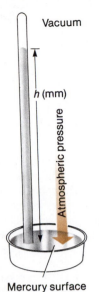

Vacuum

h (mm)

Atmospheric pressure

Mercury surface

FIGURE 11–1 The barometer. At the level of the lower mercury surface, the pressure both inside and outside the tube must be that of the atmosphere. Inside the tube, the pressure is exerted by the mercury column *h* mm high. Hence, the atmospheric pressure must equal *h* mm Hg.

For Evangelista Torricelli (1608–1647), who invented the mercury barometer.

Graham (1831), laid an experimental foundation upon which our present understanding of gases is based. For example, their investigations showed that

1. Gases can be compressed into smaller volumes; that is, their densities can be increased by applying increased pressure.
2. Gases exert pressure on their surroundings, and in turn pressure must be exerted to confine gases.
3. Gases expand without limit so that gas samples completely and uniformly occupy the volume of any container.
4. Gases diffuse into each other (mix) so that two samples of gas placed in the same container mix completely almost immediately. Conversely, different gases in a mixture like air do not separate on standing.
5. The properties of gases are described fully in terms of temperature, pressure, the volume they occupy, and the amount of gas present (i.e., the mass or number of moles). For example, a cold sample of gas occupies a much smaller volume than it does when heated to a higher temperature at the same pressure, even though the amount of gas present does not change.

11–3 Pressure

Pressure is defined as force per unit area (lb/in², commonly known as *psi*, for example). Pressure may be expressed in many different units, as we shall see. The mercury **barometer** is a simple device for measuring atmospheric pressures. A glass tube (about 800 mm long) is sealed at one end, filled with mercury, and then carefully inverted into a dish of mercury without allowing air to enter. The mercury in the tube falls to the level at which the pressure of the air on the surface of the mercury in the dish supports the column of mercury in the tube. The air pressure is measured in terms of the height of the mercury column, i.e., the vertical distance between the surface of the mercury in the open dish and that inside the closed tube. The pressure exerted by the atmosphere is equal to the pressure exerted by the column of mercury. Figure 11–1 illustrates the "heart" of a mercury barometer. Since mercury barometers are simple and well-known, gas pressures are frequently expressed in terms of millimeters of mercury (mm Hg). In recent years the unit **torr** has been used to indicate mm Hg pressure, i.e., 1 mm Hg = 1 torr.

A mercury **manometer** consists of a glass tube partially filled with mercury. One arm is open to the atmosphere and the other is connected to a container of gas. The pressure exerted by the gas in the container is equal to the atmospheric pressure plus or minus the difference in mercury levels, Δh, as shown in Figure 11–2.

Atmospheric pressure varies with atmospheric conditions and distance above sea level. It is greater at low elevations because the air at low levels is compressed by the air above it. For example, one half of the quantity of matter in the atmosphere is below 20,000 feet above sea level, and therefore atmospheric pressure is only one half as great at 20,000 feet as it is at sea level.

At sea level, at a latitude of 45° north (or south), the average atmospheric pressure supports a column of mercury 760 mm high in a simple mercury barometer when the mercury is at 0°C. This average sea level pressure of 760 torr is called **one atmosphere** of pressure:

1 atmosphere (atm) = 760 mm Hg = 760 torr = 1 bar

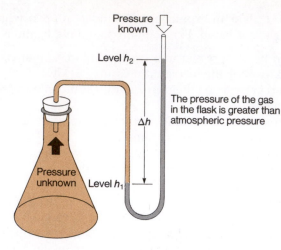

Pressure known

Level h_2

The pressure of the gas in the flask is greater than atmospheric pressure

Δh

Pressure unknown   Level h_1

FIGURE 11–2 The two-arm mercury barometer is called a manometer. At the level h_1, the total pressure on the mercury in the left arm must equal the total pressure on the mercury in the right arm.

The SI unit of pressure is the pascal (Pa). A **pascal** is defined as the pressure exerted by a force of one newton acting on an area of one square meter. By definition, one newton (N) is the force required to give a mass of one kilogram an acceleration of one meter per second per second. Symbolically, we represent one newton and one pascal as

Acceleration is the change in velocity (m/s) per unit time (s), m/s².

$$1 \text{ N} = 1 \text{ kg} \cdot \text{m/s}^2 \qquad 1 \text{ Pa} = 1 \text{ N/m}^2 = 1 \text{ kg/(m} \cdot \text{s}^2)$$

One atmosphere of pressure is 1.013×10^5 pascals or 101.3 kilopascals:

$$1 \text{ atm} = 1.013 \times 10^5 \text{ Pa} = 101.3 \text{ kPa}$$

11–4 Boyle's Law: The Relation of Volume to Pressure at Constant Temperature

In 1662 Robert Boyle summarized the results of experiments on various samples of gases. He found that *at constant temperature, the volume occupied by a definite mass of a gas is inversely proportional to the applied pressure.* In a typical experiment (Figure 11–3) a sample of a gas was trapped in a U-tube and allowed to come to constant temperature; then its volume and the difference in heights of the two mercury columns were recorded. This difference in height plus the pressure of the atmosphere represents the pressure on the gas. The results of several such experiments on a single sample of gas are given in Table 11–3.

TABLE 11–3

Pressure	Volume	Pressure × Volume*	(1/P)
5.0	40.0	200	0.20
10.0	20.0	200	0.10
15.0	13.3	200	0.067
17.0	11.8	201	0.059
20.0	10.0	200	0.050
22.0	9.10	200	0.045
30.0	6.70	201	0.033

* Both pressure and volume are in arbitrary units in this example.

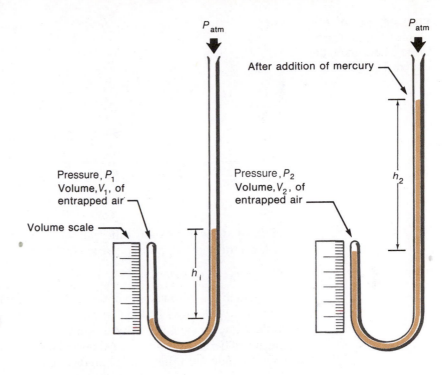

FIGURE 11-3 An artist's representation of Boyle's experiment. A sample of air is entrapped in a tube, such that the pressure on the air can be changed and the volume of the air measured. P_{atm} is the atmospheric pressure, measured with a barometer. $P_1 = h_1 + P_{atm}$, $P_2 = h_2 + P_{atm}$.

When the pressure of a sample of gas is plotted against its volume, a curve called a hyperbola* results (Figure 11-4a). This doesn't tell us much, since there are many possible mathematical descriptions for curves that look similar to this one. However, if we plot the *reciprocal* of the pressure (that is, $1/P$) versus the volume, the result is a straight line (Figure 11-4b). Each straight line has a unique equation, and it can be shown that the equation for this one is

n is the number of moles of gas in the sample, and *T* is its temperature. The units of *k* are determined by those chosen for volume, *V*, and pressure, *P*.

$$V = k\left(\frac{1}{P}\right) \quad \text{(constant } n, T) \qquad \text{[Eq. 11-1]}$$

where k is a constant number that is the slope of the line. A relationship of this kind can also be stated as "volume varies inversely with pressure" or "volume is inversely proportional to pressure," and symbolically as

$$V \propto \frac{1}{P} \quad \text{(constant } n, T)$$

where the symbol $\propto$ means "is proportional to." Rearranging Eq. 11-1 gives

Note that *k* is a constant for a given sample of gas at a fixed temperature, but varies with the mass and temperature of the sample.

$$PV = k \quad \text{(constant } n, T)$$

which is a mathematical expression of **Boyle's Law:** *At a given temperature, the product of the pressure and volume of a definite mass of gas is a constant.*

The significance of the straight-line relation can be seen from the following argument. If we denote the pressure and volume in any row of Table 11-3 by P_1 and V_1, then Eq. 11-1 says that

$$P_1 V_1 = k_1 \quad \text{(constant } n, T)$$

* If the pressure-volume relation is described exactly by Eq. 11-1, then the curve in Figure 11-4a is one branch of a hyperbola. Since neither pressure nor volume can be negative, the other branch of the hyperbola has no physical significance.

FIGURE 11–4 A graphical illustration of Boyle's Law, using the data of Table 11–3.

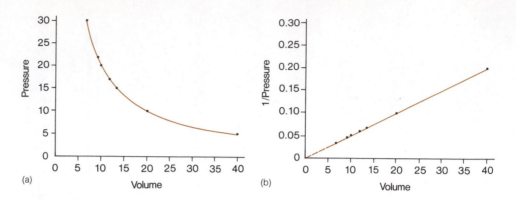

Similarly, we can pick any other pair of pressure and volume values from the table and, calling them P_2 and V_2, write Eq. 11–1 as

$$P_2V_2 = k_2 \qquad \text{(constant } n, T)$$

The data in the third column of the table show that (within roundoff error) $k_1 = k_2 = 200$ for *any* two rows (P-V pairs). For a relationship that is accurately described by a straight line (or its mathematical equation), this is true in general, even for points on the graph between the measured ones. This leads to the most useful form of Boyle's Law, the one with the greatest predictive power:

$$P_1V_1 = P_2V_2 \qquad \text{(for a given mass of a gas at constant temperature)} \qquad \text{[Eq. 11–2]}$$

This says that we can measure the pressure and volume of a gas sample *once*; then, under the assumption that the mass and temperature of the gas remain the same, we can *predict* the value of the volume at *any* pressure, or vice versa.

At normal temperatures and pressures, most gases obey Boyle's Law rather well (deviations from it will be discussed in Section 11–15). We call this *ideal behavior*. Examples 11–1 and 11–2 show how useful Boyle's Law is.

Example 11–1

A sample of gas occupies 10 liters under a pressure of 1 atmosphere. What will its volume be if the pressure is increased to 2 atmospheres? Assume that the temperature of the gas sample does not change.

Solution

Boyle's Law tells us that $P_1V_1 = P_2V_2$. Solving for V_2 gives

$$V_2 = \frac{P_1V_1}{P_2}$$

and substitution yields

$$V_2 = \frac{(1 \text{ atm})(10 \text{ L})}{(2 \text{ atm})} = \underline{5 \text{ L}}$$

Pressure and volume are inversely proportional, so doubling the pressure halves the volume of a sample of gas at constant temperature.

Another approach to the problem is to multiply the original volume by a "Boyle's Law factor." The pressure increases from 1 atmosphere to 2 atmospheres, and therefore the volume *decreases* by the factor (1 atm/2 atm). The solution then becomes

$$? \text{ L} = 10 \text{ L} \times \text{(Boyle's Law factor that represents a decrease in volume)}$$

$$= 10 \text{ L} \times \left(\frac{1 \text{ atm}}{2 \text{ atm}}\right) = \underline{5 \text{ L}}$$

Note that this "Boyle's Law factor" is *not* a unit factor (it cannot be evaluated as 1).

Example 11–2
A sample of oxygen occupies 10.0 liters under a pressure of 790 torr (105 kPa). At what pressure will it occupy 13.4 liters if the temperature does not change?

Solution
We know that $P_1V_1 = P_2V_2$, and solving for P_2 gives

$$P_2 = \frac{P_1V_1}{V_2} = \frac{(790 \text{ torr})(10.0 \text{ L})}{13.4 \text{ L}}$$

$$= \underline{590 \text{ torr}} \quad (78.6 \text{ kPa})$$

The statement of the problem tells us that volume *increases* from 10.0 liters to 13.4 liters (at constant temperature), and therefore the pressure must *decrease*. We can also solve the problem by multiplying the original pressure by a Boyle's Law factor less than unity, i.e., 10.0 L/13.4 L.

$$\underline{?} \text{ torr} = 790 \text{ torr} \times \frac{10.0 \text{ L}}{13.4 \text{ L}}$$

$$= \underline{590 \text{ torr}} \quad (78.6 \text{ kPa})$$

11–5   The Kelvin (Absolute) Temperature Scale

The absolute temperature scale was discussed in Section 1–11, and the fact that this temperature scale is based on the observed behavior of gases was emphasized. Let us consider the origin and utility of the scale in more detail.

In his pressure-volume studies on gases, Robert Boyle noticed that heating a sample of gas caused some volume change, but he failed to follow up on this observation. Shortly before 1800, two French scientists, Jacques Charles and Joseph Gay-Lussac, pioneer balloonists at the time, began studying the expansion of gases with increasing temperature. Their studies led them to conclude that the rate of expansion with increased (Celsius scale) temperature was constant, and the same for all the gases they studied as long as pressure remained constant. The implications of their discovery were not fully realized until nearly a century later, when scientists recognized that this behavior of gases could become the basis of a new temperature scale, the absolute temperature scale.

Experiments have shown that when a 273 mL sample of gas at 0°C is heated to 1°C, its volume increases by 1 mL to 274 mL. At 10°C the volume increases (by 10 mL) to 283 mL if the pressure remains constant in both cases. If a 273 mL sample of gas at 0°C is cooled to −1°C, its volume decreases to 272 mL, while at −10°C the volume decreases to 263 mL if the pressure remains constant. For each °C *increase* in temperature, the volume of a sample of gas *increases* by the fraction 1/273 of its volume at 0°C. For each °C *decrease* in temperature, the volume of a sample of gas *decreases* by the fraction 1/273 of its volume at 0°C. Thus, if a 273 mL sample of gas at 0°C is heated to 273°C at constant pressure, its volume doubles to 546 mL. On the other hand, if the temperature of a 273 mL sample of gas at 0°C should be lowered from 0°C to −273°C, the gas "should" have no volume at all because the volume should decrease at the rate of 1/273 of its volume at 0°C for each degree by which temperature is decreased, and so become zero. However, all gases liquefy, and solidify, before the temperature reaches −273°C. This temperature, −273°C (to be more exact, −273.15°C), is *absolute zero* and is taken as the zero point on the Kelvin temperature scale. As pointed out in Section 1–11, the relationship between the Celsius and Kelvin temperature scales is

Recall that degrees Kelvin is called simply kelvins and is represented by K, not °K.

$$K = °C + 273°$$

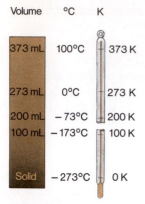

Volume	°C	K
373 mL	100°C	373 K
273 mL	0°C	273 K
200 mL	−73°C	200 K
100 mL	−173°C	100 K
Solid	−273°C	0 K

FIGURE 11–5
Volume-temperature relationship for a sample of gas at constant pressure.

Absolute zero may be thought of as the limit of thermal contraction for an ideal gas.

Figure 11–5 shows the relationship between temperature and volume for a 273 mL sample of gas originally at 0°C (273 K) at constant pressure.

11–6 Charles' Law: The Relation of Volume to Temperature at Constant Pressure

A summary of observations on the effect of temperature changes on the volumes of samples of gases at constant pressure is known as **Charles' Law:** *At constant pressure, the volume occupied by a definite mass of a gas is directly proportional to its absolute temperature.* The temperature-volume relationship, at constant pressure, is illustrated in Figures 11–6a and 11–6b.

The solid line portions of Figure 11–6b illustrate the behavior observed by Charles and Gay-Lussac. The lines represent the same sample under different pressures. Lord Kelvin, a British physicist, noticed that an extension of the different temperature-volume lines back to zero volume (dashed line) yields a common intercept. This intercept is −273.15°C on the temperature axis, and Kelvin named this temperature **absolute zero.** Since the degrees are the same size over the entire scale, 0°C becomes 273.15 degrees above absolute zero. In honor of Lord Kelvin's work, this temperature scale is called the Kelvin temperature scale.

Charles' Law, since it involves a *direct* proportionality between the volume, V, and the temperature, T, of a definite mass of gas at constant pressure, can be written

$$V = kT \qquad \text{(constant } n, P\text{)}$$

[Eq. 11–3]

Rearranging the expression gives $V/T = k$. If the volume and temperature of a sample of gas are measured, a numerical value can be calculated for k (for that sample at that pressure). Call it k_1.

$$\frac{V_1}{T_1} = k_1 \qquad (P \text{ is constant})$$

Changing the temperature causes the volume to change, and new values are obtained for V and T:

$$\frac{V_2}{T_2} = k_2 \qquad (P \text{ is constant})$$

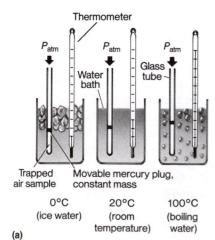

Thermometer

P_{atm} P_{atm} P_{atm}

Water bath

Glass tube

Trapped air sample   Movable mercury plug, constant mass

0°C (ice water)   20°C (room temperature)   100°C (boiling water)

(a)

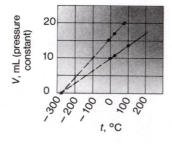

V, mL (pressure constant)

20

10

0

−300 −200 −100 0 100 200

t, °C

(b)

FIGURE 11–6 (a) An illustration showing that the volume of a gas increases as the temperature is increased at constant pressure. A mercury plug of constant mass plus atmospheric pressure (P_{atm}) maintains a constant pressure on the trapped air. (b) A plot of volume versus temperature shows that gases expand when heated at constant pressure. The two lines represent the same mass of the same gas at different pressures.

In Figure 11–5, for example, the initial volume was chosen so that the V/T ratio was unity, so $k_1 = k_2$.

Experiments show that for a given sample of gas at constant pressure, $k_1 = k_2$. See Figure 11–6. Since things equal to the same thing are equal to each other, we may write

$$\frac{V_1}{T_1} = \frac{V_2}{T_2} \qquad (n \text{ and } P \text{ are constant}) \qquad \text{[Eq. 11–4]}$$

which is the most useful form of Charles' Law.

The following examples illustrate the use of Charles' Law. Always remember that this relationship is valid *only* when temperature, T, is expressed on an absolute (usually the Kelvin) scale.

Example 11–3

A 250 mL sample of gas is confined under one atmosphere pressure at 25°C. What volume will the gas occupy at 50°C if the pressure remains constant?

Solution

First we write all of the known and unknown variables:

$$V_1 = 250 \text{ mL} \qquad V_2 = \text{?}$$
$$\begin{aligned} T_1 &= 25°C + 273° & T_2 &= 50°C + 273° \\ &= 298 \text{ K} & &= 323 \text{ K} \end{aligned}$$

Charles' Law (Eq. 11–4) is solved for the unknown quantity, V_2:

$$\frac{V_1}{T_1} = \frac{V_2}{T_2} \qquad \text{so} \qquad V_2 = \frac{T_2 V_1}{T_1}$$

$$V_2 = \frac{(323 \text{ K})(250 \text{ mL})}{(298 \text{ K})} = \underline{271 \text{ mL}}$$

Note that the temperature doubles on the Celsius scale (25°C to 50°C), but on the Kelvin scale the fractional absolute temperature increase is much smaller (298 K to 323 K). Another approach to the problem is to multiply the original volume by a Charles' Law factor. The temperature *increases* from 298 K to 323 K, which causes the volume to *increase* by the factor 323 K/298 K,

$$\text{? mL} = 250 \text{ mL} \times \frac{323 \text{ K}}{298 \text{ K}} = \underline{271 \text{ mL}}$$

Example 11–4

A sample of nitrogen occupies 400 milliliters at 100°C. At what temperature will it occupy 200 milliliters if the pressure does not change?

Solution

$$\begin{aligned} V_1 &= 400 \text{ mL} & V_2 &= 200 \text{ mL} \\ T_1 &= 100°C + 273° = 373 \text{ K} & T_2 &= \text{?} \end{aligned}$$

$$\frac{V_1}{T_1} = \frac{V_2}{T_2} \qquad \text{so} \qquad T_2 = \frac{V_2 T_1}{V_1}$$

$$T_2 = \frac{(200 \text{ mL})(373 \text{ K})}{400 \text{ mL}} = 186 \text{ K}$$

$$°C = 186 \text{ K} - 273° = \underline{-87°C}$$

Recall that according to Boyle's Law, if the temperature had remained constant, the volume decrease would have caused an increase in pressure. We see that the temperature must be reduced to $-87°C$ to reduce the pressure to its original value. As in Example 11–3, this problem can also be solved using a Charles' Law factor. A decrease in volume causes the temperature to decrease at constant pressure, so the initial temperature must be multiplied by the volume factor less than one, 200 mL/400 mL.

$$\text{? K} = 373 \text{ K} \times \frac{200 \text{ mL}}{400 \text{ mL}} = 186 \text{ K}, \text{ which is } \underline{-87°C}$$

11–7   Standard Temperature and Pressure

We have demonstrated that both temperature and pressure affect the volumes (and therefore densities) of gases. It is convenient to choose some "standard" temperature and pressure as a reference point in discussing gases. **Standard conditions of temperature and pressure** (STP or SC) are, by international agreement, 0°C (273 K) and one atmosphere of pressure (760 torr).

11–8   The Combined Gas Laws

Boyle's Law relates the pressure and volume of a sample of gas at constant temperature, while Charles' Law relates the volume and temperature at constant pressure, i.e., $P_1V_1 = P_2V_2$ and $V_1/T_1 = V_2/T_2$. Combination of the essence of Boyle's Law and Charles' Law into a single expression gives the **combined gas law equation.**

$$\frac{P_1V_1}{T_1} = \frac{P_2V_2}{T_2} \qquad \text{(constant } n\text{)}$$

[Eq. 11–5]

The following examples illustrate applications of the combined gas laws.

Example 11–5

A sample of neon occupies 100 liters at 27°C under a pressure of 1000 torr. What volume would it occupy at standard conditions?

Solution

$V_1 = 100$ L $V_2 = ?$
$P_1 = 1000$ torr $P_2 = 760$ torr
$T_1 = 27°C + 273° = 300$ K $T_2 = 273$ K

$$\frac{P_1V_1}{T_1} = \frac{P_2V_2}{T_2}$$

Solving the combined gas law expression for V_2 gives

$$V_2 = \frac{P_1V_1T_2}{P_2T_1} = \frac{(1000 \text{ torr})(100 \text{ L})(273 \text{ K})}{(760 \text{ torr})(300 \text{ K})}$$
$$= 120 \text{ L}$$

Alternatively, we can multiply the original volume by a Boyle's Law factor *and* a Charles' Law factor. The pressure decreases from 1000 torr to 760 torr (volume will increase) so the Boyle's Law factor is 1000 torr/760 torr. The temperature decreases from 300 K to 273 K (volume will decrease) so the Charles' Law factor is 273 K/300 K. Multiplication of the original volume by these factors gives the same result obtained above.

$$? \text{L} = 100 \text{ L} \times \frac{1000 \text{ torr}}{760 \text{ torr}} \times \frac{273 \text{ K}}{300 \text{ K}} = 120 \text{ L}$$

Note that the pressure decrease (from 1000 torr to 760 torr) alone would result in a significant *increase* in the volume of neon. The temperature decrease (from 300 K to 273 K) alone would give only a small *decrease* in the volume of neon. The net result of the two changes is that the volume increases from 100 liters to 120 liters.

Example 11-6

A sample of gas occupies 10.0 liters at 240°C under a pressure of 80.0 kPa. At what temperature will the gas occupy 20.0 liters if the pressure is increased to 107 kPa (a little more than 1 atm)?

Solution

$V_1 = 10.0$ L

$P_1 = 80.0$ kPa

$T_1 = 240°C + 273° = 513$ K

$V_2 = 20.0$ L

$P_2 = 107$ kPa

$T_2 = ?$

Solving the combined gas law expression for T_2 gives

$$\frac{P_1 V_1}{T_1} = \frac{P_2 V_2}{T_2} \quad \text{so} \quad T_2 = \frac{P_2 V_2 T_1}{P_1 V_1}$$

$$T_2 = \frac{(107 \text{ kPa})(20.0 \text{ L})(513 \text{ K})}{(80.0 \text{ kPa})(10.0 \text{ L})}$$

$$= \underline{1.37 \times 10^3 \text{ K}}$$

Since K = °C + 273°,

$$°C = 1.37 \times 10^3 \text{ K} - 273°$$

$$= \underline{1.10 \times 10^3 \text{ °C}}$$

The combined gas law equation is useful when both the temperature and pressure of a sample of gas are changed. Or, when any five of the variables in the equation are known, the sixth variable can be calculated.

11-9 Gas Densities and the Standard Molar Volume

In 1811, Amedeo Avogadro postulated that *at the same temperature and pressure, equal volumes of all gases contain the same number of molecules.* Numerous experiments have demonstrated that Avogadro's hypothesis was correct to within about ±2%, and the statement is now known as **Avogadro's Law.** Thus, the number of moles of molecules of the gas in a sample of a given volume, at a given temperature and pressure, does not depend on the identity of the gas.

Gas densities depend on pressure and temperature. Recall that density is defined as mass per unit volume, or density = mass/volume. Since pressure changes affect volumes of gases according to Boyle's Law, while temperature changes affect volumes of gases according to Charles' Law, gas densities measured at various temperatures and pressures can be used to calculate the value of the gas's density at *standard temperature and pressure* by application of Boyle's and Charles' Laws. Table 11-4 gives the densities of several gases at standard conditions. The volume occupied by a mole of gas at STP (found by dividing the molecular weight by the

TABLE 11-4 Densities and Standard Molar Volumes for Some Common Gases

Gas	Formula	Mol. Wt. (g/mol)	Density at STP (g/L)	Standard Molar Volume (L/mol)
hydrogen	H_2	2.02	0.090	22.428
helium	He	4.003	0.178	22.426
neon	Ne	20.18	0.900	22.425
nitrogen	N_2	28.01	1.250	22.404
oxygen	O_2	32.00	1.429	22.394
argon	Ar	39.95	1.784	22.393
carbon dioxide	CO_2	44.01	1.977	22.256
ammonia	NH_3	17.03	0.771	22.094
chlorine	Cl_2	70.91	3.214	22.063

density at STP) is very nearly the same for all gases and is referred to as the **standard molar volume.** For most gases the standard molar volume can be taken to be 22.4 liters.

    The following examples illustrate the usefulness of the standard molar volume idea.

Example 11–7

If 1.00 mole of a gas occupies 22.4 liters at STP, what volume will 2.65 moles of gas occupy at STP?

Solution

Since the standard molar volume is about 22.4 L for all gases (± about 2%), we can use the unit factor 22.4 L_{STP}/mol of gas.

$$? \text{ L}_{STP} = 2.65 \text{ mol gas} \times \frac{22.4 \text{ L}_{STP}}{1 \text{ mol gas}}$$

$$= \underline{59.4 \text{ L at STP}}$$

Example 11–8

A mole of a given gas occupies 27.0 liters and its density is 1.41 g/L at a particular temperature and pressure. What is its molecular weight? What is the density of the gas at STP?

Solution

We multiply the density by the unit factor constructed from the equality 27.0 L = 1 mol, in order to generate the appropriate units, g/mol.

$$? \frac{g}{mol} = \frac{1.41 \text{ g}}{L} \times \frac{27.0 \text{ L}}{mol} = 38.1 \text{ g/mol}$$

Hence, the molecular weight is $\underline{38.1 \text{ amu}}$. At STP, one mole of the gas (38.1 g) occupies 22.4 L, so its density is

$$\text{Density} = \frac{38.1 \text{ g}}{22.4 \text{ L}} = \underline{1.70 \text{ g/L at STP}}$$

11–10   A Summary of the Gas Laws — The Ideal Gas Equation

Let us summarize what we have learned about gases. A sample of gas can be described in terms of its pressure, temperature (Kelvin), volume, and the number of moles, n, present. Any three of these variables determine the fourth. Most gases behave quite similarly, so that the behavior of an ideal gas is illustrative of gases in general. An **ideal gas** is one that obeys the gas laws exactly. Many real gases show slight deviations from ideality, but at normal temperatures and pressures the deviations are small enough to be ignored in most cases. We'll do so for the present, and discuss deviations later.

    Summarizing the behavior of gases in a general way, we have

The symbol $\propto$ means "proportional to."

Boyle's Law	$V \propto \dfrac{1}{P}$	(at constant T and n)
Charles' Law	$V \propto T$	(at constant P and n)
Avogadro's Law	$V \propto n$	(at constant T and P)

and therefore, $V \propto nT/P$ or, rearranging, $PV \propto nT$.

The proportionality, $PV \propto nT$, can be converted into an equality by introducing a proportionality constant (for which it is traditional to use R):

$$PV = nRT$$

[Eq. 11–6]

This relationship is called the **ideal gas equation** or the **ideal gas law.** The numerical value of R, the proportionality constant, depends upon the choices of the units for P, V, and T. Recall that 1.00 mole of an ideal gas occupies 22.4 liters at 1.00 atmosphere and 273 K (STP). Solving the ideal gas law for R gives

$$R = \frac{PV}{nT} = \frac{(1.00 \text{ atm})(22.4 \text{ L})}{(1.00 \text{ mol})(273 \text{ K})} = 0.0821 \frac{\text{L} \cdot \text{atm}}{\text{mol} \cdot \text{K}}$$

R is called the **universal gas constant,** and its more exact value is 0.08206 L·atm/mol·K. It is a constant of nature and may be expressed in a variety of units.

Example 11–9

R can have any *energy* units per mole per kelvin. Calculate R in terms of joules per mole per kelvin and in SI units of kPa·dm³/mol·K.

Solution

Appendix C shows that 1 L·atm = 101.32 joules.

$$R = \frac{0.08206 \text{ L} \cdot \text{atm}}{\text{mol} \cdot \text{K}} \times \frac{101.32 \text{ J}}{1 \text{ L} \cdot \text{atm}}$$

$$= 8.314 \text{ J/mol} \cdot \text{K}$$

Let us now evaluate R in SI units. One atmosphere pressure is 101.3 kilopascals and the molar volume at STP is 22.4 dm³. (Recall that 1 dm³ = 1 L.)

$$R = \frac{PV}{nT} = \frac{101.3 \text{ kPa} \times 22.4 \text{ dm}^3}{1 \text{ mol} \times 273 \text{ K}}$$

$$= 8.31 \frac{\text{kPa} \cdot \text{dm}^3}{\text{mol} \cdot \text{K}}$$

We have now evaluated R, the universal gas constant, in three different sets of units, so we can write

$$R = 0.0821 \frac{\text{L} \cdot \text{atm}}{\text{mol} \cdot \text{K}} = 8.314 \frac{\text{J}}{\text{mol} \cdot \text{K}}$$

$$= 8.31 \frac{\text{kPa} \cdot \text{dm}^3}{\text{mol} \cdot \text{K}}$$

Often, Boyle's, Charles', and Avogadro's Laws are sufficient to calculate a change for a sample of gas. The usefulness of the ideal gas equation lies in the fact that it relates the four variables , P, V, n, and T, that describe a sample of gas. If three variables are known, the fourth can be calculated, as the following examples show.

Example 11–10

What is the volume of a gas balloon filled with 4.00 moles of helium when the atmospheric pressure is 748 torr and the temperature is 30°C?

Solution

We always look at the variables first and arrange them in some order with the proper units.

$$? \text{ atm} = 748 \text{ torr} \times \frac{1 \text{ atm}}{760 \text{ torr}} = 0.984 \text{ atm } (P)$$

$$? \text{ mol} = 4.00 \text{ mol } (n)$$

$$? \text{ K} = 30°\text{C} + 273° = 303 \text{ K } (T)$$

Solving the ideal gas equation for V and substituting these values gives

$$PV = nRT \qquad \text{so} \qquad V = \frac{nRT}{P}$$

$$V = \frac{(4.00 \text{ mol}) \left(0.0821 \dfrac{\text{L} \cdot \text{atm}}{\text{mol} \cdot \text{K}} \right) (303 \text{ K})}{0.984 \text{ atm}}$$

$$= 101 \text{ L}$$

You may wonder whether pressures are given in torr or mm Hg and temperatures in °C just to confuse the issue. This is not the case. Pressures are often measured with Torricellian (mercury) barometers, while temperatures are measured with Celsius thermometers.

Example 11–11

How many moles of hydrogen are present in a 5.00 liter container at a pressure of 400 torr at 26°C?

Solution

$P = 400 \text{ torr} \times \dfrac{1 \text{ atm}}{760 \text{ torr}} = 0.526 \text{ atm}$

$V = 5.00 \text{ L}$

$T = 26°C + 273° = 299 \text{ K}$

Solving $PV = nRT$ for n and substituting gives

$$n = \frac{PV}{RT} = \frac{(0.526 \text{ atm})(5.00 \text{ L})}{\left(0.0821 \dfrac{\text{L} \cdot \text{atm}}{\text{mol} \cdot \text{K}}\right)(299 \text{ K})}$$

$$= \underline{0.107 \text{ mol}}$$

Example 11–12

What pressure in kilopascals is exerted by 54.0 grams of xenon, Xe, in a 1.00 liter flask at 20°C?

Solution

$V = 1.00 \text{ L} = 1.00 \text{ dm}^3$

$T = 20°C + 273° = 293 \text{ K}$

$n = 54.0 \text{ g Xe} \times \dfrac{1 \text{ mol}}{131.3 \text{ g Xe}} = 0.411 \text{ mol}$

Solving $PV = nRT$ for P and substituting gives

$$P = \frac{nRT}{V}$$

$$= \frac{(0.411 \text{ mol})\left(\dfrac{8.31 \text{ kPa} \cdot \text{dm}^3}{\text{mol} \cdot \text{K}}\right)(293 \text{ K})}{1.00 \text{ dm}^3}$$

$$= \underline{1.00 \times 10^3 \text{ kPa}} \quad (9.87 \text{ atm})$$

11–11   Determination of Molecular Weights and Molecular Formulas of Gaseous Compounds

In Section 2–5, we distinguished between the simplest formula and the molecular formula of a covalent compound, and illustrated how simplest formulas can be calculated from the percent compositions of compounds. The formula (molecular) weight must be known in order to determine the molecular formula for a compound. For compounds that are gases at reasonable temperatures and pressures, the ideal gas law provides a basis for calculating molecular weights, as Example 11–13 illustrates.

Example 11–13

A 0.109 gram sample of gas occupies 112 milliliters at 100°C and 750 torr. What is the molecular weight of the compound?

Solution

We can use the ideal gas law, $PV = nRT$, to calculate the number of moles of gas:

$P = 750 \text{ torr} \times \dfrac{1 \text{ atm}}{760 \text{ torr}} = 0.987 \text{ atm}$

$V = 0.112 \text{ L}$

$T = 100°C + 273° = 373 \text{ K}$

$$n = \frac{PV}{RT} = \frac{(0.987 \text{ atm})(0.112 \text{ L})}{\left(0.0821 \dfrac{\text{L} \cdot \text{atm}}{\text{mol} \cdot \text{K}}\right)(373 \text{ K})}$$

$$= 0.00361 \text{ mol}$$

We now know that 0.109 g of the gas is 0.00361 mole, so we can calculate the mass of one mole, which is the molecular weight of the gas.

$$? \frac{\text{g}}{\text{mol}} = \frac{0.109 \text{ g}}{0.00361 \text{ mol}} = \underline{30.2 \text{ g/mol}}$$

The molecular weight of the gas is 30.2 amu.

Example 11–14

Analysis of a sample of a gaseous compound shows that it contains 85.7% carbon and 14.3% hydrogen by mass. At standard conditions, 100 milliliters of the compound weighs 0.188 gram. What is the molecular formula for the compound?

Solution

Let us determine the empirical formula for the compound as we did in Section 2–5.

$$? \text{ mol C atoms} = 85.7 \text{ g C} \times \frac{1 \text{ mol C atoms}}{12.0 \text{ g C}}$$

$$= 7.14 \text{ mol C atoms}$$

$$? \text{ mol H atoms} = 14.3 \text{ g H} \times \frac{1 \text{ mol H atoms}}{1.01 \text{ g H}}$$

$$= 14.2 \text{ mol H atoms}$$

The ratio of moles of carbon atoms to moles of hydrogen atoms is

$$\frac{7.14}{7.14} = 1 \text{ C}$$

simplest formula = CH_2 and formula weight = 14 amu

$$\frac{14.2}{7.14} = 2 \text{ H}$$

Thus, the molecular formula of the compound is some multiple of CH_2. Let us calculate the number of moles of molecules in the sample of gas by substituting into the ideal gas law, $PV = nRT$.

$$V = 0.100 \text{ L} \qquad P = 1.00 \text{ atm}$$

$$T = 0°C = 273 \text{ K}$$

$$n = \frac{PV}{RT} = \frac{(1.00 \text{ atm})(0.100 \text{ L})}{\left(0.0821 \dfrac{\text{L} \cdot \text{atm}}{\text{mol} \cdot \text{K}}\right)(273 \text{ K})}$$

$$= 0.00446 \text{ mol of gas}$$

This result tells us that 0.00446 mole has a mass of 0.188 g, which enables us to calculate the molecular weight.

$$? \frac{\text{g}}{\text{mol}} = \frac{0.188 \text{ g}}{0.00446 \text{ mol}} = 42.2 \text{ g/mol}$$

We now know that one mole of the compound has a mass of 42.2 grams, and we can determine its molecular formula. We divide the molecular weight of the compound (which corresponds to its molecular formula) by the simplest formula weight to obtain the nearest integer.

$$\frac{\text{molecular weight}}{\text{simplest formula weight}} = \frac{42.2 \text{ amu}}{14 \text{ amu}} = 3$$

This tells us that the molecular formula is three times the simplest formula. Therefore, the molecular formula is $(CH_2)_3 = C_3H_6$, and the gas is either propene or cyclopropane, both of which have the formula C_3H_6.

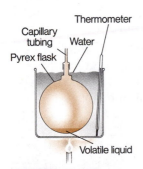

Thermometer
Capillary tubing   Water
Pyrex flask

Volatile liquid

FIGURE 11–7
Determination of molecular weight by the Dumas method. A volatile liquid is vaporized in the boiling water bath, and excess vapor is driven from the flask. The remaining vapor, at known *P*, *T*, and *V*, is condensed and weighed.

The kind of calculation illustrated in Example 11–14 can be extended to volatile liquids (those with low boiling points). If the boiling point of a liquid compound is below 100°C, the molecular weight of the compound may be determined by the **Dumas method.** In the Dumas experiment, a sample of volatile liquid is placed in a previously weighed flask, the flask is placed in a boiling water bath and the liquid is allowed to evaporate, and the excess vapor escapes into the air. As the last bit of liquid evaporates, the container is filled completely with vapor of the volatile liquid at 100°C. At this point the container is removed from the boiling water bath, cooled quickly, and weighed again. Rapid cooling condenses the vapor that filled the flask at 100°C, and the flask is filled with air again. Recall that the volume of a liquid is extremely small in comparison to the volume occupied by the same mass of the gaseous compound at atmospheric pressure. The difference between the mass of the flask containing the condensed liquid and the empty flask is the mass of the condensed liquid and, therefore, the mass of the vapor that filled the flask at 100°C and atmospheric pressure (Figure 11–7). The "empty" flask contains air, of course, but it also contained air when weighed previously, so the mass of air remains constant and does not affect the result. Example 11–15 illustrates the determination of the molecular weight of a volatile liquid by the Dumas method. Although the compound is a liquid at room temperature, it is a gas at 100°C, and therefore can be treated by the appropriate gas laws. If the water surrounding the flask is replaced by a liquid with a higher boiling point, the Dumas method can be applied to liquids that vaporize at or below the boiling point of the liquid surrounding the flask (Figure 11–7).

Example 11–15

The molecular weight of a volatile liquid was determined by the Dumas method. A 120 milliliter flask contained 0.345 gram of vapor at 100°C and 1.00 atmosphere pressure. What is the molecular weight of the compound?

Solution

We substitute into the ideal gas law, $PV = nRT$, to determine the number of moles of vapor that filled the flask.

$V = 0.120$ L $\qquad P = 1.00$ atm

$T = 100°C + 273° = 373$ K

$$n = \frac{PV}{RT} = \frac{(1.00 \text{ atm})(0.120 \text{ L})}{\left(0.0821 \dfrac{\text{L} \cdot \text{atm}}{\text{mol} \cdot \text{K}}\right)(373 \text{ K})}$$

$$= 0.00392 \text{ mol}$$

The mass of 0.00392 mole of vapor is 0.345 g, so we can calculate the mass of one mole.

$$\underset{=}{?} \frac{\text{g}}{\text{mol}} = \frac{0.345 \text{ g}}{0.00392 \text{ mol}} = \underline{88.0 \text{ g/mol}}$$

The molecular weight of the volatile liquid is 88.0 amu.

Let's carry the calculation one step further in Example 11–16.

Example 11–16

Analysis of the volatile liquid in Example 11–15 showed that it contained 54.5% carbon, 9.1% hydrogen, and 36.4% oxygen by mass. What is the molecular formula for the volatile liquid?

Solution

Let's calculate the simplest formula for the compound.

$$\underset{=}{?} \text{ mol C atoms} = 54.5 \text{ g C} \times \frac{1 \text{ mol C atoms}}{12.0 \text{ g C}}$$

$$= 4.54 \text{ mol C atoms}$$

$$\underset{=}{?} \text{ mol H atoms} = 9.1 \text{ g H} \times \frac{1 \text{ mol H atoms}}{1.0 \text{ g H}}$$

$$= 9.1 \text{ mol H atoms}$$

$$\underset{=}{?} \text{ mol O atoms} = 36.4 \text{ g O} \times \frac{1 \text{ mol O atoms}}{16.0 \text{ g O}}$$

$$= 2.28 \text{ mol O atoms}$$

$$\frac{4.54}{2.28} = 2 \text{ C}$$

$$\frac{9.1}{2.28} = 4 \text{ H}$$

$$\frac{2.28}{2.28} = 1 \text{ O}$$

simplest formula is C_2H_4O
formula weight = 44 amu

Division of the molecular weight by the simplest formula weight gives:

$$\frac{\text{molecular weight}}{\text{simplest formula weight}} = \frac{88 \text{ amu}}{44 \text{ amu}} = 2$$

Therefore, the molecular formula is $(C_2H_4O)_2 = \underline{C_4H_8O_2}$. One compound that has this composition is ethyl acetate, a common solvent used in nail polishes and polish removers.

11–12 Dalton's Law of Partial Pressures

Many gases, including our atmosphere, are mixtures of gases that consist of different kinds of gas molecules. Since the number of molecules of each type in a mixture can be expressed as a number of moles, the total number of moles in the gas mixture is given by

$$n_{\text{total}} = n_A + n_B + n_C + \ldots + n_x \qquad \text{[Eq. 11–7]}$$

where n_A, n_B, and so on, represent the number of moles of each of the x types of gas present.

Solving the ideal gas equation (Eq. 11–6) for n gives

$$n = \frac{PV}{RT}$$

All the gas molecules in a mixture occupy the same container at the same temperature and pressure. Therefore, we can write an ideal gas law expression for the number of moles of each component in a mixture of gases,

$$n_A = \left(\frac{V}{RT}\right) P_A, \qquad n_B = \left(\frac{V}{RT}\right) P_B, \qquad n_C = \left(\frac{V}{RT}\right) P_C \qquad \text{and so on}$$

The pressure exerted by the A molecules is proportional to n_A, the number of moles of A, and so on. The pressure term P_A is called the **partial pressure** of gas A.

Substituting these equations into the equation for n_{total} gives an interesting result.

$$n_{\text{total}} = \frac{V}{RT}(P_{\text{total}}) = \frac{V}{RT}(P_A) + \frac{V}{RT}(P_B) + \frac{V}{RT}(P_C) + \dots$$

$$\frac{V}{RT}(P_{\text{total}}) = \frac{V}{RT}(P_A + P_B + P_C + \dots)$$

Cancelling common terms gives the total pressure of a mixture of gases in terms of the partial pressures of the individual components.

$$P_{\text{total}} = P_A + P_B + P_C + \dots \qquad\qquad \text{[Eq. 11–8]}$$

The total pressure exerted by a mixture of gases is the sum of the partial pressures of those gases. This is known as **Dalton's Law of Partial Pressures,** since John Dalton was the first to notice this effect in 1807 while studying the composition of moist and dry air.

Two gases at one atmosphere pressure, in vessels of equal volume connected by a closed stopcock, are depicted in Figure 11–8. If the stopcock is opened, both gases diffuse to occupy both vessels equally; each exhibits a partial pressure of one-half atmosphere and the total pressure is one atmosphere after the gases are mixed.

Dalton's Law is useful in describing gaseous mixtures because it allows us to

FIGURE 11–8

Representation of diffusion of gases. The space between the molecules allows for ease of admission of one gas into another. Collisions of molecules with the walls of the container are responsible for the pressure of the gas.

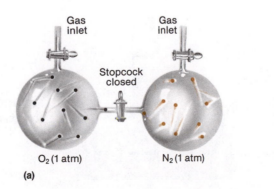

Gas inlet   Gas inlet

Stopcock closed

O$_2$ (1 atm)   N$_2$ (1 atm)

(a)

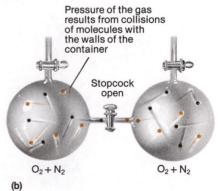

Pressure of the gas results from collisions of molecules with the walls of the container

Stopcock open

O$_2$ + N$_2$ O$_2$ + N$_2$

(b)

relate total measured pressures to the composition of mixtures. Consider the following example.

Example 11–17

A 10.0 liter flask contains 0.200 mole of methane, 0.300 mole of hydrogen, and 0.400 mole of nitrogen at 25°C. (a) What is the pressure in atmospheres inside the flask? (b) What is the partial pressure of each component of the mixture of gases?

Solution

(a) We are given the numbers of moles of each component, and the ideal gas law is used to calculate the total pressure.

$n = 0.200$ mol $CH_4 + 0.300$ mol H_2
$\qquad\qquad\qquad + 0.400$ mol N_2

$\quad = 0.900$ mol of gas

$V = 10.0$ L

$T = 25°C + 273° = 298$ K

Solving $PV = nRT$ for P gives $P = nRT/V$. Substitution gives

$$P = \frac{(0.900 \text{ mol})\left(0.0821 \dfrac{\text{L}\cdot\text{atm}}{\text{mol}\cdot\text{K}}\right)(298 \text{ K})}{10.0 \text{ L}}$$

$\quad = \underline{2.20 \text{ atm}}$

The total pressure exerted by the mixture of gases is 2.20 atmospheres.

(b) The partial pressure of each gas in the mixture can be calculated by substituting the number of moles of each gas into $PV = nRT$ individually; for CH_4, $n = 0.200$ mole and the values for V and T are the same as above.

$$P_{CH_4} = \frac{(n_{CH_4})RT}{V}$$

$$= \frac{(0.200 \text{ mol})\left(0.0821 \dfrac{\text{L}\cdot\text{atm}}{\text{mol}\cdot\text{K}}\right)(298 \text{ K})}{10.0 \text{ L}}$$

$P_{CH_4} = \underline{0.489 \text{ atm}}$

Similar calculations for the partial pressures of hydrogen and nitrogen give

$P_{H_2} = \underline{0.734 \text{ atm}}$ and $P_{N_2} = \underline{0.979 \text{ atm}}$

Dalton's Law states, "The total pressure of a mixture of gases is the sum of the partial pressures of the component gases." Addition of the partial pressures of this mixture of gases should give the total pressure, that is,

$P_{CH_4} + P_{H_2} + P_{N_2} = P_{total}$

Substitution gives the expected result,

$0.489 \text{ atm} + 0.734 \text{ atm} + 0.979 \text{ atm} = \underline{2.20 \text{ atm}}$

Frequently gases, or mixtures of gases, are conveniently collected over water. Figure 11–9 illustrates the collection of a sample of hydrogen over water. A gas produced in a reaction will displace the more dense water from the inverted water-filled jar. Gases that are soluble in water or that react with water are not collected by this method. The pressure on the gas inside the collection jar can be made equal to atmospheric pressure by raising or lowering the jar until the water

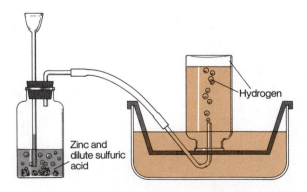

Zinc and dilute sulfuric acid

Hydrogen

FIGURE 11–9 An apparatus for preparing hydrogen from zinc and dilute sulfuric acid (Zn (s) + 2H+ (aq) → Zn²⁺ (aq) + H₂ (g)). The hydrogen is collected over water.

TABLE 11–5   Vapor Pressure of Water Near Room Temperature

Temperature (°C)	Vapor Pressure of Water (torr)	Temperature (°C)	Vapor Pressure of Water (torr)
15	12.79	23	21.07
16	13.63	24	22.38
17	14.53	25	23.76
18	15.48	26	25.21
19	16.48	27	26.74
20	17.54	28	28.35
21	18.65	29	30.04
22	19.83	30	31.82

level inside the jar is the same as that outside. The atmospheric pressure is then measured by an ordinary mercury barometer.

However, one complication arises. A gas in contact with water soon becomes saturated with water vapor; i.e., the pressure inside the jar is the sum of the partial pressure of the gas itself *plus* the partial pressure exerted by the water molecules in the gas (that is, the **vapor pressure** of water). Every liquid shows characteristic vapor pressures that vary with temperature. Table 11–5 displays vapor pressures of water near room temperature.

Thus, if we measure the temperature of the collecting jar, we can look up the vapor pressure of the water and substitute it into the Dalton's Law expression (where we set the total pressure equal to atmospheric pressure):

$$P_{atm} = P_{gas} + P_{H_2O} \qquad \text{or} \qquad P_{gas} = P_{atm} - P_{H_2O}$$

Example 11–18 provides a detailed illustration.

Example 11–18

A 300 milliliter sample of hydrogen was collected over water at 21°C on a day when the atmospheric pressure was 748 torr. (a) What volume will the dry hydrogen occupy at STP? (b) What is the mass of the sample of dry hydrogen?

Solution

(a) We can correct the volume of hydrogen to STP as in Section 11–8, but we must also correct for the vapor pressure of water at 21°C (see Table 11–5).

$V_1 = 300$ mL $V_2 = ?$
$P_1 = P_{atm} - P_{H_2O}$ $P_2 = 760$ torr
  $= 748$ torr $- 19$ torr
$P_1 = 729$ torr

$T_1 = 21°C + 273°$ $T_2 = 273$ K
  $= 294$ K

$$V_2 = 300 \text{ mL} \times \frac{729 \text{ torr}}{760 \text{ torr}} \times \frac{273 \text{ K}}{294 \text{ K}}$$

  $= 267$ mL of H_2 at STP

(b) Now that we know the volume of dry hydrogen at STP, we can calculate its mass. One mole of dry hydrogen occupies 22,400 mL at STP and has a mass of 2.02 g.

$$? \text{ g } H_2 = 267 \text{ mL} \times \frac{2.02 \text{ g } H_2}{22,400 \text{ mL } H_2}$$

  $= 0.0241$ g H_2

11–13   The Kinetic-Molecular Theory

As early as 1738, Johann Bernoulli envisioned gaseous molecules in ceaseless motion striking the walls of their container and thereby exerting pressure. In 1857, Rudolf Clausius published a theory that attempted to explain various experimental

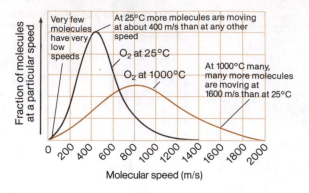

FIGURE 11-11 The Maxwellian distribution function for molecular speeds. This graph shows the relative numbers of O_2 molecules having any given speed at 25°C and at 1000°C. At 25°C most O_2 molecules have speeds between 200 and 600 m/s.

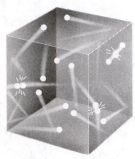

FIGURE 11-10
Visualization of molecular motion.

observations that had been summarized by Boyle's, Dalton's, Charles', and Avogadro's Laws. The basic assumptions of his **kinetic-molecular theory** are

1. Gases consist of discrete molecules. The individual molecules are relatively far apart, and they exert very little attraction for each other except near the liquefaction point (the conditions of temperature and pressure at which a gas liquefies). The volume occupied by the gas molecules themselves is insignificant compared to the volume occupied by the gas at ordinary temperatures and pressures (see Section 11-1).

2. Gaseous molecules are in continuous random, straight-line motion with various velocities (see Figure 11-10). Collisions between gas molecules and with the walls of the container are elastic; i.e., there is no net gain or loss of energy in these collisions. At any given instant only a small number of molecules are involved in collisions relative to the number present.

3. The average kinetic energies of molecules of different gases are equal at a given temperature. Kinetic energies of gases increase with increasing temperature and decrease as temperature decreases. Kinetic energy is the energy a body possesses by virtue of its motion. It is $\frac{1}{2} mv^2$, where m, the body's mass, can be expressed in grams, and v, its velocity, can be expressed in meters per second, m/s. Thus, small, light molecules like hydrogen and helium have much higher average velocities than heavier molecules like carbon dioxide and sulfur dioxide at the same temperature. All gases possess the same average kinetic energy at a given temperature. We may summarize: The average kinetic energy of gaseous molecules is proportional to the absolute temperature. Figure 11-11 shows the distribution of velocities of gaseous molecules at two temperatures.

The utility of the kinetic-molecular theory is that it satisfactorily explains most of the observed behavior of gases. Numerous experiments have indicated its basic validity. Let's look at the gas laws in light of the kinetic-molecular theory.

Boyle's and Dalton's Laws

The pressure exerted by a gas upon the walls of its container is caused by gas molecules striking the walls. Clearly, pressure depends upon two factors: (1) the number of molecules striking the walls per unit time, and (2) how vigorously the molecules strike the walls. Halving the volume of a given sample of gas doubles the pressure at constant temperature because twice as many molecules strike a given area on the container walls per unit time. Likewise, doubling the volume of a sample

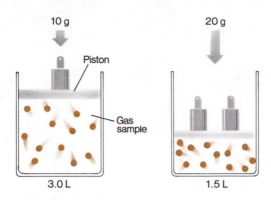

FIGURE 11-12 An illustration of the change in volume of a gas with changes in pressure (temperature constant). The entire apparatus is enclosed in a vacuum.

of gas halves the pressure because only half as many gas molecules strike a given area on the container walls per unit time (Figure 11-12).

Charles' Law

Recall that average kinetic energy is directly proportional to the absolute temperature. Doubling the *absolute* temperature of a sample of gas at constant pressure doubles the average kinetic energy of the gaseous molecules, and the increased vigor of collision doubles the volume at constant pressure. Similarly, halving the absolute temperature at constant pressure decreases kinetic energy to one-half its original value and the volume decreases by one-half because of reduced vigor of collision with the container walls (Figure 11-13).

11-14   Graham's Law: Rates of Effusion (Diffusion) of Gases

Because gas molecules are in constant, rapid, random motion, they diffuse quickly throughout any container. For example, if hydrogen sulfide (the essence of rotten eggs) is released in a large room, the odor can be detected throughout the room in a very short time. If a mixture of gases is placed in a container with porous walls, the molecules effuse through the walls. Lighter gas molecules always effuse through the tiny openings of porous materials faster than heavier molecules (Figure 11-14) because they move faster.

Chemists use *"effusion"* to describe the escape of a gas through a tiny hole, and *"diffusion"* to describe movement of a gas into a space or the mixing of one gas with another.

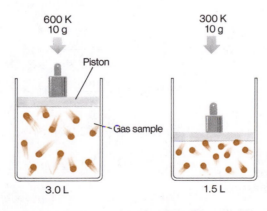

FIGURE 11-13 An illustration of the change in volume of gas with changes in temperature (pressure constant). The entire apparatus is enclosed in a vacuum.

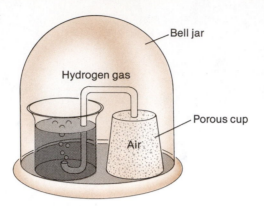

Bell jar

Hydrogen gas

Porous cup

Air

FIGURE 11-14 Effusion of gases. If a bell jar full of hydrogen is brought down over a porous cup full of air, hydrogen will effuse into the cup faster than the oxygen and nitrogen in the air can effuse out of the cup. This causes an increase in pressure in the cup sufficient to produce bubbles in the water in the beaker.

The rates of effusion of different gases were studied by Thomas Graham in 1832, and he demonstrated that *the rates of effusion of gases are inversely proportional to the square roots of their molecular weights* (or densities). This statement is known as Graham's Law. Symbolically, it may be represented as

$$\frac{\text{Rate of effusion of gas A}}{\text{Rate of effusion of gas B}} = \sqrt{\frac{\text{Molecular weight of gas B}}{\text{Molecular weight of gas A}}} \qquad \text{[Eq. 11-9]}$$

Or, to use simpler notation,

$$\frac{\text{Rate}_A}{\text{Rate}_B} = \sqrt{\frac{M_B}{M_A}} \qquad \text{where: Rate refers to rates of effusion and } M \text{ refers to molecular weights.}$$

Example 11-19 demonstrates Graham's Law for two common gases, methane, CH_4, and sulfur dioxide, SO_2.

Example 11-19

Calculate the ratio of the rate of effusion of methane to that of sulfur dioxide.

Solution

Substitution into the Graham's Law equation gives

$$\frac{\text{Rate}_{CH_4}}{\text{Rate}_{SO_2}} = \sqrt{\frac{M_{SO_2}}{M_{CH_4}}} = \sqrt{\frac{64 \text{ amu}}{16 \text{ amu}}} = \sqrt{\frac{4}{1}} = \frac{2}{1}$$

Since $\text{Rate}_{CH_4}/\text{Rate}_{SO_2} = 2/1$, we can write $\text{Rate}_{CH_4} = 2 \text{ Rate}_{SO_2}$, which tells us that the rate of effusion of CH_4 is twice that of SO_2. At a given temperature the average velocity of CH_4 molecules is twice that of SO_2 molecules.

Example 11-20

A 100 milliliter sample of hydrogen effuses through a porous container four times as rapidly as an unknown gas. Calculate the molecular weight of the unknown gas.

Solution

The Graham's Law equation

$$\frac{\text{Rate}_{H_2}}{\text{Rate}_{Unk}} = \sqrt{\frac{M_{Unk}}{M_{H_2}}}$$

can be solved for M_{Unk}. Let's square both sides of the equation, solve the resulting expression for

M_{Unk}, and substitute the known values into the expression.

$$\left(\frac{\text{Rate}_{H_2}}{\text{Rate}_{Unk}}\right)^2 = \frac{M_{Unk}}{M_{H_2}} \quad \text{and}$$

$$M_{Unk} = \left(\frac{\text{Rate}_{H_2}}{\text{Rate}_{Unk}}\right)^2 M_{H_2}$$

$$M_{Unk} = \left(\frac{4}{1}\right)^2 (2.0 \text{ amu})$$

$$= \left(\frac{16}{1}\right) 2.0 \text{ amu} = \underline{32 \text{ amu}}$$

The unknown gas is probably oxygen.

Interestingly, Graham's Law can be derived from the part of the kinetic-molecular theory which says that *at a given temperature the average kinetic energies of molecules of different gases are equal.* If we represent the average kinetic energy of gas molecules A as $KE_A = \frac{1}{2} m_A v_A^2$, and the average kinetic energy of gas molecules B as $KE_B = \frac{1}{2} m_B v_B^2$, we can write

$$KE_A = KE_B \qquad \text{or} \qquad \frac{1}{2} m_A v_A^2 = \frac{1}{2} m_B v_B^2 \qquad \text{[Eq. 11-10]}$$

Multiplying the latter expression by 2 gives $m_A v_A^2 = m_B v_B^2$, and rearranging the expression we obtain

$$\frac{v_A^2}{v_B^2} = \frac{m_B}{m_A} \qquad \text{so} \qquad \frac{v_A}{v_B} = \sqrt{\frac{m_B}{m_A}}$$

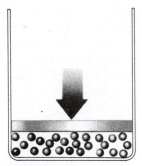

FIGURE 11-15 A sample of gas under high pressure. The molecules are quite close together.

which says that the ratio of the velocity of molecule A to that of molecule B is the square root of the ratio of the mass of molecule B to the mass of molecule A. Rates of effusion are directly proportional to velocities, and molecular weights are numerically equal to masses of molecules. Therefore, we may write

$$\frac{\text{Rate}_A}{\text{Rate}_B} = \sqrt{\frac{M_B}{M_A}}$$

which is the usual form of Graham's Law (Eq. 11-9). We can also show that Equation 11-9 is equivalent to

$$\frac{\text{Rate}_A}{\text{Rate}_B} = \sqrt{\frac{D_B}{D_A}} \qquad \text{[Eq. 11-11]}$$

where D refers to gas densities measured at the same pressure and temperature.

11-15   Real Gases—Deviations from Ideality

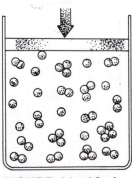

FIGURE 11-16 A sample of gas at a low temperature. Each sphere represents a molecule.

The van der Waals equation is known as an *equation of state,* i.e., an equation that describes a state of matter.

Under ordinary conditions most real gases behave like ideal gases, in that they obey the postulates of the kinetic-molecular theory and the ideal gas law. Recall that according to the kinetic-molecular model most of the volume of a gas sample is empty space and gases are very compressible. Also, the molecules of *ideal* gases exhibit no forces of attraction for each other because they are so far apart and moving so rapidly.

But, at *low temperatures* and/or *high pressures,* near the liquefaction point, a real gas deviates significantly from ideality because the postulates do not describe the behavior of matter accurately. Under high pressures a gas is compressed to the point that the volume of the molecules themselves becomes a significant fraction of the total volume occupied by the gas. At low temperatures the molecules travel more slowly than at higher temperatures and forces of attraction can overcome the motion of some of the molecules. Both of these factors result in a greater tendency for molecules to "stick together", at least briefly, as they collide (Figures 11-15 and 11-16). In 1867, after studying deviations of real gases from ideal behavior, Johannes van der Waals empirically adjusted the ideal gas equation (Eq. 11-6)

$$(P)(V) = nRT$$

to take into account two complicating factors. The **van der Waals equation** is

$$\left(P + \frac{n^2 a}{V^2}\right)(V - nb) = nRT \qquad \text{[Eq. 11-12]}$$

TABLE 11–6
van der Waals
Constants

Gas	a $\left(\dfrac{L^2 \cdot atm}{mol^2}\right)$	b $\left(\dfrac{L}{mol}\right)$
H_2	0.244	0.0266
He	0.034	0.0237
N_2	1.39	0.0391
NH_3	4.17	0.0371
CO_2	3.59	0.0427
CH_4	2.25	0.0428

in which P, V, T, and n represent the same variables as in the ideal gas law, but a and b are experimentally measured constants that differ for different gases. The a term corrects for the fact that molecules do exert attractive forces upon each other. Large values of a indicate strong attractive forces between molecules. The b factor corrects for the volume occupied by the molecules themselves. Note that if both a and b are zero, the van der Waals equation reduces to the ideal gas equation. Table 11–6 gives van der Waals constants for some common gases.

The tendency of the molecules of a gas to "stick together" when they collide depends upon the forces of attraction between them. Note that a for helium is very small. This is the case for all noble gases and other nonpolar molecules, since only very weak attractive forces, called London forces, exist among them. London forces result from short-lived electrical dipoles produced by the attraction of one atom's nucleus for an adjacent atom's electrons. These forces exist for all molecules but are important only in slightly polar and nonpolar molecules, which would never liquefy if these forces did not exist. Polar molecules, like ammonia, NH_3, have permanent charge separations (dipoles) and therefore exhibit greater forces of attraction for each other. Note the high value of a for ammonia. London forces and permanent dipole forces of attraction are discussed in more detail in Chapter 12.

The following example illustrates the deviation of methane, CH_4, from ideal gas behavior under a typical set of conditions.

Example 11–21

Calculate the pressure exerted by 1.00 mole of methane, CH_4, in a 10.0 liter vessel at 25°C assuming (a) ideal behavior and (b) nonideal behavior, as described by the van der Waals equation.

Solution

(a) Ideal gases obey the ideal gas law, and we'll assume that methane does.

$$PV = nRT$$

$$P = \frac{nRT}{V} = \frac{(1.00 \text{ mol})\left(0.0821 \dfrac{L \cdot atm}{mol \cdot K}\right)(298 \text{ K})}{10.0 \text{ L}}$$

$$= \underline{2.45 \text{ atm}}$$

(b) As a real gas methane obeys the van der Waals equation.

$$\left(P + \frac{n^2 a}{V^2}\right)(V - nb) = nRT$$

For CH_4, $a = 2.25$ $L^2 \cdot atm/mol^2$ and $b = 0.0428$ L/mol (Table 11–6).

$$\left[P + \frac{(1.00 \text{ mol})^2 (2.25 \text{ L}^2 \cdot atm/mol^2)}{(10.0 \text{ L})^2}\right]\left[10.0 \text{ L} - (1.00 \text{ mol})\left(0.0428 \frac{L}{mol}\right)\right]$$

$$= (1.00 \text{ mol})\left(0.0821 \frac{L \cdot atm}{mol \cdot K}\right)(298 \text{ K})$$

Combining terms and canceling units we get

$$[P + 0.0225 \text{ atm}][9.957 \text{ L}] = 24.47 \text{ L} \cdot atm$$

$$P + 0.0225 \text{ atm} = 2.46 \text{ atm}$$

$$P = \underline{2.44 \text{ atm}}$$

This value is 0.01 atmosphere (0.4%) less than the pressure calculated from the ideal gas law. We see that almost no error is introduced by assuming that methane behaves as an ideal gas under these conditions. However, methane behaves much less ideally at very high pressures. Repeating the calculations of this example with $V = 1.00$ L gives pressures (ideal and nonideal, respectively) of 24.5 atm and 23.3 atm, a difference of 4.9%.

Stoichiometry in Reactions Involving Gases

Thus far we have considered the behavior of gases with no emphasis on stoichiometry. Let us add a dimension to our study of stoichiometry.

11−16   Gay-Lussac's Law: The Law of Combining Volumes

Gases react in simple, definite proportions by volume. For example, *one* volume of hydrogen always combines (reacts) with *one* volume of chlorine to form *two* volumes of hydrogen chloride. It is understood that all volumes are measured at the same temperature and pressure.

$$H_2 \text{ (gas)} + Cl_2 \text{ (gas)} \longrightarrow 2HCl \text{ (gas)}$$

1 volume + 1 volume $\longrightarrow$ 2 volumes

Volumes may be expressed in any units as long as the same unit is used for all. Gay-Lussac (1788–1850) summarized several experimental observations on combining volumes of gases. The summary is known as **Gay-Lussac's Law** or the **Law of Combining Volumes:** *At constant temperature and pressure, the volumes of reacting gases can be expressed as a ratio of simple whole numbers.* The ratio is obtained from the coefficients in the balanced equation for the reaction. Clearly, the law applies only to *gaseous* substances at the same temperature and pressure. No generalizations can be made about the volumes of solids and liquids as they undergo chemical reactions. Consider the following examples, based on experimental observations at constant temperature and pressure. Hundreds more could be cited.

1. One volume of nitrogen reacts with three volumes of hydrogen to form two volumes of ammonia.

   $$N_2 \text{ (g)} + 3H_2 \text{ (g)} \longrightarrow 2NH_3 \text{ (g)}$$

   1 volume + 3 volumes   2 volumes

2. Sulfur (a solid) reacts with one volume of oxygen to form one volume of sulfur dioxide.

   $$S \text{ (s)} + O_2 \text{ (g)} \longrightarrow SO_2 \text{ (g)}$$

   1 volume   1 volume

3. Four volumes of ammonia burn in five volumes of oxygen to produce four volumes of nitric oxide and six volumes of steam.

   $$4NH_3 \text{ (g)} + 5O_2 \text{ (g)} \longrightarrow 4NO \text{ (g)} + 6H_2O \text{ (g)}$$

   4 volumes + 5 volumes   4 volumes + 6 volumes

Avogadro's Law (Section 11–9) provides an explanation of Gay-Lussac's Law. Consider the reaction of fluorine and hydrogen to produce hydrogen fluoride. All volumes are measured at the same temperature and pressure. Experiments show that:

1 volume of fluorine + 1 volume of hydrogen $\longrightarrow$ 2 volumes of hydrogen fluoride

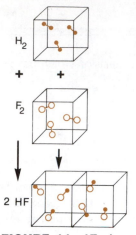

FIGURE 11–17 A representation of the combination of fluorine and hydrogen.

Figure 11–17 illustrates the volumetric relationships in the combination of fluorine and hydrogen to form hydrogen fluoride. Avogadro's Law tells us that equal volumes of fluorine, hydrogen, and hydrogen fluoride contain equal numbers of molecules. The experimentally observed facts and deductions based on these facts are:

1. Two molecules of hydrogen fluoride are formed from one molecule of fluorine and one molecule of hydrogen.
2. Each hydrogen fluoride molecule must contain *at least* one fluorine atom and one hydrogen atom.
3. Therefore, two hydrogen fluoride molecules contain at least two fluorine atoms and two hydrogen atoms, which must have been present in one fluorine molecule and one hydrogen molecule, i.e.,

$$1 \text{ volume of } F_2 + 1 \text{ volume of } H_2 \longrightarrow 2 \text{ volumes of HF}$$

or   1 molecule   1 molecule   2 molecules

4. There are no known reactions in which one mole of fluorine or one mole of hydrogen contains enough atoms to form more than two moles of product. Therefore, we may safely assume that each fluorine molecule and each hydrogen molecule contains exactly two atoms, i.e., that the formulas for the molecules are F_2 and H_2.

Similar experiments and reasoning show that the other common diatomic elements (2 atoms per molecule) are oxygen, O_2; nitrogen, N_2; chlorine, Cl_2; bromine, Br_2; and iodine, I_2.

Other reactions involving gases can be explained by Avogadro's and Gay-Lussac's Laws. It is interesting to note that Dalton, in the first decade of the nineteenth century, considered but then quite properly rejected the notion that equal volumes of gases contain the same numbers of *atoms*. The idea of the existence of diatomic (and more complex) molecules had not yet evolved, and it didn't occur to Dalton.

Application of Gay-Lussac's Law to gas phase reactions accurately predicts the volumes of gases involved.

Example 11–22

What volume of hydrogen will combine with 40 liters of oxygen to form steam at 750°C and atmospheric pressure?

Solution

The balanced equation for the reaction

$$2H_2 \text{ (g)} + O_2 \text{ (g)} \longrightarrow 2H_2O \text{ (g)}$$

tells us that two hydrogen molecules combine with one oxygen molecule to form two water molecules. Applying Gay-Lussac's Law, we can write

$$2H_2 \quad + \quad O_2 \quad \longrightarrow \quad 2H_2O$$

2 volumes   1 volume   2 volumes

and we have

$$\underline{?} \text{ L } H_2 = 40 \text{ L } O_2 \times \frac{2 \text{ volumes } H_2}{1 \text{ volume } O_2} = \underline{80 \text{ L } H_2}$$

11–17   Mass-Volume Relationships in Reactions Involving Gases

Small amounts of oxygen are produced in the laboratory by heating potassium chlorate, $KClO_3$, in the presence of a small amount of a catalyst, manganese dioxide,

MnO_2. As in Section 2–7, we can represent the stoichiometry from the balanced equation as

$$2KClO_3 \text{ (s)} \longrightarrow 2KCl \text{ (s)} + 3O_2 \text{ (g)}$$

2 mol	2 mol	3 mol
2(122.6 g)	2(74.6 g)	3(32.0 g)

But we now know that one mole of gas, measured at STP, occupies 22.4 liters, and this information can be used. We can write the equation as

$$2KClO_3 \text{ (s)} \longrightarrow 2KCl \text{ (s)} + \qquad 3O_2 \text{ (g)}$$

2 mol	2 mol	3 mol
245.2 g	149.2 g	96.0 g
		$3(22.4 \text{ L}_{STP}) = 67.2 \text{ L}_{STP}$

Unit factors can be constructed using any two of these quantities just as in Section 2–7.

Example 11–23

What volume of O_2 (STP) can be produced by heating 100.0 grams of $KClO_3$?

Solution

Reference to the preceding equation shows that 2 moles of $KClO_3$ produce 3 moles of O_2.

$$\underset{=}{?} \text{ L}_{STP} \text{ } O_2 = 100.0 \text{ g } KClO_3 \times \frac{1 \text{ mol } KClO_3}{122.6 \text{ g } KClO_3}$$

$$\times \frac{3 \text{ mol } O_2}{2 \text{ mol } KClO_3} \times \frac{22.4 \text{ L}_{STP} \text{ } O_2}{1 \text{ mol } O_2}$$

$$= 27.4 \text{ L}_{STP} \text{ } O_2$$

Alternatively, we could have solved this problem using a single unit factor, $67.2 \text{ L}_{STP} \text{ } O_2/245.2 \text{ g } KClO_3$, read directly from the balanced equation.

$$\underset{=}{?} \text{ L}_{STP} \text{ } O_2 = 100 \text{ g } KClO_3 \times \frac{67.2 \text{ L}_{STP} \text{ } O_2}{245.2 \text{ g } KClO_3}$$

$$= 27.4 \text{ L}_{STP} \text{ } O_2$$

Example 11–24

A 1.80 gram mixture of potassium chlorate, $KClO_3$, and potassium chloride, KCl, is heated until all the $KClO_3$ has decomposed. The liberated oxygen occupies 400 milliliters collected over water at 25°C when the barometric pressure is 750 torr. What percentage of the original mixture is $KClO_3$?

Solution

Note that heating KCl produces no oxygen, and therefore we are concerned with the same reaction as before,

$$2KClO_3 \text{ (s)} \longrightarrow 2KCl \text{ (s)} + 3O_2 \text{ (g)}$$

The O_2 is collected over water at 25°C and 750 torr. Therefore, we must subtract the vapor pressure of water (Table 11–5) to obtain the partial pressure exerted by O_2,

$$P_{O_2} = 750 \text{ torr} - 24 \text{ torr} = 726 \text{ torr}$$

We can now use the ideal gas law to calculate the number of moles of O_2 produced:

$$V = 0.400 \text{ L}$$

$$P = 726 \text{ torr} \times \frac{1 \text{ atm}}{760 \text{ torr}} = 0.955 \text{ atm}$$

$$T = 25°C + 273° = 298 \text{ K}$$

$$n = \frac{PV}{RT} = \frac{(0.955 \text{ atm})(0.400 \text{ L})}{\left(0.0821 \dfrac{\text{L} \cdot \text{atm}}{\text{mol} \cdot \text{K}}\right)(298 \text{ K})}$$

$$= 0.0156 \text{ mol } O_2$$

We now know that 0.0156 mole of O_2 was produced, so we can calculate the mass of $KClO_3$ that decomposed to produce it.

$$\underset{=}{?} \text{ g } KClO_3 = 0.0156 \text{ mol } O_2$$

$$\times \frac{2 \text{ mol } KClO_3}{3 \text{ mol } O_2} \times \frac{122.6 \text{ g } KClO_3}{1 \text{ mol } KClO_3}$$

$$= 1.28 \text{ g } KClO_3$$

The sample contained 1.28 g of $KClO_3$. The percentage of $KClO_3$ in the sample is

$$\% \ KClO_3 = \frac{g \ KClO_3}{g \ sample} \times 100\% = \frac{1.28 \ g}{1.80 \ g} \times 100\%$$
$$= \underline{71.1\% \ KClO_3}$$

The sample contains 71.1% $KClO_3$ and 28.9% KCl (obtained by difference).

Key Terms

Absolute zero   zero degrees on the absolute temperature scale; $-273.15°C$ or 0 K; theoretically, the temperature at which molecular motion ceases.

Atmosphere   a unit of pressure; the pressure that will support a column of mercury 760 mm high at 0°C.

Avogadro's Law   at the same temperature and pressure, equal volumes of all gases contain the same number of molecules.

Barometer   a simple device for measuring pressure. See Figure 11–1.

Boyle's Law   at constant temperature the volume occupied by a definite mass of a gas is inversely proportional to the applied pressure.

Charles' Law   at constant pressure the volume occupied by a definite mass of a gas is directly proportional to the absolute temperature.

Condensed states   the solid and liquid states.

Dalton's Law   see *Law of Partial Pressures*.

Dumas method   a method used to determine the molecular weights of volatile liquids. See Figure 11–7.

Equation of state   an equation that describes the behavior of matter in a given state; the van der Waals equation describes the behavior of the gaseous state.

Fluids   substances that flow freely; gases and liquids.

Gay-Lussac's Law   see *Law of Combining Volumes*.

Graham's Law   The rates of effusion of gases are inversely proportional to the square roots of their molecular weights (or densities).

Ideal gas   a hypothetical gas that obeys exactly all postulates of the kinetic-molecular theory.

Ideal Gas Law   the product of the pressure and volume of an ideal gas is directly proportional to the number of moles of the gas and the absolute temperature.

Kinetic-Molecular Theory   a theory, originally published by Clausius in 1857, that attempts to explain macroscopic observations on gases in microscopic or molecular terms.

Law of Combining Volumes   at constant temperature and pressure, the volumes of reacting gases (and any gaseous products) can be expressed as ratios of small whole numbers; also known as Gay-Lussac's Law.

Law of Partial Pressures   the total pressure exerted by a mixture of gases is the sum of the partial pressures of the individual gases; also called Dalton's Law.

Manometer   a two-armed barometer. See Figure 11–2.

Partial pressure   the pressure exerted by one gas in a mixture of gases.

Pressure   force per unit area.

Real gases   gases that deviate from ideal gas behavior.

Standard conditions (STP or SC)   standard temperature, 0°C, and standard pressure, one atmosphere, are standard conditions for gases.

Standard molar volume   the volume occupied by one mole of an ideal gas under standard conditions; 22.4 liters.

Universal gas constant R, the proportionality constant in the ideal gas equation, $PV = nRT$.

Vapor   a gas formed by boiling or evaporating a liquid.

Vapor pressure   the pressure exerted by a vapor at the surface of its mother liquid.

van der Waals equation   an equation of state that extends the ideal gas law to real gases by inclusion of two empirically determined parameters, which are different for different gases.

Exercises

General Ideas

1. What are the three states of matter? Compare and contrast them.
2. What evidence is there for the statement that the individual molecules are quite far apart in gases?
3. What is the composition of the atmosphere?
4. On what experimental foundation does our current understanding of gases rest?

Pressure

5. (a) What is pressure?
   (b) Describe the mercurial barometer. How does it work?
6. What is a manometer? How does it work?
7. List some units of pressure. Express one atmosphere of pressure in several different units.
8. Does a tire gauge measure the absolute pressure inside a tire? Why?
9. What does the statement that an automobile tire has 28 pounds per square inch of air in it mean?

Boyle's Law: The Pressure-Volume Relationship

10. On what kinds of observations (measurements) is Boyle's Law based? State the law.
11. Could the words "a fixed number of moles" be substituted for "a definite mass of" in the statement of Boyle's Law? Why?
12. Why is a plot of pressure versus volume at constant temperature (Figure 11–4) one branch of a hyperbola?
13. Use the statement of Boyle's Law to derive a simple mathematical expression for Boyle's Law.
14. A sample of nitrogen occupies 11.2 liters under a pressure of 580 torr at 32°C. What volume would it occupy at 32°C if the pressure were increased to 840 torr?
15. A sample of oxygen occupies 47.2 liters under a pressure of 730 torr at 25°C.
    (a) What volume would it occupy at 25°C if the pressure were increased to 1240 torr?
    (b) By what percent does the pressure increase?
    (c) By what percent does the volume decrease?
    (d) Is answer (b) the same as answer (c)? Why?
16. Under what pressure would the sample of oxygen described in Exercise 15 occupy 100 liters at 25°C? Express the answer in torr, atm, Pa, and kPa.
17. If the pressure on a 100 liter sample of hydrogen is doubled at constant temperature, by what fraction and by what percent does the volume change?
18. If the pressure on a 100 liter sample of hydrogen is halved at constant temperature, by what multiple and by what percent does the volume change?

The Kelvin Temperature Scale

19. Describe the experiments that led to the evolution of the absolute temperature scale. What is the relationship between the Celsius and Kelvin temperature scales?
    (a) What does "absolute temperature scale" mean?
    (b) What does "absolute zero" mean?
20. (a) Can an absolute temperature scale based on "Fahrenheit" rather than "Celsius" degrees be evolved? Why?
    (b) Can an absolute temperature scale that is based on a "degree" twice as large as a Celsius degree be developed? Why?

Charles' Law

21. On what kind of observations (measurements) is Charles' Law based? State the law.
22. Why is a plot of volume versus temperature at constant pressure (Figure 11–6b) a straight line?
23. Use the statement of Charles' Law to derive a simple mathematical expression for Charles' Law.
24. A 400 mL sample of hydrogen is confined under a pressure of 0.500 atm at 50.0°C. What volume would it occupy at 100°C under the same pressure?
    (a) By what fraction and by what percent does the temperature increase on the Celsius scale? On the Kelvin scale?
    (b) By what fraction and by what percent does the volume increase? Why?
25. Refer to Exercise 24. If the pressure were expressed as 380 torr or 50.65 kilopascals would the same answer be obtained? Why?
26. A sample of methane, CH_4, occupies 800 mL at

150°C. At what temperature will it occupy 400 mL if the pressure does not change?

27. Which of the following statements are true? Which are false? Why is each true or false? *Assume constant pressure* in each case.
    (a) If a sample of gas is heated from 100°C to 200°C the volume will double.
    (b) If a sample of gas is heated from 0°C to 273°C the volume will double.
    (c) If a sample of gas is cooled from 400°C to 200°C the volume will decrease by a factor of two.
    (d) If a sample of gas is cooled from 1000°C to 200°C the volume will decrease by a factor of five.
    (e) If a sample of gas is heated from 200°C to 2000°C the volume will increase by a factor of ten.

Standard Conditions

28. (a) What are standard conditions?
    (b) Express standard temperature on three scales.
    (c) Express standard pressure in torr, mm Hg, atm, Pa, and kPa.
29. Why do we have a standard temperature and a standard pressure?

The Combined Gas Laws

30. (a) To what does "the combined gas laws" refer?
    (b) What is the mathematical expression for the combined gas laws?
    (c) Outline the logic used to obtain the mathematical expression for the combined gas laws.
31. A sample of nitrogen occupies a volume of 300 mL under a pressure of 380 torr at 177°C. What volume would the gas occupy at STP?
32. A sample of gas occupies 250 mL at 100°C under a pressure of one atmosphere. If the pressure were increased to 1900 torr, to what temperature would the sample have to be heated to occupy a volume of 500 mL?
33. A sample of gas occupies 300 mL at STP. Under what pressure would this sample occupy 150 mL if the temperature were increased to 546°C?
34. A sample of hydrogen occupies 250 mL at STP. If the temperature were increased to 546°C,

what final pressure would be necessary to keep the volume constant at 250 mL?

35. A sample of nitrogen occupies 400 mL at 400°C under a pressure of 4.00 atm. If the pressure were increased to 12.00 atm, what final temperature would be necessary to keep the volume constant at 400 mL?

Gas Densities and the Standard Molar Volume

36. What is Avogadro's Law? What does it mean?
37. What does standard molar volume mean?
38. A mole of oxygen occupies 22.4 L at STP. What is its density in g/L and g/mL at STP?
39. Refer to Exercise 38. What is the volume of 3.85 moles of oxygen at STP?
40. (a) How many moles of oxygen are contained in 83.7 liters at STP?
    (b) What is the mass of this sample of oxygen?
41. At a given temperature and pressure one mole of a gas occupies 38.3 liters. Its density is 1.37 g/L under these conditions. What is the mass of one mole of the gas?
42. If 4.00 g of a gas occupies a volume of 4.48 liters at STP, what is the molecular weight of the gas?
43. Calculate the mass of a 6.72 liter sample of carbon monoxide, CO, measured at standard conditions.
44. If 4.00 grams of a gas occupies 1.12 liters at STP, what is the mass of 4.00 moles of the gas?

The Ideal Gas Equation

45. What is an "ideal gas"?
46. (a) What is the ideal gas equation?
    (b) Outline the logic used to obtain the ideal gas equation.
    (c) What is R? How is it obtained?
47. Calculate R in $L \cdot atm/mol \cdot K$, in $kPa \cdot dm^3/mol \cdot K$, in $J/mol \cdot K$, and in $kJ/mol \cdot K$.
48. Calculate the volume occupied by 64.0 g of CH_4 at 127°C under a pressure of 1520 torr.
49. Calculate the pressure exerted by 60.0 g of C_2H_6 in a 30.0 liter vessel at 27.0°C.
50. How many moles of nitrogen are contained in 328 mL of the gas under a pressure of 3040 torr at 527°C? How many nitrogen atoms does the sample contain?
51. What is the mass of 8.21 liters of CH_4 at 227°C under a pressure of 1520 torr?
52. What volume would 90.0 grams of C_2H_6 occupy at 27°C under a pressure of 380 torr?

53. What pressure would a mixture of 3.2 g of O_2, 6.4 g of CH_4, and 6.4 g of SO_2 exert if the gases were placed in a 40.0 L container at 127°C?

54. A student was given a container of ethane, C_2H_6, that had been sealed at STP. By making appropriate measurements, the student found that the mass of the sample of ethane was 0.218 gram and the volume of the container was 165 mL. Use the student's data to calculate the molecular weight of ethane. What percent error is obtained? Suggest some possible sources of the error.

55. A student was given a sample of an unknown gas that occupied 41.5 mL at 25°C and 754 torr. The mass of the sample of gas was 0.0761 g. The student was asked to calculate the molecular weight of the gas. What value did she obtain? The student was then told that the sample of gas was propane, C_3H_8, but that the sample was contaminated with either ethane, C_2H_6, or n-butane, C_4H_{10}. What was the contaminant? Justify your answer.

Molecular Weights and Molecular Formulas for Gaseous Compounds

56. Distinguish between simplest (empirical) formulas and molecular formulas.

57. A compound containing only carbon and hydrogen is 80.0% C and 20.0% H by mass. At STP, 1.12 liters of the gas has a mass of 1.50 g. What is the molecular formula for the compound?

58. A 0.580 g sample of a compound containing only carbon and hydrogen contains 0.480 g of carbon and 0.100 g of hydrogen. At STP, 33.6 mL of the gas has a mass of 0.087 g. What is the molecular formula for the compound?

59. Analysis of a volatile liquid shows that it contains 62.04% carbon, 10.41% hydrogen, and 27.54% oxygen by mass. At 150°C and 1.00 atm, 500 mL of the vapor has a mass of 1.673 g.
    (a) What is the molecular weight of the compound?
    (b) What is its molecular formula?

Dalton's Law

60. (a) State Dalton's Law. Express it symbolically.
    (b) What are partial pressures of gases?

61. One-liter samples of N_2, H_2, He, and O_2 are collected in four different vessels at the same temperature under a pressure of one atmosphere. If the four gases are then forced into one of the containers at the original temperature, what will the resulting pressure be?

62. If 2.00 liters of nitrogen under a pressure of 2.00 atmospheres at 127°C and 3.00 liters of nitrogen under a pressure of 1.00 atmosphere at 127°C are forced into a 4.00 liter container at 127°C, what will the final pressure be?

63. Suppose that 0.005 mole of CO_2 at STP and 0.005 mole of CO at STP are forced into a container that has a volume of 11.2 liters at a pressure of one atmosphere. What will the temperature of the mixture be in the new container?

64. A 5.42 liter sample of a gas was collected over water on a day when the temperature was 24°C and the barometric pressure was 706 torr. The dry sample of gas had a mass of 5.60 g. What is the mass of three moles of the dry gas? At 24°C the vapor pressure of water is 22 torr.

65. A 296 mL sample of oxygen is collected over water at 23°C on a day when the barometric pressure is 753 torr. What volume would the dry oxygen occupy at 487°C under a pressure of 1.00 atmosphere?

66. Calculate the mass of dry hydrogen in 750 mL of moist hydrogen collected over water at 25.0°C and 755 torr. The density of hydrogen is 0.08987 g/L at STP. The vapor pressure of water is 24 torr at 25°C.

The Kinetic-Molecular Theory

67. Outline the kinetic-molecular theory.

68. How do average velocities of gaseous molecules vary with temperature?

69. How does the kinetic-molecular theory explain (a) Boyle's Law? (b) Dalton's Law? (c) Charles' Law?

Graham's Law: Effusion of Gases

70. State Graham's Law. What does it mean?

71. Show how the kinetic-molecular theory leads to the mathematical expression for Graham's Law.

72. (a) Explain how the hydrogen bubbler (Figure 11–14) works.
    (b) Could sulfur hexafluoride, SF_6, be used in this bubbler? Why?

73. At a given temperature and pressure, the average velocity of O_2 molecules would be _____

times as great as the average velocity of HI molecules.

74. At 500°C the average velocity of HI molecules should be _____ times the average velocity of CH_3OH molecules.

75. Calculate the ratio of the rate of effusion of CH_4 to that of SO_2.

76. If the average velocity of helium atoms is 0.707 mile per second at room temperature, what will be the average velocity of oxygen molecules at the same temperature?

Real Gases and Deviations from Ideality

77. How do "real" and "ideal" gases differ?

78. Under what kind of conditions are deviations from ideality most important? Why?

79. What is the van der Waals equation? How does it differ from the ideal gas equation?

80. (a) How do the van der Waals constants (Table 11–6) for He and H_2 differ? Why?
    (b) Compare the van der Waals constants for H_2 and NH_3. Explain why those for NH_3 would be expected to be larger.

81. Suppose you have samples of each of the gases listed in Table 11–6 under a pressure of 100 atmospheres, each at a temperature 10°C above its liquefaction point (at 100 atm). For which gas would you expect deviation from ideal behavior to be (a) greatest and (b) least? Justify your answers.

82. Use both the ideal gas law and the van der Waals equation to calculate the pressure exerted by a 150 g sample of ammonia in a 60.0 liter container at 1000°C. By what percentage do the two results differ?

83. (a) Use both the ideal gas law and the van der Waals equation to calculate the pressure exerted by a 150 g sample of ammonia in a 600 liter container at 1000°C. By how much do these results differ?
    (b) Explain the difference between the results obtained in Exercises 82 and 83a.

Stoichiometry in Reactions Involving Gases

84. (a) What is Gay-Lussac's Law?
    (b) What does it mean?
    (c) What is the Law of Combining Volumes?

85. What volume of chlorine, Cl_2, is necessary to prepare 38 liters of hydrogen chloride, HCl, if an excess of hydrogen, H_2, is used? All gases are measured at 100°C at 1 atm.

86. (a) What volume of ammonia, NH_3, can be prepared from 50 liters of nitrogen, N_2, and excess hydrogen, H_2, if all gases are measured at the same temperature and pressure?
    (b) What volume of hydrogen would be required in (a)?

87. What volume of hydrogen (STP) would be required to react with 0.100 mole of nitrogen?

$$N_2 + 3H_2 \xrightarrow{\Delta} 2NH_3$$

88. What volume of oxygen (STP) would be produced by the thermal decomposition of 0.600 mole of potassium nitrate?

$$2KNO_3 \text{ (s)} \xrightarrow{\Delta} 2KNO_2 \text{ (s)} + O_2 \text{ (g)}$$

89. What mass of $KClO_3$ would have to be decomposed to produce 13.4 liters of oxygen gas measured at STP?

$$2KClO_3 \text{ (s)} \xrightarrow{\Delta} 2KCl \text{ (s)} + 3O_2 \text{ (g)}$$

90. An impure sample of $KClO_3$ that had a mass of 30.0 g was heated until all of the $KClO_3$ had decomposed. The liberated oxygen occupied 6.72 liters at STP. What percentage of the sample was $KClO_3$? Refer to Exercise 89 for the balanced equation.

91. What volume of dry NO (g) at STP could be prepared by reacting 6.35 grams of Cu with an excess of HNO_3?

$$3Cu \text{ (s)} + 8HNO_3 \text{ (aq)} \longrightarrow$$
$$3Cu(NO_3)_2 \text{ (aq)} + 2NO \text{ (g)} + 4H_2O \text{ (}\ell\text{)}$$

92. If sufficient acid is used to react completely with 4.862 g of magnesium, what volume of hydrogen (STP) would be produced? What volume would the gas occupy if it were collected over water on a day when the temperature was 75°F and the barometric pressure was 740 torr?

$$Mg \text{ (s)} + 2HCl \text{ (aq)} \longrightarrow MgCl_2 \text{ (aq)} + H_2 \text{ (g)}$$

93. If sufficient acid is used to react completely with 13.5 g of aluminum, what volume of hydrogen (STP) would be produced? What volume would the hydrogen occupy if it were collected over water on a day when the temperature was 80°F and the barometric pressure was 750 torr?

$$2Al \text{ (s)} + 6HCl \text{ (aq)} \longrightarrow$$
$$2AlCl_3 \text{ (aq)} + 3H_2 \text{ (g)}$$

94. What volume of hydrogen (STP) would be required to produce 0.400 mole of HCl?

$$H_2 \text{ (g)} + Cl_2 \text{ (g)} \longrightarrow 2HCl \text{ (g)}$$

95. If 0.500 mole of carbon disulfide reacts with oxygen completely according to the following equation, what volume (*total*) would the products occupy if they were measured at STP?

$$CS_2 \text{ (}\ell\text{)} + 3O_2 \text{ (g)} \longrightarrow CO_2 \text{ (g)} + 2SO_2 \text{ (g)}$$

96. Based on the following equation, what volume of hydrogen (STP) would be required to react with 10.4 g of C_2H_2? If a chemist obtained 18.0 g of C_2H_6 by this reaction, what total volume (STP) of hydrogen and C_2H_2 did he use?

$$2H_2 \text{ (g)} + C_2H_2 \text{ (g)} \xrightarrow{\text{cat.}} C_2H_6 \text{ (g)}$$

Liquids and Solids

12

FIGURE 12–1 A representation of the three states of matter at the molecular level.

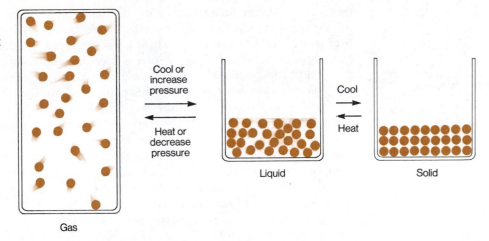

The molecules of most gases are so widely separated at ordinary temperatures and pressures that they do not interact with each other significantly. Consequently, the physical properties of gases are reasonably well described by the relatively simple mathematical relationships described in Chapter 11. In liquids and solids, the so-called **condensed phases,** the particles are closely spaced and interact strongly. Although the properties of liquids and solids can be described, they cannot be adequately explained by simple mathematical relationships. Figure 12–1 and Table 12–1 summarize some of the distinguishing characteristics of gases, liquids, and solids.

These properties, as well as those that we shall introduce later in this chapter, such as melting points, boiling points, and heats of transformation, can be accounted for, at least in a general way, by considering the types and magnitudes of attractive forces that exist among the particles that make up the sample of interest.

12–1 Kinetic-Molecular Description of Liquids and Solids

Let us begin by extending the kinetic-molecular theory (Section 11–13) to liquids and solids. The properties listed in Table 12–1 can be qualitatively explained in

TABLE 12–1 Some Common Characteristics of Gases, Liquids, and Solids

Gases	Liquids	Solids
1. No definite shape (fill containers completely)	1. No definite shape (assume shapes of containers)	1. Definite shape (resist deformation)
2. Compressible	2. Only very slightly compressible	2. Nearly incompressible
3. Low density	3. Intermediate density	3. High density
4. Fluid	4. Fluid	4. Not fluid
5. Diffuse rapidly	5. Diffuse through other liquids	5. Diffuse only very slowly through solids
6. Extremely disordered particles; much empty space; rapid, random motion in three dimensions	6. Disordered clusters of particles; quite close together; random motion in three dimensions	6. Ordered arrangement of particles; vibrational motion only; particles very close together

terms of the kinetic-molecular theory of Chapter 11. We saw in Section 11–13 that the average kinetic energy of a collection of gas molecules decreases as the temperature is lowered. As the temperature falls, the rapid, random motion of gaseous molecules decreases. If a sample of gas is cooled and compressed sufficiently, the molecules approach each other closely enough for intermolecular attractions to overcome the kinetic energies. At this point condensation, or liquefaction, occurs. Since different kinds of molecules have different attractive forces, the temperatures and pressures required for condensation vary from gas to gas.

In the liquid state, the forces of attraction among particles are great enough that disordered clustering occurs. The particles are so close together that very little of the total volume occupied by a liquid is empty space. As a result, liquids are nearly incompressible. Liquid particles do not have sufficient energy of motion to overcome the attractive forces between them, but they are able to slide past each other, so that the liquid assumes the shape of its container. They are also able to diffuse quite rapidly through other liquids with which they are *miscible.* For example, a drop of red food coloring added to a glass of water causes the water to become homogeneously red after diffusion is complete. However, the natural diffusion rate is slow enough to be aided by swirling. Since the average separations among liquid particles are far less than in gases, the densities of liquids are much higher than densities of gases.

Further cooling a liquid lowers its molecular kinetic energy and causes the motion of its molecules to decrease more. If the temperature is lowered sufficiently, at ordinary pressures, stronger but shorter-range attractive interactions overcome the decreasing kinetic energies of the molecules to cause solidification. The temperature required for crystallization at a given pressure depends upon the nature of short-range interactions among the particles and is characteristic of each substance.

Most solids are characterized by ordered arrangements of particles with very restricted motion. Particles in the solid state are unable to move freely past one another, and only vibrate about fixed positions. Consequently, solids have definite shapes. Since the particles are so close together, solids are nearly incompressible,

*Inter*molecular attractions refer to attractions between different molecules. *Intra*molecular refers to attractions between atoms within a single molecule.

The miscibility of two liquids refers to their ability to mix and produce a homogeneous solution.

FIGURE 12–2 A representation of diffusion in solids. When blocks of two different metals are clamped together for a long time, a few atoms of each metal diffuse into the other metal.

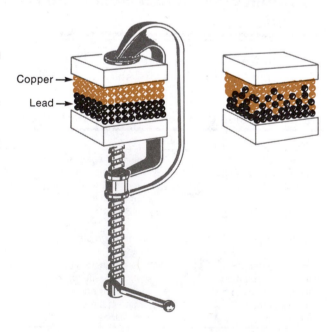

Copper

Lead

and are very dense relative to gases. Solid particles do not diffuse readily into other solids. However, analysis of two blocks of different solids such as copper and lead, after they have been pressed together for a period of years, shows that each block contains some atoms of the other element. This demonstrates that solids do diffuse, but very slowly (Figure 12–2).

The Liquid State

In the next several sections, we shall briefly describe several of the properties of the liquid state. These properties vary markedly among various liquids. Such variations depend on the nature and strength of the attractive forces between the particles (atoms, molecules, ions) making up the liquid. In Section 12–8, we shall discuss the nature of the intermolecular forces in more detail.

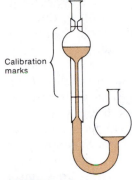

Calibration marks

FIGURE 12–3 The Ostwald viscometer, a device used to measure viscosity.

12–2 Viscosity

Viscosity is the resistance to flow of a liquid, and is related to how strongly the molecules "stick together," and to the sizes and shapes of the constituent particles. Freely flowing diethyl ether has a low viscosity, while molasses is very viscous at room temperature. Viscosity can be measured by using a viscometer such as the one depicted in Figure 12–3, in which the time it takes for a known volume of a liquid to flow through a small neck of known size is measured. As the temperature of a liquid increases, the kinetic energies of its molecules increase, and therefore they more easily overcome the intermolecular forces of attraction. As a result, viscosity decreases with increasing temperature, so long as no change in composition occurs.

12–3 Surface Tension

Molecules below the surface of a liquid are influenced by intermolecular attractions from all directions, but those on the surface are attracted only toward the interior, as shown in Figure 12–4. The attractions pull the surface layer toward the center. From an energetic point of view, the most stable situation is one in which the surface area is minimal. This is the reason that drops of liquid tend to assume spherical shapes. **Surface tension** is a measure of the inward forces that must be overcome in order to expand the surface area of a liquid.

Fire polishing removes the rough edges of laboratory glass tubing because of the tendency of the heated (liquid) glass to form a rounded surface. Surface tension is also responsible for the ability of water to support the mass of some insects that are more dense than water itself. Perhaps you've heard of floating a needle in a glass of water.

FIGURE 12–4 Molecular level view of the attractive forces experienced by molecules at and below the surface of a liquid.

12–4 Capillary Action

All the forces holding a liquid together are called **cohesive forces.** The forces of attraction between a liquid and another surface are **adhesive forces.** Water wets

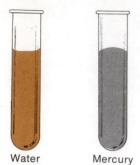

Water   Mercury

FIGURE 12–5 The meniscus, as observed in glass tubes with water and mercury.

glass and increases its surface area as it creeps up the side of a glass tube due to the strong adhesive forces between water and glass, which have similar structures. The surface of the water, its **meniscus,** takes on a concave shape (Figure 12–5). On the other hand, mercury does not wet glass because its cohesive forces are much stronger than its attraction to glass. Thus, its meniscus is convex. **Capillary action** occurs when one end of a capillary tube, a glass tube with small bore (diameter), is immersed in a liquid. If adhesive forces exceed cohesive forces, the liquid creeps up the sides of the tube until a balance is reached between adhesive forces and the weight of liquid. The smaller the bore, the higher the liquid climbs. Plant roots take up water and dissolved nutrients from the soil and transmit them up the stems by capillary action. The roots, like glass, exhibit strong adhesive forces for water.

12–5   Evaporation

Kinetic energies of liquids show the same type of temperature dependence as those of gases. The distribution of kinetic energies among liquid molecules at two different temperatures is shown in Figure 12–6. **Evaporation** or **vaporization** is the process by which molecules on the surface of the liquid break away and go into the gas phase (Figure 12–7). In order to break away, the molecules must possess a minimum kinetic energy, one half the product of mass times the square of the escape velocity. Figure 12–6 shows that the higher the temperature, the greater the fraction of molecules possessing at least that minimum energy. The rate of evaporation increases as temperature increases. Further, it is only the higher energy molecules that can escape from the liquid phase. The average molecular kinetic energy of the molecules remaining in the liquid state is thereby reduced, so that a lower temperature in the liquid results. Since the liquid would then be cooler than its surroundings, it absorbs heat from its surroundings. You are familiar with a common example of

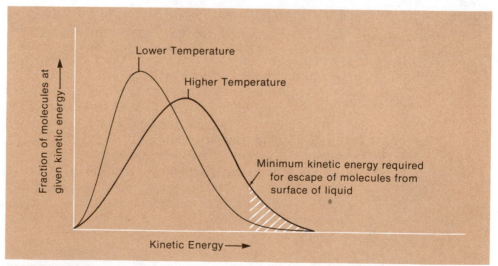

FIGURE 12–6 Distribution of kinetic energies of molecules in a liquid at different temperatures. Note that at the lower temperature, a smaller fraction of the molecules have the energy required to escape from the liquid, so evaporation is slower at the lower temperature.

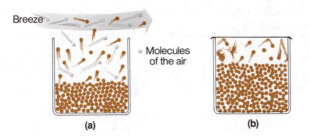

FIGURE 12-7 (a) Liquid continually evaporates from an open vessel; (b) equilibrium between liquid and vapor is established in a closed container in which molecules return to the liquid at the same rate as they leave it.

the cooling effect of evaporation on the surroundings of a liquid, the cooling of the body by evaporation of perspiration.

A molecule in the vapor may collide with a molecule of air, bounce back to the liquid surface, and be captured there. This process is the reverse of evaporation and is called **condensation.** As evaporation proceeds in a closed container, the volume of liquid decreases and the number of gas molecules above the surface increases. Since more collisions will then occur in the gas phase, the rate of condensation increases. The system composed of the liquid and gas molecules (of the same substance) eventually achieves a **dynamic equilibrium** in which the rate of evaporation equals the rate of condensation in the closed container.

The two opposing rates counterbalance each other but are not individually equal to zero; hence the use of the term "dynamic" rather than "static" equilibrium.

$$\text{Liquid} \underset{\text{condensation}}{\overset{\text{evaporation}}{\rightleftharpoons}} \text{Vapor}$$

However, if the vessel were left open to the air this equilibrium could not be established, since molecules would diffuse away and the slightest air currents would also sweep some gas molecules away from the liquid surface, allowing more evaporation to occur in order to replace the lost vapor molecules. Consequently, a liquid eventually evaporates entirely if it is left uncovered. This situation illustrates **LeChatelier's Principle,** which states that a system at equilibrium, or striving to attain equilibrium, responds in the way that best accommodates or "undoes" any stress placed upon it. In this example, the stress is the removal of molecules in the vapor phase, and the response is the continued evaporation of the liquid.

This is one of the guiding principles that allows us to understand chemical equilibrium. It is discussed further in Chapter 16.

12-6   Vapor Pressure

Vapor molecules cannot escape when vaporization of a liquid occurs in a closed container. As more molecules leave the liquid, more collisions between gaseous molecules and the walls of the container occur and more condensation occurs. This is responsible for the formation of liquid droplets that adhere to the sides of the vessel above a liquid surface and for the eventual establishment of equilibrium between liquid and vapor.

The partial pressure of vapor molecules above the surface of a liquid at equilibrium is the **vapor pressure** of the liquid. Since the rate of evaporation increases and the rate of condensation decreases with increasing temperature, vapor pressures of liquids always increase as temperature increases. At any given temperature the vapor pressures of different liquids are different (Table 12-2) because their cohesive forces are different. Easily vaporized liquids are called **volatile** liquids, and they have relatively high vapor pressures. The most volatile liquid listed

TABLE 12–2 Vapor Pressures (in Torr) of Some Liquids at Different Temperatures

	0°C	25°C	50°C	75°C	100°C
water	4.6	23.8	92.5	300	760
methyl alcohol	29.7	122	404	1126	
benzene	27.1	94.4	271	644	1360
diethyl ether	185	470	1325	2680	4859

in Table 12–2 is diethyl ether, and water is the least volatile. Vapor pressures can be measured with manometers such as the one shown in Figure 12–8.

12–7   Boiling Points and Distillation

Heating a liquid always increases its vapor pressure. When a liquid is heated to a sufficiently high temperature under a given applied (atmospheric) pressure, bubbles of vapor begin to form below the surface. They rise to the surface and burst, releasing the vapor into the air. This process is called boiling, and is distinctly different from evaporation.* If the vapor pressure inside the bubbles is less than the applied pressure on the surface of the liquid, the bubbles collapse as soon as they form and no boiling occurs. The **boiling point** of a liquid under a given pressure is the temperature at which its vapor pressure is just equal to the applied pressure. The **normal boiling point** is the temperature at which the vapor pressure of a liquid is equal to exactly one atmosphere (760 torr) pressure. The vapor pressure of water is 760 torr at 100°C, its normal boiling point. If the applied pressure is reduced below 760 torr, water boils below 100°C. It takes longer to cook food in boiling water at high altitudes than at or below sea level because of the reduction in atmospheric

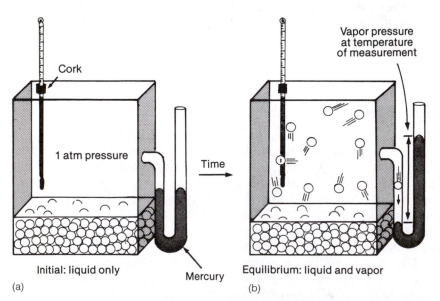

FIGURE 12–8 (a) A simplified representation of the measurement of vapor pressure of a liquid at a given temperature. At the instant the liquid is added to the container, the space above the liquid is occupied by air only. (b) As time passes, some of the liquid vaporizes until equilibrium is established. The difference between the heights of the mercury columns initially and at equilibrium is a measure of the vapor pressure of the liquid at that temperature.

* When water is heated, but before it boils, small bubbles may begin to adhere to the container, and then rise to the surface and burst. This is not boiling of water, but rather the formation of bubbles of previously dissolved gases such as carbon dioxide or oxygen, whose solubilities in water decrease with increasing temperature.

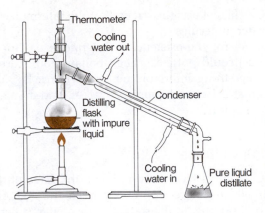

Thermometer

Cooling
water out

Condenser

Distilling
flask
with impure
liquid

Cooling
water in

Pure liquid
distillate

FIGURE 12–9 Laboratory set-up for distillation. During distillation of an impure liquid, nonvolatile substances remain in the distilling flask. The liquid is vaporized and condensed before being collected in the receiving flask.

pressure. A pressure cooker is efficient in cooking food rapidly because water boils at higher temperatures under high pressures.

Since different liquids have different cohesive forces, they have different vapor pressures, and boil at different temperatures. A mixture of liquids with different enough boiling points can often be separated into its components by **distillation.** In this process the mixture is heated slowly, until the temperature reaches the point at which the most volatile component boils off. If this component is a liquid under ordinary conditions, it is subsequently recondensed in a water-cooled condensing column (Figure 12–9) and collected as a distillate. After enough heat has been added to vaporize all of the most volatile liquid, the temperature again rises slowly until the boiling point of the next substance is reached and the process continues. Any nonvolatile substances dissolved in the liquid do not boil, but remain in the distilling flask. Impure water can be purified and separated from its dissolved salts by distillation. Compounds with similar boiling points, especially those that interact very strongly with each other, are not well separated by simple distillation but require a modification called fractional distillation, which is discussed in Section 13–11.

Intermolecular Interactions

12–8   Intermolecular Forces of Attraction and Phase Changes

We have referred to the influence of the strengths of intermolecular forces of attraction on such properties of liquids as boiling point, vapor pressure, and ease of evaporation. Such forces are also directly related to the properties of solids, such as melting point and heat of fusion, which will be discussed in Section 12–12. *Inter*molecular forces refer to the forces between particles (atoms, molecules, ions) of a substance. Intermolecular forces are quite weak relative to *intra*molecular forces, i.e., covalent and ionic bonds within compounds. For example, 920 kJ (220 kcal) of heat are required to decompose one mole of water vapor into hydrogen and oxygen atoms. This reflects the strength of intramolecular forces (chemical bonds). But it requires only 40.7 kJ to convert one mole of liquid water into steam (gaseous water) at 100°C. This reflects the strength of the intermolecular forces of attraction between the water molecules.

If it were not for the existence of intermolecular forces, condensed phases (liquid and solids) could not exist. High boiling points are associated with compounds exhibiting strong intermolecular attractions. Let us consider the general types of forces that exist among ionic, covalent, and monatomic species and their effects on boiling points.

1   Ion-Ion Interactions

According to Coulomb's Law (Section 6–9), the force of attraction between two oppositely charged ions is directly proportional to the charges on the ions, q^+ and q^-, and inversely proportional to the square of the distance between them, d.

$$F \propto \frac{q^+q^-}{d^2}$$

[Eq. 12–1]

Energy has the units of force × distance, $F \times d$, so the energy of attraction between two oppositely charged ions is directly proportional to the charges on the ions and inversely proportional to the distance of separation, rather than the square of the distance.

$$E \propto \frac{q^+q^-}{d}$$

[Eq. 12–2]

Since ionic compounds such as $NaCl$, $CaBr_2$, and K_2SO_4 exist as extended arrays of discrete ions in the solid state, ionic bonding may be thought of as both *inter-* and *intramolecular* bonding. Most ionic bonding is strong, and as a result most ionic compounds have high melting points. At high enough temperatures ionic solids will melt, as the added heat energy overcomes the energy of attraction of oppositely charged ions that holds the solid together. The ions in the resulting *molten* samples are free to move about more or less randomly, accounting for the ability of molten ionic substances to conduct electricity well. Melting of solids also produces greater average separations among the ions, which means that the forces (and energies) of attraction among the ions are less than in the solid state because d is greater in the melt. These energies of attraction are still generally considerably greater than those among the molecules of covalent or monatomic liquids. This accounts for the very high boiling points of molten ionic substances.

Since the product q^+q^- increases as the charges on the ions increase, ionic substances containing multiply charged ions such as Al^{3+}, Mg^{2+}, O^{2-}, and S^{2-} usually have higher melting and boiling points than salts containing only univalent ions such as Na^+, K^+, F^-, and Cl^-.

2   Dipole-Dipole Interactions

Permanent dipole-dipole interactions occur between polar covalent molecules, because of the attraction of the $\delta+$ atoms of one molecule to the $\delta-$ atoms of another molecule (Section 7–10). In contrast to the electrostatic forces between two ions, dipole-dipole forces vary as $1/d^4$ and, in order to be at all effective, must operate over only very short distances. In addition, such forces are weaker than in the ion-ion cases because q^+ and q^- are only "partial charges." Average dipole-dipole interaction energies are approximately 4 kJ per mole of "bonds," whereas an

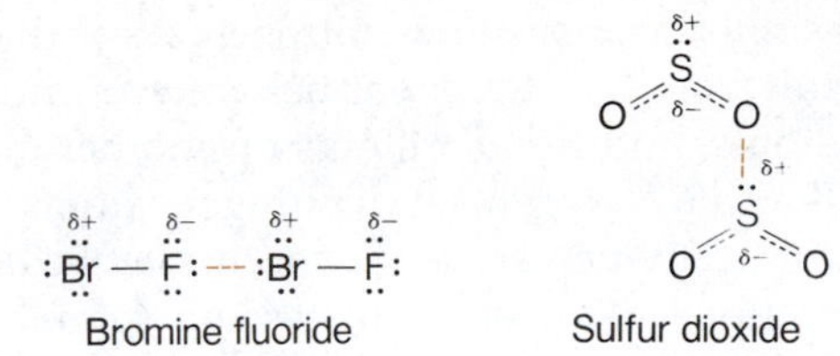

FIGURE 12-10
Dipole-dipole interactions
(---) in bromine fluoride
and sulfur dioxide.

Bromine fluoride

Sulfur dioxide

average value for typical ionic and covalent bond energies is approximately 400 kJ per mole of bonds. Substances in which permanent dipole-dipole interaction affects physical properties include bromine fluoride, BrF, and sulfur dioxide, SO_2. Dipole-dipole interactions are illustrated in Figure 12-10.

All dipole-dipole interactions, including hydrogen bonding (discussed below), are somewhat directional. An increase in temperature causes an increase in the random motion of molecules. This produces more randomness of orientation of molecules relative to each other, and a consequent decrease in the strength of dipole-dipole interactions. This fact, together with the weakness of their interactions, makes compounds having only dipole-dipole interactions more volatile than ionic compounds.

3 Hydrogen Bonding

Ordinarily, we restrict use of the term hydrogen bonding to compounds in which H is directly bonded to F, O, or N.

Hydrogen bonds are not really chemical bonds in the formal sense, but are rather a very strong, special case of dipole-dipole interaction that occurs among polar covalent molecules containing hydrogen and one of the small, highly electronega-

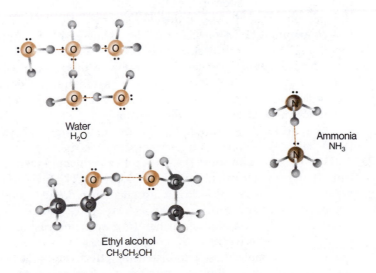

Water
H_2O

Ammonia
NH_3

Ethyl alcohol
CH_3CH_2OH

FIGURE 12-11 Hydrogen bonding (---) in water, ethyl alcohol, and ammonia. This is a special case of very strong dipole-dipole interaction.

tive elements, such as fluorine, oxygen, or nitrogen. As with ordinary dipole-dipole interactions, they result from the attractions between $\delta+$ atoms of one molecule, in this case hydrogen atoms, and the $\delta-$ atoms of another molecule. However, the small sizes of the fluorine, oxygen, and nitrogen atoms, combined with their extremely high electronegativities, serve to concentrate the electrons of these molecules around the $\delta-$ atoms. This effect forces the connected hydrogen atom to behave, in some ways, as a bare proton. This, in turn, has a dramatic effect on the hydrogen atom's attraction to the $\delta-$ pole of adjacent molecules.

Hydrogen bonding is responsible for the abnormally high melting and boiling points of compounds such as water, ethyl alcohol, and ammonia relative to other compounds of similar molecular weight and molecular geometry (refer to Figure 12–11). Hydrogen bonding in hydrogen fluoride is discussed and illustrated in Section 13–4. Hydrogen bonding between carboxylate groups ($-CO_2-$) and amino groups ($-NH_2$) of amino acid subunits is very important in establishing the three-dimensional structures of proteins.

4   London Forces

London forces are weak attractive forces that are important only over *extremely* short distances, since they vary as $1/d^7$. They exist for all types of molecules in condensed phases but are very weak for small molecules. Nevertheless, London forces are the only kind of intermolecular forces present among symmetrical nonpolar substances such as SO_3, CO_2, O_2, N_2, Br_2, and H_2 and monatomic species such as the noble gases. Thus London forces are responsible for the liquefaction of these substances, which in some instances occurs only at very low temperatures and/or high pressures.

London forces result from the attraction of the positively charged nucleus of one atom for the electron cloud of an adjacent atom in another molecule. This induces temporary dipoles in the atoms or molecules, which increase in strength as the polarizability of their electron clouds increases. Polarizability increases with increasing atomic or molecular size, because as electron clouds become larger and more diffuse they are attracted less strongly by their own (positively charged) nuclei; thus they are more easily distorted, or polarized, by adjacent nuclei. London forces are depicted in Figure 12–12 for monatomic argon. They exist in polar covalent substances as well as in nonpolar substances.

Figure 12–13 shows that polar covalent compounds with hydrogen bonding (H_2O, HF, NH_3) boil at higher temperatures than analogous polar compounds without hydrogen bonding (H_2S, HCl, PH_3) which, in turn, boil at higher temperatures than symmetrical, nonpolar compounds (CH_4, SiH_4) of comparable molecular weight. It also shows that in the absence of hydrogen bonding, boiling points of analogous substances (CH_4, SiH_4, GeH_4, SnH_4) increase fairly regularly with increasing molecular size, and therefore, with increasing molecular weight, due to increasing effectiveness of London forces. This is true even in the case of polar covalent molecules. The increasing effectiveness of London forces accounts for the increase in boiling points in the sequences $HCl < HBr < HI$ and $H_2S < H_2Se < H_2Te$, which involve non-hydrogen bonded polar covalent molecules. Since the differences in electronegativities between hydrogen and the other nonmetals *decrease* in these sequences, the effectiveness of *permanent* dipole-dipole interactions decreases. This effect alone would lead to *decreasing* boiling points. Since the boiling points actually *increase* in the order of increasing molecular size,

Ar Ar Ar Ar

No dipoles

$\delta+$ $\delta-$

Ar   Ar Ar Ar

Temporary dipole
in one atom

$\delta+\delta-$ $\delta+\delta-$ $\delta+\delta-$ $\delta+\delta-$

Ar   Ar   Ar   Ar

Induced dipoles
in other atoms

FIGURE 12–12

Simplified picture of the generation of London forces in symmetrical (nonpolar) argon atoms. An instantaneous nonspherical electronic distribution which gives a dipole in one atom induces a temporary dipole in an adjacent atom. The resulting pair of transient dipoles causes a weak attraction between the two atoms.

For comparison, the boiling points of the noble gases are:

Helium	−269°C
Neon	−246°C
Argon	−186°C
Krypton	−153°C
Xenon	−108°C
Radon	−62°C

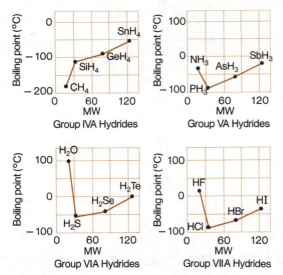

FIGURE 12–13 Boiling points of some hydrides as a function of molecular weights. The high boiling points of NH_3, H_2O, and HF are due to hydrogen bonding. The electronegativity difference between H and C is so small that CH_4, which has no lone pairs, is not hydrogen bonded.

this tells us that the increasing London forces actually override the decreasing permanent dipole-dipole forces in these cases.

It is interesting to compare the magnitudes of the various contributions to the total energy of interactions in a group of simple molecules. Table 12–3 shows the permanent dipole moments and the energy contributions for five simple molecules. Note that the contribution from London forces is large in all cases. The variations of these total energies of interaction are closely related to boiling points, as we might expect.

Example 12–1
Predict the order of increasing boiling points for the following: H_2S, H_2O, CH_4, Ne, and KBr.

Solution
Since KBr is ionic, it boils at the highest temperature. Water exhibits hydrogen bonding and boils at the next highest temperature. Hydrogen sulfide is the only other polar covalent substance, so it boils below H_2O. Nonpolar CH_4 boils at a higher temperature than the noble gas Ne.

$$\underline{Ne < CH_4 < H_2S < H_2O < KBr}$$
increasing boiling points

The general effects of intermolecular attractions on the physical properties of liquids are summarized in Table 12–4. Please keep in mind the fact that high and low are relative terms. Table 12–4 is included to show only very general trends.

The lack of *exact* correlation of total energy with boiling points for Ar and CO is probably due to approximations made in calculating total energies, and the fact that some kinds of interaction that make very small contributions have been omitted.

TABLE 12–3   Approximate Contributions to the Total Energy of Interaction Between Molecules (kJ/mol)

Molecule	Permanent Dipole Moment (D)	Permanent Dipole-Dipole Energy	London Energy	Total Energy	Boiling Point (K)
Ar	0	0	8.5	8.5	87
CO	0.1	0.0	8.7	8.7	81
HCl	1.03	3.3	17.8	21	188
NH_3	1.5	13	16.3	29	240
H_2O	1.8	36	10.9	47	373

TABLE 12–4 General Effects of Intermolecular Attractions on Physical Properties

Property	Volatile Liquids: Weak Intermolecular Attractions	Nonvolatile Liquids: Strong Intermolecular Attractions
cohesive forces	low	high
viscosity	low	high
surface tension	low	high
specific heat	low	high
vapor pressure	high	low
rate of evaporation	high	low
boiling point	low	high
heat of vaporization	low	high

Heat of vaporization will be discussed in Section 12–12.2.

The Solid State

12–9 Amorphous Solids and Crystalline Solids

We have already seen that solids have definite shapes and volumes, are not very compressible, are dense, diffuse only very slowly into other solids, and are generally characterized by compact ordered arrangements of particles that vibrate about fixed positions in their crystal lattices.

However, a small class of noncrystalline solids, *amorphous solids,* has no well-defined, ordered structure. Examples are rubber, some kinds of plastics, and amorphous sulfur.

Glasses are sometimes called amorphous solids and sometimes called **supercooled liquids.** The justification for calling them supercooled liquids is that, like liquids, they flow, although very slowly. The irregular structures of glasses are really intermediate between those of freely flowing liquids and crystalline solids; there is only short-range order. Unlike crystalline solids, glasses and other amorphous solids do not exhibit sharp melting points. Instead, they soften over a temperature range. Since particles in amorphous solids are irregularly arranged, intermolecular forces among particles vary in strength within the sample. Melting occurs at different temperatures for various portions of the same sample as the intermolecular forces are overcome.

One test for the purity of a crystalline solid is the sharpness of its melting point. Impurities disrupt the intermolecular forces and cause melting to occur over a considerable temperature range.

The shattering of a crystalline solid produces fragments having the same structural characteristics as the original sample. The shattering of a large cube of rock salt produces several small cubes of rock salt. This symmetrical shattering of crystals occurs along planes between which the interionic or intermolecular forces of attraction are weakest. Amorphous solids like fused glasses, with irregular structures, shatter irregularly to yield pieces with jagged edges and irregular angles.

X-Ray Diffraction

Atoms, molecules, and ions are much too small to be seen with the eye. The arrangements of particles in crystalline solids are determined indirectly by x-ray diffraction (scattering) or by x-ray reflection. In 1912 the German physicist Max von Laue showed mathematically that any crystal could serve as a three-dimensional diffraction grating for incident electromagnetic radiation with wavelengths approximating the internuclear separations ($\sim 10^{-10}$ m) of atoms in the crystal. Such radiation is in the x-ray region of the

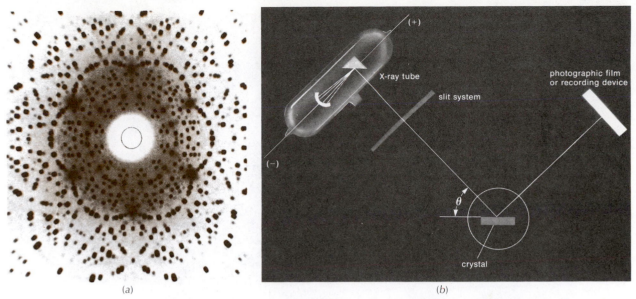

FIGURE 12-14   (a) Laue photograph of a crystal. The beam of x-rays was focused perpendicular to one of the lattice planes. (b) X-ray diffraction of crystals (schematic).

electromagnetic spectrum. Using an apparatus similar to that shown in Figure 12-14, a monochromatic (single wavelength) x-ray beam is formed by a system of slits and directed onto the surface of a slowly rotated crystal so as to vary the angle of incidence θ. At various angles, strong beams of deflected x-rays strike a photographic plate which, upon development, shows a central spot due to the primary beam and a set of symmetrically disposed spots due to deflected x-rays. Different kinds of crystals produce different arrangements of spots.

In 1913, the English scientists William and Lawrence Bragg found that the Laue photographs are more easily interpreted by treating the crystal as a reflection grating rather than a diffraction grating. The analysis of the spots is really quite complicated, but an experienced crystallographer can determine the separations between atoms within identical layers and the distances between layers of atoms. Moreover, it is possible not only to determine the relative positions of particles but also to determine unambiguously the identities of individual atoms if they are sufficiently different in atomic weight. Heavier atoms (with more electrons) deflect x-rays more strongly than light atoms.

Figure 12-15 illustrates the determination of spacings between layers of atoms. The x-ray beam strikes the surface of the crystal at an angle θ. Those rays colliding with atoms in the first layer are reflected at the same angle θ. Those passing through the first layer may be reflected from the second layer, third layer, and so forth. A strong reflected beam results only if all rays are in phase. In order for the waves to be in phase (interfere constructively), the following condition must be met.

$$n\lambda = 2d \sin\theta \qquad \text{or} \qquad \sin\theta = \frac{n\lambda}{2d} \qquad\qquad \text{[Eq. 12-3]}$$

This is known as the Bragg equation. It tells us that for x-rays of a given wavelength, λ, atoms in planes separated by distances d give rise to intense reflections for angles of incidence θ. The angles θ increase with increasing order, $n = 1, 2, 3$, and so forth.

FIGURE 12–15
Reflection of a mono-chromatic beam of x-rays by two planes of a crystal.

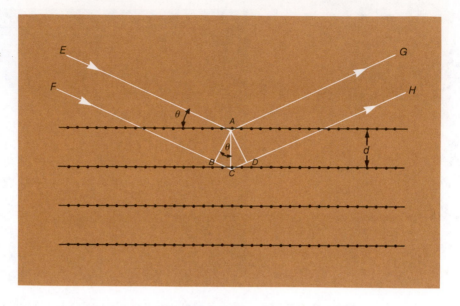

12–10 Structures of Crystals

All crystals are polyhedra consisting of regularly repeating arrays of atoms, molecules, or ions. Each type of crystal is described by a **lattice,** which is a set of mathematical points with identical surroundings. The smallest volume of a crystal showing all the characteristics of its lattice is called a **unit cell.** The unit cell is analogous to the repeating unit of a wallpaper pattern, but in three dimensions. Once we identify the pattern of a wallpaper, we can repeat it in two dimensions to cover a wall. Similarly, unit cells are stacked in three dimensions to build crystals.

It can be proved that every unit cell belongs to one of only seven crystal systems (Table 12–5), which are distinguished by the relative lengths of their axes a, b, and c (which are related to the spacing between layers, d), the angles between the axes α, β, and γ (Figure 12–16), and the symmetry of the resulting three-dimensional patterns.

Crystals have the same symmetry as their constituent unit cells, since all crystals are repetitive multiples of such cells. The corners of unit cells of simple compounds are occupied by atoms, molecules, or ions. Equivalent sets of particles may occupy certain other positions, resulting in different modifications of the same

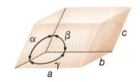

FIGURE 12–16
Representation of a unit cell.

TABLE 12–5 The Unit Cell Dimensions for the Seven Crystal Systems

System	Axis Lengths	Angles	Example (Common Name)
cubic	$a = b = c$	$\alpha = \beta = \gamma = 90°$	NaCl (rock salt)
tetragonal	$a = b \neq c$	$\alpha = \beta = \gamma = 90°$	TiO_2 (rutile)
orthorhombic	$a \neq b \neq c$	$\alpha = \beta = \gamma = 90°$	$MgSO_4 \cdot 7H_2O$ (epsomite)
monoclinic	$a \neq b \neq c$	$\alpha = \gamma = 90°$; $\beta \neq 90°$	$CaSO_4 \cdot 2H_2O$ (gypsum)
triclinic	$a \neq b \neq c$	$\alpha \neq \beta \neq \gamma \neq 90°$	$K_2Cr_2O_7$ (potassium dichromate)
hexagonal	$a = b \neq c$	$\alpha = \beta = 90°$; $\gamma = 120°$	SiO_2 (silica)
rhombohedral	$a = b = c$	$\alpha = \beta = \gamma \neq 90°$	$CaCO_3$ (calcite)

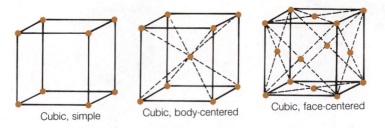

FIGURE 12–17 The cubic crystal lattices.

Cubic, simple   Cubic, body-centered   Cubic, face-centered

unit cell, giving rise to a total of fourteen crystal lattices. The unit cells of three of the simplest ones are shown in Figure 12–17. Different substances that crystallize with the same atomic arrangement in the same type of lattice are said to be **isomorphous.** When a single substance can crystallize in more than one form, it is said to be **polymorphous.** In the simple cubic lattice, equivalent particles occupy only the eight corners of the cubic unit cell. In the body-centered cubic (bcc) lattice, an additional particle is located at the center of the unit cell. The face-centered cubic (fcc) structure involves eight particles at the corners and an additional six particles, one in the middle of each of the six square faces of the cube.

In certain types of crystals, particles other than those forming the framework of the unit cell may occupy extra positions within the unit cell. For example, sodium chloride crystallizes in a face-centered cubic lattice. The unit cell can be visualized as consisting of chloride ions at the corners and middles of the faces, and sodium ions occupying spaces on the edges between the chloride ions and a space at the center (Figure 12–18a). By simple translation of the outline of the unit cell by half a unit cell in any direction within the lattice, we could visualize an alternative unit cell in which sodium and chloride ions have exchanged positions. Such an exchange is not possible for most types of crystals. Most particles of a unit cell are not entirely within the unit cell. Those at corners are shared by the eight unit cells that meet at the center of the particle. Those lying on edges, but not on corners, are shared by four unit cells; and those on faces are shared by two unit cells (Figure 12–19). While a unit cell of NaCl *may appear* to consist of fourteen chloride ions and twelve sodium ions (or the reverse), it really consists of $8\,(\frac{1}{8}) + 6\,(\frac{1}{2}) = 4$ chloride ions and $12\,(\frac{1}{4}) + 1 = 4$ sodium ions (or the reverse). The unit cell reflects the symmetry of the crystal and the stoichiometric proportions of its chemical formula.

We have discussed the structure of NaCl in some detail because it is quite easy

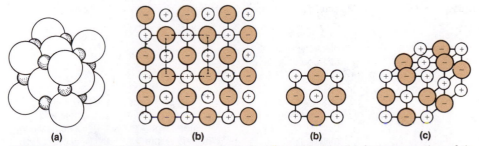

(a)   (b)   (b)   (c)

FIGURE 12–18 (a) The crystal structure of sodium chloride. (b) A cross-section of the lattice of NaCl, showing the repeating pattern of its unit cell at the right. The dashed lines outline an alternative choice of the unit cell. Note that the entire pattern is generated by repeating either unit cell (and its contents) in all three directions. Several such choices of unit cells are usually possible. (c) The three-dimensional representation of (b).

FIGURE 12–19 (a) The sharing of a corner atom by eight unit cells. (b) The sharing of a face-centered atom by two unit cells. (c) A representation of the unit cell of sodium chloride that indicates the relative sizes of the Na^+ and Cl^- ions as well as how ions are shared between unit cells. Particles at the corners and faces of unit cells are shared by other unit cells.

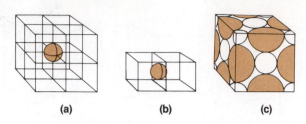

(a)   (b)   (c)

to visualize. However, it should be noted that many compounds with complicated molecules or complicated formula units are described by lattices with unit cells that are much more difficult to visualize. Experimental determinations of the crystal structures of such solids are usually correspondingly more complex as well.

12–11   Structure and Bonding in Crystalline Solids

We can now classify crystalline solids in four categories depending upon the types of particles making up the crystal and the bonding or interactions among them. The four categories are (1) molecular solids, (2) covalent solids, (3) ionic solids, and (4) metallic solids. Table 12–6 summarizes some of their important characteristics.

1   Molecular Solids

The lattice positions that describe unit cells of molecular solids are occupied by molecules or monatomic elements (sometimes referred to as monatomic molecules). While the bonds *within* the molecules are covalent, the forces of attraction *between* molecules are quite weak. They range from hydrogen bonds and weaker dipole-dipole interactions in polar molecules like H_2O and SO_2 to very weak London forces in symmetrical, nonpolar molecules such as CH_4, CO_2, and O_2 and monatomic elements such as the noble gases. Because of the relatively weak intermolecular forces of attraction, molecules can be easily displaced; molecular solids are characteristically soft substances with low melting points. Because electrons do not move

London forces are also present among polar molecules.

TABLE 12–6   Characteristics of Types of Solids

	Molecular	Covalent	Ionic	Metallic
Particles of unit cell	Molecules (or atoms)	Atoms	Anions, cations	Metal ions in "electron gas"
Strongest interparticle forces	London, dipole-dipole, and/or hydrogen bonds	Covalent bonds	Electrostatic	Metallic bonds (electrical attraction between cations and electrons)
Properties	Soft; poor heat and electrical conductors; low melting points	Very hard; poor heat and electrical conductors; high melting points	Hard; brittle; poor heat and electrical conductors; high melting points	Soft to very hard; good heat and electrical conductors; wide range of melting points
Examples	CH_4 (methane), P_4, O_2, Ar, CO_2, H_2O, S_8	C (diamond), SiO_2 (quartz)	NaCl, $CaBr_2$, K_2SO_4 (typical salts)	Li, K, Ca, Cu, Cr, Ni (metals)

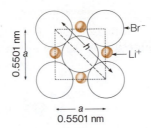

0.5501 nm
a

Br⁻
Li⁺

a
0.5501 nm

LiBr

FIGURE 12–20 One face of the face-centered cubic lattice of lithium bromide.

from one molecule to another under ordinary conditions, molecular solids are poor electrical conductors and good insulators.

2   Covalent Solids

Although atoms occupy the lattice points of covalent solids, these solids should not be confused with molecular solids that consist of *monatomic elements*, such as argon. There are no covalent bonds in monatomic molecular solids (sometimes classified separately as "atomic solids"). Covalent solids (or "network solids") are really giant molecules consisting of covalently bonded atoms in an extended, rigid crystalline network. Diamond (one crystalline form of carbon, Section 22–18) and quartz, SiO_2 (Section 22–24), are examples of covalent solids. Because of their rigid, strongly bonded structures, most covalent solids are very hard and melt at high temperatures. Since the valence electrons are localized in covalent bonds, they are not freely mobile, so covalent solids are usually poor thermal and electrical conductors at ordinary temperatures.

3   Ionic Solids

Most salts crystallize as ionic solids, with ions occupying lattice sites. Sodium chloride has been cited as an example. Since the ions in an ionic solid are not free to move, other than to vibrate about their fixed positions, ionic solids are poor electrical and thermal conductors. However, molten salts are excellent conductors because their ions are freely mobile.

Ionic radii like those in Figure 6–5 are obtained from x-ray crystallographic determinations of unit cell dimensions, assuming that adjacent ions are in contact with each other. Examples 12–2 and 12–3 illustrate such calculations.

Example 12–2

Lithium bromide, LiBr, crystallizes in the NaCl face-centered cubic lattice with a unit cell edge length of $a = b = c = 0.5501$ nm. Assume that the bromide ions at the corners of the unit cell are in contact with those at the centers of the faces. Determine the ionic radius of the bromide ion. One face of the unit cell is depicted in Figure 12–20.

Solution

We may visualize the face as consisting of two right isosceles triangles sharing a common hypotenuse, h, and having sides of length $a = 0.5501$ nm. The

hypotenuse is equal to four times the radius of the bromide ion.

$$h = 4r_{Br^-}$$

The hypotenuse can be calculated from the Pythagorean theorem, $h^2 = a^2 + a^2$.

$$h = \sqrt{a^2 + a^2} = \sqrt{2a^2} = \sqrt{2(0.5501 \text{ nm})^2}$$
$$= 0.7780 \text{ nm}$$

The radius of the bromide ion is one-fourth of h:

$$r_{Br^-} = \frac{0.7780 \text{ nm}}{4} = \underline{0.1945 \text{ nm}}$$

This value agrees well with the ionic radius of the bromide ion listed in Figure 6–5. However, the tabulated value is the *average* value obtained from a number of crystal structures of compounds containing bromide ions. Calculations of ionic radii are based on the assumption that anion-anion contact exists. This may not always be true. Therefore, calculated radii vary from structure to structure. We should not place too much emphasis on an ionic radius value obtained from any single structure determination.

Example 12–3

Refer to Example 12–2. Calculate the ionic radius of Li$^+$ in LiBr, assuming anion-cation contact along an edge of the unit cell.

Solution

The edge length, $a = 0.5501$ nm, is two times the radius of the bromide ion plus two times the radius of the lithium ion. We know from Example 12–2 that the radius of the bromide ion is 0.1945 nm. Then

$$0.5501 \text{ nm} = 2r_{Br^-} + 2r_{Li^+}$$
$$2r_{Li^+} = 0.5501 \text{ nm} - 2(0.1945 \text{ nm})$$
$$2r_{Li^+} = 0.1611 \text{ nm}$$
$$r_{Li^+} = \underline{0.0806 \text{ nm}}$$

The radius of the Li$^+$ ion is calculated to be 0.0806 nm, but Figure 6–5 lists it as 0.060 nm. The discrepancy lies in the fallacy of the assumption that the lithium ion is in contact with both bromide ions simultaneously. It is simply too small, and is free to vibrate about a fixed center position between two large Br$^-$ ions. We now see that there is occasionally some difficulty in determining precise values of ionic radii. Similar difficulties can arise in the determination of atomic radii from study of molecular and covalent solids, or metallic radii from solid metals.

An ionic compound of the MX type contains one cation and one anion per formula unit.

a Coordination Number   Most salts of the MX type crystallize in one of the three kinds of lattices shown in Figure 12–21. A convenient way of describing the environment of each cation is to specify its **coordination number** (the number of anions that are its nearest neighbors) and the *symmetry* of the position that it occupies. In the cesium chloride structure (Figure 12–21a), the positions of the anions describe a simple cube. The coordination number of the Cs$^+$ ion is 8 in the CsCl lattice (simple cubic). The eight neighboring Cl$^-$ ions are at the corners of a cube surrounding Cs$^+$, so this cation is said to occupy a *cubic hole.*

As we have already seen, the sodium chloride structure is face-centered cubic with respect to the Cl$^-$ ions. By counting the number of Cl$^-$ ions that are closest to the Na$^+$ ion in the center of the unit cell (Figure 12–21b) we can see that the coordination number of Na$^+$ is 6 and that this ion sits in an *octahedral hole* defined by the six nearest Cl$^-$ ions.

The sulfide ions in the zinc sulfide (or zinc blende) lattice form a face-centered cubic arrangement (Figure 12–21c). Each Zn^{2+} ion is surrounded tetrahedrally by four sulfide ions. So Zn^{2+} has a coordination number of 4 and sits in a *tetrahedral hole.*

It is important to re-emphasize that the structures shown in Figure 12–21 have been expanded to show spatial arrangements more clearly. Actually there is

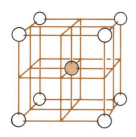

(a) Cesium chloride
CsCl; ● = Cs$^+$, ○ = Cl$^-$

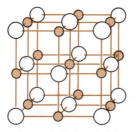

(b) Sodium chloride
NaCl; ● = Na$^+$, ○ = Cl$^-$

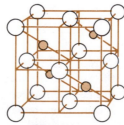

(c) Zinc blende
ZnS; ● = Zn^{2+}, ○ = S^{2-}

FIGURE 12–21 The coordination number of a cation in an ionic solid is the number of anions that are its nearest neighbors. It is 8 for Cs$^+$, 6 for Na$^+$, and 4 for Zn^{2+} in these structures.

anion-anion contact in most ionic crystals, and sometimes anion-cation contact. This situation occurs because most cations are smaller than most anions (Section 6–7). The structure of an ionic crystal depends not only upon the stoichiometry of the compound (electroneutrality must be maintained) but also upon the relative sizes of cation and anion. A relatively large cation can fit more anions of a particular kind around it than a smaller cation can. Similarly, a particular cation can accommodate more small anions around it than it can large anions. Thus cations tend to exhibit high coordination numbers when they are large compared to other cations, and when the anions are small compared to other anions.

b Crystal Lattice Energy The energy holding the particles together in a crystalline arrangement is called the **crystal lattice energy, ΔH_{xtal}**. It is a measure of the stability of an ionic solid. The more negative its value, the more energy is released in the hypothetical reaction in which a mole of ionic solid is formed from its constituent ions in the gaseous state. The definition is illustrated below for sodium chloride.

The symbol U, rather than ΔH_{xtal}, is often used to represent crystal lattice energy. The abbreviation "xtal" is commonly used for "crystal."

$$Na^+ (g) + Cl^- (g) \longrightarrow NaCl (s) \qquad \Delta H_{rxn} = \Delta H_{xtal} \qquad \text{[Eq. 12–4]}$$

We can apply Eq. 12–2 to determine the energy of attraction of the ions in a crystal. When the ions are in contact, d is equal to the sum of the radii of the ions. We can convert this proportionality into an equality by inserting a proportionality constant, C, whose value depends upon the kind of crystal lattice.

$$E = \Delta H_{xtal} = C \frac{q^+ q^-}{d} \qquad \text{(crystal lattice energy)} \qquad \text{[Eq. 12–5]}$$

This much energy must be supplied in order to convert a mole of ionic solid completely into gaseous ions. In order to melt the ionic solid, *some* of this energy must be supplied but not all of it since there are significant interactions among ions in the molten state.

If this were not so, the melt would vaporize to form gaseous ions spontaneously.

Equation 12–5 tells us that the crystal lattice energy increases in magnitude (becomes more negative) as the charges on the ions increase. Thus most ionic solids containing multiply charged ions have more negative crystal lattice energies than those containing only univalent ions. To illustrate, ΔH_{xtal} is -1.14×10^4 kJ/mol for SnO_2 and -3.89×10^3 kJ/mol for MgO, but only -781 kJ/mol for LiBr and -682 kJ/mol for RbCl. The equation also shows that ΔH_{xtal} gets larger (more negative) as the separation between anion and cation decreases, that is, as r_+ and r_- decrease. For instance, ΔH_{xtal} is -585 kJ/mol for CsI ($r = 0.169$ nm for Cs^+, 0.216 nm for I^-), but it is considerably larger when smaller ions of the same charges are involved, such as -1.03×10^3 kJ/mol for LiF ($r = 0.060$ nm for Li^+, 0.136 nm for F^-).

The symbols r_+ and r_- represent the radii of cations and anions, respectively.

c Born-Haber Cycle Crystal lattice energy *cannot* be measured directly in most cases. However, by application of Hess' Law (Section 3–7) to a series of reactions beginning with consumption of the constituent elements in their atomic or molecular form and ending with the production of the ionic solid, ΔH_{xtal} can be evaluated. Such a sequence of reactions for ionic substances is called a **Born-Haber cycle,** after Max Born and Fritz Haber, who developed the general procedure. Its use is illustrated below for sodium chloride.

The heat of reaction can easily be determined:

$$Na (s) + \tfrac{1}{2}Cl_2 (g) \longrightarrow NaCl (s) \qquad \Delta H_{rxn} = \Delta H_{f\,NaCl\,(s)} = -411 \text{ kJ}$$

The following reactions and experimental data can be summed to give the same equation:

Sublimation is the process by which a solid is converted directly to a gas (Section 12–13).

(1) Sublimation (vaporization) of sodium metal. The heat change is ΔH_f for Na (g) $= \Delta H_{subl}$.

$$\text{Na (s)} \longrightarrow \text{Na (g)} \qquad \Delta H_{subl} = 109 \text{ kJ}$$

(2) Dissociation of one-half mole of Cl_2 (g) into gaseous chlorine atoms. The heat change is ΔH_f for Cl (g) $= \frac{1}{2}\Delta H_{diss}$ (also $\frac{1}{2}$ the bond energy for Cl_2 (g)).

$$\tfrac{1}{2}Cl_2 \text{ (g)} \longrightarrow \text{Cl (g)} \qquad \tfrac{1}{2}\Delta H_{diss} = 122 \text{ kJ}$$

(3) Ionization of a mole of sodium atoms. The heat change is the ionization energy for sodium expressed in kilojoules,

$$\text{Na (g)} \longrightarrow \text{Na}^+ \text{ (g)} + e^- \qquad \Delta H_{ie} = 496 \text{ kJ}$$

(4) Addition of one mole of electrons to one mole of gaseous chlorine atoms. This heat change is the electron affinity of chlorine expressed in kilojoules,

$$\text{Cl (g)} + e^- \longrightarrow \text{Cl}^- \text{ (g)} \qquad \Delta H_{ea} = -348 \text{ kJ}$$

(5) Condensation of the gaseous ions to form a mole of solid NaCl. This heat change cannot be measured directly.

$$\text{Na}^+ \text{ (g)} + \text{Cl}^- \text{ (g)} \longrightarrow \text{NaCl (s)} \qquad \Delta H_{xtal} = ?$$

Summation of these five reactions and their ΔH values allows us to calculate ΔH_{xtal}.

(1)	Na (s) $\longrightarrow$ Na (g)	$\Delta H_{subl} = 109$ kJ
(2)	$\tfrac{1}{2}Cl_2$ (g) $\longrightarrow$ Cl (g)	$\tfrac{1}{2}\Delta H_{diss} = 122$ kJ
(3)	Na (g) $\longrightarrow$ Na$^+$ (g) $+ e^-$	$\Delta H_{ie} = 496$ kJ
(4)	Cl (g) $+ e^- \longrightarrow$ Cl$^-$ (g)	$\Delta H_{ea} = -348$ kJ
(5)	Na$^+$ (g) $+$ Cl$^-$ (g) $\longrightarrow$ NaCl (s)	$\Delta H_{xtal} = ?$
	Na (s) $+ \tfrac{1}{2}Cl_2$ (g) $\longrightarrow$ NaCl (s)	$\Delta H_{rxn} =$ sum of five ΔH's given above $= -411$ kJ

We showed by Hess' Law (Section 3–7) that for a series of reactions, the overall ΔH_{rxn} value can be related to the enthalpy changes for the individual reactions. In this case it is

$$\Delta H_{rxn} = \Delta H_{subl} + \tfrac{1}{2}\Delta H_{diss} + \Delta H_{ie} + \Delta H_{ea} + \Delta H_{xtal}$$
$$-411 \text{ kJ} = 109 \text{ kJ} + 122 \text{ kJ} + 496 \text{ kJ} + (-348 \text{ kJ}) + \Delta H_{xtal}$$

Solving for ΔH_{xtal},

$$\Delta H_{xtal} = (-411 - 109 - 122 - 496 + 348)\text{kJ}$$
$$= -790 \text{ kJ per mole of NaCl (s)}$$

The entire Born-Haber cycle is summarized in Figure 12–22. The value of -790 kJ per mole of NaCl (s) suggests that this is a very stable solid.

4   Metallic Solids

Metals crystallize as metallic solids in which metal ions occupy the lattice sites and are embedded in a cloud of delocalized valence electrons. Practically all metals

Crystal lattice energies cannot be measured directly, but they can be calculated from other known enthalpy values.

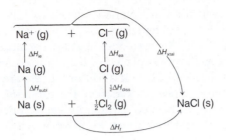

FIGURE 12-22 The Born-Haber cycle for NaCl (s).

(a)

(b)

crystallize in one of three types of lattices: (1) body-centered cubic, (2) face-centered cubic (also called cubic close-packed), and (3) hexagonal close-packed. The latter two types are called close-packed structures because the particles (in this case metal atoms) are packed together as closely as possible. The differences between the two close-packed structures are illustrated in Figures 12-23 and 12-24. We shall let spheres of equal size represent the identical metal atoms, or any other particles, that form close-packed structures. Consider a layer of spheres packed in a plane, A, as closely as possible (Figure 12-24a). An identical plane of spheres, B, is placed in the

FIGURE 12-23 (a) Spheres in the same plane, packed as closely as possible; each sphere touches six others. (b) Spheres in two planes, packed as closely as possible. In real crystals, there are far more than two planes. Each sphere touches six others in its own layer, three others in the layer below it, and three in the layer above it, resulting in a total contact (coordination number) for each sphere of 12 other spheres.

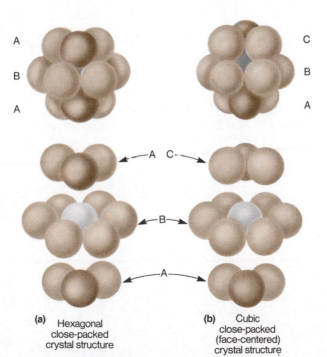

(a) Hexagonal close-packed crystal structure

(b) Cubic close-packed (face-centered) crystal structure

FIGURE 12-24 There are two crystal structures in which atoms are packed together as compactly as possible. The diagrams show the structures expanded to clarify the difference between them. In the hexagonal close-packed structure (left), the first and third layers are oriented in the same direction. In the cubic close-packed structure (right), the layers are oriented in opposite directions. In both cases, every atom is surrounded by 12 other atoms if the structure is extended infinitely, and so has a coordination number of 12. Although it is not obvious from this figure, the cubic close-packed structure is also face-centered cubic. To see this we would have to include additional atoms and rotate the resulting cluster of atoms.

depressions of plane *A*. If the third plane is placed with its spheres directly above those in plane *A*, the *ABA* arrangement results. This is the hexagonal close-packed structure (Figure 12–24a). The extended pattern of arrangement of planes is *ABABAB*. . . . If the third layer is placed in the alternate set of depressions in the second layer so that spheres in the first and third layers are *not* directly above and below each other, the cubic close-packed (fcc) structure, *ABCABCABC* . . . , results (Figure 12–24b). In closest-packed structures each sphere has a coordination number of 12, i.e., 12 nearest neighbors. In all (ideal) close-packed structures it is found that 76% of a given volume is due to spheres and 24% is empty space. The body-centered cubic structure is less efficient in packing; each sphere has only eight nearest neighbors, and there is more empty space.

Example 12–4

The unit cell of metallic silver is face-centered cubic (or cubic close-packed) with $a = b = c = 0.4086$ nm. Calculate (a) the radius of a silver atom, (b) the volume of a silver atom in cm^3 given that for a sphere $V = (4/3) \pi r^3$, and (c) the volume of a unit cell in cm^3.

Solution

(a) One face of the face-centered cubic (fcc) unit cell can be visualized as consisting of (parts of) five silver atoms describing two right triangles sharing a hypotenuse. (This can be visualized by picturing the Br^- ions in Figure 12–20 as Ag atoms and removing the Li^+ ions.) The hypotenuse, h, is four times the radius of the silver atom. By the Pythagorean theorem we can evaluate r_{Ag}.

$$h = \sqrt{a^2 + a^2} = \sqrt{2a^2}$$
$$= \sqrt{2(0.4086 \text{ nm})^2}$$
$$= 0.5778 \text{ nm} = 4r_{Ag}$$
$$r_{Ag} = \frac{0.5778 \text{ nm}}{4} = \underline{0.1444 \text{ nm}}$$

(b) The volume of a silver atom is $(4/3)\pi r_{Ag}^3$, or

$$V = (4/3)\pi(0.1444 \text{ nm})^3 = 0.01261 \text{ nm}^3$$

We now convert nm^3 to cm^3 (using $1 \text{ nm} = 10^{-7}$ cm):

$$\underline{?} \text{ cm}^3 = 0.01261 \text{ nm}^3 \times \left(\frac{10^{-7} \text{ cm}}{\text{nm}}\right)^3$$
$$= \underline{1.261 \times 10^{-23} \text{ cm}^3}$$

(c) Since the unit cell is cubic, its entire volume is a^3.

$$V_{\text{unit cell}} = (0.4086 \text{ nm})^3 = 6.822 \times 10^{-2} \text{ nm}^3$$

We now convert this to cm^3.

$$\underline{?} \text{ cm}^3 = 6.822 \times 10^{-2} \text{ nm}^3 \times \left(\frac{1 \times 10^{-7} \text{ cm}}{\text{nm}}\right)^3$$
$$= \underline{6.822 \times 10^{-23} \text{ cm}^3}$$

Example 12–5

From data in Example 12–4, calculate the density of metallic silver. Use 6.022×10^{23} as Avogadro's number.

Solution

We first determine the mass of a unit cell, that is, the mass of four atoms of silver.

$$\underline{?} \text{ g Ag} = 4 \text{ Ag atoms} \times \frac{1 \text{ mol Ag}}{6.022 \times 10^{23} \text{ Ag atoms}}$$
$$\times \frac{107.87 \text{ g Ag}}{\text{mol Ag}}$$
$$= 7.165 \times 10^{-22} \text{ g Ag}$$

The density of the unit cell, and therefore of bulk silver, is its mass divided by its volume.

$$\text{Density} = \frac{7.165 \times 10^{-22} \text{ g}}{6.822 \times 10^{-23} \text{ cm}^3} = \underline{10.50 \text{ g/cm}^3}$$

A handbook gives the density of silver at 20°C as 10.5 g/cm^3.

Phase Changes and Associated Enthalpy Changes

12–12 Changes of Physical State and Heats of Transition

1 Heat Transfer in Transitions Between Solid and Liquid States

The **melting point** of a solid substance, which is the same as the **freezing point** of its liquid, *is the temperature at which the rate of melting of a solid is the same as the rate of freezing of its liquid under a given applied pressure,* usually atmospheric pressure. Stated differently, the melting point of a solid or freezing point of the liquid is *the temperature at which the solid and liquid phases of a substance exist in equilibrium.* The melting point of a substance at one atmosphere pressure is called its *normal* melting point.

The double half-arrows signify that both forward and reverse processes occur at this temperature.

$$\text{Liquid} \underset{\text{melting}}{\overset{\text{freezing}}{\rightleftarrows}} \text{Solid}$$

When heat is added to a solid below its melting point, its temperature rise is determined by its specific heat (Section 3–2). After enough heat has been added to reach the melting point, the addition of more heat is not accompanied by a further temperature rise; instead, the solid begins to melt. Only after all the solid has melted

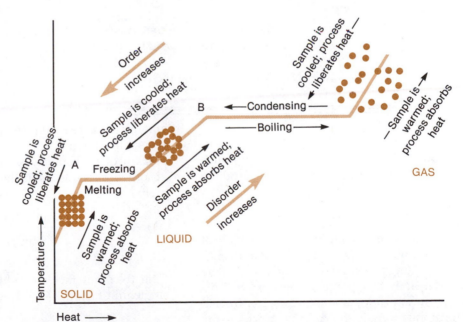

FIGURE 12–25 This heating curve shows that when heat energy is added to a solid below its melting point, the temperature of the solid rises until its melting point is reached (point A). If the solid is heated at its melting point, its temperature remains constant until the solid has melted, because the melting process requires energy. When all of the solid has melted, heating the liquid raises its temperature until its boiling point is reached (point B). If heat is added to the liquid at its boiling point, the added heat energy is absorbed as the liquid boils. When all the liquid has been converted to a gas (vapor), the addition of more heat raises the temperature of the gas. Each step in the process can be reversed by removing heat.

Specific heats and molar heat capacities have units of energy/g·°C and energy/mol·°C, respectively, but heats of transition have no temperature units because no temperature changes accompany the processes to which they apply.

TABLE 12–7   Melting Points and Heats of Fusion of Some Common Substances

Substance	Melting Point (°C)	Heat of Fusion (J/g)	Heat of Fusion (kJ/mol)
methane	−182	58.6	0.92
ethyl alcohol	−117	109	5.02
water	0	334	6.02
aluminum	658	395	10.6
sodium chloride	801	519	30.3

will the continued addition of heat result in an increase in the temperature of the liquid. This is shown on the left side of Figure 12–25, a typical heating curve.

Notice that melting is always endothermic. The term "fusion" means melting.

The amount of heat that must be absorbed to melt one gram of a solid at its melting point is its **enthalpy of fusion** *or* **heat of fusion**. The **molar heat of fusion** (ΔH_{fus}, kJ/mol) *is the amount of heat that must be absorbed to melt one mole of a solid at its melting point.* The heat of fusion is a measure of the *inter*molecular forces of attraction in the solid state. Heats of fusion are usually larger for substances with high melting points than for those with low melting points. Melting points and heats of fusion of some common substances are shown in Table 12–7.

The **enthalpy of solidification** or **heat of solidification**, ΔH_{sol}, of a liquid is *equal in magnitude, but opposite in sign, to the* **enthalpy (heat) of fusion**, ΔH_{fus}.

$$\Delta H_{sol} = -\Delta H_{fus}$$

[Eq. 12–6]

It represents the amount of heat that must be *removed* from a given amount of liquid to solidify the liquid at its freezing point. For water,

$$\text{Water} \underset{+334 \text{ J/g}}{\overset{-334 \text{ J/g}}{\rightleftharpoons}} \text{Ice} \qquad (\text{at } 0°C)$$

Example 12–6

The molar heat of fusion, ΔH_{fus}, of sodium at its melting point, 97.5°C, is 2.6 kJ/mol. How much heat must be absorbed by 5.0 grams of solid sodium at 97.5°C to convert it to molten sodium?

Solution

The molar heat of fusion tells us that every mole of Na, 23 grams, will absorb 2.6 kJ of heat at 97.5°C during the melting process. We are interested in the amount of heat that 5.0 grams would absorb. Thus we choose the appropriate unit factors, constructed from the atomic weight and ΔH_{fus}, to calculate the

amount of heat absorbed by 5.0 g of Na. The heat of fusion, like other heats of transition, is an *extensive property* of a substance, because the amount of heat absorbed depends upon the amount of solid that melts.

$$\underline{?}\text{ kJ} = 5.0 \text{ g Na} \times \frac{1 \text{ mol Na}}{23 \text{ g Na}} \times \frac{2.6 \text{ kJ}}{1 \text{ mol Na}}$$

$$= \underline{0.57 \text{ kJ}}$$

(The two unit factors could be combined into a single unit factor, 2.6 kJ/23 g Na.)

Example 12–7

Calculate the amount of heat that must be absorbed by 50.0 grams of ice at −12.0°C to convert it to water at 20.0°C.

Solution

We must determine the heat absorbed during three processes: (1) raising the temperature of 50.0 g of ice from −12.0°C to its melting point, 0.0°C, which is governed by the specific heat of ice, 2.09 J/g·°C; (2) melting the ice with no change in temperature at

0.0°C, which is governed by the heat of fusion of ice, 334 J/g; and (3) raising the temperature of the resulting water from 0.0°C to 20.0°C, which is governed by the specific heat of water, 4.18 J/g·°C.

$$(1) \quad 50.0 \text{ g} \times \frac{2.09 \text{ J}}{\text{g} \cdot {}^{\circ}\text{C}} \times [0.0 - (-12.0)]{}^{\circ}\text{C} = 1.25 \times 10^3 \text{ J} = 0.125 \times 10^4 \text{ J}$$

$$(2) \quad 50.0 \text{ g} \times \frac{334 \text{ J}}{\text{g}} \qquad\qquad\qquad\qquad = 1.67 \times 10^4 \text{ J}$$

$$(3) \quad 50.0 \text{ g} \times \frac{4.18 \text{ J}}{\text{g} \cdot {}^{\circ}\text{C}} \times (20.0 - 0.0){}^{\circ}\text{C} \quad = 4.18 \times 10^3 \text{ J} = 0.418 \times 10^4 \text{ J}$$

$$\text{Total heat absorbed} \quad = 2.21 \times 10^4 \text{ J} = 22.1 \text{ kJ}$$

Note that most of the heat absorbed was required in Step 2, melting the ice.

2   Heat Transfer in Transitions Between Liquid and Vapor States

Just as for solids, heat must be added to a liquid in proportion to its specific heat in order to raise its temperature. If heat is supplied to a liquid at a constant rate, the temperature rises at a constant rate until the boiling point of the liquid is reached. At that point the temperature remains constant until enough heat has been added to boil away all the liquid (Figure 12–25). As defined in Section 12–7, the boiling point of a liquid is the temperature at which the vapor pressure of the liquid equals the applied pressure, and the *normal* boiling point is the temperature at which the vapor pressure equals one atmosphere.

Only *molar* enthalpy of vaporization is properly designated by ΔH_{vap}, and only *molar* enthalpy of fusion is designated by ΔH_{fus}.

The **molar enthalpy (heat) of vaporization** (ΔH_{vap}) *of a liquid is the amount of heat that must be added to one mole of the liquid at its normal boiling point to convert it to vapor with no change in temperature.* Heats or enthalpies of vaporization can also be expressed in joules per gram. The heat of vaporization for water is 2.26×10^3 J/g or, as shown below, 40.7 kJ/mol.

$$\underline{?} \frac{\text{kJ}}{\text{mol}} = \frac{2.26 \times 10^3 \text{ J}}{\text{g}} \times \frac{1 \text{ kJ}}{1000 \text{ J}} \times \frac{18.02 \text{ g}}{\text{mol}} = 40.7 \frac{\text{kJ}}{\text{mol}}$$

As with many other properties of liquids, their heats of vaporization and boiling points reflect the strengths of *inter*molecular forces that "hold the molecules together" as a liquid. Comparison of the data in Tables 12–7 and 12–8 shows that for water and ethyl alcohol (or for any other substance, for that matter), ΔH_{vap} is considerably larger than ΔH_{fus}. As we have seen, this is true because the *intermo-*

Boiling points are much more sensitive to pressure than are melting points.

TABLE 12–8   Heats of Vaporization and Boiling Points of Some Common Liquids

Liquid	Boiling Point at 1 Atmosphere (°C)	Heat of Vaporization at Boiling Point (J/g)	(kJ/mol)
water	100	2.26×10^3	40.7
ethyl alcohol	78.3	855	39.3
benzene	80	395	30.8
diethyl ether	34.6	351	26.0

lecular forces in liquids are much stronger than those in gases. Heats of vaporization are generally larger for substances whose boiling points (and intermolecular forces) are greater, as the data in Table 12–8 show. The unusually high heat of vaporization for water, particularly on a joule-per-gram basis, makes it very efficient as a coolant and, in the form of steam, as a source of heat.

Liquids can be converted to vapors even below their boiling points by *evaporation* (Section 12–5). The water in perspiration is an effective coolant for your body because as it evaporates, each gram absorbs 2.41 kilojoules of heat, mostly from your body, and carries it into the air in the form of water vapor. You feel even cooler in a breeze because the evaporation of perspiration is faster and heat is removed more rapidly.

Condensation is the reverse of vaporization. *The amount of heat that must be removed from a vapor to condense it, without change in temperature, is called the* **enthalpy (heat) of condensation,** ΔH_{cond}.

$$\text{Liquid} + \text{Heat} \underset{\text{condensation}}{\overset{\text{vaporization}}{\rightleftharpoons}} \text{Vapor}$$

The heat of condensation of a liquid is equal in magnitude, but opposite in sign, to the heat of vaporization, and is released to the surroundings by the vapor during condensation.

$$\Delta H_{cond} = -\Delta H_{vap} \qquad\qquad \text{[Eq. 12–7]}$$

Since 2.26 kJ/g or 40.7 kJ/mol must be absorbed in order to vaporize water at 100°C, it follows that 2.26 kJ/g or 40.7 kJ/mol must be released to the environment when steam at 100°C condenses to form liquid water at 100°C. In steam-heated radiators, steam condenses and releases 2.26 kJ of heat per gram as its molecules collide with the cooler radiator walls and condense there. The metallic walls are good heat conductors, and they transfer the heat to molecules in the air that bounce off the outside walls of the radiator. Other substances such as benzene and diethyl ether are much less effective as heating and cooling agents because of their lower heats of vaporization and condensation (see Table 12–8).

Many large buildings are heated by steam; i.e., the condensation of steam liberates the heat that warms the buildings.

The heat of vaporization of water at normal body temperature, about 37°C, is 2.41 kJ/g. It is 2.26 kJ/g at 100°C.

The molecules in air are mostly nitrogen, N_2, and oxygen, O_2.

Example 12–8

Calculate the amount of heat in joules required to convert 180 grams of water at 10.0°C to steam at 105.0°C.

Solution

The total amount of heat absorbed is the sum of the amounts required to (1) raise the temperature of the liquid water from 10.0°C to 100.0°C, (2) convert the liquid water to steam at 100°C, and (3) raise the temperature of the steam from 100.0°C to 105.0°C. Steps 1 and 3 involve the specific heats of water and steam, 4.18 J/g·°C and 2.03 J/g·°C, respectively (see Appendix E), while Step 2 involves the heat of vaporization of water (2.26 × 10³ J/g).

(1)　$? \text{ J} = 180 \text{ g} \times \dfrac{4.18 \text{ J}}{\text{g} \cdot °\text{C}} \times (100.0°\text{C} - 10.0°\text{C})$　$= 6.77 \times 10^4 \text{ J} = 0.677 \times 10^5 \text{ J}$

(2)　$? \text{ J} = 180 \text{ g} \times \dfrac{2.26 \times 10^3 \text{ J}}{\text{g}}$　$\qquad\qquad\qquad\qquad = 4.07 \times 10^5 \text{ J}$

(3)　$? \text{ J} = 180 \text{ g} \times \dfrac{2.03 \text{ J}}{\text{g} \cdot °\text{C}} \times (105.0°\text{C} - 100.0°\text{C}) = 1.83 \times 10^3 \text{ J} = 0.0183 \times 10^5 \text{ J}$

Heat absorbed $= \underline{4.77 \times 10^5 \text{ J}}$

Example 12-9

Compare the amount of "cooling" experienced by an individual who drinks 400 mL of ice-water (0°C) with the amount of "cooling" experienced by an individual who "sweats out" 400 mL of water.

Solution

The density of water is very nearly 1.00 g/mL at both 0°C and 37°C, average body temperature. The heat of vaporization of water is 2.41 kJ/g at 37°C. The amount of heat required to raise the temperature of 400 g of water from 0°C to 37°C is

$$? J = (400 \text{ g})(4.18 \text{ J/g} \cdot °C)(37°C)$$
$$= 6.19 \times 10^4 \text{ J or } 61.9 \text{ kJ}$$

Evaporating, or "sweating out," 400 mL of water at 37°C requires

$$? J = (400 \text{ g})(2.41 \times 10^3 \text{ J/g})$$
$$= 9.64 \times 10^5 \text{ J or } 964 \text{ kJ}$$

Thus, "sweating out" 400 mL of water removes 964 kJ of heat from one's body, while drinking 400 mL of ice-water removes only 61.9 kJ. Stated differently, "sweating" is 15.6 times (964/61.9) more efficient than drinking ice-water!

12-13 Sublimation and the Vapor Pressure of Solids

This behavior led to the trademark for solid carbon dioxide, Dry Ice. It is used for shipping items, such as biological samples, that must be refrigerated; the "ice" does not drip. The sublimation temperature of CO_2 is −78.5°C.

Some solids, such as iodine and solid carbon dioxide, vaporize without passing through the liquid state at atmospheric pressure. We say they **sublime.** Solids exhibit vapor pressures just as liquids do, but generally have much lower vapor pressures. Solids with high vapor pressures sublime easily. The characteristic odors of the common household solids naphthalene (moth balls) and paradichlorobenzene (bathroom deodorizer) are due to sublimation. The reverse process, by which a vapor solidifies without passing through the liquid phase, is called **deposition.**

$$\text{Solid} \xrightleftharpoons[\text{deposition}]{\text{sublimation}} \text{Gas}$$

Some impure solids can be purified by sublimation and subsequent deposition of the vapor (as a solid) onto the surface of a cooler object. A sublimation apparatus is illustrated in Figure 12–26. Iodine is commonly purified by sublimation.

12-14 Phase Diagrams

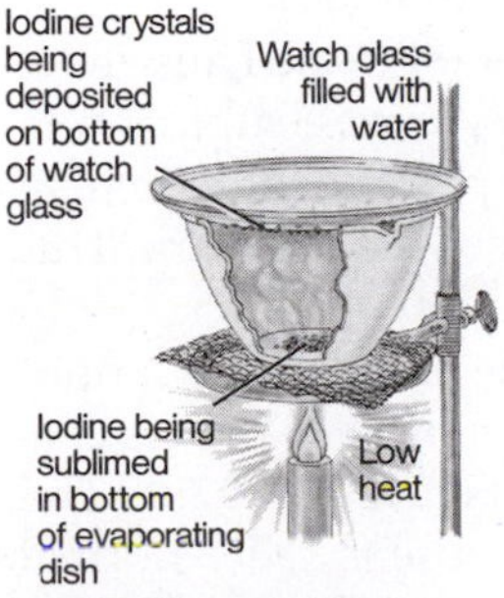

Phase diagrams show the equilibrium pressure-temperature relationships between the different phases of a given substance. Figure 12–27 shows a portion of the phase diagrams for water and carbon dioxide. (The shapes of the curves are exaggerated to allow us to describe the changes of state accompanying pressure or temperature changes using one diagram rather than several.) The curved line from A to B in Figure 12–27a is a vapor pressure curve obtained experimentally by measuring the vapor pressures of water at various temperatures (see Table 12–9).

Iodine crystals being deposited on bottom of watch glass

Watch glass filled with water

Iodine being sublimed in bottom of evaporating dish

Low heat

FIGURE 12-26
Apparatus for the recrystallization of iodine by sublimation.

TABLE 12-9   Points on the Vapor Pressure Curve for Water

temperature (°C)	−10	0	20	30	50	70	90	95	100	101
vapor pressure (torr)	2.1	4.6	17.5	31.8	92.5	234	526	634	760	788

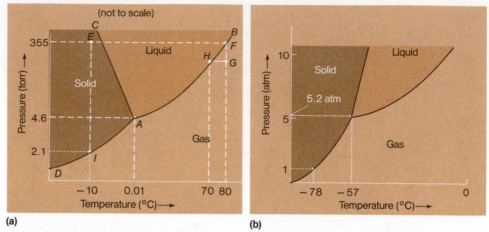

FIGURE 12–27 Phase diagrams. (a) Diagram for water (not to scale). For the few such substances for which the solid is less dense than the liquid, the solid-liquid equilibrium line (*AC*) has negative slope. (b) Diagram for carbon dioxide (not to scale), a substance for which the solid is more dense than the liquid. Note that the solid-liquid equilibrium line has positive slope. This is true for most substances.

The points along this curve represent the temperature-pressure combinations for which liquid and gas (vapor) coexist in equilibrium. In the region directly above *AB* the stable form of water is liquid; below the curve it is vapor.

Line *AC* represents the liquid-solid equilibrium conditions. Note that it has a negative slope. Water is one of the very few substances for which this is the case. The slope up and to the left indicates that increasing the pressure sufficiently on the surface of ice causes it to melt. This is because ice is *less dense* than liquid water in the vicinity of the liquid-solid equilibrium. This causes ice to float in liquid water and is due to the fact that the network of hydrogen bonding in ice, which is more extensive than that in liquid water, requires a greater separation of H_2O molecules than in the liquid. It is this property that helps make ice-skating possible. Frictional heating and the pressure exerted by the skate blade on the ice melt the ice just below it, thus producing a very slick surface which refreezes after the skater has passed. One could not skate on most other solids, even at temperatures and pressures near the solid-liquid equilibrium, because most solids are more compact than their liquids. Most substances, therefore, have positive slopes associated with line *AC*.

The stable form of water at points to the left of *AC* is solid (ice). Thus *AC* is called a melting curve. At pressures and temperatures along *AD*, the sublimation curve, solid and vapor exist in equilibrium. There is only one point, *A*, at which all three phases—ice, liquid water, and water vapor—can coexist at equilibrium. This is called the **triple point**, and for water it occurs at 4.6 torr and 0.01°C.

Note that at pressures below the triple point pressure, the liquid phase cannot exist; rather the substance goes directly from solid to gas (sublimes) or the reverse (crystals are formed directly from the gas). Consider CO_2 (Figure 12–27b). The triple point is at 5.2 atmospheres and −57°C. Since this pressure is *above* normal atmospheric pressure, liquid CO_2 cannot exist at atmospheric pressure.

To illustrate the use of a phase diagram in determining the physical state or states of a system under different sets of pressures and temperatures, let's consider a sample of water at point *E* in Figure 12–27a (355 torr and −10°C). At this point all

Compare the description here with that accompanying Figure 12–25. Each horizontal line on a phase diagram contains the information in one heating curve, obtained at a particular pressure. Phase diagrams are obtained by combining the results of heating curves measured experimentally at different pressures.

the water is in the form of ice. Suppose we hold the pressure constant and gradually increase the temperature; in other words, trace a path from left to right along EF. When we reach the temperature at which EF intersects AC, the melting curve, some of the ice melts; if we stopped here, equilibrium between ice and liquid water would eventually be established. That is, both phases would be present. If we continue to add heat, all the ice melts with no temperature change. Recall that all phase changes of pure substances occur at constant temperature. Once the ice is completely melted, additional heat causes the temperature to rise again. Eventually at point F (355 torr and 80°C) some of the liquid begins to boil; liquid and vapor coexist at equilibrium. Adding more heat at constant pressure vaporizes the rest of the water with no temperature change. Complete vaporization also occurs if, at point F and before all the liquid had vaporized, the temperature were held constant and the pressure were decreased to, say, 234 torr at point G. If we then held the pressure at 234 torr and desired to condense some of the vapor, it would be necessary to cool the vapor to 70°C, point H, at which we again arrive at the vapor pressure curve, AB. To state this in another way, the vapor pressure of water at 70°C is 234 torr.

Suppose we move back to ice at point E (355 torr and $-10°C$) again. If we now hold the temperature at $-10°C$ and reduce the pressure, we move vertically down along EI. At a pressure of 2.1 torr we reach the sublimation curve, at which point some of the solid passes directly into the gas phase (sublimes). Decreasing the pressure further vaporizes all the ice. An important application of this phenomenon is in the freezing-drying of foods. In this process a water-containing food is cooled below the freezing point of water to form ice, which is then removed as a vapor by decreasing the pressure.

The kinetic energies of molecules in the vapor phase oppose the intermolecular attractions during the condensation process. As the temperature of a sample of a gas decreases, the average kinetic energy of the molecules decreases and there is a greater likelihood that the intermolecular attractions can cause liquefaction (particularly under a high pressure, which forces the molecules close together). Each gas has a certain temperature above which no amount of pressure will cause liquefaction. This is called the **critical temperature.** The minimum pressure required to condense a vapor at its critical temperature is called the **critical pressure.**

Critical temperatures increase as intermolecular attractions increase, because more thermal kinetic energy is required to prevent liquefaction. Thus critical temperatures generally increase as boiling points and heats of vaporization increase. Some representative values are shown in Table 12–10.

TABLE 12–10 Critical Temperatures and Normal Boiling Points of Some Common Substances

Substance	Critical Temperature (°C)	Normal Boiling Point (°C)
H_2	306	-253
N_2	399	-196
O_2	427	-183
CO_2	577	-78*
NH_3	673	-33
H_2O	920	100

* Sublimes.

Special Topics in Solid State Properties

12–15   Metallic Bonding and Band Theory

The vast majority of metals crystallize in close-packed lattices (Section 12–11.4). The close packing suggests strong attractive interactions among the eight to 12 nearest neighbors. This is somewhat difficult to rationalize if we recall that each Group IA and Group IIA metal atom has only one or two valence electrons available for bonding, and this is too few to participate in bonds localized between it and each of its nearest neighbors.

Bonding in metals, called **metallic bonding,** results from the electrical attractions among positively charged metal ions and mobile, delocalized electrons belonging to the crystal as a whole. The properties associated with metallic bonding (metallic luster, high thermal and electrical conductivity, and others) are well explained by the **band theory** of metals, which we shall now describe.

The interaction of two atomic orbitals, say the $3s$ orbitals of two sodium atoms, produces two molecular orbitals (one bonding orbital and one antibonding orbital; see Chapter 8). If N atomic orbitals interact, N molecular orbitals are formed. In a single metallic crystal containing one mole of sodium atoms, for example, the interaction of 6.022×10^{23} $3s$ atomic orbitals produces 6.022×10^{23} molecular orbitals. Atoms interact more strongly with those nearby than with those farther away. As discussed in Chapter 8, the energy separating bonding and antibonding molecular orbitals resulting from two given atomic orbitals decreases as the interaction (overlap) between the atomic orbitals decreases. Thus, when all possible interactions among the mole of sodium atoms are considered, there results a series of very closely spaced molecular orbitals (formally σ_{3s} and σ_{3s}^{*}) that form a nearly continuous band of orbitals belonging to the crystal as a whole. The 6.022×10^{23} orbitals in the band are half filled, since one mole of sodium atoms contributes 6.022×10^{23} valence electrons (Figure 12–28).

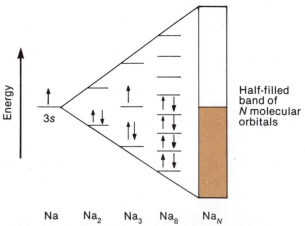

FIGURE 12–28 The band of orbitals resulting from interaction of the $3s$ orbitals in a crystal of sodium.

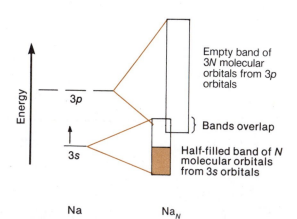

FIGURE 12–29 Overlapping of half-filled "$3s$" band with empty "$3p$" band of Na_N crystal.

The empty $3p$ atomic orbitals of the sodium atoms also interact to form a wide band of $3 \times 6.022 \times 10^{23}$ orbitals. Since the $3s$ and $3p$ atomic orbitals are quite close in energy, the two bands of molecular orbitals actually overlap as shown in Figure 12–29. The two overlapping bands contain $4 \times 6.022 \times 10^{23}$ orbitals and only 6.022×10^{23} electrons. Since each orbital can accommodate two electrons, the resulting combination of bands is only one-eighth full.

The ability of metallic sodium to conduct electricity is due to the ability of any of the highest energy electrons in the "$3s$" band to jump to a vacant orbital of slightly higher energy in the same band when an electric field is applied. As this occurs, the net flow of electrons through the crystal is in the direction of the applied field.

The fact that the "$3s$" and "$3p$" bands overlap is of no consequence in explaining the ability of sodium (or any other alkali metal) to conduct electricity, since it could do so by using only the half filled "$3s$" band. However, such overlap is of great significance for the alkaline earth metals. Consider a crystal of magnesium as an example. The $3s$ atomic orbital of an isolated magnesium atom is filled with two electrons. Thus, in the absence of this overlap, the "$3s$" band in a crystal of magnesium also would be filled. Magnesium would not be able to conduct electricity at room temperature if the highest energy electrons were not able to move readily into vacant orbitals in the "$3p$" band (Figure 12–30).

According to band theory, the highest energy electrons of metallic crystals occupy either a partially filled band or a filled band that overlaps an empty band. A band within which (or into which) electrons must move to allow electrical conduction is called a **conduction band.** Crystalline nonmetals, like diamond and phosphorus, are **insulators** in that they do not conduct electricity (or conduct heat or have metallic luster). This is because their highest energy electrons occupy filled bands of molecular orbitals that are separated from the lowest empty band (conduction band) by a large energy gap called a **forbidden zone,** which is too large an energy difference for electrons to jump in order to get to the conduction band. This is illustrated in Figure 12–31.

Elements that are semiconductors have filled bands that are only slightly below, but do not overlap with, empty bands. They do not conduct electricity at room temperature, but a small increase in temperature is sufficient to excite some of the highest energy electrons into the empty conduction band. We shall discuss semiconductors in more detail in the next section.

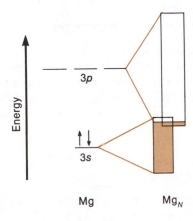

FIGURE 12–30 Overlapping of filled "$3s$" band with empty "$3p$" band of Mg_N crystal. The higher energy electrons are able to move into the "$3p$" band due to overlap.

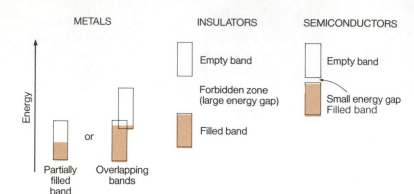

FIGURE 12−31 Distinction between metals, insulators, and semiconductors. In each case □ represents a conduction band.

Let us now explain some of the properties of metals, originally summarized in Section 6−3, in terms of the band theory of metallic bonding.

1. We have just accounted for the *ability of metals to conduct electricity.*
2. Metals also are *good conductors of heat.* They can absorb heat as electrons become thermally excited to low-lying vacant orbitals in a conduction band. The reverse process accompanies the release of heat.
3. Metals have a *lustrous appearance* because the mobile electrons can absorb a wide range of wavelengths of radiant energy and jump to higher energy levels. They then immediately emit photons of visible light (the cause of the luster) as they fall back to lower levels within the conduction band.
4. Metals are *malleable and/or ductile.* A crystal of a metal is easily deformed when a mechanical stress is applied to it. Since all the metal ions are identical and since they are imbedded in a "sea of electrons," as bonds are broken new ones are readily formed with adjacent metal ions as soon as the deformation is complete. The features of the lattice remain unchanged and the environment of each metal ion is essentially the same as before the deformation occurred (Figure 12−32). The breakage of bonds involves the promotion of electrons to higher energy levels. The formation of bonds is accompanied by the return of the electrons to the original energy levels.

A malleable substance can be rolled or pounded into sheets. A ductile substance can be drawn into wires.

FIGURE 12−32
When a metal is distorted (e.g., rolled into sheets or drawn into wires), new metallic bonds are formed and the environment around each atom is essentially unchanged. This explains why sheet metal and metal wires are strong.

12−16   Semiconductors

Present-day technology provides us with many solid state devices, such as calculators, color television sets, and solar cells. In electronics the term "solid state" refers to the use of minute devices constructed from semiconducting elements for the purposes of generating and regulating electrical signals. These devices replace the much larger and less reliable vacuum tubes that were formerly used for these purposes.

1   Intrinsic Semiconductors

As described in the preceding section, the semiconducting elements are those that do not conduct electricity at room temperature but do conduct at slightly elevated

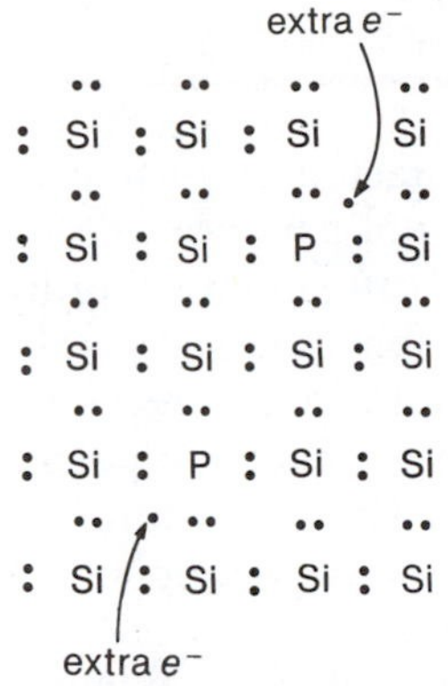

FIGURE 12–33

When a few · P̈: atoms are substituted for · S̈i· atoms in ultrapure silicon, the "extra" electron on each P atom can exchange positions with other electrons throughout the crystal. This effect is called an *n*-type defect.

temperatures. The absorption of small amounts of heat energy is sufficient to cause excitation of valence electrons from a filled band, over a very small energy gap, into a conduction band. A major way in which metals are distinguished from semiconducting elements is that the electrical conductivity of a semiconductor increases with increasing temperature, while that of a metal decreases. (Presumably, increasing the temperature of a metal causes sufficient thermal agitation of the metal ions in the crystal to impede the flow of electrons when an electrical field is applied.)

In Sections 6–3 and 12–15 we distinguished between metals and nonmetals (which are insulators) and indicated that the metalloids, the elements along the stepwise division in the periodic table, have properties intermediate between them. With this in mind, it should not be surprising that many of the metalloids are semiconducting elements. We usually designate seven of them as semiconductors: B, Si, Ge, As, Sb, Se, and Te. Generally they resemble metals in appearance and can crystallize in metal-like lattices, but are more like nonmetals in chemical properties.

2 Extrinsic Semiconductors

The pictures we have used to represent crystal structures imply perfect ordering of particles. In fact, virtually all crystals are imperfect to at least a small degree, in that atoms, molecules, or ions may be missing, displaced, or replaced by other particles. It has been estimated, for example, that a typical single crystal of "pure" sodium chloride weighing 0.0584 gram (0.00100 mole) contains about 10^{15} imperfections. This may seem like a lot, but it represents only about one imperfection per million ions.

It is possible to increase the conductivity of a semiconductor by purposely introducing impurities, a process called **doping.** The resulting defect crystals are called **extrinsic semiconductors.**

The most commonly used semiconducting element is silicon. Pure silicon is a poor conductor at room temperature. Silicon is a Group IVA element, and each atom has four valence electrons, which can form a total of four covalent bonds with other silicon atoms in the crystal lattice. If four or five out of every million silicon atoms are replaced by phosphorus atoms, the lattice framework remains essentially unchanged, but a great increase in conductivity results. Since phosphorus is a Group VA element, it has five valence electrons rather than four; each phosphorus atom contributes an extra electron, a **negative defect** or *n*-type defect (Figure 12–33). These extra electrons are free to exchange positions with other electrons and, thus, to transport charge throughout the lattice when an electrical field is applied.

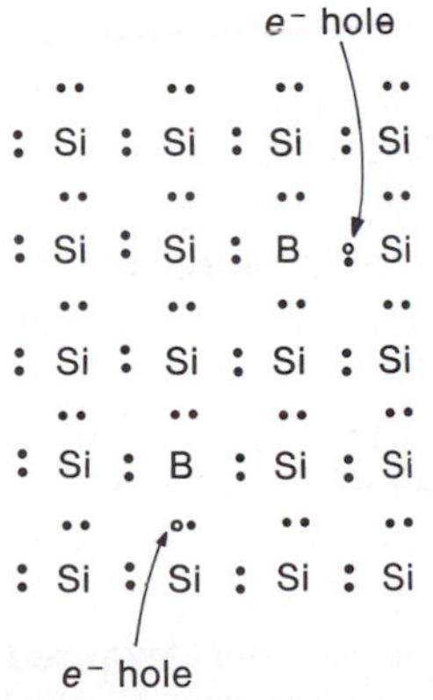

FIGURE 12–34

When a few · B̈ · atoms are substituted for · S̈i· atoms in ultrapure silicon, an "electron hole" is created at the site of each B atom. This effect is called a *p*-type defect.

Alternatively, the conductivity of silicon can be enhanced by doping it with trace quantities of a Group IIIA element, like boron, which has only three valence electrons. This leaves the lattice with an "electron hole" for every boron atom. These vacancies tend to attract electrons and so are called **positive defects** or *p*-type defects (Figure 12–34). The application of an electrical field causes electrons to jump into these holes and move in a direction governed by the polarity of the field. As an electron leaves one location to go to an electron hole, it creates a new electron hole at its original location. In this way charge is carried across the lattice of a *p*-type semiconductor.

Key Terms

Adhesive forces forces of attraction between a liquid and another surface.

Amorphous solid a noncrystalline solid with no well-defined ordered structure.

Band a series of very closely spaced, nearly continuous molecular orbitals that belong to a metal crystal as a whole.

Band theory of metals theory that accounts for the bonding and properties of metallic solids.

Boiling point the temperature at which the vapor pressure of a liquid is equal to the applied pressure; also the condensation point.

Born-Haber cycle a series of reactions (and accompanying enthalpy changes) which, when summed, represents the hypothetical one-step reaction by which elements in their standard states are converted into crystals of ionic compounds (and the accompanying enthalpy change).

Capillary action the drawing of a liquid up the inside of a small bore tube when adhesive forces exceed cohesive forces, or the depression of the surface of the liquid when cohesive forces exceed adhesive forces.

Cohesive forces all the forces of attraction among particles of a liquid.

Condensation liquefaction of vapor.

Condensed phases the liquid and solid phases; phases in which particles interact strongly.

Conduction band a partially filled band or a band of vacant energy levels slightly higher in energy than a filled band; a band within which, or into which, electrons must be promoted to allow electrical conduction to occur in a solid.

Coordination number in describing crystals, the number of nearest neighbors of an atom or ion.

Critical pressure minimum pressure required to condense a vapor at its critical temperature.

Critical temperature temperature above which a vapor cannot be condensed.

Crystal defect an impurity or imperfection in a crystal.

Crystal lattice pattern of arrangement of particles in a crystal.

Crystal lattice energy the amount of energy required to convert one mole of an ionic solid to its gaseous ions.

Crystalline solid a solid characterized by a regular, ordered arrangement of particles.

Deposition the direct solidification of a vapor by cooling; the reverse of sublimation.

Dipole-dipole interactions attractive interactions between molecules with permanent dipoles.

Distillation the separation of a liquid mixture into its components on the basis of differences in boiling points.

Doping the intentional incorporation of an impurity into a crystal.

Dynamic equilibrium a situation in which two (or more) processes occur at the same rate so that no net change occurs.

Evaporation vaporization of a liquid at a temperature below its boiling point.

Extrinsic semiconductor semiconductor whose conductivity properties have been enhanced by doping.

Forbidden zone a relatively large energy separation between an insulator's highest filled electron energy band and the vacant next higher energy band.

Heat of condensation the amount of heat that must be removed from one gram of a vapor at its condensation point to condense the vapor with no change in temperature.

Heat of crystallization the amount of heat that must be removed from one gram of a liquid at its freezing point to freeze it with no change in temperature.

Heat of fusion the amount of heat required to melt one gram of a solid at its melting point with no change in temperature. Usually expressed in J/g. The *molar heat of fusion* is the amount of heat required to melt one mole of a solid at its melting point with no change in temperature and is usually expressed in kJ/mol.

Heat of vaporization the amount of heat required to vaporize one gram of a liquid at its boiling point with no change in temperature. Usually expressed in J/g. The *molar heat of vaporization* is the amount of heat required to vaporize one mole of a liquid at its boiling point with no change in temperature and is usually expressed in kJ/mol.

Hydrogen bond a fairly strong dipole-dipole interaction (but still considerably weaker than covalent or ionic bonds) between molecules containing hydrogen directly bonded to a small, highly electronegative atom such as nitrogen, oxygen, or fluorine.

Insulator   very poor conductor or nonconductor of heat and electricity

Intermolecular forces   forces between particles (atoms, molecules, ions) of a substance.

Intramolecular forces   bonds between atoms (or ions) within molecules (or formula units).

Intrinsic semiconductor   a substance that is a semiconductor in the pure (undoped) form.

Isomorphous   refers to crystals having the same atomic arrangement.

LeChatelier's Principle   states that a system at equilibrium, or striving to attain equilibrium, responds in such a way as to "counteract" any stress placed upon it.

London forces   very weak and very short-range attractive forces between temporary (induced) dipoles.

Melting point   the temperature at which liquid and solid coexist in equilibrium; also the freezing point.

Meniscus   the shape assumed by the surface of a liquid in a cylindrical container.

Metallic bonding   bonding within metals due to the electrical attraction of positively charged metal ions for mobile electrons that belong to the crystal as a whole.

Negative (*n*-type) defect   an "extra" electron introduced in a semiconductor by doping.

Normal boiling point   the temperature at which the vapor pressure of a liquid is equal to one atmosphere.

Normal melting point   the melting (freezing) point at one atmosphere pressure.

Phase diagram   graph that shows the equilibrium pressure-temperature relationships between the various phases of a substance.

Polymorphous   refers to solids that can crystallize in more than one arrangement of atoms.

Positive (*p*-type) defect   an "electron hole" produced in a semiconductor by doping.

Semiconductor   a substance that does not conduct electricity well at room temperature but does at higher temperatures.

Sublimation   the direct vaporization of a solid by heating without passing through the liquid state.

Surface tension   a measure of the intermolecular forces of attraction among liquid particles that must be overcome in order to expand the surface area.

Triple point   point on a phase diagram for a substance corresponding to the only pressure and temperature at which solid, liquid, and gas can coexist at equilibrium.

Unit cell   smallest repeating unit showing all the structural characteristics of a crystal.

Vapor pressure   the partial pressure of a vapor at the surface of its parent liquid.

Viscosity   the tendency of a liquid to resist flow; a measure of its fluidity.

Volatility   the ease with which a liquid vaporizes.

Exercises

General Concepts

1. List several characteristics that distinguish solids from liquids and liquids from gases.
2. Why are the properties of solids and liquids more difficult to describe in general mathematical relationships than those of gases?
3. Arrange the following in order of increasing strength of interaction: hydrogen bonds, covalent bonds, London forces, permanent dipole-dipole forces.
4. Which of the following substances have permanent dipole-dipole forces? (a) HCl, (b) Cl_2, (c) BrF, (d) BrF_5, (e) SF_6, (f) CO_2, (g) CO, (h) Kr, (i) CCl_4, (j) $CHBr_3$.

5. For which of the substances in Exercise 4 are London forces the only important ones in determining boiling points? How do London forces operate?
6. Which of the following substances exhibit strong hydrogen bonding in the liquid and solid states? (a) CH_3OH, methyl alcohol, (b) PH_3, (c) CH_4, (d) CH_2Cl_2, (e) H_2S, (f) NH_3, (g) SiH_4, (h) HF, (i) HCl, (j) CH_3NH_2.

The Liquid State

7. Describe the behavior of liquids with changing temperature in terms of kinetic-molecular theory.

8. Why are liquids more dense than gases?

9. We don't see "holes" in liquids even though there are spaces between particles or clusters of particles. Why? Atoms and ions are mostly empty space. We don't see or feel spaces in solids. Why?

10. Give two examples of highly viscous liquids and two examples of only slightly viscous liquids. What is responsible for high viscosity?

11. Distinguish between cohesion and adhesion. How are they related to the shape of the meniscus of a liquid in a cylindrical container?

12. Explain how plants are able to obtain nutrients from fertilizers applied as solids.

13. Distinguish between evaporation and boiling. Explain the dependence of rate of evaporation on temperature in terms of kinetic-molecular theory.

14. Support or criticize the statement that liquids with high normal boiling points have low vapor pressures. Give examples of three liquids that have relatively high vapor pressures at 25°C and three that have low vapor pressures at 25°C.

15. The normal boiling point of H_2 is $-259°C$ and that of F_2 is $-188°C$. Why does H_2 have the lower boiling point?

16. Using Table 12–2, draw a graph of vapor pressure vs. temperature for the four listed liquids. From your graph estimate the normal boiling point of each liquid. Give an explanation for the fact that the normal boiling point of water is higher than those of the other three liquids.

17. Consider a liquid in a beaker being heated with a bunsen burner. Bubbles begin to form as it begins to boil. Where do they form and why? What are the bubbles?

18. Explain the effectiveness of a pressure cooker in cooking.

19. Explain how liquids can be separated from each other by distillation. Is the separation more effective if boiling points of liquids are close together or far apart? Why should the temperature of the vapor near the entrance to the condenser be monitored during distillation in an apparatus like that in Figure 12–9?

20. Arrange each of the following lists in order of increasing magnitude of heat of vaporization. (You may consult a handbook.) (a) H_2S, H_2O, H_2Te, H_2Se; (b) HBr, HF, HI, HCl; (c) PH_3, AsH_3, NH_3, SbH_3, BiH_3.

21. Arrange the lists in Exercise 20 in order of increasing boiling points and explain the orders. (You may consult a handbook.)

22. Arrange the following substances in order of increasing (a) melting and (b) boiling points. (You may consult a handbook.) $CH_3CH_2CH_3$, Na_2S, CH_3CH_3, H_2O. Why is this order observed?

The Solid State

23. Comment on the statement: "The only perfectly ordered state of matter is the crystalline state."

24. Ice floats in water. Why? Would you expect solid mercury to float in liquid mercury at its freezing point? Explain why or why not.

25. How do amorphous and crystalline solids differ? Give three examples of each.

26. We normally think of glasses as being solid. What is the rationale for describing glasses as supercooled liquids?

27. Since even the largest atoms are so small that they cannot be seen even under very powerful microscopes, how are the relative arrangements of atoms in solids determined?

28. Sodium chloride, NaCl, and magnesium sulfide, MgS, are isomorphous. Which has the higher melting point? Why?

29. Covalent bonding occurs in both molecular solids and covalent solids, yet molecular solids are relatively soft and have low melting points, while covalent solids are hard and have high melting points. Why is this so?

30. How are the enthalpy of sublimation and the crystal lattice energy of a molecular solid related?

31. Classify each of the following into one of the four categories of solids: molecular, ionic, covalent (network) or metallic.

	Melting Point (°C)	Boiling Point (°C)	Electrical Conductor	
			Solid	Liquid
Ar	−189.3	−185.6	no	no
Ag	960.8	1950	yes	yes
Sc	1541	2831	yes	yes
Ge	937	2830	poor	poor
$MgCl_2$	708	1412	no	yes
PBr_3	−40	173	no	no
OF_2	−223.8	−144.8	no	no

32. Determine the number of sodium ions and

chloride ions that belong to the unit cell shown in Figure 12–18.

33. Determine the number of atoms per unit cell for the cells shown in Figure 12–17.

34. Determine the number of ions present in each unit cell shown in Figure 12–21.

35. Illustrate and describe the significance of the following equation to the determination of the structure of crystals.

$$\sin \theta = \frac{n\lambda}{2d}$$

36. What is a unit cell? Sketch a unit cell and label the dimensions a, b, c, and angles α, β, and γ. How do cubic, tetragonal, and hexagonal crystal systems differ?

37. Distinguish among and sketch simple cubic, body-centered cubic, and face-centered cubic lattices. Use CsCl, sodium, and nickel as examples of solids existing in simple cubic, bcc, and fcc lattices, respectively.

38. What are the two types of closest-packed crystal structures? How do they differ from other types of structures?

39. Sketch the arrangement of atoms in the crystal structures of cadmium and copper, which have cubic close-packed and hexagonal close-packed structures, respectively. Represent the atoms as spheres.

40. Distinguish among and compare the characteristics of molecular, covalent, ionic, and metallic solids. Give two examples of each kind of solid.

41. Distinguish between isomorphous and polymorphous solids.

Unit Cell Data and Atomic and Ionic Sizes

42. Metallic rhodium crystallizes in a face-centered cubic lattice with a unit cell edge length of 0.3803 nm. Calculate the molar volume of rhodium, i.e., the volume of one mole of rhodium (including the empty spaces).

43. Zinc selenide, ZnSe, crystallizes in a face-centered cubic unit cell and has a density of 5.267 g/cm³. Determine the edge length of the unit cell.

44. The atomic radius of palladium is 0.1375 nm. The unit cell of palladium is a face-centered cube. Calculate the density of palladium.

45. A certain metal is found to have a specific gravity of 10.200 at 25°C. It is found to crystallize in a body-centered cubic lattice with a unit

cell edge length of 0.3147 nm. Determine the atomic weight and identify the metal.

46. A certain metal crystallizes in the hexagonal closest-packed structure and has a density of 1.737 g/cm³. Its atomic radius is 0.160 nm. There are four atoms per unit cell. Determine its atomic weight. What is the metal?

47. Nickel crystallizes in a face-centered cubic unit cell with an edge length of 0.3524 nm. Its density is 8.902 g/cm³ and its atomic weight is 58.70. From these data calculate Avogadro's number.

48. Barium crystallizes in a body-centered cubic unit cell with an edge length of 0.5025 nm. Calculate the fraction of empty space in the body-centered cubic lattice. How does this compare with the fraction of empty space in close-packed lattices? (Hint: the atom in the center touches the atoms at the corners of the cube. Atoms along edges of the cube do not touch.)

49. Zirconium crystallizes in a hexagonal closest-packed lattice. Its density is 6.51 g/cm³, and its atomic radius is 0.157 nm. Estimate the molar volume of zirconium and the volume occupied by a mole of zirconium atoms (no empty space).

50. The unit cell of nickel is a face-centered cube having a volume of 4.3756×10^{-2} nm³. The atom at the center of each face just touches the atoms at the corners. Determine the atomic radius and atomic volume of nickel.

51. The spacing between successive planes of platinum atoms comprising the parallel faces of face-centered unit cells is 0.2256 nm. When x-radiation emitted by molybdenum strikes a crystal of platinum metal, the minimum diffraction angle of x-rays is 9.045°. What is the wavelength of the Mo radiation?

52. Gold crystallizes in an fcc lattice. When x-radiation of 0.15404 nm wavelength from copper is used for structure determination of metallic gold, the minimum diffraction angle of x-rays by the gold is 19.14°. Calculate the spacing between parallel layers of gold atoms.

Crystal Lattice Energy and Born-Haber Cycles

53. Determine the crystal lattice energy for LiF (s), given the following: sublimation energy for Li = 155 kJ/mol, ΔH_f^o for F (g) = 78.99 kJ/mol, the first ionization energy of Li = 520 kJ/mol, the electron affinity of fluorine = -322 kJ/

mol, and the standard molar enthalpy of formation of LiF (s) = −589.5 kJ/mol.

54. From the following information, determine the crystal lattice energy of KI: sublimation energy for potassium = 90 kJ/mol, the first ionization energy for potassium = 419 kJ/mol, the sublimation energy for iodine = 62 kJ/mol, the dissociation energy for iodine = 151 kJ mol, the electron affinity for iodine = −295 kJ/mol, and the heat of formation of KI (s) = −328 kJ/mol.

55. Construct a Born-Haber cycle for the production of calcium oxide, CaO (s), from Ca (s) and O_2 (g). It is not necessary to include the values of enthalpy changes for the various steps.

56. Calculate the crystal lattice energy for $CaBr_2$ from the following information: ΔH_{subl} for calcium = 193 kJ/mol, ΔH_{ie1} for calcium = 590 kJ/mol, ΔH_{ie2} for calcium = 1145 kJ/mol, ΔH_{vap} for Br_2 is 315 kJ/mol [Br_2 (ℓ) is standard state for bromine], ΔH_{diss} for Br_2 (g) = 193 kJ/mol, ΔH_{ea} for bromine = −324 kJ/mol, and ΔH_f^o for $CaBr_2$ (g) = −675 kJ/mol.

57. Calculate the electron affinity for chlorine from the following information: ΔH_f^o for $MgCl_2$ (s) = −641.8 kJ/mol, ΔH_{subl} for Mg = 151 kJ/mol, the first and second ionization energies for Mg are 738 and 1451 kJ/mol, ΔH_{diss} for Cl_2 = 243.4 kJ/mol, and the crystal lattice energy for $MgCl_2$ (s) = −2529 kJ/mol.

58. The value of ΔH_{ie1} is *always* positive, yet the ionization of a metal atom to form a metal ion with a noble gas electron configuration, M (g) → M^{n+} (g) + ne^-, occurs readily in the formation of many ionic solids from their elements. What is the primary factor that makes the overall process favorable?

Phase Changes and Associated Enthalpy Changes

59. Explain why perspiration has a cooling effect on the skin.

60. Write the equation for a reaction whose enthalpy change is equal to
  (a) $\Delta H_{sublimation}$ for carbon dioxide.
  (b) ΔH_{fusion} for ice.
  (c) $\Delta H_{vaporization}$ for pentane, C_5H_{12}.
  (d) $\Delta H_{condensation}$ for sulfur, S_8.

The following values will be useful in working Exercises 61 to 71.

Specific heat of ice	2.09 J/g·°C
Heat of fusion of ice at 0°C	333 J/g
Specific heat for liquid H_2O	4.18 J/g·°C
Heat of vaporization of liquid H_2O at 100°C	2.26×10^3 J/g
Specific heat for steam	2.03 J/g·°C

61. Calculate the amount of heat required to raise the temperature of 10.0 grams of water at 10.0°C to 50.0°C.

62. If 100 grams of liquid water at 100°C and 200 grams of water at 20.0°C are mixed in an insulated container, what is the final temperature?

63. Calculate the amount of heat required to convert 10.0 grams of ice at 0.0°C to liquid water at 50.0°C.

64. Calculate the amount of heat required to convert 10.0 grams of ice at 0.0°C to liquid water from 100°C.

65. Calculate the amount of heat required to convert 10.0 grams of ice at −20.0°C to steam at 120.0°C.

66. Calculate the amount of heat given up when 10.0 grams of steam at 100°C is condensed and cooled to 20.0°C.

67. If 10.0 grams of ice at −10.0°C and 10.0 grams of liquid water at 100°C are mixed in an insulated container, what will the final temperature be?

68. If 105 grams of liquid water at 0.0°C and 10.5 grams of steam at 110.0°C are mixed in an insulated container, what will the final temperature be?

69. If 10.0 grams of steam at 110.0°C are bubbled slowly into 50.0 grams of liquid water at 0.0°C in an insulated container, will all of the steam be condensed?

70. If 100 g of iron at 100.0°C is placed in 200 g of water at 20.0°C in an insulated container, what will the temperature of the iron and water be when both are the same? The specific heat of iron is 0.444 J/g·°C. Compare this result with the one obtained in Exercise 62 when 100 g of water at 100°C was mixed with 200 g of water at 20.0°C.

71. (a) Calculate the amount of heat required to raise the temperature of 100 g of mercury from 20.0°C to 90.0°C. The specific heat of mercury is 0.138 J/g·°C.
  (b) Calculate the amount of heat required to raise the temperature of 100 g of water from 20.0°C to 90.0°C.

(c) What is the ratio of the specific heat of liquid water to that of liquid mercury?

(d) What does the answer to (c) tell us?

Phase Diagrams

72. How can a substance be purified by sublimation?

Refer to the phase diagram of CO_2 in Figure 12–27b to answer Exercises 73 to 76.

73. What phase of CO_2 exists at 2 atm pressure and a temperature of $-78°C$? $-50°C$? $0°C$?

74. What phases of CO_2 are present at (a) a temperature of $-78°C$ and a pressure of 1 atm? (b) at $-57°C$ and a pressure of 5.2 atm?

75. List the phases that would be observed if a sample of CO_2 at 6 atm pressure were heated from $-100°C$ to $0°C$.

76. How does the melting point of CO_2 change with pressure? What does this indicate about the relative density of solid CO_2 versus liquid CO_2?

Metallic Bonding and Semiconductors

77. Compare the temperature dependence of electrical conductivity of a metal with that of a typical metalloid. Explain the difference.

78. How do intrinsic and extrinsic semiconductors differ?

79. Propose a reasonable combination of an element and a dopant (impurity) that could be used to prepare a semiconductor with an n-type defect. Choose an example other than the one described in the text.

80. Repeat Exercise 79 for a p-type defect.

81. What single factor accounts for the ability of metals to conduct both heat and electricity in the solid state? Why are ionic solids poor conductors of heat and electricity even though they are composed of charged particles?

82. In general, metallic solids are ductile and malleable, whereas ionic salts are brittle and shatter readily (although they are hard). Refer to Figures 12–18c and 12–32 and explain this observation.

Physical Properties of Mixtures: Solutions and Colloids

13

Until now, our study of the relationships between the structures of substances and their physical properties and behavior have dealt almost entirely with individual substances. However, we very rarely encounter pure substances; almost all important applications of chemistry involve mixtures. In this chapter, we shall be concerned with the properties of several types of mixtures, and with the relationship of their properties to those of the individual components.

Many familiar mixtures are *solutions*—air, a soft drink, brass, a martini, vinegar, a glucose solution such as those used in intravenous feeding, stainless steel, dental fillings, brine (salt water), and pancake syrup are some examples. Biochemical processes in all living organisms take place in solution. Most important industrial processes, such as those which produce medicinal chemicals, plastics, adhesives, and petroleum products, involve solutions. A solution is a *homogeneous mixture* of substances in which no settling occurs. A true solution consists of a solvent and one or more solutes whose proportions can vary from solution to solution. The *essential feature* of a solution is that the solute is dispersed in the solvent as individual molecules or ions, so that a true solution consists of a single phase.

Solutions may involve many different combinations in which a solid, liquid, or gas acts as either *solvent* or *solute*. The most commonly encountered kinds are those in which the solvent is a liquid. For instance, sea water is an aqueous solution of many salts and some gases (such as carbon dioxide and oxygen). However, examples of solutions in which the solvent is not a liquid also are numerous and quite common. Air is a solution of gases (with variable composition). Dental fillings are solid amalgams, or solutions of liquid mercury dissolved in solid metals. Alloys are solid solutions of one or more solid metals dissolved in another; "yellow gold" is a solid solution of copper in gold, and "white gold" is a solution of nickel, palladium, or zinc in gold.

Some mixtures, however, are *heterogeneous,* containing portions that are clearly distinguishable from other portions. Many important mixtures such as fog, smoke, blood, milk, whipped cream, and paint are actually heterogeneous mixtures in which solute-like particles are suspended, but not dissolved, in a solvent-like phase. Such mixtures are known as **colloids.** No settling occurs in a true colloid, but it is possible to distinguish between the suspended particles and the suspending medium. The particles are large enough to make the solution appear cloudy, so they are clearly not individual ions or molecules.

In this chapter, we shall study the physical properties of solutions and colloids, beginning with the factors that influence the ability of solutes to dissolve in solvents. We shall place special emphasis on aqueous solutions, because they are so important. These factors will help us to understand why certain substances are water-soluble, according to the solubility rules presented in Section 9-4.5, and why others are quite insoluble. Then we shall study how the physical properties of solvents change as solutes are dissolved into them. These are called **colligative properties.** Finally, we shall classify colloids into various types and study some of their properties.

Before continuing with this chapter, you may wish to review the following sections, which are directly related to the study of solutions: Sections 2-12 to 2-14, 9-4.1 and 9-4.5, 10-4.1 and 10-4.2, and 10-5.2a and b.

Solutions—The Dissolving Process

13-1   Mixing and Spontaneity of the Dissolving Process

In Section 9-4.5 we covered the solubility rules that apply to common solutes in water. Now we shall investigate the major factors that influence solubility.

A substance may dissolve with or without reaction with the solvent. Metallic sodium "dissolves" in water with the evolution of bubbles of hydrogen. A chemical change occurs in which hydrogen and soluble, ionic sodium hydroxide, NaOH, are produced:

$$2Na\ (s) + 2H_2O\ (\ell) \longrightarrow 2[Na^+\ (aq) + OH^-\ (aq)] + H_2\ (g)$$

Strictly speaking, even solutes that don't "react" with the solvent undergo solvation, a kind of reaction in which molecules of solvent are attached in oriented clusters to the solute particles.

Solid sodium chloride, NaCl, on the other hand, dissolves in water with no evidence of chemical reaction.

$$NaCl\ (s) \xrightarrow{H_2O} Na^+\ (aq) + Cl^-\ (aq)$$

If the first solution is evaporated to dryness, solid sodium hydroxide, NaOH, is obtained rather than metallic sodium. This, along with the production of bubbles of hydrogen, is evidence for a reaction with the solvent. But evaporation of the sodium chloride solution yields the original NaCl. We shall consider only dissolving of the latter type, in which no irreversible reaction between components occurs.

The ease of the dissolving process depends upon two factors, (1) the change in enthalpy (exothermicity or endothermicity), and (2) the change in disorder (called entropy change, Sections 14–5 and 14–6) accompanying the process. In Chapter 14, we shall study both of these factors in detail for many kinds of physical and chemical changes. For now, you need to know that the spontaneity of a process is *favored* by a decrease in the enthalpy of the system, which corresponds to an *exothermic process,* and by an *increase in the disorder,* or randomness, of the system.

"Spontaneity" translates roughly into "ability to occur."

Let us concentrate first on the factors that determine the change in heat content (enthalpy). This change is called the **heat of solution,** $\Delta H_{solution}$, which is defined as the difference between the enthalpy of the solution and the total enthalpy of its pure components before mixing.

$$\Delta H_{solution} = H_{solution} - H_{components} \qquad \text{[Eq. 13–1]}$$

As with other thermodynamic quantities, we cannot determine the absolute enthalpies, $H_{solution}$ and $H_{components}$, but we *can* measure the difference, which corresponds to the enthalpy *change* accompanying the dissolving process. When $\Delta H_{solution}$ is negative, the process is exothermic; when it is positive, the process is endothermic.

In a pure liquid to be used as a solvent, the intermolecular forces are all between like molecules; when the liquid and a solute are mixed, each molecule then experiences forces from molecules (or ions) unlike it as well as from like molecules. The relative strengths of such interactions help to determine the extent of solubility of substances in a liquid. The interactions that must be considered to assess $\Delta H_{solution}$ for the dissolving of a specific solute in a specific solvent are:

1. solute-solute attractions
2. solvent-solvent attractions
3. solvent-solute attractions

Dissolution is favored when the first two of these interactions are relatively small and the third is relatively large, as summarized in Figure 13–1. The intermolecular or interionic attractions among solute particles in the pure solute must be overcome (Step a) to dissolve the solute. This part of the process requires an *input* of energy. Likewise, separating the solvent molecules from each other (Step b) to "make room" for the solute particles also requires the *input* of energy. However,

FIGURE 13-1 The enthalpy changes associated with the hypothetical three-step sequence in a dissolution process. The diagram shows a solid solute dissolving in a liquid solvent, but similar considerations would apply to other combinations.
(a) An exothermic process: $\Delta H_a + \Delta H_b < \Delta H_c$, so $\Delta H_{solution}$ is negative.
(b) An endothermic process: $\Delta H_a + \Delta H_b > \Delta H_c$, so $\Delta H_{solution}$ is positive.

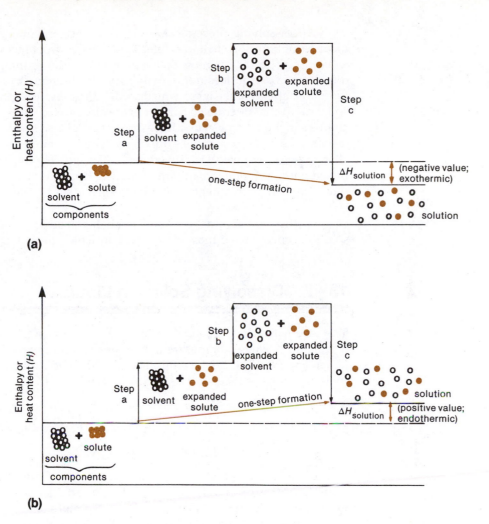

energy is *released* as the solute particles and solvent molecules interact in the solution (Step c). Thus, the dissolving process is endothermic (and disfavored with respect to $\Delta H_{solution}$) if the amount of heat absorbed in (hypothetical) Steps a and b is greater than the amount of heat released in Step c. It is exothermic (and favored) if the amount of heat absorbed in Steps a and b is less than the amount of heat released in Step c.

Many solids dissolve in liquids by endothermic processes. The reason such solids are soluble in liquids is that the endothermicity is outweighed by a great increase in disorder of the solute accompanying the dissolving process. The solute particles are very highly ordered in a solid crystal, but are free to move about randomly in liquid solutions. Likewise, the solvent particles increase in their degree of disorder as the solution is formed, since they then are in a more random environment (Figure 13-1).

All dissolving processes are obviously accompanied by an increase in the disorder of both solvent and solute. Thus, this disorder factor is *always* favorable to solubility. The determining factor is whether the enthalpy change also favors dissolution or, if it does not, whether it is small enough to be outweighed by the favorable effects of the increasing disorder. In gases, for instance, the molecules are so far apart that intermolecular forces are quite weak. Thus, when gases are mixed, changes in the intermolecular forces are very slight, so the very favorable increase in disorder which accompanies mixing always is more important than possible changes in intermolecular attractions (enthalpy). Therefore, gases can always be mixed with each other in any proportions. (This statement does not apply to gases that react with each other chemically. In such cases the gases are converted to other substances.)

In the next several sections, we shall consider in more detail the types of solutions encountered most often, those in which the solvent is a liquid.

13–2   Dissolving Solids in Liquids

The ease with which a solid goes into solution depends to a large extent upon the crystal lattice energy, or the strength of interactions among the particles holding the lattice together. The crystal lattice energy is defined as the enthalpy change for the following process (Section 12–11.3b), which is invariably exothermic.

$$M^+ (g) + X^- (g) \longrightarrow MX (s) + energy \qquad \Delta H = \Delta H_{xtal} \qquad \text{[Eq. 13–2]}$$

The reverse of the crystal formation reaction, the separation of the crystal into ions,

$$MX (s) + energy \longrightarrow M^+ (g) + X^- (g) \qquad \text{[Eq. 13–3]}$$

can be considered as the hypothetical first step (a in Figure 13–1a and b) in forming a solution. It is invariably endothermic. The smaller the crystal lattice energy (a measure of the solute-solute interactions), the more readily dissolving occurs. That is, less energy must be supplied to start the dissolving process.

If the solvent is water, the energy that must be supplied to expand the solvent (Step b in Figure 13–1) includes that required to break up some of the hydrogen bonding between water molecules.

The third major factor contributing to $\Delta H_{solution}$ is the extent to which the solvent molecules interact with particles of the solid. The process by which solvent molecules surround and interact with solute ions or molecules is called **solvation.** When the solvent is water, the more specific term is **hydration.** The **molar hydration energy** (related to Step c in Figure 13–1) is defined as the amount of energy involved in the (exothermic) hydration of one mole of gaseous ions:

The standard molar enthalpy of hydration is often represented as ΔH^0_{hyd}.

$$M^{n+} (g) + xH_2O \longrightarrow M(OH_2)_x^{n+} + energy \qquad \Delta H = \Delta H_{hyd} \; M^{n+} \qquad \text{[Eq. 13–4]}$$

$$X^{y-} (g) + rH_2O \longrightarrow X(H_2O)_r^{y-} + energy \qquad \Delta H = \Delta H_{hyd} \; X^{y-} \qquad \text{[Eq. 13–5]}$$

Hydration energy is generally very negative for ionic or polar covalent compounds, since the polar water molecules interact very strongly with ions and polar molecules. In fact, the only solutes that are appreciably soluble in water either undergo dissociation or ionization or are able to hydrogen-bond with water. Nonpolar solids like naphthalene, $C_{10}H_8$, do not dissolve appreciably in polar solvents like water because the two substances do not attract each other significantly. This is true

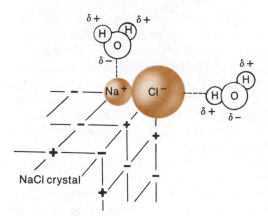

FIGURE 13–2 Electrostatic attraction in the dissolution of NaCl in water.

despite the fact that crystal lattice energies of solids consisting of nonpolar molecules are much smaller than those of ionic solids. However, naphthalene dissolves readily in nonpolar solvents like benzene because there are no strong attractive forces between solute molecules or between solvent molecules. These facts provide the basis for explaining the observation that "like dissolves like."

Let us consider what happens when a cube of sodium chloride, a typical ionic solid, is placed in water. The Cl^- ions located at the corners of the cube (Figure 13–2) are held less strongly by the lattice than any other Cl^- ions because they have only three nearest neighbor Na^+ ions. They go into solution most readily. Ions on the edges are attached to four nearest neighbors of opposite charge. Those on faces are held by electrostatic attraction to five such nearest neighbors, while those within the cube have six nearest neighbors of opposite charge.

The $\delta+$ ends of water molecules associate with the negative chloride ions. As chloride ions at the corners go into solution, sodium ions, now surrounded by only three nearest neighbor chloride ions, are exposed to water molecules whose $\delta-$ ends orient themselves toward the sodium ions and solvate them. This is also shown in Figures 13–2 and 13–3.

When we write Na^+ (aq) and Cl^- (aq) we refer to hydrated ions. The number of water molecules directly attached to an ion is different for different ions. Sodium ions are thought to be hexahydrated; that is, Na^+ (aq) probably represents $Na(OH_2)_6^+$. Most cations in aqueous solution are surrounded by four to nine water

For simplicity we frequently omit the (aq) designations from dissolved ions. But we must remember that all ions in aqueous solution are hydrated, whether this is indicated or not.

FIGURE 13–3 A two-dimensional representation of the dissolution of an ionic solid, such as NaCl, in water. The hydrated ions that form as the solid dissolves are shown on the right.

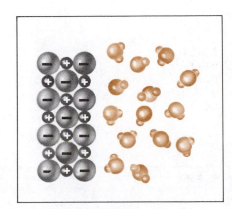
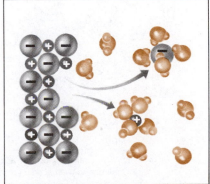

molecules, with six being the most common number. Generally, larger cations can accommodate more water molecules around them than smaller cations.

Many solids that are appreciably soluble in water are ionic in nature. Crystal lattice energies generally increase with increasing charge and decreasing size of ions. That is, lattice energies increase as the ionic charge densities increase, and therefore as electrostatic attractions within the lattice increase (Section 12–11.3b). Hydration energies also increase in the same order (Table 13–1). As we indicated earlier, crystal lattice energies and hydration energies are generally much smaller for molecular solids than for ionic solids.

The effects of lattice energies and hydration energies oppose each other in the dissolving process, but since lattice energies and hydration energies are usually of approximately the same magnitude for low-charge species, they tend to cancel each other. As a result, the dissolving process is slightly endothermic for many ionic substances. Ammonium nitrate, NH_4NO_3, is typical of salts that dissolve endothermically. This property is exploited in the "instant ice packs" used by athletic trainers in the treatment of sprains and other minor injuries. Ammonium nitrate and water are packaged in a plastic bag in which they are kept separate by a partition that is easily broken when squeezed. As the ammonium nitrate reaches water and dissolves, it absorbs heat from its surroundings and the bag becomes cold to the touch.

There are some ionic solids that dissolve with the release of heat. Examples are anhydrous sodium sulfate, Na_2SO_4, calcium chloride, $CaCl_2$, and lithium sulfate monohydrate, $Li_2SO_4 \cdot H_2O$.

As the charge-to-size ratio (charge density) increases for ions in ionic solids, the crystal lattice energy usually increases faster than the hydration energy. This makes dissolving of solids like aluminum fluoride, AlF_3, magnesium oxide, MgO, and chromium(III) oxide, Cr_2O_3, very endothermic. As we have already noted, high

TABLE 13–1   Ionic Radii, Charge/Radius Ratios, and Heats of Hydration for Some Cations

Ion	Ionic Radius, nm	Charge/Radius Ratio*	Heat of Hydration†
K^+	0.133	7.5	−351
Na^+	0.095	10.5	−435
Li^+	0.060	16.7	−544
Ca^{2+}	0.099	20.2	−1650
Mn^{2+}	0.078	25.6	−1900
Fe^{2+}	0.076	26.3	−1980
Zn^{2+}	0.074	27.0	−2100
Co^{2+}	0.074	27.0	−2110
Cu^{2+}	0.072	27.8	−2160
Ni^{2+}	0.072	27.8	−2170
Fe^{3+}	0.064	46.9	−4340
Cr^{3+}	0.062	48.4	−4370
Al^{3+}	0.052	57.7	−4750

* The charge/radius ratio is the ionic charge divided by the ionic radius in nanometers. The cations are listed in order of increasing charge/radius ratio, which is a measure of the *charge density* around the ion.
† Heats of hydration are in kJ/mol.

endothermicity usually results in very *low* solubility. Chromium(III) oxide is quite insoluble in water.

13-3   Dissolving Liquids in Liquids (Miscibility)

Miscibility refers to the ability of one liquid to mix with (dissolve in) another. The three kinds of attractive interactions listed in Section 13-1 must be considered for liquid-liquid solutions just as they were for solid-liquid solutions. Since solute-solute attractions are usually much lower for liquid solutes than for solids (as measured by crystal lattice energy), this factor is less important, so that the liquid-liquid mixing process is often exothermic for miscible liquids.

As we have seen, polar liquids tend to interact strongly with and dissolve readily in polar solvents. Methanol, CH_3OH, ethanol, CH_3CH_2OH, acetonitrile, CH_3CN, and sulfuric acid, H_2SO_4, are all polar liquids that are soluble in most polar solvents such as water. The hydrogen bonding between methanol and water molecules, and the dipolar interaction between acetonitrile and water molecules, are both depicted in Figure 13-4.

In the case of concentrated aqueous sulfuric acid, H_2SO_4, the sulfate ion is able to undergo strong hydrogen bonding with the water molecules. X-ray determinations of crystal structures of many solid hydrated sulfate compounds, like copper(II) sulfate pentahydrate, $CuSO_4 \cdot 5H_2O$, reveal that the sulfate anion interacts very strongly with water molecules. This is also presumed to be the case in aqueous solutions containing sulfate ions, especially in concentrated solutions. The 109°28′ O—S—O bond angles in the tetrahedral sulfate ion are the right size for strong attractive interactions with both hydrogens of the $\delta+$ end of a water molecule, which has a bond angle of about 105° (Figure 13-5).

> This compound is more appropriately formulated as $[Cu(OH_2)_4]SO_4 \cdot H_2O$ and named tetraaquacopper(II) sulfate hydrate.

Because this interaction is so strong, large amounts of heat are released when concentrated sulfuric acid is diluted with water, often enough to cause the solution to boil and spatter. For this reason, *sulfuric acid* (as well as all other mineral acids) *is always diluted by adding the acid slowly and carefully to water. Water is never added to the acid.* If spattering does occur when the acid is added to water, it is mainly water that spatters, not the corrosive concentrated acid.

> The potential danger when water is added to concentrated acid is due more to the spattering of the acid itself than to the steam from boiling water.

The interaction between sulfate ion and water is also enhanced by the presence of the 2− charge on the ion. The perchlorate ion, ClO_4^-, is also tetrahedral and interacts with water molecules. Dissolution and dilution of perchloric acid, $HClO_4$, in water is exothermic, but less so than for H_2SO_4; the ClO_4^- ion has only a 1− charge and does not interact with water nearly so strongly as does the SO_4^{2-} ion with its 2− charge.

Nonpolar liquids that do not react with the solvent generally are not very soluble in polar liquids, because of the mismatch of forces of interaction. Nonpolar liquids are, however, often quite soluble in other nonpolar liquids.

FIGURE 13-4 (a) Hydrogen bonding in methanol/water solution. (b) Dipolar interaction in acetonitrile/water solution.

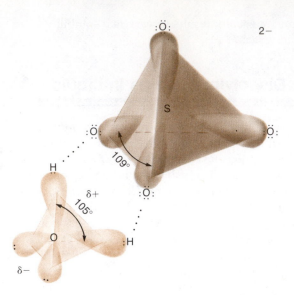

FIGURE 13–5 Interaction of a sulfate ion and a water molecule.

13–4   Dissolving Gases in Liquids

Based on what you have read in Section 12–8 and the foregoing discussion, it should come as no surprise that polar gases are generally most soluble in polar solvents, and nonpolar gases are most soluble in nonpolar liquids. The hydrogen halides, HF, HCl, HBr, and HI, are all polar covalent gases. In the gas phase the interactions among the widely separated molecules are not very strong, so solute-solute attractions are minimal and the dissolution processes are exothermic. The resulting solutions, called hydrohalic acids, contain predominantly ionized HX (X = Cl, Br, I). The ionization involves protonation of a water molecule by HX to form a hydrated hydrogen ion and halide ion X^- (which is also hydrated). HCl is used as an example below.

Protonation describes reactions in which H^+ (a bare proton) combines with another species.

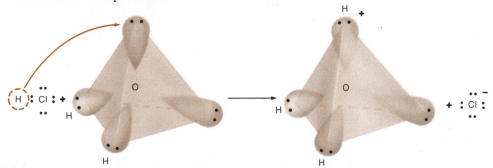

HF is only slightly ionized in aqueous solution because of the difficulty of breaking the strong covalent bond between highly electronegative fluorine and hydrogen. In addition, the more polar bond between H and the small F atom leads to very strong hydrogen bonding between H_2O and the largely intact HF molecules.

$$\overset{\delta+}{H}-\overset{\cdot\cdot}{\underset{\underset{\underset{\delta+}{H}}{|}}{\overset{\delta-}{O}}}\!\!:\text{---}\overset{\delta+}{H}-\overset{\delta-}{\underset{\cdot\cdot}{F}}\!:\qquad \text{as well as}\qquad \overset{\delta+}{H}-\overset{\delta-}{\underset{\cdot\cdot}{F}}\!:\text{---}\overset{\delta+}{H}-\overset{\delta-}{\underset{\underset{\underset{\delta+}{H}}{|}}{O}}\!:$$

Hydrogen fluoride also hydrogen bonds to itself in aqueous solutions (as well as in the gas phase), forming species that can be represented as $(HF)_n$, where n may be as large as five.

Although carbon dioxide, CO_2, and oxygen, O_2, are nonpolar gases, they do dissolve to limited extents in water. Carbon dioxide is somewhat more soluble because it reacts with water to form carbonic acid, H_2CO_3, which in turn ionizes slightly in two steps to give hydrogen ions, bicarbonate ions, and carbonate ions.

$$CO_2\ (g) + H_2O\ (\ell) \rightleftharpoons H_2CO_3\ (aq) \qquad \text{carbonic acid (exists only in solution)}$$

$$H_2CO_3\ (aq) \rightleftharpoons H^+\ (aq) + HCO_3^-\ (aq)$$

$$HCO_{3-}\ (aq) \rightleftharpoons H^+\ (aq) + CO_3^{2-}\ (aq)$$

Approximately 1.45 grams of CO_2 (0.033 mole) dissolves in a liter of water at 25°C and one atmosphere pressure. But only about 0.0045 gram of O_2 (1.4×10^{-4} mole) dissolves in a liter of water under the same conditions. Yet this is sufficient to support aquatic life.

To summarize, the only gases that dissolve appreciably in water are those that are capable of hydrogen bonding (like HF), those that ionize (like HCl, HBr, and HI), and those that react with water (like CO_2).

13-5 Rates of Dissolving and Saturation

At a given temperature, the rate at which a solid dissolves increases if large crystals are ground to a powder. This increases the surface area which, in turn, increases the number of solute ions or molecules in contact with the solvent. Pulverization also increases the number of corners and edges, where the ions or molecules are less firmly held. When an ionic solid is placed in water, some of its ions solvate and dissolve. The rate of this process slows as time passes because the surface area of each crystallite gets smaller and smaller. At the same time, the number of ions in solution increases and collisions between dissolved ions and solid occur more frequently. Some such collisions result in recrystallization or precipitation. After some time the rates of the two opposing processes become equal, and the solid and dissolved ions are then said to be in equilibrium with each other.

Solid $\overset{\text{dissolution}}{\underset{\text{crystallization}}{\rightleftharpoons}}$ Dissolved ions

The double arrows ($\rightleftharpoons$) signify that both processes occur simultaneously, and the equilibrium is called a dynamic rather than a static equilibrium. The slow "patching" of defects on the surfaces of crystals without mass increase when imperfect crystals are placed in saturated solutions of their ions provides evidence for the dynamic character of the equilibrium. After equilibrium is established, no more solid dissolves without the simultaneous crystallization of an equal mass of dissolved ions. Such a solution is said to be **saturated.** Saturation occurs at very low concentrations of dissolved species for slightly soluble substances and at high concentrations for very soluble substances.

Carbon dioxide is called an acid anhydride or an "acid without water." As noted in Section 9-5.2, many other oxides of nonmetals such as N_2O_3, SO_2, and P_4O_{10} are also acid anhydrides, and most are soluble in water.

Dynamic equilibria occur in all saturated solutions; for instance, there is a continuous exchange of oxygen molecules across the surface of water in an open container. This is fortunate for fish, which "breathe" dissolved oxygen.

The solubility of many solids increases at higher temperatures, and **supersaturated** solutions, which actually contain higher than saturation concentrations of solute, can sometimes be prepared by saturating a solution at a high temperature. The saturated solution is cooled slowly, without agitation, to a temperature at which the solute is less soluble. The resulting solution is *metastable,* and produces crystals rapidly if it is slightly disturbed or if it is "seeded" with a dust particle or a crystallite. Under such conditions enough solid crystallizes to leave a solution that is just saturated (Figure 13–6).

13–6   Effect of Temperature on Solubility

In Section 12–5 we introduced LeChatelier's Principle, which states that *when a stress is applied to a system at equilibrium, the system responds in a way that best relieves the stress.* Recall that exothermic processes release heat and endothermic processes absorb heat.

Exothermic: Reactants $\longrightarrow$ Products + Heat

Endothermic: Reactants + Heat $\longrightarrow$ Products

Since many solids dissolve by endothermic processes, their solubilities in water usually *increase* as heat is added and the temperature increases. For example, KCl dissolves endothermically.

$$KCl\ (s) + 17.0\ kJ \xrightarrow{\text{H}_2\text{O}} K^+\ (aq) + Cl^-\ (aq)$$

Figure 13–7 shows that the solubility of KCl increases as the temperature increases; this happens because more heat is available to drive the dissolving process.

FIGURE 13–6
Seeding a supersaturated solution of sodium acetate (top); the excess solute precipitates in the form of needle-like crystals (middle). The end result is a saturated solution of sodium acetate in equilibrium with excess solid.

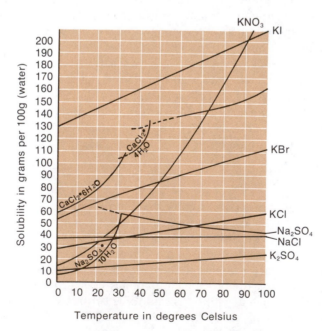

FIGURE 13–7 A graph illustrating the effect of temperature on the solubilities of some salts.

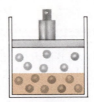

Increase
pressure

FIGURE 13-8
Illustration of Henry's
Law. The solubility of a
gas (that does not react
with the solvent)
increases with increasing
pressure of the gas
above the solution.

In contrast, some solids (and many liquids and gases) dissolve by endothermic processes, and their solubilities usually *decrease* as temperature increases. Anhydrous Na_2SO_4 is an example of such a solid compound. The major reason for the exothermicity of the dissolving of liquids and gases is that solute-solute interactions are much weaker than in solids; there is no crystal lattice energy to be overcome.

The solubility of O_2 in water decreases (by 22%) from 0.0045 gram per liter of water at 25°C to 0.0035 gram per liter at 50°C. The **thermal pollution** of rivers and lakes by heated waste water from industrial plants and nuclear power plants refers to the increased heat content of the water. A slight temperature increase causes a small but significant decrease in the amount of dissolved oxygen, so the water can no longer adequately support the marine life it ordinarily could.

13-7 Effect of Pressure on Solubility

Changing the pressure has no appreciable effect on the solubilities of either solids or liquids in liquids. However, the solubilities of gases in all solvents increase as the partial pressures of the gases increase. Carbonated water is a saturated solution of carbon dioxide in water under pressure. When a can or bottle of a carbonated beverage is opened, the pressure on the surface of the beverage is reduced to atmospheric pressure and much of the CO_2 bubbles out of solution. If the container is left open, the beverage becomes ''flat'' faster than it would if tightly stoppered, because CO_2 is released and escapes faster from the open container.

Henry's Law is applicable to gases that do not react with the solvent in which they dissolve (as well as to some cases in which gases react slightly but some gas remains unreacted): *the concentration of a gas in a solution is proportional to the pressure of the gas above the surface of the solution.* Henry's Law can be represented symbolically as

Dissolving some gas in a liquid reduces the vapor pressure of the liquid at its surface.

$$M_{gas} = kP_{gas} \qquad \text{[Eq. 13-6]}$$

in which M_{gas} is the molar concentration of the gas in solution, P_{gas} is the partial pressure of the gas above the solution, and k is a constant for a particular gas and solvent at a particular temperature. The relationship is valid at low concentrations and low pressures.

Colligative Properties of Solutions

Physical properties of solutions that depend upon the *number*, but not the *kind*, of solute particles in a given amount of solution are called **colligative properties.** The four most important colligative properties of a solution are (1) lowering of the vapor pressure of the solvent, (2) elevation of its boiling point, (3) depression of its freezing point, and (4) its osmotic pressure. Each will be discussed in turn, after we introduce the expression of concentrations of solutions in terms of mole fraction and molality, which often are used in describing colligative properties of solutions.

13-8   Mole Fraction

The **mole fraction** of component A in a solution composed of A and B is defined as the fraction of the total number of moles present that is due to A. The symbol X_A is used for the mole fraction of A:

$$X_A = \frac{\text{number of moles of A}}{\text{number of moles of A} + \text{number of moles of B}} \qquad \text{[Eq. 13-7]}$$

Example 13-1

What are the mole fractions of NH_3 and H_2O in commercial aqueous ammonia, which is 28.0% NH_3 by mass?

Solution

Consider, for convenience, a 100.0 g sample of the solution; it must contain 28.0 g NH_3 and 72.0 g H_2O. To calculate the mole fractions we must first calculate the number of *moles* of each component in these masses.

$$? \text{ mol } NH_3 = 28.0 \text{ g } NH_3 \times \frac{1 \text{ mol } NH_3}{17.0 \text{ g } NH_3}$$

$$= 1.65 \text{ mol } NH_3$$

$$? \text{ mol } H_2O = 72.0 \text{ g } H_2O \times \frac{1 \text{ mol } H_2O}{18.0 \text{ g } H_2O}$$

$$= 4.00 \text{ mol } H_2O$$

Thus,

$$X_{NH_3} = \frac{\text{mol } NH_3}{\text{mol } NH_3 + \text{mol } H_2O}$$

$$= \frac{1.65 \text{ mol}}{1.65 \text{ mol} + 4.00 \text{ mol}} = \underline{0.292}$$

$$X_{H_2O} = \frac{4.00 \text{ mol}}{1.65 \text{ mol} + 4.00 \text{ mol}} = \underline{0.708}$$

Alternatively, since the sum of the fractions must always be unity, $X_{NH_3} + X_{H_2O} = 1$, and we could compute the mole fraction of water by difference:

$$X_{H_2O} = 1.000 - 0.292 = \underline{0.708}$$

13-9   Molality

The mathematical treatment of colligative properties is often made easier by expressing concentrations in molality units. Other concentration units are more useful in describing *chemical* properties. The **molality,** *m,* of a solute in solution is the number of moles of solute *per kilogram of solvent* (not solution).

$$\text{molality} = \frac{\text{number of moles solute}}{\text{number of kilograms solvent}} \qquad \text{[Eq. 13-8]}$$

Example 13-2

Determine the molality of a solution containing 45.0 grams of potassium chloride in 1100 grams of water.

Solution

$$? \frac{\text{mol KCl}}{\text{kg } H_2O} = \frac{45.0 \text{ g KCl}}{1.100 \text{ kg } H_2O} \times \frac{1 \text{ mol KCl}}{74.6 \text{ g KCl}}$$

$$= \underline{0.548 \text{ mol KCl/kg } H_2O}$$

Example 13-3

How many grams of water must be used to dissolve 50.0 grams of sucrose, $C_{12}H_{22}O_{11}$, to prepare a 0.100 m solution of sucrose?

Solution

The molecular weight of sucrose is $12(12) + 22(1) + 11(16) = 342$ amu, so

$$? \text{ mol sucrose} = 50.0 \text{ g sucrose} \times \frac{1 \text{ mol sucrose}}{342 \text{ g sucrose}}$$

$$= 0.146 \text{ mol sucrose}$$

Since

$$\text{molality of solution} = \frac{\text{number of mol sucrose}}{\text{number of kg } H_2O},$$

rearranging gives

$$\text{number of kg } H_2O = \frac{\text{number of mol sucrose}}{\text{molality of solution}}$$

$$= \frac{0.146 \text{ mol sucrose}}{0.100 \text{ } m}$$

$$= 1.46 \text{ kg } H_2O$$

$$= 1.46 \times 10^3 \text{ g } H_2O$$

13-10   Lowering of Vapor Pressure and Raoult's Law

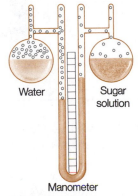

Water   Sugar solution

Manometer

FIGURE 13-9

Vapor pressure lowering. If no air is present in the apparatus, the pressure above the two liquids is due to water vapor. This pressure is less over the solution of water and sugar, because there are fewer water molecules on the surface to evaporate.

We could also represent P_{solvent} as P_{solution} since the solute is nonvolatile.

Numerous experiments have shown that solutions containing nonvolatile liquids or solids as solutes always have lower vapor pressures than the pure solvents (Figure 13-9). The vapor pressure of a liquid depends upon the ease with which the molecules are able to escape from the surface of the liquid. When a solute is dissolved in a liquid, some of the total volume of the solution is occupied by solute molecules, and there are therefore fewer solvent molecules per unit area at the surface. As a result, solvent molecules vaporize at a slower rate than if no solute were present. This results in lowering the vapor pressure of the solvent, which is a colligative property; it is a function of the number, and not the kind, of solute particles in solution. We should point out that solutions of gases or low boiling (volatile) liquids dissolved in liquids have *higher* total vapor pressures than the pure solvent, so this discussion does not apply to them.

The lowering of vapor pressure associated with nonvolatile, nonionizing solutes is summarized by **Raoult's Law:** *The vapor pressure of a solvent in a solution decreases as its mole fraction decreases.* The relationship can be expressed mathematically as

$$P_{\text{solvent}} = X_{\text{solvent}} P^0_{\text{solvent}} \qquad \text{[Eq. 13-9]}$$

in which X_{solvent} represents the mole fraction of the solvent in a solution, P^0_{solvent} is the vapor pressure of the *pure* solvent, and P_{solvent} is the vapor pressure of the solvent *in the solution.*

The lowering of vapor pressure, $\Delta P_{\text{solvent}}$, is defined as

$$\Delta P_{\text{solvent}} = P^0_{\text{solvent}} - P_{\text{solvent}} \qquad \text{[Eq. 13-10]}$$

Thus,

$$\Delta P_{\text{solvent}} = P^0_{\text{solvent}} - (X_{\text{solvent}} P^0_{\text{solvent}}) \qquad \text{[Eq. 13-11]}$$

$$= (1 - X_{\text{solvent}})P^0_{\text{solvent}} \qquad \text{[Eq. 13-12]}$$

Since $X_{\text{solvent}} + X_{\text{solute}} = 1$, then it must be true that $1 - X_{\text{solvent}} = X_{\text{solute}}$. So we can express the lowering of vapor pressure in terms of the mole fraction of solute.

$$\Delta P_{solvent} = X_{solute} P^0_{solvent}$$

[Eq. 13–13]

A hypothetical solution that obeys this relationship exactly is called an **ideal solution.** (For the present discussion, we shall consider only solutions of nonelectrolytes that behave ideally, or very nearly ideally.)

Example 13–4

By how much is the vapor pressure of water lowered by dissolving 35.0 grams of glucose, $C_6H_{12}O_6$, in 100 grams of water at 25°C? The vapor pressure of pure water at 25°C is 23.8 torr.

Solution

We first calculate the number of moles of $C_6H_{12}O_6$ and H_2O present.

$$? \text{ mol } C_6H_{12}O_6 = 35.0 \text{ g } C_6H_{12}O_6 \times \frac{1 \text{ mol } C_6H_{12}O_6}{180 \text{ g } C_6H_{12}O_6}$$

$$= 0.194 \text{ mol } C_6H_{12}O_6$$

$$? \text{ mol } H_2O = 100 \text{ g } H_2O \times \frac{1 \text{ mol } H_2O}{18.0 \text{ g } H_2O}$$

$$= 5.56 \text{ mol } H_2O$$

Now we calculate the mole fraction of $C_6H_{12}O_6$.

$$X_{C_6H_{12}O_6} = \frac{0.194 \text{ mol}}{0.194 \text{ mol} + 5.56 \text{ mol}}$$

$$= 0.0337$$

Finally, we calculate the lowering of vapor pressure.

$$\Delta P_{H_2O} = X_{C_6H_{12}O_6} P^0_{H_2O}$$

$$= (0.0337)(23.8 \text{ torr})$$

$$= \underline{0.802 \text{ torr}}$$

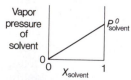

FIGURE 13–10
Raoult's Law for an ideal solution of a solute in a volatile liquid. The vapor pressure exerted by the liquid is proportional to its mole fraction in the solution.

Raoult's Law, for the case of a nonvolatile solute, is illustrated in Figure 13–10, which shows that the vapor pressure of a volatile solvent increases in direct proportion to the mole fraction of the solvent.

When a solution consists of two components that are very similar, each component behaves essentially as it would if it were pure. For example, the two liquids n-heptane, C_7H_{16}, and n-octane, C_8H_{18}, are so similar that each heptane molecule experiences nearly the same intermolecular forces whether it is near another heptane molecule or near an n-octane molecule, and similarly for each n-octane molecule. The properties of such a solution can be predicted from a knowledge of its composition and the properties of each component. Such a solution is very nearly ideal.

Consider an ideal solution of two volatile components, A and B. The vapor pressure of each component, A and B, above the solution is proportional to its mole fraction in the solution:

$$P_A = X_A P^0_A \qquad \text{and} \qquad P_B = X_B P^0_B$$

[Eq. 13–14]

The total vapor pressure of the solution is, by Dalton's Law of Partial Pressures (Section 11–12), equal to the sum of the vapor pressures:

$$P_{total} = P_A + P_B \qquad P_{total} = X_A P^0_A + X_B P^0_B$$

[Eq. 13–15]

This is shown graphically in Figure 13–11. Equation 13–15 can be used to predict the total vapor pressure of an ideal solution, as Example 13–5 illustrates.

Example 13–5

At 40°C, the vapor pressure of pure n-heptane is 92.0 torr, while the vapor pressure of pure n-octane is 31.0 torr. Consider a solution that contains 1.00 mole of n-heptane and 4.00 moles of n-octane.

Calculate the vapor pressure of each component and the total vapor pressure above the solution.

Solution

We first calculate the mole fraction of each component in the liquid solution.

$$X_{heptane} = \frac{1.00 \text{ mol heptane}}{(1.00 \text{ mol heptane}) + (4.00 \text{ mol octane})}$$

$$= 0.200$$

$$X_{octane} = 1 - X_{heptane} = 0.800$$

Then, applying Raoult's Law for two volatile components:

$$P_{heptane} = X_{heptane}P^0_{heptane}$$
$$= (0.200)(92.0 \text{ torr}) = 18.4 \text{ torr}$$

$$P_{octane} = X_{octane}P^0_{octane}$$
$$= (0.800)(31.0 \text{ torr}) = 24.8 \text{ torr}$$

$$P_{total} = P_{heptane} + P_{octane}$$
$$= 18.4 \text{ torr} + 24.8 \text{ torr}$$
$$= 43.2 \text{ torr}$$

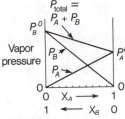

FIGURE 13–11
Raoult's Law for an ideal solution of two volatile components. Note that the left-hand side of the plot corresponds to pure B ($X_A = 0$, $X_B = 1$), and the right-hand side corresponds to pure A ($X_A = 1$, $X_B = 0$). Of these hypothetical liquids, pure B is more volatile than pure A ($P_B^0 > P_A^0$).

Many solutions do not behave ideally. For some solutions, the observed vapor pressure is greater than that predicted from Raoult's Law, as shown in Figure 13–12a. This kind of deviation, known as a *positive deviation*, is due to differences in polarity of the two components. On the molecular level, the two substances do not mix completely randomly, so that there is self-association of each component with local regions consisting of clumps of individual types of molecules. In a clump of A molecules, substance A acts as if its mole fraction were greater than it is in the solution as a whole, and the vapor pressure due to A is greater than if the solution were ideal. A similar description applies to component B. The total vapor pressure is then greater than if the solution were behaving ideally. For instance, a solution of ethanol, C_2H_5OH, and water behaves in this fashion.

In some other (more common) solutions, which are said to exhibit *negative deviations* from Raoult's Law, the observed vapor pressure of each component, and hence the total vapor pressure, is less than that predicted (Figure 13–12 b). Such an effect is due to unusually strong attractions (such as hydrogen bonding) between unlike molecules. This attraction holds the species in the liquid phase unusually tightly, so that fewer of them escape to the vapor phase. An example of such a solution is hydrogen chloride in water.

13–11   Fractional Distillation

In Chapter 12 we described simple distillation as a process by which a liquid solution can be separated into its volatile and nonvolatile components. However, a solution consisting of two volatile components cannot be completely separated by

FIGURE 13–12
Deviations from Raoult's Law. (a) Positive deviation. (b) Negative deviation. See text for explanation.

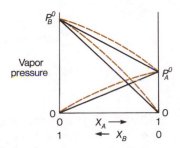

(a)

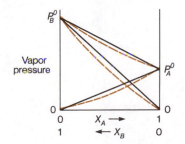

(b)

this method. As the temperature is slowly raised, such a solution will begin to boil when the sum of the vapor pressures of the components reaches the applied pressure on the surface of the liquid. If the solution obeys Raoult's Law, then the partial pressure of each component in the vapor above the boiling liquid can be found from Eqs. 13–14 and 13–15 (as in Example 13–5). Since the total pressure is known (it is equal to the external pressure, say, 1 atm), the mole fractions of the components *in the vapor* can be found:

$$X_{A,vapor} = \frac{P_A}{P_{total}} \quad \text{and} \quad X_{B,vapor} = \frac{P_B}{P_{total}}$$

Since both components exert vapor pressures, both will be present in the vapor. The more volatile component (that is, the one with the greater vapor pressure in the pure state, P^0) will have a *greater* mole fraction in the vapor than it has in the original liquid solution. If the vapor is collected in several batches (called *fractions*), condensed, and redistilled, this enrichment of the vapor in the more volatile component is compounded. By repeated application of this method of **fractional distillation**, the two components can be obtained in nearly pure form.

An apparatus called a *fractionating column* has been devised to avoid the necessity of repeated collection and distillation (Figure 13–13). The column is packed with small glass beads, which provide surfaces on which condensation can occur. As the less volatile component condenses, heat is released, which keeps the more volatile component in the vapor phase. By the time the vapor reaches the top of the column, practically all of the less volatile component has condensed and fallen back into the flask. The more volatile component goes into the condenser, where it is liquefied and delivered into the collection flask. The longer the column or the greater the surface area of the packing, the better is the separation.

Solutions that exhibit positive deviations from Raoult's Law cannot be separated even by fractional distillation. Because the vapor pressure of the mixture is greater than that of either pure component, there is a fixed composition that boils at a temperature lower than the boiling points of the components. Such a solution is called a **minimum boiling azeotrope.** The best separation that can be achieved is one pure component and the azeotrope. A solution that exhibits negative deviations from Raoult's Law, similarly, produces a **maximum boiling azeotrope** that boils at a temperature above the boiling points of the pure components.

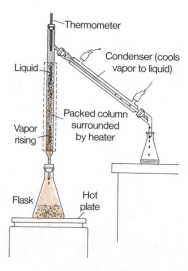

FIGURE 13–13 A fractional distillation apparatus. The vapor phase rising in the column is in equilibrium with the liquid phase condensed out and flowing slowly back down the column.

TABLE 13-2 Properties of Some Common Solvents

Solvent	Boiling Point (°C)	K_b (°C/m)	Freezing Point (°C)	K_f (°C/m)
water	100	0.512	0	1.86
benzene	80.1	2.53	5.48	5.12
acetic acid	118.1	3.07	16.6	3.90
nitrobenzene	210.88	5.24	5.7	7.00
phenol	182	3.56	43	7.40

13-12 Boiling Point Elevation

The vapor pressure of a solvent at a given temperature is lowered by the presence of a *nonvolatile* solute in it, and such a solution must be heated to a higher temperature than the pure solvent in order for its vapor pressure to equal atmospheric pressure (Figure 13-14). Recall that the boiling point of a liquid is the temperature at which its vapor pressure just equals the applied pressure on its surface or, for liquids in open containers, atmospheric pressure. Therefore, in accord with Raoult's Law, the elevation of the boiling point of a solvent caused by the presence of a nonvolatile, nonionizing solute is proportional to the number of moles of solute dissolved in a given mass of solvent. Mathematically, this is usually expressed as follows:

$$\Delta T_b = K_b m \qquad \text{[Eq. 13-16]}$$

Note that ΔT_b is always positive for solutions that contain *nonvolatile* solutes, because the boiling points of such solutions are always higher than the boiling points of the pure solvents.

The term ΔT_b represents the elevation of boiling point of the solvent, i.e., the boiling point of the solution minus the boiling point of the pure solvent. The m is the molality of the solute, and K_b is a proportionality constant called the *molal boiling point elevation constant,* which is different for different solvents and does not depend on the solute. Values of K_b for various solvents are tabulated in Table 13-2. They correspond numerically to the change in boiling point produced by one molal solutions of nonvolatile nonelectrolytes. Elevations of boiling points, and depressions of freezing points which are discussed below, are usually quite small for solutions of typical concentrations. As a result they can be measured accurately only with specially constructed (and expensive) differential thermometers that measure small temperature changes accurately to the nearest 0.001°C.

Example 13-6

Calculate the boiling point of a solution that contains 100 g of sucrose, $C_{12}H_{22}O_{11}$, in 500 g of water.

Solution

We must relate the increase of boiling point to the molality of the solution. First we determine that molality:

$$\underline{?}\ \frac{\text{mol sucrose}}{\text{kg H}_2\text{O}} = \frac{100\ \text{g sucrose}}{0.500\ \text{kg H}_2\text{O}} \times \frac{1\ \text{mol sucrose}}{342\ \text{g sucrose}}$$
$$= 0.585\ m\ \text{sucrose solution}$$

Using Eq. 13-16,

$$\Delta T_b = K_b m$$

From Table 13-2, K_b for water is 0.512°C/m, so

$$\Delta T_b = (0.512°C/m)(0.585\ m) = 0.300°C$$

This is the boiling point elevation. The normal boiling point of pure water is 100.000°C, so the boiling point of the sucrose solution is 100.300°C.

13-13   Freezing Point Depression

Ethylene glycol, CH_2OHCH_2OH, the major component of "permanent" antifreeze, effectively depresses the freezing point of water in an automobile radiator and raises its boiling point so that the solution remains in the liquid state over a wider temperature range than does pure water.

Molecules of most liquids approach each other more closely as the temperature is lowered, because their kinetic energies decrease and collisions become less frequent and less vigorous. The freezing point of a liquid is the temperature at which the forces of attraction among molecules are just great enough to cause a phase change from the liquid to the solid state. Strictly speaking, the freezing (melting) point of a substance is the temperature at which the liquid and solid phases exist in equilibrium. Clearly, the solvent molecules in a solution are somewhat more separated from each other (because of solute particles) than they are in the pure solvent. Consequently, the temperature of a solution must be lowered below the freezing point of the pure solvent in order to freeze it (Figure 13-14).

The freezing point depression of solutions of nonelectrolytes has been found to be equal to the molality of the solute times a proportionality constant called the *molal freezing point depression constant*, K_f.

ΔT_f is the *depression* of freezing point:
$\Delta T_f = T_{f(solvent)} - T_{f(solution)}$; it is always *positive*.

$$\Delta T_f = K_f m \qquad \text{[Eq. 13-17]}$$

The values of K_f for different solvents were given in Table 13-2. They are numerically equal to the freezing point depression caused by dissolving one mole of a nonelectrolyte in one kilogram of solvent.

Example 13-7

When 15.0 g of ethanol, C_2H_5OH, are dissolved in 750 g of formic acid, the freezing point of the solution is 7.20°C. The freezing point of pure formic acid is 8.40°C. Calculate K_f for formic acid.

Solution

The relationship $\Delta T_f = K_f m$ can be solved for K_f, and values for ΔT_f and m can be substituted into the resulting expression:

$$K_f = \frac{\Delta T_f}{m}$$

The molality and the depression of the freezing point are calculated first:

$$? \frac{\text{mol ethanol}}{\text{kg formic acid}} = \frac{15.0 \text{ g ethanol}}{0.750 \text{ kg formic acid}}$$

$$\times \frac{1 \text{ mol ethanol}}{46.0 \text{ g ethanol}}$$

$$= 0.435 \; m$$

$$\Delta T_f = (T_{f, \text{ formic acid}}) - (T_{f, \text{ solution}})$$

$$= 8.40°C - 7.20°C = 1.20°C \text{ depression}$$

Then

$$K_f = \frac{1.20°C}{0.435 \; m} = \underline{2.76°C/m}$$

Recall that the molal freezing point depression constant is the depression for a one molal solution of a nonelectrolyte.

Example 13-8

What is the freezing point of a solution of 23.0 g of ethanol in 600 g of water?

Solution

For water, $K_f = 1.86°C/m$. First we must calculate the molality of the solution:

$$? \frac{\text{mol ethanol}}{\text{kg H}_2\text{O}} = \frac{23.0 \text{ g ethanol}}{0.600 \text{ kg H}_2\text{O}} \times \frac{1 \text{ mol ethanol}}{46.0 \text{ g ethanol}}$$

$$= 0.833 \; m \text{ ethanol}$$

Then

$$\Delta T_f = (1.86°C/m)(0.833 \; m) = 1.55°C$$

The temperature at which the solution freezes (at one atmosphere pressure) is 1.55°C *below* the freezing point of pure water, or

$$T_{f,\text{solution}} = 0.00°C - 1.55°C = -1.55°C$$

13–14   Determination of Molecular Weight by Freezing Point Depression

The colligative properties of freezing point depression and, to a lesser extent, boiling point elevation are useful in determining the molecular weights of solutes. The solutes must be nonvolatile in the temperature range of the experiment if boiling point elevations are used. They must also be nonelectrolytes, since partial dissociation of the solute would produce more solute particles than would be expected if no dissociation occurred.

Example 13–9
A sample of an unknown covalent organic compound with a mass of 1.20 g is dissolved in 50.0 g of benzene. The resulting solution freezes at 4.92°C. Determine the molecular weight of the solute.

Solution
From Table 13–2, the freezing point of benzene is 5.48°C and its K_f is 5.12°C/m. Thus, the freezing point depression is

$$\Delta T_f = 5.48°C - 4.92°C = 0.56°C$$

so the molality of the solution is calculated to be

$$m = \frac{\Delta T_f}{K_f} = \frac{0.56°C}{5.12°C/m} = 0.11 \ m$$

The molality is the number of moles of solute per kilogram of benzene, so the number of moles in 0.0500 kg of benzene can be calculated.

$$0.11 \ m = \frac{? \ \text{mol solute}}{0.0500 \ \text{kg } C_6H_6}$$

$$? \ \text{mol solute} = (0.11 \ m)(0.0500 \ \text{kg})$$
$$= 0.0055 \ \text{mol solute}$$

$$\text{mass of 1.0 mol} = \frac{\text{number of g solute}}{\text{number of mol solute}}$$

$$= \frac{1.20 \ \text{g solute}}{0.0055 \ \text{mol solute}}$$

$$= 2.2 \times 10^2 \ \text{g/mol}$$

$$\text{molecular weight} = \underline{2.2 \times 10^2 \ \text{amu}}$$

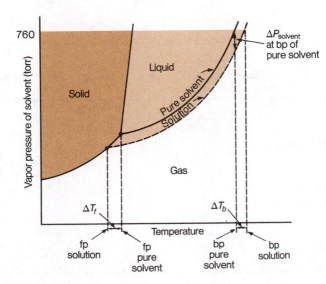

FIGURE 13–14 Because a *nonvolatile* solute lowers the vapor pressure of a solvent, the boiling point of a solution is higher and the freezing point lower than the corresponding points for the pure solvent. The magnitude of boiling point elevation, ΔT_b, is less than the magnitude of freezing point depression, ΔT_f.

13–15   Ion Association in Solutions of Strong Electrolytes

$\Delta T_f = K_f m$
$= (1.86°C/m) \times (0.100\ m)$
$= 0.186°C$

As we said earlier, colligative properties depend upon the *number* of solute particles in a given amount of solvent. A 0.100 molal *aqueous* solution of a covalent compound that does not ionize gives a freezing point depression of 0.186°C. We might predict that a 0.100 molal solution of a 1:1 strong electrolyte like KBr would have a freezing point depression of $2 \times 0.186°C$, or 0.372°C, since 0.100 *m* KBr would have an *effective* molality of 0.200 *m*, or 0.100 *m* K^+ + 0.100 *m* Br^-, assuming dissociation were complete. In fact, the *observed* depression is only 0.349°C. The discrepancy is due to the fact that some of the ions undergo association in solution. At any given instant some K^+ and Br^- ions collide and "stick together." During the brief time that they are in contact they behave as a single particle, and this tends to reduce the effective molality, and therefore the freezing point depression (as well as boiling point elevation and lowering of vapor pressure).

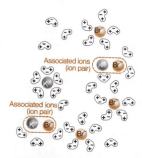

FIGURE 13–15
Some of the various species thought to be present in a solution of KBr in water, which would explain unexpected values for its colligative properties, such as freezing point depression.

A (more concentrated) 1.00 *m* solution of KBr might be expected to have a freezing point depression of $2 \times 1.86°C = 3.72°C$, but the observed depression is 3.29°C. There is a greater deviation from the predicted depression in the more concentrated solution. This is due to the greater frequency of collisions of ions, and consequent increase in ionic association, in the more concentrated solution.

To allow for dissociation and for ionic association, we use modified forms of Equations 13–17 and 13–16 when we consider solutions of electrolytes:

$$\Delta T_f = iK_f m \qquad \text{or} \qquad \Delta T_f = K_f(im) \qquad \text{[Eq. 13–18]}$$

and

$$\Delta T_b = iK_b m \qquad \text{or} \qquad \Delta T_b = K_b(im) \qquad \text{[Eq. 13–19]}$$

In these equations *i*, called the **van't Hoff factor,** is a multiplier that converts the stated molality to the *effective* molality. Thus we can also say that

$$m_{effective} = im \qquad \text{[Eq. 13–20]}$$

When $i = 1$, Eqs. 13–18 and 13–19 reduce to Eqs. 13–17 and 13–16, respectively.

For solutions of a nonelectrolyte, $i = 1$ and $m_{effective}$ is equal to *m* because no dissociation occurs.

For a hypothetical solution of a 1:1 strong electrolyte, like KBr, that is so dilute that *no* ion association occurs, $i = 2$; this is the limiting or maximum value for such a

Note that *i* values vary with both the solute involved and its concentration.

TABLE 13–3   Observed Van't Hoff Factors (*i*) for Aqueous Solutions of Sucrose and Strong Electrolytes

Compound	0.100 *m*	0.00100 *m*
sucrose	1.00	1.00
nonelectrolytes	1.00	1.00
KBr	1.88	1.97
HCl	1.89	1.98
if 2 ions/formula unit with no ion association	2.00	2.00
K_2SO_4	2.32	2.84
K_2CO_3	2.45	—
if 3 ions/formula unit with no ion association	3.00	3.00

salt. The value of i for a given electrolyte decreases with increasing concentration to reflect greater ionic association. The limiting value of i is 3 for salts like $MgCl_2$ (3 ions per formula unit) and 4 for salts like $K_3[Fe(CN)_6]$ (4 ions per formula unit; $3K^+$ and $1[Fe(CN)_6]^{3-}$). The values of i for solutions of sucrose (table sugar, which is a nonelectrolyte) and a few strong electrolytes are given in Table 13-3.

Experimentally, van't Hoff factors can be determined by dividing the *observed* change in a colligative property by the change calculated *as if* the solute were a nonelectrolyte. So,

$$i = \frac{\Delta T_{f(observed)}}{\Delta T_{f(if\ nonelectrolyte)}}$$

[Eq. 13-21]

Example 13-10

A 0.0311 m solution of iron(III) chloride, $FeCl_3$, in water freezes at $-0.206°C$. Determine the van't Hoff factor for $FeCl_3$ in this solution. K_f for water = $1.86°C/m$.

Solution

We must first calculate the freezing point that would be observed if $FeCl_3$ were a nonelectrolyte.

$$\Delta T_{f(if\ nonelectrolyte)} = K_f m = (1.86°C/m)(0.0311\ m)$$
$$= 0.0578°C$$

The observed freezing point is $-0.206°C$, so the observed freezing point depression is

$$\Delta T_{f(observed)} = 0.000°C - (-0.206°C) = 0.206°C$$

Now we can calculate i, using Eq. 13-21.

$$i = \frac{\Delta T_{f(observed)}}{\Delta T_{f(if\ nonelectrolyte)}} = \frac{0.206°C}{0.0578°C} = \underline{3.56}$$

compared to the limiting value of 4 that would be expected if the solution were infinitely dilute.

13-16 Activities and Activity Coefficients

As we said earlier, i acts as a multiplier that converts the stated molality to an effective molality, or the sum of the molalities of all the solute particles (Eq. 13-20):

$$m_{effective} = im$$

For the 0.0311 m solution of $FeCl_3$ of Example 13-10,

$$m_{effective} = (3.56)(0.0311\ m) = 0.111\ m$$

In other words, the solution *behaves* as if the concentration of all solute particles together is 0.111 m. If no ion association occurred, the effective molality would be $(4.00)(0.0311\ m) = 0.124\ m$. We could also view the situation *as if* some of the $FeCl_3$ did not dissociate in the first place, and we could then calculate the apparent percentage dissociation.

If we let x represent the molality of the $FeCl_3$ that behaves as if it were dissociated, the molalities of Fe^{3+} and Cl^- ions would be x molal and $3x$ molal, respectively. The molality of $FeCl_3$ that would remain undissociated would be $(0.0311 - x)\ m$.

$$FeCl_3 \longrightarrow Fe^{3+} + 3Cl^-$$
$$(0.0311 - x)\ m \qquad x\ m \qquad 3x\ m$$

The effective molality, 0.111 m (see above), would have to equal the sum of the molalities of all three kinds of solute particles.

$$m_{\text{effective}} = 0.111 \, m = (0.0311 - x) \, m + x \, m + 3x \, m$$

$$0.111 \, m = (0.0311 + 3x) \, m$$

$$0.080 \, m = 3x \, m$$

$$x = 0.027$$

Thus, according to the definition of x, it *appears as if* 0.027 m of the original 0.0311 m $FeCl_3$ dissociates. We can now calculate the *apparent* percentage dissociation.

$$\% \text{ dissociation} = \frac{m \, FeCl_3 \text{ that appears to be dissociated}}{m \, FeCl_3 \text{ original}} \times 100\%$$

$$= \frac{0.027 \, m}{0.0311 \, m} \times 100\% = 87\%$$

The "effective concentrations" or "apparent concentrations" of ions in solution as discussed above are referred to as **activities** of the ions. According to our calculations, the concentration of Fe^{3+} ions in 0.0311 m $FeCl_3$ solution appears to be only 87% of what we might expect. So the activity, a, of Fe^{3+} is 0.027 m.

$$a_{Fe^{3+}} = 0.87(0.0311 \, m) = 0.027 \, m$$

The activity of the Cl^- ions is

$$a_{Cl^-} = 0.87(3 \times 0.0311 \, m) = 0.081 \, m$$

The factor 0.87 for this solution (87% expressed as a decimal) is called the **activity coefficient, γ.** In general, the activity of an ion equals the activity coefficient of the ion times its expected concentration assuming 100% dissociation. Activity coefficients decrease with increasing concentration, and are different for different compounds; they must be determined experimentally.

$$a = \gamma \times \text{expected concentration if 100\% dissociation} \qquad \text{[Eq. 13-22]}$$

13-17   Percent Ionization of Weak Electrolytes

Most weak electrolytes, such as weak acids and weak bases, are quite soluble in water, but they dissociate only slightly. Example 13-11 illustrates how the percentage dissociation of a weak electrolyte can be determined from freezing point depression data.

Example 13-11

Lactic acid, C_2H_5OCOOH, is a weak acid (and therefore a weak electrolyte) found in sour milk. It is also formed in muscles during intense physical activity and is responsible for the pain felt during strenuous exercise. The freezing point of a 0.0100 m aqueous solution of lactic acid is $-0.0206°C$. Calculate the van't Hoff factor, i, and the percentage ionization in this solution.

Solution

We first calculate $\Delta T_{f\text{(if nonelectrolyte)}}$ and then i.

$$\Delta T_{f \text{ (if nonelectrolyte)}} = K_f \, m$$
$$= (1.86°C/m)(0.0100 \, m)$$
$$= 0.0186°C$$

$$\Delta T_{f \text{ (observed)}} = 0.0000°C - (-0.0206°C)$$
$$= 0.0206°C$$

$$i = \frac{\Delta T_{f \text{ (observed)}}}{\Delta T_{f \text{ (if nonelectrolyte)}}}$$

$$= \frac{0.0206°C}{0.0186°C} = 1.11$$

The effective molality is

$$m_{effective} = im = (1.11)(0.0100 \ m)$$
$$= 0.0111 \ m$$

The equation for the ionization of lactic acid and the concentrations of various species in the solution are (where x = molality of lactic acid ionized):

$$C_2H_5OCOOH \longrightarrow H^+ + C_2H_5OCOO^-$$
$$(0.0100 - x) \ m \qquad x \ m \qquad x \ m$$

The effective molality of the solution, $0.0111 \ m$, is also equal to the sum of the molalities of the species in solution, $[(0.0100 - x) + (x + x)] \ m$, or

$$m_{effective} = (0.0100 + x) \ m = 0.0111 \ m$$
$$x = 0.0011 \ m \qquad \text{(molality ionized)}$$

Hence the percentage ionization for $0.0100 \ m$ lactic acid is

$$\% \text{ lactic acid ionized} = \frac{0.0011 \ m}{0.0100 \ m} \times 100\% = \underline{11\%}$$

This calculation tells us that in $0.0100 \ m$ solution, 89% of the lactic acid exists as nonionized molecules and 11% exists as H^+ and $C_2H_5OCOO^-$ ions.

13–18 Membrane Osmotic Pressure

Osmosis is the spontaneous process by which solvent molecules pass through a semipermeable membrane from a solution of lower concentration into a solution of higher concentration. A **semipermeable membrane** (such as cellophane) is used as a thin film or partition between two solutions. Solvent molecules may pass through the membrane in either direction, but the rate at which they pass into the more concentrated solution is found to be greater than the rate in the reverse direction. This is because more solvent molecules hit the membrane, per unit time, on the dilute side than on the concentrated side. The initial difference between the two rates is directly proportional to the difference in concentration between the two solutions.

The net movement of solvent particles from the dilute solution into the concentrated solution continues until the two sides reach equal concentrations (a dynamic equilibrium). Another way to cause dynamic equilibrium to occur (without equalizing the concentrations) is to apply a physical force to the concentrated solution that makes the rate of reverse passage equal to the forward rate. Figure 13–16 shows an arrangement in which this force is exerted by the hydrostatic pressure of the rising column of the more concentrated solution. The pressure exerted when equilibrium is reached is called the *osmotic pressure* of the solution.

> The kinetic-molecular theory (Section 11–13) explained pressure in terms of frequency and momentum of collisions of molecules.

The osmotic pressure of a given aqueous solution can be measured with an apparatus such as that depicted in Figure 13–16. The solution of interest is placed inside an inverted thistle tube, which has a cellophane film firmly fastened across the bottom. This part of the thistle tube and its cellophane are then immersed into a container of pure water. As time passes, the height of the solution in the neck rises until the weight of the column of solution is counterbalanced by the osmotic pressure.

> The greater the number of solute particles, the greater height to which the column rises, and the greater the osmotic pressure.

Osmotic pressure is a measure of the forces that bind solvent molecules together, thereby causing molecules of the pure solvent to pass through the membrane into the solution in order to replace those that have been tied up by interactions with the solute. Osmotic pressure is dependent upon the number, and not the kind, of solute particles in solution; it is therefore a colligative property.

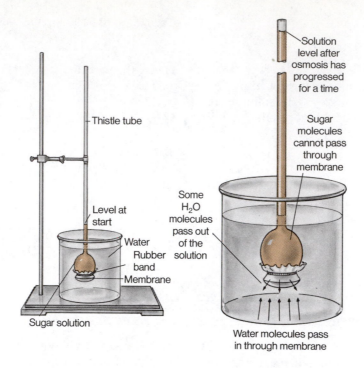

FIGURE 13-16
Laboratory apparatus for demonstrating osmosis. The picture at the right gives some details of the process.

Solute particles are quite widely separated in very dilute solutions and do not interact significantly with each other. For very dilute solutions, osmotic pressure, represented by π, is found to follow the equation

$$\pi = \frac{nRT}{V}$$ [Eq. 13-23]

In this equation, n is the number of moles of solute in volume V (in liters) of solution. The other quantities have the same meaning as in the ideal gas law. The term (n/V) is a concentration term. If concentration is expressed in terms of molarity (M), we obtain

$$\pi = MRT$$ [Eq. 13-24]

Osmotic pressure is proportional to T because T affects the number of solvent-membrane collisions per unit time, and π is proportional to M because M affects the difference in the number of solvent molecules hitting the membrane from each side. For dilute aqueous solutions the molarity is approximately equal to the molality (because the density is nearly 1 kg/liter), and the relationship becomes

For a dilute solution of an electrolyte, $\pi = imRT$.

$$\pi = mRT$$ [Eq. 13-25]

Osmotic pressures represent very significant forces. For example, a 1.0 molal solution of a nonelectrolyte in water at 0°C produces an equilibrium osmotic pressure of approximately 22.4 atmospheres.

The use of measurements of osmotic pressure for the determination of molecular weights has both advantages and disadvantages. An important advantage is that even very dilute solutions give rise to easily measurable osmotic pressures. This is in contrast to the vanishingly small freezing point depressions and boiling point elevations associated with the same solutions. This method therefore finds application in determination of the molecular weight of very expensive substances, or substances that can be prepared in only very small amounts, or substances of very high molecular weight such as polymers and many biological compounds. Because many biological materials are difficult, and in some cases impossible, to obtain in a high state of purity, determinations of their molecular weights are not as accurate as we might like. Nonetheless, osmotic pressures provide a very useful method of estimating molecular weights, as Example 13–12 illustrates.

Example 13–12

Pepsin is an enzyme present in the human digestive tract. An enzyme is a protein that acts as a biological catalyst. Pepsin catalyzes the metabolic cleavage of amino acid chains (called peptide chains) in other proteins. A solution of a 0.500 g sample of purified pepsin in 30.0 mL of benzene exhibits an osmotic pressure of 8.92 torr at 27.0°C. Estimate the molecular weight of pepsin.

Solution

$$\pi = MRT = \left(\frac{n}{V}\right)RT$$

We solve for n, the number of moles of solute (pepsin), and convert 8.92 torr to atmospheres to be consistent with the units of R.

$$n = \frac{\pi V}{RT} = \frac{\left(8.92 \text{ torr} \times \dfrac{1 \text{ atm}}{760 \text{ torr}}\right)(0.0300 \text{ L})}{\left(0.0821 \dfrac{\text{L} \cdot \text{atm}}{\text{mol} \cdot \text{K}}\right)(300 \text{ K})}$$

$$n = 1.43 \times 10^{-5} \text{ mol pepsin}$$

Thus 0.500 g of pepsin is 1.43×10^{-5} mol. We can now estimate its molecular weight.

$$\underline{?} \text{ g/mol} = \frac{0.500 \text{ g}}{1.43 \times 10^{-5} \text{ mol}}$$

$$= 3.50 \times 10^4 \text{ g/mol}$$

The molecular weight of pepsin is approximately 35,000 amu. This is typical for medium-sized proteins.

The freezing point of this very dilute solution would be depressed by only about 0.003°C, which would be extremely hard to measure accurately. The osmotic pressure of 8.92 torr, on the other hand, is easily measured.

Colloids

At the beginning of the chapter, a solution was defined as a homogeneous mixture in which no settling occurs and in which solute particles are at the molecular or ionic state of subdivision. This represents one extreme of mixtures. The other extreme would be a suspension, a clearly heterogeneous mixture in which solute-like particles immediately settle out after mixing with a solvent-like phase. Such a situation results when a handful of sand is dropped into water. **Colloids, colloidal suspensions,** or **colloidal dispersions** represent an intermediate kind of mixture in which the solute-like particles, or **dispersed phase,** are suspended in the solvent-like phase, or **dispersing medium.** The particles of the dispersed phase are small

TABLE 13–4   Types of Colloids and Examples

Dispersed (Solute-like) Phase		Dispersing (Solvent-like) Medium	Common Name	Examples
solid	in	solid	solid sol	many alloys (such as steel and duralumin), colored gems, reinforced rubber, porcelain, pigmented plastics
liquid	in	solid	solid emulsion	cheese, butter, jellies
gas	in	solid	solid foam	sponge, rubber, pumice, Styrofoam
solid	in	liquid	sols and gels	milk of magnesia, paints, mud, puddings
liquid	in	liquid	emulsion	milk, face cream, salad dressings, mayonnaise
gas	in	liquid	foam	shaving cream, whipped cream, foam on beer
solid	in	gas	solid aerosol	smoke, airborne viruses and particulate matter, auto exhaust
liquid	in	gas	liquid aerosol	fog, mist, aerosol spray, clouds

enough that no settling occurs, yet large enough to make the mixture appear cloudy (and in many cases, opaque), owing to scattering of light as it passes through the colloid.

Table 13–4 indicates that all combinations of solids, liquids, and gases are candidates for colloids except mixtures of gases in gases (all of which are homogeneous and, therefore, true solutions). Whether a given mixture forms a solution, a colloidal dispersion, or a suspension depends upon the size of the solute-like particles (Table 13–5), as well as solubility and miscibility.

13–19   The Tyndall Effect

The scattering of light by colloidal particles is called the **Tyndall effect** (Figure 13–17). In order to scatter visible light, a particle must have at least one dimension of approximately 1 nm. Solute particles in solutions are below this limit. The maximum dimension of a colloidal particle is about 1000 nm.

The scattering of light from automobile headlights by fogs and mists is an example of the Tyndall effect, as is the reflection of a light beam from a movie projector by dust particles in the air.

TABLE 13–5   Dispersed Particle Sizes

Dispersion	Example	Particle Size
suspension	sand in water	larger than 1000 nm
colloidal dispersion	starch in water	1–1000 nm
solution	sugar in water	0.1–1 nm

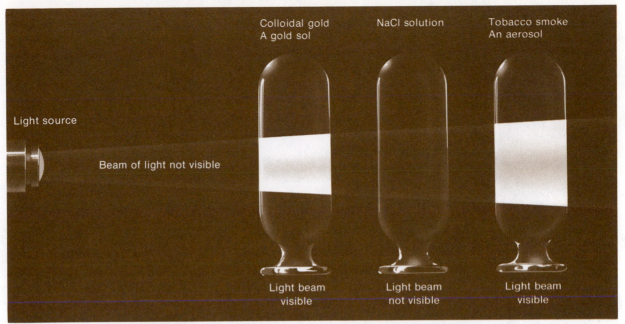

FIGURE 13-17 The presence of colloidal particles is easily detected. The dispersion of a beam of light by colloidal particles is called the Tyndall effect.

13-20 The Adsorption Phenomenon

Much of the chemistry of everyday life is the chemistry of colloids, as one can tell from a quick glance at Table 13-4. Since colloidal particles are so finely divided, they have enormous surface areas in relation to their volumes. It is not surprising, therefore, that an understanding of colloidal behavior requires an understanding of surface phenomena.

Atoms on the surface of a colloidal particle are bonded to other atoms of the particle in only two dimensions. Since these atoms, like all others, are capable of bonding in three dimensions, they have a tendency to interact with whatever comes in contact with the surface. Colloidal particles often adsorb ions or other charged particles, as well as gases and liquids. The process of **adsorption** involves adhesion of any such species onto the surfaces of particles. For example, a bright red sol (solid dispersed in liquid) is formed by the addition of hot water to a concentrated aqueous solution of iron(III) chloride.

$$2x[\text{Fe}^{3+}(aq) + 3\text{Cl}^-(aq)] + x(3 + y)\text{H}_2\text{O} \longrightarrow [\text{Fe}_2\text{O}_3 \cdot y\text{H}_2\text{O}]_x(s) + 6x[\text{H}^+ + \text{Cl}^-]$$

yellow solution   bright red sol

The surfaces of hydrated Fe_2O_3 particles tend to adsorb ions that fit readily into the crystal structure, so Fe^{3+} is preferentially adsorbed rather than Cl^-.

Each colloidal particle of this sol consists of a cluster of many hydrated Fe_2O_3 "molecules," which tends to attract positively charged Fe^{3+} ions to its surface. Since each particle is then surrounded by a shell of positively charged ions, the particles repel each other and cannot combine to the extent necessary to cause actual precipitation (see Figure 13-18).

FIGURE 13–18
Stabilization of a colloid
by electrostatic forces.

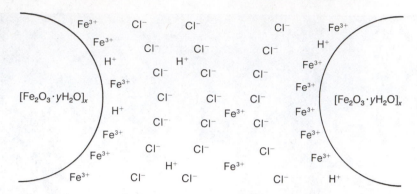

13–21   Hydrophilic and Hydrophobic Colloids

Colloids can be classified as **hydrophilic** ("water loving") or **hydrophobic** ("water hating") on the basis of the surface characteristics of the dispersed particles.

Proteins like the oxygen-carrier hemoglobin form *hydrophilic* sols when they are suspended in saline aqueous body fluids like blood plasma. Such proteins are macromolecules (giant molecules) that fold and twist in an aqueous environment so that polar groups are exposed to the fluid, while nonpolar groups are encased (see Figure 13–19). Protoplasm and human cells are examples of gels, which are special types of sols in which the solid particles (in this case mainly proteins and carbohydrates) join together in a semirigid network structure that encloses the dispersing medium. Other examples of gels are gelatin and jellies, and gelatinous precipitates such as $Al(OH)_3$.

Hydrophobic colloids cannot exist without the presence of **emulsifying agents,** or **emulsifiers,** which coat the particles of the dispersed phase to prevent their coagulation into a separate phase. Milk and mayonnaise are good examples of hydrophobic phases (milk fat in milk and vegetable oil in mayonnaise) that stay suspended with the aid of emulsifying agents (casein in milk and egg yolk in mayonnaise).

Consider the mixture resulting from vigorous shaking of oil (nonpolar) and water (polar), or salad oil and vinegar. Droplets of hydrophobic oil are temporarily suspended in the water. However, in a short time the polar water molecules force the oil molecules away from themselves. The oil floats on the surface of the water,

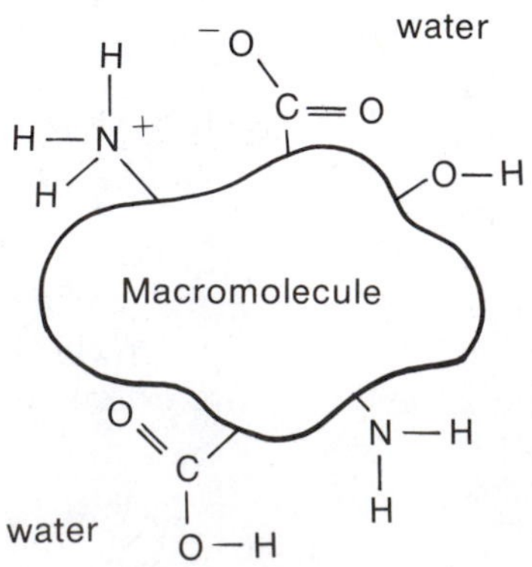

FIGURE 13–19 Examples of hydrophilic groups at the surface of a giant molecule (macromolecule) that help keep the macromolecule suspended in water.

since it is less dense. If we now add an emulsifying agent such as egg yolk, and shake or beat the mixture, a stable emulsion called mayonnaise results.

Soaps and other detergents are emulsifying agents. Solid soaps are usually sodium salts of long-chain organic acids called fatty acids. They are said to have a polar "head" and a nonpolar "hydrocarbon tail." A typical soap is sodium stearate, the salt of sodium hydroxide and stearic acid, shown below.

Sodium stearate is also a major component of most stick deodorants.

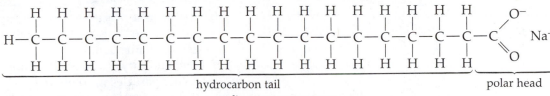

hydrocarbon tail   polar head
sodium stearate, a soap

If soap is added to an oil-water mixture and the mixture is vigorously shaken, a true emulsion forms. The polar heads of soap are attracted into the water. The hydrocarbon tails are soluble in the nonpolar oil. Droplets of the oil are surrounded by a shell of soap molecules with their polar heads in the water. As a result, the oil droplets are unable to coalesce and they remain suspended in the water (Figure 13–20). This is the mechanism of the cleansing action by which soaps and other detergents remove dirt, grease, or stains.

"Hard" water contains Fe^{3+}, Ca^{2+}, and/or Mg^{2+} ions, all of which displace Na^+ from soap molecules to form precipitates. This removes the soap from the water and puts an undesirable coating on the bathtub or on the fabric being laundered. **Synthetic detergents** are soap-like emulsifiers that contain sulfonate, $-SO_3^-$, sulfate, $-OSO_3^-$, or phosphate groups instead of carboxylate groups, $-COO^-$. They do not form precipitates with the ions of hard water, so they can be used in hard water as soap substitutes without forming undesirable scum.

However, the use of phosphate detergents is now discouraged because of their tendency to cause **eutrophication** in rivers and streams that receive sewage. This is a condition (not related to colloids) in which there is an unsightly overgrowth of vegetation caused by the high concentration of phosphorus, which is a plant nutrient. This overgrowth is accompanied by a decreased content of dissolved oxygen in the water, which causes the gradual elimination of marine life. There is also a foaming problem associated with alkylbenzenesulfonate (ABS) detergents in streams and pipes, tanks, and pumps of sewage treatment plants. Such detergents are not **biodegradable;** that is, they cannot be broken down by bacteria. However, currently used linear-chain alkyl sulfonate (LAS) detergents are biodegradable and do not cause such foaming.

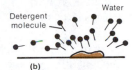

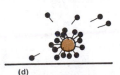

FIGURE 13–20
Representation of the method by which soaps and detergents act on dirt. The circles represent the polar "heads" of soap molecules, and the lines represent their nonpolar "tails."

$CH_3(CH_2)_{10}CH_2$ ⟨benzene ring⟩ $-SO_3^-Na^+$

sodium lauryl benzenesulfonate
a linear alkyl sulfonate (LAS)
a biodegradable detergent

Alkyl is a term used in organic chemistry to designate non-cyclic hydrocarbon structures.

ABS's have branched chain substituents.

$$CH_3CHCH_2CHCH_2CHCH_2CH$$ ⟨benzene ring⟩ $-SO_3^-Na^+$
with CH_3 groups

a sodium alkylbenzenesulfonate (ABS)
a nonbiodegradable detergent

Key Terms

The following terms were listed at the end of Chapter 2: **concentration, dilution, mass percent, molarity, solute, solution, and solvent.**

Activity of a dissolved species, concentration multiplied by a fraction called an activity coefficient that corrects for deviation from ideality. The "effective concentration" of a dissolved species.

Activity coefficient of a dissolved species, the decimal fraction that when multiplied by the expected concentration of a dissolved species gives its "effective" concentration.

Adsorption adhesion of species onto surfaces of particles.

Aerosol colloidal suspension of a solid or liquid in a gas.

Apparent percentage dissociation of strong electrolytes, the percentage of formula units of solute that appear to be dissociated into discrete ions in a solution of a given concentration.

Associated ions short-lived species formed by the chance collision of dissolved ions of opposite charge.

Azeotrope a solution of two or more volatile liquids that has a definite composition and a boiling point either higher or lower than that of any of its components.

Biodegradability the ability of a substance to be broken down into simpler substances by bacteria.

Boiling point elevation the increase in boiling point of a solvent caused by dissolution of a nonvolatile solute.

Coagulation combination of colloidal particles into particles that are large enough to precipitate.

Colligative properties physical properties of solutions that depend upon the number but not the kind of solute particles present.

Colloid a heterogeneous mixture in which solute-like particles do not settle out after introduction into a solvent-like phase.

Detergent a soap-like emulsifier that contains a sulfonate, $-SO_3^-$, sulfate, $-OSO_3^-$, or phosphate group instead of a carboxylate group.

Dispersed phase the solute-like species in a colloid.

Dispersing medium the solvent-like phase in a colloid.

Distillation the process in which components of a mixture are separated by boiling away the more volatile liquid.

Effective molality the sum of the molalities of all solute particles in solution.

Emulsifying agent a substance that coats the particles of a dispersed phase and prevents coagulation of colloidal particles; an emulsifier.

Emulsion colloidal suspension of a liquid in a liquid.

Eutrophication the undesirable overgrowth of vegetation caused by high concentration of plant nutrients in water.

Foam colloidal suspension of a gas in a liquid.

Fractional distillation repeated distillation of batches (fractions) of a mixture of volatile liquids to obtain the nearly pure components.

Freezing point depression the decrease in the freezing point of a solvent caused by the dissolution of a solute.

Gel colloidal suspension of a solid dispersed in a liquid; a semirigid sol.

Hard water water containing Fe^{3+}, Ca^{2+}, and/or Mg^{2+} ions, which form precipitates with soaps.

Heat of solution the difference between the enthalpies of a solution and its pure components.

Henry's Law the concentration of a gas in a solution is proportional to the pressure of the gas above the surface of the solution.

Hydration the interaction (surrounding) of an ion with water molecules.

Hydration energy (molar) of an ion, the amount of energy released in the hydration of a mole of gaseous ions.

Hydrophilic colloids colloidal particles that attract water molecules.

Hydrophobic colloids colloidal particles that repel water molecules.

Ideal solution a solution that obeys Raoult's Law exactly.

Ion pair see *Associated ions.*

Miscibility the ability of one liquid to mix with (dissolve in) another liquid.

Molality (*m*) concentration expressed as number of moles of solute per kilogram of solvent.

Mole fraction of a component in solution, the number of moles of the component divided by the sum of the numbers of moles of all components.

Osmosis the process by which solvent molecules pass through a semipermeable membrane from a dilute solution into a more concentrated solution.

Osmotic pressure the hydrostatic pressure produced on the surface of a semipermeable membrane by osmosis.

Percentage ionization of weak electrolytes, the percentage of the weak electrolyte that actually ionizes in a solution of given concentration.

Raoult's Law the vapor pressure of a solvent in a solution decreases as its mole fraction decreases.

Saturated solution solution in which no more solute will dissolve.

Semipermeable membrane a thin partition between two solutions through which certain molecules can pass but others cannot.

Soap an emulsifier that can disperse nonpolar substances in water; the sodium salt of a long chain organic acid; consists of a long hydrocarbon chain attached to a carboxylate group, $-CO_2^-Na^+$.

Sol colloidal suspension of a solid dispersed in a liquid.

Solvation the process by which solvent molecules surround and interact with solute ions or molecules.

Supersaturated solution a (metastable) solution that contains a higher than saturation concentration of solute; slight disturbance causes crystallization of excess solute.

Suspension a heterogeneous mixture in which solute-like particles settle out of a solvent-like phase some time after their introduction.

Thermal pollution introduction of heated waste water into natural waters.

Tyndall effect the scattering of light by colloidal particles.

van't Hoff factor (*i*) a multiplier that converts the stated molality of a solution to the effective molality.

Exercises

General Concepts — The Dissolving Process

1. There are no true solutions in which the solvent is gaseous and the solute is either liquid or solid. Why?
2. Explain why (a) solute-solute, (b) solvent-solvent, and (c) solute-solvent interactions are important in determining the ease of dissolution of a solute in a solvent.
3. What is the relative importance of each factor listed in Exercise 2 when (a) solids, (b) liquids, and (c) gases dissolve in water?
4. Why do many solids dissolve in water in endothermic processes, while most miscible liquids mix with each other in exothermic processes?
5. Define and distinguish between solvation and hydration.
6. The amount of heat released or absorbed in the dissolution process is important in determining whether or not the dissolution process is spontaneous, i.e., whether or not it can occur. What is the other important factor? How does it influence solubility?
7. Consider the following solutions. In each case predict whether the solubility of the solute should be high or low. Justify your answer. (a) KCl in H_2O, (b) KCl in CCl_4, (c) NH_4Cl in pentane, C_5H_{12}, (d) H_2O in CH_3OH, (e) CH_3OH in H_2O, (f) CCl_4 in H_2O, (g) HCl in H_2O, (h) HF in H_2O, (i) N_2 in H_2O, (j) Fe_2O_3 in H_2O.
8. For those solutions that can be prepared in "reasonable" concentrations in Exercise 7, classify the solutes as nonelectrolytes, weak electrolytes, or strong electrolytes.
9. Give a reasonable explanation for the fact that the metal hydroxides of Group IA and those of the lower members of Group IIA are considered water-soluble, while other metal hydroxides are insoluble in water.
10. Refer to Figure 13–7 to answer the following questions.

(a) A solution is prepared by dissolving 80 g of potassium iodide, KI, in 100 g of water at 60°C. Is the solution supersaturated, saturated, or unsaturated?

(b) The solution of (a) is slowly cooled to 35.0°C with no crystallization. Is the solution supersaturated, saturated, or unsaturated?

11. Is it possible for a saturated solution to be a dilute solution also? If so, give an example.

12. Describe the preparation of a saturated aqueous solution of calcium hydroxide, $Ca(OH)_2$. Describe what happens at the molecular or ionic level as saturation is achieved.

13. Supersaturated solutions do not involve equilibrium between a solid and its dissolved ions or molecules. Why?

14. Describe and explain the effect of changing the temperature on the solubility of (a) a solid that dissolves by an exothermic process, and (b) a solid that dissolves by an endothermic process.

15. What is the effect of raising the temperature on the solubility of most gases in water?

16. Describe the effect of increasing the pressure on the solubility of gases in liquids.

Mole Fraction

17. In a certain solution of ethanol and water, the mole fraction of ethanol is 0.36. What is the mole fraction of water?

18. Calculate the mole fraction of carbon tetrachloride, CCl_4, in a solution prepared by mixing 36.0 g of CCl_4 with 64.0 g of benzene, C_6H_6.

19. What is the mole fraction of ethanol, C_2H_5OH, in an alcoholic beverage that is 50% ethanol and 50% water by mass?

20. What mass of glucose, $C_6H_{12}O_6$, should be dissolved in 100 mL of water so that the mole fraction of $C_6H_{12}O_6$ is 0.100? The density of water is 1.00 g/mL.

21. What is the percentage by mass of benzene, C_6H_6, and carbon tetrachloride, CCl_4, in a solution in which each has a mole fraction of 0.500?

Raoult's Law and Vapor Pressure

22. In your own words, explain briefly *why* the vapor pressure of a solvent is lowered by dissolving a nonvolatile solute in it.

23. Calculate (a) the lowering of vapor pressure and (b) the vapor pressure of a solution pre- pared by dissolving 60.0 g of naphthalene, $C_{10}H_8$ (a nonvolatile nonelectrolyte), in 100.0 g of benzene, C_6H_6 at 20°C. Assume the solution is an ideal solution. The vapor pressure of ben- zene is 74.6 torr at 20°C.

24. (a) Calculate the lowering of vapor pressure associated with dissolving 75.0 grams of (solid) sucrose, $C_{12}H_{22}O_{11}$, in 300 grams of water at 25.0°C. (b) What is the vapor pressure of the solution? Assume the solution is ideal. The vapor pressure of water at 25°C is 23.76 torr.

25. What is the vapor pressure of the solution of Exercise 24 at 100°C?

26. Nearly all solutions of two volatile components that have highly negative heats of solution show negative deviations from ideality, i.e., $P_{total} < P_{ideal}$. Why should this be so?

27. Use the following diagram to estimate (a) the partial pressure of chloroform, (b) the partial pressure of acetone, and (c) the total vapor pressure of a solution in which the mole frac- tion of $CHCl_3$ is 0.2, assuming *ideal* behavior.

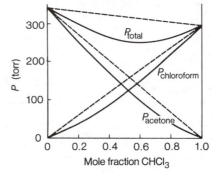

28. Answer Exercise 27 for the *real* solution of acetone and chloroform.

29. What is the vapor pressure of a solution con- taining 50.0 g of chloroform, $CHCl_3$, and 50.0 g of carbon tetrachloride, CCl_4, at 25°C? The vapor pressures at 25°C are 172.0 torr for pure $CHCl_3$ and 98.3 torr for pure CCl_4. As- sume the solution behaves ideally.

30. A solution is prepared by mixing 30.0 g of dichloromethane, CH_2Cl_2, and 50.0 g of dibro- momethane, CH_2Br_2, at 0°C. The vapor pres- sure at 0°C of pure CH_2Cl_2 is 0.175 atm and that of CH_2Br_2 is 0.015 atm. Assuming ideal behavior, calculate the total vapor pressure of the solution.

31. Refer to Exercise 30. Calculate the mole frac- tions of CH_2Cl_2 and of CH_2Br_2 in the *vapor*

above the liquid. Assume that both the vapor and the solution behave ideally.

Molality

32. What is the molality of a solution prepared by dissolving 6.00 g of benzoic acid, C_6H_5COOH, in 150 g of benzene, C_6H_6?

33. Calculate the molality of a solution that contains 60 g of methyl alcohol, CH_3OH, in 350 g of water.

34. Calculate the molality of a solution that contains 150 g of ethyl alcohol, C_2H_5OH, in 500 mL of benzene. The density of benzene is 0.879 g/mL.

35. What is the molality of ethanol, C_2H_5OH, in a solution prepared by mixing 50.0 mL of C_2H_5OH with 100 mL of water at 20°C? The density of C_2H_5OH is 0.789 g/mL at 20°C.

36. A student wishes to prepare an aqueous solution of sucrose, $C_{12}H_{22}O_{11}$, that is 0.250 m. What mass of sucrose must be dissolved in 300 g of water?

37. Urea, N_2H_4CO, is a product of metabolism of proteins. An aqueous solution is 20.0% urea by mass and has a density of 1.05 g/mL. Calculate the molality of urea in the solution.

Boiling Point Elevation and Freezing Point Depression — Solutions of Nonelectrolytes

38. Refer to Table 13–2. Suppose you had 0.100 molal solutions of a nonvolatile nonelectrolyte in each of the solvents listed there. Which one would have (a) the greatest freezing point depression, (b) the lowest freezing point, (c) the greatest boiling point elevation, and (d) the highest boiling point?

39. Explain qualitatively why boiling points are elevated, whereas freezing points are depressed, by dissolving nonvolatile solutes in solvents.

40. What is the significance of (a) the molal freezing point depression constant, K_f, and (b) the molal boiling point elevation constant, K_b? How could their values for given solvents be determined in the laboratory?

41. Explain why automobile radiator antifreeze also protects against radiator boilover.

42. Calculate the freezing point and the boiling point of a solution that contains 30.0 g of urea, N_2H_4CO, in 250 g of water. Urea is a nonvolatile nonelectrolyte.

43. Calculate the freezing point and boiling point of a solution that contains 3.42 g of ordinary sugar (sucrose, $C_{12}H_{22}O_{11}$) in 20.0 grams of water. Sucrose is a nonvolatile nonelectrolyte.

44. Using data from Table 13–2, calculate the freezing point and the boiling point of a solution that contains 12.2 g of benzoic acid (C_6H_5COOH) in 100 g of benzene (C_6H_6). Benzoic acid is a nonvolatile nonelectrolyte in benzene.

45. When 2.00 g of an unknown nonelectrolyte is dissolved in 10.0 g of water, the resulting solution freezes at −3.72°C. What is the molecular weight of the unknown compound?

46. When 0.400 g of an unknown nonelectrolyte is dissolved in 40.0 g of water, the resulting solution freezes at −0.465°C. What is the molecular weight of the compound?

47. What minimum mass of ethylene glycol, CH_2OHCH_2OH (an antifreeze component), must be mixed with 6.00 gallons of water to prevent it from freezing at −10.0°F? One gallon is 3.785 liters.

48. If ethylene glycol is treated as a nonvolatile solute, at what temperature would the solution of Exercise 47 boil?

Boiling Point Elevation and Freezing Point Depression — Solutions of Electrolytes

49. What is ion association in solution? Can you suggest why the term "ion pairing" is sometimes used to describe this phenomenon?

50. What is the significance of the van't Hoff factor, i? What is its value for the following strong electrolytes at infinite dilution? (a) Na_2SO_4, (b) KOH, (c) $Al_2(SO_4)_3$, (d) $Ba(OH)_2$

51. Why do the percentage ionization of a weak acid and the activity of a strong electrolyte increase with increasing dilution?

52. What is an activity coefficient?

53. Acetic acid (CH_3COOH) dissolves in water according to

$$CH_3COOH \rightleftharpoons H^+ + CH_3COO^-$$

A 0.0100 m acetic acid solution freezes at −0.01938°C. Calculate the percentage ionization of CH_3COOH in this solution.

54. A 0.100 m acetic acid solution in water freezes at −0.1884°C. Calculate the percentage ionization of CH_3COOH in this solution.

55. A weak acid, HX, ionizes in water according to

$$HX \rightleftharpoons H^+ + X^-$$

In a 0.100 m solution, HX is 15.0% ionized. Calculate the freezing point of this solution.

56. NaCl dissolves in water according to

$$NaCl \longrightarrow Na^+ + Cl^-$$

A 0.100 m solution of NaCl freezes at $-0.348°C$. Calculate i and the apparent percentage dissociation of NaCl in this solution.

57. K_2SO_4 dissolves in water according to

$$K_2SO_4 \longrightarrow 2K^+ + SO_4{}^{2-}$$

A 0.100 m solution of K_2SO_4 freezes at $-0.432°C$. Calculate i and the apparent percentage dissociation of K_2SO_4 in this solution.

58. The complex compound $K_3[Fe(CN)_6]$ dissolves in water according to

$$K_3[Fe(CN)_6] \longrightarrow 3K^+ + [Fe(CN)_6{}^{3-}]$$

A 0.100 m solution of $K_3[Fe(CN)_6]$ freezes at $-0.530°C$. Calculate i and the apparent percentage dissociation of $K_3[Fe(CN)_6]$ in this solution.

59. A solution of 23.8 grams of mercury(II) chloride, $HgCl_2$, in 500 g of water boils at 100.085°C at one atmosphere pressure. (a) Calculate i for this solution. (b) Would you classify $HgCl_2$ as a nonelectrolyte, a weak electrolyte, or a strong electrolyte?

Osmotic Pressure

60. What are osmosis and osmotic pressure?

61. Show how the expression

$$\pi = MRT$$

where π is osmotic pressure, is similar to the ideal gas law. Rationalize qualitatively why this should be so.

62. Show numerically that the molality and molarity of 1.00×10^{-13} M aqueous sodium chloride are nearly equal. Why is this true? Would this

be true if another solvent, say acetonitrile (methyl cyanide), CH_3CN, replaced water? Why? The density of CH_3CN is 0.786 g/mL at 20°C.

63. Estimate the molecular weight of a biological macromolecule if a 0.250 gram sample dissolved in 75.0 mL of benzene has an osmotic pressure of 12.17 torr at 25.0°C. The density of benzene is 0.879 g/mL.

64. Estimate the osmotic pressure associated with 30.0 grams of an enzyme of molecular weight 1.2×10^5 dissolved in 1700 mL of benzene at 40.0°C.

65. Calculate the freezing point depression and boiling point elevation associated with the solution of Exercise 64.

66. A solution of 0.900 g of an unknown nonelectrolyte in 300 mL of water at 27.0°C has an osmotic pressure of 38.4 torr. What is the molecular weight of the compound?

Colloids

67. Distinguish between solutions and colloids. Give examples of each.

68. Distinguish among (a) sol, (b) gel, (c) emulsion, (d) foam, (e) solid sol, (f) solid emulsion, (g) solid foam, (h) solid aerosol, (i) liquid aerosol. Try to give an example of each that is not already listed in Table 13–4.

69. What is the Tyndall effect, and how is it caused?

70. Distinguish between hydrophilic and hydrophobic colloids.

71. What is an emulsifier?

72. What is coagulation? Coagulation does not occur readily in colloidal aqueous ferric oxide. Why?

73. Distinguish between soaps and detergents. How do they interact with hard water? Write an equation to show the interaction between a soap and hard water.

74. What is the disadvantage of alkylbenzenesulfonate (ABS) detergents compared to linear alkyl sulfonate (LAS) detergents?

Chemical Thermodynamics: The Driving Force for Changes

14

We shall use the term *process* to mean either a chemical or physical change, and in some cases both.

At many stages in our study of chemistry, we have been concerned with the concept of energy. Recall that one of the ways in which the energy of a system can change is by the absorption or release of heat. Thus, one of the ways we can study energy changes is by observing heat flow, as by calorimetry (Section 3–5). We have studied the heat transfer that accompanies chemical reactions (Chapter 3) and phase transitions (Section 12–12). In these discussions, we used the quantity ΔH, the *change in enthalpy* of the system under study. In addition, we used many of the same ideas in our study of chemical bonding (Section 7–11). In all of these studies we applied the branch of science known as **thermodynamics,** which is *the study of the energy changes that accompany physical and chemical changes.*

In earlier chapters, we were concerned primarily with the aspect of thermodynamics that deals with *observing, describing, and predicting* the magnitudes of the energy changes that accompany physical or chemical changes. Another major concern of thermodynamics, however, is predicting *whether* a particular process *can* occur under specified conditions. We may summarize this concern in the question, "Which would be more stable at the given conditions — the collection of reactants or the collection of products?" A change for which the collection of products has *greater thermodynamic stability* than the collection of reactants under the given conditions *can* occur, and it is said to be **spontaneous** under those conditions. A change for which the products have *less thermodynamic stability* than the reactants under the given conditions *cannot* occur, and is described as **nonspontaneous** under those conditions. Some changes are spontaneous under all conditions, while others are nonspontaneous under all conditions. The great majority of changes, however, are spontaneous under some conditions but not under others. One goal of the study of thermodynamics is the prediction of conditions for which the last group is spontaneous.

Before you study this chapter, you should review Sections 3–1 through 3–3.

In this chapter, we shall develop the ideas of chemical thermodynamics with the aim of understanding and predicting the spontaneity of chemical and physical processes. In order to do this, we shall first look more closely at quantities such as *heat flow*, *state functions*, and *enthalpy*, which we have already used.

14–1   Spontaneity

Let us specify clearly just what is meant by the term "spontaneous" as it is used in thermodynamics. The fact that a process is spontaneous does not mean that it *will* occur at an observable rate. It may occur rapidly, at a moderate rate, or so slowly that we cannot detect its progress. The study of the rate at which a spontaneous reaction occurs is called *kinetics* (Chapter 15). The rate at which a reaction occurs is *not* a thermodynamic quantity. The only absolute statement that we shall be able to make on the basis of thermodynamic data is: *under a given set of conditions, a nonspontaneous reaction will not occur.*

Some familiar examples of processes that we know to be spontaneous, or nonspontaneous, will be helpful. We know that a truck can *roll down* a hill (Figure 14–1), but that a truck at the bottom of a hill would not, of its own accord, *roll up* the hill. In thermodynamic terms, the reason is this: at the top of the hill, the truck is at a position of higher potential energy than it is at the bottom. One of the factors that make a process (in this example, rolling down the hill) spontaneous is the *lowering* of potential energy. We note however, that all we can tell is (1) that the truck would have lower potential energy at point B than at point A (and by how much, if we know the difference in heights, the mass of the truck and the gravitational constant), and (2) that the process *can* occur. We cannot guarantee that it will occur, since other factors (such as whether the parking brake is set) may come into play. Nor can we tell how fast it would roll down the hill unless we know, for example, the frictional forces that would slow it, or which way the wind is blowing. We are quite sure, however, that a truck at point B would *not* roll *up* the hill in the absence of some outside force, such as that applied by the engine or by someone pushing the truck.

As we saw in Chapter 12, the freezing of one mole of ice at 0°C must be accompanied by the release of 6.02 kilojoules of energy.

$H_2O\,(\ell) \longrightarrow$
        $H_2O\,(s) + 6.02$ kJ

As a second example, consider the *physical change* involving the freezing of water to make ice at $-20°C$, a temperature below the normal freezing point of water. Our experience is that this is a spontaneous process. Further, we observe (Section 12–12.1) that this process is *exothermic*, i.e., the sample gives off heat to its surroundings when the water freezes. (A refrigerator produces ice cubes by removing a sufficient amount of heat from the water in the freezing compartment.) This spontaneous process, too, is accompanied by a lowering of the internal energy, which is accomplished by the removal of heat. Thus, the lowering of energy again appears to be favorable to spontaneity.

Many spontaneous *chemical reactions* also proceed with the release of energy in the form of heat (exothermic reactions). The combustion reactions of fossil fuels

FIGURE 14–1 The truck can roll spontaneously down the hill from point *A* to point *B*, reaching a position of lower (gravitational) potential energy. The truck will not spontaneously roll from point *B* to point *A*.

A hydrocarbon is a binary compound of hydrogen and carbon. Hydrocarbons may be gaseous, liquid, or solid; all burn.

are familiar examples. Hydrocarbons (including methane, the principal component of natural gas, and octane, one of the minor components of gasoline) undergo combustion to yield carbon dioxide and water with the release of energy, as shown on the product side of each equation.

$$CH_4 \text{ (g)} + 2O_2 \text{ (g)} \longrightarrow CO_2 \text{ (g)} + 2H_2O \text{ } (\ell) + 890 \text{ kJ}$$

$$2C_8H_{18} \text{ } (\ell) + 25O_2 \text{ (g)} \longrightarrow 16CO_2 \text{ (g)} + 18H_2O \text{ } (\ell) + 1.090 \times 10^4 \text{ kJ}$$

The amounts of heat shown are the amounts liberated by the reaction of one mole of methane, CH_4, and the reaction of two moles of octane, C_8H_{18}, respectively.

In such reactions the total energy of the products is lower than that of the reactants by the amount of energy released, most of which is heat. *Spontaneous processes tend toward a state of lower energy.*

The process $H_2O \text{ (s)} \longrightarrow H_2O \text{ } (\ell)$ is *endothermic* by 6.02 kJ/mol.

Entropy is a measure of the disorder in a system (Section 14–5).

The lowering of energy is *not* the only factor that determines spontaneity, however. Consider the melting of ice to form liquid water at +20°C. Experience tells us that this process is spontaneous but *endothermic. It is not necessary that spontaneous processes be exothermic or that exothermic processes be spontaneous.* Another factor, related to the disorder of reactants and products and called **entropy,** must also be considered. As the ice melts, the water molecules become less orderly than they were in crystalline ice. The effect of entropy outweighs the effect of heat energy in this case. *An increase in disorder of the system favors the spontaneity of a process.*

14–2 Changes in Internal Energy, Δ*E*: Heat, Work, and the First Law of Thermodynamics

The **internal energy, *E*,** of a specific amount of a substance represents all the energy contained within the substance. It includes such forms as kinetic energies of the molecules, energies of attraction and repulsion among subatomic particles, atoms, ions, or molecules, as well as other forms of energy. The internal energy of a collection of molecules is a state function (Section 3–3). Therefore, the difference between the internal energy of the products and the internal energy of reactants of a chemical reaction or physical change, Δ*E*, is given by the equation:

$$\Delta E = E_{\text{final}} - E_{\text{initial}} \qquad \text{or} \qquad \Delta E = E_{\text{products}} - E_{\text{reactants}} \qquad \text{[Eq. 14–1]}$$

Heat is a form of energy, so one obvious way of changing the internal energy of a system is by putting in or withdrawing heat. We shall denote the amount of heat by the symbol q. The sign of q is determined by the following convention:

q is positive: heat is *absorbed* by the system from the surroundings

q is negative: heat is *released* by the system to the surroundings

Recall that a fundamental definition of energy is the capacity to do work. When a system does work on its surroundings (for example, a compressed spring moves a weight, or an expanding gas pushes a piston), the energy of the system decreases by the amount of work done. So another way to change the internal energy of a system is by doing work on it or by allowing it to do work on its surroundings.

Other than electrical work, which will be covered in Chapter 19, the only type of work we need to be concerned with in most chemical and physical changes is **pressure-volume work.** From dimensional analysis we can see that the product of pressure and volume is work. Pressure is the force, f, exerted per unit area, where area has the dimensions of distance squared, d^2; volume has the dimensions of

FIGURE 14-2 (a)
Steam engine (diagram-
matic). (b) Steam
produced from an *open*
beaker does work
against atmospheric
pressure. (c) Internal
combustion engine;
gases produced by
burned gasoline do work
against the piston.

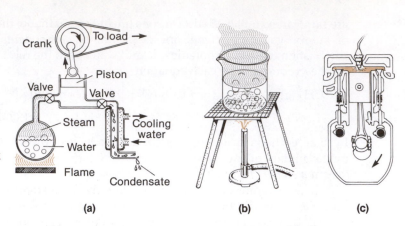

(a)   (b)   (c)

distance cubed, d^3. Thus, the product of pressure and volume is force times distance, or work:

pressure × volume = work

$$\frac{f}{d^2} \times \quad d^3 = fd$$

When a gas expands against an external pressure, it does work. An example of a physical process involving such work is the expansion of vapor from a boiling liquid, which can be used (as in a steam engine) to do work (Figure 14–2a.) Similarly, steam expanding from an open vessel (Figure 14–2b) also does work as it expands against the pressure of the atmosphere. Some chemical reactions produce gases. When this occurs in an open vessel at atmospheric pressure, the gas does work as its molecules push against the pressure of the atmosphere. Such reactions can also be used for useful work, as in an internal combustion engine (Figure 14–2c). In such cases, the work done by the system (or conversely, work done on the system by its surroundings) can result in a change in the energy of the molecules of the system. We denote the work done by (or on) the system by the symbol w, and its sign is given by the following convention.

> w is positive: work is done *by* the system on the surroundings

> w is negative: work is done *on* the system by the surroundings

The work done on or by a system equals the change in the product of pressure and volume, $\Delta(PV)$. When the pressure is constant during the change, this is the same as the pressure times the change in volume.

$$w = P\Delta V \qquad\qquad \text{[Eq. 14–2]}$$

The quantitative use of this idea will be illustrated later in this section. Be sure to note that the symbol P here represents the *external* pressure *against which* the work is done, e.g., atmospheric pressure. In constant-volume changes, no work can be done. Although the internal pressure varies, the absence of a change in volume means that nothing "moves through a distance," so $d = 0$ and $fd = 0$.

Now let us ask the question, "What can result when we put heat into a sample of matter (in thermodynamic jargon, a system)?" Several results are possible:

According to the kinetic-molecular theory, the temperature of a sample is a measure of the average kinetic energy of its molecules or ions (Section 11–13).

1. The energy of the system can increase. This may cause the temperature to increase, or it may cause a phase change such as melting or boiling to occur.
2. The system can do work on its surroundings.
3. Some combination of (1) and (2) can occur.

Suppose that we ask the more fundamental question, "How does the internal energy of a system change if the system absorbs or releases heat and/or does work or has work done on it?" This question is answered by the **First Law of Thermodynamics,** also known as the *Law of Conservation of Energy:* Energy is neither created nor destroyed in ordinary chemical or physical changes (Section 3–1). This means that any energy gained by the system, whether in the form of heat or as a result of having work done on it, must come from the surroundings. Conversely, any energy lost by the system, either by release of heat or by the system doing work, does not just disappear, but must go to its surroundings. The energy change in the system, ΔE, is given by the relationship

Careful attention must be paid to the sign conventions for q and w given earlier.

$$\Delta E = q - w \qquad \text{[Eq. 14–3]}$$

This equation, which represents another statement of the First Law of Thermodynamics, simply says that the increase in energy of a system is the amount of energy it gains in the form of heat *minus* the amount of energy it expends in doing work. Examples 14–1 and 14–2 illustrate the use of Eq. 14–3. Heat and work are *not* state functions of the system, but the internal energy is a state function.

Example 14–1

A sample of a gas is heated by the addition of 3600 kJ of heat. (a) In the first case, the volume it occupies is not allowed to increase or decrease. What is the change in internal energy, ΔE, of this sample? (b) Suppose that in addition to putting heat into the sample, we also do 800 kJ of work on the sample (e.g., by compressing it). What is the ΔE of the sample? (c) Suppose, instead, that as the original sample is heated, it is allowed to expand against an external pressure so that it does 5200 kJ of work on its surroundings. What is the ΔE of the sample?

Solution

(a) Since heat is being put *into* the sample, q is positive; that is, $q = +3600$ kJ. Since no expansion or contraction takes place, no work is done either on or by the system, so $w = 0$. According to Eq. 14–3,

$$\Delta E = q - w = (+3600 - 0)\text{kJ}$$
$$= +3600 \text{ kJ}$$

The positive sign of ΔE indicates an *increase* in the internal energy of the system. All of the heat energy put into the system appears as internal energy. This results in an increase in temperature of the sample.

(b) Again, since 3600 kJ of heat energy is put into the system, $q = +3600$ kJ. However, 800 kJ of work is also done on the system, as it is compressed by its surroundings. By the convention, we denote this as $w = -800$ kJ. The change in internal energy is then calculated as

$$\Delta E = q - w = [+3600 - (-800)]\text{kJ}$$
$$= +4400 \text{ kJ}$$

Again the positive sign indicates an increase in internal energy, but by more than in part (a), since energy was also put into the system by doing work on it.

(c) Just as before, the input of 3600 kJ of heat is represented by $q = +3600$ kJ. Now, the system does 5200 kJ of work by expanding against the pressure of its surroundings. This work done *by* the system is represented as $w = +5200$ kJ. We calculate ΔE as before:

$$\Delta E = q - w = [+3600 - (+5200)]\text{kJ}$$
$$= -1600 \text{ kJ}$$

Now ΔE has a negative sign. This indicates that the final energy of the system is lower than the initial energy (see Eq. 14–1), or that the system uses up more energy in doing work than it absorbs as heat.

Example 14–2

Calculate q, w, and ΔE for the vaporization of 1.00 g of liquid water at 1.00 atm at 100°C, to form 1.00 g of steam at 1.00 atm at 100°C (see Figure 14–3). Make the following assumptions: (a) the density of liquid water at 100°C is 1.00 g/mL, (b) water vapor is adequately described by the ideal gas equation, and (c) the external pressure is constant at 1.00 atm. From Table 12–8, the heat of vaporization of water is 2.26 kJ/g.

Solution

From this information, the heat required to vaporize 1.00 g of water is 2.26 kJ. Since this heat is *added to* the system from its surroundings, we write

$$q = 2.26 \text{ kJ}$$

The work done by the system may be calculated from the relationship

$$w = P\Delta V = P(V_f - V_i)$$

where V_f and V_i are the final and initial volumes, respectively. The initial volume of water is

$$V_i = 1.00 \text{ g} \times \frac{0.00100 \text{ L}}{1.00 \text{ g}} = 0.00100 \text{ L}$$

(from the density of liquid water)

The final volume can be calculated by assuming the ideal gas law holds [note that $n = 1.00$ g H_2O/ (18 g H_2O/mol) = 0.0556 mol]:

$$V_f = \frac{nRT}{P} = \frac{(0.0556 \text{ mol})\left(0.0821 \dfrac{\text{L}\cdot\text{atm.}}{\text{mol}\cdot\text{K}}\right)(373\text{K})}{1.00 \text{ atm}}$$

$$= 1.70 \text{ L}$$

Then the volume change on vaporization is given by

$$\Delta V = V_f - V_i = (1.70 - 0.00100)\text{L}$$
$$= +1.70 \text{ L}$$

Note that the volume of liquid H_2O is insignificant when the rules for significant figures are applied. Since the volume of gas is so much larger than the volume of liquid, any error in the density of the liquid would have very little effect on the resulting change in volume.

We can calculate the work done:

$$w = P\Delta V = (1.00 \text{ atm})(+1.70 \text{ L}) = +1.70 \text{ L}\cdot\text{atm}$$

Converting this to joules (see Appendix C) gives

$$w = (+1.70 \text{ L}\cdot\text{atm})\left(101.3 \frac{\text{J}}{\text{L}\cdot\text{atm}}\right) = +172 \text{ J}$$

Now we can calculate the change in internal energy of the system.

$$\Delta E = q - w = [2.26 - (0.172)]\text{kJ}$$
$$= 2.09 \text{ kJ}$$

We may interpret these results as follows. Of the total 2.26 kJ required to vaporize one gram of water under these conditions, 0.172 kJ is used to do work on the surroundings; the other 2.09 kJ goes to increase the internal energy of the sample.

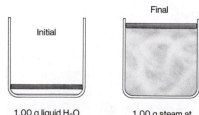

FIGURE 14–3 Vaporization of 1.00 g of liquid water at 1.00 atm and 100°C to form 1.00 g of steam at 1.00 atm and 100°C.

Initial

1.00 g liquid H_2O
at 100°C, 1.00 atm
$V_1 = 1.00$ mL $= 0.00100$ L

Final

1.00 g steam at
100°C, 1.00 atm
$V_2 = 1.70$ L

The positive sign of w in Example 14–2 tells us that the system is doing work on its surroundings. We are describing a primitive steam engine, which has a *very* low efficiency of converting heat into work.

These concepts also may be applied to chemical reactions, as we can illustrate by considering the complete reaction of a 2 : 1 mole ratio of hydrogen and oxygen to produce steam at some constant temperature above 100°C and at one atmosphere pressure, as depicted in Figure 14–4. Assuming that the constant temperature bath surrounding the cylinder in which the reaction occurs completely absorbs all the evolved heat, the temperature of the gaseous system does not change. The volume change is then due to the combination of gaseous substances. The volume of the system decreases by one third, because three moles of gases react to form two moles of gas.

$$2H_2\,(g) + O_2\,(g) \longrightarrow 2H_2O\,(g)$$

3 mol gaseous reactants   2 mol gaseous products

The surroundings exert a constant pressure of one atmosphere, and do work on the system by compressing it. The internal energy of the system simultaneously increases by an amount equal to the amount of work done on it.

Because liquids and solids have much higher densities than gases, their presence makes a negligible contribution to the total volume in processes in which gases are involved. Furthermore, since solids and liquids do not expand or contract significantly as the pressure changes, their production or consumption involves only very little work (because $\Delta V \approx 0$). In reactions in which equal numbers of moles of gases are produced and consumed at constant temperature and pressure, no work is done. Thus, the work term, w, has a significant value *only (1) when there are different numbers of moles of gaseous products and reactants* and (2) *the volume of the system changes significantly.* Many reactions take place at constant volume, either because the reaction takes place in a rigid container or because there are no gaseous reactants or products. In such constant volume processes, the change of internal energy is just the amount of heat absorbed or released at constant volume, q_v; that is,

$$\Delta E = q_v \qquad \text{(constant volume process)} \qquad \text{[Eq. 14–4]}$$

A high boiling liquid such as mineral oil is used in constant temperature baths above the boiling point of water.

FIGURE 14–4 An illustration of the decrease in volume by a factor of one-third that accompanies the reaction

$2H_2\,(g) + O_2\,(g) \longrightarrow$
$\qquad\qquad\qquad 2H_2O\,(g)$

(Note: The temperature is above 100°C, so H_2O is a gas.)

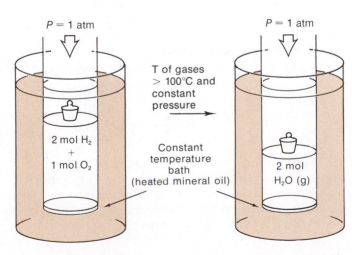

You might review the calculations of Section 3–5. Note that ΔH is the quantity that can be measured directly in a "coffee-cup" calorimeter.

14–3   Bomb Calorimeters

We saw in Section 3–5 how the ideas of heat transfer can be used to determine experimentally the amount of heat absorbed or released by a chemical or physical change. At that point, and throughout the subsequent studies of heat transfer, only changes taking place at constant *pressure* were considered. We saw that a simple "coffee-cup" calorimeter could provide a means of measuring heats of reaction in aqueous solution.

A **bomb calorimeter** is a device that measures the heat evolved or absorbed by a reaction occurring at constant *volume* (Figure 14–5). In this type of calorimeter, a strong steel vessel (the bomb) is immersed in a large volume of water. As heat is produced or absorbed by a reaction inside the steel vessel, heat is transferred to or from the large volume of water, so only very small temperature changes occur. For all practical purposes, the energy changes associated with the reactions are measured at constant volume *and* constant temperature. No work is done when a reaction is carried out in a bomb calorimeter even if gases are involved, because $\Delta V = 0$, and therefore

$$\Delta E = q_v \qquad \text{(constant volume)}$$

For exothermic reactions, we may write

amount of heat lost by system = (amount of heat gained by calorimeter)
    + (amount of heat gained by water)

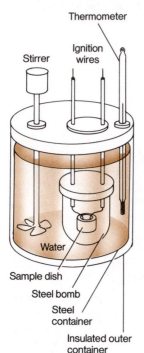

Thermometer

Ignition wires

Stirrer

Water

Sample dish

Steel bomb

Steel container

Insulated outer container

FIGURE 14–5 A bomb calorimeter measures the amount of heat evolved or absorbed by a reaction occurring at constant volume.

Benzoic acid, C_6H_5COOH, is usually used to determine water equivalents. It is a solid that can be compressed into pellets, and its heat of combustion is 3227 kJ/mol or 26.43 kJ/g.

To simplify calculations, the amount of heat absorbed by the calorimeter is usually expressed as its **water equivalent,** which is the amount of water that would absorb the same amount of heat as the calorimeter per degree temperature change. The water equivalent of a calorimeter is determined by burning a sample of a compound that produces a known amount of heat, and measuring the temperature rise of the calorimeter. For example, suppose we observe that burning 0.3630 gram of benzoic acid raises the temperature of a calorimeter and its 3000 grams of water by 0.629°C. The heat lost by the burning benzoic acid is

$$\underline{?}\text{ kJ lost by sample} = 0.3630\text{ g} \times \frac{26.43\text{ kJ}}{\text{g}} = 9.594\text{ kJ} \quad (9594\text{ J})$$

so

heat gained by calorimeter = (heat lost by sample) − (heat gained by water)
$$= 9594\text{ J} - (3000\text{ g})(4.184\text{ J/g} \cdot °\text{C})(0.629°\text{C})$$
$$= (9594 - 7895)\text{ J}$$
$$= 1699\text{ J} \quad (1.699\text{ kJ})$$

temp. change
sp. ht. of H_2O
amount of water

We can now calculate the amount of water that would absorb the same amount of heat in going through the same temperature change. The calculation is equivalent to the following problem: If you have a sample of water of unknown mass, but you know how much energy is required to raise its temperature by a given amount, what is its mass? The answer is

An alternative to measuring the water equivalent of the calorimeter is to calibrate the calorimeter prior to use by electrical heating, to provide the "calorimeter constant" in joules per degree.

$$\underline{?}\text{ g }H_2O = \frac{1699\text{ J}}{0.629°\text{C}} \times \frac{\text{g} \cdot °\text{C}}{4.184\text{ J}} = 646\text{ g }H_2O$$

In order to measure temperatures to the nearest thousandth of a degree, very sensitive differential thermometers are required.

Thus, the water equivalent of the calorimeter is 646 g. This means that the amount of heat required to raise the temperature of the steel vessel 1.000°C is the same as the amount of heat required to raise the temperature of 646 grams of water 1.000°C.

Example 14-3

A 1.000 gram sample of ethanol, C_2H_5OH, was burned in the sealed bomb calorimeter described above, which was surrounded by 3000 grams of water. The temperature of the water rose from 24.284°C to 26.225°C. Determine ΔE for the reaction in kilojoules per gram of ethanol, and then in kilojoules per mole of ethanol. The specific heat of water is 4.184 J/g·°C. The combustion reaction is

$$C_2H_5OH\ (\ell) + 3O_2\ (g) \longrightarrow 2CO_2\ (g) + 3H_2O\ (\ell)$$

Solution

Heat from the sealed compartment (the system) raises the temperature of the calorimeter and water. Recall that the water equivalent of this calorimeter is 646 grams, so we have 3000 + 646 = 3646 g of water as the effective total. The temperature increase is

$$?°C = 26.225°C - 24.284°C = 1.941°C \text{ rise}$$

The number of kilojoules of heat responsible for the increase in the temperature of the water is

$$?\ kJ = 3646\ g \times \frac{4.184\ J}{g\cdot°C} \times 1.941°C$$
$$= 2.961 \times 10^4\ J = 29.61\ kJ$$

One gram of C_2H_5OH *liberates* 29.61 kJ of energy in the form of heat. That is,

$$\Delta E = q_v = -29.61\ kJ/g\ C_2H_5OH$$

Now we may evaluate ΔE in kilojoules per mole by converting grams of C_2H_5OH to moles.

$$\Delta E = \frac{?\ kJ}{mol} = \frac{-29.61\ kJ}{g} \times \frac{46.07\ g\ C_2H_5OH}{1\ mol\ C_2H_5OH}$$
$$= -1364\ kJ/mol$$

This calculation shows that for the combustion of ethanol at constant temperature and constant volume, the change in internal energy, ΔE, is -1364 kJ/mol. The negative sign indicates that energy is released to the surroundings.

14-4 Enthalpy

Many common processes take place at constant (atmospheric) pressure, under usual laboratory conditions. It is often convenient to measure the heat released or absorbed by a reaction at *constant pressure conditions* rather than at constant volume. We have seen that the First Law of Thermodynamics can be written in the form of Eq. 14-3:

$$\Delta E = q - w \qquad \text{or} \qquad q = \Delta E + w$$

For changes carried out at *constant pressure*, and in which only expansion work is done, this may be written as

The subscript p reminds us that we are referring to a constant pressure process.

$$q_p = \Delta E + P\Delta V \qquad \text{(constant pressure)} \qquad \text{[Eq. 14-5]}$$

We find it convenient to define a new state function, **enthalpy, H,** as

$$H = E + PV \qquad\qquad \text{[Eq. 14-6]}$$

In a practical sense, we are concerned only with *changes* in state functions. For a change occurring at constant pressure, this definition leads to

$$\Delta H = \Delta E + P\Delta V \qquad \text{(constant pressure)} \qquad \text{[Eq. 14-7]}$$

Comparing this with Eq. 14−5, we see that for changes occurring at constant pressure, the heat added to or removed from the system is a direct measure of the change in this state function, enthalpy:

$$\Delta H = q_p \qquad \text{(constant pressure)} \qquad\qquad \text{[Eq. 14−8]}$$

For this reason, *enthalpy* is often referred to as *heat content*. Of course, we have made much use of the expression represented in Eq. 14−8, even before this formal introduction of enthalpy in the context of chemical thermodynamics.

For *constant temperature* changes involving *ideal gases,* it can be shown that the work, $P\Delta V$, can be represented by $(\Delta n)RT$, where Δn is the change in the number of moles of ideal gaseous substances. In such cases, the relationship in Eq. 14−7 between ΔE and ΔH can be written as

> In Eq. 14−9, Δn is the number of moles of *gaseous products* minus the number of moles of *gaseous reactants.*

$$\Delta H = \Delta E + (\Delta n)RT \qquad \text{(ideal gas, constant } T) \qquad\qquad \text{[Eq. 14−9]}$$

Recalling that the heat change at constant volume, q_v, measures ΔE, and that the heat change at constant pressure, q_p, measures ΔH, we can use Eq. 14−9 to find one of these if we have measured the other (or calculated it from tabulated values), as shown below.

Combustion of one mole of ethanol at 298 K and constant *pressure* releases 1367 kilojoules of heat, and therefore we can say that

$$\Delta H = -1367 \frac{\text{kilojoules}}{\text{mole ethanol}}$$

In Example 14−3 we found that the change in internal energy, ΔE, for the combustion of ethanol is −1364 kilojoules per mole at 298 K.

$$C_2H_5OH \text{ (}\ell\text{)} + 3O_2 \text{ (g)} \longrightarrow 2CO_2 \text{ (g)} + 3H_2O \text{ (}\ell\text{)} \qquad \Delta E = -1364 \text{ kJ}$$

The difference, 3 kJ, is due to the work term, $P\Delta V$ or $(\Delta n)RT$. In this reaction there are fewer moles of gaseous products than gaseous reactants: $\Delta n = 2 - 3 = -1$. Thus, the atmosphere does work on the reacting system and the system is compressed. Let us evaluate the work term, $(\Delta n)RT$.

> The negative sign is consistent with the fact that work is done on the system.

$$w = (\Delta n)RT = (-1 \text{ mol}) \left(\frac{8.314 \times 10^{-3} \text{ kJ}}{\text{mol} \cdot \text{K}} \right) (298 \text{ K}) = -2.48 \text{ kJ}$$

We can use Eq. 14−9 to calculate ΔE for the reaction from ΔH and $(\Delta n)RT$ values:

$$\Delta E = \Delta H - (\Delta n)RT = [-1367 - (-2.48)] \text{ kJ} = -1365 \text{ kJ}$$

This value agrees with the result of Example 14−3, within round-off error. Note also that the work term (−2.48 kJ) is almost insignificant in comparison to ΔH (−1367 kJ). This is true for many reactions. Of course, $\Delta H = \Delta E$ when $\Delta n = 0$. Measurements of ΔH for reactions are usually simpler to make than measurements of ΔE, and are entirely adequate substitutes in most cases.

By convention, enthalpy changes are usually reported at one atmosphere pressure. This is standard pressure for thermodynamic measurements, and the superscript in ΔH^0 indicates standard pressure. Also, unless otherwise stated, thermodynamic values refer to 25°C (298 K). See Appendix K for ΔH^0_{f298} values.

14−5   Entropy: A Measure of Disorder

Early in this chapter we said that two factors must be considered in determining whether or not a process is spontaneous under a given set of conditions. One of these factors, the change in energy, is most conveniently measured at constant pressure conditions as ΔH, the enthalpy change. The effect of this factor is that spontaneity is *favored* (but not required) by *exothermicity*, while nonspontaneity is favored by endothermicity.

The other factor is the change in the degree of disorder that accompanies the process. The thermodynamic quantity that measures disorder is **entropy, *S*,** which is defined in such a way that a state of *greater disorder* is said to be at *higher entropy*. The effect of this disorder factor is summarized in the **Second Law of Thermodynamics** which states that *in spontaneous changes the universe tends toward a state of greater disorder (higher entropy)*. It should be noted that the Second Law is based on our experiences, but it cannot be proven rigorously.

The general applicability of this law in the macroscopic world is illustrated by a few common examples. When mirrors are dropped, they shatter; a cracked mirror is quite unlikely to mend itself. When a drop of food coloring is added to a glass of water, it diffuses until a homogeneously colored solution results. When a truck drives through the wall of a building, the wall crumbles and the truck becomes more disordered. On the other hand, we would be very surprised to drop the pieces of a jigsaw puzzle to the floor and find that the puzzle had spontaneously assembled itself.

Consider three different arrangements of a collection of 20 coins, with each arrangement held three feet above the floor (Figure 14–6). Since each sample is the same height above the floor, they all have the same gravitational potential energy. However, from the idea of entropy as a measure of disorder, you will see that arrangement (c), in which the coins are in a row, all arranged with heads in the same direction, has the lowest entropy (least disorder), while arrangement (a), in which the coins are randomly placed and randomly heads and tails, has the highest entropy (greatest disorder).

Suppose that the 20 coins were lying, randomly arranged, on the floor, and you were required to put them on the table in one of these three arrangements. You will probably agree that you would have to expend more effort to arrive at arrangement (b) than at arrangement (a), since (b) would require that you place the coins in a row in addition to lifting them to the table. The additional effort required goes to overcome the natural tendency of the system to be in the most disordered state—not aligned neatly, but randomly situated. In thermodynamic terms, you would do work to lower the entropy of the system. Obviously, even more effort would be required to arrive at arrangement (c), in which the coins are not only arranged in a row, but also all have heads facing in the same direction.

An important distinction must be made here, however, between the work done to raise the coins to the table and that done to attain a required arrangement. If we were to drop the coins to the floor, each arrangement could do the same amount of work, since the gravitational potential energy of each arrangement would decrease by the same amount. Thus, the energy required to raise the coins to the

Please note, though, that you would need to do the same amount of work in case (c) whether you chose to turn the coins heads up on the floor and then raise them to three feet or, instead, to lift them individually to the required height and then turn them heads up.

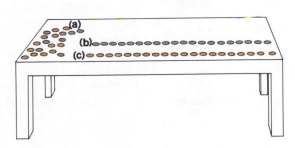

FIGURE 14–6 Three different ways of arranging a collection of 20 coins, all on a table 3 feet above the floor. (a) 20 coins randomly placed and randomly heads or tails. (b) 20 coins in a row, randomly heads and tails. (c) 20 coins in a row, all with "heads up." Arrangement (a) has highest disorder, while arrangement (c) has least disorder.

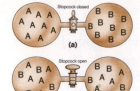

FIGURE 14–7 (a) A sample of gas in which all molecules of gas A are in left-hand bulb and all molecules of gas B are in the left-hand bulb and all molecules of gas B are in the right-hand bulb. (b) A sample of gas, containing the same numbers of A and B molecules as in part (a), but randomly mixed in the two bulbs. Sample (b) has greater disorder (higher entropy), and is thus more probable.

table is "recoverable" energy, while that expended in arranging the coins cannot be put to any useful purpose or recovered in any way. We shall return to this idea of "useful work" in Section 14–7 when we discuss changes in free energy.

We can apply this same idea of entropy to describe the degree of disorder in examples of chemical interest. For instance, of two gas samples, the more ordered arrangement in Figure 14–7a has lower entropy than the randomly mixed arrangement in Figure 14–7b. If we open the stopcock between the two bulbs in Figure 14–7a, we expect the gases to mix spontaneously. Our experience tells us, however, that the homogeneous sample in Figure 14–7b should not be expected to "unmix" to form the arrangement in Figure 14–7a. Likewise, a sample of ice (an ordered crystalline form) at 0°C is at lower entropy than the same mass of liquid water at 0°C, which in turn is a system lower in entropy than the same mass of water vapor. For any particular substance, the particles in the solid state are more highly ordered than those in the liquid state which, in turn, are more highly ordered than those in the gaseous state.

In similar fashion, we can see that a given sample is at higher entropy at a higher temperature than at a lower one; at the higher temperature, the more rapidly moving molecules are more disordered. As an analogy, suppose that you are required to arrange the 20 coins of Figure 14–8 so that they are all "heads up." To do so would require that you expend a certain amount of effort to attain this special ordered arrangement. If a mischievous colleague shakes the tray vigorously while you are working (in analogy to the greater molecular motion at higher temperature), your job is made even more difficult. In thermodynamic terms, the entropy of the bouncing collection of coins is higher than that of the stationary ones. The *amount of change in order* required to go from one arrangement to another is designated as ΔS, **the entropy change,** which is proportional to the number of coins that must be flipped to achieve the "all heads" or "all tails" arrangement. However, in a collection of molecules, the *effort* (energy, perhaps in the form of heat, q) required to bring about an entropy change is proportional to the kinetic energy of the molecules, and thus to the absolute temperature.

The product $T\Delta S$ is an energy term. For reversible changes (i.e., those that occur in such a way that all of the heat put into the system may be recovered), the entropy change of the system is given by

The units of ΔS are the same as those of q/T, that is, J/K.

$$\Delta S = \frac{q}{T}$$

However, reversible heat flow can take place only when the system and its surroundings are in thermal equilibrium (at the same temperature). This situation would result in no heat flow (another way of saying that an infinite time would be required for the heat q to flow). Any real change must take place *irreversibly*, in which case the entropy increases by more than q/T; that is, $\Delta S > q_{irr}/T$. The more slowly we carry out the change, the less will ΔS exceed q/T.

Let us now describe the scale on which we measure the entropy of a pure substance. The **Third Law of Thermodynamics** states that *the entropy of a pure, perfect crystalline substance (infinitely ordered) is zero at absolute zero (Kelvin).* Of course, no such perfect crystalline substance exists, in part because we are unable to attain this temperature. Even the particles making up the most perfectly ordered crystals are able to vibrate about fixed positions in the crystal lattice at all temperatures even infinitesimally above absolute zero (Figure 14–9). As the temperature of a substance is increased, the particles vibrate more vigorously, so the entropy

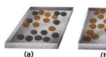

FIGURE 14–8 (a) 20 coins in a tray, some heads and some tails. (b) 20 coins in a tray, all heads. Which arrangement is more ordered?

FIGURE 14–9 (a) A simplified side view of a perfect crystal of a diatomic compound such as HCl at 0 K. Note the perfect alignment of the dipoles in all molecules in a "perfect" crystal, which causes its entropy to be zero at 0 K. However, there are no perfect crystals because even the purest substances that scientists have prepared to date are contaminated by trace impurities that occupy a few of the lattice positions. Additionally, there are some vacancies in the crystals of even very highly purified substances, such as those used in semiconductors (Section 12–16). (b) A simplified representation of the same "perfect" crystal at a temperature slightly above 0 K. Vibrations of the individual molecules within the crystal lattice cause some dipoles to be oriented in directions other than those in a "perfect" crystal at 0 K. The entropy of such a crystalline solid is greater than zero because there is some disorder in the crystal.

Enthalpies are defined only as *differences* with respect to an arbitrary standard state, so the standard enthalpy of formation of a substance is written as ΔH_f^0. Entropies, in contrast, are defined relative to an absolute zero level, so the Δ is omitted when referring to the entropy of a single substance. In either case, however, it is only differences, $\Delta X = X_{final} - X_{initial}$, with which we shall be concerned.

increases. Further heat input causes either increased temperature that results in higher entropy, or phase transitions (melting, sublimation, or boiling) that are also characterized by higher entropy. The total increase in the entropy of a substance as it goes from 0 K to the temperature in question is called the **absolute entropy** of the substance. The absolute entropy of a substance in its standard state at 25°C (298 K) is called its **standard entropy,** designated by S^0. Sometimes the standard entropy is designated as S_{298}^0, as a reminder that it refers to 298 K. Values of S_{298}^0 for various substances are tabulated in Appendix K. Please note that the standard entropy of *any* substance, even an element, in its standard state is greater than zero. The more ordered a substance is, the lower is its absolute entropy at that temperature. Consider as an example the standard entropies listed in Table 14–1.

14–6 Entropy Change: A Contributor to Spontaneity

Entropy, S, like enthalpy, H, is a state function. According to the Second Law of Thermodynamics, a characteristic of *any* spontaneous process is that the *entropy of the universe increases.* Of course, we can't measure directly the entropy of the universe (!), so we need to find out what this means in terms of the entropy change of the *system* under consideration.

If we recognize that the entropy of the universe is the sum of the entropy of the system plus that of its surroundings, then we can represent changes in entropies as

$$\Delta S_{universe} = \Delta S_{system} + \Delta S_{surroundings} > 0 \qquad \text{[Eq. 14–10]}$$

So we can see that, even though the entropy of the system may *decrease* during a change (ΔS_{system} is *negative*), the process still may be spontaneous ($\Delta S_{universe}$ is *positive*) if the *increase* in entropy of the surroundings is large enough ($\Delta S_{surroundings}$ *positive*). A good illustration of this kind of process is an ordinary refrigerator, which removes heat from the inside of the box (the system) and ejects that heat, *plus* the heat generated by the refrigerator compressor, into the room (the surroundings).

TABLE 14–1 Absolute Entropies for a Few Common Substances at 298 K

	S^0(J/mol·K)
C (diamond)	2.38
C (g)	158.0
H_2O (ℓ)	69.91
H_2O (g)	188.7
I_2 (s)	116.1
I_2 (g)	260.6

This should let you see the fallacy of trying to cool the kitchen by leaving the refrigerator door open!

Although the entropy of the system decreases (the air molecules inside the box move more slowly), the increase in the entropy of the surroundings more than makes up for it. The entropy of the universe (system plus surroundings) increases, so the process is spontaneous.

On the other hand, even though the entropy of the system may increase during a process, that process will be nonspontaneous if the entropy of the surroundings decreases by a greater amount. Thus, for *the system,* we can express the important aspect of disorder as follows: If the degree of disorder, or entropy, of a system increases during a process, spontaneity of the process is *favored,* but not required.

Similar arguments apply for condensing a gas at its condensation point (boiling point).

As another example for which you can predict the results, let us consider the entropy changes that occur when a liquid solidifies at a temperature below its freezing (melting) point. (See line 2(c) of Table 14–2.) We all know that this is a spontaneous process, yet ΔS_{system} is negative, since a solid forms from its liquid. However, $\Delta S_{\text{surroundings}}$ is positive and of larger magnitude than ΔS_{system}. The reason is that a liquid releases heat to its surroundings (atmosphere) as it crystallizes, and the released heat increases the motion (randomness, disorder) of the molecules of the surroundings. As a result, $\Delta S_{\text{universe}}$ is positive. As the temperature decreases, the $\Delta S_{\text{surroundings}}$ contribution becomes more important, $\Delta S_{\text{universe}}$ becomes more positive, and the process becomes more spontaneous.

The situation is reversed when a liquid is boiled or a solid is melted. For example, at temperatures above its melting point (see line 1(a) of Table 14–2), a solid spontaneously melts and ΔS_{system} is positive. The heat absorbed as the solid (system) melts is removed from the surroundings, and thus decreases the motion of the molecules of the surroundings. Therefore, $\Delta S_{\text{surroundings}}$ is negative (the surroundings become less disordered). However, the positive ΔS_{system} is larger than the negative $\Delta S_{\text{surroundings}}$, so that $\Delta S_{\text{universe}}$ is positive.

Above the melting point, $\Delta S_{\text{universe}}$ is positive for melting. Below the melting point, $\Delta S_{\text{universe}}$ is positive for freezing. At the melting point, (lines 1(b) and 2(b) of Table 14–2), $\Delta S_{\text{surroundings}}$ is equal in magnitude and opposite in sign to ΔS_{system}. Then $\Delta S_{\text{universe}}$ is zero for both melting and freezing, and the system is at equilibrium. Table 14–2 summarizes the entropy effects associated with phase changes.

The subscript "system" is commonly omitted; the symbol ΔS usually refers to the change in entropy of the reacting system, just as ΔH refers to the change in enthalpy of the reacting system.

In summary, if ΔS_{system} is positive, at least this is in the direction that is *favorable* to spontaneity. Such a process could be nonspontaneous only if $\Delta S_{\text{surroundings}}$ were quite negative. Consequently, entropy changes that accompany physical and chemical changes are reported in terms of ΔS_{system}.

TABLE 14–2   Entropy Effects Associated with Melting and Freezing

Change	Temperature	Sign of ΔS_{sys}	Sign of ΔS_{surr}	(Magnitude of ΔS_{sys}) compared to (Magnitude of ΔS_{surr})	$\Delta S_{\text{univ}} = \Delta S_{\text{sys}} + \Delta S_{\text{surr}}$	Spontaneity
1. Melting (solid $\rightarrow$ liquid)	(a) $>$mp	$+$	$-$	$>$	>0	spontaneous
	(b) $=$mp	$+$	$-$	$=$	$=0$	equilibrium
	(c) $<$mp	$+$	$-$	$<$	<0	nonspontaneous
2. Freezing (liquid $\rightarrow$ solid)	(a) $>$mp	$-$	$+$	$>$	<0	nonspontaneous
	(b) $=$mp	$-$	$+$	$=$	$=0$	equilibrium
	(c) $<$mp	$-$	$+$	$<$	>0	spontaneous

The Σ means that the S^0 values of all products and reactants must be summed; each S^0 value must be multiplied by its coefficient, n, in the balanced chemical equation.

The **standard entropy change,** ΔS^0, of any process can be determined from the tabulated absolute entropies of the reactants and products in the same way that an enthalpy change can be determined from values of standard enthalpies of formation of reactants and products. The relationship is similar to Hess' Law:

$$\Delta S^0_{rxn} = \Sigma n S^0_{products} - \Sigma n S^0_{reactants}$$

[Eq. 14–11]

Example 14–4

Using the values of standard molar entropies in Appendix K, calculate the entropy change at 25°C and one atmosphere pressure for the reaction of hydrazine with hydrogen peroxide. The mixture is explosive, and has been used for rocket propulsion. Do you think the reaction is spontaneous? The balanced equation for the reaction is

$$N_2H_4 \,(\ell) + 2H_2O_2 \,(\ell) \longrightarrow$$
$$N_2 \,(g) + 4H_2O \,(g) + 642.2 \text{ kJ}$$

Solution

From Eq. 14–11,

We have multiplied the absolute entropy of each species by the number of moles of that species and *included the unit moles*. It is common practice to omit the unit "mol" in such calculations for conciseness, and we shall do so in later examples.

Although it may not appear to be, $+605.9$ J/K is a relatively large value of ΔS (for the system). The entropy change is positive and favors spontaneity. In this case, the reaction is also exothermic, $\Delta H^0 = -642.2$ kilojoules. Therefore, the reaction *must* be spontaneous, as we shall demonstrate in the next section, because ΔS^0 is positive and ΔH^0 is negative.

$$\Delta S^0_{rxn} = [S^0_{N_2 \,(g)} + 4S^0_{H_2O \,(g)}] - [S^0_{N_2H_4 \,(\ell)} + 2S^0_{H_2O_2 \,(\ell)}]$$
$$= (1 \text{ mol})(191.5 \text{ J/mol·K}) + (4 \text{ mol})(188.7 \text{ J/mol·K}) - (1 \text{ mol})(121.2 \text{ J/mol·K}) - (2 \text{ mol})(109.6 \text{ J/mol·K})$$

$$\Delta S^0_{rxn} = +605.9 \text{ J/K} \qquad (+0.6059 \text{ kJ/K})$$

Realizing that changes in the thermodynamic quantity *entropy* may be understood in terms of changes in *molecular disorder*, we can often predict the sign of ΔS_{system}. The following illustrations emphasize several common types of processes that result in predictable entropy changes.

(a) *Phase changes.* For instance, when melting occurs, the molecules or ions are taken from their quite ordered crystalline arrangement to a more disordered one in which they are able to move past one another in the liquid. Thus a melting process is always accompanied by an entropy increase ($\Delta S_{system} > 0$). Likewise, vaporization and sublimation both take place with large increases in disorder, and hence with increases in entropy. Entropy decreases, since order increases, for the reverse processes of freezing, condensation, and deposition.

(b) *Temperature changes,* for example, warming a gas from 25°C to 50°C. As any sample is warmed, the molecules undergo more (random) motion; hence entropy increases ($\Delta S_{system} > 0$) as temperature increases. Likewise, as the temperature of a solid increases, the particles vibrate more vigorously about their lattice positions, so that at any instant, there is a larger average displacement from their mean positions; this results in an entropy increase.

(c) *Volume changes.* When the volume of a sample of gas increases, the molecules can occupy more positions, and hence are more randomly arranged than if they were close together in a smaller volume. Hence an expansion is accompanied by an increase in entropy ($\Delta S_{system} > 0$). Conversely, as a sample is compressed, the molecules are more restricted in their locations, and a situation of greater order (lower entropy) results.

(d) *Mixing of substances* without chemical reaction. Situations in which the molecules are more "mixed up" are more disordered, and hence are at higher entropy. As an analogy to this, consider a modification of Figure 14–8 in which you are required to sort a mixture of pennies and dimes so that they are at opposite ends of the box. The more mixed up they are to begin with, the more difficult it is for you to order them. Likewise, we pointed out that the two gases of Figure 14–7b were more disordered than the separated gas molecules of Figure 14–7a, and hence that the former was a situation of higher entropy. We see, then, that mixing of gases by diffusion is a process for which $\Delta S_{system} > 0$; we know from experience that it is always spontaneous. We have already pointed out (Section 13–1) that the increase in disorder (entropy increase) that accompanies mixing often provides the driving force for solubility of one substance in another. For example, when one mole of solid NaCl dissolves in water, NaCl (s) $\rightarrow$ NaCl (aq), the entropy (Appendix K) increases from 72.4 J/K to 115.5 J/K, or $\Delta S^0 = +43.1$ J/K. The term "mixing" can be applied rather liberally. For example, the process H_2 (g) + Cl_2 (g) $\rightarrow$ 2HCl (g) has $\Delta S^0 > 0$; in the reactants, each atom is paired with like atoms, a less "mixed-up" situation than in the products where unlike atoms are bonded together.

(e) *Increase in the number of particles,* as in the dissociation of a diatomic gas like F_2 (g) $\rightarrow$ 2F (g). Any process in which the number of particles increases results in an increase in entropy, $\Delta S_{system} > 0$. Values of ΔS^0 for several reactions of this type, as calculated from Equation 14–11, are given in Table 14–3. As you can see, the ΔS^0 values for the dissociation process $X_2 \rightarrow 2X$ are all similar for X = H, F, Cl, and N. Why, then, is the value given in Table 14–3 so much larger for X = Br? The answer lies in the fact that the value is for the process starting with *liquid* Br_2. The total process Br_2 (ℓ) $\rightarrow$ 2Br (g), for which $\Delta S^0 = 197.5$ J/K, can be treated as the result of *two* processes. The first of these is *vaporization,* Br_2 (ℓ) $\rightarrow$ Br_2 (g), for which $\Delta S^0 = 93.1$ J/K. The second step is the dissociation of gaseous bromine, Br_2 (g) $\rightarrow$ 2Br (g), for which $\Delta S^0 = 104.4$ J/K; note that this entropy increase is about the same as for the other processes that involve *only* dissociation of a gaseous diatomic species.

(f) *Changes in the number of moles of gaseous substances.* Processes that result in an increase in the number of moles of gaseous substances have $\Delta S_{system} > 0$. Example 14–4 may be taken as an illustration of this. The reactants of the balanced chemical equation include no gaseous substances, while the products include 5 moles of gas. Conversely, the process $2H_2$ (g) + O_2 (g) $\rightarrow$ $2H_2O$ (g) has a negative ΔS^0 value; in the balanced equation, three moles of gas are consumed while only two moles are produced, for a net decrease in the number of moles in the gas phase. You

Can you rationalize, in comparable terms, the even higher value given in the table for I_2 (s) $\rightarrow$ 2I (g)?

Do you think that the process $2H_2$ (g) + O_2 (g) $\rightarrow$ $2H_2O$ (ℓ) would have a higher or a lower value of ΔS^0 than when the water is in the gas phase? Confirm this by calculation.

TABLE 14–3 Entropy Changes for Some Processes $X_2 \rightarrow 2X$

Reaction	ΔS^0 (J/K)
H_2 (g) $\rightarrow$ 2H (g)	98.0
N_2 (g) $\rightarrow$ 2N (g)	114.9
F_2 (g) $\rightarrow$ 2F (g)	114.5
Cl_2 (g) $\rightarrow$ 2Cl (g)	107.2
Br_2 (ℓ) $\rightarrow$ 2Br (g)	197.5
I_2 (s) $\rightarrow$ 2I (g)	245.3

TABLE 14–4 Thermodynamic Classes of Reactions

Class	ΔH	$T\Delta S$	Comments
1	−	+	Both factors favorable to spontaneity; process is spontaneous at all temperatures.
2	−	−	Opposing factors; at lower temperatures, $T\Delta S$ becomes less important, so spontaneity is favored; at higher temperatures, the decrease in entropy could overwhelm the exothermicity, and process could become nonspontaneous.
3	+	+	Opposing factors; at lower temperatures $T\Delta S$ matters little, so the endothermic reaction is nonspontaneous; at higher temperatures, the favorable positive $T\Delta S$ term swamps the effect of ΔH, and process becomes spontaneous.
4	+	−	Neither factor favorable to spontaneity; process is nonspontaneous at all temperatures.

should be able to calculate the value of ΔS^0 for this reaction from the values in Appendix K.

14–7 Free Energy Change: A Criterion for Spontaneity

> Remember that the units of $T\Delta S$ are units of energy. Since T (temperature on the Kelvin scale) is always positive, the sign of $T\Delta S$ is determined by the sign of ΔS.

We have seen that there are two factors that *favor* the spontaneity of a process. (1) A lowering of the energy, as measured at constant pressure by ΔH of the system, helps to make a process spontaneous. (2) An increase in the disorder of the system, ΔS, indicates a tendency toward spontaneity. Since disorder of any system increases with temperature, the contribution of disorder to spontaneity is more usefully described by the product $T\Delta S$. Since both ΔH and ΔS can be either positive or negative, we can classify reactions into four categories with respect to spontaneity, as shown in Table 14–4.

Table 14–5 includes examples of each of these classes of chemical reactions (or physical changes), as well as temperatures at which each reaction listed is spontaneous, where appropriate.

TABLE 14–5 Examples of Thermodynamic Classes of Reactions

Class*	Reaction	ΔH (kJ)	ΔS (J/K)	Temperature Range of Spontaneity
1	$2H_2O_2 (\ell) \rightarrow 2H_2O (\ell) + O_2 (g)$	−196	+126	all temps.
	$H_2 (g) + Br_2 (\ell) \rightarrow 2HBr (g)$	−73	+114	all temps.
2	$NH_3 (g) + HCl (g) \rightarrow NH_4Cl (s)$	−176	−285	lower temps. (<619K)
	$2H_2S (g) + SO_2 (g) \rightarrow 3S (s) + 2H_2O (\ell)$	−223	−424	lower temps. (<550K)
3	$NH_4Cl (s) \rightarrow NH_3 (g) + HCl (g)$	+176	+285	higher temps. (>619K)
	$CCl_4 (\ell) \rightarrow C (graphite) + 2Cl_2 (g)$	+136	+235	higher temps. (>517K)
4	$2H_2O (\ell) + O_2 (g) \rightarrow 2H_2O_2 (\ell)$	+196	−126	nonspon. all temps.
	$3O_2 (g) \rightarrow 2O_3 (g)$	+285	−137	nonspon. all temps.

* The numbering corresponds to that in Table 14–4.

We shall now develop a single criterion that simultaneously takes into account the effects on spontaneity of both enthalpy and entropy.

If heat is released in a chemical reaction (ΔH is negative), some of the heat may be converted into useful work, but some of it may be expended to increase the order of the system (if ΔS for the change is negative), so that less energy than that indicated by ΔH actually would become available for doing work. Conversely, if a system becomes more disordered (ΔS is positive), then more useful energy becomes available than indicated by the enthalpy change alone. J. Willard Gibbs, a prominent nineteenth-century professor of mathematics and physics, suggested the use of another state function, now called the **Gibbs free energy,** defined by $G = H - TS$. The relationship between the enthalpy change and entropy change for a process is then expressed in terms of the change in this state function, ΔG, which we call the **Gibbs free energy change.** The relationship at *constant temperature* is given by the Gibbs-Helmholtz equation,

$$\Delta G = \Delta H - T\Delta S \qquad\qquad \text{[Eq. 14–12]}$$

where T is the absolute temperature. We can interpret this equation as follows:

$$\begin{pmatrix} \text{change in} \\ \text{available} \\ \text{energy} \end{pmatrix} = \begin{pmatrix} \text{change in} \\ \text{heat} \\ \text{content} \end{pmatrix} - \begin{pmatrix} \text{change in energy} \\ \text{due to change in} \\ \text{order/disorder} \end{pmatrix}$$

The change in Gibbs free energy is the *maximum useful energy* obtainable from a given process at constant temperature. It is also the *indicator of spontaneity of a reaction or physical change.* If there is a net release of useful energy, ΔG is negative and the process is spontaneous. Note from Eq. 14–12 that ΔG tends to become more negative as ΔH becomes more negative (exothermic) and as ΔS becomes more positive (increase in disorder). If there is a net absorption of free energy by the system during a process, ΔG is positive and the process is nonspontaneous. This means, of course, that the reverse process is spontaneous under the given conditions; that is, reversing the process reverses the signs of ΔH, ΔS, and ΔG.

When $\Delta G = 0$, there is no net transfer of free energy, so neither the forward nor the reverse process is favored over the other; and the system is then at a state of **equilibrium.** Thus, $\Delta G = 0$ describes a system at equilibrium, a point that we shall discuss shortly.

The relationship between ΔG and the spontaneity of a reaction is summarized in Table 14–6.

The free energy content of a system depends on temperature and pressure, (and, for mixtures, on concentrations). Thus, the change in free energy for a process depends on the states and the concentrations of the various substances involved. Just as in the case of ΔH_f^0 (Section 3–6), we choose some set of conditions for a reference point, usually one atmosphere and 25°C. The **standard state** of a pure

> You should be able to see how the sign of ΔG is related to the various thermodynamic classes of reactions given in Table 14–4.

TABLE 14–6   The Relationship Between ΔG and the Spontaneity of a Reaction

ΔG	Spontaneity of Reaction
ΔG is positive	reaction is nonspontaneous
ΔG is zero	system is at equilibrium
ΔG is negative	reaction is spontaneous

substance is the state (solid, liquid, gas) in which it exists in its most stable form at the standard conditions. For ideal gaseous mixtures, each substance is considered to be in its standard state when its partial pressure is one atmosphere; for a solute in liquid solution, the standard state is taken as a one molar concentration of the solute.

The change in Gibbs free energy at the standard state conditions, ΔG^0, is given by the appropriate form of the Gibbs-Helmholtz equation:

$$\Delta G^0 = \Delta H^0 - T\Delta S^0 \qquad \text{(constant temperature)} \qquad \text{[Eq. 14-13]}$$

The value of ΔH^0 can be obtained from tabulated values of ΔH_f^0, as we learned in Section 3-6, while ΔS^0 can be evaluated from tables of absolute entropies, as we saw in Section 14-6. The value and sign of ΔG^0 may then be used, as specified in Table 14-6, as a criterion to determine the spontaneity of reactions at specified standard conditions. In order to predict the spontaneity at nonstandard conditions (e.g., concentrations other than 1 M, gas partial pressures other than 1 atm) we shall have to calculate ΔG. This quantity can be calculated from ΔG^0, although we shall postpone that treatment until Chapter 16.

It is important to distinguish among the various notations for free energy change: (1) ΔG^0 represents the *standard free energy change* for the process at some temperature, T. This is the change in free energy for the process in which the numbers of moles of reactants indicated in the balanced equation, all at their standard states, are converted to the numbers of moles of products indicated in the balanced equation, also all at their standard states.

(2) ΔG_{298}^0 represents the standard free energy change for the process at 25°C (298 K). Since ΔH and ΔS do not depend very much on temperature, we can usually apply Eq. 14-13 using the values of ΔH^0 and ΔS^0 determined at 298 K, and assume that they are valid at other temperatures, so long as the temperature difference is not great.

This is not the same as saying that *H* and *S* do not depend on temperature — only that *changes* in these quantities are not very dependent upon temperature.

(3) ΔG represents the change in free energy for *any concentrations* of reactants and products, whether or not they are at standard states or conditions.

Let us now consider several examples in which we determine the spontaneity of reactions or physical changes at standard conditions with the aid of the ΔG^0 criterion. *Remember that this criterion applies only to changes that occur at constant temperature.* Example 14-5 illustrates the use of Eq. 14-13.

Example 14-5

Evaluate ΔG^0 for the reaction in which diatomic hydrogen and oxygen gases react to form water vapor at 298 K, using values of ΔH_f^0 and S^0 from Appendix K. Is the reaction spontaneous at 298 K?

$$2H_2\,(g) + O_2\,(g) \longrightarrow 2H_2O\,(g)$$

Solution

First we must evaluate ΔH_{rxn}^0 and ΔS_{rxn}^0, using values of ΔH_f^0 and S^0 (at 298 K) tabulated in Appendix K:

$$\Delta H_{rxn}^0 = \Sigma n\Delta H_{f\,products}^0 - \Sigma n\Delta H_{f\,reactants}^0$$
$$= 2\Delta H_{f\,H_2O\,(g)}^0 - [2\Delta H_{f\,H_2\,(g)}^0 + \Delta H_{f\,O_2\,(g)}^0]$$
$$= [2(-241.8) - (0 + 0)]\ \text{kJ}$$

$$\Delta H_{rxn}^0 = -483.6\ \text{kJ}$$

$$\Delta S_{rxn}^0 = \Sigma nS_{products}^0 - \Sigma nS_{reactants}^0$$
$$= 2S_{H_2O\,(g)}^0 - [2S_{H_2\,(g)}^0 + S_{O_2\,(g)}^0]$$
$$= [2(188.7) - (2 \times 130.6 + 205.00)]\ \text{J/K}$$

$$\Delta S_{rxn}^0 = -88.8\ \text{J/K} = -0.0888\ \text{kJ/K}$$

It is reasonable that entropy decreases, since fewer moles of gas are formed than are consumed.

Now we may use the Gibbs-Helmholtz equation (Eq. 14-13) with $T = 298$ K, since we are evaluating the free energy change under standard state conditions at 298 K.

$$\Delta G^0_{rxn} = \Delta H^0_{rxn} - T\Delta S^0_{rxn}$$
$$= -483.6 \text{ kJ} - (298 \text{ K})(-0.0888 \text{ kJ/K})$$
$$= -483.6 \text{ kJ} - (-26.5 \text{ kJ})$$

$$\Delta G^0_{rxn} = \underline{-457.1 \text{ kJ}}$$

Since ΔG^0_{rxn} is negative, the reaction is spontaneous at 298 K under standard conditions. This means that 2 moles of H_2O (g) are thermodynamically more stable than a mixture of 2 moles of H_2 (g) and 1 mole of O_2 (g) at standard conditions, 1 atm and 298 K. It does *not* guarantee that the reaction will occur at an observable rate.

Notice that Example 14–5 is a case in which ΔH^0_{rxn} is favorable to spontaneity (negative) while ΔS^0_{rxn} is unfavorable to spontaneity (negative)—class 2 of Table 14–4. Thus we could not predict spontaneity by qualitative reasoning. Rather we must calculate ΔG^0_{rxn}, the composite criterion that takes into account *both* ΔH^0_{rxn} and ΔS^0_{rxn} at a specified temperature.

Values of ΔG^0 for a reaction can be calculated from experimentally determined equilibrium constants (Chapter 16), from the electrochemical potential for a process (Chapter 19), and by other techniques. Since values of ΔG and ΔH usually can be obtained experimentally with little difficulty, ΔS values can be calculated by the Gibbs relationship. There is no way of measuring ΔS for a reaction directly.

Values of standard molar free energy of formation, ΔG^0_f, for many substances are tabulated in Appendix K. Notice that for elements in their standard states, $\Delta G^0_f = 0$. The values may be used, in a manner which is now familiar to you, to calculate the standard free energy change of a reaction at 298 K by using the relationship given in Eq. 14–14, as shown in Example 14–6.

$$\Delta G^0_{rxn} = \Sigma n \Delta G^0_{f\,products} - \Sigma n \Delta G^0_{f\,reactants} \qquad \text{[Eq. 14–14]}$$

Example 14–6

Make the same determination as in Example 14–5, using standard free energies of formation from Appendix K.

Solution

$$2H_2 \text{ (g)} + O_2 \text{ (g)} \longrightarrow 2H_2O \text{ (g)}$$

$$\Delta G^0_{rxn} = 2\Delta G^0_{f\,H_2O \text{ (g)}} - [2\,\Delta G^0_{f\,H_2 \text{ (g)}} + \Delta G^0_{f\,O_2 \text{ (g)}}]$$
$$= [2(-228.6 \text{ kJ}) - (0 + 0)] \text{ kJ}$$
$$= \underline{-457.2 \text{ kJ}}$$

This is the same value obtained in Example 14–5, within rounding error.

Let us emphasize once again that both ΔH and ΔS actually vary with temperature, but not enough to introduce significant errors when the temperature changes are not too great. The value of ΔG for a reaction, however, depends strongly on temperature. Thus, Eq. 14–14 can be applied *only* for a reaction or process at 298 K, while Eq. 14–13 can give quite reliable values of ΔG^0 over a considerable range of temperatures.

The Gibbs-Helmholtz equation can be used to *estimate* the temperature at which a physical process is in equilibrium, as shown in the following example.

Example 14–7

Use the thermodynamic data in the appendices to estimate the normal boiling temperature of liquid bromine, Br_2. Assume that ΔH and ΔS do not change with temperature.

Solution

The reaction of interest is

$$Br_2 \text{ (}\ell\text{)} \longrightarrow Br_2 \text{ (g)}$$

By definition, the normal boiling point of a liquid is the temperature at which liquid and gas exist in equilibrium at one atmosphere, and therefore $\Delta G = 0$.

$$\Delta H^0_{rxn} = \Delta H^0_{f\,Br_2\,(g)} - \Delta H^0_{f\,Br_2\,(\ell)}$$
$$= 30.91\ kJ - 0\ kJ$$

$$\Delta H^0_{rxn} = 30.91\ kJ$$

$$\Delta S^0_{rxn} = S^0_{Br_2\,(g)} - S^0_{Br_2\,(\ell)}$$
$$= (245.4 - 152.2)\ J/K$$

$$\Delta S^0_{rxn} = 93.2\ J/K = 0.0932\ kJ/K$$

We can now solve for the temperature at which the system is in equilibrium, i.e., the boiling point of Br_2.

$$\Delta G^0_{rxn} = \Delta H^0_{rxn} - T\Delta S^0_{rxn} = 0$$

$$\Delta H^0_{rxn} = T\Delta S^0_{rxn}$$

$$T = \frac{\Delta H^0_{rxn}}{\Delta S^0_{rxn}} = \frac{30.91\ kJ}{0.0932\ kJ/K} = \underline{332\ K\ (59°C)}$$

The value listed in a handbook of chemistry is 58.78°C.

For this process, ΔH^0_{rxn} is positive (unfavorable to spontaneity) while ΔS^0_{rxn} is positive (favorable)— class 3 of Table 14-3. Thus, the process of boiling would be spontaneous at temperatures above the calculated temperature, while it would be nonspontaneous (the reverse process, condensation, would be spontaneous) at temperatures below this.

A similar approach allows determination of the temperature range over which a reaction or process is spontaneous, as shown in the next example.

Example 14-8

Mercury(II) sulfide is found in a dark red mineral called cinnabar. Metallic mercury is obtained by roasting the sulfide in a limited amount of air. Determine the temperature range in which the reaction is spontaneous.

$$HgS\ (s) + O_2\ (g) \longrightarrow Hg\ (\ell) + SO_2\ (g)$$

Solution

We first evaluate ΔH^0_{rxn} and ΔS^0_{rxn} and *assume* that their values are independent of temperature.

$$\Delta H^0_{rxn} = \Delta H^0_{f\,Hg\,(\ell)} + \Delta H^0_{f\,SO_2\,(g)}$$
$$- [\Delta H^0_{f\,HgS\,(s)} + \Delta H^0_{f\,O_2\,(g)}]$$
$$= (0 - 296.8 + 58.2 - 0)\ kJ$$

$$\Delta H^0_{rxn} = \underline{-238.6\ kJ}$$

$$\Delta S^0_{rxn} = S^0_{Hg\,(\ell)} + S^0_{SO_2\,(g)} - [S^0_{HgS\,(s)} + S^0_{O_2\,(g)}]$$
$$= (76.02 + 248.1 - 82.4 - 205.0)\ J/K$$

$$\Delta S^0_{rxn} = +36.7\ J/K$$

Since ΔH^0_{rxn} is negative and ΔS^0_{rxn} is positive, the reaction is spontaneous at all temperatures. The reverse reaction is, therefore, nonspontaneous at all temperatures. This illustrates a corollary of the Second Law of Thermodynamics: The reverse of a spontaneous reaction or process is nonspontaneous.

14-8 Free Energy and Chemical Equilibrium

We learned in Section 14-7 that the change in standard free energy, ΔG^0, can tell us whether a particular process or reaction is spontaneous or nonspontaneous under certain conditions. It is important to realize that such an approach, using ΔG^0, applies only to a very special case, that in which the numbers of moles of reactants indicated produce the numbers of moles of products indicated, with all substances in their standard states. This means that with the ΔG^0 criterion we can consider only changes involving reactants and products that are pure liquids, pure solids, ideal gaseous mixtures with each component present at 1 atm, or solutions in which the concentration of each solute is 1 M. Most of the time, we do not work at these very special conditions. In this section, we shall introduce the reasoning that allows us to

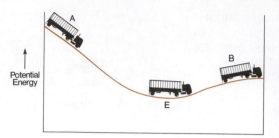

FIGURE 14–10 Potential energy of a truck. The truck has lower potential energy at point *B* than at point *A*. However, the point *E* represents the most stable position (lowest accessible potential energy), known as the equilibrium position. From any other point, movement toward point *E* would be spontaneous.

apply thermodynamics to other conditions, especially those involving the important condition of chemical equilibrium, leaving the detailed development of such applications to Chapter 16.

As an analogy, consider first the truck shown in Figure 14–10. Suppose we first ask the question, "Which is the more stable place for the truck—point A or point B?" Since point A is at higher gravitational potential energy, we are sure that point B would be preferable. Take these two possibilities as analogous to reactants (A) and products (B), so that calculation of ΔG_{rxn}^0 is analogous to considering the potential energy difference between A and B. We would say, then, that going from point A to point B is expected to be *spontaneous*, while going from point B to point A would be *nonspontaneous*.

However, it is obvious from Figure 14–10 that there is a more stable place, between A and B, for the truck to stop. As it rolls down the hill from point A, it reaches its *lowest accessible* potential energy at point E. Likewise, if it started at point B, and moved toward point A, it would reach this same lowest energy point E. In other words, the deduction that point B would be more stable than point A does not guarantee that the truck must go *all the way* to point B. Likewise, the conclusion that the process B → A is nonspontaneous does not mean that the truck would not roll *part way* from B to A.

To see how this reasoning applies to a chemical reaction, let us consider the formation of gaseous iodine chloride, ICl, at 25°C, according to the chemical equation

$$I_2 \text{ (g)} + Cl_2 \text{ (g)} \longrightarrow 2ICl \text{ (g)}$$

Equation 14–14 leads to the value $\Delta G^0 = -30.40$ kJ. By the reasoning of Section 14–7, we would classify this reaction as spontaneous at 25°C, while the reverse reaction

$$2ICl \text{ (g)} \longrightarrow I_2 \text{ (g)} + Cl_2 \text{ (g)}$$

is nonspontaneous at 25°C. Pure I_2 (g) + Cl_2 (g) is represented by point A, and pure ICl by point B, in Figure 14–11. Notice that we do not know any absolute values on the vertical axis (free energy). We do know, however, that the free energy of B is lower than that of A by 30.40 kJ. This means that two moles of ICl gas at 25°C, 1 atm, is chemically more stable than the mixture of one mole of Cl_2 gas and one mole of I_2 gas at 25°C with both I_2 and Cl_2 at 1 atm pressure (2 atm total pressure).

Does this mean that at these conditions a sample of 2 moles of ICl would not react at all, or that a mixture of 1 mole Cl_2 and 1 mole I_2 would necessarily react completely? No, this is not quite the correct interpretation.

The free energy content, *G*, of any substance in a mixture depends on its concentration (or for a gas, on its partial pressure). We have seen that mixing is

Keep in mind that we are discussing *only whether* these reactions can occur at the specified conditions, because thermodynamics tells us nothing about the *rate* at which a process, however probable, would occur.

FIGURE 14–11 Variation in total free energy for the reaction I_2 (g) + Cl_2 (g) $\longrightarrow$ 2ICl (g), and its reverse reaction, 2ICl (g) $\longrightarrow$ I_2 (g) + Cl_2 (g). The reaction is carried out at constant T. Point E represents the condition of chemical equilibrium. The *standard* free energy change, ΔG^0, represents the free energy difference between points A and B. The direction of this change (ΔG^0 negative) indicates that 2 moles of ICl at 1 atm is more stable than a mixture of 1 mole of I_2 and 1 mole of Cl_2, each at 1 atm. The minimum in the curve corresponds to a mixture of all three gases, which is even more stable (equilibrium).

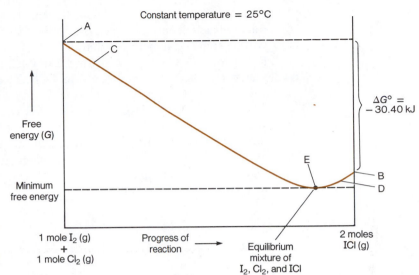

accompanied by an increase in entropy, which decreases free energy (it is spontaneous). Furthermore, the total free energy content of a mixture is the sum of the free energy contents of all substances present. Suppose we consider starting with a mixture of 1 mole I_2 and 1 mole Cl_2 (point A) at 1 atm each. If the forward reaction occurs only slightly, the partial pressures of these gases decrease somewhat, decreasing their free energy. In addition, a little ICl is formed, so that there is "mixing" of components, which also decreases free energy. The total free energy content (point C) is now a little less than at point A. Thus, $\Delta G < 0$ for this increment of reaction, and this amount of reaction is spontaneous. As the reaction proceeds, G for the mixture continues to decrease until point E is reached.

Now, suppose that we begin with 2 moles of pure ICl (g) (point B), and consider it to decompose very slightly by the reverse reaction. The free energy of the ICl decreases, because its partial pressure is now less than 1 atm; the free energy of the very small amounts of I_2 (g) and Cl_2 (g) would be quite small, so the total free energy content at point D would be less than that at point B. Now $\Delta G < 0$ for the change from point B to point D, and *that* small amount of reverse reaction, too, is seen to be spontaneous! This is what we mean when we sometimes say that "any chemical reaction is reversible." We do not mean that the reaction goes *entirely* in the reverse direction, or that it will necessarily go at a measurable rate at the specified conditions.

Starting with only I_2 and Cl_2, the reaction will cause the free energy to decrease. Likewise, starting with only ICl, the reverse reaction will also decrease the free energy. In either case, the free energy will reach a minimum value at some mixture of I_2, Cl_2, and ICl (point E). A change in either direction from point E would be "uphill" in terms of free energy. We have just described, in thermodynamic terms, a situation called **chemical equilibrium.** An *infinitesimally* small progress of the reaction from this equilibrium point would have $\Delta G = 0$, while any larger (finite) departure from this point would be nonspontaneous (G would increase). This is what we meant in Section 14–7 when we said that $\Delta G = 0$ *denotes a process at equilibrium.*

The dependence of free energy on equilibrium also lets us conclude that the equilibrium condition always lies closer to the side of the reaction with the lowest

ΔG (not ΔG^0) is proportional to the slope of the tangent to the curve connecting any initial and final states of interest. $\Delta G = 0$ at the minimum of the curve because the slope of the tangent at that point is 0.

standard free energy, G^0. Thus, for the I_2 (g) + Cl_2 (g) → 2ICl (g) reaction, the relatively small negative value of $\Delta G^0 = -30.40$ kJ tells us that at equilibrium, there will be somewhat more ICl than I_2 and Cl_2. Another way to say this is that 1 mol I_2 + 1 mol Cl_2 would react toward 2 mol ICl somewhat more fully than 2 mol ICl would react toward 1 mol I_2 + 1 mol Cl_2. By comparison, the quite large negative value of $\Delta G^0 = -457.1$ kJ found in Example 14–5 for the reaction $2H_2$ (g) + O_2 (g) → $2H_2O$ (g) tells us that the forward reaction is *much* more favored than the reverse reaction, so that the equilibrium condition would lie *much* further toward the product (H_2O) side of the reaction. Conversely, a positive value of ΔG^0 just tells us that products react to form reactants more favorably than reactants react to form products, so that the equilibrium mixture would be richer in reactants.

Key Terms

Absolute entropy (of a substance) the increase in the entropy of a substance as it goes from a perfect crystalline form at 0 K (where its entropy is zero) to the temperature in question.

Bomb calorimeter a device used to measure the heat transfer between system and surroundings at constant volume and temperature.

Chemical equilibrium see *Equilibrium.*

Endothermicity the absorption of heat by a system as a process occurs.

Enthalpy, H the heat content of a specific amount of a substance; thermodynamically defined as $E + PV$.

Entropy, S a thermodynamic state function that measures the degree of disorder or randomness of a system. The product $T\Delta S$ represents a change in energy, so it has units of energy.

Equilibrium (chemical equilibrium) a state of dynamic balance in which the rates of forward and reverse reactions are equal; the state of a system when neither the forward nor the reverse reaction is thermodynamically favored.

Exothermicity the release of heat by a system as a process occurs.

First Law of Thermodynamics the total amount of energy in the universe is constant; (also known as the Law of Conservation of Energy) energy is neither created nor destroyed in ordinary chemical reactions and physical changes.

Free energy, G (also known as Gibbs free energy) a thermodynamic state function of a system

that represents the amount of energy available for the system to do useful work.

Free energy change, ΔG the indicator of spontaneity of a process. If ΔG is negative for a process, it is spontaneous.

Gibbs free energy see *Free energy.*

Hess' Law of heat summation the enthalpy change of a reaction is the same whether it occurs in one step or a series of steps.

Internal energy, E the sum of all forms of energy associated with a specific amount of a substance.

Isothermal expansion (of a gas) expansion at constant temperature; $\Delta E = 0$ for such a process involving an ideal gas.

Pressure-volume work work done by a gas when it expands against an external pressure or work done on a system as gases are compressed or consumed in the presence of an external pressure.

Second Law of Thermodynamics the universe tends toward a state of greater disorder in spontaneous processes.

Spontaneity of a process, its property of being energetically favorable and therefore capable of proceeding in the forward direction (but not necessarily at an observable rate).

Standard entropy, S^0 (of a substance) the absolute entropy of a substance in its standard state at 298 K.

Standard molar enthalpy of formation, ΔH_f^0 the amount of heat absorbed in the formation of one mole of a substance in its standard state from its elements in their standard states.

Standard state the most stable form of a substance at 25°C and one atmosphere pressure.

State function a function that depends on the thermodynamic state of the system and is independent of the pathway by which a process occurs.

Surroundings everything in the environment of the system.

System the substances of interest in a process; the part of the universe under investigation.

Thermodynamics the study of the energy transfers accompanying physical and chemical processes.

Third Law of Thermodynamics the entropy of a hypothetical pure, perfect, crystalline substance at absolute zero temperature is zero.

Universe the system plus the surroundings.

Water equivalent (of a calorimeter) the amount of water that would absorb the same amount of heat as the calorimeter per degree temperature increase.

Work the application of a force through a distance; for chemical reactions at constant pressure, it is the pressure times the change in volume ($P\Delta V$).

Exercises

Basic Ideas

In Exercises 1 and 2, state precisely the meaning of each of the following terms that were introduced and used in Chapters 1, 3, and 7.

1. (a) energy, (b) kinetic energy, (c) potential energy, (d) heat, (e) calorie, (f) joule, (g) specific heat, (h) heat content

2. (a) enthalpy, (b) change in enthalpy, (c) system, (d) surroundings, (e) work, (f) state function, (g) thermodynamic state of a system, (h) thermodynamic standard state of a substance, (i) standard molar enthalpy of formation, (j) Hess' Law

3. Use words to state precisely and completely the meaning of the following thermochemical equations.

   (a) $CH_4 (g) + 2O_2 (g) \rightarrow CO_2 (g) + 2H_2O (\ell)$
   $$\Delta H^0 = -890.3 \text{ kJ}$$
   (b) $CH_4 (g) + 2O_2 (g) \rightarrow CO_2 (g) + 2H_2O (g)$
   $$\Delta H^0 = -802.3 \text{ kJ}$$
   (c) $N_2 (g) + 2H_2 (g) \rightarrow N_2H_4 (\ell)$
   $$\Delta H^0 = 50.63 \text{ kJ}$$
   (d) $CaO (s) \rightarrow Ca (s) + \frac{1}{2} O_2 (g)$
   $$\Delta H^0 = 635.5 \text{ kJ}$$

4. Distinguish between spontaneous and nonspontaneous processes.

5. What can be said about the sign of ΔH^0 for (a) a spontaneous process and (b) a nonspontaneous process?

6. Distinguish between endothermic and exothermic processes. If we know that a reaction is exothermic in one direction, what can be said about the reaction in the reverse direction?

7. What is entropy? What do we mean when we say that an increase in entropy favors the spontaneity of a process?

Internal Energy and Changes in Internal Energy

8. State the First Law of Thermodynamics. What does it mean?

9. What is internal energy? What is ΔE? What are the sign conventions for q, the amount of heat added to or removed from a system?

10. Show that $P\Delta V$ is equal to work, and then that it is proportional to the change in number of moles of gases in a process at a particular absolute temperature.

11. A sample of a gas is heated by the addition of 4000 kJ of heat. (a) If the volume remains constant, what is the change in internal energy, ΔE, of this sample? (b) Suppose that in addition to putting heat into the sample, we also do 1000 kJ of work on the sample. What is the change in internal energy, ΔE, of the sample? (c) Suppose, instead, that as the original sample is heated, it is allowed to expand against the atmosphere so that it does 6000 kJ of work on its surroundings. What is the change in internal energy, ΔE, of the sample?

12. Calculate q, w, and ΔE for the vaporization of 1.00 g of liquid ethanol, C_2H_5OH, at 1.00 atm at 78.0°C, to form gaseous ethanol at 1.00 atm at 78.0°C. Make the following simplifying assumptions: (a) the density of liquid ethanol at 78.0°C is 0.789 g/mL and (b) gaseous ethanol is adequately described by the ideal gas equa-

tion. The heat of vaporization of ethanol is 854 J/g.

13. Assuming the gases are ideal, calculate the amount of work done (in joules) in each of the following reactions. In each case, is the work done *on* or *by* the system?

   (a) A reaction in the Mond process for purifying nickel which involves formation of the volatile gas, nickel(0) tetracarbonyl, at $50-100°C$. Assume one mole of nickel is used and a constant temperature of 75°C is maintained.

   $$Ni (s) + 4CO (g) \longrightarrow Ni(CO)_4 (g)$$

   (b) The conversion of one mole of brown nitrogen dioxide into colorless dinitrogen tetroxide at 30.0°C.

   $$2NO_2 (g) \longrightarrow N_2O_4 (g)$$

   (c) The decomposition of one mole of an air pollutant, nitric oxide, at 300°C.

   $$2NO (g) \longrightarrow N_2 (g) + O_2 (g)$$

14. What does the term "isothermal expansion of an ideal gas" mean? Why is $\Delta E = 0$ for the isothermal expansion of an ideal gas? (Remember that internal energy, E, represents *all* forms of energy associated with a sample, and that all forms of energy can be classified as either kinetic or potential energy. Consider this in light of the Kinetic-Molecular theory as it applies to ideal gases.)

15. An ideal gas is allowed to expand isothermally from 2.00 L at 5.00 atm to 5.00 L at 2.00 atm, against a vacuum (i.e., no outside pressure). Calculate q and w for this change.

16. The sample of gas in Exercise 15 is allowed to undergo the same isothermal expansion, but now against a constant external pressure of 2.00 atm. Calculate q and w for this change.

17. The same isothermal expansion in the two preceding exercises is carried out in two steps: Step 1, against a constant external pressure of 3.00 atm, followed by step 2, against a constant external pressure of 2.00 atm. Calculate q and w.

18. What is a bomb calorimeter? How do bomb calorimeters give us useful data?

19. A 2.00 g sample of hydrazine, N_2H_4, is burned in a bomb calorimeter that contains

7.00×10^3 g of H_2O and the temperature increases from 25.00°C to 26.17°C. The water equivalent of the calorimeter is known to be 9.00×10^2 g. Calculate ΔE for the combustion of N_2H_4 in kJ/g and in kJ/mol.

20. The combustion of 1.048 grams of benzene, C_6H_6 (ℓ), in a calorimeter compartment surrounded by 826 grams of water raised the temperature of the water from 23.640°C to 33.700°C. The calorimeter's water equivalent is 216 g of H_2O.

   (a) Write the balanced equation for the combustion reaction, assuming that CO_2 (g) and H_2O (ℓ) are the only products.

   (b) Use the calorimetric data to calculate ΔE for the combustion of benzene in kJ/g and in kJ/mol.

Enthalpy and Changes in Enthalpy

You may wish to refer to Sections $3-4$ through $3-7$ and $7-11$ to review earlier discussions of enthalpy and changes in enthalpy.

21. How do enthalpy and internal energy differ?

22. When 1.00 mole of ice melts at 0°C and a constant pressure of 1.00 atm, 6.02 kJ of heat are absorbed by the system. The molar volumes (the volume occupied by 1 mole) of ice and water are 0.0196 and 0.0180 liter, respectively. Calculate ΔH and ΔE. Comment on the value(s) of ΔH and ΔE. Can you make a generalization on the relative values of ΔH and ΔE for processes in which only liquids and solids are involved?

23. Our bodies convert chemical bond energy from compounds such as carbohydrates into useful heat energy by oxidation. The combustion of cane sugar or sucrose (a carbohydrate) is shown below. Combustion is rapid oxidation accompanied by the liberation of heat and light. The same amount of energy is released in slow oxidation.

$$C_{12}H_{22}O_{11} (s) + 12O_2 (g) \longrightarrow 12CO_2 (g) + 11H_2O (\ell)$$

   (a) Calculate the work done in the oxidation of one mole of sucrose at 25°C and one atmosphere pressure.

   (b) Use data from Appendix K and Eq. $3-6$ to calculate ΔH^0_{298} for the combustion of sucrose at 25°C. ΔH_f^0 for $C_{12}H_{22}O_{11} = -2218$ kJ/mol.

(c) Calculate ΔE^0 for the oxidation of one mole of sucrose, using Eq. 14–3.

(d) Most hydrocarbons and carbohydrates (compounds containing C, H, and O) undergo exothermic and thermodynamically spontaneous oxidation. Since a large portion of our bodies is made up of carbohydrates, how can our bodies remain intact when they are in constant contact with oxygen?

(e) Compare the value of ΔH^0 with the value of ΔE^0 calculated in part (c).

24. (a) Use data from Appendix K and Eq. 3–6 to calculate ΔH^0 for the combustion of benzene, C_6H_6, at 25°C. The products are CO_2 (g) and H_2O (ℓ). (b) Compare the value of ΔH^0 with the value of ΔE^0 calculated in Exercise 20.

25. In Exercise 23 you found that $\Delta E^0 = \Delta H^0$ for the combustion of sucrose. In Exercises 20 and 24 you found that $\Delta E^0 \neq \Delta H^0$ for the combustion of benzene. Why?

Entropy and Entropy Changes

26. State the Second Law of Thermodynamics. We cannot use $\Delta S_{universe}$ directly as a measure of the spontaneity of a reaction. Why?

27. State the Third Law of Thermodynamics. What does it mean?

28. Define and distinguish among the following terms clearly and concisely. (a) entropy, (b) absolute entropy, (c) standard entropy

29. Explain why the term $T\Delta S$ is an energy term.

30. Why is it impossible for any substance to have an absolute entropy of zero at temperatures above 0 K?

31. Describe the characteristic of a reaction that its entropy change measures.

32. Which of the following processes are accompanied by an increase in entropy?
    (a) The freezing of water.
    (b) The evaporation of carbon tetrachloride, CCl_4.
    (c) The sublimation of iodine, I_2 (s) $\xrightarrow{\Delta}$ I_2 (g)
    (d) The precipitation of white silver chloride, AgCl, from a solution containing silver ions and chloride ions.
    (e) The reaction PCl_5 (g) $\rightarrow$ PCl_3 (g) + Cl_2 (g)
    (f) The reaction PCl_3 (g) + Cl_2 (g) $\rightarrow$ PCl_5 (g)

(g) Thirty-five pennies are removed from a bag and placed heads up on a table.

(h) The pennies of (g) are swept off the table and back into the bag.

33. Use the data from Appendix K to calculate the value of ΔS_{298}^0 for each reaction in Exercise 13. Compare the signs and magnitudes for these ΔS_{298}^0 values and explain your observations.

34. List six kinds of processes that result in increases in entropy and explain why each does cause an increase.

35. Explain why ΔS^0 may be referred to as a contributor to spontaneity.

Free Energy Changes and Spontaneity

36. What are the two factors that favor spontaneity of a process?

37. What is free energy? Change in free energy?

38. Most spontaneous reactions are exothermic, but some are not. Explain.

39. Explain how the signs and magnitudes of ΔH^0 and ΔS^0 are related to the spontaneity of a process and how they affect it.

40. Explain the meaning of the equation $\Delta G = \Delta H - T\Delta S$.

41. Under a specific set of conditions, why must the reverse of a spontaneous reaction be nonspontaneous?

42. It is not necessary to know the absolute values of free energies, G, of substances in order to predict the spontaneity of reactions involving the substances. Why?

43. Using values of standard free energy of formation, ΔG_f^0, from Appendix K, calculate the standard free energy change for each of the following reactions.
    (a) $2K$ (s) + $2H_2O$ (ℓ) $\rightarrow$ $2KOH$ (s) + H_2 (g)
    (b) $3Fe$ (s) + $4H_2O$ (ℓ) $\rightarrow$ Fe_3O_4 (s) + $4H_2$ (g)

44. Make the same calculations as in Exercise 43, using values of standard enthalpy of formation and absolute entropy instead of values of ΔG_f^0.

45. Use Appendix K to calculate ΔG^0 at 298 K for each of the following reactions. Which ones are spontaneous at 298 K?
    (a) CaH_2 (s) + $2H_2O$ (ℓ) $\rightarrow$
    $$Ca(OH)_2 \text{ (s)} + 2H_2 \text{ (g)}$$
    (b) Na_2CO_3 (s) + $2HCl$ (g) $\rightarrow$
    $$2NaCl \text{ (aq)} + H_2O \text{ (ℓ)} + CO_2 \text{ (g)}$$
    (c) CaO (s) + H_2O (ℓ) $\rightarrow$ $Ca(OH)_2$ (aq)

(d) $2C_2H_2$ (g) $+ 5O_2$ (g) $\rightarrow$
$$4CO_2 \text{ (g)} + 2H_2O \text{ } (\ell)$$
(e) C_2H_4 (g) $+ H_2O$ (g) $\rightarrow C_2H_5OH$ (ℓ)

Temperature Range of Spontaneity

46. What are the effects of temperature changes on the spontaneity of reactions for which:
    (a) ΔH is negative and ΔS is positive?
    (b) ΔH is negative and ΔS is negative?
    (c) ΔH is positive and ΔS is positive?
    (d) ΔH is positive and ΔS is negative?
47. Determine the temperature ranges over which the following reactions are spontaneous.
    (a) The reaction by which sulfuric acid droplets from polluted air convert water-insoluble limestone or marble (calcium carbonate) to slightly soluble calcium sulfate, which is slowly washed away by rain.

    $$CaCO_3 \text{ (s)} + H_2SO_4 \text{ } (\ell) \longrightarrow$$
    $$CaSO_4 \text{ (s)} + H_2O \text{ } (\ell) + CO_2 \text{ (g)}$$

    (b) The reaction by which Antoine Lavoisier achieved the first laboratory preparation of oxygen in the late eighteenth century: the thermal decomposition of the red-orange powder, mercury(II) oxide, to oxygen and the silvery liquid metal, mercury:

    $$2HgO \text{ (s)} \longrightarrow 2Hg \text{ } (\ell) + O_2 \text{ (g)}$$

    (c) The reaction of coke (carbon) with carbon dioxide to form the reducing agent, carbon monoxide, which is used to reduce some metal ores to the metals.

    $$CO_2 \text{ (g)} + C \text{ (s)} \longrightarrow 2CO \text{ (g)}$$

    (d) The reverse of the reaction by which iron rusts:

    $$2Fe_2O_3 \text{ (s)} \longrightarrow 4Fe \text{ (s)} + 3O_2 \text{ (g)}$$

48. (a) Determine the temperature range over which the reaction below is spontaneous.

    $$2CO \text{ (g)} + O_2 \text{ (g)} \longrightarrow 2CO_2 \text{ (g)}$$

    (b) In light of your answer to (a), how do you explain the fact that carbon monoxide from auto exhausts is a major health hazard?
49. Determine the temperature ranges over which the following reactions are spontaneous.
    (a) $2NOCl$ (g) $\rightarrow 2NO$ (g) $+ Cl_2$ (g)
    (b) $2PH_3$ (g) $\rightarrow 3H_2$ (g) $+ 2P$ (g)

50. (a) Estimate the boiling point of water (the temperature at which liquid and vapor are in equilibrium with each other) at one atmosphere pressure, using Appendix K.
    (b) Compare the temperature obtained with the known boiling point of water. Can you explain the discrepancy?
51. Estimate the normal boiling point of titanium(IV) chloride, $TiCl_4$, using Appendix K.
52. Estimate the sublimation temperature (solid to vapor) of dark violet solid iodine, I_2, using the data of Appendix K. Sublimation and subsequent deposition onto a cold surface is a common method of purification of I_2 and other solids that sublime readily.

Free Energy and Chemical Equilibrium

53. Explain the relationship between the change in Gibbs free energy (ΔG) for a process and chemical equilibrium.

Unscramble Them

54. Dissociation reactions are those in which molecules break apart. Why do high temperatures favor the spontaneity of most dissociation reactions?
55. When is it true that $\Delta S = \dfrac{\Delta H}{T}$?
56. Using your chemical intuition, decide which of the following reactions are spontaneous and occur at an observable rate at room temperature and atmospheric pressure.
    (a) H_2O (ℓ) $\rightarrow H_2O$ (s)
    (b) $2O$ (g) $\rightarrow O_2$ (g)
    (c) $2H_2O$ (ℓ) $\rightarrow 2H_2$ (g) $+ O_2$ (g)
    (d) N_2 (g) $+ O_2$ (g) $\rightarrow 2NO$ (g)
    (e) $NaOH$ (s) $+ \frac{1}{2}H_2$ (g) $\rightarrow Na$ (s) $+ H_2O$ (ℓ)
57. Determine the standard entropy changes at 298 K for each of the following reactions from ΔG_f^0 and ΔH_f^0 values.
    (a) The careful thermal decomposition of ammonium nitrate, which is a laboratory preparation of nitrous oxide (dinitrogen oxide), also known as laughing gas, a mild anesthetic often used in dental work:

    $$NH_4NO_3 \text{ (s)} \xrightarrow{\Delta} 2H_2O \text{ (g)} + N_2O \text{ (g)}$$

    (b) Reaction (a) can become dangerous if large (bulk) quantities of ammonium nitrate are used or if the salt is heated too strongly. The reaction below was partially responsible for the disastrous explosions that occurred in the port of Texas City, Texas in 1947:

$$2NH_4NO_3 (s) \longrightarrow$$
$$4H_2O (g) + 2N_2 (g) + O_2 (g)$$

58. Determine the standard entropy changes at 298 K associated with the reactions of Exercise 57 by using absolute entropies.

59. Without consulting Appendix K, decide whether or not the following reactions are spontaneous at room temperature and atmospheric pressure. Why must the reactions be spontaneous or nonspontaneous, as the case may be?
   (a) $(NH_4)_2Cr_2O_7 (s) \rightarrow$
      $Cr_2O_3 (s) + N_2 (g) + 4H_2O (g) + 71.8$ kcal
   (b) $Cs^+ (aq) + F^- (aq) + 8.4$ kJ $\rightarrow$ CsF (aq)

60. (a) Using the data in Appendix K, calculate ΔG^0 at 25°C for each of the following reactions, which are utilized in the industrial production of nitric acid, HNO_3, by the oxidation of ammonia (the Ostwald Process).

$$4NH_3 (g) + 5O_2 (g) \longrightarrow$$
$$4NO (g) + 6H_2O (g)$$

$$2NO (g) + O_2 (g) \longrightarrow 2NO_2 (g)$$

$$3NO_2 (g) + H_2O (\ell) \longrightarrow$$
$$2HNO_3 (\ell) + NO (g) \quad \text{(recycled)}$$

   (b) Calculate the standard enthalpy change accompanying complete conversion of 100 grams of ammonia into nitric acid by this process (neglect recycling).

61. If half of a given sample of ammonia is used to prepare nitric acid as in Exercise 60, the remainder can be used to neutralize the nitric acid to form ammonium nitrate, an important nitrogen fertilizer.

$$NH_3 (g) + HNO_3 (\ell) \longrightarrow NH_4NO_3 (s)$$

Calculate the standard (a) enthalpy and (b) free energy changes at 25°C for the neutralization of the amount of nitric acid prepared in part (b) of Exercise 60. (c) Is the entropy change positive or negative?

62. Standard entropy changes cannot be measured directly in the laboratory. They are calculated from experimentally obtained values of ΔG^0 and ΔH^0. Calculate ΔS^0 at 298 K for each of the following reactions.

	ΔH^0	ΔG^0
(a) $3NO_2 (g) + H_2O (\ell) \rightarrow 2HNO_3 (\ell) + NO (g)$	-71.71 kJ	8.28 kJ
(b) $OF_2 (g) + H_2O (g) \rightarrow O_2 (g) + 2HF (g)$	-323 kJ	-358.4 kJ
(c) $CaC_2 (s) + 2H_2O (\ell) \rightarrow Ca(OH)_2 (s) + C_2H_2 (g)$	-125.4 kJ	-145.4 kJ
(d) $2PbS (s) + 3O_2 (g) \rightarrow 2PbO (s) (red) + 2SO_2 (g)$	-830.8 kJ	-780.9 kJ
(e) $SnO_2 (s) + 2CO (g) \rightarrow 2CO_2 (g) + Sn (s)$	14.8 kJ	5.23 kJ

63. Many organic compounds containing nitro ($-NO_2$) and nitrate ($-O-NO_2$) groups are explosives. An example is nitroglycerin, which undergoes the following explosive decomposition.

$$4C_3H_5(NO_3)_3 (\ell) \longrightarrow$$
$$12CO_2 (g) + 10H_2O (g) + 6N_2 (g) + O_2 (g)$$

(Note the similarity to the explosion reaction of Exercise 57.) At 298 K this reaction releases about 6.28 kJ per gram of nitroglycerin.
   (a) What is ΔH^0 for the reaction in kilojoules per mole of nitroglycerin?
   (b) What is ΔH^0 in kilojoules for the reaction as written?

(c) Using Appendix K, calculate ΔH_f^o for nitroglycerin.

(d) What is suggested by the sign and magnitude of ΔH_f^o for nitroglycerin?

(e) What can you say about the entropy change of the reaction?

(f) What can you say about the temperature range over which the explosion is spontaneous?

(g) How much work in (joules) does the system do on the environment as it expands against the atmosphere if 50.0 grams of nitroglycerin explode at 25°C?

(Most of the damage caused by such an explosion is due to the shock wave resulting from the sudden expansion of the gases produced. The expansion is accelerated by the evolved heat.)

64. Carbon tetrachloride was formerly used widely as a dry cleaning solvent for removal of greasy or oily stains because of its nonpolar character. Its use has been curtailed because it is now a suspected carcinogen. It is still used (with care!) as a reactant or solvent in many organic reactions. It can be prepared by the action of chlorine on chloroform.

$$CHCl_3 \, (\ell) + Cl_2 \, (g) \longrightarrow CCl_4 \, (\ell) + HCl \, (g)$$

(a) Evaluate ΔG^0 at 25°C for this reaction using ΔH_f^o and S^0 values from Appendix K.

(b) Evaluate ΔG for the same reaction at 500°C when all species are in the gas phase, assuming ΔH and ΔS have the same values as at 25°C.

(c) Evaluate ΔG for the gas phase reaction at 1000°C.

(d) At which of these temperatures is the gas phase reaction thermodynamically most favorable?

Chemical Kinetics

In our study of thermodynamics we addressed the question of whether a reaction *can* occur. If a reaction is thermodynamically unfavorable at given conditions, it *will not* occur appreciably under those conditions. If a reaction is thermodynamically favorable, it *can* occur, but this does not guarantee that it *will* occur at an appreciable rate.

Our experience is that different chemical reactions occur at very different rates. For instance, combustion reactions such as the burning of the methane, CH_4, in natural gas, and combustion of *iso*-octane, C_8H_{18}, in gasoline take place very rapidly, sometimes even explosively.

$$CH_4\ (g) + 2O_2\ (g) \longrightarrow CO_2\ (g) + 2H_2O\ (g)$$

$$2C_8H_{18}\ (\ell) + 25O_2\ (g) \longrightarrow 16CO_2\ (g) + 18H_2O\ (g)$$

The rusting of iron is a complicated process that may be represented in simplified form as

$$4Fe\ (s) + 3O_2\ (g) \longrightarrow 2Fe_2O_3\ (s)$$

This is the major reaction that occurs in one's digestive system when an antacid, containing relatively insoluble magnesium hydroxide, neutralizes excess stomach acid. Fortunately, the reaction is spontaneous and also rapid.

On the other hand, a chemical change such as the rusting of iron takes place only very slowly.

The reactions of acids with bases are thermodynamically favorable, and they also occur at very rapid rates. Consider, for example, the reaction of 1 *M* hydrochloric acid with solid magnesium hydroxide. It is thermodynamically spontaneous under standard state conditions, as indicated by the negative ΔG^0 value, and it occurs rapidly.

$$2HCl\ (aq) + Mg(OH)_2\ (s) \longrightarrow MgCl_2\ (aq) + 2H_2O\ (\ell) \qquad \Delta G^0 = -97\ kJ$$

The reaction of diamond, a form of solid carbon, with oxygen is also spontaneous.

$$C(diamond) + O_2\ (g) \longrightarrow CO_2\ (g) \qquad \Delta G^0 = -396\ kJ$$

However, we know from experience that diamonds exposed to air do not disappear by forming carbon dioxide, even over long periods of time. The reaction does not occur at an observable rate at room temperature, or even at body temperature.

We also know that chemical reactions can be slowed down or speeded up by altering the conditions under which they occur. Everyday examples include the preservation of food by refrigeration, the cooking of food at high temperatures (including the use of pressure cookers at higher altitudes), and the much slower rusting of iron in an extremely dry climate.

In this chapter we shall study the general topic of how fast chemical reactions go, and the factors that affect these *rates* — the subject known as **chemical kinetics.** We have two principal reasons for studying reaction rates. The first is a practical reason. If we can identify the factors that affect the rate of a particular reaction, we can use this information *to predict* its rate at some specified conditions. We may even be able to alter the conditions to make the reaction go at the rate we desire — either more rapidly or more slowly. These applications will be more useful if we can describe them with quantitative relationships, so that we can calculate rates or changes in rates for the reactions. So we shall attempt, whenever possible, to express what we observe about reaction rates in the form of mathematical equations. The second reason arises from the connection that has been established between our macroscopic observations about rates of reactions and our theories about what happens at the atomic and molecular level. We shall consider the collision theory of reaction rates and the transition state theory — complementary theories that attempt to provide descriptions of how reactions occur. A study of kinetics from this theoretical viewpoint can give us important information about the molecular and atomic details of reactions. Such theoretical results then may be helpful in understanding or predicting results at the macroscopic level.

15-1   Rates of Reactions

We are all familiar with processes in which something changes with time. As examples, an automobile travels at 40 mi/h, a water tap delivers water at 3 gal/min, or a factory produces 32,000 tires/day. Each of these quantities is called a rate. Each can be numerically expressed as the ratio of the amount of change in some quantity (ΔX) to the time (Δt) required to produce that change:

The term ΔX means $X_{final} - X_{initial}$, just as we have used it previously. In common terms, Δt is described as the amount of time elapsed.

$$\text{Rate} = \frac{\text{change in } X}{\text{change in time}} = \frac{\Delta X}{\Delta t}$$

[Eq. 15-1]

FIGURE 15-1 A driver notes that his odometer reads 21383 miles at 8:05 a.m. After traveling (at a steady speed) until 8:20, he sees that the odometer reads 21393 miles. His rate of travel is calculated as

$$\text{rate} = \frac{\Delta(\text{distance})}{\Delta(\text{time})}$$

$$= \frac{\text{distance}_f - \text{distance}_i}{\text{time}_f - \text{time}_i} = \frac{21393 \text{ mi} - 21383 \text{ mi}}{(8:20 - 8:05)}$$

$$= \frac{10 \text{ mi}}{15 \text{ min}} \times \frac{60 \text{ min}}{1 \text{ hr}} = 40 \text{ mi/h}$$

For example, the rate of travel of an automobile stands for (change in distance)/(change in time). A driver could determine his rate of travel as shown in Figure 15–1. In that illustration, the car travels at a steady speed (constant rate). What if it had stopped at a few traffic signals, and had been driven sometimes faster, sometimes slower, but still covered the same 10 miles in 15 minutes? Then all we could say would be that its *average rate* was 40 mi/h, while its *instantaneous rate* (the rate at which it is traveling at any instant) was changeable. As we shall see, the rate of a chemical reaction usually changes as the reaction progresses, so that we sometimes need to make the distinction between *average rate* and *instantaneous rate*.

To emphasize another point about our numerical description of rates, let us consider the situation in Figure 15–2. Here the meaning of rate is (change in volume of water)/(change in time). As the gardener allows water to run out of the reservoir into the marked bucket, she observes that the bucket, originally empty ($V_i = 0$ gal), fills up to the 5.0 gallon mark ($V_f = 5.0$ gal) in 100 seconds. By the approach already illustrated, we would calculate the rate of water flow through the hose as

$$\text{Rate} = \frac{V_f - V_i}{\text{time elapsed}} = \frac{5.0 \text{ gal} - 0 \text{ gal}}{(100 \text{ s})(1 \text{ min}/60 \text{ s})} = 3.0 \text{ gal/min}$$

Suppose, though, that we find it more convenient to determine the rate by observing the volume of water that is removed from the reservoir in a given time. In that case, as shown in Figure 15–2, the value of V_i is 285.0 gal, while the value of V_f after 100 s is 280.0 gal. Applying the approach we have used before, $V_f - V_i = (280.0 \text{ gal} - 285.0 \text{ gal}) = -5.0 \text{ gal}$, and the rate we calculate is a negative number, -3.0 gal/min. What is wrong here? Of course, there is nothing "wrong" — it is just that what we have actually expressed in this second case is the rate of change of the amount of water *in the reservoir*. Since the amount of water in the reservoir decreases, the rate is a negative number. What we are really trying to express is the rate of water flow through the hose, but we can only measure this in terms of the amount of water present in the bucket (increasing) or in the reservoir (decreasing). If we wish to have the rate of water flow through the hose expressed as a positive number (a reasonable goal), but wish to base this on the reservoir level, then we must do something special, like taking the *absolute value* of the change in the water volume, wherever it

Taking the absolute value of a (real) number means to multiply it by $+1$ if it is a positive number or by -1 if it is a negative number, so that it will come out the same size, but always positive. Thus, the absolute value of 3.56 is 3.56, while the absolute value of -7.4 is 7.4.

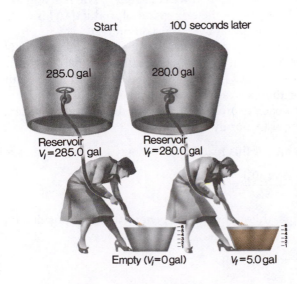

Start   100 seconds later

285.0 gal   280.0 gal

Reservoir   Reservoir
V_i = 285.0 gal V_f = 280.0 gal

Empty (V_i = 0 gal) V_f = 5.0 gal

FIGURE 15–2 Determination of the rate of water flow through a garden hose. At some time, t_i, the bucket is empty, and the reservoir contains 285.0 gallons of water; 100 seconds later, the bucket contains 5.0 gallons, and the reservoir contains 280.0 gallons.

has been. We shall find very similar circumstances when we deal with rates of chemical reactions.

Now let us see how we can express numerically the **rate of a chemical reaction.** In general terms, the question "What is the rate of this reaction" means "How much does this reaction proceed in a given time?" To be more specific, we describe the progress of a reaction either in terms of the amount of reactant consumed or the amount of product that has been formed; usually the amount is expressed as a molar concentration, but sometimes other units are used. The units of reaction rates are thus seen to be moles liter^{-1} time^{-1}; time is often expressed in seconds (mol L^{-1} s^{-1}), though other time units may be desirable for very fast or very slow reactions (e.g., mol L^{-1} ms^{-1} or mol L^{-1} h^{-1}). The rate of a reaction may be determined by measuring the change in concentration of any product or reactant that can be detected conveniently by quantitative means. Methods for making such measurements will be discussed in Section 15–2. Changes in concentrations of products and reactants are related to each other by balanced chemical equations.

We introduce here a notation for concentration that will be used throughout the next several chapters. The symbol [X] stands for the molar concentration of substance X (moles/liter). Whenever we use this notation, the units mol/L are implied.

Let us illustrate the approach first with the gas phase decomposition of dinitrogen tetroxide, N_2O_4:

$$N_2O_4 \text{ (g)} \longrightarrow 2NO_2 \text{ (g)}$$

The rate of the reaction will be expressed in terms of the change in some concentration per unit time. Two special features must be accounted for:

1. Suppose we use the measured change in concentration of product, NO_2, to express the rate. Then, since the amount of product increases with elapsed time, $[NO_2]_f$ is greater than $[NO_2]_i$, and the rate would be a positive number. If, however, we choose to use the measured change in concentration of reactant to express the rate, then $[N_2O_4]_f$ would be *less than* $[N_2O_4]_i$ (its concentration is *decreasing*), and the rate would appear to come out negative. In order to be consistent, we multiply the expression $\Delta[\text{reactant}]/\Delta t$ by -1. This makes it a positive number, so that we could say, "The reaction is progressing by so many mol L^{-1} s^{-1}." Since the expression $\Delta[\text{product}]/\Delta t$ is already positive, we leave it alone (or multiply it by $+1$, if you prefer).

2. In this reaction, the coefficients of the balanced equation tell us that for every *one* mole of N_2O_4 that disappears, *two* moles of NO_2 are formed. Thus, if we use the expression $\Delta[NO_2]/\Delta t$ for the rate, we would have a number *twice as great* as if we use the expression $-\Delta[N_2O_4]/\Delta t$. We take care of this inconsistency by dividing each expression by the stoichiometric coefficient of that substance (i.e., by 1 for N_2O_4, or by 2 for NO_2).

The rate of this reaction, then, may be expressed as follows:

$$\text{Rate} = -\frac{\Delta[N_2O_4]}{\Delta t} = \frac{\Delta[NO_2]}{2\Delta t}$$

To put this approach in general terms, consider the general hypothetical reaction

$$a\text{A} + b\text{B} \longrightarrow c\text{C} + d\text{D}$$

In reactions involving gases, rates of reaction may be related to rates of change of partial pressures, since pressures and concentrations of gases are directly proportional.

When we write $[CO_2] =$ 1.75 M, we mean that the concentration of CO_2 is 1.75 mol/L.

Compare the reasoning here to the earlier illustration involving the garden hose. Reactants are like the water in the reservoir, while products are like the water in the gardener's bucket.

In such a general representation of a reaction, the capital letters represent substances, while the lower-case letters represent their stoichiometric coefficients.

FIGURE 15–3 A plot of concentrations of all reactants and products in the reaction of 2.000 M ICl with 1.000 M H_2 vs. time.

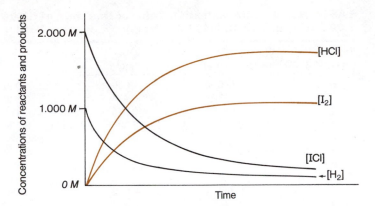

The rate at which the reaction proceeds can be measured in terms of the rate at which one of the reactants disappears or one of the products appears.

$$\text{Rate of reaction} = \underbrace{\frac{-\Delta[A]}{a\Delta t} = \frac{-\Delta[B]}{b\Delta t}}_{\substack{\text{Rates of decrease of} \\ \text{conc'n of reactants}}} = \underbrace{\frac{\Delta[C]}{c\Delta t} = \frac{\Delta[D]}{d\Delta t}}_{\substack{\text{Rates of increase of} \\ \text{conc'n of products}}} \qquad \text{[Eq. 15–2]}$$

This representation gives several equalities, any one of which can be used to relate changes in observed concentrations to the rate of reaction.

As a more detailed example of the application of these ideas, consider the gas-phase reaction carried out starting with 2.000 moles of iodine chloride, ICl, and 1.000 mole of hydrogen in a closed 1.000 liter container at 230°C.

$$2ICl\,(g) + H_2\,(g) \longrightarrow I_2\,(g) + 2HCl\,(g)$$

$$\text{Rate of reaction} = \frac{-\Delta[H_2]}{\Delta t} = \frac{-\Delta[ICl]}{2\Delta t} = \frac{\Delta[I_2]}{\Delta t} = \frac{\Delta[HCl]}{2\Delta t} \qquad \text{[Eq. 15–3]}$$

Since the balanced chemical equation tells us that two moles of ICl are consumed per one mole of H_2 consumed, we must divide the rate of disappearance of ICl by 2 in order to equate it to the rate of disappearance of H_2. Similarly, the equal coefficients of H_2 and I_2 tell us that I_2 appears at the same molar rate that H_2 is consumed. Finally, the coefficient of HCl in the balanced equation tells us that for each mole of H_2 that disappears, two moles of HCl will be formed; thus, we need to divide the rate of appearance of HCl by 2.

Table 15–1 lists the concentrations of reactants remaining and products present at one-second intervals, beginning with the time of mixing ($t = 0$ seconds). Be sure to note that all quantities in this table *could* have been observed. In fact, it would be necessary only to keep track of the time (first column) and measure the concentration of *one* of the reactants or products, if we knew the starting concentrations. The starting concentrations were $[ICl] = 2.000\ M$, $[H_2] = 1.000\ M$, $[I_2] = [HCl] = 0\ M$. Figure 15–3 shows graphically how the concentrations of all substances involved change with time.

Now we may determine the rate of the reaction in terms of any one of the reactants or products. If we choose to express the rate as $-\Delta[H_2]/\Delta t$, we can work out the numbers in the "Rate" column of Table 15–2. For instance, you might see if you can express the rate in terms of the concentration of HCl.

You should verify for yourself, using the numerical data of Table 15–2, that the rate of loss of ICl is twice that for H_2, and that the same numerical values for the rate can be determined from any of the equalities of Eq. 15–3.

TABLE 15–1   Concentration Data for Reaction of 2.000 *M* ICl and 1.000 *M* H$_2$ at 230°C: 2ICl (g) + H$_2$ (g) $\longrightarrow$ I$_2$ (g) + 2HCl (g)

Time (*t*) (seconds)	[ICl] (mol/L)	[H$_2$] (mol/L)	[I$_2$] (mol/L)	[HCl] (mol/L)
0	2.000	1.000	0.000	0.000
1	1.348	0.674	0.326	0.652
2	1.052	0.526	0.474	0.948
3	0.872	0.436	0.564	1.128
4	0.748	0.374	0.626	1.252
5	0.656	0.328	0.672	1.344
6	0.586	0.293	0.707	1.414
7	0.530	0.265	0.735	1.470
8	0.484	0.242	0.758	1.516
9	0.444	0.222	0.778	1.556
10	0.412	0.206	0.794	1.588
11	0.384	0.192	0.808	1.616
12	0.360	0.180	0.820	1.640
13	0.338	0.169	0.831	1.662
14	0.320	0.160	0.840	1.680
15	0.304	0.152	0.848	1.696
16	0.288	0.144	0.856	1.712

Let us look more closely at the variation of the concentration of hydrogen versus time, as shown graphically in Figure 15–4. The rate of reaction at any time *t* is the negative of the slope of the tangent to the curve at time *t*.

> The *slope* of a line is the change in the quantity plotted along the vertical (*y*) axis for a specified segment of the line, divided by the corresponding change along the horizontal (*x*) axis; that is, $\Delta y / \Delta x$ = slope. We must be careful to determine the change in the same direction for both variables. For any curve that is not a straight line, the slope of the curve at any point is defined to be the slope of the straight line that is tangent to the curve at that point. If you've had calculus, you may recognize this as the derivative of the concentration function, or $d[H_2]/dt$. In this case, the slope of the line (the tangent) is negative because [H$_2$] decreases as time passes (increases).

The slope of the tangent to the curve at *t* = 3.5 seconds (for example) can be approximated by dividing the change in [H$_2$] in the interval between 3.0 and 4.0 seconds by 1.0 second, the corresponding change in time. Since the rate is changing, even slightly, during this period, what we would calculate is the *average rate* for this time period. As the interval used for this calculation gets shorter, the average rate for that time period becomes a better approximation of the *instantaneous rate*.

Refer back to the discussion of average vs. instantaneous rate in the automobile example earlier in this section.

The reaction rate is seen to decrease as time goes on. This is because the set of concentrations of reactants is one of the factors that affects reaction rate. As the reaction proceeds, some ICl and H$_2$ are used up, so that lower concentrations of ICl and H$_2$ result in fewer collisions and less chance for reaction. We shall consider this concentration effect, both experimentally and theoretically, in more detail in Sections 15–4 and 15–5.

The rate of the reaction at the instant of mixing the reactants is called the **initial rate,** and it is equal to the negative of the slope of the [H$_2$] curve at *t* = 0. Had

TABLE 15-2 [H₂] Data and Derived Rate Data for the Reaction in Table 15-1

[H₂] (mol/L)	Rate = $\dfrac{-\Delta[H_2]}{\Delta t}$ (mol/L · s)	Time (t) (seconds)
1.000		0
	0.326	
0.674		1
	0.148	
0.526		2
	0.090	
0.436		3
	0.062	
0.374		4
	0.046	
0.328		5
	0.035	
0.293		6
	0.028	
0.265		7
	0.023	
0.242		8
	0.020	
0.222		9
	0.016	
0.206		10
	0.014	
0.192		11
	0.012	
0.180		12
	0.011	
0.169		13
	0.009	
0.160		14
	0.008	
0.152		15
	0.008	
0.144		16

FIGURE 15-4 A plot of H₂ concentration versus time for the reaction of 2.000 M ICl with 1.000 M H₂ at 230°C from data in Table 15-1. The rate of reaction, which decreases with time, equals the negative of the slope of the tangent to the curve.

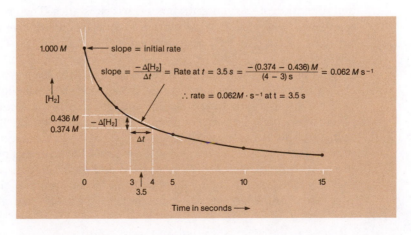

slope = initial rate

$$\text{slope} = \frac{-\Delta[H_2]}{\Delta t} = \text{Rate at } t = 3.5\,s = \frac{-(0.374 - 0.436)\,M}{(4-3)\,s} = 0.062\,M\,s^{-1}$$

$$\therefore \text{rate} = 0.062\,M \cdot s^{-1} \text{ at } t = 3.5\,s$$

1.000 M

[H₂]

0.436 M
0.374 M

$-\Delta[H_2]$

Δt

0   3   4   5   10   15
         3.5

Time in seconds ⟶

we looked at a plot of the concentration of a product, I_2 or HCl, the rate would have been related (how?) to the (positive) slope of the tangent at $t = 0$.

15–2   Methods of Following the Progress of a Reaction

As we saw in Section 15–1, two kinds of observations must be made in order to describe numerically the rate at which a reaction proceeds. One of these is the familiar measurement of time. Experimental measurement of the other quantity, concentration of one or more of the substances involved in the reaction, may be accomplished in a wide variety of ways, some of which we shall mention in this section.

Many methods involve observations of *physical properties* of the reaction mixture. As a simple example of this, we could determine the rate of burning of a candle simply by measuring the height of the candle at various times during the course of the reaction.

Some reactions involve the net production or consumption of gases. The progress of such reactions may be followed by measuring the pressure of the reaction mixture at constant volume (or, alternatively, the volume of the reaction mixture at constant pressure). For example, this approach might be used to follow the reaction N_2O_4 (g) $\rightarrow$ $2NO_2$ (g), because *one* mole of gaseous reactant produces *two* moles of gaseous product. Thus, as the reaction proceeds (at constant volume) the pressure would increase; observing this pressure could provide a measure of the progress of the reaction, and partial pressures thus obtained (Chapter 11) could allow calculation of concentrations of all reactants and products. This approach depends on the fact that, in the balanced chemical equation, the *total number of moles* of gaseous reactants is *different* from the *total number of moles* of gaseous products. Such a method could not be used to monitor the progress of a reaction such as H_2 (g) $+$ Cl_2 (g) $\rightarrow$ $2HCl$ (g), in which two moles of gaseous products are formed for every two moles of gaseous reactants consumed.

Sometimes one of the substances involved in a reaction has a distinctive color. The increase in intensity of the color that is characteristic of the particular product (or decrease in color due to a reactant) can then provide a method of following the progress of the reaction. This is a special case of the widely used *spectroscopic* class of methods. These involve the measurement of the amount of electromagnetic radiation (visible, infrared, ultraviolet) absorbed by a reactant or product, which allows determination of the concentration of that substance. The use of lasers has permitted detection of very low concentration levels; some of these methods can detect substances that may be present for extremely short times — nanoseconds or even picoseconds.

Other physical methods that have been used to study reaction kinetics involve electrical measurements. For instance, a reaction mixture in which ions are produced or used up in solution may exhibit changes in electrical conductivity. Since cell potentials depend on concentrations of substances (Section 19–20), the change in voltage of an electrochemical cell may be used as a measure of the amount of a particular substance present (or of the ratio of the amounts of two substances).

Physical methods, such as those mentioned here, have the advantage that they can be continuously monitored without interrupting or disturbing the progress of the reaction. They do, however, require that some correlation be established

An interesting twist on this has been to use the height of the candle (after calibration) as a means of measuring *time*.

$1 \text{ ns} = 1 \times 10^{-9}$ s,
$1 \text{ ps} = 1 \times 10^{-12}$ s

between the quantity being measured and the concentration of one or more substances involved in a reaction.

The progress of a reaction also may be followed by *chemical methods.* For instance, it may be possible to withdraw a small sample of the reaction mixture and then to "quench" (stop) the reaction by cooling it rapidly (Section 15–7). The concentrations present in this small sample are then characteristic of those in the larger reaction mixture at the time of quenching. Alternatively, the reaction can be effectively stopped by diluting a very small sample (to change all concentrations by a known amount—Section 15–4), and then the sample can be analyzed for the substances of interest. Such chemical methods have the disadvantage that they do not provide a means of continuously monitoring concentrations because some time elapses between withdrawal of the sample and completion of the analysis.

Factors Affecting Reaction Rates

There are four factors that have marked effects on the rates of chemical reactions. These are (1) nature of reactants, (2) concentrations of reactants, (3) temperature, and (4) presence of a catalyst. A description of the effects of these four factors on reaction rates provides information that can help us control the rate of a reaction. In addition, a study of these factors can give considerable insight into the molecular processes by which a reaction occurs, and is the basis for development of theories of chemical kinetics. The remainder of this chapter will be devoted to the study of these factors and presentation of the resulting theories—collision theory and transition state theory.

15–3   Nature of Reactants

The rate at which a reaction occurs depends on the chemical identities and physical states of the reactants. As we have seen, solutions of an acid and a base react very rapidly when they are mixed, as does metallic sodium when it contacts water. However, metallic calcium reacts slowly with water at room temperature.

The physical states of reacting substances are important in determining their reactivities. White phosphorus and red phosphorus are different solid forms (allotropes) of elemental phosphorus. White phosphorus ignites when exposed to oxygen of the air. By contrast, red phosphorus may be kept in open containers for long periods of time.

The state of subdivision of solids can be crucial in determining reaction rates. Large chunks of most metals do not burn, but many powdered metals with larger surface areas, and consequently more atoms exposed to the oxygen of the air, burn easily. The effects of large surface area on rates of combustion are violently exemplified by the dust explosions that can occur in grain elevators, coal mines, and chemical plants in which large amounts of powdered oxidizable substances are produced. Similarly, we find that the reaction between substances in two solutions is much faster if the solutions are well mixed than if the solids themselves are mixed. Samples of dry solid potassium sulfate, K_2SO_4, and dry solid barium nitrate, $Ba(NO_3)_2$, can be mixed with no appreciable reaction occurring over a period of several years. But, if solutions of the two in reasonable concentrations are mixed, a reaction occurs rapidly, as evidenced by the immediate formation of a white precipitate, barium sulfate:

$$Ba^{2+} (aq) + SO_4^{2-} (aq) \longrightarrow BaSO_4 (s)$$

Additionally, it is observed that spontaneous reactions involving ions in solutions are generally much more rapid than reactions in which covalent bonds must be broken.

Only general remarks are possible regarding the effects of "nature of reactants" on reaction rates, but such considerations are of great importance in actually carrying out a reaction at a desirable rate, whether in the laboratory, in an industrial setting, or in nature.

15–4   Concentration: The Rate-Law Expression

A second factor that has a marked effect on the rate of a reaction is the concentrations of reactants. We have already seen in our discussion of the reaction of ICl with H_2 that the reaction slows down as time passes, because the reactants are being used up. In terms of molecular behavior, we can rationalize this dependence by observing that reactant molecules will undergo the collisions necessary for reaction much more often if there are more molecules in a given volume (higher concentration). This idea, which is part of the *theoretical implication* of kinetics, will be more fully discussed in Section 15–6. At this point we must emphasize that we can study the rates *experimentally* without understanding the theory. Indeed, our theories of kinetic behavior in terms of molecules are, like all other theories in chemistry, nothing more than our attempts to explain what we have already observed. With this in mind, let us now see how we can first observe and then describe how the rate of a reaction changes when we change the concentrations of reactants.

We observe that the rate decreases as concentrations of reactants decrease. The quantitative expression of this relationship is that *rate is proportional to reactant concentrations raised to some powers.* For the general hypothetical reaction

$$aA + bB \longrightarrow cC + dD$$

this expression has the general form

Rate $= k[A]^x[B]^y$ [Eq. 15–4]

This is called the **rate-law expression** for this reaction. As before, [A] and [B] represent the concentrations of reactants A and B at some particular time. After the reaction has begun, these will be smaller than the initial concentrations. The powers to which the concentrations are raised, x and y, are usually integers or zero, but are occasionally fractions. *The values of these exponents must be determined experimentally, and they have no necessary relationship to the coefficients in the balanced chemical equation for the reaction.* An exponent of 1 means that the rate is directly proportional to the concentration of that substance. For example, for the reaction

$$N_2O_4 (g) \longrightarrow 2NO_2 (g)$$

the rate-law expression *is found experimentally* to be

Rate $= k[N_2O_4]$

(i.e., exponent of 1). This means, for instance, that when $[N_2O_4] = 3.00$ mol/L,

As in mathematics, a quantity appearing without an exponent is understood to have an exponent of 1. When the concentration of N_2O_4 is doubled, the rate increases by a factor of 2^1, or 2.

there would be twice as many moles per liter of NO_2 being formed per second as when $[N_2O_4] = 1.50$ mol/L. On the other hand, for the reaction

$$3NO\ (g) \longrightarrow N_2O\ (g) + NO_2\ (g)$$

the experimentally determined rate-law expression is

$$\text{Rate} = k[NO]^2$$

For this reaction, when the concentration of NO is doubled, the rate goes up by a factor of 2^2, or 4.

(i.e., exponent of 2). Now the rate increases as the *square* of the reactant concentration. For instance, when $[NO] = 3.00$ mol/L, the rate would be *four times* as great (*four times* as many moles per liter of product being formed per second) as when $[NO] = 1.50$ mol/L, i.e., $2^2 = 4$.

Referring to Eq. 15–4, the value of x is said to be the **order** of the reaction with respect to A, while y is the order of the reaction with respect to B. The **overall order** of the reaction is $x + y$. Some examples follow.

$$3NO\ (g) \longrightarrow N_2O\ (g) + NO_2\ (g)$$

$$\text{Rate} = k[NO]^2 \qquad \text{second order in NO; second order overall}$$

$$2NO_2\ (g) + F_2\ (g) \longrightarrow 2NO_2F\ (g)$$

$$\text{Rate} = k[NO_2][F_2] \qquad \text{first order in } NO_2 \text{ and } F_2; \text{ second order overall}$$

$$H_2O_2\ (aq) + 3I^-\ (aq) + 2H^+\ (aq) \longrightarrow 2H_2O\ (\ell) + I_3^-\ (aq)$$

Any number raised to the *zero* power is equal to 1. When a reaction is "zero order" in a reactant, its rate is independent of the concentration of that reactant.

$$\text{Rate} = k[H_2O_2][I^-] \qquad \text{first order in } H_2O_2 \text{ and } I^-; \text{ zero order in } H^+; \text{ second order overall}$$

The quantity k is called the **specific rate constant** (or sometimes just the rate constant) for the reaction at a particular temperature. Be sure to remember the following points about the meaning of k: (1) The value of k is for a *specific reaction*. Thus, the symbol k, as it is used in the three examples above, refers to three *different numerical values*. (2) The value of k for a reaction refers to that reaction at a *particular temperature*. If we were to change the temperature of the reaction, the value of k would change, as we shall see in Section 15–7. (3) The value of k for a reaction *does not change with time*. (4) The value of k for a reaction *does not change with concentrations* of either reactants or products. (5) The value of k *does* depend on whether a *catalyst* is present, as we shall discuss in Section 15–9. (6) The value of k for a reaction *must be determined experimentally* for that reaction at those particular conditions. (7) The units of k depend on the *order* of the reaction. These units can always be determined from the rate-law expression, since k must convert the concentration units (M), raised to the power equal to the overall order, into the units of rate ($M \cdot \text{time}^{-1}$). For example, a reaction that is second order overall would be described by a rate-law expression such as

The slowing of a reaction as it proceeds is obviously due, then, to the decrease in concentrations of reactants.

$$\text{Rate} = k(\text{concentration})^2$$

Putting in the units for all except the specific rate constant,

$$M \cdot \text{time}^{-1} = k\ M^2$$

Thus, the units of k must be $M^{-1} \cdot \text{time}^{-1}$, so that

$$M \cdot \text{time}^{-1} = (M^{-1} \cdot \text{time}^{-1})(M^2)$$

The use of a known rate-law expression is illustrated in the following example.

Example 15–1

The decomposition of N_2O_5 in carbon tetrachloride, CCl_4, solution,

$$2N_2O_5 \longrightarrow 4NO_2 + O_2$$

at 45°C is found experimentally to follow the rate-law expression

$$\text{Rate} = k[N_2O_5]$$

The value of the specific rate constant at this temperature is found to be 6.2×10^{-4} s^{-1}. (a) What is the order of the reaction? (b) What would be the rate of the reaction when $[N_2O_5] = 1.50$ M? (c) What would be the rate of the reaction when $[N_2O_5] = 0.50$ M? (d) What concentration of N_2O_5 would be necessary if we wish to have the reaction rate be 1.3×10^{-3} $M \cdot$s^{-1}?

Solution

(a) Since the exponent of $[N_2O_5]$ is 1, the reaction is said to be <u>first order in N_2O_5</u>. No other substance appears in the rate expression, so the reaction is said to be <u>first order overall.</u> (Even if the rate-law expression had not been given, we could have deduced that the reaction was first order overall from the units, s^{-1}, of the rate constant.)

(b) Substituting the known values for k and $[N_2O_5]$ into the known rate-law expression, we obtain

$$\text{Rate} = (6.2 \times 10^{-4}\ \text{s}^{-1})(1.50\ M)$$
$$= \underline{9.3 \times 10^{-4}\ M \cdot \text{s}^{-1}}$$

(c) Remember that the rate constant doesn't depend on reactant concentrations. Substituting the value for k and the new value for $[N_2O_5]$, we obtain

$$\text{Rate} = (6.2 \times 10^{-4}\ \text{s}^{-1})(0.50\ M)$$
$$= \underline{3.1 \times 10^{-4}\ M \cdot \text{s}^{-1}}$$

Since we know that the reaction is first order, we should expect that multiplying the concentration of the reactant by $\frac{1}{3}$ would multiply the rate by $(\frac{1}{3})^1$, as confirmed by our calculations.

(d) Solving the rate-law expression for concentration, we obtain

$$[N_2O_5] = \frac{\text{Rate}}{k}$$

Using the known value for k and the desired value for rate, we can calculate the required concentration:

$$[N_2O_5] = \frac{(1.3 \times 10^{-3}\ M \cdot \text{s}^{-1})}{(6.2 \times 10^{-4}\ \text{s}^{-1})} = \underline{2.1\ M}$$

We should emphasize again here that the *only way* that we can *obtain* a rate-law expression is from *experimental data*; this expression can *never* be predicted just on the basis of an overall balanced chemical equation.

Let us now consider some examples that show how the rate-law expression could be derived from experimental data. A common approach is called the **method of initial rates.** In this method, a reaction is carried out several times, each time with different starting concentrations of reactants. For each run, the observed concentration vs. time data are analyzed as was illustrated in Section 15–1, to determine the *initial rate* of the reaction for those concentrations. Using the initial rate data for these trial runs, we can calculate the order and rate constant of the reaction, as we next demonstrate.

The tabulated data below represent typical experimentally measured kinetic data and refer to the hypothetical reaction

$$A_2 + 2B \longrightarrow 2AB$$

at a specific temperature. The brackets refer to the concentrations of the reacting

Trial Run	Initial $[A_2]$	Initial $[B]$	Initial Rate of Formation of AB ($M \cdot$s^{-1})
1	1.0×10^{-2} M	1.0×10^{-2} M	4.0×10^{-4}
2	1.0×10^{-2} M	2.0×10^{-2} M	4.0×10^{-4}
3	2.0×10^{-2} M	3.0×10^{-2} M	8.0×10^{-4}

species *at the beginning* of the experimental runs listed in the first column, i.e., the initial concentrations for each trial reaction.

Since we are describing the same reaction in each trial run, each is governed by the same rate law,

$$Rate = k[A_2]^x[B]^y$$

We observe that the initial concentration of A_2 in trials 1 and 2 is the same, so for these trials any change in reaction rate is due to different initial concentrations of B. We can evaluate y by solving the ratio of the rate expressions for the two trial runs for y. Note that the operation involves dividing both sides of an equation by terms that are equal.

$$\frac{Rate_{(1)}}{Rate_{(2)}} = \frac{k[A_2]^x_{(1)}[B]^y_{(1)}}{k[A_2]^x_{(2)}[B]^y_{(2)}}$$

The value of k always cancels from such ratios, since it is constant at any particular temperature. For trials 1 and 2, the initial concentrations of A_2 are equal, so they too cancel. Thus, the expression simplifies to

$$\frac{Rate_{(1)}}{Rate_{(2)}} = \left(\frac{[B]_{(1)}}{[B]_{(2)}}\right)^y$$

The only unknown in this equation is y. Substituting the appropriate values from the data for trials 1 and 2 into the equation, we get

$$\frac{4.0 \times 10^{-4}}{4.0 \times 10^{-4}} = \left(\frac{1.0 \times 10^{-2}}{2.0 \times 10^{-2}}\right)^y$$

$n^0 = 1$ for any value of n.

$$1.0 = (0.5)^y \quad \text{and} \quad \underline{y = 0}$$

Since the units of $Rate_{(1)}$ and $Rate_{(2)}$ are identical, they obviously cancel. Similarly, the units of $[B]_{(1)}$ and $[B]_{(2)}$ are identical, and they too cancel. Both sets of units have been omitted to simplify the mathematical expression. Because any number raised to the zero power equals one, y must be zero. This result indicates that the initial concentration of B is raised to the zero power in the rate expression, and that the rate at which the reaction occurs is *independent* of the concentration of B. Thus far we know that the rate expression is

$$Rate = k[A_2]^x[B]^0 \quad \text{or} \quad Rate = k[A_2]^x$$

Next we shall evaluate x. In trial runs 1 and 3, the initial concentration of A_2 is doubled and the rate doubles. We may neglect the initial concentration of B, since we have just shown that it does not affect the rate of the reaction.

$$\frac{Rate_{(3)}}{Rate_{(1)}} = \frac{k[A_2]^x_{(3)}}{k[A_2]^x_{(1)}} = \left(\frac{[A_2]_{(3)}}{[A_2]_{(1)}}\right)^x$$

$$\frac{8.0 \times 10^{-4}}{4.0 \times 10^{-4}} = \left(\frac{2.0 \times 10^{-2}}{1.0 \times 10^{-2}}\right)^x$$

$n^1 = n$ for any value of n.

$$2.0 = (2.0)^x \quad \text{and} \quad \underline{x = 1}$$

The power to which $[A_2]$ is raised in the rate-law expression is one, and we know that the rate law for this reaction is

$$Rate = k[A_2]^1 = \underline{k[A_2]}$$

The specific rate constant, k, can be evaluated by substituting any of the three sets of data into the rate-law expression. Substitution of the data from trial run 1 gives

$$\text{Rate}_{(1)} = k[A_2]_{(1)}$$

$$4.0 \times 10^{-4}\ M \cdot s^{-1} = k(1.0 \times 10^{-2}\ M)$$

$$k = \frac{4.0 \times 10^{-4}}{1.0 \times 10^{-2}}\ s^{-1} = \underline{4.0 \times 10^{-2}\ s^{-1}}$$

At the temperature at which the measurements were made, the rate-law expression for this reaction is

$$\text{Rate} = k[A_2] \quad \text{or} \quad \text{Rate} = \underline{4.0 \times 10^{-2}\ s^{-1}\ [A_2]}$$

We may check our calculations by evaluating k from one of the other sets of data. Using the data from trial run 3 in the rate expression gives

$$\text{Rate}_{(3)} = k[A_2]_{(3)}$$

$$8.0 \times 10^{-4}\ M \cdot s^{-1} = k(2.0 \times 10^{-2}\ M)$$

$$k = \frac{8.0 \times 10^{-4}}{2.0 \times 10^{-2}}\ s^{-1} = 4.0 \times 10^{-2}\ s^{-1}$$

This is the same value for k obtained above, and verifies the earlier calculation.

Once we have determined the rate-law expression, including the value of the specific rate constant, k, we can use it to predict rates of reaction at other concentrations. We must realize, though, that the value of k applies only to that reaction at that temperature. Such a use of the rate law is shown in Example 15–2.

Example 15–2

For the following reaction the rate-law expression has been determined experimentally to be Rate = $k[A]^2[C]$, with $k = 3.0 \times 10^{-4}\ M^{-2} \cdot min^{-1}$. For this reaction (a) determine the initial rate of the reaction if we start with concentrations $[A] = 0.10\ M$, $[B_2] = 0.35\ M$, $[C] = 0.25\ M$, and (b) determine the rate after 0.04 mol/L of A has reacted.

$$2A + B_2 + C \longrightarrow A_2B + BC$$

Solution

(a) Rate = $k[A]^2[C]$

$$= (3.0 \times 10^{-4}\ M^{-2} \cdot min^{-1}) \times$$
$$(0.10\ M)^2(0.25\ M)$$
$$= \underline{7.5 \times 10^{-7}\ M \cdot min^{-1}}$$

(b) Looking at the balanced chemical equation, we can see that when 0.04 mol/L of A react, 0.02 mol/L of B_2 and 0.02 mol/L of C must also react. (This is a 2:1:1 mole ratio.) This allows us to calculate the concentrations of all three reactants at this later time:

	2A	+ B_2	+ C → prod
Starting concentrations:	0.10 M	0.35 M	0.25 M
Change:	−0.04 M	−0.02 M	−0.02 M
Remaining:	0.06 M	0.33 M	0.23 M

Putting these concentrations into the rate expression:

$$\text{Rate} = k[A]^2[C]$$

$$= (3.0 \times 10^{-4}\ M^{-2} \cdot min^{-1})\ (0.06\ M)^2(0.23\ M)$$

$$= \underline{2 \times 10^{-7}\ M \cdot min^{-1}}$$

As the reaction proceeds, the concentrations of reactants decrease, and this causes the rate of the reaction to decrease.

The value of $[B_2]$ does not enter into the rate expression. Of course, some B_2 must be present for the reaction to proceed at all, but as long as any B_2 is there, changing its concentration would not cause the reaction to speed up or slow down.

In Section 15–8 we shall see how experimentally determined rate-law expressions can help us to postulate some of the detailed molecular processes by which reactions occur. Let us now describe another way to represent experimental information about rates.

15–5 Concentration: The Integrated Rate Equation

As we saw in Section 15–4, the rate-law expression for a reaction gives a relationship between the rate of a reaction and the concentrations of its reactants. Sometimes it is desirable to calculate the concentration of one of the reactants that would remain after some specified time, or how long it would take for a specified fraction of the reactants to be used up.

It is possible to reformulate the rate-law expression for a reaction into an equation that relates reactant concentration and time. This equation, called the **integrated rate equation** (or sometimes just the **rate equation**), allows the kinds of calculations just mentioned. The mathematical form of the integrated rate equation is different for reactions of different order. In this section, we shall consider two of the most common types of reactions, those that are overall first order, and those that are second order with respect to a particular reactant and second order overall.

For a reaction that is first order with respect to a particular reactant, A, and first order overall, the rate-law expression is

Rate = $k[A]$

The integrated rate equation, which can be derived from this rate law, is

For interested students with experience in calculus: Eq. 15–5 can be derived by integrating a differential form of the rate law expression.

$$\log\left(\frac{[A]_0}{[A]}\right) = \frac{kt}{2.303} \qquad \text{(First order)} \qquad \text{[Eq. 15–5]}$$

In this equation, $[A]$ is the concentration of A at some time, t, after the reaction begins, and $[A]_0$ is the initial concentration of A. (The logarithm is to the base 10.*) If we solve this relationship for t we get

$$t = \log\left(\frac{[A]_0}{[A]}\right) \times \frac{2.303}{k} \qquad \text{(First order)} \qquad \text{[Eq. 15–6]}$$

For reactions known to be first order, and for which the value of k has been determined, the integrated rate equation can be used to calculate the amounts of reactants remaining (and products formed) after some specified time, t, has elapsed. Example 15–3 illustrates such a calculation.

---

* If the *natural logarithm* (base e) is used, the factor 2.303 is omitted, i.e., $\ln\left(\frac{[A]_0}{[A]}\right) = kt$

Example 15–3

The gas phase reaction

$2N_2O_5 (g) \longrightarrow 4NO_2 (g) + O_2 (g)$

obeys the first order rate law

Rate = $k[N_2O_5]$

in which the specific rate constant is $1.68 \times 10^{-2}\,s^{-1}$ at a certain temperature. If 2.50 moles of N_2O_5 (g) are placed in a 5.00 liter container at that temperature, how many moles of N_2O_5 would remain after 1.00 minute? How much O_2 would have been produced?

Solution

We may apply the relationship given in Eq. 15–5:

$$\log\left(\frac{[N_2O_5]_0}{[N_2O_5]}\right) = \frac{kt}{2.303}$$

First we must determine the original concentration of N_2O_5:

$$[N_2O_5]_0 = \frac{2.50 \text{ mol}}{5.00 \text{ L}} = 0.500 \ M$$

The only unknown is $[N_2O_5]$ after 1.00 minute. Let us solve for the unknown. Since $\log x/y = \log x - \log y$,

$$\log [N_2O_5]_0 - \log [N_2O_5] = \frac{kt}{2.303}$$

$$\log [N_2O_5] = \log [N_2O_5]_0 - \frac{kt}{2.303}$$

$$= \log (0.500) - \frac{(1.68 \times 10^{-2} \text{ s}^{-1})(60.0 \text{ s})}{2.303}$$

$$= -0.301 - 0.438$$

$$\log [N_2O_5] = -0.739$$

Taking the antilogarithm of both sides,

$$[N_2O_5] = 10^{-0.739} = 10^{+0.261-1.000}$$

$$= 10^{+0.261} \times 10^{-1.000}$$

$$[N_2O_5] = 1.82 \times 10^{-1} \ M$$

Thus, after 1.00 minute of reaction, the concentration of the reactant is 0.182 M N_2O_5. Now we convert this to moles of unreacted N_2O_5 and compute the amount of O_2 produced.

unreacted:

$$? \text{ mol } N_2O_5 = 5.00 \text{ L} \times \frac{1.82 \times 10^{-1} \text{ mol}}{L}$$

$$= 0.910 \text{ mol } N_2O_5$$

reacted:

$$? \text{ mol } N_2O_5 = 2.50 \text{ mol} - 0.91 \text{ mol}$$

$$= 1.59 \text{ mol } N_2O_5$$

produced:

$$? \text{ mol } O_2 = 1.59 \text{ mol } N_2O_5 \times \frac{1 \text{ mol } O_2}{2 \text{ mol } N_2O_5}$$

$$= 0.795 \text{ mol } O_2$$

We have solved the integrated rate equation, Eq. 15–5, for [A]. It is also possible to solve Eq. 15–5 for $[A]_0$, or for t, or for k, provided that the other quantities are known or can be calculated from given information. Because of possible experimental inaccuracies in measuring concentrations, however, it is usually better to use concentrations measured at several times to derive the value of k by a graphical approach, to be illustrated later in this section.

A very convenient way to express the rate of a reaction is in terms of the **half-life** (sometimes called the half-time), defined as *the time it takes for half of a specified amount of reactant to be consumed.* That is, when $t = t_{1/2}$, the concentration of A will be half its original concentration, or $[A] = \frac{1}{2}[A]_0$. This means that we can substitute $t_{1/2}$ for t and $\frac{1}{2}[A]_0$ for [A] in Eq. 15–6:

$$t_{1/2} = \log\left(\frac{[A]_0}{\frac{1}{2}[A]_0}\right) \times \frac{2.303}{k}$$

$$= \log (2) \times \frac{2.303}{k}$$

$$= 0.301 \times \frac{2.303}{k}$$

$$t_{1/2} = \frac{0.693}{k} \qquad \text{(First order)} \qquad \text{[Eq. 15–7]}$$

This is the relationship between the half-life of a reactant in an overall first order reaction and its rate constant k. Note that in such reactions the half-life is independent of the initial concentration of A. But this is not the case in all types of reactions, as we shall see.

Example 15-4

Compound A decomposes to form B and C in a reaction that is first order with respect to A and first order overall. At 25°C the specific rate constant for the reaction is 0.0450 s⁻¹. What is the half-life of A at 25°C?

$$A \longrightarrow B + C$$

Solution

We substitute the given data into Eq. 15-7:

$$t_{1/2} = \frac{0.693}{k} = \frac{0.693}{0.0450 \text{ s}^{-1}} = 15.4 \text{ s}$$

After 15.4 seconds of reaction, half of the original reactant remains.

For a reaction that is second order with respect to reactant A and second order overall, we know that the rate-law expression (Section 15-4) is

$$\text{Rate} = k[A]^2$$

Again, using calculus, we can derive from this the rate equation that relates concentration and time:

$$kt = \frac{1}{[A]} - \frac{1}{[A]_0} \qquad \text{(Second order)} \qquad \text{[Eq. 15-8]}$$

This equation can be rearranged to solve for time:

$$t = \frac{1}{k}\left(\frac{1}{[A]} - \frac{1}{[A]_0}\right) \qquad \text{(Second order)} \qquad \text{[Eq. 15-9]}$$

Example 15-5

Compounds A and B react to form C and D in a reaction that is found to be second order overall and second order in B. The rate constant at 30°C is 0.622 liter per mole per minute. How many minutes does it take to use up 7.0×10^{-3} M of B if 4.00×10^{-2} M of B is mixed with excess A?

$$A + B \longrightarrow C + D$$

Solution

We first determine how much B remains after 7.0×10^{-3} M of B is used up.

$$\underline{?} \, M \text{ B remaining} = 0.0400 \, M - 0.0070 \, M$$
$$= 0.0330 \, M = [B]$$

We may apply the relationship given in Eq. 15-9 (the reaction is second order overall and second order in B):

$$t = \frac{1}{k}\left(\frac{1}{[B]} - \frac{1}{[B]_0}\right)$$
$$= \left(\frac{1}{0.622 \, M^{-1} \cdot \text{min}^{-1}}\right)\left(\frac{1}{0.0330 \, M} - \frac{1}{0.0400 \, M}\right)$$
$$= (1.61 \, M \cdot \text{min})(30.3 \, M^{-1} - 25.0 \, M^{-1})$$
$$= \underline{8.5 \text{ min}}$$

Example 15-6

Consider the reaction of Example 15-5 at 30°C. What concentration of B will remain after 10.0 minutes of reaction, starting with 2.40×10^{-2} M of B and excess A?

Solution

We start with Eq. 15-8.

$$kt = \frac{1}{[B]} - \frac{1}{[B]_0}$$

Inserting the known values:

$$(0.622 \, M^{-1} \cdot \text{min}^{-1})(10.0 \text{ min}) = \frac{1}{[B]} - \frac{1}{0.0240 \, M}$$

Solving for $1/[B]$:

$$\frac{1}{[B]} = (0.622 \ M^{-1} \cdot min^{-1})(10.0 \ min) + \frac{1}{0.0240 \ M}$$

$$= 6.22 \ M^{-1} + 41.7 \ M^{-1}$$

$$\frac{1}{[B]} = 47.9 \ M^{-1}$$

Inverting, $[B] = \dfrac{1}{47.9 \ M^{-1}} = 2.09 \times 10^{-2} \ M$

To set up the equation for half-life for a second order reaction, we again note that at $t = t_{1/2}$, we have $[A] = \frac{1}{2}[A]_0$ and substitute into Eq. 15–8.

$$kt_{1/2} = \frac{1}{\frac{1}{2}[A]_0} - \frac{1}{[A]_0}$$

Rearranging and simplifying, we obtain the relationship between the rate constant and $t_{1/2}$ for a reaction that is *second order overall and second order in A,*

$$t_{1/2} = \frac{1}{k[A]_0} \qquad \text{(Second order)} \qquad\qquad \text{[Eq. 15–10]}$$

Note that, in this case, $t_{1/2}$ *depends upon the initial concentration of A.* This equation also may be used for reactions that are second order overall and first order with respect to each of two reactants that are initially present in equal concentrations.

Example 15–7

Consider the reaction of Example 15–5 at 30°C. What is the half-life of B, if $4.10 \times 10^{-2} \ M$ B reacts with excess A?

$$t_{1/2} = \frac{1}{k[B]_0} = \frac{1}{(0.622 \ M^{-1} \cdot min^{-1})(4.10 \times 10^{-2} \ M)}$$

$$= 39.2 \ min$$

Solution

As long as A is present in stoichiometric excess, only the concentration of B affects the rate. The half-life is given by Eq. 15–10 (in terms of $[B]_0$):

Example 15–8

What is the half-life of B in the reaction of Example 15–7 when an initial $3.50 \times 10^{-4} \ M$ concentration of B reacts with excess A?

For a second order reaction the half-life does depend upon the initial concentration of reactant. The half-life of this reaction is now 4.59×10^3 minutes, or 3.19 days.

Solution

$$t_{1/2} = \frac{1}{k[B]_0} = \frac{1}{(0.622 \ M^{-1} \cdot min^{-1})(3.50 \times 10^{-4} \ M)}$$

$$= 4.59 \times 10^3 \ min$$

Integrated rate equations can be derived for other cases, such as reactions that are first order in A, first order in B, and second order overall, or such as third order reactions, but those are beyond the scope of this text.

Let us close this section by taking a graphical look at the two integrated rate equations that we have used. We can rearrange the first order integrated rate equation (Eq. 15–5)

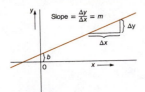

FIGURE 15-5 Plot of the equation $y = mx + b$, where m and b are constants. The slope of the line (positive in this case) is equal to m, while the intercept on the y-axis is equal to b.

$$\log\left(\frac{[A_0]}{[A]}\right) = \frac{kt}{2.303} \qquad \text{(First order)}$$

as follows. Remembering that the logarithm of a quotient, $\log(x/y)$, is equal to the difference of the logarithms, $\log x - \log y$, we can write

$$\log [A]_0 - \log [A] = \frac{kt}{2.303} \qquad \text{or} \qquad \log [A] = \frac{-k}{2.303}(t) + \log [A]_0 \qquad \text{[Eq. 15-11]}$$

Recall that the equation of a straight line may be written

$$y = mx + b \qquad \text{[Eq. 15-12]}$$

where x is the variable plotted along the abscissa (horizontal axis), y is the variable plotted along the ordinate (vertical), m is the slope of the line, and b is the intercept of the line with the y-axis (Figure 15-5). If we compare Eq. 15-11 with Eq. 15-12, we find that $\log [A]$ can be interpreted as y, and t is in the place of x.

$$\underbrace{\log [A]}_{\text{variable}} = \underbrace{\frac{-k}{2.303}}_{\text{constant}} \underbrace{(t)}_{\text{variable}} + \underbrace{\log [A]_0}_{\text{constant}}$$

$$ y = m x + b$$

The expression $-k/2.303$ is a constant as the reaction proceeds, so it can be interpreted as m, while the log of the initial concentration, $\log [A]_0$, is also constant for a particular run of the reaction, so it can be interpreted as b. Thus, a plot of $\log [A]$ vs. time would be expected to give a straight line (Figure 15-6b) with the slope of the line equal to $-k/2.303$ and the intercept equal to $\log [A]_0$.

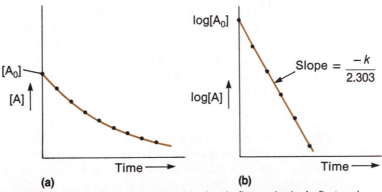

(a)   (b)

FIGURE 15-6 Plots for a reaction that is first-order in A, first-order overall. (a) Plot of concentration of reactant A vs. time (compare to Figure 15-4). (b) Plot of log [A] vs. time. The observation that this plot is a straight line confirms that the reaction is first-order in A and first-order overall, i.e., Rate = k[A]. The slope is equal to $(-k/2.303)$. Since k is a positive number, the slope of the line is always negative. Note that logarithms are dimensionless, so the slope has the units $(\text{time})^{-1}$. If the concentrations were less than 1 molar, the entire plot would be shifted below the time-axis.

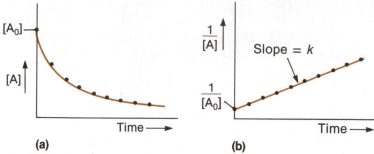

FIGURE 15–7 Plots for a reaction that is second-order in A and second-order overall. (a) Plot of concentration of reactant A vs. time (compare to Figure 15–4). (b) Plot of $1/[A]$ vs. time. The observation that this plot is a straight line confirms that the reaction is second-order in A, second-order overall, i.e., Rate $= k[A]^2$. The slope is equal to k. Since k is a positive number, the slope of the line is always positive. Note that the units of the slope are M^{-1} time^{-1}. Since concentrations cannot be negative, $1/[A]$ is always positive, and the line is always *above* the time-axis.

We may work in similar fashion with the integrated rate equation (Eq. 15–8) for a reaction that is second order in A and second order overall,

$$\frac{1}{[A]} - \frac{1}{[A]_0} = kt, \quad \text{simply rearranging it to read} \quad \frac{1}{[A]} = kt + \frac{1}{[A]_0} \qquad \text{[Eq. 15–13]}$$

Again comparing this with the equation for a straight line (Eq. 15–12), we see that a plot of $1/[A]$ vs. time would give a straight line, with slope equal to the specific rate constant, k, and intercept equal to $1/[A]_0$, as shown in Figure 15–7b.

Sometimes a convenient way to deduce an unknown rate-law expression from experimental data is to plot the data in various ways as suggested above. This approach is particularly useful for decomposition reactions (reactions involving only one reactant):

A $\longrightarrow$ products

A plot of concentration vs. time would be a curve with shape similar to that in Figure 15–4, no matter what the order of the reaction, since rate always slows as concentration diminishes. However, *if* the reaction follows first order kinetics, *then* a plot of log [A] vs. t would give a straight line (Eq. 15–11), whose slope could be interpreted to give the value of k. If, on the other hand, the reaction is second order in A and second order overall, *then* that log plot would not give a straight line, but a plot of $1/[A]$ vs. t would (Eq. 15–13). If neither of these plots gives a straight line (within expected scatter due to experimental error), we would know that neither of these is the correct order (rate law) for the reaction. Plots to test for other orders can be devised, as can graphical tests for rate-law expressions involving more than one reactant, but those are subjects for more advanced texts. The graphical approach that we have described is illustrated in Example 15–9.

It would not be possible for both plots to yield straight lines.

Example 15-9

We carry out the reaction A → B + C at a particular temperature. As the reaction proceeds, we measure the molarity of reactant A at various times, observing the following results:

Time (min)	[A] (mol/L)
0.00	2.00
1.00	1.46
2.00	1.07
3.00	0.78
4.00	0.57
5.00	0.42
6.00	0.31
7.00	0.22
8.00	0.16
9.00	0.12
10.00	0.088

(a) Plot [A] vs. *t*. (b) Plot log [A] vs. *t*. (c) Plot 1/[A] vs. *t*. (d) What is the order of the reaction? (e) Write the rate law expression for the reaction. (f) What is the value of *k* at this temperature? (Use seconds as the units of time.)

Solution

(a) Plot of [A] vs. *t*:

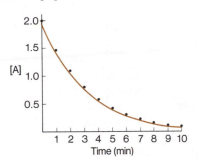

(b) In order to plot log [A] vs. *t*, we must calculate log [A] for each experiment reading:

t (min)	[A] (mol/L)	log [A]
0.00	2.00	0.301
1.00	1.46	0.166
2.00	1.07	0.030
3.00	0.78	−0.105
4.00	0.57	−0.241
5.00	0.42	−0.376
6.00	0.31	−0.512
7.00	0.22	−0.647
8.00	0.16	−0.783
9.00	0.12	−0.918
10.00	0.088	−1.054

Now, we plot log [A] vs. *t*:

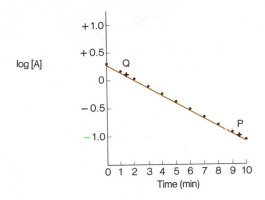

(c) In order to plot 1/[A] vs. *t*, we calculate 1/[A] for each of the experiment points:

t(min)	[A] (mol/L)	1/[A] (L/mol)
0.00	2.00	0.50
1.00	1.46	0.68
2.00	1.07	0.93
3.00	0.78	1.27
4.00	0.57	1.74
5.00	0.42	2.38
6.00	0.31	3.25
7.00	0.22	4.44
8.00	0.16	6.07
9.00	0.12	8.29
10.00	0.088	11.32

Now we plot 1/[A] vs. *t*:

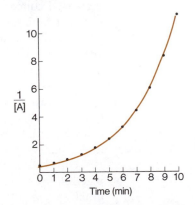

(d) It is clear from the answer to part (b) that the plot of log [A] vs. *t* gives a straight line. As we see in the answer to part (c), the plot of 1/[A] vs. *t* does *not* give a straight line (although the nonlinearity of the plot does not become very obvious until some time has elapsed). Thus the reaction is *first order* in [A].

(e) Putting our answer to part (d) into the form of an equation, we can write the rate law for the reaction:

Rate = $k[A]$

(f) We can determine the value of the rate constant for this first order reaction from the relationship

slope = $-k/2.303$ or $k = -2.303$ (slope)

In order to determine the slope of the line, we may pick any two points like P and Q in the plot of log [A] vs. t. From their coordinates, we calculate

slope = (change in ordinate)/(change in abscissa)

$$= \frac{(-1.00) - (0.10)}{(9.50 \text{ min} - 1.40 \text{ min})\left(60 \ \frac{\text{s}}{\text{min}}\right)}$$

$$= -0.00226 \text{ s}^{-1}$$

$$k = (-2.303)(-0.00226 \text{ s}^{-1})$$
$$= +0.0052 \text{ s}^{-1}$$

(Remember that the ordinate is the vertical axis, while the abscissa is the horizontal one. If you aren't careful to keep the points in the same order in the numerator and denominator, you'll get the wrong sign for the slope.)

15–6   Collision Theory and Transition State Theory

Two related theories help us understand the rates of chemical reactions in terms of molecular detail. They attempt to explain what the *atoms* and *molecules* are doing as we observe *substances* being consumed and formed at particular rates. The first theory we will examine is the collision theory.

1   Collision Theory

In its full form, collision theory is quite mathematical, so we shall present only a qualitative description of some of its main features. The fundamental notion of the **collision theory** of reaction rates is that *in order for reaction to occur between atoms, ions, or molecules, they must first collide.* This idea accounts for the observed increase in the rate of any reaction when the concentrations of its reactants are increased. Increased concentrations of *reacting species* result in greater numbers of collisions per unit time, and the reaction rate is correspondingly higher. This idea may be developed quantitatively using the mathematical form of the collision theory. Such an approach leads to a connection between *observed rates* and the *number* and *type* of molecules that must collide in the critical step of the reaction. Application of this idea will be more fully discussed in Section 15–8.

    An important conclusion of collision theory is that most collisions are ineffective. The collision theory equations may be used to calculate the total number of collisions that occur per unit time. For instance, in a mixture of gases (with a molecular weight of 100) that has a concentration of 0.01 M, there would be about 10^{30} collisions per liter per second at room temperature. If each collision resulted in reaction, the rate would be of the order of 10^6 moles per liter per second. This is much greater than most observed reaction rates. Many reactions in the gas phase take place only at rates of about 10^{-4} M s^{-1}, or 10^{10} times slower than would be predicted from the total number of collisions. The conclusion that we must draw from this is that *not all collisions result in reaction,* i.e., not all collisions are **effective collisions.** In order for a collision to be effective, the reacting species must (1) have the proper orientations toward each other at the time of collision, and (2) possess at least a certain minimum energy necessary to rearrange outer electrons in breaking bonds and forming new ones.

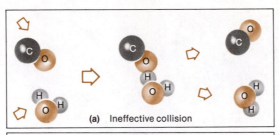

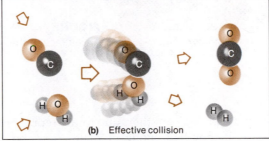

(a) Ineffective collision

(b) Effective collision

FIGURE 15–8 (a) Ineffective and (b) effective collisions between carbon monoxide and water molecules in the gas phase. The molecules must possess both sufficient energy to react and also the proper orientation relative to each other.

This is the final reaction in an important commercial method of preparing hydrogen. The carbon monoxide is produced by reaction of methane, CH_4, or some other hydrocarbon with steam in the presence of catalysts.

If colliding molecules have improper orientations, they do not react even though they may possess sufficient energy. Figure 15–8 depicts ineffective and effective collisions between molecules of carbon monoxide and water vapor, each having sufficient energy to react according to the equation

$$CO\ (g) + H_2O\ (g) \longrightarrow CO_2\ (g) + H_2\ (g)$$

Recall from Chapter 11 that the average kinetic energy of a collection of molecules is proportional to the absolute temperature. At higher temperatures, a greater fraction of the molecules possess enough energy to react. We shall discuss this temperature dependence further in Section 15–7. The fraction of energetically favorable collisions that have the proper orientation can be increased for many reactions by the introduction of a heterogeneous catalyst, discussed in Section 15–9.

2 Transition State Theory

In contrast with collision theory, this second theory dealing with reaction rates focuses on the energy changes that occur in the course of the reaction.

Chemical reactions involve the making and breaking of chemical bonds by shifting electrons, and are accompanied by changes in potential energy. Many reactions occur with release of energy to the surroundings. (Recall that if the energy is released in the form of heat, the reaction is said to be exothermic.) Consider the following hypothetical, one-step exothermic reaction at a certain temperature:

$$A_2 + B_2 \longrightarrow 2AB + energy$$

Figure 15–9 shows a plot of potential energy (which is related to bond energy) versus progress of reaction. The ground state energy of the reactants, A_2 and B_2, is higher than the ground state energy of the products, $2AB$. The energy released in the reaction is the difference between these two energies ΔE, which is related to the change in enthalpy or heat content.

It is clear that some covalent bonds must be broken, and others formed, in order for the molecules to react. If the molecules approached one another too gently,

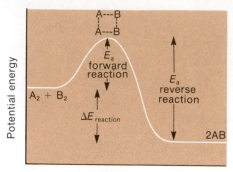

FIGURE 15–9 Potential energy diagram for a reaction that releases energy.

they could simply repel one another, due to the repulsions between their electron clouds. However, if they had enough kinetic energy to overcome this repulsion, they could collide in such a way that their kinetic energy becomes available to break some bonds so that new ones can form. According to the transition state theory, the reactants pass through a short-lived, high-energy state called a **transition state,** before the products are formed.

The **activation energy,** E_a, is the energy that must be absorbed by the reactants in their ground states to allow them to reach the transition state. If A_2 and B_2 do not possess the necessary amount of energy, E_a, above their ground states when they collide, no reaction will occur. If they do possess sufficient energy to "climb the energy barrier" to the transition state, the reaction can proceed to completion with the release of an amount of energy equal to the energy of activation *plus* ΔE. Thus, the activation energy must be supplied to the system from its environment, but is subsequently released to the environment along with additional energy in reactions that release energy. The net release of energy is ΔE.

The reverse of any exothermic reaction requires a net absorption of energy from its surroundings. (The reaction is endothermic if the energy is absorbed as heat.) A typical example, $2AB \longrightarrow A_2 + B_2$, is represented by Figure 15–10. In this case ΔE is the *net* energy *absorbed* by the system from its surroundings as the reaction occurs. Not all the energy of activation is returned to the environment when the products are formed. Note that the activation energy for the reverse reaction is greater than for the (original) forward reaction, and the magnitude, but not the sign, of ΔE is the same.

As a particular example of the ideas of this section, consider the reaction of iodide ions with chloroform.

The CH$_3$Cl and CH$_3$I molecules are (distorted) tetrahedra.

$$I^- + CH_3Cl \longrightarrow CH_3I + Cl^-$$

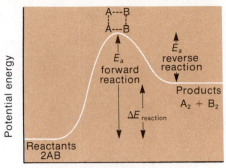

FIGURE 15-10 Potential energy diagram for a reaction that absorbs energy.

Extensive studies have shown that this reaction proceeds as shown in Figure 15-11a. The I^- ion must approach the CH_3Cl molecule from the "back side" of the C—Cl bond, through the middle of the three hydrogen atoms. A collision of the I^- ion with the CH_3Cl molecule from any other angle (Figure 15-11b) does not cause electrons to be shifted so as to break the C—Cl bond and allow the new I—C bond to form. A collision in the appropriate orientation can allow the new I—C bond to form at the same time that the three C—H bonds are shifting positions and the C—Cl bond is breaking. This collection of atoms, which we might write as

$$\text{I----C----Cl}$$

with H atoms

FIGURE 15-11 (a) A collision that could lead to reaction of I^- + CH_3Cl to give CH_3I + Cl^-. The I^- must approach along the "backside" of the C—Cl bond. (b) Two collisions that are not at the "correct" orientation.

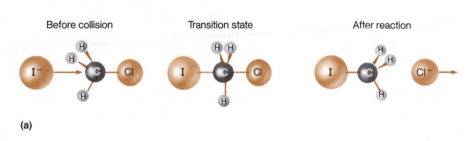

Before collision   Transition state   After reaction

(a)

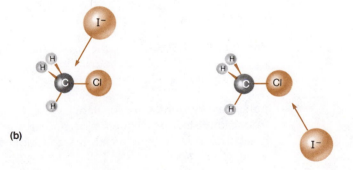

(b)

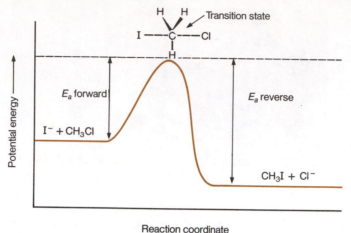

I —— C —— Cl
Transition state

Potential energy

E_a forward

$I^- + CH_3Cl$

E_a reverse

$CH_3I + Cl^-$

Reaction coordinate

FIGURE 15–12
Potential energy diagram (not to scale) for the reaction $I^- + CH_3Cl \rightarrow CH_3I + Cl^-$.

is what we mean by the transition state of this reaction (see Figure 15–12). From this state, either of two things could happen: (1) the I—C bond could finish forming, and the C—Cl bond could finish breaking, leading to products, or (2) the I—C bond could fall apart with I^- leaving, and the C—Cl bond could re-form, leading back to reactants.

15–7   Temperature

The average kinetic energy of a collection of molecules is proportional to the absolute temperature. At any particular temperature, T_1, a definite fraction of the molecules possess kinetic energies above the average for that temperature. Therefore, at a given temperature, a definite fraction of the reactant molecules have at least enough energy, E_a, to react upon collision. At a higher temperature, T_2, a greater percentage of the molecules possess the necessary activation energy, and the reaction proceeds at a faster rate. This situation is depicted in Figure 15–13.

From experimental observations, Svante Arrhenius developed the mathematical relationship among activation energy, absolute temperature, and the specific rate constant of a reaction, k, at that temperature. The relationship, known as the **Arrhenius equation,** is

This is the same Arrhenius who proposed a theory of acids and bases (Chapter 9).

$$k = Ae^{-E_a/RT}$$

[Eq. 15–14]

or, in natural logarithmic form,

$e \approx 2.718$ is the base of the natural logarithms.

$$\ln k = \ln A - \frac{E_a}{RT}$$

[Eq. 15–15]

Since $\ln (x) = 2.303 \log (x)$, the Arrhenius equation can also be written as

$$\log k = \log A - \frac{E_a}{2.303\ RT}$$

[Eq. 15–16]

In these expressions A is a constant having the same units as the rate constant, and is proportional to the frequency of properly oriented collisions between reacting molecules; R is the universal gas constant, with the same energy units in its

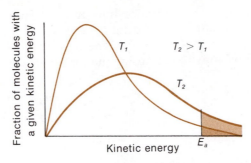

FIGURE 15–13 The effect of temperature on the number of molecules that have kinetic energies greater than E_a, the activation energy. Note that at T_2, a larger fraction of the molecules possesses at least E_a.

numerator as are used for E_a. The important point to note is that at a given temperature, a reaction with a larger E_a would have a smaller k (all other factors being equal), so its rate of reaction would be less at the same concentrations.

Let's look at how the rate constant varies with temperature for a single reaction. If we assume that the activation energy and the factor A do not depend on temperature, we can write Eq. 15–16 twice for two different temperatures, subtract one equation from the other, and rearrange the result to obtain the very useful form of the Arrhenius equation:

$$\log \frac{k_2}{k_1} = \frac{E_a}{2.303\,R}\left(\frac{T_2 - T_1}{T_1 T_2}\right) \qquad \text{[Eq. 15–17]}$$

Let us illustrate the use of this equation by substituting some representative values. The activation energy for many reactions is about 50 kJ/mol, and R is 8.314 J/mol·K; so for an increase in temperature from 300 K to 310 K we find

300 K is 27°C, slightly above room temperature.

$$\log \frac{k_2}{k_1} = \frac{50,000\ \text{J/mol}}{(2.303)(8.314\ \text{J/mol·K})}\left(\frac{310\ \text{K} - 300\ \text{K}}{(300\ \text{K})(310\ \text{K})}\right) = 0.28$$

$$\frac{k_2}{k_1} = 1.9 \approx 2$$

This is summarized by the rule of thumb that the rates of many reactions approximately double with a 10°C rise in temperature. *Such a rule must be used with care, however, since it depends critically on the activation energy.*

The following example further emphasizes the dependence of reaction rates on temperature. It also illustrates a useful application of the "two-temperature" form of the Arrhenius equation (Eq. 15–17) to determine the value of the activation energy E_a from measurements of rate constant data at various temperatures.

Example 15–10

Ethyl iodide decomposes to give ethylene and hydrogen iodide in the first order gas phase reaction

$$C_2H_5I \longrightarrow C_2H_4 + HI$$

At 600 K, the value of k is determined to be $1.60 \times 10^{-5}\ \text{s}^{-1}$. When the temperature is raised to 700 K, the value of k increases to $6.36 \times 10^{-3}\ \text{s}^{-1}$.

(a) Suppose we start with a concentration of C_2H_5I of 1.00 mol L^{-1}. How long would it take at 600 K for the C_2H_5I concentration to decrease to half this value?

(b) With the same starting concentration as in part (a), how long would it take at 700 K for half the reactant to disappear?

(c) Determine the activation energy for this reaction.

Solution

(a) The time for half of a given amount of reactant to disappear is given by the half-life, $t_{1/2}$. For a first-order reaction, this does not depend on concentration, and is given by Eq. 15–7,

$$t_{1/2} = \frac{0.693}{k}$$

Putting in the value of k, 1.60×10^{-5} s^{-1}, at 600 K, we obtain

$$t_{1/2} = \frac{0.693}{1.60 \times 10^{-5} \text{ s}^{-1}}$$

$$= 4.33 \times 10^4 \text{ s} \qquad \text{(about 12 hours)}$$

(b) Similarly, using the data given at 700 K, we obtain

$$t_{1/2} = \frac{0.693}{6.36 \times 10^{-3} \text{ s}^{-1}}$$

$$= 1.09 \times 10^2 \text{ s} \qquad \text{(less than 2 minutes)}$$

Notice how much faster the reaction goes at a temperature 100 degrees higher.

(c) The activation energy may be determined by using the Arrhenius equation in the form of Eq. 15–17:

$$\log \frac{k_2}{k_1} = \frac{E_a}{2.303 \, R} \left(\frac{T_2 - T_1}{T_1 T_2} \right)$$

$$k_1 = 1.60 \times 10^{-5} \text{ s}^{-1} \text{ at } T_1 = 600 \text{ K}$$
$$k_2 = 6.36 \times 10^{-3} \text{ s}^{-1} \text{ at } T_2 = 700 \text{ K}$$

Rearranging:

$$E_a = (2.303 \, R) \left(\frac{T_1 T_2}{T_2 - T_1} \right) \left[\log \frac{k_2}{k_1} \right]$$

$$= (2.303 \, R) \left(\frac{(600 \text{ K})(700 \text{ K})}{700 \text{ K} - 600 \text{ K}} \right)$$

$$\times \left[\log \left(\frac{6.36 \times 10^{-3} \text{ s}^{-1}}{1.60 \times 10^{-5} \text{ s}^{-1}} \right) \right]$$

$$= (2.303) \left(8.314 \frac{\text{J}}{\text{mol} \cdot \text{K}} \right) (4.20 \times 10^3 \text{ K})(2.60)$$

$$= 2.09 \times 10^5 \text{ J/mol} \qquad \text{or} \qquad \underline{209 \text{ kJ/mol}}$$

The determination of E_a in the manner illustrated in Example 15–10 may be subject to considerable error, since it depends on the measurement of k at only two different temperatures. Any error in either of these k values would be strongly reflected in the resulting value of E_a. A more trustworthy method is based on a graphical approach, as follows. Let us rearrange Eq. 15–16 and compare it with Eq. 15–12, the equation of a straight line.

$$\log k = \left(-\frac{E_a}{2.303 \, R} \right) \left(\frac{1}{T} \right) + \log A$$
$$ \quad y \quad = \quad m \quad \quad x \quad + \quad b$$

Compare this approach to that described in Section 15–5 for the determination of k.

Since the value of A is (very nearly) constant over moderate temperature changes, $\log A$ can be interpreted as the constant term in the equation (the intercept). The slope of the straight line obtained by plotting $\log k$ vs. $1/T$ then may be interpreted as $(-E_a/2.303 \, R)$, allowing us to determine the value of the activation energy from this slope.

Once we have determined the activation energy for a reaction, either as in Example 15–10 or by the graphical method just described, we can predict the rate constant for the reaction at any temperature. Example 15–11 illustrates this useful application of the Arrhenius equation.

Example 15–11

(a) Using the results of Example 15–10, predict the value of k at room temperature (25°C). (b) What would be the half-life of the reaction at 25°C?

Solution

(a) We may use Eq. 15–17 to calculate the value of k_2 at $T_2 = 25°C$ (298 K). We may use the known value for k at any temperature—e.g., let us take

$T_1 = 600$ K, at which temperature $k_1 = 1.60 \times 10^{-5}$ s^{-1}. Then

$$\log \frac{k_2}{k_1} = \frac{2.09 \times 10^5 \text{ J/mol}}{2.303 \left(8.314 \dfrac{\text{J}}{\text{mol} \cdot \text{K}}\right)} \left(\frac{298 \text{ K} - 600 \text{ K}}{(600 \text{ K})(298 \text{ K})}\right)$$

$$= (1.09 \times 10^4 \text{ K})(-1.69 \times 10^{-3} \text{ K}^{-1})$$

$$\log \frac{k_2}{k_1} = -18.4$$

Taking antilogs:

$$\frac{k_2}{k_1} = 4 \times 10^{-19}$$

$$k_2 = (4 \times 10^{-19}) \, k_1$$
$$= (4 \times 10^{-19})(1.60 \times 10^{-5} \text{ s}^{-1})$$
$$= \underline{6 \times 10^{-24} \text{ s}^{-1}}$$

(b) At 298 K,

$$t_{1/2} = \frac{0.693}{k} = \frac{0.693}{6 \times 10^{-24} \text{ s}^{-1}}$$

$$= \underline{1 \times 10^{23} \text{ s}} \qquad \text{(or } 3 \times 10^{15} \text{ years!)}$$

Although we can use the Arrhenius equation to treat activation energy in a quantitative way, as we have just seen, we shall deal with it only qualitatively in the sections that follow.

15–8 Reaction Mechanisms and the Rate-Law Expression

Many of the applications encountered so far in our study of chemical kinetics have been quite practical ones—how we can change the concentrations to make a reaction go faster or slower, how we can make a reaction proceed at some desired rate by modifying the temperature at which it is carried out, and so forth. Such practical aspects of chemical kinetics can be of great significance, whether in the laboratory, in industrial settings, or in everyday life. For instance, the burning of a fuel in an internal combustion engine must be at the proper rate to provide power, without reaching explosive speed. The chemical reaction that causes an epoxy adhesive to "set" may be far too slow at temperatures lower than those for which the product was designed. Some laboratory reactions in which useful substances are synthesized must be carried out at low concentrations with the temperature carefully controlled, so that the reaction will not occur too rapidly and explode or proceed to give some unwanted side-products.

In this section, we shall see another very important reason for studying chemical kinetics. This reason involves the insight that such studies can provide into the details of molecular behavior. Some reactions occur in a single step, in which a single collision could result in the formation of products. However, most reactions occur in a series of steps, some of which may be quite fast, while others may be slower. Such a series of steps is called the **mechanism** of the reaction. Much of our knowledge about such molecular details of reactions has come from studies of the rates of reactions, especially their concentration dependence as described by rate-law expressions.

Using a combination of experimental data and chemical intuition, we can postulate a mechanism by which a reaction occurs. There is no way to verify absolutely the validity of a proposed mechanism. All we can do is propose a mechanism that is consistent with experimental data. If further investigations produce information inconsistent with a postulated mechanism, such as the identification of intermediate species that are not part of the proposed mechanism, then the proposed mechanism must be modified to conform to all observations.

In many mechanisms one step is much slower than the others. A reaction can never occur faster than its slowest step. That is, the speed at which the slow step occurs determines the rate at which the overall reaction occurs. For this reason, the slow step of such a reaction is called the **rate-determining step.** For the general overall reaction

$$a A + b B \longrightarrow c C + d D$$

the experimentally determined rate-law expression has the form Rate $= k[A]^x[B]^y$. The values of x and y are influenced by the coefficients of the reactants in the slowest step (*not* the overall balanced equation, which is the sum of the individual steps) or, in some cases, in steps preceding the slowest step.

An important result of collision theory (Section 15–6) is that *for a reaction that takes place in a single step, the observed orders in the rate-law expression match the coefficients of the reacting substances.* For instance, consider the hypothetical reaction in which one molecule of AB collides with one molecule of C and reacts to give A + BC *in a single step.*

$$AB + C \longrightarrow A + BC \qquad \text{(single step)}$$

Such a reaction step, involving the collision of two molecules, is said to be *bimolecular.* According to collision theory, for this single-step reaction, the predicted rate-law expression would be

$$\text{Rate} = k[AB][C]$$

Be sure to notice that the relation between stoichiometric coefficients and rate-law exponents is *valid only for individual reaction steps.* In general, for the *individual step*

$$a A + b B \longrightarrow \text{products}$$

the rate-law expression would be

$$\text{Rate} = k[A]^a[B]^b \qquad \text{(for an individual step)}$$

As an example of the use of the rate-law expression and this prediction of collision theory in the study of reaction mechanisms, let us consider some specific examples.

One of the earliest kinetic studies undertaken involved the gas-phase reaction of hydrogen and iodine to form hydrogen iodide. The reaction was found to be first order in both hydrogen and iodine:

$$H_2\,(g) + I_2\,(g) \longrightarrow 2HI\,(g)$$

$$\text{Rate} = k[H_2][I_2]$$

The mechanism originally postulated involved collision of single molecules of H_2 and I_2 in a simple one-step reaction. However, present evidence indicates a more complex process. Most kineticists now accept the following mechanism:

(1)	I_2	$\longrightarrow 2I$	(fast)
(2)	$I \ + H_2$	$\longrightarrow H_2I$	(fast)
(3)	$H_2I + I$	$\longrightarrow 2HI$	(slow)
	$H_2 \ + I_2$	$\longrightarrow 2HI$	overall

In this case, neither of the original reactants appears in the rate-determining step, but both appear in the rate-law expression. Since each step is a reaction in itself, it

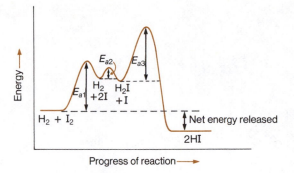

FIGURE 15–14 The relative energies of activation for a postulated mechanism for the gas phase reaction H_2 (g) + I_2 (g) $\rightleftharpoons$ 2HI (g).

When several steps have similar activation energies, the analysis of experimental data is complex.

follows that (according to transition-state theory) each step has its own activation energy. Since Step 3 is the slowest, its activation energy is the highest, as shown in Figure 15–14.

The gas-phase reaction of nitrogen oxide and bromine is known to be second order in NO and first order in Br_2:

$$2NO \text{ (g)} + Br_2 \text{ (g)} \longrightarrow 2NOBr \text{ (g)}$$

$$Rate = k[NO]^2[Br_2]$$

A one-step, three-molecule collision involving two NO molecules and one Br_2 molecule would be consistent with the experimentally determined rate-law expression. However, the likelihood of all three molecules colliding simultaneously is far less than the likelihood that two will collide. Therefore, more favorable *routes involving only bimolecular collisions* (*or unimolecular decompositions*) *are thought to be important in reaction mechanisms.* The mechanism is believed to be

$$
\begin{array}{lll}
(1) & NO + Br_2 \longrightarrow NOBr_2 & \text{(fast)} \\
(2) & NOBr_2 + NO \longrightarrow 2NOBr & \text{(slow)} \\
\hline
 & 2NO + Br_2 \longrightarrow 2NOBr & \text{overall}
\end{array}
$$

The first step involves the collision of one NO and one Br_2 to produce the intermediate species, $NOBr_2$, which then reacts relatively slowly with one NO molecule to produce two NOBr molecules. The rate-determining step involves one molecule of NO and one of $NOBr_2$. We could express the rate as

$$Rate = k'[NOBr_2][NO]$$

However, the $NOBr_2$ is a reaction intermediate, and its concentration at the beginning of the second step cannot be directly measured. Since the $NOBr_2$ is produced from one NO molecule and one Br_2 molecule, its concentration is proportional to the concentrations of both.

$$[NOBr_2] = k''[NO][Br_2]$$

Thus, if we substitute the right side of the equation above for $[NOBr_2]$ in the rate expression, we arrive at the original experimentally determined expression.

$$Rate = k'k''[NO][Br_2][NO]$$

$$Rate = k[NO]^2[Br_2]$$

Similar situations apply to most other overall third or higher order reactions, as well as some lower order reactions, such as that of H_2 and I_2 above.

FIGURE 15–15 The two steps in the reaction $2NO$ (g) $+ O_2$ (g) $\rightarrow 2NO_2$ (g). The first step is fast, and the second is slow.

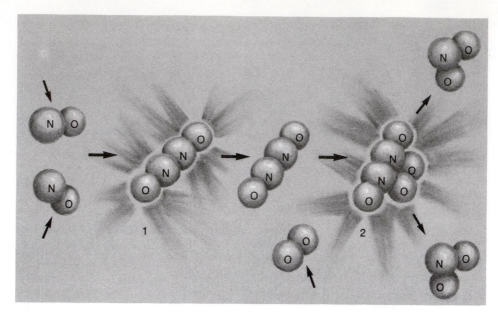

The mechanism by which nitrogen oxide from auto exhaust reacts with oxygen to form another pollutant, NO_2, is thought to be a two-step mechanism. An intermediate, N_2O_2, is formed in the first step. This then reacts with O_2 to form two NO_2 molecules, as depicted in Figure 15–15.

$$
\begin{array}{llll}
(1) & 2NO \text{ (g)} & \longrightarrow N_2O_2 \text{ (g)} & \text{(fast)} \\
(2) & N_2O_2 \text{ (g)} + O_2 \text{ (g)} \longrightarrow 2NO_2 \text{ (g)} & & \text{(slow)} \\
\hline
 & 2NO \text{ (g)} + O_2 \text{ (g)} \longrightarrow 2NO_2 \text{ (g)} & & \text{overall}
\end{array}
$$

15–9 Catalysts

Catalysts are substances that can be added to reacting systems either to increase or, rarely, to decrease the rate of reaction. They allow reactions to occur via alternate pathways, and these affect reaction rates by altering activation energies. The activation energy is lowered in most catalyzed reactions, as depicted in Figures 15–16 and 15–17. However, in the case of inhibitory catalysis, the activation energy is raised as the reaction is forced to occur by a less favorable route. Although a catalyst may enter into a reaction, it does not appear in the balanced equation for the reaction. If a catalyst does react, it is regenerated in subsequent steps; if it is not regenerated, it is not a catalyst but a reactant.

Catalysts may be classified in two categories: (1) homogeneous catalysts and (2) heterogeneous catalysts or contact catalysts. A **homogeneous catalyst** exists in the same phase as the reactants. Strong acids function as homogeneous catalysts in the acid-catalyzed hydrolysis of esters (a class of organic compounds). Using ethyl acetate (a component of nail polish removers) as an example of an ester, the overall reaction can be written

"Hydrolysis" means reaction with water and is discussed in detail in Chapter 17.

$$
\underset{\text{ethyl acetate}}{CH_3-\overset{\overset{\textstyle O}{\|}}{C}-OCH_2CH_3 \text{ (aq)}} + H_2O \xrightarrow{H^+} \underset{\text{acetic acid}}{CH_3-\overset{\overset{\textstyle O}{\|}}{C}-OH \text{ (aq)}} + \underset{\text{ethanol}}{CH_3CH_2OH \text{ (aq)}}
$$

This is a thermodynamically favorable reaction, but because of its high energy of activation it occurs only very, very slowly when no catalyst is present. However, in the presence of strong acids, the reaction occurs rapidly. In acid-catalyzed hydrolysis, different intermediates with lower activation energies are formed. The *postulated* sequence of steps is shown below.

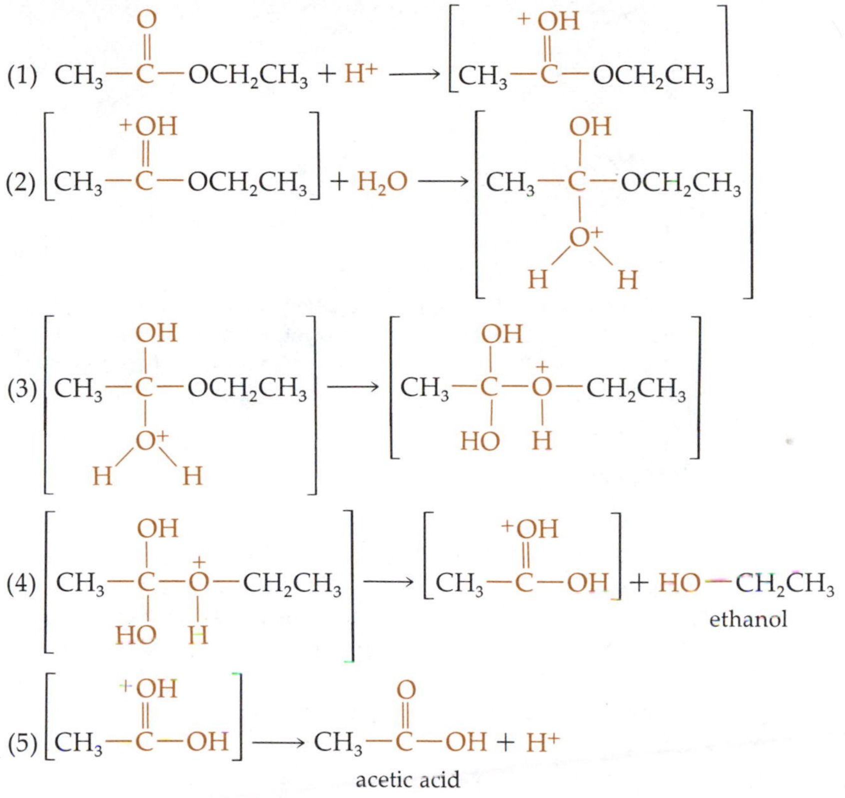

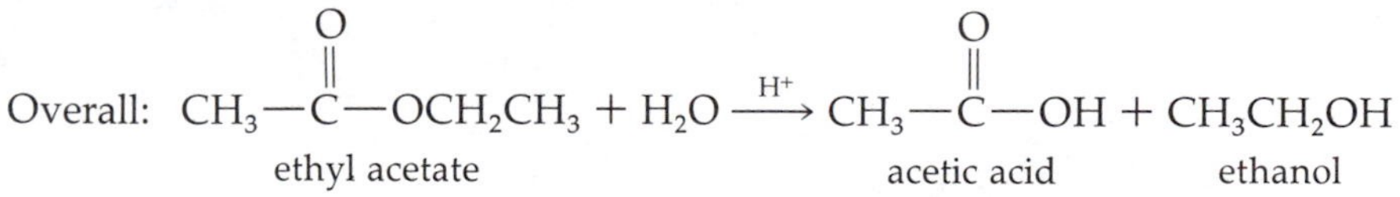

(1) $CH_3-\overset{\overset{O}{\|}}{C}-OCH_2CH_3 + H^+ \longrightarrow \left[CH_3-\overset{\overset{+OH}{\|}}{C}-OCH_2CH_3\right]$

(2) $\left[CH_3-\overset{\overset{+OH}{\|}}{C}-OCH_2CH_3\right] + H_2O \longrightarrow CH_3-\underset{\underset{\underset{H \quad H}{\diagdown \diagup}}{O^+}}{\overset{\overset{OH}{|}}{C}}-OCH_2CH_3$

(3) $\left[CH_3-\underset{\underset{\underset{H \quad H}{\diagdown \diagup}}{O^+}}{\overset{\overset{OH}{|}}{C}}-OCH_2CH_3\right] \longrightarrow \left[CH_3-\underset{\underset{HO \quad H}{|}}{\overset{\overset{OH}{|}}{C}}-\overset{+}{O}-CH_2CH_3\right]$

(4) $\left[CH_3-\underset{\underset{HO \quad H}{|}}{\overset{\overset{OH}{|}}{C}}-\overset{+}{O}-CH_2CH_3\right] \longrightarrow \left[CH_3-\overset{\overset{+OH}{\|}}{C}-OH\right] + HO-CH_2CH_3$

ethanol

(5) $\left[CH_3-\overset{\overset{+OH}{\|}}{C}-OH\right] \longrightarrow CH_3-\overset{\overset{O}{\|}}{C}-OH + H^+$

acetic acid

Overall: $CH_3-\overset{\overset{O}{\|}}{C}-OCH_2CH_3 + H_2O \xrightarrow{H^+} CH_3-\overset{\overset{O}{\|}}{C}-OH + CH_3CH_2OH$

ethyl acetate   acetic acid   ethanol

Some reaction intermediates are so unstable and short-lived that it is very difficult to prove experimentally that they exist. Others are so stable that they can be sold commercially.

Note that H^+ is a reactant in equation (1) and a product in equation (5) and is, therefore, a catalyst. The charged species (shown in brackets) produced in equations (1) through (4) are intermediates; they are not considered catalysts because they were not present in the original reaction mixture. Of course, ethyl acetate and water are reactants and acetic acid and ethanol are products of the overall catalyzed reaction.

Heterogeneous catalysts exist in a different phase than the reactants. They are usually solids, and they lower activation energies by providing surfaces upon which reactions can occur. One or more of the reactants may be preferentially adsorbed onto catalytic surfaces in particular orientations, resulting in more effective collisions between reactants than when a catalyst is not present. Most contact catalysts are more effective as finely divided powders, with large surface areas.

The reaction between hydrogen and oxygen to produce water is a thermodynamically favorable reaction,

$$2H_2 \text{ (g)} + O_2 \text{ (g)} \longrightarrow 2H_2O \text{ (}\ell\text{)} \qquad \Delta G^0 = -474 \text{ kJ}$$

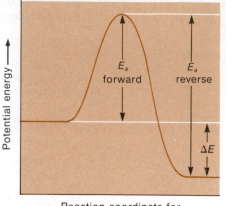 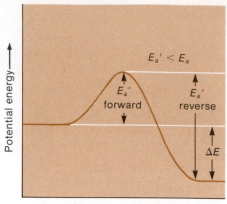

Reaction coordinate for
uncatalyzed reaction

Reaction coordinate for
catalyzed reaction

FIGURE 15–16 The effect a catalyst has on the potential energy diagram. The reaction mechanism is changed by the catalyst, which provides a low-energy mechanism for the formation of the products. ΔE has the same value for each path, since the value of ΔE depends on the state of the reactants and products only.

but it does not proceed significantly at room temperature in the absence of energy input or a catalyst. When finely divided noble metals (unreactive metals such as platinum or palladium) are used as catalysts, the reaction can occur explosively at room temperature. The effect of this catalysis is depicted in Figure 15–18.

The catalytic mufflers built into today's automobiles contain two types of heterogeneous catalysts, powdered noble metals and powdered transition metal oxides. They catalyze the oxidation of unburned fuel and of partial combustion products such as carbon monoxide (see Figure 15–19).

$$2C_8H_{18} \text{ (g)} + 25O_2 \text{ (g)} \xrightarrow[\text{NiO}]{\text{Pt}} 16CO_2 \text{ (g)} + 18H_2O \text{ (g)}$$

iso-octane (a component of gasoline)

$$2CO \text{ (g)} + O_2 \text{ (g)} \xrightarrow[\text{NiO}]{\text{Pt}} 2CO_2 \text{ (g)}$$

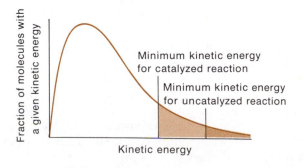

Minimum kinetic energy
for catalyzed reaction

Minimum kinetic energy
for uncatalyzed reaction

FIGURE 15–17 When a catalyst is present, more molecules possess the minimum kinetic energy necessary for reaction.

FIGURE 15-18
Catalysis of
$2H_2 + O_2 \rightarrow 2H_2O$ on
metal surface.

The same catalysts also catalyze the decomposition of nitrogen oxide, NO, one of the by-products of combustion of any fuel, into harmless nitrogen and oxygen:

$$2NO \text{ (g)} \xrightarrow[\text{NiO}]{\text{Pt}} N_2 \text{ (g)} + O_2 \text{ (g)}$$

Nitrogen oxide is a serious air pollutant because it is oxidized to nitrogen dioxide, NO_2, which reacts with water to form nitric acid and with alcohols to form nitrites (eye irritants).

These three reactions, catalyzed in catalytic mufflers, are all exothermic and thermodynamically favorable. Unfortunately, other energetically favored reactions are also accelerated by the mixed catalysts. All fossil fuels contain sulfur impurities, which are oxidized to sulfur dioxide during combustion. Sulfur dioxide, itself an air pollutant, undergoes further oxidation to form sulfur trioxide as it passes through the catalytic bed:

$$2SO_2 \text{ (g)} + O_2 \text{ (g)} \xrightarrow[\text{NiO}]{\text{Pt}} 2SO_3 \text{ (g)}$$

Sulfur trioxide is probably a worse pollutant than sulfur dioxide, because SO_3 is the acid anhydride of strong, corrosive sulfuric acid. Sulfur trioxide reacts with water vapor in the air, as well as in auto exhausts, to form sulfuric acid droplets. This is a major problem that must be overcome if the current type of catalytic converter is to see continued use. These same catalysts also suffer from the problem of being "poisoned" (made inactive) by lead. Leaded fuels, which contain tetraethyl lead, $Pb(C_2H_5)_4$, and tetramethyl lead, $Pb(CH_3)_4$, are not suitable for automobiles equipped with catalytic converters and are excluded by law from use in such cars.

The Haber process for the preparation of ammonia, an extremely important industrial chemical, involves the use of iron as a catalyst at $450°-500°C$ and high pressures. The reaction is thermodynamically spontaneous, but very slow.

$$N_2 \text{ (g)} + 3H_2 \text{ (g)} \xrightarrow{\text{Fe}} 2NH_3 \text{ (g)}$$

In the absence of a catalyst the reaction does not occur at room temperature, and even with a catalyst is not efficient at atmospheric pressure. It is discussed in detail in Section 16-5.

Enzymes are proteins that act as catalysts in living systems for specific biochemical reactions. The reaction between nitrogen and hydrogen to form ammonia is catalyzed at room temperature and atmospheric pressure by a class of enzymes, called nitrogenases, that are present in some bacteria. Most of the essential nutrients for both plants and animals contain nitrogen. Legumes are plants that support certain bacteria, which are able to obtain nitrogen as N_2 from the atmos-

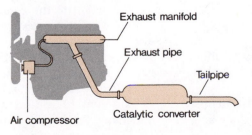

FIGURE 15-19 The arrangement of a catalytic converter.

phere and convert it to ammonia. This can then be used in the synthesis of proteins, nucleic acids, and other nitrogen-containing biological compounds. The process is called nitrogen fixation.

Currently, one of the most active areas of chemical research involves attempts to discover or synthesize catalysts that are nearly as efficient as naturally occurring enzymes such as nitrogenases. Such a development would be industrially important because it would eliminate the costs of high temperature and high pressure currently necessary in the Haber process, and so could decrease the cost of food grown with the aid of ammonia-based fertilizer. Ultimately this would aid greatly in feeding the world's growing population.

In comparison to man-made catalysts, most enzymes are tremendously efficient under very mild conditions. If chemists and biochemists could develop catalysts with but a small fraction of the efficiency of enzymes, it could be a great boon to the world's health and economy.

Key Terms

Activation energy energy that must be absorbed by reactants in their ground states in order to reach the transition state so that a reaction can occur.

Arrhenius equation equation that expresses relationship among activation energy, absolute temperature, and specific rate constant for a reaction.

Catalyst a substance that alters (usually increases) the rate at which a reaction occurs.

Chemical kinetics the study of rates and mechanisms of chemical reactions.

Collision theory theory of reaction rates that states that effective collisions between reactant molecules must occur in order for reaction to occur.

Contact catalyst see *Heterogeneous catalyst.*

Effective collision collision between molecules resulting in reaction; one in which molecules collide with proper relative orientations and sufficient energy to react.

Enzyme a protein that acts as a catalyst in biological systems.

Half-life of a reactant, the time required for half of a given reactant to be converted into product(s).

Heterogeneous catalyst a catalyst that exists in a different phase (solid, liquid, or gas) from the reactants; a contact catalyst.

Homogeneous catalyst a catalyst that exists in the same phase (solid, liquid, or gas) as the reactants.

Inhibitory catalyst inhibitor; a catalyst that decreases the rate of a reaction.

Integrated rate equation equation (obtained by integrating the rate-law expression) that relates reactant concentrations to time; different for reactions of different order.

Intermediate a (usually short-lived) species produced and consumed during a reaction.

Mechanism the sequence of steps by which reactants are converted into products.

Method of initial rates method of determining the rate-law expression by carrying out a reaction with different initial concentrations and analyzing concentration vs. rate data.

Order of a reactant, the power to which the reactant's concentration is raised in the rate-law expression; of a reaction, the sum of the powers to which all concentrations are raised in the rate-law expression (also called overall order).

Rate-determining step the slowest step in a mechanism; the step that determines the overall rate of reaction.

Rate-law expression equation relating the rate of a reaction to the concentrations of the reactants and the specific rate constant.

Rate of reaction change in concentration of a reactant or product per unit time.

Specific rate constant an experimentally determined (proportionality) constant, which is different for different reactions and which changes only with temperature; k in the rate-law expression: Rate $= k[A]^x[B]^y$.

Thermodynamically favorable (spontaneous) reaction   reaction that occurs with a net release of free energy, G; reaction for which ΔG is negative (see Section 14–7).

Transition state   relatively high energy state, in which bonds in reactant molecules are partially broken and new ones are partially formed.

Transition state theory   theory of reaction rates that states that reactants pass through high-energy transition states before forming products.

Exercises

Basic Concepts

1. Summarize briefly the effects of each of the four factors affecting rates of reactions.
2. Describe the basic features of collision theory and transition state theory.
3. What is a rate-law expression? Describe how it is determined for a particular reaction.
4. Distinguish between reactions that are thermodynamically favorable and reactions that are kinetically favorable. What can be said about relationships between the two?
5. Draw typical reaction-energy diagrams for one-step reactions that release energy and that absorb energy. Distinguish between the net energy change for each kind of reaction and the activation energy. Indicate potential energies of products and reactants for both kinds of reactions.
6. Describe and illustrate graphically how the presence of a catalyst can affect the rate of a reaction.
7. What is meant by the order of a reaction?
8. Define reaction mechanism. Why do we assume that only bimolecular collisions and unimolecular decompositions are important in most reaction mechanisms?
9. What, if anything, can be said about the relationship between the coefficients of the balanced *overall* equation for a reaction and the powers to which concentrations are raised in the rate-law expression? To what are these powers related?
10. How do homogeneous catalysts and heterogeneous catalysts differ? What is an inhibitor?

Rate-Law Expression

11. If tripling the initial concentration of a reactant triples the initial rate of reaction, what is the order of the reaction with respect to the reactant? If the rate increases by a factor of nine, what is the order? If the rate remains the same, what is the order?

12. What are the units of the rate constant for reactions that are overall: (a) first order? (b) second order? (c) third order? (d) of order $1\frac{1}{2}$?

13. The following rate data were obtained at 25°C for the following reaction. What is the rate-law expression for this reaction?

$$2A + B \longrightarrow 3C$$

Experiment	[A](mol/L)	[B](mol/L)	Initial Rate of Formation of C
1	0.10	0.10	4.0×10^{-4} M/min
2	0.30	0.30	1.2×10^{-3} M/min
3	0.10	0.30	4.0×10^{-4} M/min
4	0.20	0.40	8.0×10^{-4} M/min

14. The following data were obtained for the following reaction at 25°C. What is the rate-law expression for the reaction?

$$2A + B + 2C \longrightarrow 3D$$

Experiment	Initial [A]	Initial [B]	Initial [C]	Initial Rate of Formation of D
1	0.20 M	0.10 M	0.10 M	4.0×10^{-4} M/min
2	0.20 M	0.30 M	0.20 M	1.2×10^{-3} M/min
3	0.20 M	0.10 M	0.30 M	4.0×10^{-4} M/min
4	0.60 M	0.30 M	0.40 M	3.6×10^{-3} M/min

15. The following data were collected for the following reaction at a particular temperature. What is the rate-law expression for this reaction?

$$A + B \longrightarrow C$$

Experiment	Initial [A]	Initial [B]	Initial Rate of Formation of C
1	0.10 M	0.10 M	4.0×10^{-4} M/min
2	0.20 M	0.20 M	3.2×10^{-3} M/min
3	0.10 M	0.20 M	1.6×10^{-3} M/min

16. The rate-law expression for the following reaction is found to be of the form: Rate = $k[N_2O_5]$. What is the overall reaction order?

$$2N_2O_5 \text{ (g)} \longrightarrow 4NO_2 \text{ (g)} + O_2 \text{ (g)}$$

17. We know that the rate expression for the following reaction at a certain temperature is Rate = $k[NO]^2[O_2]$. If two experiments involving this reaction are carried out at the same temperature, but if in the second experiment the initial concentration of NO is doubled while the initial concentration of O_2 is halved, the initial rate in the second experiment will be ____ times that of the first.

$$2NO + O_2 \longrightarrow 2NO_2$$

18. We know that the rate expression for the following reaction is Rate = $k[A][B_2]^2$. According to this expression, if during a reaction the concentrations of both A and B_2 are suddenly tripled, the rate of the reaction will ____ by a factor of ____.

$$A + B_2 \longrightarrow \text{Products}$$

19. Nitric oxide reacts with hydrogen to produce nitrogen and water vapor according to the following equation:

$$2NO \text{ (g)} + 2H_2 \text{ (g)} \longrightarrow N_2 \text{ (g)} + 2H_2O \text{ (g)}$$

This reaction is thought to proceed by the following two-step mechanism:

$$2NO + H_2 \longrightarrow N_2O + H_2O \quad \text{(slow)}$$
$$N_2O + H_2 \longrightarrow N_2 \ + H_2O \quad \text{(fast)}$$

(a) According to this mechanism, what is the rate-law expression for this reaction?
(b) What is the overall reaction order?

20. Given the following data for the reaction $A + B \rightarrow C$, write the rate-law expression.

Experiment	Initial [A]	Initial [B]	Initial Rate of Formation of C
1	0.20 M	0.10 M	5.0×10^{-6} M/s
2	0.30 M	0.10 M	7.5×10^{-6} M/s
3	0.40 M	0.20 M	4.0×10^{-5} M/s

21. Given the following data for the reaction $A + B \rightarrow C$, write the rate-law expression.

Experiment	Initial [A]	Initial [B]	Initial Rate of Formation of C
1	0.10 M	0.10 M	8.0×10^{-5} M/s
2	0.20 M	0.10 M	3.2×10^{-4} M/s
3	0.20 M	0.20 M	1.28×10^{-3} M/s

22. Given the following data for the reaction $A + B \rightarrow C$, write the rate-law expression.

Experiment	Initial [A]	Initial [B]	Initial Rate of Formation of C
1	0.10 M	0.10 M	2.0×10^{-4} M/s
2	0.20 M	0.10 M	8.0×10^{-4} M/s
3	0.40 M	0.20 M	2.56×10^{-2} M/s

23. Consider a chemical reaction between compounds A and B that is first order in A and first order in B. From the information given below, fill in the blanks.

Experiment	Rate ($M \cdot s^{-1}$)	[A]	[B]
1	0.10	0.20 M	0.050 M
2	0.40	___ M	0.050 M
3	0.80	0.40 M	____ M

24. Consider a chemical reaction of compounds A and B that was found to be first order in A and second order in B. From the information given below, fill in the blanks.

Experiment	Rate ($M \cdot s^{-1}$)	[A]	[B]
1	0.10	1.0 M	0.20 M
2	_____	2.0 M	0.20 M
3	_____	2.0 M	0.40 M

25. Consider the reaction below, for which the activation energy of the forward reaction is 69.9 kJ/mol and that of the reverse reaction is 82.0 kJ/mol at 25°C.

$$A + 2B \longrightarrow C + D$$

Sketch a potential energy versus progress of reaction diagram for the reaction. Does the reaction release energy or absorb energy?

Integrated Rate-Law Equation

26. What is meant by the half-life of a reactant?

27. We carry out the reaction $A \longrightarrow B + C_2$ at a particular temperature, starting with $[A]_0 = 2.00\ M$. As the reaction proceeds, we measure the molar concentration of reactant A at various times, observing the following results:

Time (min)	[A](mol/L)
0.00	2.00
1.00	1.23
2.00	0.89
3.00	0.70
4.00	0.57
5.00	0.48
6.00	0.42
7.00	0.37
8.00	0.33
9.00	0.30
10.00	0.28

(a) Plot [A] vs. t. (b) Plot log [A] vs. t. (c) Plot $1/[A]$ vs. t. (d) What is the order of the reaction? (e) Write the rate equation for the reaction. (f) What is the value of k at this temperature? (Use seconds as the units of time.) (g) Compare the results of this problem with those of Example 15–9. Compare the numerical values of k. Compare the behavior of the three curves. (You may find it helpful to re-plot the results of Example 15–9 on the same scale as your solution to this question.)

28. In the gas phase, acetaldehyde decomposes to give methane and carbon monoxide

$$CH_3CHO \longrightarrow CH_4 + CO$$

This decomposition reaction is carried out at 760 K, starting with an acetaldehyde concentration of 0.100 M. As the reaction proceeds, we measure the concentration of acetaldehyde remaining, with the following results:

Time (min)	[CH₃CHO](mol/L)
0.00	0.100
1.00	0.061
2.00	0.044
3.00	0.035
4.00	0.028
5.00	0.024
6.00	0.021
7.00	0.018
8.00	0.016
9.00	0.015
10.00	0.014

(a) Write the rate equation for this reaction. (b) What is the value, with units, of the specific rate constant for this reaction at 760 K? (c) What is the half-life of this reaction at 760 K, starting with an acetaldehyde concentration of 0.100 M? (d) Suppose we start with an acetaldehyde concentration of 0.400 M. What would be the time required for the concentration to decrease to 0.100 M?

29. The decomposition of SO_2Cl_2 in the gas phase,

$$SO_2Cl_2 \longrightarrow SO_2 + Cl_2$$

can be studied by measuring the concentration of Cl_2 gas as the reaction proceeds. We begin with $[SO_2Cl_2]_0 = 0.250\ M$. Holding the temperature constant at 320°C, we monitor the Cl_2 concentration at intervals of two hours, with the following results:

Time (hours)	[Cl₂](mol/L)
0.00	0.000
2.00	0.037
4.00	0.068
6.00	0.095
8.00	0.117
10.00	0.137
12.00	0.153
14.00	0.168
16.00	0.180
18.00	0.190
20.00	0.199

(a) Plot $[Cl_2]$ vs. t. (b) Plot $[SO_2Cl_2]$ vs. t. (c) Determine the rate law for this reaction. (d) What is the value, with units, for the specific rate constant at 320°C? (e) How long would it take for 90 percent of the original SO_2Cl_2 to react?

30. Cyclopropane rearranges to form propene

cyclopropane   propene

in a reaction that follows first order kinetics. At 800 K, the specific rate constant for this reaction is $2.74 \times 10^{-3}\ s^{-1}$. Suppose we start with a cyclopropane concentration of 0.150 M. How long will it take for 99.0 percent of the cyclopropane to disappear according to this reaction?

31. The first order rate constant for the conversion of cyclobutane to ethylene at 1000°C is 87 s^{-1}.

cyclobutane   ethylene

(a) What is the half-life of this reaction at 1000°C?
(b) If one started with 1.00 gram of cyclobutane, how long would it take to consume 0.70 gram of it?
(c) How much of an initial 1.00 gram sample of cyclobutane would remain after 1.00×10^{-3} second?

32. The decomposition of carbon disulfide, CS_2, to carbon monosulfide, CS, and sulfur is first order with $k = 2.8 \times 10^{-7}$ s^{-1} at 1000°C.

$$CS_2 \longrightarrow CS + S$$

(a) What is the half-life of this reaction at 1000°C?
(b) How many days would pass before a 1.00 gram sample of CS_2 had decomposed to the extent that 0.60 gram of CS_2 remained?
(c) Refer to (b). How many grams of CS would be present after this length of time?
(d) How much of a 1.00 gram sample of CS_2 would remain after 35.0 days?

33. For the reaction

$$2NO_2 \longrightarrow 2NO + O_2$$

the rate law at 25°C is
Rate = 1.4×10^{-10} $M^{-1} \cdot s^{-1}[NO_2]^2$.
(a) If 2.50 moles of NO_2 are initially present in a sealed 1.00 liter vessel at 25°C, what is the half-life of the reaction?
(b) Refer to (a). What concentration and how many grams of NO_2 remain after 100 years?
(c) Refer to (b). What concentration of NO would have been produced during the same period of time?

Arrhenius Equation

34. Derive Eq. 15–17 from the Arrhenius equation, Eq. 15–16.

35. Consider the forward reaction of Exercise 25. If its second order rate constant, k, is 6.0×10^{-5} $M^{-1} \cdot s^{-1}$ at 25°C, what is its rate constant at 50°C?

36. The rearrangement reaction

$$CH_3NC \longrightarrow CH_3CN$$

follows first order kinetics. In a table of kinetic data we find the following values listed for this reaction: $A = 3.98 \times 10^{13}$ s^{-1}, $E_a = 160$ kJ·mol^{-1}. (a) Calculate the value of the specific rate constant at room temperature, 25°C. (b) At room temperature, how long would it take for half of a given amount of CH_3NC to rearrange to CH_3CN? (c) Calculate the value of the specific rate constant at 327°C. (d) How long would it take for half of a given amount of CH_3NC to rearrange at 327°C?

37. Use the data given in Exercise 36 to calculate values of the specific rate constant, k, at the following temperatures (all °C): 25, 77, 127, 177, 227, 277, 327, 377. Then plot log(k) vs. $1/T$, to verify that Eq. 15–16 does yield a straight line. (Reminder: Convert temperatures to the Kelvin scale.)

38. The rearrangement of cyclopropane to propene described in Exercise 30 has been studied at various temperatures. The following values for the specific rate constant have been determined experimentally:

T (K)	k (s^{-1})
600	3.30×10^{-9}
650	2.19×10^{-7}
700	7.96×10^{-6}
750	1.80×10^{-4}
800	2.74×10^{-3}
850	3.04×10^{-2}
900	2.58×10^{-1}
950	1.75
1000	9.78

From the appropriate plot of these data, determine the value of the activation energy for this reaction.

39. The gas phase reaction

$$N_2O_5 \longrightarrow NO_2 + NO_3$$

follows first order kinetics. The activation energy of this reaction is measured to be 88 kJ·mol^{-1}. The value of k at 0°C is determined to be 9.16×10^{-3} s^{-1}. What would be the value of k for this reaction at room temperature, 25°C?

40. In the chapter, it was calculated that the rate constant, k, for a reaction whose activation energy is 50 kJ/mol approximately doubles (factor of 1.9) when the temperature is increased from 300 K to 310 K. Over this same temperature range, determine the factor by which k increases for a reaction whose activation energy is 100 kJ/mol. Can you rationalize this result?

41. At temperatures below about 500°C, the dimerization of tetrafluoroethylene to octafluorocyclobutane

$$2 \quad \begin{matrix} F \\ \diagdown \\ F \end{matrix} C=C \begin{matrix} \diagup F \\ \\ \diagdown F \end{matrix} \longrightarrow \begin{matrix} F & F \\ | & | \\ F-C-C-F \\ | & | \\ F-C-C-F \\ | & | \\ F & F \end{matrix}$$

follows second order kinetics. At 140°C, the specific rate constant for this reaction is 3.42×10^{-3} $M^{-1} \cdot$ s^{-1}. All parts of this question refer to the reaction at 140°C. (a) Suppose we start with a C$_2$F$_4$ concentration of 0.200 M. What fraction of the original C$_2$F$_4$ would remain after 1.00 hour? (b) What would be the concentration of octafluorocyclobutane at the end of the 1.00 hour period? (c) With the starting concentration of part (a), how long would be required for 95.0 percent of the tetrafluoroethylene to dimerize? (d) Starting with $[C_2F_4]_0 = 0.500$ M, how long would be required for 95.0 percent of the tetrafluoroethylene to dimerize?

Reaction Mechanisms

42. The rate equation for the reaction

$$Cl_2 \text{ (aq)} + H_2S \text{ (aq)} \longrightarrow S \text{ (s)} + 2HCl \text{ (aq)}$$

is found to be Rate $= k[Cl_2][H_2S]$. Which of the following mechanisms are consistent with the rate law?

(a) $Cl_2 \longrightarrow Cl^+ + Cl^-$ slow
$Cl^- + H_2S \longrightarrow HCl + HS^-$ fast
$Cl^+ + HS^- \longrightarrow HCl + S$ fast
$Cl_2 + H_2S \longrightarrow S + 2HCl$ overall

(b) $Cl_2 + H_2S \longrightarrow HCl + Cl^+ + HS^-$ slow
$Cl^+ + HS^- \longrightarrow HCl + S$ fast
$Cl_2 + H_2S \longrightarrow S + 2HCl$ overall

(c) $Cl_2 \longrightarrow Cl + Cl$ fast
$Cl + H_2S \longrightarrow HCl + HS$ fast
$HS + Cl \longrightarrow HCl + S$ slow
$Cl_2 + H_2S \longrightarrow S + 2HCl$ overall

43. The ozone, O$_3$, of the stratosphere can be decomposed by reaction with nitrogen oxide (or, more commonly, nitric oxide), NO, from high-flying jet aircraft.

$$O_3 \text{ (g)} + NO \text{ (g)} \longrightarrow NO_2 \text{ (g)} + O_2 \text{ (g)}$$

The rate expression is Rate $= k[O_3][NO]$. Which of the following mechanisms are consistent with the observed rate expression?

(a) $NO + O_3 \longrightarrow NO_3 + O$ slow
$NO_3 + O \longrightarrow NO_2 + O_2$ fast
$O_3 + NO \longrightarrow NO_2 + O_2$ overall

(b) $NO + O_3 \longrightarrow NO_2 + O_2$ slow (one step)

(c) $O_3 \longrightarrow O_2 + O$ slow
$O + NO \longrightarrow NO_2$ fast
$O_3 + NO \longrightarrow NO_2 + O_2$ overall

(d) $NO \longrightarrow N + O$ slow
$O + O_3 \longrightarrow 2O_2$ fast
$O_2 + N \longrightarrow NO_2$ fast
$O_3 + NO \longrightarrow NO_2 + O_2$ overall

(e) $NO \longrightarrow N + O$ fast
$O + O_3 \longrightarrow 2O_2$ slow
$O_2 + N \longrightarrow NO_2$ fast
$O_3 + NO \longrightarrow NO_2 + O_2$ overall

44. Propose a possible mechanism consistent with the observed rate-law expression, Rate $= k[NO_2Cl]$, for the reaction below.

$$2NO_2Cl \longrightarrow 2NO_2 + Cl_2$$

45. Propose a possible mechanism for the reaction of iodide ion, I$^-$, with hypochlorite ion, OCl$^-$.

$$I^- + OCl^- \longrightarrow OI^- + Cl^-$$

The observed rate expression is Rate $= k[I^-][OCl^-]$.

46. Propose a mechanism consistent with the rate expression Rate $= k[NO]^2[O_2]$ for the reaction

$$2NO + O_2 \longrightarrow 2NO_2$$

Chemical Equilibrium

16

16 – 1   Basic Concepts

Most chemical reactions do not go to completion. That is, when reactants are mixed in stoichiometric quantities, they are not completely converted to products. Reactions that do not go to completion *and* that can occur in either direction are called *reversible reactions.* In Section 9 – 4.4, we discussed the ionization of acetic acid in water as a specific example of a reversible reaction.

$$CH_3COOH \text{ (aq)} + H_2O \text{ (ℓ)} \rightleftharpoons H_3O^+ \text{ (aq)} + CH_3COO^- \text{ (aq)}$$

Reversible reactions may be represented in general terms as shown below, where the capital letters represent formulas and the lower case letters represent coefficients in the balanced equation. (In the equation for the ionization of acetic acid, *a*, *b*, *c*, and *d* are all ones.)

$$a\text{A} + b\text{B} \rightleftharpoons c\text{C} + d\text{D}$$

The double arrow ($\rightleftharpoons$) indicates that the reaction is reversible and that both the forward (left-to-right) and reverse (right-to-left) reactions can occur simultaneously. When A and B react to form C and D at the same rate at which C and D react to form A and B, the system is at *equilibrium*. **Chemical equilibrium** exists when two opposing reactions occur simultaneously at the same rate. Chemical equilibria are **dynamic equilibria** in the sense that there are always individual molecules reacting, even though the overall composition of the reaction mixture does not change. In a system at equilibrium, the equilibrium is said to lie toward the right if more C and D are present than A and B, and to lie toward the left if more A and B are present.

The dynamic nature of chemical equilibrium can be proved experimentally by "tagging" a small percentage of reactant molecules with radioactive atoms, and following them through the reaction. Even when the initial mixture is at equilibrium, radioactive atoms will appear in product molecules.

Consider the simple case in which the coefficients of the equation for a reaction are all one. If substances A and B are mixed, the rate of the forward reaction decreases as time passes because the concentrations of A and B decrease.

$$A + B \longrightarrow C + D \qquad (1)$$

As C and D are formed, these species react to form A and B.

$$C + D \longrightarrow A + B \qquad (2)$$

The rate of reaction between C and D increases with time because as more C and D molecules are formed, more collide and react. Eventually, the two reactions occur at the same rate and the system is at equilibrium. See Figure 16–1. If a reaction begins with only C and D present, reaction (2) proceeds at a rate that decreases with time, and the rate of reaction (1) increases with time, until the two rates are equal.

In previous chapters we have dealt almost exclusively with reactions that were assumed to occur in only one direction. Those that go essentially to completion are commonly written with only a single arrow, signifying that when equilibrium is established only insignificantly small amounts of reactants remain.

The square brackets, [], represent concentrations of the species enclosed within them in moles per liter.

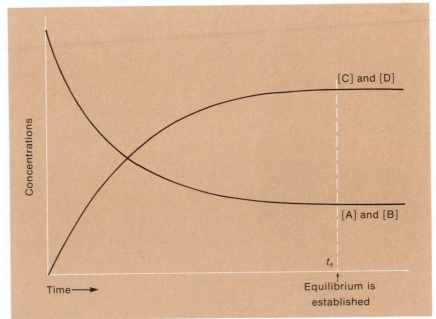

FIGURE 16–1 Variation in the amounts of species present in the $A + B \rightleftharpoons C + D$ system, as equilibrium is approached beginning with equal amounts of A and B only. This equilibrium favors the products, C and D.

The SO_2, O_2, SO_3 System

Let us examine the reversible reaction of sulfur dioxide with oxygen to form sulfur trioxide at 1500 K. The numbers in the following discussion were determined experimentally.

$$2SO_2 \text{ (g)} + O_2 \text{ (g)} \rightleftharpoons 2SO_3 \text{ (g)}$$

Suppose 0.400 mole of SO_2 and 0.200 mole of O_2 are injected into a closed container of one liter capacity. When equilibrium is established (at time t_e, Figure 16–2), 0.056 mole of SO_3 has formed and 0.344 mole of SO_2 and 0.172 mole of O_2 remain unreacted; *or*, 0.056 mole of SO_2 and 0.028 mole of O_2 have reacted to form 0.056 mole of SO_3. The reaction ratio is a 2 : 1 : 2 molar ratio as required by the coefficients of the balanced equation; that is, 2(0.028) mole of SO_2 and 1(0.028) mole of O_2 have been consumed and 2(0.028) mole of SO_3 has been produced. *But* the reaction does not go to completion. The situation is summarized below, using molarity units rather than moles. (They are numerically identical since the volume of the reaction vessel is 1.00 liter.) The *net reaction* is represented by the *changes* in concentrations. A positive (+) change represents an increase, while a negative (−) change indicates a decrease in concentration.

	$2SO_2$ (g) +	O_2 (g) $\rightleftharpoons$	$2SO_3$ (g)
initial conc'n	0.400 *M*	0.200 *M*	0
change in conc'n due to reaction (ratio *set* by coefficients)	−0.056 *M*	−0.028 *M*	+0.056 *M*
equilibrium conc'n	0.344 *M*	0.172 *M*	0.056 *M*

In another experiment, only 0.400 mole of SO_3 is introduced into a closed one liter container. The same total numbers of sulfur and oxygen atoms are present as in the previously described experiment. When equilibrium is established (time t_e in Figure 16–3), 0.056 mole of SO_3, 0.172 mole O_2, and 0.344 mole of SO_2 are present. These are the same amounts found at equilibrium in the previous case. Note that the net reaction proceeds from *right to left* as the equation is written, and that the changes in concentration are in the same 2 : 1 : 2 ratio as in the previous case, as required by the coefficients of the balanced equation.

	$2SO_2$ (g) +	O_2 (g) $\rightleftharpoons$	$2SO_3$ (g)
initial conc'n	0	0	0.400 *M*
change in conc'n due to reaction (ratio *set* by coefficients)	+0.344 *M*	+0.172 *M*	−0.344 *M*
equilibrium conc'n	0.344 *M*	0.172 *M*	0.056 *M*

The data tabulated below summarize the results of these two experiments.

	Initial Concentrations			Equilibrium Concentrations		
	$[SO_2]$	$[O_2]$	$[SO_3]$	$[SO_2]$	$[O_2]$	$[SO_3]$
Experiment 1	0.400 *M*	0.200 *M*	0 *M*	0.344 *M*	0.172 *M*	0.056 *M*
Experiment 2	0 *M*	0 *M*	0.400 *M*	0.344 *M*	0.172 *M*	0.056 *M*

These results are displayed graphically in Figures 16–2 and 16–3.

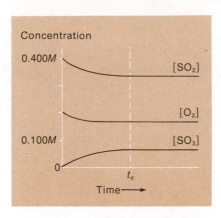

FIGURE 16-2 Establishment of equilibrium in the $2SO_2 + O_2 \rightleftharpoons 2SO_3$ system, beginning with only SO_2 and O_2.

16-2 The Equilibrium Constant

Recall that the simultaneous collision of three bodies is an unlikely occurrence from a statistical point of view (Section 15-8).

Assume that the reversible reaction below occurs via a *simple one-step mechanism.*

$$2A + B \rightleftharpoons A_2B$$

The rate of the forward reaction is $\text{Rate}_f = k_f[A]^2[B]$, while the rate of the reverse reaction is $\text{Rate}_r = k_r[A_2B]$. In these expressions k_f and k_r are the *specific rate constants* of the forward and reverse reactions, respectively. By definition the two rates are equal at equilibrium, and since $\text{Rate}_f = \text{Rate}_r$ we may write

$$k_f[A]^2[B] = k_r[A_2B]$$

Rearranging the expression by dividing both sides of the equation by $k_r[A]^2[B]$ gives

$$\frac{k_f}{k_r} = \frac{[A_2B]}{[A]^2[B]}$$

At any specific temperature, k_f/k_r is a constant, since both k_f and k_r are constants. This ratio is given a special name and symbol, the **equilibrium constant,** K_c or simply K, where the subscript refers to concentrations. In this reaction,

$$K_c = \frac{[A_2B]}{[A]^2[B]}$$

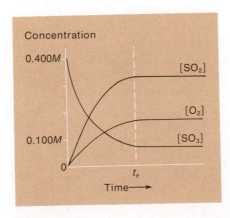

FIGURE 16-3 Establishment of equilibrium in the $2SO_2 + O_2 \rightleftharpoons 2SO_3$ system, beginning with only SO_3. Greater changes in concentrations occur to establish equilibrium starting with SO_3 than starting with SO_2 and O_2. The equilibrium favors SO_2 and O_2.

The square brackets, [ ], in an equilibrium constant expression indicate *equilibrium concentrations* in moles per liter.

We have dealt with a simple one-step reaction. Suppose, instead, that a reaction involves a two-step mechanism with the rate constants shown. An equilibrium constant expression can be written for each step:

$$(1) \quad 2A \underset{k_{1r}}{\overset{k_{1f}}{\rightleftharpoons}} A_2$$

$$(2) \quad A_2 + B \underset{k_{2r}}{\overset{k_{2f}}{\rightleftharpoons}} A_2B$$

The overall reaction is
$$2A + B \rightleftharpoons A_2B$$

$$K_1 = \frac{k_{1f}}{k_{1r}} = \frac{[A_2]}{[A]^2}$$

$$K_2 = \frac{k_{2f}}{k_{2r}} = \frac{[A_2B]}{[A_2][B]}$$

By multiplying K_1 and K_2 the $[A_2]$ term may be eliminated. Since K_1 and K_2 are both constants, K_c is also a constant and we obtain

$$K_1 \times K_2 = \frac{[A_2]}{[A]^2} \times \frac{[A_2B]}{[A_2][B]} = \frac{[A_2B]}{[A]^2[B]} = K_c$$

Regardless of the mechanism by which a reaction occurs, the same expression for the equilibrium constant is obtained. For a reaction in general terms,

$$\underbrace{aA + bB}_{\text{Reactants}} \rightleftharpoons \underbrace{cC + dD}_{\text{Products}}$$

the equilibrium constant expression is given by

$$K_c = \frac{[C]^c[D]^d}{[A]^a[B]^b} \quad \begin{matrix} \longleftarrow \text{Products} \\ \longleftarrow \text{Reactants} \end{matrix}$$

[Eq. 16-1]

By convention we refer to the substances on the right side of the balanced equation as "products" and those on the left as "reactants," even if the reaction is initiated by products, with no "reactants" initially present. The equilibrium constant, K_c, is defined as the product of the *equilibrium concentrations* (moles per liter) of the products, each raised to the power that corresponds to its coefficient in the balanced equation, divided by the product of the *equilibrium concentrations* of reactants, each raised to the power that corresponds to its coefficient in the balanced equation.

The equilibrium constant expressions are given for the following reversible reactions to illustrate some common types.

$$N_2\,(g) + O_2\,(g) \rightleftharpoons 2NO\,(g), \qquad\qquad K_c = \frac{[NO]^2}{[N_2][O_2]}$$

$$CH_4\,(g) + Cl_2\,(g) \rightleftharpoons CH_3Cl\,(g) + HCl\,(g), \qquad K_c = \frac{[CH_3Cl][HCl]}{[CH_4][Cl_2]}$$

$$N_2\,(g) + 3H_2\,(g) \rightleftharpoons 2NH_3\,(g), \qquad\qquad K_c = \frac{[NH_3]^2}{[N_2][H_2]^3}$$

Numerical values for equilibrium constants must be determined experimentally. Let us reconsider the $SO_2/O_2/SO_3$ equilibrium described earlier. The equilibrium concentrations were the same in the two experiments. We may use these equilibrium concentrations to calculate the value of the equilibrium constant for the reaction. The equilibrium concentrations were $[SO_3] = 0.056\ M$, $[SO_2] = 0.344\ M$, and $[O_2] = 0.172\ M$.

$$2SO_2\ (g) + O_2\ (g) \rightleftharpoons 2SO_3\ (g)$$

equil. conc'n $\quad$ 0.344 M $\quad$ 0.172 M $\quad\quad$ 0.056 M

Substitution of these values into the equilibrium expression allows evaluation of the equilibrium constant:

$$K_c = \frac{[SO_3]^2}{[SO_2]^2[O_2]} = \frac{(0.056\ M)^2}{(0.344\ M)^2(0.172\ M)} = 1.5 \times 10^{-1}\ M^{-1}$$

For the reaction written *as it is*, K_c is 0.15 M^{-1} at 1500 K. The value and units of K_c (in this case M^{-1}) depend upon the form of the balanced equation for the reaction. The units are commonly omitted. The value of K_c, for any particular equation, varies only with the temperature, is constant for the reaction at a given temperature, and is independent of initial concentrations.

> The magnitude of K_c is a measure of the extent to which reaction occurs. A very large value of K_c indicates that at equilibrium most of the species on the left have been converted into the species on the right side of the balanced chemical equation. If K_c is much smaller than 1, equilibrium is established when most of the reactants (left side) remain unreacted and relatively small amounts of products (right side) have been formed.

Example 16-1

Some nitrogen and hydrogen gases are pumped into an empty 5.00 L vessel at 500°C. When equilibrium is established, 3.00 moles of N_2, 2.10 moles of H_2, and 0.298 mole of NH_3 are present. Evaluate K_c for the reaction of nitrogen and hydrogen at 500°C.

$$N_2\ (g) + 3H_2\ (g) \rightleftharpoons 2NH_3\ (g)$$

Solution

The equilibrium concentrations are obtained by dividing the number of moles of each reactant and product by the volume, 5.00 liters.

$$[N_2] = 3.00\ mol/5.00\ L = 0.600\ M$$

$$[H_2] = 2.10\ mol/5.00\ L = 0.420\ M$$

$$[NH_3] = 0.298\ mol/5.00\ L = 0.0596\ M$$

The resulting concentrations (not numbers of moles) are then substituted into the expression for the equilibrium constant:

$$K_c = \frac{[NH_3]^2}{[N_2][H_2]^3} = \frac{(0.0596)^2}{(0.600)(0.420)^3} = 0.080$$

Thus, for the reaction of hydrogen and nitrogen to form ammonia at 500°C, we can write

$$K_c = \frac{[NH_3]^2}{[N_2][H_2]^3} = \underline{0.080}$$

The small value of K_c indicates that the equilibrium lies far to the left.

Variation of K_c with the Form of the Balanced Equation

We wrote the equation for the reaction of SO_2 with O_2 to form SO_3 as

$$2SO_2\ (g) + O_2\ (g) \rightleftharpoons 2SO_3\ (g)$$

and showed the equilibrium constant to be

$$K_c = \frac{[SO_3]^2}{[SO_2]^2[O_2]} = \frac{(0.0560)^2}{(0.344)^2(0.172)} = 0.15$$

Suppose we had written the equation for the same reaction in reverse,

$$2SO_3\ (g) \rightleftharpoons 2SO_2\ (g) + O_2\ (g)$$

The equilibrium constant, K_c', for the reverse reaction is

$$K_c' = \frac{[SO_2]^2[O_2]}{[SO_3]^2} = \frac{(0.344)^2(0.172)}{(0.056)^2} = 6.5$$

We see that K_c', the equilibrium constant for the reverse reaction, is *the reciprocal* of K_c, the equilibrium constant for the forward reaction. If the equation for the reaction is written as

$$SO_2\ (g) + \tfrac{1}{2}O_2\ (g) \rightleftharpoons SO_3\ (g)$$

then

$$K_c'' = \frac{[SO_3]}{[SO_2][O_2]^{1/2}} = \frac{0.056}{(0.344)(0.172)^{1/2}} = 0.39 = K_c^{1/2}$$

where K_c'' is the square root of K_c.

To generalize, if an *equation* for a reaction is *multiplied* by a positive or negative integer or fraction, n, then the *original value of* K_c is raised to the nth *power*. (Reversing an equation is the same as multiplying all coefficients by -1.)

16-3   Uses of the Equilibrium Constant

The equilibrium constant for a given reaction at a given temperature is a type of *ratio* of product concentrations to reactant concentrations, and expresses the extent of the reaction at equilibrium. We have seen how to calculate its value from one set of equilibrium concentrations; once that value is obtained, the computational process can be turned around to calculate equilibrium *concentrations* from the equilibrium *constant*. (Remember that the equilibrium constant is a "constant" only if the temperature does not change. We will see in Section 16-11 how to compute a new equilibrium constant for a different temperature.)

As in any algebraic equation, one of the quantities involved can be calculated if values are specified for all of the others. In Example 16-1, the equation involved three concentrations and the equilibrium constant; we were given values of the three equilibrium concentrations, and we calculated the equilibrium constant. In the next example, given all the *initial* concentrations and the equilibrium constant, we shall calculate the equilibrium concentrations.

Example 16-2

For the reaction

$$A + B \rightleftharpoons C + D$$

the equilibrium constant is 144 at a certain temperature. If 0.400 mole each of A and B are placed in a 2.00 liter container at the temperature, what concentrations of all species are present at equilibrium?

Solution

In practically all equilibrium calculations it is helpful to write down the values, or symbols for the values, of initial concentrations, changes in concentrations (due to the reaction), and the concentrations at equilibrium as shown below. Since the coefficients of the equation are all ones, the reaction ratio must be $1:1:1:1$.

Let x = moles per liter of both A and B that have reacted, and then

x = moles per liter of C and D that have been formed

	A	+	B	⇌	C	+	D
in:	$0.200\ M$		$0.200\ M$		$0\ M$		$0\ M$
ch:	$-x\ M$		$-x\ M$		$+x\ M$		$+x\ M$
eq:	$(0.200 - x)\ M$		$(0.200 - x)\ M$		$x\ M$		$x\ M$

(We use *in* for *initial concentration,* *ch* for *changes in concentration,* and *eq* for *equilibrium concentration.*) Now substitute the equilibrium concentrations (*not* the initial concentrations) into the equation for the equilibrium constant:

$$K_c = \frac{[C][D]}{[A][B]} = 144$$

$$\frac{(x)(x)}{(0.200 - x)(0.200 - x)} = \frac{x^2}{(0.200 - x)^2} = 144$$

We have a quadratic equation that is also a perfect square, and it can be solved by extracting the square root of both sides of the equation and then solving for x.

$$\frac{x}{0.200 - x} = 12.0$$

$$x = 2.40 - 12.0x$$

$$13.0x = 2.40$$

$$x = \frac{2.40}{13.0} = 0.185\ M$$

Now that the value of x is known, the equilibrium concentrations may be easily obtained:

$$[A] = [B] = (0.200 - x)\ M = \underline{0.015\ M}$$

$$[C] = [D] = x = \underline{0.185\ M}$$

These values may be substituted into the equilibrium expression to provide a quick check on the calculation. Recall that $K_c = 144$.

$$K_c = \frac{[C][D]}{[A][B]} = \frac{(0.185)^2}{(0.015)^2} = 1.5 \times 10^2$$

Note that the value 1.5×10^2 differs somewhat from 144. However, its calculation involved squaring two numbers, each of which contains considerable round-off error, and then dividing one of these numbers into the other. You might like to play with the number combinations generated by your calculator until you are convinced that the statement is correct.

Example 16–3 is similar to Example 16–2, except that A and B are mixed in nonstoichiometric amounts.

Example 16–3

Consider the same system as in Example 16–2 at the same temperature. If 0.600 mole of A and 0.200 mole of B are mixed in a 2.00 liter container and allowed to reach equilibrium, what are the equilibrium concentrations of all species?

Solution

Let x = concentrations of A and B that have reacted, and

x = concentrations of C and D formed

	A	+	B	⇌	C	+	D
in:	$0.300\ M$		$0.100\ M$		$0\ M$		$0\ M$
ch:	$-x\ M$		$-x\ M$		$+x\ M$		$+x\ M$
eq:	$(0.300 - x)\ M$		$(0.100 - x)\ M$		$x\ M$		$x\ M$

Note that initial concentrations are governed only by the amounts of reactants mixed together. But *changes in concentrations* due to reaction must occur in a $1:1:1:1$ ratio as required by the coefficients in the balanced equation. Now, the equilibrium constant is still

$$K_c = \frac{[C][D]}{[A][B]} = 144$$

or, substituting,

$$\frac{(x)(x)}{(0.300 - x)(0.100 - x)} = 144$$

This equation is not a perfect square but it is a quadratic equation, and we can rearrange it into the standard form of a quadratic equation:

$$\frac{x^2}{0.0300 - 0.400x + x^2} = 144$$

$$x^2 = 4.32 - 57.6x + 144x^2$$

$$143x^2 - 57.6x + 4.32 = 0$$

All quadratic equations of the form

$$ax^2 + bx + c = 0$$

can be solved by use of the quadratic formula,

$$x = \frac{-b \pm \sqrt{b^2 - 4ac}}{2a} \qquad \text{(see Appendix A)}$$

In this case, $a = 143$, $b = -57.6$, and $c = 4.32$. Substituting in the appropriate values,

$$x = \frac{-(-57.6) \pm \sqrt{(-57.6)^2 - 4(143)(4.32)}}{2(143)}$$

$$= \frac{57.6 \pm \sqrt{3318 - 2471}}{286}$$

$$= \frac{57.6 \pm \sqrt{847}}{286} = \frac{57.6 \pm 29.1}{286}$$

$$x = 0.303 \quad \text{or} \quad 0.0997$$

Solutions of quadratic equations always yield two roots. One root has physical meaning (the answer) and the other root is extraneous; *i.e.*, it has no physical meaning. The maximum acceptable value of x is 0.100 M (based on the chemical equation). The value of x is defined as the number of moles of A per liter and the number of moles of B per liter that react, and no more B can be consumed than was initially present, as would be the case if $x = 0.303$. Thus $x = 0.0997$ M is the root in which we are interested, and the extraneous root is 0.303. (The two roots for quadratic equations that are also perfect squares are equal in magnitude but opposite in sign. Clearly, concentrations can never be less than zero.)

The equilibrium concentrations are:

$$[A] = 0.300 - x = \underline{0.200 \ M}$$

$$[B] = 0.100 - x = \underline{0.0003 \ M}$$

$$[C] = [D] = x = \underline{0.0997 \ M}$$

The following table summarizes the last two examples. Recall that $K_c = 144$.

	Initial Concentration (*M*)				Equilibrium Concentration (*M*)			
	[A]	[B]	[C]	[D]	[A]	[B]	[C]	[D]
Example 16-2	0.200	0.200	0	0	0.015	0.015	0.185	0.185
Example 16-3	0.300	0.100	0	0	0.200	0.0003	0.0997	0.0997

Very large round-off errors are generated because numbers with one or two significant figures cannot be used to calculate numbers with three significant figures.

The data from the table can be substituted into the equilibrium constant expression to show that even though the reaction is initiated by different relative amounts of reactants in the two cases, the equilibrium ratios of concentrations of products to reactants (each raised to the first power) are identical within round-off error. Note that B is the limiting reagent in Example 16-3, and that less B remains at equilibrium than in Example 16-2, where reactants are present in stoichiometric amounts.

16-4 Reaction Quotient

"The law of mass action" is another name for LeChatelier's Principle (Section 12-5).

We define the **mass action expression** or **reaction quotient,** Q, for the general reaction

$$aA + bB \rightleftharpoons cC + dD$$

as

$$Q = \frac{[C]^c[D]^d}{[A]^a[B]^b} \quad \begin{array}{l} \text{not necessarily} \\ \text{equilibrium concentrations} \end{array}$$

[Eq. 16–2]

The reaction quotient has the same *form* as the equilibrium constant, except that it involves specific concentrations that are *not necessarily* equilibrium concentrations. However, if they are equilibrium concentrations, then $Q = K_c$. As we shall see, the concept of reaction quotient is extremely valuable because if we can compare the magnitude of Q with that of K for any particular reaction under a given set of conditions, we can decide whether the forward or reverse reaction will proceed to a greater extent to establish equilibrium.

If at any time during an experiment $Q < K$, the forward reaction proceeds to a greater extent than the reverse reaction until equilibrium is established. This is because when $Q < K$, the numerator of the reaction quotient is too small and the denominator is too large. The only way to reduce the denominator and increase the numerator is by A and B reacting to produce C and D. Conversely, if $Q > K$, the reverse reaction proceeds to a greater extent than the forward reaction until equilibrium is established. When $Q = K$, the system is at equilibrium and both forward and reverse reactions occur at the same rate. In summary:

$Q < K$ forward reaction occurs to a greater extent than reverse reaction until equilibrium is established

$Q = K$ system is at equilibrium

$Q > K$ reverse reaction occurs to a greater extent than forward reaction until equilibrium is established

---

Example 16–4

At a very high temperature, $K_c = 1 \times 10^{-13}$ for

$$2HF (g) \rightleftharpoons H_2 (g) + F_2 (g)$$

At a certain time the following concentrations were detected: $[HF] = 0.500\ M$, $[H_2] = 1.00 \times 10^{-3}\ M$, and $[F_2] = 4.00 \times 10^{-3}\ M$. Is the system at equilibrium? If not, what must occur in order for equilibrium to be established?

Solution

We substitute these concentrations into the mass action expression,

$$Q = \frac{[H_2][F_2]}{[HF]^2} = \frac{(1.00 \times 10^{-3})(4.00 \times 10^{-3})}{(0.500)^2}$$

$$= 1.60 \times 10^{-5}$$

and find that $Q = 1.60 \times 10^{-5}$ and therefore $Q > K$. In order for equilibrium to be established, the value of Q must become smaller until it equals K_c. This can occur only if the numerator decreases and the denominator increases. Thus, the right-to-left reaction must occur to a greater extent than the forward reaction; i.e., H_2 and F_2 react to form more HF.

16–5 Factors That Affect Equilibria

Once a reacting system has achieved equilibrium, it remains at equilibrium until it is disturbed by some change of conditions. We shall now consider the different types of changes, remembering that the *value* of an equilibrium constant changes only with temperature. The guiding principle, known as **LeChatelier's Principle,** was introduced in Section 12–5. It states: *If a change of conditions (stress) is applied to a system at equilibrium, the system will respond in the way that best reduces the stress in*

For reactions involving gases at constant temperature, pressure changes cause volume changes and vice-versa.

reaching a new state of equilibrium. There are four types of changes to consider: (1) concentration changes, (2) pressure changes (which can also be treated as volume changes for gas phase reactions), (3) temperature changes, and (4) introduction of catalysts.

In this section we shall look at the effects of these types of stresses from a purely qualitative or descriptive point of view. In Section 16–6 we shall illustrate the validity of our discussion with several quantitative examples.

1   Changes in Concentration

Consider the following system at equilibrium:

$$A + B \rightleftharpoons C + D \qquad K_c = \frac{[C][D]}{[A][B]}$$

If an additional amount of any reactant or product is *added* to the system, the stress is relieved as the reaction that consumes the added substance occurs more rapidly than its reverse reaction. Let us compare the mass action expressions for Q and K. If more A or more B is added, then $Q < K$ and the forward reaction proceeds to a greater extent than the reverse reaction until equilibrium is re-established. If more C or D is added, $Q > K$ and the reverse reaction occurs more rapidly until equilibrium is re-established.

If a reactant or product is *removed* from a system at equilibrium, the reaction that produces the substance occurs more rapidly than its reverse. For example, if some C is removed, then $Q < K$, and the forward reaction is favored until equilibrium is re-established. If A is removed, the reverse reaction is favored. To summarize:

Stress	Q	Direction of Faster Reaction
Increase concentration of A or B	$Q < K$	→ right
Increase concentration of C or D	$Q > K$	left ←
Decrease concentration of A or B	$Q > K$	left ←
Decrease concentration of C or D	$Q < K$	→ right

2   Changes in Volume and Pressure

Changes in pressure do not significantly affect the concentrations of solids or liquids, since they are nearly incompressible. However, the concentrations of gases are significantly altered by changes in pressure. Therefore, such changes affect the value of Q for reactions in which the number of moles of gaseous reactants differs from the number of moles of gaseous products. For an ideal gas

$$PV = nRT \qquad \text{and} \qquad P = (n/V)(RT)$$

The term n/V is concentration in moles per liter. At constant temperature n, R, and T are constants. Thus, if the volume occupied by a gas decreases, its partial pressure increases and its concentration (n/V) increases. If the volume increases, both its partial pressure and its concentration decrease.

Consider the following gaseous system at equilibrium:

$$A (g) \rightleftharpoons 2D (g) \qquad K = \frac{[D]^2}{[A]}$$

At constant temperature, an increase in pressure (decrease in volume) increases the concentrations of both A and D. In the mass action expression, the concentration of D is squared and the concentration of A is raised to the first power. As a result, the numerator of Q increases more dramatically with a given increase in pressure than does the denominator. Thus, $Q > K$ and the reverse reaction is favored. Conversely, a decrease in pressure (increase in volume) causes the reaction to the right to be favored until equilibrium is re-established, since $Q < K$. Thus, for the system shown above (and any system in which there are more moles of gases on the right side of the equation than on the left side) initially at equilibrium, we may summarize the effect of pressure (volume) changes:

Stress	Q (for this reaction)	Direction of Faster Reaction
Pressure increase (Volume decrease)	$Q > K$	Toward smaller number of moles of gas (left for *this* reaction)
Pressure decrease (Volume increase)	$Q < K$	Toward larger number of moles of gas (right for *this* reaction)

In general, an increase in pressure (decrease in volume) applied to a system at equilibrium favors the reaction in the direction that produces the smaller number of moles of gases, and a decrease in pressure favors the opposite reaction. If there is no change in number of moles of gases in a reaction, a pressure (volume) change does not affect the equilibrium.

It should be noted that the foregoing argument applies only when pressure changes are due to volume changes. It *does not apply* if the total pressure of a gaseous system is raised by pumping in an inert gas, which has no effect on the equilibrium. In such a situation, the *partial* pressure of each reacting gas remains constant, so the system remains at equilibrium.

3 Changes in Temperature

Consider the following system at equilibrium:

$$A + B \rightleftharpoons C + D + \text{heat}$$

Heat is a product of the forward (exothermic) reaction. Increasing the temperature at constant pressure and volume increases the amount of heat in the system. This increases the rate of the reaction to the left, consuming some of the extra heat. Lowering the temperature favors the reaction to the right as the system replenishes some of the heat removed. Likewise, for the following system at equilibrium,

$$W + X + \text{heat} \rightleftharpoons Y + Z$$

an increase in temperature at constant pressure and volume favors the reaction to the right, and a decrease in temperature favors the reaction to the left. In fact, the *values* of equilibrium constants change as temperature changes. No other stresses affect the value of K.

4 Introduction of a Catalyst

A catalyst may be added to a system to change the rate of a reaction (Section 15–9), but it *cannot* favor either the forward or the reverse reaction. Since a catalyst affects the activation energy of both forward and reverse reactions, it changes both equally, and only the time required to establish equilibrium is altered.

Not all reactions attain equilibrium; they simply occur too slowly, or else products or reactants are continually added or removed. Such is the case with most reactions in biological systems. On the other hand, some reactions, such as typical acid-base neutralizations, achieve equilibrium very rapidly.

The Haber Process

The Haber process for the industrial production of ammonia is interesting because it provides insight into both the kinetic and thermodynamic factors that influence reaction rates. Ammonia is produced by the combination of nitrogen and hydrogen, an exothermic, reversible reaction that in fact never reaches a state of equilibrium, but moves toward it.

For uses of NH_3 see Section 9–2.2 and Table 9–5.

$$N_2 \text{ (g)} + 3H_2 \text{ (g)} \rightleftharpoons 2NH_3 \text{ (g)} + 92.22 \text{ kJ}$$

$$K_c = \frac{[NH_3]^2}{[N_2][H_2]^3} = 4.0 \times 10^8 \text{ at } 25°C$$

The process is diagrammed in Figure 16–4. The reaction is run at about 450°C under pressures ranging from 200 to 1000 atmospheres. Hydrogen is obtained from coal gas or petroleum refining, and nitrogen from liquefied air.

The very large value of K_c indicates that *at equilibrium* virtually all of the nitrogen and hydrogen (mixed in a 1:3 ratio) are converted to ammonia. However, the reaction occurs so slowly at 25°C that no measurable amount of ammonia is produced within a reasonable period of time. Thus, the large equilibrium constant (a thermodynamic factor) indicates that the reaction proceeds toward to the right almost completely, *but* it tells us nothing about how fast the reaction occurs (a kinetic factor).

Since there are four moles of gases on the left side of the balanced equation and only two moles of gas on the right side, increasing the pressure on this system favors the production of ammonia. Therefore, the Haber process is carried out at very high pressures, as high as the equipment will safely stand.

FIGURE 16–4 A simplified representation of the Haber process for synthesizing ammonia.

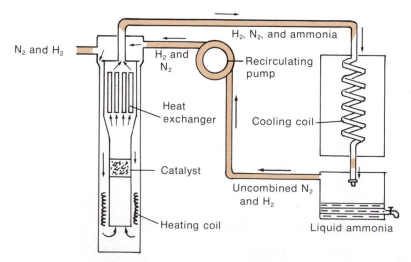

TABLE 16–1   Effect of Temperature and Pressure on Yield of Ammonia

°C	K_c	Mole % NH₃ in Equilibrium Mixture		
		10 atm	100 atm	1000 atm
200	650	51	82	98
400	0.5	4	25	80
600	0.014	0.5	5	13

The reaction is exothermic ($\Delta H^0_{rxn} = -92.22$ kJ), so increasing the temperature favors the decomposition of ammonia. The rates of both forward and reverse reactions increase as temperature increases. Therefore the reaction is carried out at high temperatures *to speed up the reaction*, even though somewhat less ammonia is produced. The economic advantage of the increase in rate outweighs the disadvantage of decreasing the yield of ammonia.

The addition of a catalyst consisting of finely divided iron and small amounts of selected oxides speeds up both the forward and reverse reactions because activation energies are lowered. This enables the reaction to proceed at a lower temperature, which is more favorable for yield and for life of the equipment.

Table 16–1 shows the effects of increases in temperature and pressure on the yield of ammonia from the Haber process. Note that K_c decreases from 4.0×10^8 at 25°C to only 1.4×10^{-2} at 600°C, which tells us that the reaction proceeds *very far to the left* as the temperature is raised. Casual examination of the data in Table 16–1 might indicate that the reaction should be run at 200°C, since a high percentage of the nitrogen and hydrogen mixture is converted into ammonia. However, the reaction occurs so slowly, even in the presence of a catalyst, that it cannot be run economically until the temperature reaches about 450°C, at as high a pressure as possible. In practice, the mixed reactants are compressed by special pumps and injected into the heated reaction vessel.

The emerging reaction mixture is cooled and ammonia (bp = −33.43°C) is removed periodically by liquefaction. This, of course, favors the forward reaction. The unreacted nitrogen and hydrogen are recycled. Excess nitrogen is used to favor the reaction to the right.

The Haber process is economically the most important process by which atmospheric nitrogen is converted to a soluble, reactive form. (N_2 is extremely stable and therefore very unreactive.) Many dyes, plastics, explosives, fertilizers, and synthetic fibers are nitrogen-containing compounds produced from ammonia.

16–6   Application of Stress to a System at Equilibrium

Example 16–5 illustrates how an equilibrium constant can be used to determine a set of new equilibrium concentrations resulting from addition of one or more species to a system at equilibrium.

Example 16–5

Some hydrogen and iodine are mixed at 400°C in a 1.00 liter container, and when equilibrium is established the following concentrations are present: [HI] = 0.490 M, [H_2] = 0.080 M, and [I_2] = 0.060 M. If an additional 0.300 mole of HI is added, what concentrations will be present when the new equilibrium is established?

Solution

We can calculate the equilibrium constant from the first set of equilibrium concentrations:

$$H_2 (g) + I_2 (g) \rightleftharpoons 2HI (g)$$

$$K_c = \frac{[HI]^2}{[H_2][I_2]} = \frac{(0.490)^2}{(0.080)(0.060)} = 50$$

When 0.300 mole of HI is added, the concentration of HI instantaneously increases by $0.300\ M$ (the volume is 1.00 liter).

	$H_2 (g)$	$+$	$I_2 (g)$	$\rightleftharpoons$	$2HI (g)$
eq:	$0.080\ M$		$0.060\ M$		$0.490\ M$
added	$0\ M$		$0\ M$		$+0.300\ M$
new in:	$0.080\ M$		$0.060\ M$		$0.790\ M$

Substitution of these concentrations into the mass action expression shows that $Q > K$.

$$Q = \frac{[HI]^2}{[H_2][I_2]} = \frac{(0.790)^2}{(0.080)(0.060)} = 130$$

Since $Q > K$, the reaction to the left occurs more rapidly to establish a new equilibrium, and the new equilibrium concentrations can be determined as follows. Let $x =$ concentration of H_2 and I_2 formed, and $2x =$ concentration of HI consumed. Then

	$H_2 (g)$	$+$	$I_2 (g)$	$\rightleftharpoons$	$2HI (g)$
new in:	$0.080\ M$		$0.060\ M$		$0.790\ M$
ch:	$+x\ M$		$+x\ M$		$-2x\ M$
new eq:	$(0.080 + x)\ M$		$(0.060 + x)\ M$		$(0.790 - 2x)\ M$

Substitution of these values into K_c allows us to evaluate x:

$$K_c = 50 = \frac{(0.790 - 2x)^2}{(0.080 + x)(0.060 + x)}$$

$$50 = \frac{0.624 - 3.16x + 4x^2}{0.0048 + 0.14x + x^2}$$

$$0.24 + 7.0x + 50x^2 = 0.624 - 3.16x + 4x^2$$

$$46x^2 + 10.2x - 0.38 = 0$$

$$x^2 + 0.22x - 0.0083 = 0$$

Solution by the quadratic formula gives $x = 0.033$ or -0.25. Clearly $x = -0.25$ is the extraneous root, because this would lead to negative equilibrium concentrations for H_2 and I_2, a physically impossible situation. Since $x = 0.033\ M$ is the root with physical meaning, the new equilibrium concentrations are

$$[H_2] = (0.080 + x)\ M = (0.080 + 0.033)\ M$$
$$= \underline{0.113\ M}$$

$$[I_2] = (0.060 + x)\ M = (0.060 + 0.033)\ M$$
$$= \underline{0.093\ M}$$

$$[HI] = (0.790 - 2x)\ M = (0.790 - 0.066)\ M$$
$$= \underline{0.724\ M}$$

The data below summarize this example.

Original Equilibrium	Stress Applied	New Equilibrium
$[H_2] = 0.080\ M$		$[H_2] = 0.113\ M$
$[I_2] = 0.060\ M$	Add	$[I_2] = 0.093\ M$
$[HI] = 0.490\ M$	$0.300\ M$ HI	$[HI] = 0.724\ M$

Knowing the value of the equilibrium constant is crucial to the determination of the new equilibrium concentrations. We see that some of the additional HI is consumed, but not all of it. More HI remains after the new equilibrium is established than was present before the stress was imposed. However, the new equilibrium concentrations of H_2 and I_2 are substantially larger than the original equilibrium concentrations. At the new equilibrium conditions, the equilibrium constant, K_c, must still be satisfied (within round-off error).

Example 16–6 shows how equilibrium constants can be used to calculate new equilibrium concentrations resulting from decreasing the volume (increasing the pressure) of a gaseous system that was initially at equilibrium.

Example 16-6

At 25°C the equilibrium constant, K_c, for the following reaction is 4.66×10^{-3}. If 0.800 mole of N_2O_4 is injected into a closed 1.00 liter glass container at 25°C, what will the equilibrium concentrations of the two gases be? What will the concentrations be at equilibrium if the volume is suddenly halved at constant temperature?

$$N_2O_4 (g) \rightleftharpoons 2NO_2 (g)$$

$$K_c = \frac{[NO_2]^2}{[N_2O_4]} = 4.66 \times 10^{-3}$$

Solution

$$N_2O_4 (g) \rightleftharpoons 2NO_2 (g)$$

in:	0.800 M	0 M
ch:	−x M	+2x M
eq:	(0.800 − x) M	2x M

$$K_c = 4.66 \times 10^{-3} = \frac{(2x)^2}{0.800 - x} = \frac{4x^2}{0.800 - x}$$

$$3.73 \times 10^{-3} - 4.66 \times 10^{-3}x = 4x^2$$

$$4x^2 + 4.66 \times 10^{-3}x - 3.73 \times 10^{-3} = 0$$

$$x^2 + 1.16 \times 10^{-3}x - 9.32 \times 10^{-4} = 0$$

Solving by the quadratic formula gives $x = 3.00 \times 10^{-2}$ and $x = -3.11 \times 10^{-2}$. The value of x must be between $0\ M$ and $0.800\ M$, and the acceptable value of x is

$$x = 3.00 \times 10^{-2}\ M$$

The original equilibrium concentrations are

$$[NO_2] = 2x\ M = 6.00 \times 10^{-2}\ M$$

$$\begin{aligned}[N_2O_4] &= (0.800 - x)\ M \\ &= (0.800 - 3.00 \times 10^{-2})\ M \\ &= 0.770\ M\end{aligned}$$

When the volume of the reaction vessel is halved (the pressure is doubled), the concentrations are doubled instantaneously, so the new initial concentrations of N_2O_4 and NO_2 are $2(0.770\ M) = 1.54\ M$ and $2(6.00 \times 10^{-2}\ M) = 0.120\ M$, respectively. A decrease in volume (increase in pressure) favors the production of N_2O_4.

$$Q = \frac{(0.120)^2}{1.54} = 9.35 \times 10^{-3}$$

$Q > K$, so reaction to left is favored

$$N_2O_4 (g) \rightleftharpoons 2NO_2 (g)$$

new in:	1.54 M	0.120 M
ch:	+x M	−2x M
new eq:	(1.54 + x) M	(0.120 − 2x) M

$$K_c = 4.66 \times 10^{-3} = \frac{(0.120 - 2x)^2}{1.54 + x}$$

$$7.18 \times 10^{-3} + 4.66 \times 10^{-3}x = 1.44 \times 10^{-2} - 0.480x + 4x^2$$

Rearranging into the general form of a quadratic equation:

$$4x^2 - 0.485x + 7.22 \times 10^{-3} = 0$$

$$x^2 - 0.121x + 1.81 \times 10^{-3} = 0$$

Solving by the quadratic formula gives $x = 0.104$ or 0.017. The maximum value of x is $0.060\ M$ (or $0.120\ M/2$), since $2x$ may not exceed the concentration of NO_2 initially present when the volume is halved. Thus, $x = 0.104$ is the extraneous root and $x = 0.017\ M$ is the root with physical significance. The new equilibrium concentrations in the 0.500 liter container are

$$\begin{aligned}[NO_2] &= (0.120 - 2x)\ M = (0.120 - 0.034)\ M \\ &= 0.086\ M\end{aligned}$$

$$\begin{aligned}[N_2O_4] &= (1.54 + x)\ M = (1.54 + 0.017)\ M \\ &= 1.56\ M\end{aligned}$$

The data below summarize this example.

First Equilibrium	Stress	New Equilibrium
$[N_2O_4] = 0.770\ M$ $[NO_2] = 6.00 \times 10^{-2}\ M$	Decrease volume from 1.00 L to 0.500 L	$[N_2O_4] = 1.56\ M$ $[NO_2] = 0.086\ M$

The fact that the concentrations of both N_2O_4 and NO_2 *increase* is a consequence of the large decrease in volume. But you may verify that the *number of moles* of N_2O_4 increases and the *number of moles* of NO_2 decreases, as one would predict by applying the principle of LeChatelier.

16–7 Partial Pressures and the Equilibrium Constant

It is often more convenient to measure pressures rather than concentrations of gases. Recall that if the ideal gas law is solved for pressure, the result is

$$P = \frac{n}{V}(RT)$$

Thus, the pressure of a gas is directly proportional to its concentration (n/V). For equilibria involving gases, the equilibrium constant is often expressed in terms of partial pressures in atmospheres (K_p) rather than in terms of concentrations (K_c). For instance, for the following reversible reaction,

$$N_2 \text{ (g)} + 3H_2 \text{ (g)} \rightleftharpoons 2NH_3 \text{ (g)}$$

K_p is defined as

$$K_p = \frac{(P_{NH_3})^2}{(P_{N_2})(P_{H_2})^3}$$

Example 16–7

The following partial pressures were measured at equilibrium at 500°C: $P_{NH_3} = 0.147$ atm, $P_{N_2} = 5.00$ atm, and $P_{H_2} = 6.00$ atm. Evaluate K_p at 500°C for the reaction

$$N_2 \text{ (g)} + 3H_2 \text{ (g)} \rightleftharpoons 2NH_3 \text{ (g)}$$

Solution

$$K_p = \frac{(P_{NH_3})^2}{(P_{N_2})(P_{H_2})^3} = \frac{(0.147 \text{ atm})^2}{(5.00 \text{ atm})(6.00 \text{ atm})^3}$$

$$K_p = 2.0 \times 10^{-5} \text{ atm}^{-2}$$

In general, for a reaction involving gases,

$$aA(g) + bB(g) \rightleftharpoons cC(g) + dD(g)$$

K_p is defined as

$$K_p = \frac{(P_C)^c(P_D)^d}{(P_A)^a(P_B)^b} \qquad \text{[Eq. 16–3]}$$

16–8 Relationship Between K_p and K_c

The term n/V is a concentration term (moles per liter). If the ideal gas law is rearranged, the concentration of a gas is

$$\frac{n}{V} = \frac{P}{RT} \qquad \text{or} \qquad [\text{conc'n}] = \frac{P}{RT}$$

Substituting P/RT for n/V into the K_c expression for the ammonia equilibrium gives the relationship between K_c and K_p.

$$K_c = \frac{[NH_3]^2}{[N_2][H_2]^3} = \frac{\left(\dfrac{P_{NH_3}}{RT}\right)^2}{\left(\dfrac{P_{N_2}}{RT}\right)\left(\dfrac{P_{H_2}}{RT}\right)^3} = \frac{(P_{NH_3})^2}{(P_{N_2})(P_{H_2})^3} \times \frac{\left(\dfrac{1}{RT}\right)^2}{\left(\dfrac{1}{RT}\right)^4}$$

$$K_c = K_p(RT)^2 \qquad \text{and} \qquad K_p = K_c(RT)^{-2}$$

In general, the realtionship between K_c and K_p is

$$K_p = K_c(RT)^{\Delta n} \qquad \text{[Eq. 16–4]}$$

where Δn is the number of moles of gaseous products minus the number of moles of gaseous reactants in the balanced equation,

$$\Delta n = (n_{\text{gas prod}}) - (n_{\text{gas react}}) \qquad \text{[Eq. 16–5]}$$

For reactions in which there are the same number of moles of gases on both sides of the balanced equation, $\Delta n = 0$ and $K_p = K_c$. For the ammonia equilibrium, N_2 (g) + $3H_2$ (g) $\rightleftharpoons$ $2NH_3$ (g), we find $\Delta n = -2$, so at 500°C,

$$K_p = K_c(RT)^{\Delta n} \qquad \text{and} \qquad K_c = K_p(RT)^{-\Delta n}$$

$$K_c = (2.0 \times 10^{-5} \text{ atm}^{-2})[(0.0821 \text{ L} \cdot \text{atm/mol} \cdot \text{K})(773 \text{ K})]^{-(-2)}$$
$$= (2.0 \times 10^{-5})(4028) \text{ L}^2/\text{mol}^2 = 0.081 \text{ } M^{-2}$$

which agrees with the value at 500°C calculated in Example 16–1, within the range of round-off error.

16–9 Heterogeneous Equilibria

Thus far we have considered only equilibria involving species in a single phase, i.e., homogeneous equilibria. **Heterogeneous equilibria** are equilibria involving species in more than one phase. Consider the following reversible reaction at 25°C:

$$2HgO \text{ (s)} \rightleftharpoons 2Hg \text{ (ℓ)} + O_2 \text{ (g)}$$

When equilibrium is established for this system, a solid, a liquid, and a gas are present. Neither solids nor liquids are affected significantly by changes in pressure unless the changes are extreme. The concentrations of solids and pure liquids are directly proportional to their densities, which vary only with temperature.* Therefore, the concentrations and partial pressures of pure solids and liquids remain constant in any reaction as long as any solid or liquid is present and the temperature remains constant. As a result, we can simplify the equilibrium expressions for the above reaction, which we represent as

$$K_c' = \frac{[Hg]^2[O_2]}{[HgO]^2} \qquad \text{and} \qquad K_p' = \frac{(P_{Hg})^2(P_{O_2})}{(P_{HgO})^2},$$

by including the concentrations or partial pressures of Hg and HgO (which are constants) in the values of the equilibrium constants. We do this by multiplying both sides of the K_c' equation by $[HgO]^2/[Hg]^2$, and by multiplying both sides of the K_p' equation by $(P_{HgO})^2/(P_{Hg})^2$. The new equilibrium constants are K_c and K_p.

$$\underbrace{K_c' \frac{[HgO]^2}{[Hg]^2}}_{\text{constants}} = [O_2] = K_c \qquad \underbrace{K_p' \frac{(P_{HgO})^2}{(P_{Hg})^2}}_{\text{constants}} = P_{O_2} = K_p$$

---

* $\underbrace{\dfrac{\text{mol}}{\text{L}}}_{} = \underbrace{\dfrac{\text{g}}{\text{mL}}}_{} \times \underbrace{\dfrac{1000 \text{ mL}}{\text{L}}}_{} \times \underbrace{\dfrac{\text{mol}}{\text{g}}}_{}$

conc'n = density $\times$ constant $\times \dfrac{1}{\text{formula weight}}$

= constant $\times$ constant $\times$ constant

The equilibrium constant expressions indicate that equilibrium exists at a given temperature for *one and only one* concentration and one partial pressure of oxygen in contact with liquid mercury and solid mercury(II) oxide.

By convention, tabulated values of equilibrium constants for reactions involving solids and/or pure liquids include the concentrations (partial pressures) of the solids and liquids as part of the constants. Therefore, concentration or partial pressure terms for pure liquids and solids are omitted from equilibrium constant expressions.

Example 16–8

Write both K_c and K_p expressions for the following reversible reactions.

Solution

(a) $2SO_2$ (g) $+ O_2$ (g) $\rightleftharpoons 2SO_3$ (g)

$$K_c = \frac{[SO_3]^2}{[SO_2]^2[O_2]} \qquad\qquad K_p = \frac{(P_{SO_3})^2}{(P_{SO_2})^2(P_{O_2})}$$

(b) $CaCO_3$ (s) $\rightleftharpoons CaO$ (s) $+ CO_2$ (g)

$$K_c = [CO_2] \qquad\qquad K_p = P_{CO_2}$$

(c) H_2O (ℓ) $+ CO_2$ (g) $\rightleftharpoons H_2CO_3$ (aq)

$$K_c = \frac{[H_2CO_3]}{[CO_2]} \qquad\qquad K_p = \frac{1}{P_{CO_2}} = (P_{CO_2})^{-1}$$

(d) $2NH_3$ (g) $+ H_2SO_4$ (ℓ) $\rightleftharpoons (NH_4)_2SO_4$ (s)

$$K_c = \frac{1}{[NH_3]^2} = [NH_3]^{-2} \qquad\qquad K_p = \frac{1}{(P_{NH_3})^2} = (P_{NH_3})^{-2}$$

(e) S (s) $+ H_2SO_3$ (aq) $\rightleftharpoons H_2S_2O_3$ (aq)

$$K_c = \frac{[H_2S_2O_3]}{[H_2SO_3]} \qquad\qquad K_p \text{ undefined;}$$
$$\qquad\qquad\qquad\qquad\qquad \text{no gases involved}$$

16–10   Relationship Between ΔG^0 and the Equilibrium Constant

The standard free energy change for a reaction is ΔG^0. This is the free energy change that accompanies complete conversion of reactants to products at standard conditions (*all* present in 1 M concentrations in solution, or as pure solids or pure liquids, or gases at 1 atm partial pressure.) The free energy change under any specific set of conditions is ΔG (no superscript zero). The two quantities are related by the equation

Demonstration of the validity of this relationship is beyond the scope of this text.

$$\Delta G = \Delta G^0 + RT \ln Q \qquad\qquad\qquad\qquad \text{[Eq. 16–6]}$$

or, after converting natural log ($\log_e$ or ln) to $\log_{10}$ (common log),

$$\Delta G = \Delta G^0 + 2.303RT \log Q \qquad\qquad\qquad \text{[Eq. 16–7]}$$

where R is the universal gas constant, T is the absolute temperature, and Q is the reaction quotient. Recall from Section 16–4 that the reaction quotient has the same form as the equilibrium expression for a reaction, but does not (necessarily) involve concentrations at equilibrium.

When a system is at equilibrium, $\Delta G = 0$ (see Section 14–8), and

$$0 = \Delta G^0 + 2.303RT \log Q$$

or

$$\Delta G^0 = -2.303RT \log Q$$

At equilibrium $Q = K$, and therefore

$$\Delta G^0 = -2.303RT \log K \qquad \text{[Eq. 16–8]}$$

This equation shows the relationship between the standard free energy change and the equilibrium constant for a reaction. Strictly speaking, the equilibrium constant in this equation is the **thermodynamic equilibrium constant.** For the generalized reaction

$$a\text{A} + b\text{B} \rightleftharpoons c\text{C} + d\text{D}$$

this thermodynamic equilibrium constant is defined in terms of the activities or effective concentrations of the species involved:

$$K = \frac{(a_\text{C})^c (a_\text{D})^d}{(a_\text{A})^a (a_\text{B})^b} \qquad \text{[Eq. 16–9]}$$

where a_A is the activity of substance A, and so on. (Section 13–16 discussed the activities of species in electrolyte solutions.) For our present purposes, we can consider the activity of each species to be a dimensionless quantity whose numerical value may be determined as follows: (a) For pure liquids and solids, the activity is taken as 1. (b) For components in ideal solution, the activity of each component is taken to be equal to its molar concentration. (c) For gases in an ideal mixture, the activity of each component is taken to be equal to its partial pressure in atm.

Making these substitutions, we can see the thermodynamic origin of Eq. 16–1 (Section 16–2) for K_c, Eq. 16–3 (Section 16–7) for K_p, and the relationships for equilibrium constants in heterogeneous equilibria (Section 16–9). Further, we can see that, in thermodynamic terms, the equilibrium constant (Eq. 16–9) is dimensionless. This thermodynamic approach helps to explain why we frequently omit units for K_c and K_p, as long as their evaluation involves molar concentrations and partial pressures in atmospheres, respectively.

To summarize, we apply Eq. 16–8 as follows. For reactions involving only gases (or only gases and pure solids or pure liquids), the equilibrium constant related to ΔG^0 is K_p. For reactions involving substances in solution, it is K_c whose value is given by this relationship.

How can we rationalize the fact that ΔG^0 is the free energy change that occurs during a reaction *beginning* with all reactants and products at unit molarities or partial pressures, while K involves equilibrium concentrations? This idea was first presented in Section 14–8, and we now relate that discussion to the equilibrium constant.

When ΔG^0 is negative, the reaction is spontaneous; i.e., the forward reaction may proceed to a greater extent than the reverse reaction *until* equilibrium is established. At that instant ($\Delta G = 0$, and $Q = K$) and thereafter, equilibrium concentrations are present. Conversely, if ΔG^0 for a reaction is positive, the *reverse reaction* is spontaneous and is favored over the forward reaction until equilibrium is

You may wish to refer back to the discussion of the relation between the value of ΔG^0 and the position of equilibrium in Section 14–8.

In thermodynamic terms, this is why "concentrations" of pure liquids and solids do not appear in the equilibrium constant expression.

established. In the special case where the system is already at equilibrium (when all species are present in unit concentrations or at unit partial pressures), $\Delta G^0 = 0$ and $Q = K = 1$.

Thus, the sign of ΔG^0 indicates whether the forward or reverse reaction is spontaneous at unit concentrations or partial pressures, and the magnitude indicates how spontaneous or nonspontaneous a reaction is when all reactants and products are at unit concentrations or partial pressures. When ΔG^0 is large and negative, K is very large, indicating that *after equilibrium is established,* the ratio of products to reactants will be very large. When ΔG^0 is large and positive, K is very small, and *at equilibrium* the ratio of products to reactants will be very small. In other words, ΔG^0 tells us *how far away from equilibrium* the starting reaction mixture is and whether the reaction occurs to the right or left to achieve equilibrium. To summarize:

ΔG^0 is $-$, $K > 1$	forward reaction is spontaneous at unit concentrations or partial pressures
ΔG^0 is 0, $K = 1$	system is at equilibrium at unit concentrations or partial pressures (very rare case)
ΔG^0 is $+$, $K < 1$	reverse reaction is spontaneous at unit concentrations or partial pressures

Always keep in mind that if a reaction is spontaneous, the reverse reaction is nonspontaneous. Also, $\Delta G = 0$ when a system is at equilibrium, but ΔG^0 is usually not equal to zero.

Example 16–9

Using the data in Appendix K, evaluate the equilibrium constant for the following reaction at 25°C.

$$2C_2H_2 (g) + 5O_2 (g) \rightleftharpoons 4CO_2 (g) + 2H_2O (g)$$

Solution

We may evaluate ΔG^0_{rxn} for the forward reaction from ΔG^0_f values.

$$\Delta G^0_{rxn} = 4\Delta G^0_{f\,CO_2 (g)} + 2\Delta G^0_{f\,H_2O (g)}$$
$$- [2\Delta G^0_{f\,C_2H_2 (g)} + 5\Delta G^0_{f\,O_2 (g)}]$$
$$= [4(-394.4) + 2(-228.6)]\ kJ$$
$$- [2(209.2) + 5(0)]\ kJ$$
$$= -2.45 \times 10^3\ kJ$$
$$\text{or} -2.45 \times 10^6\ J/\text{mol of reaction}$$

This is a gas-phase reaction, and ΔG^0_{rxn} is related to K_p by

$$\Delta G^0_{rxn} = -2.303RT \log K_p$$

$$\log K_p = \frac{-2.45 \times 10^6\ J/\text{mol}}{-2.303(8.314\ J/\text{mol} \cdot K)(298\ K)}$$

$$\log K_p = 429$$

$$K_p = \text{antilog } 429 = 10^{429}$$

The very large value of K_p tells us that the equilibrium lies very far to the right.

In Example 16–9 we have used units on all numbers. Note that ΔG^0_{rxn} is expressed in J/mol of reaction; this means that the free energy of the system changes by -2.45×10^6 J for every two moles of C_2H_2 and five moles of O_2 that react. Units are often omitted from calculations of this type.

Example 16–10

Diatomic nitrogen and oxygen molecules make up about 99% of all the molecules in reasonably "unpolluted" air (Section 9–7.4). For the following reaction, ΔG^0 is 173.1 kJ or 1.731×10^5 J. Calculate K_p at 25°C.

$$N_2 (g) + O_2 (g) \rightleftharpoons 2NO (g)$$

Solution

$$\Delta G^0 = -2.303RT \log K_p$$

$$\log K_p = \frac{1.731 \times 10^5}{-2.303(8.314)(298)}$$

$$\log K_p = -30.34$$
$$K_p = 10^{0.66} \times 10^{-31} = \underline{4.6 \times 10^{-31}}$$

Clearly, this very small number indicates that at equilibrium, almost no N_2 and O_2 are converted to NO at 25°C. For all practical purposes, the reaction does not occur at 25°C.

Example 16–11

The equilibrium constant for the following reaction is 5.04×10^{17} atm^{-1} at 25°C. Calculate ΔG^0.

$$C_2H_4 \text{ (g)} + H_2 \text{ (g)} \Longrightarrow C_2H_6 \text{ (g)}$$
ethylene   ethane

Solution

$$\begin{aligned}
\Delta G^0 &= -2.303RT \log K_p \\
&= -2.303(8.314)(298) \log (5.04 \times 10^{17}) \\
&= -5706 \log (5.04 \times 10^{17}) \\
&= -5706(17.702) = -1.010 \times 10^5 \text{ J}
\end{aligned}$$
$$\Delta G^0 = -\underline{101 \text{ kJ}}$$

16–11   Evaluation of Equilibrium Constants at Different Temperatures

Chemists have determined equilibrium constants for thousands of reactions. It would be an impossibly large task to catalog such constants for each reaction at every temperature of interest. Fortunately, there is no need to do this. If we determine the equilibrium constant K_{T_1} for a reaction at one temperature, T_1, and also its ΔH^0, then we can estimate the equilibrium constant K_{T_2} at a second temperature T_2 by using the **van't Hoff equation:**

$$\frac{\Delta H^0(T_2 - T_1)}{2.303RT_2T_1} = \log \left(\frac{K_{T_2}}{K_{T_1}} \right) \qquad \text{[Eq. 16–10]}$$

Thus, if we know ΔH^0 for a reaction and K (or ΔG^0, from which we can calculate K) at a given temperature (say 298 K), we can use the equation to calculate the value of K at any other temperature.

Note that Eq. 16–10 has the same form as Eq. 15–17, the equation that relates k, E_a, and T in kinetics. Consequently, ΔH^0 can also be determined graphically from a plot of $\log K$ vs. $1/T$, which provides an alternate means of estimating ΔH^0_{rxn}.

Example 16–12

We found in Example 16–10 that $K_p = 4.6 \times 10^{-31}$ at 25°C and, from Appendix K, $\Delta H^0 = 180.5$ kJ for the reaction

$$N_2 \text{ (g)} + O_2 \text{ (g)} \Longrightarrow 2NO \text{ (g)}$$

Evaluate the equilibrium constant for this reaction at 2400 K, and then let us compare $K_{2400 \text{ K}}$ with $K_{298 \text{ K}}$. (2400 K is a typical temperature inside the combustion chambers of automobile engines. Large quantities of N_2 and O_2 are present during gasoline combustion, since the gasoline is mixed with air.)

Solution

Let $T_1 = 298$ K and $T_2 = 2400$ K. Then

$$\frac{\Delta H^0(T_2 - T_1)}{2.303RT_2T_1} = \log \left(\frac{K_{T_2}}{K_{T_1}} \right)$$

We know all of the terms except K_{T_2}. Let us solve for $\log K_{T_2}$. First we use the general relation, $\log (x/y) = \log x - \log y$.

$$\frac{\Delta H^0(T_2 - T_1)}{2.303RT_2T_1} = \log K_{T_2} - \log K_{T_1}$$

$$\frac{\Delta H^0(T_2 - T_1)}{2.303RT_2T_1} + \log K_{T_1} = \log K_{T_2}$$

If we express R in $J/mol \cdot K$, then ΔH^0 must be expressed in J/mol.

$$\log K_{T_2} = \frac{(1.805 \times 10^5)(2400 - 298)}{(2.303)(8.314)(2400)(298)}$$
$$+ \log (4.6 \times 10^{-31})$$

$$\log K_{T_2} = 27.71 + \log (4.6 \times 10^{-31})$$
$$= 27.71 + (-30.34) = -2.63$$

$$K_{T_2} = 10^{-2.63} = 10^{0.37} \times 10^{-3}$$

$$K_{T_2} = 2.34 \times 10^{-3}$$

We see that K_{T_2} (K_p at 2400 K) is quite small, and that the equilibrium favors N_2 and O_2 rather than NO. However, K_{T_2} is very much larger than K_{T_1}, which is 4.6×10^{-31}. At 2400 K, significantly more NO is present at equilibrium, relative to N_2 and O_2, than at 298 K. Thus, small but significant amounts of NO are exhausted into the atmosphere by automobiles. Only small amounts of NO are necessary to cause severe air pollution problems. Catalytic converters are designed to catalyze the forward and reverse reactions (and others) involving breakdown of nitrogen oxide into nitrogen and oxygen.

$$2NO\ (g) \rightleftharpoons N_2\ (g) + O_2\ (g)$$

The forward reaction as written here is spontaneous. Remember that catalysts do not affect the equilibrium. That is, they favor neither the consumption nor production of nitrogen oxide. They merely allow the system to reach equilibrium more rapidly.

Key Terms

Chemical equilibrium a state of dynamic balance in which the rates of forward and reverse reactions are equal; there is no net change in concentrations of reactants or products while a system is at equilibrium.

Dynamic equilibrium an equilibrium in which processes occur continuously, with no *net* change.

Equilibrium constant a quantity that characterizes the position of equilibrium for a reversible reaction; its magnitude is equal to the mass action expression at equilibrium. K varies with temperature.

Heterogeneous equilibria equilibria involving species in more than one phase.

Homogeneous equilibria equilibria involving only species in a single phase.

LeChatelier's Principle if a stress (change of conditions) is applied to a system at equilibrium, the system will respond in the way that best reduces the stress in reaching a new state of equilibrium.

Mass action expression for a reversible reaction, the product of the concentrations of the products (species on the right), each raised to the power that corresponds to its coefficient in the balanced chemical equation, divided by the product of the concentrations of the reactants (species on the left), each raised to the power that corresponds to its coefficient in the balanced chemical equation. At equilibrium the mass action expression is equal to K; at other conditions, it is Q.

$$aA + bB \rightleftharpoons cC + dD$$

$$\frac{[C]^c[D]^d}{[A]^a[B]^b} = Q, \text{ or at equilibrium, } K$$

Reaction quotient, Q, the mass action expression under any set of conditions (not necessarily equilibrium); its magnitude relative to K determines the direction in which reaction must occur to establish equilibrium.

Reversible reactions reactions that do not go to completion and occur in both the forward and reverse directions.

van't Hoff equation the relationship between ΔH^0 for a reaction and its equilibrium constants at two different temperatures.

Exercises

Basic Concepts

1. Define and illustrate the following terms: (a) reversible reaction, (b) chemical equilibrium, (c) equilibrium constant, K.

2. Why are chemical equilibria called dynamic equilibria?

3. Write the expression for K for each of the following reactions in terms of concentrations.
   (a) $2NO_2 (g) \rightleftharpoons 2NO (g) + O_2 (g)$
   (b) $CO (g) + H_2O (g) \rightleftharpoons CO_2 (g) + H_2 (g)$
   (c) $2H_2S (g) + 3O_2 (g) \rightleftharpoons$
   $\qquad\qquad\qquad 2H_2O (g) + 2SO_2 (g)$
   (d) $PCl_5 (g) \rightleftharpoons PCl_3 (g) + Cl_2 (g)$
   (e) $MgCO_3 (s) \rightleftharpoons MgO (s) + CO_2 (g)$
   (f) $P_4 (g) + 3O_2 (g) \rightleftharpoons P_4O_6 (s)$
   (g) $H_2 (g) + I_2 (s) \rightleftharpoons 2HI (g)$

4. Explain the significance of (a) a very large value of K, (b) a very small value of K, (c) a value of K of about 1.0.

5. Why may we omit concentrations of pure solids and pure liquids from equilibrium constant expressions?

6. Arrange the following in order of increasing tendency of the forward reactions to proceed toward completion at 298 K and one atmosphere pressure.
   (a) $H_2O (g) \rightleftharpoons H_2O (\ell)$
   $\qquad K_c = 782$
   (b) $F_2 (g) \rightleftharpoons 2F (g) \qquad K_c = 4.9 \times 10^{-21}$
   (c) $C \text{ (graphite)} + O_2 (g) \rightleftharpoons CO_2 (g)$
   $\qquad\qquad\qquad\qquad K_c = 1.3 \times 10^{69}$
   (d) $H_2 (g) + C_2H_4 (g) \rightleftharpoons C_2H_6 (g)$
   $\qquad\qquad\qquad\qquad K_c = 9.8 \times 10^{18}$
   (e) $N_2O_4 (g) \rightleftharpoons 2NO_2 (g)$
   $\qquad\qquad\qquad\qquad K_c = 4.6 \times 10^{-3}$

7. Define the reaction quotient Q. Distinguish between Q and K.

8. Why is it useful to compare Q with K? What is the situation in the case of (a) $Q = K$, (b) $Q < K$, (c) $Q > K$?

9. Given the equilibrium constants for the following two reactions at a particular temperature:
   $$NiO (s) + H_2 (g) \rightleftharpoons Ni (s) + H_2O (g)$$
   $$K_c = 25$$
   $$NiO (s) + CO (g) \rightleftharpoons Ni (s) + CO_2 (g)$$
   $$K_c = 500$$

   Calculate the value for the equilibrium constant, K_c, for the reaction
   $$CO_2 (g) + H_2 (g) \rightleftharpoons CO (g) + H_2O (g)$$
   at the same temperature.

10. The concentration equilibrium constant for the gas phase reaction
    $$HCHO \rightleftharpoons H_2 + CO$$
    has the numerical value 0.50 at 600°C. A mixture of HCHO, H_2, and CO is introduced into a flask at 600°C. After a short time, analysis of a small sample of the reaction mixture shows the concentrations to be $[HCHO] = 0.50\ M$, $[H_2] = 1.50\ M$, and $[CO] = 0.25\ M$. Classify each of the following statements about this reaction mixture as true or false.
    (a) The reaction mixture is at equilibrium.
    (b) The reaction mixture is not at equilibrium, but no further reaction will occur.
    (c) The reaction mixture is not at equilibrium, but will move toward equilibrium by using up more HCHO.
    (d) The forward rate of this reaction is the same as the reverse rate.

11. At some temperature, the reaction
    $$N_2 (g) + 3H_2 (g) \rightleftharpoons 2NH_3 (g)$$
    has an equilibrium constant K_c numerically equal to 1. For such a situation, state whether each of the following statements is true or false, and explain why.
    (a) An equilibrium mixture must have the H_2 concentration three times that of N_2 and the NH_3 concentration twice that of H_2.
    (b) An equilibrium mixture must have the H_2 concentration three times that of N_2.
    (c) A mixture in which the H_2 concentration is three times that of N_2 *and* the NH_3 concentration is twice that of N_2 could be an equilibrium mixture.
    (d) A mixture in which the concentration of each reactant and each product is 1 M is an equilibrium mixture.

(e) Any mixture in which the concentrations of all reactants and products are equal is an equilibrium mixture.

(f) An equilibrium mixture must have equal concentrations of all reactants and products.

Calculation and Uses of *K*

12. Given:

$$A (g) + B (g) \rightleftharpoons C (g) + 2D (g)$$

1.00 mole of A and 1.00 mole of B are placed in a 2.00 liter container. After equilibrium has been established, 0.10 mole of C is present in the container. Calculate the equilibrium constant for the reaction.

13. For the reaction

$$CO (g) + H_2O (g) \rightleftharpoons CO_2 (g) + H_2 (g)$$

the value of the equilibrium constant K_c is 1.845 at 600°C. We place 0.500 mole CO and 0.500 mole H_2O in a 1.00 L container at 600°C, and allow the reaction to reach equilibrium. What will be the equilibrium concentrations of all substances present?

14. We place 0.500 mole CO, 0.500 mole H_2O, 0.500 mole CO_2, and 0.500 mole H_2 in a 1.00 L container under the conditions of Exercise 13 and allow the reaction to reach equilibrium. What will be the equilibrium concentrations of all substances present?

15. At 800°C (1073 K), K_c values for gas phase dissociation reactions of two of the hydrogen halides are known to be

$$2HBr (g) \rightleftharpoons H_2 (g) + Br_2 (g)$$
$$K_c = 7.00 \times 10^{-7}$$

$$2HI (g) \rightleftharpoons H_2 (g) + I_2 (g)$$
$$K_c = 2.31 \times 10^{-3}$$

(a) We start with [HBr] = 0.100 M at 800°C. What would be the concentration of H_2 at equilibrium?

(b) Calculate the percentage dissociation of HBr in part (a).

(c) We start with [HI] = 0.100 M at 800°C. What would be the concentration of H_2 at equilibrium?

(d) Calculate the percentage dissociation of HI in part (c).

16. Initially 0.84 mole of PCl_5 and 0.18 mole of PCl_3 are mixed in a 1.0 L vessel at some temperature. It is later found that 0.72 mole PCl_5 is present at equilibrium (plus some amounts of PCl_3 and Cl_2, of course).

(a) What is the numerical value of K_c for the following reaction?

$$PCl_5 (g) \rightleftharpoons PCl_3 (g) + Cl_2 (g)$$

(b) Calculate the equilibrium concentrations of PCl_5, PCl_3 and Cl_2 if we put 0.60 mole of PCl_5 in a closed vessel of volume 4.0 liters, at the same temperature as in part (a), and then allow the reaction to proceed to equilibrium.

17. Given

$$A (g) + 3B (g) \rightleftharpoons C (g) + 2D (g)$$

1.00 mole of A and 1.00 mole of B are placed in a 5.00 L container. After equilibrium has been established, 0.50 mole of D is present in the container. Calculate the equilibrium constant for the reaction.

18. Antimony pentachloride decomposes in a gas phase reaction at 448°C as follows:

$$SbCl_5 (g) \rightleftharpoons SbCl_3 (g) + Cl_2 (g)$$

An equilibrium mixture in a 5.00 liter vessel is found to contain 3.84 grams of $SbCl_5$, 9.14 grams of $SbCl_3$ and 2.84 grams of Cl_2. Evaluate K_c at 448°C.

19. If 10.0 grams of $SbCl_5$ are placed in a 5.00 liter container at 448°C and allowed to establish equilibrium as in Exercise 18, what will be the equilibrium concentrations of all species? For this reaction, $K_c = 2.51 \times 10^{-2}$.

20. If 5.00 grams of $SbCl_5$ and 5.00 grams of $SbCl_3$ are placed in a 5.00 liter container and allowed to establish equilibrium as in Exercise 18, what will the equilibrium concentrations be? ($K_c = 2.51 \times 10^{-2}$)

21. At 600°C, K_c for the reaction below is 0.542.

$$CO_2 (g) + H_2 (g) \rightleftharpoons CO (g) + H_2O (g)$$

(a) If 0.100 mole of CO_2 and 0.100 mole of H_2 are placed into a closed 2.00 liter container, what equilibrium concentrations of all species will result?

(b) If 0.200 mole of CO_2 and 0.100 mole of H_2 are placed into a closed 2.00 liter container, what equilibrium concentrations of all species will result?

22. At a certain temperature, K_c for the reaction below is 0.300. If 0.600 mole of $POCl_3$ is placed in a closed 3.00 liter vessel at this temperature, what percentage of it will be dissociated when equilibrium is established?

$$POCl_3 \text{ (g)} \rightleftharpoons POCl \text{ (g)} + Cl_2 \text{ (g)}$$

23. Bromine chloride, BrCl, a reddish covalent gas with properties similar to those of Cl_2, may eventually replace Cl_2 as a water disinfectant. One mole of chlorine and one mole of bromine are enclosed in a 5.00 liter flask and allowed to reach equilibrium at a certain temperature.

$$Cl_2 \text{ (g)} + Br_2 \text{ (g)} \rightleftharpoons 2BrCl \text{ (g)}$$
$$K_c = 4.7 \times 10^{-2}$$

(a) What percentage of the chlorine has reacted at equilibrium?
(b) What mass of each species is present at equilibrium?
(c) How would a decrease in volume affect the equilibrium, if at all?

The Equilibrium Constant in Terms of Partial Pressures

24. Write the K_p expression for each of the reactions in Exercise 3.
25. Under what conditions are no units associated with (a) K_c and (b) K_p?
26. Under what conditions are K_c and K_p for a reaction numerically equal?
27. Given:

$$2Cl_2 \text{ (g)} + 2H_2O \text{ (g)} \rightleftharpoons 4HCl \text{ (g)} + O_2 \text{ (g)}$$
$$K_p = 4.6 \times 10^{-14}$$

Write the equilibrium constant expressions for K_c and K_p and calculate their values for the reaction written in the following forms.
(a) $4HCl \text{ (g)} + O_2 \text{ (g)} \rightleftharpoons 2Cl_2 \text{ (g)} + 2H_2O \text{ (g)}$
(b) $2HCl \text{ (g)} + \frac{1}{2}O_2 \text{ (g)} \rightleftharpoons Cl_2 \text{ (g)} + H_2O \text{ (g)}$
(c) $Cl_2 \text{ (g)} + H_2O \text{ (g)} \rightleftharpoons 2HCl \text{ (g)} + \frac{1}{2}O_2 \text{ (g)}$
(d) $4Cl_2 \text{ (g)} + 4H_2O \text{ (g)} \rightleftharpoons 8HCl \text{ (g)} + 2O_2 \text{ (g)}$
(e) $\frac{1}{2}Cl_2 \text{ (g)} + \frac{1}{2}H_2O \text{ (g)} \rightleftharpoons HCl \text{ (g)} + \frac{1}{4}O_2 \text{ (g)}$

28. A sample of SO_3 (g) is allowed to decompose into SO_2 (g) and O_2 (g) at 900 K.
(a) Write the balanced chemical equation for this reaction, using the smallest whole number coefficients.
(b) At equilibrium, we find that the molar concentrations are $[SO_3] = 0.262$, $[O_2] = 0.0162$, and $[SO_2] = 0.0324$. What is the value of K_c for the reaction at this temperature?
(c) What is the value of K_p for the reaction at 900 K?

29. In the Mond process for refining nickel, the very poisonous gaseous compound nickel carbonyl, $Ni(CO)_4$, is heated to cause it to decompose into solid nickel and gaseous carbon monoxide, CO.
(a) Write the balanced chemical equation for this reaction, using the smallest whole number coefficients.
(b) Write the expression for the equilibrium constant, K_p, for this reaction.

30. Solid ammonium carbamate dissociates on heating into ammonia and carbon dioxide:

$$NH_4CO_2NH_2 \text{ (s)} \rightleftharpoons 2NH_3 \text{ (g)} + CO_2 \text{ (g)}$$

When pure ammonium carbamate is put into a closed container and allowed to come to equilibrium with the gaseous products at a constant temperature of 35°C, the total pressure is found to be 0.30 atm. What is the value of K_p for this reaction at 35°C?

31. At 25°C the equilibrium constant K_c for the reaction

$$N_2O_4 \text{ (g)} \rightleftharpoons 2NO_2 \text{ (g)}$$

is 4.66×10^{-3}. We introduce 4.60 g of NO_2 into a 1.0 L vessel.
(a) What is the total pressure in the vessel before any reaction takes place?
(b) The reaction proceeds to establish equilibrium. What will be the equilibrium concentrations of all substances present?
(c) What is the total pressure in the vessel at equilibrium?

32. The dark red solid interhalogen compound iodine chloride, ICl, decomposes to the dark violet solid, I_2, and the greenish-yellow gas, Cl_2. At 25°C, K_p for the reaction is 0.24 atm.

$$2ICl \text{ (s)} \rightleftharpoons I_2 \text{ (s)} + Cl_2 \text{ (g)}$$

(a) If 1.0 mole of ICl is placed in a closed vessel, what will be the pressure of Cl_2 at equilibrium? Note that it is not necessary to specify the volume of the vessel in this case.

(b) If the 1.0 mole of ICl is placed in a 1.0 liter container, what concentration of Cl_2 is present at equilibrium?

(c) How many grams of ICl will remain at equilibrium in the 1.0 liter container?

33. Given:

$$PCl_5 \text{ (g)} \rightleftharpoons PCl_3 \text{ (g)} + Cl_2 \text{ (g)}$$

At 250°C, a sample of PCl_5 was placed in a 24 liter evacuated reaction vessel and allowed to come to equilibrium. Analysis showed that at equilibrium 0.42 mole of PCl_5, 0.64 mole of PCl_3, and 0.64 mole of Cl_2 were present in the vessel. Calculate K_c and K_p for this system at 250°C.

34. Given:

$$H_2 \text{ (g)} + I_2 \text{ (g)} \rightleftharpoons 2HI \text{ (g)}$$

1.00 g of H_2 and 127 g of I_2 are injected into a 10.0 liter evacuated reaction vessel at 448°C. K_c for this reaction is 50.

(a) What is K_p at 448°C?

(b) What are the partial pressure of each substance in the equilibrium mixture and the total pressure in the vessel at equilibrium?

(c) How many moles and grams of I_2 remain unreacted at equilibrium?

Factors Affecting K and Stresses on Equilibria

35. What would be the effect of decreasing the temperature on each of the following systems at equilibrium?

(a) H_2 (g) + I_2 (g) $\rightleftharpoons$ 2HI (g) + 9.45 kJ

(b) PCl_5 (g) + 92.5 kJ $\rightleftharpoons$ PCl_3 (g) + Cl_2 (g)

(c) $2SO_2$ (g) + O_2 (g) $\rightleftharpoons$ $2SO_3$ (g)
$$\Delta H^0 = -198 \text{ kJ}$$

(d) 2NOCl (g) $\rightleftharpoons$ 2NO (g) + Cl_2 (g)
$$\Delta H^0 = 75 \text{ kJ}$$

(e) C (s) + H_2O (g) + 131 kJ $\rightleftharpoons$
$$CO \text{ (g)} + H_2 \text{ (g)}$$

36. What would be the effect of increasing the volume of each of the following systems at equilibrium?

(a) 2CO (g) + O_2 (g) $\rightleftharpoons$ $2CO_2$ (g)

(b) 2NO (g) $\rightleftharpoons$ N_2 (g) + O_2 (g)

(c) N_2O_4 (g) $\rightleftharpoons$ $2NO_2$ (g)

(d) Ni (s) + 4CO (g) $\rightleftharpoons$ $Ni(CO)_4$ (g)

(e) N_2 (g) + $3H_2$ (g) $\rightleftharpoons$ $2NH_3$ (g)

37. What would be the effect of increasing the pressure by decreasing the volume of each of the systems of Exercise 36 at equilibrium?

38. What would be the effect of each of the following concentration changes on the systems at equilibrium?

(a) CO (g) + Cl_2 (g) $\rightleftharpoons$ $COCl_2$ (g)
add Cl_2 (g)

(b) CO (g) + Cl_2 (g) $\rightleftharpoons$ $COCl_2$ (g)
remove $COCl_2$ (g)

(c) PCl_3 (g) + Cl_2 (g) $\rightleftharpoons$ PCl_5 (g)
remove PCl_3 (g)

(d) CaO (s) + SO_3 (g) $\rightleftharpoons$ $CaSO_4$ (s)
add SO_3 (g)

(e) CaO (s) + SO_3 (g) $\rightleftharpoons$ $CaSO_4$(s)
add $CaSO_4$ (s)

(f) 2PbS (s) + $3O_2$ (g) $\rightleftharpoons$
$$2PbO \text{ (s)} + 2SO_2 \text{ (g)}$$
add O_2 (g)

(g) 2PbS (s) + $3O_2$ (g) $\rightleftharpoons$
$$2PbO \text{ (s)} + 2SO_2 \text{ (g)}$$
remove PbS (s)

39. What effect does the presence of a catalyst have on the equilibrium for a reversible reaction? Why?

40. Given: PCl_5 (g) $\rightleftharpoons$ PCl_3 (g) + Cl_2 (g)
$$\Delta H^0 = 92.5 \text{ kJ}$$
How will each of the following influence this system at equilibrium? (a) increased temperature, (b) increased pressure, (c) increased concentration of Cl_2, (d) increased concentration of PCl_5, (e) adding a catalyst.

41. Given: N_2 (g) + $3H_2$ (g) $\rightleftharpoons$ $2NH_3$ (g)
$$\Delta H = -92 \text{ kJ}$$
How will each of the following influence this system at equilibrium? (a) increased temperature, (b) increased pressure, (c) increased concentration of NH_3, (d) increased concentration of N_2, (e) adding a catalyst.

42. Given: A (g) + B (g) $\rightleftharpoons$ C (g) + D (g)

(a) At equilibrium a 1.0 liter container was found to contain 1.6 mole of C, 1.6 mole of D, 0.20 mole of A and 0.20 mole of B. Calculate the equilibrium constant for this reaction.

(b) If 0.20 mole of A and 0.20 mole of B are added to this system, what will the new equilibrium concentration of A be?

43. Given: A (g) + B (g) $\rightleftharpoons$ C (g) + D (g)
    When 1.0 mole of A and 1.0 mole of B are mixed and allowed to reach equilibrium at room temperature the mixture is found to contain $\frac{2}{3}$ mole of C and $\frac{2}{3}$ mole of D.
    (a) Calculate the equilibrium constant.
    (b) If three moles of A were mixed with one mole of B and allowed to reach equilibrium, how many moles of C would be present at equilibrium?

44. Given: A (g) $\rightleftharpoons$ B (g) + C (g)
    (a) When the system is at equilibrium at 200°C, the concentrations are found to be [A] = 0.20 M, [B] = 0.30 M = [C]. Calculate K_c.
    (b) If the volume of the container in which the system is at equilibrium is suddenly doubled at 200°C, what will the new equilibrium concentrations be?
    (c) Refer to part (a). If the volume of the container is suddenly halved at 200°C, what will the new equilibrium concentrations be?

45. The equilibrium constant, K_c, for the dissociation of phosphorus pentachloride is 4.0×10^{-2} at 250°C. How many moles and grams of PCl_5 must be added to a 3.0 liter flask to obtain a Cl_2 concentration of 0.15 M? Refer to Exercise 40 for the equation.

46. (a) What is K_p for the reaction of Exercise 45 at 250°C?
    (b) What are the equilibrium partial pressures of each of the species?
    (c) What is the total pressure of the system at equilibrium?
    (d) What was the partial pressure of PCl_5 before any of it decomposed?

47. Consider the following system *at equilibrium* at 1800°C.

    $$CO_2 \text{ (g)} + H_2 \text{ (g)} \rightleftharpoons H_2O \text{ (g)} + CO \text{ (g)}$$

    A 2.0 liter vessel contains 0.48 mole of CO_2, 0.48 mole of H_2, 0.96 mole of H_2O, and 0.96 mole of CO.

(a) How many moles and how many grams of H_2 must be added to bring the concentration of CO to 0.60 M?
(b) How many moles and how many grams of CO_2 must be added to bring the CO concentration to 0.60 M?
(c) How many moles of H_2O must be removed to bring the CO concentration to 0.60 M?

48. At 25°C, K_c is 4.66×10^{-3} for the dissociation of dinitrogen tetroxide to nitrogen dioxide.

    $$N_2O_4 \text{ (g)} \rightleftharpoons 2NO_2 \text{ (g)}$$

    (a) Calculate the equilibrium concentrations of both gases when 2.50 grams of N_2O_4 are placed in a 2.00 liter flask at 25°C.
    (b) What will be the new equilibrium concentrations if the volume of the system is suddenly increased to 4.00 liters at 25°C?
    (c) What will be the new equilibrium concentrations if the volume is decreased to 1.00 liter at 25°C?

49. Consider the equilibrium described in part (a) of Exercise 48.
    (a) What will be the new equilibrium concentrations if 0.20 gram of NO_2 is removed?
    (b) What will be the new equilibrium concentrations if 0.20 gram of NO_2 is added?
    (c) What will be the new equilibrium concentrations if 0.20 gram of N_2O_4 is added?

Relationships Among K, ΔG^0, ΔH^0, and T

50. What must be true of the value of ΔG^0 for a reaction if (a) $K \gg 1$; (b) $K \ll 1$?
51. What kind of equilibrium constant can be calculated from a ΔG^0 value for a reaction involving only gases?
52. The air pollutant sulfur dioxide can be partially removed from stack gases in industrial processes and converted to sulfur trioxide, the acid anhydride of commercially important sulfuric acid. Write the equation for the reaction using the smallest whole number coefficients. Calculate the value of the equilibrium constant for this reaction at 25°C, from values of ΔG_f^0 in Appendix K.
53. In Chapter 3, the value of ΔH^0 for the reaction in Exercise 52 was calculated to be -197.6 kJ.
    (a) Predict qualitatively (i.e., without calculation) whether the value of K_p for this reaction at 500°C would be greater than, the

same as, or less than the value at room temperature (25°C).

(b) Now calculate the value of K_p at 500°C.

54. For the gas phase reaction

$$4HCl + O_2 \rightleftharpoons 2Cl_2 + 2H_2O$$

K_p has the value 7.77×10^5 at 400°C. The reaction is exothermic by 114 kJ.

(a) Calculate the value of K_p for this reaction at 600 K and at 800 K.

(b) At what temperature would this reaction have an equilibrium constant equal to 1?

55. A mixture of 3.00 moles of Cl_2 and 3.00 moles of CO is enclosed in a 5.00 liter flask at 600°C. At equilibrium, 3.3% of the Cl_2 has been consumed.

$$CO\ (g) + Cl_2\ (g) \rightleftharpoons COCl_2\ (g)$$

(a) Calculate K_c for the reaction at 600°C.

(b) Calculate ΔG^0 for the reaction at this temperature.

56. Given that K_p is 4.6×10^{-14} at 25°C for the reaction below:

$$2Cl_2\ (g) + 2H_2O\ (g) \rightleftharpoons 4HCl\ (g) + O_2\ (g)$$
$$\Delta H^0 = +115\ kJ$$

Calculate K_p and K_c for the reaction at 500°C and at 1000°C.

57. (a) Use the tabulated thermodynamic values of ΔH_f^0 and S^0 to calculate the value of K at 25°C for the gas phase reaction

$$CO + H_2O \rightleftharpoons CO_2 + H_2$$

(b) Calculate the value of K for this reaction at 200°C, by the same method as in part (a).

(c) Repeat the calculation of part (a), using tabulated values of ΔG_f^0.

(d) Can you calculate the value of K for this reaction at 200°C using just tabulated values of ΔG_f^0?

58. The following is an example of an alkylation reaction important in the production of iso-octane (2,2,4-trimethylpentane) from two components of crude oil, isobutane and isobutene. Iso-octane is an anti-knock additive for gasoline.

isobutane   isobutene

iso-octane

The equilibrium constant, K, for this reaction at 25°C is 4.3×10^6, and ΔH^0 is -78.58 kJ.

(a) Calculate ΔG^0 at 25°C.

(b) Calculate K at 700°C.

(c) Calculate ΔG^0 at 700°C.

(d) How does the spontaneity of the forward reaction at 700°C compare with that at 25°C?

(e) Why do you think the reaction mixture is heated in the industrial preparation of iso-octane?

(f) What is the purpose of the catalyst? Does it affect the forward reaction more than the reverse reaction?

59. At sufficiently high temperatures, chlorine gas dissociates, according to

$$Cl_2\ (g) \rightleftharpoons 2Cl\ (g)$$

At 800°C, K_p for this reaction is 5.63×10^{-7}.

(a) A sample originally contained Cl_2 at 1 atm and 800°C. Calculate the percentage dissociation of Cl_2 when this reaction has reached equilibrium.

(b) At what temperature would Cl_2 (originally at 1 atm pressure) be 1% dissociated into Cl atoms?

Ionic Equilibria—I: Acids and Bases

17

17-1   Introduction and Review of Basic Ideas

Of all of the classes of systems that involve equilibria among ionic substances, one of the largest and best studied is the class of acidic and basic solutions in water. Now that we have the idea of the equilibrium constant available to us, we can present acid and base equilibria in a quantitative manner, complementing the qualitative discussion in Chapter 9. Knowing the initial concentrations of acidic or basic substances in an aqueous solution, we will be able to calculate the equilibrium concentrations of the species that give such a solution its characteristic properties.

In Section 9–4 we introduced some general ideas about acids and bases. Acids were defined as substances that increase the concentration of hydrogen ions, H^+, in aqueous solution. Bases were defined as substances that increase the concentration of hydroxide ions, OH^-, in aqueous solution. We distinguish between *strong* acids and bases, which are nearly completely ionized or dissociated in dilute aqueous solutions, and *weak* acids and bases, which are only slightly ionized in dilute aqueous solutions. The common strong acids and the strong soluble bases are listed in the margin. Most other acids are weak acids and most other bases are weak bases

Common Strong Acids
HCl HNO_3
HBr $HClO_4$
HI $HClO_3$
        H_2SO_4

Strong Soluble Bases
LiOH
NaOH
KOH $Ca(OH)_2$
RbOH $Sr(OH)_2$
CsOH $Ba(OH)_2$

The common inorganic acids and low molecular weight organic acids are soluble in water.

or are insoluble in water. The strong acids, strong soluble bases, and soluble ionic salts are *strong electrolytes*. (See the solubility rules in Section 9–4.5.) Most other solutes are *weak electrolytes* or *nonelectrolytes*.

We have also described the general properties of acids and bases (Section 9–4.3), as well as methods of preparing them (Sections 9–5.1 and 9–5.2), and reactions between acids and bases (Section 9–6). You may wish to review this material before continuing with this chapter.

17–2   The Hydrated Hydrogen Ion

The most common isotope of hydrogen, $^1_1$H, has no neutrons. Thus $^1_1$H$^+$ would be a bare proton if it were not hydrated.

Although Arrhenius described H$^+$ ions in water as bare ions in 1884, we now know that they are hydrated in aqueous solution, and exist as $H^+(H_2O)_n$ in which n is some small integer. This is due to the attraction of the protons for the oxygen end ($\delta-$) of water molecules. While we do not know the extent of hydration of H$^+$ in most solutions, we often represent the hydrated hydrogen ion as the hydronium ion, H_3O^+, or $H^+(H_2O)_n$ in which $n = 1$. The hydrated hydrogen ion is the species that gives aqueous solutions of acids their characteristic acidic properties.

$$H^+ + :\overset{..}{\underset{..}{O}}:H \longrightarrow H:\overset{..}{\underset{H}{O}}:H^+$$

17–3   The Brønsted-Lowry Theory

In 1923 Brønsted and Lowry independently presented logical extensions of the Arrhenius theory. Brønsted's contribution was more thorough than Lowry's, and the result is known as the Brønsted theory or the **Brønsted-Lowry theory.**

An **acid** is defined in this theory as a *proton donor,* and a **base** is defined as a *proton acceptor.* The definitions are sufficiently broad that any hydrogen-containing molecule or ion capable of releasing a proton, H$^+$, is an acid, while any molecule or ion that can accept a proton is a base according to the Brønsted-Lowry theory.

According to this theory *an acid-base reaction is the transfer of a proton from an acid to a base.* Thus the complete ionization of hydrogen chloride, HCl, a *strong* acid, in water is an acid-base reaction in which water acts as a base, a proton acceptor.

Step 1:	$HCl(g) \xrightarrow{H_2O} H^+(aq) + Cl^-(aq)$	(Arrhenius description)
Step 2:	$H^+(aq) + H_2O(\ell) \longrightarrow H_3O^+$	
Overall:	$HCl(g) + H_2O(\ell) \longrightarrow H_3O^+ + Cl^-(aq)$	(Brønsted-Lowry description)

$$\text{(H)}:\overset{..}{\underset{..}{Cl}}: + H:\overset{..}{\underset{H}{O}}: \longrightarrow H:\overset{H}{\underset{H}{\overset{..}{O}}}:^+ + :\overset{..}{\underset{..}{Cl}}:^-$$

acid   base

The ionization of hydrogen fluoride, a *weak* acid, is similar except that it occurs to only a slight extent, so we use double arrows to indicate that the reaction is reversible.

$$HF \text{ (g)} + H_2O \text{ (}\ell\text{)} \rightleftharpoons H_3O^+ + F^- \text{ (aq)}$$

acid$_1$ base$_2$ acid$_2$ base$_1$

Various measurements (electrical conductivity, freezing point depression, etc.) indicate that HF is only *slightly* ionized in water.

We can describe Brønsted-Lowry acid-base reactions in terms of **conjugate acid-base pairs,** *which are species on opposite sides of an equation that differ by a proton.* In the above equation, HF (acid$_1$) and F$^-$ (base$_1$) are one conjugate acid-base pair and H$_2$O (base$_2$) and H$_3$O$^+$ (acid$_2$) are the other pair. The members of each conjugate pair are designated by the same numerical subscript, although it makes no difference which pair is assigned the subscript 1 or 2.

At equilibrium, which is established very rapidly for most acids and bases in water, both forward and reverse reactions occur simultaneously at the same rate. In the forward reaction HF and H$_2$O act as acid and base, respectively. In the reverse reaction H$_3$O$^+$ functions as the acid, or proton donor, and F$^-$ functions as the base, or proton acceptor. At equilibrium there are far more nonionized HF molecules than H$_3$O$^+$ or F$^-$ ions, which tells us that F$^-$ is stronger as a base than HF is as an acid.

On the other hand, in dilute hydrochloric acid solutions H$_3$O$^+$ and Cl$^-$ ions are the predominant species, and there are (practically) no nonionized HCl molecules. This tells us that HCl is a strong acid, but that the Cl$^-$ ion has virtually no tendency to accept a proton to form an HCl molecule in aqueous solution and is, therefore, an *extremely weak* base.

We can generalize: *The weaker an acid is, the greater is the base strength of its conjugate base. Likewise, the weaker a base is, the stronger is its conjugate acid.* Strong and weak, like many other adjectives, are used in a relative sense. For example, we do not mean to imply that the fluoride ion, F$^-$, is a strong base compared to species such as the hydroxide ion, OH$^-$. We mean that *relative to the anions of strong acids,* which are extremely weak bases, F$^-$ is a much stronger base.

In 0.10 *M* solution, NH$_3$ is only 1.3% ionized and 98.7% nonionized.

Ammonia acts as a weak Brønsted-Lowry base and water acts as an acid in the ionization of aqueous ammonia.

$$NH_3 \text{ (g)} + H_2O \text{ (}\ell\text{)} \rightleftharpoons NH_4^+ \text{ (aq)} + OH^- \text{ (aq)}$$

base$_1$ acid$_2$ acid$_1$ base$_2$

In three dimensions, the molecular structures are:

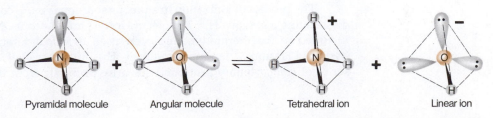

Pyramidal molecule   Angular molecule   Tetrahedral ion   Linear ion

Ammonium ion, NH_4^+, is the conjugate acid of NH_3, and hydroxide ion, OH^-, is the conjugate base of water.

17–4   A Closer Look at Acidic and Basic Character

What are the major factors that determine the relative strengths of acids and bases? As we shall see presently, the relative strengths of acids and bases can be correlated with the position in the periodic table of the characteristic nonmetal or metal in the acid or base, and especially with trends in electronegativities (Table 6–8).

1   Strengths of Binary Acids (Hydrides of Groups VIIA and VIA Elements)

In comparing strengths of a series of binary acids, we must compare *similar* compounds; i.e., we must *not* compare "apples with oranges." We shall compare binary acids from the *same family* in the periodic table.

Recall that *protonic* acids contain one or more H atoms that are capable of ionization.

The relative strengths of a series of *similar binary protonic acids* can be predicted from electronegativity differences. The ionization of a binary acid involves breaking H—X bonds; as the electronegativity difference between H and element X *decreases*, so does bond strength, which makes ionization easier. The electronegativity differences for the hydrogen halides are (see Table 6–8):

	H——F	H——Cl	H——Br	H——I
EN	2.1 4.0	2.1 3.0	2.1 2.8	2.1 2.5
ΔEN	1.9	0.9	0.7	0.4

The electronegativity difference for HF is much greater than that for HCl, and the HF bond is much stronger than the HCl bond. Thus, hydrogen fluoride ionizes only slightly in dilute aqueous solutions because the HF bond is so strong.

H^+ (aq) is often used to represent H_3O^+.

$$HF \text{ (g)} \xrightleftharpoons{H_2O} H^+ \text{ (aq)} + F^- \text{ (aq)}$$

The ionization of HF is complicated by strong hydrogen bonding that produces species such as $(HF)_n$, where n ranges from 2 to 5, as well as species such as HF_2^- and H_2F^+. However, HCl, HBr, and HI ionize completely in dilute aqueous solutions because their H—X bonds are much weaker.

The trends in binary acid strengths *across* a period (e.g., $CH_4 < NH_3 < H_2O < HF$) are the reverse of those predicted from trends in bond energies and electronegativity differences. The correlations used for *vertical* trends *cannot* be used for *horizontal* trends because a "horizontal" series of compounds has different stoichiometries and different numbers of lone pairs of electrons on their central atoms. (Remember, we can't compare apples with oranges.)

$$HX \text{ (g)} \xrightarrow{H_2O} H^+ \text{ (aq)} + X^- \text{ (aq)} \qquad X = Cl, Br, I$$

The order of *bond strengths* for the hydrogen halides is

$$HF \gg HCl > HBr > HI \text{ (weakest bond)}$$

while, as we shall see, the order of *acid strengths* is

$$HF \ll HCl < HBr < HI \text{ (strongest acid)}$$

The process of ionization in aqueous solution is more complicated than simply dissociating a gaseous H—X molecule into H and X atoms. Many other factors are included in the thermodynamic cycle used to calculate acid strength, in which hydrated H^+ and X^- ions are produced. The differences between the strengths of a series of *similar* binary protonic acids arise primarily from the large differences in their bond energies and the energies of hydration of their anions. The bond energy differences are the dominant factor, and energies of hydration are the second most important factor. The differences among the other thermodynamic terms are so small that they are not very important in the overall (Born-Haber type) cycle.

In dilute aqueous solutions, hydrochloric, hydrobromic, and hydroiodic acids all show the same apparent acid strength because they are completely ionized. Recall that water can act as either an acid or as a base. Water is sufficiently basic that it does not distinguish among the acid strengths of HCl, HBr, and HI, and therefore it is referred to as a **leveling solvent.** It is not possible to determine the order of increasing strengths of these three acids in water (because they are completely ionized). When these compounds are dissolved in anhydrous acetic acid, or other solvents less basic than water, however, significant differences in their acid strengths are observed. The order of increasing acid strengths is the order expected from the decreasing differences in electronegativities and bond strengths:

HCl < HBr < HI

Similar statements can be made for other series of similar binary protonic acids. For those of the Group VIA elements, the order of increasing acid strengths is

$H_2O \ll H_2S < H_2Se < H_2Te$ (strongest acid)

Table 17-1 displays relative acid and base strengths of a number of conjugate acid-base pairs.

The hydrated hydrogen ion is the strongest acid that can exist in aqueous solution.

Acids stronger than H^+ (aq) or H_3O^+ react with water to produce H^+ (aq) and their conjugate bases. For example, $HClO_4$ (see Table 17-1) reacts with H_2O completely to form H^+ (aq) and ClO_4^- (aq). Similar observations can be made for

This phenomenon is known as the **leveling effect.** When several acids show the same strength in a solvent (i.e., ionize to the same extent), their acid strengths are said to be *leveled* by the solvent.

TABLE 17-1   Relative Strengths of Conjugate Acid-Base Pairs

Acid			Base
$HClO_4$ HI HBr HCl HNO_3	100% ionized in dil. aq. soln. No molecules of nonionized acid.	Negligible base strength in water	ClO_4^- I^- Br^- Cl^- NO_3^-

$$\xrightarrow{-H^+} \atop \xleftarrow{+H^+}$$

Acid			Base
H_3O^+ HF CH_3COOH HCN NH_4^+ H_2O NH_3	Equilibrium mixture of nonionized molecules of acid, conjugate base, and H^+ (aq).	Reacts completely with H_2O. Cannot exist in aqueous solution.	H_2O F^- CH_3COO^- CN^- NH_3 OH^- NH_2^-

Acid Strength Increases

Base Strength Increases

aqueous solutions of strong soluble bases such as NaOH and KOH. Both are completely dissociated in dilute aqueous solutions.

$$Na^+OH^- \ (s) \xrightarrow{H_2O} Na^+ \ (aq) + OH^- \ (aq)$$

The hydroxide ion is the strongest base that can exist in aqueous solution.

Bases stronger than OH^- react with H_2O to produce OH^- and their conjugate acids. For instance, when metal amides such as sodium amide, $NaNH_2$, are placed in water, the amide ion, NH_2^-, reacts with H_2O completely:

$$NH_2^- + H_2O \longrightarrow NH_3 \ (aq) + OH^- \ (aq)$$

Thus, we see that H_2O is a leveling solvent for all bases stronger than OH^-.

2 Strengths of Ternary Acids

bond that breaks to form H^+ and NO_3^-

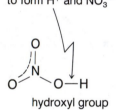

hydroxyl group

Ternary acids ionize to produce H^+ (aq); however, they can be considered to be *hydroxides of nonmetals*. The formula for nitric acid is commonly written HNO_3 to emphasize the presence of an acidic hydrogen atom, but it could also be written as $NO_2(OH)$, as its structure shows.

We usually reserve the term hydroxide for substances that produce basic solutions, and call the other "hydroxides" acids because they ionize to produce H^+ (aq). In ternary acids the hydroxyl oxygen is bonded to a fairly electronegative element such as nitrogen. In nitric acid the nitrogen draws the electrons of the N—O (hydroxyl) bond more closely toward itself than would a less electronegative element such as sodium. The oxygen pulls the electrons of the O—H bond close enough so that the hydrogen atom ionizes as H^+, leaving NO_3^-.

$$HNO_3 \ (\ell) \xrightarrow{H_2O} H^+ \ (aq) + NO_3^- \ (aq)$$

In contrast, let us consider hydroxides of metals. Oxygen is so much more electronegative than most metals, such as sodium, that it draws the electrons of the sodium-oxygen bond in NaOH (a strong soluble base) close enough to itself that the bonding is ionic, and therefore NaOH exists as Na^+ and OH^- ions, even in the solid state.

We usually write the formula for sulfuric acid as H_2SO_4 to emphasize the fact that it is an acid. However, the formula can also be written as $(HO)_2SO_2$, because the structure of sulfuric acid shows clearly that H_2SO_4 contains two —O—H groups bound to a sulfur atom. Since the O—H bonds are easier to break than the S—O bonds, sulfuric acid ionizes as an acid when it is dissolved in water.

Sulfuric acid is called a *polyprotic acid* because it has more than one ionizable hydrogen atom per molecule.

Keep in mind the fact that H_2SO_4 is the only *common strong* polyprotic acid (Table 9–2) and that most others are weak (Table 9–3 and Appendix F).

1st Step: $$H_2SO_4 \ (\ell) \xrightarrow{H_2O} H^+ \ (aq) + HSO_4^- \ (aq)$$

2nd Step: $$HSO_4^- \ (aq) \xrightarrow{H_2O} H^+ \ (aq) + SO_4^{2-} \ (aq)$$

The first step in the ionization of H_2SO_4 is complete in dilute aqueous solution. The second step is nearly complete only in very dilute solutions. As we shall see in Section 17–8, the first step in the ionization of a polyprotic acid always occurs to a greater extent than the second step.

Sulfurous acid, H_2SO_3, is a polyprotic acid that contains the same elements as H_2SO_4. However, H_2SO_3 is a weak acid, which tells us that the H—O bonds in

H_2SO_3 are stronger than those in H_2SO_4. Also, comparison of the acid strengths of nitric acid, HNO_3, and nitrous acid, HNO_2, shows that HNO_3 is a much stronger acid than HNO_2. Similar observations show that *acid strengths of most ternary acids containing the same central element increase with the oxidation state of the central element and with increasing numbers of oxygen atoms per central nonmetal atom.* The following orders of increasing strength are typical.

$$H_2SO_3 < H_2SO_4$$
$$HNO_2 < HNO_3$$
$$HClO < HClO_2 < HClO_3 < HClO_4$$

strongest acids are on the right

Acid strengths of most ternary acids containing different elements in the same oxidation state from the same group in the periodic table increase with increasing electronegativity of the central element, as in the following examples.

$$H_2SeO_4 < H_2SO_4 \qquad H_2SeO_3 < H_2SO_3$$
$$H_3PO_4 < HNO_3$$
$$HBrO_4 < HClO_4 \qquad HBrO_3 < HClO_3$$

Contrary to what we might expect, H_3PO_3 is a stronger acid than HNO_2.

A word of caution is necessary. Comparisons such as the one just presented are valid only for acids with similar structures. For example, H_3PO_2, which contains two H atoms bonded to the P atom, is a stronger acid than H_3PO_3, which contains one H atom bonded to the P atom. H_3PO_3 is a stronger acid than H_3PO_4, which has no H atoms bonded to the P atom.

17–5 The Autoionization of Water

Amphoterism was introduced in Section 9–6.2.

We have seen that water acts as an acid (H^+ donor) in its reaction with NH_3, whereas it acts as a base (H^+ acceptor) in its reactions with HCl and HF. Whether water acts as an acid or as a base depends on its environment, i.e., the other species present. Therefore, water is **amphoteric** (that is, it may act as either an acid or a base). Careful measurements have shown that pure water ionizes ever-so-slightly to produce equal numbers (concentrations) of H_3O^+ and OH^- ions.

The ionization of water is often represented in simplified notation as $H_2O(\ell) \rightleftharpoons H^+(aq) + OH^-(aq)$, where $H^+(aq)$ is understood to refer to H_3O^+.

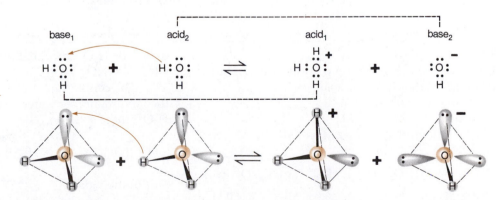

The prefix "auto-" means "self," as in autobiography. Autoionization refers to a substance ionizing by reacting with itself.

Water also exhibits amphoterism in its **autoionization.** One molecule acts as a Brønsted-Lowry acid and another acts as a Brønsted-Lowry base. This reaction occurs only *very* slightly. Evidence for this is the fact that the concentration of each of its ions is only 1.0×10^{-7} M in pure water at 25°C, as shown by measurements of electrical conductivity. This is consistent with the observation that the combination of H^+ (aq) ions (or H_3O^+) with OH^- ions to form nonionized water molecules occurs to completion in neutralization reactions of strong acids with strong soluble bases. The combination of H_3O^+ with OH^- is the reverse of the autoionization of water, which *must*, therefore, occur only very slightly.

$$H_2O \ (\ell) + H_2O \ (\ell) \rightleftharpoons H_3O^+ \ (aq) + OH^- \ (aq)$$

The equilibrium constant expression for the autoionization of water is

$$K_c = \frac{[H_3O^+][OH^-]}{[H_2O]^2} \qquad \text{[Eq. 17–1]}$$

Since the concentration of water is constant in pure water, it can be included in the equilibrium constant.

$$K_c[H_2O]^2 = [H_3O^+][OH^-] \qquad \text{[Eq. 17–2]}$$

In pure H_2O at 25°C, as we have seen,

$$[H_3O^+] = [OH^-] = 1.0 \times 10^{-7} \ \text{mol/L} \qquad \text{[Eq. 17–3]}$$

Since the concentrations of both H_3O^+ and OH^- are known for pure water, substitution into Eq. 17–2 gives

$$K_c[H_2O]^2 = [H_3O^+][OH^-] = (1.0 \times 10^{-7})(1.0 \times 10^{-7}) = 1.0 \times 10^{-14}$$

or, in a more useful form,

$$[H_3O^+][OH^-] = 1.0 \times 10^{-14} = K_w \qquad \text{[Eq. 17–4]}$$

The quantity 1.0×10^{-14} is known as the **ion product** for water and is usually represented as K_w.

Although the expression $K_w = [H_3O^+][OH^-] = 1.0 \times 10^{-14}$ was obtained for pure water, *it is valid for dilute aqueous solutions at 25°C*, because the concentration of water remains nearly constant in all such solutions and is very close to that in pure water. Indeed, this is one of the most useful relationships chemists have discovered. It is useful because it gives a simple (inverse) relationship between H_3O^+ and OH^- concentrations in *all* dilute aqueous solutions.

Solutions in which the concentration of solute is less than 1 mole/liter are usually called dilute solutions.

Example 17–1

Calculate the concentration of H_3O^+ and OH^- in a 0.050 M solution of nitric acid.

Solution

The equation for the ionization of HNO_3, a strong acid,

$$HNO_3 \ (\ell) + H_2O \ (\ell) \longrightarrow H_3O^+ \ (aq) + NO_3^- \ (aq)$$

| 0.050 M | | 0.050 M | 0.050 M |

shows that 0.050 mol/L of HNO_3 produces 0.050 mol/L of H_3O^+ *and* 0.050 mol/L of NO_3^-. Thus, $[H_3O^+] = [NO_3^-] = \underline{0.050 \ M}$.

The concentration of OH^- is calculated from the ion product for water and the concentration of H_3O^+ in the solution.

$$[H_3O^+][OH^-] = 1.0 \times 10^{-14}; \quad [OH^-] = \frac{1.0 \times 10^{-14}}{[H_3O^+]}$$

Substitution gives

$$[OH^-] = \frac{1.0 \times 10^{-14}}{5.0 \times 10^{-2}} = \underline{2.0 \times 10^{-13} \ M}$$

In solving this problem, we have assumed that *all* of the H_3O^+ comes from the ionization of HNO_3 and have neglected the H_3O^+ formed by the ionization of water. The ionization of water produces only 2.0×10^{-13} M H_3O^+ and 2.0×10^{-13} M OH^- in this solution. The 2.0×10^{-13} M H_3O^+ produced by the ionization of water is so small compared to the concentration of H_3O^+ formed by the ionization of HNO_3 (5.0×10^{-2} M) that it is neglected. We are justified in assuming that the H_3O^+ concentration is derived solely from HNO_3.

When nitric acid is added to water, large numbers of H_3O^+ ions are produced instantaneously. The large increase in H_3O^+ concentration causes the reaction to the left to predominate (LeChatelier's Principle) and the concentration of OH^- is decreased.

$$H_2O\ (\ell) + H_2O\ (\ell) \rightleftharpoons H_3O^+\ (aq) + OH^-\ (aq)$$

In acidic solutions the H_3O^+ concentration is always greater than the OH^- concentration. We should not conclude that acidic solutions contain no OH^- ions, but rather that the concentration of OH^- is always less than 1.0×10^{-7} M in acidic solutions. The converse is true for basic solutions, where the concentration of OH^- is always greater than 1.0×10^{-7} M. By definition, "neutral" aqueous solutions are solutions in which $[H_3O^+] = [OH^-] = 1.0 \times 10^{-7}$ M.

$[H_3O^+] = [OH^-]$ in neutral solution.
$[H_3O^+] > [OH^-]$ in acidic solution.
$[H_3O^+] < [OH^-]$ in basic solution.

17-6   The pH Scale

The pH scale provides a convenient way of expressing the acidity and basicity of dilute aqueous solutions. The pH of a solution is defined as

This is the base-10 (common) logarithm, *not* the base-*e* (natural) logarithm.

$$pH = \log \frac{1}{[H_3O^+]} \qquad or \qquad pH = -\log [H_3O^+] \qquad \text{[Eq. 17-5]}$$

Note that we use pH rather than pH_3O. At the time the pH concept was developed, it was believed that all aqueous solutions of acids contained H^+ ions. A number of "p" terms are currently used. For example, $pOH = -\log [OH^-]$. In general, a lower case **"p"** before a symbol is read "negative logarithm of the symbol." Thus, **pH** is the negative logarithm of H_3O^+ concentration, pOH is the negative logarithm of OH^- concentration, pAg refers to the negative logarithm of Ag^+ concentration, and pK refers to the negative logarithm of an equilibrium constant.

If H_3O^+ and OH^- concentrations are known for a particular solution, the pH and pOH can be calculated by taking the negative logarithms of these concentrations, as Examples 17-2 and 17-3 illustrate.

Example 17-2

Calculate the pH of a solution in which the concentration of H_3O^+ is 0.010 M.

Solution

We are given $[H_3O^+] = 0.010\ M$, or more conveniently, $[H_3O^+] = 1.0 \times 10^{-2}\ M$.

$$pH = -\log [H_3O^+] = -\log [1.0 \times 10^{-2}]$$
$$= -[0.00 + (-2)] = \underline{2.00}$$

Note that pH = 2.00 contains two significant figures because the 2 only indicates decimal position in the number from which pH was calculated.

Example 17–3

Calculate the pH of a solution in which the H_3O^+ concentration is 0.050 mol/L.

Solution

We are given $[H_3O^+] = 0.050\ M$ or $[H_3O^+] = 5.0 \times 10^{-2}\ M$.

$$pH = -\log [H_3O^+] = -\log [5.0 \times 10^{-2}]$$
$$= -[0.699 + (-2)] = 1.301$$

Since we are justified in using only two significant figures, pH = 1.30. The "1" in 1.30 is derived from the exponential term and is *not* a significant figure.

Example 17–4

The pH of a solution is 3.301. What is the concentration of H_3O^+ in this solution?

Solution

Recall that $pH = -\log [H_3O^+]$. Therefore, we can write

$$-\log [H_3O^+] = 3.301$$

or $$\log [H_3O^+] = -3.301$$

Taking the antilogarithm of both sides of the equation gives

$[H_3O^+] = 10^{-3.301}$,   or   in   general   terms, $[H_3O^+] = 10^{-pH}$

Since logarithm tables contain only positive mantissas (fractional powers of the base 10; e.g., $10^{0.699} = 5.00$), we express the negative exponent in terms of a positive fractional exponent *and* a negative integral exponent.

$$[H_3O^+] = 10^{-3.301} = 10^{0.699} \times 10^{-4}$$

We look up 0.699 in the table of mantissas, which gives

$$[H_3O^+] = 5.00 \times 10^{-4}\ M$$

Scientific calculators can handle negative logarithms directly.

A convenient relationship between pH and pOH can be derived easily. Recall that *for all dilute aqueous solutions,*

$$[H_3O^+][OH^-] = 1.0 \times 10^{-14}$$

Taking the logarithm of both sides of this equation gives

$$\log [H_3O^+] + \log [OH^-] = \log (1.0 \times 10^{-14})$$

Multiplying both sides of this equation by -1 gives

$$(-\log [H_3O^+]) + (-\log (OH^-)) = -\log (1.0 \times 10^{-14})$$

or, in slightly different form

$$pH + pOH = 14 \qquad \text{[Eq. 17–6]}$$

This is an extraordinarily useful expression because it gives a simple relationship between pH and pOH in dilute aqueous solutions. We now have expressions for relating $[H_3O^+]$ and $[OH^-]$ as well as pH and pOH, *i.e.,*

$$[H_3O^+][OH^-] = 1.0 \times 10^{-14} \qquad \text{[Eq. 17–4]}$$

and

$$pH + pOH = 14 \qquad \text{[Eq. 17–6]}$$

These two expressions are used frequently enough to warrant a small location in your memory!

From the relationship pH + pOH = 14, we see that positive values of both pH and pOH are limited to the range *0 to 14*. If either pH or pOH is greater than 14, the other is obviously negative. As a matter of convention, we do not normally express acidities and basicities on the pH and pOH scale outside the range 0 to 14. This does not mean that pH and pOH cannot have negative values or values greater than 14, but just that they are seldom used. Example 17–5 illustrates the relationships among $[H_3O^+]$, pH, $[OH^-]$, and pOH.

Example 17–5
Calculate $[H_3O^+]$, pH, $[OH^-]$, and pOH for 0.010 M HNO_3 solution.

$$HNO_3 + H_2O \rightarrow H_3O^+ + NO_3^-$$

Solution
Because nitric acid is a strong acid (it ionizes completely), we know that

$[H_3O^+] = 1.0 \times 10^{-2} M$

$$pH = -\log [H_3O^+] = -\log [1.0 \times 10^{-2}]$$
$$= -[0.00 + (-2)] = 2.00$$

We also know that pH + pOH = 14. Therefore,

$$pOH = 14 - pH = 14 - 2.00 = \underline{12.00}$$

Since $[H_3O^+][OH^-] = 1.0 \times 10^{-14}$, $[OH^-]$ is easily calculated.

$$[OH^-] = \frac{1.0 \times 10^{-14}}{1.0 \times 10^{-2}} = \underline{1.0 \times 10^{-12} M}$$

Also, since pOH = 12.00, we have $-\log [OH^-] = 12$ or $[OH^-] = 1.0 \times 10^{-12} M$.

To develop familiarity with the pH scale, consider a series of solutions in which the H_3O^+ concentration varies from 1.0 M to 1.0×10^{-14} M. Obviously the OH^- concentration will vary from 1.0×10^{-14} M to 1.0 M in these solutions. Let us tabulate $[H_3O^+]$, $[OH^-]$, pH, and pOH for this series of solutions, so the relationships become obvious (Table 17–2).

TABLE 17–2   Relationship Among $[H_3O^+]$, $[OH^-]$, pH, and pOH at 25°C

$[H_3O^+]$ (M)	$[OH^-]$ (M)	pH	pOH	
1.0	1.0×10^{-14}	0	14	increasing acidity
1.0×10^{-1}	1.0×10^{-13}	1	13	
1.0×10^{-2}	1.0×10^{-12}	2	12	
1.0×10^{-3}	1.0×10^{-11}	3	11	
1.0×10^{-4}	1.0×10^{-10}	4	10	
1.0×10^{-5}	1.0×10^{-9}	5	9	
1.0×10^{-6}	1.0×10^{-8}	6	8	
1.0×10^{-7}	1.0×10^{-7}	7	7	neutral
1.0×10^{-8}	1.0×10^{-6}	8	6	increasing basicity
1.0×10^{-9}	1.0×10^{-5}	9	5	
1.0×10^{-10}	1.0×10^{-4}	10	4	
1.0×10^{-11}	1.0×10^{-3}	11	3	
1.0×10^{-12}	1.0×10^{-2}	12	2	
1.0×10^{-13}	1.0×10^{-1}	13	1	
1.0×10^{-14}	1.0	14	0	

TABLE 17–3 pH Ranges for Some Common Substances

Substances	pH Range
gastric contents (human)	1.0–3.0
limes	1.8–2.0
soft drinks	2.0–4.0
lemons	2.2–2.4
vinegar	2.4–3.4
apples	2.9–3.3
tomatoes	4.0–4.4
beer	4.0–5.0
bananas	4.5–4.7
urine (human)	4.8–8.4
carrots	4.9–5.3
milk (cow's)	6.3–6.6
saliva (human)	6.5–7.5
blood plasma (human)	7.3–7.5
egg white	7.6–8.0
milk of magnesia	10.5
household ammonia	11–12

Table 17–3 shows the pH ranges for few common substances. Note that fruits and vegetables are acidic; some are strongly acidic.

The pH of an aqueous solution may be determined by using indicators (Section 18–4) or by using pH meters, which are electrical instruments constructed for that purpose.

In the indicator method, a series of solutions of known pH, called *standard solutions*, is prepared and the appropriate mixture of indicators is added to each so that solutions of different acidities have different colors. The same mixture of indicators is added to the unknown solution, and its color is then compared to those of the standard solutions. Solutions that have the same pH develop the same color. Alternatively, "universal indicator" papers may be used to determine the pH of a solution. These are pieces of paper that are impregnated with a mixture of indicators; a color chart is attached to each container of paper. A piece of paper is dipped into a solution, and its color is then compared to the color chart to establish the pH of the solution.

These are similar to the standard solutions used for acid-base titrations (Section 10–4).

The pH meter (Figure 17–1) is based on the glass electrode, a sensing device that generates a voltage. This voltage is proportional to the pH of the solution in which the electrode is placed. The rest of the instrument consists of an electrical circuit to amplify the voltage from the electrode, and a meter that relates the voltage to the pH. Like other scientific instruments, pH meters must be calibrated. This is accomplished by using a series of solutions of known pH. Once calibrated, the pH meter can then be used to determine the pH of an unknown solution.

17–7 Weak Molecular Acids and Bases

Recall that the common strong electrolytes are (1) strong acids, (2) strong soluble bases, and (3) soluble ionic salts.

Since *strong electrolytes* are completely (or nearly completely) ionized in dilute aqueous solution, the concentrations of ions in solutions of such compounds can be determined easily from the concentrations of the strong electrolytes themselves. As examples, an aqueous 0.10 M HCl solution is 0.10 M in H_3O^+ and 0.10 M in Cl^-

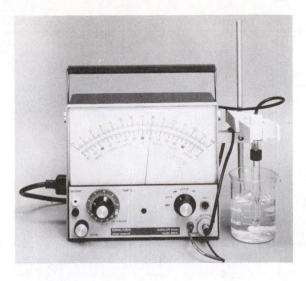

FIGURE 17-1 A pH meter is an instrument that reads the pH of a solution directly. The electrode is dipped into the solution of interest, and the meter shows the pH of the solution. A pH meter must be calibrated frequently with a series of solutions of known pH.

ions; $0.10\ M$ Ba(OH)$_2$ solution is $0.10\ M$ in Ba^{2+} and $0.20\ M$ in OH$^-$; $0.10\ M$ Ca(NO$_3$)$_2$ is $0.10\ M$ in Ca^{2+} and $0.20\ M$ in NO$_3^-$ ions.

Weak acids and weak bases are often called **molecular acids** or **molecular bases** because their dilute aqueous solutions contain mostly nonionized molecules. When we consider specified molarities of *weak* electrolytes as solutes we must use **ionization constants,** *equilibrium constants that describe ionization reactions,* to determine equilibrium concentrations of the ions and nonionized molecules in solution. Tables 17-4 and 17-5 list ionization constants for a few common weak molecular acids and bases, as well as the reactions to which they apply. Appendices F and G contain more comprehensive lists. K_a's refer to ionization of *acids* and K_b's refer to ionization of *bases.*

We can think of several weak acids with which we are familiar because of their common occurrence and uses. Vinegar is a 5% solution of acetic acid, CH$_3$COOH. Carbonated beverages are saturated solutions of carbon dioxide dissolved in water to produce carbonic acid (CO$_2$ + H$_2$O $\rightleftharpoons$ H$_2$CO$_3$). Citrus fruits contain citric acid, C$_3$H$_5$O(COOH)$_3$, and many of the ointments and powders used for medicinal purposes contain boric acid, H$_3$BO$_3$. These everyday uses of weak acids indicate that there is indeed a fundamental difference between the properties of strong and weak acids. The difference is that strong acids ionize completely in dilute aqueous solution while weak acids ionize only slightly.

Let us consider the reaction that occurs when a weak acid, such as acetic acid, is dissolved in water. The equation for the ionization of acetic acid is

$$CH_3COOH + H_2O \rightleftharpoons H_3O^+ + CH_3COO^-$$

How do you *know* that carbonated beverages are *saturated* solutions?

In the remainder of this chapter, we'll discuss compounds that are soluble in water. All ions are hydrated in aqueous solution, and we'll omit the designation (ℓ), (g), (s), (aq), etc. to simplify writing equations.

TABLE 17-4 Ionization Constants for Some Weak Monoprotic Molecular Acids at 25°C

Acid	Ionization Reaction	K_a at 25°C
hydrofluoric acid	HF + H$_2$O $\rightleftharpoons$ H$_3$O$^+$ + F$^-$	7.2×10^{-4}
nitrous acid	HNO$_2$ + H$_2$O $\rightleftharpoons$ H$_3$O$^+$ + NO$_2^-$	4.5×10^{-4}
acetic acid	CH$_3$COOH + H$_2$O $\rightleftharpoons$ H$_3$O$^+$ + CH$_3$COO$^-$	1.8×10^{-5}
hypochlorous acid	HClO + H$_2$O $\rightleftharpoons$ H$_3$O$^+$ + ClO$^-$	3.5×10^{-8}
hydrocyanic acid	HCN + H$_2$O $\rightleftharpoons$ H$_3$O$^+$ + CN$^-$	4.0×10^{-10}

TABLE 17–5   Ionization Constants for Some Weak Molecular Bases at 25°C

Base	Ionization Reaction	K_b at 25°C
ammonia	$NH_3 + H_2O \rightleftharpoons NH_4^+ + OH^-$	1.8×10^{-5}
methylamine	$(CH_3)NH_2 + H_2O \rightleftharpoons (CH_3)NH_3^+ + OH^-$	5.0×10^{-4}
dimethylamine	$(CH_3)_2NH + H_2O \rightleftharpoons (CH_3)_2NH_2^+ + OH^-$	7.4×10^{-4}
trimethylamine	$(CH_3)_3N + H_2O \rightleftharpoons (CH_3)_3NH^+ + OH^-$	7.4×10^{-5}
pyridine	$C_5H_5N + H_2O \rightleftharpoons C_5H_5NH^+ + OH^-$	1.5×10^{-9}

The eqilibrium constant for this reaction is

$$K_c = \frac{[H_3O^+][CH_3COO^-]}{[CH_3COOH][H_2O]}$$

Note that this expression contains the concentration of water. If we restrict our discussion to *dilute* aqueous solutions, the concentration of water is very high. There are 55.6 moles of water in one liter of pure water. In dilute aqueous solutions, the concentration of water is *essentially* constant; if we assume that it is constant, we can rearrange the above expression to give

$$K_c[H_2O] = \frac{[H_3O^+][CH_3COO^-]}{[CH_3COOH]}$$

Since K_c is a constant (at a specified temperature), and the concentration of water is treated as a constant, we define the **ionization constant** of a weak acid as the product $K_c[H_2O]$:

$$K_a = K_c[H_2O] \tag{Eq. 17–7}$$

For acetic acid at 25°C,

$$K_a = \frac{[H_3O^+][CH_3COO^-]}{[CH_3COOH]} = 1.8 \times 10^{-5} \tag{Eq. 17–8}$$

This expression tells us that in dilute aqueous solutions of acetic acid, the concentration of H_3O^+ multiplied by the concentration of CH_3COO^- and then divided by the concentration of *nonionized* acetic acid is equal to 1.8×10^{-5}.

Ionization constants for weak acids and bases *must* be calculated from data that are determined experimentally. Conductivity, depression of freezing point, and measurements of pH provide data from which ionization constants can be calculated.

Example 17–6

In a freezing point depression experiment, it is found that acetic acid is 4.2% ionized in 0.0100 M solution. Calculate its ionization constant from this information.

Solution

We have seen that the equation for the ionization of CH_3COOH is

$$CH_3COOH + H_2O \rightleftharpoons H_3O^+ + CH_3COO^-$$

and that the equilibrium constant expression is

$$K_a = \frac{[H_3O^+][CH_3COO^-]}{[CH_3COOH]}$$

The balanced equation indicates that the concentrations of H_3O^+ and CH_3COO^- are equal, and the percentage ionization means that each is 4.2% of the total acid concentration:

$$[H_3O^+] = [CH_3COO^-] = 0.042 \times 0.0100 \ M$$
$$= 4.2 \times 10^{-4} \ M$$

The concentration of nonionized CH_3COOH is 95.8% (or 100.0% − 4.2%) of the total acid concentration:

$$[CH_3COOH] = 0.958 \times 0.0100\ M$$
$$= 9.58 \times 10^{-3}\ M$$

Substitution of these values into the K_a expression enables us to calculate K_a for CH_3COOH.

$$K_a = \frac{[H_3O^+][CH_3COO^-]}{[CH_3COOH]}$$
$$= \frac{(4.2 \times 10^{-4})(4.2 \times 10^{-4})}{9.58 \times 10^{-3}}$$
$$= \underline{1.8 \times 10^{-5}}$$

Example 17–7 illustrates how K_a for a weak acid can be determined from experimental data such as the pH of a solution of the weak monoprotic acid of known concentration.

Example 17–7

The pH of a 0.100 M solution of a weak monoprotic acid is 1.47. Calculate K_a for the acid.

Solution

The ionization of the weak monoprotic acid (call it HA) may be represented as

$$HA + H_2O \rightleftharpoons H_3O^+ + A^-$$

and its ionization constant expression is

$$K_a = \frac{[H_3O^+][A^-]}{[HA]}$$

We can calculate $[H_3O^+]$ from the definition of pH.

$$pH = -\log [H_3O^+]$$
$$[H_3O^+] = 10^{-pH} = 10^{-1.47}$$
$$[H_3O^+] = 10^{0.53} \times 10^{-2} = 3.4 \times 10^{-2}\ M$$

The ionization equation tells us that the concentrations of H_3O^+ and A^- must be equal, since they are formed in a 1:1 ratio. [Recall that the ionization of

pure water produces only $1.0 \times 10^{-7}\ M\ H_3O^+$. The concentration of H_3O^+ *from* water is even less when an acid is dissolved in water (see the discussion following Example 17–1).]

$$[H_3O^+] = [A^-] = 3.4 \times 10^{-2}\ M$$

Since each molecule of HA that ionizes produces one H_3O^+ (and one A^-), the concentration of nonionized HA is

$$[HA]_{nonionized} = [HA]_{total} - [HA]_{ionized}$$
$$= [HA]_{total} - [H_3O^+]$$
$$[HA]_{nonionized} = (0.100 - 0.034)\ M = 0.066\ M$$

Now that all concentrations are known, the ionization constant can be calculated.

$$K_a = \frac{[H_3O^+][A^-]}{[HA]}$$
$$= \frac{(3.4 \times 10^{-2})(3.4 \times 10^{-2})}{6.6 \times 10^{-2}}$$
$$= \underline{1.8 \times 10^{-2}}$$

Since ionization constants are equilibrium constants for ionization reactions, their values indicate the extents to which weak electrolytes ionize. Acids with larger ionization constants ionize to greater extents (and are therefore stronger acids) than acids with smaller ionization constants. If we look back at Table 17–4, we see that for the five weak acids listed there, the order of decreasing acid strength is

$$HF > HNO_2 > CH_3COOH > HClO > HCN$$

Conversely, we note that in terms of Brønsted-Lowry terminology, the order of increasing base strengths of the anions of these acids is

$$F^- < NO_2^- < CH_3COO^- < ClO^- < CN^-$$

Knowing the value of the ionization constant for a weak acid enables us to calculate the concentrations of the various species in solutions of the weak acid of known concentrations, as Example 17–8 illustrates.

Example 17–8

Calculate the concentrations of the various species in $0.10\ M$ hypochlorous acid, for which $K_a = 3.5 \times 10^{-8}$.

Solution

The equation for the ionization of HOCl (Table 17–4) is

$$HOCl + H_2O \rightleftharpoons H_3O^+ + OCl^-$$

and the expression for the ionization constant is

$$K_a = \frac{[H_3O^+][OCl^-]}{[HOCl]} = 3.5 \times 10^{-8}$$

(We have written the formula for hypochlorous acid as HOCl rather than HClO to emphasize that the structure is H—O—Cl.) Having written *both* the equation for the ionization of HOCl and the expression for the ionization constant, we are ready to attack the problem. We would like to know the concentrations of H_3O^+, OCl^-, and nonionized HOCl in the solution. An algebraic representation of concentrations is required, since there is no other obvious way to obtain the desired concentrations. If we let $x = [H_3O^+]$, we can also let $x = [OCl^-]$ because the ionization of one HOCl molecule produces one H_3O^+ and one OCl^-. If $x = [H_3O^+]$, then the equilibrium concentration of *nonionized* HOCl is $(0.10 - x)$. We will neglect the $1.0 \times 10^{-7}\ M$ of H_3O^+ produced by the ionization of water. We have now represented the concentrations of the various species algebraically, i.e.,

	$HOCl + H_2O$	$\rightleftharpoons$	H_3O^+	$+ OCl^-$
Initial:	$0.10\ M$		$\sim 0\ M$	$0\ M$
Change:	$-x\ M$		$+x\ M$	$+x\ M$
Equil:	$(0.10 - x)M$		$x\ M$	$x\ M$

Substituting the algebraic representations into the equilibrium constant expression gives

$$\frac{[H_3O^+][OCl^-]}{[HOCl]} = \frac{(x)(x)}{(0.10 - x)} = 3.5 \times 10^{-8}$$

This is a quadratic equation, but it is not necessary to solve it by the quadratic formula. If we assume that $(0.10 - x)$ is very nearly equal to 0.10 (see the box that follows), the equation becomes

$$\frac{x^2}{0.10} \approx 3.5 \times 10^{-8}$$

$$x^2 \approx 3.5 \times 10^{-9}$$

$$x \approx 5.9 \times 10^{-5}$$

In our algebraic representation, we let

$$[H_3O^+] = x = \underline{5.9 \times 10^{-5}\ M}$$
$$[OCl^-] = x = \underline{5.9 \times 10^{-5}\ M}$$
$$[HOCl] = 0.10 - x = 0.10 - 0.000059$$
$$= \underline{0.10\ M}$$

Simplifying Quadratic Equations

A property of quadratic equations is that when the linear variable (x) is added to or subtracted from a much larger number, the linear variable may be disregarded if it is sufficiently small. A reasonable rule-of-thumb for determining whether the variable can be disregarded is: If the exponent of 10 in the equilibrium constant is -5 or less ($-5, -6, -7$, etc.), then the variable can be disregarded when it is added to or subtracted from a number greater than 0.05.

Let's examine the assumption as it applies to Example 17–8 in detail. Our quadratic equation is

$$\frac{(x)(x)}{(0.10 - x)} = 3.5 \times 10^{-8}$$

If we convert this to the standard quadratic equation form, $ax^2 + bx + c = 0$, it becomes

$$x^2 + (3.5 \times 10^{-8})x - 3.5 \times 10^{-9} = 0$$

Since x is obviously a small number, then $(3.5 \times 10^{-8})x$ is so small that it is of no significance, and we are justified in assuming that $(0.10 - x) = 0.10$. By making the simplifying assumption, our quadratic equation becomes $x^2 = 3.5 \times 10^{-9}$, which is easily solved by taking the square root of both sides. You may wish to use the quadratic formula to verify that the answer obtained this way is correct within round-off error.

Although the above argument is purely algebraic, we could use our chemical intuition to arrive at exactly the same conclusion. A small ionization constant (10^{-5} or less) tells us that the extent of ionization is very small. If the extent of ionization is small, most of the weak acid exists as nonionized molecules, and the amount that ionizes is not significant when compared to the concentration of nonionized weak acid.

From our calculations, we can make some observations. In a solution containing only a weak monoprotic acid, the concentration of H_3O^+ is equal to the concentration of the anion of the weak monoprotic acid. Unless the weak acid is *very* dilute, say less than 0.050 M, the concentration of nonionized acid is approximately equal to the molarity of the solution. Only if the ionization constant for the weak acid is much greater than 10^{-5} will the extent of ionization be sufficiently large so that there will be a significant difference between the concentration of nonionized acid and the molarity of the solution. Example 17-9 further illustrates these statements.

Example 17-9

Calculate the percentage ionization of a 0.10 M solution of acetic acid.

Solution

The equation for the ionization of CH_3COOH is

$$CH_3COOH + H_2O \rightleftharpoons H_3O^+ + CH_3COO^-$$

Having written the appropriate equation, let us decide how to proceed. Since percentage is defined as (part/whole) $\times 100\%$, we represent the percentage ionization of acetic acid as

$$\% \text{ ionization} = \frac{[CH_3COOH]_{\text{ionized}}}{[CH_3COOH]_{\text{total}}} \times 100\%$$

Examination of the chemical equation shows that each molecule of CH_3COOH that ionizes produces one H_3O^+. Therefore, we can represent the concentration of CH_3COOH that ionizes by the equilibrium concentration of H_3O^+, and the above expression becomes

$$\% \text{ ionization} = \frac{[H_3O^+]}{[CH_3COOH]_{\text{total}}} \times 100\%$$

We can solve for $[H_3O^+]$ as we did in Example 17-8. Let $x = [H_3O^+] = [CH_3COO^-]$, and $[CH_3COOH] = (0.10 - x)$ at equilibrium.

	CH_3COOH	$+ H_2O \rightleftharpoons$	H_3O^+	$+ CH_3COO^-$
in:	0.10 M		~0 M	0 M
ch:	$-x$ M		$+x$ M	$+x$ M
eq:	$(0.10 - x)$ M		x M	x M

Substituting these values into the ionization constant expression (Eq. 17-8) gives

$$K_a = \frac{[H_3O^+][CH_3COO^-]}{[CH_3COOH]} = \frac{(x)(x)}{(0.10 - x)}$$
$$= 1.8 \times 10^{-5}$$

If we make the simplifying assumption as in Example 17-8, namely that $(0.10 - x)$ is approximately equal to 0.10, we have

$$\frac{x^2}{0.10} = 1.8 \times 10^{-5} \quad \text{and} \quad x^2 = 1.8 \times 10^{-6}$$

which gives $x = 1.3 \times 10^{-3}$ mol/L $= [H_3O^+]$. Now that we know $[H_3O^+]$, we can calculate the percentage ionization for 0.10 M CH_3COOH solution.

$$\% \text{ ionization} = \frac{[CH_3COOH]_{\text{ionized}}}{[CH_3COOH]_{\text{total}}} \times 100\%$$
$$= \frac{[H_3O^+]}{[CH_3COOH]_{\text{total}}} \times 100\%$$
$$= \frac{1.3 \times 10^{-3}}{0.10} \times 100\% = 1.3\%$$

Note that our assumption that $(0.10 - x)$ is approximately 0.10 is reasonable because $(0.10 - x) = (0.10 - 0.0013)$. If we follow the rules for significant figures, $(0.10 - 0.0013)$ is 0.10. However, if the ionization constant for a weak acid is considerably greater than 10^{-5}, this assumption would introduce a noticeable error into our calculations.

In dilute solutions, acetic acid exists primarily as nonionized molecules (as do all weak acids), and there are relatively few hydronium and acetate ions. To be specific, in 0.10 M solution, CH_3COOH is 1.3% ionized; for each 1000 formula units of CH_3COOH in the solution, there are 13 H_3O^+ ions, 13 CH_3COO^- ions, and 987 nonionized CH_3COOH molecules. For weaker acids, the number of molecules of nonionized acid is even larger.

By now you should be developing some "feel" for the strength of an acid by simply looking at its ionization constant. Let's consider 0.10 M solutions of HCl (a strong acid), CH_3COOH (Example 17–9), and HOCl (Example 17–8). If we calculate the percentage ionization for 0.10 M HOCl (as we did for 0.10 M CH_3COOH in Example 17–9), we find that it is 0.059% ionized. In 0.10 M solution, HCl is very nearly completely ionized. The data in Table 17–6 show that the concentration of H_3O^+ in 0.10 M HCl is approximately 77 times greater than that in 0.10 M CH_3COOH, and approximately 1700 times greater than that in 0.10 M HOCl.

Thus far we have focused our attention primarily on acids. Very few common weak bases are soluble in water. Aqueous ammonia is the most frequently encountered example. The organic derivatives of ammonia are called *amines,* and many are also soluble weak bases. Hundreds of amines are known, and some are very important in the structures of proteins. From our earlier discussion of bonding in covalent compounds (Section 7–5.3) we recall that there is one unshared pair of electrons on the nitrogen atom in ammonia (frequently written NH_3 for emphasis), and that when ammonia dissolves in water it accepts a proton from a water molecule as it ionizes slightly. (See Section 17–3.) The low molecular weight amines show similar behavior.

Let us now consider the behavior of ammonia in aqueous solutions. As indicated earlier, the reaction of ammonia with water is

$$NH_3 + H_2O \rightleftharpoons NH_4^+ + OH^-$$

and the ionization constant expression is

> The subscript b indicates that the substance ionizes as a base.

$$K_b = \frac{[NH_4^+][OH^-]}{[NH_3]} = 1.8 \times 10^{-5} \qquad \text{[Eq. 17–9]}$$

Because the concentration of water is constant, or very nearly so, it is included in the ionization constant as it was for weak acids. The fact that the ionization constant for aqueous ammonia has the same value as the ionization constant for acetic acid is pure coincidence. However, it does tell us that in aqueous solutions of the same concentration, CH_3COOH and NH_3 are ionized to the same extent.

> Solutions of ammonia in water are sometimes called "ammonium hydroxide." There is no physical evidence to indicate the existence of "ammonium hydroxide," but the name is still fairly widely used. We prefer the term *aqueous ammonia.*

TABLE 17–6   Comparison of Ionizations of Acids

Acid Solution	Ionization Constant	[H_3O^+]	pH	% Ionization
0.10 M HCl	very large	0.10 M	1.00	~100
0.10 M CH_3COOH	1.8×10^{-5}	0.0013 M	2.89	1.3
0.10 M HClO	3.5×10^{-8}	0.000059 M	4.23	0.059

Ionization constants for weak bases are used in the same way as ionization constants for weak acids, as the following examples illustrate.

Example 17-10

Calculate the concentration of OH^-, the pH, and the % ionization for a 0.20 M solution of aqueous ammonia.

Solution

The equation for the ionization of aqueous ammonia and the algebraic representation of equilibrium concentrations are

$$NH_3 \qquad + H_2O \rightleftharpoons NH_4^+ + OH^-$$

in:	0.20 M	0 M	~0 M
ch:	$-x\ M$	$+x\ M$	$+x\ M$
eq:	$(0.20 - x)\ M$	$x\ M$	$x\ M$

Substitution into the ionization constant expression gives

$$K_b = \frac{[NH_4^+][OH^-]}{[NH_3]} = 1.8 \times 10^{-5} = \frac{(x)(x)}{(0.20 - x)}$$

If we assume that $(0.20 - x) \approx 0.20$, we have

$$\frac{x^2}{0.20} = 1.8 \times 10^{-5} \qquad \text{and} \qquad x^2 = 3.6 \times 10^{-6}$$

which gives $x = 1.9 \times 10^{-3}\ M = [OH^-]$.

Since $[OH^-] = 1.9 \times 10^{-3}\ M$, pOH $= 2.72$ and pH $= 11.28$.

The percentage ionization may be represented as

$$\% \text{ ionization} = \frac{[NH_3]_{ionized}}{[NH_3]_{total}} \times 100\%$$

$$= \frac{[OH^-]}{[NH_3]_{total}} \times 100\%$$

Each NH_3 molecule that ionizes produces one OH^- ion, and therefore we represent $[NH_3]_{ionized}$ by $[OH^-]$. Substitution into the above relationship gives

$$\frac{[OH^-]}{[NH_3]_{total}} \times 100\% = \frac{1.9 \times 10^{-3}}{0.20} \times 100\%$$

$$= 0.95\% \text{ ionized}$$

Example 17-11

The pH of an aqueous solution of ammonia is 11.50. Calculate the molarity of the solution.

Solution

Let's examine the problem to determine what we are given and what we would like to know. Since pH $= 11.50$, we know that pOH $= 2.50$, and

$$[OH^-] = 10^{-2.50} = 10^{0.50} \times 10^{-3}$$
$$= 3.2 \times 10^{-3}\ M$$

Now that we know $[OH^-] = 3.2 \times 10^{-3}\ M$, we also know that $[NH_4^+] = 3.2 \times 10^{-3}\ M$, and we can represent the equilibrium concentrations by letting $x\ M$ represent the original concentration of NH_3.

Since x mol/L of NH_3 is present in the solution and 3.2×10^{-3} mol/L of NH_3 ionizes, the equilibrium concentration of NH_3 is $(x - 3.2 \times 10^{-3})$ mol/L. Substituting these values into the equilibrium expression for aqueous ammonia gives

$$K_b = \frac{[NH_4^+][OH^-]}{[NH_3]}$$

$$= \frac{(3.2 \times 10^{-3})(3.2 \times 10^{-3})}{(x - 3.2 \times 10^{-3})} = 1.8 \times 10^{-5}$$

Examination of this equation suggests that $(x - 3.2 \times 10^{-3})$ is approximately equal to x (that is, 3.2×10^{-3} is very small compared to x). Making this assumption simplifies the calculation.

$$\frac{(3.2 \times 10^{-3})(3.2 \times 10^{-3})}{x} = 1.8 \times 10^{-5},$$

$$\text{and} \quad x = 0.57\ M$$

Therefore, the solution is 0.57 molar in ammonia. Our assumption that $(x - 3.2 \times 10^{-3}) \approx x$ was justified.

$$NH_3 \qquad + H_2O \rightleftharpoons \qquad NH_4^+ \quad + \qquad OH^-$$

in:	$x\ M$	0 M	~0 M
ch:	$-3.2 \times 10^{-3}\ M$	$+3.2 \times 10^{-3}\ M$	$+3.2 \times 10^{-3}\ M$
eq:	$(x - 3.2 \times 10^{-3})\ M$	$3.2 \times 10^{-3}\ M$	$3.2 \times 10^{-3}\ M$

17–8   Polyprotic Acids

Thus far we have restricted our discussion of weak acids to *monoprotic* acids, i.e., acids that furnish only one hydronium ion per molecule. Many weak acids furnish two or more hydronium ions per molecule, and are called **polyprotic** acids. The ionizations of polyprotic acids occur stepwise, that is, one proton at a time. An ionization constant expression can be written for each step in the ionization of a polyprotic acid, as the following example illustrates. Consider phosphoric acid as a typical polyprotic acid. It contains three acidic hydrogen atoms and ionizes in three steps:

$$H_3PO_4 + H_2O \rightleftharpoons H_3O^+ + H_2PO_4^- \qquad K_1 = \frac{[H_3O^+][H_2PO_4^-]}{[H_3PO_4]} = 7.5 \times 10^{-3}$$

$$H_2PO_4^- + H_2O \rightleftharpoons H_3O^+ + HPO_4^{2-} \qquad K_2 = \frac{[H_3O^+][HPO_4^{2-}]}{[H_2PO_4^-]} = 6.2 \times 10^{-8}$$

$$HPO_4^{2-} + H_2O \rightleftharpoons H_3O^+ + PO_4^{3-} \qquad K_3 = \frac{[H_3O^+][PO_4^{3-}]}{[HPO_4^{-2}]} = 3.6 \times 10^{-13}$$

Examination of the three ionization constants for H_3PO_4 reveals that K_1 is much larger than K_2, and that K_2 is much larger than K_3. This is generally true for polyprotic *inorganic* acids. (Refer to Appendix F.) Successive ionization constants commonly decrease by a factor of approximately 10^4 to 10^6, although some differences are outside this range. A large decrease in the value of each successive ionization constant means that each step in the ionization of a polyprotic acid occurs to a much smaller extent than the previous step. The extent of ionization in the second and third steps is even less than the values of K_2 and K_3 indicate, because the concentration of H_3O^+ produced in the first step is very large compared to the concentration of H_3O^+ produced in the second and third steps. Since the expressions for K_2 and K_3 include $[H_3O^+]$, they are satisfied by the *total concentration of H_3O^+ in the solution.* As a matter of fact, in all except extremely dilute solutions of H_3PO_4, the concentration of H_3O^+ may be assumed to be that furnished by the first step in the ionization alone.

Example 17–12

Calculate the concentrations of all species present in 0.100 M H_3PO_4.

Solution

Since K_1 is so much larger than K_2, we assume that the amount of H_3O^+ formed in the second and third steps of the ionization of H_3PO_4 and by the ionization of water is insignificant.

$$H_3PO_4 \quad + H_2O \rightleftharpoons H_3O^+ + H_2PO_4^-$$
$$(0.100 - x)\,M \qquad\qquad x\,M \quad\ x\,M$$

Substitution into the expression for K_1 gives

$$\frac{[H_3O^+][H_2PO_4^-]}{[H_3PO_4]} = \frac{(x)(x)}{(0.100 - x)} = 7.5 \times 10^{-3}$$

This equation must be solved by the quadratic formula. Solving the quadratic equation gives the positive root $x = 2.4 \times 10^{-2}$; $x = -3.1 \times 10^{-2}$ is the extraneous root, and is discarded because concentrations can't possibly be less than zero. Thus, from the first step in the ionization of H_3PO_4

$$x = \underline{[H_3O^+]} = \underline{[H_2PO_4^-]} = 2.4 \times 10^{-2}\,M$$

$$(0.100 - x) = \underline{[H_3PO_4]} = 7.6 \times 10^{-2}\,M$$

For the second step in the ionization of H_3PO_4, we use the concentrations H_3O^+ and

$H_2PO_4^-$ obtained from the first step, and let $y =$ [H_3O^+] produced in the second step as well as [HPO_4^{2-}].

$$H_2PO_4^- + H_2O \rightleftharpoons H_3O^+ + HPO_4^{2-}$$

$$(2.4 \times 10^{-2} - y)\,M \qquad (2.4 \times 10^{-2} + y)\,M \qquad y\,M$$

from 1st step

Substitution into the expression for K_2 gives

$$\frac{[H_3O^+][HPO_4^{2-}]}{[H_2PO_4^-]} = \frac{(2.4 \times 10^{-2} + y)(y)}{(2.4 \times 10^{-2} - y)}$$
$$= 6.2 \times 10^{-8}$$

Examination of this equation indicates that 2.4×10^{-2} is much greater than y, and making the usual simplifying assumption gives

$$\frac{(2.4 \times 10^{-2})(y)}{(2.4 \times 10^{-2})} = 6.2 \times 10^{-8},$$

$$y = [HPO_4^{2-}] = 6.2 \times 10^{-8}\,M$$

Also, $y =$ [H_3O^+] produced in the second step. Thus, we find that the concentration of HPO_4^{2-} is equal to K_2.

A general statement can be made for solutions of weak polyprotic acids for which $K_1 \gg K_2$ and that

contain no other electrolytes: **the concentration of the anion produced in the second step ionization is always equal to K_2 in solutions of reasonable concentration.** Note that our assumption that the [H_3O^+] produced in the first step (2.4×10^{-2} M) is much greater than the [H_3O^+] produced in the second step ionization (6.2×10^{-8} M) is valid.

We have now calculated the concentrations of all the species in 0.10 M H_3PO_4 solution except PO_4^{3-}, the anion produced in the third step ionization, and OH^-, which is present in all aqueous solutions. To calculate the concentration of PO_4^{3-} we use the equation for the third step ionization and let $z =$ [PO_4^{3-}] and [H_3O^+] produced in the third step.

$$HPO_4^{2-} + H_2O \rightleftharpoons H_3O^+ + PO_4^{3-}$$

$$(6.2 \times 10^{-8} - z)\,M \qquad (2.4 \times 10^{-2} + z)\,M \qquad z\,M$$

from 1st step

$$\frac{[H_3O^+][PO_4^{3-}]}{[HPO_4^{2-}]} = \frac{(2.4 \times 10^{-2} + z)(z)}{(6.2 \times 10^{-8} - z)}$$
$$= 3.6 \times 10^{-13}$$

We can make our simplifying assumption, and find that $z =$ [PO_4^{3-}] $= 9.3 \times 10^{-19}$ M.

We have now calculated the concentration of each of the species formed by the ionization of 0.100 M H_3PO_4. For easy comparison these concentrations are listed in Table 17–7. As a point of interest, the concentration of OH^- in 0.100 M H_3PO_4 is also included in the tabulation; it is calculated from the known value for [H_3O^+] using the ion product for water, [H_3O^+][OH^-] $= 1.0 \times 10^{-14}$.

Note that nonionized H_3PO_4 is present in greater concentration than any other species in 0.100 M H_3PO_4 solution, and that the only other species present in significant concentrations are H_3O^+ and $H_2PO_4^-$. Analogous statements can be made for other weak polyprotic acids for which the last K is very small.

TABLE 17–7 Concentrations of the Species in 0.100 M H_3PO_4

Species	Concentration (mol/L)
H_3PO_4	0.076
H_3O^+	0.024
$H_2PO_4^-$	0.024
HPO_4^{2-}	0.000000062
OH^-	0.00000000000042
PO_4^{3-}	0.00000000000000000093

17 – 9   Anions and Cations of Salts as Weak Bases or Acids

In Section 17 – 3 we learned that the anion of a weak acid is the conjugate base of that acid. The weaker a molecular acid is, the stronger its anion is as a base. Likewise, the cation of a weak base is the conjugate acid of that base. The weaker a molecular base is, the stronger its cation is as an acid. Consequently, *aqueous solutions of soluble ionic salts are usually acidic or basic if one (or more) of the ions is related to a weak acid or a weak base.*

The reactions of anions or cations of salts with water (or with its ions, H_3O^+ or OH^-) are called **hydrolysis** reactions.* We shall classify soluble salts into six general categories, based on whether their ions are related to strong acids, weak monoprotic acids, polyprotic acids, strong soluble bases, weak bases, or insoluble bases. For each class we shall describe the kind of hydrolysis reactions that can occur, if any, and whether the resulting solutions are acidic, basic, or neutral.

1   Salts of Strong Soluble Bases and Strong Acids (Neutral Solutions)

Solutions of salts derived from strong soluble bases and strong acids are neutral because neither the cation nor the anion of these salts reacts with water. For example, consider an aqueous solution of NaCl, the salt of NaOH and HCl. Sodium chloride is ionic even in the solid state and dissociates into hydrated ions in water. The water also ionizes slightly to produce equal numbers of H_3O^+ and OH^- ions.

$$NaCl \text{ (solid)} \longrightarrow Na^+ + Cl^-$$

$$H_2O + H_2O \rightleftharpoons OH^- + H_3O^+$$

Thus, aqueous solutions of NaCl contain four ions, Na^+, Cl^-, H_3O^+ and OH^-. However, the cation of the salt, Na^+, is such a weak acid that it does not react with the anion of water, OH^-. Likewise, the anion of the salt, Cl^-, is such a weak base that it does not react with the cation of water, H_3O^+. Therefore, solutions of salts of strong bases and strong acids are neutral because neither the cation nor the anion of such salts reacts to upset the H_3O^+/OH^- balance in water.

2   Salts of Strong Soluble Bases and Weak Acids (Basic Solutions)

When salts derived from strong soluble bases and weak acids are dissolved in water, the resulting solutions are always found to be basic. This happens because anions of weak acids react with water to form hydroxide ions, as the following example illustrates. Consider a solution of sodium acetate, $NaCH_3COO$, the salt of NaOH and CH_3COOH. It is soluble and dissociates completely in water.

$$NaCH_3COO \text{ (s)} \xrightarrow{H_2O} Na^+ + CH_3COO^-$$

$$H_2O + H_2O \rightleftharpoons OH^- + H_3O^+$$

$$\text{excess}$$

$$CH_3COOH + H_2O$$

* The reaction of a substance with *any* solvent in which it is dissolved (say, liquid ammonia) is called **solvolysis.** Because it is most common, hydrolysis is given a special name. "Lysis" means *splitting* (of the ions of the solute).

LeChatelier's Principle applies to equilibria in aqueous solution, and it enables us to make accurate predictions.

Acetate ion from $NaCH_3COO$ is somewhat basic since it is the conjugate base of a *weak* acid, CH_3COOH. Thus it combines with H_3O^+ and causes more water to ionize. As H_3O^+ is removed from the solution, an excess of OH^- builds up, so the solution becomes basic. The above equations can be combined into a single equation that represents the reaction much more simply, namely

$$CH_3COO^- + H_2O \rightleftharpoons CH_3COOH + OH^-$$

Similar equations can be written for other salts derived from strong bases and weak acids.

The equilibrium constant for this reaction is called a **hydrolysis constant,** or K_b for CH_3COO^-, and the equilibrium expression is written

$$K_b = \frac{[CH_3COOH][OH^-]}{[CH_3COO^-]}$$

Hydrolysis constants, or K_b's for anions of weak acids, can be determined experimentally and the values obtained from experiments agree with the calculated values. Please note that the K_b we use here refers to a reaction in which the anion of a weak acid acts as a base.

For the first time, we are able to calculate equilibrium constants using other known expressions. (Previously we have considered equilibrium constants that were determined experimentally, or derived from changes in free energy.) If we multiply the above expression by $[H_3O^+]/[H_3O^+]$, an algebraic manipulation that doesn't change the value of the expression, we have:

$$K_b = \frac{[CH_3COOH][OH^-]}{[CH_3COO^-]} \times \frac{[H_3O^+]}{[H_3O^+]} = \frac{[CH_3COOH]}{[H_3O^+][CH_3COO^-]} \times \frac{[H_3O^+][OH^-]}{1}$$

Comparing the last factors with Eqs. 17–4 and 17–8, we recognize that

$$K_b = \frac{1}{K_{a(CH_3COOH)}} \times \frac{K_w}{1} = \frac{1.0 \times 10^{-14}}{1.8 \times 10^{-5}}$$

or

$$K_b = \frac{[CH_3COOH][OH^-]}{[CH_3COO^-]} = 5.6 \times 10^{-10} \qquad \text{[Eq. 17–10]}$$

We have calculated the hydrolysis constant for the acetate ion, CH_3COO^-. We can perform the same operations for the anion of any weak monoprotic acid, and find that $K_b = K_w/K_a$ where K_a refers to the ionization constant for the weak monoprotic acid from which the anion under consideration is derived, and K_b refers to the hydrolysis constant for the anion.

Note that the equation can be rearranged to

$$K_a K_b = K_w \qquad \text{[Eq. 17–11]}$$

This relationship is valid for *all conjugate acid-base pairs* in aqueous solution. If either K_a or K_b is known, the other can be calculated.

The general equation for the hydrolysis of the anion of a weak monoprotic acid (from a salt) is

$$\underset{\substack{\text{Anion of} \\ \text{weak acid}}}{A^-} + H_2O \rightleftharpoons \underset{\text{weak acid}}{HA} + OH^-$$

The following examples illustrate the very different base strengths of different anions.

Example 17–13

Calculate $[OH^-]$ and pH for 0.10 M solutions of $NaCH_3COO$, sodium acetate, and NaCN, sodium cyanide.

Solution

Since we have two distinct problems, let's work them one at a time. The $NaCH_3COO$ is completely dissociated in $0.10\ M$ solution, and the overall equation for its reaction as a base with water is

$$CH_3COO^- + H_2O \rightleftharpoons CH_3COOH + OH^-$$

The equilibrium expression is Eq. 17–10,

$$K_b = \frac{[CH_3COOH][OH^-]}{[CH_3COO^-]} = 5.6 \times 10^{-10}$$

From the balanced equation, we see that $[CH_3COOH] = [OH^-]$, and we represent these concentrations by x.

$$CH_3COO^- + H_2O \rightleftharpoons CH_3COOH + OH^-$$
$$(0.10 - x)\ M \qquad\qquad x\ M \qquad x\ M$$

Substitution into the equilibrium constant expression gives

$$\frac{[CH_3COOH][OH^-]}{[CH_3COO^-]} = \frac{(x)(x)}{(0.10 - x)} = 5.6 \times 10^{-10}$$

Making our usual simplifying assumption gives $x = 7.5 \times 10^{-6}\ M = [OH^-]$. Since $[OH^-] = 7.5 \times 10^{-6}\ M$, we find pOH = 5.12 and pH = 8.88. The $0.10\ M$ NaCH$_3$COO solution is distinctly basic.

If we perform the same kind of calculation for $0.10\ M$ NaCN, a soluble salt that is completely dissociated in water, we have:

$$CN^- + H_2O \rightleftharpoons HCN + OH^-$$
$$(0.10 - y)\ M \qquad\qquad y\ M \qquad y\ M$$

$$K_b = \frac{[HCN][OH^-]}{[CN^-]} = \frac{K_w}{K_a} = \frac{1.0 \times 10^{-14}}{4.0 \times 10^{-10}}$$
$$= 2.5 \times 10^{-5}$$

Substitution into this expression gives

$$\frac{(y)(y)}{(0.10 - y)} = 2.5 \times 10^{-5}$$

and $\qquad y = 1.6 \times 10^{-3}\ M = [OH^-]$,

from which we find pOH = 2.80 and pH = 11.20.

The $0.10\ M$ solution of NaCN is much more basic than the $0.10\ M$ solution of NaCH$_3$COO because CN^- is a much stronger base than CH_3COO^-. This is entirely predictable because HCN is a much weaker acid than CH_3COOH.

To emphasize further the difference in the basicities of CH_3COO^- and CN^-, let's calculate the percentage hydrolysis of these anions in $0.10\ M$ solutions. For the hydrolysis of $0.10\ M\ CH_3COOH$, we found that $[OH^-] = 7.5 \times 10^{-6}\ M$.

$$\%\ \text{hydrolysis} = \frac{[CH_3COO^-]_{\text{hydrolyzed}}}{[CH_3COO^-]_{\text{total}}} \times 100\%$$

Since each CH_3COO^- that hydrolyzes produces one OH^-, the concentration of OH^- is the same as the concentration of CH_3COO^- that *hydrolyzed*. Therefore, we can write

$$\%\ \text{hydrolysis} = \frac{[CH_3COO^-]_{\text{hydrolyzed}}}{[CH_3COO^-]_{\text{total}}} \times 100\% = \frac{[OH^-]}{[CH_3COO^-]_{\text{total}}} \times 100\%$$

Substituting $7.5 \times 10^{-6}\ M$ for $[OH^-]$ and $0.10\ M$ for $[CH_3COO^-]_{\text{total}}$ gives

$$\frac{7.5 \times 10^{-6}}{0.10} \times 100\% = 0.0075\%\ \text{hydrolysis} \qquad (0.10\ M\ \text{NaCH}_3\text{COO})$$

For the $0.10\ M$ solution of NaCN, we have an analogous situation in which each CN^- ion that hydrolyzes produces one OH^- ion. Therefore, we can represent the percentage hydrolysis as

$$\%\ \text{hydrolysis} = \frac{[CN^-]_{\text{hydrolyzed}}}{[CN^-]_{\text{total}}} \times 100\% = \frac{[OH^-]}{[CN^-]_{\text{total}}} \times 100\%$$

$$= \frac{1.6 \times 10^{-3}}{0.10} \times 100\% = 1.6\%\ \text{hydrolysis} \qquad (0.10\ M\ \text{NaCN})$$

TABLE 17–8   Comparison of 0.10 M Solutions of NaCH$_3$COO, NaCN, and NH$_3$

	0.10 M NaCH$_3$COO	0.10 M NaCN	0.10 M aq NH$_3$
K_a for parent acid	1.8×10^{-5}	4.0×10^{-10}	
K_b for anion	5.6×10^{-10}	2.5×10^{-5}	K_b for NH$_3$ is 1.8×10^{-5}
[OH$^-$]	7.5×10^{-6} M	1.6×10^{-3} M	1.3×10^{-3} M
% hydrolysis	0.0075%	1.6%	1.3% ionized
pH	8.88	11.20	11.11

We have just shown that the % hydrolysis for 0.10 M NaCN (1.6%) is approximately 213 times greater than the % hydrolysis for 0.10 M NaCH$_3$COO (0.0075%). Table 17–8 summarizes the two calculations. Data for a 0.10 M solution of aqueous ammonia are included to provide a reference point. Note that the cyanide ion, CN$^-$, is a stronger base than aqueous NH$_3$.

3 Salts of Weak Bases and Strong Acids (Acidic Solutions)

The second common kind of hydrolysis reaction involves the combination of the cation (the conjugate acid) of a weak base with OH$^-$ from water to form nonionized molecules of the weak base. The removal of OH$^-$ likewise upsets the H$_3$O$^+$/OH$^-$ balance in water, but it produces *acidic* solutions:

$$\underset{\text{cation}^+}{BH^+} + OH^- \longrightarrow \underset{\text{weak base}}{B} + H_2O$$

This reaction is more frequently represented as

$$\underset{\text{cation}^+}{BH^+} + H_2O \longrightarrow \underset{\text{weak base}}{B} + H_3O^+$$

because the removal of OH$^-$ causes more H$_2$O to ionize to give an excess of H$_3$O$^+$.

Aqueous solutions of salts of weak bases and strong acids are acidic because cations derived from weak bases react with water to produce hydronium ions. Consider a solution of ammonium chloride, NH$_4$Cl, the salt of aqueous ammonia and hydrochloric acid.

Ammonium chloride is an ionic solid that is soluble in water.

$$NH_4Cl\,(s) \xrightarrow{\;H_2O\;} \boxed{NH_4^+} + Cl^-$$
$$H_2O + H_2O \rightleftharpoons \boxed{OH^-} + H_3O^+ \quad \text{excess}$$
$$\Updownarrow$$
$$NH_3\;+\;H_2O$$

Ammonium ions, from NH$_4$Cl, combine with OH$^-$ to form nonionized ammonia and water molecules. Since this reaction removes OH$^-$ from the system, it causes more water to ionize and produces an excess of H$_3$O$^+$ in the solution. We can conveniently combine the above equations into a single equation that represents the reaction accurately.

Analogous equations can be written for cations derived from other weak bases.

$$NH_4^+ + H_2O \rightleftharpoons NH_3 + H_3O^+$$

The equilibrium expression for this reaction is written

$$K_a = \frac{[NH_3][H_3O^+]}{[NH_4^+]}$$

Like hydrolysis constants for anions of weak acids, hydrolysis constants for cations of weak bases can be calculated. If we multiply the above expression by $[OH^-]/[OH^-]$, we have

$$K_a = \frac{[NH_3][H_3O^+]}{[NH_4^+]} \times \frac{[OH^-]}{[OH^-]} = \frac{[NH_3]}{[NH_4^+][OH^-]} \times \frac{[H_3O^+][OH^-]}{1}$$

We recognize that

$$K_a = \frac{1}{K_{b(NH_3)}} \times \frac{K_w}{1} = \frac{1.0 \times 10^{-14}}{1.8 \times 10^{-5}} = 5.6 \times 10^{-10}$$

This manipulation gives the hydrolysis constant,

$$K_a = \frac{[NH_3][H_3O^+]}{[NH_4^+]} = 5.6 \times 10^{-10} \qquad \text{[Eq. 17–12]}$$

which enables us the calculate pH values for solutions of salts of aqueous ammonia and strong acids.

In calculating the equilibrium constant for NH_4^+, we found that $K_a = K_w/K_b$. The process by which we obtained this hydrolysis constant can be used to obtain equilibrium constants for cations of other weak bases. Alternatively, we can use Eq. 17–11,

$$K_a \times K_b = K_w \qquad \text{(any conjugate acid-base pair)}$$

and find

$$K_a = \frac{K_w}{K_b}$$

the same expression obtained above.

The fact that K_a for the ammonium ion, NH_4^+, is the same as K_b for the acetate ion, CH_3COO^-, should not be surprising. Recall that the ionization constants for CH_3COOH and aqueous NH_3 are equal. Thus, we might expect the anion of CH_3COOH to hydrolyze to exactly the same extent as the cation of aqueous NH_3, and it does.

Example 17–14 illustrates how the appropriate equilibrium constant can be used to calculate concentrations of the various species in solutions containing soluble salts of weak bases and strong acids.

Example 17 – 14

Calculate the pH of a 0.20 M solution of ammonium nitrate, NH_4NO_3.

Solution

We recognize that NH_4NO_3 is the salt of aqueous ammonia and nitric acid, a strong acid. Therefore, the cation of the weak base reacts with water.

$$NH_4^+ + H_2O \rightleftharpoons NH_3 + H_3O^+$$
$$(0.20 - x)\,M \qquad\qquad x\,M \quad x\,M$$

The equation tells us that $[NH_3] = [H_3O^+]$ and, since we don't know either, we represent both as x mol/L. The equilibrium concentration of NH_4^+ is represented as $(0.20 - x)$ mol/L. In our earlier discussion we found that K_a for NH_4^+, Eq. 17–12, is

$$K_a = \frac{[NH_3][H_3O^+]}{[NH_4^+]} = 5.6 \times 10^{-10}$$

Substituting into this expression gives

$$\frac{(x)(x)}{(0.20 - x)} = 5.6 \times 10^{-10}$$

Making the usual simplifying assumption gives $x = 1.1 \times 10^{-5}\,M = [H_3O^+]$, and pH = 4.96. Note that 0.20 M NH_4NO_3 solution is distinctly acidic. Ammonium nitrate is widely used as a fertilizer. As this example shows, it contributes significantly to soil acidity.

4 Salts of Weak Bases and Weak Acids (Acidic, Basic, or Neutral Solutions)

Soluble salts of weak bases and weak acids make up the fourth class of salts we wish to examine. In our earlier discussion we pointed out the fact that anions of weak acids react with water to give basic solutions, and cations of weak bases give acidic solutions. Since salts of weak bases and weak acids contain cations that give acidic solutions and anions that give basic solutions, will solutions of salts of weak bases and weak acids be neutral, basic, or acidic? The answer is that they may be any one of the three depending on the relative strengths of the basic or acidic species involved. Thus, salts of weak bases and weak acids may be divided into three types that depend on the values of the ionization constants of the parent weak bases and weak acids.

a. Salts of Weak Bases and Weak Acids for which $K_{b(\text{parent base})} = K_{a(\text{parent acid})}$ (Neutral Solutions) These cases are very rare. Ammonium acetate, NH_4CH_3COO, is the only common example. Both cation and anion hydrolyze to the *same* extent (Eqs. 17–12 and 17–10):

$$NH_4^+ + H_2O \rightleftharpoons NH_3 + H_3O^+$$

$$K_a = \frac{K_w}{K_{b(NH_3)}} = \frac{1.0 \times 10^{-14}}{1.8 \times 10^{-5}} = 5.6 \times 10^{-10}$$

$$CH_3COO^- + H_2O \rightleftharpoons CH_3COOH + OH^-$$

$$K_b = \frac{K_w}{K_{a(CH_3COOH)}} = \frac{1.0 \times 10^{-14}}{1.8 \times 10^{-5}} = 5.6 \times 10^{-10}$$

Note that K_a for NH_4^+ is equal to K_b for CH_3COO^- because K_b for NH_3 is equal to K_a for CH_3COOH. Thus, H_3O^+ and OH^- are produced in equal concentrations and the solution remains neutral. Most of these ions combine to form H_2O. This increases the extents of both forward reactions, leading to greater percentage hydrolysis than for the salts covered previously, but such solutions remain neutral.

b. Salts of Weak Bases and Weak Acids for which $K_{b(\text{parent base})} > K_{a(\text{parent acid})}$ (Basic Solutions) Salts of weak bases and weak acids for which K_{base} is greater than K_{acid} are always basic because the anion of the weaker acid hydrolyzes to a greater extent than the cation of the stronger base, as the following example illustrates.

Consider NH_4CN, ammonium cyanide, as a typical salt of a weak base and a weak acid for which the ionization constant of the parent weak base is greater than the ionization constant of the parent weak acid. Since K_a for HCN (4.0×10^{-10}) is much smaller than K_b for NH_3 (1.8×10^{-5}), then K_b for CN^- (2.5×10^{-5}) is much larger than K_a for NH_4^+ (5.6×10^{-10}). This tells us that the CN^- ion hydrolyzes to a much greater extent than NH_4^+ ion, and so solutions of NH_4CN are distinctly basic. Stated differently, CN^- is much stronger as a base than NH_4^+ is as an acid.

$$NH_4^+ + H_2O \rightleftharpoons NH_3 + H_3O^+$$
$$CN^- + H_2O \rightleftharpoons HCN + OH^-$$

$\longrightarrow 2H_2O$

This reaction occurs to greater extent, ∴ solution is basic

c. Salts of Weak Bases and Weak Acids for which $K_{b(\text{parent base})} <$ $K_{a(\text{parent acid})}$ (Acidic Solutions)

Salts of weak bases and weak acids for which K_{base} is less than K_{acid} are acidic because the cation of the weaker base hydrolyzes to a greater extent than the anion of the stronger acid. Consider ammonium fluoride, NH_4F, as a typical example. It is the salt of aqueous ammonia and hydrofluoric acid.

Recall that K_b for aqueous NH_3 is 1.8×10^{-5} and K_a for HF is 7.2×10^{-4}. The K_a value for NH_4^+ (5.6×10^{-10}) is slightly larger than the K_b value for F^- (1.4×10^{-11}), which tells us that the NH_4^+ is slightly stronger as an acid than F^- is as a base, so solutions of ammonium fluoride are slightly acidic.

$$NH_4^+ + H_2O \rightleftharpoons NH_3 + H_3O^+$$
$$F^- \; + H_2O \rightleftharpoons HF \; + OH^-$$

$\rightarrow 2H_2O$ This reaction occurs to greater extent, $\therefore$ solution is acidic

5   Salts Containing Metal Ions Related to Insoluble Bases (Acidic Solutions)

Solutions of many common salts of strong acids are acidic. For this reason, many homeowners apply iron(II) sulfate, $FeSO_4 \cdot 7H_2O$, or aluminum sulfate, $Al_2(SO_4)_3 \cdot 18H_2O$, to the soil around the "acid-loving" plants such as azaleas, camellias, and hollies. You are probably familiar with the sour "acid" taste of alum, $KAl(SO_4)_2 \cdot 12H_2O$, a substance that is frequently added to pickles.

> All the common metal ions except those derived from strong soluble bases fall into this category.

Solutions of salts that contain small, **highly charged metal ions** together with anions of strong acids are acidic, because such cations hydrolyze to produce excess hydronium ions. Consider aluminum chloride, $AlCl_3$, as a typical example. When solid anhydrous $AlCl_3$ is added to water, the water becomes very warm. In fact, if a large quantity of anhydrous aluminum chloride is added rapidly to a small amount of water, a small "explosion" occurs! The reaction of $AlCl_3$ with water is so highly exothermic that steam is produced, and the escaping steam causes severe spattering of the mixture. As we have pointed out earlier, ions are always hydrated in solution. In many cases the interaction between positively charged ions and the negative end of polar water molecules is sufficiently strong that salts of such cations crystallize from aqueous solution in combination with definite numbers of water molecules.

> Non-coordinating anions do not form coordinate co-valent bonds with metal ions in aqueous solution.

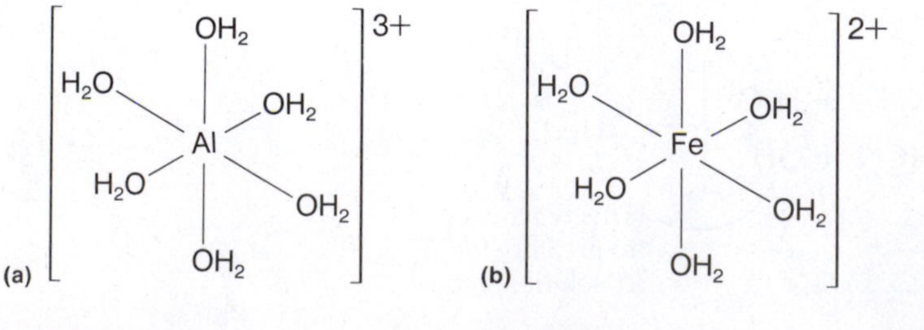

Salts containing Al^{3+}, Fe^{2+}, Fe^{3+}, and Cr^{3+} ions combined with non-coordinating anions usually crystallize from aqueous solutions with six water molecules associated with each metal ion. Many common salts of these cations contain hydrated cations, such as $[Al(OH_2)_6]^{3+}$, $[Fe(OH_2)_6]^{2+}$, $[Fe(OH_2)_6]^{3+}$, and $[Cr(OH_2)_6]^{3+}$, in the solid state. All of these species are octahedral; i.e., the metal ion (M^{n+}) is located at the center of a regular octahedron and the six water molecules are located at the corners, as illustrated in Figure 17–2. In the metal-oxygen bonds of these hydrated

FIGURE 17–2 Structures of (a) the hydrated aluminum ion, $[Al(OH_2)_6]^{3+}$, which is often represented as Al^{3+} (aq) or more simply as Al^{3+}, and (b) the hydrated iron(II) ion, $[Fe(OH_2)_6]^{2+}$, which is represented as Fe^{2+} (aq) or simply as Fe^{2+}.

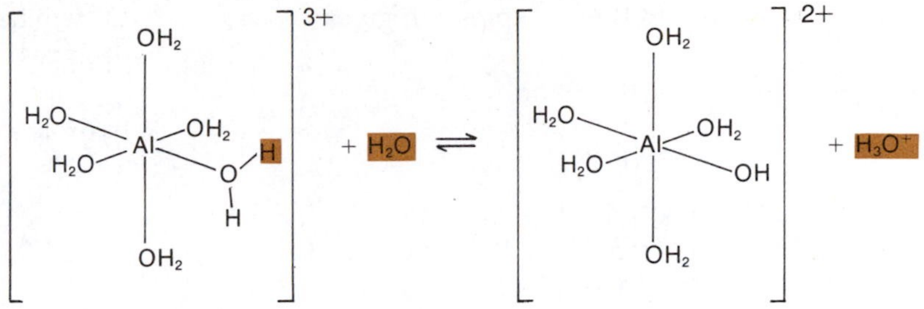

FIGURE 17-3 Hydrolysis of hydrated aluminum ion to produce H_3O^+, that is, the abstraction of a proton from a coordinated H_2O molecule by a non-coordinated H_2O molecule.

"Abstract" means to pull off.

cations, electron density is decreased around the oxygen end of the water molecule by the positively charged metal ions. This decrease in electron density weakens the hydrogen-oxygen bonds in coordinated water molecules relative to the hydrogen-oxygen bonds in non-coordinated water molecules. Consequently, non-coordinated water molecules can "abstract hydrogen ions" from the coordinated water molecules to form hydronium ions and produce acidic solutions as shown in Figure 17-3. This reaction is written as

$$[Al(OH_2)_6]^{3+} + H_2O \rightleftharpoons [Al(OH)(OH_2)_5]^{2+} + H_3O^+$$

or even more simply as

$$Al^{3+} + 2H_2O \rightleftharpoons Al(OH)^{2+} + H_3O^+$$

Note that abstraction of a hydrogen ion converts a coordinated water molecule to a coordinated hydroxide ion, and decreases the positive charge on the hydrated species. We write equilibrium expressions for the hydrolysis of small, highly charged cations just as we have for other reactions. Thus, the equilibrium constant expression for the hydrolysis of the hydrated aluminum ion is

$$K_a = \frac{[[Al(OH)(OH_2)_5]^{2+}][H_3O^+]}{[[Al(OH_2)_6]^{3+}]} = 1.2 \times 10^{-5}$$

or more simply

$$K_a = \frac{[Al(OH)^{2+}][H_3O^+]}{[Al^{3+}]} = 1.2 \times 10^{-5} \qquad \text{[Eq. 17-13]}$$

Hydrolysis of small, highly charged cations may occur beyond the first step, and in many cases the reactions are quite complex. They may involve two or more cations reacting with each other to form large polymeric species. However, for most common cations, consideration of the first hydrolysis constant is adequate for the calculations we will perform.

Example 17-15

Calculate the pH and percentage hydrolysis in a 0.10 M solution of $Al(NO_3)_3$.

Solution

As we have just observed, the reaction can be written as

$$Al^{3+} + 2H_2O \rightleftharpoons Al(OH)^{2+} + H_3O^+$$

and the hydrolysis constant is

$$K_a = \frac{[Al(OH)^{2+}][H_3O^+]}{[Al^{3+}]} = 1.2 \times 10^{-5}$$

The usual algebraic representation of equilibrium concentrations gives

$$\underset{(0.10 - x)\,M}{Al^{3+}} + 2H_2O \rightleftharpoons \underset{x\,M}{Al(OH)^{2+}} + \underset{x\,M}{H_3O^+}$$

and substitution into the K_a expression gives

$$\frac{(x)(x)}{(0.10 - x)} = 1.2 \times 10^{-5} \quad \text{and}$$

$$x = 1.1 \times 10^{-3} \, M = [H_3O^+]$$

Since $[H_3O^+] = 1.1 \times 10^{-3} \, M$, we find pH = 2.96, so the solution is quite acidic. It may be helpful to recall that in $0.10 \, M \, CH_3COOH$, $[H_3O^+] = 1.3 \times 10^{-3} \, M$ and pH = 2.89. Thus, a $0.10 \, M$ acetic acid solution is only slightly more acidic than a $0.10 \, M$ aluminum nitrate solution.

To calculate the percentage hydrolysis, we observe that each Al^{3+} that hydrolyzes produces one H_3O^+. Thus, the equilibrium concentration of H_3O^+ will be equal to the concentration of Al^{3+} that hydrolyzes.

$$\% \text{ hydrolysis} = \frac{[Al^{3+}]_{\text{hydrolyzed}}}{[Al^{3+}]_{\text{total}}} \times 100\%$$

$$= \frac{[H_3O^+]}{[Al^{3+}]_{\text{total}}} \times 100\%$$

$$= \frac{1.1 \times 10^{-3}}{0.10} \times 100\% = 1.1\%$$

Again, recall (Example 17–9) that in $0.10 \, M$ solution, CH_3COOH is 1.3% ionized; we have just found that in $0.10 \, M$ solution, $Al(NO_3)_3$ is 1.1% hydrolyzed, so the acidities of the two solutions are almost equal.

We might expect that smaller, more highly charged metal ions are stronger acids than larger, less highly charged cations because the smaller, more highly charged cations interact with coordinated water molecules more strongly. Indeed, this is usually the case, as Table 17–9 shows.

If we compare isoelectronic cations in the same period in the periodic table, the smaller, more highly charged metal ion is the stronger acid. (Compare K_a's for Li^+ and Be^{2+} and for Na^+, Mg^{2+}, Al^{3+}.) For cations with the same charge from the same group in the periodic table, the smaller cation hydrolyzes to a greater extent. (Compare K_a's for Be^{2+} and Mg^{2+}.) If we compare cations of the same element in different oxidation states, the smaller, more highly charged cation is the stronger acid. (Compare K_a's for Fe^{2+} and Fe^{3+}, and for Co^{2+} and Co^{3+}.) Although there are additional considerations that enable us to predict relatively accurately the extent to which various kinds of cations hydrolyze, the previous statements are adequate for our purposes.

TABLE 17–9 Ionic Radii and Hydrolysis Constants for Some Common Cations

Cation	Ionic Radius (nm)	Hydrated Cation	K_a
Li^+	0.060	$[Li(OH_2)_4]^+$	1×10^{-14}
Be^{2+}	0.031	$[Be(OH_2)_4]^{2+}$	1.0×10^{-5}
Na^+	0.095	$[Na(OH_2)_6]^+$ (?)	10^{-14}
Mg^{2+}	0.065	$[Mg(OH_2)_6]^{2+}$	3.0×10^{-12}
Al^{3+}	0.050	$[Al(OH_2)_6]^{3+}$	1.2×10^{-5}
Fe^{2+}	0.076	$[Fe(OH_2)_6]^{2+}$	3.0×10^{-10}
Fe^{3+}	0.064	$[Fe(OH_2)_6]^{3+}$	4.0×10^{-3}
Co^{2+}	0.074	$[Co(OH_2)_6]^{2+}$	5.0×10^{-10}
Co^{3+}	0.063	$[Co(OH_2)_6]^{3+}$	1.7×10^{-2}
Cu^{2+}	0.096	$[Cu(OH_2)_4]^{2+}$	1.0×10^{-8}
Zn^{2+}	0.074	$[Zn(OH_2)_6]^{2+}$	2.5×10^{-10}
Hg^{2+}	0.110	$[Hg(OH_2)_6]^{2+}$	8.3×10^{-7}
Bi^{3+}	0.074	$[Bi(OH_2)_6]^{3+}$	1.0×10^{-2}

6 Salts Containing Anions with Ionizable Hydrogen Atoms (Acidic or Basic Solutions)

Acidic salts, those with hydrogen-containing anions such as hydrogen carbonate ion, HCO_3^-, hydrogen sulfate ion, HSO_4^-, and dihydrogen phosphate ion, $H_2PO_4^-$, can produce acidic or basic solutions. Such anions can either ionize as acids or act as bases like other anions of weak acids in hydrolysis reactions. The kind of reaction that predominates can be determined by comparing the magnitudes of the equilibrium constants for the two possible reactions.

First, let us use sodium hydrogen sulfate, $NaHSO_4$, as an example of such salts. To understand the character of the anion we must first consider the ionization of the related diprotic acid, H_2SO_4, which is a strong acid with respect to its first step ionization.

Step 1: $H_2SO_4 (\ell) + H_2O (\ell) \longrightarrow H_3O^+ + HSO_4^- (aq) \qquad K_1 =$ very large

Step 2: $HSO_4^- (aq) + H_2O (\ell) \rightleftharpoons H_3O^+ + SO_4^{2-} (aq) \qquad K_2 = 1.2 \times 10^{-2}$

In very dilute solutions the second step ionization also is nearly complete, but in more concentrated solutions it is not. Thus we can consider the HSO_4^- ion as a relatively strong acid compared to other weak acids. Note that K_2 is large compared to K_a for CH_3COOH, which is 1.8×10^{-5}. The large value of K_2 for H_2SO_4, which is K_a for HSO_4^-, tells us that solutions of $NaHSO_4$ and similar salts are distinctly acidic.

> HSO_4^- does not hydrolyze to give a basic solution because it is the anion (conjugate base) of the very strong acid, H_2SO_4. K_b for HSO_4^- is extremely small.

Let us study one other example, sodium hydrogen carbonate, $NaHCO_3$. Let us first consider the ionization of carbonic acid, H_2CO_3, a very weak acid.

Step 1: $H_2CO_3 (aq) + H_2O (\ell) \rightleftharpoons H_3O^+ + HCO_3^- (aq) \qquad K_1 = 4.2 \times 10^{-7}$

Step 2: $HCO_3^- (aq) + H_2O (\ell) \rightleftharpoons H_3O^+ + CO_3^{2-} (aq) \qquad K_2 = 4.8 \times 10^{-11}$

The possible reactions that HCO_3^- ion can undergo in aqueous $NaHCO_3$ are the step 2 ionization as an acid (above), for which K_2 (or K_a for HCO_3^-) = 4.8×10^{-11}, and hydrolysis as the anion of a weak acid. Since HCO_3^- is the conjugate base of H_2CO_3, K_b for HCO_3^- is K_w / K_1 for H_2CO_3.

$HCO_3^- (aq) + H_2O (\ell) \rightleftharpoons H_2CO_3 (aq) + OH^- (aq)$

$$K_b = \frac{K_w}{K_1} = \frac{1.0 \times 10^{-14}}{4.2 \times 10^{-7}} = 2.4 \times 10^{-8}$$

> These arguments also apply to other hydrogen carbonate salts that are derived from strong soluble bases, such as $KHCO_3$, $Ca(HCO_3)_2$, and $Ba(HCO_3)_2$, all of which produce basic solutions.

Since K_b for HCO_3^-, 2.4×10^{-8}, is greater than K_2 (or K_a for HCO_3^-), 4.8×10^{-11}, the hydrolysis as the anion of a weak acid predominates and solutions of $NaHCO_3$ are *basic*.

We have now considered solutions of salts in each of six different categories, defined by the characteristics of the ions of the salt. We see that we can systematically determine whether salt solutions are acidic, basic, or neutral by (1) identifying the parent acids and bases to which the ions are related, (2) obtaining the K_a and K_b values for the parent acids and bases (if they are weak); and (3) calculating the hydrolysis constants for the ions, K_a or K_b, from these values.

The explanations used in this chapter have been based on the Brønsted-Lowry theory of acid-base behavior. Let us finally consider a more general and very useful theory of acids and bases, the Lewis theory.

17–10   The Lewis Theory

In 1923 Professor G. N. Lewis presented the most comprehensive of the classical acid-base theories. In the Lewis view, an **acid** *is defined as any species that can accept a share in an electron pair.* A **base** *is any species that can make available or "donate" a share in an electron pair.* The definitions do *not* specify that an electron pair must be transferred from one atom to another—only that an electron pair, residing originally on one atom, be shared between two atoms. *Neutralization* is defined as the formation of a **coordinate covalent bond,** which is a covalent bond in which both electrons were furnished by one atom.

The reaction of boron trifluoride with ammonia is a typical Lewis acid-base reaction.

$$BF_3\,(g) + NH_3\,(g) \longrightarrow F_3B\!:\!NH_3(s)$$

acid   base   product

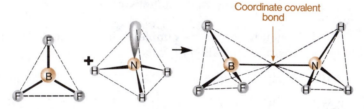

The Lewis theory is sufficiently general that it covers *all* acid-base reactions that the other theories include, plus many additional reactions such as complex formation (Chapter 23).

The autoionization of water (Section 17–5) was described in terms of the Brønsted-Lowry theory as

$$H_2O + H_2O \rightleftharpoons H_3O^+ + OH^-$$

base₁   acid₂   acid₁   base₂

In Lewis theory terminology, this is also an acid-base reaction:

$$H\!:\!\ddot{O}\!: + H\!:\!\ddot{O}\!: \rightleftharpoons H\!:\!\ddot{O}\!:^+ + H\!:\!\ddot{O}\!:^-$$

base   acid

The acceptance of a proton, H^+, by a base involves the formation of a coordinate covalent bond. Here we shift our viewpoint to consider the reaction as "donation" of the electron pair rather than "acceptance" of a proton.

Theoretically, any species that contains an unshared electron pair is a base. In fact, most ions and molecules that contain unshared electron pairs undergo some reactions by sharing their electron pairs. Conversely, many Lewis acids contain only six electrons in the highest occupied energy level of the central element, and react by accepting a share in an additional pair of electrons. These species are said to have an **open sextet.** Many compounds of the Group IIIA elements are Lewis acids, as illustrated by the reaction of boron trifluoride with ammonia presented above.

Anhydrous aluminum chloride is a common Lewis acid. The dissolution of $AlCl_3$ in hydrochloric acid gives a solution that contains $AlCl_4^-$ ions.

$$AlCl_3 \text{ (s)} + Cl^- \text{ (aq)} \longrightarrow AlCl_4^- \text{ (aq)}$$
$$\quad\text{acid} \qquad\quad \text{base} \qquad\qquad\qquad \text{product}$$

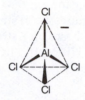

Al is sp^3 hybridized

Experienced chemists find the Lewis theory to be very useful because so many chemical reactions are covered by it. The less experienced usually find the theory less useful, but as their knowledge expands so does the utility of the Lewis theory.

Key Terms

Acidic salt a salt containing an ionizable hydrogen atom; does not necessarily produce acidic solutions.

Amines derivatives of ammonia in which one or more hydrogen atoms have been replaced by organic groups.

Amphoterism ability of a substance to act as either an acid or a base.

Autoionization an ionization reaction between identical molecules.

Brønsted-Lowry acid a proton donor.

Brønsted-Lowry base a proton acceptor.

Conjugate acid-base pair in Brønsted-Lowry terminology, a reactant and product that differ by a proton, H^+.

Coordinate covalent bond covalent bond in which both shared electrons were furnished by the same species; bond between a Lewis acid and a Lewis base.

Hydration the process by which water molecules bind to ions or molecules in the liquid or solid state or in solution.

Hydrolysis the reaction of a substance with water or its ions.

Hydrolysis constant an equilibrium constant for a hydrolysis reaction.

Ionization the breaking up of a compound into separate ions.

Ionization constant equilibrium constant for the ionization of a weak electrolyte.

Ion product for water equilibrium constant for the ionization of water, $K_w = [H^+][OH^-] = 1.00 \times 10^{-14}$ at 25°C

Lewis acid any species that can accept a share in an electron pair.

Lewis base any species that can make available a share in an electron pair.

Leveling effect effect by which all acids stronger than the acid that is characteristic of the solvent react with solvent to produce that acid; similar statement applies to bases. The strongest acid (base) that can exist in a given solvent is the acid (base) characteristic of that solvent.

Monoprotic acid an acid that contains one ionizable hydrogen atom per formula unit.

Open sextet refers to species that have only six electrons in the highest energy level of the central element (many Lewis acids).

pH the negative logarithm of the concentration (mol/L) of the H^+ (H_3O^+) ion; scale is commonly used over the range 0 to 14.

p[ ] the negative logarithm of the concentration (mol/L) of the indicated species.

Polyprotic acid an acid that contains more than one ionizable hydrogen atom per formula unit.

Protonic acid an Arrhenius (classical) acid.

Solvolysis the reaction of a substance with the solvent in which it is dissolved.

Exercises

Basic Ideas

1. Define and illustrate the following terms clearly and concisely, and provide a specific example of each.
   - (a) strong electrolyte
   - (b) weak electrolyte
   - (c) nonelectrolyte
   - (d) strong acid
   - (e) strong soluble base
   - (f) weak acid
   - (g) weak base

2. Distinguish between the following pairs of terms and provide a specific example of each.
   - (a) strong acid and weak acid
   - (b) strong soluble base and weak base
   - (c) strong soluble base and insoluble base

3. Write formulas and names for
   - (a) the common strong acids
   - (b) three weak acids
   - (c) the common strong soluble bases
   - (d) the most common weak base
   - (e) three soluble ionic salts
   - (f) three insoluble salts

The Hydrated Hydrogen Ion

4. (a) Describe the hydrated hydrogen ion in words and with formulas.
   - (b) Why is the hydrated hydrogen ion important?
   - (c) Criticize the following statement. "The hydrated hydrogen ion should always be represented as H_3O^+."

Brønsted-Lowry Theory

5. State the basic ideas of the Brønsted-Lowry Theory.

6. Use Brønsted-Lowry terminology to define the following terms. Illustrate each with a specific example.
   - (a) acid
   - (b) conjugate base
   - (c) base
   - (d) conjugate acid
   - (e) conjugate acid-base pair

7. Write balanced equations that describe the ionization of the following acids in dilute aqueous solution. Use a single arrow ($\rightarrow$) to represent complete, or nearly complete ionization, and a double arrow ($\rightleftharpoons$) to represent a small extent of ionization.
   - (a) HNO_3
   - (b) CH_3COOH
   - (c) HBr
   - (d) HCN
   - (e) HF
   - (f) $HClO_4$

8. Use words and equations to describe how ammonia can act as a base in (a) aqueous solution, and (b) the pure state, i.e., as gaseous ammonia molecules when it reacts with gaseous hydrogen chloride or a similar anhydrous acid.

Acidic and Basic Character

9. What does "acid strength" mean?

10. What does "base strength" mean?

11. Classify the following compounds as (a) strong soluble bases, (b) insoluble bases, (c) strong acids, or (d) weak acids: HBr, HF, KOH, $Cu(OH)_2$, CH_3COOH, $Ba(OH)_2$, $HClO_4$, HNO_3, HCN, $Fe(OH)_3$.

12. (a) What are binary protonic acids?
    - (b) Write names and formulas for four binary protonic acids.

13. (a) How can the order of increasing acid strength in a series of similar binary protonic acids be explained?
    - (b) Illustrate your answer for the series HF, HCl, HBr, and HI.
    - (c) What is the order of increasing base strength of the conjugate bases of the acids in (b)? Why?
    - (d) Is your explanation applicable to the series H_2O, H_2S, H_2Se, and H_2Te? Why?

14. (a) What are ternary acids? Write names and formulas for four of them.
    - (b) Why can we describe nitric and sulfuric acids as "hydroxides" of nonmetals?

15. Explain the order of increasing acid strength for the following groups of acids and the order of increasing base strength for their conjugate bases.
    - (a) H_2SO_3, H_2SO_4
    - (b) HNO_2, HNO_3
    - (c) H_3PO_3, H_3PO_4
    - (d) $HClO$, $HClO_2$, $HClO_3$, $HClO_4$

16. What does the term "leveling effect" mean? Illustrate your answer with three specific examples.

17. (a) Write a generalization that describes the order of acid strengths for a series of ternary acids that contain different elements in the same oxidation state from the same group in the periodic table.

(b) Indicate the order of acid strengths for the following:
   (1) HNO_3, H_3PO_4 (2) H_3PO_4, H_3AsO_4
   (3) H_2SO_4, H_2SeO_4 (4) $HClO_3$, $HBrO_3$

The Autoionization of Water

18. (a) What is meant by the autoionization of water?
   (b) Write the balanced equation for the autoionization of water and draw Lewis formulas for all species in the equation.
   (c) Pure anhydrous ammonia can be liquified at $-31°C$ to form a clear colorless liquid whose autoionization is similar to that of water. Write the balanced equation and draw Lewis formulas for all species in the equation.

19. (a) What is the meaning of the statement, ''Water is amphoteric''?
   (b) Based on the equation you wrote in Exercise 18c, can you say that liquid ammonia is amphoteric? Why?

20. Calculate the pH of solutions that have the following concentrations of H_3O^+. (a) 0.10 M, (b) 0.37 M, (c) 1.4×10^{-3} M, (d) 8.4×10^{-9} M

21. Calculate the concentrations of H_3O^+ in solutions that have the following pH's: (a) 1.23, (b) 4.92, (c) 8.30, (d) 10.92

22. What are the relationships (a) between $[H_3O^+]$ and $[OH^-]$ and (b) between pH and pOH in aqueous solutions?

23. Complete the following table.

Solution	$[H_3O^+]$	$[OH^-]$	pH	pOH
(a) 0.25 M HCl	_____	_____	_____	_____
(b) 0.25 M NaOH	_____	_____	_____	_____
(c) 0.030 M Ca(OH)$_2$	_____	_____	_____	_____

24. Given one of the four values in each row of the following table, calculate the other three.

Solution	$[H_3O^+]$	$[OH^-]$	pH	pOH
(a)	1.0×10^{-4} M	_____	_____	_____
(b)	_____	5.8×10^{-6} M	_____	_____
(c)	_____	_____	8.32	_____
(d)	_____	_____	_____	3.71

25. What is a pH meter?

Strong Electrolytes

26. Calculate the concentrations of the constituent ions in solutions of the following compounds in the indicated concentrations. The salts are soluble ionic compounds. Ignore hydrolysis.
   (a) 0.10 M HI   (b) 0.050 M KOH
   (c) 0.010 M Sr(OH)$_2$ (d) 0.0020 M Ba(NO$_3$)$_2$
   (e) 0.00030 M H$_2$SO$_4$ (f) 0.0035 M Fe$_2$(SO$_4$)$_3$

27. Calculate the concentrations of the constituent ions in the following solutions. The salts are soluble ionic compounds.
   (a) 4.0 g NaOH in 2.0 liters of solution
   (b) 0.74 g Ca(OH)$_2$ in 200 mL of solution
   (c) 2.64 g (NH$_4$)$_2$SO$_4$ in 100 mL of solution
   (d) 3.42 g Al$_2$(SO$_4$)$_3$ in 300 mL of solution
   (e) 43.8 g CaCl$_2$ · 6H$_2$O in 4.00 liters of solution

Weak Molecular Acids and Bases

28. In a solution of a weak acid,

$$HA + H_2O \rightleftharpoons H_3O^+ + A^-,$$

the following equilibrium concentrations are found: $[H_3O^+] = 0.0020$ M and $[HA] = 0.098$ M. Calculate the ionization constant for the weak acid, HA.

29. A 0.050 M aqueous solution of a weak, monoprotic acid is 1.2% ionized. Calculate the ionization constant for the acid.

30. In 0.10 M solution a weak acid, $HX + H_2O \rightleftharpoons H_3O^+ + X^-$, HX is 1.6% ionized. Calculate the ionization constant for the weak acid.

31. A 0.0100 molal solution of acetic acid $(CH_3COOH + H_2O \rightleftharpoons H_3O^+ + CH_3COO^-)$ freezes at $-0.01938°C$. Calculate the ionization constant for acetic acid. A 0.0100 molal solution is sufficiently dilute that it may be assumed to be 0.0100 molar without introducing a significant error.

32. Calculate the concentrations of the various species in 0.20 M hypochlorous acid, for which $K_a = 3.5 \times 10^{-8}$.

33. Calculate $[H_3O^+]$, $[OH^-]$, pH, pOH, and % ionization for 0.050 M CH_3COOH solution, for which $K_a = 1.8 \times 10^{-5}$.

34. Calculate $[H_3O^+]$, $[OH^-]$, pH, pOH, and % ionization for 0.050 M HCN solution, for which $K_a = 4.0 \times 10^{-10}$.

35. (a) In 0.050 M CH_3COOH the H_3O^+ concentration is approximately _____ times greater than in 0.050 M HCN. Refer to Exercises 33 and 34.
    (b) In 0.050 M HCN solution the OH^- concentration is approximately _____ times greater than in 0.050 M CH_3COOH solution. Refer to Exercises 33 and 34.
    (c) In 0.050 M CH_3COOH solution the % ionization is approximately _____ times greater than in 0.050 M HCN solution. Refer to Exercises 33 and 34.

36. Calculate the concentration of H_3O^+, pH, and % ionization of 0.10 M solutions of the following weak, monoprotic acids. (a) HCN, (b) HF, (c) HBrO, (d) HN_3

37. The pH of a hydrofluoric acid solution is 3.00. Calculate the molarity of the solution. Can you make a simplifying assumption in this case?

38. A 0.10 M solution of a newly discovered monoprotic acid is found to have a pH of 3.22. What is the value of the ionization constant for the acid? Represent the new acid by the general formula HY.

39. Answer the following questions for *0.10 M solutions* of the weak acids listed in Table 17–4.
    (a) Which solution contains
        1. the highest concentration of H_3O^+?
        2. the highest concentration of OH^-?
        3. the lowest concentration of H_3O^+?
        4. the lowest concentration of OH^-?
        5. the highest concentration of nonionized acid molecules?
        6. the lowest concentration of nonionized acid molecules?
    (b) In which solution is
        1. the pH highest?
        2. the pH lowest?
        3. the pOH highest?
        4. the pOH lowest?
    (c) Which solution contains
        1. the highest concentration of the anion of the weak acid?
        2. the lowest concentration of the anion of the weak acid?

40. Calculate $[H_3O^+]$, $[OH^-]$, pH, pOH, and % ionization for 0.050 M aqueous ammonia solution, $K_b = 1.8 \times 10^{-5}$. How do these numbers compare with those obtained for 0.050 M CH_3COOH solution in Exercise 33? Why?

41. Calculate $[OH^-]$, % ionization, and pH for (a) 0.10 M aqueous ammonia, and (b) 0.10 M methylamine solution. (See Table 17–5).

42. Refer to Table 17–5 and answer the following questions for *0.10 M solutions* of the weak bases listed there.
    (a) Which solution contains
        1. the highest concentration of OH^-?
        2. the highest concentration of H_3O^+?
        3. the lowest concentration of OH^-?
        4. the lowest concentration of H_3O^+?
        5. the highest concentration of nonionized molecules of weak base?
        6. the lowest concentration of nonionized molecules of weak base?
        7. the highest concentration of the cation of the weak base?
        8. the lowest concentration of the cation of the weak base?
    (b) In which solution is
        1. the pH highest?
        2. the pH lowest?
        3. the pOH highest?
        4. the pOH lowest?

43. Vinegar is a dilute aqueous solution of acetic acid, and is one of the most common acidic materials found in most homes. Household ammonia is a dilute aqueous solution of ammo-

nia that also contains a little detergent. It is one of the most common basic materials in most homes.

(a) Write the balanced molecular equation for the reaction that occurs when vinegar and household ammonia are mixed.

(b) Would you expect 1.00 L of vinegar that is 5.00% CH_3COOH by mass to neutralize 1.00 L of household ammonia that is 5.00% NH_3 by mass? Why?

(c) The density of a sample of household ammonia that is 5.00% NH_3 by mass is 0.980 g/mL. What is the molarity of this solution?

(d) The density of a sample of white vinegar that is 5.00% CH_3COOH by mass is 1.01 g/mL. What is the molarity of this solution?

(e) What volume of the vinegar described in (d) would be required to neutralize 1.00 L of the household ammonia described in (c)?

(f) What volume of the household ammonia would be required to neutralize 1.00 L of the vinegar?

Polyprotic Acids

44. Calculate the concentrations of H_3O^+, OH^-, HCO_3^-, and CO_3^{2-} in 0.050 M H_2CO_3 solution.

45. Calculate the concentrations of H_3O^+, OH^-, $H(COO)_2^-$, and $(COO)_2^{2-}$ in 0.050 M $(COOH)_2$ solution. Compare the concentrations with those obtained in Exercise 44. How can you explain the difference between the concentrations of HCO_3^- and $H(COO)_2^-$? Between CO_3^{2-} and $(COO)_2^{2-}$?

46. Calculate the concentrations of the various species in 0.100 M H_3AsO_4 solution. Compare the concentrations with those of the analogous species in 0.100 M H_3PO_4 solution (see Example 17–12).

47. Calculate the concentration of each species in each of the following solutions. Include the concentration of OH^-.
   (a) 0.020 M H_2CO_3 (b) 0.20 M H_3AsO_4
   (c) 0.15 M H_3PO_4 (d) 0.10 M H_2S

Anions and Cations as Weak Bases or Acids

48. Define and illustrate the following terms clearly and concisely: (a) solvolysis, (b) hydrolysis.

49. Why do salts of strong soluble bases and strong acids give neutral aqueous solutions? Use KNO_3 to illustrate. Write names and formulas

for three other salts of strong soluble bases and strong acids.

50. Why do salts of strong soluble bases and weak acids give basic aqueous solutions? Use sodium hypochlorite, $NaClO$, to illustrate. (Chlorox, Purex, and the other "chlorine bleaches" are 5% $NaClO$.)

51. Write names and formulas for three salts of strong soluble bases and weak acids (other than $NaClO$).

52. Calculate hydrolysis constants for the following anions of weak acids: (a) F^-, (b) ClO^-, (c) CN^-. Is there a relationship between K_a, the ionization constant for a weak acid, and K_b, the hydrolysis constant for the anion of the weak acid? What is it?

53. Calculate the pH of 0.15 M solutions of the following salts: (a) NaF, (b) $NaClO$, (c) $NaCN$. (Refer to Exercise 52 for K's.)

54. What is the percentage hydrolysis in each of the solutions in Exercise 53?

55. Why do salts of weak bases and strong acids give acidic aqueous solutions? Illustrate with NH_4NO_3, a common fertilizer.

56. Write names and formulas for four salts of weak bases and strong acids.

57. Calculate hydrolysis constants for the following cations of weak bases: (a) NH_4^+, (b) $(CH_3)NH_3^+$, methylammonium ion, (c) $C_5H_5NH^+$, pyridinium ion.

58. Calculate the pH of 0.15 M solutions of (a) NH_4NO_3, (b) $(CH_3)NH_3NO_3$, (c) $C_5H_5NHNO_3$. (Refer to Exercise 57 for K's.)

59. Why are some aqueous solutions of weak acids and weak bases neutral, while others are acidic, and still others are basic?

60. Write the name and formula for a salt of a weak acid and a weak base that gives (a) neutral, (b) acidic, and (c) basic aqueous solutions.

61. Why do salts that contain cations related to insoluble bases (metal hydroxides) and anions related to strong acids give acidic aqueous solutions? Use $Cr(NO_3)_3$ to illustrate.

62. Calculate the pH and percentage hydrolysis for the following.
   (a) 0.050 M $Al(NO_3)_3$, aluminum nitrate
   (b) 0.20 M $Co(ClO_4)_2$, cobalt(II) perchlorate
   (c) 0.15 M $MgCl_2$, magnesium chloride

63. Given pH values for solutions of the following concentrations, calculate hydrolysis constants for the cations.
    (a) 0.00050 M $CeCl_3$, cerium(III) chloride, pH = 5.99
    (b) 0.10 M $Cu(NO_3)_2$, copper(II) nitrate, pH = 4.50
    (c) 0.10 M $Sc(ClO_4)_3$, scandium perchlorate, pH = 3.44

64. Use words and chemical equations to explain why a 0.10 M solution of $KHSO_4$ is acidic while a 0.10 M solution of $KHCO_3$ is basic.

65. Are 0.10 M solutions of the following salts acidic or basic? Justify your answers.
    (a) $NaHSeO_4$, (b) NaH_2PO_4, (c) Na_2HPO_4.

The Lewis Theory

66. Define the following terms clearly and concisely. Write an equation to illustrate the meaning of each. (a) Lewis acid, (b) Lewis base, (c) neutralization (according to Lewis theory).

67. Compare and contrast the advantages and disadvantages of the Brønsted-Lowry theory and the Lewis theory.

68. Draw a Lewis formula for each species in the following equations. Label the acids and bases using Lewis theory terminology.
    (a) $H_2O + H_2O \rightleftharpoons H_3O^+ + OH^-$
    (b) $HBr\ (g) + H_2O \longrightarrow H_3O^+ + Br^-$
    (c) $NH_3\ (g) + H_2O \rightleftharpoons NH_4^+ + OH^-$
    (d) $NH_3\ (g) + HBr\ (g) \longrightarrow NH_4Br\ (s)$
    (e) $NH_3\ (g) + BF_3\ (g) \longrightarrow H_3N:BF_3\ (s)$

69. Draw a Lewis formula for each species in the following equations. Label the acids and bases using Lewis theory terminology.
    (a) $AlCl_3 + Cl^- \longrightarrow AlCl_4^-$
    (b) $SnCl_2 + 2Cl^- \longrightarrow SnCl_4^{2-}$
    (c) $SnCl_4 + 2Cl^- \longrightarrow SnCl_6^{2-}$
    (d) $SO_2 + MgO \longrightarrow MgSO_3$
    (e) $CO_2 + CaO \longrightarrow CaCO_3$
    (f) $SO_3 + K_2O \longrightarrow K_2SO_4$

Ionic Equilibria — II: Buffers and Acid-Base Titrations; Slightly Soluble Compounds

18

Buffer Solutions and Acid-Base Titrations

All the solutions considered in Chapter 17 were prepared by dissolving a *single* solute in water. In the first part of this chapter we shall consider acid-base equilibria in which more than one solute is present.

18–1   The Common Ion Effect and Buffer Solutions

It is often useful in laboratory reactions, industrial processes, and the bodies of plants and animals to keep the pH nearly constant despite the addition of considerable amounts of acids or bases. For instance, the oxygen-carrying capacity of the hemoglobin in your blood and the activity of the enzymes in your cells depend very strongly on the pH of your body fluids. In order to keep the pH within a narrow range, the body uses a combination of compounds known as a *buffer solution,* whose operation depends on the *common ion effect,* a special case of LeChatelier's Principle. This effect can be observed in the laboratory.

Buffer systems resist changes in pH.

The term **common ion effect** is used to describe the behavior of solutions in which the same ion is produced by two different compounds. Although there are many types of solutions that exhibit this effect, two of the most frequently encountered kinds are

1. a solution of a weak acid *plus* a soluble ionic salt of the weak acid;
2. a solution of a weak base *plus* a soluble ionic salt of the weak base.

Consider a solution that contains acetic acid, CH_3COOH, *and* sodium acetate, $NaCH_3COO$, a soluble ionic salt of CH_3COOH. The $NaCH_3COO$ is completely dissociated into its constituent ions, but the CH_3COOH is only slightly ionized, as indicated by the following equations.

$$NaCH_3COO \xrightarrow{H_2O} Na^+ + CH_3COO^-$$

$$CH_3COOH + H_2O \rightleftharpoons H_3O^+ + CH_3COO^-$$

LeChatelier's Principle (Section 16–5) is applicable to equilibria in aqueous solution.

In solutions containing both CH_3COOH and $NaCH_3COO$ in reasonable concentrations, both compounds serve as sources of CH_3COO^- ions. The completely dissociated $NaCH_3COO$ provides a high concentration of CH_3COO^- ions, which strongly favors the combination of CH_3COO^- with H_3O^+ to form nonionized CH_3COOH and H_2O. This results in a drastic decrease in the concentration of H_3O^+ in the solution. Solutions containing a weak acid *plus a salt of the weak acid* are always less acidic than solutions containing the same concentration of the weak acid alone, as Example 18–1 illustrates.

Example 18–1

Calculate the concentration of H_3O^+ and the pH of a solution that is 0.10 *M* in CH_3COOH and 0.20 *M* in $NaCH_3COO$.

Solution

The appropriate equations are shown above, and the ionization constant expression for CH_3COOH (Eq. 17–8) is

$$K_a = \frac{[H_3O^+][CH_3COO^-]}{[CH_3COOH]} = 1.8 \times 10^{-5}$$

This ionization constant expression is valid for all solutions that contain CH_3COOH; the fact that the solution also contains $NaCH_3COO$ does not affect its validity. In solutions that contain both CH_3COOH and $NaCH_3COO$, acetate ions are derived from two sources, and the ionization constant

is satisfied by the total CH_3COO^- concentration in the solution.

Since $NaCH_3COO$ is completely dissociated, the concentration of CH_3COO^- from $NaCH_3COO$ will be 0.20 mol/L. If we let $x = [H_3O^+]$, then x is also equal to the concentration of CH_3COO^- from CH_3COOH, and the concentration of nonionized CH_3COOH is $(0.10 - x)$ *M*. The *total* concentration of CH_3COO^- is $(0.20 + x)$ *M*. This may be represented as

$$NaCH_3COO \longrightarrow Na^+ + CH_3COO^-$$
$$\quad 0.20\ M \qquad\qquad 0.20\ M \quad 0.20\ M$$

$$CH_3COOH + H_2O \rightleftharpoons H_3O^+ + CH_3COO^-$$
$$(0.10 - x)\ M \qquad\qquad x\ M \qquad x\ M$$

Substitution into the ionization expression for CH_3COOH gives

$$\frac{(x)(0.20 + x)}{(0.10 - x)} = 1.8 \times 10^{-5}$$

Examination of this equation indicates that x is a very small number, and we are able to make two simplifying assumptions, namely

$$(0.20 + x) \approx 0.20$$

and

$$(0.10 - x) \approx 0.10$$

Making these assumptions gives

$$\frac{0.20 \, x}{0.10} = 1.8 \times 10^{-5}$$

and $x = 9.0 \times 10^{-6} \, M = [H_3O^+]$

Since $[H_3O^+] = 9.0 \times 10^{-6} \, M$, we find pH = 5.05.

You may verify the validity of the assumptions by substituting $x = 9.0 \times 10^{-6}$, into the original equation.

To gain an appreciation for just how much the acidity of the 0.10 M CH_3COOH solution is reduced by making the solution 0.20 M in $NaCH_3COO$ also, refer back to Example 17–9, where we found that the H_3O^+ concentration in 0.10 M CH_3COOH is 1.3×10^{-3} mol/L and the pH of the solution is 2.89. Table 18–1 makes the point clearly.

TABLE 18–1   Comparison of $[H_3O^+]$ and pH in Acetic Acid and Sodium Acetate–Acetic Acid Solutions

Solution	$[H_3O^+]$	pH
0.10 M CH_3COOH	$1.3 \times 10^{-3} \, M$	2.89
0.10 M CH_3COOH and 0.20 M $NaCH_3COO$	$9.0 \times 10^{-6} \, M$	5.05

ΔpH = 2.16

Comparison of $[H_3O^+]$ in the two solutions shows that $[H_3O^+]$ is 144 times greater in 0.10 M CH_3COOH than in the solution that is both 0.10 M in CH_3COOH and 0.20 M in $NaCH_3COO$, which demonstrates LeChatelier's Principle.

The calculation of $[H_3O^+]$ in solutions containing both a weak acid and a salt of the weak acid can be simplified greatly. If we write the equation for the ionization of a weak monoprotic acid in the following way, a useful relationship is easily evolved.

$$H(Anion) + H_2O \rightleftharpoons H_3O^+ + Anion^-$$

The ionization constant expression can be written as

$$\frac{[H_3O^+][Anion^-]}{[H(Anion)]} = K_a$$

Solving this expression for $[H_3O^+]$ gives

$$[H_3O^+] = \frac{[H(Anion)]}{[Anion^-]} \times K_a \qquad \text{[Eq. 18–1]}$$

We now impose two conditions: (1) The concentrations of both the weak acid and its salt are some reasonable value, say greater than 0.050 M; (2) the salt contains a univalent cation. Under these conditions, the concentration of the anion, $[Anion^-]$, in the solution can be assumed to be the same as the concentration of the salt. With these restrictions, the above expression for $[H_3O^+]$ becomes

$$[H_3O^+] = \frac{[Acid]}{[Salt]} \times K_a \qquad \text{[Eq. 18–2]}$$

where [Acid] is the concentration of nonionized acid (for all practical purposes, the total acid concentration) and [Salt] is the concentration of the salt of the weak acid.

If we take the logarithm of Eq. 18–2, we obtain

$$\log [H_3O^+] = \log \frac{[Acid]}{[Salt]} + \log K_a$$

Multiplying by -1 gives

$$-\log [H_3O^+] = -\log \frac{[Acid]}{[Salt]} - \log K_a$$

and rearrangement gives

$$pH = pK_a + \log \frac{[Salt]}{[Acid]} \qquad \text{where } pK_a = -\log K_a \qquad \text{[Eq. 18–3]}$$

This relationship is valid only for solutions that contain a weak monoprotic acid and a soluble, ionic salt of the weak acid with a univalent cation, both in reasonable concentrations.

This equation is known as the **Henderson-Hasselbalch equation,** and workers in the biological sciences use it frequently.

Let's consider the second common kind of buffer solution, containing a weak base and its salt. A solution that contains aqueous NH_3 and ammonium chloride, NH_4Cl, a soluble ionic salt of NH_3, is typical. The NH_4Cl is completely dissociated into its constituent ions, but aqueous NH_3 is only slightly ionized, as indicated by the following equations.

$$NH_4Cl \xrightarrow{\,H_2O\,} NH_4^+ + Cl^-$$
$$NH_3 + H_2O \rightleftharpoons NH_4^+ + OH^-$$

In this solution both NH_4Cl and aqueous NH_3 produce NH_4^+ ions. The completely dissociated NH_4Cl provides a high concentration of NH_4^+, which favors the combination of NH_4^+ ions with OH^- ions to form nonionized NH_3 and H_2O. Thus, the concentration of OH^- is decreased significantly. Solutions containing a weak base plus a salt of the weak base are less basic than solutions containing the same concentration of weak base alone, as Example 18–2 illustrates.

Example 18–2

Calculate the concentration of OH^- and the pH of a solution that is 0.20 M in aqueous NH_3 *and* 0.10 M in NH_4Cl.

Solution

The appropriate equations and algebraic representation of concentrations are

$$NH_4Cl \longrightarrow NH_4^+ + Cl^-$$
$$0.10\,M \qquad\qquad 0.10\,M \quad 0.10\,M$$

$$NH_3 + H_2O \rightleftharpoons NH_4^+ + OH^-$$
$$(0.20 - x)\,M \qquad\qquad x\,M \quad x\,M$$
$$\overline{\text{Total } [NH_4^+] = (0.10 + x)\,M}$$

Substitution into the ionization expression for aqueous NH_3 gives

$$K_b = \frac{[NH_4^+][OH^-]}{[NH_3]} = 1.8 \times 10^{-5} = \frac{(0.10 + x)(x)}{(0.20 - x)}$$

Again, we can assume that $(0.10 + x) \approx 0.10$ and $(0.20 - x) \approx 0.20$, which gives

$$\frac{0.10x}{0.20} = 1.8 \times 10^{-5}$$

and $\qquad x = 3.6 \times 10^{-5}\,M = [OH^-]$

Since $[OH^-] = 3.6 \times 10^{-5}\,M$, we find $pOH = 4.44$ and pH = 9.56.

TABLE 18-2 Comparison of [OH⁻] and pH in Ammonia and Ammonium Chloride–Ammonia Solutions

Solution	[OH⁻]	pH	
0.20 M aq. NH_3	1.9×10^{-3} M	11.28	$\Delta pH = -1.72$
0.20 M aq. NH_3 and 0.10 M aq. NH_4Cl	3.6×10^{-5} M	9.56	

Recall that in Example 17-10 we calculated [OH⁻] and pH for 0.20 M aqueous NH_3. Comparison of that result with the results obtained here is instructive (Table 18-2).

Note that the concentration of OH⁻ is 53 times greater in the solution containing only 0.20 M aqueous NH_3 than in the solution containing both 0.20 M aqueous NH_3 and 0.10 M NH_4Cl, another demonstration of LeChatelier's Principle.

We can derive a relationship for concentration of OH⁻ in solutions containing weak bases *plus* salts of the weak bases, just as we did for weak acids. The equation for the ionization of a monoprotic weak base, in general terms, is

$$Base + H_2O \rightleftharpoons Cation^+ + OH^-$$

The ionization constant expression then becomes

$$\frac{[Cation^+][OH^-]}{[Base]} = K_b \qquad \text{[Eq. 18-4]}$$

Solving this expression for [OH⁻] gives

$$[OH^-] = \frac{[Base]}{[Cation^+]} \times K_b$$

Taking the logarithm of both sides of the equation gives

$$\log [OH^-] = \log \frac{[Base]}{[Cation^+]} + \log K_b$$

As long as we consider salts of weak bases that contain univalent anions, [Cation⁺] = [Salt]. Multiplication by −1 and rearrangement gives the Henderson-Hasselbalch equation for solutions containing a weak base plus a salt of the weak base:

$$pOH = pK_b + \log \frac{[Salt]}{[Base]} \qquad \text{where } pK_b = -\log K_b \qquad \text{[Eq. 18-5]}$$

If we consider other salts, such as $(NH_4)_2SO_4$, that contain divalent anions, then [Cation⁺] = 2[Salt]. The above relationship is valid for solutions of weak bases plus salts of weak bases in reasonable concentrations, as Example 18-3 illustrates.

Example 18-3

Calculate the concentration of OH⁻ and the pH of a solution that is 0.15 M in aqueous NH_3 and 0.075 M in $(NH_4)_2SO_4$.

$$\begin{array}{ccc} (NH_4)_2SO_4 & \longrightarrow & 2NH_4^+ + SO_4^{2-} \\ 0.075\ M & & 0.15\ M \quad 0.075\ M \end{array}$$

$$\begin{array}{ccc} NH_3 + H_2O \rightleftharpoons & NH_4^+ + OH^- \\ (0.15 - x)\ M & x\ M \quad\quad x\ M \end{array}$$

Solution

The equations and algebraic representations are

We can substitute into the ionization constant expression for aqueous NH_3, as we did in Example 18-2:

$$K_b = \frac{[NH_4^+][OH^-]}{[NH_3]}$$

$$= 1.8 \times 10^{-5} = \frac{(0.15 + x)(x)}{(0.15 - x)}$$

which gives $x = 1.8 \times 10^{-5} M = [OH^-]$, pOH = 4.74, and pH = 9.26.

Alternatively, we can use the ionization constant expression for aqueous NH_3 in the general form

$$K_b = \frac{[Cation^+][OH^-]}{[Base]} = 1.8 \times 10^{-5}$$

$$[OH^-] = \frac{[Base]}{[Cation^+]} \times 1.8 \times 10^{-5}$$

$$= \frac{(0.15)}{(0.15)} \times 1.8 \times 10^{-5}$$

which gives $[OH^-] = 1.8 \times 10^{-5} M$, pOH = 4.74, and pH = 9.26.

As another alternative, substitution into the Henderson-Hasselbalch equation gives the same result. $K_b = 1.8 \times 10^{-5}$ so $pK_b = 4.74$

$$pOH = pK_b + \log \frac{2[Salt]}{[Base]}$$

$$= 4.74 + \log \frac{0.15}{0.15}$$

$$pOH = 4.74 + 0 = 4.74, \quad pH = 9.26.$$

18-2 Buffering Action

As we stated earlier, buffer solutions resist changes in pH. By this statement, we mean that we can add small amounts of strong base or strong acid to buffer solutions and the pH will change very little. These statements imply that buffer solutions are able to react with both H_3O^+ ions and OH^- ions, and indeed this is the case. The two common kinds of buffer solutions are the ones we have just discussed, namely, solutions containing (1) a weak acid plus a salt of the weak acid and (2) a weak base plus a salt of the weak base.

All buffer solutions contain conjugate acid-base pairs. In all cases *the more acidic component is the one that reacts with (and tends to neutralize) added strong base, and the more basic component reacts with added strong acid.*

1 Solutions of a Weak Acid and a Salt of the Weak Acid

Let us consider a solution containing acetic acid, CH_3COOH, and sodium acetate, $NaCH_3COO$, as a common example of this kind of buffer solution. The more acidic component is CH_3COOH. The more basic component is $NaCH_3COO$ because the acetate ion, CH_3COO^-, is the conjugate base of CH_3COOH. Thus the following reaction tends to neutralize added strong acid, such as HCl.

$$HCl + NaCH_3COO \xrightarrow{\sim 100\%} CH_3COOH + NaCl$$

or as a net ionic equation,

The net effect is to neutralize most of the H_3O^+ from HCl by forming nonionized CH_3COOH molecules, and to decrease slightly the ratio $[CH_3COO^-]/[CH_3COOH]$, which governs the pH of the solution (see Eq. 18-3).

$$\underset{\substack{\text{added strong} \\ \text{acid}}}{H_3O^+} + \underset{\text{base}}{CH_3COO^-} \xrightarrow{\sim 100\%} \underset{\text{weak acid}}{CH_3COOH} + H_2O$$

This reaction goes nearly to completion because CH_3COOH is a *weak* acid; its constituent ions, when mixed from separate sources, have a strong tendency to combine rather than remain separate.

On the other hand, when a strong soluble base such as NaOH is added to the $CH_3COOH/NaCH_3COO$ buffer solution, it is consumed by the *acidic component*, CH_3COOH.

The net effect is to neutralize most of the OH^- from NaOH, and to increase slightly the ratio of $[CH_3COO^-]/[CH_3COOH]$, which governs the pH of the solution.

$$NaOH \ + \ CH_3COOH \xrightarrow{\sim 100\%} NaCH_3COO + H_2O$$
$$\text{added base} \qquad \text{acid} \qquad\qquad \text{salt} \qquad \text{water}$$

or as a net ionic equation,

$$OH^- + CH_3COOH \xrightarrow{\sim 100\%} CH_3COO^- + H_2O$$

Example 18-4

If 0.010 mole of solid NaOH is added to one liter of a buffer solution that is 0.10 M in CH_3COOH and 0.10 M in $NaCH_3COO$, how much will $[H_3O^+]$ and pH change? Assume no volume change due to the addition of solid NaOH.

Solution

From our earlier discussion, for the buffer solution we can write

$$[H_3O^+] = \frac{[Acid]}{[Salt]} \times K_a = \frac{0.10}{0.10} \times 1.8 \times 10^{-5}$$

which gives $[H_3O^+] = 1.8 \times 10^{-5} \ M$ and pH = 4.74 before the NaOH is added. When solid NaOH is added to the solution, it reacts with CH_3COOH to form more $NaCH_3COO$.

	NaOH	+ CH_3COOH	$\rightarrow NaCH_3COO$	+ H_2O
in:	0.01 mol	0.10 mol	0.10 mol	
ch:	−0.01 mol	−0.01 mol	+0.01 mol	
eq:	0 mol	0.09 mol	0.11 mol	

Since the volume of the solution is one liter, we now have a solution that is 0.09 M in CH_3COOH and 0.11 M in $NaCH_3COO$. In this solution

$$[H_3O^+] = \frac{[Acid]}{[Salt]} \times K_a = \frac{0.09}{0.11} \times 1.8 \times 10^{-5}$$

which gives $[H_3O^+] = 1.5 \times 10^{-5} \ M$ and pH = 4.82.

The calculation shows that the addition of 0.010 mole of solid NaOH to one liter of the buffer solution (which is enough NaOH to neutralize 10% of the acid) decreases $[H_3O^+]$ from $1.8 \times 10^{-5} \ M$ to $1.5 \times 10^{-5} \ M$ and increases pH from 4.74 to 4.82, a change of 0.08 pH unit. These are very slight changes.

By contrast to the case in Example 18-4, adding 0.010 mole of NaOH to enough pure water to make one liter of solution gives a 0.010 M NaOH solution in which $[OH^-] = 1.0 \times 10^{-2} \ M$, and pOH = 2.00. Since pH + pOH = 14, the pH of this solution is 12.00, an increase of 5 pH units above that of pure water.

Addition of 0.010 mole of solid NaOH to one liter of 0.10 M CH_3COOH (pH = 2.89 from Table 18-1) gives a solution that is 0.09 M in CH_3COOH and 0.01 M in $NaCH_3COO$. The pH of this solution is 3.80, which is 0.91 pH units higher than that of the 0.10 M CH_3COOH solution. Table 18-3 summarizes these data.

2 Solutions of a Weak Base and a Salt of the Weak Base

A common example of a buffer solution in this category is one containing the weak base aqueous ammonia, NH_3, and its soluble ionic salt ammonium chloride, NH_4Cl. If strong acid, such as HCl, is added to such a buffer solution, the more basic component, NH_3, neutralizes most of the H_3O^+ from HCl.

TABLE 18–3   Changes in pH Caused by Addition of 0.010 Mole of Solid NaOH to One Liter of Solution

Solution	Change in pH
0.10 M CH$_3$COOH ⎫ 0.10 M NaCH$_3$COO ⎭	0.08
0.10 M CH$_3$COOH alone	0.91
pure H$_2$O	5.00

$$\text{HCl} + \text{NH}_3 + \text{H}_2\text{O} \xrightarrow{\sim 100\%} \text{NH}_4\text{Cl} + \text{H}_2\text{O}$$
$$\text{acid} \quad \text{base} \qquad\qquad \text{salt} \quad \text{water}$$

or in net ionic form,

$$\text{H}_3\text{O}^+ + \text{NH}_3 \xrightarrow{\sim 100\%} \text{NH}_4^+ + \text{H}_2\text{O}$$

> The net effect is to neutralize most of the H_3O^+ from the HCl and to increase the ratio [NH$_4^+$]/[NH$_3$], which governs the pH of the solution (see Eq. 18–5).

If a strong soluble base, such as NaOH, is added to the original buffer solution, it is neutralized by the more acidic component, NH_4^+ (from NH_4Cl), which is the conjugate acid of aqueous ammonia.

$$\text{NaOH} + \text{NH}_4\text{Cl} \xrightarrow{\sim 100\%} \text{NH}_3 + \text{H}_2\text{O} + \text{NaCl}$$

or in net ionic form,

$$\text{OH}^- \quad + \text{NH}_4^+ \longrightarrow \quad \text{NH}_3 \quad + \text{H}_2\text{O}$$
$$\text{added strong} \quad \text{acid} \qquad \text{weak base}$$
$$\text{base}$$

> The net effect is to neutralize most of the OH$^-$ from NaOH, and to decrease the ratio [NH$_4^+$]/[NH$_3$], which governs the pH of the solution.

Thus we see that changes in pH (that is, changes in [H$_3$O$^+$] and [OH$^-$]) are minimized in buffer solutions because one (basic) component can react with excess H$_3$O$^+$ ions and another (acidic) component can react with excess OH$^-$ ions.

18–3   Preparation of Buffer Solutions

Buffer solutions are frequently prepared by mixing other solutions. When solutions are mixed, the volume in which each solute is contained increases and the concentrations of the dissolved species change. This change in concentration must be taken into consideration, as Example 18–5 illustrates.

Example 18–5

Calculate the concentration of H$_3$O$^+$ in a buffer solution prepared by mixing 200 mL of 0.10 M NaF and 100 mL of 0.050 M HF. $K_a = 7.2 \times 10^{-4}$ for HF.

Solution

Since two dilute solutions are mixed, we may assume that the volumes are additive, and the volume of the new solution will be 300 mL. Mixing a solution of a weak acid with a solution of a salt of the weak acid does not result in any new chemical species being formed, and we have a straightforward buffer calculation. Therefore, we shall calculate the number of millimoles of each compound and the molarities in the new solution.

$$? \text{ mmol NaF} = 200 \text{ mL} \times \frac{0.10 \text{ mmol NaF}}{\text{mL}}$$

$$= 20 \text{ mmol NaF}$$

$$? \text{ mmol HF} = 100 \text{ mL} \times \frac{0.050 \text{ mmol HF}}{\text{mL}}$$

$$= 5.0 \text{ mmol HF}$$

Since the 300 mL of solution contain 20 mmol of NaF and 5.0 mmol of HF, the molarities of NaF and HF are

$$\frac{20 \text{ mmol NaF}}{300 \text{ mL}} = 0.067 \text{ M NaF} \quad \text{and}$$

$$\frac{5.0 \text{ mmol HF}}{300 \text{ mL}} = 0.017 \text{ M HF}$$

The appropriate equations and algebraic representations of concentrations are

$$\text{NaF} \longrightarrow \text{Na}^+ + \text{F}^-$$
$$0.067 \text{ M} \qquad\qquad 0.067 \text{ M} \quad 0.067 \text{ M}$$
$$\text{HF} + \text{H}_2\text{O} \rightleftharpoons \text{H}_3\text{O}^+ + \text{F}^-$$
$$(0.017 - x) \text{ M} \qquad\qquad x \text{ M} \qquad x \text{ M}$$

The ionization constant expression for hydrofluoric acid is

$$K_a = \frac{[\text{H}_3\text{O}^+][\text{F}^-]}{[\text{HF}]} = 7.2 \times 10^{-4}$$

Substitution gives

$$\frac{(x)(0.067 + x)}{(0.017 - x)} = 7.2 \times 10^{-4}$$

Some question may arise as to whether we can assume that x is negligible compared to 0.067 and 0.017 in this expression. When in doubt, solve the equation using the simplifying assumption and then determine whether the assumption is valid. Assuming that $(0.067 + x) \approx 0.067$ and that $(0.017 - x) \approx 0.017$ gives

$$\frac{0.067x}{0.017} = 7.2 \times 10^{-4},$$

and $\quad x = 1.8 \times 10^{-4} \text{ M} = [\text{H}_3\text{O}^+]$

Our assumption is valid.

Calculated quantities of solutions of weak base (or weak acid) and strong acid (or strong soluble base) also could be mixed as in Example 18-5 to produce a buffer solution of a designated pH.

On occasion it is desirable to prepare a buffer solution of a particular pH. Example 18-6 illustrates one method by which such solutions can be prepared. This involves adding a solid salt of a weak base (or weak acid) to a solution of the weak base (or weak acid) itself.

Example 18-6

Calculate the number of moles of solid ammonium chloride, NH_4Cl, that must be used to prepare 2.0 liters of a buffer solution that is 0.10 M in aqueous ammonia and that has a pH of 9.15.

Solution

Since pH = 9.15, we must have pOH = 14.00 − 9.15 = 4.85, and

$$[\text{OH}^-] = \text{antilog}(-4.85)$$
$$= 10^{-4.85} = 1.4 \times 10^{-5} \text{ M}$$

Let us define the necessary molarity of NH_4Cl to be x M. Since $[\text{OH}^-] = 1.4 \times 10^{-5}$ M, this must be the concentration of OH^- ions produced by ionization of NH_3. The ionization equations and representations of equilibrium concentrations are shown below.

$$\text{NH}_4\text{Cl} \xrightarrow{100\%} \text{NH}_4^+ + \text{Cl}^-$$
$$x \text{ M} \qquad x \text{ M} \qquad x \text{ M}$$
$$\text{NH}_3 + \text{H}_2\text{O} \rightleftharpoons \text{NH}_4^+ + \text{OH}^-$$
$$(0.10 - 1.4 \times 10^{-5}) \text{ M} \qquad 1.4 \times 10^{-5} \text{ M} \quad 1.4 \times 10^{-5} \text{ M}$$

Substitution into the ionization constant expression for aqueous ammonia gives the following.

$$K_b = \frac{[NH_4^+][OH^-]}{[NH_3]} = 1.8 \times 10^{-5}$$

$$= \frac{(x + 1.4 \times 10^{-5})(1.4 \times 10^{-5})}{0.10 - 1.4 \times 10^{-5}}$$

Since NH_4Cl is completely dissociated, we may assume that $x \gg 1.4 \times 10^{-5}$ so that $(x + 1.4 \times 10^{-5}) \approx x$. The expression then becomes

$$\frac{(x)(1.4 \times 10^{-5})}{0.10} = 1.8 \times 10^{-5}$$

$$x = 0.13\ M = [NH_4^+] = M_{NH_4Cl}$$

Now we calculate the number of moles of NH_4Cl that must be added to prepare 2.0 L of buffer solution.

$$\underline{?}\ mol\ NH_4Cl = 2.0\ L \times \frac{0.13\ mol\ NH_4Cl}{L}$$

$$= 0.26\ mol\ NH_4Cl$$

This corresponds to 0.26 mol $\times$ 53.5 g/mol = 14 g NH_4Cl.

So far in this chapter we have seen (1) how the presence of a "common ion" from a strong electrolyte represses the ionization of a weak electrolyte by LeChatelier's Principle, (2) how buffer solutions resist changes in pH as strong acid or strong soluble base is added, and (3) how we can prepare buffer solutions. The next sections will be devoted to studying acid-base indicators, how pH changes during the course of acid-base titrations, and how these changes are influenced by the strengths of the acids and bases involved.

18–4 Acid-Base Indicators

In Section 10–4 we defined *titration* as the procedure in which a solution of accurately known concentration, a standard solution, is added slowly to a solution of unknown concentration until the chemical reaction between the two solutes is complete. Suppose we have a standard solution of sodium hydroxide and a solution of hydrochloric acid of unknown concentration (Figure 10–1). Clearly the acid and base react in a 1 : 1 mole ratio to form ordinary salt and water.

$$NaOH\ (aq) + HCl\ (aq) \longrightarrow NaCl\ (aq) + H_2O$$

How can we determine the point at which the reaction is complete? Some organic compounds exhibit different colors in solutions of different acidities, and change colors when the hydrogen ion concentration in the solution reaches a definite value. They are called **indicators** and are used to determine (as nearly as possible) the point at which the reaction between the acid and base is complete. We introduced indicators in Section 10–4.

Phenolphthalein is a common acid-base indicator. It is colorless in aqueous solutions of pH less than 8 (that is, $[H_3O^+] > 10^{-8}\ M$), and turns reddish violet as the pH approaches 10.

Most acid-base indicators are weak organic acids, HIn, or weak organic bases, InOH, where "In" represents complex organic groups. Bromthymol blue is typical of the acid-base indicators that are weak organic acids. Its ionization constant is about 1×10^{-7}. We can represent its dissociation in dilute aqueous solution as:

HIn $\quad$ + $H_2O \rightleftharpoons H_3O^+ \quad$ In$^-$

color 1 $\qquad\qquad\qquad\qquad\qquad$ color 2
yellow (for bromthymol blue) $\quad$ blue

Phenolphthalein is also the active component of the laxative "Exlax," and is sometimes added to laboratory ethyl alcohol to discourage its consumption.

where HIn represents nonionized acid molecules and In⁻ represents the anion (conjugate base) of the weak acid. The essential characteristic of acid-base indicators is that HIn and In⁻ *must* have different colors. The relative amounts of the two species determine the colors of solutions. The ionization constant expression for such indicators is

$$K_a = \frac{[H_3O^+][In^-]}{[HIn]}$$
[Eq. 18–6]

Addition of an acid causes the formation of more HIn molecules, while addition of a base causes the formation of more In⁻ ions.

The ionization constant expression can be rearranged to

$$\frac{[In^-]}{[HIn]} = \frac{K_a}{[H_3O^+]}$$
[Eq. 18–7]

which shows clearly the dependence of the $[In^-]/[HIn]$ ratio on $[H_3O^+]$ (or on pH) and the value of the ionization constant for the indicator. As a rule-of-thumb, when $[HIn]/[In^-] \leq 0.10$, color 2 is observed; conversely, when $[HIn]/[In^-] \geq 10$, color 1 is observed. The point at which the color change for the indicator occurs in a titration, known as the **end point,** is determined by the value of the ionization constant for the indicator. Table 18–4 shows a few acid-base indicators and the pH ranges over which their colors change. Typically, color changes occur over a range of 1.5 to 2.0 pH units.

The **equivalence point** (Section 10–4) is the point at which chemically equivalent amounts of acid and base have reacted. Ideally, the end point and the equivalence point should coincide in a titration. In practice, the best we can do is to select an indicator whose range of color change spans the equivalence point of the reaction of interest. By choosing an indicator with the proper end point and using the same procedures in both standardization and analysis, any error arising from a difference between end point and equivalence point can be minimized.

18–5 Acid-Base Titration Curves

1 Titrations of Strong Acids with Strong Soluble Bases

A **titration curve** is a plot of pH versus the amount (volume, usually) of acid or base added. It displays graphically the change in pH as acid or base is added to a solution, and indicates clearly how pH changes near the equivalence point.

Consider the titration of 100.0 mL of a 0.100 *M* solution of HCl with a 0.100 *M* solution of NaOH. We shall calculate the pH of the solution as NaOH is added. As

Titrations are usually done with 50 mL, or smaller, burets. We have chosen to use 100 mL of solution in this illustrative example to simplify the arithmetic.

TABLE 18–4 Range and Color Changes of Some Common Acid-Base Indicators

	pH Scale
Indicators	1 2 3 4 5 6 7 8 9 10 11 12 13

Methyl orange	←— red —→ 3.1 – 4 4 ←————————— yellow ————————→
Methyl red	←———— red ———→ 4.4 ———— 6.2 ←——————— yellow ————→
Bromthymol blue	←——— yellow ———→ 6.2 — 7.6 ←———— blue ————→
Neutral red	←———— red ————→ 6.8—— 8.0 ←——— yellow ————→
Phenolphthalein	←——— colorless ————→ 8.0 ——— 10.0 ←— red-violet —→ colorless beyond 13.0

pointed out earlier, NaOH and HCl react in a 1:1 ratio. Before any NaOH is added to the 0.100 M HCl solution, its pH is

$$HCl + H_2O \longrightarrow H_3O^+ + Cl^-$$

$$[H_3O^+] = 0.100 \ M$$

$$\underline{pH = 1.00}$$

After 20.0 mL of 0.100 M NaOH has been added, the pH is:

	HCl	+	NaOH	$\longrightarrow$	NaCl	+ H$_2$O
Start:	10.0 mmol		2.00 mmol		0 mmol	
Change:	-2.00 mmol		-2.00 mmol		$+2.00$ mmol	
After rxn:	8.00 mmol		0 mmol		2.00 mmol	

$$M_{HCl} = \frac{8.00 \text{ mmol}}{120 \text{ mL}} = 0.067 \ M$$

At this point the total volume is 100.0 + 20.00 = 120.0 mL.

$[H_3O^+] = 6.7 \times 10^{-2} \ M$, so $\underline{pH = 1.17}$

The pH at any other point before the equivalence point can be calculated in similar fashion.

After 100.0 mL of 0.100 M NaOH has been added (the equivalence point), the pH is $\underline{7.00}$ because the NaOH exactly neutralizes the HCl.

Beyond the equivalence point, the pH is governed by the excess strong soluble base.

After 110.0 mL of 0.100 M NaOH has been added, the pH is determined by the excess NaOH:

	HCl	+	NaOH	$\longrightarrow$	NaCl	+ H$_2$O
Start:	10.0 mmol		11.0 mmol		0 mmol	
Change:	-10.0 mmol		-10.0 mmol		$+10.0$ mmol	
After rxn:	0 mmol		1.0 mmol		10.0 mmol	

$$M_{NaOH} = \frac{1.0 \text{ mmol}}{210.0 \text{ mL}} = 0.0048 \ M \text{ NaOH}$$

$[OH^-] = 4.8 \times 10^{-3} \ M$, so pOH $= 2.32$ and $\underline{pH = 11.68}$

Table 18–5 displays the data for the titration of 100.0 mL of 0.100 M HCl by 0.100 M NaOH solution. A few additional points have been included to show the shape of the curve definitively. The data are plotted in Figure 18–1.

TABLE 18–5   Titration Data for HCl vs. NaOH

The change in pH from 3.60 to 10.40 corresponds to a change in [H$_3$O$^+$] of about 10^7. This is due to the logarithmic nature of pH.

mL of 0.100 M NaOH Added	mmol NaOH Added	mmol Excess Acid or Base	pH
0.0	0.00	10.0 H$_3$O$^+$	1.00
20.0	2.00	8.0	1.17
50.0	5.00	5.0	1.48
90.0	9.00	1.0	2.28
99.0	9.90	0.10	3.30
99.5	9.95	0.05	3.60
100.0	10.00	0.00	7.00
100.5	10.05	0.05 OH$^-$	10.40
110.0	11.00	1.00	11.68
120.0	12.00	2.00	11.96

FIGURE 18–1

Titration curve for 100 mL of 0.100 *M* HCl vs. 0.100 *M* NaOH. Note that the "vertical section" of the curve is quite long. The titration curves for other strong acids and bases are identical with this one *if* the same concentrations of acid and base are used.

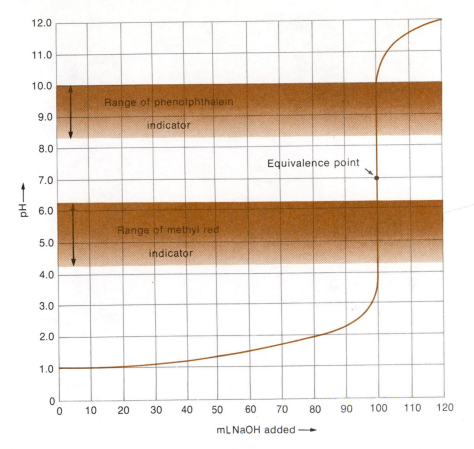

Note that the titration curve has a long "vertical section" over which pH changes very rapidly with the addition of very small amounts of base. The pH changes from 3.60 (99.5 mL NaOH added) to 10.40 (100.5 mL of NaOH added) in the vicinity of the equivalence point (100.0 mL NaOH added). The midpoint of the vertical section (pH = 7.00) is the equivalence point.

Indicators with color changes in the range pH 4 to 10 can be used in the titration of strong acids with strong bases, although ideally the color change should occur at pH = 7.00. Figure 18–1 shows the pH ranges over which color changes occur for methyl red and phenolphthalein, two widely used indicators. Both fall within the vertical section of the NaOH/HCl titration curve, which tells us that either could be used in this titration.

When a strong acid is added to a solution of a strong base, the titration curve is inverted, but its essential characteristics are the same, as Figure 18–2 shows.

2 Titrations Involving a Weak Monoprotic Acid or Base

When a weak acid is titrated with a strong base, the curve is dramatically different from Figure 18–1 before and at the equivalence point. The solution is buffered before the equivalence point, and it is basic at the equivalence point because salts of weak acids and strong bases hydrolyze to give basic solutions.

Consider the titration of 100.0 mL of 0.100 *M* CH_3COOH with a 0.100 *M* solution of NaOH. (The strong electrolyte is added to the weak electrolyte.) Before any base is added the pH is 2.89 (see Example 17–9 and Table 18–1).

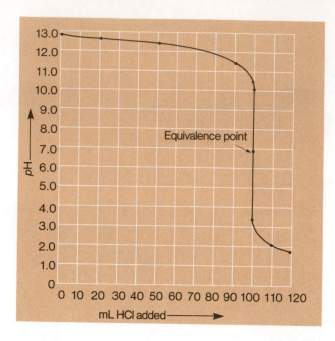

FIGURE 18–2 Titration curve for 100 mL of 0.100 M NaOH vs. 0.100 M HCl. Notice that this titration curve is similar to that in Figure 18–1, but inverted.

As soon as some NaOH is added, and until the equivalence point is reached, the solution is buffered because it contains $NaCH_3COO$ and (unreacted) CH_3COOH.

$$NaOH + CH_3COOH \longrightarrow NaCH_3COO + H_2O$$

lim. amt.   excess

After 20.0 mL of NaOH solution has been added, we have

	NaOH	+ CH₃COOH	⟶ NaCH₃COO + H₂O
Start:	2.00 mmol	10.00 mmol	0
Change:	−2.00	−2.00	+2.00
After rxn:	0	8.00 mmol	2.00 mmol

These amounts are present in 120 mL of solution, so the concentrations are

$$M_{CH_3COOH} = \frac{8.00 \text{ mmol}}{120 \text{ mL}} = 0.0667 \ M \text{ CH}_3\text{COOH}$$

$$M_{NaCH_3COO} = \frac{2.00 \text{ mmol}}{120 \text{ mL}} = 0.0167 \ M \text{ NaCH}_3\text{COO}$$

The pH of this solution is calculated as we did in Example 18–1.

$$\frac{[H^+][CH_3COO^-]}{[CH_3COOH]} = 1.8 \times 10^{-5}$$

$$[H^+] = 1.8 \times 10^{-5} \times \frac{[CH_3COOH]}{[CH_3COO^-]} = 1.8 \times 10^{-5} \times \frac{0.0667}{0.0167}$$

$$[H^+] = \underline{7.2 \times 10^{-5} \ M} \qquad \text{and} \qquad pH = \underline{4.14}$$

TABLE 18–6 Titration Data for CH₃COOH vs. NaOH

mL 0.100 M NaOH Added	mmol Base Added	mmol Excess Acid or Base	pH
0.0 mL	0	10.0 CH₃COOH	2.89
20.0 mL	2.00	8.00	4.14
50.0 mL	5.00	5.00	4.74
75.0 mL	7.50	2.50	5.22
90.0 mL	9.00	1.00	5.70
95.0 mL	9.50	0.50	6.02
99.0 mL	9.90	0.10	6.74
100.0 mL	10.0	0	8.72
101.0 mL	10.1	0.10 OH⁻	10.70
110.0 mL	11.0	1.0	11.68
120.0 mL	12.0	2.0	11.96

Just before the equivalence point, the solution contains relatively high concentrations of NaCH₃COO and relatively low concentrations of CH₃COOH. Just after the equivalence point, the solution contains relatively high concentrations of NaCH₃COO and relatively low concentrations of NaOH, both basic compounds. In both regions our calculations are only *approximations*. Exact calculations of pH in these regions are beyond the scope of this text.

All points before the equivalence point are calculated in the same way. After some NaOH has been added, the solution contains both $NaCH_3COO$ and CH_3COOH, and so it is buffered until the equivalence point is reached. At the equivalence point, the solution is 0.0500 M in $NaCH_3COO$.

$$NaOH + CH_3COOH \longrightarrow NaCH_3COO + H_2O$$

10.0 mmol   10.0 mmol   10.0 mmol

$$M_{NaCH_3COO} = \frac{10.0 \text{ mmol}}{200 \text{ mL}} = 0.0500 \ M \text{ NaCH}_3\text{COO}$$

The pH of a 0.0500 M solution of $NaCH_3COO$ is 8.72. (See Example 17–13 for a similar calculation.) The solution is distinctly basic at the equivalence point, because of the hydrolysis of the acetate ion.

Beyond the equivalence point, the excess NaOH determines the pH of the solution just as it did in the titration of a strong acid.

Table 18–6 lists several points on the titration curve and Figure 18–3 shows the titration curve for 100.0 mL of 0.100 M CH_3COOH titrated with a 0.100 M solution of NaOH. This titration curve has a short vertical section (pH = 8 to 10) and the indicator range is limited. Phenolphthalein is the indicator commonly used when weak acids are titrated with strong bases (see Table 18–4).

The titration curves for weak bases and strong acids are similar to those for weak acids and strong bases except that they are inverted (recall that strong is added to weak). Figure 18–4 displays the titration curve for 100.0 mL of 0.100 M aqueous ammonia titrated with 0.100 M HCl solution.

When solutions of *weak acids* are titrated with *weak bases*, the vertical portion of the curve is much too short for color indicators to be used. The solution is buffered both before and after the equivalence point. Figure 18–5 shows the titration curve for 100.0 mL of 0.100 M CH_3COOH solution titrated with 0.100 M aqueous ammonia.

Slightly Soluble Compounds and the Solubility Product Principle

In our discussion of ionic equilibrium thus far we have considered only compounds that are readily soluble in water. Many compounds are only slightly soluble in water, and such substances are usually referred to as "insoluble compounds."

FIGURE 18–3
Titration curve for 100 mL of 0.100 *M* CH₃COOH vs. 0.100 *M* NaOH. Notice that the vertical section of this curve is much shorter than the vertical sections in Figures 18–1 and 18–2 because the solution is buffered before the equivalence point.

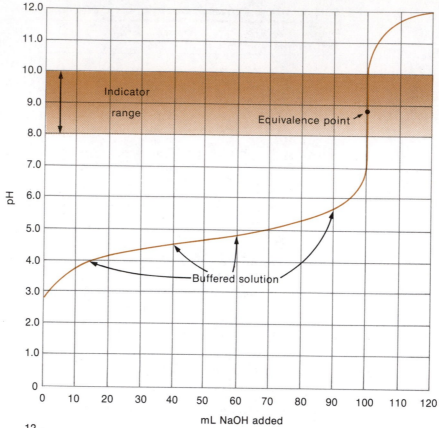

FIGURE 18–4
Titration curve for 100 mL of 0.100 *M* aqueous ammonia vs. 0.100 *M* HCl. Notice that the vertical section of the curve is relatively short because the solution is buffered before the equivalence point. Notice also that the curve is similar to Figure 18–3, but inverted.

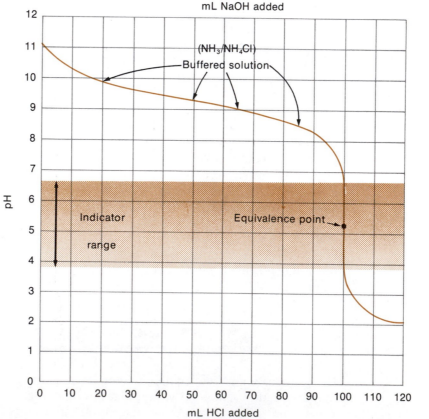

FIGURE 18-5

Titration curve for 100 mL of 0.100 M CH_3COOH vs. 0.100 M aqueous NH_3. Because the solution is buffered before and after the equivalence point, the vertical section of the curve is too short to be noticed. Therefore, color indicators cannot be used in such titrations.

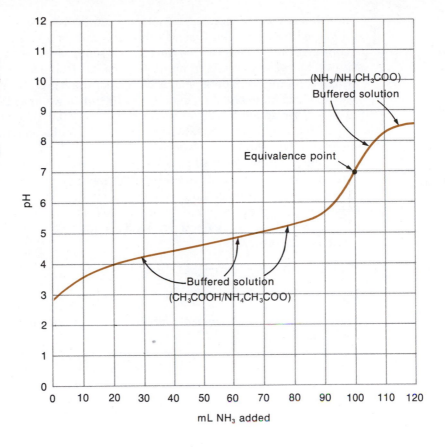

Barium sulfate is the "insoluble" substance taken orally before intestinal x-ray photos are taken, because the heavy barium atoms absorb x-rays well. Even though barium ions are quite toxic, barium sulfate can still be taken orally without danger. The compound is so insoluble that most of it passes through the digestive system unchanged, and only an insignificant amount of barium ions dissolves.

Nearly all compounds dissolve in water to some extent, however, and we shall now consider those that are only very slightly soluble. As a rough rule-of-thumb, compounds that dissolve in water to the extent of at least 0.020 mole/liter are classified as soluble. Refer to the solubility rules (Section 9–4.5) as necessary.

18-6 Solubility Product Constants

Let's consider what happens when powdered barium sulfate, $BaSO_4$, is placed in water and stirred vigorously. Careful experiments show that a very small amount of barium sulfate dissolves in the water to give a saturated solution, and the $BaSO_4$ that dissolves is completely dissociated into its constituent ions:

$$BaSO_4 \text{ (s)} \rightleftharpoons Ba^{2+} \text{ (aq)} + SO_4^{2-} \text{ (aq)}$$

The existence of a substance in the solid state is indicated a number of ways. For example, $BaSO_4$ (s), $\underline{BaSO_4}$, and $BaSO_4 \downarrow$ are three different ways of representing solid $BaSO_4$. The last two are frequently used to indicate the formation of solids (precipitates) in reactions in aqueous solution. As a matter of consistency, we shall use the notation in the preceding equation for formulas of solid substances in equilibrium with dissolved species in aqueous solution. The equilibrium expression for the dissolution of $BaSO_4$ in water may be written

$$K_c = \frac{[Ba^{2+}][SO_4^{2-}]}{[BaSO_4 \text{ (s)}]}$$

[Eq. 18–8]

Note that this expression contains the concentration of solid $BaSO_4$, [$BaSO_4$ (s)]. In an equilibrium between a solid substance and its ions in aqueous solution, the "concentration of the solid" has no physical significance, as the following discussion shows. Suppose that one gram of solid $BaSO_4$ is added to 100 mL of water and stirred until the solution is *saturated*. Most of the $BaSO_4$ does not dissolve. In fact, only 0.00025 gram of $BaSO_4$ dissolves in 100 mL of water at 25°C. Even if ten grams of $BaSO_4$ are added to 100 mL of water, still only 0.00025 gram dissolves at 25°C. No matter how much $BaSO_4$ we add to the 100 mL of water, the same small amount dissolves. The expression "concentration of solid" has no meaning *in equilibria involving solids and their saturated solutions*, since the amount of solid remaining does not affect the concentrations of the ions. In many equilibria involving the dissolution of slightly soluble compounds in water, it is convenient to determine experimentally the concentrations of *dissolved ions* in saturated solutions. Therefore, for such equilibria, we define a new kind of equilibrium constant called a **solubility product constant,** K_{sp}:

> Strictly speaking, the activity of pure crystalline $BaSO_4$ is the same as its activity in the standard state, *i.e.*, $a_{BaSO_4}/a^0_{BaSO_4} = 1$. Therefore, the denominator of this and other solubility product constants is unity. One may also argue that the concentration of a solid is fixed by its density (Section 16–9).

$$BaSO_4 \text{ (s)} \rightleftharpoons Ba^{2+} \text{ (aq)} + SO_4^{2-} \text{ (aq)}$$

$$K_{sp} = [Ba^{2+}][SO_4^{2-}] \qquad \text{[Eq. 18–9]}$$

The solubility product constant for $BaSO_4$ is the product of the concentrations of the constituent ions.

In general terms, *the* **solubility product expression** *for a compound is the product of the concentrations of the constituent ions, each raised to the power that corresponds to the number of ions in one formula unit of the compound.* This statement is the **solubility product principle.** The following examples illustrate the point.

Consider the dissolution of the slightly soluble compound calcium fluoride, CaF_2, in water.

> Unless otherwise indicated, solubility product constants and solubility data are given for 25°C.

$$CaF_2 \text{ (s)} \rightleftharpoons Ca^{2+} \text{ (aq)} + 2F^- \text{ (aq)}$$

Its solubility product expression is $K_{sp} = [Ca^{2+}][F^-]^2$. On the other hand, the dissolution of solid bismuth sulfide, Bi_2S_3, gives two bismuth ions and three sulfide ions per formula unit.

$$Bi_2S_3 \text{ (s)} \rightleftharpoons 2Bi^{3+} \text{ (aq)} + 3S^{2-} \text{ (aq)}$$

so $K_{sp} = [Bi^{3+}]^2[S^{2-}]^3$. Generally, we may represent the dissolution of a slightly soluble compound as

$$M_yX_z \text{ (s)} \rightleftharpoons y\,M^{z+} \text{ (aq)} + z\,X^{y-} \text{ (aq)}$$

and its solubility product expression is

> The units of all solubility product constants are mol/L raised to the appropriate power. Units are commonly omitted from K_{sp}'s.

$$K_{sp} = [M^{z+}]^y[X^{y-}]^z \qquad \text{[Eq. 18–10]}$$

If there are more than two kinds of ions in the formula for a compound, the above generalization still applies. The dissolution of the slightly soluble compound magnesium ammonium phosphate, $MgNH_4PO_4$, in water is represented as

$$MgNH_4PO_4 \text{ (s)} \rightleftharpoons Mg^{2+} \text{ (aq)} + NH_4^+ \text{ (aq)} + PO_4^{3-} \text{ (aq)}$$

and its solubility product expression is

$$K_{sp} = [Mg^{2+}][NH_4^+][PO_4^{3-}] \qquad \text{[Eq. 18–11]}$$

A word about conventions. We frequently shorten the term "solubility product expression" to "solubility product." Thus, the solubility product for barium sulfate, $BaSO_4$, is written

$$K_{sp} = [Ba^{2+}][SO_4^{2-}] = 1.1 \times 10^{-10}$$

and the solubility product for calcium fluoride, CaF_2, is written

$$K_{sp} = [Ca^{2+}][F^-]^2 = 3.9 \times 10^{-11} \qquad \text{[Eq. 18–12]}$$

18–7 Experimental Determination of Solubility Product Constants

Although we frequently use statements like "the solution contains 0.0025 g of dissolved $BaSO_4$," what we mean is: 0.0025 g of solid $BaSO_4$ dissolves to give a saturated solution that contains equal numbers of Ba^{2+} and SO_4^{2-} ions.

Solubility product constants can be determined in a number of ways. For example, careful measurements of conductivity show that one liter of a saturated solution of barium sulfate contains 0.0025 gram of dissolved $BaSO_4$. If the solubility of a compound is known, its solubility product can be calculated as in Example 18–7.

Example 18–7

One liter of saturated barium sulfate contains 0.0025 gram of dissolved $BaSO_4$. Calculate the solubility product constant for $BaSO_4$.

Solution

As with previous problems, we write the appropriate (1) chemical equation and (2) equilibrium constant expression. Recall that in a saturated solution, equilibrium exists between solid and dissolved solute.

$$BaSO_4 \text{ (s)} \rightleftharpoons Ba^{2+} \text{ (aq)} + SO_4^{2-} \text{ (aq)}$$

$$K_{sp} = [Ba^{2+}][SO_4^{2-}]$$

From the solubility of barium sulfate in water we can calculate its *molar solubility*, the number of moles that dissolve to give one liter of saturated solution.

$$\frac{? \text{ mol } BaSO_4}{L} = \frac{0.0025 \text{ g } BaSO_4}{L} \times \frac{1 \text{ mol } BaSO_4}{233 \text{ g } BaSO_4}$$

$$= 1.1 \times 10^{-5} \text{ mol } BaSO_4/L \quad \text{molar solubility}$$

(We are not suggesting that the concentration of solid $BaSO_4$ is 1.1×10^{-5} *M*, only that its molar solubility is 1.1×10^{-5} *M*.) Now that the molar solubility of $BaSO_4$ is known, reference to the dissolution equation shows each formula unit of $BaSO_4$ dissolves to produce one Ba^{2+} and one SO_4^{2-} ion.

$$BaSO_4 \text{ (s)} \rightleftharpoons Ba^{2+} \text{ (aq)} + SO_4^{2-} \text{ (aq)}$$
$$1.1 \times 10^{-5} \text{ M} \qquad 1.1 \times 10^{-5} \text{ M} \qquad 1.1 \times 10^{-5} \text{ M}$$

The molar solubility of $BaSO_4$ is 1.1×10^{-5} mol/L, and in a saturated solution $[Ba^{2+}] = [SO_4^{2-}] = 1.1 \times 10^{-5}$ *M*. Substituting these values into the solubility product constant expression for $BaSO_4$ gives its solubility product constant.

$$K_{sp} = [Ba^{2+}][SO_4^{2-}] = (1.1 \times 10^{-5})(1.1 \times 10^{-5})$$
$$= 1.2 \times 10^{-10}$$

The difference between this value and the tabulated 1.1×10^{-10} is due to round-off error.

This expression is particularly useful because it is applicable *to all saturated solutions* of $BaSO_4$ at 25°C. The *origin* of the Ba^{2+} and SO_4^{2-} ions is of no importance. For example, suppose a solution of barium chloride (which contains Ba^{2+} and Cl^- ions) and a solution of sodium sulfate (which contains Na^+ and SO_4^{2-} ions) are mixed at 25°C. (Both of these compounds are very soluble.) As soon as the concentration of Ba^{2+} ions multiplied by the concentration of SO_4^{2-} ions exceeds 1.1×10^{-10}, solid $BaSO_4$ begins to precipitate from the solution. *Or*, if solid $BaSO_4$ is placed in water at 25°C, it dissolves until the concentration of Ba^{2+} multiplied by the concentration of SO_4^{2-} ions just equals 1.1×10^{-10}.

Example 18–8 is similar to Example 18–7, but there are some differences, due to the 2:1 ion stoichiometry of Ag_2CrO_4.

Example 18–8

One liter of a saturated solution of silver chromate at 25°C contains 0.0435 g of Ag_2CrO_4. Calculate its solubility product constant.

Solution

The equation for the dissolution of silver chromate in water and its solubility product expression are

$$Ag_2CrO_4 \text{ (s)} \rightleftharpoons 2Ag^+ \text{ (aq)} + CrO_4^{2-} \text{ (aq)}$$

$$K_{sp} = [Ag^+]^2[CrO_4^{2-}]$$

The molar solubility of silver chromate is calculated as

$$\frac{? \text{ mol } Ag_2CrO_4}{L} = \frac{0.0435 \text{ g } Ag_2CrO_4}{L}$$

$$\times \frac{1 \text{ mol } Ag_2CrO_4}{332 \text{ g } Ag_2CrO_4}$$

$$= \underline{1.31 \times 10^{-4} \text{ mol/L}}$$

The equation for dissolution of Ag_2CrO_4 and its molar solubility give the concentrations of Ag^+ and CrO_4^{2-} in a saturated solution.

$$Ag_2CrO_4 \text{ (s)} \rightleftharpoons 2Ag^+ \text{ (aq)} + CrO_4^{2-} \text{ (aq)}$$

$$1.31 \times 10^{-4} \, M \qquad 2.62 \times 10^{-4} \, M \qquad 1.31 \times 10^{-4} \, M$$

Substitution into the solubility product expression for Ag_2CrO_4 gives its solubility product.

$$K_{sp} = [Ag^+]^2[CrO_4^{2-}] = (2.62 \times 10^{-4})^2(1.31 \times 10^{-4})$$
$$= \underline{9.0 \times 10^{-12}}$$

The values obtained in Examples 18–7 and 18–8 are compared in Table 18–7. These data show that the molar solubility of Ag_2CrO_4 is greater than that of $BaSO_4$. However, the solubility product for $BaSO_4$ is greater than the solubility product for Ag_2CrO_4 because the solubility product expression for Ag_2CrO_4 contains a *squared* term, $[Ag^+]^2$. Thus, we see that care must be exercised in comparing solubility products for different compounds.

TABLE 18–7   Comparison of Solubilities of $BaSO_4$ and Ag_2CrO_4

Compound	Molar Solubility	K_{sp}
$BaSO_4$	1.1×10^{-5} mol/L	$[Ba^{2+}][SO_4^{2-}] = 1.1 \times 10^{-10}$
Ag_2CrO_4	1.3×10^{-4} mol/L	$[Ag^+]^2[CrO_4^{2-}] = 9.0 \times 10^{-12}$

18–8   Uses of Solubility Product Constants

If the solubility product for a compound is known, the solubility of the compound in water at 25°C can be calculated conveniently, as Example 18–9 illustrates.

Solubility products for some common compounds are tabulated in Appendix H. Refer to this appendix as needed.

Example 18-9

Calculate molar solubilities, concentrations of the constituent ions, and solubilities in g/L for the following compounds: (a) silver chloride, AgCl, and (b) zinc hydroxide, $Zn(OH)_2$.

Solution

(a) The equation for the dissolution reaction of silver chloride and its solubility product expression (obtained from Appendix H) are

$$AgCl (s) \rightleftharpoons Ag^+ (aq) + Cl^- (aq)$$

$$K_{sp} = [Ag^+][Cl^-] = 1.8 \times 10^{-10}$$

Each formula unit of AgCl that dissolves produces one Ag^+ and one Cl^-, and we can let x = the molar solubility, which gives

$$AgCl (s) \rightleftharpoons Ag^+ (aq) + Cl^- (aq)$$
$$x \text{ mol/L} \qquad x \, M \qquad x \, M$$

Substitution into the solubility product expression gives

$$[Ag^+][Cl^-] = (x)(x) = 1.8 \times 10^{-10}$$
$$x^2 = 1.8 \times 10^{-10}$$
$$x = 1.3 \times 10^{-5} \text{ mol AgCl/L}$$

One liter of saturated AgCl contains 1.3×10^{-5} mole of AgCl at 25°C. Now that we know the value of x, the molar solubility, we also know the concentrations of the constituent ions,

$$x = [Ag^+] = [Cl^-] = 1.3 \times 10^{-5} \text{ mol/L}$$

To calculate the mass of AgCl in one liter of saturated solution, we convert 1.3×10^{-5} mol AgCl/L to g AgCl/L.

$$\frac{? \text{ g AgCl}}{L} = \frac{1.3 \times 10^{-5} \text{ mol AgCl}}{L} \times \frac{143 \text{ g AgCl}}{1 \text{ mol AgCl}}$$
$$= 1.9 \times 10^{-3} \text{ g AgCl/L}$$

A liter of saturated AgCl solution contains only 0.0019 g of AgCl, and we are justified in referring to AgCl as a slightly soluble compound.

(b) The dissolution of zinc hydroxide, $Zn(OH)_2$, in water and its solubility product expression are represented as

$$Zn(OH)_2 (s) \rightleftharpoons Zn^{2+} (aq) + 2OH^- (aq)$$

$$K_{sp} = [Zn^{2+}][OH^-]^2 = 4.5 \times 10^{-17}$$

If, as before, we let x = molar solubility, then $[Zn^{2+}] = x$ and $[OH^-] = 2x$, and we have

$$Zn(OH)_2 (s) \rightleftharpoons Zn^{2+} (aq) + 2OH^- (aq)$$
$$x \text{ mol/L} \qquad x \, M \qquad 2x \, M$$

Substitution into the solubility product expression gives

$$[Zn^{2+}][OH^-]^2 = (x)(2x)^2 = 4.5 \times 10^{-17}$$
$$4x^3 = 4.5 \times 10^{-17}$$
$$x^3 = 11 \times 10^{-18}$$
$$x = 2.2 \times 10^{-6} \text{ mol } Zn(OH)_2/L$$

so

$$x = [Zn^{2+}] = 2.2 \times 10^{-6} \, M$$
$$2x = [OH^-] = 4.4 \times 10^{-6} \, M$$

We can now calculate the mass of $Zn(OH)_2$ in one liter of saturated solution,

$$\frac{? \text{ g } Zn(OH)_2}{L} = \frac{2.2 \times 10^{-6} \text{ mol } Zn(OH)_2}{L}$$
$$\times \frac{99 \text{ g } Zn(OH)_2}{1 \text{ mol } Zn(OH)_2}$$
$$= 2.2 \times 10^{-4} \text{ g } Zn(OH)_2/L$$

The common ion effect applies to solubility equilibria, just as it does to ionic equilibria in general. The solubility of any compound is less in a solution containing an ion in common with the compound than in pure water, as long as no other reaction is caused by the presence of the common ion (such as dissolution of an amphoteric metal hydroxide in the presence of a large excess of hydroxide ions). Example 18–10 illustrates of the common ion effect.

Example 18–10

The molar solubility of magnesium fluoride, MgF_2, in pure water at 25°C is 1.2×10^{-3} mol/L. Calculate the molar solubility of MgF_2 in 0.10 M sodium fluoride, NaF, solution at 25°C. $K_{sp} = 6.4 \times 10^{-9}$ for MgF_2.

Solution

NaF is a soluble ionic salt, and therefore 0.10 M F$^-$ is produced by

$$NaF\ (s) \xrightarrow{\ H_2O\ } Na^+\ (aq)\ +\ F^-\ (aq)$$
$$\qquad\qquad\quad 0.10\ M \qquad 0.10\ M$$

For MgF_2, a slightly soluble salt, letting x represent the molar solubility of MgF_2,

$$MgF_2\ (s) \rightleftharpoons Mg^{2+}\ (aq)\ +\ 2F^-\ (aq)$$
$$x\ \text{mol/L} \qquad\quad x\ M \qquad\quad 2x\ M$$

$$K_{sp} = [Mg^{2+}][F^-]^2 = 6.4 \times 10^{-9}$$

$$(x)(0.10 + 2x)^2 = 6.4 \times 10^{-9}$$

We make the simplifying assumption that $2x \ll 0.10$, so

$$(x)(0.10)^2 = 6.4 \times 10^{-9}$$
$$x = \underline{6.4 \times 10^{-7}\ M}$$

This is the molar solubility of MgF_2 in 0.10 M NaF. (Note that the assumption is valid.) The molar solubility of MgF_2 in 0.10 M NaF ($6.4 \times 10^{-7}\ M$) is much less than in pure water ($1.2 \times 10^{-3}\ M$).

Another application of the solubility product principle is the calculation of the maximum concentrations of ions that can exist in solution, and from these the determination of whether or not a precipitate will form in a given solution. This involves comparison of Q_{sp}, the reaction quotient (Section 16–4) applied to solubility equilibria, with K_{sp}.

If $Q_{sp} < K_{sp}$ forward process is favored; no precipitation occurs, more solid will dissolve

$Q_{sp} = K_{sp}$ solution is just saturated; neither forward nor reverse process is favored

$Q_{sp} > K_{sp}$ reverse process is favored; precipitation occurs

Example 18–11 illustrates this idea.

Example 18–11

If 100 mL of 0.0010 M sodium sulfate, Na_2SO_4, and 100 mL of 0.010 M barium chloride, $BaCl_2$, solutions are mixed, will a precipitate form?

Solution

First consider the kinds of compounds mixed and determine whether a reaction *can* occur. Both Na_2SO_4 and $BaCl_2$ are soluble ionic salts; therefore, at the moment of mixing the new solution contains a mixture of Na$^+$, SO_4^{2-}, Ba^{2+}, and Cl$^-$ ions in the following concentrations. (Since the volumes of dilute aqueous solutions are very nearly additive, the volume doubles to 200 mL and the concentrations are halved.)

$$Na_2SO_4\ (s) \xrightarrow{\ H_2O\ } 2Na^+\ (aq)\ +\ SO_4^{2-}\ (aq)$$
$$0.00050\ \text{mol/L} \qquad 2(0.00050\ M) \qquad 0.00050\ M$$
$$= 0.0010\ M$$

$$BaCl_2\ (s) \xrightarrow{\ H_2O\ } Ba^{2+}\ (aq)\ +\ 2Cl^-\ (aq)$$
$$0.0050\ \text{mol/L} \qquad\quad 0.0050\ M \qquad 2(0.0050\ M)$$
$$= 0.010\ M$$

The possibility of forming two new compounds, NaCl and $BaSO_4$, exists. Sodium chloride is a soluble ionic compound; Na^+ and Cl^- do not react in dilute aqueous solutions. However, $BaSO_4$ is only very slightly soluble, and solid $BaSO_4$ will precipitate from the solution *if* the product of the concentrations of Ba^{2+} and SO_4^{2-} exceeds the solubility product for $BaSO_4$. Having determined that a precipitate may form, we perform calculations to determine whether solid $BaSO_4$ precipitates from this particular solution. The solubility product for $BaSO_4$ is 1.1×10^{-10} (Appendix H). If $[Ba^{2+}][SO_4^{2-}]$ exceeds 1.1×10^{-10}, then solid $BaSO_4$ will precipitate. Substituting $[Ba^{2+}] = 0.0050\ M$ and $[SO_4^{2-}] =$ 0.00050 M into the solubility product expression for $BaSO_4$ we calculate Q_{sp}:

$$Q_{sp} = [Ba^{2+}][SO_4^{2-}]$$
$$= (5.0 \times 10^{-3})(5.0 \times 10^{-4})$$
$$= 2.5 \times 10^{-6}$$

The product of the actual concentrations of the constituent ions, Q_{sp}, exceeds the solubility product for $BaSO_4$, 1.1×10^{-10}. Since $Q_{sp} > K_{sp}$, solid $BaSO_4$ will precipitate until the concentrations of the constituent ions just satisfy K_{sp} for $BaSO_4$.

Since the actual concentrations of Ba^{2+} and SO_4^{2-} multiplied together give 2.5×10^{-6}, K_{sp} for $BaSO_4$ is exceeded by a factor of approximately 2×10^4, and there will be no difficulty in seeing some precipitated $BaSO_4$.

The human eye is not a particularly sensitive detection device; when two solutions are mixed, enough solid must form so that we can see it if we are to be aware of precipitate formation. As a general rule of thumb, a precipitate can be seen with the naked eye if the solubility product of the compound is exceeded by a factor of 1000. If the solubility product is exceeded by a factor less than 1000, the amount of solid will likely be so small that it cannot be seen.

Solubility products enable us to calculate the concentration of an ion necessary to initiate precipitation in a particular solution, as Example 18–12 demonstrates.

Example 18–12

What concentration of barium ion (from a soluble compound such as barium chloride) is necessary to start the precipitation of barium sulfate in a solution that is 0.0015 M in sodium sulfate?

Solution

We are dealing with the same compounds we described in Example 18–11. Since Na_2SO_4 is a soluble ionic compound, we know that $[SO_4^{2-}] = 0.0015\ M$. Therefore, we can use K_{sp} for $BaSO_4$ to calculate the concentration of barium ions required to exceed K_{sp}.

$$[Ba^{2+}][SO_4^{2-}] = 1.1 \times 10^{-10}$$

$$[Ba^{2+}] = \frac{1.1 \times 10^{-10}}{[SO_4^{2-}]} = \frac{1.1 \times 10^{-10}}{1.5 \times 10^{-3}}$$
$$= \underline{7.3 \times 10^{-8}\ M}$$

Addition of enough $BaCl_2$ to give a barium ion concentration of $7.3 \times 10^{-8}\ M$ *just satisfies* K_{sp} for $BaSO_4$. Ever-so-slightly more $BaCl_2$ would be required to exceed K_{sp} and initiate precipitation of $BaSO_4$. Therefore, we say

$$[Ba^{2+}] \geq \underline{7.3 \times 10^{-8}\ M} \qquad \text{to initiate precipitation of } BaSO_4$$

Recall that K_{sp} must be exceeded by a factor of approximately 1000 in order for a precipitate to be visible.

Frequently, it is desirable to remove a particular ion from solution by forming an insoluble compound (as in water purification). Calculations based on solubility products enable us to calculate the concentrations of ions remaining in solution *after* precipitation has occurred.

Example 18–13

Suppose we wish to recover silver from an aqueous solution containing a soluble silver compound such as silver nitrate, $AgNO_3$, by precipitating silver ions as the insoluble compound silver chloride, AgCl. What concentration of chloride ion is necessary to reduce the concentration of silver ions to 1.0×10^{-9} M? A soluble ionic compound such as NaCl is used as a source of chloride ions.

Solution

The reaction is

$$Ag^+ (aq) + Cl^- (aq) \rightleftharpoons AgCl (s)$$

and the solubility product for AgCl is

$$[Ag^+][Cl^-] = 1.8 \times 10^{-10}$$

To determine the concentration of Cl^- required to reduce the concentration of Ag^+ to 1.0×10^{-9} M, we simply solve the solubility product expression for $[Cl^-]$,

$$[Cl^-] = \frac{1.8 \times 10^{-10}}{[Ag^+]} = \frac{1.8 \times 10^{-10}}{1.0 \times 10^{-9}}$$
$$= \underline{0.18\ M\ Cl^-}$$

Thus, to reduce $[Ag^+]$ to 1.0×10^{-9} M, NaCl would be added until the final concentration of Cl^- in the solution is 0.18 M.

The recovery of silver from solutions used in developing and fixing photographic film and prints presents just such a problem. Silver is an expensive metal, and the recovery is profitable. Moreover, if not recovered, the silver ions would constitute an undesirable pollutant in water supplies.

18–9   Fractional Precipitation

Compound	Solubility Product
AgCl	1.8×10^{-10}
AgBr	3.3×10^{-13}
AgI	1.5×10^{-16}

The increasing polarizability of the halide ions (Section 21–5) in going from F^- to I^- leads to greatest covalent character in AgI.

On occasion it is desirable to remove some ions from solution by precipitation while leaving other ions with very similar properties in solution. The process is called **fractional precipitation.** Consider a solution that contains Cl^-, Br^- and I^- ions. Since these halide ions are derived from elements in the same family in the periodic table, we should expect that there will be similarities in their properties. But we should also expect some differences in properties, and indeed this is what we find when we examine the solubility products for these silver halides. The solubility products show that AgI is less soluble than AgBr, and AgBr is less soluble than AgCl. This should not be surprising, since the iodide ion is larger and more polarizable than the bromide ion, and the bromide ion in turn is larger and more polarizable than the chloride ion. As a matter of interest, silver fluoride is quite soluble in water because the small fluoride ion is not easily polarized.

Example 18–14

Solid silver nitrate is slowly added to a solution that is 0.0010 M each in NaCl, NaBr, and NaI. Calculate the concentration of Ag^+ required to initiate the precipitation of each of these silver halides. (Recall that NaCl, NaBr, NaI, and $AgNO_3$ are soluble ionic compounds that are completely dissociated in dilute aqueous solution. Sodium nitrate, $NaNO_3$, is also a soluble ionic compound, so it cannot precipitate from this solution.)

Solution

Let us calculate the concentration of Ag^+ necessary to begin to precipitate each of the halide ions as we

did in Example 18–12. The solubility product for AgI is

$$[Ag^+][I^-] = 1.5 \times 10^{-16}$$

Since $[I^-] = 1.0 \times 10^{-3}$ M, the $[Ag^+]$ that must be exceeded to initiate precipitation of AgI is

$$[Ag^+] = \frac{1.5 \times 10^{-16}}{[I^-]} = \frac{1.5 \times 10^{-16}}{1.0 \times 10^{-3}}$$
$$= \underline{1.5 \times 10^{-13}\ M}$$

Therefore, $[Ag^+] \geq 1.5 \times 10^{-13}$ M is required to start precipitation of silver iodide. Repeating this kind of calculation for silver bromide gives

$$[Ag^+] = \frac{3.3 \times 10^{-13}}{[Br^-]} = \frac{3.3 \times 10^{-13}}{1.0 \times 10^{-3}}$$

$$= 3.3 \times 10^{-10} \, M$$

Thus, $[Ag^+] \geq 3.3 \times 10^{-10} \, M$ is needed to start precipitation of silver bromide. For the precipitation of silver chloride to begin,

$$[Ag^+] = \frac{1.8 \times 10^{-10}}{[Cl^-]} = \frac{1.8 \times 10^{-10}}{1.0 \times 10^{-3}}$$

$$= 1.8 \times 10^{-7} \, M$$

For silver chloride to precipitate, we must have $[Ag^+] \geq 1.8 \times 10^{-7} \, M$. We have demonstrated that

to precipitate AgI, $[Ag^+] \geq 1.5 \times 10^{-13} \, M$,
to precipitate AgBr, $[Ag^+] \geq 3.3 \times 10^{-10} \, M$,
to precipitate AgCl, $[Ag^+] \geq 1.8 \times 10^{-7} \, M$.

This calculation tells us that when silver nitrate is added slowly to a solution that is 0.0010 M in each of NaI, NaBr, and NaCl, silver iodide precipitates first, silver bromide precipitates second, and silver chloride precipitates last. The calculation of the amount of I^- precipitated before Br^- begins to precipitate and the amounts of I^- and Br^- precipitated before Cl^- begins to precipitate is illustrated in Example 18–15.

Example 18–15

(a) Calculate the percentage of I^- precipitated before AgBr precipitates. Refer to Example 18–14 as necessary for data. (b) Calculate the percentage of I^- and Br^- precipitated before Cl^- precipitates.

Solution

(a) In Example 18–14 we found that $[Ag^+] \geq 3.3 \times 10^{-10} \, M$ is needed to begin precipitation of AgBr. This value for $[Ag^+]$ can be substituted into the solubility product expression for silver iodide to determine $[I^-]$ remaining *in solution* just before AgBr begins to precipitate. (In Example 18–13 we did a similar calculation, but stopped short of expressing the result in terms of percentage of an ion precipitated.)

$$[Ag^+][I^-] = 1.5 \times 10^{-16}$$

$$[I^-]_{unppt'd} = \frac{1.5 \times 10^{-16}}{[Ag^+]} = \frac{1.5 \times 10^{-16}}{3.3 \times 10^{-10}}$$

$$= 4.5 \times 10^{-7} \, M$$

The percentage of I^- unprecipitated is

$$\% \, I^-_{unppt'd} = \frac{[I^-]_{unppt'd}}{[I^-]_{orig}} \times 100\%$$

$$= \frac{4.5 \times 10^{-7}}{1.0 \times 10^{-3}} \times 100\%$$

$$= 0.045\% \, I^- \text{ unprecipitated}$$

Therefore, 99.955% of the I^- precipitates *before* AgBr begins to precipitate. (We have subtracted 0.045% from *exactly* 100%, and therefore we have not violated the rules for significant figures.)

(b) Similar calculations show that *just before* AgCl begins to precipitate, $[Ag^+] = 1.8 \times 10^{-7} \, M$ and the $[I^-]$ unprecipitated is calculated as in (a).

$$[Ag^+][I^-] = 1.5 \times 10^{-16}$$

$$[I^-]_{unppt'd} = \frac{1.5 \times 10^{-16}}{[Ag^+]} = \frac{1.5 \times 10^{-16}}{1.8 \times 10^{-7}}$$

$$= 8.3 \times 10^{-10} \, M$$

The percentage of I^- unprecipitated just before AgCl precipitates is

$$\% \, I^-_{unppt'd} = \frac{[I^-]_{unppt'd}}{[I^-]_{orig}} \times 100\%$$

$$= \frac{8.3 \times 10^{-10}}{1.0 \times 10^{-3}} \times 100\%$$

$$= 0.000083\% \, I^- \text{ unprecipitated}$$

Therefore, 99.999917% of the I^- precipitates before AgCl begins to precipitate.

A similar calculation for the amount of Br^- precipitated before AgCl begins to precipitate gives

$$[Ag^+][Br^-] = 3.3 \times 10^{-13}$$

$$[Br^-]_{unppt'd} = \frac{3.3 \times 10^{-13}}{[Ag^+]} = \frac{3.3 \times 10^{-13}}{1.8 \times 10^{-7}}$$

$$= 1.8 \times 10^{-6} \, M$$

$$\% \, Br^-_{unppt'd} = \frac{[Br^-]_{unppt'd}}{[Br^-]_{orig}} \times 100\%$$

$$= \frac{1.8 \times 10^{-6}}{1.0 \times 10^{-3}} \times 100\%$$

$$= 0.18\% \, Br^- \text{ unprecipitated}$$

Thus, 99.82% of the Br^- precipitates before AgCl begins to precipitate.

We have described the series of reactions that occurs when solid $AgNO_3$ is added slowly to a solution that is $0.0010 \, M$ in Cl^-, Br^-, and I^-. Silver iodide begins to precipitate first, and 99.955% of the I^- precipitates before any solid AgBr is formed. Silver bromide begins to precipitate next, and 99.82% of the Br^- and 99.999917% of the I^- precipitate before any solid AgCl forms.

18–10 Simultaneous Equilibria Involving Slightly Soluble Compounds

Reactions between weak acids or bases and metal ions to form insoluble compounds are very common. Example 18–16 provides insight into the reaction of a metal ion with aqueous ammonia to form an insoluble metal hydroxide.

Example 18–16

If a solution is made $0.10 \, M$ in magnesium nitrate, $Mg(NO_3)_2$, *and* $0.10 \, M$ in aqueous ammonia, a weak base, will magnesium hydroxide, $Mg(OH)_2$, precipitate? The solubility product for $Mg(OH)_2$ is 1.5×10^{-11} and the ionization constant for aqueous NH_3 is 1.8×10^{-5}.

Solution

Two equilibria must be considered, and we must determine whether or not K_{sp} for $Mg(OH)_2$ is exceeded in the solution. That is, is the ion product in the solution, Q_{sp}, greater than K_{sp}, $[Mg^{2+}][OH^-]^2 = 1.5 \times 10^{-11}$? To answer the question the concentrations of both ions are calculated.

Magnesium nitrate is a soluble ionic compound and $[Mg^{2+}] = 0.10 \, M$ in a $0.10 \, M$ solution of $Mg(NO_3)_2$.

The concentration of OH^- in $0.10 \, M$ aqueous ammonia is calculated as in Example 17–10.

$$NH_3 \, (aq) \; + H_2O \rightleftharpoons NH_4^+ \, (aq) + OH^- \, (aq)$$
$$(0.10 - x) \, M \qquad\qquad x \, M \qquad\quad x \, M$$

$$\frac{[NH_4^+][OH^-]}{[NH_3]} = 1.8 \times 10^{-5}$$

Substitution into the ionization constant expression for aqueous NH_3 gives

$$\frac{(x)(x)}{(0.10 - x)} = 1.8 \times 10^{-5}$$

$$x = 1.3 \times 10^{-3} \, M = [OH^-]$$

Now that the concentrations of both Mg^{2+} and OH^- are known, they can be substituted into the solubility product for $Mg(OH)_2$ to determine whether $Mg(OH)_2$ will precipitate at these concentrations of Mg^{2+} and OH^-.

$$Q_{sp} = [Mg^{2+}][OH^-]^2$$
$$= (0.10)(1.3 \times 10^{-3})^2 = 1.7 \times 10^{-7}$$

$$K_{sp} = [Mg^{2+}][OH^-]^2 = 1.5 \times 10^{-11}$$

The ion product, $Q_{sp} = 1.7 \times 10^{-7}$, is greater than K_{sp} for $Mg(OH)_2$, and therefore $Mg(OH)_2$ precipitates until its K_{sp} is just satisfied.

Example 18–17 demonstrates how the amount of a weak base required to initiate precipitation of an insoluble metal hydroxide may be calculated.

Example 18–17

What concentration of aqueous ammonia is necessary to just initiate precipitation of $Mg(OH)_2$ from a 0.10 M solution of $Mg(NO_3)_2$? Refer to Example 18–16.

Solution

Two equilibria must be considered:

$Mg(OH)_2$ (s) $\rightleftharpoons$ Mg^{2+} (aq) $+ 2OH^-$ (aq)
$K_{sp} = 1.5 \times 10^{-11}$

NH_3 (aq) $+ H_2O \rightleftharpoons NH_4^+$ (aq) $+ OH^-$ (aq)
$K_b = 1.8 \times 10^{-5}$

First we determine the $[OH^-]$ necessary to initiate precipitation of $Mg(OH)_2$ from 0.10 M $Mg(NO_3)_2$. Since $Mg(NO_3)_2$ is completely dissociated, $[Mg^{2+}] = 0.10 \, M$.

$$K_{sp} = 1.5 \times 10^{-11} = [Mg^{2+}][OH^-]^2$$

$$[OH^-]^2 = \frac{1.5 \times 10^{-11}}{[Mg^{2+}]} = \frac{1.5 \times 10^{-11}}{0.10}$$

$$= 1.5 \times 10^{-10}$$

$$[OH^-] = 1.2 \times 10^{-5} \, M$$

Now we determine the concentration of NH_3 that will supply $1.2 \times 10^{-5} \, M$ OH^-. If we let x be the original concentration of NH_3, the equilibrium concentrations are

NH_3 (aq) $+ H_2O \rightleftharpoons NH_4^+$ (aq) $+ OH^-$ (aq)
$(x - 1.2 \times 10^{-5}) \, M$ $\qquad 1.2 \times 10^{-5} \, M \quad 1.2 \times 10^{-5} \, M$

$$K_b = \frac{[NH_4^+][OH^-]}{[NH_3]}$$

$$1.8 \times 10^{-5} = \frac{(1.2 \times 10^{-5})(1.2 \times 10^{-5})}{x - 1.2 \times 10^{-5}}$$

$$1.44 \times 10^{-10} = 1.8 \times 10^{-5} \, x - 2.16 \times 10^{-10}$$

$$1.8 \times 10^{-5} \, x = 3.6 \times 10^{-10}$$

$$x = 2.0 \times 10^{-5} \, M = \text{initial } [NH_3]$$

Thus, the solution must be ever so slightly greater than $2.0 \times 10^{-5} \, M$ in NH_3 to initiate precipitation of $Mg(OH)_2$ from a 0.10 M solution of $Mg(NO_3)_2$.

On occasion, solutions containing a weak base are buffered to decrease their basicity so that a significant concentration of a metal ion may exist in the solution. Example 18–18 illustrates this.

Example 18–18

How many moles of ammonium chloride, NH_4Cl, must be added to 1.00 liter of solution that is to be made 0.10 M in $Mg(NO_3)_2$ *and* 0.10 M in ammonia to prevent precipitation of $Mg(OH)_2$?

Solution

Note that the concentrations of both $Mg(NO_3)_2$ and NH_3 are the same as in Example 18–16, and we demonstrated that $Mg(OH)_2$ precipitates from that solution. The buffering action of NH_4Cl in the presence of NH_3 decreases the concentration of OH^-. Again, two equilibria must be considered:

$Mg(OH)_2$ (s) $\rightleftharpoons$ Mg^{2+} (aq) $+ 2OH^-$ (aq)

NH_3 (aq) $+ H_2O \rightleftharpoons NH_4^+$ (aq) $+ OH^-$ (aq)

The concentration of Mg^{2+} cannot be varied. From Example 18–17 we know that $[OH^-]$ cannot exceed $1.2 \times 10^{-5} \, M$ in this solution.

K_b for aqueous NH_3 is used to calculate the number of moles of NH_4Cl necessary to buffer 0.10 M aqueous NH_3 so that $[OH^-] = 1.2 \times 10^{-5} \, M$. (You may wish to refer back to Section 18–3 to refresh your memory on buffer solutions.) Let $x = $ number of moles of NH_4Cl required:

$$NH_4Cl \ (aq) \longrightarrow NH_4^+ \ (aq) \ + Cl^- \ (aq)$$
$$ x \ M x \ M x \ M$$

$$NH_3 \ (aq) + H_2O \rightleftharpoons NH_4^+ \ (aq) \ + \ OH^- \ (aq)$$
$$(0.10 - 1.2 \times 10^{-5} \ M) 1.2 \times 10^{-5} \ M 1.2 \times 10^{-5} \ M$$

$$K_b = \frac{[NH_4^+][OH^-]}{[NH_3]}$$

$$1.8 \times 10^{-5} = \frac{(x + 1.2 \times 10^{-5})(1.2 \times 10^{-5})}{(0.10 - 1.2 \times 10^{-5})}$$

Two simplifying assumptions, $(x + 1.2 \times 10^{-5}) \approx x$ and $(0.10 - 1.2 \times 10^{-5}) \approx 0.10$, can be made to give

$$\frac{(x)(1.2 \times 10^{-5})}{0.10} = 1.8 \times 10^{-5}$$

$x = 0.15$ mol of NH_4^+ per liter of solution

Addition of 0.15 mol/L of NH_4Cl to 0.10 M aqueous NH_3 decreases $[OH^-]$ to $1.2 \times 10^{-5} \ M$, so that K_{sp} for $Mg(OH)_2$ is not exceeded in this solution.

We have just demonstrated (Examples 18–16, 18–17, 18–18) the fact that when more than one equilibrium must be considered to describe a solution, *all relevant equilibria must be satisfied.* Suppose we are asked to calculate $[H^+]$ and the pH of the solution described in Example 18–18. From the solution of the problem we know that

$$[OH^-] = 1.2 \times 10^{-5} \ M$$

We can use the ion product for water (another equilibrium constant that must be satisfied) to calculate $[H^+]$ and pH.

$$[H^+][OH^-] = 1.0 \times 10^{-14}$$

$$[H^+] = \frac{1.0 \times 10^{-14}}{[OH^-]} = \frac{1.0 \times 10^{-14}}{1.2 \times 10^{-5}} = 8.3 \times 10^{-10} \ M \ H^+$$

Since $[H^+] = 8.3 \times 10^{-10} \ M$, then pH = 9.08.

18–11   Dissolution of Precipitates

Precipitates can be dissolved by one or more of the three types of reactions outlined below. Dissolution of a precipitate is accomplished by reducing the concentrations of the constituent ions so that K_{sp} is no longer exceeded; i.e., when $Q_{sp} < K_{sp}$ then a precipitate dissolves until the concentrations of the ions just satisfy K_{sp}.

Recall that solubility products, like other equilibrium constants, are thermodynamic quantities. They tell us nothing about how fast a given reaction occurs, only that it can, or cannot, occur under the specified conditions.

1   Converting an Ion to a Weak Electrolyte

Typical examples are described below:

a. Acidification of insoluble aluminum hydroxide in contact with its saturated solution converts OH^- ions to the weak electrolyte, water, until $[Al^{3+}][OH^-]^3 < K_{sp}$, and dissolution occurs.

$$Al(OH)_3 \ (s) \rightleftharpoons Al^{3+} \ (aq) + 3OH^- \ (aq)$$
$$\underline{3H^+ \ (aq) + 3OH^- \ (aq) \longrightarrow 3H_2O}$$
overall reaction $\quad Al(OH)_3 \ (s) + 3H^+ \ (aq) \longrightarrow Al^{3+} \ (aq) + 3H_2O$

b. Treatment of magnesium hydroxide, in contact with its saturated solution, with ammonium ion from a salt, such as ammonium chloride, NH_4Cl, converts OH^- ions to the weak electrolytes, ammonia and water, with the result that $[Mg^{2+}][OH^-]^2 < K_{sp}$, and dissolution occurs.

$$Mg(OH)_2 \text{ (s)} \rightleftharpoons Mg^{2+} \text{ (aq)} + 2OH^- \text{(aq)}$$
$$2NH_4^+ \text{ (aq)} + 2OH^- \text{ (aq)} \longrightarrow 2NH_3 \text{ (aq)} + 2H_2O$$

overall reaction $\overline{Mg(OH)_2 \text{ (s)} + 2NH_4^+ \text{ (aq)} \longrightarrow Mg^{2+} \text{ (aq)} + 2NH_3 \text{ (aq)} + 2H_2O}$

Observe that this process, the dissolution of $Mg(OH)_2$ in an NH_4Cl solution, is the reverse of the reaction we considered in Example 18–16, the precipitation of $Mg(OH)_2$ from a solution of aqueous ammonia.

c. Acidification of some metal sulfides, for example manganese(II) sulfide, $K_{sp} = 5.1 \times 10^{-15}$, converts sulfide ions from the saturated solution to hydrogen sulfide, a weak electrolyte, with the result that $[Mn^{2+}][S^{2-}] < K_{sp}$ and MnS dissolves.

$$MnS \text{ (s)} \rightleftharpoons Mn^{2+} \text{ (aq)} + S^{2-} \text{ (aq)}$$
$$2H^+ \text{ (aq)} + S^{2-} \text{ (aq)} \longrightarrow H_2S \text{ (g)}$$

overall reaction $\overline{MnS \text{ (s)} + 2H^+ \text{ (aq)} \longrightarrow Mn^{2+} \text{ (aq)} + H_2S \text{ (g)}}$

The latter technique is effective only for the more soluble of the "insoluble" metal sulfides, since the very insoluble sulfides do not produce sufficiently high concentrations of sulfide ions in their saturated solutions to react with even the strongest acids.

2   Converting an Ion to Another Species by Oxidation-Reduction Reaction

Most insoluble metal sulfides can be dissolved in hot dilute nitric acid because nitrate ions oxidize sulfide ions to elemental sulfur, thereby removing sulfide ions from the solution.

$$3S^{2-} \text{ (aq)} + 2NO_3^- \text{ (aq)} + 8H^+ \text{ (aq)} \longrightarrow 3S \text{ (s)} + 2NO \text{ (g)} + 4H_2O \text{ (}\ell\text{)}$$

Consider copper(II) sulfide, CuS, as a typical metal sulfide in equilibrium with its ions. This equilibrium strongly favors solid CuS, since $K_{sp} = 8.7 \times 10^{-36}$. However, removal of the sulfide ions by oxidation to elemental sulfur causes CuS (s) to dissolve in hot dilute HNO_3.

$$CuS \text{ (s)} \rightleftharpoons Cu^{2+} \text{ (aq)} + S^{2-} \text{ (aq)}$$

$$3S^{2-} \text{ (aq)} + 2NO_3^- \text{ (aq)} + 8H^+ \text{ (aq)} \longrightarrow 3S \text{ (s)} + 2NO \text{ (g)} + 4H_2O \text{ (}\ell\text{)}$$

Multiplication of the first equation by 3, followed by addition of these two equations, gives the net ionic equation for the dissolution of CuS (s) in hot dilute HNO_3.

$$3CuS \text{ (s)} + 2NO_3^- \text{ (aq)} + 8H^+ \text{ (aq)} \longrightarrow 3Cu^{2+} \text{ (aq)} + 3S \text{ (s)} + 2NO \text{ (g)} + 4H_2O \text{ (}\ell\text{)}$$

3   Complex Ion Formation

Ligand is the name given to an atom or a group of atoms bonded to the central element in complex ions.

Many slightly soluble compounds contain cations that are capable of forming soluble compounds that contain complex ions. They do this by accepting shares in electron pairs from such molecules or ions as NH_3, CN^-, OH^-, SCN^-, F^-, Cl^-, Br^-, I^-, and so on. Coordinate covalent bonds are formed as these *ligands* replace water molecules from hydrated metal ions (Section 13–2). Some complex ions and their dissociation constants, K_d, are given in Appendix I.

The smaller K_d is, the more effectively the ligand competes with water for a coordination site on the metal ion; stated differently, the smaller the K_d, the more stable the complex ion is. Many copper(II) compounds react with excess aqueous ammonia to form the deep blue colored complex ion $Cu(NH_3)_4^{2+}$ (aq). The dissociation reaction of this ion is:

$$Cu(NH_3)_4^{2+} \rightleftharpoons Cu^{2+} (aq) + 4NH_3 (aq)$$

$$K_d = \frac{[Cu^{2+}][NH_3]^4}{[Cu(NH_3)_4^{2+}]} = 8.5 \times 10^{-13}$$

Recall that Cu^{2+} (aq) is really a hydrated ion, $Cu(OH_2)_4^{2+}$. We could write the above reaction more accurately as

$$Cu(NH_3)_4^{2+} + 4H_2O \rightleftharpoons Cu(OH_2)_4^{2+} + 4NH_3$$

$$K_d = \frac{[Cu(OH_2)_4^{2+}][NH_3]^4}{[Cu(NH_3)_4^{2+}]} = 8.5 \times 10^{-13}$$

Example 18–19

What are the concentrations of hydrated Cu^{2+}, NH_3, and $Cu(NH_3)_4^{2+}$ in a 0.20 M solution of $Cu(NH_3)_4SO_4$?

Solution

The soluble deep blue complex salt, $Cu(NH_3)_4SO_4$, dissociates completely to produce tetraamminecopper(II) ions and sulfate ions:

$$Cu(NH_3)_4SO_4 \xrightarrow{\;100\%\;} Cu(NH_3)_4^{2+} + SO_4^{2-}$$
$$\quad 0.20\ M \qquad\qquad 0.20\ M \qquad 0.20\ M$$

Some of the $Cu(NH_3)_4^{2+}$ ions then dissociate. Let x be the concentration of $Cu(NH_3)_4^{2+}$ that dissociates:

$$Cu(NH_3)_4^{2+} \rightleftharpoons Cu^{2+} (aq) + 4NH_3$$
$$(0.20 - x)\ M \qquad x\ M \qquad 4x\ M$$

$$K_d = \frac{[Cu^{2+}][NH_3]^4}{[Cu(NH_3)_4^{2+}]} = 8.5 \times 10^{-13}$$

$$8.5 \times 10^{-13} = \frac{x(4x)^4}{0.20 - x} \qquad \text{Assume } (0.20 - x) \approx 0.20.$$

$$= \frac{256\ x^5}{0.20}$$

$$x^5 = 6.6 \times 10^{-16}$$

Taking the fifth root of both sides of this equation gives:

$$x = 9.2 \times 10^{-4}\ M = [Cu^{2+}]$$

$$[NH_3] = 4x = 3.7 \times 10^{-3}\ M$$

$$[Cu(NH_3)_4^{2+}] = (0.20 - x)M = 0.20\ M$$

Our assumption, $(0.20 - x) \approx 0.20$, is valid since 9.2×10^{-4} is much less than 0.20.

The dissolution of solid zinc hydroxide in excess sodium hydroxide solution to form the complex ion $Zn(OH)_4^{2-}$ illustrates the amphoteric nature of $Zn(OH)_2$ (Section 9–6.2).

$$Zn(OH)_2 (s) + 2OH^- \rightleftharpoons Zn(OH)_4^{2-}$$

Example 18–20

Some solid $Zn(OH)_2$ is suspended in a saturated solution of $Zn(OH)_2$. A solution of sodium hydroxide is added until all the $Zn(OH)_2$ just dissolves. The pH of the solution is 11.80. What are the concentrations of Zn^{2+} and $Zn(OH)_4^{2-}$ in the solution? K_{sp} for $Zn(OH)_2 = 4.5 \times 10^{-17}$, and K_d for $Zn(OH)_4^{2-} = 3.5 \times 10^{-16}$.

Solution

The equilibria of interest are

$$Zn(OH)_2 (s) \rightleftharpoons Zn^{2+} (aq) + 2OH^- (aq)$$
$$K_{sp} = 4.5 \times 10^{-17}$$

$$Zn(OH)_4^{2-} \text{ (aq)} \rightleftharpoons Zn^{2+} \text{ (aq)} + 4OH^- \text{ (aq)}$$
$$K_d = 3.5 \times 10^{-16}$$

Since we know the pH, we can calculate $[OH^-]$ from pH + pOH = 14.00.

$$pOH = 14.00 - pH = 14.00 - 11.80 = 2.20$$

$$[OH^-] = 10^{-pOH} = 10^{-2.20} = 10^{+0.80} \times 10^{-3}$$
$$= 6.3 \times 10^{-3} M = [OH^-]$$

We can use the solubility product relationship to calculate $[Zn^{2+}]$.

$$K_{sp} = 4.5 \times 10^{-17} = [Zn^{2+}][OH^-]^2$$
$$4.5 \times 10^{-17} = [Zn^{2+}](6.3 \times 10^{-3})^2$$

$$[Zn^{2+}] = \frac{4.5 \times 10^{-17}}{(6.3 \times 10^{-3})^2} = \underline{1.1 \times 10^{-12} M}$$

Since both equilibria are established in the same solution, the same $[Zn^{2+}]$ and $[OH^-]$ also satisfy the complex ion dissociation equilibrium.

$$K_d = 3.5 \times 10^{-16} = \frac{[Zn^{2+}][OH^-]^4}{[Zn(OH)_4^{2-}]}$$

$$[Zn(OH)_4^{2-}] = \frac{[Zn^{2+}][OH^-]^4}{3.5 \times 10^{-16}}$$

$$= \frac{(1.1 \times 10^{-12})(6.3 \times 10^{-3})^4}{3.5 \times 10^{-16}}$$

$$\underline{[Zn(OH)_4^{2-}] = 5.0 \times 10^{-6} M}$$

Key Terms

Buffer solution solution that resists change in pH; contains either a weak acid and soluble ionic salt of the acid, or a weak base and soluble ionic salt of the base.

Common ion effect suppression of ionization of a weak electrolyte by the presence in the same solution of a strong electrolyte containing one of the same ions as the weak electrolyte.

Complex ion ion resulting from the formation of coordinate covalent bonds between simple cations and other ions or molecules.

Dissociation constant equilibrium constant that applies to the dissociation of a complex ion into a simple ion and coordinating species (ligands).

End point the point at which an indicator changes color and a titration is stopped.

Equivalence point the point at which chemically equivalent amounts of reactants have reacted.

Fractional precipitation removal of some ions from solution by precipitation while leaving other ions with similar properties in solution.

Indicators for acid-base titrations, organic compounds that exhibit different colors in solutions of different acidities; used to determine the point at which reaction between two solutes is complete.

Insoluble compound actually a very slightly soluble compound.

Molar solubility number of moles of a solute that dissolve to produce a liter of saturated solution.

Precipitate solid formed by mixing in solution the constituent ions of an insoluble compound.

Solubility product constant equilibrium constant that applies to the dissolution of a slightly soluble compound.

Solubility product principle the solubility product constant expression for a slightly soluble compound is the product of the concentrations of the constituent ions, each raised to the power that corresponds to the number of ions in one formula unit.

Standard solution a solution of accurately known concentration.

Titration procedure in which a solution of known concentration, a standard solution, is added to a solution of unknown concentration until the chemical reaction between the two solutes is complete; yields concentration of unknown solution.

Titration curve for acid-base titration, a plot of pH versus volume of acid or base added.

Exercises

The Common Ion Effect and Buffer Solutions

1. Define and illustrate the following terms clearly: (a) common ion effect, (b) buffer solution.

2. List two frequently encountered kinds of buffer solutions. List two specific examples of each kind.

3. Consider solutions that contain the indicated concentrations of the following pairs of compounds. Which ones are buffer solutions? Why?
   (a) 0.10 M HF and 0.10 M NaF
   (b) 0.10 M HCl and 0.10 M NaCl
   (c) 0.050 M HClO and 0.10 M NaClO
   (d) 0.10 M HClO$_4$ and 0.050 M KClO$_4$
   (e) 0.10 M NH$_3$ and 0.10 M NH$_4$Cl
   (f) 0.10 M NH$_3$ and 0.50 M NH$_4$NO$_3$

4. Calculate the hydronium ion concentration and the pH of the following solutions. After you have done the calculations, refer to Exercise 36 of Chapter 17 and compare [H$_3$O$^+$] and pH for solutions containing only the acids.
   (a) 0.10 M HCN and 0.10 M NaCN
   (b) 0.10 M HF and 0.10 M NaF

5. (a) Calculate the ratio of concentrations [CH$_3$COOH]/[NaCH$_3$COO] that gives solutions with pH = 5.00.
   (b) Calculate the ratio of concentrations [CH$_3$COOH]/[NaCH$_3$COO] that gives solutions with pH = 4.74.

6. Calculate pH for each of the following buffer solutions:
   (a) 0.10 M HCNO and 0.20 M KCNO
   (b) 0.050 M HCN and 0.025 M Ba(CN)$_2$

7. Calculate the concentration of OH$^-$ and the pH of each of the following solutions:
   (a) 0.10 M aq. NH$_3$ and 0.10 M NH$_4$NO$_3$
   (b) 0.20 M aq. NH$_3$ and 0.050 M (NH$_4$)$_2$SO$_4$
   (c) 0.10 M pyridine and 0.10 M pyridinium chloride, C$_5$H$_5$NHCl.

8. Calculate the concentration of OH$^-$ and pH for the following buffer solutions:
   (a) 0.10 M aq. NH$_3$ and 0.20 M NH$_4$NO$_3$
   (b) 0.10 M aq. NH$_3$ and 0.10 M (NH$_4$)$_2$SO$_4$

9. Calculate the ratio of concentrations [NH$_3$]/[NH$_4$$^+$] that gives
   (a) solutions of pH = 9.00.
   (b) solutions of pH = 9.25.

10. Calculate the change in [H$_3$O$^+$] and pH for the following buffer solutions:
    (a) 0.010 mole of HCl is added to one liter of a solution that is 0.10 M in CH$_3$COOH and 0.10 M in NaCH$_3$COO. Assume no volume change.
    (b) 0.010 mole of HNO$_3$ is added to 500 mL of a solution that is 0.10 M in aqueous ammonia and 0.20 M in NH$_4$NO$_3$. Assume no volume change.
    (c) 0.010 mole of NaOH is added to 500 mL of a solution that is 0.10 M in aqueous ammonia and 0.20 M in NH$_4$Cl. Assume no volume change.

11. (a) What is the pH of a 0.20 M formic acid solution, HCO$_2$H? $K_a = 1.8 \times 10^{-4}$.
    (b) What is the pH of a solution containing 0.20 M formic acid and 0.10 M sodium formate, NaCO$_2$H?
    (c) What is the pH of a solution prepared by adding 50 mL of 0.010 M NaOH to 100 mL of 0.20 M formic acid?
    (d) What is the pH of a solution prepared by adding 50 mL of 0.010 M NaOH to 100 mL of a solution containing 0.20 M HCO$_2$H and 0.10 M NaCO$_2$H?
    (e) What is the pH of a solution prepared by adding 50 mL of 0.10 M HCl to 100 mL of a solution containing 0.20 M HCO$_2$H and 0.10 M NaCO$_2$H?

12. If 100 mL of 0.020 M HCl and 100 mL of 0.10 M NaCN solutions are mixed, what will the pH of the resulting solution be? Would this be wise? Why?

13. Calculate the pH of each of the following solutions:
    (a) 100 mL of 0.100 M HCl mixed with 50.0 mL of 0.100 M KOH.
    (b) 100 mL of 0.100 M HCl mixed with 100 mL of 0.100 M KOH.
    (c) 100 mL of 0.100 M HCl mixed with 150 mL of 0.100 M KOH.
    (d) 100 mL of 0.100 M CH$_3$COOH mixed with 50.0 mL of 0.100 M NaOH.

14. Calculate [H$_3$O$^+$], [OH$^-$], pH, and pOH for the following solutions:
    (a) 50 mL of 2.0 M aqueous NH$_3$.
    (b) 50 mL of a solution containing 2.0 M aqueous NH$_3$ and 1.0 M NH$_4$Cl.
    (c) 50 mL of a solution containing 2.0 M aqueous NH$_3$ and 1.0 M (NH$_4$)$_2$SO$_4$.

(d) A solution prepared by adding 10 mL of 1.0 M HNO_3 to the solution in (a).

(e) A solution prepared by adding 10 mL of 1.0 M HNO_3 to the solution in (b).

(f) A solution prepared by adding 10 mL of 1.0 M KOH to the solution in (c).

15. How many grams of NH_4Cl must be added to 100 mL of 4.0 M aqueous NH_3 in order to prepare a buffer solution having a pH of 10.20?

16. How many grams of sodium acetate must be added to 200 mL of 6.0 M CH_3COOH to prepare a buffer solution having a pH of 3.62?

17. How much 0.20 M NaF solution would have to be added to 100 mL of 0.10 M HF to give a solution of pH = 3.00?

18. How much 0.10 M NaCNO solution would have to be added to 100 mL of 0.10 M HCNO to give a solution of pH = 4.00?

Titration and Titration Curves

19. What is a titration?

20. (a) What are acid-base indicators? (b) What is the essential characteristic of acid-base indicators? (c) What determines the color of an acid-base indicator in an aqueous solution?

21. Define and distinguish between end point and equivalence point in an acid-base titration.

22. What colors do the following indicators exhibit in solutions of the indicated pH?

	pH = 2.00	pH = 11.00
methyl red	_____	_____
neutral red	_____	_____
phenolphthalein	_____	_____

23. Demonstrate mathematically that bromthymol blue is yellow in solutions of pH 4.00, while it is blue in solutions of pH 9.00. Its ionization constant is 1×10^{-7}.

24. What are titration curves?

25. Construct a table similar to Table 18–5 for the titration of 25.0 mL of 0.100 M HNO_3 solution by 0.100 M KOH solution. Calculate the pH of the solution before any KOH is added and after the addition of the following amounts of 0.100 M KOH: (a) 5.0 mL, (b) 10.0 mL, (c) 12.5 mL, (d) 20.0 mL, (e) 24.0 mL, (f) 24.9 mL, (g) 25.0 mL, (h) 25.1 mL, (i) 27.0 mL, and (j) 30.0 mL. (1) Plot pH vs. mL of KOH added. (2) Does this titration curve resemble the curve in Figure 18–1? Why? (3) Over what pH range could indicators change colors for this titration?

26. Construct a table similar to Table 18–5 for the titration of 25.0 mL of 0.100 M NaOH solution by 0.100 M $HClO_4$ solution. Calculate the pH of the solution before any $HClO_4$ is added and after the addition of the following amounts of 0.100 M $HClO_4$: (a) 5.0 mL, (b) 10.0 mL, (c) 12.5 mL, (d) 20.0 mL, (e) 24.0 mL, (f) 24.9 mL, (g) 25.0 mL, (h) 25.1 mL, (i) 27.0 mL, and (j) 30.0 mL. (1) Plot pH vs. mL of $HClO_4$ added. (2) Does this titration curve resemble the curve in Figure 18–2? Why? (3) Over what pH range could indicators change colors for this titration?

For Exercises 27–30, calculate $[H_3O^+]$, $[OH^-]$, pH, and pOH at the indicated points. In each case assume that pure acid (or base) is added to exactly one liter of 0.0100 molar solution of the indicated base (or acid). This simplifies the arithmetic because we may assume that the volume of each solution is constant throughout the titration. Plot each titration curve with pH on the vertical axis and moles of base (or acid) added on the horizontal axis.

27. *Titration Curves—I: HCl vs. NaOH*

Solid NaOH is added to one liter of 0.0100 M HCl solution. Consult Table 18–4 and list the indicators that could be used satisfactorily in this titration.

Total Moles NaOH Added	$[H^+]$	$[OH^-]$	pH	pOH
none	_____	_____	_____	_____
0.00100	_____	_____	_____	_____
0.00300	_____	_____	_____	_____
0.00500 (50%)	_____	_____	_____	_____
0.00700	_____	_____	_____	_____
0.00900	_____	_____	_____	_____
0.00950	_____	_____	_____	_____
0.0100 (100%)	_____	_____	_____	_____
0.0105	_____	_____	_____	_____
0.0120	_____	_____	_____	_____
0.0150 (50% excess NaOH)	_____	_____	_____	_____

28. *Titration Curves — II: CH₃COOH vs. NaOH*

    Solid NaOH is added to one liter of 0.0100 M CH₃COOH solution. Consult Table 18–4 and list the indicators that could be used satisfactorily in this titration.

Total Moles NaOH Added	[H⁺]	[OH⁻]	pH	pOH
none	___	___	___	___
0.00200	___	___	___	___
0.00400	___	___	___	___
0.00500 (50%)	___	___	___	___
0.00700	___	___	___	___
0.00900	___	___	___	___
0.00950	___	___	___	___
0.0100 (100%)	___	___	___	___
0.0105	___	___	___	___
0.0120	___	___	___	___
0.0150 (50% excess NaOH)	___	___	___	___

29. *Titration Curves — III: HCl vs. aqueous NH₃*

    Gaseous HCl is added to one liter of 0.0100 M aqueous ammonia solution. Consult Table 18–4 and list the indicators that could be used satisfactorily in this titration.

Total Moles HCl Added	[H⁺]	[OH⁻]	pH	pOH
none	___	___	___	___
0.00100	___	___	___	___
0.00300	___	___	___	___
0.00500 (50%)	___	___	___	___
0.00700	___	___	___	___
0.00900	___	___	___	___
0.00950	___	___	___	___
0.0100 (100%)	___	___	___	___
0.0105	___	___	___	___
0.0120	___	___	___	___
0.0150 (50% excess HCl)	___	___	___	___

30. *Titration Curves — IV: NH₃ vs. CH₃COOH*

    Gaseous NH₃ is added to one liter of 0.0100 M CH₃COOH solution.

Total Moles NH₃ Added	[H⁺]	[OH⁻]	pH	pOH
none	___	___	___	___
0.00100	___	___	___	___
0.00400	___	___	___	___
0.00500 (50%)	___	___	___	___
0.00900	___	___	___	___
0.00950	___	___	___	___
0.0100 (100%)	___	___	___	___
0.0105	___	___	___	___
0.0130	___	___	___	___

(a) What is the major difference between the titration curve for the reaction of CH₃COOH and NH₃ and the other curves that you have plotted?

(b) Consult Table 18–4. Can you suggest a satisfactory indicator for this titration? Why?

Slightly Soluble Compounds

31. (a) Are "insoluble" substances really insoluble?
    (b) What do we mean when we refer to insoluble substances?
32. State the solubility product principle. What does it mean?

Solubility Product Constants

33. (a) Why are solubility product constant expressions written as products of concentrations of ions raised to appropriate powers?
    (b) What does the term "concentration of a solid" mean when we discuss an equilibrium involving a slightly soluble compound such as $BaSO_4$ dissolving in water to form a saturated solution?
34. Write an equation for the dissolution of, and the solubility product constant expression for, each of the following slightly soluble compounds. (a) AgCl, (b) $CaCO_3$, (c) $AlPO_4$, (d) Ag_2S, (e) BaF_2, (f) $Sn(OH)_2$, (g) $Fe(OH)_3$, (h) Sb_2S_3
35. What do we mean when we refer to the molar solubility of a compound?

Experimental Determination of K_{sp}'s

36. From the solubility data given for the following compounds, calculate their solubility product constants. Your calculated values may not agree exactly with the solubility products given in Appendix H because round-off errors become large in calculations to two significant figures. Also, there is some disagreement within the scientific community over the exact values for some solubility products. These solubility data were taken from a handbook of chemistry and physics.
    (a) AgBr, silver bromide, 2.7×10^{-6} g/L
    (b) $BaCrO_4$, barium chromate, 0.00041 g/100 mL
    (c) $CaCO_3$, calcium carbonate, 0.0014 g/10 mL
    (d) CaF_2, calcium fluoride, 0.016 g/L
37. Calculate molar solubilities, concentrations of constituent ions, and solubilities in g/L for the following compounds at 25°C. (Refer to Appendix H for solubility product constants.)
    (a) CuI, copper(I) iodide
    (b) BaF_2, barium fluoride
    (c) Ag_2SO_4, silver sulfate
    (d) $Ag_4Fe(CN)_6$, silver hexacyanoferrate(II)

38. Construct a table similar to Table 18–7 for the compounds listed in Exercise 37. Which compound has (a) the highest molar solubility, (b) the lowest molar solubility, (c) the highest solubility expressed in g/L, and (d) the lowest solubility expressed in g/L?

Uses of K_{sp}'s

39. Calculate the number of moles and the mass of:
    (a) CuBr that will dissolve in one liter of 0.010 M HBr.
    (b) Ag_2CrO_4 that will dissolve in one liter of 0.10 M $AgNO_3$.
    (c) Ag_2CrO_4 that will dissolve in one liter of 0.10 M Na_2CrO_4.
40. If the following volumes of the indicated solutions are mixed, will a precipitate form? If so, do you expect to be able to see the solid?
    (a) 100 mL of 0.20 M KCl and 100 mL of 0.50 M Na_2S
    (b) 100 mL of 0.00050 M $AgNO_3$ and 100 mL of 0.0020 M NaCl
    (c) 200 mL of 0.015 M $Pb(NO_3)_2$ and 100 mL of 0.030 M KI
    (d) 20 mL of 0.0015 M $AgNO_3$ and 10 mL of 0.0030 M Na_3PO_4
41. Suppose you have beakers that contain 100 mL of the following solutions:
    (a) 0.0015 M KOH   (b) 0.0015 M K_2CO_3
    (c) 0.0015 M K_2S
    If solid copper(II) nitrate is added to each beaker, what concentration of Cu^{2+} is required to initiate precipitation? If solid copper(II) nitrate were added to each beaker until $[Cu^{2+}] = 0.0015\ M$, what concentrations of OH^-, CO_3^{2-} and S^{2-} would remain in solution, i.e., unprecipitated?
42. Suppose we wish to recover silver from an aqueous solution that contains a soluble silver compound such as $AgNO_3$ by precipitating silver ions in the form of an insoluble compound such as AgCl.
    (a) How much sodium chloride should we add to reduce the silver ion concentration to $1.0 \times 10^{-6}\ M$?
    (b) What mass of Ag^+ remains in one liter of solution?
    (c) What volume of the solution contains one gram of Ag^+?
43. If silver ions are to be removed from solution by precipitation of Ag_2S, what final concentration of sulfide ions is required to reduce the Ag^+

concentration to 1.0×10^{-10} molar? What volume of this solution contains 1.0 g of Ag^+?

44. If calcium ions are to be removed from a city water supply by the addition of soda ash, Na_2CO_3, what final concentration of CO_3^{2-} is required to reduce $[Ca^{2+}]$ in the water to 0.000010 M? What volume of this solution contains 1.0 g of Ca^{2+}? Assume no dilution due to the addition of solid soda ash.

45. If solid NaOH is added to 0.020 M $Mg(NO_3)_2$ until the pH reaches 11.00, what concentration of Mg^{2+} will remain in solution? What percentage of the Mg^{2+} will be precipitated? Assume no volume change due to the addition of solid NaOH.

Fractional Precipitation

46. What is fractional precipitation?

47. Solid Na_2SO_4 is added slowly to a solution that is 0.10 M in $Pb(NO_3)_2$ and 0.10 M in $Ba(NO_3)_2$. In what order will solid $PbSO_4$ and $BaSO_4$ form? Calculate the percentage of Ba^{2+} that precipitates just before $PbSO_4$ begins to precipitate.

48. To a solution that is 0.010 M in Cu^+, 0.010 M Ag^+, and 0.010 M in Au^+, *solid* NaCl is added slowly. Assume no volume change due to the addition of solid NaCl.
    (a) Which compound will begin to precipitate first?
    (b) Calculate $[Au^+]$ when AgCl just begins to precipitate. What percentage of the Au^+ has precipitated at this point?
    (c) Calculate $[Au^+]$ and $[Ag^+]$ when CuCl just begins to precipitate.

49. Solid $Pb(NO_3)_2$ is added slowly to a solution that is 0.010 M each in NaOH, K_2CO_3, and Na_2SO_4. In what order will solid $Pb(OH)_2$, $PbCO_3$ and $PbSO_4$ form? Calculate the percentage of OH^- and CO_3^{2-} that precipitate just before $PbSO_4$ begins to precipitate.

Simultaneous Equilibria

50. If a solution is made 0.10 M in $Mg(NO_3)_2$, 1.0 M in aqueous ammonia, and 3.0 M in NH_4NO_3, will $Mg(OH)_2$ precipitate? What is the pH of this solution?

51. If a solution is made 0.10 M in $Mg(NO_3)_2$, 1.0 M in aqueous ammonia, and 0.10 M in NH_4NO_3, will $Mg(OH)_2$ precipitate? What is the pH of this solution?

52. (a) What is the minimum concentration of ammonium nitrate necessary to prevent precipitation of $Mg(OH)_2$ in the solution described in Exercise 51?
    (b) What mass of ammonium nitrate is this in one liter of solution?
    (c) What is the pH of this solution if it contains the minimum concentration of NH_4NO_3 necessary to prevent the precipitation of $Mg(OH)_2$?

53. If a solution is 0.010 M in manganese(II) nitrate, $Mn(NO_3)_2$, and 0.10 M in aqueous ammonia, will manganese(II) hydroxide, $Mn(OH)_2$, precipitate?

54. If a solution is 1.0×10^{-5} M in $Mn(NO_3)_2$ and 1.0×10^{-3} M in aqueous ammonia, will $Mn(OH)_2$ precipitate?

55. What concentration of NH_4NO_3 is necessary to prevent precipitation of $Mn(OH)_2$ in the solution of Exercise 53?

Dissolution of Precipitates

56. Explain, by writing appropriate equations, how the following insoluble compounds can be dissolved by the addition of a solution of nitric acid. (Carbonates dissolve in strong acids to form carbon dioxide, which is evolved as a gas, and water.) What is the "driving force" for each reaction? (a) $Fe(OH)_2$, (b) $Fe(OH)_3$, (c) $Co(OH)_2$, (d) $PbCO_3$, (e) $(CuOH)_2CO_3$

57. Explain, by writing equations, how the following insoluble compounds can be dissolved by the addition of a solution of ammonium nitrate or ammonium chloride.
    (a) $Mg(OH)_2$, (b) $Mn(OH)_2$, (c) $Ni(OH)_2$

58. The following insoluble sulfides can be dissolved in 3 M hydrochloric acid. Explain how this is possible and write the appropriate equations.
    (a) FeS, (b) ZnS

59. The following sulfides are less soluble than those listed in Exercise 58. These sulfides can be dissolved in hot 6 M nitric acid, an oxidizing acid. Explain how, and write the appropriate balanced equations.
    (a) PbS, (b) CuS, (c) Bi_2S_3

60. Which of the following insoluble metal hydroxides can be dissolved in 6 M sodium hydroxide? (You may wish to refer to Table 9–6.)
    (a) $Cd(OH)_2$, (b) $Mg(OH)_2$, (c) $Zn(OH)_2$, (d)

$Al(OH)_3$, (e) $Cr(OH)_3$, (f) $Ni(OH)_2$, (g) $Cu(OH)_2$

61. Calculate the concentration of complex ion, metal ion, and ammonia in the following solutions. The complex ions are enclosed in brackets. All these compounds are both soluble and ionic.

    (a) 0.10 M [Ag(NH$_3$)$_2$]Cl
    (b) 0.10 M [Cu(NH$_3$)$_4$]SO$_4$
    (c) 0.10 M [Co(NH$_3$)$_6$](ClO$_4$)$_3$

Electrochemistry

19

Electrochemistry deals with chemical changes produced by an electric current, and with the production of electricity by chemical reactions. By its nature, electrochemistry requires some method of introducing a stream of electrons into a reacting chemical system, and some means of withdrawing electrons. The reacting system is contained in a **cell** and an electric current is led in or out by **electrodes.**

There are two kinds of electrochemical cells, electrolytic cells and voltaic cells.

1. **Electrolytic cells** are those in which electrical energy causes *nonspontaneous* chemical reactions to occur.
2. **Voltaic cells** are those in which *spontaneous* chemical reactions produce electricity and supply it to an external circuit.

The study of electrochemistry has provided much of our knowledge about chemical reactions. The amount of electrical energy consumed or produced can be measured quite accurately. Additionally, the parts of an electrochemical reaction are separated physically, so that one half of a reaction occurs at one electrode while the other half occurs at the other electrode. All electrochemical reactions involve the transfer of electrons and are therefore *oxidation-reduction* reactions.

19–1 Electrical Conduction

Electrical current can be conducted through pure liquid electrolytes or solutions containing electrolytes, and through metal wires or along metallic surfaces. The latter type of conduction is called **metallic conduction.** It involves the flow of electrons along a metal surface with no similar movement of the atoms of the metal and no obvious changes in the metal. **Ionic** or **electrolytic conduction** refers to the conduction of electrical current by the motion of ions through a solution or a pure liquid. The positively charged ions (cations) migrate spontaneously toward the negative electrode, while the negatively charged ions (anions) move toward the positive electrode. Both kinds of conduction, ionic and metallic, occur in electrochemical cells (Figure 19–1).

19–2 Electrodes

Electrodes are usually metal surfaces upon which oxidation or reduction half-reactions occur. They may or may not participate in the reactions. Those that do not react

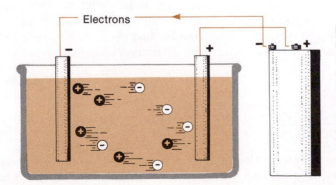

Electrons

FIGURE 19–1 The motion of ions through a solution is an electric current, and accounts for ionic (electrolytic) conduction. Positively charged ions move toward the negative electrode, while negatively charged ions move toward the positive electrode.

with the reactants or products are called **inert electrodes.** Regardless of the kind of cell, electrolytic or voltaic, the **cathode** is defined as the electrode at which *reduction* occurs as electrons are gained by some species. The **anode** is defined as the electrode at which *oxidation* occurs as electrons are lost by some species.

Electrolytic Cells

"Lysis" means "splitting apart." In many electrolytic cells, compounds are split into their constituent elements.

Electrolytic cells are electrochemical cells in which *nonspontaneous* chemical reactions are made to occur by the forced input of electrical energy. This process is called **electrolysis.** An electrolytic cell consists of a container for the reaction material, plus electrodes immersed in the reaction material and connected to a source of direct current. Inert electrodes are usually used, so that the electrodes are not involved in the chemical reaction.

We shall discuss several electrochemical cells in this chapter. For most of them, experimental observations will be presented from which we can deduce the electrode reactions and, from them, the overall reactions. We can then construct simplified diagrams of the cells.

19-3   The Electrolysis of Molten Sodium Chloride (The Downs Cell)

Electrons cannot migrate through the NaCl crystal the way they do in metals (Section 12–15).

Solid sodium chloride does not conduct electricity because, although its ions vibrate about fixed positions, they are not free to move throughout the lattice. However, molten (melted) sodium chloride (melting point, 801°C; a clear, colorless liquid that looks like water) is an excellent conductor because its ions are freely mobile. Consider a cell in which a source of direct current is connected by a wire to two inert graphite electrodes that are immersed in a container of sodium chloride heated above its melting point. When the current is flowing, we observe that:

1. A pale green gas, which is chlorine, Cl_2, is liberated at one electrode.
2. Molten, silvery-white metallic sodium, Na, forms at the other electrode and rises to the top of the molten sodium chloride.

From this information we can determine the essential features of the cell. Since chlorine is produced, it must be produced by oxidation of chloride ions, and the electrode at which this happens must be the anode. Metallic sodium is produced by reduction of sodium ions at the cathode, where electrons are being forced into the cell. The metal remains liquid because its melting point is only 97.8°C, and it floats because it is less dense than the molten sodium chloride.

$$2Cl^- \longrightarrow Cl_2\,(g) + 2e^- \quad \text{(oxidation, anode half-reaction)}$$
$$2(Na^+ + e^- \longrightarrow Na\,(\ell)) \quad \text{(reduction, cathode half-reaction)}$$
$$\underline{2Na^+ + 2Cl^- \longrightarrow 2Na\,(\ell) + Cl_2\,(g)} \quad \text{(overall cell reaction)}$$
$$2NaCl\,(\ell)$$

The formation of metallic sodium and gaseous chlorine from sodium chloride is nonspontaneous except at temperatures very much higher than 801°C, so the

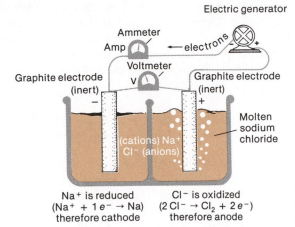

FIGURE 19–2 Apparatus for electrolysis of molten sodium chloride.

generator must supply considerable electrical energy to cause this reaction to occur. Electrons are produced at the anode (oxidation) and consumed at the cathode (reduction) and, therefore, they travel through the wire from *anode to cathode*. Since the reaction is nonspontaneous, the generator forces electrons to flow, nonspontaneously, from the positive electrode to the negative electrode. Thus, the anode is the positive electrode and the cathode the negative electrode *in all electrolytic cells*. Figure 19–2 shows a simplified diagram of the cell.

The direction of spontaneous flow for negatively charged particles is from negative to positive.

The Downs cell used for industrial electrolysis of NaCl (Figure 19–3) is more sophisticated than that depicted in Figure 19–2. Sodium and chlorine must not be allowed to come in contact with each other because they react spontaneously, rapidly, and explosively to form sodium chloride (ΔG^0 is very negative for this reaction at 801°C). The Downs cell is expensive to use mainly because of the cost of construction, the cost of the electricity, and the cost of heating the sodium chloride to 801°C. However, electrolysis of a molten sodium salt is the only means by which metallic sodium can be obtained, because of its extremely high reactivity. Once liberated by the electrolysis, the liquid sodium is drained off, cooled, and cast into blocks. These must be stored in an inert environment, such as below the surface of mineral oil, to prevent reaction with oxygen or other components of the atmosphere.

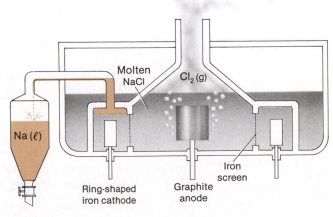

FIGURE 19–3 The Downs cell, the apparatus in which molten sodium chloride is commercially electrolyzed to produce sodium metal and chlorine gas.

Electrolysis of molten compounds is also the common method of obtaining other Group IA and IIA metals (except barium) and aluminum (Chapter 20). Most chlorine is produced by the less expensive electrolysis of aqueous sodium chloride, but that produced in the Downs cell is cooled, compressed, and marketed. This partially offsets the expense of producing metallic sodium.

19–4   The Electrolysis of Aqueous Sodium Chloride

We have indicated many times that all ions in aqueous solutions are hydrated. We will not use the notation that indicates states of substances, (s), (ℓ), (g), and that ions are hydrated in solution, (aq), in the remainder of the text except in those cases where such information is not obvious. This greatly simplifies writing chemical equations.

Consider the electrolysis of a solution of sodium chloride in water, using inert electrodes. The following experimental observations can be made when a sufficiently high voltage is applied across the electrodes of a suitable cell:

1. Gaseous hydrogen is liberated at one electrode, and the solution becomes basic around that electrode.
2. Gaseous chlorine is liberated at the other electrode.

Chloride ions are obviously being oxidized to chlorine in this cell, as they were in the electrolysis of molten sodium chloride. But sodium ions are not reduced to metallic sodium. Instead, gaseous hydrogen and aqueous hydroxide ions are produced by reduction of water molecules at the cathode. It must be that water is more easily reduced than Na^+ ions. This illustrates an important principle of electrochemistry and oxidation-reduction reactions in general:

The most easily oxidized species is oxidized and the most easily reduced species is reduced in preference to other species.

The half-reactions and overall cell reaction for the electrolysis of aqueous sodium chloride solution are

$$2Cl^- \longrightarrow Cl_2 + 2e^- \qquad \text{(at anode)}$$
$$2H_2O + 2e^- \longrightarrow 2OH^- + H_2 \qquad \text{(at cathode)}$$

$$\underbrace{2H_2O + 2Cl^-}_{} \longrightarrow \underbrace{2OH^- + H_2 + Cl_2}_{} \qquad \text{(overall cell reaction)}$$
$$\underbrace{+\ 2Na^+}_{2NaCl} \qquad \underbrace{+\ 2Na^+}_{2NaOH} \qquad \text{(spectator ions)}$$

The cell is illustrated in Figure 19–4. As before, the electrons flow from the anode (+) through the wire to the cathode (−) under the influence of the source of electrical energy that causes the nonspontaneous cell reaction to occur.

The overall cell reaction produces gaseous H_2 and Cl_2 and an aqueous solution of sodium hydroxide, sometimes called caustic soda. Solid sodium hydroxide is obtained by evaporation of the residual solution. This process represents the most important commercial preparation of each of these substances, and is much less expensive than the electrolysis of molten sodium chloride since it is not necessary to heat the solution.

Not surprisingly, the fluctuations in the commercial prices of these widely used industrial products nearly always parallel each other.

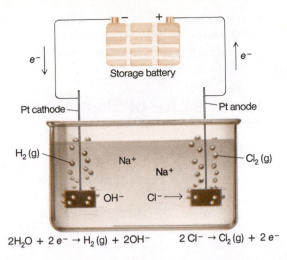

FIGURE 19-4 Electrolysis of aqueous NaCl solution. Although several reactions occur at both the anode and the cathode, the end result is the production of H_2 (g) and NaOH (aq) at the cathode and Cl_2 (g) at the anode.

$$2H_2O + 2\,e^- \rightarrow H_2\,(g) + 2OH^- \qquad 2\,Cl^- \rightarrow Cl_2\,(g) + 2\,e^-$$

19-5   The Electrolysis of Aqueous Sodium Sulfate

The following observations can be made for the electrolysis of aqueous sodium sulfate using inert electrodes.

1. Gaseous hydrogen is produced at one electrode, and the solution becomes basic around that electrode.
2. Gaseous oxygen is produced at the other electrode, and the solution becomes acidic around that electrode.

As in the previous example, water is reduced in preference to Na^+ at the cathode. Observation 2 suggests that water is also preferentially oxidized, relative to the sulfate ion, SO_4^{2-}, at the anode:

$$2(2H_2O + 2e^- \longrightarrow H_2 + 2OH^-) \qquad \text{(cathode)}$$
$$2H_2O \longrightarrow O_2 + 4H^+ + 4e^- \qquad \text{(anode)}$$

$$6H_2O \longrightarrow 2H_2 + O_2 + \underbrace{4H^+ + 4OH^-}_{4H_2O} \qquad \text{(overall cell reaction)}$$

$$2H_2O \longrightarrow 2H_2 + O_2 \qquad \text{(net reaction)}$$

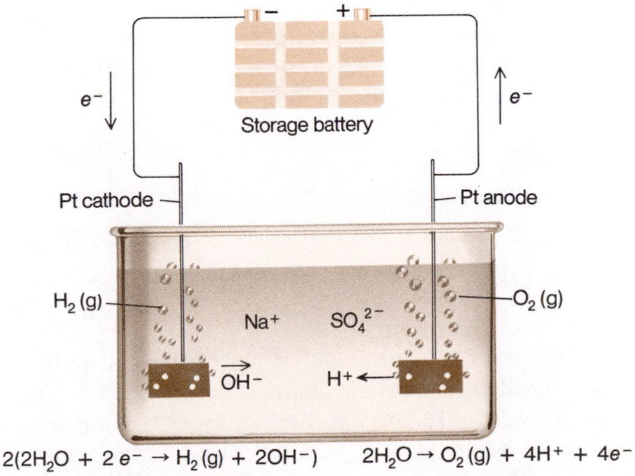

FIGURE 19-5 Electrolysis of aqueous Na_2SO_4 solution to produce H_2 (g) at the cathode and O_2 (g) at the anode.

$$2(2H_2O + 2\,e^- \rightarrow H_2\,(g) + 2OH^-) \qquad 2H_2O \rightarrow O_2\,(g) + 4H^+ + 4e^-$$

The net reaction describes the electrolysis of pure water, which happens because water is more easily reduced than Na^+ and more easily oxidized than SO_4^{2-}. The ions of Na_2SO_4 conduct the current through the solution, and take no part in the reaction. The cell is diagrammed in Figure 19–5.

19–6   Faraday's Law of Electrolysis

In 1832–33 Michael Faraday observed that the amount of substance undergoing oxidation or reduction at each electrode during electrolysis is directly proportional to the amount of electricity that passes through the cell. This statement is known as **Faraday's Law of Electrolysis.** The quantitative unit of electricity, now called the **faraday,** is the amount of electricity that reduces one equivalent weight of a substance at the cathode and oxidizes one equivalent weight of a substance at the anode. This corresponds to the gain or loss, and therefore the passage, of 6.022×10^{23} electrons.

A smaller electrical unit, commonly used in physics and electronics, is the **coulomb.** One coulomb is formally defined to be the amount of charge that passes a given point when one ampere of electrical current flows for one second. It is also equal to the amount of electricity that will deposit 0.001118 gram of silver at the cathode during the electrolysis of an aqueous solution containing silver ions, Ag^+, without limit of time. One ampere of current equals one coulomb per second. One faraday is found to be equal to 96,487 coulombs of charge.

$$1 \text{ ampere} = 1 \frac{\text{coulomb}}{\text{second}} \qquad \text{[Eq. 19–1]}$$

$$1 \text{ faraday} = 6.022 \times 10^{23} e^- = 96,487 \text{ coul} \qquad \text{[Eq. 19–2]}$$

We may restate Faraday's Law: During electrolysis one faraday of electricity (96,487 coulombs) reduces and oxidizes respectively one equivalent weight of the oxidizing and reducing agents. This corresponds to the passage of 6.022×10^{23} electrons through the cell. Table 19–1 shows the amounts of several elements produced during electrolysis by the passage of one faraday of electricity.

Example 19–1

Calculate the mass of copper produced by the reduction of copper(II) ions during the passage of 2.50 amperes of current through a solution of copper(II) sulfate for 45.0 minutes.

Solution

The equation for the reduction of copper(II) ions is

$$Cu^{2+} + \quad 2e^- \quad \longrightarrow \quad Cu \qquad \text{(at cathode)}$$

1 mol	$2(6.02 \times 10^{23})e^-$	1 mol
63.5 g	2(96,500 coul)	63.5 g

(Because most of our calculations will be done to three significant figures, 96,487 coulombs will be rounded off to 96,500 coulombs.) From this information we see that 63.5 grams of copper "plate out" for every 2(96,500 coulombs) of electronic charge.

We first calculate the number of coulombs passing through the cell.

$$? \text{ coul} = 45.0 \text{ min} \times \frac{60 \text{ s}}{1 \text{ min}} \times \frac{2.50 \text{ coul}}{s}$$
$$= 6.75 \times 10^3 \text{ coul}$$

We now calculate the mass of copper produced by the passage of 6.75×10^3 coulombs.

$$? \text{ g Cu} = 6750 \text{ coul} \times \frac{63.5 \text{ g Cu}}{2 \text{ mol } e^- \left(96,500 \frac{\text{coul}}{\text{mol } e^-}\right)}$$
$$= 2.22 \text{ g Cu}$$

Notice how little copper is deposited by this considerable current in 45 minutes.

TABLE 19–1 Elements Produced at One Electrode in Electrolysis by One Faraday of Electricity

Half-Reaction	Product (Electrode)	Amount (= 1 Eq. Wt.)
$Ag^+ (aq) + e^- \longrightarrow Ag (s)$	Ag (cathode)	1 mol = 107.868 g
$2H^+ (aq) + 2e^- \longrightarrow H_2 (g)$	H_2 (cathode)	$\frac{1}{2}$ mol = 1.008 g
$Cu^{2+} (aq) + 2e^- \longrightarrow Cu (s)$	Cu (cathode)	$\frac{1}{2}$ mol = 31.77 g
$Au^{3+} (aq) + 3e^- \longrightarrow Au (s)$	Au (cathode)	$\frac{1}{3}$ mol = 65.6555 g
$2Cl^- (aq) \longrightarrow Cl_2 (g) + 2e^-$	Cl_2 (anode)	$\frac{1}{2}$ mol = 35.453 g = 11.2 L_{STP}
$2H_2O (\ell) \longrightarrow O_2 (g) + 4H^+ (aq) + 4e^-$	O_2 (anode)	$\frac{1}{4}$ mol = 7.9997 g = 5.60 L_{STP}

Example 19–2

What volume of oxygen (measured at STP) is produced by the oxidation of water in the electrolysis of copper(II) sulfate in Example 19–1?

Solution

The equation for the oxidation of water, and the equivalence between coulombs and the volume of oxygen produced at STP, are given below:

$$2H_2O \longrightarrow \quad O_2 \quad + 4H^+ + \quad 4e^- \quad \text{(at anode)}$$

$$\begin{array}{cc} & 1 \text{ mol} & 4(6.02 \times 10^{23})e^- \\ & 22.4 \text{ } L_{STP} & 4(96,500 \text{ coul}) \end{array}$$

The number of coulombs passing through the cell is the same as in Example 19–1, 6750 coulombs. For every 4(96,500 coulombs) passing through the cell, 22.4 liters of O_2 at STP are produced.

$$? \text{ } L_{STP} \text{ } O_2 = 6750 \text{ coul} \times \frac{22.4 \text{ } L_{STP} \text{ } O_2}{4 \text{ mol } e^- \left(96,500 \dfrac{\text{coul}}{\text{mol } e^-}\right)}$$

$$= \underline{0.392 \text{ } L_{STP} \text{ } O_2}$$

Note again how little oxygen is produced by what seems to be a lot of electricity. This shows why electrolytic production of gases and metals is so expensive.

19–7 Electrolytic Refining and Electroplating of Metals

Electrolytic reduction is the only means by which many active metals can be obtained (Section 20–9). Other less active metals that can be obtained from their ores by less expensive chemical reduction, and metals that occur in nature in the native (uncombined) state, are frequently purified or refined by electrolysis. The electrolytic method used for refining metals is also called *electroplating* when applied to plating a metal onto a surface. For example, impure metallic copper obtained from the chemical reduction of Cu_2S and CuS (Section 20–11.3) is purified by using an electrolytic cell like the one shown in Figure 19–6. In this cell, thin sheets of very pure copper are made cathodes by connecting them to the negative terminal of a direct-current generator. Impure chunks of copper are connected to the positive terminal and function as anodes. The electrodes are immersed in a solution of copper(II) sulfate and sulfuric acid. When the circuit is closed, copper from the impure anode is oxidized and goes into solution as Cu^{2+} ions, while Cu^{2+} ions from the solution are reduced and plate out as metallic copper on the pure copper cathodes. Other active metals from the impure bars also go into solution after oxidation, but they do not plate out onto the bars of pure copper because of the far greater concentration of more easily reduced copper ions already in solution.

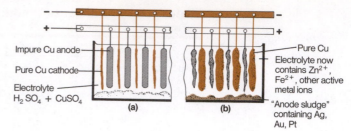

FIGURE 19−6 A schematic of the electrolytic cell used for refining copper. (a) Its appearance before electrolysis; (b) after electrolysis.

Overall, there is no net reaction, merely a transfer of copper from anode to solution and from solution to cathode:

$$\text{(impure)} \quad Cu \longrightarrow Cu^{2+} + 2e^- \qquad \text{(anode)}$$
$$\underline{\qquad\qquad Cu^{2+} + 2e^- \longrightarrow Cu \text{ (pure)} \qquad \text{(cathode)}}$$
$$\text{no net reaction}$$

Although there is no net reaction, the net effect is that small bars of very pure copper and large bars of impure copper are converted to large bars of very pure copper and small bars of impure copper. The energy provided by the electric generator is used to force a decrease in the entropy of the system by separating the copper from the other ions in the impure bars.

A sludge called anode mud collects under the anodes. It contains such valuable and difficult-to-oxidize elements as gold, platinum, silver, selenium, and tellurium; the separation, purification, and sale of these elements reduces the cost of refined copper. Copper can be plated onto other objects by the same mechanism (Figure 19−7).

Examples of metal-plated articles are common in our society. Jewelry and tableware are often plated with silver. Gold is plated on jewelry and electrical contacts. Automobiles have steel bumpers with thin films of chromium. The plating process for a typical bumper requires approximately three seconds of electrolysis to produce a smooth, shiny surface. Rapid plating of metal results in rough, grainy, black surfaces due to the many discontinuities of the metallic lattices produced, because the atoms are deposited from ions so rapidly that they are not able to form extended lattices. Slower plating produces smooth surfaces. "Tin cans" are steel cans plated electrolytically with tin; these may now be replaced by cans plated in one-third of a second with a chromium film only a few atoms thick.

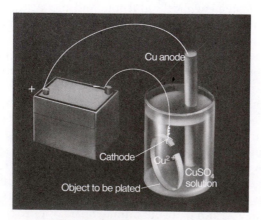

FIGURE 19−7 Electroplating with copper. The anode is made of pure copper, which dissolves during the electroplating process, keeping the concentration of Cu^{2+} constant.

Voltaic or Galvanic Cells

Named for Alessandro Volta and Luigi Galvani, two Italian physicists of the eighteenth century.

Voltaic or **galvanic cells** are electrochemical cells in which *spontaneous* oxidation-reduction reactions produce electrical energy. The two halves of the redox reaction are separated, so that electron transfer is forced to occur by passage of the electrons through an external circuit. In this way, useful electrical energy is obtained. Everyone is familiar with some voltaic cells. The dry cells commonly used in such things as flashlights and transistor radios are voltaic cells. Automobile batteries are also voltaic cells. We shall consider first some simple laboratory cells used to measure the potential difference, or voltage, as it is sometimes called, of a reaction under study, and then look at some more common types of galvanic cells.

19-8 The Construction of Simple Voltaic Cells

A **half-cell** contains the oxidized and reduced forms of an element, or other more complex species, in contact with each other. A common kind of half-cell consists of a piece of metal (the electrode) immersed in a solution of its ions. Consider two such half-cells in separate beakers, using two different elements (Figure 19–8). Electrical contact between the two solutions is made by a **salt bridge.** A salt bridge can be any medium through which ions can slowly pass; the most useful one may be prepared by bending a piece of glass tubing into the shape of a "U", filling it with hot saturated potassium chloride/5% agar solution, and allowing it to cool. The cooled mixture "sets" to the consistency of firm gelatin, so the solution does not run out when the tube is inverted (see Figure 19–8). A salt bridge serves two functions; it avoids mixing the electrode solutions, and it makes electrical contact between the two solutions, thereby enabling an electrical circuit to be completed.

Agar is a gelatinous material obtained from algae.

A cell in which the reactants and products are in their thermodynamic standard states (one molar concentration for dissolved ions and one atmosphere partial pressure for gases) is called a **standard cell.**

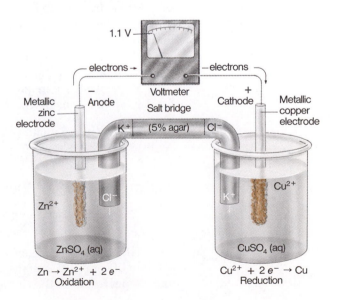

FIGURE 19-8 The zinc/copper voltaic cell uses the reaction Zn (s) + Cu²⁺ (aq) → Zn²⁺ (aq) + Cu (s). The potential of this cell is 1.1 volts.

19–9   The Zinc-Copper Cell

Consider a standard cell that consists of two half-cells, one with a strip of metallic copper immersed in 1.0 M copper(II) sulfate solution, and the other with a strip of zinc immersed in 1.0 M zinc sulfate solution. The electrodes are connected by a wire, and the solutions by a salt bridge. A voltmeter can be inserted into the circuit to measure the potential difference between the two electrodes, or an ammeter can be inserted to measure the flow of electricity. The electrical current is the direct result of the spontaneous redox reaction that occurs, and we seek to measure the potential difference under standard state conditions.

When the copper-zinc cell is constructed as described, the following experimental observations can be made:

1. The initial voltage is 1.10 volts.
2. The mass of the copper electrode increases, and the concentration of Cu^{2+} decreases in the solution around the copper electrode as the cell operates.
3. The mass of the zinc electrode decreases, and the concentration of Zn^{2+} increases in the solution around the zinc electrode as the cell operates.

From these observations we deduce that the half-reaction at the cathode is the reduction of copper(II) ions to copper metal, which plates out on the copper electrode. The zinc electrode is the anode. It loses mass because the zinc metal is oxidized to zinc ions, which go into solution.

$$Zn \longrightarrow Zn^{2+} + 2e^- \qquad \text{(anode)}$$
$$\underline{Cu^{2+} + 2e^- \longrightarrow Cu \qquad \text{(cathode)}}$$
$$Cu^{2+} + Zn \longrightarrow Cu + Zn^{2+} \qquad \text{(overall cell reaction)}$$

This is the cell that was shown in Figure 19–8. Electrons are released at the anode and consumed at the cathode. Therefore they flow through the wire from anode to cathode, as is the case in all electrochemical cells. Since the electrons flow spontaneously in all voltaic cells, they go from the negative electrode to the positive electrode. So, in contrast to electrolytic cells, the anode must be negative and the cathode must be positive. In order to maintain electroneutrality and complete the circuit, two Cl^- ions from the salt bridge migrate into the anode solution for every Zn^{2+} ion formed. Simultaneously, two K^+ ions migrate into the cathode solution to replace every Cu^{2+} ion reduced. Neither K^+ nor Cl^- ions are oxidized or reduced in preference to the zinc metal or Cu^{2+} ions.

Voltaic cells are frequently represented in shorthand form, as illustrated below for the cell just considered.

species (and concentrations)
in contact with electrode surfaces

salt bridge

$$Zn/Zn^{2+} \ (1.0 \ M) \ \| \ Cu^{2+} \ (1.0 \ M)/Cu$$

electrode
surfaces

The same net reaction occurs when a piece of zinc is dropped into a blue solution of copper(II) sulfate. The zinc dissolves; the blue color of Cu^{2+} disappears; and copper forms on the zinc and then settles to the bottom of the container. But no electricity flows, since the two half-reactions are not physically separated as they are in the cell we have just described.

19–10 The Copper-Silver Cell

Let us now consider a similar standard voltaic cell consisting of a strip of copper immersed in 1.0 M copper(II) sulfate and a strip of silver immersed in 1.0 M silver nitrate, $AgNO_3$, solution. A wire and a salt bridge complete the circuit. The following experimental observations can be made.

1. The initial voltage of the cell is 0.46 volt.
2. The mass of the copper electrode decreases, and the Cu^{2+} ion concentration increases in the solution around the copper electrode.
3. The mass of the silver electrode increases, and the Ag^+ ion concentration decreases in the solution around the silver electrode.

We can tell that the copper electrode is the anode in this cell because copper metal is oxidized to Cu^{2+}. The silver electrode is the cathode because Ag^+ ions are reduced to metallic silver.

$$Cu \longrightarrow Cu^{2+} + 2e^- \quad \text{(anode)}$$
$$\underline{2(Ag^+ + e^- \longrightarrow Ag) \quad \text{(cathode)}}$$
$$Cu + 2Ag^+ \longrightarrow Cu^{2+} + 2Ag \quad \text{(overall cell reaction)}$$

Chloride and potassium ions from the salt bridge migrate into the anode and cathode solutions, respectively, to maintain electroneutrality and to complete the circuit. Some NO_3^- ions (from the cathode vessel) and some Cu^{2+} ions (from the anode vessel) also migrate into the salt bridge. The cell is diagrammed in Figure 19–9.

Recall that in the zinc-copper cell the *copper electrode* is the *cathode,* while in the copper-silver cell the *copper electrode* is the *anode.* Whether a particular electrode

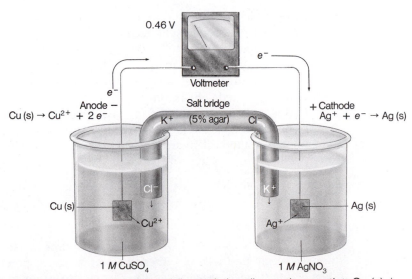

FIGURE 19–9 The copper/silver voltaic cell uses the reaction Cu (s) + 2Ag⁺ (aq) → Cu²⁺ (aq) + 2Ag (s). The potential of this cell is 0.46 volt.

behaves as an anode or a cathode depends on what the other electrode of the cell is. The two cells we have described demonstrate that the Cu^{2+} ion is a stronger oxidizing agent than Zn^{2+}, and Cu^{2+} oxidizes metallic zinc to Zn^{2+}. By contrast, silver ion is a stronger oxidizing agent than Cu^{2+} ion, and Ag^+ oxidizes copper atoms to Cu^{2+}. Conversely, metallic zinc is a stronger reducing agent than metallic copper, and metallic copper is a stronger reducing agent than metallic silver. We can now arrange the species we have studied thus far in order of increasing strength as oxidizing agents and as reducing agents.

Oxidizing Agents	Reducing Agents
$Zn^{2+} < Cu^{2+} < Ag^+$	$Ag < Cu < Zn$
$\longrightarrow$	$\longrightarrow$
increasing strength	increasing strength

Standard Electrode Potentials

The potentials of the standard zinc-copper and copper-silver voltaic cells are 1.10 volts and 0.46 volt, respectively. The magnitude of a cell's potential is a direct measure of the spontaneity of its redox reaction. Higher (more positive) cell potentials indicate greater spontaneity. Under standard conditions the oxidation of metallic zinc by Cu^{2+} ions has a greater tendency to go toward completion than does the oxidation of metallic copper by Ag^+ ions. It is desirable to separate the individual contributions to the total cell potential of the two electrode half-reactions. In this way we can determine the relative tendencies of particular oxidation or reduction half-reactions to occur. Such information gives us a quantitative basis for specifying strengths of oxidizing and reducing agents.

19–11   The Standard Hydrogen Electrode

FIGURE 19–10 The standard hydrogen electrode.

The superscript in E^0 indicates that the potential is under standard conditions.

It is not possible to determine experimentally the potential of a single electrode, because every oxidation must be accompanied by a reduction (that is, the electrons must have somewhere to go). Therefore, it is necessary to establish some arbitrary standard. By international agreement, the reference electrode is the **standard hydrogen electrode (SHE).** A standard hydrogen electrode contains a piece of metal electrolytically coated with a grainy black surface of inert platinum metal, immersed in a 1.0 M H^+ solution. Hydrogen, H_2, is bubbled through a glass envelope over the platinized electrode at one atmosphere pressure (Figure 19–10). By international agreement the standard hydrogen electrode is arbitrarily assigned a potential of exactly 0.0000 . . . volt.

SHE half-reaction	E^0 (standard electrode potential)
$H_2 \longrightarrow 2H^+ + 2e^-$	0.0000 . . . V (SHE as anode)
$2H^+ + 2e^- \longrightarrow H_2$	0.0000 . . . V (SHE as cathode)

When a cell is constructed of a standard hydrogen electrode and some other standard electrode (half-cell), then the standard cell potential, E^0_{cell}, is arbitrarily taken as the *standard potential of the other electrode.* Let us now consider two cells, one consisting of the standard zinc electrode versus the SHE, and the other consisting of the standard copper electrode versus the SHE.

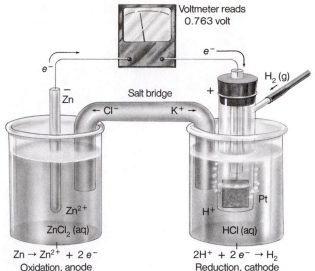

FIGURE 19–11 The Zn/Zn^{2+} (1.0 M)$\|H^+$ (1.0 M); H_2 (1 atm)/Pt cell, in which the reaction Zn (s) + $2H^+$ (aq) → Zn^{2+} (aq) + H_2 (g) occurs.

$\underset{\text{Oxidation, anode}}{Zn \longrightarrow Zn^{2+} + 2\,e^-}$

$\underset{\text{Reduction, cathode}}{2H^+ + 2\,e^- \rightarrow H_2}$

19–12 The Zinc-SHE Cell

This cell consists of an SHE in one beaker and a strip of zinc immersed in 1.0 M zinc sulfate solution in another beaker. A wire and salt bridge complete the circuit. When the circuit is closed, the following observations can be made.

1. As the cell operates, the mass of the zinc electrode decreases and the concentration of Zn^{2+} ion increases in the solution around the zinc electrode.
2. The H^+ concentration decreases in the SHE, and hydrogen is produced.
3. The initial potential of the cell is 0.763 volt.

We can conclude from these observations that the following half-reactions and cell reaction occur.

		E^0
(anode)	$Zn \longrightarrow Zn^{2+} + 2e^-$	0.763 V
(cathode)	$2H^+ + 2e^- \longrightarrow H_2$	0.000 V
(cell reaction)	$Zn + 2H^+ \longrightarrow Zn^{2+} + H_2$	$E^0_{cell} = 0.763$ V

The standard potential at the anode *plus* the standard potential at the cathode gives the standard cell potential. Since the standard cell potential is found to be 0.763 volt, and the potential of the SHE is 0.000 volt, the standard potential of the zinc anode is 0.763 volt. The Zn/Zn^{2+} (1.0 M)$\|H^+$ (1.0 M); H_2 (1 atm)/Pt cell is depicted in Figure 19–11.

Note that in this cell the SHE is the *cathode,* and metallic zinc reduces H^+ to H_2. The zinc electrode is the anode in this cell as well as in the zinc-copper cell we examined earlier.

19–13 The Copper-SHE Cell

Consider a cell that consists of an SHE in one beaker and a strip of metallic copper immersed in a 1.0 M solution of copper(II) sulfate in another beaker. A wire and a

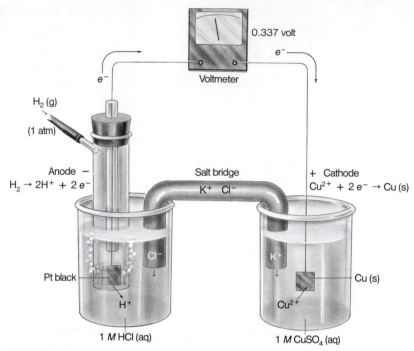

FIGURE 19–12 The standard Cu/SHE cell.

salt bridge complete the circuit. The following observations can be made when the cell is in operation:

1. The mass of the copper electrode increases, and the concentration of Cu^{2+} ions decreases in the solution around the copper electrode.
2. Gaseous hydrogen is used up and the H^+ concentration increases in the solution of the SHE.
3. The initial cell potential is 0.337 volt.

Thus the following half-reactions and cell reaction occur.

		E^0
(anode)	$H_2 \longrightarrow 2H^+ + 2e^-$	0.000 V
(cathode)	$Cu^{2+} + 2e^- \longrightarrow Cu$	0.337 V
(cell reaction)	$H_2 + Cu^{2+} \longrightarrow 2H^+ + Cu$ $\quad E^0_{cell} = 0.337$ V	

The cell is shown in Figure 19–12.

The SHE functions as the *anode* in this cell, and Cu^{2+} ions oxidize H_2 to H^+ ions. (In the Zn-SHE cell, the SHE functions as the *cathode*.) The standard electrode potential of the copper electrode is 0.337 volt as a *cathode* in the Cu-SHE cell.

19–14 The Electromotive Series or Activity Series of the Elements

By measuring the potentials of several other standard electrodes versus the SHE in the same manner as described for the standard Zn-SHE and standard Cu-SHE voltaic cells, a series of standard electrode potentials can be established. When the electrodes involve metals or nonmetals in contact with their ions, the resulting series is called the **electromotive series** or **activity series** of the elements. In Section 19–12 we deduced that the standard zinc electrode behaves as the anode versus the SHE and that the standard *oxidation* potential for the zinc electrode is 0.763 volt.

$$E^0_{\text{oxidation}}$$

(as anode) $Zn \longrightarrow Zn^{2+} + 2e^-$ $+0.763$ V

in general: reduced form $\longrightarrow$ oxidized form $+ ne^-$ standard oxidation potential

Therefore, the *reduction* potential for the standard zinc electrode (to act *as a cathode* relative to the SHE) is the negative of this, or -0.763 volt.

$$E^0_{\text{reduction}}$$

(as cathode) $Zn^{2+} + 2e^- \longrightarrow Zn$ -0.763 V

in general: oxidized form $+ ne^- \longrightarrow$ reduced form   standard reduction potential

By international convention, the standard potentials of electrodes are tabulated for *reduction half-reactions,* indicating the tendencies of the electrodes to behave as cathodes toward the SHE. Those with positive E^0 values for reduction half-reactions do in fact act as cathodes versus the SHE, while those with negative E^0 values for reduction half-reactions behave instead as anodes versus the SHE. Stated differ-

TABLE 19–2   Standard Aqueous Electrode Potentials at 25°C — The Electromotive Series*

Element	Electrode Reaction	Standard Reduction Potential E^0, volts
Li†	$Li^+ + e^- \rightleftharpoons Li$	-3.045
K	$K^+ + e^- \rightleftharpoons K$	-2.925
Ca	$Ca^{2+} + 2e^- \rightleftharpoons Ca$	-2.87
Na	$Na^+ + e^- \rightleftharpoons Na$	-2.714
Mg	$Mg^{2+} + 2e^- \rightleftharpoons Mg$	-2.37
Al	$Al^{3+} + 3e^- \rightleftharpoons Al$	-1.66
Zn	$Zn^{2+} + 2e^- \rightleftharpoons Zn$	-0.7628
Cr	$Cr^{3+} + 3e^- \rightleftharpoons Cr$	-0.74
Fe	$Fe^{2+} + 2e^- \rightleftharpoons Fe$	-0.44
Cd	$Cd^{2+} + 2e^- \rightleftharpoons Cd$	-0.403
Ni	$Ni^{2+} + 2e^- \rightleftharpoons Ni$	-0.25
Sn	$Sn^{2+} + 2e^- \rightleftharpoons Sn$	-0.14
Pb	$Pb^{2+} + 2e^- \rightleftharpoons Pb$	-0.126
H₂	$2H^+ + 2e^- \rightleftharpoons H_2$	(reference electrode) 0.000
Cu	$Cu^{2+} + 2e^- \rightleftharpoons Cu$	$+0.337$
I₂	$I_2 + 2e^- \rightleftharpoons 2I^-$	$+0.535$
Ag	$Ag^+ + e^- \rightleftharpoons Ag$	$+0.7994$
Hg	$Hg^{2+} + 2e^- \rightleftharpoons Hg$	$+0.885$
Br₂	$Br_2 + 2e^- \rightleftharpoons 2Br^-$	$+1.08$
Cl₂	$Cl_2 + 2e^- \rightleftharpoons 2Cl^-$	$+1.360$
Au	$Au^{3+} + 3e^- \rightleftharpoons Au$	$+1.50$
F₂	$F_2 + 2e^- \rightleftharpoons 2F^-$	$+2.87$

(left vertical label: Increasing Strength as Oxidizing Agent; right vertical label: Increasing Strength as Reducing Agent)

* Defined in accord with Josiah Gibbs, one of the founders of chemical thermodynamics, and the International Union of Pure and Applied Chemistry, IUPAC-GIBBS STOCKHOLM Sign Convention.
† For pure metals that react violently with water, an amalgam saturated with metal is used as the electrode.

ently, *the more positive the E^0 value for a half-reaction is, the greater the tendency for the half-reaction to occur in the forward direction as written.* Conversely, the more *negative* the E^0 value for a half-reaction is, the greater the tendency for the reaction to occur in the *reverse* direction as written.

The electromotive series is shown in Table 19–2. The more positive the reduction potential is, the stronger the species on the left is as an oxidizing agent and the weaker the species on the right is as a reducing agent.

19–15   Uses of the Electromotive Series

The electromotive series can be used in many ways. One important application is in the determination of the spontaneity of redox reactions. It is not necessary that the series be used only with respect to reactions in electrochemical cells. Standard electrode potentials can be used to determine the spontaneity of redox reactions in general, and whether or not the half-reactions are physically separated is of no importance. Before this application is illustrated, we must consider two further points about conventions used in the series.

1. The species on the *left* side are all either cations of metals or hydrogen, or elemental nonmetals. These are all *oxidizing agents* (or *oxidized forms* of the elements). Their strengths as oxidizing agents increase from top to bottom, i.e., as the $E^0_{\text{reduction}}$ values become more positive. Fluorine is the strongest oxidizing agent, and Li^+ is a very weak oxidizing agent. (As we shall see in Section 19–16, it is possible to expand a table such as this one to include oxidized and reduced forms, neither of which involve free elements. Such an expanded version is given in Appendix J.)
2. The species on the *right* side are free metals, hydrogen, or anions of nonmetals. These are all reducing agents (reduced forms of the elements). Their strengths as reducing agents increase from bottom to top, i.e., as the $E^0_{\text{reduction}}$ values become more negative. Metallic Li is a very strong reducing agent, and F^- is a very weak reducing agent.

The half-reaction for the standard lithium electrode is

$$Li^+ + e^- \longrightarrow Li \qquad E^0 = -3.045 \text{ V}$$

The very negative E^0 value tells us that this half-reaction does not occur except under extreme circumstances. As a matter of fact, electrolytic methods in nonaqueous cells are required to reduce lithium ions to metallic lithium (Section 19–3). The half-reaction for the standard fluorine electrode is

$$F_2 + 2e^- \longrightarrow 2F^- \qquad E^0 = +2.87 \text{ V}$$

The very positive E^0 value tells us that this half-reaction occurs readily and fluorine, F_2, is the strongest of all common oxidizing agents.

Note that the elements with high ionization energies and highly negative electron affinities have the greatest tendencies to exist as anions. The elements with low ionization energies and low electron affinities have the greatest tendencies to exist as cations.

The following examples will illustrate how E^0 values can be used to predict the spontaneity of a redox reaction.

Example 19–3

Will copper(II) ions oxidize metallic zinc to zinc ions, or will Zn^{2+} ions oxidize metallic copper to Cu^{2+}?

Solution

One of the two possible reactions is spontaneous, and the reverse reaction is nonspontaneous. We must determine which one is spontaneous. We already know the answer to this question from experimental results (Section 19–9), but let us demonstrate the procedure for determining the spontaneous reaction.

1. Choose the appropriate half-reactions from a table of standard reduction potentials.
2. Write the equation for the reaction with the more positive E^0 value for reduction first, along with its potential.
3. Then write the equation for the other half-reaction *as an oxidation* and write its *oxidation potential*; that is, reverse the tabulated reduction half-reaction and change the sign of E^0. (Reversing a half-reaction or a complete reaction also changes the sign of its potential.)
4. Balance the electron transfer.
5. Add the reduction and oxidation half-reactions and add the reduction and oxidation potentials. E^0_{cell} *will be positive* for the resulting overall cell reaction. *This indicates that the forward reaction is spontaneous.* (Of course, any overall cell reaction for which E^0_{cell} is negative is nonspontaneous.)

Following these steps, we obtain the equation for the spontaneous reaction.

E^0

Zn^{2+}/Zn couple has the less positive reduction potential $\longrightarrow$

$1(Cu^{2+} + 2e^- \longrightarrow Cu)$ reduction $+0.337$ V $\overset{\text{oxidation}}{\searrow}$

$1(Zn \longrightarrow Zn^{2+} + 2e^-)$ oxidation $+0.763$ V   potential

spontaneous reaction $Cu^{2+} + Zn \longrightarrow Cu + Zn^{2+}$ $E^0_{cell} = +1.10$ V

The fact that E^0_{cell} is positive tells us that the forward reaction is spontaneous. Reference to Section 19–9 shows that 1.10 volts is, indeed, the potential of the standard zinc-copper voltaic cell and that this is the spontaneous reaction that occurs. Copper(II) ions oxidize metallic zinc to zinc(II) ions and in turn are reduced to metallic copper.

In order to make the reverse reaction (nonspontaneous, negative E^0) occur, electrical energy would have to be supplied with a potential difference greater than 1.10 volts.

nonspontaneous reaction $Cu + Zn^{2+} \longrightarrow Cu^{2+} + Zn$ $E^0_{cell} = -1.10$ volts

---

Example 19–4

Will chromium(III) ions, Cr^{3+}, oxidize metallic copper to copper(II) ions, Cu^{2+}, or will Cu^{2+} oxidize metallic chromium to Cr^{3+} ions?

Solution

As in Example 19–3, we refer to the table of standard reduction potentials and choose the two appropriate half-reactions. The copper half-reaction has the more positive (less negative) reduction potential, so we write it first. Then we write the chromium half-reaction as an oxidation, balance the electron transfer, and add the second half-reaction, and its oxidation potential, to the first. *Note that the potentials are not multiplied by the numbers (2 and 3) used to balance the electron transfer!*

E^0

$3(Cu^{2+} + 2e^- \longrightarrow Cu)$ reduction $+0.337$ V

$2(Cr \longrightarrow Cr^{3+} + 3e^-)$ oxidation $+0.74$ V

$2Cr + 3Cu^{2+} \longrightarrow 2Cr^{3+} + 3Cu$ $E^0_{cell} = +1.08$ V

This is the spontaneous reaction because E^0_{cell} is positive. Copper(II) ions spontaneously oxidize metallic chromium to chromium(III) ions and are reduced to metallic copper.

It is worth emphasizing again a point made in Example 19–4: the standard electrode potentials of the half-cells are *not* multiplied by the numbers used to balance the electron transfer. This is in distinct opposition to the rule used in Hess' Law calculations, and comes about because electrode potentials, *unlike* enthalpy changes, are *intensive* properties of electrochemical reactions. That is, an electrode potential does not depend on "how much" reaction occurs.

19–16   Electrode Potentials for More Complex Half-Reactions

It is possible to construct half-cells in which both oxidized and reduced species are in solution as ions in contact with inert electrodes. For example, the standard iron(III) ion/iron(II) ion half-cell contains 1.0 M concentrations of the two ions and involves the following half-reaction:

$$Fe^{3+} + e^- \longrightarrow Fe^{2+} \qquad E^0 = +0.771 \text{ V}$$

Another example is the standard dichromate ($Cr_2O_7^{2-}$) ion/chromium(III) ion half-cell, which consists of 1.0 M concentrations of the two ions in contact with an inert electrode. The balanced half-reaction in acidic solution (1.0 M H$^+$) is

$$Cr_2O_7^{2-} + 14H^+ + 6e^- \longrightarrow 2Cr^{3+} + 7H_2O \qquad E^0 = +1.33 \text{ V}$$

Other examples of more complex standard electrode potentials are given in Table 19–3 and in Appendix J.

These electrode potentials are analogous to those of the electromotive series, and can be used in similar fashion, as Example 19–5 illustrates.

TABLE 19–3   Standard Electrode Potentials for Selected Half-Cells

Electrode Reaction (Reduction)	Standard Electrode Potential E^0, volts
$Zn(OH)_4^{2-} + 2e^- \rightleftharpoons Zn + 4OH^-$	−1.22
$Fe(OH)_2 + 2e^- \rightleftharpoons Fe + 2OH^-$	−0.877
$2H_2O + 2e^- \rightleftharpoons H_2 + 2OH^-$	−0.8277
$PbSO_4 + 2e^- \rightleftharpoons Pb + SO_4^{2-}$	−0.356
$NO_3^- + H_2O + 2e^- \rightleftharpoons NO_2^- + 2OH^-$	+0.01
$Sn^{4+} + 2e^- \rightleftharpoons Sn^{2+}$	+0.15
$AgCl + e^- \rightleftharpoons Ag + Cl^-$	+0.222
$Hg_2Cl_2 + 2e^- \rightleftharpoons 2Hg + 2Cl^-$	+0.27
$O_2 + 2H_2O + 4e^- \rightleftharpoons 4OH^-$	+0.40
$NiO_2 + 2H_2O + 2e^- \rightleftharpoons Ni(OH)_2 + 2OH^-$	+0.49
$Fe^{3+} + e^- \rightleftharpoons Fe^{2+}$	+0.771
$ClO^- + H_2O + 2e^- \rightleftharpoons Cl^- + 2OH^-$	+0.89
$NO_3^- + 4H^+ + 3e^- \longrightarrow NO + 2H_2O$	+0.96
$Cr_2O_7^{2-} + 14H^+ + 6e^- \rightleftharpoons 2Cr^{3+} + 7H_2O$	+1.33
$MnO_4^- + 8H^+ + 5e^- \rightleftharpoons Mn^{2+} + 4H_2O$	+1.51
$PbO_2 + SO_4^{2-} + 4H^+ + 2e^- \rightleftharpoons PbSO_4 + 2H_2O$	+1.685

Example 19-5

Will tin(IV) ions, Sn^{4+}, oxidize gaseous nitrogen oxide, NO, to nitrate ions, NO_3^-, in acidic solution, or will NO_3^- oxidize Sn^{2+} to Sn^{4+}?

Solution

The nitrate/nitrogen oxide half-reaction has the more positive E^0 value, so we write the tin(IV)/tin(II) half-reaction as an oxidation. We then balance the electron transfer and add the reduction and oxidation half-reactions to obtain the equation for the spontaneous reaction.

$$
\begin{array}{lr}
 & E^0 \\
\hline
2(NO_3^- + 4H^+ + 3e^- \longrightarrow NO + 2H_2O) & +0.96 \text{ V} \\
3(Sn^{2+} \longrightarrow Sn^{4+} + 2e^-) & -0.15 \text{ V} \\
\hline
2NO_3^- + 8H^+ + 3Sn^{2+} \longrightarrow 2NO + 4H_2O + 3Sn^{4+} \quad E^0_{cell} = & +0.81 \text{ V}
\end{array}
$$

Since E^0_{cell} is positive for this reaction, nitrate ions spontaneously oxidize tin(II) ions to tin(IV) ions and are reduced to nitrogen oxide in acidic solution, when all species are present at unit activities.

Now that we are familiar with the use of standard reduction potentials, let us apply them to explain the reaction that occurs in the electrolysis of aqueous solutions of sodium chloride. Recall that the first two electrolytic cells we considered involved molten NaCl and aqueous NaCl. In the first case there was no doubt that metallic sodium would be produced by reduction of Na^+ and gaseous Cl_2 would be produced by oxidation of Cl^-. But in aqueous NaCl we found that H_2O, rather than Na^+, was reduced. This result is consistent with the considerably less negative reduction potential of H_2O as compared to Na^+.

$$
\begin{array}{lr}
 & E^0 \\
\hline
2H_2O + 2e^- \longrightarrow H_2 + 2OH^- & -0.828 \text{ V} \\
Na^+ + e^- \longrightarrow Na & -2.714 \text{ V}
\end{array}
$$

The more easily reduced species, H_2O, is reduced preferentially.

19-17 Corrosion

Ordinary **corrosion** is the redox process by which metals are oxidized by oxygen, O_2, in the presence of moisture. There are other kinds, but this is the most common. The problems of corrosion and its prevention are of both theoretical and practical interest, because they are responsible for the loss of billions of dollars annually in metal products. The mechanism of corrosion has been studied extensively, and it is now known that the oxidation of metals occurs most readily at points of strain (where the metals are most "active"). Thus a steel nail, which is mostly iron (Section 20-6), first corrodes at the tip and the head (Figure 19-13). A bent nail is also susceptible to corrosion at the bend.

In the formation of iron rust, Fe_2O_3, iron(II) is ultimately further oxidized to iron(III) ions.

The end of a steel nail acts as an anode where the iron is oxidized to iron(II) ions.

$$Fe \longrightarrow Fe^{2+} + 2e^- \qquad \text{(anode)}$$

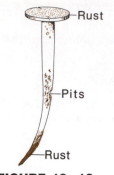

FIGURE 19–13
Corrosion of a bent nail at points of strain and "active" metal atoms.

The electrons produced then flow along the nail to areas containing impurities or weak points, which act as cathodes where oxygen is reduced to hydroxide ions, OH^-, and pits are formed.

$$O_2 + 2H_2O + 4e^- \longrightarrow 4OH^- \qquad \text{(cathode)}$$

The overall reaction is obtained by balancing the electron transfer and adding the two half-reactions.

$$
\begin{array}{ll}
2(Fe \longrightarrow Fe^{2+} + 2e^-) & \text{(oxidation occurs at anode)} \\
O_2 + 2H_2O + 4e^- \longrightarrow 4OH^- & \text{(reduction occurs at cathode)} \\
\hline
2Fe + O_2 + 2H_2O \longrightarrow \underbrace{2Fe^{2+} + 4OH^-}_{2Fe(OH)_2} & \text{(net reaction)}
\end{array}
$$

The $Fe(OH)_2$ may be dehydrated to iron(II) oxide, FeO, or further oxidized to $Fe(OH)_3$ and then dehydrated to iron rust, Fe_2O_3.

19–18   Corrosion Protection

Several methods for protection of metals against corrosion have been developed. The most widely used methods are (1) plating the metal with a thin layer of a less easily oxidized metal, (2) connecting the metal directly to a "sacrificial anode," a piece of another metal that is more active and therefore preferentially oxidized, (3) allowing a protective film, such as a metal oxide, to form naturally on the surface of the metal, and (4) galvanizing, in which steel is coated with zinc, a more active metal.

The thin layer of tin in tin-plated steel cans is not easily oxidized, and it protects the steel underneath from corrosion. It is deposited by dipping the can into molten tin, or by electroplating. Copper is also less active than iron (see Table 19–2) and is sometimes used for protecting metals when food is not involved. It is deposited by electroplating. Whenever the layer of tin or copper is breached, the iron beneath it corrodes even more rapidly than it would without the coating, because of the adverse electrochemical cell set up.

Figure 19–14 shows an iron or steel pipe connected to a bar of magnesium, a more active metal, to protect the iron from oxidation. The magnesium is preferentially oxidized to magnesium ions and is called a "sacrificial anode." Zinc could be used for the same purpose. Similar methods are used to protect bridges and ships' hulls from corrosion.

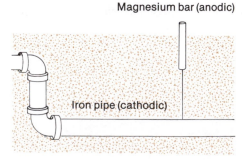

Magnesium bar (anodic)

Iron pipe (cathodic)

FIGURE 19–14 Cathodic protection of buried iron pipe. A magnesium or zinc bar is oxidized instead of the iron.

Aluminum is a very active metal, so active that it reacts rapidly with oxygen from the air to form a surface layer of aluminum oxide, Al_2O_3, which is so thin that it is transparent. This is a very tough, hard substance that is inert toward oxygen, water, and most other corrosive agents in the environment. For this reason, objects made of aluminum form their own protective layer and need not be treated further to inhibit corrosion.

Effect of Concentrations (or Partial Pressures) on Electrode Potentials

19–19   The Nernst Equation

If we connect (by a salt bridge and a voltmeter) two half-cells consisting of strips of copper immersed in 1.0 M and 0.010 M Cu^{2+} ions, we observe a voltage of 0.059 volt. Of course, there would have been a potential of zero if both solutions had been 1.0 M in Cu^{2+} ion (or if both had been 0.010 M). Deviations of concentrations from 1.0 M or of partial pressures from 1.0 atm cause the corresponding potentials to deviate from standard electrode potentials.

Standard electrode potentials, designated E^0, refer to standard conditions. These standard conditions are one molar solutions for ions, partial pressures of one atmosphere for gases, and solids and liquids in their standard states at 25°C. It is important to remember that we refer to *thermodynamic* standard state conditions, and not standard conditions as in gas law calculations.

The **Nernst equation** was derived to calculate electrode potentials for concentrations and partial pressures other than the standard condition values. The Nernst equation is

$$E = E^0 - \frac{2.303RT}{nF} \log Q \qquad \text{[Eq. 19–3]}$$

in which E = the electrode potential under the *nonstandard* conditions
E^0 = the *standard* electrode potential
R = the gas constant, 8.314 J/mol·K or 1.987 cal/mol·K
T = the absolute temperature, in K
n = the number of electrons transferred in the half-reaction
F = the faraday, (96,487 coul/mol e^-) × (1 J/V·coul)
  = 96,487 J/V·mol e^- or 23,060 cal/V·mol e^-
Q = a term analogous to the reaction quotient (Section 16–4) in which dissolved species are expressed in instantaneous molar concentrations and gases in instantaneous partial pressures (in atm) in the same expression

Substituting the values given above into the Nernst equation, with $T = 25°C = 298$ K, simplifies the equation to

$$E = E^0 - \frac{0.0592}{n} \log Q \qquad \text{(at 298 K)} \qquad \text{[Eq. 19–4]}$$

Recall that the reaction quotient, Q, has the same form as the equilibrium constant, K, except that the concentrations or partial pressures are not necessarily

Developed by Walther Nernst (1864–1941).

In this equation, the expression to the right of the minus sign represents the deviation of electrode potential from the E^0 value due to the *nonstandard* conditions.

those at equilibrium. For the familiar half-reaction involving metallic zinc and zinc ions,

$$Zn + 2e^- \rightleftharpoons Zn^{2+} \qquad E^0 = +0.763 \text{ volt}$$

the corresponding Nernst equation is

Since metallic Zn is a solid, its concentration does not appear in Q.

$$E = E^0 - \frac{0.0592}{2} \log \frac{[Zn^{2+}]}{1} \qquad \text{(for reduction)}$$

Substituting the E^0 value into the equation:

$$E = +0.763 \text{ V} - \frac{0.0592}{2} \log \frac{[Zn^{2+}]}{1}$$

Note that under standard conditions of $[Zn^{2+}] = 1.0 \ M$, the electrode potential, E, is equal to the standard electrode potential, E^0, because the correction factor is equal to zero.

$$E = +0.763 \text{ V} - \frac{0.0592}{2} \log 1.0 \qquad \log (1.0) = 0$$

$$E = +0.763 \text{ V} - 0 \text{ V}$$

$$E = E^0$$

Recall that half-reactions for standard reduction potentials are written

$$Ox + ne^- \rightleftharpoons Red$$

where Red refers to the reduced species and Ox refers to the oxidized species. The Nernst equation for any cathode half-cell, where the reduction half-reaction occurs, is

$$E = E^0 - \frac{0.0592}{n} \log \frac{[Red]}{[Ox]} \qquad \text{[Eq. 19-5]}$$

Example 19-6

Calculate the (reduction) potential for the Fe^{3+}/Fe^{2+} electrode if the concentration of Fe^{2+} is five times that of Fe^{3+}.

Solution

The reduction half-reaction is

$$Fe^{3+} + e^- \longrightarrow Fe^{2+} \qquad E^0 = +0.771 \text{ V}$$

First we use the given information to calculate the value of Q.

$$Q = \frac{[Fe^{2+}]}{[Fe^{3+}]} = \frac{5[Fe^{3+}]}{[Fe^{3+}]} = 5$$

Now we substitute this into the Nernst equation with $n = 1$:

$$E = E^0 - \frac{0.0592}{n} \log \frac{[Red]}{[Ox]}$$

$$= +0.771 - \frac{0.0592}{1} \log 5$$

$$= +0.771 - 0.0592(0.699)$$
$$= (+0.771 - 0.041) \text{ V}$$
$$= +0.730 \text{ V}$$

Example 19-7

Calculate E for the Fe^{3+}/Fe^{2+} electrode when the Fe^{3+} concentration is five times the Fe^{2+} concentration.

Solution

$$Q = \frac{[Fe^{2+}]}{[Fe^{3+}]} = \frac{1}{5}$$

$$E = +0.771 - \frac{0.0592}{1} \log \left(\frac{1}{5}\right)$$

$$= +0.771 - 0.0592\,(-0.699)$$
$$= (+0.771 + 0.041)\ \mathrm{V}$$
$$= +\underline{0.812\ \mathrm{V}} \quad \text{(for reduction)}$$

Note that the correction factor, $+0.041$ volt, differs from that in Example 19–6 only in sign.

Example 19–8

Calculate the potential of the chlorine/chloride ion, Cl_2/Cl^-, electrode when the partial pressure of Cl_2 is 10.0 atm and $[Cl^-] = 1.00 \times 10^{-3}\ M$.

Solution

The half-reaction and standard reduction potential are

$$Cl_2 + 2e^- \rightleftharpoons 2Cl^- \qquad E^0 = +1.360\ \mathrm{V}$$

Remember that in evaluating Q in the Nernst equation, (1) molar concentrations are used for dissolved species, and (2) partial pressures of gases are expressed in atmospheres. Also, for this reaction $n = 2$. Thus, the appropriate Nernst equation is

$$E = +1.360 - \frac{0.0592}{2} \log \frac{[Cl^-]^2}{P_{Cl_2}}$$

Substituting $[Cl^-] = 1.00 \times 10^{-3}\ M$ and $P_{Cl_2} = 10.0$ atm into the equation,

$$E = +1.360 - \frac{0.0592}{2} \log \frac{(1.00 \times 10^{-3})^2}{10.0}$$

$$= +1.360 - \frac{0.0592}{2} \log (1.00 \times 10^{-7})$$

$$= +1.360 - \frac{0.0592}{2} (-7.00)$$

$$= [\,+1.360 - (-0.207)\,]\ \mathrm{V}$$

$$= +\underline{1.567\ \mathrm{V}}$$

Example 19–9 illustrates how the Nernst equation can be applied to balanced equations for redox reactions, just as we applied it to equations for half-reactions.

Example 19–9

Calculate the cell potential for the following cell: one electrode consists of the Fe^{3+}/Fe^{2+} couple in which $[Fe^{3+}] = 1.00\ M$ and $[Fe^{2+}] = 0.100\ M$; the other involves the MnO_4^-/Mn^{2+} couple in acidic solution in which $[MnO_4^-] = 1.00 \times 10^{-2}\ M$, $[Mn^{2+}] = 1.00 \times 10^{-4}\ M$, and $[H^+] = 1.00 \times 10^{-3}\ M$.

Solution

We first determine the reaction that occurs and the standard cell potential, E^0_{cell}, in the usual way.

$$
\begin{array}{ll}
 & E^0 \\
\hline
MnO_4^- + 8H^+ + 5e^- \longrightarrow Mn^{2+} + 4H_2O & +1.51\ \mathrm{V} \\
5(Fe^{2+} \longrightarrow Fe^{3+} + e^-) & -0.771\ \mathrm{V} \\
\hline
MnO_4^- + 8H^+ + 5Fe^{2+} \longrightarrow Mn^{2+} + 4H_2O + 5Fe^{3+} & E^0_{cell} = +0.74\ \mathrm{V}
\end{array}
$$

Now we apply the Nernst equation to this reaction, with $n = 5$.

$$E_{cell} = E^0_{cell} - \frac{0.0592}{5} \log \frac{[Mn^{2+}][Fe^{3+}]^5}{[MnO_4^-][H^+]^8[Fe^{2+}]^5}$$

Substituting the appropriate values, we get

$$E_{cell} = +0.74 \text{ V} - \frac{0.0592}{5} \log \frac{(1.00 \times 10^{-4})(1.00)^5}{(1.00 \times 10^{-2})(1.00 \times 10^{-3})^8(0.100)^5}$$

$$= +0.74 \text{ V} - \frac{0.0592}{5} \log (1.00 \times 10^{+27})$$

$$= +0.74 \text{ V} - \frac{0.0592}{5} (27.0) = (+0.74 - 0.32) \text{ V}$$

$$= +0.42 \text{ V}$$

The reaction shown is spontaneous under the stated conditions, with a potential of $+0.42$ volt.

19–20   Relationship of E^0_{cell} to ΔG^0 and K

In Section 16–10 the relationship between the standard Gibbs free energy change, ΔG^0, and the equilibrium constant, K, was studied.

$$\Delta G^0 = -2.303RT \log K$$

There also exists a simple relationship between ΔG^0 and the standard cell potential, E^0_{cell}, for a redox reaction:

$$\Delta G^0 = -nFE^0_{cell} \qquad \text{[Eq. 19–6]}$$

ΔG^0 can also be thought of as the *maximum electrical work* that can be obtained from a redox reaction. In this equation, n is the number of electrons involved in the overall process, and F is the faraday, 96,487 J/V·mol e^-.

Combining these two relationships for ΔG^0 enables us to relate E^0_{cell} values to equilibrium constants.

$$\underbrace{-nFE^0_{cell}}_{\Delta G^0} = \underbrace{-2.303RT \log K}_{\Delta G^0} \qquad \text{[Eq. 19–7]}$$

Solving for E^0_{cell},

$$E^0_{cell} = \frac{2.303RT \log K}{nF} \qquad \text{[Eq. 19–7a]}$$

or for K,

$$\log K = \frac{nFE^0_{cell}}{2.303RT} \qquad K = \text{antilog}\left(\frac{nFE^0_{cell}}{2.303RT}\right) \qquad \text{[Eq. 19–7b]}$$

If any one of the three quantities ΔG^0, K, or E^0_{cell} is known, each of the other two can be calculated using one or more of these relationships. You should keep in mind the following correspondences for all redox reactions *under standard conditions*.

Spontaneity of Forward Reaction	ΔG^0	K	E_{cell}	
spontaneous	−	>1	+	
at equilibrium	0	1	0	(Standard conditions)
nonspontaneous	+	<1	−	

Example 19-10

Calculate the standard Gibbs free energy change, ΔG^0, in J/mol at 25°C from standard electrode potentials for the following reaction.

$$3Sn^{4+} + 2Cr \longrightarrow 3Sn^{2+} + 2Cr^{3+}$$

Recall (Example 16-9) that ΔG^0 is expressed in joules per mole of occurrences of the entire reaction *as written*. Here, we are asking for the number of joules of free energy change that would accompany the reaction of 3 moles of tin(IV) ions and 2 moles of chromium atoms.

Solution

In order to evaluate the potential of this cell, we simply look up the two appropriate half-reactions in a table of standard reduction potentials; then we reverse the one that corresponds to the oxidation and change the sign of its E^0 value. The standard reduction potential for the Sn^{4+}/Sn^{2+} couple is +0.15 volt, while that of the Cr^{3+}/Cr couple is −0.74 volt. Since the equation for the reaction shows Cr being oxidized to Cr^{3+}, the sign of the E^0 value for the Cr^{3+}/Cr couple is reversed. The overall reaction, the sum of the oxidation and reduction half-reactions, has a cell potential equal to the sum of the potentials of the two half-reactions, as shown below.

	E^0
$3(Sn^{4+} + 2e^- \longrightarrow Sn^{2+})$	+0.15 V
$2(Cr \longrightarrow Cr^{3+} + 3e^-)$	+0.74 V
$3Sn^{4+} + 2Cr \longrightarrow 3Sn^{2+} + 2Cr^{3+}$ $E^0_{cell} =$	+0.89 V

The positive value of E^0_{cell} indicates that the forward reaction is spontaneous. We now calculate ΔG^0 using the relationship

$$\Delta G^0 = -nFE^0_{cell}$$
$$= -(6)(96,500 \text{ J/V} \cdot \text{mol})(+0.89 \text{ V})$$
$$= -5.2 \times 10^5 \text{ J/mol} \quad \text{or} -5.2 \times 10^2 \text{ kJ/mol}$$

Example 19-11

Calculate the equilibrium constant, K, for the reaction in Example 19-10 at 25°C by (a) relating it to ΔG^0, and (b) relating it to E^0_{cell}.

Solution

(a) The relationship between ΔG^0 and K is

$$\Delta G^0 = -2.303RT \log K$$

$$-5.2 \times 10^5 \text{ J/mol}$$
$$= (-2.303)(8.314 \text{ J/mol} \cdot \text{K})(298 \text{ K}) \log K$$

$$\log K = \frac{-5.2 \times 10^5}{(-2.303)(8.314)(298)} = +91$$

$$K = 1 \times 10^{91}$$

This very large value of K is consistent with the fact that the reaction is spontaneous.

(b) The relationship between E^0_{cell} and K is

$$\log K = \frac{nFE^0_{cell}}{2.303RT}$$

$$= \frac{(6)(96,500 \text{ J/V} \cdot \text{mol})(+0.89 \text{ V})}{(2.303)(8.314 \text{ J/mol} \cdot \text{K})(298 \text{ K})}$$

$$= +90$$
$$K = 1 \times 10^{90}$$

This result agrees with that obtained in (a) within round-off error. Both 1×10^{91} and 1×10^{90} are so very large that the difference between them is insignificant. Therefore, at equilibrium, only very little Sn^{4+} and Cr are present, and we say that for all practical purposes the reaction goes to completion.

$$K = \frac{[Cr^{3+}]^2[Sn^{2+}]^3}{[Sn^{4+}]^3} = 1 \times 10^{91}$$

Under nonstandard conditions the relationship between the Gibbs free energy change, ΔG, and the cell potential, E_{cell}, is similar to that at standard conditions.

$$\Delta G = -nFE_{cell} \quad \text{(at nonstandard conditions)} \quad \text{[Eq. 19-8]}$$

Example 19-12

Calculate the equilibrium constant, K, at 25°C for the following reaction.

$$2Cu + PtCl_6^{2-} \longrightarrow 2Cu^+ + PtCl_4^{2-} + 2Cl^-$$

Solution

First we find the appropriate half-reactions. The Cu^+/Cu couple occurs as an oxidation in this reaction, so it is written as an oxidation and the sign of its tabulated E^0 value is changed. The electron transfer is balanced and the oxidation and reduction half-reactions are added. The resulting E^0_{cell} value can be used to calculate the equilibrium constant, K, which *does not change* with changing concentrations.

$$
\begin{array}{lr}
 & E^0 \\
2(Cu \longrightarrow Cu^+ + e^-) & -0.521 \text{ V} \\
PtCl_6^{2-} + 2e^- \longrightarrow PtCl_4^{2-} + 2Cl^- & +0.68 \text{ V} \\
\hline
2Cu + PtCl_6^{2-} \longrightarrow 2Cu^+ + PtCl_4^{2-} + 2Cl^- \quad E^0_{cell} = & +0.16 \text{ V}
\end{array}
$$

The equilibrium constant is calculated using Eq. 19–7b:

$$\log K = \frac{nFE^0_{cell}}{2.303RT} = \frac{(2)(96,500)(+0.16)}{(2.303)(8.314)(298)} = 5.4; \qquad K = \underline{2.5 \times 10^5}$$

At equilibrium:

$$K = \frac{[Cu^+]^2[PtCl_4^{2-}][Cl^-]^2}{[PtCl_6^{2-}]} = 2.5 \times 10^5$$

The forward reaction is spontaneous.

Example 19–13

Calculate ΔG at 25°C for the reaction of Example 19–12 at the following concentrations:

$$[PtCl_6^{2-}] = 1.00 \times 10^{-2} \ M$$

$$[Cu^+] = 1.00 \times 10^{-3} \ M$$

$$[PtCl_4^{2-}] = 2.00 \times 10^{-5} \ M$$

$$[Cl^-] = 1.00 \times 10^{-3} \ M$$

Solution

The Gibbs free energy change at nonstandard conditions, ΔG, is related to E_{cell} (not E^0_{cell}) by Eq. 19–8. We must evaluate E_{cell}. Recall (Eq. 19–3) that for a *half*-reaction, n is the number of electrons transferred in the half-reaction, not in the overall reaction. For the Cu^+/Cu couple written *as a reduction*, $n = 1$ and the Nernst equation is

$$E = +0.521 - \frac{0.0592}{1} \log \frac{1}{[Cu^+]}$$

$$= +0.521 - \frac{0.0592}{1} \log \frac{1}{1.00 \times 10^{-3}}$$

$$= [+0.521 - 0.0592(3.00)] \text{ V}$$

$$= +0.343 \text{ V} \qquad \textit{as a reduction}$$

For the other (reduction) half-reaction,

$$E = +0.68 - \frac{0.0592}{2} \log \frac{[PtCl_4^{2-}][Cl^-]^2}{[PtCl_6^{2-}]}$$

$$= +0.68 - \frac{0.0592}{2} \log \frac{(2.00 \times 10^{-5})(1.00 \times 10^{-3})^2}{1.00 \times 10^{-2}}$$

$$= (+0.68 + 0.26) \text{ V}$$

$$= +0.94 \text{ V} \qquad \textit{as a reduction}$$

Subtracting a reduction potential is equivalent to adding an oxidation potential. Since both potentials are reduction potentials, then

$$
\begin{aligned}
E_{cell} &= (E_{PtCl_6^{2-}/PtCl_4^{2-}, \, Cl^-}) - (E_{Cu^+/Cu}) \\
&= (+0.94 \text{ V}) - (+0.343 \text{ V}) \\
&= +0.60 \text{ V}
\end{aligned}
$$

We saw earlier that the forward reaction

$$2Cu + PtCl_6^{2-} \longrightarrow 2Cu^+ + PtCl_4^{2-} + 2Cl^-$$

is spontaneous when all ions are present in 1.0 M concentrations ($E^0_{cell} = +0.16$ V). Now we see that it is even more spontaneous ($E_{cell} = +0.60$ V) under the conditions stated. The Gibbs free energy change under these conditions can now be calculated.

$$
\begin{aligned}
\Delta G &= nFE_{cell} \\
&= -(2)(96,500 \text{ J/V·mol})(0.60 \text{ V}) \\
&= \underline{1.2 \times 10^5 \text{ J/mol}}
\end{aligned}
$$

Primary Voltaic Cells

Primary voltaic cells are cells in which, once the chemicals have been consumed, further chemical action is not possible. The properties of the electrolytes and/or electrodes are such that they cannot be regenerated by reversing the current flow through the cell using an external direct current source of electrical energy. The most familiar examples of primary voltaic cells are ordinary "dry" cells that are used as energy sources in flashlights and other small appliances.

19-21 The Dry Cell (Leclanché Cell)

The dry cell was patented by Georges Leclanché in 1866 (Figure 19-15). The container of a dry cell is made of zinc, which also serves as one of the electrodes. The other electrode is a carbon rod in the center of the cell. The zinc container is lined with porous paper to separate it from the other materials of the cell. A moist mixture (the cell is not really dry) of ammonium chloride, NH_4Cl, manganese(IV) oxide, MnO_2, zinc chloride, $ZnCl_2$, and a porous, inert filler occupies the space between the paper-lined zinc container and the graphite rod. Dry cells are sealed with wax or other sealant to keep the moisture from evaporating. As the cell operates, after the electrodes are connected, the metallic zinc is oxidized to Zn^{2+} and the electrons liberated are left on the container. Thus the zinc electrode is the negative electrode (anode).

$$Zn \longrightarrow Zn^{2+} + 2e^- \quad \text{(anode)}$$

The carbon rod is the cathode, at which ammonium ion is reduced.

$$2NH_4^+ + 2e^- \longrightarrow 2NH_3 + H_2 \quad \text{(cathode)}$$

Addition of the half-reactions gives the overall cell reaction.

$$Zn + 2NH_4^+ \longrightarrow Zn^{2+} + 2NH_3 + H_2 \qquad E_{cell} = 1.6 \text{ V}$$

As hydrogen is formed, it is oxidized by the manganese(IV) oxide in the cell. This prevents collection of hydrogen on the cathode, which would stop the reaction, a

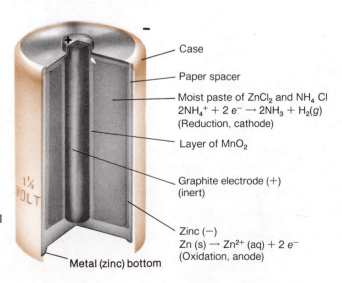

FIGURE 19-15 A cross-section of a typical dry cell that generates a potential difference of about 1.5 volts.

Case

Paper spacer

Moist paste of $ZnCl_2$ and NH_4Cl
$2NH_4^+ + 2e^- \rightarrow 2NH_3 + H_2(g)$
(Reduction, cathode)

Layer of MnO_2

Graphite electrode (+)
(inert)

Zinc (−)
$Zn(s) \rightarrow Zn^{2+}(aq) + 2e^-$
(Oxidation, anode)

Metal (zinc) bottom

situation referred to as **polarization** of the electrode.

$$H_2 + 2MnO_2 \longrightarrow 2MnO(OH)$$

The ammonia produced at the cathode combines with zinc ion and forms a soluble compound containing the complex ion $Zn(NH_3)_4^{2+}$.

$$Zn^{2+} + 4NH_3 \longrightarrow Zn(NH_3)_4^{2+}$$

This reaction prevents polarization due to build-up of ammonia, and it prevents the concentration of Zn^{2+} from increasing substantially, which would decrease the cell potential.

Alkaline dry cells are similar to ordinary dry cells, with two exceptions. (1) The electrolyte is basic (alkaline) because it contains potassium hydroxide. (2) The interior surface of the zinc container is rough, which gives a larger surface area than if it were smooth. The reactions that occur in an alkaline dry cell during discharge are:

anode: $Zn\ (s) + 2OH^-\ (aq) \longrightarrow Zn(OH)_2\ (s) + 2e^-$

cathode: $2MnO_2\ (s) + 2H_2O\ (\ell) + 2e^- \longrightarrow 2MnO(OH)\ (s) + 2OH^-\ (aq)$

overall: $Zn\ (s) + 2MnO_2\ (s) + 2H_2O\ (\ell) \longrightarrow Zn(OH)_2\ (s) + 2MnO(OH)\ (s)$

Alkaline batteries have longer shelf life than ordinary dry cells, and they stand up better under heavy use. The voltage of an alkaline battery is approximately 1.5 V.

Secondary Voltaic Cells

Secondary cells, or *reversible cells,* are cells in which the original reactants can be regenerated by passing a direct current through the cell in the direction opposite to the current flow when the cell is *discharging,* or producing electrical current. This process is referred to as *charging* or recharging a cell or battery. The lead storage battery, used in most automobiles, is the most common example of a secondary voltaic cell.

19–22   The Lead Storage Battery

The lead storage battery is depicted in Figure 19–16. The battery consists of a group of lead plates bearing compressed spongy lead, alternating with a group of lead plates bearing lead (IV) oxide, PbO_2. These electrodes are immersed in a solution of about 30% sulfuric acid. When the cell discharges, it operates as a voltaic cell. The spongy lead is oxidized to lead ions and the lead plates accumulate a negative charge:

$$Pb \longrightarrow Pb^{2+} + 2e^- \qquad \text{(oxidation)}$$

The lead ions then combine with sulfate ions from the sulfuric acid to form insoluble lead(II) sulfate, which begins to coat the lead electrode.

$$Pb^{2+} + SO_4^{2-} \longrightarrow PbSO_4\ (s) \qquad \text{(precipitation)}$$

Thus the net process at the anode *when the cell delivers current* is

$$Pb + SO_4^{2-} \longrightarrow PbSO_4\ (s) + 2e^- \qquad \text{(anode during discharge)}$$

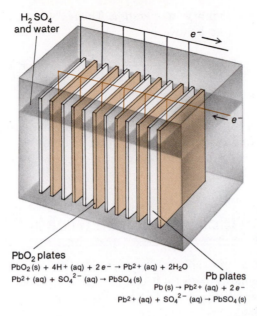

H₂SO₄ and water

e⁻ →

← e⁻

PbO₂ plates
$PbO_2(s) + 4H^+(aq) + 2e^- \rightarrow Pb^{2+}(aq) + 2H_2O$
$Pb^{2+}(aq) + SO_4^{2-}(aq) \rightarrow PbSO_4(s)$

Pb plates
$Pb(s) \rightarrow Pb^{2+}(aq) + 2e^-$
$Pb^{2+}(aq) + SO_4^{2-}(aq) \rightarrow PbSO_4(s)$

FIGURE 19–16 A schematic representation of a twelve-volt lead storage battery. Alternate lead grids are packed with spongy lead and lead(IV) oxide, and then connected in series. The grids are immersed in a solution of sulfuric acid, which serves as the electrolyte.

The electrons produced at the lead electrode travel through the external circuit and re-enter the cell at the lead(IV) oxide, PbO_2, electrode, which is the cathode during discharge. Here, in the presence of hydrogen ions, the lead(IV) oxide is reduced to lead(II) ions, Pb^{2+}. These ions, of course, also combine with sulfate ions from the sulfuric acid to form additional insoluble lead sulfate, which coats the lead(IV) oxide electrode.

$$\begin{aligned} PbO_2 + 4H^+ + 2e^- &\longrightarrow Pb^{2+} + 2H_2O & \text{(reduction)} \\ Pb^{2+} + SO_4^{2-} &\longrightarrow PbSO_4\,(s) & \text{(precipitation)} \\ \hline PbO_2 + 4H^+ + SO_4^{2-} + 2e^- &\longrightarrow PbSO_4\,(s) + 2H_2O & \text{(cathode during discharge)} \end{aligned}$$

The net cell reaction and its standard potential during discharge are obtained by adding the net anode and cathode half-reactions and their tabulated potentials. The tabulated E^0 value for the anode half-reaction is reversed in sign, since it occurs as oxidation during discharge.

$$\begin{array}{lr} & E^0 \\ \hline Pb + SO_4^{2-} \longrightarrow PbSO_4\,(s) + 2e^- & +0.356 \text{ V} \\ PbO_2 + 4H^+ + SO_4^{2-} + 2e^- \longrightarrow PbSO_4\,(s) + 2H_2O & +1.685 \text{ V} \\ \hline Pb + PbO_2 + \underbrace{4H^+ + 2SO_4^{2-}}_{2H_2SO_4} \longrightarrow 2PbSO_4\,(s) + 2H_2O \quad E^0_{cell} = +2.041 \text{ V} \end{array}$$

<aside>The decrease in the concentration of sulfuric acid provides an easy method for measuring the degree of discharge, because the density of the solution decreases proportionally; just measure the density of the solution with a hydrometer.</aside>

One cell creates a potential of about 2 volts. Most automobile batteries are 12 volt batteries that involve six cells connected in series. The potential declines only slightly in use, since solid reagents are being consumed. Eventually the sulfuric acid is consumed, leaving only water. (It is for this reason that a discharged lead-acid battery can freeze, rupturing its case.)

When a potential slightly greater than the potential the battery can generate is imposed across the electrodes, the current flow can be reversed and the battery can be recharged by reversal of all reactions. The alternator or generator applies this

An *alternator* supplies alternating current, so a rectifier (an electronic device) is required to supply the necessary direct current to the battery. A *generator* supplies direct current.

potential when the engine is in operation. The reactions that occur in a lead storage battery are summarized below.

$$Pb + PbO_2 + 2[2H^+ + SO_4^{2-}] \underset{\text{charge}}{\overset{\text{discharge}}{\rightleftharpoons}} 2PbSO_4 \text{ (s)} + 2H_2O$$

Note that sulfuric acid and the electrodes are consumed during discharge, and lead sulfate is deposited. After many repeated charge-discharge cycles, some of the lead sulfate falls to the bottom of the container, the sulfuric acid concentration remains correspondingly low, and the battery cannot be recharged fully. It is then traded in on a new one, and the lead is recovered and re-used to make new storage batteries.

This is one of the oldest and most successful examples of recycling.

19–23 The Nickel-Cadmium (Nicad) Cell

In recent years a new kind of dry cell, the nickel-cadmium cell, has gained widespread popularity because it can be recharged and thus has a much longer useful life than ordinary (Leclanché) dry cells. Nicad batteries are commonly used in electronic calculators and photographic equipment.

The anode is made of cadmium, while the cathode is made of nickel(IV) oxide. The electrolytic solution is basic. The reactions that occur during discharge can be represented as:

anode: $Cd \text{ (s)} + 2OH^- \text{ (aq)} \longrightarrow Cd(OH)_2 \text{ (s)} + 2e^-$
cathode: $NiO_2 \text{ (s)} + 2H_2O \text{ (}\ell\text{)} + 2e^- \longrightarrow Ni(OH)_2 \text{ (s)} + 2OH^- \text{ (aq)}$
overall: $Cd \text{ (s)} + NiO_2 \text{ (s)} + 2H_2O \text{ (}\ell\text{)} \longrightarrow Cd(OH)_2 \text{ (s)} + Ni(OH)_2 \text{ (s)}$

As in the lead storage battery, the solid reaction product at each electrode adheres to the electrode surface. Therefore, a nicad battery can be recharged by an external source of electricity (i.e., the electrode reactions can be reversed). Because no gases are produced by the reactions in a nicad battery, the unit can be sealed. The voltage of a nicad cell is about 1.4 volts, slightly less than that of a Leclanché cell.

19–24 The Hydrogen-Oxygen Fuel Cell

Fuel cells are voltaic cells in which the reactants are continuously supplied to the cell. The hydrogen-oxygen fuel cell (Figure 19–17) already has many applications and is expected to have even more in the future. It has been widely used in spacecraft to supplement the energy obtained from solar cells. Since liquid hydrogen is carried on board as a propellant, the vapor that ordinarily would be lost is used in a fuel cell to generate electrical power, and incidentally, to provide drinking water.

Hydrogen (the fuel) is supplied to one compartment, the anode, where it is oxidized. Oxygen (the oxidizer) is fed into the cathode compartment, where it is reduced. The diffusion rates of the gases into the cell must be carefully regulated to achieve maximum efficiency. The net reaction is the same as the burning of hydrogen in oxygen to form water, but combustion does not actually occur. Most of the chemical energy from the formation of H—O bonds is converted directly into electrical energy, rather than into heat energy as in combustion.

The efficiency of energy conversion of the fuel cell operation is only 60 to 70 percent of the theoretical maximum (based on ΔG), but this still represents about twice the efficiency that can be realized from burning hydrogen in a heat engine coupled to a generator.

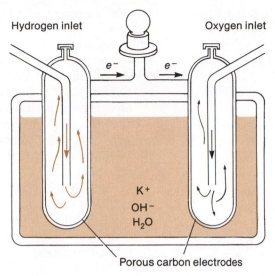

FIGURE 19–17 Schematic drawing of a hydrogen-oxygen fuel cell.

Oxygen undergoes reduction at the cathode, which consists of porous carbon impregnated with finely divided platinum or palladium catalyst.

$$O_2 + 2H_2O + 4e^- \xrightarrow{\text{Pt}} 4OH^- \qquad \text{(cathode)}$$

The hydroxide ions migrate through the electrolyte, an aqueous solution of a base, to the anode. The anode is also porous carbon containing a small amount of catalyst (Pt, Ag, or CoO). Here the hydrogen is oxidized to water.

$$H_2 + 2OH^- \longrightarrow 2H_2O + 2e^- \qquad \text{(anode)}$$

The net reaction is obtained by balancing the electron transfer and adding the two half-reactions:

$$
\begin{array}{ll}
O_2 + 2H_2O + 4e^- \longrightarrow 4OH^- & \text{(cathode)} \\
\underline{2(H_2 + 2OH^- \longrightarrow 2H_2O + 2e^-)} & \text{(anode)} \\
2H_2 + O_2 \longrightarrow 2H_2O & \text{net cell reaction}
\end{array}
$$

Many cells are usually connected together to obtain kilowatts of power.

Current research is aimed at modifying the design of fuel cells to lower their cost. This includes the development of better catalysts to speed the reactions and allow more rapid generation of electricity, and to get more power per unit volume. The use of fuel cells offers an advantage over other methods of obtaining energy in that the H_2/O_2 cell itself is entirely nonpolluting; the only substance released is water.

Key Terms

Activity series the relative order of tendencies for elements and their simple ions to act as oxidizing or reducing agents; also called the electromotive series.

Alkaline battery a dry cell in which the electrolyte contains KOH.

Ampere SI base unit of electrical current.

Anode electrode at which oxidation occurs.

Cathode electrode at which reduction occurs.

Cathodic protection protection of a metal against corrosion by attaching it to a sacrificial anode of a more easily oxidized metal.

Cell potential potential difference, E_{cell}, between oxidation and reduction half-cells under (possibly) nonstandard conditions.

Corrosion oxidation of metals in the presence of moisture.

Coulomb unit of electrical charge, equal to 1 ampere·second.

Downs cell electrolytic cell for the commercial electrolysis of molten sodium chloride.

Dry cells ordinary batteries (voltaic cells); Leclanché cells.

Electrochemistry study of chemical changes produced by electrical current and production of electricity by chemical reactions.

Electrode potentials potentials, E, of half-reactions as reductions versus the standard hydrogen electrode.

Electrodes surfaces upon which oxidation and reduction half-reactions occur in electrochemical cells.

Electrolysis process that occurs in electrolytic cells.

Electrolytic cell electrochemical cell in which electrical energy causes nonspontaneous redox reactions to occur.

Electrolytic conduction conduction of electrical current by ions through a solution or pure liquid.

Electromotive series see *Activity series.*

Electroplating depositing a metal onto a (cathode) surface by electrolysis.

Faraday the charge on 6.022×10^{23} electrons, or 96,487 coulombs.

Faraday's Law of Electrolysis one equivalent weight of a substance is produced at each electrode during the passage of 96,487 coulombs of charge through an electrolytic cell.

Fuel cell a voltaic cell in which the reactants (usually gases) are supplied continuously.

Galvanic cell see *Voltaic cell.*

Half-cell compartment in which the oxidation or reduction half-reaction occurs in a voltaic cell.

Lead storage battery secondary voltaic cell used in most automobiles.

Leclanché cell a type of dry cell.

Metallic conduction conduction of electric current through a metal or along a metallic surface.

Nernst equation corrects standard electrode potentials for nonstandard conditions.

Nickel-cadmium cell (battery) a dry cell in which the anode is Cd, the cathode is NiO_2, and the electrolyte is basic.

Polarization of an electrode, build-up of a product of the cell reaction at an electrode, preventing further reaction.

Primary voltaic cell a voltaic cell in which no further chemical action is possible once the reactants are consumed; cannot be recharged.

Sacrificial anode a more active metal that is attached to a less active metal to protect it against corrosion.

Salt bridge tube containing electrolyte, which connects two half-cells of a voltaic cell.

Secondary voltaic cell a voltaic cell that can be recharged; original reactants can be regenerated by reversing the direction of current flow.

Standard cell potential potential difference, E^0_{cell}, between standard reduction and oxidation half-cells.

Standard electrochemical conditions 1.0 M concentration of each dissolved ion, 1.0 atm partial pressure of gases, and pure solids and liquids.

Standard electrode potential the potential, E^0, of a half-reaction as a reduction relative to the standard hydrogen electrode when all species are present at unit activity.

Standard electrodes half-cells in which the oxidized and reduced forms of a species are present at standard electrochemical conditions.

Standard hydrogen electrode (SHE) the standard electrode for oxidation (or reduction) of hydrogen (or hydrogen ions); arbitrarily assigned an E^0 of exactly zero volts.

Voltage measure of the chemical potential for a redox reaction to occur; potential difference between two electrodes.

Voltaic cell electrochemical cell in which spontaneous chemical reactions produce electricity; also called galvanic cell.

Exercises

General Concepts

1. Define and distinguish among electrochemical cells, electrolytic cells, voltaic cells, anode, and cathode.
2. Why do all electrochemical cells involve redox reactions?
3. Compare and contrast ionic conduction and metallic conduction.
4. Support or refute the statement that the positive electrode in any electrochemical cell is the one toward which the electrons flow through the wire.

Electrolytic Cells

5. Why do solids like potassium bromide, KBr, and sodium nitrate, $NaNO_3$, not conduct electrical current even though they are ionic? Can these solids be electrolyzed?
6. Why can metallic potassium not be obtained by electrolysis of aqueous potassium chloride, KCl?
7. Support or refute the statement that the Gibbs free energy change, ΔG, for any electrolysis reaction is positive.
8. The following compounds are natural sources of the metals contained within them. For which ones can the pure metal be obtained only by electrolysis of the compound or some other compound prepared from it? Explain. $CaCO_3$, Al_2O_3, NiS, KCl, PbS, Fe_2O_3, $MgCO_3$, Cr_2O_3, Cu_2S.
9. Why must the products of many electrolysis reactions be kept separate once they are produced?
10. Why are there no sodium ions in the overall cell reaction for the electrolysis of aqueous sodium chloride?
11. Consider the electrolysis of molten magnesium chloride with inert electrodes. The following experimental observations can be made when current is supplied.
    (a) Bubbles of pale green chlorine gas, Cl_2, are produced at one electrode.
    (b) Silvery white molten metallic magnesium is produced at the other electrode.
    Diagram the cell, indicating the anode, the cathode, the positive and negative electrodes, the half-reactions occurring at the electrodes,

the overall cell reaction, and the direction of electron flow through the wire.

12. Do the same as in Exercise 11 for the electrolysis of molten aluminum oxide, Al_2O_3, dissolved in cryolite, Na_3AlF_6. This is the Hall process for commercial production of aluminum (Section 20–11.1). The observations are:
    (a) Silvery metallic aluminum is produced at one electrode.
    (b) Oxygen, O_2, bubbles off at the other electrode.
13. Do the same as in Exercise 11 for the electrolysis of aqueous hydrogen chloride, that is, hydrochloric acid, HCl. The observations are:
    (a) Bubbles of gaseous hydrogen are produced at one electrode and the solution becomes less acidic around that electrode.
    (b) Bubbles of chlorine gas are produced at the other electrode.
14. Do the same as in Exercise 11 for the electrolysis of aqueous calcium sulfate, $CaSO_4$. The observations are:
    (a) Bubbles of gaseous hydrogen are produced at one electrode and the solution becomes more basic around that electrode.
    (b) Bubbles of gaseous oxygen are produced at the other electrode and the solution becomes more acidic around that electrode.

Faraday's Law

15. What are (a) a coulomb, (b) electrical current, (c) an ampere, (d) a faraday?
16. If separate solutions of the following salts are electrolyzed with the same current for the same time period, which metal would be produced in the greatest mass? What will be the ratio of numbers of moles of metal plated out? $AgNO_3$, CuCl, $CuCl_2$, $CrCl_3$, $Hg(NO_3)_2$, $Hg_2(NO_3)_2$.
17. Calculate the charge in coulombs on a single electron.
18. Calculate the number of electrons that have a total charge of 1.00 coulomb.
19. How many faradays would be required to reduce 1.00 mole of each of the following cations to the free metal?
    (a) Fe^{3+}, (b) Cu^{2+}, (c) Cu^+, (d) Ag^+, (e) Al^{3+}
20. How many coulombs would be necessary to

plate out 1.00 gram of each of the metals from the ions listed in Exercise 19?

21. Calculate the mass of gold plated out during the electrolysis of gold(III) sulfate with a 0.150 ampere current for 52.0 hours.

22. Refer to Exercise 21. How many grams and how many milliliters of dry O_2, measured at STP, would be produced simultaneously at the anode?

23. How long would electrolysis of aqueous gold(III) sulfate have to occur to produce 5.00 grams of gold at the cathode at a current of 0.150 ampere?

24. How long must the electrolysis of Exercise 21 occur to produce 5.00 liters of dry O_2 measured at STP?

25. How long must the electrolysis of Exercise 21 occur to produce 5.00 liters of O_2, collected over water at 32.0°C and 740 torr pressure? The vapor pressure of water at 32.0°C is 35.7 torr.

26. How many amperes of electrical current would be required for 10.0 hours to produce 1.00 gram of
    (a) Fe from Fe^{2+}?   (b) Cr from Cr^{3+}?
    (c) Ag from Ag^+?   (d) Br_2 from Br^-?

27. Describe how the charge on an ion may be determined by electrolysis.

28. The total charge of electricity required to plate out 15.54 grams of a metal from a solution of its dipositive ions is 14,475 coulombs. What is the metal?

29. What is the charge on an ion of tin if 14.84 grams of metallic tin are plated out from the passage of 24,125 coulombs through a solution containing the ion?

30. How many grams of H_2 and O_2 are produced during the electrolysis of water by a 1.30 ampere current for 5.00 hours? What volumes of dry gases are produced at STP?

31. How many minutes would be required to
    (a) produce 3.00 faradays of electricity at a current of 2.00 amperes?
    (b) pass 67,400 coulombs of charge through a cell at a current of 1.50 amperes?
    (c) reduce 0.606 gram of Au^{3+} to metallic gold at a current of 2.50 amperes?

32. What mass of silver could be plated onto a spoon from electrolysis of silver nitrate at a 1.00 ampere current for 10.0 minutes?

33. What mass of platinum could be plated onto a ring from the electrolysis of a platinum(II) salt at a 0.100 ampere current for 30.0 seconds?

34. How many coulombs of electricity would be required to reduce the iron in 36.0 grams of potassium hexacyanoferrate(III), $K_3[Fe(CN)_6]$, to metallic iron?

35. Describe and illustrate the electrolytic purification of impure silver.

36. How does the smoothness or coarseness of a thin film of an electroplated metal depend on the time period of electrolysis?

Voltaic Cells

37. What kind of energy is converted into electrical energy in voltaic cells?

38. What are the functions of a salt bridge?

39. Describe and sketch the following electrodes:
    (a) the standard magnesium electrode, Mg^{2+}/Mg
    (b) the standard Fe^{3+}/Fe^{2+} electrode
    (c) the standard chlorine electrode, Cl_2/Cl^-

40. A voltaic cell containing a standard Co^{2+}/Co electrode and a standard Au^{3+}/Au electrode is constructed and the circuit is closed. Without consulting the table of standard reduction potentials, diagram and completely describe the cell from the following experimental observations.
    (a) Metallic gold plates out on one electrode and the gold ion concentration decreases around that electrode.
    (b) The mass of the cobalt electrode decreases and the cobalt(II) ion concentration increases around the same electrode.

41. Repeat Exercise 40 for a voltaic cell containing standard Fe^{3+}/Fe^{2+} and Ga^{3+}/Ga electrodes. The observations are
    (a) The mass of the gallium electrode decreases and the gallium ion concentration around that electrode increases.
    (b) The ferrous ion, Fe^{2+}, concentration increases in the other electrode solution.

42. What are standard electrochemical conditions?

43. What does voltage measure? How does it vary with time in a primary voltaic cell? Why?

44. When metallic copper is placed into aqueous silver nitrate, a spontaneous redox reaction occurs. Why is electricity not produced?

45. Distinguish among (a) primary voltaic cells, (b) secondary voltaic cells, and (c) fuel cells.

46. Sketch and describe the operation of (a) a dry cell, (b) a lead storage battery, and (c) a hydrogen-oxygen fuel cell.

Standard Electrode Potentials

47. Why are we permitted to assign arbitrarily an electrode potential of 0.000 volt to the standard hydrogen electrode?
48. What does the sign of the standard reduction potential of a half-reaction indicate? What does the magnitude indicate?
49. What is the electromotive series? What information does it contain? How was the electromotive series constructed?
50. In general, how do ionization energies and electron affinities of elements vary with their position in the electromotive series?
51. Diagram the following cells. In each case write the balanced equation for the reaction that occurs spontaneously, and calculate the cell potential. Indicate the direction of electron flow, the anode, the cathode, and the polarity (+ or −) of each electrode. In each case assume that the circuit is completed by a wire and a salt bridge.
    (a) A strip of copper is immersed in a solution that is 1.0 M in Cu^{2+} and a strip of silver is immersed in a solution that is 1.0 M in Ag^+.
    (b) A strip of aluminum is immersed in a solution that is 1.0 M in Al^{3+} and a strip of copper is immersed in a solution that is 1.0 M in Cu^{2+}.
    (c) A strip of magnesium is immersed in a solution that is 1.0 M in Mg^{2+} and a strip of silver is immersed in a solution that is 1.0 M in Ag^+.
52. Why are E^0 values for half-reactions *not* multiplied by the factors necessary to balance electron transfer in generating net cell reactions and standard cell potentials, E^0_{cell}?
53. Describe the process of corrosion. How can corrosion of an oxidizable metal be prevented if it must be exposed to the weather?

In answering the following questions, **justify your answer by appropriate calculations.**

54. Will Fe^{3+} oxidize Sn^{2+} to Sn^{4+} in acidic solution?
55. Will dichromate ions oxidize fluoride ions (F^-) to free fluorine (F_2) in acidic solution?

56. Will permanganate ions oxidize arsenous acid (H_3AsO_3) to arsenic acid (H_3AsO_4) in acid solution?
57. Will permanganate ions oxidize hydrogen peroxide (H_2O_2) to free oxygen (O_2) in acidic solution?
58. Will dichromate ions oxidize Mn^{2+} to MnO_4^- in acidic solution?
59. Calculate the standard cell potential, E^0_{cell}, for each of the cells described in Exercises 40 and 41.
60. By consulting a table of standard reduction potentials, determine which of the following reactions are spontaneous under standard electrochemical conditions.
    (a) Cr^{3+} (aq) + Fe^{2+} (aq) →
    $$Cr^{2+} (aq) + Fe^{3+} (aq)$$
    (b) Mn (s) + $2H^+$ (aq) → H_2 (s) + Mn^{2+} (aq)
    (c) $2Al^{3+}$ (aq) + $3H_2$ (g) → 2Al (s) + $6H^+$ (aq)
    (d) H_2 (g) → H^+ (aq) + H^- (aq)
    (e) Cl_2 (g) + $2Br^-$ (aq) → Br_2 (ℓ) + $2Cl^-$ (aq)
61. Which of the following reactions are spontaneous in voltaic cells under standard conditions?
    (a) Si (s) + $6OH^-$ (aq) + $2Cu^{2+}$ (aq) →
    $$SiO_3^{2-} (aq) + 3H_2O (ℓ) + 2Cu (s)$$
    (b) Zn (s) + $4CN^-$ (aq) + Ag_2CrO_4 (s) →
    $$Zn(CN)_4^{2-} (aq) + 2Ag (s) + CrO_4^{2-} (aq)$$
    (c) MnO_2 (s) + $4H^+$ (aq) + Sr (s) →
    $$Mn^{2+} (aq) + 2H_2O (ℓ) + Sr^{2+} (aq)$$
    (d) $2Cr(OH)_3$ (s) + $6F^-$ (aq) →
    $$2Cr (s) + 6OH^- (aq) + 3F_2 (g)$$
    (e) Cl_2 (g) + $2H_2O$ (ℓ) + ZnS (s) →
    $$2HClO (aq) + H_2S (aq) + Zn (s)$$
62. Which of each pair is the better oxidizing agent?
    (a) Cd^{2+} or Al^{3+} (b) Sn^{2+} or Sn^{4+}
    (c) H^+ or Cr^{2+} (d) Cl_2 or Br_2
    (e) MnO_4^- in acidic solution or MnO_4^- in basic solution
    (f) F_2 or Pb^{2+}
63. Which of each pair is the better oxidizing agent?
    (a) H^+ or Cl_2
    (b) Ni^{2+} or Se in contact with acidic solution
    (c) Fe^{2+} or Zn^{2+} (d) Al^{3+} or H^+
    (e) HgS or Au^+ (f) $Cr_2O_7^{2-}$ or Br_2
64. Which of each pair is the better reducing agent?
    (a) Ag or H_2 (b) Pt or Co^{2+}
    (c) Hg or Au
    (d) Cl^- in acidic solution or Cl^- in basic solution
    (e) Ce^{3+} or Cu   (f) Rb or H_2

65. Which of each pair is the better reducing agent?
    (a) H^- or K   (b) V or Fe^{2+}
    (c) Zn or H^-
    (d) $Mn(OH)_2$ in basic solution or Mn^{2+} in acidic solution
    (e) H_2S or Zr   (f) Ga or Cr

The Nernst Equation

66. How is the Nernst equation of value in electrochemistry?

67. Identify all of the terms in the Nernst equation. What part of the Nernst equation represents the correction factor for nonstandard electrochemical conditions?

68. Calculate the reduction potentials for the following electrodes under the stated conditions. (You will need to write the balanced half-reactions to evaluate Q.)
    (a) Fe^{2+}/Fe when $[Fe^{2+}] = 2.00\ M$
    (b) SnF_6^{2-}/Sn, F^- when $[SnF_6^{2-}] = 0.500\ M$ and $[F^-] = 1.00 \times 10^{-4}\ M$
    (c) CdS/Cd, S^{2-} when $[S^{2-}] = 1.00 \times 10^{-19}\ M$
    (d) $Cr(OH)_3/Cr$, OH^- when $[OH^-] = 5.80 \times 10^{-4}\ M$
    (e) TeO_2, H^+/Te when $[H^+] = 3.50 \times 10^{-3}\ M$
    (f) MnO_4^-, H^+/Mn^{2+} when $[MnO_4^-] = 5.00 \times 10^{-2}\ M$, $[H^+] = 1.00 \times 10^{-3}\ M$, and $[Mn^{2+}] = 6.62 \times 10^{-1}\ M$

69. Calculate the reduction potentials for the following electrodes under the stated conditions. (You will need to write the balanced half-reactions to evaluate Q.)
    (a) Cl_2/Cl^- when $P_{Cl_2} = 2.50$ atm and $[Cl^-] = 1.00\ M$
    (b) Cl_2/Cl^- when $P_{Cl_2} = 1900$ torr and $[Cl^-] = 1.00\ M$
    (c) O_2, H^+/H_2O_2 when $P_{O_2} = 10.0$ atm, $[H^+] = 2.00 \times 10^{-2}\ M$, and $[H_2O_2] = 1.00\ M$
    (d) NO_3^-, H^+/NO when $[NO_3^-] = [H^+] = 3.00 \times 10^{-5}\ M$ and $P_{NO} = 0.00350$ atm
    (e) H^+/H_2 when $[H^+] = 5.00\ M$ and $P_{H_2} = 1.00 \times 10^{-5}$ atm

70. Write the equation for the spontaneous reaction and calculate the cell potential, E_{cell}, for the following combinations of electrodes from Exercises 68 and 69.
    (a) 68(a) and 68(b)   (b) 68(c) and 68(f)
    (c) 69(c) and 68(e)   (d) 69(d) and 69(e)

71. Determine the cell potential for each of the following reactions under the specified conditions at 25°C. Concentrations of ions and par-

tial pressures of gases are indicated in parentheses.
    (a) $Zn\ (s) + 2H^+\ (1.0 \times 10^{-3}\ M) \longrightarrow$
        $\qquad Zn^{2+}\ (3.0\ M) + H_2\ (g)\ (5.0$ atm$)$
    (b) $Cu\ (s) + 2Ag^+\ (1.0 \times 10^{-6}\ M) \longrightarrow$
        $\qquad Cu^{2+}\ (5.0 \times 10^{-2}\ M) + 2Ag\ (s)$
    (c) $Sn^{2+}\ (0.010\ M) + Ni\ (s) \longrightarrow$
        $\qquad Sn\ (s) + Ni^{2+}\ (6.0\ M)$

72. Calculate E_{cell} for each of the following reactions under the specified conditions at 25°C.
    (a) $PbO_2\ (s) + SO_4^{2-}\ (1.0 \times 10^{-2}\ M) +$
        $4H^+\ (0.10\ M) + Zn\ (s) \longrightarrow$
        $\qquad PbSO_4\ (s) + 2H_2O + Zn^{2+}\ (1.0 \times 10^{-5}\ M)$
    (b) $2MnO_4^-\ (0.10\ M) + 16H^+\ (3.0\ M) +$
        $10Cl^-\ (3.0\ M) \longrightarrow$
        $2Mn^{2+}\ (1.0 \times 10^{-2}\ M)$
        $\qquad\qquad + 5Cl_2\ (g)\ (0.50$ atm$) + 8H_2O$

Relationships Among ΔG^0, E^0_{cell}, and K

73. How are the signs and magnitudes of E^0_{cell}, ΔG^0, and K related for a particular reaction?

74. If E_{cell} is related to ΔG and E^0_{cell} is related to ΔG^0, why is the equilibrium constant K related only to E^0_{cell} and not E_{cell}?

75. In light of your answer to Exercise 52, how do you explain the fact that ΔG^0 for a redox reaction *does* depend on the number of electrons transferred, according to $\Delta G^0 = -nFE^0_{cell}$?

76. Calculate E^0_{cell} from the tabulated standard reduction potentials for each of the following reactions in aqueous solution. Then calculate ΔG^0 and K at 25°C from E^0_{cell}. Which reactions are spontaneous as written?
    (a) $Sn^{4+} + 2Fe^{2+} \longrightarrow Sn^{2+} + 2Fe^{3+}$
    (b) $2Cu^+ \longrightarrow Cu^{2+} + Cu\ (s)$
    (c) $3Zn + 2MnO_4^- + 4H_2O \longrightarrow$
        $\qquad 2MnO_2\ (s) + 3Zn(OH)_2\ (s) + 2OH^-$

77. Calculate ΔG^0 (overall) and ΔG^0 per mole of metal for each of the following reactions from E^0 values.
    (a) Zinc dissolves in dilute hydrochloric acid to produce a solution that contains Zn^{2+}, and hydrogen gas is evolved.
    (b) Aluminum dissolves in dilute hydrochloric acid to produce a solution that contains Al^{3+}, and hydrogen gas is evolved.
    (c) Silver dissolves in dilute nitric acid to form a solution that contains Ag^+, and NO is liberated as a gas.
    (d) Lead dissolves in dilute nitric acid to form a

solution that contains Pb^{2+}, and NO is liberated as a gas.

78. Write balanced equations for the following spontaneous reactions. Calculate the standard cell potential for each cell, ΔG^0 (overall), and ΔG^0 per mole of each reactant.
   (a) In acidic solution, dichromate ions (from $K_2Cr_2O_7$) oxidize iodide ions (from NaI) to free iodine.
   (b) In acidic solution, permanganate ions (from $KMnO_4$) oxidize sulfurous acid (H_2SO_3) to sulfate ions.
   (c) In acidic solution, permanganate ions (from $KMnO_4$) oxidize oxalic acid $(COOH)_2$ to carbon dioxide.

79. Calculate the standard Gibbs free energy change, ΔG^0, at 25°C for each of the cells of Exercise 71.

80. Calculate the thermodynamic equilibrium constant, K, at 25°C for each of the voltaic cell reactions in Exercise 71.

81. Calculate the free energy change, ΔG, for each reaction of Exercise 71 under the stated conditions at 25°C.

82. Given ΔG^0 at 25°C for each of the following reactions, calculate E^0_{cell} and the thermodynamic equilibrium constant, K, at 25°C.
   (a) $MnO_4^- + 8H^+ + 5Fe^{2+} \longrightarrow$
   $$Mn^{2+} + 5Fe^{3+} + 4H_2O$$
   $\Delta G^0 = -357$ kJ
   (b) $Cr_2O_7^{2-} + 14H^+ + 6I^- \longrightarrow$
   $$2Cr^{3+} + 3I_2 (s) + 7H_2O$$
   $\Delta G^0 = -460$ kJ

(c) $3Zn (s) + 8H^+ + 2NO_3^- \longrightarrow$
$$3Zn^{2+} + 2NO (g) + 4H_2O$$
$\Delta G^0 = -997$ kJ

83. Given the thermodynamic equilibrium constant, K, at 25°C for each reaction below, calculate E^0_{cell} at 25°C.
   (a) $Cl_2 (g) + 2Br^- (aq) \rightleftharpoons Br_2 (\ell) + 2Cl^- (aq)$
   $$K = \frac{[Cl^-]^2}{(P_{Cl_2})[Br^-]^2} = 2.9 \times 10^9$$
   (b) $4ClO_3^- (aq) \rightleftharpoons Cl^- (aq) + 3ClO_4^- (aq)$
   (in basic solution)
   $$K = K_c = \frac{[Cl^-][ClO_4^-]^3}{[ClO_3^-]^4} = 2.4 \times 10^{26}$$
   (c) $Cl_2 (g) + 2OH^- (aq) \rightleftharpoons$
   $$ClO^- (aq) + Cl^- (aq) + H_2O (\ell)$$
   $$K = \frac{[ClO^-][Cl^-]}{(P_{Cl_2})[OH^-]^2} = 2.9 \times 10^{32}$$

84. Given the following E^0 values at 25°C, calculate K_{sp} for cadmium sulfide, CdS.
   $Cd^{2+} (aq) + 2e^- \longrightarrow Cd (s)$
   $E^0 = -0.403$ V
   $CdS (s) + 2e^- \longrightarrow Cd (s) + S^{2-} (aq)$
   $E^0 = -1.21$ V

85. Refer to Exercise 84. Evaluate ΔG^0 at 25°C for the process
   $$Cd (S) \rightleftharpoons Cd^{2+} (aq) + S^{2-} (aq)$$

The Metals and Metallurgy

20

20–1 Introduction

We are about to study the properties of the metals and how they vary with position in the periodic table. We shall begin with the representative metals, whose valence electrons are in *s* and/or *p* orbitals in the *outermost* shell. These include the alkali metals (IA), the alkaline earth metals (IIA), and the post-transition metals, which are the heavier members of Groups IIIA, IVA, and VA. Then we shall describe the *d*-transition metals, which have valence electrons in *d* orbitals *one shell inside* the outermost occupied shell. See Figure 20–1.

We shall conclude the chapter with a description of the kinds of processes used to extract metals from their ores, known as metallurgy. We shall see that the amount of energy required to obtain a metal from its ore, and thus the cost, increases as the reactivity of the metal increases.

We described metallic bonding and related its effectiveness to the characteristic properties of metals in Section 12–15. In our study of periodicity we learned that metallic character increases toward the left and toward the bottom of the periodic table (Sections 6–3 through 6–8), and that oxides of most metals are basic (Section 9–5). The oxides of some metals (and the metalloids) are amphoteric (Section 9–6).

You may wish to review some of these sections.

The nonmetals C, H, O, N, P, S and Cl are present in large quantities in biological systems. Several other nonmetals are also known to be essential to life.

We usually think of metals as substances used in buildings, railroads, ships, cars, and trucks. Metals are also used for containers, and to conduct heat and electricity. In addition, metal ions serve many important biological functions; medical and nutritional research have provided most of our insight into these processes. Sodium, potassium, calcium, and magnesium are all present in substantial quantities in humans. Table 20–1 shows the dietary "trace elements," known to

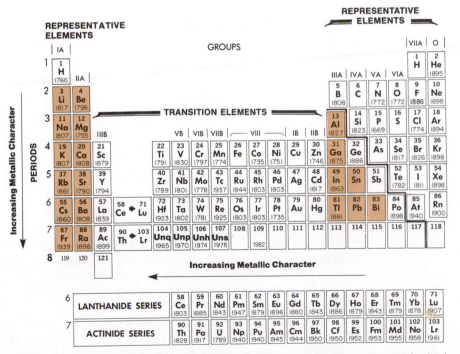

FIGURE 20–1 Metals include about 80% of all known elements. Except for hydrogen, the elements located below and to the left of the stepwise division are metals. The representative metals are shaded.

TABLE 20–1   Dietary Trace Elements Shown to be Essential in Humans and Animals

Element	Date Discovered to Be an Essential Dietary Trace Element in Animals	Dietary Sources	Functions in Humans and/or Animals
GROUP 1 (*Need in Humans Established and Quantified*)			
Iron	17th century	Meat, liver, fish, poultry, beans, peas, raisins, prunes	Component of hemoglobin and myoglobin; oxidative enzymes
Iodine	1850	Iodized table salt, shellfish, kelp	Needed to make the thyroid hormones thyroxine and triiodothyronine; prevents iodine-deficiency goiter
Zinc	1934	Meat, liver, eggs, shellfish	Present in at least 90 enzymes and in the hormone insulin
GROUP 2 (*Need in Humans Established But Not Yet Quantified*)			
Copper	1928	Nuts, liver, shellfish	Component of oxidative enzymes; involved in absorption and mobilization of iron needed for making hemoglobin
Manganese	1931	Nuts, fruits, vegetables, whole-grain cereals	Involved in formation of enzymes, bone
Cobalt	1935	Meat, dairy products	Component of vitamin B_{12}
Molybdenum	1953	Organ meats, green leafy vegetables, legumes	Involved in formation of enzymes, proteins
Selenium	1957	Meat, seafood	Involved in enzyme formation, fat metabolism
Chromium	1959	Meat, beer, unrefined wheat flour	Required for glucose metabolism
GROUP 3 (*Need Established in Animals But Not Yet in Humans*)			
Tin	1970	*	Essential for normal growth of rats
Vanadium	1971	*	Needed for optimum growth of chicks and rats
Fluorine	1972	Fluoridated water	Essential for normal growth of rats
Silicon	1972	*	Needed for growth and bone development in chicks; needed for growth of rats
Nickel	1974	*	Needed for normal growth and for formation of red blood cells in rats; needed by chicks and swine
Arsenic	1975	*	Required for normal growth of rats, goats, minipigs; shortage of element can impair reproductive ability of goats, minipigs

* Present in trace quantities in many foods and in the environment.

be essential in humans and animals; ten of the fifteen listed are metals. The major difficulty in investigations of these trace elements has been the measurement of the extremely small amounts in which they occur in foods. As an example, the vanadium content of fresh peas is usually less than 4.0×10^{-10} gram per gram of peas. Based on this figure, 2700 tons of peas contain only 1.0 gram of vanadium. Obviously, very sensitive techniques are required to detect elements in such low concentrations; these techniques have been developed only recently.

Let us now turn our attention to more traditional aspects of the metallic elements, beginning with the representative metals. They show the greatest variations from one group to the next because their valence electrons are all in the outermost occupied shell.

The Representative Metals

20-2   The Alkali Metals (Group IA)

1   Properties and Occurrence

See the discussion of electrolysis of sodium chloride in Section 19-3.

Because they are so easily oxidized, the alkali metals are not found free in nature. They are obtained by electrolysis of their molten salts. Sodium (2.6% abundance by mass) and potassium (2.4% abundance) are very common in the earth's crust, but the other IA metals are quite rare. Francium consists only of short-lived radioactive isotopes formed by alpha-particle emission (Section 24-3) from actinium. Both potassium and cesium also have radioisotopes. The properties of the alkali metals vary regularly as the group is descended, and are summarized in Table 20-2.

Hydrogen, a *nonmetal*, is included in Group IA in the periodic table. However, it does not share properties with the metals.

The free metals except lithium are soft, silvery corrosive metals that can be cut with a knife; lithium is harder. Cesium is slightly golden and melts in the hand (wrapped in plastic since it is so corrosive). The relatively low melting and boiling points of the alkali metals are consequences of the fairly weak bonding forces, since each atom can furnish only one electron for metallic bonding (Section 12-15). Because their outer electrons are so loosely held, the metals are excellent electrical and thermal conductors. They exhibit the photoelectric effect with low-energy radiation, i.e., they ionize when irradiated with light. These effects become more pronounced with increasing atomic size. For this reason, cesium is used in photoelectric cells.

The ionization energies reveal that the single electron in the outer shell of the IA metals is very easily removed. Cesium and francium are the least electronegative; lithium the most. In all alkali metal compounds the metals exhibit the +1 oxidation state, and virtually all are ionic. The extremely high second ionization energies show that removal of an electron from the next (filled) shell is impossible by ordinary chemical means.

TABLE 20-2   Properties of the Group IA Metals

Property	Lithium	Sodium	Potassium	Rubidium	Cesium	Francium
outer electrons	$2s^1$	$3s^1$	$4s^1$	$5s^1$	$6s^1$	$7s^1$
melting point (°C)	186	97.5	63.6	38.9	28.5	27
boiling point (°C)	1326	889	774	688	690	677
density (g/cm³)	0.534	0.971	0.862	1.53	1.87	—
atomic radius (nm)	0.152	0.186	0.231	0.244	0.262	—
ionic radius, M^+ (nm)	0.060	0.095	0.133	0.148	0.169	—
electronegativity	1.0	1.0	0.9	0.9	0.8	0.8
standard reduction potential (V) $M^+ (aq) + e^- \rightarrow M (s)$	−3.05	−2.71	−2.93	−2.93	−2.92	—
ionization energy (kJ/mol) $M (g) \rightarrow M^+ (g) + e^-$	520	496	419	403	376	—
$M^+ (g) \rightarrow M^{2+} (g) + e^-$	7298	4562	3051	2632	2420	—
heat of hydration of gaseous ion (kJ/mol) $M^+ (g) + xH_2O \rightarrow M^+ (aq)$	−515	−406	−322	−293	−264	—

2 Reactions

Many of the reactions of the alkali metals are summarized in Table 20–3. All are characterized by the loss of one electron per metal atom. These metals are very strong reducing agents. Reactions of the alkali metals with hydrogen and oxygen were discussed in Sections 9–2 and 9–3, those with the halogens in Section 6–10, and those with water in Section 9–7.

The high reactivities of the alkali metals are exemplified by their vigorous reactions with water. Lithium reacts readily, sodium reacts so vigorously that it may ignite, and potassium, rubidium, and cesium burst into flames when dropped into water. The large amounts of heat released provide the activation energy to ignite the evolved hydrogen. The elements also react with water vapor in the air, or moisture from the skin.

Alkali metals should never be handled with bare hands.

$$2K + 2H_2O \longrightarrow 2[K^+ + OH^-] + H_2 \qquad \Delta H^0 = -390.8 \text{ kJ}$$

For this reason, alkali metals are stored under anhydrous nonpolar liquids such as mineral oil.

As is often true for elements of the second period, lithium differs in many ways from the other members of its family. Its ionic charge density and electronegativity are close to those of magnesium, so lithium compounds resemble those of magnesium in some ways. This is illustrative of the **diagonal similarities** that exist between elements in successive groups near the top of the periodic table.

IA	IIA	IIIA	IVA
Li	Be	B	C
Na	Mg	Al	Si

For example, lithium is the only IA metal that combines with nitrogen to form a nitride, Li_3N. Magnesium readily forms magnesium nitride, Mg_3N_2. Both metals readily combine with carbon to form carbides, whereas the other alkali metals do not react readily with carbon. The solubilities of lithium compounds are closer to those of magnesium compounds than to those of other IA compounds. The fluorides, phosphates, and carbonates of both lithium and magnesium are only slightly soluble, but their chlorides, bromides, and iodides are very soluble. Both lithium and

TABLE 20–3 Reactions of the IA Metals

Reaction	Remarks
$4M + O_2 \rightarrow 2M_2O$	limited O_2
$4Li + O_2 \rightarrow 2Li_2O$	excess O_2 (lithium oxide)
$2Na + O_2 \rightarrow Na_2O_2$	(sodium peroxide)
$M + O_2 \rightarrow MO_2$	M = K, Rb, Cs; excess O_2 (superoxides)
$2M + H_2 \rightarrow 2MH$	molten metals
$6Li + N_2 \rightarrow 2Li_3N$	at high temperature
$2M + X_2 \rightarrow 2MX$	X = halogen (Group VIIA)
$2M + S \rightarrow M_2S$	also with Se, Te of Group VIA
$3M + P \rightarrow M_3P$	also with As, Sb of Group VA
$2M + 2H_2O \rightarrow 2MOH + H_2$	K, Rb, and Cs react explosively
$2M + 2NH_3 \rightarrow 2MNH_2 + H_2$	with NH_3 (ℓ) in presence of catalyst; with NH_3 (g) at high temperature (solutions also contain M^+ + solvated e^-)

magnesium form normal oxides, Li_2O and MgO, when burned in air at one atmosphere pressure, rather than the peroxides or superoxides formed by other alkali metals.

The IA metal oxides are all basic and react with water to produce strong soluble bases.

$$Na_2O \text{ (s)} + H_2O \longrightarrow 2[Na^+ + OH^-]$$

$$K_2O \text{ (s)} + H_2O \longrightarrow 2[K^+ + OH^-]$$

Since the IA cations are all related to strong soluble bases, they do not hydrolyze in their salt solutions.

3   Uses of the Alkali Metals and Their Compounds

Commercially, **sodium** is by far the most widely used alkali metal because it is so abundant. Its salts are absolutely essential for life. The metal itself is used as a reducing agent in the manufacture of drugs and dyes and in the metallurgy of metals such as titanium and zirconium, which are obtained from their chlorides.

$$TiCl_4 \text{ (g)} + 4Na \text{ (}\ell\text{)} \xrightarrow{\Delta} 4NaCl \text{ (s)} + Ti \text{ (s)}$$

Sodium has been used in the form of sodium-lead alloys in the production of leaded gasoline. The metal is also used as heat-transfer liquid in some experimental nuclear fast-breeder reactors. Highway lamps often contain sodium vapor, which produces a bright yellow glow when electricity excites its electrons. Just a few examples of the wide variety of uses for sodium compounds are $NaOH$, called caustic soda or — when in solution — lye or soda lye (production of rayon, cleansers, textiles, soap, paper, petroleum products); Na_2CO_3, soda or soda ash, and $Na_2CO_3 \cdot 10H_2O$, washing soda (and as a substitute for $NaOH$ when a weaker base is required); $NaHCO_3$, baking soda (baking and other household uses); $NaCl$ (table salt and source of all other compounds of sodium and chlorine); $NaNO_3$, Chile saltpeter (a nitrogen fertilizer); $NaHSO_4$ (production of HCl from NaCl); Na_2SO_4, salt cake, a byproduct of HCl manufacture (production of brown wrapping paper and corrugated boxes); and NaH (synthesis of $NaBH_4$, which is used to recover silver and mercury from waste water).

Like sodium, metallic **lithium** is used as a heat transfer medium in experimental nuclear reactors because it has the highest heat capacity of any element. Lithium is used as a reducing agent for the synthesis of many organic compounds. Lithium compounds are components of some dry cells and storage batteries as well as some glasses and ceramics. $LiCl$ and $LiBr$ are very hygroscopic (they absorb water vapor from the air) and are used in industrial drying processes and air-conditioning. Lithium compounds are also used for the treatment of some types of mental disorders such as manic depression.

Like salts of sodium (and probably lithium), those of **potassium** are also absolutely essential for life. KNO_3, commonly known as nitre or saltpeter, is used as a potassium and nitrogen fertilizer. There are few major industrial uses for potassium that cannot be satisfied as well with the more abundant and cheaper sodium.

There are very few practical uses for the rare metals **rubidium, cesium,** and **francium.** Metallic cesium is, however, used in some photoelectric cells (Section 5–2).

Leaded gasoline is now being phased out of use due to the air pollution problems it causes.

The highly corrosive nature of both lithium and sodium is a major drawback to applications of the pure metals.

20–3   The Alkaline Earth Metals (Group IIA)

1   Properties and Occurrence

The alkaline earth metals are very silvery-white, malleable, ductile, and somewhat harder than their immediate predecessors in Group IA. Activity increases from top to bottom within the group, with calcium, strontium, and barium being quite active. They all have two electrons in the highest energy level, which are both lost in ionic compound formation, though not as easily as the outer electron of an alkali metal. (Compare their ionization energies in Tables 20–2 and 20–4.) While nearly all other IIA compounds are ionic, those of beryllium exhibit a great deal of covalent character. This is due to the extremely high charge density of Be^{2+}. Its 2+ charge is spread over only a very small volume, resulting in high polarizing power. Beryllium compounds therefore resemble those of aluminum in Group IIIA. The IIA elements exhibit the +2 oxidation state in all their compounds. The tendency to react by forming 2+ ions increases from beryllium to radium.

Recall the principle of diagonal similarities mentioned in Section 20–2.

The alkaline earth metals show a wider range of chemical properties than the alkali metals. Although the IIA metals are not quite as reactive as the IA metals, they are much too reactive to occur free in nature, and are obtained by electrolysis of their molten chlorides. In order to increase the electrical conductivity of molten anhydrous beryllium chloride, which is covalent and polymeric

In Section 7–3, we found that gaseous $BeCl_2$ is linear. However, the Be atoms in $BeCl_2$ molecules act as Lewis acids and, in the solid state, accept shares in electron pairs from Cl atoms in other molecules to form polymers.

$$\left(\begin{array}{ccc} & \text{Cl} & \quad\quad \text{Cl} & \quad\quad \text{Cl} \\ & \diagup\ \ \diagdown & \diagup\ \ \diagdown & \diagup \\ & \text{Be} & \quad \text{Be} & \quad \text{Be} \\ & \diagup\ \ \diagdown & \diagup\ \ \diagdown & \diagdown \\ & \text{Cl} & \quad\quad \text{Cl} & \quad\quad \text{Cl} \end{array} \right),\ \text{small amounts of NaCl are added to the melt.}$$

Calcium and magnesium are quite abundant in the earth's crust, especially as carbonates and sulfates. Beryllium, strontium, and barium are less abundant. All known radium isotopes are radioactive and are extremely rare.

TABLE 20–4   Properties of the Group IIA Metals

Property	Beryllium	Magnesium	Calcium	Strontium	Barium	Radium
outer electrons	$2s^2$	$3s^2$	$4s^2$	$5s^2$	$6s^2$	$7s^2$
melting point (°C)	1283	650	845	770	725	700
boiling point (°C)	2970	1120	1420	1380	1640	1140
density (g/cm³)	1.85	1.74	1.55	2.60	3.51	5
atomic radius (nm)	0.111	0.160	0.197	0.215	0.217	0.220
ionic radius, M^{2+} (nm)	0.031	0.065	0.099	0.113	0.135	—
electronegativity	1.5	1.2	1.0	1.0	1.0	1.0
standard reduction potential (V)						
$M^{2+}(aq) + 2e^- \longrightarrow M(s)$	−1.85	−2.37	−2.87	−2.89	−2.90	−2.92
ionization energy (kJ/mol)						
$M(g) \longrightarrow M^+(g) + e^-$	899	738	590	549	503	509
$M^+(g) \longrightarrow M^{2+}(g) + e^-$	1757	1451	1145	1064	965	(979)
heat of hydration of gaseous ion (kJ/mol) (approx.)						
$M^{2+}(g) + xH_2O \longrightarrow M^{2+}(aq)$	—	−1925	−1652	−1485	−1276	—

TABLE 20-5   **Reactions of the Group IIA Metals**

Reaction	Remarks
$2M + O_2 \longrightarrow 2MO$	very exothermic (except Be)
$Ba + O_2 \longrightarrow BaO_2$	almost exclusively
$M + H_2 \longrightarrow MH_2$	M = Ca, Sr, Ba at high temperatures
$3M + N_2 \longrightarrow M_3N_2$	at high temperatures
$6M + P_4 \longrightarrow 2M_3P_2$	at high temperatures
$M + X_2 \longrightarrow MX_2$	X = halogen (Group VIIA)
$M + S \longrightarrow MS$	also with Se, Te of Group VIA
$M + 2H_2O \longrightarrow M(OH)_2 + H_2$	M = Ca, Sr, Ba at room temperature; Mg gives MgO at high temperatures
$M + 2NH_3 \longrightarrow M(NH_2)_2 + H_2$	M = Ca, Sr, Ba in NH_3 (ℓ) in presence of catalyst; NH_3 (g) with heat.
$3M + 2NH_3 \text{ (g)} \longrightarrow M_3N_2 + 3H_2$	at high temperatures
$Be + 2OH^- + 2H_2O \longrightarrow Be(OH)_4^{2-} + H_2$	only with Be

2   Reactions

Table 20-5 summarizes many of the reactions of the alkaline earth metals, which (except for stoichiometry) are quite similar to the corresponding reactions of the alkali metals. Reactions with hydrogen and oxygen were discussed in Sections 9-2 and 9-3, those with the halogens in Section 6-10, and those with water in Section 9-7.

Except for beryllium, all the alkaline earth metals are oxidized to oxides in air. As is true of the oxides of the alkali metals, the IIA oxides (except BeO) are basic and react with water to give hydroxides. Beryllium hydroxide, $Be(OH)_2$, is quite insoluble in water and is amphoteric. Magnesium hydroxide, $Mg(OH)_2$, is only slightly soluble in water ($K_{sp} = 1.5 \times 10^{-11}$). The hydroxides of calcium, strontium, and barium are considered water-soluble and strong bases.

When one considers the position of beryllium at the top of Group IIA, it is not surprising to find that its oxide is amphoteric while the lower members of the group have basic oxides. Metallic character increases from top to bottom within a group and from right to left across a period. This results in increasing basicity and decreasing acidity of the oxides in the same directions.

Group IA	Group IIA	Group IIIA
Li_2O	BeO	B_2O_3
basic	amphoteric	acidic
Na_2O	MgO	Al_2O_3
basic	basic	amphoteric
K_2O	CaO	Ga_2O_3
basic	basic	amphoteric
		In_2O_3
		basic

Decreasing Acidity of Oxides

Increasing Basicity of Oxides

Increasing Metallic Character of Elements

(left margin, vertical) Decreasing Acidity of Oxides   Increasing Basicity of Oxides   Increasing Metallic Character of Elements

3 Uses of the Alkaline Earth Metals and Their Compounds

The shells of most mollusks are composed primarily of calcium carbonate.

Calcium and its compounds are widely used commercially. The element is used as a reducing agent in the metallurgy of uranium, thorium, and other metals, as a scavenger to remove dissolved impurities such as oxygen, sulfur, and carbon in molten metals, and to remove residual gases in vacuum tubes. It is also a component of many alloys. Heating limestone produces *quicklime*, CaO, which can then be treated with water to form *slaked lime*, $Ca(OH)_2$, a cheap base for which industry finds many uses. When slaked lime is mixed with sand and exposed to the CO_2 of the air it hardens to form mortar and, with a binder, lime plaster for coating walls and ceilings. Careful heating of gypsum, $CaSO_4 \cdot 2H_2O$, produces plaster of Paris, $2CaSO_4 \cdot H_2O$. Calcium phosphate and carbonate occur in seashells and animal bones.

Metallic **magnesium** burns in air with such a brilliant white light that it is used in photographic flashbulbs, fireworks, and incendiary bombs. It is very lightweight, and is used in many alloys for structural purposes. Given its inexhaustible supply in the oceans, it is likely to find many more structural uses as the reserves of iron ore dwindle and as the energy shortage dictates the construction of automobiles, trucks, and airplanes from lighter materials. The metal is used as a reagent for many important organic syntheses. When magnesite, $MgCO_3$, is thermally decomposed, **magnesia,** MgO, is produced. Magnesia is an excellent heat insulator which is used in making furnaces, ovens, and crucibles. It can be converted to $Mg(OH)_2$ by reaction with aqueous ammonium salts (it does not slake readily). A milky white aqueous suspension of finely divided $Mg(OH)_2$, called milk of magnesia, is used as a stomach antacid and as a laxative. Anhydrous $MgSO_4$ and $Mg(ClO_4)_2$ are used as drying agents. A number of other magnesium compounds are commercially important.

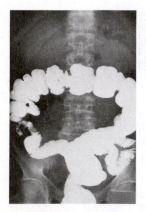

A suspension of barium sulfate blocks x-rays, making the intestines opaque.

Because of its rarity **beryllium** has only a few practical uses; many of its compounds are quite toxic. **Strontium** salts are used in fireworks and flares that show the characteristic red glow of strontium in a flame. The metal itself has no practical uses. **Barium** is a constituent of alloys used for spark-plugs because of the ease with which it emits electrons when heated. A slurry of finely divided barium sulfate from barite, $BaSO_4$, is used to coat the gastrointestinal tract for obtaining x-ray photographs because it absorbs x-rays so well. It is so insoluble that it is not poisonous; all soluble barium salts are very toxic.

20-4 The Post-Transition Metals

The metals below the stepwise division of the periodic table in Groups IIIA through VA are the post-transition metals. These include aluminum, gallium, indium, and thallium from Group IIIA; tin and lead from Group IVA; and bismuth from Group VA. In Section 6-3 we classified the elements *along* the stepwise division as metalloids; however, we shall discuss aluminum here because so many of its properties are characteristic of metals. Aluminum is the only post-transition metal that is considered very reactive.

| | | VIIA | 0 |
| | | | |

				VIIA	0
				1 **H**	2 **He**
IIIA	IVA	VA	VIA		
5 **B**	6 **C**	7 **N**	8 **O**	9 **F**	10 **Ne**
13 **Al**	14 **Si**	15 **P**	16 **S**	17 **Cl**	18 **Ar**
31 **Ga**	32 **Ge**	33 **As**	34 **Se**	35 **Br**	36 **Kr**
49 **In**	50 **Sn**	51 **Sb**	52 **Te**	53 **I**	54 **Xe**
81 **Tl**	82 **Pb**	83 **Bi**	84 **Po**	85 **At**	86 **Rn**

Post-transition metals are shaded

1 Periodic Trends — Groups IIIA and IVA

The properties of the Groups IIIA and IVA elements, listed in Tables 20-6 and 20-7, respectively, vary less regularly down the groups than those of the IA and IIA metals do.

TABLE 20-6   Properties of the Group IIIA Elements

Property	Boron	Aluminum	Gallium	Indium	Thallium
outer electrons	$2s^2 2p^1$	$3s^2 3p^1$	$4s^2 4p^1$	$5s^2 5p^1$	$6s^2 6p^1$
melting point (°C)	2300	660	29.8	156.6	303.5
boiling point (°C)	2550	2327	2403	2000	1457
density (g/cm^3)	2.34	2.70	5.91	7.31	11.85
atomic radius (nm)	0.088	0.143	0.122	0.162	0.171
ionic radius, M^{3+} (nm)	0.020*	0.050	0.062	0.081	0.095
electronegativity	2.0	1.5	1.7	1.6	1.6
standard reduction potential (V) $M^{3+} (aq) + 3e^- \longrightarrow M (s)$	-0.90*	-1.66	-0.53	-0.34	-0.34†
oxidation states	-3 to $+3$	$+3$	$+1, +3$	$+1, +3$	$+1, +3$
heat of hydration of gaseous ion (kJ/mol) $M^{3+} (g) + xH_2O \longrightarrow M^{3+} (aq)$	—	-4749	-4703	-4159	-4117

* For covalent $+3$ oxidation state.
† For Tl$^+$ (aq) $+ e^- \longrightarrow$ Tl (s); for Tl^{3+} (aq) $+ 2e^- \longrightarrow$ Tl$^+$ (aq), $E^0 = 1.28$ V.

The Group IIIA elements are all solids. *Boron*, at the top of the group, is a nonmetal. Its melting point, 2300°C, is very high because it crystallizes as a covalent network solid (Sections 12–11 and 22–28). The other elements, aluminum through thallium, form metallic crystals and consequently have considerably lower melting points.

Gallium is unusual in that it melts when held in the hand, and has the largest liquid-state temperature range of any element. It is used in transistors and high-temperature thermometers.

Indium is a soft bluish metal that is used in some alloys with silver and lead to make good heat conductors. Most indium is used in electronics.

Thallium is a soft, heavy metal that resembles lead. It is quite toxic and is not used in any important practical way.

Ga, In, and Tl are all quite rare.

The elements at the top of Group IVA, carbon, silicon, and germanium, all have relatively high melting points (especially C), like those at the top of Group IIIA, because they crystallize in covalent network lattices, while the metals, tin and lead, have lower melting points.

The Group IIIA elements all have the $ns^2 np^1$ outer electronic configuration. Aluminum shows only the $+3$ oxidation state in its compounds, but the heavier

TABLE 20-7   Properties of the Group IVA Elements

Property	Carbon	Silicon	Germanium	Tin	Lead
outer electrons	$2s^2 2p^2$	$3s^2 3p^2$	$4s^2 4p^2$	$5s^2 5p^2$	$6s^2 6p^2$
melting point (°C)	3570	1414	937	232	328
boiling point (°C)	sublimes	2355	2830	2270	1750
density (g/cm^3)	2.25 (graphite)	2.33	5.35	7.30	11.35
atomic radius (nm)	0.077	0.117	0.122	0.140	0.175
ionic radius, M^{2+} (nm)	—	—	0.073	0.093	0.121
electronegativity	2.5	1.8	1.9	1.8	1.7
standard reduction potential (V) $M^{2+} (aq) + 2e^- \longrightarrow M (s)$	$+0.39$*	$+0.10$*	-0.3	-0.14	-0.126
oxidation states	$\pm 2, \pm 4$	± 4	$+2, +4$	$+2, +4$	$+2, +4$

* For covalent $+2$ oxidation state.

metals (Ga, In, Tl) can lose or share either the single p valence electron or else the p and both s electrons, to exhibit the $+1$ and $+3$ oxidation states, respectively. The IVA elements have the ns^2np^2 configuration. The IVA metals can lose or share the two p electrons to exhibit the $+2$ oxidation state, or they may share all four valence electrons to produce the $+4$ oxidation state. In general, then, the post-transition metals can exhibit oxidation states of $(g-2)+$ and $g+$ where $g =$ periodic group number. As examples, TlCl and $TlCl_3$ both exist, as do $SnCl_2$ and $SnCl_4$. The stability of the lower oxidation state increases upon descending the groups. This is called the **inert (s) pair effect,** since the two s electrons remain nonionized or unshared for the $(g-2)+$ oxidation state. To illustrate, $AlCl_3$ exists but not AlCl; TlCl is more stable than $TlCl_3$. Likewise $GeCl_4$ is more stable than $GeCl_2$, but $PbCl_2$ is more stable than $PbCl_4$. However, for the IIIA metals the $+3$ compounds are stabilized (relative to $+1$) in aqueous solution by the very high heats of hydration of the positive ions.

As is generally true, for each pair of compounds the covalent character is greater for the higher (more polarizing) oxidation state.

Bi from Group VA exhibits the $+3$ and $+5$ oxidation states, but the $+5$ state is quite rare.

Let us now look more closely at some of the more important post-transition metals — aluminum, tin, lead, and bismuth.

2 Aluminum

The Alcoa Bulding is sheathed in aluminum, which resists corrosion.

Aluminum is the most abundant metal in the earth's crust (7.5%) and the third most abundant element. Most is contained in aluminosilicate minerals (Section 22–26) such as clays and micas. Unfortunately, it is not economical to extract aluminum from these sources. The metal itself is produced by electrolysis of Al_2O_3 obtained from its ore, *bauxite,* by the Hall process (Section 20–11.1). Aluminum is quite inexpensive compared to most other metals. It is soft and can be extruded into wires or rolled, pressed, or cast into shapes very readily.

Because of its relatively low density, aluminum is often used as a lightweight structural metal. It is often alloyed with Mg and some Cu and Si to increase its strength.

Although pure aluminum conducts only about two-thirds as much electrical current per unit volume as copper, it is only one-third as dense. As a result, aluminum can conduct twice as much current as the same mass of copper. It is now used in electrical transmission lines and has been used in wiring in homes. However, the latter use has been implicated as a fire hazard due to the heat that can be generated during high current flow at the junction of the aluminum wire and fixtures of other metals.

The metal is a good reducing agent.

$$Al^{3+} (aq) + 3e^- \longrightarrow Al (s) \qquad E^0 = -1.66 \text{ V}$$

Although aluminum is quite reactive, a thin transparent film of Al_2O_3 forms when aluminum comes into contact with air, and this protects it from further oxidation. It is even passive toward nitric acid, HNO_3, a strong oxidizing agent, for this reason. However, if the oxide coating is brushed off with steel wool, aluminum reacts vigorously with HNO_3.

$$Al (s) + 4HNO_3 (aq) \longrightarrow Al(NO_3)_3 (aq) + NO (g) + 2H_2O (\ell)$$

Recall that HNO_3 is an oxidizing acid that does not produce hydrogen when it reacts with active metals.

The very negative enthalpy of formation of aluminum oxide makes aluminum an especially good reducing agent for other metal oxides. The **thermite reaction** is a spectacular example that generates enough heat to produce molten iron for welding steel.

$$2Al \, (s) + Fe_2O_3 \, (s) \longrightarrow 2Fe \, (s) + Al_2O_3 \, (s) \qquad \Delta H^0 = -852 \text{ kJ}$$

Anhydrous Al_2O_3 occurs naturally as the extremely hard, high melting mineral *corundum,* which has a network structure. It is colorless when pure, but becomes colored when a few transition metal ions replace Al^{3+} in the crystal lattice. *Sapphire* is blue and contains some iron and titanium. *Ruby* is red due to the presence of small amounts of chromium.

Aluminum also occurs naturally (and was first discovered) in a series of double salts known as *alums.* Common **alums** are hydrated sulfates of the general formula $M^+M^{3+}(SO_4)_2 \cdot 12H_2O$. The most common alum is $KAl(SO_4)_2 \cdot 12H_2O$, but Li^+, Na^+ and NH_4^+ easily substitute for K^+, and Cr^{3+} and Fe^{3+} can substitute for Al^{3+} to form a large variety of alums. All the alums crystallize in the same lattice; the large, octahedral crystals are easily grown in the laboratory. Alums are used in large quantities in the dye industry and for sizing in the paper industry. Sizing helps paper to retain its shape and to become water-repellent.

The amphoteric nature of aluminum hydroxide, $Al(OH)_3$, has been discussed in Section 9–6.2. Metallic aluminum, itself, is also amphoteric. Freshly brushed aluminum (brushing removes the protective coating of Al_2O_3) will dissolve in non-oxidizing acids to form aqueous aluminum salts and hydrogen,

$$Al \, (s) + 6H^+ \, (aq) \longrightarrow Al^{3+} \, (aq) + 3H_2 \, (g)$$

and in solutions of strong soluble bases to form aqueous aluminate salts — or, more properly, tetrahydroxoaluminate salts — and hydrogen.

$$2Al \, (s) + 2OH^- \, (aq) + 6H_2O \, (\ell) \longrightarrow \quad 2Al(OH)_4^- \quad + 3H_2 \, (g)$$
$$\text{``aluminate ion''}$$

Most anhydrous compounds of aluminum have considerable covalent character, even Al_2O_3. Anhydrous AlF_3 is the only common exception. However, most aluminum salts are ionized in aqueous solution and crystallize from such solutions as hydrated ionic salts because of the high hydration energy of Al^{3+} ion (Table 13–1). For example, aluminum chloride crystallizes from aqueous solution as $AlCl_3 \cdot 6H_2O$, which is $Al(OH_2)_6Cl_3$.

By contrast, anhydrous aluminum chloride, like $BeCl_2$, is a three-dimensional polymeric solid in which chlorine atoms form bridges between adjacent aluminum atoms via coordinate covalent bonds. In the vapor phase it exists primarily as **dimers** of $AlCl_3$, that is, Al_2Cl_6. In both the solid and vapor phases the aluminum atoms are tetrahedrally surrounded by chlorine atoms, and are sp^3 hybridized. In this way, the aluminum atoms attain an octet of electrons in their valence shells.

Al_2Cl_6

As discussed in Section 17–9.5, solutions of aluminum salts are acidic because Al^{3+} (aq) is a small, highly charged cation that hydrolyzes extensively, with $K_a = 1.2 \times 10^{-5}$.

$$Al^{3+} \, (aq) + H_2O \, (\ell) \rightleftharpoons Al(OH)^{2+} \, (aq) + H^+ \, (aq)$$

A mixture of powdered aluminum and iron(III) oxide, called thermite, is used to weld steel rods in construction.

Most oven cleaners contain NaOH (lye). Can you see why they should not be used to clean aluminum pots and pans?

Recall that Al^{3+} (aq) is $Al(OH_2)_6^{3+}$.

A dimer is a molecule formed by the combination of two identical smaller molecules. Compare this with BCl_3 and BF_3 (Section 7–4), which are trigonal planar molecules with sp^2 hybridization at the boron atom.

Some antiperspirants contain Al^{3+} salts. The H_3O^+ produced in such hydrolysis reactions kills bacteria in perspiration. Perhaps you have heard of "aluminum chlorhydrate."

3   Tin and Lead (Group IVA) and Bismuth (Group VA)

Tin is obtained mainly from cassiterite, SnO_2. It is used in some alloys such as solder (with lead), bronze (with copper), and "white metal." Its widest use is in pewter and as tin plate, which is produced by dipping sheet iron or steel into molten tin or by electroplating the tin onto the sheet. Tin exists in three allotropic forms, gray tin, malleable tin, and brittle tin. Malleable tin is most common. It is silver-white and resistant to air oxidation.

Lead is bluish-white and malleable, but is more dense than tin. It is most often obtained from galena, PbS. It is used as a protective absorber of x-rays, in battery plates, and in alloys such as solder. Like most heavy metals, lead is toxic. Its use in gasoline as tetraethyllead, $Pb(C_2H_5)_4$, an anti-knock additive, is decreasing since federal regulations now prohibit its use in the United States in newly manufactured cars. Lead "poisons" the catalysts in catalytic converters. $Pb_3(OH)_2(CO_3)_2$ was formerly used as a pigment in white paints. Pb_3O_4 is still used in red corrosion-resistant outdoor paints.

These Group IVA metals, tin and lead, exhibit the $+2$ and $+4$ oxidation states. All four oxides, SnO, SnO_2, PbO, and PbO_2, are amphoteric. PbO_2 is thermally unstable and decomposes to PbO. Tin(II) oxide, SnO, and lead(II) oxide, PbO, are more basic and have greater ionic character than tin(IV) oxide, SnO_2, and lead(IV) oxide, PbO_2. This is consistent with the general trend that oxides and hydroxides of an element in a higher oxidation state are more acidic and covalent than those of the same element in lower oxidation states.

Both Sn^{2+} and Pb^{2+} ions from *salts* hydrolyze (Sn^{2+} quite extensively) to produce acidic solutions.

$$[Sn(OH_2)_6]^{2+} + H_2O \rightleftharpoons [Sn(OH_2)_5(OH)]^+ + H_3O^+ \qquad K_a \approx 10^{-2}$$

$$[Pb(OH_2)_6]^{2+} + H_2O \rightleftharpoons [Pb(OH_2)_5(OH)]^+ + H_3O^+ \qquad K_a \approx 10^{-8}$$

Aqueous solutions of Sn^{2+} salts are acidified to inhibit hydrolysis and subsequent precipitation of various basic salts. This, of course, is an application of LeChatelier's Principle.

Lead(IV) compounds can act as oxidizing agents, as can be seen by the very positive standard reduction potential.

$$Pb^{4+} (aq) + 2e^- \longrightarrow Pb^{2+} (aq) \qquad E^0 = +1.8 \text{ V}$$

Bismuth (Group VA) is a dense metal with a yellowish tinge sometimes found in the uncombined form in nature. It is also found as bismuth(III) oxide, Bi_2O_3, and as bismuth(III) sulfide, Bi_2S_3. It is incorporated into many low-melting alloys to take advantage of its unusual ability to expand upon freezing, thus giving sharp impressions of mold-cast objects. Low-melting bismuth alloys are used in such things as fire alarms, automatic sprinkler systems, and electrical fuses. Bismuth subcarbonate, $(BiO)_2CO_3$, and bismuth subnitrate, $(BiO)NO_3$, are used medically in the treatment of stomach and skin disorders.

Bismuth exists primarily in the $+3$ oxidation state, and only rarely in the $+5$ oxidation state. The most common oxide, Bi_2O_3, is distinctly basic. It dissolves in acids to produce bismuth(III) salts but does not dissolve in bases.

$$Bi_2O_3 (s) + 6H^+ (aq) \longrightarrow 2Bi^{3+} (aq) + 3H_2O$$

Solutions of bismuth(III) salts hydrolyze so extensively that a precipitate of a basic

Heavy metals tend to accumulate in the body; they often inhibit enzymes that catalyze reactions in the body.

Sn(II) = stannous;
Sn(IV) = stannic;
Pb(II) = plumbous;
Pb(IV) = plumbic

salt of bismuth forms. For solutions of $BiCl_3$, the reaction is

The first step hydrolysis constant is 1.0×10^{-2}.

$$Bi^{3+} (aq) + 2H_2O + Cl^- \longrightarrow Bi(OH)_2Cl (s) + 2H^+ (aq)$$

$$Bi(OH)_2Cl (s) \longrightarrow BiOCl (s) + H_2O$$

Other Bi^{3+} salts react similarly. Hydrolysis can be prevented by acidification. Bismuth(V) in sodium bismuthate, $NaBiO_3$, is a powerful oxidizing agent.

$$NaBiO_3 (s) + 6H^+ (aq) + 2e^- \longrightarrow Bi^{3+} (aq) + Na^+ (aq) + 3H_2O \qquad E^0 = \sim 1.6 \text{ V}$$

The Transition Metals

Remember that oxides of most nonmetals are acidic and oxides of most metals are basic.

The term "transition elements" was originally coined to denote the elements in the middle of the periodic table that provide a transition between the "base formers" on the left and the "acid formers" on the right. It actually applies to both the d- and f-transition elements, all of which are metals, but we commonly use the term "transition metals" to refer to only the more frequently encountered d-transition metals.

20-5 General Properties of the Transition Metals

These metals are located between Groups IIA and IIIA in the periodic table. Strictly speaking, in order for an element to be classified a d-transition metal, it must have a partially filled set of d orbitals. Zinc, cadmium, and mercury (the Group IIB elements) and their cations have completely filled sets of d orbitals, and therefore are not really d-transition metals. However, they are often discussed with them because of the similarities of their properties to those of the transition metals. All of the other elements in this region have partially filled sets, except copper, silver, and gold (the IB elements) and palladium, all of which have completely filled sets. However, some of the cations of these latter elements have only partially filled sets of d orbitals.

The following general properties are characteristic of the transition elements.

1. All are metals.
2. Most are harder, more brittle, and have higher melting points and boiling points and higher heats of vaporization than non-transition metals.
3. Their ions and compounds are often colored because they have one or more unpaired electrons (see Chapter 23).
4. They form many complex ions (see Chapter 23).
5. With few exceptions, they exhibit multiple oxidation states.
6. Many of them are paramagnetic, as are many of their compounds.
7. Many of the metals and their compounds are good catalysts.

Some specific properties of the $3d$-transition metals are tabulated in Table 20-8. Properties of the transition metals vary somewhat more irregularly than do those of the representative elements because the differentiating electrons are one shell *inside* the outermost occupied shell. Let us consider some of these properties.

TABLE 20–8 Properties of the First Transition Series

Property	Scandium	Titanium	Vanadium	Chromium	Manganese	Iron	Cobalt	Nickel	Copper	Zinc
melting point (°C)	1541	1660	1890	1900	1244	1535	1495	1453	1083	420
boiling point (°C)	2831	3287	3380	2672	1962	2750	2870	2732	2567	907
density (g/cm³)	3.0	4.51	6.11	7.9	7.2	7.87	8.7	8.91	8.94	7.13
atomic radius (nm)	0.162	0.147	0.134	0.125	0.129	0.126	0.125	0.124	0.128	0.138
ionic radius, M^{2+} (nm)	—	0.094	0.088	0.089	0.080	0.074	0.072	0.069	0.070	0.074
electronegativity	1.3	1.4	1.5	1.6	1.6	1.7	1.8	1.8	1.8	1.6
standard reduction potential (V) M^{2+} (aq) $+ 2e^- \longrightarrow$ M (s)	−2.08*	−1.63	−1.2	−0.91	−1.18	−0.44	−0.28	−0.25	+0.34	−0.76
ionization energy (kJ/mol)										
first	631	658	650	653	717	759	758	640	745	906
second	1235	1310	1414	1592	1509	1561	1646	1753	1958	1733

* For Sc^{3+} (aq) $+ 3e^- \longrightarrow$ Sc (s).

1 Electronic Configurations and Oxidation States

The electronic configurations of the three d-transition series are given in Table 20–9. The properties of the transition metals can be correlated roughly with either the total number of d electrons or the number of unpaired electrons.

One of the characteristics of most transition metals is the ability to exhibit more than one oxidation state. The *maximum* oxidation state is given by a metal's group number, but this is usually not the most stable oxidation state.

The outer s electrons, of course, lie outside the d electrons and are always the first ones lost in ionization. All the metals of the first transition series except scandium and zinc exhibit more than one oxidation state. Scandium loses its two $4s$ electrons and its only $3d$ electron to form Sc^{3+}, its single oxidation state. Zinc loses its two $4s$ electrons to give only Zn^{2+}.

Recall that in "building" atoms by the Aufbau Principle, the outer s orbitals are filled before the inner d orbitals.

$$
\begin{array}{ccc}
 & 3d & 4s \\
\end{array}
$$

$_{21}$Sc [Ar] $\uparrow$ $\quad$ $\uparrow\downarrow$ $\xrightarrow{-3e^-}$ $_{21}Sc^{3+}$ [Ar]

$_{30}$Zn [Ar] $\uparrow\downarrow$ $\uparrow\downarrow$ $\uparrow\downarrow$ $\uparrow\downarrow$ $\uparrow\downarrow$ $\uparrow\downarrow$ $\xrightarrow{-2e^-}$ $_{30}Zn^{2+}$ [Ar]$3d^{10}$

All of the other $3d$-transition metals exhibit at least two oxidation states. For example, cobalt can form Co^{2+} and Co^{3+} ions.

TABLE 20–9 Electronic Configurations of the d-Transition Metals

Period 4	Period 5	Period 6
$_{21}$Sc [Ar]$3d^1 4s^2$	$_{39}$Y [Kr]$4d^1 5s^2$	$_{57}$La [Xe]$5d^1 6s^2$
$_{22}$Ti [Ar]$3d^2 4s^2$	$_{40}$Zr [Kr]$4d^2 5s^2$	$_{72}$Hf [Xe]$4f^{14} 5d^2 6s^2$
$_{23}$V [Ar]$3d^3 4s^2$	$_{41}$Nb [Kr]$4d^4 5s^1$	$_{73}$Ta [Xe]$4f^{14} 5d^3 6s^2$
$_{24}$Cr [Ar]$3d^5 4s^1$	$_{42}$Mo [Kr]$4d^5 5s^1$	$_{74}$W [Xe]$4f^{14} 5d^4 6s^2$
$_{25}$Mn [Ar]$3d^5 4s^2$	$_{43}$Tc [Kr]$4d^5 5s^2$	$_{75}$Re [Xe]$4f^{14} 5d^5 6s^2$
$_{26}$Fe [Ar]$3d^6 4s^2$	$_{44}$Ru [Kr]$4d^7 5s^1$	$_{76}$Os [Xe]$4f^{14} 5d^6 6s^2$
$_{27}$Co [Ar]$3d^7 4s^2$	$_{45}$Rh [Kr]$4d^8 5s^1$	$_{77}$Ir [Xe]$4f^{14} 5d^7 6s^2$
$_{28}$Ni [Ar]$3d^8 4s^2$	$_{46}$Pd [Kr]$4d^{10}$	$_{78}$Pt [Xe]$4f^{14} 5d^9 6s^1$
$_{29}$Cu [Ar]$3d^{10} 4s^1$	$_{47}$Ag [Kr]$4d^{10} 5s^1$	$_{79}$Au [Xe]$4f^{14} 5d^{10} 6s^1$
$_{30}$Zn [Ar]$3d^{10} 4s^2$	$_{48}$Cd [Kr]$4d^{10} 5s^2$	$_{80}$Hg [Xe]$4f^{14} 5d^{10} 6s^2$

$$
\begin{array}{ccc}
& 3d \qquad\qquad 4s & \qquad\qquad 3d \qquad\quad 4s \\
{}_{27}\text{Co [Ar]}\ \uparrow\downarrow\ \uparrow\downarrow\ \uparrow\ \uparrow\ \uparrow\ \uparrow\downarrow & \xrightarrow{-2e^-} & {}_{27}\text{Co}^{2+}\text{[Ar]}\ \uparrow\downarrow\ \uparrow\downarrow\ \uparrow\ \uparrow\ \uparrow\ \underline{\quad} \\
{}_{27}\text{Co [Ar]}\ \uparrow\downarrow\ \uparrow\downarrow\ \uparrow\ \uparrow\ \uparrow\ \uparrow\downarrow & \xrightarrow{-3e^-} & {}_{27}\text{Co}^{3+}\text{[Ar]}\ \uparrow\downarrow\ \uparrow\ \uparrow\ \uparrow\ \uparrow\ \underline{\quad}
\end{array}
$$

Simple aqueous salts of cobalt(III) oxidize water and are reduced to cobalt(II) salts.

2 Atomic Radii

The atomic radii of the d-transition (and IIB) metals are given in Table 20–10. Like the representative elements, the d-transition metals generally decrease in atomic radii from left to right across a given period as the nuclear charge (and therefore the nuclear attraction for electrons in the outer shell) increases. But these decreases are not as consistent as for the representative elements. The radii of elements at the end of a transition series increase because the increases in effective nuclear charge are outweighed by greater repulsions among d electrons in a given set of orbitals.

The lanthanide contraction has important effects on the properties of the sixth period transition and post-transition elements.

The radii of representative elements increase substantially upon descending a group, as electrons occupy energy levels farther from the nucleus. Likewise, the transition metals of the fifth period have larger radii than those of the fourth period. But the transition metals of the sixth period have nearly the same radii as the metals above them. This phenomenon is known as the **lanthanide contraction,** following lanthanum (atomic number 57). The insertion of fourteen lanthanides, with fourteen poorly-shielding f electrons two shells inside the outermost shell (and fourteen positive charges in the nucleus), preceding hafnium (atomic number 72) results in higher effective nuclear charges felt by the outermost electrons than would be predicted if f electrons were good charge shielders. The higher effective nuclear charges tend to pull the outer electrons closer to the nucleus, which produces smaller radii than would be anticipated otherwise. The radii of the d-transition metals are plotted against periodic position in Figure 20–2.

The f electrons shield outer electrons less effectively than s or p electrons.

3 Densities

As expected, densities vary inversely with the atomic radii (Tables 20–8 and 20–10, and Figure 20–3). Densities increase as the radii decrease within given periods, and increase upon descending a group. In going from the fifth period to the sixth period, the metals have considerably more protons and neutrons but nearly the same atomic volume. Thus, the sixth-period transition metals are very dense because of the lanthanide contraction.

Osmium and iridium, with densities of 22.6 g/cm³, are the densest elements known.

TABLE 20–10 Atomic Radii of the *d*-Transition Metals (nm)

Period 4	Sc	Ti	V	Cr	Mn	Fe	Co	Ni	Cu	Zn
	0.162	0.147	0.134	0.125	0.129	0.126	0.125	0.124	0.128	0.138
Period 5	Y	Zr	Nb	Mo	Tc	Ru	Rh	Pd	Ag	Cd
	0.180	0.160	0.146	0.139	0.136	0.134	0.134	0.137	0.144	0.154
Period 6	La	Hf	Ta	W	Re	Os	Ir	Pt	Au	Hg
	0.187	0.158	0.146	0.139	0.137	0.135	0.136	0.138	0.144	0.157

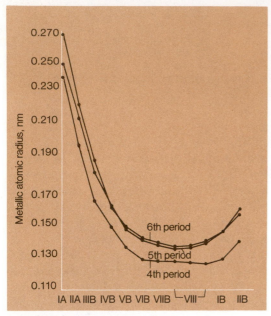

FIGURE 20-2 Metallic atomic radii of alkali, alkaline earth, and transition metals of the 4th, 5th, and 6th periods.

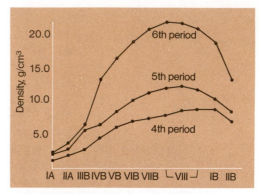

FIGURE 20-3 Densities of the alkali, alkaline earth, and transition metals of the 4th, 5th, and 6th periods.

4   Paramagnetism

Many transition metals and metal ions have one or more unpaired electrons and are paramagnetic (Section 5–10). Figure 20–4 shows the good agreement between atomic theory and experiment in correlating the number of unpaired electrons with the observed degree of paramagnetism. Magnetic measurements are also important in explaining the colors and bonding of the transition metal compounds, as will be seen in Chapter 23.

5   Melting Points

Other properties of transition metals that are roughly related to the number of unpaired electrons are melting point and boiling point. The transition metals, of course, form metallic crystals. The strength of metallic bonding increases with the availability of electrons to participate in the bonding by delocalization over the lattice (Section 12–15). Thus, note the similarity of the shapes of the plots in Figures 20–4 and 20–5. Since the alkali and alkaline earth metals have only one or two outer electrons available, their melting points are relatively low in comparison to other metals. But the increasing availability of d electrons, especially unpaired electrons, causes an increase in melting point of the transition metals from the beginning toward the middle of each transition series. As electron pairing increases toward the right of the transition series, melting points decrease.

Although most transition metals are hard with fairly high melting points, the elements of Group IIB [zinc, cadmium, and mercury (a liquid)] are quite soft with low melting points. This might be expected, since they have pseudo-noble gas electron configurations and no unpaired electrons available for metallic bonding.

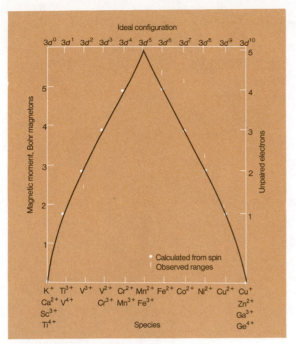

FIGURE 20–4 Magnetic properties of ions in the first transition series and a few transition metal ions.

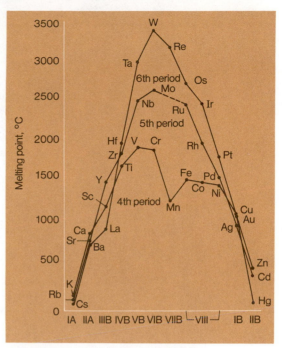

FIGURE 20–5 Melting points of alkali, alkaline earth, and transition metals of the 4th, 5th, and 6th periods.

20–6 Some Transition Metals of Period 4

Let us now describe the properties and chemistries of the more common "first row" transition metals, the elements that are filling the $3d$ sublevel.

1 Chromium ([Ar] $3d^5 4s^1$)

Chromium is a lustrous, hard metal that is very resistant to corrosion because it forms a transparent, protective oxide coat. We are familiar with the use of metallic chromium as an electrolytically deposited coating (chrome) for other metal objects, such as steel automobile bumpers.

The most important ore of chromium is *chromite* (also called chrome-iron ore), $FeCr_2O_4$. The ore is reduced by carbon in an electric furnace to form an alloy called ferrochrome.

$$FeCr_2O_4 + 4C \longrightarrow \underbrace{Fe + 2Cr}_{\text{ferrochrome}} + 4CO$$

Ferrochrome is used in making chromium steels. Stainless steel contains 14 to 18% Cr and is very resistant to corrosion. Chrome-vanadium steel contains 1 to 10% Cr together with about 0.15% vanadium. These steels are very strong, and are used to make axles that must withstand constant strains as well as frequent shock and vibration.

TABLE 20–11   Some Common Chromium Compounds

Oxidation State	Oxide	Hydroxide	Name	Acid or Base Character of Oxide or Hydroxide	Related Salt	Name
+2	CrO black	Cr(OH)$_2$	chromous or chromium(II) hydroxide	basic	CrCl$_2$ anhydrous, colorless; aqueous, lt. blue	chromous or chromium(II) chloride
+3	Cr$_2$O$_3$ green	Cr(OH)$_3$	chromic or chromium(III) hydroxide	amphoteric	CrCl$_3$ anhydrous, violet; aqueous, green	chromic or chromium(III) chloride (acid sol'n)
					KCrO$_2$ green	potassium chromite (basic solution)
+6	CrO$_3$ dk. red	H$_2$CrO$_4$ [CrO$_2$(OH)$_2$]	chromic acid	weakly acidic	K$_2$CrO$_4$ yellow	potassium chromate
		H$_2$Cr$_2$O$_7$ [Cr$_2$O$_5$(OH)$_2$]	dichromic acid	acidic	K$_2$Cr$_2$O$_7$ orange	potassium dichromate

Pure chromium is prepared from Cr$_2$O$_3$ by a thermite reaction.

$$Cr_2O_3 \text{ (s)} + 2Al \text{ (s)} \longrightarrow Al_2O_3 \text{ (s)} + 2Cr \text{ (s)} \qquad \Delta H^0 = -536 \text{ kJ}$$

Typical of the metals near the middle of each transition series, chromium shows a number of oxidation states, of which +2, +3, and +6 are the most common. The +3 state is most stable. The most common oxides are CrO (basic), Cr$_2$O$_3$ (amphoteric), and CrO$_3$ (acidic), as shown in Table 20–11.

Chromium is a strong reducing agent, as are solutions of blue Cr(II) salts, which are air-oxidized to gray-green Cr^{3+} (aq).

$$Cr^{2+} \text{ (aq)} + 2e^- \longrightarrow Cr \text{ (s)} \qquad E^0 = -0.91V$$

$$Cr^{3+} \text{ (aq)} + e^- \longrightarrow Cr^{2+} \text{ (aq)} \qquad E^0 = -0.41V$$

Chromium(VI) species are oxidizing agents. Basic solutions containing the chromate ion, CrO$_4^{2-}$, are weakly oxidizing; but acidification produces the dichromate ion, Cr$_2$O$_7^{2-}$ (also chromium(VI) oxide), a powerful oxidizing agent.

$$Cr_2O_7^{2-} \text{ (aq)} + 14H^+ \text{ (aq)} + 6e^- \longrightarrow 2Cr^{3+} \text{ (aq)} + 7H_2O \text{ (}\ell\text{)} \qquad E^0 = +1.33 \text{ volts}$$

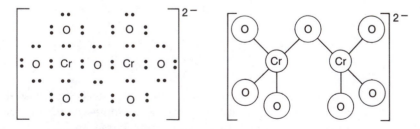

Red chromium(VI) oxide, CrO$_3$, is the acid anhydride of two acids, chromic acid, H$_2$CrO$_4$, and dichromic acid, H$_2$Cr$_2$O$_7$. Neither acid has ever been isolated in the pure form, although chromate and dichromate salts are common. CrO_3 dissolves in water to produce a strongly acidic solution containing hydrogen ions and (predominantly) dichromate ions.

$$2CrO_3 + H_2O \longrightarrow \underbrace{[2H^+ + Cr_2O_7{}^{2-}]}_{\substack{\text{dichromic acid} \\ \text{(red-orange)}}}$$

From such solutions orange dichromate salts can be crystallized after adding a stoichiometric amount of base. Addition of excess base produces a yellow solution from which only yellow chromate salts can be obtained. The two Cr(VI) anions exist in solution in a pH-dependent equilibrium as follows:

$$2CrO_4{}^{2-} + 2H^+ \rightleftharpoons Cr_2O_7{}^{2-} + H_2O$$

yellow   orange

$$K_c = \frac{[Cr_2O_7{}^{2-}]}{[CrO_4{}^{2-}]^2[H^+]^2} = 4.2 \times 10^{14}$$

We see that decreasing the concentration of H^+ by addition of OH^- produces more $CrO_4{}^{2-}$, while acidification favors the production of $Cr_2O_7{}^{2-}$.

Dehydration of a chromate or dichromate salt with concentrated sulfuric acid, an excellent dehydrating agent, produces chromium(VI) oxide, CrO_3, a strong oxidizing agent. A powerful "cleaning solution" used for removing greasy stains and coatings from laboratory glassware is made by adding concentrated H_2SO_4 to a concentrated solution of potassium dichromate. The active ingredients are CrO_3, an oxidizing agent, and sulfuric acid, an excellent solvent.

Sodium dichromate, $Na_2Cr_2O_7$, is used as an oxidizing agent in the chrome tanning process in which hides are converted to leather, which is tough, pliable, and resistant to biological decay. Lead chromate, $PbCrO_4$, which is yellow, and green Cr_2O_3 are used commercially as paint pigments.

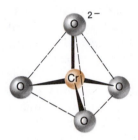

Chromate ion, $CrO_4{}^{2-}$

This cleaning solution should be handled with **extreme care,** as dangerous amounts of heat can be generated, e.g., when organic solvents are attacked by this mixture.

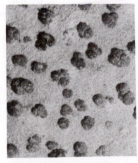

Manganese nodules on the sea floor. They also contain copper, nickel, and cobalt.

2 Manganese ([Ar] $3d^5 4s^2$)

Manganese is a brittle gray metal with a reddish tinge, which corrodes in moist air. It is the eleventh most abundant element in the earth's crust. The ores of manganese are oxides, with *pyrolusite*, $MnO_2 \cdot xH_2O$, being the most important. It also occurs as MnO_2 in "manganese nodules" in the sea. Since manganese is used primarily to alloy with iron, the ore is usually reduced with coke in a blast furnace (Section 20–11), which also reduces iron oxides, and the impure mixture of metals is used to make steel.

$$MnO_2 + 2C \longrightarrow Mn \text{ (impure)} + 2CO$$

Steel that contains 10 to 18% manganese is hard, tough, and resistant to wear. It is used to make rock crushing machinery, armor plate, railroad rails, and similar items.

Manganese exhibits more oxidation states than any other $3d$ transition metal, from +2 to +7, with the +2 oxidation state being most stable under ordinary conditions. Potassium permanganate, in which manganese is in the +7 oxidation state, is a very strong oxidizing agent and common laboratory reagent.

$$MnO_4{}^- \text{ (aq)} + 8H^+ \text{ (aq)} + 5e^- \longrightarrow Mn^{2+} \text{ (aq)} + 4H_2O \text{ }(\ell) \qquad E^0 = +1.51 \text{ V}$$

Manganese(II) ion in aqueous solution, $Mn(OH_2)_6{}^{2+}$, is very pale pink, nearly colorless.

MnO₂ is a component of dry cell batteries (Section 19–21).

It is often used in redox titrations, in which its deep purple color serves as its own indicator (Section 10–5.2).

As an oxidizing agent in basic solution, $MnO_4{}^-$ is readily reduced to MnO_2 rather than Mn^{2+}.

$$MnO_4{}^- \text{ (aq)} + 2H_2O + 3e^- \longrightarrow MnO_2 \text{ (s)} + 4OH^- \qquad E^0 = +0.588 \text{ V}$$

3   Iron ([Ar] $3d^6 4s^2$)

Iron is the second most abundant metal (4.7%) in the earth's crust after aluminum. It is the most widely used of all metals. Iron was known as early as 4000 BC, because of its wide distribution and ease of reduction to the free metal. The red color of many rocks, clays, and soils is due to the presence of Fe_2O_3, the compound formed when iron rusts.

Iron is a silvery-white metal of high tensile strength that takes a high polish. *Pure* iron is ductile and relatively soft compared to other metals; it has little practical use except when alloyed. The primary ores are *hematite*, Fe_2O_3, *magnetite*, Fe_3O_4, *siderite*, $FeCO_3$, and *taconites*, which are iron oxides containing silicates. The metallurgy of iron will be discussed in Section 20–11.

Iron is widely used because its properties can be modified by alloying with a variety of elements. Additionally, it can be protected from corrosion by many different kinds of coatings. For example, iron can be coated with (1) other metals such as Zn, Sn, Cu, Ni, Cr, Cd, and Pb; (2) ceramic materials like those used in bathtubs and refrigerators; or (3) organic materials such as paint, lacquer, or asphalt.

An interesting property of pure iron, as well as cobalt and nickel, is **ferromagnetism,** which is related to paramagnetism but is thousands of times stronger. Ferromagnetic substances become permanently magnetized when placed in a magnetic field. This effect is the result of the alignment of spins of unpaired electrons of *different* iron atoms at certain optimum distances from each other to form regions called **domains** containing a million or more atoms. Certain alloys, not necessarily containing Fe, Co, or Ni, also exhibit ferromagnetism.

The common oxidation states are +2 and +3. Solutions of Fe^{2+} (aq) are readily air-oxidized to Fe^{3+} (aq). Solutions of Fe^{3+} (aq) salts hydrolyze strongly (Section 17–9.5) to produce acidic solutions.

Iron dissolves in nonoxidizing acids to form iron(II) salts and liberate hydrogen. Iron(II) sulfate, $FeSO_4$, is used as a weed killer and wood preservative.

$$Fe \text{ (s)} + 2H^+ \text{ (aq)} \longrightarrow Fe^{2+} \text{ (aq)} + H_2 \text{ (g)} \qquad E^0 = +0.44 \text{ V}$$

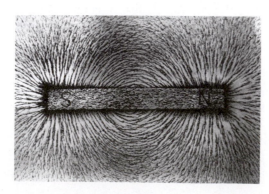

Iron filings aligned in the field of a bar magnet demonstrate ferromagnetism.

However, iron(II) ions are oxidized to iron(III) ions by mild oxidizing agents, such as the oxygen in the air.

$$4Fe^{2+} (aq) + 4H^+ (aq) + O_2 (g) \longrightarrow 4Fe^{3+} (aq) + 2H_2O (\ell) \qquad E^0 = +0.47 \text{ V}$$

Interestingly, iron is passive toward concentrated nitric acid because it quickly builds up a thin protective oxide coat. The corrosion or rusting of iron and corrosion protection are described in Sections 19–17 and 19–18.

4 Copper ([Ar] $3d^{10}4s^1$)

Copper is a relatively soft, reddish-yellow metal that is an excellent conductor of heat and electricity. Copper was known to many ancient civilizations, and it was probably the first metal used to fabricate tools and utensils.

Copper sometimes occurs as the free metal. The largest known deposits are near Houghton, Michigan, and the largest piece found to date weighs more than 400 tons.

The commercially important ores of copper are copper(I) sulfide (Cu_2S, *chalcocite*) and a mixed sulfide ($CuFeS_2$, *chalcopyrite*), which contains less than 10% copper. Most known sources of the high grade oxide ores *cuprite*, Cu_2O, and *melaconite*, CuO, have been exhausted. The metallurgy of copper will be discussed in Section 20–11. The metal is refined electrolytically (Section 19–7).

Copper is considered second in importance to iron, the most widely used metal, because of its excellent thermal and electrical conductivity, chemical inertness, and usefulness in alloying elements. Copper is used extensively in alloys such as brass (Cu-Zn), bronze (Cu-Zn-Sn), sterling silver (Cu-Ag), aluminum bronze (Cu-Al), and German silver (Cu-Zn-Ni).

Copper pipes are used extensively in plumbing because it is easy to "work" and because it does not react with hot or cold water at an appreciable rate. Copper statues and roofs turn brown as a thin adherent film of copper oxide or copper sulfide forms. On prolonged exposure to the atmosphere, such items eventually turn green due to the formation of the basic carbonate, $Cu(OH)_2 \cdot CuCO_3$.

Trace amounts of copper are found in plants that grow near deposits of copper ores, and a copper compound (hemocyanin) serves the same oxygen-carrying function in the fluids of lobsters and oysters that the iron compound hemoglobin serves in the blood of higher animals.

Copper is the only one of the $3d$ transition metals that will not dissolve in nonoxidizing acids; it is below hydrogen in the activity series (Section 19–14). However, it will dissolve readily in nitric acid to form sky blue Cu^{2+} (aq), which is $Cu(OH_2)_4^{2+}$. The common oxidation states are $+1$ and $+2$, with $+2$ the more stable state. The cuprous ion, Cu^+, disproportionates in water to copper metal and cupric ion, Cu^{2+}.

		E^0
reduction:	$Cu^+ (aq) + e^- \longrightarrow Cu (s)$	$+0.521V$
oxidation:	$Cu^+ (aq) \longrightarrow Cu^{2+} (aq) + e^-$	$-0.153V$
overall:	$2Cu^+ (aq) \longrightarrow Cu (s) + Cu^{2+} (aq)$	$E^0_{cell} = +0.368V$

Copper(II) is toxic in greater than trace amounts to bacteria, fungi, and algae, so it is used in the form of copper(II) sulfate, $CuSO_4$, and a few other compounds to control such pests.

The fact that copper utensils were used by American Indians in the southern part of the country is taken as evidence for trade among widely separated tribes.

Copper is used in the manufacture of U.S. coins, both in pennies and in copper-clad dimes and quarters.

5 Zinc ([Ar] $3d^{10}4s^2$)

Zinc is a Group IIB element, and strictly speaking it is not a d-transition metal. It is a reactive metal, but considerably less so than aluminum. Its only nonzero oxidation state is +2. Most compounds of Zn^{2+} are colorless.

$$Zn^{2+} (aq) + 2e^- \longrightarrow Zn (s) \qquad E^0 = -0.76 \text{ V}$$

Like aluminum, zinc becomes coated with a protective oxide layer. Galvanizing is the process of coating a metal object (usually iron) with zinc. The zinc is then converted to the blue-gray basic carbonate, $Zn(OH)_2 \cdot ZnCO_3$, on exposure to moist air. Even if the basic carbonate is scratched away, the zinc will protect the metal object by undergoing corrosion (oxidation) itself.

Zinc hydroxide is amphoteric (Table 9–6). The metal dissolves in both nonoxidizing acids and strong soluble bases with the evolution of hydrogen.

The main ore of zinc contains zinc sulfide, ZnS, and is called *zinc blende* or *sphalerite.* Cadmium, iron, lead, and arsenic sulfides usually occur in zinc sulfide ores.

Zinc is used in many alloys and in dry cell batteries (Section 19–21). It also provides cathodic protection for many underground steel pipelines (Section 19–18). Ointments containing white zinc oxide, ZnO, act as sunscreens and antiseptics. Along with white zinc sulfide, ZnO serves as a white paint pigment.

The outer casings of dry cell batteries are made of zinc.

Metallurgy

Mixtures of metals frequently have properties that are more desirable for a particular purpose than are those of the free metal. In such cases metals are alloyed.

Metallurgy is the term applied to the commercial extraction of metals from their ores and the preparation of the metals for use. It is usually divided into a sequence of steps: (1) mining, (2) pretreatment of the ore, (3) reduction of the ore to the free metal, (4) refining or purifying the metal, and (5) alloying, or mixing with other elements, if necessary. Before describing each individually, let us first see the predominant forms in which the metals occur naturally.

20–7 Occurrence of Metals

Metals with negative standard reduction potentials (that is, active metals; those that are easily oxidized) are found only in the combined state in nature. Those with

The most widespread minerals are silicates, but because extraction of metals from silicates is very difficult, metals are obtained from silicate minerals only when there is no other more economical alternative.

TABLE 20–12 Common Classes of Ores

Anion	Examples and Name of Mineral
none (Native ores)	Au, Ag, Pt, Os, Ir, Ru, Rh, Pd, As, Sb, Bi
oxide	hematite, Fe_2O_3; magnetite, Fe_3O_4; bauxite, Al_2O_3; cassiterite, SnO_2; periclase, MgO; silica, SiO_2
sulfide	chalcopyrite, $CuFeS_2$; chalcocite, Cu_2S; sphalerite, ZnS; galena, PbS; iron pyrites, FeS_2; cinnabar, HgS
chloride	rock salt, NaCl; sylvite, KCl; carnallite, $KCl \cdot MgCl_2$
carbonate	limestone, $CaCO_3$; magnesite, $MgCO_3$; dolomite, $MgCO_3 \cdot CaCO_3$
sulfate	gypsum, $CaSO_4 \cdot 2H_2O$; epsom salts, $MgSO_4 \cdot 7H_2O$; barite, $BaSO_4$
silicate	beryl, $Be_3Al_2Si_6O_{18}$; kaolinite, $Al_2(Si_2O_8)(OH)_4$; spodumene, $LiAl(SiO_3)_2$

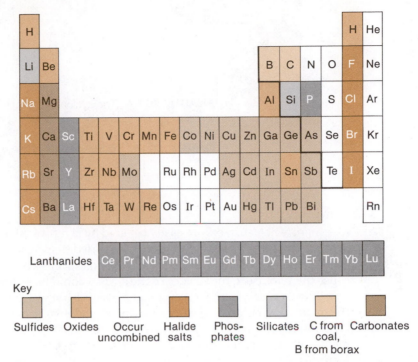

Key

| Sulfides | Oxides | Occur uncombined | Halide salts | Phos-phates | Silicates | C from coal, B from borax | Carbonates |

FIGURE 20–6 Natural sources of the elements. The soluble halide salts are found in the ocean or in solid deposits. Most of the rare gases are obtained from air.

positive reduction potentials, the less active metals, may occur in the uncombined free state or **native ores.** Examples of the latter are copper, silver, gold, and the less abundant platinum, osmium, iridium, ruthenium, rhodium, and palladium. Copper, silver, and gold are also found in the combined state.

Many rather insoluble compounds of the metals are found in the earth's crust. Solids containing these compounds are the **ores** from which metals are extracted. Ores contain **minerals,** comparatively pure compounds of the metal of interest, mixed with relatively large amounts of sand, soil, clay, rock and other material called **gangue,** a German term meaning mining vein. Water-soluble metal compounds are usually found dissolved in the sea or in salt beds in areas where large bodies of water have evaporated. Metal ores can be classified in terms of the anions with which the metal ion is combined. The most common types are listed with examples in Table 20–12. (See Figure 20–6.)

20–8   Pretreatment of Ores

After an ore is mined, but before it is reduced to the free metal, it must be concentrated by removal of most of the gangue. Most sulfides have relatively high densities and are more dense than gangue. After pulverization the lighter gangue particles are removed by a variety of methods. One involves blowing the lighter particles away using a cyclone separator (Figure 20–7). In some cases the lighter particles are sifted out through layers of vibrating wire mesh or inclined vibration tables.

The **flotation** method is particularly applicable to sulfides, carbonates, and silicates, which either are not "wet" by water or can be made water-repellent by

These are called hydrophobic materials.

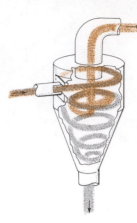

FIGURE 20–7 The cyclone separator, used for enriching metal ores. The crushed ore is blown at high velocity into the separator, where centrifugal force takes the heavier particles, with a high percentage of the desired metal, to the wall of the separator. From there, these particles spiral down to the collection bin at the bottom. Lighter particles, not as rich in the metal, move into the center of the separator, and are carried out the top in the air stream.

treatment. Their surfaces are easily covered by layers of oil or other flotation agents. When a stream of air is blown through a swirled suspension of such an ore in water and oil (or other agent), bubbles form on the oil surfaces on the mineral particles and cause them to rise to the surface of the suspension. The bubbles are prevented from breaking and escaping by a layer of oil and emulsifying agent at the surface, where a frothy ore concentrate forms. By varying the relative amounts of oil and water, the types of oil additive, the air pressure, and so on, it is even possible to separate one metal sulfide, carbonate, or silicate from another. A flotation apparatus is shown in Figure 20–8.

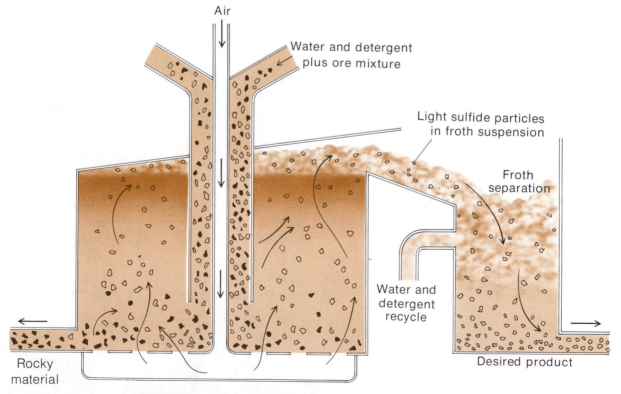

FIGURE 20–8 Flotation process for enrichment of copper sulfide ore. The relatively light sulfide particles are put into suspension in the water-oil-detergent mixture and collected as a froth. The denser material sinks to the bottom of the container.

FIGURE 20–9 Coppertown Basin (Ducktown), Tennessee, photographed in 1943. Copper ore (principally copper sulfide, Cu_2S) had been mined and smelted in this area since 1847. In the early years, large quantities of sulfur dioxide, a by-product, were discharged directly into the atmosphere and killed all vegetation for miles around the smelter. Today the sulfur is reclaimed in the exhaust stacks to make sulfuric acid, but the denuded soil remains a monument to the misuse of the atmosphere.

Another pretreatment process involves chemical modification in order to convert metal compounds to a more easily reduced form. Carbonates and hydroxides may be heated to drive off carbon dioxide and water, respectively, as shown by the following equations.

$$CaCO_3 \text{ (s)} \xrightarrow{\Delta} CaO \text{ (s)} + CO_2 \text{ (g)}$$

$$Mg(OH)_2 \text{ (s)} \xrightarrow{\Delta} MgO \text{ (s)} + H_2O \text{ (g)}$$

Some sulfides are converted to oxides by **roasting,** i.e., heating below their melting points in the presence of oxygen from air. For example,

$$2ZnS \text{ (s)} + 3O_2 \text{ (g)} \xrightarrow{\Delta} 2ZnO \text{ (s)} + 2SO_2 \text{ (g)}$$

Roasting of sulfides has caused problems of air pollution due to the escape of enormous quantities of SO_2 into the atmosphere (Section 9–7.4). Figure 20–9 shows the damage caused by SO_2 from roasting Cu_2S in Coppertown Basin, Tennessee in the early 1940's. Federal regulations now require limitations of the amount of SO_2 that escapes with the stack gases and fuel gases, so most of the SO_2 now is trapped and used in the manufacture of sulfuric acid (Section 21–17).

20–9 Reduction to Free Metals

The method used for reduction, or smelting, of metal ores to the free metals depends upon the strength of the bonds between metal ions and anions. The stronger the bonds, the more energy is required to break them (and reduce the metal) and, generally, the more expensive the reduction process. The most active metals usually form the strongest bonds.

The least reactive metals, which can occur in the free state, require no reduction. Examples are gold, silver, and platinum.

Weakly active metals, such as mercury, can be obtained from the sulfide ores by roasting alone, which reduces metal ions to the free metal by oxidation of the sulfide ions.

$$HgS\ (s)\ +\ O_2\ (g) \xrightarrow{\Delta} Hg\ (g)\ +\ SO_2\ (g)$$

cinnabar   from   obtained as vapor;
           air   later condensed

Roasting a more active metal sulfide produces metal oxide, but no free metal.

$$2NiS\ (s)\ +\ 3O_2\ (g) \xrightarrow{\Delta} 2NiO\ (s)\ +\ 2SO_2\ (g)$$

The resulting metal oxides are then reduced to free metals with coke (impure carbon) or carbon monoxide. If carbon must be avoided, other reducing agents such as hydrogen, iron, or aluminum are used.

$$SnO_2\ (s)\ +\ 2C\ (s) \xrightarrow{\Delta} Sn\ (\ell)\ +\ 2CO\ (g)$$

$$WO_3\ (s)\ +\ 3H_2\ (g) \xrightarrow{\Delta} W\ (s)\ +\ 3H_2O\ (g)$$

The very active metals, such as aluminum and sodium, can be reduced only electrochemically, usually from their anhydrous molten salts. If water is present, it is reduced preferentially. Tables 20–13 and 20–14 summarize reduction processes for some metal ions.

TABLE 20–13 Reduction Processes for Some Metals

Metal Ion	Typical Reduction Process
lithium, Li^+ potassium, K^+ calcium, Ca^{2+} sodium, Na^+ magnesium, Mg^{2+} aluminum, Al^{3+}	electrolysis of molten salt
manganese, Mn^{2+} zinc, Zn^{2+} chromium, Cr^{2+}, Cr^{3+} iron, Fe^{2+}, Fe^{3+}	reaction of oxide with coke (carbon) or carbon monoxide (CO)
lead, Pb^{2+} copper, Cu^{2+} silver, Ag^+ mercury, Hg^{2+} platinum, Pt^{2+} gold, Au^+	element occurring free, or easily obtained by roasting the sulfide or oxide ore

(increasing activity of metals)

TABLE 20-14 Some Specific Reduction Processes

Metal	Ore	Reduction Process	Comments
Mercury	HgS, cinnabar	Roast reduction; heat ore in air $HgS + O_2 \xrightarrow{\Delta} Hg + SO_2$	
Copper	Sulfides such as Cu_2S, chalcocite	Blow oxygen through purified molten Cu_2S: $Cu_2S + O_2 \xrightarrow{\Delta} 2Cu + SO_2$	Preliminary ore concentration and purification steps required to remove FeS impurities
Zinc	ZnS, sphalerite	Conversion to oxide and reduction with carbon: $2ZnS + 3O_2 \xrightarrow{\Delta} 2ZnO + 2SO_2$ $ZnO + C \xrightarrow{\Delta} Zn + CO$	Process also used for the production of lead from galena, PbS
Iron	Fe_2O_3, hematite	Reduction with carbon monoxide: $2C\ (coke) + O_2 \xrightarrow{\Delta} 2CO$ $Fe_2O_3 + 3CO \xrightarrow{\Delta} 2Fe + 3CO_2$	
Titanium	TiO_2, rutile	Conversion of oxide to halide salt and reduction with an active metal: $TiO_2 + 2Cl_2 + 2C \xrightarrow{\Delta} TiCl_4 + 2CO$ $TiCl_4 + 2Mg \xrightarrow{\Delta} Ti + 2MgCl_2$	Also used for the reduction of UF_4 obtained from UO_2, pitchblende
Tungsten	$FeWO_4$, wolframite	Reduction with hydrogen: $WO_3 + 3H_2 \xrightarrow{\Delta} W + 3H_2O$	Used for molybdenum also
Aluminum	$Al_2O_3 \cdot nH_2O$, bauxite	Electrolytic reduction (electrolysis) in molten cryolite, Na_3AlF_6, at 1000°C: $2Al_2O_3 \xrightarrow{\Delta} 4Al + 3O_2$	
Sodium	Salt beds, sea water	Electrolysis of molten chlorides: $2NaCl \xrightarrow{\Delta} 2Na + Cl_2$	Also for calcium, magnesium, and other active metals in Groups IA and IIA

20-10 Refining or Purification of Metals

Metals obtained from the procedures described above are almost always impure, and further purification (refining) is required in most cases. This may be accomplished by distillation if the metal is more volatile than its impurities, as in the case of mercury obtained by roasting HgS. Copper, silver, gold, and aluminum are among the metals purified electrolytically (Section 19-7).

Zone refining is often used when extremely pure metals are desired for such applications as solar cells and semiconductors. This method relies on the fact that many impurities will not fit into the lattice of a pure crystal. The process is illustrated in Figure 20-10. An induction heater surrounds a bar of the impure solid and passes slowly from one end to the other. As it passes, it melts portions of the bar, which slowly recrystallize as the heating element moves away. Since the impurity does not fit into the lattice as easily as the element of interest, most of it is carried along in the molten portion until it reaches the end. Repeated passes of the heating element produce a bar of high purity.

The end containing the impurities is sliced off and often recycled.

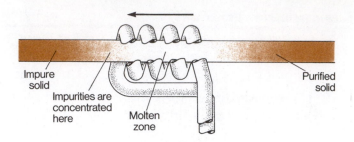

FIGURE 20–10 A zone-refining apparatus.

After purification many metals are alloyed, or mixed with other elements, to change their physical and chemical characteristics. In some cases, certain impurities are allowed to remain during refining because their presence improves the desired properties of the metal. For example, the presence of a small amount of carbon in iron greatly enhances its hardness. Some examples of alloys with improved properties are brass, bronze, duralumin, and stainless steel.

20–11   The Metallurgies of Specific Metals

The metallurgies of aluminum, iron, and copper will now be discussed as specific examples.

1   Aluminum

Aluminum is obtained from bauxite, or hydrated aluminum oxide, $Al_2O_3 \cdot xH_2O$. The aluminum ions can be reduced to metallic aluminum only by electrolysis in the absence of water. The crushed bauxite is first purified by dissolving it in a concentrated solution of sodium hydroxide to form soluble $NaAl(OH)_4$, and then precipitating $Al(OH)_3 \cdot xH_2O$ from the filtered solution by blowing in carbon dioxide to

Recall that Al_2O_3 is amphoteric.

Electrolytic cells used in the Hall process. In the foreground, molten aluminum is being poured.

neutralize the unreacted sodium hydroxide as well as one OH^- ion per formula unit of $NaAl(OH)_4$. The hydrated compound is dehydrated to Al_2O_3 by heating it. The melting point of Al_2O_3 is 2045°C; electrolysis of pure molten Al_2O_3 would have to be carried out at or above this temperature, with great expense. However, it can be done at a much lower temperature when the Al_2O_3 is mixed with much lower-melting cryolite, a mixture of NaF and AlF_3 often represented as Na_3AlF_6. The molten mixture can be electrolyzed at 1000°C with carbon electrodes. The cell used industrially for this process, called the **Hall process,** is shown in Figure 20–11.

> A mixture of compounds typically has a lower melting point than either of the pure compounds.

The inner surface of the cell is plated with carbon or carbonized iron, which functions as the cathode at which aluminum ions are reduced to the free metal. Oxide ions are oxidized to oxygen at the graphite anode, which is consumed almost quantitatively by the oxygen and must be replaced frequently. This represents one of the chief costs of aluminum production.

$$
\begin{aligned}
\text{cathode:} && 4Al^{3+} + 12e^- &\longrightarrow 4Al\,(\ell) \\
\text{anode:} && 6O^{2-} &\longrightarrow 3O_2\,(g) + 12e^- \\
\hline
\text{net reaction:} && 4Al^{3+} + 6O^{2-} &\longrightarrow 4Al\,(\ell) + 3O_2\,(g)
\end{aligned}
$$

Molten aluminum is more dense than molten cryolite, and it collects in the bottom of the cell until drawn off and cooled to a solid.

Recently a more economical process has been developed. The anhydrous bauxite is first converted to $AlCl_3$ by reaction with Cl_2 in the presence of carbon. The $AlCl_3$ is then melted and electrolyzed to aluminum and chlorine, using only about 30% as much electricity as in the Hall process.

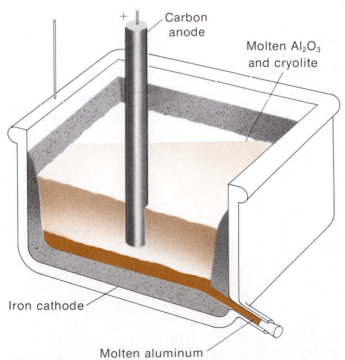

FIGURE 20–11 Schematic drawing of a cell for producing aluminum by electrolysis of a melt of Al_2O_3 in Na_3AlF_6. The molten aluminum collects in the cathode container.

Iron ore is scooped up in an open-pit mine.

2   Iron

The most desirable iron ores contain hematite, Fe_2O_3, or magnetite, Fe_3O_4. As the available supplies of these high grade ores have dwindled, taconite, which is magnetite in a very hard silica rock, has become an important source of iron. The oxide is reduced in blast furnaces (Figure 20–12) by carbon monoxide. Coke mixed with limestone ($CaCO_3$) and crushed ore is loaded at the top of the furnace as the "charge." A blast of hot air from the bottom burns the coke to carbon monoxide with the evolution of more heat.

$$2C \ (s) + O_2 \ (g) \xrightarrow{\Delta} 2CO \ (g) + heat$$

Most of the oxide is reduced to molten iron by the carbon monoxide, although some is reduced directly by coke. Several stepwise reductions occur, but the overall reactions for Fe_2O_3 may be summarized as

$$Fe_2O_3 \ (s) + 3CO \ (g) \xrightarrow{\Delta} 2Fe \ (\ell) + 3CO_2 \ (g) + heat$$

$$Fe_2O_3 \ (s) + 3C \ (s) \xrightarrow{\Delta} 2Fe \ (\ell) + 3CO \ (g) + heat$$

Much of the CO_2 reacts with excess coke to produce more CO to reduce the next incoming charge.

$$CO_2 \ (g) + C \ (s) \xrightarrow{\Delta} 2CO \ (g)$$

The limestone, called a **flux,** is added to react with the silica gangue in the ore to form a molten **slag** of calcium silicate:

$$CaCO_3 \ (s) \xrightarrow{\Delta} CaO \ (s) + CO_2 \ (g)$$

limestone

$$CaO \ (s) + SiO_2 \ (s) \xrightarrow{\Delta} CaSiO_3 \ (\ell)$$

gangue   slag

The slag is less dense than the molten iron; it floats on the surface of the iron and protects it from atmospheric oxidation. Both are drawn off periodically. Some of the slag is subsequently used in the manufacture of cement.

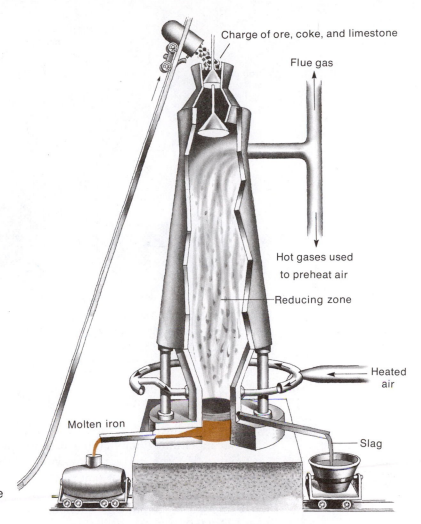

Charge of ore, coke, and limestone

Flue gas

Hot gases used
to preheat air

Reducing zone

Heated
air

Molten iron

Slag

FIGURE 20–12 Blast furnace
for reduction of iron ore.

Molten steel is poured
from a basic oxygen
furnace.

The iron obtained is still impure and contains carbon, among other things. It is called **pig iron.** If it is remelted, run into molds, and cooled, it becomes **cast iron,** which is brittle because it contains much iron carbide, Fe_3C. If some of the carbon is removed and other metals such as Mn, Cr, Ni, W, Mo, and V are added, the mixture becomes mechanically stronger and is known as **steel.** There are many types of steel containing alloyed metals and other elements in various controlled proportions. Stainless steel shows excellent resistance to corrosion and high tensile strength. The most common kind contains 14 to 18% chromium and 7 to 9% nickel. If all the carbon is removed, nearly pure iron can be produced. It is silvery in appearance, quite soft and of little use.

Pig iron can also be converted to steel by burning out most of the carbon with oxygen in a basic oxygen furnace, shown in Figure 20–13. Oxygen is blown through a heat-resistant tube inserted below the surface of the molten iron. When the iron is heated to high temperatures, the carbon burns to carbon monoxide, which subsequently escapes and burns to carbon dioxide.

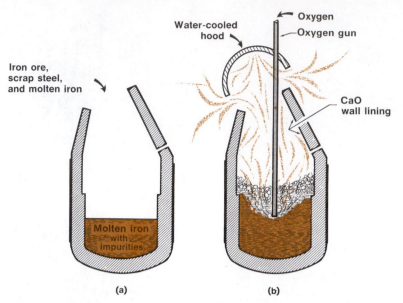

FIGURE 20–13 The basic oxygen process furnace. Much of the steel manufactured today is refined by blowing oxygen through a furnace charged with ore, scrap, and molten iron. The steel industry is one of the nation's largest consumers of oxygen.

3 Copper

Copper is so widely used, especially in its alloys such as bronze (copper and tin) and brass (copper and zinc), that it is becoming very scarce. The U.S. Bureau of Mines estimates that the currently known worldwide reserves of copper ore will be exhausted during the first half of the next century. The two principal classes of copper ores are the mixed sulfides of copper and iron, such as chalcopyrite, $CuFeS_2$, and the basic carbonates such as azurite, $Cu_3(CO_3)_2(OH)_2$, and malachite, $Cu_2CO_3(OH)_2$.

It is now profitable to mine ores containing as little as 0.25% copper.

Let us consider $CuFeS_2$ (or $CuS \cdot FeS$) as an example. After the ore is mined, the copper compound is separated by flotation and then roasted to remove volatile impurities. Enough air is used to convert iron(II) sulfide, but not the less active copper(II) sulfide, to the oxide.

$$2CuFeS_2 \text{ (s)} + 3O_2 \text{ (g)} \xrightarrow{\Delta} 2FeO \text{ (s)} + 2CuS \text{ (s)} + 2SO_2 \text{ (g)}$$

The roasted ore is then mixed with sand (silica or SiO_2), crushed limestone ($CaCO_3$), and some unroasted ore that contains copper(II) sulfide in a reverberatory furnace at 1100°C (Figure 20–14). Copper(II) sulfide, CuS, is reduced to Cu_2S, which melts. The limestone and silica form a calcium silicate glass, which dissolves the ferrous oxide to form a slag less dense than molten copper(I) sulfide upon which it floats.

$$CaCO_3 \text{ (s)} + SiO_2 \text{ (s)} \xrightarrow{\Delta} CaSiO_3 \text{ (}\ell\text{)} + CO_2 \text{ (g)}$$

$$CaSiO_3 \text{ (}\ell\text{)} + FeO \text{ (s)} + SiO_2 \text{ (s)} \xrightarrow{\Delta} CaSiO_3 \cdot FeSiO_3 \text{ (}\ell\text{)}$$

The slag is periodically drained off. The molten copper(I) sulfide is drawn off into a Bessemer converter where it is again heated and air-oxidized to free copper and

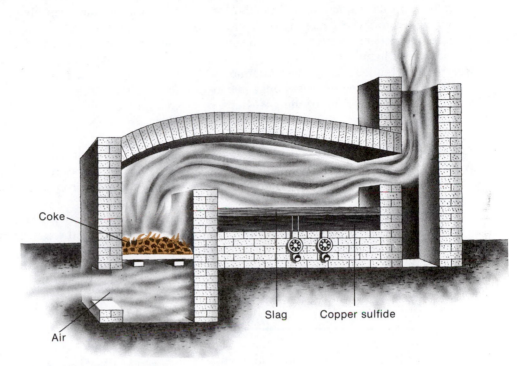

Coke

Air

Slag   Copper sulfide

FIGURE 20–14 Reverberatory furnace for the production of copper.

SO_2. This oxidizes the sulfide ions, but not copper(I) ions; they are reduced to metallic copper:

$$Cu_2S\ (\ell) + O_2\ (g) \xrightarrow{\Delta} 2Cu\ (\ell) + SO_2\ (g)$$

The impure copper is refined electrolytically as described in detail in Section 19–7.

Key Terms

Alkali metals Group IA metals.

Alkaline earth metals Group IIA metals.

Alloying mixing of a metal with other elements (usually other metals) to modify properties.

Alums hydrated sulfates of the general formula $M^+M^{3+}(SO_4)_2 \cdot 12H_2O$.

Charge a sample of crushed ore as it is admitted to a furnace for smelting.

Diagonal similarities chemical similarities of elements of Period 2 to elements of Period 3 one group to the right; especially evident toward the left of the periodic table.

Dimer molecule formed by combination of two identical molecules.

Domain a cluster of atoms in a ferromagnetic substance, all of which align in the same direction in the presence of an external magnetic field.

Double salt solid consisting of two co-crystallized salts.

Ferromagnetism permanent magnetization of a substance by exposure to an external magnetic field.

Flotation method by which hydrophobic (water-repelling) particles of an ore are separated from hydrophilic (water-attracting) particles in a metallurgical pretreatment process.

Flux a substance added to react with the charge, or

a product of its reduction, in metallurgy; usually added to lower a melting point.

Gangue sand and other rock surrounding the mineral of interest in an ore.

Inert s-pair effect tendency of the two outermost s electrons to remain nonionized or unshared in compounds; characteristic of the post-transition metals.

Lanthanide contraction a decrease in the radii of the elements following the lanthanides relative to what would be expected if there were no f-transition metals.

Metallurgy the overall processes by which metals are extracted from ores.

Native state the occurrence of an element in an uncombined or free state in nature.

Ore a natural deposit containing a mineral of an element to be extracted.

Pig iron impure iron obtained from a blast furnace.

Post-transition metals representative metals in the "p-block."

Refining purification of a substance.

Roasting heating a compound below its melting point in the presence of air.

Slag unwanted material produced during smelting.

Smelting chemical reduction of a substance at high temperature in metallurgy.

Steels alloys of iron with other elements, mostly metals.

Zone refining method of purifying a bar of a metal by passing it through an induction heater; this causes impurities to move along in the melted or softened portion.

Exercises

General Concepts

1. How do the acidities or basicities of representative metal oxides vary with oxidation number of the same metal?

2. Discuss the general differences in electronic configurations of (a) representative metals, (b) d-transition metals, and (c) f-transition metals.

3. Compare the extents to which the properties (in general) of successive elements in the periodic table differ for (a) representative metals, (b) d-transition metals, and (c) f-transition metals. Explain.

4. How do the physical properties of metals differ from those of nonmetals?

5. Compare the metals and nonmetals with respect to (a) outer shell electrons, (b) electronegativities, (c) standard reduction potentials, and (d) ionization energies.

Alkali and Alkaline Earth Metals

6. Compare the alkali metals with the alkaline earth metals with respect to (a) atomic radii, (b) densities, (c) first ionization energies, and (d) second ionization energies. Explain the comparisons.

7. Write general equations for the reactions of alkali metals with (a) hydrogen, (b) sulfur, (c) water, and (d) halogens. Represent the metal as M and the halogen as X.

8. Summarize the chemical and physical properties of the alkali metals.

9. Describe the "diagonal relationships" in the periodic table.

10. Why are the standard reduction potentials of lithium and beryllium out of line with respect to group trends?

11. Write general equations for reactions of alkaline earth metals with: (a) hydrogen, (b) chlorine, (c) water, and (d) sulfur.

12. Calculate ΔH^0 values at 25°C for the reactions of one mole quantities of each of the following metals with stoichiometric quantities of water to form metal hydroxides and hydrogen: (a) Li, (b) K, and (c) Ca. What do these values indicate?

Post-transition Metals

13. Describe the "inert-pair effect" associated with the post-transition metals. What is its cause? What are the most likely oxidation states for post-transition metals in the various groups?

14. Why are M^{3+} ions highly polarizing? What does this mean? Would high polarizing power of the metal lead to high covalent or ionic character of its bonds? For a given oxidation state, are the lighter or heavier members of a family more polarizing? Why? Answer the same questions for IVA and VA elements.

15. Write equations illustrating the amphoterism of two post-transition metal hydroxides.

16. Which hydrolyzes to the greater extent, Sn^{2+} or Pb^{2+}? Why? How can hydrolysis and precipita-

tion of $Sn(OH)_2$ be prevented in stannous salt solutions? Relate this to LeChatelier's Principle.

17. Calculate the pH of a 0.010 M aqueous solution of stannous chloride. Assume that K_a for $Sn(OH_2)_6^{2+}$ is 1×10^{-2}.

d-Transition Metals

18. Write out the entire electronic configurations for the following species: (a) V, (b) Fe, (c) Cu, (d) Zn, (e) Fe^{3+}, (f) Ni^{2+}, (g) Ag, (h) Ag^+.

19. Why do copper and chromium atoms have "unexpected" electronic configurations?

20. Discuss the similarities and differences among corresponding A and B groups in the periodic table, using IIIA and IIIB as examples.

21. Why are most transition metal compounds colored?

22. Why do we say that zinc, cadmium, and mercury have pseudo-noble gas configurations?

23. What is the lanthanide contraction? Which transition elements does it affect? How does it affect radii and densities?

24. Copper exists in the +1, +2, and +3 oxidation states. Which is the most stable? Which is a good oxidizing agent and which a good reducing agent?

25. For a given transition metal in different oxidation states, how does the acidic character of its oxides increase? How do ionic and covalent character vary? Characterize a series of metal oxides as examples.

26. Chromium(VI) oxide is the acid anhydride of what two acids? What is the oxidation state of the chromium in the acids?

27. How is the number of unpaired electrons per atom or ion experimentally determined?

28. What is ferromagnetism? What are domains? Which pure elements are ferromagnetic?

29. What single characteristic seems to be the most important in determining melting points of the d-transition metals?

30. How many grams of Co_3O_4 must react with excess aluminum to produce 300 g of metallic cobalt, assuming 72% yield? The reaction is

$$3Co_3O_4 + 8Al \xrightarrow{\Delta} 9Co + 4Al_2O_3$$

31. Scandium is a quite active metal. What volume of hydrogen, measured at STP, is produced by the reaction of 6.00 g of scandium with excess hydrochloric acid? The reaction produces $ScCl_3$.

32. What is the ratio of $[Cr_2O_7^{2-}]$ to $[CrO_4^{2-}]$ at 25°C in a solution prepared by dissolving 1.0×10^{-3} mol of sodium dichromate, $Na_2Cr_2O_7$, in enough of an aqueous solution buffered at pH = 11.00 to produce 200 mL of solution?

33. Answer Exercise 32 for a solution buffered at pH = 3.00.

34. Classify the following substances as acidic or basic: MnO, Mn_2O_3, and Mn_2O_7. Explain.

35. Arrange the substances of Exercise 34 in order of increasing covalent character.

36. Calculate ΔG^0 for the oxidation of ferrous ion to ferric ion by O_2 in acidic solution.

$$4Fe^{2+} (aq) + 4H^+ (aq) + O_2 (g) \longrightarrow 4Fe^{3+} (aq) + 2H_2O (\ell)$$

$$E^0 = +0.47 \text{ V}$$

37. Without consulting a table of electrode potentials, indicate which of the following substances should be strong oxidizing agents: Cr, $Fe(OH)_2$, Mn_2O_7, Cu_2O, CrO_3.

38. Aqueous solutions of most transition metal ions are colored. What are the colors of the following ions in aqueous solution? Cr^{2+}, Cr^{3+}, Mn^{2+}, Cu^{2+}, Zn^{2+}.

Metallurgy

39. What is a native ore? What kinds of metals are most apt to occur in the uncombined (native) state in nature?

40. List the six anions (and their formulas) that are most often combined with metals in ores. Give at least one example of an ore of each kind. What anion is the most commonly encountered?

41. Give the five general steps involved in extracting a metal from its ore. Briefly describe the importance of each.

42. Describe the flotation method of ore pretreatment. Are any chemical changes involved?

43. What kinds of ores are roasted? What kinds of compounds are converted to oxides by roasting? What kinds are converted directly to the free metals?

44. Of the following compounds, which would require electrolysis to obtain the free metals: KCl, $Cr_2(SO_4)_3$, Fe_2O_3, Al_2O_3, Ag_2S, $MgSO_4$? Why?

45. At which electrode is the free metal produced in the electrolysis of a metal compound? Why?

46. How are bond strengths related to ease of obtaining free metals from ores? All other factors

equal, would it be more economically advantageous to be able to obtain a free metal from its ore by chemical reduction (such as heating a metal oxide with coke, hydrogen, and so on) or by electrolysis? Which method requires the greater energy input? Why?

47. Suggest the best method of obtaining manganese from an ore containing manganese(III) oxide, Mn_2O_3. On what basis do you make the suggestion?

48. Describe the metallurgy of (a) aluminum and (b) iron.

49. Describe the metallurgy of copper.

50. What is the purpose of utilizing the basic oxygen furnace after the blast furnace in the production of iron?

51. What is steel? How does the hardness of iron compare with that of steel?

52. Describe and illustrate the electrolytic refining of copper.

53. (a) Calculate the weight in pounds of sulfur dioxide produced in the roasting of one ton of chalcocite ore containing 6.5% Cu_2S, 0.4% Ag_2S, and no other source of sulfur.

    (b) What weight of sulfuric acid can be prepared from the SO_2 generated, assuming 93% of it can be recovered from stack gases and 88% of that recovered can be converted to sulfuric acid?

    (c) How many pounds of pure copper can be obtained, assuming 75% efficient extraction and purification?

    (d) How many pounds of silver can be produced, assuming 82% of it can be extracted and purified?

54. Thirty-five pounds of Al_2O_3 obtained from bauxite are mixed with cryolite and electrolyzed. How long would a 0.500 ampere current have to be passed to convert all the Al^{3+} (from Al_2O_3) to aluminum metal? What volume of oxygen collected at 840 torr and 145°C, would be produced in the same period of time?

55. Calculate the percentage of iron in hematite ore containing 62.4% Fe_2O_3 by mass. How many pounds of iron would be contained in one ton of the ore?

56. Find the standard molar enthalpies of formation of Al_2O_3, Fe_2O_3 and HgS in Appendix K. Are the values in line with what might be predicted in view of the methods by which the metal ions are reduced in extractive metallurgy?

57. Using data from Appendix K, calculate ΔG^0_{298} for the following reactions:

    (a) Al_2O_3 (s) $\rightarrow$ 2Al (s) + $\frac{3}{2}O_2$ (g)
    (b) Fe_2O_3 (s) $\rightarrow$ 2Fe (s) + $\frac{3}{2}O_2$ (g)
    (c) HgS (s) $\rightarrow$ Hg (ℓ) + S (s)

Are any of the reactions spontaneous at 25°C? Are the ΔG^0_{298} values in line with what would be predicted based on the relative activities of the metal ions involved? Do increases in temperature favor these reactions?

The Nonmetals—I: Groups 0, VIIA, and VIA

21

Only about 20% of the elements are classified as nonmetals. With the exception of hydrogen, they are located in the upper right-hand corner of the periodic table, above the somewhat arbitrary stepwise division line.

In general, metallic character increases from *right to left* within periods and from *top to bottom* within groups. The elements along the stepwise division are called *metalloids* because their properties are intermediate between those of metals

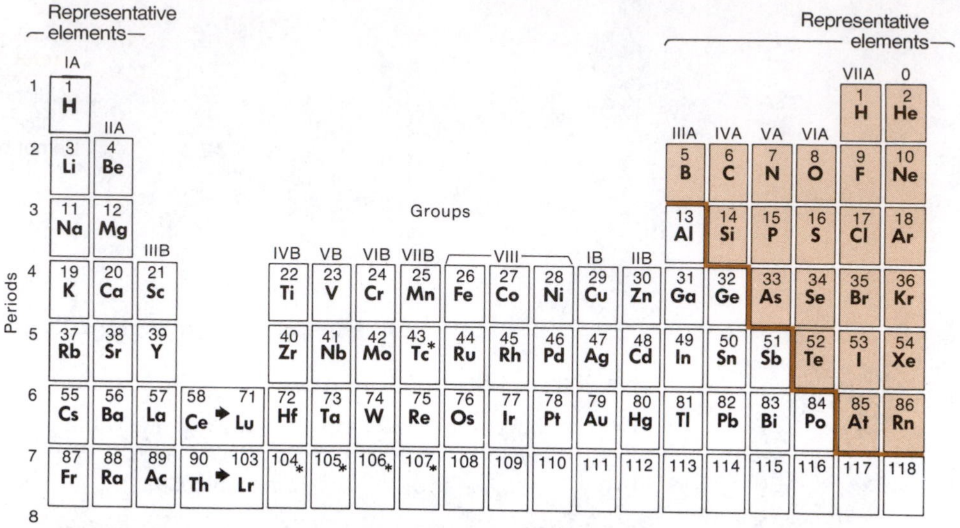

FIGURE 21–1 The periodic table with the nonmetals highlighted. The lanthanides (numbers 58–71) and the actinides (numbers 90–103) are metals, so they have been omitted.

and nonmetals. In this chapter we shall consider the chemistry and properties of the noble gases (Group 0), the halogens (Group VIIA) and the nonmetals of Group VIA. Other nonmetals will be discussed in the next chapter.

The Noble Gases (Group 0)

21–1 Occurrence, Isolation, and Uses of the Noble Gases

TABLE 21–1 Percentages of Noble Gases in the Atmosphere (by Volume)

He	0.0005%
Ne	0.0015%
Ar	0.94%
Kr	0.00011%
Xe	0.000009%
Rn	0%

The noble gases are very low boiling gases that can be isolated by fractional distillation of liquefied air (except for radon, all of whose isotopes are radioactive). Radon must be collected from the radioactive disintegration of radium salts. Table 21–1 gives the percentage of each noble gas in the atmosphere.

Helium is most economically recovered from helium-containing natural gas fields in the United States. This source was discovered in 1905 by H. P. Cady and D. F. McFarland at the University of Kansas, when they were asked to analyze a nonflammable component of natural gas from a Kansas gas well. The main uses of the noble gases are summarized in Table 21–2.

21–2 Physical Properties of the Noble Gases

The noble gases are colorless, tasteless, and odorless, and all have extremely low melting and boiling points. The only forces of attraction among the atoms in the

TABLE 21-2 Uses of the Noble Gases

Noble Gas	Use	Useful Properties or Reasons
helium	1. filling of observation balloons and other lighter-than-air craft	nonflammable; 93% of lifting power of flammable hydrogen
	2. He/O$_2$ mixtures rather than N$_2$/O$_2$ for deep sea breathing	He not blood-soluble, prevents nitrogen narcosis and "bends"
	3. diluent for gaseous anesthetics	nonflammable, nonreactive
	4. He/O$_2$ mixtures for respiratory patients	low density, flows easily through restricted passages
	5. heat transfer medium for nuclear reactors	transfers heat readily; does not become radioactive; chemically inert
	6. industrial applications, such as inert atmosphere for welding easily oxidized metals	chemically inert
	7. liquid He used to maintain very low temperatures in research (cryogenics)	extremely low boiling point
neon	neon signs	even at low Ne pressure, moderate electric current causes bright orange-red glow; color can be modified by colored glass or mixing with Ar or Hg vapor; flashes with changes in current
argon	1. inert atmosphere for metallurgical operations such as welding	chemically inert
	2. filling incandescent light bulbs (with Hg)	inert; inhibits vaporization of W and blackening of bulbs
xenon	Xe and Kr mixture in high intensity, short exposure photographic flash tubes	both have fast response to electric current
krypton	used in airport runway and approach lights	gives longer life to incandescent lights than Ar, but more expensive
radon	radiotherapy of cancerous tissues	radioactive

liquid and solid states are very weak London or van der Waals forces. Since polarizability, and therefore interatomic interaction, increases with increasing atomic size, the melting and boiling points increase with increasing atomic numbers. The forces of attraction among helium atoms are so small that helium remains liquid at one atmosphere pressure even at a temperature of 0.001 K. Table 21-3 summarizes some of the physical properties of the noble gases.

A pressure of about 26 atmospheres is required for solidification at this temperature.

TABLE 21-3 Properties of Noble Gases

Property	Helium	Neon	Argon	Krypton	Xenon	Radon
atomic number	2	10	18	36	54	86
outer electrons	$1s^2$	$2s^22p^6$	$3s^23p^6$	$4s^24p^6$	$5s^25p^6$	$6s^26p^6$
atomic radius (nm)	0.05	0.070	0.094	0.109	0.130	0.14
melting point (°C, 1 atm)	−272.2*	−248.6	−189.3	−157	−112	−71
boiling point (°C, 1 atm)	−268.9	−245.9	−185.6	−152.9	−107.1	−61.8
density (g/L at STP)	0.18	0.90	1.78	3.75	5.90	9.73
first ionization energy (kJ/mol)	2372	2081	1521	1351	1170	1037

* At 26 atm.

21–3   Chemical Properties of the Noble Gases

The consensus of opinion in the scientific community until the early 1960's was that the Group 0 elements (then called the inert gases) would not combine chemically with any elements. To do so would require an expanded valence shell for the noble gas. This opinion was strongly held even though there were many known examples of stable substances containing elements with expanded valence shells: SF_6, PCl_5, IF_7, SiF_6^{2-}, and so on.

In 1962, Neil Bartlett and his research group at the University of British Columbia were studying the powerful oxidizing agent, PtF_6. After accidentally preparing and identifying $O_2^+PtF_6^-$ by reaction of oxygen with PtF_6, Bartlett reasoned that xenon also should be oxidized by PtF_6 since the first ionization energy of molecular oxygen is actually slightly larger (1.31×10^3 kJ/mol) than that of xenon (1.17×10^3 kJ/mol). His attempts yielded a red crystalline solid initially believed to be $Xe^+PtF_6^-$ but now known to be more complex.

Since Bartlett's discovery many other noble gas compounds have been synthesized. Most are compounds of xenon, and the best characterized compounds are the xenon fluorides, though oxygen compounds are also well known. Reaction of xenon with fluorine, also an extremely strong oxidizing agent, in different stoichiometric ratios produces xenon difluoride, XeF_2, xenon tetrafluoride, XeF_4, and xenon hexafluoride, XeF_6, all colorless crystals.

All the xenon fluorides are formed in exothermic reactions, and are reasonably stable with Xe—F bond energies of about 125 kJ/mol of bonds. For comparison, strong bond energies generally range from about 170 to 500 kJ/mol, whereas bond energies of hydrogen bonds and other weak bonding interactions are typically less than 40 kJ/mol.

Oxygen is second only to fluorine in electronegativity.

The Halogens (Group VIIA)

The elements of Group VIIA are known as **halogens** (Greek: salt formers). The term **halides** is used to describe their binary compounds.

21–4   Properties of the Halogens

The elemental halogens exist as diatomic molecules containing single covalent bonds. Because many of the chemical properties of the halogens are common to the whole group, it is often convenient to let X represent a halogen atom without specifying a particular halogen.

$$:\overset{..}{\underset{..}{X}}:\overset{..}{\underset{..}{X}}:$$ Lewis dot formula for a halogen molecule

F_2 Cl_2 Br_2 I_2 At_2

TABLE 21–4 Properties of the Halogens

Property	Fluorine	Chlorine	Bromine	Iodine	Astatine
physical state (25°C, 1 atm)	gas	gas	liquid	solid	solid
color	pale yellow	yellow-green	red-brown	violet (g) black (s)	—
outer electrons	$2s^2 2p^5$	$3s^2 3p^5$	$4s^2 4p^5$	$5s^2 5p^5$	$6s^2 6p^5$
atomic radius (nm)	0.064	0.099	0.114	0.133	0.140
ionic radius, X^- (nm)	0.136	0.181	0.195	0.216	—
first ionization energy (kJ/mol)	1681	1255	1146	1017	916
electronegativity	4.0	3.0	2.8	2.5	2.1
melting point (°C, 1 atm)	−218	−101	−7.1	114	—
boiling point (°C, 1 atm)	−188	−35	59	184	—
X—X bond energy (kJ/mol)	158	243	192	151	—
standard reduction potential (V) (aq. soln)	+2.87	+1.36	+1.08	+0.54	+0.3
heat of hydration of X^- (kJ/mol)	−510	−372	−339	−301	—
solubility in water (mol/L, 20°C)	reacts	0.09	0.21	0.0013	—

Some of the important properties of the halogens are given in Table 21–4. The properties of the halogens show rather obvious trends. The high electronegativities of the halogens indicate that they attract electrons strongly. Nearly all binary compounds containing a metal and a halogen are ionic.

The high standard reduction potentials indicate that fluorine, chlorine, and bromine are strong oxidizing agents while iodine is a mild oxidizing agent. Conversely, fluoride, chloride, and bromide ions are weak reducing agents while iodide ions are mild reducing agents.

The monatomic ions are assumed to be spherical. The surface area of a sphere is $4\pi r^2$ or $12.57\, r^2$.

The small fluoride ion (radius = 0.136 nm) is not easily polarized, whereas the large iodide ion (radius = 0.216 nm) is. The unit negative charge associated with each halide ion may be thought of as distributed over the entire surface of the ion. Isolated halide ions are spherical, and it is easy to see why there should be significant differences between fluoride and iodide ions. On the average the single negative charge associated with an iodide ion (surface area = 0.59 nm²) is spread over a much larger surface area than is the negative charge associated with a fluoride ion (surface area = 0.23 nm²). Stated another way, the average charge density on the fluoride ion is 2.6 times greater than on the iodide ion. Additionally, the iodide ion has low-lying energy levels that contain large numbers of shielding electrons. Consequently, an I^- ion is much more easily polarized (that is, the electron cloud is more easily distorted from spherical geometry) when it interacts with a positively charged ion such as Ag^+. Similar reasoning leads to the conclusion that small fluoride ions should not be easily polarized. The properties of chloride and bromide ions are intermediate between those of fluoride and iodide ions.

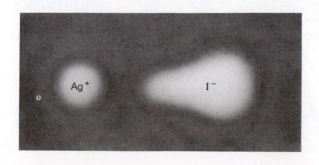

The halogens resemble each other more closely than do elements in any other periodic group, with the exception of the noble gases and possibly the Group IA metals. But their properties do differ significantly. The main reason for the rise in melting and boiling points from fluorine to iodine is the increase in atomic size and the accompanying increase in ease of polarization of outer shell electrons by adjacent nuclei. This results in greater intermolecular attractive forces as described in Section 12–8. All halogens, except the short-lived radioactive element astatine, are decidedly nonmetallic. They show the -1 oxidation number in most of their compounds although, except for fluorine, they also commonly exhibit oxidation numbers of $+1$, $+3$, $+5$, and $+7$.

21–5   Occurrence, Production, and Uses of the Halogens

Sodium iodate, $NaIO_3$, occurs as an impurity in the Chilean nitrate deposits, which are primarily $NaNO_3$.

The halogens are so reactive that they do not occur free in nature. The most abundant sources of halogens are halide salts, although a primary source of iodine is $NaIO_3$. The halogens are obtained by oxidation of the halide ions, i.e.,

$$2X^- \longrightarrow X_2 + 2e^-$$

The order of increasing ease of oxidation of halide ions is $F^- \ll Cl^- < Br^- < I^- < At^-$.

1   Fluorine

Fluorine occurs in large quantities in the minerals *fluorspar* or *fluorite*, CaF_2, *cryolite*, Na_3AlF_6, and *fluoroapatite*, $Ca_{10}(PO_4)_6F_2$. It also occurs in small amounts in seawater, teeth, bones, and blood. Since F_2 is such a strong oxidizing agent, it cannot be produced by chemical oxidation of fluoride ions. Rather, the pale yellow gas must be prepared by electrolysis of a molten mixture of $KF + HF$, or KHF_2, in a Monel cell under strictly anhydrous conditions, since water is more readily oxidized than F^-.

Monel metal is an alloy of Ni, Cu, Fe, and Al that is resistant to attack by hydrogen fluoride.

$$2KHF_2 \xrightarrow[\text{melt}]{\text{Electrolysis}} F_2\,(g) + H_2\,(g) + 2KF$$
$$\quad\quad\quad\quad\quad\quad \text{anode} \quad\;\; \text{cathode}$$

Industrially and in research, fluorine is used in the direct fluorination of inorganic compounds and as a powerful oxidizing agent. Many fluorinated organic compounds, called fluorocarbons, are stable and nonflammable. They have found use as refrigerants, lubricants, plastics, insecticides, coolants, and, until recently, as aerosol propellants. A common refrigerant is freon-12 or CCl_2F_2, and CCl_3F is an insecticide. *Teflon* is a very inert plastic consisting of polymers of $-CF_2-CF_2-$ units, used for nonstick surfaces on cooking utensils and a variety of other articles requiring smooth inert surfaces (such as artificial valves for the heart). Gaseous fluorine is also used in the production of volatile uranium hexafluoride, UF_6, which is used in separating fissionable and nonfissionable uranium isotopes for nuclear reactors. In a very different area, the fluoride ion is effective in preventing dental caries because it encourages formation of fluoroapatite, $Ca_{10}(PO_4)_6F_2$, in teeth instead of the more acid-soluble hydroxyapatite, $Ca_{10}(PO_4)_6(OH)_2$. The fluoride ion is added to public water supplies in carefully regulated, nontoxic amounts.

2 Chlorine

Chlorine (Greek: *chloros,* meaning "green") occurs in abundance in NaCl, KCl, $MgCl_2$, and $CaCl_2$ in salt water and in salt beds. It is also present as HCl in gastric juices. The toxic yellowish-green gas is prepared commercially by electrolysis of concentrated aqueous sodium chloride, in which industrially important hydrogen and caustic soda (NaOH) are also produced. The electrode reactions were given in Section 19–4.

21.12 billion pounds of chlorine were produced in the United States in 1981.

Chlorine is a very important reagent in the production of many widely used and commercially important products. Tremendous amounts are used in extractive metallurgy and in chlorinating hydrocarbons to produce a variety of compounds such as polyvinyl chloride, a plastic. Chlorine is present as Cl_2, NaClO, $Ca(ClO)_2$ or Ca(ClO)Cl in household bleaches as well as in bleaches for wood pulp and textiles. Cl_2 itself is used under carefully controlled conditions to kill bacteria in public water supplies.

ClO⁻ is the hypochlorite ion.

3 Bromine

Bromine (Greek: *bromos,* meaning "stench") is less abundant than fluorine and chlorine. In the elemental form it is a dense, freely flowing, corrosive, dark red liquid with a brick red vapor at room temperature. The element occurs mainly in NaBr, KBr, $MgBr_2$ and $CaBr_2$ in salt water, underground salt brines, and salt beds. The major source for commercial production of bromine is sea water, although its concentration there is only 0.0066%. Sea water is normally alkaline because of the presence of CO_3^{2-} from calcium carbonate. It is first acidified with H_2SO_4 to a pH of 3.5, and chlorine gas is bubbled through it.

Approximately 32 kilograms of bromine can be obtained by treatment of a million kilograms of sea water.

$$Cl_2\ (g) + 2Br^- \longrightarrow 2Cl^- + Br_2\ (\ell)$$

The bromine vapor is then swept by a stream of air into a saturated alkaline solution of Na_2CO_3, in which it is trapped as a concentrated solution of sodium bromide and sodium bromate. This reaction, the disproportionation of Br_2, occurs readily in basic solution.

This displacement reaction occurs because Cl_2 is a stronger oxidizing agent (more active) than Br_2 (Section 9–7).

$$3Br_2\ (g) + 3CO_3^{2-} \longrightarrow 5Br^- + BrO_3^- + 3CO_2\ (g)$$

Bromine vapor can then be obtained conveniently when desired by acidifying this solution with H_2SO_4.

$$5Br^- + BrO_3^- + 6H^+ \longrightarrow 3Br_2\ (\ell) + 3H_2O$$

Bromine is used in the production of 1,2-dibromoethane, $C_2H_4Br_2$, a component of leaded gasolines that reacts with lead in automobile engines to form volatile $PbBr_2$. The $PbBr_2$ is exhausted into the atmosphere and prevents buildup of nonvolatile lead compounds inside engines. However, it poses a serious health hazard when present in the air. Its use for this purpose is declining with the phasing out of leaded fuels.

Bromine is also used in the production of silver bromide for light-sensitive eyeglasses and photographic film, in the production of sodium bromide, a mild sedative, and in methyl bromide, CH_3Br, a soil fumigant.

4   Iodine

Iodine (Greek: *iodos,* meaning "purple") is a black crystalline solid with a metallic luster. It exists in equilibrium with a violet vapor at room temperature. The element can be obtained from dried seaweed or shellfish or from $NaIO_3$ impurities in Chilean nitrate ($NaNO_3$) deposits. It is contained in the growth-regulating hormone, thyroxine, produced by the thyroid gland. About 0.02% of "iodized" table salt is KI, which helps prevent goiter, a condition in which the thyroid enlarges. Iodine is also used as an antiseptic and germicide in the form of tincture of iodine, a solution in alcohol. Silver iodide is used in "cloud seeding" to cause rain.

More recently available is an aqueous solution of an iodine complex of polyvinylpyrolidone or "povidone," which does not sting when applied to open wounds.

The commercial preparation of iodine involves reduction of iodate ion from $NaIO_3$ with sodium hydrogen sulfite, $NaHSO_3$.

$$2IO_3^- + 5HSO_3^- \longrightarrow 3HSO_4^- + 2SO_4^{2-} + H_2O + I_2 \text{ (s)}$$

Iodine is then purified by sublimation (Figure 12–26). Elemental chlorine or bromine also can be used to displace iodine from iodide salts.

$$2I^- + Cl_2 \longrightarrow I_2 + 2Cl^-$$

21–6   Reactions of the Free Halogens

The free halogens react directly with most other elements and many compounds. For example, all the Group IA metals react with all the halogens to form simple binary ionic compounds, as described in Section 6–10.

Direct reactions of F_2 with other elements or compounds are dangerous because of the vigor with which F_2 oxidizes other substances. They are carried out only rarely and with *extreme* caution.

In general, the most vigorous reactions are those of F_2, which usually oxidizes other species to their highest possible oxidation states. On the other hand, I_2 is only a mild oxidizing agent (I^- is a mild reducing agent) and usually does not oxidize substances to high oxidation states. We might also say that F^- stabilizes high oxidation states of cations, while I^- stabilizes low oxidation states. As illustrations, consider the following reactions of halogens with metals that exhibit variable oxidation numbers, iron on the left and copper on the right.

$$2Fe + 3F_2 \longrightarrow 2\overset{+3}{Fe}F_3 \qquad\qquad Cu + X_2 \longrightarrow \overset{+2}{Cu}X_2 \quad (X = F, Cl, Br)$$

$$2Fe + 3Cl_2 \text{ (excess)} \longrightarrow 2\overset{+3}{Fe}Cl_3 \qquad\qquad 2Cu + I_2 \longrightarrow 2\overset{+1}{Cu}I$$

$$Fe + Cl_2 \text{ (lim. amt.)} \longrightarrow \overset{+2}{Fe}Cl_2 \qquad \overset{+2}{Cu^{2+}} + 2I^- \longrightarrow \overset{+1}{Cu}I + \tfrac{1}{2}I_2$$

$$2Fe + 3Br_2 \text{ (excess)} \longrightarrow 2\overset{+3}{Fe}Br_3$$

$$Fe + Br_2 \text{ (lim. amt.)} \longrightarrow \overset{+2}{Fe}Br_2$$

$$Fe + I_2 \longrightarrow \overset{+2}{Fe}I_2 \text{ (only)}$$

$$Fe^{3+} + I^- \longrightarrow Fe^{2+} + \tfrac{1}{2}I_2$$

Table 21–5 summarizes many of the reactions of the free halogens.

TABLE 21–5 Some Common Reactions of the Free Halogens

General Reaction	Remarks
$nX_2 + 2M \longrightarrow 2MX_n$	all X_2 with most metals (most vigorous reaction with F_2 and Group IA metals)
$X_2 + nX_2' \longrightarrow 2XX_n'$	formation of interhalogens
$X_2 + H_2 \longrightarrow 2HX$	
$3X_2 + 2P \longrightarrow 2PX_3$	with all X_2; and with As, Sb, Bi replacing P
$5X_2 + 2P \longrightarrow 2PX_5$	not with I_2; also Sb $\longrightarrow SbF_5$, $SbCl_5$; As $\longrightarrow AsF_5$; Bi $\longrightarrow BiF_5$
$X_2 + H_2S \longrightarrow S + 2HX$	with all X_2
$X_2' + 2X^- \longrightarrow 2X'^- + X_2$	$F_2 \longrightarrow Cl_2, Br_2, I_2$
	$Cl_2 \longrightarrow Br_2, I_2$
	$Br_2 \longrightarrow I_2$
$X_2 + C_nH_y \longrightarrow C_nH_{y-1}X + HX$	halogenation; substitution of many saturated hydrocarbons with Cl_2, Br_2
$X_2 + C_nH_{2n} \longrightarrow C_nH_{2n}X_2$	halogenation of hydrocarbon double bonds with Cl_2, Br_2

21–7 The Hydrogen Halides

For example, aqueous solutions of hydrogen fluoride are called hydrofluoric acid.

The hydrogen halides (Figure 21–2) are all colorless gases that dissolve in water to give acidic solutions called hydrohalic acids. The gases all have piercing, irritating odors. Some of the properties of the hydrogen halides are given in Table 21–6.

TABLE 21–6 Properties of the Hydrogen Halides

Property	HF	HCl	HBr	HI
melting point (°C)	−83.1	−114.8	−86.9	−50.7
boiling point (°C)	19.54	−84.9	−66.8	−35.4
H—X bond energy (kJ/mol)	569	431	368	297
apparent % ionization in 0.1 M aq. sol'n (18°C)	< 10 (wk. acid)	92.6 (st. acid)	93 (st. acid)	95 (st. acid)

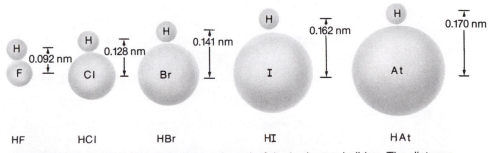

HF   HCl   HBr   HI   HAt

FIGURE 21–2 Relative sizes (approximate) of the hydrogen halides. The distance between the center of the hydrogen atom and the center of the halogen atom (internuclear distance) is indicated for each molecule.

21-8 Preparation of Hydrogen Halides

All the hydrogen halides may be prepared by direct combination of the elements.

$$H_2 + X_2 \longrightarrow 2HX \text{ (g)} \qquad X = F, Cl, Br, I$$

However, the reaction with F_2 to produce HF is explosive and very dangerous under all conditions. The reaction producing HCl does not occur significantly in the dark but occurs rapidly by a photochemical chain reaction when exposed to light. Light energy is absorbed by chlorine molecules, which break apart into very reactive chlorine **radicals** (atoms with unpaired electrons). These subsequently attack hydrogen molecules and produce HCl molecules, leaving hydrogen atoms, also radicals, behind. The hydrogen radicals, in turn, attack chlorine molecules to form HCl molecules and chlorine radicals.

A photochemical reaction is one in which some species (usually a molecule) interacts with radiant energy to produce very reactive species, which then undergo further reaction.

$$Cl_2 \xrightarrow{\ h\nu\ } 2 \; :\!\ddot{\underset{..}{Cl}}\cdot \qquad \text{initiation step}$$

$$\left. \begin{aligned} :\!\ddot{\underset{..}{Cl}}\cdot + H_2 &\longrightarrow HCl + H\cdot \\ H\cdot + Cl_2 &\longrightarrow HCl + \; :\!\ddot{\underset{..}{Cl}}\cdot \end{aligned} \right\} \text{chain propagation steps}$$

This chain reaction continues as long as there is a significant concentration of radicals. The following reaction termination steps eliminate (by combination) two radicals and eventually terminate the reaction.

$$\left. \begin{aligned} H\cdot + \; H\cdot &\longrightarrow H_2 \\ :\!\ddot{\underset{..}{Cl}}\cdot + \; :\!\ddot{\underset{..}{Cl}}\cdot &\longrightarrow Cl_2 \\ H\cdot + \; :\!\ddot{\underset{..}{Cl}}\cdot &\longrightarrow HCl \end{aligned} \right\} \text{termination steps}$$

The reaction of H_2 with Br_2 is also a photochemical reaction. The reaction of H_2 with I_2 is very slow, even at high temperatures and with illumination.

A safer way of preparing HF or HCl is the reaction of a metal halide with a nonvolatile acid, such as concentrated sulfuric or phosphoric acid. The more volatile hydrogen halide is evolved from the resulting solution as a gas.

$$\underset{bp = 338°C}{NaF \text{ (s)}} + \; H_2SO_4 \text{ (}\ell\text{)} \longrightarrow NaHSO_4 \text{ (s)} + \; \underset{bp = 20°C}{HF \text{ (g)}}$$

HBr and HI are not prepared from H_2SO_4 because they are oxidized to Br_2 or I_2 by hot concentrated H_2SO_4. However, H_3PO_4 is a nonoxidizing, nonvolatile acid that can be used to prepare HBr and HI from their salts.

$$\underset{bp = 213°C}{NaBr \text{ (s)}} + \; H_3PO_4 \text{ (}\ell\text{)} \longrightarrow NaH_2PO_4 \text{ (s)} + \; \underset{bp = -67°C}{HBr \text{ (g)}}$$

Most nonmetal halides hydrolyze to produce hydrogen halides and a ternary acid or oxide of the nonmetal:

$$BCl_3 \text{ (}\ell\text{)} + 3H_2O \longrightarrow H_3BO_3 \text{ (aq)} + 3HCl \text{ (g)}$$

$$SiCl_4 \text{ (}\ell\text{)} + 2H_2O \longrightarrow SiO_2 \text{ (s)} + 4HCl \text{ (g)}$$

Hydrogen bromide and hydrogen iodide are often prepared by hydrolysis of phosphorus tribromide and phosphorus triiodide, respectively, and are separated from the reaction mixture by distillation.

$$PX_3 + 3H_2O \longrightarrow H_3PO_3 \text{ (s)} + 3HX \text{ (g)} \qquad X = Cl, Br, I$$

A commercially important class of reactions is the *halogenation of saturated hydrocarbons.* Hydrogen halides are by-products of such reactions. A general reaction of this type is

$$C_nH_y + X_2 \longrightarrow C_nH_{y-1}X + HX \qquad X = F, Cl, Br, I$$

However, fluorination of hydrocarbons with F_2 is so dangerously explosive that such reactions are not performed. Halogenation of hydrocarbons with I_2 is not economically practical.

21-9   Aqueous Solutions of Hydrogen Halides (Hydrohalic Acids)

The hydrogen halides all dissolve readily in water, producing hydrohalic acids that ionize as shown.

$$H\!:\!\overset{\cdot\cdot}{\underset{H}{\overset{\cdot\cdot}{O}}}\!: + H\!:\!\overset{\cdot\cdot}{\underset{\cdot\cdot}{X}}\!: \rightleftharpoons H\!:\!\overset{\cdot\cdot}{\underset{H}{O}}\!:\!\overset{+}{H} + :\!\overset{\cdot\cdot}{\underset{\cdot\cdot}{X}}\!:^-$$

The reaction is essentially complete for dilute HCl (aq), HBr (aq), and HI (aq), which are classified as strong acids. However, because of the strong H—F bonds as well as hydrogen bonding between HF and H_2O (Figure 21-3), dilute HF (aq) is a weak acid with $K_a = 7.2 \times 10^{-4}$. In concentrated solution the more acidic dimeric $(HF)_2$ units are present, and they ionize as shown below.

$$(HF)_2 + H_2O \rightleftharpoons H_3O^+ + HF_2^- \qquad K \approx 5$$

Water is a sufficiently strong base that it does not discriminate among the acid strengths of hydrochloric, hydrobromic, and hydroiodic acids. However, reactions in less basic solvents, such as anhydrous acetic acid, indicate that the order of increasing acid strengths of the hydrohalic acids is the same as for the anhydrous hydrogen halides: HF (aq) ≪ HCl (aq) < HBr (aq) < HI (aq).

The only acid used to a greater extent in industry than hydrochloric acid is sulfuric acid. Hydrochloric acid is used in the production of metal chlorides, dyes, and many other commercially important products. It is also used on a large scale for

> The phosphorus trihalide is usually formed at the time of reaction by mixing red phosphorus with the free halogen.

> This is the leveling effect (Section 17-4).

> HCl (aq) is known commercially as muriatic acid.

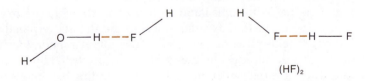

dilute solution   concentrated solution

FIGURE 21-3 Hydrogen bonding (dashed lines) in dilute and concentrated aqueous solutions of HF.

dissolving metal oxide coatings from iron and steel prior to galvanization or enameling.

Hydrofluoric acid is used in the production of fluorine-containing compounds, and for etching of glass. The acid reacts with silicates, such as calcium silicate, $CaSiO_3$, in the glass to produce a very volatile and thermodynamically stable compound, silicon tetrafluoride, SiF_4.

$$CaSiO_3 (s) + 6HF (aq) \longrightarrow CaF_2 (s) + SiF_4 (g) + 3H_2O (\ell)$$

Hydrobromic and hydroiodic acids are relatively less important, although they are used in the synthesis of bromine-containing and iodine-containing organic compounds and in chemical research.

21–10 Halides of Other Elements

The halogens combine with most other elements to form halides. They range from high melting, water-soluble, ionic electrolytes (metal halides) such as NaCl, KBr, and $CaCl_2$, to volatile, covalent nonelectrolytes (nonmetal halides) like PCl_3 and SF_6. Generally the former properties are those of halides of metals, and the latter properties characterize the halides of nonmetals.

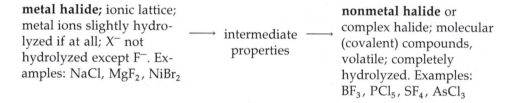

Halides (increasing nonmetallic character of other element)

metal halide; ionic lattice; metal ions slightly hydrolyzed if at all; X^- not hydrolyzed except F^-. Examples: NaCl, MgF_2, $NiBr_2$ $\longrightarrow$ intermediate properties $\longrightarrow$ **nonmetal halide** or complex halide; molecular (covalent) compounds, volatile; completely hydrolyzed. Examples: BF_3, PCl_5, SF_4, $AsCl_3$

21–11 The Oxyacids (Ternary Acids) of the Halogens and Their Salts

Table 21–7 lists the known oxyacids of the halogens and their sodium salts, and some trends in properties.

Only three oxyacids, $HClO_4$, HIO_3, and H_5IO_6, have been isolated in anhydrous form. The others are known only in aqueous solution. In all these acids the hydrogen is bonded through an oxygen.

The Lewis formulas and structures of the chlorine oxyanions are shown in Figure 21–4. The four oxyanions of chlorine all have tetrahedral *electronic geometry*, but only the perchlorate ion, ClO_4^-, also has tetrahedral *ionic geometry*. The other three have the ionic geometries that can be predicted by the (imaginary) successive removal of the oxygen atoms. The corresponding oxyanions of bromine and iodine have similar structures.

Since oxygen is so electronegative, greater numbers of oxygen atoms around the halogen cause the electron density of the H—O bonds of the acids to be shifted more toward oxygen. This results in an increase in the ease of H—O bond cleavage (Section 17–4); therefore, formation of a hydrogen ion and an anion becomes easier

TABLE 21-7   Oxyacids of the Halogens and Their Salts

Oxidation State	Acid	Name of Acid	Thermal Stability and Acid Strength	Oxidizing Power of Acid	Sodium Salt	Name of Salt	Thermal Stability	Oxidizing Power and Hydrolysis of Anion	Nature of Halogen
+1	HXO	Hypohalous acids	Increases	Increases	NaXO	sodium hypohalite	Increases	Increases	X = F*, Cl, Br, I
+3	HXO_2	Halous acids			$NaXO_2$	sodium halite			X = Cl, Br (?)
+5	HXO_3	Halic acids			$NaXO_3$	sodium halate			X = Cl, Br, I
+7	HXO_4	Perhalic acids			$NaXO_4$	sodium perhalate			X = Cl, Br, I
+7	H_5XO_6	Paraperhalic acids			several types	sodium paraperhalates			X = I

* The oxidation state of F is −1 in HOF.

as the number of oxygen atoms increases. Such behavior is observed for the series of oxyacids of any element, *if* all the H atoms are bonded to O atoms. Consistent with this is the observation that the basic strength, or degree of hydrolysis, of the anions of the salts decrease in the same order.

When concentrated, all of these acids are strong oxidizing agents; oxidizing power decreases with increasing number of oxygen atoms and increasing oxidation state of the halogen.

The only oxyacid of fluorine that has been prepared is hypofluorous acid, HOF. It can be prepared in the vapor phase by passing fluorine rapidly over ice.

> In HOF the oxidation states are: F = −1, H = +1, O = 0.

$$F_2 + H_2O \xrightarrow{0°C} HOF + HF$$

It is difficult to isolate because HOF reacts with water to produce hydrofluoric acid and hydrogen peroxide.

Aqueous *hypohalous acids* (except HOF) can be prepared by the reaction of free halogens (Cl_2, Br_2, I_2) with cold water.

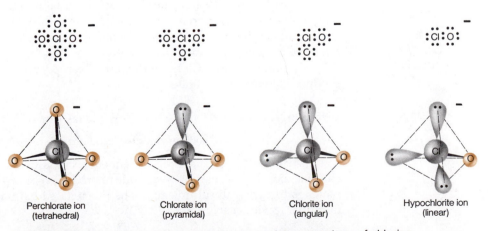

Perchlorate ion (tetrahedral)   Chlorate ion (pyramidal)   Chlorite ion (angular)   Hypochlorite ion (linear)

FIGURE 21-4 Lewis formulas and structures of the oxyanions of chlorine.

These reactions all involve disproportionation of the halogen.

$$\overset{0}{X_2} + H_2O \rightleftharpoons \underset{\substack{\text{hydrohalic} \\ \text{acid}}}{\overset{-1}{HX}} + \underset{\substack{\text{hypohalous} \\ \text{acid}}}{\overset{+1}{HOX}} \qquad\qquad X = Cl,\ Br,\ I$$

The smaller the halogen, the farther to the right the equilibrium lies.

Hypohalite salts can be prepared by reactions of the halogens with *cold* dilute bases.

Hypohalite ions can be represented as either XO⁻ or OX⁻.

$$\overset{0}{X_2} + 2NaOH \longrightarrow \underset{\substack{\text{sodium} \\ \text{halide}}}{\overset{-1}{NaX}} + \underset{\substack{\text{sodium} \\ \text{hypohalite}}}{\overset{+1}{NaOX}} + H_2O \qquad\qquad X = Cl,\ Br,\ I$$

The hypohalites are used as bleaching agents. Sodium hypochlorite is made on a huge commercial scale by electrolyzing aqueous sodium chloride and mixing two of the products, Cl_2 and NaOH (Section 19–4). Sometimes Cl_2 is used as a bleach or as a disinfectant, but the above reactions provide the hypochlorite ion when the Cl_2 is introduced into water.

Solid household bleaches are usually Ca(ClO)Cl, prepared by reaction of Cl_2 with Ca(OH)$_2$.

Ca(OH)$_2$ + Cl$_2$ $\longrightarrow$ Ca(ClO)Cl + H$_2$O

$$\underset{\substack{\text{hypohalite} \\ \text{ion}}}{OX^-} + H_2O \rightleftharpoons \underset{\substack{\text{hypohalous} \\ \text{acid}}}{HOX} + OH^-$$

The hypochlorous acid produced by this hydrolysis then undergoes partial decomposition to HCl and O radicals.

$$HOCl \longrightarrow HCl + :\overset{..}{\underset{.}{O}}\cdot$$

These oxygen radicals, which are very strong oxidizing agents, are the effective bleaching and disinfecting agent in aqueous solutions of Cl_2 or hypochlorite salts.

Upon heating, the hypohalite ions disproportionate to produce halates and halides.

$$3\overset{+1}{OX^-} \overset{\Delta}{\longrightarrow} \underset{\text{halate ion}}{\overset{+5}{XO_3^-}} + \underset{\text{halide ion}}{\overset{-1}{2X^-}} \qquad\qquad X = Cl,\ Br,\ I$$

Anhydrous *halous acids* have not been isolated. Fluorous and iodous acids apparently do not exist, even in aqueous solution, and the existence of bromous acid is questionable. Chlorite ion has long been known, and aqueous solutions of bromite ion have been prepared recently.

Chlorine dioxide, ClO_2, undergoes disproportionation in basic solution to produce chlorite and chlorate salts.

$$2\overset{+4}{ClO_2} + 2NaOH \longrightarrow \underset{\text{sodium chlorite}}{\overset{+3}{NaClO_2}} + \underset{\text{sodium chlorate}}{\overset{+5}{NaClO_3}} + H_2O$$

A convenient preparation of aqueous chlorous acid involves reaction of chlorine dioxide with barium hydroxide to produce barium chlorite, a suspension of which is treated with sulfuric acid. The precipitated barium sulfate is removed by filtration, leaving a solution of $HClO_2$.

$$\underset{\text{barium chlorite}}{Ba(ClO_2)_2} + H_2SO_4 \longrightarrow BaSO_4 + \underset{\text{chlorous acid}}{2HClO_2}$$

Aqueous solutions of chloric acid and bromic acid are prepared, similarly, by reactions of barium chlorate and bromate with sulfuric acid.

When stoichiometric amounts of reactants are mixed, Ba^{2+} and SO_4^{2-} ions combine to form insoluble $BaSO_4$, which can be separated by filtration. Fairly pure $HClO_3$ and $HBrO_3$ solutions remain.

$$Ba(XO_3)_2 + H_2SO_4 \longrightarrow \underset{\text{precipitate}}{BaSO_4} + \underset{\text{halic acid}}{2HXO_3} \qquad X = Cl, Br$$

The *halate* salts can be prepared by reaction of free halogens (except F_2) with *hot aqueous* alkali, or by thermal decomposition of hypohalites.

$$\overset{0}{3X_2} + 6NaOH \longrightarrow \overset{+5}{Na}\underset{\substack{\text{sodium} \\ \text{halate}}}{XO_3} + \overset{-1}{5NaX} + 3H_2O \qquad X = Cl, Br, I$$

All halates decompose upon heating. Potassium chlorate is used as a convenient laboratory source of small amounts of oxygen, especially when heated with a catalyst like MnO_2, which lowers the temperature of decomposition:

$$\overset{+5}{2KClO_3} \xrightarrow[MnO_2]{\Delta} \overset{-1}{2KCl} + 3O_2$$

Gentle heating without a catalyst, however, produces potassium perchlorate and potassium chloride by disproportionation:

$$\overset{+5}{4KClO_3} \longrightarrow \overset{+7}{3KClO_4} + \overset{-1}{KCl}$$

Neither perfluoric acid nor perfluorate ion is known. The other *perhalic acids* and *perhalates* are known. Anhydrous perchloric acid, $HClO_4$, a colorless oily liquid, distills under reduced pressure (10 to 20 torr) after reaction of a perchlorate salt with nonvolatile concentrated sulfuric acid.

$$NaClO_4 + H_2SO_4 \longrightarrow NaHSO_4 + \underset{\substack{\text{perchloric acid} \\ \text{distills out}}}{HClO_4}$$

Treatment of $HClO_4$ with a strong dehydrating agent like P_4O_{10} produces its explosive acid anhydride, dichlorine heptoxide, Cl_2O_7. The oxidation state of Cl is +7 in both compounds.

Hot, concentrated perchloric acid can explode in the presence of reducing agents, especially organic reducing agents. While hot, concentrated $HClO_4$ solutions are very strong oxidizing agents, cold, dilute perchloric acid is only a weak oxidizing agent. In fact, the hydrogen ions in such solutions are reduced to hydrogen with no reduction of perchlorate ions in the presence of fairly strong reducing agents, such as metallic zinc.

$$Zn + \underset{\text{cold, dilute}}{2HClO_4} \longrightarrow Zn(ClO_4)_2 + H_2$$

Perchloric acid is the strongest of all common acids with respect to ionization, exceeding even HNO_3 and HCl. It is used commercially as a substitute for sulfuric acid under conditions in which H_2SO_4 could be reduced to SO_2.

The Heavier Members of Group VIA

In the following sections the chemistry of the heavier members of Group VIA will be discussed. Particular emphasis will be placed on sulfur and its compounds. **The properties and reactions of oxygen were discussed in Chapter 9.** The Group VIA

TABLE 21–8   Properties of the Group VIA Elements

Property	Oxygen	Sulfur	Selenium	Tellurium	Polonium
physical state (25°C, 1 atm)	gas	solid	solid	solid	solid
color	colorless (very pale blue liq.)	yellow	red-gray to black	brass-like metallic luster	—
outer electrons	$2s^2 2p^4$	$3s^2 3p^4$	$4s^2 4p^4$	$5s^2 5p^4$	$6s^2 6p^4$
melting point (°C, 1 atm)	−219	112	217	450	254
boiling point (°C, 1 atm)	−183	444	685	990	962
atomic radius (nm)	0.066	0.104	0.117	0.137	0.14
ionic radius, E^{2-} (nm)	0.140	0.184	0.198	0.221	—
electronegativity	3.5	2.5	2.4	2.1	1.9
first ionization energy (kJ/mol)	1314	1000	941	869	812
common oxidation states	usually −2	−2, +2, +4, +6	−2, +2, +4, +6	−2, +2, +4, +6	−2, +6

> All that is known about the chemistry of polonium has been determined through the use of radioactive tracers.

elements show less electronegative character than the halogens. As is true of all representative groups, metallic character increases with increasing atomic weight. Oxygen and sulfur are clearly nonmetallic, but selenium is less so. Tellurium is usually classified as a metalloid and crystallizes in a metal-like crystal lattice, yet its chemistry is mostly nonmetallic. Polonium, all 29 isotopes of which are radioactive, is less familiar to us, but it is known to be more metallic than tellurium.

Irregularities in variations of the properties of elements within a given family increase from the extreme right and left of the periodic table toward the middle, when only the A group elements are considered. Thus, the properties of the Group VIA elements differ more than do the properties of the halogens. Furthermore, the properties of elements in the second period usually differ significantly from those of other elements in their families, because of the absence of low-energy d orbitals. In keeping with this, the properties of oxygen are not very similar to those of the other Group VIA elements (Table 21–8).

The outer electronic configuration of the VIA elements is $ns^2 np^4$, and all show a tendency to gain or share two additional electrons in many of their compounds. All form covalent compounds of the type H_2E in which the VIA element (E) exhibits an oxidation number of −2. The maximum number of atoms with which oxygen can bond (coordination number) is four, but sulfur, selenium, tellurium, and probably polonium can bond covalently to as many as six other atoms. This is due to the availability of vacant d orbitals in the outer shell of each of the VIA elements except oxygen. One or more of the d orbitals can be used to accommodate additional electrons donated by other atoms to form a total of up to six bonds.

> The d orbitals do not occur until the third shell.

21–12   Occurrence, Properties, and Uses of Sulfur

Sulfur makes up about 0.05% of the Earth's crust, and was one of the elements known to the ancients. It was used by the Egyptians as a yellow coloring, and it was burned in some religious ceremonies because of the unusual odor it produced; it is the "brimstone" of the Bible. Alchemists tried to incorporate its "yellowness" into other substances in attempts to produce gold.

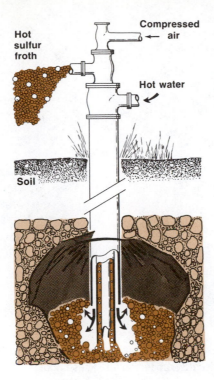

FIGURE 21–5 The Frasch process for sulfur mining. Three concentric pipes are used. Water at a temperature of about 170°C and a pressure of 7 atm is forced down the outermost pipe to melt the sulfur. Hot compressed air is pumped down the innermost pipe, and mixes with the molten sulfur to form a froth that rises through the third.

Sulfur occurs as the free element, predominantly S_8 molecules, and as metal sulfides such as galena, PbS, iron pyrite, FeS_2, and cinnabar, HgS. To a lesser extent it occurs as metal sulfates such as barite, $BaSO_4$, and gypsum, $CaSO_4 \cdot 2H_2O$, and in volcanic gases as hydrogen sulfide, H_2S, and sulfur dioxide, SO_2.

Sulfur is also a constituent of much naturally-occurring organic matter such as petroleum and coal. Its presence in fossil fuels causes environmental and health problems because many sulfur-containing compounds undergo combustion to produce sulfur dioxide (Section 9–7.4), an air pollutant, when the fuels are burned.

The elemental sulfur mined along the U.S. Gulf Coast is obtained by the **Frasch process,** or "hot water" process, described in Figure 21–5. Most of it is used in the production of sulfuric acid, H_2SO_4, probably the most important of all industrial chemicals. Sulfur is also a component of black gunpowder, and is used in the vulcanization of rubber and in the synthesis of many important sulfur-containing organic compounds. Elemental sulfur is a good electrical insulator.

In each of the three physical states, elemental sulfur exists in many forms. The stable form of oxygen is the diatomic molecule, O_2; in contrast, the two most stable forms of sulfur, the rhombic (mp 112°C) and monoclinic (mp 119°C) crystalline modifications, consist of S_8 molecules. These are puckered rings containing eight sulfur atoms (Figure 21–6) and all S—S single bonds. The S—S single bond energy (213 kJ/mol) is much greater than that for the O—O single bond (138 kJ/mol), and this accounts for the greater tendency of sulfur to bond with itself and form rings. The rings are arranged differently in the rhombic and monoclinic forms.

O_2 is doubly bonded, whereas S—S sigma bonds are apparently too long for pi bonding to be effective.

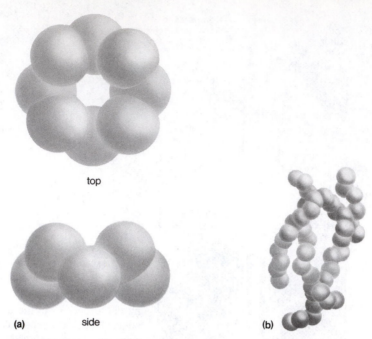

top

(a)   side   (b)

FIGURE 21–6 (*a*) Two views of the sulfur molecule, S_8. (*b*) A model of the intertwining chains of sulfur atoms in viscous, molten sulfur. These chains may be as long as 10,000 atoms.

21–13   Reactions of the Nonmetals of Group VIA

Some of the general reactions of the elements of Group VIA are summarized in Table 21–9.

TABLE 21–9   Some Reactions of the VIA Elements

General Equation	Remarks
$xE + yM \rightarrow M_yE_x$	With many metals
$zE + M_xE_y \rightarrow M_xE_{y+z}$	Especially with S, Se
$E + H_2 \rightarrow H_2E$	Decreasingly in the series O_2, S, Se, Te
$E + 3F_2 \rightarrow EF_6$	With S, Se, Te, and excess F_2
$2E + Cl_2 \rightarrow E_2Cl_2{}^*$	With S, Se (Te gives $TeCl_2$)
$E_2Cl_2 + Cl_2 \rightarrow 2ECl_2{}^*$	With S, Se
$E + 2Cl_2 \rightarrow ECl_4{}^*$	With S, Se, Te, and excess Cl_2
$E + O_2 \rightarrow EO_2$	With S (with Se, use $O_2 + NO_2$)
$3E + 4HNO_3 \rightarrow 3EO_2 + 2H_2O + 4NO$	With S, Se, Te
$3E + 6OH^- \rightarrow EO_3{}^{2-} + 2E^{2-} + 3H_2O$	With S

* Parallel reactions with Br_2.

21–14   Hydrides of the VIA Elements

HCN is the gas used in gas chambers in some states utilizing capital punishment.

All the VIA elements form covalent compounds of the type H_2E (E = O, S, Se, Te, Po) in which the VIA element is in the -2 oxidation state. All except H_2O have extremely noxious odors and are even more toxic than HCN. While H_2O is a liquid

TABLE 21–10 Properties of the Group VIA Hydrides

Property	H₂O	H₂S	H₂Se	H₂Te
melting point (°C)	0.00	−85.60	−60.4	−51
heat of fusion (kJ/mol)	6.01	2.38	—	—
boiling point (°C, 1 atm)	100	−60.75	−41.5	−1.8
heat of vaporization (kJ/mol)	40.7	18.7	19.9	24
density at boiling point (g/mL)	0.958	0.993	2.004	2.650
heat of formation at 25°C (kJ/mol)	−286	−20.6	29.7	135
free energy of formation at 25°C (kJ/mol)	−237	−33.6	15.9	130

that is absolutely essential for animal and plant life, H₂S, H₂Se, and H₂Te are colorless, noxious, poisonous gases. H₂Po is a liquid. Egg protein contains sulfur, and its decomposition forms H₂S, which is responsible for the odor of rotten eggs and for tarnishing silver. H₂Se and H₂Te smell even worse. Small amounts of H₂S will cause headaches and nausea. (If these symptoms occur while you are using this common laboratory reagent, you should leave the laboratory and get fresh air immediately.) Exposure to larger amounts can cause fainting and heart or lung failure. H₂S poses an additional problem in that, after long enough exposure, it severely retards the sense of smell so that one is no longer aware of its presence. Some properties of the VIA hydrides are summarized in Table 21–10.

Both the melting point and boiling point of water are very much higher than would be predicted from those of the heavier hydrides (Table 21–10). This is a consequence of hydrogen bonding in ice and liquid water (Section 12–8.3) caused by the strongly dipolar nature of water molecules. The energy required to overcome hydrogen bonding is reflected in the high melting and boiling points of water. Since the electronegativity differences between hydrogen and the other VIA elements are much smaller than between hydrogen and oxygen, no hydrogen bonding occurs in hydrogen sulfide, selenide, telluride or polonide. The angular, polar water molecule is described by assuming sp^3 hybridization for oxygen. The bond angle decreases, as does polarity, as the group is descended. The heavier members, H₂S, H₂Se, and H₂Te, all have bond angles of nearly 90°, and so are best described by assuming that the VIA atom utilizes p^2 bonding.

Water and hydrogen sulfide are produced from their elements in exothermic reactions, whereas the other hydrides are produced by endothermic reactions. Therefore, while water and hydrogen sulfide are stable at room temperature, hydrogen selenide and telluride slowly decompose to the elements.

Fortunately, their odors are usually ample warning of the presence of these poisonous gases.

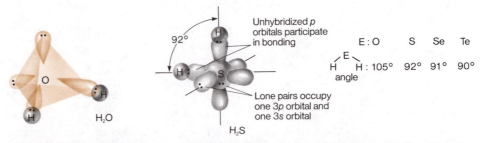

FIGURE 21–7 The Group VIA hydrides. The structures of H₂Se and H₂Te are similar to that of H₂S, but the bond angles are slightly smaller because Se and Te atoms are larger than the S atom.

H_2S is often prepared in the laboratory by hydrolysis of thioacetamide, a sulfur-containing organic compound. Heating speeds the reaction.

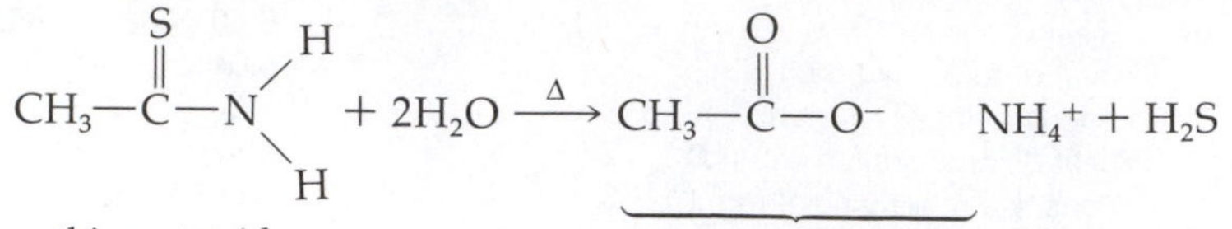

$$CH_3-\overset{\overset{\displaystyle S}{\|}}{C}-N\overset{\displaystyle H}{\underset{\displaystyle H}{\big<}} + 2H_2O \xrightarrow{\Delta} CH_3-\overset{\overset{\displaystyle O}{\|}}{C}-O^- \quad NH_4^+ + H_2S$$

thioacetamide

ammonium acetate
(NH_4CH_3COO)

21–15   Aqueous Solutions of the VIA Hydrides

Aqueous solutions of hydrogen sulfide, selenide, and telluride are acidic, with acid strength increasing upon descending the group: $H_2O < H_2S < H_2Se < H_2Te$. The same trend was observed for increasing acidity of the hydrogen halides (Section 17–4.1) and will be observed for decreasing basicity of the Group VA hydrides in Chapter 22. This is a consequence of the decrease in average bond energy in the same order. The acid ionization constants are

<table>
<tr><td></td><td></td><td>H_2S</td><td>H_2Se</td><td>H_2Te</td></tr>
<tr><td>$H_2E \rightleftharpoons H^+ + HE^-$</td><td>$K_1$:</td><td>$1.0 \times 10^{-7}$</td><td>$1.9 \times 10^{-4}$</td><td>$2.3 \times 10^{-3}$</td></tr>
<tr><td>$HE^- \rightleftharpoons H^+ + E^{2-}$</td><td>$K_2$:</td><td>$1.3 \times 10^{-13}$</td><td>$\sim 10^{-11}$</td><td>$\sim 1.6 \times 10^{-11}$</td></tr>
</table>

The solubility of H_2S in water is approximately 0.10 mol/L at 25°C.

21–16   Group VIA Oxides

Although others exist, the most important VIA oxides are the dioxides, which are acid anhydrides of sulfurous, selenous, and tellurous acids, and the trioxides, which are anhydrides of sulfuric, selenic, and telluric acids.

1   Physical States of the VIA Dioxides

The tendency toward metallic character upon descending a group is exemplified by the stable forms of the dioxides of S, Se, and Te. The very nonmetallic sulfur is covalently bonded to oxygen in SO_2, as is selenium in polymeric SeO_2, but the more metallic tellurium forms ionic TeO_2. Note the similarity of O_3, ozone (Section 9–3.1), "oxygen dioxide," to SO_2.

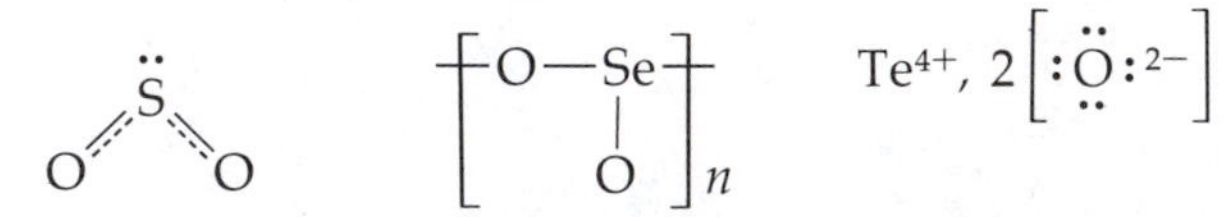

sulfur dioxide
molecular gas

selenium dioxide
polymeric solid

tellurium dioxide
ionic lattice
(also PoO_2)

Multiple bond character often exists between sulfur and oxygen atoms, which have small electronegativity differences and large electronegativity sums. Sulfur and selenium dioxides are quite soluble in water, but tellurium dioxide is only slightly soluble, owing to its high crystal lattice energy.

2 Sulfur Dioxide (An Air Pollutant)

Sulfur dioxide is a colorless, poisonous, corrosive gas with a very irritating odor. It causes coughing and nose, throat, and lung irritation when inhaled even in small quantities. It is an angular molecule with trigonal planar electronic geometry involving sp^2 hybridization at the sulfur atom and resonance stabilization. Both sulfur-oxygen bonds are of equal length, 0.143 nm, and intermediate in length between typical single and double bond distances. The resonance structures are shown in Section 7–4.3.

Sulfur dioxide is undesirably produced in a number of reactions. These primarily involve volcanic eruptions, the combustion of sulfur-containing fossil fuels, and the roasting of sulfide ores.

Metals such as copper, zinc, and lead are obtained by smelting their sulfide ores (Chapter 20). This involves heating in the presence of oxygen from the air (roasting).

$$2ZnS + 3O_2 \rightleftharpoons 2ZnO + 2SO_2$$

A waste product of the operation is sulfur dioxide which, in the past, has been almost indiscriminately released into the atmosphere along with some SO_3 produced by its reaction with O_2. However, efforts are now underway to trap gaseous sulfur dioxide and trioxide and use them to produce sulfuric acid. Some coal contains up to 5% sulfur, so both SO_2 and SO_3 are present in flue gases and should be trapped when coal is burned.

No way has been found to remove all the SO_2 from flue gases of power plants, but one way of removing most of it involves the injection of limestone, $CaCO_3$, into the combustion zone of the furnace where it undergoes thermal decomposition to lime, CaO. This then combines with SO_2 to form calcium sulfite, $CaSO_3$, an ionic solid that is then collected.

$$CaCO_3 \xrightarrow{\Delta} CaO + CO_2$$

$$CaO + SO_2 \longrightarrow CaSO_3$$

This process is called scrubbing. A disadvantage is that it creates huge quantities of solid waste ($CaSO_3$, unreacted CaO, and other substances) which must be disposed of.

Methods of catalytic oxidation are being used by the smelting industry to convert SO_2 into SO_3 and then into solutions of sulfuric acid, H_2SO_4 (up to 80% by mass). To do so, the gases containing SO_2 are passed through a series of condensers containing catalysts to speed up the desired conversions. Again, waste disposal is a major problem. In some cases the impure acid solutions can be used in other operations in the same plant, but in other cases the acid must be sold commercially. This is economically burdensome, because the solutions first must be purified and transportation costs are usually high (since most smelters are quite isolated).

Sulfur dioxide is also an important industrial, commercial, and research chemical. Sulfur dioxide can be prepared in the laboratory by treating sodium sulfite or sodium hydrogen sulfite with a nonvolatile acid such as sulfuric or phosphoric acid.

$$\underset{\substack{\text{sodium hydrogen}\\\text{sulfite}}}{2NaHSO_3} + H_2SO_4 \longrightarrow Na_2SO_4 + 2SO_2 + 2H_2O$$

It can also be prepared by the action of very weak reducing agents on hot concentrated sulfuric acid. The dissolution of metallic copper is an example.

$$Cu + 2H_2SO_4 \longrightarrow CuSO_4 + SO_2 + 2H_2O$$

3 Sulfur Trioxide

Sulfur trioxide is a liquid that boils at 44.8°C. It is the anhydride of sulfuric acid, and can be formed by the reaction of sulfur dioxide with oxygen. The ordinarily slow but very exothermic reaction is catalyzed commercially in the **contact process** by spongy platinum, silicon dioxide, or vanadium(V) oxide, V_2O_5, at high temperatures (400–700°C).

$$2SO_2 (g) + O_2 (g) \underset{\text{catalyst}}{\rightleftharpoons} 2SO_3 (g) \qquad \Delta H^0 = -197.6 \text{ kJ}$$

The high temperature favors the reverse reaction toward SO_2 and O_2, but allows equilibrium to be achieved so much more rapidly that it is economically advantageous. The SO_3 is then removed from the gaseous reaction mixture by dissolution in concentrated H_2SO_4 ($\sim$ 95% H_2SO_4 by mass) to produce polysulfuric acids, mainly pyrosulfuric acid, $H_2S_2O_7$, which is called oleum or fuming sulfuric acid. Addition of fuming sulfuric acid to water produces commercial H_2SO_4.

$$SO_3 + H_2SO_4 \longrightarrow H_2S_2O_7$$

$$H_2S_2O_7 + H_2O \longrightarrow 2H_2SO_4$$

Alternatively, SO_3 can be obtained by distillation of fuming sulfuric acid. It is used in the preparation of sulfonated oils and sulfonate detergents (Section 13–22).

Sulfur dioxide in polluted air reacts rapidly with oxygen to form sulfur trioxide in the presence of certain catalysts. Particulate matter, or suspended microparticles, such as ammonium nitrate and elemental sulfur, act as efficient catalysts for this oxidation.

The SO_3 molecule is trigonal planar, containing sp^2 hybridized sulfur, and is represented by the following resonance formulas.

The term "pyro" means heat or fire. Pyrosulfuric acid may also be obtained by heating concentrated sulfuric acid, which results in the elimination of one molecule of water per two molecules of sulfuric acid.

$$2H_2SO_4 \longrightarrow H_2S_2O_7 + H_2O$$

21–17 Oxyacids of Sulfur

1 Sulfurous Acid and Sulfites

Sulfur dioxide readily dissolves in water to produce solutions of sulfurous acid, H_2SO_3. The acid has not been isolated in anhydrous form.

$$H_2O + \overset{+4}{S}O_2 \rightleftharpoons H_2\overset{+4}{S}O_3$$

The acid ionizes in two steps in water.

$$H_2SO_3 \rightleftharpoons H^+ + \quad HSO_3^- \qquad K_1 = 1.2 \times 10^{-2}$$

sulfurous   hydrogen sulfite ion
acid   (bisulfite ion)

$$HSO_3^- \rightleftharpoons H^+ + \quad SO_3^{2-} \qquad K_2 = 6.2 \times 10^{-8}$$

sulfite ion

When SO_2 is bubbled through aqueous NaOH, sodium hydrogen sulfite, $NaHSO_3$, is produced. This acid salt can be neutralized with additional NaOH or Na_2CO_3 to produce sodium sulfite.

$$NaOH + H_2SO_3 \longrightarrow \quad NaHSO_3 \quad + H_2O$$

sodium hydrogen sulfite

$$NaOH + NaHSO_3 \longrightarrow \quad Na_2SO_3 \quad + H_2O$$

sodium sulfite

The sulfite ion is pyramidal and has tetrahedral electronic geometry.

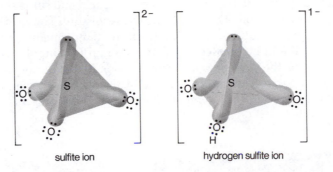

sulfite ion   hydrogen sulfite ion

2 Sulfuric Acid and Sulfates (and Acidic Rain)

More than 40 million tons of sulfuric acid are produced annually worldwide. The contact process is used for the commercial production of almost all sulfuric acid and sulfur trioxide. The solution sold commercially as "concentrated sulfuric acid" is 96–98% H_2SO_4 by mass, the rest being water, and is 18 molar H_2SO_4.

Pure sulfuric acid is a colorless oily liquid that freezes at 10.4°C and boils at 290–317°C while partially decomposing to SO_3 and water. Solid and liquid H_2SO_4 are somewhat hydrogen-bonded.

The sulfur is sp^3 hybridized in sulfuric acid and in sulfate and hydrogen sulfate ions, resulting in tetrahedral geometry around sulfur.

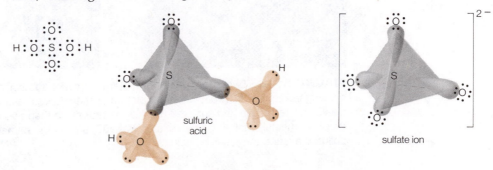

sulfuric acid   sulfate ion

Tremendous amounts of heat are evolved when concentrated sulfuric acid is diluted. This behavior illustrates the strong affinity of H_2SO_4 for water (Section 13-3). Thus, H_2SO_4 is often used as a dehydrating agent. DILUTIONS (AS WITH ALL ACIDS) SHOULD ALWAYS BE PERFORMED BY ADDING THE ACID TO WATER in order to avoid spattering the acid, a phenomenon caused by the heat evolved on mixing.

Sulfuric acid is a strong acid with respect to the first step of its ionization in water. The second ionization occurs to a somewhat lesser extent.

For selenic acid, H_2SeO_4,
K_1 = very large
$K_2 = 1.15 \times 10^{-2}$

$$H_2SO_4 \rightleftharpoons H^+ + HSO_4^- \qquad K_1 = \text{very large}$$

hydrogen sulfate
ion

$$HSO_4^- \rightleftharpoons H^+ + SO_4^{2-} \qquad K_2 = 1.2 \times 10^{-2}$$

sulfate ion

Hot concentrated sulfuric acid is also a strong oxidizing agent.

The "acidic rain" now experienced especially in highly industrialized areas is due in large part to the presence of sulfuric acid and some sulfurous acid, which are formed by reaction of SO_2 and SO_3 with moisture. The oxides are produced during combustion of sulfur-containing fossil fuels and smelting operations.

It should be pointed out that even rain in "unpolluted" air is slightly acidic because of the presence of carbon dioxide in the atmosphere, which reacts with moisture to produce carbonic acid, H_2CO_3. However, rain in some areas has been found to have a pH of about 4, due mainly to the presence of sulfuric acid.

FIGURE 21-8 Two photographs of a statue showing the corrosive power of pollutants causing acidic rain. In 1908, when the picture at the left was taken, the statue at Harten Castle in Westphalia had survived nearly intact for more than two centuries. The photograph at the right, taken 61 years later, in 1969, shows the damage done by increasingly polluted air, and the resulting acid rain.

Sulfuric acid, of course, easily reacts with structural materials such as limestone and marble (both $CaCO_3$).

$$H_2SO_4 + CaCO_3 \longrightarrow CaSO_4 + CO_2 + H_2O$$

The calcium sulfate then washes away in the rain (see Figure 21-8). The acid also breaks down structures composed of metals such as steel (which contains iron) and aluminum (which is always coated with an oxide surface when exposed to air).

$$H_2SO_4 + Fe \longrightarrow FeSO_4 + H_2$$

$$3H_2SO_4 + Al_2O_3 \longrightarrow Al_2(SO_4)_3 + 3H_2O$$

Sulfuric acid also is extremely reactive toward many organic compounds, including those in plants and human flesh; the lungs are especially susceptible to irritation.

Key Terms

Contact process industrial process by which sulfur trioxide and sulfuric acid are produced from sulfur dioxide.

Frasch process method by which elemental sulfur is mined or extracted; it is melted and forced to the surface of the earth as a slurry with superheated water.

Halogens Group VIIA elements: F, Cl, Br, I, At.

Noble (rare) gases Group 0 elements: He, Ne, Ar, Kr, Xe, Rn.

Particulate matter finely divided solid particles suspended in polluted air.

Polymer a large molecule consisting of chains or rings of linked identical units; usually characterized by high melting and boiling points.

Exercises

The Noble Gases

1. Why are the noble gases so unreactive?
2. Why were the noble gases among the last elements to be discovered?
3. List some of the uses of the noble gases and reasons for the uses.
4. Arrange the noble gases in order of increasing (a) atomic radii, (b) melting points, (c) boiling points, (d) densities, (e) first ionization energies.
5. Explain the order of increasing melting and boiling points of the noble gases in terms of polarizabilities of the atoms and forces of attraction between them.
6. What gave Neil Bartlett the idea that compounds of xenon could be synthesized?
7. Describe the bonding and geometry of XeF_2, XeF_4, and XeF_6.
8. Compare the bond energies of Xe—F bonds with other typical bonds and with hydrogen bonds.

9. Give the hybridization of Xe and draw structures of XeO_3, XeO_4, and $XeOF_4$.
10. How many grams of xenon oxide tetrafluoride, $XeOF_4$, and how many liters of HF at STP, could be prepared, assuming complete reaction of 6.50 grams of xenon tetrafluoride, XeF_4, with a stoichiometric quantity of water according to the equation below?

$$6XeF_4 (s) + 8H_2O (\ell) \longrightarrow$$
$$2XeOF_4 (\ell) + 4Xe (g) + 16HF (g) + 3O_2 (g)$$

The Halogens

11. Draw Lewis formulas for I_2 and Cl_2.
12. Describe the normal physical states and colors of the free halogens.
13. Why do the halogens not occur free in nature?
14. List the halogens in order of increasing (a) atomic radii, (b) ionic radii, (c) electronegativities, (d) melting points, (e) boiling points, (f) standard reduction potentials.
15. What is the order of increasing X—X (halogen)

bond energies? Suggest why the F—F bond energy is less than the Cl—Cl bond energy.

16. Discuss the polarizabilities of halide ions.

17. Give examples of halogen compounds in which a halogen exhibits the $+1$, $+3$, $+5$, and $+7$ oxidation states. Are these compounds ionic or covalent?

18. Why can fluorine not be prepared by electrolysis of aqueous solutions of fluoride salts?

19. Discuss the chemistry of the extraction of bromine from sea water.

20. Give two practical uses for compounds of each of the halogens.

21. Write equations describing general reactions of the free halogens, X_2, with (a) Group IA (alkali) metals, (b) Group IIA (alkaline earth) metals, (c) Group IIIA metals. Represent the metals as M.

22. The interhalogens are compounds of two different halogens, X and X', having general formulas XX', XX'_3, XX'_5, and XX'_7. X is always the larger halogen and the central atom. Draw structures and give hybridizations at the central atom for ClF_3, BrF_5, and IF_7.

23. Ions of the interhalogens (Exercise 22) also exist. An example is the ICl_4^- ion in $KICl_4$. Sketch the ion and indicate the hybridization at I in the ICl_4^- ion.

24. Distinguish between hydrogen bromide and hydrobromic acid.

25. Write equations illustrating the tendency of F^- to stabilize high oxidation states of cations and the tendency of I^- to stabilize low oxidation states. Why is this the case?

26. Give a reaction illustrating each of four general methods for preparation of hydrogen halides.

27. What is the order of decreasing melting and boiling points of the hydrogen halides? Why is HF out of line?

28. Describe the effect of hydrofluoric acid on glass.

29. Compare the general properties of metal halides with those of nonmetal halides.

30. What is the acid anhydride of perchloric acid?

31. Write the equation for the dehydration of $HClO_4$ with tetraphosphorus decoxide.

32. Name the following compounds: (a) $KBrO_3$, (b) $KBrO$, (c) $NaClO_4$, (d) $NaClO_2$, (e) $HBrO$, (f) $HBrO_3$, (g) HIO_3, (h) $HClO_4$.

33. Draw Lewis formulas and structures of the four ternary acids of chlorine.

34. Write equations describing reactions by which the following compounds can be prepared:
    (a) hypohalous acids of Cl, Br, and I (in solution with hydrohalic acids)
    (b) hypohalite salts
    (c) chlorous acid
    (d) halate salts
    (e) a perchlorate salt
    (f) perchloric acid

35. What is the order of increasing acid strength of the ternary chlorine acids? Explain the order.

36. The pseudohalogens are compounds that behave chemically much like the diatomic halogen molecules. Cyanogen, $(CN)_2$, is a pseudohalogen. How would you expect cyanogen to react with ethylene (or ethene),

37. Calculate the surface area ($4\pi r^2$) in square nanometers for all the halogen atoms and for the halide ions.

38. Calculate the ratio of rates of effusion of each of the halogen molecules to that of the fluorine molecule.

39. The volume of a sphere is given by the equation $V = \frac{4}{3}\pi r^3$. Given the radii of the halogen atoms and their atomic weights, calculate the densities of the halogen atoms in grams per milliliter. (Note: These densities do *not* include the empty space between atoms in samples of the elements.)

40. A 0.250 ampere current is applied to molten sodium chloride (see Section 19–3). The Cl_2 collected at 26°C and 774 torr pressure occupies a volume of 32,700 liters. How many moles, grams, and pounds of NaCl must have been electrolyzed? How many pounds of metallic sodium were produced? How long did the cell operate?

41. Assuming that "iodized salt" contains 0.02% potassium iodide and 99.98% sodium chloride, calculate the percentage of the total mass of a sample of "iodized salt" due to iodine alone.

42. What volume of 0.150 normal sodium hydrogen sulfite solution is required to react with sodium iodate to produce 500 grams of solid iodine? What mass of sodium iodate must react?

43. A 30.0 g sample of Cl_2 is bubbled through 500 mL of 1.50 M NaOH and 17.6 g of Cl_2 escapes. What molarity of NaClO is produced? How many grams of sodium hypochlorite and of sodium chloride would be present? How many grams of sodium hydroxide remain unreacted?

44. How many grams and how many milliliters of silicon tetrafluoride gas are generated at STP by the action of 100 milliliters of 3.00 M HF on 65.4 grams of calcium silicate?

45. What volume of oxygen collected at 727 torr and 28.2°C would be generated by the strong heating and complete decomposition of 5.271 grams of potassium chlorate in the laboratory?

46. How many grams of potassium perchlorate could be produced by *gently* heating 5.271 grams of potassium chlorate so that it all disproportionates?

47. An impure 3.00 gram sample of potassium chlorate containing some potassium chloride was heated strongly to liberate oxygen. A volume of 725 milliliters of oxygen was collected over water at 24.26°C on a day when the barometric pressure was 742.0 torr. What masses of the two salts were contained in the sample? What was the percentage of potassium chlorate in the sample?

Group VIA

48. Write out the electronic configurations for atomic oxygen, sulfur, and tellurium. Do the same for oxide, sulfide, and telluride ions.

49. Characterize the Group VIA elements with respect to color and physical state under normal conditions.

50. The Group VIA elements, except oxygen, can exhibit oxidation states ranging from -2 to $+6$. Why not -3 to $+7$?

51. Give and explain the order of increasing melting and boiling points of the Group VIA elements.

52. Is the order of decreasing first ionization energies of the Group VIA elements consistent with the order of increasing metallic character?

53. Sulfur, selenium, and tellurium are all capable of forming six-coordinate compounds such as SF_6. Give two reasons why oxygen cannot be the central atom in such six-coordinate molecules.

54. Draw diagrams that show the hybridization of atomic orbitals and three-dimensional structures showing all hybridized orbitals and outermost electrons for the following molecules: (a) H_2S, (b) SF_6, (c) SF_4, (d) SO_2, (e) SO_3.

55. Repeat Exercise 54 for (a) SeF_6, (b) SO_3^{2-}, (c) SO_4^{2-}, (d) HSO_4^-, (e) thiosulfate ion, $S_2O_3^{2-}$ (one S is the central atom).

56. Write equations for the reactions of
    (a) S, Se, or Te with excess F_2
    (b) O_2, S, Se, or Te with H_2
    (c) S, Se, or Te with O_2

57. Write equations for the reactions of
    (a) S and Te with HNO_3
    (b) S and Se with excess Cl_2
    (c) S and Se with Na, Ca, and Al

58. What is the order of increasing melting and boiling points and heats of vaporization of the Group VIA hydrides, H_2O, H_2S, H_2Se, H_2Te? What are their physical states under normal conditions? Why is H_2O out of line with the others?

59. Compare the structures of the dioxides of sulfur, selenium, tellurium, and polonium. How are they related to the metallic and nonmetallic characters of these elements?

60. Write equations for the following reactions:
    (a) the preparation of sulfur dioxide (and copper(II) sulfate and water) by reaction of copper with hot concentrated sulfuric acid.
    (b) the reaction of sulfur dioxide with oxygen.
    (c) the reaction of sulfur dioxide with water.

61. What are the acid anhydrides of sulfuric acid, selenic acid, and telluric acid?

62. Draw a two-dimensional representation of pyrosulfuric acid, $H_2S_2O_7$. Write an equation to show how it is prepared from concentrated sulfuric acid.

63. Write equations for the reactions of
    (a) sodium hydroxide with sulfuric acid (1:1 ratio).
    (b) sodium hydroxide with sulfuric acid (2:1 ratio).

64. Write equations for the reactions of
    (a) sodium hydroxide with sulfurous acid (1:1 ratio).
    (b) sodium hydroxide with sulfurous acid (2:1 ratio).

65. How much sulfur dioxide is produced from the complete combustion of one ton of coal containing 4.4% sulfur impurity?

66. A sterling silver serving piece contains 117

grams of silver. If 0.104 gram of silver sulfide (tarnish) forms by reaction of the silver with H_2S from the decomposition of eggs, how much silver must react? What percentage of the silver tarnishes?

$$4Ag + 2H_2S + O_2 \longrightarrow 2Ag_2S + 2H_2O$$

67. Calculate the concentrations of H^+, HSO_3^-, and SO_3^{2-} ions present in $0.010\ M$ sulfurous acid, H_2SO_3, solution. $K_1 = 1.2 \times 10^{-2}$ and $K_2 = 6.2 \times 10^{-8}$.

The Nonmetals—II: Group VA Nonmetals, and Carbon, Silicon, and Boron

22

In Chapter 21 we discussed the nonmetals in Groups 0, VIIA, and VIA. We are now ready to study the important remaining nonmetals: nitrogen, phosphorus, and arsenic in Group VA, carbon and silicon in Group IVA, and boron in Group IIIA. Because these nonmetals lie further to the left in the periodic table, they show more metallic character and less nonmetallic character than elements in the same periods further to the right. Nitrogen, phosphorus, and carbon are very important elements in all living systems.

Group VA Nonmetals

In the nitrogen family, nitrogen and phosphorus are nonmetals, arsenic is predominantly nonmetallic, antimony is somewhat more metallic, and bismuth is definitely metallic. Many of the properties of the Group VA elements are listed in Table 22–1.

The possible oxidation states of these elements range from -3 to $+5$, with emphasis on the odd numbers. The VA elements form only a few monatomic ions. Ions with a charge of $3-$ occur only for nitrogen and phosphorus, as in Mg_3N_2 and Ca_3P_2. Tripositive cations probably exist for antimony and bismuth in such compounds as antimony(III) sulfate, $Sb_2(SO_4)_3$, and bismuth(III) perchlorate pentahydrate, $Bi(ClO_4)_3 \cdot 5H_2O$, although in aqueous solution they are extensively hydrolyzed to SbO^+ or SbOX and BiO^+ or BiOX (X = univalent anion) to give strongly acidic solutions. There are few if any compounds containing ions with $5+$ charge, and simple cations of the Group VA elements with this charge do not exist.

All of the Group VA elements exhibit the -3 oxidation state in covalent compounds such as NH_3, PH_3, and AsH_3. The $+5$ oxidation state is exhibited only in covalent compounds, including phosphorus pentafluoride, PF_5, phosphoric acid, H_3PO_4, and in polyatomic ions such as NO_3^- and PO_4^{3-}. Nitrogen and phosphorus

Nitride and phosphide ions are N^{3-} and P^{3-}, respectively.

TABLE 22-1 Properties of the Group VA Elements

	Nitrogen	Phosphorus	Arsenic	Antimony	Bismuth
physical state (1 atm, 25°C)	gas	solid	solid	solid	solid
color	colorless	red, white, black	yellow, gray	yellow, gray	gray
outermost electrons	$2s^2 2p^3$	$3s^2 3p^3$	$4s^2 4p^3$	$5s^2 5p^3$	$6s^2 6p^3$
melting point (°C)	−210	44 (white)	814 (gray)	631 (gray)	271
boiling point (°C)	−196	280 (white)	sublimes 613	1380	1560
atomic radius (nm)	0.070	0.110	0.121	0.141	0.146
electronegativity	3.0	2.1	2.1	1.9	1.8
first ionization energy (kJ/mol)	1402	1012	947	834	703
oxidation states	−3 to +5	−3 to +5	−3 to +5	−3 to +5	−3 to +5

have many oxidation states in their compounds, but the common ones for antimony and bismuth are +3 and +5.

Oxides in which the Group VA elements exhibit the +3 oxidation state are known for all: N_2O_3, P_4O_6, As_4O_6, Sb_4O_6, and Bi_2O_3. The first two are acid anhydrides of nitrous acid, HNO_2, and phosphorous acid, H_3PO_3; both are weak acids. Saturated aqueous solutions of As_4O_6 and Sb_4O_6 are amphoteric, the former having more acidic character and the latter more basic character. Neither oxide dissolves in water to any significant extent. Bismuth(III) oxide, Bi_2O_3, is the anhydride of bismuth(III) hydroxide, $Bi(OH)_3$, an insoluble base. The trend of increasing oxide basicity on descending the group indicates increasingly metallic character of the elements.

Nitrogen

22-1 Occurrence, Properties, and Importance

Nitrogen, N_2, is a colorless, odorless, tasteless gas that makes up about 75% by mass and 78% by volume of the atmosphere. Despite the fact that nitrogen compounds form only a minor portion of the earth's crust, all living matter contains nitrogen. The primary natural inorganic deposits of nitrogen are very localized and consist primarily of potassium and sodium nitrates, KNO_3 and $NaNO_3$. Most sodium nitrate is mined in Chile.

All proteins contain nitrogen in each of their fundamental amino acid units.

The explanation for the extreme abundance of N_2 in the atmosphere but the low relative abundance of nitrogen compounds elsewhere lies in the chemical inertness of the N_2 molecule. This inertness results from the very high bond energy of the N≡N bond (946 kJ/mol). The Lewis formula shows a triple bond and no unpaired electrons, consistent with its experimentally observed diamagnetism in all three physical states.

:N⫶⫶⫶N:

22-2 The Nitrogen Cycle

Although nitrogen molecules are unreactive, nature provides mechanisms by which nitrogen atoms can be incorporated into proteins, nucleic acids, and other nitrogenous compounds. The **nitrogen cycle** is the complex series of reactions by which nitrogen is slowly but continually recycled in the atmosphere, lithosphere (earth), and hydrosphere (water).

When N_2 and O_2 molecules collide in the atmosphere (which is our nitrogen reservoir) in the vicinity of a bolt of lightning, they can absorb enough electrical energy to produce molecules of NO, nitrogen oxide (commonly called nitric oxide). Molecules of NO are quite reactive because they contain one unpaired electron, as described in Section 22–7, and NO reacts with O_2 to form nitrogen dioxide, NO_2. Most nitrogen dioxide dissolves in rainwater and falls to the earth's surface. Bacterial enzymes reduce the nitrogen in a series of reactions in which amino acids and proteins are produced. These are then used by plants, eaten by animals, metabolized, and excreted as nitrogenous compounds such as urea, $(NH_2)_2CO$, and ammonium salts like $NaNH_4HPO_4$. These products can also be enzymatically converted to ammonia, NH_3, and amino acids.

Another mechanism allows N_2 to be converted directly to ammonia. Members of a certain class of plants called legumes, such as alfalfa and clover, have nodules on their roots; within the nodules live bacteria that contain an enzyme called nitrogenase. These bacteria extract N_2 molecules directly from air trapped in the soil and convert them into ammonia. The ability of nitrogenase to catalyze such a conversion, called "nitrogen fixation," at usual temperatures and pressures with very high efficiency is a marvel to scientists, who must resort to very extreme and costly conditions to produce NH_3 from nitrogen and hydrogen (Haber process, Section 16–5). We hope to learn through research how the enzymatic conversion is carried out, so that we can accomplish the transformation much more efficiently.

Ammonia is the source of nitrogen in many nitrogen-containing fertilizers. Unfortunately, nature does not produce ammonia and related plant nutrient compounds rapidly enough to provide an adequate food supply for the world's growing population. Commercial synthetic fertilizers have helped to lessen this problem; with increasing energy costs, however, the fertilizer shortage is predicted to grow rapidly in the next few decades unless significant advances are made in the area of agricultural and fertilizer chemistry or in population control.

The suffix -ase associated with the name of a biological compound almost invariably signifies that it is an enzyme, or a protein that acts as a catalyst for a specific biological reaction.

22–3   Oxidation States of Nitrogen

No other element exhibits more oxidation states than does nitrogen. Examples are given in Table 22–2.

TABLE 22–2   Oxidation States of Nitrogen and Examples

−3	−2	−1	0	+1	+2	+3	+4	+5
NH_4^+ ammonium ion	N_2H_4 hydrazine	NH_2OH hydroxylamine	N_2 nitrogen	N_2O dinitrogen oxide	NO nitrogen oxide	N_2O_3 dinitrogen trioxide	NO_2 nitrogen dioxide	N_2O_5 dinitrogen pentoxide
NH_3 ammonia		NH_2Cl chloramine		$H_2N_2O_2$ hyponitrous acid		HNO_2 nitrous acid	N_2O_4 dinitrogen tetroxide	HNO_3 nitric acid
NH_2^- amide ion						NO_2^- nitrite ion		NO_3^- nitrate ion

Nitrogen Compounds with Hydrogen

22–4   Ammonia and Ammonium Salts

1   Production

38 billion pounds of NH_3 were produced in the U.S.A. in 1981.

Ammonia is produced commercially by the **Haber process** described in Section 16–5.

$$N_2 \text{ (g)} + 3H_2 \text{ (g)} \underset{\Delta, \text{ pressure}}{\overset{\text{iron}}{\rightleftharpoons}} 2NH_3 \text{ (g)} \qquad \Delta H^0 = -92.2 \text{ kJ}$$

Ammonia can be prepared in the laboratory by treating aqueous ammonia or ammonium salts with excess alkali and distilling. The presence of excess OH^- favors the reaction toward the right and NH_3 is evolved as a gas.

$$\underset{\text{excess}}{NH_4^+ + OH^-} \rightleftharpoons NH_3 + H_2O$$

2   Properties

Ammonia is a colorless gas with a very characteristic pungent odor. The molecules are pyramidal and very polar (Section 7–5). Ammonia is very soluble in water because it forms hydrogen bonds with water. Saturated aqueous ammonia is 15 M. The lone pair of electrons on the nitrogen atom allows NH_3 to function as a Lewis base in its reaction with water.

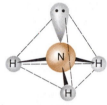

Ammonia, NH_3
mp $-77.7°C$
bp $-33.4°C$
polar, pyramidal molecule; tetrahedral (sp^3) electronic geometry

$$\underset{\text{aqueous ammonia}}{NH_3 \text{ (aq)} + H_2O \text{ (}\ell\text{)}} \rightleftharpoons NH_4^+ + OH^- \qquad K_b = 1.8 \times 10^{-5}$$

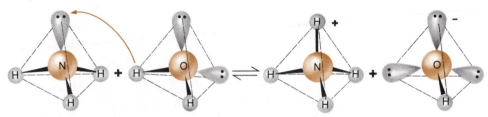

Ammonia acts as a Lewis base in its reaction with metal ions, such as aqueous Co^{2+}, or $[Co(OH_2)_6]^{2+}$, to form ammine complex ions (Section 23–2) such as $[Co(NH_3)_6]^{2+}$.

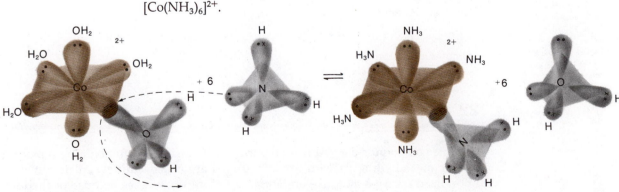

$$[Co(OH_2)_6]^{2+} + 6NH_3 \xrightarrow{H_2O} [Co(NH_3)_6]^{2+} + 6H_2O$$

hexaaquacobalt(II) ion　　　　　　　　　hexaamminecobalt(II) ion

Ammonia also acts as a Lewis base in its reactions with other strong electron acceptors to form "addition compounds." Its reaction with boron trifluoride, BF_3, is a typical example (Section 17–10).

$$BF_3 + :NH_3 \longrightarrow F_3B:NH_3$$

Ammonia burns in oxygen when heated to produce nitrogen and water.

$$4NH_3\ (g) + 3O_2\ (g) \xrightarrow{\Delta} 2N_2\ (g) + 6H_2O\ (g) \qquad \Delta H^0 = -1.27 \times 10^3\ kJ$$

But in the presence of red-hot platinum, nitrogen oxide, rather than nitrogen, is produced.

$$4NH_3\ (g) + 5O_2\ (g) \xrightarrow[Pt]{\Delta} 4NO\ (g) + 6H_2O\ (g) \qquad \Delta H^0 = -904\ kJ$$

This is an important reaction in the Ostwald process (Section 22–13) for the production of nitric acid.

Liquid ammonia (bp $-33.4°C$) is used as a solvent for some chemical reactions. It is hydrogen bonded, just as water is, but ammonia is a much more basic solvent. Its weak *autoionization* produces the ammonium ion, NH_4^+, and the amide ion, NH_2^-. Note the similarity to H_2O, which ionizes slightly to produce some H_3O^+ and OH^- ions with $K_w = 10^{-14}$.

Liquid NH_3 is a clear colorless liquid.

The autoionization of NH_3, like the autoionization of H_2O, involves the transfer of a proton.

$$NH_3\ (\ell) + NH_3\ (\ell) \rightleftharpoons NH_4^+ + NH_2^- \qquad K = 10^{-30}$$

　acid$_1$　　　 base$_2$　　　 acid$_2$　　 base$_1$

Many ammonium salts are known, and most are very soluble in water. They can be prepared by reactions of ammonia with acids, such as those shown below.

$$NH_3 + HCN \longrightarrow [NH_4^+ + CN^-] \qquad \text{ammonium cyanide}$$

$$NH_3 + [H^+ + Br^-] \longrightarrow [NH_4^+ + Br^-] \qquad \text{ammonium bromide}$$

$$NH_3 + [H^+ + NO_3^-] \longrightarrow [NH_4^+ + NO_3^-] \qquad \text{ammonium nitrate}$$

Some ammonium salts contain anions that act as oxidizing agents toward the ammonium ion; when heated, they decompose rapidly and sometimes even explosively.

Ammonium nitrate is often used as a source of nitrogen in fertilizers because of its high nitrogen content (35% by mass) and water solubility.

$$2NH_4NO_3\ (s) \xrightarrow{\Delta} 2N_2\ (g) + 4H_2O\ (g) + O_2\ (g) + heat$$

ammonium nitrate

$$(NH_4)_2Cr_2O_7\ (s) \xrightarrow{\Delta} N_2\ (g) + Cr_2O_3\ (s) + 4H_2O\ (g) + heat$$

ammonium dichromate

22–5　Amines

Organic compounds derived from or structurally related to ammonia by replacing one or more hydrogens with organic groups are called **amines**. All involve sp^3 hybridized nitrogen and, like ammonia itself, are weak bases because of the lone pair of electrons. Examples are given at the top of page 731.

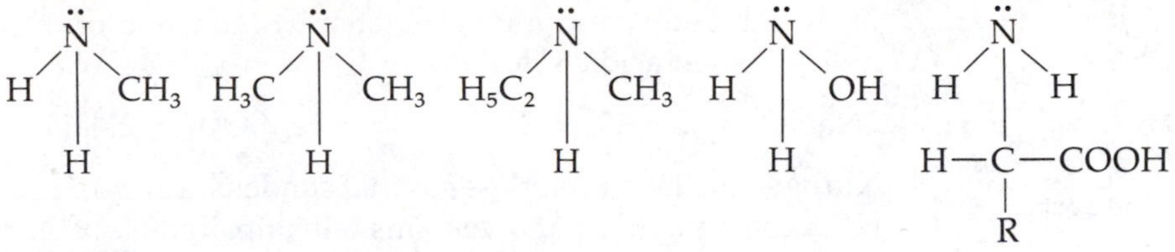

methyl-
amine

dimethyl-
amine

ethylmethyl-
amine

hydroxyl-
amine

an amino acid
R = organic
group

Nitrogen Oxides

Nitrogen forms several oxides, in which it exhibits positive oxidation states of 1 to 5 (see the right side of Table 22–2). All the oxides have positive free energies of formation, owing to the high dissociation energy of nitrogen and oxygen molecules. All those listed in Table 22–2 are gases except N_2O_5, a solid that sublimes at only 32.4°C. There is also evidence for the existence of two very unstable oxides, NO_3 and N_2O_6.

22–6 Dinitrogen Oxide (+1 Oxidation State)

Molten ammonium nitrate undergoes internal oxidation-reduction (decomposition) at 170° to 260°C to produce dinitrogen oxide, also called nitrous oxide. Above this temperature range, explosions occur, and the products are N_2, O_2, and H_2O.

Some dentists use N_2O for its mild anesthetic properties; it is also known as laughing gas because of its side effects.

$$NH_4NO_3 \xrightarrow{\Delta} N_2O + 2H_2O + heat$$

Dinitrogen oxide decomposes into nitrogen and oxygen when heated.

$$2N_2O \xrightarrow{\Delta} 2N_2 + O_2 + heat$$

Dinitrogen oxide supports combustion very well, since it produces oxygen upon heating.

The molecule is linear but nonsymmetric, with a dipole moment of 0.17 D. It is thought to have at least two important contributing resonance structures.

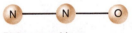

Dinitrogen oxide or nitrous oxide, N_2O
mp −90.8°C
bp −88.8°C

$$:\ddot{N}::N::\ddot{O}: \longleftrightarrow :N:::N:\ddot{O}:$$

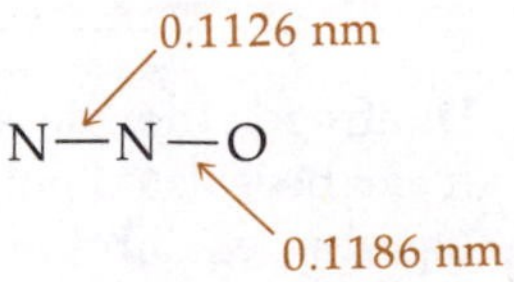

0.1126 nm

$$N \xleftarrow{} N \longrightarrow O$$

0.1186 nm

22–7 Nitrogen Oxide (+2 Oxidation State)

The first step of the Ostwald process (Section 22–13) for the production of nitric acid from ammonia is also used for the commercial preparation of nitrogen oxide.

$$4NH_3 + 5O_2 \xrightarrow[\Delta]{catalyst} 4NO + 6H_2O$$

In the laboratory it can be generated by reduction of nitrite ion by iron(II) ions or iodide ions in acidic solution.

$$2NaNO_2 + 2FeSO_4 + 3H_2SO_4 \longrightarrow Fe_2(SO_4)_3 + 2NaHSO_4 + 2H_2O + 2NO$$

Nitrogen oxide is a colorless gas that condenses at $-152°C$ to a blue liquid. Gaseous NO is paramagnetic and contains one unpaired electron per molecule.

$$:N::\ddot{O}: \longleftrightarrow :N::\dot{O}:$$

Its unpaired electron makes nitric oxide very reactive. Molecules that contain unpaired electrons are called **radicals.** It is somewhat surprising that in the gas phase NO is predominantly monomeric. However, in the solid phase it is dimerized to N_2O_2 and is diamagnetic and polar.

Nitrogen oxide, which is colorless, reacts with oxygen to form nitrogen dioxide, a brown corrosive gas.

$$2NO\ (g) + O_2\ (g) \longrightarrow 2NO_2\ (g)$$

There is concern that NO emitted in the exhaust fumes of supersonic transports (SSTs), which fly in the stratosphere, can catalyze the decomposition of ozone, O_3. In the stratosphere ozone absorbs some of the ultraviolet radiation from the sun before it reaches the surface of the earth. It is thought that the incidence of skin cancer is proportional to exposure to ultraviolet radiation. The atomic oxygen radicals in the following reactions come from the natural radiation-induced breakdown of ozone.

$$NO + O_3 \longrightarrow NO_2 + O_2$$
$$NO_2 + O \longrightarrow NO + O_2$$
$$\overline{O_3 + O \longrightarrow 2O_2\ \text{(net reaction)}}$$

Note that the NO is regenerated in the second step, so it can cause a chain reaction in which a single molecule of NO ultimately is responsible for the decomposition of several molecules of ozone.

22-8   Dinitrogen Trioxide (+3 Oxidation State)

Dinitrogen trioxide is the extremely unstable acid anhydride of nitrous acid, HNO_2. It can be isolated pure only as a solid (mp $-103°C$). The oxide is prepared as a blue liquid in equilibrium with its reactants by mixing equimolar amounts of nitrogen oxide and nitrogen dioxide at $-20°C$.

$$NO + NO_2 \rightleftharpoons N_2O_3$$

Upon boiling ($3.5°C$) it decomposes completely to a brown gaseous mixture of colorless NO and brown NO_2. Its structure is shown below and in the margin.

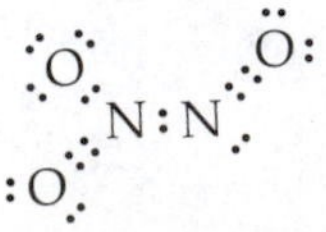

N atoms are sp^2 hybridized

Nitrogen oxide or nitric oxide, NO
mp $-163.6°C$
bp $-151.8°C$
one unpaired electron

Bond distance
(0.115 nm)
intermediate
between N≡O
(0.106 nm) and
N=O (0.12 nm)

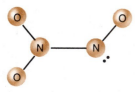

Dinitrogen trioxide, N_2O_3
mp $-103°C$
bp $3.5°C$

22-9   Nitrogen Dioxide and Dinitrogen Tetroxide (+4 Oxidation State)

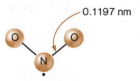

Nitrogen dioxide, NO_2
mp $-11.20°C$
bp $21.2°C$
one unpaired electron
brown gas

Nitrogen oxide is converted into brown nitrogen dioxide by reaction with oxygen. It is usually prepared in the laboratory by heating heavy metal nitrates such as lead(II) nitrate,

$$2Pb(NO_3)_2 \xrightarrow{\Delta} 2PbO + 4NO_2 + O_2$$

or by reduction of concentrated nitric acid by a metal.

Nitrogen dioxide molecules contain one unpaired electron each, and they easily dimerize to form colorless, diamagnetic dinitrogen tetroxide, N_2O_4.

$$2NO_2 \text{ (g)} \rightleftharpoons N_2O_4 \text{ (g)} \qquad \Delta H^0 = -57.2 \text{ kJ}$$

brown   colorless
paramagnetic   diamagnetic

The formation of N_2O_4 is favored at low temperatures, and NO_2 formation is favored at high temperatures, since the forward reaction is exothermic.

The NO_2 molecule is angular and is represented by resonance structures. The structures of NO_2 and N_2O_4 are represented below.

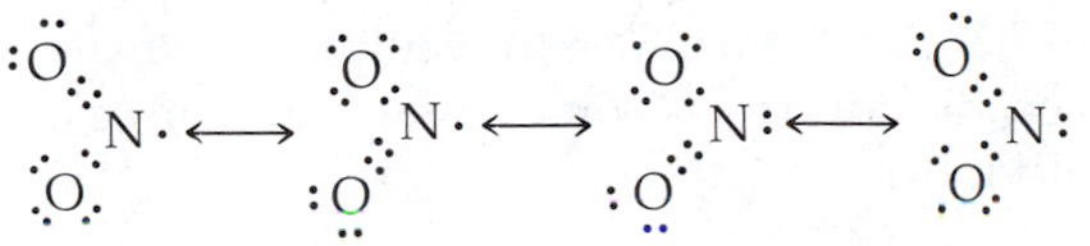

nitrogen dioxide

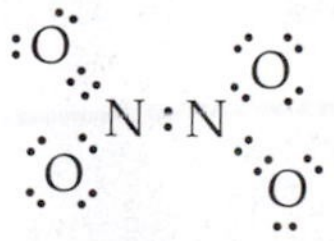

other resonance structures

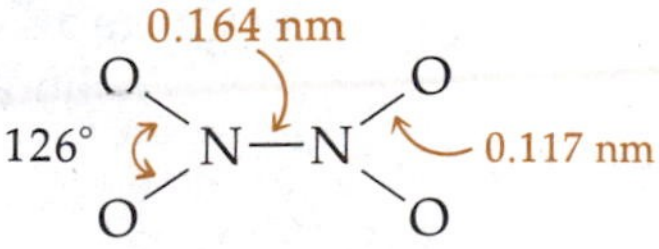

dinitrogen tetroxide

22-10   Dinitrogen Pentoxide (+5 Oxidation State)

Dinitrogen pentoxide,
N_2O_5 (gaseous)
mp $30°C$
bp $47°C$ (dec)

This is called a dehydration because it is equivalent to two occurrences of

$2HNO_3 \longrightarrow N_2O_5 + H_2O$

$H_2O + \frac{1}{2}P_4O_{10} \longrightarrow 2HPO_3$

This is not, however, the actual mechanism of the overall reaction.

Dinitrogen pentoxide, N_2O_5, is the acid anhydride of nitric acid.

$$N_2O_5 + H_2O \longrightarrow 2HNO_3$$

dinitrogen pentoxide   nitric acid

The preparation of N_2O_5 is accomplished by dehydration of nitric acid with tetraphosphorus decoxide, P_4O_{10}, a very powerful dehydrating agent.

$$4HNO_3 + P_4O_{10} \longrightarrow 4HPO_3 + 2N_2O_5$$

The white solid, N_2O_5, is removed by vacuum sublimation at $32.4°C$. Below $0°C$ it is stable, but it decomposes when heated to room temperature at atmospheric pressure. Rapid heating causes it to explode.

$$2N_2O_5 \xrightarrow{\Delta} 4NO_2 + O_2 + \text{heat}$$

22–11   Nitrogen Oxides and Photochemical Smog

Nitrogen oxides are produced in the atmosphere by natural processes and are important in the nitrogen cycle (Section 22–2). Human activities contribute only about 10% of all the oxides of nitrogen (collectively referred to as NO_x) in the atmosphere, but the human contribution occurs mostly in urban areas where the oxides may be present in concentrations a hundred times greater than in rural areas.

Just as nitrogen oxide is produced naturally by reaction of nitrogen and oxygen in electrical storms, so it is also produced by the same reaction when air is mixed with fossil fuels at high temperatures in internal combustion engines and furnaces.

$$N_2\,(g) + O_2\,(g) \rightleftharpoons 2NO\,(g) \qquad \Delta H^0 = 180\ kJ$$

The reaction does not occur to a significant extent at ordinary temperatures but, since it is endothermic, it is favored by high temperatures. At the high operating temperatures of internal combustion engines and furnaces, still only small amounts of NO are produced and released into the atmosphere. However, very small concentrations of nitrogen oxides (in the range of parts per million or parts per billion) are sufficient to cause serious problems.

The nitric oxide radical, NO, reacts with oxygen to produce nitrogen dioxide. Both NO and NO_2 are quite reactive because each possesses an unpaired electron, and they do considerable damage to plants and animals.

Nitrogen dioxide reacts with water vapor in the air to produce corrosive droplets of nitric acid and more NO.

$$3NO_2 + H_2O \longrightarrow 2HNO_3 + NO$$

The nitric acid may be washed out of the air by rainwater, or it may react with traces of ammonia in the air to form solid ammonium nitrate, a type of particulate pollutant.

$$HNO_3 + NH_3 \longrightarrow NH_4NO_3$$

This situation occurs in all urban areas, but the problem is accentuated in areas with warm, dry climates conducive to light-induced reactions, also called photochemical reactions. Here ultraviolet radiation from the sun produces damaging oxidants. The brownish hazes that often hang over such cities as Los Angeles, Denver, and Mexico City are due to the presence of brown NO_2. Problems begin in the early morning rush hour as relatively high concentrations of NO are exhausted into the air from automobiles and trucks. The NO combines with O_2 to form NO_2. Then, as the sun rises higher into the sky, NO_2 absorbs ultraviolet radiation and breaks down into NO and oxygen radicals.

$$NO_2 \xrightarrow{\ uv\ } NO + O$$

The oxygen radicals are extremely reactive, and they combine with oxygen molecules to produce ozone.

$$O + O_2 \longrightarrow O_3$$

Ozone is a powerful oxidizing agent that damages rubber, plastic materials, and all plant and animal life. Obviously, the ozone in the upper atmosphere does not pose such a problem, but that generated at the surface of the earth certainly does. It also reacts with hydrocarbons from automobile exhaust and evaporated gasoline to form secondary organic pollutants such as aldehydes and ketones (Section 25–20). The peroxyacyl nitrates (PANs) are especially damaging photochemical oxidants that are very irritating to the eyes and throat.

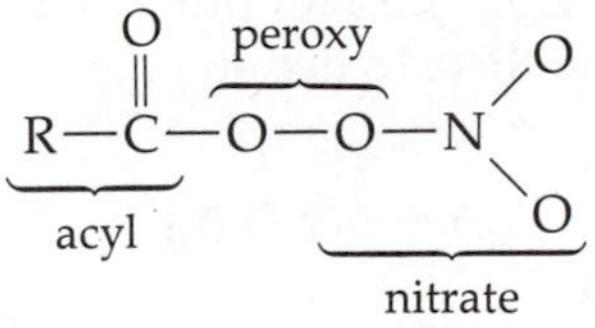

R = hydrocarbon chain or ring

The catalytic converters now being installed in automobile exhaust systems (Section 15–9) are designed to reduce emissions of oxides of nitrogen; unfortunately, evidence indicates that they also increase emission of sulfur oxides and sulfuric acid droplets, also very corrosive pollutants.

Some Oxyacids of Nitrogen and Their Salts

Nitrogen also forms hyponitrous acid, $H_2N_2O_2$, in which N is in the +1 oxidation state, as well as hyponitrite salts such as $Na_2N_2O_2$.

The important oxyacids of nitrogen are nitrous acid, HNO_2, and nitric acid, HNO_3.

22–12   Nitrous Acid and Nitrites (+3 Oxidation State)

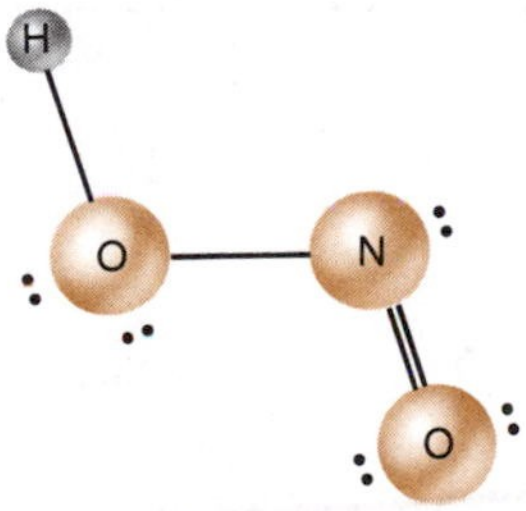

Although nitrous acid, HNO_2, is unstable and cannot be isolated in pure form, N_2O_3 can be considered its anhydride. The reaction of N_2O_3 with aqueous alkali produces nitrite salts.

$$N_2O_3 + 2[Na^+ + OH^-] \longrightarrow 2[Na^+ + NO_2^-] + H_2O$$
sodium nitrite

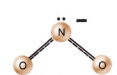

Nitrous acid, HNO_2

The acid is prepared as a pale blue solution when sulfuric acid reacts with cold aqueous sodium nitrite. Nitrous acid is a weak acid ($K_a = 4.5 \times 10^{-4}$). It acts as an oxidizing agent toward strong reducing agents and as a reducing agent toward very strong oxidizing agents.

Lewis diagrams for nitrous acid and the nitrite ion are shown below, and their structures are shown in the margin.

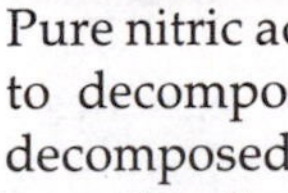

nitrous acid 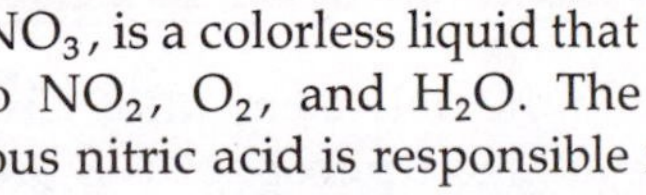   nitrite ion

22–13   Nitric Acid and Nitrates (+5 Oxidation State)

Pure nitric acid, HNO_3, is a colorless liquid that boils at 83 °C. Light or heat causes it to decompose into NO_2, O_2, and H_2O. The presence of the NO_2 in partially decomposed aqueous nitric acid is responsible for its yellow or brown tinge.

Studies on the vapor phase indicate that nitric acid has the following structure.

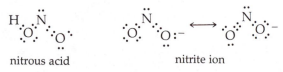

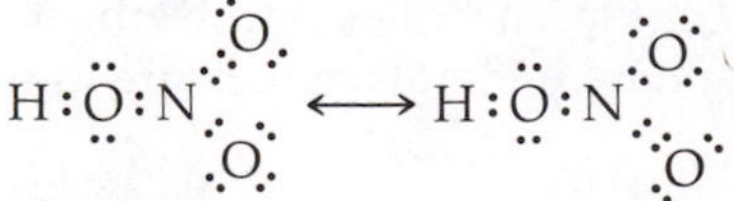

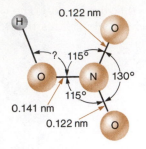

Nitric acid, HNO₃
mp − 42°C
bp 83°C

HNO₃ is commercially prepared by the **Ostwald process.** At high temperatures, ammonia is catalytically converted to nitrogen oxide, which is cooled and then air-oxidized to nitrogen dioxide. Nitrogen dioxide reacts with water to produce nitric acid and some nitrogen oxide. The NO produced in the third step is then recycled into the second step. More than 18 billion pounds of HNO₃ were produced in the U.S.A. in 1981.

$$4NH_3\ (g) + 5O_2\ (g) \xrightarrow[1000°C]{Pt} 4NO\ (g) + 6H_2O\ (g)$$

$$2NO\ (g) + O_2\ (g) \xrightarrow{cool} 2NO_2\ (g)$$

$$3NO_2\ (g) + H_2O\ (\ell) \longrightarrow 2[H^+ + NO_3^-] + NO\ (g)$$

Nitric acid is very soluble in water (∼ 16 mol/L). It is a strong acid and a strong oxidizing agent.

Nitric acid has many important industrial uses (see Table 9–5), including the production of explosives. It is interesting to note that although the Germans were cut off from their normal supply of nitrates by the Allied naval blockade during World War I, they were able to continue the production of munitions. Ammonia was obtained from putrified garbage and then converted into nitric acid, which was used to produce explosives. Thus, the war was extended considerably. (We are indebted to Professor Lloyd Jones for pointing out these facts.)

Sodium nitrite and sodium nitrate are used as food additives to retard oxidation and prevent growth of botulism bacteria. However, controversy has arisen over the possibility that nitrites combine with amines in the stomach to form carcinogenic nitrosoamines.

The nitrate ion, NO₃⁻, is a planar, resonance-stabilized ion (Section 7–4).

$$:\!\ddot{O}\!:^- \qquad \ddot{O}^- \qquad :\!\ddot{O}\!:^-$$

Neutralization of solutions of nitric acid with metal hydroxides, carbonates, or oxides produces water-soluble nitrate salts.

$$[Ca^{2+} + 2OH^-] + 2[H^+ + NO_3^-] \longrightarrow [Ca^{2+} + 2NO_3^-] + 2H_2O$$
calcium hydroxide   calcium nitrate

$$CaCO_3 + 2[H^+ + NO_3^-] \longrightarrow [Ca^{2+} + 2NO_3^-] + CO_2 + H_2O$$
calcium carbonate

Phosphorus

22–14   Occurrence, Production, and Uses

Phosphorus is the only element of Group VA that does not occur uncombined in nature. Like nitrogen, phosphorus is present in all living organisms, as organophosphates and also in calcium phosphates such as hydroxyapatite, $Ca_{10}(PO_4)_6(OH)_2$, and fluoroapatite, $Ca_{10}(PO_4)_6(F)_2$, in bones and teeth. It also occurs in these and related compounds in phosphate minerals, which are mined mostly in Florida and North Africa.

Industrially, the element is obtained from phosphate minerals by heating them to 1200–1500°C in an electric arc furnace with sand (mainly SiO_2) and coke (impure carbon). The reaction is

$$2Ca_3(PO_4)_2 \; + 6SiO_2 + 10C \xrightarrow{\Delta} \; 6CaSiO_3 \; +10CO + P_4$$

calcium phosphate   calcium silicate
(phosphate rock)   (slag)

The vaporized phosphorus is condensed to a white solid (mp 44.2°C; bp 280.3°C) under water to prevent oxidation. Even when kept under water, white phosphorus slowly converts to the more stable red phosphorus allotrope (mp 597°C; sublimes at 431°C) (Section 22–15). Red phosphorus and tetraphosphorus trisulfide, P_4S_3, are both used in ordinary matches. They do not ignite spontaneously, yet ignite easily when heated by friction. Both white and red phosphorus are insoluble in water.

The largest single user of phosphorus compounds is the fertilizer industry. Since phosphorus is an essential nutrient and since nature's phosphorus cycle is very slow, owing to the low solubility of most natural phosphates, phosphate fertilizers are absolutely essential. To increase the solubility of the natural phosphates, they are treated with sulfuric acid to produce "superphosphate of lime," a solid consisting of two salts, which is then pulverized and applied as a powder.

> This reaction represents the biggest single use of sulfuric acid, the industrial chemical produced in largest quantity.

$$Ca_3(PO_4)_2 \; + 2H_2SO_4 + 4H_2O \xrightarrow{\text{evaporate}} [Ca(H_2PO_4)_2 \; + 2(CaSO_4 \cdot 2H_2O)]$$

phosphate rock   calcium   calcium
                                        dihydrogen   sulfate
                                           phosphate   dihydrate

superphosphate of lime

22–15   Allotropes of Phosphorus

Although phosphorus exists in several allotropic forms, there are only two important forms. One has a lattice consisting of tetrahedral P_4 molecules (white phosphorus). This allotrope is more volatile, less dense, more soluble in organic solvents, more chemically active, and more toxic than the other allotrope. Red phosphorus crystallizes in a polymeric lattice consisting of bonded tetrahedra in which one bond in each P_4 tetrahedron has been broken and replaced by a bond *between* tetrahedra.

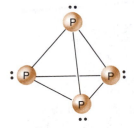

The P_4 molecule

> There are no atoms at the centers of the tetrahedra.

22–16   Oxides of Phosphorus

The most important oxides of phosphorus are P_4O_6 and P_4O_{10}. Tetraphosphorus hexoxide can be prepared by heating phosphorus under low pressures of oxygen, while high oxygen pressures produce P_4O_{10}. Both are acid anhydrides; P_4O_{10} is a very strong dehydrating agent that will remove H_2O from such compounds as HNO_3. The structure of P_4O_6 is visualized by inserting an oxygen atom between each pair of phosphorus atoms in a P_4 tetrahedron. If an additional oxygen atom is doubly-bonded to each phosphorus, the P_4O_{10} structure results.

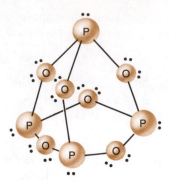

Tetraphosphorus hexoxide or
phosphorus (III) oxide, P_4O_6
mp 23.8°C
bp 175.4°C

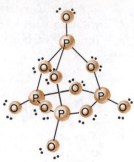

Tetraphosphorus decoxide
or
phosphorus (V) oxide,
P_4O_{10}
sublimes at 358°C

22–17 Phosphorus Oxyacids and Salts

The oxyacids of phosphorus may contain acidic hydrogens bonded to oxygen,

$$\overset{\diagdown}{\underset{\diagup}{—P}}—O—H + H_2O \rightleftharpoons \overset{\diagdown}{\underset{\diagup}{—P}}—O^- + H_3O^+$$

or nonacidic hydrogen atoms bonded directly to phosphorus, —P—H. Many such oxyacids are known. The names and formulas of the most important ones are given in Table 22–3.

1 Phosphorous Acid (+3 Oxidation State)

Tetraphosphorus hexoxide is the anhydride of (ortho)phosphorous acid, H_3PO_3.

$$P_4O_6 + 6H_2O \xrightarrow{\text{cold}} 4H_3PO_3$$

However, the acid is more easily prepared by hydrolysis of PCl_3.

$$PCl_3 + 3H_2O \longrightarrow H_3PO_3 + 3HCl$$

The acid is only diprotic ($K_1 = 1.6 \times 10^{-2}$; $K_2 = 7 \times 10^{-7}$) because one hydrogen is directly bonded to phosphorus and will not ionize, as do the two that are bonded to oxygen. Its salts are strong reducing agents in basic solution. Phosphite salts containing either $H_2PO_3^-$ or HPO_3^{2-} ions can be prepared by neutralization of the

TABLE 22–3 Important Phosphorus Oxyacids

Oxidation State	Formula	Name	No. of Acidic H	Compounds Prepared
+1	H_3PO_2	hypophosphorous acid	1	acid, salts
+3	H_3PO_3	(ortho)phosphorous acid	2	acid, salts
+5	$(HPO_3)_n$	metaphosphoric acids	n	salts ($n = 3, 4$)
+5	$H_5P_3O_{10}$	triphosphoric acid	3	salts
+5	$H_4P_2O_7$	pyrophosphoric acid	4	acid, salts
+5	H_3PO_4	(ortho)phosphoric acid	3	acid, salts

acid with the appropriate base. The Lewis diagram and structure of the acid are represented below.

Phosphorous acid, H_3PO_3
mp 73.6°C
bp 200°C (dec)

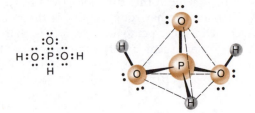

2 Phosphoric Acid (+5 Oxidation State)

Orthophosphoric acid, or simply phosphoric acid, H_3PO_4, is the most common of the phosphoric acids. Its anhydride is P_4O_{10}.

$$P_4O_{10} + 6H_2O \longrightarrow 4H_3PO_4$$

Phosphoric acid, H_3PO_4
mp 42.35°C

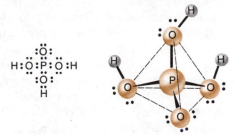

Pure phosphoric acid is a stable, colorless solid. It is a weak triprotic acid.

$$H_3PO_4 \rightleftharpoons H^+ + H_2PO_4^- \qquad K_1 = 7.5 \times 10^{-3}$$

$$H_2PO_4^- \rightleftharpoons H^+ + HPO_4^{2-} \qquad K_2 = 6.2 \times 10^{-8}$$

$$HPO_4^{2-} \rightleftharpoons H^+ + PO_4^{3-} \qquad K_3 = 3.6 \times 10^{-13}$$

Salts containing each of the three anions can be prepared by stepwise neutralization of the acid. Salts of the dihydrogen phosphate and hydrogen phosphate ions decompose upon heating.

$$n\text{NaH}_2\text{PO}_4 \xrightarrow{\Delta} \qquad (\text{NaPO}_3)_n \qquad + n\text{H}_2\text{O}$$
sodium "n" metaphosphate

$$2\text{Na}_2\text{HPO}_4 \xrightarrow{\Delta} \qquad \text{Na}_4\text{P}_2\text{O}_7 \qquad + \text{H}_2\text{O}$$
sodium pyrophosphate
or sodium diphosphate

Arsenic and Its Compounds

The most metallic element of Group VA, bismuth, was discussed in Chapter 20. Its properties are summarized in Table 22-1.

Arsenic is somewhat more metallic than phosphorus, though many of its properties, as well as those of its compounds, are similar to those of phosphorus and its compounds. It is considered a metalloid. Small amounts of arsenic occur in the free form, As_4, but most arsenic is found in the +3 oxidation state in diarsenic trisulfide, As_2S_3, and arsenopyrite, FeAsS. Roasting As_2S_3 produces As_4O_6. All arsenic compounds are poisonous, some in only trace amounts, although it has been shown that minute traces of arsenic actually stimulate the production of red blood cells.

Like phosphorus, arsenic forms trihalides ($X = F$, Cl, Br, I) as well as the pentahalide AsF_5. Tetraarsenic hexoxide, As_4O_6, is the anhydride of diprotic arsenous acid, H_3AsO_3. Diarsenic pentoxide, As_2O_5, is the anhydride of triprotic arsenic acid, H_3AsO_4. Arsenic acid is only slightly weaker than phosphoric acid, while arsenous acid ($K_1 = 6.0 \times 10^{-10}$, $K_2 = 3.0 \times 10^{-14}$) is much weaker than phosphorous acid ($K_1 = 1.6 \times 10^{-2}$, $K_2 = 7.0 \times 10^{-7}$), reflecting the more metallic character of arsenic.

> The oxidation states for As are $+3$ in As_4O_6 and H_3AsO_3 and $+5$ in As_2O_5 and H_3AsO_4.

The Nonmetals of Groups IVA and IIIA

In the sections that follow we shall examine the remaining nonmetals, carbon and silicon at the top of Group IVA, and boron at the top of Group IIIA. Many of the general properties of the elements in Groups IIIA and IVA were given in Tables 20–6 and 20–7. Carbon is a nonmetal, but silicon and boron are better classified as metalloids. Consistent with their relatively low electronegativities (for nonmetals), neither silicon nor boron forms simple negative ions. Carbon forms the carbide anion, C_2^{2-}, in saltlike compounds with highly electropositive metals, although there is undoubtedly a high degree of covalent character in these carbides. All borides and silicides are classified as covalent compounds.

Many of the chemical characteristics of silicon (Group IVA) are closer to those of boron (IIIA) than to those of carbon (IVA). This tendency for the elements of the second and third periods to exhibit "diagonal relationships" in chemical properties was introduced in Section 20–2.2. Elements within the same families tend to form compounds with similar stoichiometries, e.g., CO_2 and SiO_2; B_2O_3 and Al_2O_3. However, the properties of the compounds of the elements of the second period are often closer to those of the corresponding compounds of the third period element one group to the right. Carbon dioxide is a gas while SiO_2 and B_2O_3 are solids, and both solids are used in making glass. As another example, silicon and boron halides hydrolyze vigorously, whereas halides of carbon are very inert toward water.

Carbon

By virtue of its position in the middle of the representative elements, carbon ($1s^2 2s^2 2p^2$) has little tendency to form simple ions. The energy required to remove four outer-shell electrons (sum of the first four ionization energies) is prohibitively large for the existence of C^{4+}, and only the most metallic elements form ionic carbides containing C^{4-} or C_2^{2-} ions. In nearly all of its compounds carbon is bonded covalently.

As a consequence of the high bond energy of C—C single bonds (347 kJ/mol) and of the C—H bond (414 kJ/mol), carbon has a strong tendency to form hydrocarbon chains and rings. The self-linkage of an element to form chains or rings is called **catenation**. Doubly-bonded and triply-bonded carbon atoms,

$$\diagdown\!\!\!\!\!\!\underset{\diagup}{\overset{}{C}}\!\!=\!\!\underset{\diagdown}{\overset{}{C}}\!\!\diagup \quad , \text{ and } -C\equiv C-, \text{ are also common.}$$

Hydrocarbons fall into the domain of organic chemistry, the chemistry of most carbon-containing compounds. Carbon is the only element whose compounds comprise an entire branch of chemistry. The term *organic chemistry* arises from the fact that most compounds essential for plant and animal life contain carbon. The distinction, however, between organic carbon compounds and inorganic carbon compounds is rather arbitrary. In the present chapter we shall consider some of the compounds usually classified as inorganic compounds, most of which contain only one carbon atom per formula unit and no C—H bonds. Chapter 25 is devoted to an introduction to organic chemistry.

For many years most scientists believed that organic compounds could be obtained only from living organisms. This idea was finally proved false in 1828 when a German chemist, Friedrich Wöhler, synthesized urea (an animal metabolic product) by heating a completely inorganic compound, ammonium cyanate:

$$NH_4OCN \xrightarrow{\Delta} CO(NH_2)_2$$

ammonium cyanate $\qquad$ urea
"inorganic" $\qquad$ "organic"

22–18 Occurrence of Carbon

Carbon makes up only about 0.08% of the combined lithosphere, hydrosphere, and atmosphere. It occurs on the crust of the earth mainly in the form of calcium carbonate or magnesium carbonate rocks such as calcite, limestone, dolomite, marble, and chalk. There are also some natural deposits of elemental carbon in the form of diamond and graphite. Carbon dioxide occurs in the atmosphere, which acts as a CO_2 reservoir for photosynthesis in plants.

Carbon dioxide is incorporated by plants and bacteria into the structure of carbohydrates, which are ingested by animals and converted into other biological substances. After plants and animals die, these substances decompose to hydrocarbons and many other organic substances, which may eventually become coal, peat, natural gas, petroleum, and other **fossil fuels.** Both respiration and the complete combustion of fossil fuels produce CO_2 and water to complete the carbon cycle.

Carbon exists in two crystalline forms, graphite and diamond. The so-called amorphous forms, such as charcoal and carbon black, are really arrangements of graphite crystallites. Carbon black is obtained by burning natural gas in a deficiency of air. Charcoal is the residue from the destructive distillation of wood, in which the wood is heated in the absence of air to drive off volatile substances, and carbonaceous residues are formed at too low a temperature to permit thorough crystallization of the carbon.

Graphite. At ordinary temperatures and pressures, graphite is the most stable allotrope of carbon. It is soft, black, and slippery, and its density (2.25 g/cm³) is less than that of diamond (3.51 g/cm³). Its properties are closely related to its structure, which consists of sp^2 hybridized carbon atoms arranged in planar layers of six-membered rings (Figure 22–1). Each carbon atom is at the center of a trigonal plane. Layers easily slide horizontally over each other because the van der Waals forces that hold one layer to another are easily overcome. For this reason graphite finds use as a lubricant, as an additive for motor oil, and in pencil "lead." Certain compounds such as CO_2, H_2O, and NH_3 are able to diffuse between the layers and

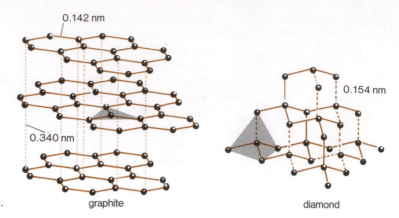

FIGURE 22-1   Allotropes of carbon.

function as cushions, improving the ability of graphite to act as a lubricant. Only three of carbon's four electrons are involved in in-plane bonding. The fourth electron of each is quite mobile, and this accounts for graphite's ability to conduct electricity.

Diamond.   In contrast to graphite, diamond is one of the hardest substances known to man. It is colorless, a nonconductor of electricity, and more dense than graphite. These properties are consistent with the structure of diamond, which consists of a network of tetrahedrally distributed sp^3 hybridized carbon atoms, each separated from its neighbors by only 0.154 nm (as compared with 0.142 nm in-plane and 0.340 nm out-of-plane separations in graphite). This structure (Figure 22-1) involves very strong bonds with no mobile electrons; this accounts for diamond's hardness, the nonconductivity, and the very high melting point of about 3570°C, which is the highest of any element.

22-19   Reactions of Carbon

1   Reactions with Oxygen

$$2C \, (s) + O_2 \, (g) \longrightarrow 2CO \, (g)$$

$$C \, (s) + O_2 \, (g) \longrightarrow CO_2 \, (g)$$

Although carbon is unreactive at room temperature, it reacts with a number of substances when heated. It reacts with oxygen from the air to produce either carbon monoxide or carbon dioxide (Section 9-3).

2   Reactions with Metal Oxides

Carbon reduces the oxides of the less reactive metals to produce the metal and carbon monoxide.

$$Fe_2O_3 + 3C \xrightarrow{\Delta} 2Fe + 3CO$$

This reaction is important in the blast furnace reduction of iron ore with coke (Section 20-11). In such a reaction with oxides of the more electropositive metals, carbides of the metals often are produced. All are very hard, high melting solids. Examples are beryllium carbide, BeC_2, and aluminum carbide, Al_4C_3. Carbide ions may be either C^{4-} or, more often, C_2^{2-} ($:C:::C:^{2-}$), as in Na_2C_2 and CaC_2. Members of the latter class are also called acetylides; e.g., Na_2C_2 is sodium acetylide. These release acetylene, $HC \equiv CH$, when treated with water or dilute acid.

3   Reactions with Nonmetal Oxides

Carbon reacts with some oxides of nonmetals to produce very hard covalent carbides. Silicon carbide, known by the trade name of carborundum, is one of the hardest substances known. It is produced by reduction of silica (SiO_2) in sand with coke in an electric furnace.

$$SiO_2 + 3C \xrightarrow{3500°C} SiC + 2CO$$

Silicon carbide is industrially important as an abrasive material in sandpaper and grinding wheels. It has the same structure as diamond, except that it contains alternating silicon and carbon atoms in the diamond lattice.

4   Reaction with Steam

Carbon reacts with steam at high temperature in an endothermic reaction to produce "water gas," a mixture of CO and H_2.

$$C + H_2O \xrightarrow{red\ heat} CO + H_2$$

Both products burn in air and constitute an important industrial fuel.

22-20   Oxides of Carbon

1   Carbon Monoxide

An oxide that has the formula C_3O_2 is known, but it is of little importance.

Carbon monoxide is a colorless, odorless, toxic gas that is formed by burning carbon or a hydrocarbon in a deficiency of oxygen. Since its carbon-oxygen bond length is intermediate between typical double and triple bond lengths, two resonance forms are thought to contribute to its structure:

$$:C:::O: \longleftrightarrow :C::O:$$

Carbon monoxide is produced commercially by the "water gas" reaction, above, or by incomplete combustion of hydrocarbons, or by reduction of CO_2 by hot carbon. Pure carbon monoxide can be obtained by dehydrating formic acid with concentrated sulfuric acid.

$$H-\overset{\overset{\displaystyle O}{\|}}{C}-O-H + H_2SO_4 \longrightarrow CO + H_2SO_4 \cdot H_2O$$

formic acid

Carbon monoxide dissolves in aqueous alkali to yield salts of formic acid, called formates.

$$CO + [Na^+ + OH^-] \xrightarrow[6\ atm]{200°C} [H-\overset{\overset{\displaystyle O}{\|}}{C}-O^- + Na^+] \qquad \text{sodium formate}$$

At high temperatures, CO reduces many metal oxides to the free metals, as described under metallurgy in Chapter 20.

$$FeO\ (s) + CO\ (g) \xrightarrow{\Delta} Fe\ (s) + CO_2\ (g)$$

It also disproportionates to more stable carbon and carbon dioxide in a reversible reaction.

$$2CO\ (g) \rightleftharpoons C\ (s) + CO_2\ (g) \qquad \Delta H^0 = -172\ kJ$$

Carbon monoxide is important industrially not only as a fuel but also as the starting material for the syntheses of many important organic substances. A few examples are

$$CO + Cl_2 \xrightarrow[\text{catalyst}]{uv} Cl{-}\overset{\overset{\displaystyle O}{\|}}{C}{-}Cl \qquad \text{phosgene or carbonyl chloride}$$

$$CO + 2H_2 \xrightarrow[\Delta,\ \text{pressure}]{\text{catalyst}} H{-}\overset{\overset{\displaystyle H}{|}}{\underset{\underset{\displaystyle H}{|}}{C}}{-}O{-}H \qquad \text{methyl alcohol or methanol}$$

$$CO + 3H_2 \xrightarrow[250°C]{\text{catalyst}} H_2O + CH_4 \qquad \text{methane}$$

Carbon monoxide is a poison because of its ability to bind strongly to the iron in the oxygen-carrier protein, hemoglobin, in red blood corpuscles, thus inhibiting the ability of the blood to carry oxygen. The hemoglobin-CO complex is bright red, so often it is not realized that the victim is suffocating. Since CO is odorless and tasteless, the victim is often unaware of danger until it is too late.

2   Carbon Dioxide (and the Greenhouse Effect)

Carbon dioxide is a colorless, odorless gas. It is the product of complete combustion of elemental carbon and of hydrocarbons. The combustion of gasoline is illustrated by complete oxidation of octane, C_8H_{18}, one of its components.

$$2C_8H_{18}\ (\ell) + 25O_2\ (g) \longrightarrow 16CO_2\ (g) + 18H_2O\ (g) + \text{heat}$$

Carbon dioxide is produced commercially as a by-product of the fermentation of sugar or starch for production of ethyl alcohol and alcoholic beverages.

$$C_6H_{12}O_6\ (s) \xrightarrow{\text{yeast}} 2C_2H_5OH\ (\ell) + 2CO_2\ (g)$$

glucose   ethyl alcohol
(a simple sugar)   or ethanol

Photosynthesis in plants involves the conversion of carbon dioxide and water into carbohydrates, such as glucose and other sugars:

$$6CO_2 + 6H_2O \xrightarrow{uv} C_6H_{12}O_6 + 6O_2$$

The CO_2 molecule is linear, with an sp hybridized carbon atom and two carbon-oxygen double bonds.

$$\ddot{O}::C::\ddot{O}$$

Carbon dioxide is a minor component of the atmosphere, in which it is present in a concentration of about 325 parts per million. Natural processes account for most of the atmospheric CO_2, which is in approximate equilibrium with the CO_2 dissolved in natural waters, that contained in the earth's crust (such as limestone, $CaCO_3$), that participating in photosynthesis, and that produced by combustion of fossil fuels.

The great increase in our use of fossil fuels over the past century has caused a significant increase in the concentration of CO_2 in the atmosphere, and this use is accelerating. It is now projected that with the anticipated development and use of synthetic fuels ("synfuels," derived from natural sources such as plants and grain) the concentration of atmospheric CO_2 could double by early in the next century.

This should result in a rise in the average global temperature by $2-3°C$ owing to the **greenhouse effect** caused by the carbon dioxide and water vapor in the atmosphere. Most of the sunlight passing through the atmosphere is in the visible region of the spectrum. Neither carbon dioxide nor water vapor absorbs visible light, so they do not prevent it from reaching the surface of the earth. However, the earth re-radiates some of the energy from sunlight in the form of lower energy infrared (heat) radiation, which is readily absorbed by both CO_2 and H_2O (and by the glass or plastic of greenhouses). Thus, some of the heat the earth must lose in order to be in thermal equilibrium is trapped in the atmosphere, and the temperature rises. While a $2-3°C$ increase in the temperature of the atmosphere may not seem like much, it is thought to be sufficient to cause such things as the partial melting of the polar icecaps, a rise in the level of the oceans that would submerge some coastal cities, and a dramatic change in climates.

Ironically, though, there is also evidence to suggest that air pollution in the form of particulate matter may partially counteract this effect, because the particles reflect visible and infrared radiation rather than absorbing it.

22-21   Carbonic Acid and the Carbonates

Carbon dioxide is moderately soluble in water, and a saturated aqueous solution is about 0.033 M. Its solubility is increased, in accord with LeChatelier's Principle, by increasing the partial pressure of CO_2. Carbonated beverages contain dissolved CO_2 under pressure. About 0.4% of the dissolved CO_2 reacts with water to form carbonic acid, H_2CO_3, which has not been isolated in pure form.

$$CO_2 \text{ (g)} + H_2O \text{ (ℓ)} \rightleftharpoons H_2CO_3 \text{ (aq)}$$

The reaction is easily reversed by heating. The structure of carbonic acid can be represented as follows.

It is a weak diprotic acid, capable of forming both hydrogen carbonate and carbonate salts upon neutralization. Carbonate salts can also be formed when basic solids absorb gaseous carbon dioxide.

$$CO_2 \text{ (g)} + 2NaOH \text{ (s)} \longrightarrow Na_2CO_3 \cdot H_2O \text{ (s)}$$

Alkali metal carbonates are soluble in water, but most other metal carbonates are quite insoluble.

Since all natural waters contain dissolved CO_2, it should not be surprising that deposits of $CaCO_3$ and $MgCO_3$ are abundant in areas formerly covered by bodies of water. Limestone caves are formed as ground water containing dissolved CO_2, or H_2CO_3, dissolves $CaCO_3$ and forms soluble $Ca(HCO_3)_2$.

Hydrogen carbonate salts are more soluble than the corresponding carbonates. Hydrogen carbonate ion, HCO_3^-, is often called bicarbonate ion.

$$CaCO_3 \text{ (s)} + H_2CO_3 \text{ (aq)} \longrightarrow [Ca^{2+} + 2HCO_3^-]$$
$$\text{calcium bicarbonate}$$

The carbonates are redeposited as stalactites and stalagmites, composed of $CaCO_3$, as water evaporates and the above reaction is reversed.

$$[Ca^{2+} + 2HCO_3^-] \longrightarrow CaCO_3 \text{ (s)} + CO_2 \text{ (g)} + H_2O \text{ (g)}$$

All metal carbonates decompose upon heating to the oxides and CO_2. The alkali metal carbonates are quite stable and require high temperatures for decomposition.

$$Li_2CO_3 \xrightarrow{1310°C} Li_2O + CO_2$$
$$PbCO_3 \xrightarrow{315°C} PbO + CO_2$$

Solutions of alkali metal carbonates are strongly basic because of the hydrolysis of carbonate ion (Section 17–9).

$$CO_3^{2-} + H_2O \rightleftharpoons HCO_3^- + OH^- \qquad K_b = 2.1 \times 10^{-4}$$

Solutions of hydrogen carbonate salts are much less basic because HCO_3^- is a weaker base than CO_3^{2-}.

$$HCO_3^- + H_2O \rightleftharpoons H_2CO_3 + OH^- \qquad K_b = 2.4 \times 10^{-8}$$

The carbonate ion is symmetrical and planar, and is represented by three resonance forms. The carbon is sp^2 hybridized and all bond angles are 120°.

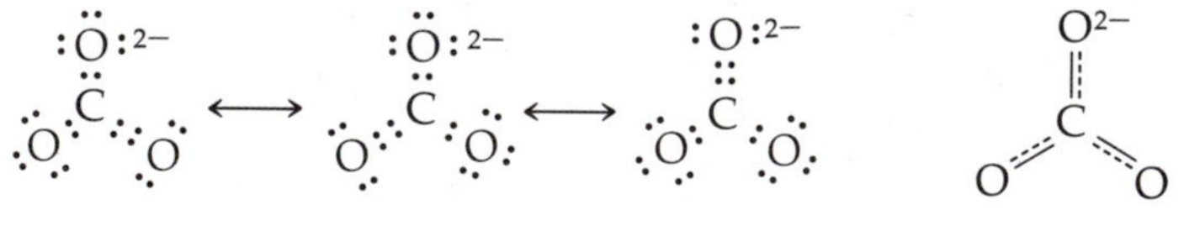

Silicon

Silicon is a shiny, blue-gray, high melting, brittle metalloid. It looks like a metal, but it is chemically more nonmetallic than metallic. It is second only to oxygen in abundance in the earth's crust, about 87% of which is composed of silica (SiO_2) and its derivatives, the silicate minerals (Section 22–26). Silicon itself constitutes 26% of the crust, compared to oxygen's 49.5%. It does not occur free in nature. Pure silicon crystallizes in a diamond-type structure, but the silicon atoms are less closely packed than carbon. Its density, therefore, is only 2.4 g/cm³ compared with 3.51 g/cm³ for diamond.

22-22 Production and Uses of Silicon

Elemental silicon is usually prepared by the high-temperature reduction of silica (silicon dioxide) from sand with coke (impure carbon). An excess of silica prevents the formation of silicon carbide.

$$SiO_2 \text{ (excess)} + 2C \xrightarrow{\Delta} Si + 2CO$$

Reduction of a mixture of silicon and iron oxides with coke produces an alloy of iron and silicon known as ferrosilicon, which is used in the production of acid-resistant steel alloys such as "duriron" and in the "deoxidation" of steel. Aluminum alloys for aircraft are strengthened with silicon.

Elemental silicon is also used as the source of silicon in silicone polymers. Its semiconducting properties (Section 12-16) are used in transistors and solar cells, for which the silicon must be ultrapure. For such applications it is best prepared by reducing a tetrahalide in the vapor phase with an active metal such as sodium or magnesium. The tetrahalide first must be distilled carefully to remove impurities of boron, aluminum, and arsenic halides.

$$SiCl_4 + 4Na \longrightarrow 4NaCl + Si$$

The sodium chloride is then dissolved in hot water, leaving behind quite pure silicon, which is melted and cast into bars. In order to obtain silicon with less than one part per billion impurity, it must be further purified by zone refining (Section 20-10).

22-23 Chemical Properties of Silicon

Si is much larger than C and therefore forms Si-Si bonds that are too long to allow effective pi bonding found in multiple bonds.

The biggest differences between silicon and carbon are that (1) silicon does not form double bonds, (2) it does not readily form Si-Si bonds unless the silicon atoms are bonded to very electronegative elements, and (3) it has vacant $3d$ orbitals in its valence shell with which to accept electrons from donor atoms. The Si-O single bond is the strongest of all silicon bonds and accounts for the stability and prominence of silica and the silicates. Although silicon does possess vacant $3d$ orbitals, it actually does not form many species in which it is bound to more than four atoms. However, the octahedral hexafluorosilicate ion, SiF_6^{2-}, exists, with sp^3d^2 hybridization at Si.

Elemental silicon is quite unreactive. It does not react with solutions of acids except a mixture of nitric acid, an oxidizing acid, and hydrofluoric acid, which dissolves the SiO_2 as it forms to produce hexafluorosilicic acid, H_2SiF_6.

$$Si + 4[H^+ + NO_3^-] + 6HF \longrightarrow [2H^+ + SiF_6^{2-}] + 4NO_2 + 4H_2O$$
$$\text{hexafluorosilicic}$$
$$\text{acid}$$

Silicon dissolves in solutions of very strong bases to produce silicate salts.

$$Si + 2[K^+ + OH^-] + H_2O \longrightarrow [2K^+ + SiO_3^{2-}] + 2H_2$$
$$\text{potassium silicate}$$

FIGURE 22–2 In SiO_2 every silicon atom (small) is bonded tetrahedrally to four oxygen atoms (large).

22–24   Silicon Dioxide (Silica)

Silicon dioxide exists in two familiar forms in nature. One is quartz, small chips of which occur in sand. The other is flint (Latin: *silex*), an uncrystallized amorphous type of silica. Silica is properly represented as $(SiO_2)_n$ since it is a polymeric solid of SiO_4 tetrahedra sharing all oxygens among surrounding tetrahedra. Its structure is represented in Figure 22–2. For comparison, solid carbon dioxide ("Dry Ice") consists of discrete $O{=}C{=}O$ molecules, just as does gaseous CO_2.

Some gems and semiprecious stones such as amethyst, opal, agate, and jasper are crystals of quartz with colored impurities. When silica is melted and rapidly cooled, it actually undercools. That is, it cools to a temperature below its normal freezing point as it forms an amorphous glass called "fused quartz," or more properly "fused silica." Since it does not absorb visible or ultraviolet radiation, fused silica is used in optical instruments and sun lamps. Because it is resistant to most chemical attack and expands or contracts only very slightly with changing temperature, it is used for the manufacture of glassware for use in laboratories. It is, however, rapidly attacked by hydrofluoric acid to form volatile silicon tetrafluoride.

$$SiO_2 + 4HF \longrightarrow 2H_2O + SiF_4$$

22–25   Silicates and Silicic Acids

Strong bases react very slowly with silica to form various soluble metal silicates.

$$SiO_2 + 2[Na^+ + OH^-] \longrightarrow [2Na^+ + SiO_3^{2-}] + H_2O$$
sodium metasilicate

$$SiO_2 + 4[Na^+ + OH^-] \longrightarrow Na_4SiO_4 \text{ (aq)} + 2H_2O$$
sodium orthosilicate

$$2SiO_2 + 6[Na^+ + OH^-] \longrightarrow Na_6Si_2O_7 \text{ (aq)} + 3H_2O$$
sodium pyrosilicate

The parent acids of these salts are unstable and cannot be isolated in anhydrous form because they revert to silica. They are metasilicic acid, H_2SiO_3, orthosilicic acid, H_4SiO_4, and pyrosilicic acid or disilicic acid, $H_6Si_2O_7$. Others are thought to exist in solution. Dehydration of any silicic acid produces SiO_2. Silica in a spongy form with 5% water content is called **silica gel.** Because of its tremendous surface area, it adsorbs water and some vapors in relatively large quantities and is used as a drying agent and dehumidifier.

22–26 Natural Silicates

Most of the crust of the earth is made up of silica and silicates. The natural silicates comprise a large variety of compounds, the structures of which are all based on SiO_4 tetrahedra with metal ions occupying spaces between the tetrahedra. The extreme stability of the silicates derives from the donation of extra electrons from oxygen into vacant $3d$ orbitals of silicon. In many common minerals, called aluminosilicates, aluminum atoms replace some silicon atoms with very little structural change. Since an aluminum atom has one less positive charge in its nucleus than silicon does, it is also necessary to introduce another ion, such as K^+ or Na^+, or to substitute, for example, Ca^{2+} for Na^+. The major classes of silicates and their structures are summarized in Table 22–4.

The physical characteristics of the silicates are often suggested by the arrangement of the SiO_4 tetrahedra. A few examples are given in the table. Diopside, $[CaMg(SiO_3)_2]_n$ (a single-chain silicate), and asbestos, $[Ca_2Mg_5(Si_4O_{11})_2(OH)_2]_n$

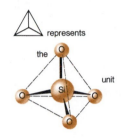

represents the unit

TABLE 22–4 Structure of Silicate Minerals

Structural Arrangement		Anion Unit	Examples
Independent tetrahedra		$(SiO_4)^{4-}$	Forsterite, Mg_2SiO_4 Olivine, $(Mg, Fe)_2SiO_4$ Fayalite, Fe_2SiO_4
Two tetrahedra sharing one oxygen		$(Si_2O_7)^{6-}$	Akermanite, $Ca_2MgSi_2O_7$
Closed rings of tetrahedra, each sharing two oxygens		$(Si_3O_9)^{6-}$, $(Si_4O_{12})^{8-}$, $(Si_6O_{18})^{12-}$	Beryl, $Al_2Be_3Si_6O_{18}$
Continuous single chains of tetrahedra, each sharing two oxygens		$(SiO_3)^{2-}$	Pyroxenes: enstatite, $MgSiO_3$, diopside, $CaMg(SiO_3)_2$
Continuous double chains of tetrahedra, sharing alternately two and three oxygens		$(Si_4O_{11})^{6-}$	Amphiboles: anthophyllite, $Mg_7(Si_4O_{11})_2(OH)_2$
Continuous sheets of tetrahedra, each sharing three oxygens		$(Si_4O_{10})^{4-}$	Talc, $Mg_3Si_4O_{10}(OH)_2$ Kaolinite, $Al_4Si_4O_{10}(OH)_8$
Continuous framework of tetrahedra, each sharing all four oxygens	See Figure 22–2	(SiO_2)	Quartz, SiO_2 Albite, $NaAlSi_3O_8$

(a double-chain silicate), occur as fibrous or needle-like crystals. Talc, $[Mg_3Si_4O_{10}(OH)_2]_n$, a silicate with a sheet-like structure, is flaky. Micas are sheet-like aluminosilicates (similar to talc) with about one of every four silicon atoms replaced by aluminum. Muscovite mica is $[KAl_2(AlSi_3O_{10})(OH)_2]_n$. Micas all occur in thin sheets that are easily peeled away from each other.

Boron

Boron is very rare, constituting only 0.0003% of the earth's crust. Much of it is localized in evaporated lake beds in the southwestern U.S. in the form of borax, $Na_2B_4O_5(OH)_4 \cdot 8H_2O$, and kernite, $Na_2B_4O_5(OH)_4 \cdot 2H_2O$, hydrated sodium salts of tetraboric acid. Borax is used mainly as a cleansing agent in laundry detergents and as a mild alkali. It is also used in soldering and welding as a flux, since it melts at low temperatures and dissolves metal-oxide films.

22–27 Boric Acids and Boric Oxide

Recrystallized mined borax reacts with sulfuric acid to produce boric acid, H_3BO_3 [or $B(OH)_3$].

$$Na_2B_4O_5(OH)_4 \cdot 8H_2O \text{ (s)} + [2H^+ + SO_4^{2-}] \longrightarrow$$
$$[2Na^+ + SO_4^{2-}] + 5H_2O + 4H_3BO_3 \text{ (aq)}$$

Boric acid itself is used as a mildly acidic antiseptic. It is a very weak acid.

$K_2 = 1.8 \times 10^{-13}$

$K_3 = 1.6 \times 10^{-14}$

$$H_3BO_3 \text{ (aq)} \rightleftharpoons H^+ + H_2BO_3^- \qquad K_1 = 7.3 \times 10^{-10}$$

The solid acid melts and loses water at high temperature, producing a colorless, transparent melt of boric oxide, B_2O_3:

$$2H_3BO_3 \xrightarrow{\Delta} B_2O_3 + 3H_2O$$

Rapid cooling produces a glasslike solid. Boric oxide is used in the fabrication of the sturdy borosilicate glasses.

Boric oxide is the anhydride of three common boric acids, two of which are obtained from boric acid (or, more properly, orthoboric acid) by dehydration.

$$4H_3BO_3 \xrightarrow[-4H_2O]{\Delta} 4HBO_2 \xrightarrow[-H_2O]{\Delta} H_2B_4O_7$$

orthoboric acid   metaboric acid   tetraboric acid
(pyroboric acid)

A number of borate salts containing anions of these acids are known.

22–28 Elemental Boron

The element can be obtained in low purity in a dark brown amorphous form by the high-temperature reduction of boric oxide with magnesium. The magnesium oxide is dissolved away by hydrochloric acid, leaving the boron behind.

$$B_2O_3 + 3Mg \xrightarrow{\Delta} 3MgO + 2B$$

Boron can be obtained in higher purity by reduction of gaseous boron tribromide or

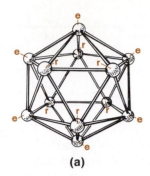

(a)

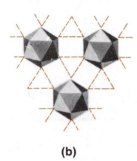

(b)

FIGURE 22–3 (a) The B_{12} icosahedron. (b) Linking of icosahedra in rhombohedral boron.

An *icosahedron* is a regular polyhedron with 20 equilateral triangular faces and 12 vertices.

trichloride with hydrogen over a hot tungsten or tantalum filament.

$$2BBr_3 + 3H_2 \xrightarrow[1000-2500°C]{W} 2B + 6HBr$$

Elemental boron exists in several allotropic modifications, of which α-rhombohedral boron is the simplest in structure. All modifications consist of large, complex polyhedral clusters of boron atoms; they are very hard and are semiconductors. The α-rhombohedral form consists of icosahedral B_{12} units. One boron atom occupies each vertex. The structure is shown in Figure 22–3.

22–29 Boron Hydrides (Boranes)

Boron forms a number of unusual binary compounds with hydrogen called **boron hydrides,** or more commonly **boranes.** They fall into two series, B_nH_{n+4} and the less stable B_nH_{n+6}. Examples are given in Table 22–5. A special system of naming is usually applied, in which the prefix indicates the number of boron atoms and the parenthetical number gives the number of hydrogen atoms. The simplest borane is diborane, B_2H_6. No BH_3 exists except as a short-lived reaction intermediate.

The higher boranes are generally prepared by controlled heating of diborane or other boranes:

$$2B_2H_6 \xrightarrow{120°C} B_4H_{10} + H_2$$

$$5B_2H_6 \xrightarrow{200-240°C} 2B_5H_9 + 6H_2$$

$$2B_9H_{15} \xrightarrow{>-30°C} B_8H_{12} + B_{10}H_{14} + 2H_2$$

Acid decomposition of magnesium boride, Mg_3B_2, also produces a number of boranes, but mainly B_2H_6.

Many boranes have structures based on fragments of icosahedra. Each borane contains unusual three-center two-electron bonds, in which three atoms are bonded together by only two electrons. Some are $\overset{B}{\underset{B \quad B}{\diagup\diagdown}}$ bonds and others are $\overset{H}{\underset{B \quad B}{\diagup\diagdown}}$ bonds involving "hydrogen bridges," as will be discussed for diborane.

The bonding in B_2H_6 is not well explained by simple Lewis formulas. Each of the six hydrogen atoms contributes one electron, and each of the two boron atoms contributes three electrons to give a total of twelve bonding electrons. If two electrons are allotted to each of the four terminal B—H (two-center) bonds, only four electrons remain to form the other bonds. Note also that the hypothetical BH_3 would not have a complete octet of bonding electrons. The structure of diborane is

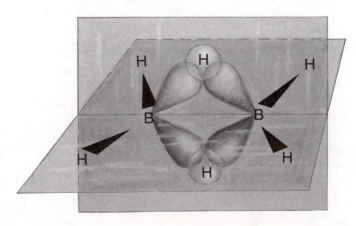

TABLE 22–5 Typical Boron Hydrides

B_nH_{n+4}	Name	Phase	B_nH_{n+6}	Name	Phase
B_2H_6	diborane	gas	B_4H_{10}	tetraborane(10)	gas
B_4H_8	tetraborane(8)	gas	B_5H_{11}	pentaborane(11)	liquid
B_5H_9	pentaborane(9)	liquid	B_6H_{12}	hexaborane(12)	liquid
B_8H_{12}	octaborane(12)	solid	$B_{10}H_{16}$	decaborane(16)	solid

It is thought that the B H B bridges are held together by two three-center two-electron bonds resulting from the overlap of approximately sp^3 hybrid orbitals on each boron atom with the $1s$ orbital of a bridging hydrogen atom. Two electrons occupy each of the resulting pair of three-center molecular orbitals.

Key Terms

Catenation bonding of atoms of the same element into chains or rings.

Greenhouse effect trapping of heat on the surface of the earth by carbon dioxide and water vapor in the atmosphere.

Haber process a process for the catalyzed industrial production of ammonia from N_2 and H_2 at high temperature and pressure.

Nitrogenases a class of enzymes found in bacteria within root nodules in some plants, which catalyze reactions by which N_2 molecules from the air are converted to ammonia.

Nitrogen cycle the complex series of reactions by which nitrogen is slowly but continually recycled in the atmosphere, lithosphere, and hydrosphere.

Ostwald process a process for the industrial production of nitrogen oxide and nitric acid from ammonia, oxygen, and water.

PANs abbreviation for peroxyacyl nitrates, photochemical oxidants in smog.

Photochemical oxidants photochemically produced oxidizing agents capable of causing severe damage to plants and animals.

Photochemical smog a brownish smog occurring in urban areas receiving large enough amounts of sunlight; caused by photochemical (light-induced) reactions among nitrogen oxides, hydrocarbons, and other components of polluted air, which produce photochemical oxidants.

Three-center two-electron bond a bond holding three atoms together but involving only two electrons; common in boron compounds.

Exercises

Group VA

1. Characterize each of the Group VA elements with respect to normal physical state and color.
2. Write out complete electron configurations for the Group VA elements as well as for the nitride ion, N^{3-}, and phosphide ion, P^{3-}.
3. Is bismuth classified as a metal or nonmetal? Why? What causes BiO^+ ions to be formed in aqueous bismuth(III) salt solutions such as $Bi(NO_3)_3$? Write an equation for the reaction that occurs. (Refer to Section 20–4.3.)
4. Describe the allotropes of solid phosphorus.
5. Compare and contrast the properties of (a) N_2 and P_4, (b) HNO_3 and H_3PO_4, (c) N_2O_3, and P_4O_6.
6. Suggest why corresponding phosphorus and

arsenic compounds more closely resemble each other than they do the corresponding nitrogen compounds.

7. List natural sources of nitrogen, phosphorus, and arsenic and at least two uses for the first two elements.

8. Describe the natural nitrogen cycle.

9. Discuss the effects of temperature, pressure, and catalysts on the Haber process for the production of ammonia. (You may wish to consult Section 16–5.)

10. Determine the oxidation states of nitrogen in the following molecules: (a) N_2, (b) NO, (c) N_2O_4, (d) HNO_3, (e) HNO_2

11. Determine the oxidation states of nitrogen in the following species: (a) NO_3^-, (b) NO_2^-, (c) N_2H_4, (d) NH_3, (e) NH_2^-

12. Calculate the bond energy of the nitrogen-nitrogen triple bond from data in Appendix K. How is this high value related to the general reactivity of N_2?

13. Draw three-dimensional structures showing all outer shell electrons and indicate hybridization (except for N^{3-}) at the central element for the following species: (a) N_2, (b) N^{3-}, (c) NH_3, (d) NH_4^+, (e) NH_2^-, amide ion, (f) N_2H_4, hydrazine.

14. Draw three-dimensional structures showing all outer shell electrons and indicate hybridization at the central element for the following species: (a) NH_2Br, bromamine, (b) HN_3, hydrazoic acid, (c) N_2O_2, (d) $NO_2^+NO_3^-$, solid nitronium nitrate, (e) HNO_3, (f) NO_2^-.

15. Draw three-dimensional structures showing all outer shell electrons for the following species: (a) P_4, (b) P_4O_{10}, (c) As_4O_6, (d) H_3PO_4, (e) AsO_4^{3-}.

16. Write molecular equations for the following reactions:
    (a) reaction of gaseous ammonia with gaseous HCl
    (b) reaction of aqueous ammonia with aqueous HCl
    (c) thermal decomposition of ammonium nitrate at temperatures above 260°C
    (d) reaction of ammonia with oxygen in the presence of red hot platinum catalyst
    (e) thermal decomposition of nitrous oxide (dinitrogen oxide), N_2O
    (f) reaction of NO_2 with water

17. Write molecular equations for the following reactions:
    (a) preparation of "superphosphate of lime"
    (b) reaction of phosphorus with limited Cl_2 (You may wish to refer to Chapter 9.)
    (c) reaction of phosphorus with excess Cl_2
    (d) preparation of phosphorous acid, H_3PO_3
    (e) preparation of phosphoric acid

18. Write two equations illustrating the ability of ammonia to function as a Lewis base.

19. In liquid ammonia would sodium amide, $NaNH_2$, be acidic, basic, or neutral? Would ammonium chloride, NH_4Cl, be acidic, basic, or neutral? Why?

20. Which of the following molecules have a non-zero dipole moment, i.e., are polar molecules? (a) NH_3, (b) NH_2Cl, (c) NO, (d) HNO_3, (e) As_4O_6

21. Describe with equations the Ostwald process for the production of nitrogen oxide (nitric oxide), NO, and nitric acid.

22. Why is NO so reactive?

23. Draw a Lewis formula for NO_2. Would you predict that it is very reactive? How about N_2O_4 (or dimerized NO_2)?

24. What are the acid anhydrides of (a) nitric acid, HNO_3, (b) nitrous acid, HNO_2, (c) phosphoric acid, H_3PO_4, (d) phosphorous acid, H_3PO_3?

25. Discuss the problem of NO_x emissions with respect to air pollution. Use equations to illustrate the important reactions.

26. Arrange the following oxides of nitrogen according to increasing oxidation state of N: N_2O_5, NO_2, N_2O, N_2O_4, NO, N_2O_3.

27. Why is calcium phosphate (phosphate rock) not applied directly as a phosphorus fertilizer?

28. Both sodium nitrate, $NaNO_3$, from Chilean nitrate deposits, and ammonium nitrate, NH_4NO_3, have been used as commercial fertilizers. Both are water-soluble and are easily absorbed by plants.
    (a) Show that ammonium nitrate contains a greater percentage of nitrogen by mass than does sodium nitrate.
    (b) The advantage gained by (a) must be weighed against the disadvantage that the ammonium ion hydrolyzes to produce acidic aqueous solutions, and thus tends to make the soil acidic. This requires application of a basic material such as limestone,

CaCO$_3$ with some MgCO$_3$, to neutralize the acidity. Calculate the pH of 0.10 M ammonium nitrate solution, using K_b for NH$_3$ = 1.8 × 10^{-5} and K_w = 1.0 × 10^{-14}.

29. Calculate the concentrations of ammonium ions and of amide ions, NH$_2^-$, in a 5.0 × 10^{-6} M solution of ammonium chloride in liquid ammonia. Assume that NH$_4$Cl is completely dissociated. K for the autoionization of liquid ammonia is 1 × 10^{-30}.

30. Using data in Appendix K, calculate the bond energies of each of the species in the following reaction.

$$N_2 \text{ (g)} + O_2 \text{ (g)} \rightleftharpoons 2NO \text{ (g)}$$

Is it surprising that nitrogen and oxygen do not react significantly at room temperature?

31. How much hydrazine could be prepared from the reaction of 100 milliliters of 0.50 M aqueous ammonia with 3.00 grams of chloroamine, NH$_2$Cl, and an excess of sodium hydroxide, assuming 68% yield with respect to the limiting reagent? The equation for the reaction is

$$NH_3 \text{ (aq)} + NH_2Cl + NaOH \text{ (aq)} \longrightarrow$$
$$N_2H_4 + NaCl \text{ (aq)} + H_2O$$

The reactants are mixed at low temperature and heated to 80–90°C in the presence of a gelatin catalyst.

32. How many grams of sodium amide must react with excess sodium nitrate at 175°C to produce 10.0 grams of sodium azide, NaN$_3$, assuming 75% yield?

$$3NaNH_2 + NaNO_3 \xrightarrow{175°C}$$
$$NaN_3 + 3NaOH + NH_3$$

33. What volume of dry N$_2$O would be liberated at STP from the complete thermal decomposition of 50.0 grams of hyponitrous acid, H$_2$N$_2$O$_2$? H$_2$O is the other product. (Write the balanced equation first.)

34. How many grams of "superphosphate of lime" could be prepared by the complete reaction of 22.7 grams of "phosphate rock" containing 62.2% calcium phosphate with an excess of sulfuric acid and water?

35. What will be the molarity of 250 milliliters of a phosphoric acid solution produced by the action of water on 20.0 grams of tetraphosphorus decoxide, P$_4$O$_{10}$, assuming complete reaction?

36. Calculate the percentage of phosphorus in P$_4$O$_{10}$ and in H$_3$PO$_4$.

37. If a household detergent contained 25% by mass sodium trimetaphosphate, Na$_3$(PO$_3$)$_3$ or Na$_3$P$_3$O$_9$, what percentage of phosphorus would it contain? How many grams of phosphorus would be contained in 10.0 grams of the detergent?

Carbon

38. Compare the properties of CO$_2$ and SiO$_2$; of B$_2$O$_3$ and SiO$_2$. How do you explain the similarities and differences?

39. Give several natural sources of carbon. Briefly summarize the natural carbon cycle.

40. Sketch the structures of graphite and diamond. Describe their properties and relate them to their structures.

41. Write equations for the reactions of carbon at high temperatures with (a) limited amounts of O$_2$, (b) excess O$_2$, (c) nickel(II) oxide, NiO, (d) silicon dioxide, SiO$_2$, (e) germanium dioxide, GeO$_2$, (f) steam

42. Draw Lewis formulas and structures for each of the following species. Also indicate the hybridization at each carbon atom. (a) CO$_2$, (b) H$_2$CO$_3$, (c) CO$_3^{2-}$

43. Draw Lewis formulas and structures for each of the following species. Also indicate the hybridization at each carbon atom. (a) CCl$_4$, (b) HCN, (c) C$_2$H$_2$

44. Why are simple (single-element) ions with a charge of 4+ or 4− so rare?

45. Write equations for reactions to show how
    (a) carbon monoxide can be obtained from formic acid, HCO$_2$H
    (b) sodium formate can be produced from CO
    (c) carbon monoxide disproportionates

46. Write equations for reactions to show how
    (a) methanol can be synthesized from CO
    (b) methane can be synthesized from CO

47. How does carbon monoxide act as a poison?

48. Describe with equations how stalactites and stalagmites are formed in caves.

49. Why is rain water generally slightly acidic, even in air that is not polluted with oxides of sulfur and nitrogen?

50. Carbon tetrachloride is effective in removal of greasy stains from clothing. Its use is now discouraged because of its toxicity. Account for its effectiveness in stain removal.

51. How many liters of methanol (a liquid) at 20°C can be produced from the complete reaction of 500 cubic feet of carbon monoxide, measured at 15 atmospheres pressure and 50°C,, with an excess of hydrogen? The specific gravity of methanol is 0.791 at 20°C.

52. A 5.631 gram mixture containing only lithium carbonate and lithium oxide was heated to drive off all the carbon dioxide from the carbonate. A total of 1.434 liters of dry CO_2 was collected at 747 torr and 32°C. What mass of lithium carbonate was contained in the original sample? What percentages of lithium carbonate and lithium oxide made up the sample?

Silicon

53. Why is it logical that if 87% of the earth's crust is composed of silicate minerals, silicates must be unreactive?

54. How is silicon obtained from natural sources? How is it purified for use in semiconductors? How does the process of zone refining work?

55. Why are pure silicon and germanium good semiconducting elements?

56. Write equations for the following reactions:
    (a) silicon with aqueous sodium hydroxide
    (b) silicon with bromine (You may wish to refer to Chapter 9.)

57. How does hydrofluoric acid attack glass?

58. Using the data of Appendix K, calculate the ΔG^0 values at 25°C for each of the reactions below for X = F, Cl. Use standard states for each of the species.

$$Si + 2X_2 \longrightarrow SiX_4$$

Which reaction is more spontaneous at 25°C?

Boron

59. How can natural borax be converted to boric acid, H_3BO_3?

60. Show how boric oxide, B_2O_3, metaboric acid, HBO_2, and tetraboric acid, $H_2B_4O_7$, can be prepared by dehydration of boric acid.

61. Write equations for the following reactions:
    (a) a preparation of elemental boron
    (b) the preparation of B_5H_9 from diborane, B_2H_6
    (c) the preparation of B_4H_{10} from diborane

62. Write equations for two reactions illustrating the ability of BCl_3 to act as a Lewis acid.

63. Is it surprising that boron participates in three-center two-electron bonds? Why?

64. Describe and illustrate the bonding in the three-center two-electron bonds of B_2H_6.

65. How much borax must react with excess sulfuric acid to produce 100 pounds of boric acid, assuming 95.2% yield based on borax?

66. Calculate the concentrations of all species and the pH in 0.10 M boric acid, H_3BO_3, a triprotic acid. $K_1 = 7.3 \times 10^{-10}$, $K_2 = 1.8 \times 10^{-13}$, $K_3 = 1.6 \times 10^{-14}$.

67. How many grams of diborane, B_2H_6, can be produced from the 88.4% efficient reaction of 10.7 grams of boron trichloride with an excess of lithium aluminum hydride in ether solution?

$$4BCl_3 + 3LiAlH_4 \longrightarrow 2B_2H_6 + 3AlCl_3 + 3LiCl$$

68. A pure sample of a gaseous boron hydride is found to be 78.4 percent boron by mass. A 0.2120 gram sample occupies 114.5 mL at STP. What is the molecular formula of the compound?

Coordination Compounds

23

23–1   Introduction

Covalent bonds in which the shared electron pair is provided by one atom are known as coordinate covalent bonds.

In Chapter 17 we discussed Lewis acid-base reactions. According to this view a *base* makes available a share in an electron pair and an *acid* accepts a share in an electron pair, to form a **coordinate covalent bond.** Such bonds are frequently represented by arrows, as shown at the top of the next page.

The chemical structure diagrams show:

ammonia (a Lewis base) + boron trifluoride (a Lewis acid) → $H_3N \rightarrow BF_3$ adduct

In the case of the second equation, the arrows do not imply that two Sn—Cl bonds are any different from the others. Once formed, all the Sn—Cl bonds in the $SnCl_6^{2-}$ ion are indistinguishable.

$2 :Cl:^-$ chloride ion (a Lewis base) + tin(IV) chloride (a Lewis acid) → hexachlorostannate(IV) ion

Most d-transition metal ions have vacant d orbitals that can accommodate shares in electron pairs, and they show a marked tendency to act as Lewis acids in forming coordinate covalent bonds in **coordination compounds (coordination complexes** or **complex ions).** Examples of transition metal ions or molecules showing coordinate covalence are $[Cr(OH_2)_6]^{3+}$, $[Co(NH_3)_6]^{3+}$, $[Ni(CN)_4]^{2-}$, $[Fe(CO)_5]$ and $[Ag(NH_3)_2]^+$. Many such complexes are very stable, as indicated by their low dissociation constants, K_d. See Section 18-11 and Appendix I.

Many biologically important substances are d-transition metal coordination compounds in which large, complicated organic species are bound to a metal ion through coordinate covalent bonds. One example is hemoglobin (Figure 23-1), a protein in blood that contains iron bound to a large, nearly planar organic group called a porphyrin ring; another is vitamin B-12, a large complex of cobalt. There are many practical applications of coordination compounds in such areas as water treatment, soil and plant treatment, protection of metal surfaces, analysis of trace amounts of metals, electroplating, and textile dyeing.

The bonding in transition metal complexes was not well understood until the pioneering research of Alfred Werner, a Swiss chemist in the 1890s and early 1900s who won the Nobel Prize in Chemistry in 1913. Although tremendous advances have been made in the field of coordination chemistry, especially in the last quarter century, Werner's classic work still remains the most important contribution by any single researcher.

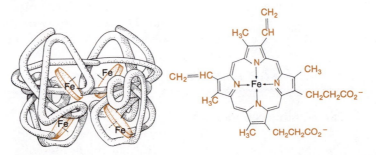

Porphyrin ring system + Iron = Heme group (disks at left)

FIGURE 23-1 The shape of the tetrameric hemoglobin molecule (mol. wt. 64,500). Individual atoms not shown. The four heme groups (one of which is shown at right) are represented by the disks (left).

TABLE 23–1 Interpretation of Experimental Data by Werner

Formula	Moles AgCl Precipitated per Formula Unit of Complex by Excess AgNO$_3$	Number of Ions per Formula Unit Indicated by Conductance	True Formula
PtCl$_4$ · 6NH$_3$	4	5	[Pt(NH$_3$)$_6$]Cl$_4$
PtCl$_4$ · 5NH$_3$	3	4	[Pt(NH$_3$)$_5$Cl]Cl$_3$
PtCl$_4$ · 4NH$_3$	2	3	[Pt(NH$_3$)$_4$Cl$_2$]Cl$_2$
PtCl$_4$ · 3NH$_3$	1	2	[Pt(NH$_3$)$_3$Cl$_3$]Cl
PtCl$_4$ · 2NH$_3$	0	0	[Pt(NH$_3$)$_2$Cl$_4$]

Double salts are ionic solids resulting from the crystallization into a single lattice of two salts from the same solution. In the example given, the solid is produced from an aqueous solution of iron(II) sulfate, FeSO$_4$, and ammonium sulfate, (NH$_4$)$_2$SO$_4$.

Prior to Werner's work the formulas of transition metal complexes were written with dots, CrCl$_3$ · 6H$_2$O, AgCl · 2NH$_3$, just like double salts such as iron(II) ammonium sulfate hexahydrate, FeSO$_4$ · (NH$_4$)$_2$SO$_4$ · 6H$_2$O. The properties of solutions of double salts are the properties expected for solutions made by mixing the individual salts. However, a solution of AgCl · 2NH$_3$, or more appropriately [Ag(NH$_3$)$_2$]Cl, behaves entirely differently from either a solution of (very insoluble) silver chloride or a solution of ammonia. The dots have been called "dots of ignorance," since they signified that the mode of bonding was unknown. The following outline illustrates the types of experiments Werner performed and interpreted to lay the foundations for modern coordination theory.

Werner was able to isolate platinum(IV) compounds with the formulas given in the first column of Table 23–1. To solutions of carefully weighed amounts of each of the five salts he added excess silver nitrate. The precipitated silver chloride was collected by filtration, dried, and weighed to determine the number of moles of silver chloride produced, and therefore the number of chloride ions precipitated per formula unit. The results appear in the second column. Werner reasoned that the precipitated chloride ions must be free (uncoordinated), while the unprecipitated chloride ions must be directly bound to platinum so they could not be removed by silver ions. He also measured the equivalent conductances of salt solutions of known concentrations to determine the number of ions per formula unit. The results are given in the third column. Piecing the evidence together, he concluded that the correct formulations are the ones listed in the last column, in which the following Pt(IV) complex ions or molecules are present (from top to bottom): [Pt(NH$_3$)$_6$]$^{4+}$,

Here the square brackets do not imply molar concentrations; instead they enclose the coordination sphere (Section 23–3).

[Pt(NH$_3$)$_5$Cl]$^{3+}$, [Pt(NH$_3$)$_4$Cl$_2$]$^{2+}$, [Pt(NH$_3$)$_3$Cl$_3$]$^+$, and [Pt(NH$_3$)$_2$Cl$_4$]. The species within the square brackets, NH$_3$ and Cl$^-$, are bonded by coordinate covalent bonds to the Lewis acid, Pt(IV) ion. The charges on the complex ions are the sums of the constituent charges. The remaining chloride ions, written outside the brackets, are bonded by ionic bonds in the solid, and dissociate in solution.

23–2 The Ammine Complexes

The **ammine complexes** are species that contain ammonia molecules bonded to metal ions by coordinate covalent bonds. They are important compounds that have been known for a long time. We shall briefly describe their preparation as an example.

Since most metal hydroxides are insoluble in water, aqueous ammonia reacts with nearly all metal ions to form insoluble metal hydroxides, or hydrated oxides. [The exceptions are the cations of the strong soluble bases — cations of the Group IA

TABLE 23-2   Common Metal Ions That Form Soluble Complexes with an Excess of Aqueous Ammonia*

Metal Ion	Insoluble Hydroxide Formed by Limited Aq. NH_3	Complex Ion Formed by Excess Aq. NH_3
Co^{2+}	$Co(OH)_2$	$[Co(NH_3)_6]^{2+}$
Co^{3+}	$Co(OH)_3$	$[Co(NH_3)_6]^{3+}$
Ni^{2+}	$Ni(OH)_2$	$[Ni(NH_3)_6]^{2+}$
Cu^+	$CuOH \longrightarrow \frac{1}{2}Cu_2O\dagger$	$[Cu(NH_3)_2]^+$
Cu^{2+}	$Cu(OH)_2$	$[Cu(NH_3)_4]^{2+}$
Ag^+	$AgOH \longrightarrow \frac{1}{2}Ag_2O\dagger$	$[Ag(NH_3)_2]^+$
Zn^{2+}	$Zn(OH)_2$	$[Zn(NH_3)_4]^{2+}$
Cd^{2+}	$Cd(OH)_2$	$[Cd(NH_3)_4]^{2+}$
Hg^{2+}	$Hg(OH)_2$	$[Hg(NH_3)_4]^{2+}$

* The ions of Rh, Ir, Pd, Pt, and Au show similar behavior, but since these ions are seldom encountered in elementary texts, they are not included here.
† CuOH and AgOH are unstable and decompose to the corresponding oxides.

metals and the heavier members of Group IIA (Ca^{2+}, Sr^{2+}, Ba^{2+}).]

$$Cu^{2+} + 2NH_3 + 2H_2O \longrightarrow Cu(OH)_2 \text{ (s)} + 2NH_4^+$$

$$Cr^{3+} + 3NH_3 + 3H_2O \longrightarrow Cr(OH)_3 \text{ (s)} + 3NH_4^+$$

In general terms, we may represent this reaction as

$$M^{n+} + nNH_3 + nH_2O \longrightarrow M(OH)_n \text{ (s)} + nNH_4^+$$

In Section 9-6 we pointed out the fact that the hydroxides of some metals and some metalloids are amphoteric; i.e., they dissolve in an excess of a strong soluble base. Aqueous ammonia is such a weak base ($K_b = 1.8 \times 10^{-5}$) that the concentration of OH^- is not high enough to dissolve the amphoteric hydroxides.

However, several metal hydroxides do dissolve in an excess of aqueous ammonia to form ammine complexes. For example, the hydroxides of copper and cobalt are readily soluble in an excess of aqueous ammonia solution.

$$Cu(OH)_2 \text{ (s)} + 4NH_3 \rightleftharpoons [Cu(NH_3)_4]^{2+} + 2OH^-$$

$$Co(OH)_2 \text{ (s)} + 6NH_3 \rightleftharpoons [Co(NH_3)_6]^{2+} + 2OH^-$$

Interestingly, all metal hydroxides that exhibit this behavior are derived from the twelve metals of the cobalt, nickel, copper, and zinc families. All the common cations of these metals except Hg_2^{2+} (which disproportionates) form soluble complexes in the presence of excess aqueous ammonia; the common ones are listed in Table 23-2.

23-3   Important Terms in Coordination Chemistry

Before we go further, let us define and illustrate a few terms that will make our discussion of coordination compounds easier. The Lewis bases in coordination compounds may be molecules, anions, or (rarely) cations, and are called **ligands,** (Latin, *ligare:* to bind). The **donor atoms** of the ligands are the atoms that actually donate electron pairs to metals. In some instances it is not possible to identify donor atoms, since the bonding electrons are not localized on specific atoms. For example, some small organic molecules such as ethylene, $H_2C{=}CH_2$, can bond to a transi-

TABLE 23–3 **Typical Simple Ligands***

Molecule	Name	Name as Ligand	Ion	Name	Name as Ligand
:$\underline{N}H_3$	ammonia	ammine	:$\ddot{\underline{C}l}$:⁻	chloride	chloro
:$\underline{O}H_2$	water	aqua	:$\ddot{\underline{F}}$:⁻	fluoride	fluoro
:$\underline{C}{\equiv}O$:	carbon monoxide	carbonyl	:$\underline{C}{\equiv}N$:⁻	cyanide	cyano
:$\underline{P}H_3$	phosphine	phosphine	:$\underline{O}H^-$	hydroxide	hydroxo
:$\underline{N}{=}\ddot{O}$	nitrogen oxide	nitrosyl	:$\underline{N}$ with $\ddot{O}$⁻ and :$\ddot{O}$:	nitrite	nitro†

* Donor atoms are underlined.

† Oxygen atoms can also function as donor atoms, in which case the ligand name is "nitrito."

tion metal through the electrons associated with the π component of their double bonds. Examples of typical simple ligands are listed in Table 23–3.

Ligands that can bond to a metal through only one donor atom at a time, such as those in Table 23–3, are **unidentate** (Latin, *dent:* tooth). Many ligands can bond simultaneously through more than one donor atom and are called **polydentate.** Of the polydentates, those that bond through two, three, four, five or six donor atoms are called bidentates, tridentates, quadridentates, quinquedentates, and sexidentates, respectively. The resulting complexes, consisting of the metal atom or ion and polydentate ligands, are called **chelate complexes** (Greek, *chele:* claw).

The **coordination number** of a metal atom or ion in a complex is the number of donor atoms to which it is coordinated, not necessarily the number of ligands. The **coordination sphere** includes the metal or metal ion and its ligands, but no uncoordinated counter-ions. For example, the coordination sphere of hexaamminecobalt(III) chloride, $[Co(NH_3)_6]Cl_3$, is the hexaamminecobalt(III) ion, $[Co(NH_3)_6]^{3+}$. Illustrations of these terms are given in Table 23–4. The naming of complexes will be discussed in the following section.

23–4 Naming Coordination Compounds

Because many thousands of rather complicated coordination compounds are known and thousands of new ones will be discovered during the next decade, the International Union of Pure and Applied Chemistry (IUPAC) has adopted a set of rules for naming such compounds. The rules are based on those originally devised by Werner. They are summarized below.

1. Consistent with the rules of nomenclature for compounds of the representative elements, cations are named before anions.
2. In naming the coordination sphere, ligands are named in alphabetical order. The prefixes di = 2, tri = 3, tetra = 4, penta = 5, hexa = 6, etc., are used to specify the number of a particular kind of coordinated ligand. For example, in dichloro, the "di" indicates that two Cl^- act as ligands, and this prefix is not considered in alphabetizing. However, when the prefix denotes the number of substituents on a single ligand, as in dimethylamine, $NH(CH_3)_2$, it *is* used to alphabetize ligands. For complicated ligands (usually chelating agents), other prefixes such as bis = 2,

TABLE 23-4 Ligands and Coordination Complexes

Ligand(s)	Classification	Complex	Oxidation Number of M	Coordination Number of M
NH_3 ammine	unidentate	$[Co(NH_3)_6]^{3+}$ hexaamminecobalt(III) ion	+3	6
$H_2N-CH_2-CH_2-NH_2$ (or N͡N) ethylenediamine (en)	bidentate	or $[Co(en)_3]^{3+}$ tris(ethylenediamine)-cobalt(III) ion	+3	6
Br^- bromo $H_2N-CH_2-CH_2-NH_2$ ethylenediamine (en)	unidentate bidentate	or $[Cu(en)Br_2]$ dibromoethylene-diaminecopper(II)	+2	4
$H_2N-CH_2-CH_2-\overset{H}{N}-CH_2-CH_2-NH_2$ (or N͡N͡N) diethylenetriamine (dien)	tridentate	or $[Fe(dien)_2]^{3+}$ bis(diethylenetriamine)-iron(III) ion	+3	6
ethylenediaminetetraacetato (edta)	sexidentate	or $[Co(edta)]^-$ ethylenediamine-tetraacetatocobaltate(III) ion	+3	6

tris = 3, tetrakis = 4, pentakis = 5, and hexakis = 6 indicate the number of ligands.

3. The names of anionic ligands end in the suffix — o. Examples are: F^-, fluoro; OH^-, hydroxo; O^{2-}, oxo; S^{2-}, sulfido; CO_3^{2-}, carbonato; CN^-, cyano; SO_4^{2-}, sulfato; NO_3^-, nitrato; $S_2O_3^{2-}$, thiosulfato.

4. The names of neutral ligands are usually unchanged. Some important exceptions are NH_3, ammine; H_2O, aqua; CO, carbonyl; NO, nitrosyl.
5. The oxidation number of the metal ion is designated by a Roman numeral in parentheses following the name of the complex ion or molecule if the metal exhibits variable oxidation states.
6. The suffix "ate" at the end of the name of the complex signifies that it is an anion. If the complex is neutral or cationic, no suffix is used. The English stem is usually used for the metal, but where the naming of an anion is awkward, the Latin stem is substituted. For example, ferrate is used rather than ironate, and plumbate rather than leadate.

The term *ammine* (two m's) which signifies the presence of ammonia as a ligand, is distinctly different from the term *amine* (one m), which describes a class of organic compounds (Section 25–18) that can be considered to be derived from ammonia.

Several examples are given to illustrate the rules.

$K_2[Cu(CN)_4]$	potassium tetracyanocuprate(II)
$[Ag(NH_3)_2]Cl$	diamminesilver chloride
$[Cr(OH_2)_6](NO_3)_3$	hexaaquachromium(III) nitrate
$[Co(en)_2Br_2]Cl$	dibromobis(ethylenediamine)cobalt(III) chloride
$[Ni(CO)_4]$	tetracarbonylnickel(0)
$[Pt(NH_3)_4][PtCl_6]$	tetraammineplatinum(II) hexachloroplatinate(IV)
$[Fe(OH_2)_5(NCS)]SO_4$	pentaaquathiocyanatoiron(III) sulfate
$[Cu(NH_3)_2(en)]Br_2$	diammine(ethylenediamine)copper(II) bromide
$Na[Al(OH)_4]$	sodium tetrahydroxoaluminate
$Na_2[CrOF_4]$	sodium tetrafluorooxochromate(IV)
$Na_2[Sn(OH)_6]$	sodium hexahydroxostannate(IV)
$[Cu(NH_3)(OH_2)Br_2]$	ammineaquadibromocopper(II)
$[Co(en)_3](NO_3)_3$	tris(ethylenediamine)cobalt(III) nitrate
$K_4[Ni(CN)_2(ox)_2]$	potassium dicyanobis(oxalato)nickelate(II)
$[Co(NH_3)_4(OH_2)Cl]Cl_2$	tetraammineaquachlorocobalt(III) chloride
$[RuCl_3\{P(C_6H_5)_3\}_2]$	trichlorobis(triphenylphosphine)ruthenium(III)

Water is written OH_2 rather than H_2O to emphasize that oxygen is the donor atom.

The oxidation state of aluminum is not given since it is always +3.

The abbreviation ox represents the oxalate ion, $(COO)_2^{2-}$. It is bidentate.

23–5 Structures of Coordination Compounds

The structures of coordination compounds are governed largely by the coordination number of the metal, and can be predicted quite accurately for most of them by application of the VSEPR method (Chapter 7). In most cases, lone pairs of electrons in *d* orbitals have only minimal influences on geometry since they are not in the outer shell. Table 23–5 summarizes the geometries and hybridizations associated with common coordination numbers.

Transition metal complexes with coordination numbers as high as seven, eight, and nine are known, but they are very rare. For coordination number five, the trigonal bipyramidal structure generally predominates over the square pyramid, although the energies associated with the two geometries are not very different. Both tetrahedral and square planar geometries are quite common for complexes with coordination number four. Note that the tabulated geometries are ideal geometries. The actual structures are sometimes distorted, especially if ligands of more than one type are present. The distortions are due to compensations for the unequal electric fields generated by the different ligands at the coordination sites.

TABLE 23-5   Geometries and Hybridizations for Various Coordination Numbers

Coordination Number	Geometry	Hybridization	Example
2	 linear	sp	$[Ag(NH_3)_2]^+$ $[Cu(CN)_2]^-$
4	 tetrahedral	sp^3	$[Zn(CN)_4]^{2-}$ $[Cd(NH_3)_4]^{2+}$
4	 square planar	dsp^2 or sp^2d	$[Cu(OH_2)_4]^{2+}$ $[Pt(NH_3)_2Cl_2]$
5	 trigonal bipyramid	dsp^3	$[Fe(CO)_5]$ $[CuCl_5]^{3-}$
5	 square pyramid	d^2sp^2	$[NiBr_3(P(C_2H_5)_3)_2]$ $[RuCl_3(P(C_6H_5)_3)_2]$
	 octahedral	d^2sp^3 or sp^3d^2	$[Fe(CN)_6]^{4-}$ $[Fe(OH_2)_6]^{2+}$

Isomerism in Coordination Compounds

The term *isomers* comes from the Greek word meaning equal parts.

Isomers are substances that have the same number and kinds of atoms, arranged differently. *Because their structures are different, isomers have different properties.* Isomers can be broadly classed into two major categories, structural isomers and stereoisomers, each of which can be further subdivided as shown below.

Structural Isomers	**Stereoisomers**
1. ionization isomers	1. geometrical (position) isomers
2. hydrate isomers	2. optical isomers
3. coordination isomers	
4. linkage isomers	

Of the two broad classes, stereoisomers are more important and noteworthy. Before considering stereoisomers, we shall give examples of the four types of structural isomers. While distinctions between stereoisomers involve only one coordination sphere, and the same ligands and donor atoms, the differences between **structural isomers** involve either more than one coordination sphere or else different donor atoms on the same ligand. They contain different *atom-to-atom bonding sequences.*

23–6 Structural Isomers

1   Ionization Isomers

These isomers result from the interchange of ions inside and outside the coordination sphere. For example, red-violet $[Co(NH_3)_5Br]SO_4$ and red $[Co(NH_3)_5SO_4]Br$ are ionization isomers.

Isomers such as those shown here may *or may not* exist in the same solution in equilibrium. This depends on the thermodynamic and kinetic properties of the reaction mechanism, if any, that interconverts a pair of isomers. Generally, such isomers are formed by *different* reactions.

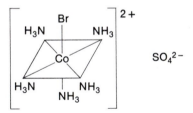

$[Co(NH_3)_5Br]SO_4$
pentaamminebromocobalt(III) sulfate

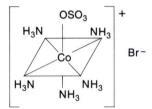

$[Co(NH_3)_5SO_4]Br$
pentaamminesulfatocobalt(III) bromide

A solution of the sulfate reacts with a solution of barium chloride to precipitate white barium sulfate, but a solution of the bromide does not. Equimolar solutions of the two also have different electrical conductivities. The sulfate solution conducts electric current better because its ions have 2+ and 2− charges rather than 1+ and 1−. Other examples of this type of isomerism include:

$[Pt(NH_3)_4Cl_2]Br_2$	and	$[Pt(NH_3)_4Br_2]Cl_2$
$[Pt(NH_3)_4SO_4](OH)_2$	and	$[Pt(NH_3)_4(OH)_2]SO_4$
$[Co(NH_3)_5NO_2]SO_4$	and	$[Co(NH_3)_5SO_4]NO_2$
$[Cr(NH_3)_5SO_4]Br$	and	$[Cr(NH_3)_5Br]SO_4$

2 Hydrate Isomers

Hydration isomerism and ionization isomerism are quite similar. In some crystalline complexes water can occur in more than one way, inside and outside the coordination sphere. For example, solutions of the three hydrate isomers given below yield three, two, and one mole of silver chloride precipitate per mole of complex when treated with excess silver nitrate.

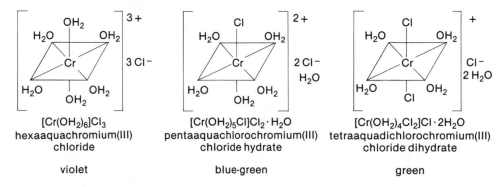

$[Cr(OH_2)_6]Cl_3$
hexaaquachromium(III)
chloride

violet

$[Cr(OH_2)_5Cl]Cl_2 \cdot H_2O$
pentaaquachlorochromium(III)
chloride hydrate

blue-green

$[Cr(OH_2)_4Cl_2]Cl \cdot 2H_2O$
tetraaquadichlorochromium(III)
chloride dihydrate

green

3 Coordination Isomers

Coordination isomerism can occur in compounds containing both complex cations and complex anions. Such isomers involve exchange of ligands between cation and anion, i.e., between coordination spheres.

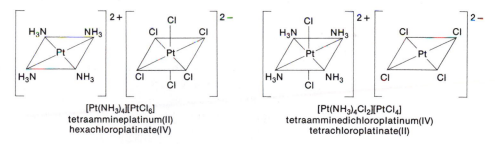

$[Pt(NH_3)_4][PtCl_6]$
tetraammineplatinum(II)
hexachloroplatinate(IV)

$[Pt(NH_3)_4Cl_2][PtCl_4]$
tetraamminedichloroplatinum(IV)
tetrachloroplatinate(II)

4 Linkage Isomers

Certain ligands can bind to a metal ion in more than one way. Examples of such ligands are cyano, $-CN^-$, and isocyano, $-NC^-$; nitro, $-NO_2^-$, and nitrito, $-ONO^-$. The donor atoms are on the left in these representations. Examples of linkage isomers are given below.

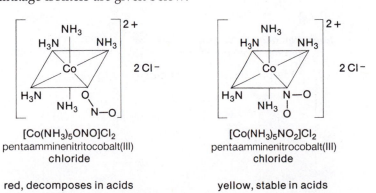

$[Co(NH_3)_5ONO]Cl_2$
pentaamminenitritocobalt(III)
chloride

red, decomposes in acids

$[Co(NH_3)_5NO_2]Cl_2$
pentaamminenitrocobalt(III)
chloride

yellow, stable in acids

23–7 Stereoisomers

Because of symmetry, a complex with coordination number 2 or 3 can have only one spatial arrangement; all apparent "isomers" are equivalent to just turning the complex around. Try building models to see this.

Compounds that contain the same atoms and the same atom-to-atom bonding sequences, but that differ only in spatial arrangements of the atoms relative to the central atom, are **stereoisomers.** Stereoisomerism can exist only in complexes with coordination number four or greater. Since the most common coordination numbers among coordination complexes are four and six, they will be used to illustrate stereoisomerism.

1 Geometrical Isomers

Stereoisomers that are not optical isomers (see p. 767) are **geometrical isomers** or position isomers. *Cis-trans* isomerism is one kind of geometrical isomerism. *Cis* means "adjacent to" and *trans* means "on the opposite side of." *Cis-* and *trans-*diamminedichloroplatinum(II) are shown below.

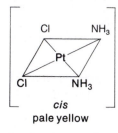

cis
pale yellow

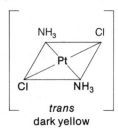

trans
dark yellow

In the *cis* complex, the chloro groups are closer to each other (on the same side of the square) than they are in the *trans* complex. The ammine groups are also closer together in the *cis* complex.

Since all ligands in tetrahedral complexes are adjacent to each other, they do not exhibit geometrical isomerism.

For octahedral complexes, geometrical isomerism can be more involved. For example, complexes of the type $MA_2B_2C_2$ can exhibit several isomeric forms. Consider as an example $[Cr(OH_2)_2(NH_3)_2Br_2]^+$. Each of the like ligands may be *trans* to each other or *cis* to each other.

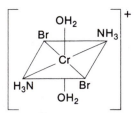

trans-diammine-*trans*-diaqua-*trans*-dibromochromium(III) ion

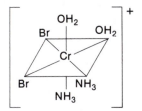

cis-diammine-*cis*-diaqua-*cis*-dibromochromium(III) ion

See whether you can discover why there is no *trans-trans-cis* isomer.

One pair of ligands may be *trans* to each other, but the others *cis.*

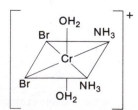

cis-diammine-*trans*-diaqua-*cis*-dibromochromium(III) ion

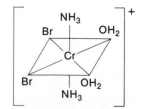

trans-diammine-*cis*-diaqua-*cis*-dibromochromium(III) ion

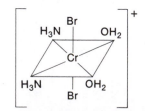

cis-diammine-*cis*-diaqua-*trans*-dibromochromium(III) ion

Interchanging the positions of the ligands further produces no new geometric isomers. However, one of the five geometric isomers can exist in two distinct forms called optical isomers.

2 Optical Isomers

The *cis*-diammine-*cis*-diaqua-*cis*-dibromochromium(III) geometric isomer exists in two forms that bear the same relationship to each other as left and right hands. They are *non-superimposable* mirror images of each other and are called **optical isomers,** or **enantiomers.**

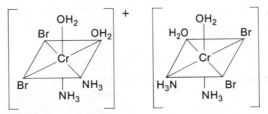

Optical isomers of *cis*-diammine-*cis*-diaqua-*cis*-dibromochromium(III) ion

Optical isomers have identical physical and chemical properties except that they interact with polarized light in different ways. Separate equimolar solutions of the two will rotate a plane of polarized light (see Figures 23–2 and 23–3) by equal amounts but in opposite directions. One solution is **dextrorotatory** (rotates to the *right*) and the other is **levorotatory** (rotates to the *left*). The optical isomers are thus called *dextro* and *levo* isomers. The phenomenon by which a plane of polarized light is rotated is called **optical activity** and can be measured by a device called a polarimeter (see Figure 23–3) or by more sophisticated instruments. A single solution containing equal amounts of the two isomers, called a **racemic mixture,** does not rotate a plane of polarized light because the equal and opposite effects of

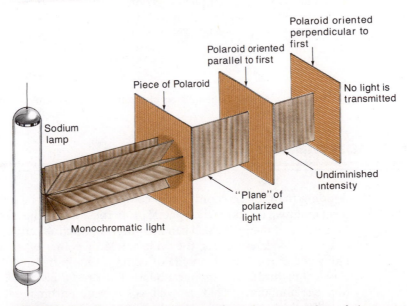

FIGURE 23–2 Light from a lamp or from the sun consists of electromagnetic waves that vibrate in all directions perpendicular to the direction of travel. "Polaroid" material absorbs all waves except those that vibrate in a single plane. The third Polaroid sheet, with a plane of polarization at right angles to the first, absorbs the polarized light completely.

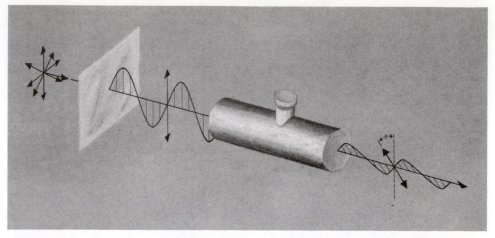

FIGURE 23–3 The plane of polarization of plane polarized light is rotated through an angle (θ) as it passes through an optically active medium. Species that rotate the plane to the right (clockwise) are dextrorotatory and those that rotate it to the left are levorotatory.

the two isomers exactly cancel. Thus, in order to exhibit optical activity, the *dextro* and *levo* isomers (sometimes designated as delta, Δ, and lambda, Λ, isomers) must be separated from each other by one of a number of chemical or physical processes broadly called optical resolution.

Another example of a pair of optical isomers is shown below. Note that they both contain ethylenediamine, a bidentate ligand. The difference is in the way the vertices of the octahedron (at the nitrogen atoms) are connected pairwise. Again, a pair of molecular models may help in investigating the nature of the isomerism.

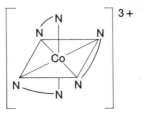

Λ-tris(ethylenediamine)cobalt(III) ion

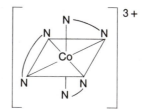

Δ-tris(ethylenediamine)cobalt(III) ion

Bonding in Coordination Compounds

Bonding theories for coordination compounds should be able to account for structural features, colors, and magnetic properties. The earliest accepted theory was the *Valence Bond* theory, which can account for structural and magnetic properties but offers no explanation for the wide range of colors of coordination compounds. However, it has the advantage of being a simple treatment of bonding using the classical picture of the chemical bond. The *Crystal Field* theory gives quite satisfactory explanations of color as well as structure and magnetic properties for many coordination compounds. We shall discuss the Valence Bond and Crystal Field theories.

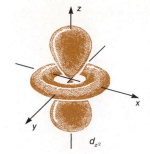

d_{z^2}

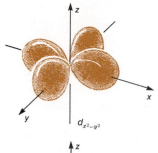

$d_{x^2-y^2}$

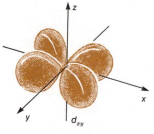

d_{xy}

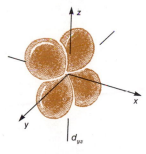

d_{yz}

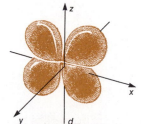

d_{xz}

FIGURE 23-4 A representation of the five d orbital boundary surfaces and their orientation in Cartesian coordinates.

23-8 Valence Bond Theory

A basic assumption of the Valence Bond theory applied to coordination compounds is that the coordinate bonds between ligands and metal are entirely covalent. Of course, this is not really true. All bonds have some degree, however slight, of both covalent and ionic character.

1 Coordination Number 6

The shapes of the d orbitals of a metal atom are depicted in Figure 23-4. The lobes of the $d_{x^2-y^2}$ and d_{z^2} orbitals are directed along the x, y, and z axes. The lobes of the d_{xy}, d_{yz}, and d_{xz} orbitals bisect the axes. The six ligand donor atoms in an octahedral complex are located at the corners of an octahedron, two along each of the three axes. Valence Bond theory postulates that they must donate electrons into a set of octahedrally hybridized d^2sp^3 or sp^3d^2 metal orbitals. (We shall distinguish between d^2sp^3 and sp^3d^2 hybridization very shortly.) Thus, the two metal d orbitals used in the hybridization must be the $d_{x^2-y^2}$ and d_{z^2} orbitals, since they are the only ones directed along the x, y, and z axes. This is illustrated in Figure 23-5.

Let us take one example and see how the Valence Bond theory accounts for its properties. Hexaaquairon(III) perchlorate, $[Fe(OH_2)_6](ClO_4)_3$, is known to be paramagnetic, with a magnetic moment corresponding to five unpaired electrons per iron atom. The valence bond description of the bonding in the $[Fe(OH_2)_6]^{3+}$ ion is as follows. Atomic iron has the electronic configuration

	$3d$	$4s$
Fe [Ar]	↑↓ ↑ ↑ ↑ ↑	↑↓

The Fe^{3+} ion is formed upon ionization of the $4s$ electrons and one of the $3d$ electrons.

	$3d$	$4s$
Fe^{3+} [Ar]	↑ ↑ ↑ ↑ ↑	__

To account for the experimentally observed fact that $[Fe(OH_2)_6]^{3+}$ has five unpaired electrons, each $3d$ orbital is assumed to have one unpaired electron in the complex. The vacant $4d_{x^2-y^2}$ and $4d_{z^2}$ orbitals are hybridized with the vacant $4s$ and $4p$ orbitals.

	$3d$	$4s$ $4p$	$4d$
Fe^{3+} [Ar]	↑ ↑ ↑ ↑ ↑	__ __ __ __	__ __ __

↓ hybridize

	$3d$	six sp^3d^2 hybrids	$4d_{xy}$ $4d_{xz}$ $4d_{yz}$
Fe^{3+} [Ar]	↑ ↑ ↑ ↑ ↑	__ __ __ __ __ __	__ __ __

Each of the six water ligands now donates two electrons into one of the six sp^3d^2 orbitals, forming the six coordinate covalent bonds. The electrons originally on the oxygen atoms are represented as "x" rather than "↑," even though, in actuality, electrons are indistinguishable. Note that these are the only *bonding* electrons, and that none of the $3d$ electrons of iron is involved in bonding.

$$[\text{Fe(OH}_2)_6]^{3+} \ [\text{Ar}] \quad \underset{3d}{\underline{\uparrow}\ \underline{\uparrow}\ \underline{\uparrow}\ \underline{\uparrow}\ \underline{\uparrow}} \quad \underset{sp^3d^2}{\underline{\text{xx}}\ \underline{\text{xx}}\ \underline{\text{xx}}\ \underline{\text{xx}}\ \underline{\text{xx}}\ \underline{\text{xx}}} \quad \underset{4d_{xy}\ 4d_{xz}\ 4d_{yz}}{\underline{\ \ }\ \underline{\ \ }\ \underline{\ \ }}$$

Since a set of outer *d* orbitals is involved in the hybridization, $[\text{Fe(OH}_2)_6]^{3+}$ is called an **outer orbital complex.**

The hexacyanoferrate(III) ion (also known commonly as the ferricyanide ion), $[\text{Fe(CN)}_6]^{3-}$, also involves Fe^{3+} (a d^5 ion), but its magnetic moment indicates only one unpaired electron per iron atom. The $3d_{x^2-y^2}$ and $3d_{z^2}$ orbitals (rather than those in the 4*d* shell) are involved in d^2sp^3 hybridization. This forces pairing of all but one of the nonbonding 3*d* electrons of Fe^{3+}.

$$\text{Fe}^{3+}\ [\text{Ar}] \quad \underset{3d}{\underline{\uparrow}\ \underline{\uparrow}\ \underline{\uparrow}\ \underline{\uparrow}\ \underline{\uparrow}} \quad \underset{4s}{\underline{\ \ }} \quad \underset{4p}{\underline{\ \ }\ \underline{\ \ }\ \underline{\ \ }}$$

$$\downarrow \text{ hybridize}$$

$$[\text{Fe(CN)}_6]^{3-}\ [\text{Ar}] \quad \underset{3d_{xy}}{\underline{\uparrow\downarrow}}\ \underset{3d_{xz}}{\underline{\uparrow\downarrow}}\ \underset{3d_{yz}}{\underline{\uparrow}} \quad \underset{\text{six } d^2sp^3 \text{ hybrids}}{\underline{\text{xx}}\ \underline{\text{xx}}\ \underline{\text{xx}}\ \underline{\text{xx}}\ \underline{\text{xx}}\ \underline{\text{xx}}}$$

Since only inner *d* orbitals are used in hybridization, $[\text{Fe(CN)}_6]^{3-}$ is called an **inner orbital complex.** As in the previous case, none of the original 3*d* electrons of the Fe^{3+} ion is involved in bonding.

In deciding whether the hybridization at the central metal ion of octahedral complexes is sp^3d^2 (outer orbital) or d^2sp^3 (inner orbital), we must know the results of magnetic measurements that indicate the number of unpaired electrons.

To account for the single unpaired electron in $[\text{Co(NH}_3)_6]^{2+}$, derived from a d^7 ion, by Valence Bond theory, it is necessary to postulate the *promotion* of a 3*d* electron to a 5*s* orbital as d^2sp^3 hybridization occurs.

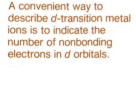

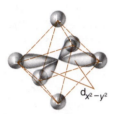

FIGURE 23–5
Orientation of $d_{x^2-y^2}$ and d_{z^2} orbitals relative to the ligands in an octahedral complex.

$$\text{Co}^{2+}\ [\text{Ar}] \quad \underset{3d}{\underline{\uparrow\downarrow}\ \underline{\uparrow\downarrow}\ \underline{\uparrow}\ \underline{\uparrow}\ \underline{\uparrow}} \quad \underset{4s}{\underline{\ \ }}\ \underset{4p}{\underline{\ \ }\ \underline{\ \ }\ \underline{\ \ }} \quad \underset{5s}{\underline{\ \ }}$$

$$\downarrow \text{ hybridize}$$

$$[\text{Co(NH}_3)_6]^{2+}\ [\text{Ar}] \quad \underset{3d_{xy}}{\underline{\uparrow\downarrow}}\ \underset{3d_{xz}}{\underline{\uparrow\downarrow}}\ \underset{3d_{yz}}{\underline{\uparrow\downarrow}} \quad \underset{d^2sp^3}{\underline{\text{xx}}\ \underline{\text{xx}}\ \underline{\text{xx}}\ \underline{\text{xx}}\ \underline{\text{xx}}\ \underline{\text{xx}}} \quad \underset{5s}{\underline{\uparrow}}$$

This structure for $[\text{Co(NH}_3)_6]^{2+}$ is consistent with the fact that $[\text{Co(NH}_3)_6]^{2+}$ is easily oxidized to $[\text{Co(NH}_3)_6]^{3+}$, an extremely stable complex ion.

The electronic configurations and hybridizations of some octahedral complexes are shown in Table 23–6.

2   Coordination Number 4

Most complexes with coordination number four are either tetrahedral or square planar. Tetrahedral geometry results from sp^3 hybridization. The IIB metals zinc, cadmium, and mercury in their +2 oxidation states are the B Group metals *commonly* involved in tetrahedral complexes. Since they all have d^{10} configurations, their inner *d* orbitals cannot participate in hybridization and it is not surprising that they exhibit

sp^3 hybridization. Tetraamminezinc ion, for instance, is diamagnetic and tetrahedral.

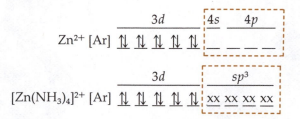

Since Cu^+ is also a d^{10} ion, this also represents the electronic configuration of diamagnetic, tetrahedral $[Cu(CN)_4]^{3-}$. (See p. 772).

TABLE 23–6 Bonding and Hybridization in Some Octahedral Complexes

Metal Ion	Outer Electron Configuration				Complex Ion	Type	Outer Electron Configuration		
	$3d$	$4s$	$4p$	$4d$			$3d$	d^2sp^3	$4d$
Co^{3+}	⇅ ↑ ↑ ↑ ↑	—	— — —	— — — —	$Co(NH_3)_6^{3+}$	inner; diamagnetic	⇅ ⇅ ⇅	xx xx xx xx xx xx	— — —
							$3d$	sp^3d^2	$4d$
					CoF_6^{3-}	outer; paramagnetic	↑ ↑ ↑ ↑ ↑	xx xx xx xx xx xx	— — —
	$3d$	$4s$	$4p$	$4d$			$3d$	sp^3d^2	$4d$
Co^{2+}	⇅ ⇅ ↑ ↑ ↑	—	— — —	— — — —	$Co(OH_2)_6^{2+}$	outer; paramagnetic	⇅ ⇅ ↑ ↑ ↑	xx xx xx xx xx xx	— — —
	$3d$	$4s$	$4p$	$4d$			$3d$	d^2sp^3	$4d$
Mn^{2+}	↑ ↑ ↑ ↑ ↑	—	— — —	— — — —	$Mn(CN)_6^{4-}$	inner; paramagnetic	⇅ ⇅ ↑	xx xx xx xx xx xx	— — — —
	$3d$	$4s$	$4p$	$4d$			$3d$	sp^3d^2	$4d$
Ni^{2+}	⇅ ⇅ ⇅ ↑ ↑	—	— — —	— — — —	$Ni(OH_2)_6^{2+}$	outer; paramagnetic	⇅ ⇅ ⇅ ↑ ↑	xx xx xx xx xx xx	— — —
	$3d$	$4s$	$4p$	$4d$			$3d$	d^2sp^3	$4d$
Cr^{3+}	↑ ↑ ↑ — —	—	— — —	— — — —	$Cr(NH_3)_6^{3+}$	inner; paramagnetic	↑ ↑ ↑	xx xx xx xx xx xx	— — — —
	$3d$	$4s$	$4p$	$4d$			$3d$	d^2sp^3	$4d$
Fe^{2+} and Co^{3+}	⇅ ↑ ↑ ↑ ↑	—	— — —	— — — —	$Fe(CN)_6^{4-}$ and $Co(CN)_6^{3-}$	inner; diamagnetic	⇅ ⇅ ⇅	xx xx xx xx xx xx	— — —

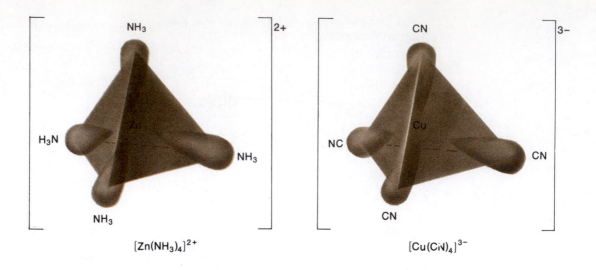

$$[Zn(NH_3)_4]^{2+}$$

$$[Cu(CN)_4]^{3-}$$

Although they are not as common, complexes of a few other transition metal ions are also tetrahedral. For example, $[NiCl_4]^{2-}$ is paramagnetic with two unpaired electrons, and is tetrahedral.

Most four-coordinate transition metal complexes are square planar and dsp^2 hybridized (utilizing the $d_{x^2-y^2}$ orbital) at the central metal ion. The metals most commonly forming square planar complexes are Cu(II), Au(III), Co(II), Ni(II), Pt(II), and Pd(II). For example, the tetracyanonickelate(II) ion, $[Ni(CN_4)^{2-}$, is square planar and diamagnetic.

We see that Valence Bond theory predicts that paramagnetic four-coordinate Ni(II) complexes are tetrahedral, whereas diamagnetic four-coordinate Ni(II) complexes are square planar.

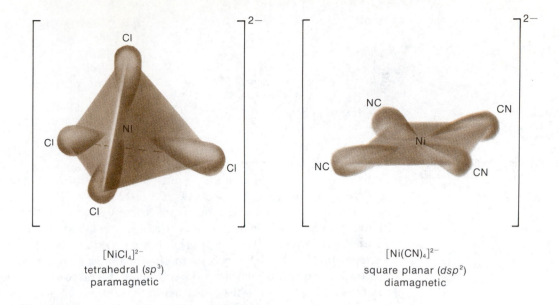

$[NiCl_4]^{2-}$
tetrahedral (sp^3)
paramagnetic

$[Ni(CN)_4]^{2-}$
square planar (dsp^2)
diamagnetic

However, the recent discovery of paramagnetic Ni(II) complexes that are square planar suggests hybridization involving the $4d_{x^2-y^2}$, $4p_x$, $4p_y$, and $4s$ orbitals to give sp^2d (outer orbital) hybrids at the Ni(II). This leaves two unpaired electrons in the $3d$ orbitals of Ni(II). Thus we see that, although Valence Bond theory accurately predicts most hybridizations from structures, it does have some flaws.

23–9 Crystal Field Theory

Hans Bethe and J. H. van Vleck originally developed the Crystal Field theory between 1919 and the early 1930s, but it was not widely used until the 1950s. In its pure form it assumes that the bonds between ligand and metal ion are completely ionic. Both ligand and metal ion are treated as infinitesimally small, nonpolarizable point charges. Recall that Valence Bond theory assumes complete covalence. A "relative" of Crystal Field theory, Ligand Field theory, attributes partial covalent and partial ionic character to the bonds.

1 Coordination Number 6

As we have seen, the $d_{x^2-y^2}$ and d_{z^2} orbitals are directed along a set of mutually perpendicular x, y, and z axes. As a group these orbitals are called the e_g orbitals. The d_{xy}, d_{yz}, and d_{xz} orbitals, or collectively the t_{2g} orbitals, bisect the axes. Since the ligand donor atoms approach the metal ion along the axes during the formation of octahedral complexes, there are greater repulsions between ligand electrons and metal ion electrons in the e_g orbitals than between ligand electrons and those in t_{2g} orbitals. Crystal Field theory proposes that the approach of the six donor atoms (point charges) along the axes sets up an electric field (the crystal field) that removes the degeneracy of the set of d orbitals and splits them into two sets, the t_{2g} set at lower energy and the e_g set at higher energy.

Recall that degenerate orbitals are orbitals of equal energy.

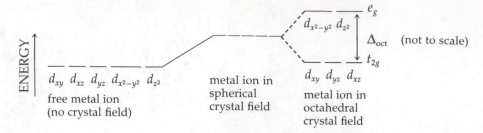

Δ_{oct} is sometimes called 10 *Dq*. Its typical values are between 100 and 400 kJ/mol.

The energy separation between the two sets, called $\Delta_{octahedral}$ or Δ_{oct}, is proportional to the *crystal field strength* of the ligands, that is, how strongly the ligand electrons repel the metal electrons. (See Section 23–10.)

The *d* electrons on the metal ion occupy the t_{2g} set in preference to the higher-energy e_g set. In fact, the t_{2g} orbitals are called **nonbonding orbitals** in octahedral complexes, but the e_g orbitals are called **antibonding orbitals** because electrons that are forced to occupy these orbitals are quite strongly repelled by the relatively close approach of ligand electrons, and tend to destabilize the octahedral complex. This will be discussed further in Section 23–11.

Let us now consider the cases of the hexafluorocobaltate(III) ion, $[CoF_6]^{3-}$, and the hexaamminecobalt(III) ion, $[Co(NH_3)_6]^{3+}$. Both contain a d^6 ion and already have been treated in terms of Valence Bond theory (Table 23–6). The former is a paramagnetic outer orbital complex; the latter is a diamagnetic inner orbital complex. We shall focus our attention on the *d* electrons only.

In the free ion, or in the absence of ligands, Co^{3+} has six electrons (four unpaired) in its $3d$ orbitals.

$$Co^{3+} \ [Ar] \ \underset{3d}{\underline{\uparrow\downarrow \ \uparrow \ \uparrow \ \uparrow \ \uparrow}}$$

Since its magnetic moment indicates that $[CoF_6]^{3-}$ also has four unpaired electrons per ion, the electrons must be arranged with four in t_{2g} orbitals and two in e_g orbitals.

This is a high spin complex; see p. 775.

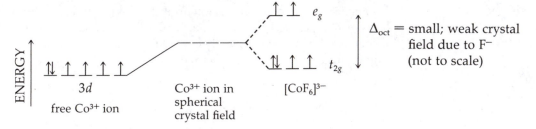

On the other hand, $[Co(NH_3)_6]^{3+}$ is diamagnetic, and all six *d* electrons must be paired in the t_{2g} orbitals.

This is a low-spin complex; see p. 775.

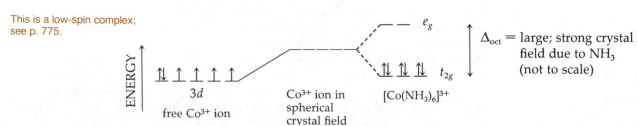

The difference in configurations is a consequence of the relative magnitudes of the crystal field splitting, Δ_{oct}, due to the different crystal field strengths of fluoride and ammonia ligands. The ammonia molecule more readily donates electrons into vacant metal orbitals than does the very electronegative fluoride ion, which holds its electrons very strongly. As a result the crystal field splitting generated by the close approach of six ammonia molecules to the metal ion is greater than that produced by the approach of six fluoride ions. That is,

$$\Delta_{oct} \text{ for } [Co(NH_3)_6]^{3+} > \Delta_{oct} \text{ for } [CoF_6]^{3-}$$

The crystal field splitting for $[CoF_6]^{3-}$ is so small that an energetically more favorable situation results if two electrons remain unpaired in the antibonding e_g orbitals rather than pairing with electrons in the lower energy, nonbonding t_{2g} orbitals. Recall that Hund's Rule requires that electrons singly occupy a set of degenerate orbitals before pairing in any one of the set, in order to avoid the unnecessary expenditure of energy in pairing electrons, or bringing two negatively charged particles into the same region of space. If the approach of a set of ligands removes the d orbital degeneracy but causes a Δ_{oct} less than the electron pairing energy, P, then the electrons will continue to occupy singly the resulting nondegenerate orbitals rather than becoming paired. After all d orbitals are half-filled, additional electrons will pair with electrons in the t_{2g} set. This is the case for $[CoF_6]^{3-}$, which is called a **high spin complex.**

The spin pairing energy is "high" compared to the crystal field splitting energy.

For $[CoF_6]^{3-}$: F^- is weak field ligand so $P > \Delta_{oct}$ for high speed complex

electron pairing energy

A high spin complex in the crystal field treatment corresponds to an outer orbital complex in the valence bond treatment.

In contrast, the $[Co(NH_3)_6]^{3+}$ ion is a **low spin complex.** The Δ_{oct} generated by the strong field ligand, ammonia, is greater than the electron pairing energy, so electrons become paired in t_{2g} orbitals before any occupy the antibonding e_g orbitals.

The spin pairing energy here is "low" compared to the crystal field splitting energy.

For $[Co(NH_3)_6]^{3+}$: NH_3 is strong field ligand so $\Delta_{oct} > P$ for low spin complex

A low spin complex corresponds to an inner orbital complex in the valence bond treatment.

Low spin configurations exist only for octahedral complexes having metal ions with d^4, d^5, d^6, and d^7 configurations. For d^1-d^3 and d^8-d^{10} ions only one possibility exists. In each case the configuration is designated as high spin. All d^n possibilities are shown in Table 23-7.

2 Coordination Number 4

Tetrahedral Complexes. Figure 23-6 shows that tetrahedrally distributed ligands are closer to and interact more strongly with the t_{2g} orbitals than with the e_g orbitals. Consequently, a tetrahedral environment splits a set of d orbitals so that the e_g orbitals are lower in energy than the t_{2g} orbitals, the reverse of the octahedral splitting.

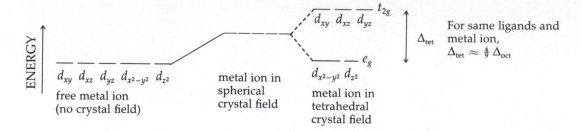

Because an octahedral splitting is caused by six ligands, while a tetrahedral splitting is caused by only four, the tetrahedral splitting caused by a given set of four ligands is smaller than the octahedral splitting caused by six of the same ligands. An approximate relationship is that $\Delta_{tet} \approx \frac{4}{9}\Delta_{oct}$. Since the tetrahedral splittings are so small, even strong field ligands like CN^- are unable to cause low spin tetrahedral complexes to form. That is, $\Delta_{tet} < P$. All tetrahedral complexes found to date are high spin.

TABLE 23–7 High and Low Spin Octahedral Configurations

d^n	Examples	High Spin	Low Spin
d^1	Ti^{3+}	$\underline{}\ \underline{}\ e_g$ $\underline{\uparrow}\ \underline{}\ \underline{}\ t_{2g}$	same as high spin
d^2	Ti^{2+}, V^{3+}, Zr^{2+}	$\underline{}\ \underline{}\ e_g$ $\underline{\uparrow}\ \underline{\uparrow}\ \underline{}\ t_{2g}$	same as high spin
d^3	V^{2+}, Cr^{3+}	$\underline{}\ \underline{}\ e_g$ $\underline{\uparrow}\ \underline{\uparrow}\ \underline{\uparrow}\ t_{2g}$	same as high spin
d^4	Mn^{3+}, Re^{3+}	$\underline{\uparrow}\ \underline{}\ e_g$ $\underline{\uparrow}\ \underline{\uparrow}\ \underline{\uparrow}\ t_{2g}$	$\underline{}\ \underline{}\ e_g$ $\underline{\uparrow\downarrow}\ \underline{\uparrow}\ \underline{\uparrow}\ t_{2g}$
d^5	Mn^{2+}, Fe^{3+}, Ru^{3+}	$\underline{\uparrow}\ \underline{\uparrow}\ e_g$ $\underline{\uparrow}\ \underline{\uparrow}\ \underline{\uparrow}\ t_{2g}$	$\underline{}\ \underline{}\ e_g$ $\underline{\uparrow\downarrow}\ \underline{\uparrow\downarrow}\ \underline{\uparrow}\ t_{2g}$
d^6	Fe^{2+}, Ru^{2+}, Pd^{4+}, Rh^{3+}, Co^{3+}	$\underline{\uparrow}\ \underline{\uparrow}\ e_g$ $\underline{\uparrow\downarrow}\ \underline{\uparrow}\ \underline{\uparrow}\ t_{2g}$	$\underline{}\ \underline{}\ e_g$ $\underline{\uparrow\downarrow}\ \underline{\uparrow\downarrow}\ \underline{\uparrow\downarrow}\ t_{2g}$
d^7	Co^{2+}, Rh^{2+}	$\underline{\uparrow}\ \underline{\uparrow}\ e_g$ $\underline{\uparrow\downarrow}\ \underline{\uparrow\downarrow}\ \underline{\uparrow}\ t_{2g}$	$\underline{\uparrow}\ \underline{}\ e_g$ $\underline{\uparrow\downarrow}\ \underline{\uparrow\downarrow}\ \underline{\uparrow\downarrow}\ t_{2g}$
d^8	Ni^{2+}, Pt^{2+}, Au^{3+}	$\underline{\uparrow}\ \underline{\uparrow}\ e_g$ $\underline{\uparrow\downarrow}\ \underline{\uparrow\downarrow}\ \underline{\uparrow\downarrow}\ t_{2g}$	same as high spin
d^9	Cu^{2+}	$\underline{\uparrow\downarrow}\ \underline{\uparrow}\ e_g$ $\underline{\uparrow\downarrow}\ \underline{\uparrow\downarrow}\ \underline{\uparrow\downarrow}\ t_{2g}$	same as high spin
d^{10}	Zn^{2+}, Ag^+, Hg^{2+}	$\underline{\uparrow\downarrow}\ \underline{\uparrow\downarrow}\ e_g$ $\underline{\uparrow\downarrow}\ \underline{\uparrow\downarrow}\ \underline{\uparrow\downarrow}\ t_{2g}$	same as high spin

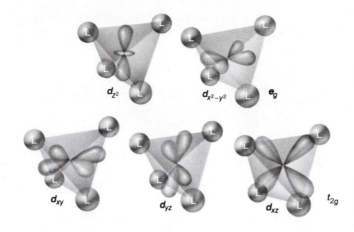

FIGURE 23–6 The five d orbitals in a tetrahedral environment.

Square Planar Complexes. Figure 23–7 shows the spatial relationship of a set of d orbitals to an environment of square planar ligands. Since the four ligands are located along the x and y axes, the $d_{x^2-y^2}$ orbital interacts most strongly with the ligand orbitals and the d_{z^2} the least strongly. Intermediate are the d_{zy}, d_{yz}, and d_{xz} orbitals; the one having both x and y components, the d_{xy} orbital, has the highest energy of the three. We may think of a square planar complex as derived from an

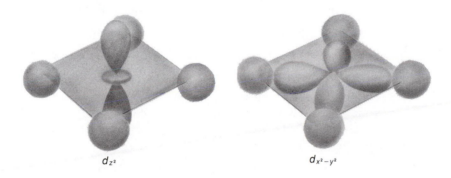

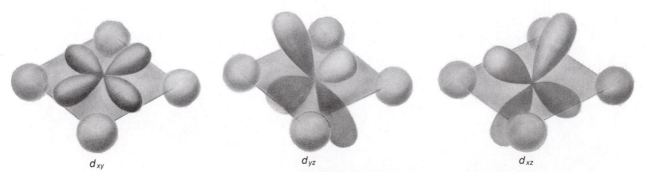

FIGURE 23–7 The five d orbitals in a square planar environment.

octahedral complex by removal of the two ligands along the z axis. The square planar splitting of orbitals is depicted below.

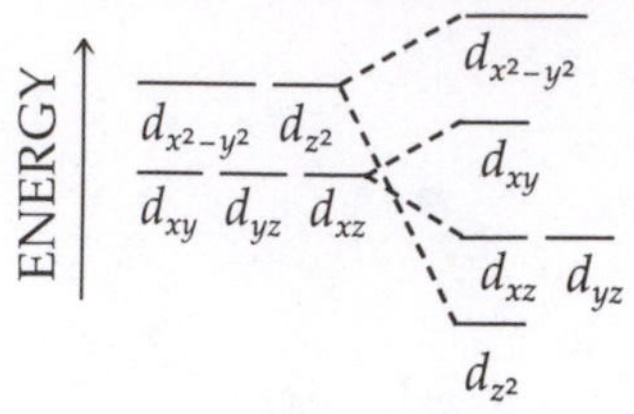

Octahedral $\rightarrow$ Square Planar

We would predict that square planar $[Cu(NH_3)_4]^{2+}$, a d^9 case, would have its only unpaired electron in the $d_{x^2-y^2}$ orbital, since it is highest in energy. Experimentally it is found that the ion is paramagnetic, with only one unpaired electron which is postulated to be in the $d_{x^2-y^2}$ orbital. Likewise $[Ni(CN)_4]^{2-}$, a d^8 case, is diamagnetic with a vacant $d_{x^2-y^2}$ orbital.

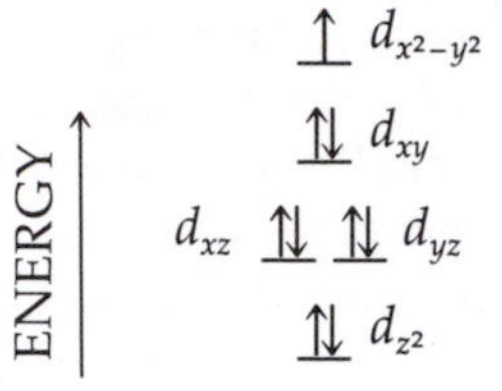

$[Cu(NH_3)_4]^{2+}$
square planar, d^9

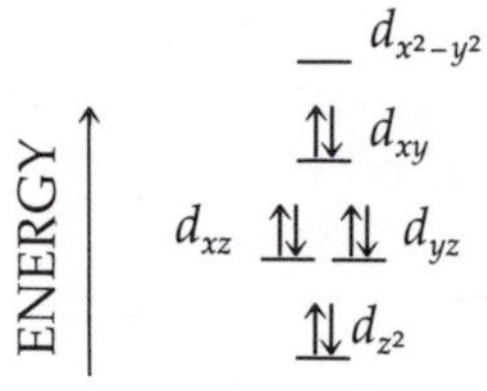

$[Ni(CN)_4]^{2-}$
square planar, d^8

23–10   Color and the Spectrochemical Series

The colors of transition metal complexes arise from the absorption of visible light as electrons undergo transitions from a t_{2g} orbital to an e_g orbital (or, in tetrahedral complexes, from e_g to t_{2g}). One transition of a high spin octahedral Co(III) complex is depicted below.

Planck's constant is $h = 6.63 \times 10^{-34}$ J · s

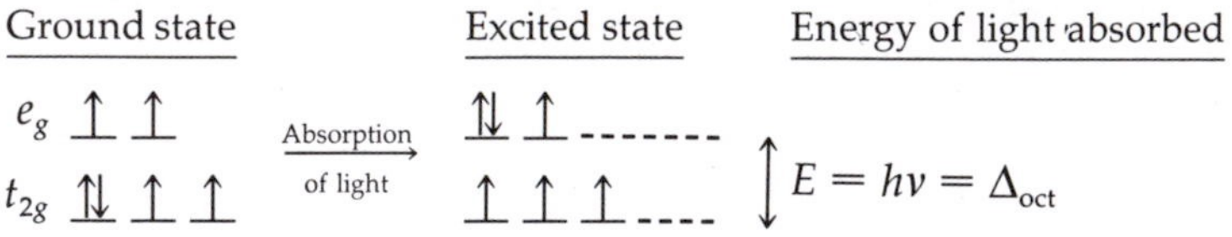

The frequency (ν), and therefore the wavelength and color, of the light absorbed is related to Δ_{oct}.* This, in turn, depends upon the crystal field strength of the ligands.

---

* The numerical relationship between Δ_{oct} and the wavelength, λ, of the absorbed light is found by combining the expressions $E = h\nu$ and $\nu = c/\lambda$, where c is the speed of light:

$$\Delta_{oct} = EN_A = \frac{hcN_A}{\lambda} \text{ where } N_A \text{ is Avogadro's number}$$

$$= \frac{(6.63 \times 10^{-34} \text{ J} \cdot \text{s})(3 \times 10^8 \text{ m} \cdot \text{s}^{-1})(6.02 \times 10^{23} \text{ mol}^{-1})}{(\lambda \text{ nm})(10^{-9} \text{ m/nm})(1000 \text{ J/kJ})}$$

$$= (1.20 \times 10^5 \text{ nm} \cdot \text{kJ/mol})/\lambda \text{ nm}$$

The approximate limits of the visible spectrum, 400 nm (violet) and 700 nm (red), thus correspond to Δ_{oct} values of 299 kJ/mol and 171 kJ/mol, respectively; the center of the visible spectrum at 550 nm (yellow) corresponds to a Δ_{oct} of 218 kJ/mol. All of these values are well within the ranges determined for common transition metal complexes of coordination number six.

TABLE 23–8 Colors of Some Chromium(III) Complexes

$[Cr(OH_2)_4Br_2]Br$	green	$[Cr(CON_2H_4)_6][SiF_6]_3$	green
$[Cr(OH_2)_6]Br_3$	bluish-gray	$[Cr(NH_3)_5Cl]Cl_2$	purple
$[Cr(OH_2)_4Cl_2]Cl$	green	$[Cr(NH_3)_4Cl_2]Cl$	violet
$[Cr(NH_3)_6]Cl_3$	yellow	$[Cr(OH_2)_4Cl_2]Cl$	green

So the colors and visible absorption spectra of transition metal complexes, as well as their magnetic properties, all yield information regarding the strengths of the ligand-metal interactions.

By interpreting the visible spectra of numerous complexes, it is possible to arrange ligands in the order of increasing crystal field strengths. Some common ligands are

$$I^- < Br^- < Cl^- < F^- < OH^- < H_2O < (COO)_2^{2-} < NH_3 < en < NO_2^- < CN^-$$

$\longrightarrow$ increasing crystal field strength

Such an arrangement is called a **spectrochemical series.** Strong field ligands, such as CN^-, usually produce low spin complexes where possible, and high crystal field splittings. Weak field ligands, like I^-, usually produce high spin complexes and small crystal field splittings. Low spin complexes generally absorb higher energy (shorter wavelength) light than do high spin complexes. The colors of several six-coordinate Cr(III) complexes are listed in Table 23–8.

In $[Cr(NH_3)_6]Cl_3$ the Cr(III) is bonded to six strong-field ammonia ligands, which produce a relatively high value of Δ_{oct} and cause the $[Cr(NH_3)_6]^{3+}$ ion to absorb relatively high-energy visible light in the blue and violet regions. The color that we see when we look at a sample of the compound (either in solid form or in solution) is the light that is *not* absorbed. Thus, we see a yellowish color, which is the complementary color of blue.

Water is a weaker field ligand than ammonia, and therefore Δ_{oct} is less for $[Cr(OH_2)_6]^{3+}$ than for $[Cr(NH_3)_6]^{3+}$. As a result, $[Cr(OH_2)_6]Br_3$ absorbs lower energy (longer wavelength) light. This causes the reflected and transmitted light to be higher-energy bluish-gray, the color we ascribed to $[Cr(OH_2)_6]Br_3$.

We see the light that is transmitted (passes through the sample) or that is reflected by the sample.

23–11 Crystal Field Stabilization Energy

We have seen that transition metals have a strong tendency to form stable coordination complexes, particularly those with coordination number six. One of the major factors contributing to the stability of complex ions relative to free, or uncomplexed, ions is the **Crystal Field Stabilization Energy** (CFSE) associated with complexes containing metal ions having certain electronic configurations. Here we will consider the CFSE's of various types of octahedral complexes. For a given set of six ligands, the t_{2g} and e_g orbitals are split by a certain amount of energy, Δ_{oct}. Each of the t_{2g} orbitals is $\frac{2}{5}\Delta_{oct}$ *below* the energy of the set of degenerate d orbitals in a spherical (homogeneous) field, and each of the e_g orbitals is $\frac{3}{5}\Delta_{oct}$ *above* the energy of the unsplit d orbitals.

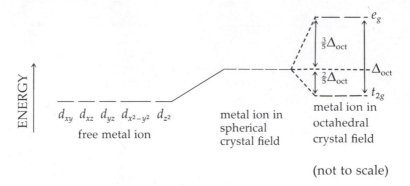

(not to scale)

There are three t_{2g} orbitals and two e_g orbitals, and the total energy of the sets of t_{2g} and e_g orbitals is the same as the energy of the set of degenerate d orbitals in a spherical field. That is, the change in total energy of the system as a result of the approach of the six ligands along the x-, y-, and z-axes is zero.

$$\Delta E = 3(-\tfrac{2}{5}\Delta_{oct}) + 2(\tfrac{3}{5}\Delta_{oct}) = 0$$
$$\quad\;\; (t_{2g}) \qquad\qquad (e_g)$$

But electrons occupying the lower energy (nonbonding) t_{2g} orbitals of an octahedral complex are lower in energy than they would be if they occupied orbitals of the metal ion in a spherical crystal field. Electrons occupying the (antibonding) e_g orbitals are higher in energy than if they occupied d orbitals of the metal ion in a spherical field. The lower the total energy of a system, the more stable it is. The CFSE of a complex is a measure of the net energy of stabilization (relative to the ion in a spherical field) of a metal ion's electrons as the complex forms.

Table 23–7 shows the electronic configurations of the common high spin and low spin octahedral complexes of d-transition metals. If an energy change of $-\frac{2}{5}\Delta_{oct}$ (stabilization) is assigned to each t_{2g} electron and an energy change of $+\frac{3}{5}\Delta_{oct}$ (destabilization) is assigned to each e_g electron, the CFSE is the algebraic sum of the energies of all the electrons in the metal atom or ion.

For example, for a high spin d^7 complex such as $[Co(OH_2)_6]^{2+}$, the CFSE is $5(-\frac{2}{5}\Delta_{oct}) + 2(+\frac{3}{5}\Delta_{oct}) = -\frac{4}{5}\Delta_{oct}$. That is, the *energy released* by the octahedral splitting of the occupied orbitals in formation of the complex ion is $\frac{4}{5}\Delta_{oct}$. The results of similar calculations for d^n configurations are given in Table 23–9. Those configurations with the most negative CFSE's are generally the ones for which large numbers of stable octahedral complexes are known. Note that no configuration can

TABLE 23–9   Crystal Field Stabilization Energies for Octahedral d^n Complexes

d^n	High Spin	Low Spin		d^n	High Spin	Low Spin
d^0	0	same as high spin				
d^1	$-\frac{2}{5}\Delta_{oct}$	same as high spin		d^6	$-\frac{2}{5}\Delta_{oct}$	$-\frac{12}{5}\Delta_{oct}$
d^2	$-\frac{4}{5}\Delta_{oct}$	same as high spin		d^7	$-\frac{4}{5}\Delta_{oct}$	$-\frac{9}{5}\Delta_{oct}$
d^3	$-\frac{6}{5}\Delta_{oct}$	same as high spin		d^8	$-\frac{6}{5}\Delta_{oct}$	same as high spin
d^4	$-\frac{3}{5}\Delta_{oct}$	$-\frac{8}{5}\Delta_{oct}$		d^9	$-\frac{3}{5}\Delta_{oct}$	same as high spin
d^5	0	$-\frac{10}{5}\Delta_{oct}$		d^{10}	0	same as high spin

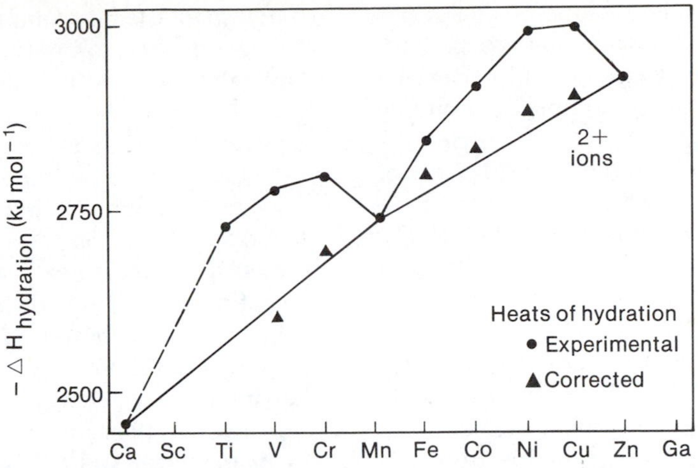

FIGURE 23-8 Heats of hydration of transition metal ions. When values of CFSE from spectroscopic Δ_{oct}'s are subtracted from experimental values of $\Delta H_{hydration}$ (·), a plot of the "corrected" values of $\Delta H_{hydration}$ (▲) versus atomic number is very nearly a straight line.

produce a CFSE greater than zero. That is, no d-transition metal ions should be less stable in an octahedral ligand environment than in a spherical crystal field.

Most magnetic and spectral properties of complex species of the d-transition elements can be accounted for by the Crystal Field theory, and this fact is strong evidence for the general validity of the theory. Consider the heats (enthalpies) of hydration for the series of 2+ ions shown in Figure 23–8. All of these ions form octahedral weak-field complexes with water $[M(OH_2)_6]^{2+}$. Recall that $\Delta H_{hydration}$ is the amount of energy absorbed when one mole of gaseous ions forms hydrated ions.

$$M^{n+} (g) + 6H_2O \longrightarrow [M(OH_2)_6]^{2+} (aq)$$

The experimental values for Ca^{2+} (d^0), Mn^{2+} (d^5) and Zn^{2+} (d^{10}) show no CFSE, because CFSE is zero for octahedral d^0, d^5, and d^{10} weak-field (high spin) complexes (see Table 23–9).

Key Terms

Ammine complexes complex species that contain ammonia molecules bonded to metal ions.

Chelate a ligand that uses two or more donor atoms in binding to metals.

***cis-trans* isomerism** a type of geometrical isomerism related to the angles between like ligands.

Coordinate covalent bond a covalent bond in which both shared electrons are donated by the same atom; a bond between a Lewis base and a Lewis acid.

Coordination compound or complex a compound containing coordinate covalent bonds.

Coordination isomers isomers involving exchange of ligands between complex cation and complex anion of the same compound.

Coordination number the number of donor atoms coordinated to a metal.

Coordination sphere the metal ion and its coordinated ligands but not any uncoordinated counter-ions.

Crystal field stabilization energy a measure of the net energy of stabilization gained by a metal ion's nonbonding d electrons as result of complex formation.

Crystal field theory theory of bonding in transition metal complexes in which ligands and metal ions are treated as point charges; a purely ionic model; ligand point charges represent the crystal (electrical) field perturbing the metal's d orbitals containing nonbonding electrons.

Δ_{oct} energy separation between e_g and t_{2g} sets of metal d orbitals caused by octahedral complexation of ligands; sometimes called $10\, Dq$.

Δ_{tet} energy separation between t_{2g} and e_g sets of metal d orbitals caused by tetrahedral complexation of ligands.

Dextrorotatory   refers to an optically active substance that rotates the plane of plane polarized light clockwise; also called dextro.

Donor atom   a ligand atom whose electrons are actually shared with a Lewis acid.

e_g **orbitals**   set of $d_{x^2-y^2}$ and d_{z^2} orbitals; those d orbitals within a set with lobes directed along the x, y, and z axes.

Enantiomers   optical isomers.

Geometrical isomers   stereoisomers that are not mirror images of each other; also known as position isomers.

High spin complex   crystal field designation for an outer orbital complex; no electrons are paired in either t_{2g} or e_g orbitals before all orbitals are singly occupied.

Hydrate isomers   isomers of crystalline complexes that differ in whether water is present inside or outside the coordination sphere.

Inner orbital complex   valence bond designation for a complex in which the metal ion utilizes d orbitals one shell inside the outermost occupied shell in its hybridization.

Ionization isomers   isomers that result from interchange of ions inside and outside the coordination sphere.

Isomers   different substances that have the same formula.

Levorotatory   refers to an optically active substance that rotates the plane of plane polarized light counterclockwise; also called levo.

Ligand   a Lewis base in a coordination compound.

Linkage isomers   isomers in which a particular ligand bonds to a metal ion through different donor atoms.

Low spin complex   crystal field designation for an

inner orbital complex; contains electrons paired in t_{2g} orbitals (octahedral complex).

Optical activity   the rotation of plane polarized light by one of a pair of optical isomers.

Optical isomers   stereoisomers that differ only by being nonsuperimposable mirror images of each other, like left and right hands; also called enantiomers.

Outer orbital complex   valence bond designation for a complex in which the metal ion utilizes d orbitals in the outermost (occupied) shell in hybridization.

Pairing energy   energy required to pair two electrons in the same orbital.

Plane polarized light   light waves in which all the electric vectors are oscillating in one plane.

Polarimeter   a device used to measure optical activity.

Polydentate   refers to ligands with more than one donor atom.

Racemic mixture   an equimolar mixture of dextro and levo optical isomers that is, therefore, optically inactive.

Spectrochemical series   arrangement of ligands in order of increasing ligand field strength.

Square planar complex   complex in which the metal is in the center of a square plane, with ligand donor atoms at each of the four corners.

Stereoisomers   isomers that differ only in the way that atoms are oriented in space; consist of geometrical and optical isomers.

Strong field ligand   ligand that exerts a strong crystal or ligand electrical field and generally forms low spin complexes with metals.

Structural isomers   (applied to coordination compounds) isomers whose differences involve more than a single coordination sphere or else different donor atoms; include ionization isomers, hydrate isomers, coordination isomers, and linkage isomers.

t_{2g} **orbitals**   set of d_{xy}, d_{yz}, and d_{xz} orbitals; those d orbitals within a set with lobes bisecting the x, y, and z axes.

Weak field ligand   ligand that exerts a weak crystal or ligand field and generally forms high spin complexes with metals.

Exercises

Basic Concepts

1. What property of transition metals allows them to form coordination complexes easily?

2. Suggest more appropriate designations for $NiSO_4 \cdot 6H_2O$, $Cu(NO_3)_2 \cdot 4NH_3$, and $Ni(NO_3)_2 \cdot 6NH_3$.

3. Describe the experiments of Alfred Werner on the compounds of the general formula $PtCl_4 \cdot nNH_3$ where $n = 2, 3, 4, 5, 6$.

4. For each of the compounds of Exercise 3, write formulas indicating the species within the coordination sphere. Also indicate the charges on the complex ions and provide a name for each compound.

5. Distinguish between the terms ligands, donor atoms, and chelates.

Naming Coordination Compounds

6. Give systematic names for the following compounds: (a) $[Ni(CO)_4]$, (b) $Na_2[Co(OH_2)_2(OH)_4]$, (c) $[Ag(NH_3)_2]Br$, (d) $[Cr(en)_3](NO_3)_3$, (e) $[Pt(NH_3)_4(NO_2)_2](NO_3)_2$, (f) $K_2[Cu(CN)_4]$.

7. Name the following compounds systematically:
   (a) $K_4[NiF_6]$
   (b) $[Mn(NH_3)_2(OH_2)_3(OH)]SO_4$
   (c) $[Mo(NCS)_2(en)_2]ClO_4$
   (d) $[Co(en)_2Cl_2]I$
   (e) $(NH_4)_3[CuCl_5]$
   (f) $[Co(NH_3)_5SO_4]NO_2$

8. Write formulas for the following compounds:
   (a) sodium tetracyanocadmate(II)
   (b) hexaamminecobalt(III) chloride
   (c) diaquadicyanocopper(II)
   (d) potassium hexachloropalladate(IV)
   (e) *cis*-diaquabis(ethylenediamine)cobalt(III) perchlorate
   (f) ammonium *trans*-dibromobis(oxalato)-chromate(III)

9. Write formulas for the following compounds:
   (a) *trans*-diamminedinitroplatinum(II) nitrate
   (b) rubidium tetracyanozincate
   (c) triaqua-*cis*-dibromochlorochromium(III)
   (d) pentacarbonyliron(0)
   (e) sodium pentacyanocobaltate(II)
   (f) hexaammineruthenium(III) tetrachloronickelate(II)

Ammine Complexes

10. Which of the following insoluble metal hydroxides will dissolve in an excess of aqueous ammonia? (a) $Zn(OH)_2$, (b) $Cr(OH)_3$, (c) $Fe(OH)_2$, (d) $Ni(OH)_2$, (e) $Cd(OH)_2$.

11. Write net ionic equations for the reactions of Exercise 10 that do occur in the dissolution process.

12. Write net ionic equations for reactions of solutions of the following transition metal salts in water with a *limited amount* of aqueous ammonia. (It is not necessary to show the ions as hydrated.) (a) $CuCl_2$, (b) $Zn(NO_3)_2$, (c) $Fe(NO_3)_3$, (d) $Hg(NO_3)_2$, (e) $MnCl_3$.

13. Write *net ionic* equations for the reactions of the insoluble products of Exercise 12 with an *excess* of aqueous ammonia, if such a reaction occurs.

Structures of Coordination Compounds

14. Write formulas and provide names for three complex cations in each of the following categories:
    (a) cations coordinated to only unidentate ligands
    (b) cations coordinated to only bidentate ligands
    (c) cations coordinated to two bidentate and two unidentate ligands
    (d) cations coordinated to one tridentate ligand, one bidentate ligand, and one unidentate ligand
    (e) cations coordinated to one tridentate ligand and three unidentate ligands

15. Provide formulas and names for three complex anions fitting each of the descriptions given in Exercise 14.

Isomerism in Coordination Compounds

16. Distinguish between structural isomers and stereoisomers.

17. Distinguish between an optically active compound and a racemic mixture.

18. Write the formula for a potential ionization isomer of each of the following compounds. Name each one. (a) $[Cr(NH_3)_4I_2]Br$, (b) $[Ni(en)_2(NO_2)_2]Cl_2$, (c) $[Fe(NH_3)_5CN]SO_4$.

19. Write the formula for a potential hydrate isomer of each of the following compounds.

Name each one. (a) $[Cu(OH_2)_4]Cl_2$, (b) $[Ni(OH_2)_5Br]Br \cdot H_2O$.

20. Write the formula for a potential coordination isomer of each of the following compounds. Name each one. (a) $[Co(NH_3)_6][Cr(CN)_6]$, (b) $[Ni(en)_3][Cu(CN)_4]$.

21. Write the formula for a potential linkage isomer for each of the following compounds. Name each one. (a) $[Co(en)_2(NO_2)_2]Cl_2$, (b) $[Cr(NH_3)_5(CN)](CN)_2$.

22. How many geometrical isomers (or forms) can exist, in theory, for the following species? Sketch and name each form.
    (a) $[Pt(NH_3)_2Cl_2]$ (square planar)
    (b) $[Pt(NH_3)Cl_3]$ (square planar)
    (c) $[Zn(NH_3)_2Cl_2]$ (tetrahedral)

23. The ion $[Co(en)_2Cl_2]^+$ exists as two geometrical isomers, one of which occurs as two optical isomers. Sketch all and name the geometrical forms. Label each according to the above description.

24. How many *geometrical* and *optical* isomers can exist, in theory, for the following species? Sketch and name them all.
    (a) $[Co(NH_3)_2Cl_4]^-$
    (b) $[Co(NH_3)_3Cl_3]$
    (c) $[Co(NH_3)(en)Cl_3]$ (The N atoms of en can bond only at adjacent positions.)

25. How many *geometrical* and *optical* isomers can exist, in theory, for the following species?
    (a) $[CoCl_2Br_2I_2]^{4-}$
    (b) $[Cr(ox)_3]^{3-}$, in which ox = oxalato,

. Represent ox as $O \frown O^{2-}$.

Valence Bond Theory

26. Give the hybridizations and sketch the structures of each of the complexes in Exercise 6. It is not necessary to distinguish between d^2sp^3 and sp^3d^2 hybridization. $[Ni(CO)_4]$ is tetrahedral.

27. Give the hybridizations and sketch the structures of each of the complexes in Exercise 8. It is not necessary to distinguish between d^2sp^3 and sp^3d^2 hybridization.

28. Why are the d_{z^2} and $d_{x^2-y^2}$ orbitals used in d^2sp^3 and sp^3d^2 hybridization rather than two of the d_{xy}, d_{yz}, and d_{xz} orbitals?

29. On the basis of the spectrochemical series, determine whether the following complexes are inner orbital or outer orbital, and diamagnetic or paramagnetic: (a) $[Cu(OH_2)_6]^{2+}$, (b) $[MnF_6]^{3-}$, (c) $[Co(CN)_6]^{3-}$, (d) $[Cr(NH_3)_6]^{3+}$, (e) $[CrCl_4Br_2]^{3-}$, (f) $[Co(en)_3]^{3+}$, (g) $[Fe(NO_2)_6]^{3-}$.

30. Draw diagrams showing outer electron configurations and hybridizations at the metal ions for each of the complexes in Exercise 29.

31. Draw diagrams showing outer electron configurations and hybridizations for the following four-coordinate complexes: (a) $[Cd(NH_3)_4]^{2+}$, (b) $[Cu(NH_3)_4]^{2+}$, (c) $[Zn(CN)_4]^{2-}$, (d) $[Pt(NH_3)_4]^{2+}$.

Crystal Field Theory

32. (a) Describe clearly what Δ is. (b) Why is Δ for a tetrahedral ligand field always less than Δ for an octahedral field? (c) How is Δ actually measured experimentally? (d) How is it related to the spectrochemical series?

33. Describe the relationship among Δ, the electron pairing energy, and whether a complex is high spin or low spin. Illustrate the relationship with Fe^{2+} in strong and weak octahedral fields.

34. On the basis of the spectrochemical series, determine whether the complexes of Exercise 29 are low spin or high spin complexes.

35. Write out the electron distributions in t_{2g} and e_g orbitals for the following ions in an octahedral field.

Metal Ions	Ligand Field Strength
V^{2+}	weak
Mn^{2+}	strong
Mn^{2+}	weak
Ni^{2+}	weak
Cu^{2+}	weak
Fe^{3+}	strong
Cu^+	weak
Ru^{3+}	strong

36. Write formulas for two complex ions that would fit into each of the categories of Exercise 35. Name the complex ions you list.

37. What is crystal field stabilization energy?

38. Given the following spectrophotometrically measured values of Δ_{oct}, calculate the CFSE's for the ions. Remember to determine first whether the complex is high spin or low spin.

Complex Ion	Δ_{oct}*
(a) $[Co(NH_3)_6]^{3+}$	22,900 cm^{-1}
(b) $[Ti(OH_2)_6]^{2+}$	20,300 cm^{-1}
(c) $[Cr(OH_2)_6]^{3+}$	17,600 cm^{-1}
(d) $[Co(CN)_6]^{3-}$	33,500 cm^{-1}
(e) $[Co(OH_2)_6]^{2+}$	10,000 cm^{-1}
(f) $[Cr(en)_3]^{3+}$	21,900 cm^{-1}
(g) $[Cu(OH_2)_6]^{2+}$	13,000 cm^{-1}
(h) $[V(OH_2)_6]^{3+}$	18,000 cm^{-1}

* The cm^{-1} is an energy unit; 1 cm^{-1} = 11.96 J/mol.

Equilibria Involving Coordination Compounds

39. Calculate (a) the molar solubility of $Zn(OH)_2$ in pure water, (b) the molar solubility of $Zn(OH)_2$ in 0.10 M NaOH solution, and (c) the concentration of $[Zn(OH)_4]^{2-}$ ions in the solution of (b). K_{sp} for $Zn(OH)_2$ is 4.5×10^{-17} and K_d for $[Zn(OH)_4]^{2-}$ is 3.5×10^{-16}.

40. Calculate the pH of a solution prepared by dissolving 0.10 mol of tetraamminecopper(II) chloride, $[Cu(NH_3)_4]Cl_2$, in enough water to make 1.0 L of solution. K_d for $[Cu(NH_3)_4]^{2+}$ is 8.5×10^{-13} and K_b for aqueous NH_3 is 1.8×10^{-5}. Ignore hydrolysis of Cu^{2+}.

41. Would a solution of 0.10 mol of tetraamminecopper(II) sulfate, $[Cu(NH_3)_4]SO_4$, in 1.0 L of aqueous solution have higher, lower, or the same pH as the solution of Exercise 40? Why? (No calculation is necessary.)

Nuclear Chemistry

24

Chemical properties are determined by electronic distributions and are only indirectly influenced by atomic nuclei. In previous chapters we have been concerned with ordinary chemical reactions and have focused attention on electronic configurations. Nuclear reactions involve changes in the composition of nuclei, and are extraordinary in the sense that they are often accompanied by the release of tremendous amounts of energy and by transmutations of elements. Some of the characteristics that differentiate between nuclear reactions and ordinary chemical reactions are summarized at the top of p. 787.

Nuclear Reaction	Ordinary Chemical Reaction
1. Elements may be converted from one to another.	1. No new elements can be produced.
2. Particles within the nucleus are involved.	2. Usually only outermost electrons participate.
3. Often accompanied by release or absorption of tremendous amounts of energy.	3. Accompanied by release or absorption of relatively small amounts of energy.
4. Rate of reaction is independent of factors such as concentration, temperature, pressure, and catalyst.	4. Rate of reaction is influenced by external factors.

Many ancient alchemists spent their lives attempting to convert other heavy metals into gold without success. Years of failure and the acceptance of Dalton's atomic theory early in the nineteenth century convinced the scientific community that elements are not interconvertible. However, Henri Becquerel's discovery of "radioactive rays" emanating from a uranium compound in 1896, and the subsequent determination of the nature of radioactive rays by Ernest Rutherford, showed that atoms of one element may indeed be converted into atoms of other elements by spontaneous nuclear disintegrations, or **natural radioactivity.** (As shown many years later, deliberate nuclear reactions initiated by bombardment of nuclei with accelerated subatomic particles or other nuclei can also transform one element into another.)

The stimulus of Becquerel's discovery led other researchers, including Marie and Pierre Curie, to discover and study new radioactive elements. Many radioactive isotopes, or **radiosotopes,** now have important medical, agricultural, and industrial uses.

Nuclear fission is the splitting of a heavy nucleus into lighter nuclei. **Nuclear fusion** is the combination of light nuclei to produce a heavier nucleus. The huge amounts of energy released per unit mass of nuclear fuels give controlled nuclear fission (Sections 24–12, 24–14) great promise for supplying a large portion of our future energy demands. Research is currently aimed at surmounting some of the technological problems associated with safe and efficient use of nuclear fission reactors and with the development of continuously controlled fusion reactors.

24–1 The Nucleus

The nucleus comprises only a minute fraction of the total volume of an atom, yet nearly all the mass of an atom resides in the nucleus. Thus nuclei are extremely dense. It has been experimentally shown that nuclei of all elements have approximately the same density, 2.44×10^{14} g/cm^3.

From an electrostatic point of view it is amazing that positively charged protons (and uncharged neutrons) can be packed so closely together. Yet nonradioactive nuclei do not spontaneously decompose, so they must be stable. In the early twentieth century, when Rutherford postulated the nuclear model of the atom, scientists were puzzled by such a situation. In attempts to explain the phenomenon, physicists have since detected many very short-lived subatomic particles (in addition to protons, neutrons, and electrons) as products of nuclear reactions. Well over a hundred have been identified, and several more are found every year. Their

If enough nuclei were gathered together to occupy one cubic centimeter, the total mass would be about 250 million tonnes!

Working out the details of nuclear structure has become a fascinating and very complex problem.

TABLE 24–1 Abundances of Naturally Occurring Nuclides

protons	even	even	odd	odd
neutrons	even	odd	even	odd
number of nuclides	157	52	50	5

functions are not entirely understood, but it is now thought that they play important roles in overcoming the coulombic proton-proton repulsions and in binding nuclear particles (**nucleons**) together. The attractive forces among nucleons appear to operate over only extremely small distances, about 10^{-13} cm.

24–2 Neutron-Proton Ratio and Nuclear Stability

The term "isotope" applies only to different forms of the same element. The term "nuclide" is used to refer to different atomic forms of all elements.

Recall that the superscript in this notation gives the mass number, and the subscript gives the atomic number (Section 4–7).

Most naturally occurring **nuclides** have even numbers of protons and even numbers of neutrons; 157 nuclides fall into this category. Nuclides with odd numbers of both are least common (there are only five), and those with odd-even combinations are intermediate in abundance (Table 24–1).

In addition to the even-even preference, there seem to be certain "magic numbers" of protons and neutrons that impart exceptional stability to nuclides, reminiscent of the 2, 8, 18, and 32 electrons that result in very stable filled electronic shells. Nuclides with a number of protons *or* a number of neutrons, *or* a sum of these two equal to 2, 8, 20, 28, 50, 82 or 126 have unusual stability. Examples are $^4_2$He, $^{16}_8$O, $^{40}_{20}$Ca, $^{88}_{38}$Sr, and $^{208}_{82}$Pb. This suggests a shell model for the nucleus similar to the shell model of electronic configurations.

Figure 24–1 shows a plot of the number of neutrons (N) versus number of protons (Z) for the stable nuclides. For low atomic numbers, below about 20, the

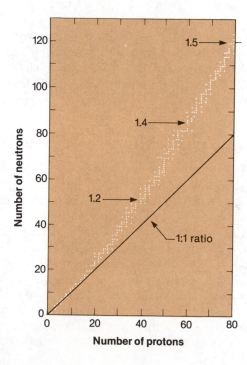

FIGURE 24–1 A plot of the number of neutrons versus the number of protons in stable nuclei. As the atomic number increases, the neutron-to-proton ratio of the stable nuclei increases. The stable nuclei are located in an area of the graph known as the band of stability. The majority of radioactive nuclei occur outside this band.

most stable nuclides have equal numbers of protons and neutrons ($N = Z$). Above atomic number 20 the most stable nuclides have more neutrons than protons. Each dot in the band represents a stable nuclide. Careful examination reveals a vague stepwise shape associated with the plot, owing to the stability of nuclides with even numbers of nucleons. There are no stable nuclides with Z greater than 82.

24–3 Radioactive Decay

Nuclei whose neutron-to-proton ratios lie outside the stable region undergo spontaneous radioactive decay by emitting one or more particles and/or electromagnetic rays. The type of decay that occurs usually depends upon whether the nucleus is above, below, or to the right of the band of stability. The most common types of radiation emitted in decay processes and their properties are summarized in Table 24–2.

The particles can be emitted at different kinetic energies, equal to the energy equivalent of the mass loss of the products relative to reactants (see discussion of mass defect and binding energy in Section 4–9) minus the energy associated with subsequently emitted gamma rays. Frequently, radioactive decay leaves a nucleus in an excited (high-energy) state. In such cases the decay is accompanied by subsequent gamma-ray emission in which the energy of the gamma ray ($h\nu$) is equal to the energy difference between the ground and excited nuclear states. This is analogous to the emissions of lower-energy electromagnetic radiation that occur when atoms in excited electronic states return to their ground states (Section 5–3). Determina-

> Recall that the energy of electromagnetic radiation is $E = h\nu$, where h is Planck's constant and ν is the frequency.

TABLE 24–2 Common Types of Radioactive Emissions

Type and Symbol*	Identity	Relative Mass (amu)	Charge	Velocity	Penetration
beta (β^-, $_{-1}^{0}\beta$, $_{-1}^{0}e$)	electron	0.00055	1−	≤90% of speed of light	low to moderate, depending on energy
positron† ($_{+1}^{0}\beta$, $_{+1}^{0}e$)	positively charged electron	0.00055	1+	≤90% of speed of light	low to moderate, depending on energy
alpha (α, $_{2}^{4}\alpha$, $_{2}^{4}He$)	helium nucleus	4.0026	2+	≤10% of speed of light	low
proton ($_{1}^{1}p$, $_{1}^{1}H$)	proton, hydrogen nucleus	1.0073	1+	≤10% of speed of light	low to moderate, depending on energy
neutron ($_{0}^{1}n$)	neutron	1.0087	0	≤10% of speed of light	very high
gamma ($_{0}^{0}\gamma$) ray	high-energy electromagnetic radiation like x-rays	0	0	speed of light	high

* The number at the upper left of the symbol is the number of nucleons, and the number at the lower left is the charge.
† On the average, a positron exists for only about a nanosecond (1×10^{-9} second) before colliding with an electron and being converted into the corresponding amount of energy.

tions of the energies of gamma radiation from decay processes strongly suggest that nuclear energy levels are quantized just as are electronic energy levels, adding further support for a shell model for the nucleus.

$$_Z^M E^* \longrightarrow _Z^M E + _0^0 \gamma$$

excited nucleus

The penetrating abilities of the particles or rays are proportional to their energies. Beta particles and positrons are about 100 times more penetrating than the heavier and slower-moving alpha particles. They can be stopped by a one-eighth-inch (0.3-cm) thick aluminum plate. They can burn skin severely but cannot reach internal organs. Alpha particles have low penetrating ability and cannot damage or penetrate skin. However, they can damage sensitive internal tissue if inhaled. The high-energy gamma rays have great penetrating power and severely damage both skin and internal organs. They travel at the speed of light and can be stopped only by thick layers of concrete or lead.

24–4   Nuclei Above the Band of Stability

Nuclei in this region have too high a ratio of neutrons to protons, and they undergo decays that decrease the ratio. The most common methods of accomplishing this are **beta emission** or, less commonly, **neutron emission.** A beta particle is really an electron ejected from the nucleus as a neutron is converted into a proton.

$$_0^1 n \longrightarrow _1^1 p + _{-1}^0 \beta$$

Beta emission results in a simultaneous increase in the number of protons and decrease in the number of neutrons by one. Examples of beta particle emission are

$$_{88}^{228}\text{Ra} \longrightarrow _{89}^{228}\text{Ac} + _{-1}^0 \beta$$

$$_{90}^{234}\text{Th} \longrightarrow _{91}^{234}\text{Pa} + _{-1}^0 \beta$$

$$_6^{14}\text{C} \longrightarrow _7^{14}\text{N} + _{-1}^0 \beta$$

Neutron emissions simply decrease the number of neutrons by one, without changing the atomic number. A lighter isotope is formed.

$$_{53}^{137}\text{I} \longrightarrow _{53}^{136}\text{I} + _0^1 n$$

$$_7^{17}\text{N} \longrightarrow _7^{16}\text{N} + _0^1 n$$

Note that in all equations for nuclear reactions,

1. the sum of the mass numbers (total number of nucleons) of the reactants equals the corresponding sum for the products, and
2. the sum of the atomic numbers (positive nuclear charges) of the reactants equals the corresponding sum for the products.

24–5   Nuclei Below the Band of Stability

These nuclei can increase their neutron-to-proton ratios by undergoing **positron emission** or **electron capture** (K capture).

A positron has the mass of an electron but a positive charge. Positrons are emitted as protons are converted to neutrons.

$$^1_1p \longrightarrow {^1_0}n + {^{\;0}_{+1}}\beta$$

By transforming a proton into a neutron, positron emission results in a decrease in atomic number and an increase in the number of neutrons by one, with no change in mass number.

$$^{38}_{19}K \longrightarrow {^{38}_{18}}Ar + {^{\;0}_{+1}}\beta$$

$$^{15}_{8}O \longrightarrow {^{15}_{7}}N + {^{\;0}_{+1}}\beta$$

The same effect can be accomplished by electron capture (*K* capture), in which an electron from the *K* shell ($n = 1$) is captured by the nucleus.

$$^{106}_{47}Ag + {^{\;0}_{-1}}e \longrightarrow {^{106}_{46}}Pd$$

$$^{37}_{18}Ar + {^{\;0}_{-1}}e \longrightarrow {^{37}_{17}}Cl$$

Some nuclides undergo both positron emission and electron capture. Sodium-22 is an example.

$$^{22}_{11}Na + {^{\;0}_{-1}}e \longrightarrow {^{22}_{10}}Ne \qquad (3\%)$$

$$^{22}_{11}Na \longrightarrow {^{22}_{10}}Ne + {^{\;0}_{+1}}\beta \qquad (97\%)$$

In addition, some nuclei, especially heavier ones, undergo **alpha emission.** Alpha particles are actually helium nuclei, ^4_2He, two protons, and two neutrons. They carry a double positive charge, but charge is usually not shown on atomic species in nuclear reactions. An example is the alpha emission of lead-204. Of course, this also results in an increase of the neutron-to-proton ratio.

$$^{204}_{82}Pb \longrightarrow {^{200}_{80}}Hg + {^4_2}\alpha$$

24-6   Nuclei with Atomic Number Greater than 82

All nuclides with atomic number greater than 82 are radioactive and are located to the right of the band of stability. Many such nuclides decay by alpha emission.

$$^{226}_{88}Ra \longrightarrow {^{222}_{86}}Rn + {^4_2}\alpha$$

$$^{210}_{84}Po \longrightarrow {^{206}_{82}}Pb + {^4_2}\alpha$$

The first of the decays shown, that of radium-226, was originally reported in 1902 by Rutherford and Soddy and was the first transmutation of an element ever observed. Heavy nuclides also decay less commonly by beta emission, positron emission, and electron capture.

Some of the isotopes of uranium ($Z = 92$) and elements of higher atomic number, the **transuranium elements,** also disintegrate via spontaneous nuclear fission, in which a heavy nuclide splits into nuclides of intermediate masses and neutrons.

$$^{252}_{98}Cf \longrightarrow {^{142}_{56}}Ba + {^{106}_{42}}Mo + 4\,{^1_0}n$$

24–7   Detection of Radiations

1   Photographic Detection

Emanations from radioactive substances affect photographic plates just as ordinary visible light does. Becquerel's discovery of radioactivity resulted from the unexpected exposure of a photographic plate wrapped in black paper by a nearby enclosed sample of a uranium-containing compound, potassium uranyl sulfate. After developing and fixing, the darkness of the spot is related to the amount of radiation that struck the plate, but detection of radiation on a quantitative basis by this method is difficult and tedious.

2   Detection by Fluorescence

The phenomenon of fluorescence can be used for other purposes. A familiar application is in fluorescent clock dials. A typical dial is coated with fluorescent zinc sulfide, ZnS, containing about one part per 100,000 of radioactive emitter, such as tritium. The constantly decaying tritium excites the ZnS so that it glows continuously.

This method is based upon the ability of **fluorescent** substances to absorb high-energy radiation such as gamma rays and subsequently emit visible light. As the gamma radiation is absorbed, the absorbing atoms jump to excited electronic states. In returning to their ground states, the excited electrons go through a series of transitions, at least some of which result in the emission of light of visible wavelengths. This method may be used for the quantitative detection of radiation, using an instrument called a **scintillation counter.**

3   Cloud Chambers

Vapors other than water can be used to saturate the air.

The original cloud chamber was devised by C. T. R. Wilson in 1911. The chamber contains air saturated with water vapor. The withdrawal of a piston expands and cools the air, causing droplets of water vapor to condense on particles emitted from a radioactive substance. The paths of the particles, which in some ways reflect their properties, can be followed by observing the fog-like tracks produced. The tracks may be photographed and studied in detail. Figures 24–2 and 24–3 show a cloud chamber and a cloud chamber photograph.

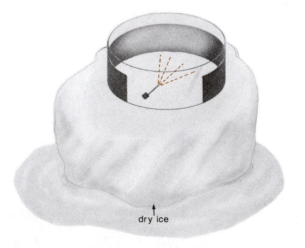

FIGURE 24–2 A cloud chamber. The emitter is glued onto a pin stuck into a stopper that is mounted onto the chamber wall. The chamber has some methyl alcohol in the bottom and rests on Dry Ice. The cool air near the bottom becomes supersaturated with methyl alcohol vapor. When an emission speeds through this supersaturated vapor, ions are produced that serve as ''seeds'' about which the vapor condenses, forming tiny droplets, or fog.

dry ice

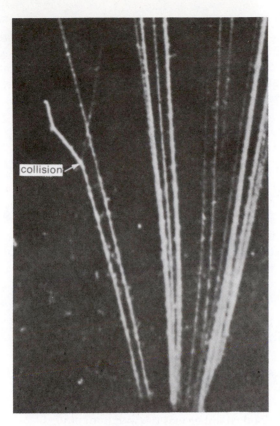

FIGURE 24–3 A historic cloud chamber photograph of alpha tracks in nitrogen gas. The forked track was shown to be due to a speeding proton (going off to the left) and an isotope of oxygen (going off to the right). It is assumed that the alpha particle struck the nucleus of a nitrogen atom at the point where the track forks.

4 Gas Ionization Counters

A common gas ionization counter is the **Geiger-Müller counter,** which is depicted in Figure 24–4. Radiation enters the tube through a thin window. Windows of different thicknesses or materials can be used to let only radiation of certain penetrating powers pass through. The tube is filled with a gas, usually argon, some molecules of which become ionized when struck by the radiation. A potential

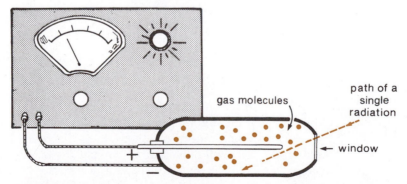

FIGURE 24–4 Principle of operation of a gas ionization counter. The center wire is positively charged, and the shell of the tube is negatively charged. When radiation enters through the window, it ionizes one or more gas atoms. The electrons are attracted to the central wire, and the positive ions are drawn to the shell. This constitutes a pulse of electric current, which is amplified and displayed on the meter or other readout.

difference of about 1200 volts is applied between the inner wire and the tube shell, which causes a current of electrons to flow toward the (+) wire and Ar^+ ions to flow toward the (−) tube shell. Amplification of the pulse from each ionization causes a clicking sound in a loudspeaker, or activates a digital counter that registers the amount of radiation entering the tube. A chart recorder may also be attached to the circuit to give a graphic display of the intensity of radiation versus time.

24 – 8   Rates of Decay and Half-Life

The rates of all radioactive decays are independent of temperature, and obey first-order kinetics. All radionuclides have different stabilities and decay at different rates. Some samples decay nearly completely in fractions of a second, and others only after millions of years or more. In Section 15 – 5 we found that *first-order* processes occur at rates proportional only to the concentration of one reacting substance.

$$\text{Rate of decay} = k[A] \qquad \text{[Eq. 24 – 1]}$$

The 2.303 is a conversion factor from natural to base-10 logarithms.

$$\log\left(\frac{A_0}{A}\right) = \frac{kt}{2.303} \quad \text{(integrated rate-law expression)} \qquad \text{[Eq. 24 – 2]}$$

Here A represents the amount of decaying radionuclide of interest remaining at some time t, and A_0 is the amount of the nuclide present at the beginning of the decay process. The k is the specific rate constant, which is different for each radioactive nuclide. If N represents the number of disintegrations per unit time, a similar relationship holds:

$$\log\left(\frac{N_0}{N}\right) = \frac{kt}{2.303} \qquad \text{[Eq. 24 – 3]}$$

Equations 24 – 2 and 24 – 4 are identical to Equations 15 – 5 and 15 – 7, which apply to chemical kinetics, except that molar concentrations are used in chemical kinetics.

We also found that the half-life, $t_{1/2}$, for a first order reaction, which is the amount of time required for half the original reacting (decaying) sample to react (time after which $A = \frac{1}{2}A_0$ or $N = \frac{1}{2}N_0$), is given by the equation

$$t_{1/2} = \frac{0.693}{k} \qquad \text{[Eq. 24 – 4]}$$

In 1963 a treaty was signed by the United States, the Soviet Union, and the United Kingdom prohibiting the further testing of nuclear weapons in the atmosphere. Since then, strontium-90 has been disappearing from the air, water, and soil according to the curve in Figure 24 – 5, so the treaty has accomplished its aim to the present time (1983).

Strontium-90 is an isotope that was introduced into the atmosphere by the atmospheric testing of nuclear weapons. Because of the chemical similarity of strontium to calcium, it now occurs with calcium in measurable quantities in milk, bones, and teeth as a result of its presence in food and water supplies. It is a radionuclide that undergoes beta emission with a half-life of 28 years, and it may cause leukemia, bone cancer, or other related disorders. If we begin with a 16 μg sample of $^{90}_{38}Sr$, 8 μg will remain after one half-life of 28 years. After 56 years 4 μg remain; after 84 years, 2 μg; after 112 years 1 μg; after 140 years 0.5 μg remain, and so on. The situation is depicted graphically in Figure 24 – 5.

Similar plots for other radionuclides all show the same shape of **exponential decay curve.** Approximately ten half-lives must pass for radionuclides to lose 99.9% of their radioactivity, or 280 years for $^{90}_{38}Sr$.

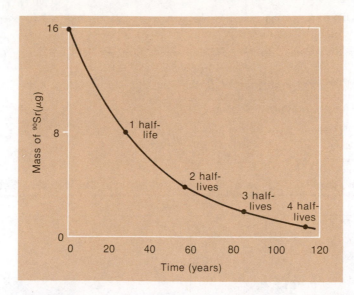

FIGURE 24–5 The decay of a 16.0μg sample of $^{90}_{38}$Sr.

Example 24–1

The "cobalt treatments" used in medicine to arrest certain types of cancer rely on the ability of gamma rays to destroy cancerous tissues. Cobalt-60 decays with the emission of beta particles and gamma rays, with a half-life of 5.27 years. How much of a 3.42 μg sample of cobalt-60 remains after 30 years?

$$^{60}_{27}\text{Co} \longrightarrow {}^{60}_{28}\text{Ni} + {}^{0}_{-1}\beta + {}^{0}_{0}\gamma$$

Solution

We first determine the value of the specific rate constant, using Eq. 24–4.

$$t_{1/2} = \frac{0.693}{k}$$

$$k = \frac{0.693}{t_{1/2}} = \frac{0.693}{5.27 \text{ yr}} = 0.131 \text{ yr}^{-1}$$

This value can now be used to determine the ratio of A_0 to A after 30 years, using Eq. 24–2.

$$\log\left(\frac{A_0}{A}\right) = \frac{kt}{2.303} = \frac{0.131 \text{ yr}^{-1} (30 \text{ yr})}{2.303} = 1.71$$

Taking the antilog of both sides,

$$\frac{A_0}{A} = \text{antilog } 1.71 = 10^{1.71} = 51.3$$

Since $A_0 = 3.42$ μg,

$$A = \frac{A_0}{51.3} = \frac{3.42 \text{ }\mu\text{g}}{51.3}$$

$$= \underline{0.0667 \text{ }\mu\text{g } {}^{60}_{27}\text{Co remains after 30 years}}$$

24–9 Disintegration Series

Many radionuclides cannot attain nuclear stability by only one nuclear reaction, so they decay in a series of disintegrations. A few such series are known to occur in nature. Two begin with isotopes of uranium, ^{238}U and ^{235}U, and one begins with ^{232}Th. All end with a stable isotope of lead (Z = 82). Uranium-238 decays by alpha emission to thorium-234 in the first step of one series. Thorium-234 subsequently emits a beta particle to produce protactinium-234 in the second step of the series, which can be summarized as shown. Emitted particles are shown over the arrows.

There are also decay series of varying lengths starting with some of the man-made radionuclides (Section 24–11).

$$^{238}_{92}U \xrightarrow{\alpha} {}^{234}_{90}Th \xrightarrow{\beta} {}^{234}_{91}Pa \xrightarrow{\beta} {}^{234}_{92}U \xrightarrow{\alpha} {}^{230}_{90}Th \xrightarrow{\alpha} {}^{226}_{88}Ra \xrightarrow{\alpha}$$

$$^{222}_{86}Rn \xrightarrow{\alpha} {}^{218}_{84}Po \xrightarrow{\alpha} {}^{214}_{82}Pb \xrightarrow{\beta} {}^{214}_{83}Bi \xrightarrow{\beta} {}^{214}_{84}Po \xrightarrow{\alpha}$$

$$^{210}_{82}Pb \xrightarrow{\beta} {}^{210}_{83}Bi \xrightarrow{\beta} {}^{210}_{84}Po \xrightarrow{\alpha} {}^{206}_{82}Pb$$

The net reaction is

$$^{238}_{92}U \longrightarrow {}^{206}_{82}Pb + 8\,{}^{4}_{2}He + 6\,{}^{0}_{-1}\beta$$

TABLE 24–3   Emissions and Half-Lives of Members of Natural Radioactive Series*

^{238}U Series	^{235}U Series	^{232}Th Series
$^{238}_{92}U \to \alpha$ 4.51 × 10⁹ y	$^{235}_{92}U \to \alpha$ 7.1 × 10⁸ y	$^{232}_{90}Th \to \alpha$ 1.41 × 10¹⁰ y
$^{234}_{90}Th \to \beta$ 24.1 d	$^{231}_{90}Th \to \beta$ 25.5 h	$^{228}_{88}Ra \to \beta$ 6.7 y
$^{234}_{91}Pa \to \beta$ 6.75 h	$^{231}_{91}Pa \to \alpha$ 3.25 × 10⁴ y	$^{228}_{89}Ac \to \beta$ 6.13 h
$^{234}_{92}U \to \alpha$ 2.47 × 10⁵ y	$\beta \leftarrow {}^{227}_{89}Ac \to \alpha$ 98.8% 21.6 y	$^{228}_{90}Th \to \alpha$ 1.91 y
$^{230}_{90}Th \to \alpha$ 8.0 × 10⁴ y	$\alpha \leftarrow {}^{227}_{90}Th \qquad {}^{223}_{87}Fr \to \beta$ 18.2 d 22 m	$^{224}_{88}Ra \to \alpha$ 3.64 d
$^{226}_{88}Ra \to \alpha$ 1.60 × 10³ y	$^{223}_{88}Ra \to \alpha$ 11.4 d	$^{220}_{86}Rn \to \alpha$ 55.3 s
$^{222}_{86}Rn \to \alpha$ 3.82 d	$^{219}_{86}Rn \to \alpha$ 4.00 s	$\beta \leftarrow {}^{216}_{84}Po \to \alpha$ 0.014% 0.14 s
$\beta \leftarrow {}^{218}_{84}Po \to \alpha$ 0.04% 3.05 m	$\beta \leftarrow {}^{215}_{84}Po \to \alpha$ 5 × 10⁻⁴% 1.78 × 10⁻³ s	$\alpha \leftarrow {}^{216}_{85}At \qquad {}^{212}_{82}Pb \to \beta$ 3 × 10⁻⁴ s 10.6 h
$\alpha \leftarrow {}^{218}_{85}At \qquad {}^{214}_{82}Pb \to \beta$ 2 s 26.8 m	$\alpha \leftarrow {}^{215}_{85}At \qquad {}^{211}_{82}Pb \to \beta$ 10⁻⁴ s 36.1 m	$\beta \leftarrow {}^{212}_{83}Bi \to \alpha$ 66.3% 60.6 m
$\beta \leftarrow {}^{214}_{83}Bi \to \alpha$ 99.96% 19.7 m	$\beta \leftarrow {}^{211}_{83}Bi \to \alpha$ 99.7% 2.16 m	$\alpha \leftarrow {}^{212}_{84}Po \qquad {}^{208}_{81}Tl \to \beta$ 3.0 × 10⁻⁷ s 3.10 m
$\alpha \leftarrow {}^{214}_{84}Po \qquad {}^{210}_{81}Tl \to \beta$ 1.6 × 10⁻⁴ s 1.32 m	$\alpha \leftarrow {}^{211}_{84}Po \qquad {}^{207}_{81}Tl \to \beta$ 0.52 s 4.79 m	$^{208}_{82}Pb$
$^{210}_{82}Pb \to \beta$ 20.4 y	$^{207}_{82}Pb$	
$\beta \leftarrow {}^{210}_{83}Bi \to \alpha$ ~100% 5.01 d		
$\alpha \leftarrow {}^{210}_{84}Po \qquad {}^{206}_{81}Tl \to \beta$ 138 d 4.19 m		
$^{206}_{82}Pb$		

* The abbreviations are y, year; d, day; m, minute; and s, second.

"Branchings" are possible at various points in the chain. That is, two successive decays may be replaced by alternate decays, but they always result in the same final product. Only the most prevalent decays were given above for the uranium-238 series. Table 24-3 outlines in detail the ^{238}U, ^{235}U, and ^{232}Th disintegration series, showing half-lives and alternate pathways. The less prevalent decays are denoted by broken arrows. For any particular decay, the decaying nuclide is called the **mother** nuclide and the product nuclide is the **daughter.**

24-10   Uses of Radionuclides

Radionuclides find many practical applications either because they decay at known rates or, in some cases, simply because they emit radiation continuously.

1   Radioactive Dating

The ages of articles of organic origin can be estimated by **radiocarbon dating.** The radioisotope carbon-14 is produced continuously in the upper atmosphere, as nitrogen atoms capture cosmic-ray neutrons.

$$^{14}_{7}N + ^{1}_{0}n \longrightarrow ^{14}_{6}C + ^{1}_{1}H$$

The carbon-14 nuclei react with oxygen molecules to form $^{14}CO_2$. This process continually supplies the atmosphere with radioactive $^{14}CO_2$, which is removed from the atmosphere by photosynthesis. Cosmic-ray intensity is related to the sun's activity. As long as cosmic-ray intensity remains constant, the amount of $^{14}CO_2$ in the atmosphere remains constant. Since it is incorporated into living plants and organisms just as ordinary $^{12}CO_2$ is, a certain fraction of all carbon atoms in living substances is carbon-14, which decays via beta emission with a half-life of 5730 years.

In recent decades, atmospheric testing of nuclear warheads has also caused fluctuations in the natural abundance of ^{14}C.

$$^{14}_{6}C \longrightarrow ^{14}_{7}N + ^{0}_{-1}\beta$$

After death the plant no longer participates in photosynthesis and no longer takes up $^{14}CO_2$. Other organisms that consume plants for food stop doing so at death. Thus, the emissions from the ^{14}C in dead tissue decrease with the passage of time, and the activity per gram of carbon is a measure of the length of time elapsed since death. Comparison of ages of ancient trees calculated from carbon-14 activity with those determined by counting rings indicates that cosmic ray intensity has varied somewhat throughout history. As a result, the calculated ages must be corrected for these variations, but this can be done. The carbon-14 technique is useful only for dating objects less than 50,000 years old. Older objects possess too little activity to be dated accurately.

Methods used for dating older objects are the **potassium-argon** and **uranium-lead methods.** Potassium-40 is present in all living organisms. It decays by electron capture to argon-40, with a half-life of 1.3 billion years.

$$^{40}_{19}K + ^{0}_{-1}e \longrightarrow ^{40}_{18}Ar$$

Gaseous argon is easily lost from minerals, and therefore measurements based on the $^{40}K/^{40}Ar$ method may or may not be reliable.

Because of its longer half-life, potassium-40 can be used to date objects up to 1 million years old by determination of the ratio of $^{40}_{19}K$ to $^{40}_{18}Ar$ in the sample. The uranium-lead method is based on the natural uranium-238 decay series, which ends with the production of stable lead-206. This method can be used for dating uranium-containing minerals several billion years old. All the lead-206 in such minerals is *assumed* to have come from uranium-238. Because of the very long half-life of $^{238}_{92}U$, 4.5 billion years, the amounts of intermediate nuclei can be neglected. The oldest object found to date is a meteorite 4.6 billion years old, which fell in Mexico in 1969. Results of $^{238}U/^{206}Pb$ studies support the "big bang" theory, which indicates that our solar system was born about 6 billion years ago.

Example 24–2

A sample of uranium ore is found to contain 4.64 mg of ^{238}U and 1.22 mg of ^{206}Pb. Estimate the age of the ore. The half-life of ^{238}U is 4.51×10^9 years.

Solution

First we calculate the amount of ^{238}U that must have decayed to produce 1.22 mg of ^{206}Pb, using the isotopic masses.

$$\underline{?} \text{ mg } ^{238}U = 1.22 \text{ mg } ^{206}Pb \times \frac{238 \text{ mg } ^{238}U}{206 \text{ mg } ^{206}Pb}$$

$$= 1.41 \text{ mg } ^{238}U$$

Thus the sample must have originally contained 4.64 mg + 1.41 mg = 6.05 mg of ^{238}U.

We must next use Eq. 24–4 to evaluate the specific rate (disintegration) constant, k, which will be used in the final calculation.

$$t_{1/2} = \frac{0.693}{k}$$

$$k = \frac{0.693}{t_{1/2}} = \frac{0.693}{4.51 \times 10^9 \text{ yr}}$$

$$= 1.54 \times 10^{-10} \text{ yr}^{-1}$$

Now we calculate the age of the sample, t, from Eq. 24–2.

$$\log\left(\frac{A_0}{A}\right) = \frac{kt}{2.303}$$

$$\log\left(\frac{6.05 \text{ mg}}{4.64 \text{ mg}}\right) = \frac{(1.54 \times 10^{-10} \text{ yr}^{-1})t}{2.303}$$

$$\log 1.30 = (6.69 \times 10^{-11} \text{ yr}^{-1})t$$

$$\frac{0.114}{6.69 \times 10^{-11} \text{ yr}^{-1}} = t$$

$$t = \underline{1.70 \times 10^9 \text{ years}}$$

The ore is approximately 1.7 billion years old.

---

2   Medical Uses of Radionuclides

The use of cobalt radiation treatments for cancerous tumors has already been described in Example 24–1. Several other nuclides are used as **radioactive tracers** in medicine. Since radioisotopes of an element have the same chemical properties as stable isotopes of the same element, they can be used to "label" an element in compounds. A radiation detector can be used to follow the path of the element throughout the body. Salt solutions containing ^{24}Na can be injected into the bloodstream in order to follow the flow of blood and locate obstructions in the circulatory system. Thallium-201 tends to concentrate in healthy heart muscle tissue, while technetium-99 concentrates in abnormal heart tissue. The two can be used in a complementary fashion in surveying the extent of damage from heart disease. Iodine-131 concentrates in the thyroid gland, the liver, and certain parts of the brain, and is used to monitor goiter and other thyroid problems, as well as liver and brain tumors. The heat generated by the decay of plutonium-238 can be converted into electrical energy in heart pacemakers. The relatively long half-life of the isotope allows the device to be used for ten years before replacement.

3   Research Applications for Radionuclides

The courses of organic, inorganic, and biochemical reactions can also be investigated using radioactive tracers. If radioactive $^{35}S^{2-}$ ion is added to a saturated solution of cobalt sulfide in equilibrium with solid cobalt sulfide, the solid becomes radioactive. This is taken as evidence that sulfide ion exchange occurs between solid and solution in the solubility equilibrium

$$CoS \text{ (s)} \rightleftharpoons Co^{2+} \text{ (aq)} + S^{2-} \text{ (aq)}$$

Photosynthesis is the process by which the carbon atoms of carbon dioxide are incorporated into glucose, $C_6H_{12}O_6$, in the presence of chlorophyll in green plants during daylight hours.

$$6CO_2 + 6H_2O \xrightarrow[\text{chlorophyll}]{\text{sunlight}} C_6H_{12}O_6 + 6O_2$$

The process is not really as simple as the net equation implies; it actually occurs in a series of steps producing a number of intermediate products. By studying the reaction with labeled $^{14}CO_2$, the intermediate molecules that are produced can be identified as the ones containing the radioactive ^{14}C atoms.

4 Agricultural Uses of Radionuclides

DDT is a polychlorinated hydrocarbon whose use as a pesticide has been criticized because of the possible toxicity to humans and animals associated with continuous exposure to it. It decomposes only very slowly, persisting in the environment for a long time, and it tends to concentrate in fatty tissues.

The use of DDT in controlling the screw-worm fly has been replaced by a radiologic technique. The irradiation of the male flies with gamma rays alters their reproductive cells so as to render them sterile. When large numbers of sterilized males are released in an infested area, they mate with females who produce no offspring, resulting in reduction and eventual disappearance of the population. The procedure works because females of this species mate only once. When an area is highly populated with sterile males, the probability of a "productive" mating is very small.

Labeled samples of fertilizers can also be used to investigate nutrient uptake by plants, and to study the growth of crops. Gamma irradiation of some foods allows them to be stored for longer periods. For example, it retards the sprouting of potatoes and onions.

24-11 Artificial Transmutations of Elements (Nuclear Bombardment)

The first artificially induced nuclear reaction was performed by Rutherford in 1915. He bombarded nitrogen-14 with alpha particles to produce an isotope of oxygen and a proton.

$$^{14}_{7}N + ^{4}_{2}He \longrightarrow ^{1}_{1}H + ^{17}_{8}O$$

Such reactions are often indicated in abbreviated form, with the bombarding particle and emitted subsidiary particles shown parenthetically between the mother and daughter nuclides.

$$^{14}_{7}N \, (^{4}_{2}\alpha, \, ^{1}_{1}p) \, ^{17}_{8}O$$

Several thousand artificially induced reactions have been carried out with such bombarding projectiles as neutrons, protons, deuterons ($^{2}_{1}H$), alpha particles, and other small nuclei.

1   Bombardment with Positive Ions

A problem arises with the use of positively charged nuclei as projectiles. In order to cause a nuclear reaction, they must actually collide with the target nuclei, which are also positively charged. Coulombic repulsion prevents collisions unless the projectiles have sufficient kinetic energy to overcome it. The required kinetic energies become greater with increasing atomic numbers of the target and of the bombarding particle. Of course, this problem does not arise when neutrons are used.

Particle accelerators called **cyclotrons** (atom smashers) and **linear accelerators** have surmounted the problem of repulsion. A cyclotron is shown in Figure 24-6. It consists of two hollow D-shaped electrodes called "dees," both of which are in an evacuated enclosure between the poles of an electromagnet. The particles to be accelerated are introduced at the center in the gap between the dees. The dees are connected to a source of high-frequency alternating current, which keeps them oppositely charged. At first, the positively charged particles are attracted toward the negative electrode. As the charge is reversed on the dees, the particles are repelled by the first dee (now positive) and attracted to the other. As this process repeats itself in synchronism with the motion of the particles, they accelerate along a spiral path and eventually emerge through an exit hole oriented in such a way that the beam of particles hits the target atoms (Figure 24-7).

In a linear accelerator, the particles are accelerated through a series of tubes within an evacuated chamber (Figure 24-8). The odd-numbered tubes are at first negatively charged, and the even ones positively charged. The positively charged particle is attracted toward the first tube. As it passes through the tube, the charges on the tubes are reversed so that the particle is repelled out of the first tube (now positive) and toward the second (negative) tube. As the particle nears the end of the second tube, the charges are again reversed, and as this process is repeated the particle is accelerated to very high velocities. Since the polarity is changed at

The first cyclotron was constructed by E. O. Lawrence and M. S. Livingston at the University of California in 1930.

The path of the particle is initially circular because of the interaction of the particle's charge with the electromagnet's field; as the particle gains energy, the path widens into a spiral.

The first linear accelerator was built in 1928 by a German physicist, Rolf Wideroe.

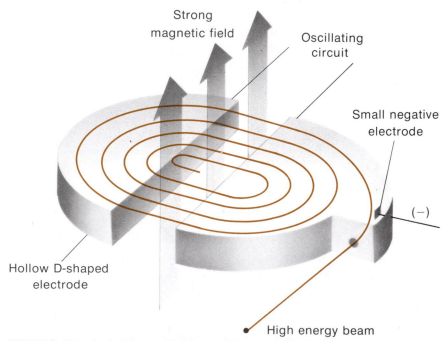

FIGURE 24-6 Schematic representation of a cyclotron.

FIGURE 24–7 A beam of deuterons (bright area) from a cyclotron at the University of California. Nuclear reactions take place when deuterons and other atomic particles strike the nuclei of atoms.

constant frequency, the lengths of subsequent tubes must be increased to accommodate the increased distance traveled by the accelerating particle per unit time. The bombardment target is located outside the last tube. If the initial polarities are reversed, negatively charged particles can also be accelerated. The longest linear accelerator is approximately 2 miles long and was completed in 1966 at Stanford University. It is capable of accelerating electrons to energies of nearly 20 GeV.

Many nuclear reactions have been induced by bombardment with positively charged particles. At the time of development of particle accelerators there were a few gaps among the first 92 elements in the periodic table. Particle accelerators were used between 1937 and 1941 to synthesize three of the four "missing" elements: number 43 (technetium), number 85 (astatine), and number 87 (francium).

One gigaelectron volt (GeV) $= 1 \times 10^9$ eV $= 1.60 \times 10^{-10}$ J. This is sometimes called 1 billion electron volts (BeV) in the United States.

$$^{96}_{42}\text{Mo} + {}^{2}_{1}\text{H} \longrightarrow {}^{97}_{43}\text{Tc} + {}^{1}_{0}n$$

$$^{209}_{83}\text{Bi} + {}^{4}_{2}\text{He} \longrightarrow {}^{210}_{85}\text{At} + 3\,{}^{1}_{0}n$$

$$^{230}_{90}\text{Th} + {}^{1}_{1}\text{H} \longrightarrow {}^{223}_{87}\text{Fr} + 2\,{}^{4}_{2}\text{He}$$

Many previously unknown and unstable isotopes of known elements were also synthesized so that their nuclear structures and behavior could be studied.

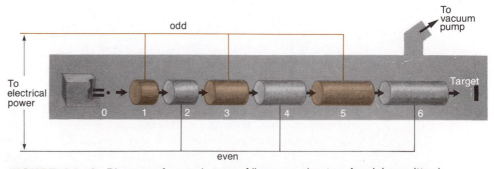

FIGURE 24–8 Diagram of an early type of linear accelerator. An alpha emitter is placed in the container at the left; only those alpha particles can escape that happen to be emitted in line with the series of accelerating tubes.

2   Neutron Bombardment

Since neutrons bear no charge, they are not repelled by nuclei as are positively charged projectiles, and so need not be accelerated to produce bombardment reactions. Neutrons can be generated in several ways. A frequently used method involves bombardment of beryllium-9 with alpha particles.

$$^{9}_{4}\text{Be} + ^{4}_{2}\text{He} \longrightarrow ^{12}_{6}\text{C} + ^{1}_{0}n$$

Nuclear reactors (Section 24–14) are also used as neutron sources. Neutrons ejected in nuclear reactions usually possess high kinetic energies and are called **fast neutrons.** When they are used as projectiles they cause reactions, such as (n, p) or (n, α) reactions, in which subsidiary particles are ejected. **Slow neutrons** are produced when fast neutrons collide with **moderators** such as hydrogen, deuterium, oxygen, or the carbon atoms in paraffin, and these neutrons are more likely to be captured by the target nuclei. Bombardments with slow neutrons cause mostly neutron-capture (n, γ) reactions.

$$^{200}_{80}\text{Hg} + ^{1}_{0}n \longrightarrow ^{201}_{80}\text{Hg} + ^{0}_{0}\gamma$$

The fourth "missing" element, number 61 (promethium), was synthesized by fast neutron bombardment of neodymium-142.

$$^{142}_{60}\text{Nd} + ^{1}_{0}n \longrightarrow ^{143}_{61}\text{Pm} + ^{0}_{-1}\beta$$

E. M. McMillan discovered the first transuranium element, neptunium, in 1940 by bombardment of uranium-238 with slow neutrons.

$$^{238}_{92}\text{U} + ^{1}_{0}n \longrightarrow ^{239}_{92}\text{U} + ^{0}_{0}\gamma$$

$$^{239}_{92}\text{U} \longrightarrow ^{239}_{93}\text{Np} + ^{0}_{-1}\beta$$

More than a dozen additional elements have since been prepared by neutron bombardment or by bombardment of nuclei so produced with positively charged particles. Some examples are

$$\left. \begin{array}{l} ^{238}_{92}\text{U} + ^{1}_{0}n \longrightarrow ^{239}_{92}\text{U} + ^{0}_{0}\gamma \\[4pt] ^{239}_{92}\text{U} \longrightarrow ^{239}_{93}\text{Np} + ^{0}_{-1}\beta \\[4pt] ^{239}_{93}\text{Np} \longrightarrow ^{239}_{94}\text{Pu} + ^{0}_{-1}\beta \end{array} \right\} \text{plutonium}$$

$$^{239}_{94}\text{Pu} + ^{4}_{2}\text{He} \longrightarrow ^{242}_{96}\text{Cm} + ^{1}_{0}n \qquad \text{curium}$$

$$^{246}_{96}\text{Cm} + ^{12}_{6}\text{C} \longrightarrow ^{254}_{102}\text{No} + 4\,^{1}_{0}n \qquad \text{nobelium}$$

Fast neutrons are moving so rapidly that they are likely to pass right through a target nucleus without reacting. Hence the probability of a reaction is low, even though the reactions may be very energetic.

24–12   Nuclear Fission and Fusion

Isotopes of some elements with atomic numbers above 80 are capable of undergoing fission, in which they split into nuclei of intermediate masses and emit one or more neutrons. Some fissions are spontaneous; others require that the activation energy be supplied by bombardment. A given nucleus can split in many different ways, liberating enormous amounts of energy. Some of the possible fissions resulting from bombardment of fissionable uranium-235 with fast neutrons are shown at the top of p. 803. The uranium-236 is a short-lived intermediate.

$$\begin{array}{c}
{}^{235}_{92}\text{U} + {}^{1}_{0}n \longrightarrow [{}^{236}_{92}\text{U}]
\begin{cases}
\nearrow {}^{160}_{62}\text{Sm} + {}^{72}_{30}\text{Zn} + 4\,{}^{1}_{0}n + \text{energy} \\
\nearrow {}^{146}_{57}\text{La} + {}^{87}_{35}\text{Br} + 3\,{}^{1}_{0}n + \text{energy} \\
\rightarrow {}^{140}_{56}\text{Ba} + {}^{93}_{36}\text{Kr} + 3\,{}^{1}_{0}n + \text{energy} \\
\searrow {}^{144}_{55}\text{Cs} + {}^{90}_{37}\text{Rb} + 2\,{}^{1}_{0}n + \text{energy} \\
\searrow {}^{144}_{54}\text{Xe} + {}^{90}_{38}\text{Sr} + 2\,{}^{1}_{0}n + \text{energy}
\end{cases}
\end{array}$$

In dealing with atomic structure (Section 4–9) we discussed mass defects and binding energies. Recall that the binding energy of an atom is the amount of energy released in forming the atom from its constituent subatomic particles, and is related by the Einstein equation, $E = mc^2$, to its mass defect, m. We could say that the binding energy represents the amount of energy that must be supplied to the atom to break it apart into subatomic particles.

Reference to Figure 24–9, a plot of binding energy per nucleon versus mass number, shows that the highest binding energies per nucleon, and therefore the greatest stabilities, are associated with atoms of intermediate mass number. The most stable atom is ${}^{56}_{26}\text{Fe}$, with a binding energy per nucleon of 8.80 MeV. Thus, fission is an energetically favorable process for heavy atoms, since atoms with intermediate masses and greater binding energies per nucleon are formed.

Which isotopes of which elements undergo fission? Experiments with particle accelerators have revealed that all elements with an atomic number of 80 or more have one or more isotopes capable of undergoing fission, provided they are bombarded at the right energy. Nuclei with atomic numbers between 89 and 98 fission spontaneously with long half-lives: 10^{17} yr to 10^4 yr. Nuclei with atomic numbers of 98 or more fission with shorter half-lives: from 60.5 days to a few milliseconds. One of the *natural* decay modes of the transuranium elements is by spontaneous fission. In fact, nuclides with mass numbers greater than 250 do this because they are just too big to be stable. Nuclides with mass numbers between 225 and 250 will not undergo fission spontaneously (except for a few with very long half-lives) but can be induced to do so when bombarded with particles of relatively low kinetic energies.

The term "nucleon" refers to a nuclear particle, either a neutron or a proton.

1 MeV = 1.60×10^{-13} J

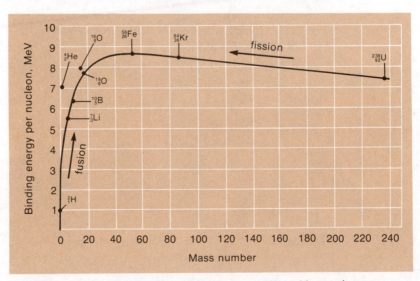

FIGURE 24–9 Variation in nuclear binding energy with atomic mass.

Particles that can supply the appropriate activation energy include slow (and fast) neutrons, protons, and alpha particles, as well as fast electrons. As we proceed toward nuclei lighter than mass number 225, the activation energy required to induce fission rises very rapidly.

In Section 24–2 we discussed the preference of nuclei to have even numbers of protons and even numbers of neutrons. Thus, when we consider the fission of nuclides with mass numbers of 225 to 250 (the isotopes of uranium, for example), we should not be surprised to learn that both ^{233}U and ^{235}U can be excited to a fissionable state by slow neutrons much more easily than ^{238}U, since they are less stable. In fact, it is so difficult to cause fission in ^{238}U that this isotope is said to be "nonfissionable."

Fusion, the joining of light nuclei to form heavier nuclei, is favorable for the very light atoms. In both fission and fusion the energy liberated is due to the loss of mass that accompanies the reactions. Much larger amounts of energy per unit mass of reacting atoms are produced in fusion than in fission.

Spectroscopic evidence indicates that the sun is a tremendous fusion reactor consisting of 73% hydrogen, 26% helium, and 1% other elements. The major fusion reaction that occurs is thought to involve the combination of a deuteron, $^{2}_{1}$H, and a triton, $^{3}_{1}$H, at tremendously high temperatures to form a helium nucleus and a neutron.

$$^{2}_{1}\text{H} + ^{3}_{1}\text{H} \longrightarrow ^{4}_{2}\text{He} + ^{1}_{0}n + \text{energy}$$

Thus solar energy is actually a form of fusion energy, which is, in fact, the only unlimited kind of energy available to us. Even our fossil fuels are just "left over" fusion energy.

The deuteron and triton are the nuclei of two isotopes of hydrogen, called deuterium and tritium. Deuterium occurs naturally in water. When the D_2O is purified as "heavy water," it can be used for several types of chemical analysis.

24–13   The Atomic Bomb (Fission)

On the average, two or three neutrons are produced per fission reaction. If sufficient fissionable material, the **critical mass,** is contained in a small enough volume, an uncontrolled explosive chain reaction can result as the neutrons ejected from the initial fission collide with other fissionable atoms to sustain and expand the process. If too few fissionable atoms are present, most of the neutrons escape the mass and no chain reaction occurs. Figure 24–10 depicts a fission chain reaction.

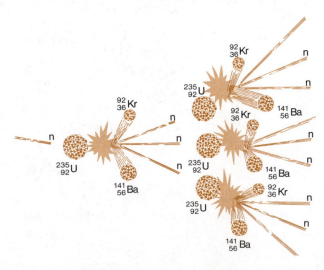

FIGURE 24–10 Self-propagating nuclear chain reaction. A stray neutron induces a single fission, liberating three more neutrons. Each of these induces another fission, each of which is accompanied by release of two or three neutrons. The chain continues to branch in this way, finally resulting in an explosive rate of fission.

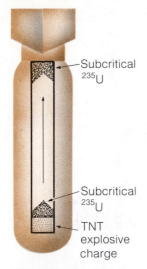

Subcritical
^{235}U

Subcritical
^{235}U

TNT
explosive
charge

FIGURE 24–11 One
design used in atomic
bombs. A conventional
explosive is used to
bring two subcritical
masses together to form
a supercritical mass.

One type of atomic bomb (Figure 24–11) contains two subcritical portions of fissionable material, one of which is driven into the other to form a supercritical mass by an ordinary chemical explosive such as trinitrotoluene (TNT). An uncontrolled nuclear fission explosion results. Tremendous amounts of heat energy are released, as well as many radionuclides whose effects are devastating to life and the environment. The radioactive dust and debris is called fallout.

24–14 Nuclear Reactors (Fission)

Controlled fission reactions in nuclear reactors are of great use and even greater potential. No possibility of nuclear explosion exists because the fuel elements of a nuclear reactor have neither the composition nor the extremely compact arrangement of the critical mass of a bomb at the instant of explosion. However, certain dangers are associated with nuclear energy generation. There is the possibility of "meltdown" if the cooling system is not working properly. This will be discussed with respect to cooling systems in light water reactors. Proper shielding precautions also must be taken to insure that the radionuclides produced are always contained within vessels from which neither they nor their radiations can escape. Long-lived radionuclides from spent fuel must be stored underground in heavy, shock-resistant containers until they have decayed to the point that they are no longer biologically harmful. As examples, strontium-90 ($t_{1/2} = 28$ years) and plutonium-239 ($t_{1/2} = 24,000$ years) must be stored for 280 years and 240,000 years, respectively, before they lose 99.9% of their activities. Critics of nuclear energy contend that the containers could fail over such long periods, or as a result of earth tremors, and that transportation and reprocessing accidents could cause environmental contamination with radionuclides. They claim that river water used for cooling is returned to the rivers with too high a heat content (thermal pollution), thus disrupting marine life. It should be noted, though, that fossil fuel electric power plants also cause the same thermal pollution, kilowatt for kilowatt. The problem of espionage also presents itself. Plutonium-239, a fissionable material, could be stolen from reprocessing plants and used to construct atomic weapons.

Proponents of the development of nuclear energy argue that the potential risks are far outweighed by the advantages. Nuclear energy plants do not pollute the air with oxides of sulfur, nitrogen, carbon, and particulate matter as do fossil fuel electric power plants. The big advantage of nuclear fuels, though, is the enormous amount of energy liberated per unit mass of fuel. With rapidly declining fossil fuel reserves, it seems likely that nuclear energy and solar energy will become increasingly important. At present, nuclear reactors provide nearly 10% of the electrical energy consumed in the United States.

Most commercial nuclear power plants in the United States are "light water" reactors, moderated and cooled by ordinary water, but the greatest potential exists for **breeder reactors,** which actually produce more fissionable material than they consume, as we shall see. Let us look at the components and operation of light water reactors first.

The Soviet Union has placed especially high priority on the development and use of breeder reactors.

1 Light Water Reactors

A schematic diagram of a light water reactor plant is shown in Figure 24–12. The main difference between this and a fossil fuel plant is that the reactor core at the left replaces the usual furnace in which coal, oil, or natural gas is burned. Such a fission

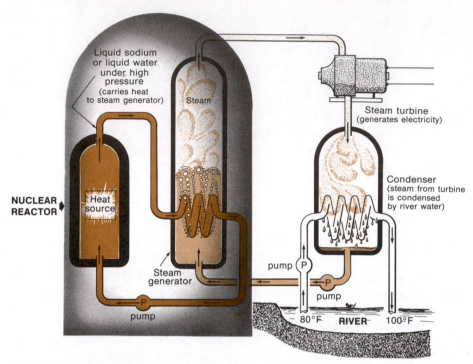

FIGURE 24–12 A schematic diagram of a light water reactor plant.

reactor has five main components: (1) fuel, (2) moderator, (3) control rods, (4) cooling system, and (5) shielding.

Fuel Rods of U_3O_8 enriched in uranium-235 serve as the fuel. Unfortunately, uranium ores contain only about 0.7% uranium-235. Most of the rest is nonfissionable uranium-238. The enrichment is accomplished in processing and reprocessing plants by separating $^{235}UF_6$ from $^{238}UF_6$, prepared from the oxide by treatment with HF and F_2. Separation by diffusion is based upon Graham's Law, which relates rate of diffusion to the molecular weights of gases (Section 11–14). Another separation procedure depends upon the ultracentrifuge.

A potentially more efficient method of enrichment involves the use of sophisticated tunable lasers to ionize uranium-235 selectively and not uranium-238. The ionized uranium-235 could then be made to react with negative ions to form another compound, easily separated from the mixture. The success of this method will lie in our ability to construct lasers capable of producing radiation monochromatic enough to excite one isotope and not the other, a very difficult challenge.

Moderator The most efficient fission reactions occur with slow neutrons. Thus the fast neutrons ejected during fission must be slowed by collisions with atoms of comparable mass that do not absorb them. Such materials are called **moderators.** The most commonly used moderator is ordinary water, although graphite is sometimes used. The most efficient moderator is helium, which slows neutrons but does not absorb them all. The next most efficient one is "heavy water" (deuterium oxide, $^2_1H_2O$ or D_2O), but this is so expensive that it has been used chiefly in research reactors. However, there is a Canadian-designed power reactor that uses heavy water and is correspondingly more neutron-efficient.

Control Rods   Control of the rate of a fission reaction is achieved by using movable control rods, usually of cadmium or boron steel. They are inserted or removed automatically from spaces between the fuel rods. Cadmium and boron are good neutron absorbers.

$$^{10}_{5}B + ^{1}_{0}n \longrightarrow ^{7}_{3}Li + ^{4}_{2}\alpha$$

The more neutrons absorbed by the control rods, the fewer fissions occur, and the less heat is produced. Hence the heat output is governed by the control system that operates the rods.

Cooling System   Actually, two cooling systems are needed. First, the moderator itself serves as a coolant for the reactor. It transfers the fission-generated heat to a second system called a steam generator, which converts water to steam. The steam then goes to the turbines to drive the generator to produce electricity. Another coolant (river water, sea water, or recirculated water) condenses the steam from the turbine, and the condensate is then recycled into the steam generator.

Note that the water that actually comes in contact with the nuclear fuel flows in a closed system and is not released to the environment.

The danger of meltdown arises if the reactor is shut down quickly. The disintegration of radioactive fission products still goes on at a furious rate, fast enough to overheat the fuel elements and to melt them. So it isn't enough to shut down the fission reaction; efficient cooling must be continued until the short-lived isotopes are gone and the heat from their disintegration is dissipated. Only then can the circulation of cooling water be stopped.

The accident that occurred in 1979 at Three Mile Island, near Harrisburg, Pennsylvania, was the result of stopping the water pumps too soon *and* the inoperability of the emergency pumps. So a combination of mechanical malfunctions, errors, and carelessness produced the overheating that damaged the fuel assembly. It did not and *could not explode,* although some melting of core material did occur.

Shielding   It is essential that the people and surrounding countryside be adequately shielded from possible exposure to radioactive nuclides. Thus the entire reactor is enclosed in a steel containment vessel, which is housed in a thick-walled concrete building. The operating personnel are protected further by a so-called biological shield, a thick layer of organic material made of compressed wood fibers. This absorbs the neutrons and beta and gamma rays that would be absorbed in the human body.

The neutrons are the worst radiation hazard. The human body contains a high percentage of H_2O, which absorbs neutrons very efficiently.

2   Breeder Reactors

It is predicted that our limited supply of uranium-235 will last only another 50 years. However, nonfissionable uranium-238 is about 100 times more plentiful and can be converted into fissionable plutonium-239. In fact, this does take place to some extent in light water reactors.

$$^{238}_{92}U + ^{1}_{0}n \longrightarrow ^{239}_{94}Pu + 2\,^{0}_{-1}\beta$$

Fissionable uranium-233 can also be produced by neutron bombardment of thorium-232.

$$^{232}_{90}Th + ^{1}_{0}n \longrightarrow ^{233}_{92}U + 2\,^{0}_{-1}\beta$$

It is possible to build reactors, called **breeder reactors,** that not only generate large quantities of heat from fission, but also generate more fuel than they use because neutrons are purposely absorbed in a thorium or uranium "blanket" to cause the above reactions to occur. However, the design of the breeder reactor has several difficulties associated with it. This type of reactor requires the use of fast neutrons, so no moderator is needed, but control is more difficult. It also must operate at higher temperatures than light water reactors. Thus water cannot be used as a coolant, and liquid sodium, which is not a neutron moderator, is used instead. Sodium is very reactive, particularly at high temperatures, and has a tendency to attack the walls of its container. Heat must be transferred very efficiently because plutonium-239 melts at a relatively low temperature of 640°C.

An additional problem is that plutonium-239 is one of the most toxic substances known, so any release would be disastrous.

Key Terms

Alpha particle (α) a helium nucleus.

Artificial transmutation an artificially induced nuclear reaction caused by bombardment of a nucleus with subatomic particles or small nuclei.

Band of stability band containing nonradioactive nuclides in a plot of number of neutrons versus atomic number.

Beta particle (β) electron emitted from the nucleus when a neutron decays to a proton and an electron.

Breeder reactor a nuclear reactor that produces more fissionable nuclear fuel than it consumes.

Chain reaction a reaction which, once initiated, sustains itself and expands.

Cloud chamber a device for observing the paths of speeding particles as vapor molecules condense on them to form foglike tracks.

Control rods rods of materials such as cadmium or boron steel that act as neutron absorbers (not merely moderators) used in nuclear reactors to control the number of free neutrons per unit volume and therefore rates of fission.

Critical mass the minimum mass of a particular fissionable nuclide in a given volume required to sustain a nuclear chain reaction.

Cyclotron a device for accelerating charged particles along a spiral path.

Daughter nuclide nuclide that is produced in a nuclear decay.

Fast neutron a neutron ejected at high kinetic energy in a nuclear reaction.

Fluorescence absorption by a substance of high-energy radiation and subsequent emission of visible light.

Gamma ray (γ) high-energy electromagnetic radiation.

Half-life of a radionuclide, the time required for half of a given sample to undergo radioactive decay.

Heavy water water containing deuterium, a heavy isotope of hydrogen, $_1^2H$.

K capture absorption of a K shell ($n = 1$) electron by a proton as it is converted to a neutron.

Linear accelerator a device used for accelerating charged particles along a straight line.

Moderator a substance such as hydrogen, deuterium, oxygen, or paraffin capable of slowing fast neutrons upon collision.

Mother nuclide nuclide that undergoes nuclear decay.

Nuclear fission the splitting of a heavy nucleus into two (or more) nuclei of intermediate masses; also releases energy and, usually, subatomic particles.

Nuclear fusion the combination of light nuclei to produce a heavier nucleus.

Nuclear reaction a change in the composition of a nucleus; can evolve or absorb an extraordinarily large amount of energy.

Nuclear reactor a system in which controlled nuclear fission reactions generate heat energy on a large scale, which is subsequently converted into electrical energy.

Nucleons particles making up the nucleus.

Nuclides different atomic forms of all elements (in contrast to "isotopes," which refers only to different atomic forms of a single element).

Positron nuclear particle with the mass of an electron but opposite charge.

Radiation high-energy particles or rays emitted in nuclear decay processes.

Radioactive dating method of dating ancient objects by determining the ratio of amounts of mother and daughter nuclides present in an object and relating the ratio to the object's age via half-life calculations.

Radioactive tracer   a small amount of radioisotope replacing a nonradioactive isotope of the element in a compound whose path (for example, in the body) or whose decomposition products are to be monitored by detection of radioactivity; also called a radioactive label.

Radioactivity   the spontaneous disintegration of atomic nuclei.

Radioisotope   a radioactive isotope of an element.

Radionuclide   a radioactive nuclide.

Scintillation counter   device used for the quantitative detection of radiation.

Slow neutron   a fast neutron slowed by collision with a moderator.

Thermonuclear energy   energy from nuclear fusion reactions.

Transuranium elements   the elements with atomic number greater than 92 (uranium); none occur naturally and all must be prepared by nuclear bombardment of other elements.

Exercises

1. How do nuclear reactions differ from ordinary chemical reactions?
2. How do nuclear fission and nuclear fusion differ? How are they similar?
3. Describe what is meant by "magic numbers" of nucleons.
4. Explain the fact that a plot of number of neutrons versus atomic number for all nuclei shows a band of stable nuclei with a somewhat step-like shape.
5. Describe the characteristics of alpha particles, beta particles, and gamma rays.
6. Fill in the missing symbols in the following nuclear reactions:
   (a) $^{23}_{11}Na + ? \longrightarrow ^{23}_{12}Mg + ^{1}_{0}n$
   (b) $^{96}_{42}Mo + ^{4}_{2}He \longrightarrow ^{100}_{43}Tc + ?$
   (c) $^{232}_{90}Th + ? \longrightarrow ^{240}_{96}Cm + 4\,^{1}_{0}n$
   (d) $? + ^{1}_{1}H \longrightarrow ^{29}_{14}Si + ^{0}_{0}\gamma$
   (e) $^{209}_{83}Bi + ? \longrightarrow ^{210}_{84}Po + ^{1}_{0}n$
   (f) $^{238}_{92}U + ^{16}_{8}O \longrightarrow ? + 5\,^{1}_{0}n$
7. Write the symbols for the daughter nuclides in the following radioactive decays:

   (a) $^{237}_{92}U \xrightarrow{-\beta}$

   (b) $^{13}C \xrightarrow{-n}$

   (c) $^{11}B \xrightarrow{-\gamma}$

   (d) $^{224}Ra \xrightarrow{-\alpha}$

   (e) $^{18}F \xrightarrow{-\frac{1}{1}p}$

   (f) $^{40}_{19}K \xrightarrow{+\beta}$ (K capture)
8. Write the symbols for the daughter nuclides in the following nuclear reactions:
   (a) $^{60}_{28}Ni(n,p)$
   (b) $^{98}_{42}Mo(n, -^{0}_{1}\beta)$
   (c) $^{35}_{17}Cl(p,\alpha)$
   (d) $^{20}_{10}Ne(\alpha,\gamma)$
   (e) $^{15}_{7}N(p,\alpha)$
   (f) $^{10}_{5}B(n,\alpha)$
9. Predict the kind of decays you would expect for the following radionuclides:
   (a) $^{60}_{27}Co$ (n/p ratio too high)
   (b) $^{20}_{11}Na$ (n/p ratio too low)
   (c) $^{224}_{88}Ra$
   (d) $^{64}_{29}Cu$ (n/p ratio too low)
   (e) $^{238}_{92}U$
   (f) $^{11}_{6}C$
10. Name and describe four methods for detection of radiation.
11. Why must all radioactive decays be first order?
12. Describe the process by which steady-state (constant) ratios of carbon-14 to (nonradioactive) carbon-12 are attained in living plants and organisms. Describe the method of radiocarbon dating.
13. Name some radionuclides that have medical uses, and give the uses.
14. Describe how radionuclides can be used in (a) research, (b) agriculture.
15. Describe how (a) cyclotrons and (b) linear accelerators work.
16. Why do both fission and fusion reactions release energy?
17. Summarize how an atomic bomb works, including how the nuclear explosion is initiated.
18. Discuss the pros and cons of the use of nuclear energy instead of other more conventional types of energy based on fossil fuels.
19. Describe and illustrate the essential features of a light water fission reactor.
20. How is fissionable uranium-235 separated from nonfissionable uranium-238?

21. Distinguish between moderators and control rods of nuclear reactors.

22. What are the major advantages and disadvantages of fusion as a potential energy source, as compared with fission?

23.

Particle	Mass
neutron	1.0087 amu
proton	1.0073 amu
electron	0.00055 amu

Given the information above and any information found in the appendices, calculate the following for $^{65}_{29}Cu$ (actual mass $= 64.9278$ amu):
   (a) mass defect in amu/atom
   (b) mass defect in g/mol
   (c) binding energy in ergs/atom
   (d) binding energy in J/atom
   (e) binding energy in kcal/mol
   (f) binding energy in kJ/mol

24. Francium-223 undergoes β decay with a half-life of 22 minutes. How much of a 26 μg sample remains after 2.0 hours?

25. Strontium-90 is one of the harmful radionuclides resulting from nuclear fission explosions. It decays by beta emission with a half-life of 28 years. How long would it take for 99.99% of a given sample released in an atmospheric test of an atomic bomb to disintegrate?

26. Carbon-14 decays by beta emission with a half-life of 5730 years. Assuming a particular object originally contained 6.50 μg of carbon-14 and now contains 0.76 μg of carbon-14, how old is the object?

27. Plutonium-239 decays by alpha particle and gamma ray emission with a half-life of 24,000 years.
   (a) Write the balanced equation for the decay.
   (b) Calculate the specific rate constant, k, for the decay.
   (c) Is ^{239}Pu or ^{235}U ($t_{1/2} = 710$ million years) more stable?

28. Potassium-42 undergoes beta decay and γ emission. If, after 50 hours, 57 minutes, a 2.000 g sample of ^{42}K has decayed to only 0.116 g, what is the half-life of ^{42}K?

Organic Chemistry: The Chemistry of Carbon Compounds

25

Organic chemistry may be described as the chemistry of substances that contain carbon-hydrogen bonds. Originally the term "organic" was used to describe compounds of plant or animal origin. However, in 1828 Friedrich Wöhler synthesized urea by boiling ammonium cyanate with water. This demonstration overthrew the "vital force" theory, which held that organic compounds could be made only by living things.

Urea, H_2N—CO—NH_2, is the principal end product of metabolism of nitrogen-containing compounds in mammals. It is eliminated in the urine. An adult man excretes about 30 g of urea in 24 hours.

$$NH_4OCN \xrightarrow[\text{boil}]{H_2O} H_2N-\overset{\overset{\textstyle O}{\|}}{C}-NH_2$$

ammonium cyanate   urea
(an inorganic compound)   (an organic compound)

Today it is commonplace to manufacture organic compounds from inorganic materials.

Of the Group IVA elements, carbon is a nonmetal, silicon and germanium are metalloids, and tin and lead are metallic elements.

Carbon is the first element in Group IVA in the periodic table, which means that it has four electrons in its outermost shell and can be expected to form single bonds with four other atoms (or with other carbon atoms, for that matter). Although carbon constitutes only about 0.027% of the earth's crust, it is unique among the elements in that it bonds to itself to form millions of different compounds. The bonding of an element to itself is known as **catenation.** Several elements exhibit catenation, but carbon does so to a much greater extent than any other element. Carbon atoms bond to each other to form long chains, branched chains, and rings that may also have chains attached to them.

Saturated Hydrocarbons

Recall that hydrocarbons contain only C and H.

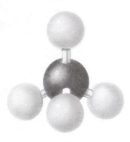

FIGURE 25–1 A molecule of methane, CH_4, formed by overlap of the four sp^3 carbon orbitals with the s orbitals of four hydrogen atoms. The resulting molecule is tetrahedral.

Saturated hydrocarbons, those containing only single bonds, compose the major portion of petroleum and natural gas. Although we normally think of petroleum and natural gas as fuel sources, we should keep in mind the fact that most synthetic organic materials are derived from these two sources. Well over half of the top 50 commercial chemicals are organic compounds derived in this way. Organic compounds are used for a variety of purposes such as clothing, fuel, plastics, paints, medicine, and foodstuffs. Many plastics and other synthetic materials find increasing uses as structural materials.

25–1   The Alkanes

The hydrocarbons are among the simplest and most common compounds of carbon. The *saturated hydrocarbons,* or **alkanes,** are compounds in which each carbon atom is bonded to four other atoms, and each hydrogen atom is, of course, bonded to only one carbon atom. The term saturated means that only single covalent bonds are present.

In Sections 7–5 and 8–3 we examined the structure of the simplest alkane, *methane* (CH_4), and found that methane molecules are tetrahedral, with sp^3 hybridization at carbon (Figure 25–1).

Ethane, C_2H_6, is the next simplest saturated hydrocarbon, and its structure is quite similar to that of methane. Two carbon atoms share a pair of electrons, and each carbon atom shares an electron pair with each of three hydrogen atoms. Both

The term "saturated" comes from early studies in which chemists tried to add hydrogen to various organic substances. Those to which no more hydrogen could be added were called saturated, by analogy with saturated solutions.

We do this *only* as an aid to visualization of the structure of ethane. It is *not* formed in this way. Refer to Section 8–6.

"Straight-chain" does not mean that the molecules are linear. It means that no carbon atom is bonded to more than two other carbon atoms. The bond angles around each carbon atom are approximately 109°28'.

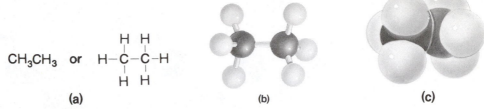

$$CH_3CH_3 \quad \text{or} \quad \begin{array}{c} H \quad H \\ | \quad | \\ H-C-C-H \\ | \quad | \\ H \quad H \end{array}$$

(a)   (b)   (c)

FIGURE 25–2 Models of ethane, C_2H_6. (a) The condensed and line formulas for ethane. (b) Ball-and-stick model and (c) space-filling model of ethane.

carbon atoms are sp^3 hybridized (Figure 25–2). One may visualize the formation of an ethane molecule from two methane molecules by removing one H atom (and its electron) from each CH_4 molecule and then joining the fragments. *Propane*, C_3H_8, is the next member of the family (Figure 25–3). Again, it can be considered as the result of removing a hydrogen atom from a methane molecule and one from an ethane molecule, and joining the fragments.

There are two compounds that have the formula C_4H_{10}. These molecules are structural **isomers** (Section 23–6). The structures of these two isomeric *butanes* are shown in Figure 25–4. These two structures correspond to the two ways in which a hydrogen atom could be removed from a propane molecule and replaced by a —CH_3 group. If the hydrogen were removed from either of the end carbon atoms and replaced by a —CH_3, the result would be *normal butane*, abbreviated as *n*-butane. "Normal" or "straight-chain" refers to hydrocarbon chains in which there is no branching. If the hydrogen were removed from the central carbon atom of propane, however, the result of replacing it with a —CH_3 group would be 2-methylpropane, or *isobutane*, the simplest *branched-chain* hydrocarbon.

Examination of the formulas of the saturated hydrocarbons discussed to this point indicates that their formulas can be written in general terms as C_nH_{2n+2}, where *n* refers to the number of carbon atoms per molecule. The first five members of the

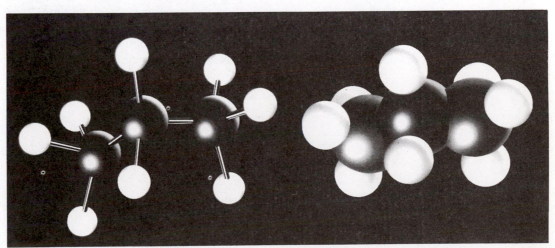

FIGURE 25–3 Ball-and-stick and space-filling models of propane, C_3H_8. Dark spheres represent C atoms; light spheres represent H atoms.

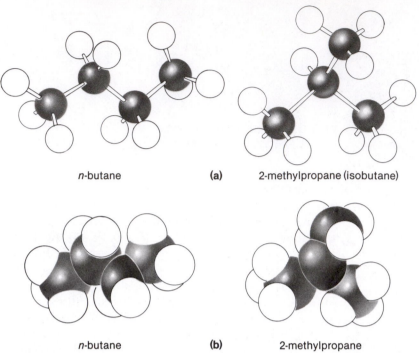

FIGURE 25-4 Models of *n*-butane, $CH_3CH_2CH_2CH_3$, and isobutane, $CH_3CH(CH_3)CH_3$.
(a) Ball-and-stick and (b) space-filling models.

series are

	CH_4	C_2H_6	C_3H_8	C_4H_{10}	C_5H_{12}
$n =$	1	2	3	4	5
$2n + 2 =$	4	6	8	10	12

Each saturated hydrocarbon differs from the next by CH_2, a **methylene group.** A series of compounds in which each member differs from the next member by a specific number and kind of atoms is referred to as a **homologous series.** The chemical and physical properties of members of a homologous series are usually closely related. For example, the boiling points of the lighter members of the saturated hydrocarbon series are shown in Figure 25–5. Note that there is a regular

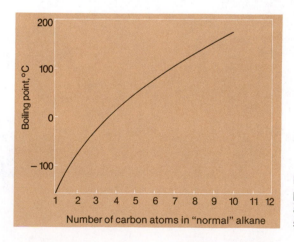

FIGURE 25-5 A plot of boiling point versus the number of carbon atoms in normal (i.e., straight-chain) saturated hydrocarbons.

TABLE 25–1   Some Normal Hydrocarbons (Alkanes)

Molecular Formula	Name	Boiling Point (°C)	Melting Point (°C)	State at Room Temp.
CH_4	methane	−161	−184	
C_2H_6	ethane	−88	−183	Gas
C_3H_8	propane	−42	−188	
C_4H_{10}	n-butane	+0.6	−138	
C_5H_{12}	n-pentane	36	−130	
C_6H_{14}	n-hexane	69	−94	
C_7H_{16}	n-heptane	98	−91	
C_8H_{18}	n-octane	126	−57	
C_9H_{20}	n-nonane	150	−54	
$C_{10}H_{22}$	n-decane	174	−30	
$C_{11}H_{24}$	n-undecane	194.5	−25.6	Liquid
$C_{12}H_{26}$	n-dodecane	214.5	−9.6	
$C_{13}H_{28}$	n-tridecane	234	−6.2	
$C_{14}H_{30}$	n-tetradecane	252.5	+5.5	
$C_{15}H_{32}$	n-pentadecane	270.5	10	
$C_{16}H_{34}$	n-hexadecane	287.5	18	
$C_{17}H_{36}$	n-heptadecane	303	22.5	
$C_{18}H_{38}$	n-octadecane	317	28	
$C_{19}H_{40}$	n-nonadecane	330	32	
$C_{20}H_{42}$	n-eicosane	205 (at 15 torr)	36.7	Solid

increase in boiling points with molecular weight. This is due to the increase in effectiveness of London forces (Section 12–8).

Because there are so many "straight-chain" hydrocarbons, some system for naming them is necessary. The names of the first 20 members of the family, as prescribed by the International Union of Pure and Applied Chemistry (IUPAC), are listed in Table 25–1. Note that the first four members' names must be memorized, but the names of the later members have prefixes (from Greek) that give the number of carbon atoms in the molecules. All names have the "-ane" ending, characteristic of all alkanes.

Branched-chain alkanes will be named in the next section.

We have already noted that there are two saturated hydrocarbons that have the formula C_4H_{10}. When we come to the C_5 hydrocarbons, we find there are three

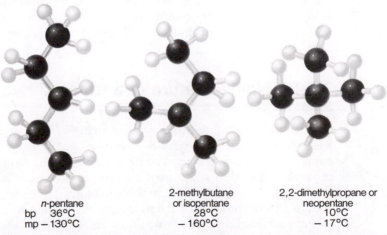

n-pentane
bp   36°C
mp −130°C

2-methylbutane or isopentane
28°C
−160°C

2,2-dimethylpropane or neopentane
10°C
−17°C

FIGURE 25–6   Three-dimensional models of the three isomeric pentanes, each containing five C atoms and twelve H atoms.

TABLE 25–2   Isomeric Hexanes, C_6H_{14}

Name	Formula	Boiling Point (°C)	Melting Point (°C)
n-hexane	$CH_3CH_2CH_2CH_2CH_2CH_3$	68.7	−94
2-methylpentane	$CH_3CH_2CH_2CHCH_3$ $\mid$ CH_3	60.3	−153.7
3-methylpentane	$CH_3CH_2CHCH_2CH_3$ $\mid$ CH_3	63.3	−118
2,2-dimethylbutane	CH_3 $\mid$ $CH_3CH_2CCH_3$ $\mid$ CH_3	49.7	−99.7
2,3-dimethylbutane	$CH_3CH—CHCH_3$ $\mid$ $\mid$ CH_3 CH_3	58.0	−128.4

TABLE 25–3   Number of Possible Isomers of Alkanes

Carbon Content	Isomers
C_8	18
C_9	35
C_{10}	75
C_{11}	159
C_{12}	355
C_{13}	802
C_{14}	1,858
C_{15}	4,347
C_{20}	366,319
C_{25}	36,797,588
C_{30}	4,111,846,763

possible arrangements of the atoms, and indeed three different *pentanes* are known (Figure 25–6).

Note that the boiling points of the pentanes decrease as branching increases (Figure 25–6). For isomers of a given molecular weight, boiling points and melting points generally decrease with increased branching. Properties such as boiling point depend on the forces between molecules, which in turn depend on the "contact area" presented by each molecule to its neighbors. As the degree of branching increases for a series of molecules of the same molecular weight, the molecules become more compact (as seen in Figure 25–6) and the "contact area" becomes smaller.

The number of structural isomers increases rapidly as the number of carbon atoms in saturated hydrocarbons increases. There are five isomeric *hexanes* (Table 25–2).

Table 25–3 displays the number of isomers of some saturated hydrocarbons (alkanes) containing increasing numbers of carbon atoms. Clearly most of the isomers have not been, and probably never will be, prepared and isolated.

25–2   Nomenclature of Saturated Hydrocarbons

The names of the lighter straight-chain alkanes were listed in Table 25–1. You may wish to refer to these to refresh your memory. Branched-chain hydrocarbons are named by naming the longest continuous chain of carbon atoms and indicating the position and kind of substituent attached to the chain. **Alkyl group substituents** attached to the longest chain are thought of as fragments of hydrocarbon molecules obtained by the removal of one hydrogen atom. They are given names related to the parent hydrocarbons from which they are derived, as shown in the following list. Other alkyl groups are named similarly.

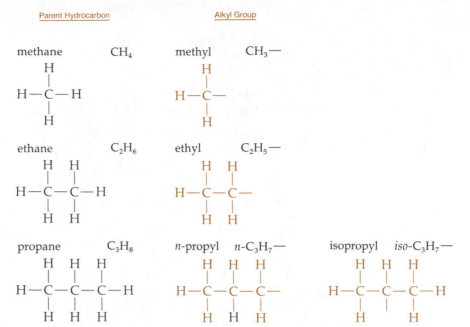

We now illustrate the naming of some saturated branched hydrocarbons. The rules are: (1) Find the longest continuous chain of carbon atoms. (2) Number the carbon atoms in the chain beginning at the end nearest the branching. (3) Assign the names and (position) numbers that indicate the substituents. If more than one of the same substituent appear, prefixes (di-, tri-, etc.) are used; the position number is given for each (see below). (4) The longest continuous chain is always numbered so that the substituents have the lowest possible numbers.

Consider the following compound.

$$H-\underset{\displaystyle H}{\overset{\displaystyle H}{C}}-\underset{\displaystyle H}{\overset{\displaystyle H-\overset{\displaystyle H}{\underset{\displaystyle H}{C}}-H}{C}}-\underset{\displaystyle H}{\overset{\displaystyle H}{C}}-\underset{\displaystyle H}{\overset{\displaystyle H}{C}}-\underset{\displaystyle H}{\overset{\displaystyle H}{C}}-\underset{\displaystyle H}{\overset{\displaystyle H}{C}}-H \quad or \quad CH_3CH(CH_3)CH_2CH_2CH_2CH_3$$

Parentheses are used to conserve space. Formulas written with parentheses must indicate unambiguously the structure of the compound.

The parentheses indicate that the CH₃ group is attached to the C that precedes it.

Following rules 1 and 2 allows us to number the carbon atoms in the longest chain.

$$\underset{1}{CH_3}-\underset{2}{\overset{\displaystyle CH_3}{\underset{\displaystyle |}{CH}}}-\underset{3}{CH_2}-\underset{4}{CH_2}-\underset{5}{CH_2}-\underset{6}{CH_3}$$

It is incorrect to name the compound as 5-methylhexane because that violates rules 2 and 4.

The methyl group is attached to the *second* carbon atom in a *six-carbon* chain, so the compound is named 2-methylhexane.

The following examples further illustrate the rules of nomenclature.

Remember that line (dash) formulas indicate atoms that are bonded to each other. They do *not* show molecular geometry.

$$\underset{1}{CH_3}-\overset{2}{\underset{\displaystyle CH_3}{\overset{\displaystyle CH_3}{\underset{\displaystyle |}{\overset{\displaystyle |}{C}}}}}-\overset{3}{CH_2}-\overset{4}{CH_2}-\overset{5}{CH_3} \quad or \quad CH_3C(CH_3)_2(CH_2)_2CH_3$$

2,2-dimethylpentane

$$\overset{1}{CH_3}-\overset{2}{CH_2}-\overset{3}{CH}-CH_3 \quad or \quad \overset{1}{CH_3}-\overset{2}{CH_2}-\overset{3}{CH}-\overset{4}{CH_2}-\overset{5}{CH_3}$$

with the substituent:
$$\underset{5}{\overset{4}{CH_2}}$$
$$\overset{5}{CH_3}$$
$$CH_3$$ (above C3)

3-methylpentane   better written so as to emphasize the 5-C chain

In general terms, we use the symbol R *to represent any alkyl group.* For example, the saturated hydrocarbons may be represented by the general formula R—H.

25–3 Conformations

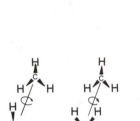

Eclipsed
conformation   Staggered
conformation

FIGURE 25–7 Two possible conformations of ethane. Rotation from one to the other is easily possible as shown by the curved arrows. The lower formulas show how the hydrogens would be arranged if the molecule were viewed end on along the carbon-carbon bond axis.

A **conformation** is one specific geometry of a molecule. Conformations of compounds differ from one another by the *extent of rotation about a single bond.* The distance between two singly-bonded carbon atoms is relatively independent of the structure of the rest of the molecule and is very nearly equal to 0.154 nm. Rotation about single carbon-carbon bonds is possible; in fact, it occurs rapidly. It might appear that there is an infinite number of kinds of ethane molecules, depending on the rotation of one carbon atom with respect to the other. However, at room temperature ethane molecules possess sufficient thermal energy to cause rapid rotation about the single carbon-carbon bond from one conformation to another. Therefore there is only one kind of ethane molecule, and no isomers are known. There is, however, a small preference for the staggered conformation over the eclipsed conformation (see Figure 25–7) because there is less repulsive interaction between hydrogen atoms on adjacent carbon atoms in the former.

Let us consider two conformations of *n*-butane (Figure 25–8). Again, there is a slight preference for the second (staggered) conformation, but at room temperature many conformations are present in a sample of pure *n*-butane.

25–4 Chemical Properties of the Alkanes

The saturated hydrocarbons are chemically relatively inert materials. For many years they were known as **paraffin hydrocarbons** because they undergo relatively few reactions. The alkanes do not react with such powerful oxidizing agents as

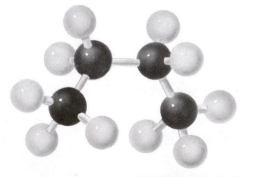
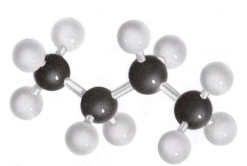

FIGURE 25–8 Two conformations of *n*-butane, *n*-C_4H_{10}.

Paraffin is a mixture of high-molecular-weight alkanes, and is chemically quite unreactive.

potassium permanganate and potassium dichromate. However, they do react with the halogens, with oxygen when ignited, and with concentrated nitric acid. As expected, the members of a homologous series show similar chemical properties. If we study the chemistry of one of these compounds, we can make predictions about the others with a fairly high degree of certainty.

25–5 Substitution Reactions

A reaction in which an atom (or a group of atoms) replaces another atom (or group of atoms) in an organic compound is called a *substitution reaction*.

Since the saturated hydrocarbons contain only single covalent bonds, they can react without a big disruption of the molecular structure only by *displacement or substitution of one atom for another*. At room temperature, chlorine and bromine react very slowly with saturated straight-chain hydrocarbons. At higher temperatures, and particularly in the presence of sunlight or other source of ultraviolet light, hydrogen atoms in the hydrocarbon can be replaced easily by halogen atoms. The reactions are called **halogenation** reactions. The mechanism of reaction of chlorine with methane (a *chlorination* reaction) may be represented as

A free radical is an atom or group of atoms that contains at least one unpaired electron. Some carry an electrical charge, others do not.

$$:\ddot{Cl}:\ddot{Cl}: \xrightarrow[\text{sunlight}]{\text{heat or}} 2:\ddot{Cl}\cdot$$

chlorine molecule   chlorine atoms (free radicals)

$$\overset{H}{\underset{H}{H:\overset{\cdot\cdot}{C}:H}} + :\ddot{Cl}\cdot \longrightarrow \overset{H}{\underset{H}{H:\overset{\cdot\cdot}{C}\cdot}} + H:\ddot{Cl}:$$

methane   chlorine atom   methyl radical   hydrogen chloride

$$\overset{H}{\underset{H}{H:\overset{\cdot\cdot}{C}\cdot}} + :\ddot{Cl}:\ddot{Cl}: \longrightarrow \overset{H}{\underset{H}{H:\overset{\cdot\cdot}{C}:\ddot{Cl}:}} + :\ddot{Cl}\cdot, \text{ etc.}$$

methyl radical   chlorine molecule   methyl chloride   chlorine atom

This reaction is called a **free radical chain reaction.** It is quite similar to the reaction of chlorine with hydrogen (Section 21–8). Chlorine molecules absorb light energy and split into very reactive chlorine atoms. Some chlorine atoms may recombine to form chlorine molecules, but others attack a hydrocarbon molecule, removing one of the hydrogen atoms to form hydrogen chloride. The methyl radical, in turn, reacts with chlorine molecules to form chloromethane (also called methyl chloride). The overall reaction is usually represented as

Note that only one half of the chlorine atoms occurs in the organic product. The other half of the chlorine atoms forms hydrogen chloride, a commercially valuable compound.

$$Cl-Cl + H-\overset{\displaystyle H}{\underset{\displaystyle H}{\overset{|}{\underset{|}{C}}}}-H \xrightarrow[\text{uv}]{\text{heat or}} H-\overset{\displaystyle H}{\underset{\displaystyle H}{\overset{|}{\underset{|}{C}}}}-Cl + HCl$$

chlorine   methane   chloromethane, bp = 23.8°C (methyl chloride)

Many organic reactions are complex, and produce more than a single product. For example, the chlorination of methane may produce several other products in addition to chloromethane, as the following equations show.

$$\text{Cl—Cl} + \text{H—}\underset{\underset{\displaystyle H}{|}}{\overset{\overset{\displaystyle H}{|}}{\text{C}}}\text{—Cl} \longrightarrow \text{Cl—}\underset{\underset{\displaystyle H}{|}}{\overset{\overset{\displaystyle H}{|}}{\text{C}}}\text{—Cl} \quad + \text{HCl}$$

dichloromethane, bp = 40.2°C
(methylene chloride)

$$\text{Cl—Cl} + \text{Cl—}\underset{\underset{\displaystyle H}{|}}{\overset{\overset{\displaystyle H}{|}}{\text{C}}}\text{—Cl} \longrightarrow \text{Cl—}\underset{\underset{\displaystyle Cl}{|}}{\overset{\overset{\displaystyle H}{|}}{\text{C}}}\text{—Cl} \quad + \text{HCl}$$

trichloromethane, bp = 61°C
(chloroform)

$$\text{Cl—Cl} + \text{Cl—}\underset{\underset{\displaystyle Cl}{|}}{\overset{\overset{\displaystyle H}{|}}{\text{C}}}\text{—Cl} \longrightarrow \text{Cl—}\underset{\underset{\displaystyle Cl}{|}}{\overset{\overset{\displaystyle Cl}{|}}{\text{C}}}\text{—Cl} \quad + \text{HCl}$$

tetrachloromethane, bp = 76.8°C
(carbon tetrachloride)

The mixture of products formed in an organic reaction can be separated by physical methods such as fractional distillation, which relies on differences in the boiling points of different compounds.

When a hydrocarbon has more than one carbon atom, the reaction with chlorine is considerably more complex than the chlorination of methane. The first step in the chlorination of ethane gives a product that contains one chlorine atom per molecule.

Ethyl chloride is widely used by athletic trainers as a spray-on pain killer.

$$\text{Cl—Cl} + \text{H—}\underset{\underset{\displaystyle H}{|}}{\overset{\overset{\displaystyle H}{|}}{\text{C}}}\text{—}\underset{\underset{\displaystyle H}{|}}{\overset{\overset{\displaystyle H}{|}}{\text{C}}}\text{—H} \xrightarrow{\text{heat or uv}} \text{H—}\underset{\underset{\displaystyle H}{|}}{\overset{\overset{\displaystyle H}{|}}{\text{C}}}\text{—}\underset{\underset{\displaystyle H}{|}}{\overset{\overset{\displaystyle H}{|}}{\text{C}}}\text{—Cl} \quad + \text{HCl}$$

ethane   chloroethane, bp = 13.1°C
(ethyl chloride)

When a second hydrogen atom is replaced, two products are possible, and indeed a mixture of both is obtained.

The product mixture does not contain equal numbers of moles of the dichloro-ethanes, so we do not show a stoichiometrically balanced equation. Since reactions of saturated hydrocarbons with chlorine produce a multiplicity of products, the reactions are not always as useful as might be desired.

$$\text{Cl—Cl} + \text{H—}\underset{\underset{\displaystyle H}{|}}{\overset{\overset{\displaystyle H}{|}}{\text{C}}}\text{—}\underset{\underset{\displaystyle H}{|}}{\overset{\overset{\displaystyle H}{|}}{\text{C}}}\text{—Cl} \xrightarrow{\text{heat or uv}} \left\{ \begin{array}{c} \text{H—}\underset{\underset{\displaystyle H}{|}}{\overset{\overset{\displaystyle H}{|}}{\text{C}}}\text{—}\underset{\underset{\displaystyle H}{|}}{\overset{\overset{\displaystyle Cl}{|}}{\text{C}}}\text{—Cl} \\ \text{1,1-dichloroethane} \\ \text{bp = 57°C} \\ \text{Cl—}\underset{\underset{\displaystyle H}{|}}{\overset{\overset{\displaystyle H}{|}}{\text{C}}}\text{—}\underset{\underset{\displaystyle H}{|}}{\overset{\overset{\displaystyle H}{|}}{\text{C}}}\text{—Cl} \\ \text{1,2-dichloroethane} \\ \text{bp = 84°C} \end{array} \right\} + \text{HCl}$$

The cyclic saturated hydrocarbons, or **cycloalkanes,** are more symmetrical and can be converted into single monosubstituted products in good yield. The first four cycloalkanes are indicated below. Note that the general formula for cycloalkanes is C_nH_{2n}.

$$
\begin{array}{cccc}
\text{cyclopropane} & \text{cyclobutane} & \text{cyclopentane} & \text{cyclohexane}
\end{array}
$$

The reaction of cyclopentane with chlorine produces only a single product, chlorocyclopentane or cyclopentyl chloride.

cyclopentane $+ Cl_2 \xrightarrow{\text{heat or}}{\text{uv}}$ chlorocyclopentane (cyclopentyl chloride) $+ HCl$

25-6 Oxidation of Alkanes

The principal use of hydrocarbons is as fuels such as gasoline, diesel fuel, heating oil, and "natural gas." Hydrocarbons burn in excess oxygen to produce carbon dioxide and water in highly exothermic processes (Section 9-7).

methane: $CH_4 + 2O_2 \longrightarrow CO_2 + 2H_2O + 891 \text{ kJ}$

n-octane: $2C_8H_{18} + 25O_2 \longrightarrow 16CO_2 + 18H_2O + 1.090 \times 10^4 \text{ kJ}$

Recall that ΔH^0 is negative for an exothermic process.

The **heat of combustion** is the amount of energy *liberated* per mole of hydrocarbon burned. Heats of combustion are assigned positive values (Table 25-4), and are therefore equal in magnitude but opposite in sign to ΔH^0 values for combustion reactions. The combustion of hydrocarbons produces large volumes of gases in addition to large amounts of heat. The sudden expansion of these gases drives the pistons or turbine blades in internal combustion engines.

TABLE 25-4 Heats of Combustion of Some Alkanes

Hydrocarbon		Heat of Combustion	
		kJ/mol	J/g
methane	CH_4	891	55.7
propane	C_3H_8	2220	50.5
n-pentane	n-C_5H_{12}	3507	48.7
n-octane	n-C_8H_{18}	5450	47.8
n-decane	n-$C_{10}H_{22}$	6737	47.4
ethanol*	C_2H_5OH	1372	29.8

* Not an alkane; included for comparison only.

> The heat of combustion for ethanol (Table 25–4), even when converted to a per gram basis, is low compared with those for saturated hydrocarbons. Clearly, the complete combustion of a given mass of ethanol produces considerably less energy than complete burning of the same mass of hydrocarbon. Ethanol is usually produced by fermentation of grain and other agricultural products. Considerable amounts of petroleum products are required to produce fertilizers to grow crops and to produce fuels to operate farm equipment. Additionally, energy is required to distill alcohol out of the fermentation mixture. Grain that is converted to alcohol must be taken from the world's supply of food. Until other sources of alcohol are found, ethanol is not likely to become an important fuel.

In the absence of sufficient oxygen, partial combustion of hydrocarbons occurs. The products may be carbon monoxide (a very poisonous gas) or carbon (which forms deposits on spark plugs, in the cylinder head, and on the pistons of automobile engines). All hydrocarbons undergo similar reactions. The reactions of methane with limited oxygen are

$$2CH_4 + 3O_2 \longrightarrow 2CO + 4H_2O$$

$$CH_4 + O_2 \longrightarrow C + 2H_2O$$

25–7   Petroleum

Petroleum or crude oil consists mainly of hydrocarbons, but small amounts of organic compounds containing nitrogen, sulfur, and oxygen are present. Each oil field produces petroleum with a particular set of characteristics. Distillation of petroleum produces several fractions, as shown in Table 25–5.

The process is called **thermal cracking.**

Since gasoline is so much in demand, higher hydrocarbons are "cracked" to increase the amount of gasoline that can be made from a barrel of petroleum. Higher-molecular-weight hydrocarbons (C_{15} to C_{18}) are heated, in the absence of air and in the presence of a catalyst, to produce a mixture of lower-molecular-weight hydrocarbons (C_5 to C_{12}) that can be used in gasoline.

Engine knock is caused by premature detonation of fuel in the combustion chamber.

The **octane number** (rating) of a gasoline refers to how smoothly the gasoline burns and how much engine "knock" it produces. Isoöctane, or 2,2,4-trimethyl-

TABLE 25–5   Petroleum Fractions

Fraction*	Principal Composition	Distillation Range
natural gas	C_1–C_4	below 20°C
bottled gas	C_5–C_6	20–60°
gasoline	C_4–C_{12}	40–200°
kerosene	C_{10}–C_{16}	175–275°
fuel oil, diesel oil	C_{15}–C_{20}	250–400°
lubricating oils	C_{18}–C_{22}	above 300°
paraffin	C_{23}–C_{29}	mp 50–60°
asphalt		viscous liquid ("bottoms fraction")
coke		solid

* Other descriptions and distillation ranges have been used, but all are similar.

pentane, has excellent combustion properties and was assigned an arbitrary octane number of 100. Normal heptane, $CH_3—(CH_2)_5—CH_3$, has very poor combustion properties and was assigned an octane number of zero.

$$CH_3—\underset{\underset{\displaystyle CH_3}{|}}{\overset{\overset{\displaystyle CH_3}{|}}{C}}—CH_2—\overset{\overset{\displaystyle CH_3}{|}}{CH}—CH_3 \qquad\qquad CH_3—CH_2—CH_2—CH_2—CH_2—CH_2—CH_3$$

isoöctane
octane number = 100

n-heptane
octane number = 0

Mixtures of isoöctane and *n*-heptane were prepared and burned in test engines to establish the "octane" scale. The octane number of such a mixture is the percentage of isoöctane in it. When gasolines are burned in standard test engines, they can be assigned octane numbers based on the compression ratio at which they begin to knock. A 90-octane gasoline produces the same amount of knock as the 90% isoöctane–10% *n*-heptane mixture. Branched-chain compounds produce less knock than straight-chain compounds. The octane numbers of two isomeric hexanes illustrate the point.

$$CH_3—(CH_2)_4—CH_3 \qquad (CH_3)_3C—CH_2—CH_3$$

n-hexane
octane number = 25

2,2-dimethylbutane
octane number = 92

Gasoline containing tetraethyllead is known as "leaded" gasoline.

For many years tetraethyllead, $Pb(C_2H_5)_4$, was used as a gasoline additive to improve smoothness of burning and increase octane rating. However, the discharge of lead into the atmosphere is believed to be hazardous to health, and the lead compounds "poison" the catalyst in catalytic converters. Therefore, "leaded" gasoline is being phased out of use.

Unsaturated Hydrocarbons

There are three classes of unsaturated hydrocarbons: (1) the alkenes and their cyclic counterparts, the cycloalkenes, (2) the alkynes and the cycloalkynes, and (3) the aromatic hydrocarbons. We shall consider each group separately.

25–8 The Alkenes (Olefins)

Recall that the cycloalkanes may also be represented by the general formula C_nH_{2n}.

The general formula for the simplest **alkenes** is C_nH_{2n} because these compounds contain one carbon-carbon double bond per molecule. The roots for their names are derived from the alkanes having the same number of carbon atoms as the longest chain containing the double bond. In the common (sometimes called trivial) system of nomenclature, the suffix -ylene is added to the characteristic root. In systematic (IUPAC) nomenclature, the suffix -ene is added to the characteristic root. In chains of four or more carbon atoms, the position of the double bond is indicated by a numerical prefix that shows the *lowest numbered* doubly bonded carbon atom.

$CH_2{=}CH_2$ $CH_3{-}CH{=}CH_2$ $\overset{4}{C}H_3\overset{3}{C}H_2{-}\overset{2}{C}H{=}\overset{1}{C}H_2$

trivial: ethylene   propylene   *n*-butylene

systematic: ethene   propene   1-butene

$\overset{1}{C}H_3\overset{2}{C}H{=}\overset{3}{C}H\overset{4}{C}H_3$

$CH_3{-}\underset{\underset{CH_3}{|}}{C}{=}CH_2$

2-butene   isobutylene

2-methylpropene

In naming more complex alkenes, the double bond takes (positional) preference over substituents on the carbon chain and is always assigned the lowest possible number.

$\overset{4}{C}H_3{-}\overset{3}{C}H_2{-}\underset{\underset{CH_3}{|}}{\overset{2}{C}}{=}\overset{1}{C}H_2$

$\overset{1}{C}H_3{-}\overset{2}{C}H{=}\overset{3}{C}H{-}\underset{\underset{CH_3}{|}}{\overset{4}{C}H}{-}\overset{5}{C}H_3$

2-methyl-1-butene   4-methyl-2-pentene

Some alkenes have two or more carbon-carbon double bonds per molecule. The suffixes -adiene, -atriene, etc., are used to indicate the number of (C=C) double bonds in a molecule.

$\overset{1}{C}H_2{=}\overset{2}{C}H{-}\overset{3}{C}H{=}\overset{4}{C}H_2$ $\overset{4}{C}H_3{-}\overset{3}{C}H{=}\overset{2}{C}{=}\overset{1}{C}H_2$

1,3-butadiene   1,2-butadiene

1,3-Butadiene and similar molecules that contain alternating single and double bonds are described as having **conjugated double bonds.** Such compounds are of special interest because of the polymerization reactions they undergo (Section 25–9).

The **cycloalkenes** are represented by the general formula C_nH_{2n-2}.

$\begin{array}{c}CH_2{-}CH_2\\|\qquad\;|\\CH\quad CH_2\\\diagdown\!{=}\!\diagup\\CH\end{array}$

$\begin{array}{c}\quad CH_2\\\diagup\quad\diagdown\\CH_2\qquad CH_2\\|\qquad\qquad|\\CH\qquad CH_2\\\diagdown\!{=}\!\diagup\\CH\end{array}$

cyclopentene   cyclohexene

The bonding in ethylene was described in detail in Section 8–7. The hybridization (sp^2) and bonding at other double-bonded carbon atoms is similar. Recall that both carbon atoms in C_2H_4 are located at the centers of trigonal planes.

Rotation about C=C double bonds at room temperature is *not* possible because it would require breaking the pi bond. (In Section 25–3, we pointed out that rotation about C—C single bonds is possible.) Therefore, compounds that have the general formulas (XY)C=C(XY) exist as a pair of *cis-trans* isomers. Figure 25–9 shows the *cis-trans* isomers of dichloroethene.

Carbon-carbon double bonds are shorter than carbon-carbon single bonds, 0.134 nm versus 0.154 nm, because the two shared electron pairs draw the carbon nuclei closer than does a single electron pair. Although the physical properties of the alkenes are similar to those of the alkanes, their chemical properties are quite different.

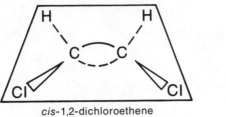

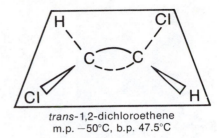

cis-1,2-dichloroethene
m.p. −80.5°C, b.p. 60.3°C

trans-1,2-dichloroethene
m.p. −50°C, b.p. 47.5°C

FIGURE 25–9 Two isomers of 1,2-dichloroethene are possible because rotation around the double bond is restricted. This is an example of geometric isomerism.

25–9 Reactions of the Alkenes

Most reactions of the *alkanes* that do not disrupt the carbon skeleton are **substitution reactions,** but the *alkenes* are characterized by **additions** to the double bond. Consider the reactions of ethane and ethene with chlorine.

ethane $CH_3{-}CH_3 + Cl_2 \longrightarrow CH_3{-}CH_2Cl + HCl$ (substitution)

ethene $CH_2{=}CH_2 + Cl_2 \longrightarrow$ $\underset{\underset{Cl}{|}}{CH_2}{-}\underset{\underset{Cl}{|}}{CH_2}$ (addition)

More functional groups will be considered later.

Carbon-carbon double bonds represent **reaction sites** and so are called **functional groups.** Most addition reactions proceed rapidly at room temperature. By contrast, the substitution reactions of the alkanes usually require catalysts and elevated temperatures.

Bromine adds readily to the alkenes to give dibromides. The reaction with ethene is

Ethylene dibromide is added to gasoline that contains tetraethyllead, $Pb(C_2H_5)_4$, to prevent the accumulation of lead in engines. It forms $PbBr_2$, which is volatile at engine-operating temperatures and is expelled through the exhaust system. The lead bromide "poisons" the catalytic converters in some modern automobiles, so the use of lead-containing gasoline in these automobiles is prohibited by law.

$Br_2 + CH_2{=}CH_2 \longrightarrow$ $\underset{\underset{Br}{|}}{CH_2}{-}\underset{\underset{Br}{|}}{CH_2}$

1,2-dibromoethane
(ethylene dibromide)

The addition of bromine to alkenes is used as a simple qualitative test for unsaturation. Bromine, a dark red liquid, is dissolved in a nonpolar solvent such as methylene chloride, CH_2Cl_2. When an alkene is added, the solution becomes colorless as the bromine reacts with the alkene to form a colorless compound. This reaction may be used to distinguish between alkanes and alkenes.

Hydrogenation is an extremely important reaction of the alkenes. At elevated temperatures, under high pressures, and in the presence of an appropriate catalyst (finely divided Pt, Pd, or Ni), hydrogen adds across double bonds.

Note that hydrogenation corresponds to reduction.

$CH_2{=}CH_2 + H_2 \xrightarrow[\text{heat}]{\text{catalyst}} CH_3{-}CH_3$

ethane

Unsaturated hydrocarbons are converted to saturated hydrocarbons in the manufacture of high octane gasoline and aviation fuels. Unsaturated vegetable oils may also be converted to solid cooking fats (shortening) by hydrogenation (Figure 25–10).

$$
\begin{array}{ccc}
\underset{\displaystyle\parallel}{\overset{\displaystyle O}{}} & & \underset{\displaystyle\parallel}{\overset{\displaystyle O}{}} \\
CH_2OC(CH_2)_7CH{=}CH(CH_2)_7CH_3 & & CH_2OC(CH_2)_{16}CH_3 \\
\mid & & \mid \\
\underset{\displaystyle\parallel}{\overset{\displaystyle O}{}} & & \underset{\displaystyle\parallel}{\overset{\displaystyle O}{}} \\
CHOC(CH_2)_7CH{=}CH(CH_2)_7CH_3 & \xrightarrow[\text{Ni catalyst}\atop\text{heat}]{3H_2} & CHOC(CH_2)_{16}CH_3 \\
\mid & & \mid \\
\underset{\displaystyle\parallel}{\overset{\displaystyle O}{}} & & \underset{\displaystyle\parallel}{\overset{\displaystyle O}{}} \\
CH_2OC(CH_2)_7CH{=}CH(CH_2)_7CH_3 & & CH_2OC(CH_2)_{16}CH_3 \\
\text{olein} & & \text{stearin} \\
\text{an oil, liquid} & & \text{a fat, solid}
\end{array}
$$

FIGURE 25–10
Fats can be obtained by hydrogenation of the olefinic double bonds in vegetable oils. The beaker in the top photo contains clear oil before hydrogenation. Below, the same oil is shown hardened by hydrogenation.

Polymerization, the combination of many small molecules to form large molecules (polymers), is another important reaction of the alkenes. Polyethylene provides a common example. In the presence of appropriate catalysts (a mixture of aluminum trialkyls, R_3Al, and titanium tetrachloride, $TiCl_4$), ethylene polymerizes into chains containing 800 or more carbon atoms.

$$n CH_2{=}CH_2 \xrightarrow{\text{catalyst}} {+}CH_2{-}CH_2{+}_n$$
$$\text{ethylene} \qquad\qquad \text{polyethylene}$$

The polymer may be represented as $CH_3(CH_2{-}CH_2)_nCH_3$ where n is approximately 400. Polyethylene is a tough, flexible plastic widely used as an electrical insulator and for the fabrication of such things as unbreakable refrigerator dishes, plastic cups, and squeeze bottles. Polypropylene is made by polymerizing propylene, $CH_3{-}CH{=}CH_2$, in much the same way that polyethylene is produced.

"Teflon" is made by polymerizing tetrafluoroethene in a reaction that is similar to the formation of polyethylene.

$$n CF_2{=}CF_2 \xrightarrow[\text{heat}]{\text{catalyst}} {+}CF_2{-}CF_2{+}_n$$
$$\text{tetrafluoroethene} \qquad\qquad \text{"Teflon"}$$

The molecular weight of Teflon is between 1×10^6 and 2×10^6; i.e., approximately 20,000 $CF_2{=}CF_2$ molecules polymerize to form a single giant molecule. Teflon is a very useful polymer in that it does *not* react with concentrated acids and bases or with most oxidizing agents, nor does it dissolve in most organic solvents.

Natural rubber is obtained from the sap of the rubber tree, a sticky liquid called latex. Rubber is a polymeric hydrocarbon formed (in the sap) by the combination of about 2000 molecules of 2-methyl-1,3-butadiene, commonly called isoprene. The molecular weight of rubber is about 136,000.

Teflon has many interesting properties, including low friction, stability at high temperatures, and the ability to be machined to close tolerances.

$$2n CH_2{=}\underset{\underset{\displaystyle CH_3}{\displaystyle |}}{C}{-}CH{=}CH_2 \longrightarrow {+}CH_2{-}\underset{\underset{\displaystyle CH_3}{\displaystyle |}}{C}{=}CH{-}CH_2{-}CH_2{-}\underset{\underset{\displaystyle CH_3}{\displaystyle |}}{C}{=}CH{-}CH_2{+}_n$$
$$\text{isoprene}$$

When natural rubber becomes warm, it flows and becomes sticky. To eliminate this problem, **vulcanization** is used. This is the process in which sulfur is added to rubber and the mixture is heated to approximately 140°C. Sulfur atoms combine with some of the double bonds in the linear polymer molecules to form bridges that bond one rubber molecule to another. This cross-linking by sulfur atoms converts the linear polymer into a three-dimensional polymer. Fillers and reinforcing agents are added during the mixing process to improve the wearing qualities of rubber and to form colored rubber. Carbon black is the most common reinforcing agent, and zinc oxide, barium sulfate, titanium dioxide, and antimony(V) sulfide are common fillers.

Some synthetic rubbers are superior to natural rubber in some ways. Neoprene is a synthetic elastomer (an elastic polymer) with properties quite similar to those of natural rubber. The basic structural unit is chloroprene. Note that chloroprene differs from isoprene in having a chlorine atom rather than a methyl group as a substituent on the 1,3-butadiene chain.

$$n\text{CH}_2\text{=CH}-\overset{\overset{\displaystyle\text{Cl}}{|}}{\text{C}}\text{=CH}_2 \xrightarrow{\text{polymerization}} \text{(CH}_2-\text{CH=}\overset{\overset{\displaystyle\text{Cl}}{|}}{\text{C}}-\text{CH}_2\text{)}_n$$

chloroprene   neoprene

Numerous other polymers are elastic enough to come under the generic name of rubber.

Neoprene is less affected by gasoline and oil, and is more elastic than natural rubber. It resists abrasion well, and is not swollen or dissolved by hydrocarbons. It is widely used to make hoses for oil and gasoline, electrical insulation, and automobile and refrigerator parts.

25-10   Oxidation of Alkenes

The alkenes, like the alkanes, burn in excess oxygen to form carbon dioxide and water in exothermic reactions.

$$\text{CH}_2\text{=CH}_2 + 3\text{O}_2 \text{ (excess)} \longrightarrow 2\text{CO}_2 + 2\text{H}_2\text{O} + 1387 \text{ kJ}$$

When an alkene is burned in air, a yellow luminous flame is observed and considerable amounts of soot, unburned carbon, are obtained. This reaction provides a qualitative test for unsaturation, because saturated hydrocarbons burn in air without the production of significant amounts of soot.

Another qualitative test for unsaturation involves the reaction of alkenes (and, as we shall see presently, the alkynes) with cold dilute aqueous solutions of potassium permanganate, $KMnO_4$. Potassium permanganate solutions are purple to pink, depending on their concentrations. The addition of an unsaturated hydrocarbon results in the disappearance of the characteristic permanganate color and the formation of manganese dioxide, MnO_2, a brown solid. With ethene the reaction is

$$3\text{CH}_2\text{=CH}_2 + 2\text{KMnO}_4 + 4\text{H}_2\text{O} \longrightarrow 3\text{CH}_2-\text{CH}_2 + 2\text{MnO}_2 + 2\text{KOH}$$
$$\qquad\qquad\qquad\qquad\qquad\qquad\qquad\quad \overset{|}{\text{OH}}\quad\overset{|}{\text{OH}}$$

ethylene glycol

Other alkenes undergo similar reactions. Hot or concentrated $KMnO_4$ solutions oxidize ethene all the way to carbon dioxide and water.

25-11   The Alkynes

The **alkynes,** or acetylenic hydrocarbons, contain carbon-carbon triple bonds. Those with one triple bond per molecule have the general formula C_nH_{2n-2}. They

FIGURE 25-11 Models of acetylene, H—C≡C—H.

are named like the alkenes except that the suffix -yne is added to the characteristic root. The first member of the series is commonly called acetylene, and experiments show that its molecular formula is C_2H_2. It may be transformed into ethene and then to ethane by the addition of hydrogen. These reactions suggest that the formula for acetylene is H—C≡C—H.

CH≡CH CH_3—C≡CH CH_3—CH_2—C≡CH CH_3—C≡C—CH_3 CH_3—CH—C≡CH
ethyne   propyne   1-butyne   2-butyne |
acetylene CH_3
3-methyl-1-butyne

The triple bond takes positional preference over substituents on the carbon chain and is assigned the lowest possible number in naming.

The bonding in alkynes is similar to that described specifically for acetylene in Section 8-8. The triply bonded carbon atoms are all *sp* hybridized, and they and the adjacent atoms lie on a straight line (Figure 25-11).

Acetylene lamps are charged with calcium carbide. Very slow addition of water produces acetylene, which is burned as it is produced. Acetylene is also used in the oxyacetylene torch for welding and cutting metals. When it is burned with oxygen, the flame reaches temperatures of 3000°C.

$CaC_2 + 2H_2O \longrightarrow CH≡CH + Ca(OH)_2$

$$2CH≡CH + 5O_2 \longrightarrow 4CO_2 + 2H_2O + 2611 \text{ kJ}$$

Since the alkynes contain two pi bonds, both of which are concentrated sources of electrons, they are more reactive than the alkenes. The most common reaction of the alkynes is addition across the triple bond. The reactions with hydrogen and with bromine are typical.

These reactions are thought of as stepwise reactions, but stopping them after the first step is difficult because the products still contain a reactive double bond.

H—C≡C—H $\xrightarrow{H_2}$ (H)C=C(H) $\xrightarrow{H_2}$ H_3C—CH_3

H—C≡C—H $\xrightarrow{Br_2}$ (H/Br)C=C(H/Br) $\xrightarrow{Br_2}$ H—C(Br)(Br)—C(Br)(Br)—H

1,2-dibromoethene   1,1,2,2-tetrabromoethane

Aromatic Hydrocarbons

Originally the word **aromatic** was applied to pleasant-smelling substances. We now use the word to describe benzene, its derivatives, and certain other compounds that exhibit similar chemical properties. Some have very foul odors because of substituents on the benzene ring. On the other hand, many fragrant compounds do not contain benzene rings.

In the nineteenth century the development of the coal tar industry in Germany provided an enormous stimulus to the systematic study of organic chemistry. The production of steel requires large amounts of coke, which is prepared by heating bituminous coal to high temperatures in the absence of air. The production of coke also results in the formation of *coal gas* and *coal tar*; the latter serves as a source of aromatic compounds.

When a ton of coal is converted to coke, about 5 kg of aromatic compounds can be obtained from it. Because of the enormous amount of coal converted to coke, coal tar is produced in large quantities.

The principal components of coal gas are hydrogen (~50%) and methane (~30%). Small amounts of NH_3, N_2, CO, CO_2, H_2O, and some low-molecular-weight hydrocarbons are also present.

Distillation of coal tar produces a variety of aromatic compounds, as we shall see shortly.

25–12   Benzene

Benzene is the simplest aromatic hydrocarbon, and by studying its reactions we can learn a great deal about aromatic hydrocarbons. Benzene was discovered in 1825 by Michael Faraday when he fractionally distilled a by-product oil obtained in the manufacture of illuminating gas from whale oil.

Elemental analysis and determination of its molecular weight show that the molecular formula for benzene is C_6H_6. The formula suggests that it is a highly unsaturated substance, but its properties are quite different from those of open-chain unsaturated hydrocarbons. It reacts with neither bromine water (aqueous Br_2) nor solutions of potassium permanganate, reagents that invariably react with unsaturated *aliphatic* hydrocarbons.

Aliphatic hydrocarbons contain no aromatic rings.

The facts that only one monosubstitution product is obtained in numerous reactions and that no addition products can be prepared indicate conclusively that benzene has a *symmetrical ring structure*. Stated differently, every hydrogen atom is equivalent to every other hydrogen atom, and this is possible only in a symmetrical ring structure:

(skeleton)

Only the *positions* of the atoms are indicated in this structure. The debate over the structure and bonding in benzene raged for about thirty years. In 1865, Friedrich Kekulé suggested that the structure of benzene was intermediate between two structures that we now call resonance structures. He actually proposed an ill-defined equilibrium between them.

The structure of benzene is described in detail in Section 8–14.2 in terms of MO theory. Figure 8–17 shows the structure.

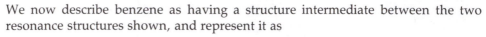

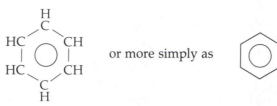

FIGURE 25-12 sp^2 hybridized carbons with p orbitals parallel.

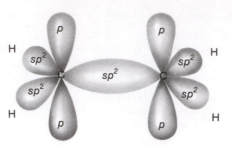

FIGURE 25-14 The symbolic formula and a model of toluene, $C_6H_5CH_3$, which may be thought of as a derivative of benzene in which one H atom has been replaced by a —CH₃ group.

We now describe benzene as having a structure intermediate between the two resonance structures shown, and represent it as

$$\text{HC} \underset{\text{HC}}{\overset{\text{H}}{\underset{}{\begin{array}{c} \text{C} \\ \bigcirc \\ \text{C} \end{array}}}} \overset{\text{CH}}{\underset{\text{CH}}{}} \qquad \text{or more simply as} \qquad \bigcirc$$

Benzene molecules are planar (all 12 atoms lie in a plane). This suggests sp^2 hybridization of each carbon, with one electron occupying the unhybridized p orbital (Figure 25-12), as in ethene (Section 8-7). The six sp^2 hybridized carbon atoms lie in a plane, and the unhybridized p orbitals extend above and below the plane. Side-by-side overlap between the p orbitals leads them to merge into pi orbitals.

The electrons associated with the pi bonds are *delocalized* over the entire benzene ring (recall that all C's are chemically equivalent, and so are all the H's); this is depicted in Figure 25-13. In Figure 25-14 we show the structure of **toluene** (methylbenzene), which is derived by substituting a methyl group for one of the hydrogens of benzene.

25-13   Other Aromatic Hydrocarbons

The distillation of coal tar provides four volatile fractions, as well as pitch that is used for surfacing roads and in the manufacture of "asphalt" roofing (Figure 25-15).

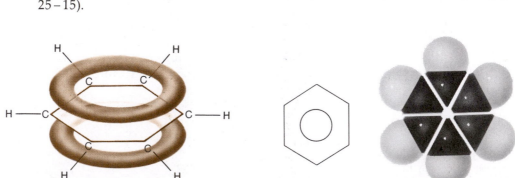

FIGURE 25-13 The electron distribution of the benzene molecule, C_6H_6, its symbol, and a space-filling model.

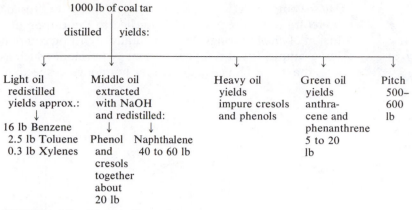

1000 lb of coal tar

distilled | yields:

| Light oil redistilled yields approx.: | Middle oil extracted with NaOH and redistilled: | | Heavy oil yields impure cresols and phenols | Green oil yields anthracene and phenanthrene 5 to 20 lb | Pitch 500– 600 lb |

↓ | | | | |

16 lb Benzene
2.5 lb Toluene
0.3 lb Xylenes

Phenol and cresols together about 20 lb

Naphthalene 40 to 60 lb

FIGURE 25–15 Fractions obtained from coal tar.

Eight aromatic hydrocarbons are obtained in significant amounts by efficient fractional distillation of the "light oil" fraction. They are called "coal-tar crudes" (Table 25–6).

The structures for benzene and toluene were given earlier. Note that three isomeric **xylenes** are listed in Table 25–6. The *o*-, *m*-, and *p*- refer to relative positions of substituents on the benzene ring. The xylenes are dimethylbenzenes, of which there are three distinctly different compounds that have the formula $C_6H_4(CH_3)_2$. The *ortho*- prefix refers to two substituents located on *adjacent* carbon atoms; i.e., 1,2-dimethylbenzene is *o*-xylene. The *meta*- prefex refers to substituents located on carbon atoms 1 and 3. Thus, 1,3-dimethylbenzene is *m*-xylene. The *para*-prefix refers to substituents located on carbon atoms 1 and 4, and 1,4-dimethylbenzene is *p*-xylene.

The structures of naphthalene, anthracene, and phenanthrene are

ortho-xylene
bp = 144°C
mp = −27°C

meta-xylene
bp = 139°C
mp = −54°C

para-xylene
bp = 138°C
mp = 13°C

Anthracene and phenanthrene are isomers.

naphthalene, $C_{10}H_8$

anthracene, $C_{14}H_{10}$

phenanthrene, $C_{14}H_{10}$

TABLE 25–6 The Coal-Tar Crudes

Name	Formula	Boiling Point (°C)	Melting Point (°C)	Solubility
benzene	C_6H_6	80	+6	
toluene	$C_6H_5CH_3$	111	−95	
o-xylene	$C_6H_4(CH_3)_2$	144	−27	All
m-xylene	$C_6H_4(CH_3)_2$	139	−54	insoluble
p-xylene	$C_6H_4(CH_3)_2$	138	+13	in
naphthalene	$C_{10}H_8$	218	+80	water
anthracene	$C_{14}H_{10}$	342	+218	
phenanthrene	$C_{14}H_{10}$	340	+101	

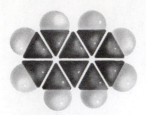

FIGURE 25–16 A model of naphthalene, $C_{10}H_8$, the simplest "condensed" aromatic ring system.

Note that the rings in the major sex hormones are mostly cycloalkane rings rather than aromatic rings.

These compounds are examples of "condensed" or "fused" ring systems. Note that there are no hydrogen atoms attached to the carbon atoms that are involved in fusion of *aromatic* rings. Many naturally occurring compounds contain fused rings. For example, fused rings occur in many hormones such as estradiol and testosterone, the major female and male sex hormones, respectively.

estradiol   testosterone

25–14   Reactions of Aromatic Hydrocarbons

Early research on the reactions of the aromatic hydrocarbons was richly rewarded as chemists learned to prepare a great variety of dyes, drugs, flavors and perfumes, and explosives. More recently, large numbers of polymeric materials such as plastics and fabrics have been prepared from these "coal-tar crudes."

The most common kind of reaction of the aromatic rings is substitution, i.e., replacement of a hydrogen atom by another atom or group of atoms.

When iron is used as a catalyst, it reacts with chlorine to form iron(III) chloride.

Halogenation, with chlorine or bromine, occurs readily in the presence of iron or anhydrous iron(III) chloride (a Lewis acid) catalyst.

benzene   chlorobenzene

The equation is usually written in condensed form as

Aromatic rings can undergo **nitration** in a mixture of concentrated nitric and sulfuric acid at low temperatures. The sulfuric acid is a catalyst and a dehydrating agent.

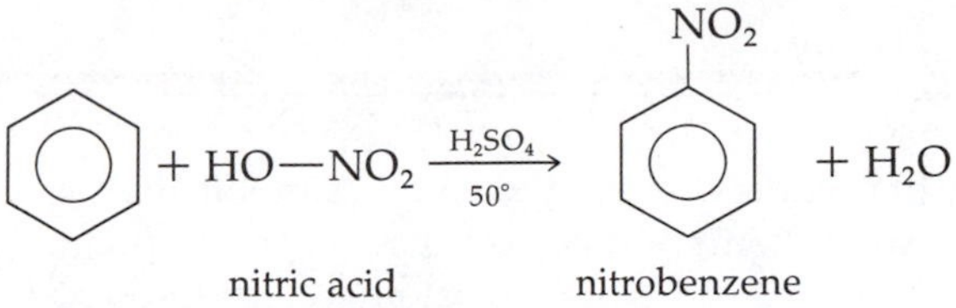

$$\text{benzene} + HO{-}NO_2 \xrightarrow[50°]{H_2SO_4} \text{nitrobenzene} + H_2O$$

nitric acid   nitrobenzene

TNT (2,4,6-trinitrotoluene) is manufactured by the nitration of toluene in several steps.

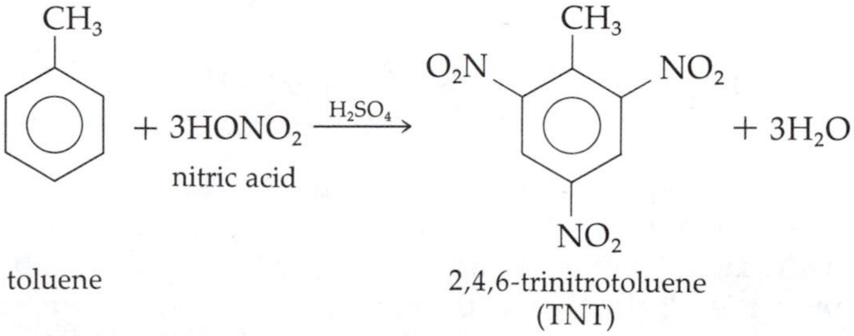

$$\text{toluene} + 3HONO_2 \xrightarrow{H_2SO_4} \text{2,4,6-trinitrotoluene (TNT)} + 3H_2O$$

nitric acid

toluene   2,4,6-trinitrotoluene
                                (TNT)

Like other hydrocarbons, aromatic hydrocarbons burn in air to release large amounts of energy.

$$2C_6H_6 + 15O_2 \longrightarrow 12CO_2 + 6H_2O + 6548 \text{ kJ}$$

benzene

The reactions of strong oxidizing agents with alkylbenzenes, which are benzene molecules containing alkyl side chains, illustrate strikingly the stability of the benzene ring system. Heating toluene with a basic solution of potassium permanganate results in a nearly quantitative yield of benzoic acid. The ring itself remains intact; only the nonaromatic portion of the molecule is oxidized.

Frequently only the product of interest is shown in organic reactions. The reduction product is MnO_2, which is removed by filtration of the basic solution. Acidification with a strong inorganic acid results in the precipitation of benzoic acid, an insoluble weak acid.

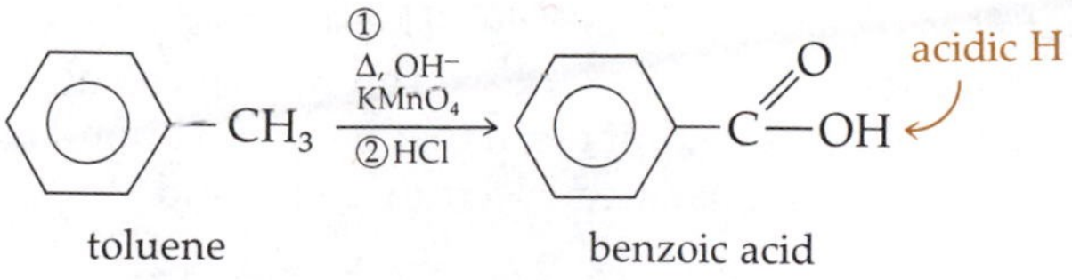

$$\text{toluene}{-}CH_3 \xrightarrow[\text{②HCl}]{\overset{①}{\underset{KMnO_4}{\Delta,\ OH^-}}} \text{benzoic acid}\ \overset{O}{\underset{}{C}}{-}OH \quad \text{acidic H}$$

toluene   benzoic acid

Two or more alkyl groups on an aromatic ring are oxidized to yield a diprotic acid, as the following example illustrates.

Organic reactions are sometimes written in extremely abbreviated form. This is often the case when a variety of common oxidizing agents will accomplish the desired conversion.

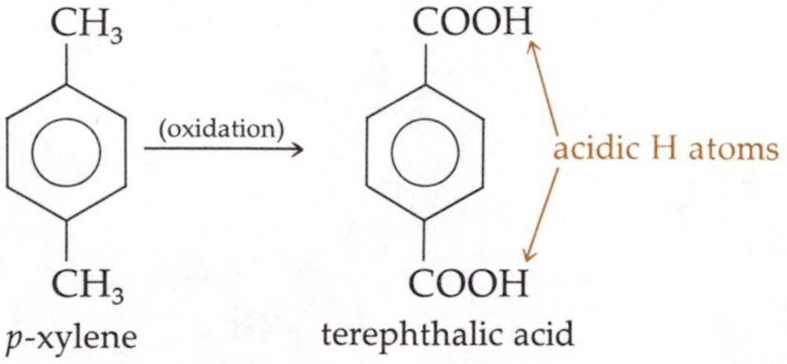

$$\text{p-xylene} \xrightarrow{\text{(oxidation)}} \text{terephthalic acid} \quad \text{acidic H atoms}$$

p-xylene   terephthalic acid

As we shall see, terephthalic acid is used to make "polyesters," an important class of polymers.

Functional Groups

The study of organic chemistry is greatly simplified by assuming that hydrocarbons represent parent compounds, and that other compounds are derived from them. **Functional groups** are groups of atoms that represent potential reaction sites in organic compounds. Because the reactions of a given functional group are similar in most compounds, we are able to systematize large amounts of information easily by studying functional groups and their characteristic reactions.

25–15   Alcohols and Phenols

Originally the term *aliphatic* meant "fat-like." Present use of the term is: the parent hydrocarbon was an alkane, an alkene, an alkyne, or one of their cyclic counterparts, rather than an aromatic compound.

The simplest phenol is called phenol. The most common member of a class of compounds is frequently called by the class name. Salt, sugar, alcohol, and phenol are common examples.

Alcohols and phenols contain the hydroxyl group ($-O-H$) as their functional group. **Alcohols** may be considered to be derived from saturated or unsaturated hydrocarbons by replacing at least one hydrogen atom by a hydroxyl group. The properties of alcohols are the properties of a hydroxyl group attached to an *aliphatic* carbon atom, $-C-O-H$. Ethanol, or ethyl alcohol, is the most common example.

When a hydrogen atom attached to an aromatic ring is replaced by a hydroxyl group, the resulting compound is known as a **phenol.** Such a compound behaves more like an acid than like an alcohol.

Alternatively, we may view alcohols and phenols as derivatives of water in which one H atom has been replaced by an organic group:

H—O—H   ethanol   phenol

water

Indeed, this is by far the better view. The structure of water was discussed in detail in Sections 7–5 and 8–3. The hydroxyl group in an alcohol or a phenol is covalently bonded to a carbon atom, but recall that the O—H bond is quite polar. The oxygen atom retains its two unshared electron pairs, and the C—O—H bond angle is nearly $104.5°$.

A group derived from an *aromatic* hydrocarbon by removing a hydrogen atom is called an aryl (Ar) group.

Clearly the presence of a bonded alkyl or aryl group changes the properties of the $-O-H$ group. *Alcohols* are so very weakly acidic that they are thought of as neutral compounds. *Phenols* are weakly acidic. For phenol, $K_a = 10^{-10}$.

Many reactions of alcohols depend on whether the hydroxyl group is attached

FIGURE 25–17 Models of ethanol (also called ethyl alcohol or grain alcohol), CH_3CH_2OH.

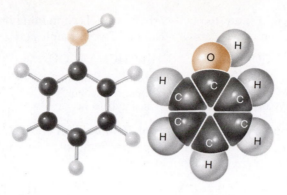

FIGURE 25–18 Models of phenol, C_6H_5OH.

to a carbon that is bonded to one, two, or three other carbon atoms. If we represent alkyl groups as R, we can illustrate the three classes of alcohols.

Recall that R represents an alkyl group.

$$\begin{array}{ccc} \overset{\displaystyle H}{\underset{\displaystyle H}{R-C-O-H}} & \overset{\displaystyle R}{\underset{\displaystyle H}{R-C-O-H}} & \overset{\displaystyle R}{\underset{\displaystyle R}{R-C-O-H}} \end{array}$$

a primary (1°) alcohol   a secondary (2°) alcohol   a tertiary (3°) alcohol

Note that primary alcohols contain one R group, secondary alcohols contain two R groups, and tertiary alcohols contain three R groups bonded to the carbon atom to which the —OH group is attached. The R groups may be the same or different.

1   Nomenclature of Alcohols and Phenols

The systematic names of alcohols consist of the characteristic stem plus an -ol ending. A numeric prefix indicates the position of the —OH group in chains of three or more carbon atoms.

The systematic name is given first; the others are common (trivial) names.

$CH_3—OH$ $CH_3CH_2—OH$ $CH_3CH_2CH_2—OH$ $CH_3CH(OH)CH_3$

methanol   ethanol   1-propanol   2-propanol
methyl alcohol   ethyl alcohol   n-propyl alcohol   isopropyl alcohol
wood alcohol   grain alcohol   a primary alcohol   rubbing alcohol
                                                                            a secondary alcohol

There are four four-carbon alcohols containing one —OH per molecule:

$CH_3CH_2CH_2CH_2OH$ $CH_3—\overset{\displaystyle }{\underset{\displaystyle CH_3}{CH}}—CH_2OH$

                                1°   1°

1-butanol   2-methyl-1-propanol
normal butyl alcohol   isobutyl alcohol

$CH_3CH_2\overset{\displaystyle }{\underset{\displaystyle OH}{CH}}CH_3$ $CH_3—\overset{\displaystyle CH_3}{\underset{\displaystyle CH_3}{C}}—OH$

                                2°   3°

2-butanol   2-methyl-2-propanol
secondary butyl alcohol   tertiary butyl alcohol

There are eight saturated five-carbon alcohols containing one —OH per molecule. They are often called "amyl" alcohols. Two examples are

$$CH_3CH_2CH_2CH_2CH_2OH \qquad (CH_3)_2CHCH_2CH_2OH$$

1-pentanol
n-pentyl alcohol
n-amyl alcohol

3-methyl-1-butanol
isopentyl alcohol
isoamyl alcohol

The **polyhydric alcohols** contain more than one —OH group per molecule. Those containing two OH groups per molecule are called **glycols.** Important examples of polyhydric alcohols include

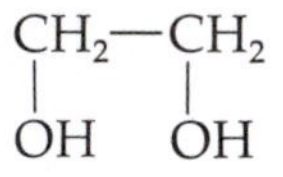

1,2-ethanediol
ethylene glycol
(permanent antifreeze)

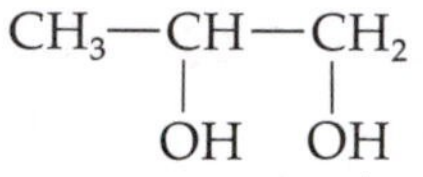

1,2-propanediol
propylene glycol

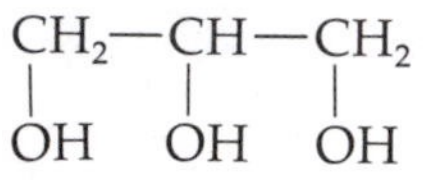

1,2,3-propanetriol
glycerine or glycerol
(the moisturizer in cosmetics)

Phenols are nearly always referred to by their common names. Some examples are

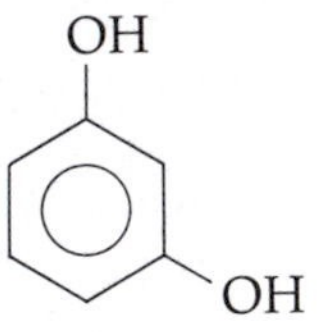

resorcinol 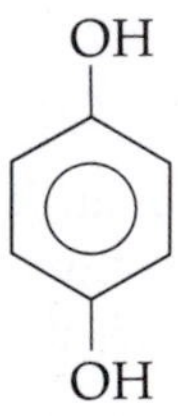   hydroquinone 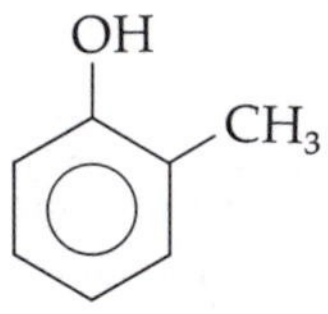   *o*-cresol 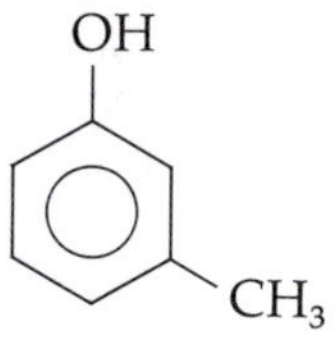   *m*-cresol

As you might guess, the cresols occur in "creosote," a wood preservative that has been widely used. There is a third isomer, *p*-cresol, in which the —OH and —CH$_3$ groups occur on opposite corners of the aromatic ring.

2   Physical Properties of Alcohols and Phenols

Since the hydroxyl group is quite polar while alkyl groups are nonpolar, the properties of alcohols depend on two factors: (1) the number of hydroxyl groups per molecule, and (2) the size of the organic portion of the molecule. We may think of alcohols as consisting of a polar part, —OH, and a nonpolar part, R.

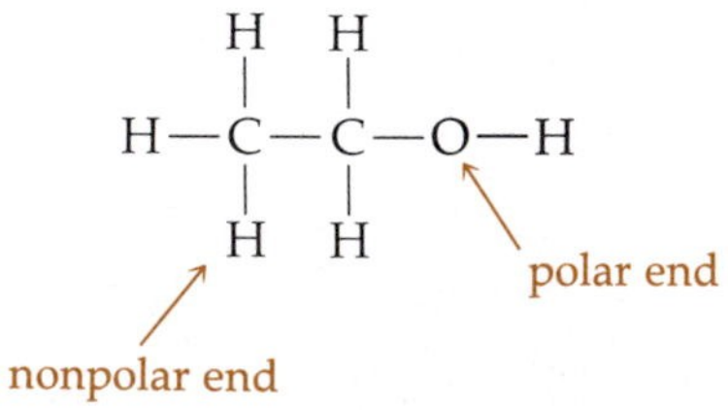

The low-molecular-weight monohydric alcohols are miscible with water in all proportions. Beginning with the four butyl alcohols, solubility in water decreases rapidly with increasing molecular weight because the nonpolar part of the molecules becomes larger and larger. Many polyhydric alcohols are very soluble in water because they contain two or more polar hydroxyl groups per molecule.

TABLE 25–7 Physical Properties of Normal Primary Alcohols

Name	Formula	Boiling Point (°C)	Solubility in H_2O, g/100 g at 20°C
methanol	CH_3OH	65	completely miscible
ethanol	CH_3CH_2OH	78.5	completely miscible
1-propanol	$CH_3CH_2CH_2OH$	97	completely miscible
1-butanol	$CH_3CH_2CH_2CH_2OH$	117.7	7.9
1-pentanol	$CH_3CH_2CH_2CH_2CH_2OH$	137.9	2.7
1-hexanol	$CH_3CH_2CH_2CH_2CH_2CH_2OH$	155.8	0.59

Table 25–7 shows that the boiling points of normal primary alcohols increase, while the solubilities in water decrease, with increasing molecular weight. The boiling points of the alcohols are much higher than those of the corresponding alkanes (Table 25–1) because of the hydrogen bonding of the hydroxyl groups.

Unless other functional groups that interact with water are present in the molecules, phenols are only slightly soluble in water. Most phenols are solids at room temperature. Dilute aqueous solutions of phenols are frequently used as antiseptics and disinfectants.

3 Preparation of Some Alcohols

Methanol, or methyl alcohol, was produced for many years by the destructive distillation of wood, and it is often called wood alcohol. Most methanol is now prepared from carbon monoxide and hydrogen at high temperatures and pressures in the presence of a mixed oxide catalyst (oxides of Zn, Cu, Cr).

$$CO + 2H_2 \xrightarrow[\substack{400°C \\ cat.}]{150\ atm} CH_3OH$$

Methanol is used as a temporary antifreeze (bp = 65°C), as a solvent for varnishes and shellacs, and as the starting material in the manufacture of formaldehyde (Section 25–20). It is very toxic and causes permanent blindness when taken internally.

Fermentation is an enzymatic process carried out by certain kinds of bacteria.

Ethanol, or ethyl alcohol, was first prepared by fermentation *a long time ago* — the most ancient literature contains references to beverages that were obviously alcoholic! The fermentation of blackstrap molasses, the residue from the purification of cane sugar, is one important source of ethanol.

$$C_{12}H_{22}O_{11} + H_2O \xrightarrow{yeast} 4CH_3CH_2OH + 4CO_2$$

cane sugar   ethanol

The starch in grain, potatoes, and similar substances can be converted into sugar by malt; this is followed by fermentation to produce ethanol, which is the most important industrial alcohol. The industrial preparation involves the hydration of ethene from petroleum, using H_2SO_4 as a catalyst.

$$
\begin{array}{c}
\quad\ \ \text{H}\ \ \text{H} \\
\quad\ \ | \quad | \\
\text{H}-\text{C}=\text{C}-\text{H} + \text{HOSO}_3\text{H} \xrightarrow{cold}
\end{array}
\quad
\begin{array}{c}
\text{H}\ \ \text{H} \\
| \quad | \\
\text{H}-\text{C}-\text{C}-\text{OSO}_3\text{H} \\
| \quad | \\
\text{H}\ \ \text{H}
\end{array}
$$

ethene   sulfuric acid   ethyl hydrogen sulfate

Note that these reactions amount to the addition of water to ethene, since H_2SO_4 is regenerated in the second reaction.

$$H-\overset{\overset{\displaystyle H}{|}}{\underset{\underset{\displaystyle H}{|}}{C}}-\overset{\overset{\displaystyle H}{|}}{\underset{\underset{\displaystyle H}{|}}{C}}-OSO_3H + H-OH \longrightarrow H-\overset{\overset{\displaystyle H}{|}}{\underset{\underset{\displaystyle H}{|}}{C}}-\overset{\overset{\displaystyle H}{|}}{\underset{\underset{\displaystyle H}{|}}{C}}-OH + HOSO_3H$$

steam   ethanol

Ethylene glycol (systematic name 1,2-ethanediol) is prepared by the reaction of ethene with the hypochlorous acid present in chlorine water, and the subsequent hydrolysis of the product in an aqueous solution of sodium carbonate.

Ethylene glycol is completely miscible with water and is widely used as a permanent antifreeze (bp = 197°C).

$$CH_2{=}CH_2 + \quad HOCl \quad \longrightarrow \overset{\displaystyle CH_2-CH_2}{\underset{\displaystyle \underset{OH}{|}\ \underset{Cl}{|}}{}} \xrightarrow[2Na^+,CO_3{}^{2-}]{H_2O} \overset{\displaystyle CH_2-CH_2}{\underset{\displaystyle \underset{OH}{|}\ \underset{OH}{|}}{}}$$

hypochlorous acid   ethylene   1,2-ethanediol
(chlorine + water)   chlorohydrin   (ethylene glycol)

4   Reactions of Alcohols and Phenols

The alcohols are *very weakly acidic* compounds. However, they do not react with strong soluble bases. They are much weaker acids than water (see Table 25 – 10), but some of their reactions are analogous to those of water.

The very reactive metals react with alcohols to form **alkoxides** with the liberation of hydrogen.

$$2CH_3-OH + 2Na \longrightarrow \quad 2Na^{+\,-}OCH_3 + H_2$$

sodium methoxide (an alkoxide)

$$2CH_3-CH_2-OH + 2Na \longrightarrow \quad 2Na^{+\,-}OCH_2CH_3 + H_2$$

sodium ethoxide (an alkoxide)

Note the similarity between these reactions and the reaction of water with active metals.

$$2H-OH + 2Na \longrightarrow 2Na^+OH^- + H_2 \quad \text{(can occur explosively)}$$

The alkoxides of low molecular weight are strong bases that react with water (i.e., hydrolyze) to form the parent alcohol and a strong soluble base.

$$Na^{+\,-}OCH_2CH_3 + H-OH \longrightarrow CH_3CH_2OH + Na^+OH^-$$

sodium ethoxide   ethyl alcohol

Phenols also react with metallic sodium to produce **phenoxides** in reactions that are analogous to those of alcohols. Since phenols are more acidic than alcohols, their reactions are more vigorous.

$$2\left[\bigcirc\!\!-O-H\right] + 2Na \longrightarrow 2\left[\bigcirc\!\!-O^-\ Na^+\right] + H_2$$

phenol   sodium phenoxide

The hydroxyl groups of alcohols can be replaced by atoms or by groups of atoms. The hydrohalic acids react with alcohols in the presence of certain Lewis acid catalysts to form **alkyl halides.** Concentrated hydrochloric acid reacts with primary alcohols very slowly at elevated temperatures.

$$H^+Cl^- + CH_3CH_2OH \xrightarrow{\underset{\Delta}{ZnCl_2}} CH_3CH_2Cl + H_2O$$

ethyl chloride

Alcohols react with common ternary inorganic oxyacids to produce **inorganic esters.** For instance, nitric acid reacts with alcohols to produce nitrates.

$$CH_3CH_2OH + HONO_2 \longrightarrow CH_3{-}CH_2{-}O{-}NO_2 + H_2O$$

nitric acid   ethyl nitrate

The reaction of nitric acid with glycerol produces the principal explosive ingredient of dynamite.

$$
\begin{array}{l}
CH_2{-}OH \\
|\\
CH{-}OH \\
|\\
CH_2{-}OH
\end{array}
+ 3HONO_2 \xrightarrow{H_2SO_4}
\begin{array}{l}
CH_2{-}ONO_2 \\
|\\
CH{-}ONO_2 \\
|\\
CH_2{-}ONO_2
\end{array}
+ 3H_2O
$$

glycerol   glyceryl trinitrate
(nitroglycerine)

Cold concentrated sulfuric acid reacts with alcohols to form **alkyl hydrogen sulfates.** The reaction with lauryl alcohol is an important industrial reaction.

$$CH_3CH_2CH_2CH_2CH_2CH_2CH_2CH_2CH_2CH_2CH_2CH_2{-}OH + H_2SO_4 \longrightarrow CH_3(CH_2)_{10}CH_2{-}O{-}SO_3H + H_2O$$

lauryl alcohol   lauryl hydrogen sulfate
1-dodecanol

The reaction of an alkyl hydrogen sulfate with sodium hydroxide produces the sodium salt of the alkyl hydrogen sulfate.

$$CH_3(CH_2)_{10}CH_2{-}O{-}SO_3H + Na^+OH^- \longrightarrow CH_3(CH_2)_{10}{-}CH_2{-}OSO_3{}^-Na^+ + H_2O$$

sodium lauryl sulfate
a detergent

Sodium salts of the alkyl hydrogen sulfates that contain approximately 12 carbon atoms are excellent detergents, and they are also biodegradable. Soaps and detergents were discussed in Section 13–22.

25–16 Carboxylic Acids

Compounds that contain the **carboxyl group,** $-\overset{\overset{\displaystyle O}{\|}}{C}-O-H$, are found to be acidic. The general formula is R—COOH, and most are *weak acids* compared to strong inorganic acids such as HCl. However, they are much stronger acids than most phenols. These acids are named systematically by dropping the terminal $-e$ from the name of the parent hydrocarbon and adding *-oic acid.* However, old habits are hard to break, and organic acids are usually called by common names.

In aromatic acids the carboxyl group is attached to the aromatic ring as shown in Figure 25–19.

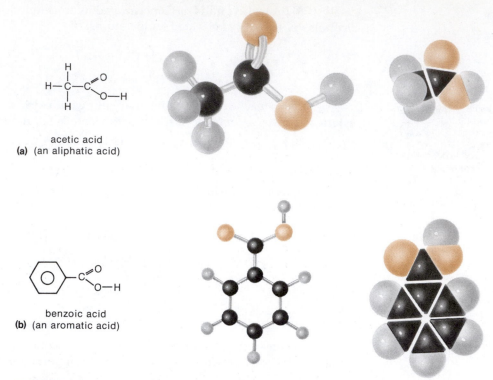

FIGURE 25–19   (a) Models of acetic acid, (b) models of benzoic acid.

Organic acids occur widely in natural products, and many have been known since ancient times. As Table 25–8 indicates, the common names are often derived from a Greek or Latin word that indicates the original source.

The names of derivatives of carboxylic acids are often derived from the trivial names of acids. Positions of substituents may be indicated by lower-case Greek letters beginning with the carbon *adjacent* to the carboxyl carbon, rather than by numbering the carbon chain (again, old habits are hard to break).

α-bromopropionic acid
(2-bromopropanoic acid)

β-methylbutyric acid
(3-methylbutanoic acid)

Formic acid was obtained by distillation of ants (L., *formica*, ant); acetic acid occurs in vinegar (L., *acetum*, vinegar); butyric acid in rancid butter (L., *butyrum*, butter); stearic acid in animal fats (Gr., *stear*, beef suet).

n-Caproic acid is one of the so-called "goat acids." Its odor is responsible for the name.

TABLE 25–8   Aliphatic Carboxylic Acids

Formula	Common Name	IUPAC Name
HCOOH	formic acid	methanoic acid
CH_3COOH	acetic acid	ethanoic acid
CH_3CH_2COOH	propionic acid	propanoic acid
$CH_3CH_2CH_2COOH$	*n*-butyric acid	butanoic acid
$CH_3CH_2CH_2CH_2CH_2COOH$	*n*-caproic acid	hexanoic acid
$CH_3(CH_2)_{10}COOH$	lauric acid	dodecanoic acid
$CH_3(CH_2)_{14}COOH$	palmitic acid	hexadecanoic acid
$CH_3(CH_2)_{16}COOH$	stearic acid	octadecanoic acid

TABLE 25–9 Aliphatic Dicarboxylic Acids

Formula	Name
HOOC—COOH	oxalic acid
HOOC—CH$_2$—COOH	malonic acid
HOOC—CH$_2$CH$_2$—COOH	succinic acid
HOOC—CH$_2$CH$_2$CH$_2$—COOH	glutaric acid
HOOC—CH$_2$CH$_2$CH$_2$CH$_2$—COOH	adipic acid

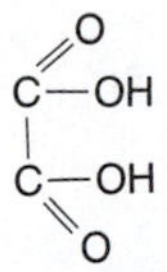

oxalic acid
(an aliphatic acid)

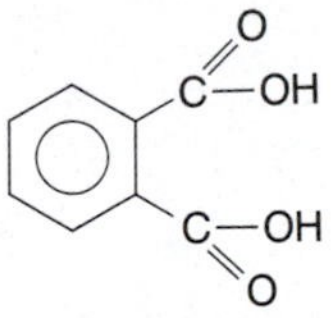

phthalic acid
(an aromatic acid)

Some carboxylic acids contain more than one carboxyl group per molecule. These acids are known almost exclusively by their common names. Oxalic acid is an aliphatic **dicarboxylic acid,** while phthalic acid is a typical aromatic dicarboxylic acid.

Aromatic acids are called by their common names or named as derivatives of benzoic acid, which is considered the "parent" aromatic acid.

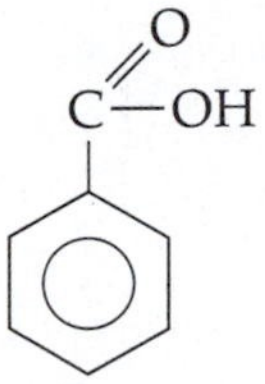

benzoic acid 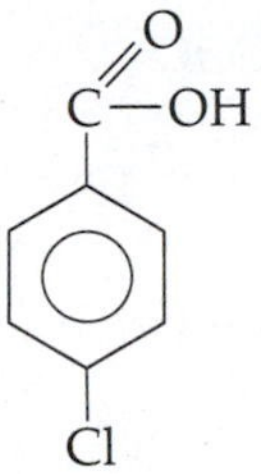   p-chlorobenzoic acid 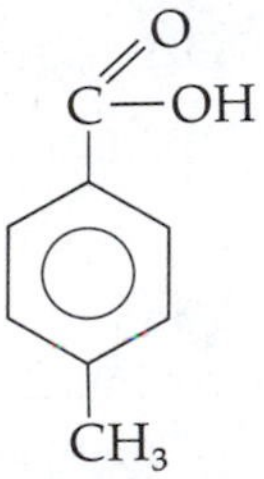   p-toluic acid

Since many reactions of carboxylic acids involve displacement of the —OH group by another atom or group of atoms, we find it useful to name the non-OH portion of acid molecules because they occur in numerous compounds.

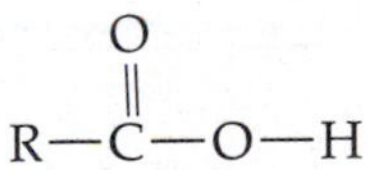

R—C—O—H   R—C—   Ar—C—O—H   Ar—C—

an aliphatic
carboxylic acid

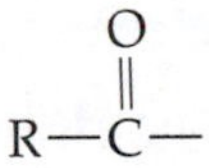

an aliphatic
acyl group

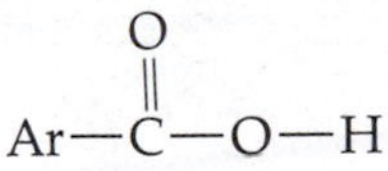

an aromatic
carboxylic acid

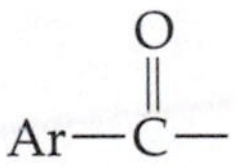

an aromatic
acyl group

Acyl groups are named as derivatives of the parent acid by dropping -ic acid and adding -yl to the characteristic stem. Some examples are:

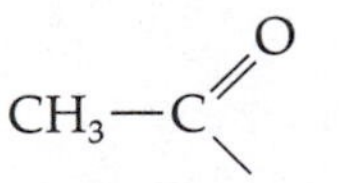

CH$_3$—C   CH$_3$CH$_2$—C   C

acetyl group 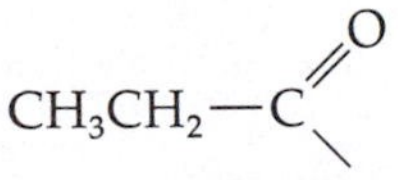   propionyl group 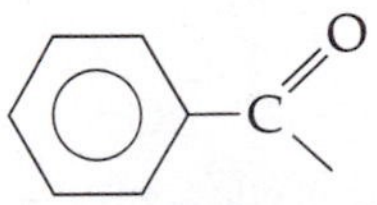   benzoyl group

Four important classes of acid derivatives are formed by the displacement of the hydroxyl group by another atom or group of atoms to form compounds that contain an acyl group.

Aromatic compounds of the same types are encountered frequently.

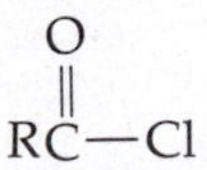

RC—Cl   RC—OR'   RC—O—CR   RC—NH$_2$

an acyl chloride
(an acid chloride)

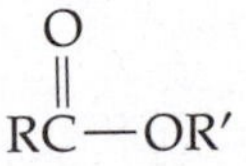

an ester

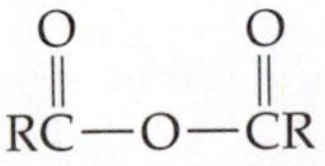

an acid anhydride

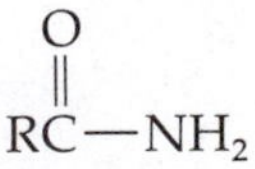

an amide

TABLE 25–10 Ionization Constants of Some Carboxylic Acids

Name	Formula	K_a
formic acid	HCOOH	2.1×10^{-4}
acetic acid	CH_3COOH	1.8×10^{-5}
propionic acid	CH_3CH_2COOH	1.4×10^{-5}
monochloroacetic acid	$ClCH_2COOH$	1.5×10^{-3}
dichloroacetic acid	$Cl_2CHCOOH$	5.0×10^{-2}
trichloroacetic acid	Cl_3CCOOH	2.0×10^{-1}
benzoic acid	C_6H_5COOH	6.3×10^{-5}
phenol	C_6H_5OH	1.3×10^{-10}
ethanol	CH_3CH_2OH	$\sim 10^{-18}$

Phenol and ethanol are not carboxylic acids. They are included in Table 25–10 to show that they are only weakly acidic compared to carboxylic acids.

In Section 17–7 we discussed the extent of ionization of acetic acid.

$$CH_3COOH \rightleftharpoons H^+ + CH_3COO^-$$

$$K_a = \frac{[H^+][CH_3COO^-]}{[CH_3COOH]} = 1.8 \times 10^{-5}$$

It is 1.3% ionized in 0.10 M solution. The acid strengths of the monocarboxylic acids are approximately the same, regardless of the length of the chain. Their acid strengths increase dramatically when highly electronegative substituents are present on the α-carbon atom, i.e., the carbon attached to the —COOH group. (See acetic acid and the three substituted acetic acids listed in Table 25–10.)

25–17 Derivatives of Carboxylic Acids

1 The Acyl Halides (Acid Halides)

The **acyl halides,** sometimes called **acid halides,** are much more reactive than their parent acids, and consequently they are often used in reactions to introduce an acyl group into another molecule. They are usually prepared by treating acids with PCl_3, PCl_5, or $SOCl_2$ (thionyl chloride).

$$\underset{\text{acetic acid}}{CH_3-\overset{\overset{\textstyle O}{\|}}{C}-O-H} + \underset{\substack{\text{phosphorus} \\ \text{pentachloride}}}{PCl_5} \longrightarrow \underset{\substack{\text{acetyl chloride} \\ \text{(an acyl chloride or} \\ \text{an acid chloride)}}}{CH_3-\overset{\overset{\textstyle O}{\|}}{C}-Cl} + HCl\,(g) + \underset{\substack{\text{phosphorus} \\ \text{oxychloride}}}{POCl_3}$$

In general terms, the reaction of acids with PCl_5 may be represented as

$$\underset{\text{acid}}{R-\overset{\overset{\textstyle O}{\|}}{C}-OH} + PCl_5 \longrightarrow \underset{\substack{\text{an acyl chloride or} \\ \text{an acid chloride}}}{R-\overset{\overset{\textstyle O}{\|}}{C}-Cl} + HCl\,(g) + POCl_3$$

2 Esters

When an organic acid is heated with an alcohol, an equilibrium is established with the resulting **ester** and water. The reaction is catalyzed by traces of strong inorganic acids, such as a few drops of concentrated H_2SO_4.

$$\underset{\text{acetic acid}}{CH_3-\overset{\overset{\displaystyle O}{\|}}{C}-OH} + \underset{\text{ethyl alcohol}}{CH_3CH_2-OH} \underset{}{\overset{H^+, \Delta}{\rightleftharpoons}} \underset{\text{ethyl acetate, an ester}}{CH_3\overset{\overset{\displaystyle O}{\|}}{C}-O-CH_2CH_3} + H_2O$$

In general terms the reaction (where R and R′ may be the same or different alkyl groups) may be represented as

Numerous experiments have demonstrated conclusively that the —OH group from the acid and —H from the alcohol are the atoms that form water molecules.

$$\underset{\text{acid}}{R-\overset{\overset{\displaystyle O}{\|}}{C}-O-H} + \underset{\text{alcohol}}{R'-OH} \rightleftharpoons \underset{\text{ester}}{R-\overset{\overset{\displaystyle O}{\|}}{C}-O-R'} + H_2O$$

Esters are nearly always called by their common names, which consist of the name of the alkyl group in the alcohol first, then the name of the anion derived from the acid. This statement does *not* imply that both oxygen atoms come from the acid; they don't.

Reactions between acids and alcohols are usually quite slow and require prolonged boiling (refluxing). However, the reactions between most acyl halides and most alcohols occur very rapidly.

$$\underset{\text{acetyl chloride}}{CH_3-\overset{\overset{\displaystyle O}{\|}}{C}-Cl} + \underset{\text{ethyl alcohol}}{CH_3-CH_2-OH} \longrightarrow \underset{\text{ethyl acetate}}{CH_3-\overset{\overset{\displaystyle O}{\|}}{C}-O-CH_2CH_3} + HCl$$

Most simple esters are pleasant-smelling substances. They are responsible for the flavors and fragrances of most fruits and flowers as well as many of the artificial fruit flavors that are used in cakes, candies, and ice cream (Table 25–11).

Esters of low molecular weight are excellent solvents for nonpolar compounds. Many nail polish removers contain ethyl acetate, an excellent solvent that also gives nail polish removers their characteristic odor.

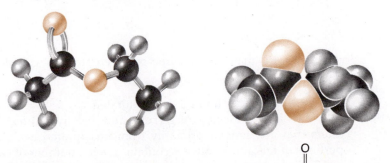

FIGURE 25–20 Models of ethyl acetate, $CH_3-\overset{\overset{\displaystyle O}{\|}}{C}-O-CH_2-CH_3$, an ester. The

$CH_3-\overset{\overset{\displaystyle O}{\|}}{C}-$ fragment is derived from acetic acid, the parent acid, while the —O—CH_2—CH_3 fragment is derived from ethanol, the parent alcohol.

TABLE 25–11   Some Common Esters

	Formula	Odor of
n-butyl acetate	$CH_3COOC_4H_9$	bananas
ethyl butyrate	$C_3H_7COOC_2H_5$	pineapples
n-amyl butyrate	$C_3H_7COOC_5H_{11}$	apricots
n-octyl acetate	$CH_3COOC_8H_{17}$	oranges
isoamyl isovalerate	$C_4H_9COOC_5H_{11}$	apples
methyl salicylate	$C_6H_4(OH)(COOCH_3)$	oil of wintergreen
methyl anthranilate	$C_6H_4(NH_2)(COOCH_3)$	grapes

Most esters are not very reactive, and strong reagents are required for many of their reactions. Esters can be hydrolyzed by refluxing with solutions of strong bases.

$$CH_3-\overset{\overset{\displaystyle O}{\|}}{C}-O-CH_2-CH_3 + Na^+OH^- \xrightarrow{\Delta} CH_3\overset{\overset{\displaystyle O}{\|}}{C}-O^-Na^+ + CH_3CH_2OH$$

ethyl acetate   sodium acetate   ethanol

In general terms the hydrolysis of esters may be represented as

$$R-\overset{\overset{\displaystyle O}{\|}}{C}-O-R' + Na^+OH^- \xrightarrow{\Delta} R-\overset{\overset{\displaystyle O}{\|}}{C}-O^-Na^+ + R'OH$$

ester   salt of an acid   alcohol

The hydrolysis of esters in the presence of strong soluble bases is called **saponification** because the hydrolysis of fats and oils produces soaps.

Fats (solids) and **oils** (liquids) are **triesters** (compounds with three ester groups) formed by the combination of glycerol and aliphatic acids of high molecular weight. "Fatty acids" are any and all organic acids that occur in fats and oils (as esters). Fats and oils may be represented by the general formula

$$CH_2-O-\overset{\overset{\displaystyle O}{\|}}{C}-R$$
$$CH-O-\overset{\overset{\displaystyle O}{\|}}{C}-R$$
$$CH_2-O-\overset{\overset{\displaystyle O}{\|}}{C}-R$$

where the R's may be the same or different groups, and the fatty acid portions

$(-\overset{\overset{\displaystyle O}{\|}}{C}-R)$ may be saturated or unsaturated.

Fats are solid esters of glycerol and (mostly) saturated acids. Oils are liquid esters that are derived primarily from unsaturated acids and glycerol. The acid portions of fats almost always contain an even number of carbon atoms, usually 16 or 18. The acids that occur in fats and oils most frequently are:

Butyric	$CH_3CH_2CH_2COOH$
Lauric	$CH_3(CH_2)_{10}COOH$
Myristic	$CH_3(CH_2)_{12}COOH$
Palmitic	$CH_3(CH_2)_{14}COOH$

Stearic	$CH_3(CH_2)_{16}COOH$
Oleic	$CH_3(CH_2)_7CH{=}CH(CH_2)_7COOH$
Linolenic	$CH_3CH_2CH{=}CHCH_2CH{=}CHCH_2CH{=}CH(CH_2)_7COOH$
Ricinoleic	$CH_3(CH_2)_5CH(OH)CH_2CH{=}CH(CH_2)_7COOH$

Figure 25–21 is a scale model of stearic acid, a long-chain saturated fatty acid.

Naturally occurring fats and oils are mixtures of many different esters. Milk fat, lard, and tallow are familiar important fats. Soybean oil, cottonseed oil, linseed oil, palm oil, and coconut oil are familiar examples of important oils.

The triesters of glycerol are called glycerides. Simple glycerides are esters in which all three R groups are identical. Two examples are

$$\begin{array}{cc}
\overset{\displaystyle O}{\overset{\|}{CH_2OC(CH_2)_{14}CH_3}} & \overset{\displaystyle O}{\overset{\|}{CH_2OC(CH_2)_{16}CH_3}} \\
\overset{\displaystyle O}{\overset{\|}{\underset{|}{CHOC(CH_2)_{14}CH_3}}} & \overset{\displaystyle O}{\overset{\|}{\underset{|}{CHOC(CH_2)_{16}CH_3}}} \\
\overset{\displaystyle O}{\overset{\|}{CH_2OC(CH_2)_{14}CH_3}} & \overset{\displaystyle O}{\overset{\|}{CH_2OC(CH_2)_{16}CH_3}}
\end{array}$$

glyceryl tripalmitate   glyceryl tristearate
(palmitin)   (stearin)

Glycerides are frequently called by their common names, which are indicated in parentheses in the examples above. The common name is the characteristic stem for the parent acid plus an -in ending.

Like other esters, fats and oils can be hydrolyzed in strongly basic solution to produce salts of the acids and the alcohol, glycerol. The resulting salts of long-chain fatty acids are soaps.

$$\begin{array}{l}
\overset{\displaystyle O}{\overset{\|}{CH_2-O-C-(CH_2)_{16}CH_3}} \\
\overset{\displaystyle O}{\overset{\|}{\underset{|}{CH-O-C-(CH_2)_{16}CH_3}}} + 3Na^+OH^- \xrightarrow{\ \Delta\ } 3CH_3-(CH_2)_{16}\overset{\displaystyle O}{\overset{\|}{C}}-O^-Na^+ + \begin{array}{l} CH_2-OH \\ CH\ -OH \\ CH_2-OH \end{array} \\
\overset{\displaystyle O}{\overset{\|}{CH_2-O-C-(CH_2)_{16}CH_3}}
\end{array}$$

glyceryl tristearate,   sodium stearate, a soap   glycerol
a fat

In Section 13–22 we described the cleansing action of soaps and detergents. In Section 25–9 we mentioned the hydrogenation of oils to convert them into fats.

Waxes are esters of fatty acids and alcohols other than glycerol. Most are derived from long-chain fatty acids and monohydric alcohols, both of which usually

FIGURE 25–21 Model of a long chain fatty acid, stearic acid, $CH_3(CH_2)_{16}COOH$.

contain even numbers of carbon atoms. Beeswax is largely $C_{15}H_{31}COOC_{30}H_{61}$, while carnauba wax contains $C_{25}H_{51}COOC_{30}H_{61}$. Both are esters of myricyl alcohol, $C_{30}H_{61}OH$.

3 Polyesters

> Dihydric alcohols contain two —OH groups per molecule.

When *dihydric alcohols* react with *dicarboxylic acids,* ester linkages may be formed at each end of each molecule to build up large molecules containing many ester linkages. The resulting *polymeric esters* are called **polyesters.** The familiar synthetic fiber Dacron is a polyester prepared from ethylene glycol and terephthalic acid.

terephthalic acid　　ethylene glycol　　terephthalic acid　　ethylene glycol

polyethylene terephthalate
(Dacron)

Dacron, the fiber produced from this polyester, absorbs very little moisture, and its properties are very nearly the same when wet or dry. Additionally, it possesses exceptional elastic recovery properties, so it is used to make "no-iron" or "permanent press" fabrics. This polyester can also be made into films, such as Mylar, of great strength. There are many other polyesters.

25 – 18　Amines

The **amines** can be considered to be derivatives of ammonia in which one or more hydrogen atoms have been replaced by alkyl or aryl groups. Amines are basic compounds, and their basicity depends on the nature of the organic substituents (Table 25 – 12).

The odors of amines are quite unpleasant; the malodorous compounds released as fish decays are simple anions. Amines of high molecular weight are

> The common names of tetramethylenediamine, *putrescine,* and pentamethylenediamine, *cadaverine,* are indeed suggestive.

TABLE 25 – 12　Some Properties of Amines

Name	Formula	Boiling Point (°C)	Ionization Constant, K_b
ammonia	NH_3	−33.4	1.8×10^{-5}
methylamine	CH_3NH_2	−6.5	50×10^{-5}
dimethylamine	$(CH_3)_2NH$	7.4	74×10^{-5}
trimethylamine	$(CH_3)_3N$	3.5	7.4×10^{-5}
ethylamine	$CH_3CH_2NH_2$	16.6	47×10^{-5}
aniline	$C_6H_5NH_2$	184	4.2×10^{-10}
ethylenediamine	$H_2NCH_2CH_2NH_2$	116.5	8.5×10^{-5}
pyridine	C_5H_5N	115.3	15×10^{-10}

nonvolatile, and they have little odor. Many aromatic amines are used to prepare organic dyes that are widely used in industrial societies. Amines are also used to produce many medicinal products, including local anesthetics and sulfa drugs.

1 Structure and Nomenclature of Amines

Recall that ammonia is a Lewis base (Section 17–10) because there is one unshared pair of electrons on the nitrogen atom. As a reference point, a $0.10\ M$ solution of aqueous ammonia is 1.3% ionized. (See Section 17–7.)

There are three classes of amines, depending on whether one, two, or three hydrogen atoms have been replaced by organic groups. They are called primary, secondary, and tertiary amines respectively.

Note that the use of the terms primary, secondary, and tertiary for amines differs from that for alcohols (Section 25–15).

NH_3	RNH_2	R_2NH	R_3N
H—N—H $\mid$ H	CH₃—N—H $\mid$ H	CH₃—N—CH₃ $\mid$ H	CH₃—N—CH₃ $\mid$ CH₃
ammonia	methylamine (primary)	dimethylamine (secondary)	trimethylamine (tertiary)

Models of ammonia, methylamine, dimethylamine, and trimethylamine are shown in Figure 25–22.

Aniline is the simplest aromatic amine, and many aromatic amines are named as derivatives of aniline.

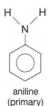

aniline
(primary)

Heterocyclic amines contain nitrogen as a part of a ring, bound to two carbon atoms. Many heterocyclic amines are found in coal tar and a variety of natural products.

pyridine
(tertiary)

pyrrole
(secondary)

quinoline
(tertiary)

purine

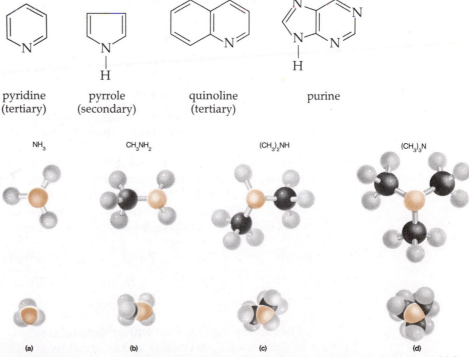

FIGURE 25–22 Models of (a) ammonia, (b) methylamine, (c) dimethylamine, and (d) trimethylamine.

Table 25–12 shows that aliphatic amines are much stronger bases than aromatic and heterocyclic amines. Most low-molecular-weight aliphatic amines are somewhat stronger bases than ammonia.

The aliphatic amines of low molecular weight are soluble in water. Even aliphatic diamines of fairly high molecular weight are soluble in water, because each molecule contains highly polar $-NH_2$ groups. The $-NH_2$ group is called the **amino** group in amines.

Reactions of amines with water are similar to that of ammonia with water (see Section 17–3):

$$NH_3 + H_2O \rightleftharpoons NH_4^+ + OH^-$$
ammonium ion

$$CH_3NH_2 + H_2O \rightleftharpoons CH_3NH_3^+ + OH^-$$
methylammonium ion

$$(CH_3)_2NH + H_2O \rightleftharpoons (CH_3)_2NH_2^+ + OH^-$$
dimethylammonium ion

There are no general methods by which all types of amines can be prepared. We shall leave the preparation of amines for the organic chemistry course.

2 Reactions of Amines

All classes of amines form salts of inorganic acids in reactions similar to the reaction of ammonia and hydrochloric acid (or any inorganic acid, as well as many organic acids).

$$NH_3 + H^+Cl^- \longrightarrow NH_4^+Cl^-$$
ammonium chloride

$$CH_3NH_2 + H^+Cl^- \longrightarrow CH_3NH_3^+Cl^-$$
methylammonium chloride
(methylamine hydrochloride)

Most salts of amines are soluble in water, but insoluble in hydrocarbon solvents. As the above example illustrates, the common names for salts of amines consist of (name of amine) + hydro + (name of anion of acid).

Just as the reaction between ammonium salts and strong bases liberates ammonia, the reaction of strong bases with salts of amines liberates the free amine.

$$NH_4^+Cl^- + Na^+OH^- \longrightarrow NH_3\,(g) + Na^+Cl^- + H_2O$$

$$CH_3NH_3^+Cl^- + Na^+OH^- \longrightarrow CH_3NH_2 + Na^+Cl^- + H_2O$$
methylamine

The fact that amines react with acids to form salts is frequently used to extract amines (for example, strychnine and nicotine) from their natural sources by treatment with acids. The resulting solutions are then made basic with sodium hydroxide, which liberates the free amine.

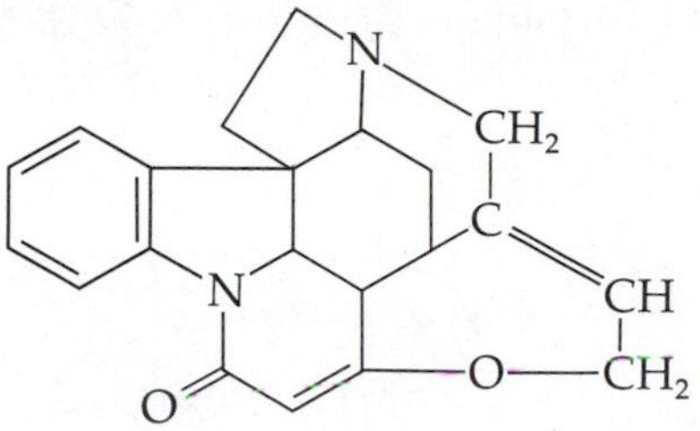

strychnine 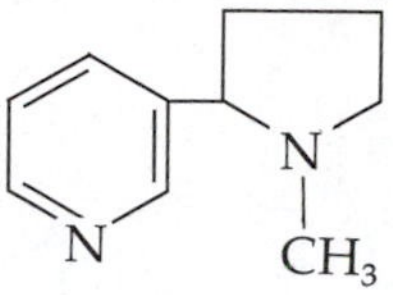   nicotine

25–19 Amides

Amides may be thought of as derivatives of primary or secondary amines and organic acids. Amides contain the $-\overset{O}{\overset{\|}{C}}-N\!\!<$ grouping of atoms. However, they are usually *not* prepared by the reaction of an amine with an organic acid. Acyl halides (acid halides) and acid anhydrides react with primary and secondary amines to produce amides readily. The reaction of an acyl halide with two molecules of a primary or secondary amine produces an amide and a salt of the amine. One molecule of the amine is incorporated in the amide and the other molecule forms a salt.

$$2CH_3NH_2 \;+\; CH_3-\overset{O}{\overset{\|}{C}}-Cl \longrightarrow CH_3-\overset{O}{\overset{\|}{C}}-N\overset{H}{\underset{CH_3}{<}} \;+\; CH_3NH_3{}^+Cl^-$$

methylamine   acetyl chloride   N-methylacetamide   methylammonium
a primary amine   an acyl halide   chloride (a salt)

The reaction of acid anhydrides (see the following box) with primary or secondary amines produces an amide and a salt of the amine. Note that only one molecule of the amine is converted to an amide.

$$2(CH_3)_2NH \;+\; CH_3\overset{O}{\overset{\|}{C}}-O-\overset{O}{\overset{\|}{C}}-CH_3 \longrightarrow CH_3-\overset{O}{\overset{\|}{C}}-N\overset{CH_3}{\underset{CH_3}{<}} \;+\; CH_3-\overset{O}{\overset{\|}{C}}-O^-\;{}^+H_2N(CH_3)_2$$

dimethylamine   acetic anhydride   N,N-dimethyl-   dimethylammonium
a secondary amine   an acid anhydride   acetamide   acetate (a salt)

Acid anhydrides are usually prepared by indirect methods, but this illustration shows the structural relationship that is important for our purposes.

The relationship between a monocarboxylic acid and its anhydrides is:

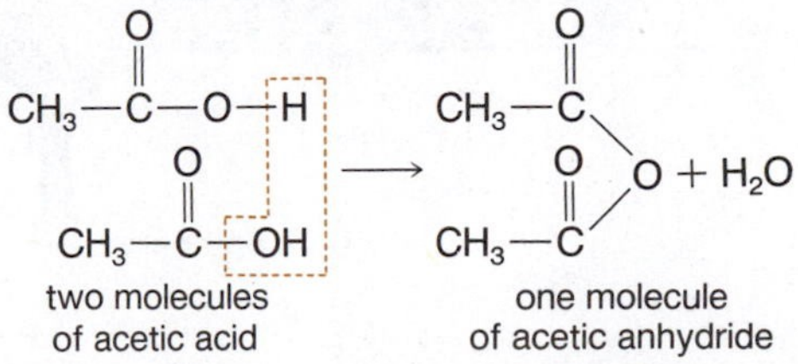

$$CH_3-\overset{O}{\overset{\|}{C}}-O-H$$
$$CH_3-\overset{O}{\overset{\|}{C}}-OH$$
two molecules
of acetic acid
$$\longrightarrow$$
$$CH_3-\overset{O}{\overset{\|}{C}}\!\!\diagdown_{\!\!O}\!\!\diagup^{\!\!O}\!\!\overset{O}{\overset{\|}{C}}-CH_3\;\; O + H_2O$$
one molecule
of acetic anhydride

Acetanilide, the amide of acetic acid and aniline (which is sometimes called antifebrin), is used to treat headaches, neuralgia, and mild fevers. N,N-diethyl-*m-*

toluamide, the amide of metatoluic acid and N,N-diethylamine, is the active ingredient in some insect repellents.

$$CH_3\overset{\overset{\displaystyle O}{\|}}{C}-NH-\hspace{-0.5em}\bigcirc\hspace{2em}\underset{\text{N,N-diethyl-}m\text{-toluamide}}{\overset{\displaystyle CH_3}{\bigcirc}-\overset{\overset{\displaystyle O}{\|}}{C}-N(CH_2CH_3)_2}$$

acetanilide

The polymeric amides are an especially important class of compounds, also called **polyamides. Nylon** is the best known polymeric amide. It is prepared by heating anhydrous hexamethylenediamine with anhydrous adipic acid, a dibasic acid.

$$HO-\overset{\overset{\displaystyle O}{\|}}{C}-(CH_2)_4-\overset{\overset{\displaystyle O}{\|}}{C}-OH + H_2N-(CH_2)_6-NH_2 \xrightarrow[-H_2O]{\text{heat}}$$

adipic acid   hexamethylenediamine

$$-NH\left(\overset{\overset{\displaystyle O}{\|}}{C}-(CH_2)_4-\overset{\overset{\displaystyle O}{\|}}{C}-NH-(CH_2)_6-NH\right)_n\overset{\overset{\displaystyle O}{\|}}{C}-$$

Nylon 6-6
(a polyamide)

The molecular weight of the polymer varies from about 10,000 to about 25,000. It melts at approximately 260° to 270°C.

This substance is known as Nylon 6-6 because the parent diamine and dicarboxylic acid each contain six carbon atoms.

Molten Nylon is drawn into threads that, after cooling to room temperature, can be stretched to about four times their original length (Figure 25–23). The "cold drawing" process orients the polymer molecules so that their longer axes are parallel to the fiber axis. At regular intervals there are N—H----O hydrogen bonds that crosslink oxygen and nitrogen atoms on adjacent chains to give strength to the fiber.

We might note in passing that petroleum is the ultimate source of both adipic acid and hexamethylenediamine. This statement does not imply that these compounds are present in petroleum, only that they are made from petroleum. Similar

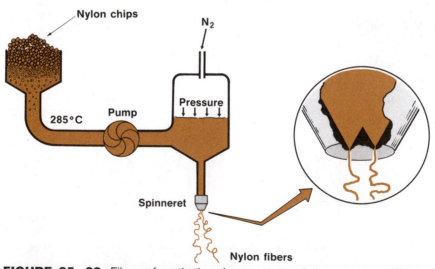

FIGURE 25–23 Fibers of synthetic polymers are made by extrusion of the molten material through tiny holes, called spinnerets. After cooling, they are stretched to about four times their original length.

statements can be made for many industrial chemicals. The price of petroleum is an important factor in our economy because so many different products are derived from petroleum.

25-20 Aldehydes and Ketones

1 Structure and Nonmenclature of Aldehydes and Ketones

Aldehydes and ketones contain the carbonyl group, $\diagdown C{=}O$. **Aldehydes** are compounds in which one alkyl or aryl group and one hydrogen atom are bonded to the carbonyl group. **Ketones** have two alkyl or aryl groups bonded to a carbonyl group.

$$
\underset{\text{aliphatic aldehyde}}{R{-}\overset{\overset{\displaystyle O}{\|}}{C}{-}H} \qquad
\underset{\text{aromatic aldehyde}}{Ar{-}\overset{\overset{\displaystyle O}{\|}}{C}{-}H} \qquad
\underset{\text{aliphatic ketone}}{R{-}\overset{\overset{\displaystyle O}{\|}}{C}{-}R} \qquad
\underset{\text{aromatic ketone}}{Ar{-}\overset{\overset{\displaystyle O}{\|}}{C}{-}Ar} \qquad
\underset{\text{mixed ketone}}{Ar{-}\overset{\overset{\displaystyle O}{\|}}{C}{-}R}
$$

Models of formaldehyde, the simplest aldehyde, and acetone, the simplest ketone, are shown in Figure 25-24.

Aldehydes are usually called by their common names, which are derived from the name of the acid formed when the aldehyde is oxidized (Table 25-13). The systematic name is derived from the name of the parent hydrocarbon. The suffix -al is added to the characteristic stem. The carbonyl group takes positional preference over other substituents.

Simple, commonly encountered ketones are usually called by their common names, which are derived by naming the alkyl or aryl groups attached to the carbonyl group.

$$
\underset{\text{acetone}}{CH_3\overset{\overset{\displaystyle O}{\|}}{C}CH_3} \qquad
\underset{\text{methyl ethyl ketone}}{CH_3\overset{\overset{\displaystyle O}{\|}}{C}CH_2CH_3} \qquad
\underset{\text{diethyl ketone}}{CH_3CH_2\overset{\overset{\displaystyle O}{\|}}{C}CH_2CH_3}
$$

When the benzene ring ($C_6H_5{-}$) is a substituent, it can be called a phenyl group.

cyclohexanone

acetophenone
(methyl phenyl ketone)

benzophenone
(diphenyl ketone)

TABLE 25-13 Properties of Some Simple Aldehydes

Common Name (IUPAC Name)	Formula	Boiling Point (°C)
formaldehyde (methanal)	$H{-}\overset{\overset{\displaystyle O}{\|}}{C}{-}H$	-21
acetaldehyde (ethanal)	$CH_3{-}\overset{\overset{\displaystyle O}{\|}}{C}{-}H$	20.2
propionaldehyde (propanal)	$CH_3CH_2\overset{\overset{\displaystyle O}{\|}}{C}{-}H$	48.8
benzaldehyde	$C_6H_5{-}\overset{\overset{\displaystyle O}{\|}}{C}{-}H$	179.5

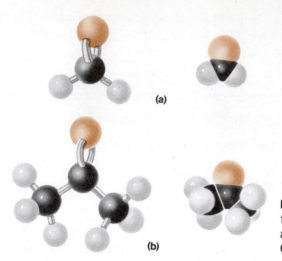

FIGURE 25–24 (a) Models of formaldehyde, HCHO, the simplest aldehyde, and (b) models of acetone, $CH_3—CO—CH_3$, the simplest ketone.

In some cases trivial names that give few clues about structure are used. Acetone, acetophenone, and benzophenone are three examples given above.

The systematic names for ketones are derived from their parent hydrocarbons. The suffix -one is added to the characteristic stem.

$$\overset{O}{\overset{\|}{\underset{1}{CH_3}—\underset{2}{C}—\underset{3}{CH_2}—\underset{4}{CH_3}}}$$

2-butanone

$$\overset{\quad\quad O \quad CH_3}{\overset{\quad\quad\|\quad\;|}{\underset{1}{CH_3}—\underset{2}{CH_2}—\underset{3}{C}—\underset{4}{CH}—\underset{5}{CH_2}—\underset{6}{CH_3}}}$$

4-methyl-3-hexanone

Formaldehyde has been used as a disinfectant and as a preservative for biological specimens (this includes embalming fluid) for many years. Its most important use is in the production of certain plastics and in binders for plywood. Many important natural substances are aldehydes and ketones. Examples include sex hormones, some vitamins, camphor, and the flavors from almonds and cinnamon. Since aldehydes contain a carbon-oxygen double bond, they are very reactive compounds. Reaction usually occurs by addition to the carbonyl group.

The ketones are excellent solvents, and acetone is particularly useful because it dissolves most organic compounds yet is completely miscible with water. Acetone is widely used as a solvent in the manufacture of lacquers, paint removers, explosives, plastics, drugs, and disinfectants. Some ketones of high molecular weight are used extensively in the blending of perfumes. Structures of some naturally occurring aldehydes and ketones are given below.

benzaldehyde
(almonds)

cinnamaldehyde
(cinnamon)

vanillin (vanilla)

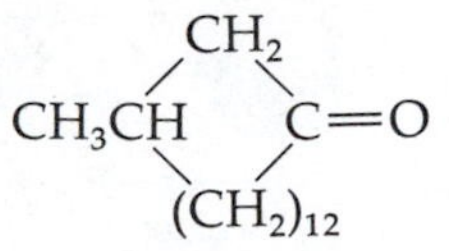

muscone
(musk deer, used
in perfumes)

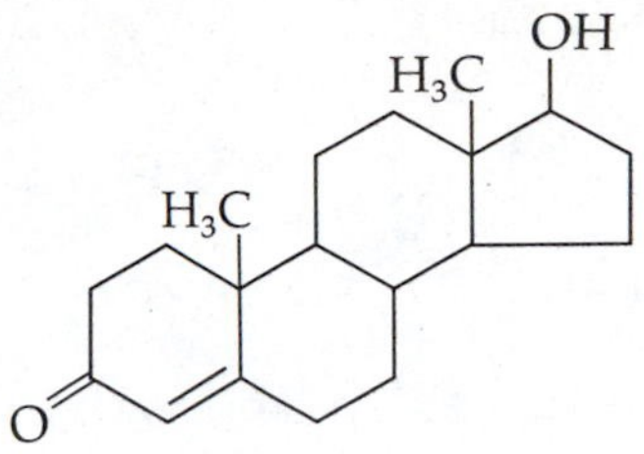

testosterone
(male sex hormone)

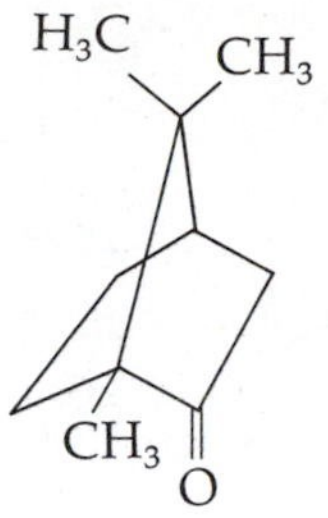

camphor

2 Preparation of Aldehydes and Ketones

Since aldehydes are easily oxidized to acids, they must be removed from the reaction mixture as rapidly as they are formed. Aldehydes have lower boiling points than the alcohols from which they are formed.

Aldehydes may be prepared by the oxidation of *primary* alcohols. The reaction mixture is heated to a temperature slightly above the boiling point of the aldehyde so that it distills out as soon as it is formed. Potassium dichromate, in the presence of dilute sulfuric acid, is the comon oxidizing agent.

$$CH_3OH \xrightarrow[\text{dil. } H_2SO_4]{K_2Cr_2O_7} H-\overset{\overset{\displaystyle O}{\|}}{C}-H$$

methanol
bp = 65°C

methanal (formaldehyde)
bp = -21°C

Ketones may be prepared by the oxidation of *secondary* alcohols. Ketones are not as susceptible to oxidation as are aldehydes, and alkaline solutions of potassium permanganate may be used as the oxidizing agent.

$$CH_3-\overset{\overset{\displaystyle OH}{|}}{CH}-CH_3 \xrightarrow[\text{OH}^-]{KMnO_4} CH_3-\overset{\overset{\displaystyle O}{\|}}{C}-CH_3$$

isopropyl alcohol
2-propanol

acetone

$$CH_3-CH_2-\overset{\overset{\displaystyle OH}{|}}{CH}-CH_3 \xrightarrow[\text{OH}^-]{KMnO_4} CH_3-CH_2-\overset{\overset{\displaystyle O}{\|}}{C}-CH_3$$

2-butanol

2-butanone
methyl ethyl ketone

Aldehydes and ketones may be prepared commercially by a catalytic process that involves passing alcohol vapors and air over a copper gauze or powder at approximately 300°C.

$$2CH_3OH + O_2 \xrightarrow[300°C]{Cu} 2H-\overset{\overset{\displaystyle O}{\|}}{C}-H + 2H_2O$$

methanol

formaldehyde

Formaldehyde is quite soluble in water; the gaseous compound may be dissolved in water to give a 40% solution.

Acetaldehyde may be prepared by the similar oxidation of ethanol.

$$2CH_3CH_2OH + O_2 \xrightarrow[300°C]{Cu} 2CH_3-\overset{\overset{\displaystyle O}{\|}}{C}-H + 2H_2O$$

H—O—H
water

R—O—H
alcohol

R—O—R
ether

FIGURE 25–25 Models showing the structural relationship among water, alcohols, and ethers.

25–21   Ethers

When the word ether is mentioned, most people think of the well-known anesthetic, diethyl ether. There are many ethers, and they are used for a variety of other purposes such as artificial flavors, refrigerants, and a very important class of solvents. **Ethers** are compounds in which an oxygen atom is bonded to two organic groups.

$$-\underset{|}{\overset{|}{C}}-O-\underset{|}{\overset{|}{C}}-$$

Note the difference between an ether and an ester. An ether does not have a carbonyl group.

Remember that alcohols are considered derivatives of water in which one hydrogen atom has been replaced by an organic group. Ethers may be considered derivatives of water in which both hydrogen atoms have been replaced by organic groups (Figure 25–25). However, the similarity is only structural because ethers are not very polar and are chemically quite unreactive.

Three kinds of ethers are known: (1) aliphatic, (2) aromatic, and (3) mixed ethers. Common names are used for ethers in most cases.

$CH_3—O—CH_3$
methoxymethane
dimethyl ether

$CH_3—O—CH_2CH_3$
methoxyethane
methyl ethyl ether

—O—CH_3
methoxybenzene
methyl phenyl ether
anisole
(a mixed ether)

—O—
phenoxybenzene
diphenyl ether
(an aromatic ether)

Diethyl either is a very low boiling liquid, bp = 35°C. Dimethyl ether is a gas that is used as a refrigerant. The aliphatic ethers of higher molecular weights are liquids, and the aromatic ethers are liquids or solids.

Key Terms

Acid anhydride   compound produced by dehydration of a carboxylic acid; general formula is

$$R-\overset{\overset{\displaystyle O}{\|}}{C}-O-\overset{\overset{\displaystyle O}{\|}}{C}-R.$$

Acyl group   group of atoms remaining after removal of an —OH group of a carboxylic acid.

Addition reaction   a reaction in which two atoms are added to a molecule, one on each side of a double or triple bond.

Alcohol   hydrocarbon derivative containing an —OH group attached to a carbon atom that is not in an aromatic ring.

Aldehyde   compound in which an alkyl or aryl group and a hydrogen atom are attached to a carbonyl group; general formula, $R-\overset{\overset{\displaystyle O}{\|}}{C}-H$.

Alkanes see *Saturated hydrocarbons*.

Alkenes (olefins) unsaturated hydrocarbons that contain one or more carbon-carbon double bonds.

Alkylbenzene a compound containing an alkyl group bonded to a benzene ring.

Alkyl group a group of atoms derived from an alkane by the removal of one hydrogen atom.

Alkynes unsaturated hydrocarbons that contain one or more carbon-carbon triple bonds.

Amide compound containing the $-\overset{\overset{\displaystyle O}{\|}}{C}-N\diagup$ group.

Amine compound that can be considered a derivative of ammonia in which one or more hydrogens are replaced by alkyl or aryl groups.

Amino group the $-NH_2$ group.

Aromatic hydrocarbons benzene and its derivatives.

Aryl group group of atoms remaining after a hydrogen atom is removed from an aromatic system.

Carbonyl group the $-\overset{\overset{\displaystyle O}{\|}}{C}-$ group.

Carboxylic acid compound containing a $-\overset{\overset{\displaystyle O}{\|}}{C}-O-H$ group.

Catenation the bonding of an element to itself.

Conformations structures of a compound that differ by the extent of rotation about a single bond.

Conjugated double bonds double bonds that are separated from each other by one single bond, as in $C{=}C-C{=}C$.

Cycloalkanes cyclic saturated hydrocarbons.

Cycloalkenes cyclic hydrocarbons that contain one or more double bonds.

Ester compound of the general formula $R-\overset{\overset{\displaystyle O}{\|}}{C}-O-R'$ where R and R' may be the same or different, and may be either aliphatic or aromatic.

Ether compound in which an oxygen atom is bonded to two alkyl or two aryl groups, or one alkyl and one aryl group.

Fat solid triester of glycerol and (mostly) saturated fatty acids.

Fatty acid an aliphatic acid; many can be obtained from animal fats.

Functional group generally, a group of atoms that represents a potential reaction site in an organic compound.

Glyceride triester of glycerol.

Heterocyclic amine amine in which the nitrogen is part of a ring.

Homologous series a series of compounds in which each member differs from the next by a specific number and kind of atoms.

Hydrocarbons compounds that contain only carbon and hydrogen.

Hydrogenation the reaction in which hydrogen adds across a double or triple bond.

Ketone compound in which a carbonyl group is bound to two alkyl or two aryl groups, or to one alkyl and one aryl group.

Phenol hydrocarbon derivative containing an $-OH$ group bound to an aromatic ring.

Polyamide a polymeric amide.

Polyester a polymeric ester.

Polyhydric alcohol an alcohol containing more than one $-OH$ group.

Polymerization the combination of many small molecules to form large molecules.

Polymers large molecules formed by the combination of many small molecules.

Octane number a number that indicates how smoothly a gasoline burns.

Oil liquid triester of glycerol and unsaturated fatty acids.

Organic chemistry the chemistry of substances that contain carbon-hydrogen bonds.

Saponification hydrolysis of esters in the presence of strong soluble bases.

Saturated hydrocarbons hydrocarbons that contain only single bonds. They are also called *alkanes* or *paraffin hydrocarbons*.

Soap sodium salt of long chain fatty acid.

Structural isomers compounds that contain the same number of the same kinds of atoms in different geometric arrangements.

Substitution reaction a reaction in which an atom or a group of atoms is replaced by another atom or group of atoms.

Triester compound containing three ester groups.

Unsaturated hydrocarbons hydrocarbons that contain double or triple carbon-carbon bonds.

Vulcanization the process in which sulfur is added to rubber and heated to 140°C.

Exercises

Basic Ideas

1. (a) What is organic chemistry? (b) What was the "vital force" theory? (c) What happened to the "vital force" theory?
2. (a) What is catenation? (b) How is carbon unique among the elements?
3. How many "everyday" uses of organic compounds can you think of? List them.
4. (a) What are the principal sources of organic compounds? (b) Some chemists argue that the ultimate source of all naturally occurring organic compounds is carbon dioxide. Could this be possible? Hint: Think about the origin of coal, natural gas, and petroleum.

Alkanes (Saturated Hydrocarbons)

5. (a) What are hydrocarbons? (b) What are saturated hydrocarbons? (c) What are the alkanes?
6. Describe the bonding in and the geometry of molecules of the following alkanes. (a) methane, (b) ethane, (c) propane, (d) n-butane. How are the formulas for these compounds similar? Different?
7. (a) What are "normal" hydrocarbons? (b) What are branched-chain hydrocarbons? (c) Cite three examples of each.
8. (a) What is a homologous series? (b) Provide specific examples of compounds that are members of a homologous series. (c) What is a methylene group?
9. (a) How do the melting points and boiling points of the normal alkanes vary with molecular weight? (b) Do you expect them to vary in this order? (c) Why?

Isomerism and Nomenclature

10. (a) What are structural isomers? (b) Draw all possible structural isomers for compounds that have the following formulas: C_4H_{10}, C_5H_{12}, C_6H_{14}. (c) Why are there no structural isomers for C_3H_8?
11. Draw structural formulas for and write the names of the first ten straight-chain alkanes.
12. (a) What are alkyl groups? (b) Draw structures for and write the names of five alkyl groups. (c) What is the origin of the name for an alkyl group?
13. Write the name for each structural isomer you drew in answering Exercise 10.
14. What are conformations?

Reactions of the Alkanes

15. Why are the alkanes also called paraffin hydrocarbons?
16. (a) What is a substitution reaction? (b) What is a halogenation reaction?
17. (a) Describe the reaction of methane with chlorine in the presence of ultraviolet light. (b) Write equations that show formulas for all compounds that can be formed by reaction (a). (c) Write names for all compounds in these equations.
18. (a) Describe the reaction of ethane with chlorine in the presence of ultraviolet light. (b) Write equations that show formulas for all compounds that can be formed by reaction (a). (c) Write names for all compounds in these equations.
19. Why are the halogenation reactions of the alkanes of limited value?
20. What is a free radical chain reaction?
21. (a) What are the cycloalkanes? (b) How do they differ from other alkanes?
22. Why does halogenation of cycloalkanes give a single product, whereas halogenation of normal alkanes give a multiplicity of products?

Oxidation of Alkanes

23. (a) What are the principal uses of hydrocarbons? (b) Why are hydrocarbons used for these purposes? (c) What does heat of combustion mean? Illustrate.
24. How does the heat of combustion of ethyl alcohol compare with the heats of combustion of saturated hydrocarbons on a per mole as well as per gram basis?

Petroleum

25. (a) What is petroleum? (b) What are the principal fractions obtained by the distillation of petroleum? (c) For what is each fraction used?
26. (a) What do we mean when we say that higher-molecular-weight hydrocarbons can be cracked? (b) Why is "cracking" important?
27. (a) To what does the octane rating of a gasoline refer? (b) Why is octane rating important?
28. What kinds of hydrocarbons have low octane ratings? High octane ratings?

Unsaturated Hydrocarbons

29. (a) What are alkenes? (b) What other names are used to describe alkenes?

30. (a) How does the general formula for the alkenes differ from the general formula for the alkanes? (b) Why are the general formulas identical for alkenes and cycloalkanes that contain the same number of carbon atoms?
31. Write names and formulas for six alkenes.
32. (a) What are dienes? (b)Provide some examples.
33. (a) What are cycloalkenes? (b) What is their general formula? (c) Provide three examples.

Bonding in the Alkenes

34. Describe the bonding at each carbon atom in (a) ethene, (b) propene, (c) 1-butene, and (d) 2-butene.
35. (a) What are geometric (structural) isomers? (b) Why is rotation around a double bond not possible at room temperature? (c) What do *cis* and *trans* mean? (d) Draw structures for *cis*- and *trans*-1,2-dichloroethene. How do their melting and boiling points compare?
36. Distinguish between conformations and isomers.
37. How do carbon-carbon single bond lengths and carbon-carbon double bond lengths compare? Why?
38. Most reactions of the alkanes that do not disrupt the carbon skeleton are substitution reactions, while the alkenes are characterized by addition to the double bond. What does this statement mean?
39. (a) What are functional groups? (b) What is the functional group that occurs in all alkenes?
40. Write equations for two reactions in which halogens undergo addition reactions with alkenes. Name all compounds.
41. How can bromination be used to distinguish between alkenes and alkanes?
42. (a) What is hydrogenation? (b) Why is it important? (c) Write equations for two reactions that involve hydrogenation of alkenes. (d) Name all compounds in (c).
43. What is the difference between a vegetable oil and a shortening? Illustrate.

Polymers

44. (a) What is polymerization? (b) Write equations for three polymerization reactions.
45. (a) What is rubber? (b) What is vulcanization? (c) What is the purpose of vulcanizing rubber? (d) What are fillers and reinforcing agents? (e) What is their purpose?

46. (a) What is an elastomer? (b) Cite a specific example. (c) What are some of the advantages of neoprene compared to natural rubber?

Oxidation of Alkenes

47. (a) Describe two qualitative tests that can be used to distinguish between alkenes and alkanes. (b) Cite some specific examples. (c) What functional group is the basis for the qualitative distinction between alkanes and alkenes?

The Alkynes

48. (a) What are alkynes? (b) What other name is used to describe them? (c) What is the general formula for alkynes? (d) How does the general formula for alkynes compare with the general formula for cycloalkenes? Why?
49. (a) What is the most familiar alkyne? (b) What is its common name? (c) For what is it commonly used? (d) How is it prepared?
50. (a) Write names and formulas for four alkynes. (b) What is the functional group in alkynes?

Bonding in the Alkynes

51. Describe the bonding in and the geometry of the alkynes you listed in Exercise 50.

Reactions of the Alkynes

52. (a) Why are alkynes more reactive than alkenes? (b) What is the most common kind of reaction that alkynes undergo? (c) Write equations for four such reactions. (d) Name all compounds in part (c).

Aromatic Hydrocarbons

53. (a) What are aromatic hydrocarbons? (b) What is the principal source of aromatic hydrocarbons?
54. (a) What is the simplest aromatic hydrocarbon? (b) From what source was it first isolated? When?
55. What is coal tar? Coal gas?

Bonding in Benzene

56. (a) Describe the structure of and the bonding in benzene. (b) What evidence indicates conclusively that benzene has a symmetrical ring structure?
57. (a) Draw resonance structures for benzene. (b) What are resonance structures? (c) What do we mean when we say that the electrons asso-

ciated with the pi bonds in benzene are delocalized over the entire ring?

Other Aromatic Hydrocarbons

58. (a) Write names and draw structural formulas for seven aromatic hydrocarbons (other than benzene). (b) How is each structurally related to benzene?

59. What are coal-tar crudes?

60. How are the prefixes *ortho-*, *meta-*, and *para-* used in naming aromatic compounds?

Reactions of Aromatic Hydrocarbons

61. (a) What is the most common kind of reaction that the benzene ring undergoes? (b) Write equations for the reaction of benzene with chlorine in the presence of an iron catalyst and for the analogous reaction with bromine.

62. Write equations to illustrate both aromatic and aliphatic substitution reactions of toluene using (a) chlorine and (b) bromine.

63. Write equations to illustrate the nitration of benzene and toluene.

64. (a) Do you expect aromatic hydrocarbons to produce soot as they burn? Why? (b) Would you expect the flames to be blue or yellow?

65. Write equations to illustrate the oxidation of the following aromatic hydrocarbons by potassium permanganate in basic solution. (a) toluene, (b) ethylbenzene, (c) 1,2-dimethylbenzene.

Alcohols and Phenols — Structure and Nomenclature

66. (a) What are alcohols and phenols? (b) How do they differ? (c) Why can alcohols and phenols be viewed as derivatives of hydrocarbons? As derivatives of water?

67. Distinguish between alkyl and aryl groups.

68. Draw structural formulas and write names for six alcohols and three phenols.

69. (a) Distinguish among primary, secondary, and tertiary alcohols. (b) Write names and formulas for three alcohols of each type.

70. (a) Draw structural formulas for and write the names of the four (saturated) alcohols that contain four carbon atoms and one —OH group per molecule. (b) Draw structural formulas for and write the names of the eight (saturated) alcohols that contain five carbon atoms and one —OH group per molecule.

Which ones may be classified as primary alcohols? Secondary alcohols? Tertiary alcohols?

71. What are glycols?

Physical Properties of Alcohols and Phenols

72. Refer to Table 25–7 and explain the trends in boiling points and solubilities of alcohols in water listed there.

73. Why are glycols more soluble in water than monohydric alcohols that contain the same number of carbon atoms?

74. Why are most phenols only slightly soluble in water?

Preparation of Alcohols

75. Why are methyl alcohol and ethyl alcohol called wood alcohol and grain alcohol respectively?

76. (a) How is methyl alchol prepared commercially? (b) List some uses for methyl alcohol.

77. (a) What is the important source of ethyl alcohol obtained by fermentation? (b) Write the overall equation for this reaction. (c) What is blackstrap molasses?

78. Describe the industrial preparation of ethyl alcohol and write balanced equations for the reactions.

79. Describe the industrial preparation of ethylene glycol and write balanced equations for the reactions.

Reactions of Alcohols

80. (a) Write equations for the reactions of three alcohols with metallic sodium. (b) Name all compounds in these equations. (c) Are these reactions similar to the reaction of metallic sodium with water? How?

81. (a) What are alkoxides? (b) What do we mean when we say that the low molecular weight alkoxides are strong bases?

82. (a) Write equations for the reaction of nitric acid with the following alcohols: methanol, ethanol, *n*-propanal, *n*-butanol. (b) Name the inorganic ester formed in each case. (c) What are inorganic esters?

83. (a) What is nitroglycerine? (b) How is it produced? (c) List two important uses for nitroglycerine. Are they similar?

84. (a) What are alkyl hydrogen sulfates? (b) How are they prepared? (c) Can alkyl hydrogen sulfates be classified as acids? Why? (d) What are detergents? (e) What is sodium lauryl sul-

fate? (f) What is the common use of sodium lauryl sulfate?

Carboxylic Acids — Names and Formulas

85. (a) What are carboxylic acids? (b) Draw structural formulas for and write the names of seven carboxylic acids. (c) Why are aliphatic carboxylic acids sometimes called fatty acids? Cite two examples.

Reactions of Carboxylic Acids

86. (a) What are acyl chlorides or acid chlorides? (b) How are they prepared? (c) Write equations for the preparation of four acid chlorides. Name all compounds.
87. (a) What are esters? (b) Draw structures for four esters and write their names.
88. Write equations for the formation of three different esters, starting with an acid and an alcohol in each example. Name all compounds.
89. Write equations for the formation of three esters, starting with an acid chloride and an alcohol in each case. Name all compounds.

Occurrence and Hydrolysis of Esters

90. List six naturally occurring esters and their sources.
91. (a) What is saponification? (b) Why is this kind of reaction called saponification?
92. Write equations for the hydrolysis of (a) methyl acetate, (b) ethyl formate, (c) n-butyl acetate, and (d) n-octyl acetate. Name all products.
93. (a) What are fats? Oils? (b) Write the general formula for fats and oils.
94. What are glycerides? Distinguish between simple glycerides and mixed glycerides.
95. Write the names and formulas of some acids that occur in fats and oils (as esters, of course).
96. Write names and formulas for three simple glycerides.
97. Write equations for the hydrolysis (saponification) of glyceryl tristearate and glyceryl tripalmitate.
98. (a) What are soaps? (b) What is the difference between a soap and a detergent?
99. What are waxes?

Polyester

100. (a) What are polyesters? (b) What is Dacron? (c) How is Dacron prepared?

Amines

101. (a) What are amines? (b) Why are amines thought of as derivatives of ammonia?
102. Distinguish among primary, secondary, and tertiary amines.
103. Write names and formulas for three amines in each of the following classes without using any compound more than once: (a) aliphatic amines, (b) aromatic amines, (c) primary aliphatic amines, (d) secondary aliphatic amines, and (e) tertiary aliphatic amines.
104. Why are aqueous solutions of amines basic?
105. Show that the reactions of amines with inorganic acids such as HCl are similar to the reactions of ammonia with inorganic acids.

Amides

106. (a) What are amides? (b) Draw structural formulas for and write the names of three amides.
107. Write equations for the preparation of an amide starting with
    (a) a primary amine and an acid chloride
    (b) a secondary amine and an acid chloride
    (c) a primary amine and an acid anhydride
    (d) a secondary amine and an acid anhydride
108. What are polyamides? Why are they an important class of compounds?
109. (a) What is Nylon? (b) How is it prepared?
110. Common Nylon is called Nylon 6-6. (a) What does this mean? (b) Can you visualize other Nylons?

Aldehydes and Ketones

111. (a) Distinguish between aldehydes and ketones. (b) Cite three examples (each) of aliphatic and aromatic aldehydes and ketones by drawing structural formulas and naming the compounds.
112. (a) List several naturally occurring aldehydes and ketones. (b) What are their sources? (c) What are some uses of these compounds?
113. Describe the preparation of three aldehydes from alcohols and write appropriate equations. Name all reactants and products.
114. Describe the preparation of three ketones from alcohols and write appropriate equations. Name all reactants and products.
115. (a) Describe the commercial preparation of aldehydes and ketones. (b) Write equations for

the preparation of two aldehydes and two ketones. (c) Name all reactants and products.

116. What is dehydrogenation? Does it correspond to oxidation? Why?

117. How can aldehydes be converted to alcohols? To acids?

Ethers

118. (a) What are ethers? (b) Draw structural formulas for four ethers. Name them.

119. Distinguish between simple ethers and mixed ethers.

120. List some uses for some ethers.

Appendices

Appendix A   Some Mathematical Operations

We frequently use very large or very small numbers in chemistry, and such numbers are conveniently expressed in *scientific* or *exponential notation.*

A.1   Scientific Notation

In scientific notation a number is expressed as the *product of two numbers.* By convention, the first number, called the digit term, is between 1 and 10. The second number, called the *exponential term,* is an integer power of 10. Some examples follow.

$$10000 = 1 \times 10^4 \qquad\qquad 24327 = 2.4327 \times 10^4$$
$$1000 = 1 \times 10^3 \qquad\qquad 7958 = 7.958 \ \times 10^3$$
$$100 = 1 \times 10^2 \qquad\qquad 594 = 5.94 \ \ \times 10^2$$
$$10 = 1 \times 10^1 \qquad\qquad 98 = 9.8 \ \ \ \times 10^1$$
$$1 = 1 \times 10^0 \qquad \text{Recall, (any base)}^0 = 1 \text{ (by definition)}$$
$$1/10 = 0.1 = 1 \times 10^{-1} \qquad\qquad 0.32 = 3.2 \ \ \times 10^{-1}$$
$$1/100 = 0.01 = 1 \times 10^{-2} \qquad\qquad 0.067 = 6.7 \ \ \times 10^{-2}$$
$$1/1000 = 0.001 = 1 \times 10^{-3} \qquad\qquad 0.0049 = 4.9 \ \ \times 10^{-3}$$
$$1/10000 = 0.0001 = 1 \times 10^{-4} \qquad\qquad 0.00017 = 1.7 \ \ \times 10^{-4}$$

The exponent of 10 (the power to which 10 is raised) tells us how many places the decimal point must be shifted to give the number in long form. A *positive exponent* indicates that the decimal point is *shifted right* that number of places to give the long form, and a *negative exponent* indicates that the decimal point is *shifted left*, as the following examples illustrate.

$$7.3 \times 10^3 = 73 \times 10^2 \quad = 730 \times 10^1 \quad = 7300$$

$$8.6 \times 10^{-3} = 86 \times 10^{-4} \quad = 860 \times 10^{-5}$$

$$0.0086 = 0.086 \times 10^{-1} = 0.86 \times 10^{-2} = 8.6 \times 10^{-3}$$

When numbers are written in *standard scientific notation* there is one nonzero digit to the left of the decimal point.

1. Addition and Subtraction In addition and subtraction, all numbers are converted to the same power of 10 and the digit terms are added or subtracted.

$$(4.21 \times 10^{-3}) + (1.4 \times 10^{-4}) = (4.21 \times 10^{-3}) + (0.14 \times 10^{-3})$$
$$= \underline{4.35 \times 10^{-3}}$$

$$(8.97 \times 10^4) - (2.31 \times 10^3) = (8.97 \times 10^4) - (0.231 \times 10^4)$$
$$= \underline{8.74 \times 10^4}$$

2. Multiplication The digit terms are multiplied in the usual way, the exponents are added algebraically, and the product is written with one nonzero digit to the left of the decimal.

$$(4.7 \times 10^7)(1.6 \times 10^2) = (4.7)(1.6) \times 10^{7+2}$$
$$= 7.52 \times 10^9 = \underline{7.5 \times 10^9} \text{ (two sig. figs.)}$$

$$(8.3 \times 10^4)(9.3 \times 10^{-9}) = (8.3)(9.3) \times 10^{4-9}$$
$$= 77.19 \times 10^{-5} = \underline{7.7 \times 10^{-4}} \text{ (two sig. figs.)}$$

3. Division The digit term of the numerator is divided by the digit term of the denominator, the exponents are subtracted algebraically, and the quotient is written with one nonzero digit to the left of the decimal.

$$\frac{8.4 \times 10^7}{2.0 \times 10^3} = \frac{8.4}{2.0} \times 10^{7-3}$$
$$= \underline{4.2 \times 10^4}$$

$$\frac{3.9 \times 10^9}{8.4 \times 10^{-3}} = \frac{3.9}{8.4} \times 10^{9-(-3)}$$
$$= 0.4643 \times 10^{12} = \underline{4.6 \times 10^{11}} \text{ (two sig. figs.)}$$

4. Powers of Exponentials The digit term is raised to the indicated power in the usual way, and the exponent is multiplied by the number that indicates the power.

$$(1.2 \times 10^3)^2 = (1.2)^2 \times 10^{3 \times 2}$$
$$= 1.44 \times 10^6 = \underline{1.4 \times 10^6} \text{ (two sig. figs.)}$$

$$(3.0 \times 10^{-3})^4 = (3.0)^4 \times 10^{-3 \times 4}$$
$$= 81 \times 10^{-12} = \underline{8.1 \times 10^{-11}}$$

Electronic Calculators: *To square a number:* (1) enter the number, (2) touch the (x^2) button.

$(7.3)^2 = 53.29 = \underline{53}$ (two sig. figs.)

To raise a number to a higher power: (1) enter the number, (2) touch the (y^x) button, (3) enter the power, and (4) touch the $(=)$ button.

$(7.3)^4 = 2839.8241 = \underline{2.8 \times 10^3}$ (two sig. figs.)

$(7.30 \times 10^2)^5 = 2.0730716 \times 10^{14} = \underline{2.07 \times 10^{14}}$ (three sig. figs.)

5. Roots of Exponentials The exponent must be divisible by the desired root if a calculator is not used. The root of the digit term is extracted in the usual way and the exponent is divided by the desired root.

$$\sqrt{2.5 \times 10^5} = \sqrt{25 \times 10^4} = \sqrt{25} \times \sqrt{10^4} = \underline{5.0 \times 10^2}$$

$$\sqrt[3]{2.7 \times 10^{-8}} = \sqrt[3]{27 \times 10^{-9}} = \sqrt[3]{27} \times \sqrt[3]{10^{-9}} = \underline{3.0 \times 10^{-3}}$$

Electronic Calculators: *To extract the square root of a number:* (1) enter the number, and (2) touch the $(\sqrt{x})$ button.

$\sqrt{23} = 4.7958315 = \underline{4.8}$ (two sig. figs.)

To extract a higher order root: (1) enter the number, (2) touch the $(\sqrt[x]{y})$ button, (3) enter the root to be extracted, and (4) touch the $(=)$ button.

$\sqrt[3]{12.0} = 2.2894285 = \underline{2.29}$ (three sig. figs.)

$\sqrt[4]{1.2 \times 10^{-9}} = 5.8856619 \times 10^{-3} = \underline{5.9 \times 10^{-3}}$ (two sig. figs.)

A.2 Logarithms

The power to which the base 10 must be raised to equal a number is called the *logarithm* (log) of the number. In the simplest view a logarithm is an exponent. The number 10 must be raised to the third power to equal 1000. Therefore the logarithm of 1000 is 3, written as $\log 1000 = 3$. Some examples are listed below.

Number	Exponential Expression	Logarithm
1000	10^3	3
100	10^2	2
10	10^1	1
1	10^0	0
1/10 = 0.1	10^{-1}	-1
1/100 = 0.01	10^{-2}	-2
1/1000 = 0.001	10^{-3}	-3

Most numbers are neither exact multiples nor exact fractions of 10. A table of common (base 10) logarithms is a tabulation of *fractional powers of the base 10*. These fractional powers are called *mantissas*. The four-digit numbers in Appendix B are mantissas or fractional powers of 10.

The following examples are illustrative:

$10^{0.0414} = 1.1,$ so $\log 1.1 = 0.0414$ $\qquad 10^{0.3222} = 2.1,$ so $\log 2.1 = 0.3222$

$10^{0.6435} = 4.4,$ so $\log 4.4 = 0.6435$ $\qquad 10^{0.9956} = 9.9,$ so $\log 9.9 = 0.9956$

Note that the first two digits of a number are tabulated in the left-hand vertical column, and the third digit is listed in the top horizontal row. For example, we find the log of 3.33 by locating 3.3 in the first vertical column and moving right until we come to the mantissa under the 3 column, where we find .5224. Thus,

$$\log 3.33 = 0.5224 \qquad \text{or} \qquad 10^{0.5224} = 3.33$$

The above procedure can be reversed. Suppose that we know that $\log x = 0.7259$. How do we obtain x? First, the terminology. The number x is called the *antilogarithm* of 0.7259. We locate 7259 in table of mantissas, look to the left and see that 7259 is located in row that starts with 5.3. Then we look to the top of column and see that 7259 is located in the column that is headed by 2. Therefore,

$$\text{antilog } 0.7259 = 5.32 \qquad \text{or} \qquad 10^{0.7259} = 5.32$$

For numbers much larger or much smaller than those in these examples, logarithms may be obtained by expressing the number in exponential form. Remember: logarithms are powers of 10 and they are treated like all other exponents, *i.e.*, when two numbers are multiplied together their exponents are added.

$$\log 37 = \log (3.7 \times 10^1) = \log 3.7 + \log 10^1 = 0.5682 + 1 = \underline{1.5682}$$

so

$$\log 37 = 1.5682 \qquad \text{or} \qquad 10^{1.5682} = 37$$

Recall that $\log 10 = 1$ and $\log 100 = 2$, so log 37 must be greater than 1 but less than 2.

$$\log 0.037 = \log (3.7 \times 10^{-2}) = \log 3.7 + \log 10^{-2} = 0.5682 + (-2) = -\underline{1.4318}$$

so

$$\log 0.037 = -1.4318 \qquad \text{or} \qquad 10^{-1.4318} = 0.037$$

Recall that $\log 0.01 = -2$ and $\log 0.1 = -1$, so log 0.037 must be greater than -2 but less than -1.

Electronic Calculators: To obtain the logarithm of a number from an electronic calculator, (1) enter the number, and (2) touch the (log x) button.

$$\log 7.4 = 0.86923172 = \underline{0.869} \text{ (three sig. figs.)}$$

$$\log 7.4 \times 10^{-3} = -2.1307683 = -\underline{2.13} \text{ (three sig. figs.)}$$

To obtain an antilogarithm, (1) enter the logarithm, (2) touch the (INV) button, and then (3) touch the (log x) button.

$$\text{Antilog } 1.301 = 19.998619 = \underline{20.0} \text{ (three sig. figs.)}$$

1. Multiplication Using Logarithms $\quad \log xy = \log x + \log y$

To multiply two numbers, we (1) add their logarithms and then (2) find the antilogarithm (number) that corresponds to the logarithm of the product. Suppose we wish to multiply 213 by 42.6.

$$
\begin{array}{l}
\log 213 = 2.3284 \\
\underline{+\log 42.6 = 1.6294} \\
\log \text{product} = 3.9578 \approx 3 + 0.9576 \text{ (nearest mantissa listed} \\
\qquad\qquad\qquad\qquad\qquad\qquad \text{in Appendix B)}
\end{array}
$$

$$\text{antilog } 3.9576 = 9.07 \times 10^3$$

Interpolation is a process for approximating values between those listed in a table. If you need to use it, you can find an explanation in any common algebra book.

We are justified in reporting only 3 significant figures, since we have not interpolated. We've simply taken the antilog of the mantissa nearest to 0.9578.

Thus, we write $(213)(42.6) = 9.07 \times 10^3$ (three sig. figs.)

2. Division Using Logarithms $\log \dfrac{x}{y} = \log x - \log y$

Division is just the reverse of multiplication. Suppose we wish to divide 213 by 42.6.

$$\begin{array}{r} \log 213 = 2.3284 \\ -\log 42.6 = 1.6294 \\ \hline \log \text{quotient} = 0.6990 \end{array}$$

antilog $0.6990 = 5.00$

Or, we may write $213/42.6 = 5.00$

3. Extraction of Higher Order Roots Using Logarithms

Although most electronic calculators have the capability of extracting higher order roots (p. 863), your instructor may insist that you use logarithms to demonstrate that you understand the process.

Recall that $\sqrt[5]{y} = (y)^{1/5}$ or in general terms $\sqrt[n]{y} = (y)^{1/n}$. To evaluate an expression such as

$$(2x)^2(3x)^3 = 1.6 \times 10^{-72}$$

we perform the indicated multiplication and obtain:

$$108\, x^5 = 1.6 \times 10^{-72} \qquad \text{or} \qquad x^5 = 1.48 \times 10^{-74}$$

Since we wish to extract the fifth root, the exponent must be divisible by five if a calculator is not used. Therefore, we decrease the exponent to -75 by shifting the decimal point one place to the right,

$$x^5 = 14.8 \times 10^{-75}$$

and extract the fifth root of the exponential term.

$$x = \sqrt[5]{14.8 \times 10^{-75}} = \sqrt[5]{14.8} \times \sqrt[5]{10^{-75}} = \sqrt[5]{14.8} \times 10^{-15}$$

To obtain the fifth root of 14.8, we (1) look up its logarithm, (2) divide the logarithm by five, and (3) take the antilog of the logarithm obtained in step (2).

(1) $\log 14.8 = 1.1703$

(2) $\dfrac{\log 14.8}{5} = \dfrac{1.1703}{5} = 0.2341$

(3) antilog $0.2341 = 1.71$

Thus, $\sqrt[5]{14.8} = 1.71$ and $\sqrt[5]{14.8 \times 10^{-75}} = 1.71 \times 10^{-15} = x$ which should be reported as

$$x = 1.7 \times 10^{-15}$$

because the original equation, $108\, x^5 = 1.6 \times 10^{-72}$, contains only two significant figures on the right side.

A.3 Quadratic Equations

Algebraic expressions of the form

$$ax^2 + bx + c = 0$$

are called **quadratic equations.** Each of the constant terms (a, b, and c) may be either positive or negative. All quadratic equations may be solved by the **quadratic formula:**

$$x = \frac{-b \pm \sqrt{b^2 - 4ac}}{2a}$$

If we wish to solve the quadratic equation $3x^2 - 4x - 8 = 0$, we observe that $a = 3$, $b = -4$, and $c = -8$. Substitution of these values into the quadratic formula gives

$$x = \frac{-(-4) \pm \sqrt{(-4)^2 - 4(3)(-8)}}{2(3)} = \frac{4 \pm \sqrt{16 + 96}}{6} = \frac{4 \pm \sqrt{112}}{6} = \frac{4 \pm 10.6}{6}$$

The two roots of this quadratic equation are

$$x = 2.4 \qquad \text{and} \qquad x = -1.1$$

As you construct and solve quadratic equations based on the observed behavior of matter, you must decide which root has physical significance. Examination of the *equation that defines x* always gives clues about possible values for x. In this way you can tell which is extraneous (has no physical significance). Negative roots are *usually* extraneous.

When you have solved a quadratic equation you should always check the values you obtain by substitution into the original equation. In the above example we obtained $x = 2.4$ and $x = -1.1$. Substitution of these values into the original equation, $3x^2 - 4x - 8 = 0$, shows that both roots are correct. Note, however, that such substitutions often don't give a perfect check because some round-off error has been introduced.

Appendix B

Four-Place Table of Logarithms

	0	1	2	3	4	5	6	7	8	9
1.0	.0000	.0043	.0086	.0128	.0170	.0212	.0253	.0294	.0334	.0374
1.1	.0414	.0453	.0492	.0531	.0569	.0607	.0645	.0682	.0719	.0755
1.2	.0792	.0828	.0864	.0899	.0934	.0969	.1004	.1038	.1072	.1106
1.3	.1139	.1173	.1206	.1239	.1271	.1303	.1335	.1367	.1399	.1430
1.4	.1461	.1492	.1523	.1553	.1584	.1614	.1644	.1673	.1703	.1732
1.5	.1761	.1790	.1818	.1847	.1875	.1903	.1931	.1959	.1987	.2014
1.6	.2041	.2068	.2095	.2122	.2148	.2175	.2201	.2227	.2253	.2279
1.7	.2304	.2330	.2355	.2380	.2405	.2430	.2455	.2480	.2504	.2529
1.8	.2553	.2577	.2601	.2625	.2648	.2672	.2695	.2718	.2742	.2765
1.9	.2788	.2810	.2833	.2856	.2878	.2900	.2923	.2945	.2967	.2989

Appendix B Continued

Four-Place Table of Logarithms

	0	1	2	3	4	5	6	7	8	9
2.0	.3010	.3032	.3054	.3075	.3096	.3118	.3139	.3160	.3181	.3201
2.1	.3222	.3243	.3263	.3284	.3304	.3324	.3345	.3365	.3385	.3404
2.2	.3424	.3444	.3464	.3483	.3502	.3522	.3541	.3560	.3579	.3598
2.3	.3617	.3636	.3655	.3674	.3692	.3711	.3729	.3747	.3766	.3784
2.4	.3802	.3820	.3838	.3856	.3874	.3892	.3909	.3927	.3945	.3962
2.5	.3979	.3997	.4014	.4031	.4048	.4065	.4082	.4099	.4116	.4133
2.6	.4150	.4166	.4183	.4200	.4216	.4232	.4249	.4265	.4281	.4298
2.7	.4314	.4330	.4346	.4362	.4378	.4393	.4409	.4425	.4440	.4456
2.8	.4472	.4487	.4502	.4518	.4533	.4548	.4564	.4579	.4594	.4609
2.9	.4624	.4639	.4654	.4669	.4683	.4698	.4713	.4728	.4742	.4757
3.0	.4771	.4786	.4800	.4814	.4829	.4843	.4857	.4871	.4886	.4900
3.1	.4914	.4928	.4942	.4955	.4969	.4983	.4997	.5011	.5024	.5038
3.2	.5051	.5065	.5079	.5092	.5105	.5119	.5132	.5145	.5159	.5172
3.3	.5185	.5198	.5211	.5224	.5237	.5250	.5263	.5276	.5289	.5302
3.4	.5315	.5328	.5340	.5353	.5366	.5378	.5391	.5403	.5416	.5428
3.5	.5441	.5453	.5465	.5478	.5490	.5502	.5514	.5527	.5539	.5551
3.6	.5563	.5575	.5587	.5599	.5611	.5623	.5635	.5647	.5658	.5670
3.7	.5682	.5694	.5705	.5717	.5729	.5740	.5752	.5763	.5775	.5786
3.8	.5798	.5809	.5821	.5832	.5843	.5855	.5866	.5877	.5888	.5899
3.9	.5911	.5922	.5933	.5944	.5955	.5966	.5977	.5988	.5999	.6010
4.0	.6021	.6031	.6042	.6053	.6064	.6075	.6085	.6096	.6107	.6117
4.1	.6128	.6138	.6149	.6160	.6170	.6180	.6191	.6201	.6212	.6222
4.2	.6232	.6243	.6253	.6263	.6274	.6284	.6294	.6304	.6314	.6325
4.3	.6335	.6345	.6355	.6365	.6375	.6385	.6395	.6405	.6415	.6425
4.4	.6435	.6444	.6454	.6464	.6474	.6484	.6493	.6503	.6513	.6522
4.5	.6532	.6542	.6551	.6561	.6571	.6580	.6590	.6599	.6609	.6618
4.6	.6628	.6637	.6646	.6656	.6665	.6675	.6684	.6693	.6702	.6712
4.7	.6721	.6730	.6739	.6749	.6758	.6767	.6776	.6785	.6794	.6803
4.8	.6812	.6821	.6830	.6839	.6848	.6857	.6866	.6875	.6884	.6893
4.9	.6902	.6911	.6920	.6928	.6937	.6946	.6955	.6964	.6972	.6981
5.0	.6990	.6998	.7007	.7016	.7024	.7033	.7042	.7050	.7059	.7067
5.1	.7076	.7084	.7093	.7101	.7110	.7118	.7126	.7135	.7143	.7152
5.2	.7160	.7168	.7177	.7185	.7193	.7202	.7210	.7218	.7226	.7235
5.3	.7243	.7251	.7259	.7267	.7275	.7284	.7292	.7300	.7308	.7316
5.4	.7324	.7332	.7340	.7348	.7356	.7364	.7372	.7380	.7388	.7396
5.5	.7404	.7412	.7419	.7427	.7435	.7443	.7451	.7459	.7466	.7474
5.6	.7482	.7490	.7497	.7505	.7513	.7520	.7528	.7536	.7543	.7551
5.7	.7559	.7566	.7574	.7582	.7589	.7597	.7604	.7612	.7619	.7627
5.8	.7634	.7642	.7649	.7657	.7664	.7672	.7679	.7686	.7694	.7701
5.9	.7709	.7716	.7723	.7731	.7738	.7745	.7752	.7760	.7767	.7774
6.0	.7782	.7789	.7796	.7803	.7810	.7818	.7825	.7832	.7839	.7846
6.1	.7853	.7860	.7868	.7875	.7882	.7889	.7896	.7903	.7910	.7917
6.2	.7924	.7931	.7938	.7945	.7952	.7959	.7966	.7973	.7980	.7987
6.3	.7993	.8000	.8007	.8014	.8021	.8028	.8035	.8041	.8048	.8055
6.4	.8062	.8069	.8075	.8082	.8089	.8096	.8102	.8109	.8116	.8122
6.5	.8129	.8136	.8142	.8149	.8156	.8162	.8169	.8176	.8182	.8189
6.6	.8195	.8202	.8209	.8215	.8222	.8228	.8235	.8241	.8248	.8254
6.7	.8261	.8267	.8274	.8280	.8287	.8293	.8299	.8306	.8312	.8319
6.8	.8325	.8331	.8338	.8344	.8351	.8357	.8363	.8370	.8376	.8382
6.9	.8388	.8395	.8401	.8407	.8414	.8420	.8426	.8432	.8439	.8445

Table continued on p. 868

Appendix B Continued

Four-Place Table of Logarithms

	0	1	2	3	4	5	6	7	8	9
7.0	.8451	.8457	.8463	.8470	.8476	.8482	.8488	.8494	.8500	.8506
7.1	.8513	.8519	.8525	.8531	.8537	.8543	.8549	.8555	.8561	.8567
7.2	.8573	.8579	.8585	.8591	.8597	.8603	.8609	.8615	.8621	.8627
7.3	.8633	.8639	.8645	.8651	.8657	.8663	.8669	.8675	.8681	.8686
7.4	.8692	.8698	.8704	.8710	.8716	.8722	.8727	.8733	.8739	.8745
7.5	.8751	.8756	.8762	.8768	.8774	.8779	.8785	.8791	.8797	.8802
7.6	.8808	.8814	.8820	.8825	.8831	.8837	.8842	.8848	.8854	.8859
7.7	.8865	.8871	.8876	.8882	.8887	.8893	.8899	.8904	.8910	.8915
7.8	.8921	.8927	.8932	.8938	.8943	.8949	.8954	.8960	.8965	.8971
7.9	.8976	.8982	.8987	.8993	.8998	.9004	.9009	.9015	.9020	.9026
8.0	.9031	.9036	.9042	.9047	.9053	.9058	.9063	.9069	.9074	.9079
8.1	.9085	.9090	.9096	.9101	.9106	.9112	.9117	.9122	.9128	.9133
8.2	.9138	.9143	.9149	.9154	.9159	.9165	.9170	.9175	.9180	.9186
8.3	.9191	.9196	.9201	.9206	.9212	.9217	.9222	.9227	.9232	.9238
8.4	.9243	.9248	.9253	.9258	.9263	.9269	.9274	.9279	.9284	.9289
8.5	.9294	.9299	.9304	.9309	.9315	.9320	.9325	.9330	.9335	.9340
8.6	.9345	.9350	.9355	.9360	.9365	.9370	.9375	.9380	.9385	.9390
8.7	.9395	.9400	.9405	.9410	.9415	.9420	.9425	.9430	.9435	.9440
8.8	.9445	.9450	.9455	.9460	.9465	.9469	.9474	.9479	.9484	.9489
8.9	.9494	.9499	.9504	.9509	.9513	.9518	.9523	.9528	.9533	.9538
9.0	.9542	.9547	.9552	.9557	.9562	.9566	.9571	.9576	.9581	.9586
9.1	.9590	.9595	.9600	.9605	.9609	.9614	.9619	.9624	.9628	.9633
9.2	.9638	.9643	.9647	.9652	.9657	.9661	.9666	.9671	.9675	.9680
9.3	.9685	.9689	.9694	.9699	.9703	.9708	.9713	.9717	.9722	.9727
9.4	.9731	.9736	.9741	.9745	.9750	.9754	.9759	.9763	.9768	.9773
9.5	.9777	.9782	.9786	.9791	.9795	.9800	.9805	.9809	.9814	.9818
9.6	.9823	.9827	.9832	.9836	.9841	.9845	.9850	.9854	.9859	.9863
9.7	.9868	.9872	.9877	.9881	.9886	.9890	.9894	.9899	.9903	.9908
9.8	.9912	.9917	.9921	.9926	.9930	.9934	.9939	.9943	.9948	.9952
9.9	.9956	.9961	.9965	.9969	.9974	.9978	.9983	.9987	.9991	.9996

Appendix C Common Units, Equivalences, and Conversion Factors

Fundamental Units of the SI System

The metric system was begun by the French National Assembly in 1790 and has undergone many modifications. The International System of Units or *Système International* (SI), which represents an extension of the metric system, was adopted by the 11th General Conference of Weights and Measures in 1960. It is constructed from seven fundamental units, each of which represents a particular physical quantity (Table I).

TABLE I   SI Fundamental Units

Physical Quantity	Name of Unit	Symbol
length	metre	m
mass	kilogram	kg
time	second	s
temperature	kelvin	K
amount of substance	mole	mol
electric current	ampere	A
luminous intensity	candela	cd

The first five units listed in Table I are particularly useful in general chemistry. They are defined as follows.

1. The *metre* was redefined in 1960 to be equal to 1,650,763.73 wavelengths of a certain line in the emission spectrum of krypton-86.
2. The *kilogram* represents the mass of a platinum-iridium block kept at the International Bureau of Weights and Measures at Sevres, France.
3. The *second* was redefined in 1967 as the duration of 9,192,631,770 periods of a certain line in the microwave spectrum of cesium-133.
4. The *kelvin* is 1/273.16 of the temperature interval between absolute zero and the triple point of water.
5. The *mole* is the amount of substance that contains as many entities as there are atoms in exactly 0.012 kg of carbon-12 (12 g of ^{12}C atoms).

Prefixes Used with Traditional Metric Units and SI Units

Decimal fractions and multiples of metric and SI units are designated by using the prefixes listed in Table II. Those most commonly used in general chemistry are underlined.

TABLE II   Traditional Metric and SI Prefixes

Factor	Prefix	Symbol	Factor	Prefix	Symbol
10^{12}	tera	T	10^{-1}	deci	d
10^{9}	giga	G	10^{-2}	centi	c
10^{6}	mega	M	10^{-3}	milli	m
10^{3}	kilo	k	10^{-6}	micro	μ
10^{2}	hecto	h	10^{-9}	nano	n
10^{1}	deka	da	10^{-12}	pico	p
			10^{-15}	femto	f
			10^{-18}	atto	a

Derived SI Units

In the International System of Units, all physical quantities are represented by appropriate combinations of the fundamental units listed in Table I. A list of the derived units frequently used in general chemistry is given in Table III.

TABLE III   Derived SI Units

Physical Quantity	Name of Unit	Symbol	Definition
area	square metre	m^2	
volume	cubic metre	m^3	
density	kilogram per cubic metre	kg/m^3	
force	newton	N	$kg\ m/s^2$
pressure	pascal	Pa	N/m^2
energy	joule	J	$kg\ m^2/s^2$
electric charge	coulomb	C	A s
electric potential difference	volt	V	J/(A s)

Common Units of Mass and Weight

1 pound = 453.59 grams

1 pound = 453.59 g = 0.45359 kilogram
1 kilogram = 1000 grams = 2.205 pounds
1 gram = 10 decigrams = 100 centigrams = 1000 milligrams
1 gram = 6.022×10^{23} atomic mass units
1 atomic mass unit = 1.6606×10^{-24} gram
1 short ton = 2000 pounds = 907.2 kilograms
1 long ton = 2240 pounds
1 metric tonne = 1000 kilograms = 2205 pounds

Common Units of Length

1 inch = 2.54 centimeters (exactly)

1 mile = 5280 feet = 1.609 kilometers
1 yard = 36 inches = 0.9144 meter
1 meter = 100 centimeters = 39.37 inches = 3.281 feet
     = 1.094 yards
1 kilometer = 1000 meters = 1094 yards = 0.6215 mile
1 Ångstrom = 1.0×10^{-8} centimeter = 0.10 nanometer
         = 1.0×10^{-10} meter = 3.937×10^{-9} inch

Common Units of Volume

1 quart = 0.9463 liter
1 liter = 1.056 quarts

1 liter = 1 cubic decimeter = 1000 cubic centimeters
     = 0.001 cubic meter
1 milliliter = 1 cubic centimeter = 0.001 liter = 1.056×10^{-3} quart
1 cubic foot = 28.316 liters = 29.902 quarts = 7.475 gallons

Common Units of Force* and Pressure

1 atmosphere = 760 millimeters of mercury = 1.013×10^5 pascals
     = 14.70 pounds per square inch
1 bar = 10^5 pascals
1 torr = 1 millimeter of mercury
1 pascal = 1 kg/m s^2 = 1 N/m^2
*Force: 1 newton (N) = 1 $kg\ m/s^2$, *i.e.,* the force that when applied
        for one second gives a one kilogram mass a velocity of
        one meter per second.

Common Units of Energy

1 joule = 1 × 10⁷ ergs

1 thermochemical calorie* = 4.184 joules = 4.184 × 10⁷ergs = 4.129 × 10⁻² liter-atmospheres
$$= 2.612 × 10^{19} \text{ electron volts}$$

1 erg = 1 × 10⁻⁷ joule = 2.3901 × 10⁻⁸ calorie
1 electron volt = 1.6022 × 10⁻¹⁹ joule = 1.6022 × 10⁻¹² erg = 96.487 kJ/mol†
1 liter-atmosphere = 24.217 calories = 101.32 joules = 1.0132 × 10⁹ ergs
1 British thermal unit = 1055.06 joules = 1.05506 × 10¹⁰ ergs = 252.2 calories

* The amount of heat required to raise the temperature of one gram of water from 14.5°C to 15.5°C.
† Note that the other units are per particle and must be multiplied by 6.022 × 10²³ to be strictly comparable.

Appendix D   Physical Constants

Quantity	Symbol	Traditional Units	SI Units
acceleration of gravity	g	980.6 cm/s	9.806 m/s
atomic mass unit (1/12 the mass of ¹²C atom)	amu or u	1.6606 × 10⁻²⁴ g	1.6606 × 10⁻²⁷ kg
Avogadro's number	N	6.022 × 10²³ particles/mol	6.022 × 10²³ particles/mol
Bohr radius	a_0	0.52918 Å 5.2918 × 10⁻⁹ cm	5.2918 × 10⁻¹¹ m
Boltzmann constant	k	1.3807 × 10⁻¹⁶ erg/K	1.3807 × 10⁻²³ J/K
charge-to-mass ratio of electron	e/m	1.7588 × 10⁸ coulomb/g	1.7588 × 10¹¹ C/kg
electronic charge	e	1.6022 × 10⁻¹⁹ coulomb 4.8033 × 10⁻¹⁰ esu	1.6022 × 10⁻¹⁹ C
electron rest mass	m_e	9.1095 × 10⁻²⁸ g 0.00054859 amu	9.1095 × 10⁻³¹ kg
Faraday constant	F	96,487 coulombs/eq 23.06 kcal/volt eq	96,487 C/mol e^- 96,487 J/V mol e^-
gas constant	R	$0.08206 \dfrac{\text{L atm}}{\text{mol K}}$ $1.987 \dfrac{\text{cal}}{\text{mol K}}$	$8.3145 \dfrac{\text{Pa dm}^3}{\text{mol K}}$ 8.3145 J/mol K
molar volume (STP)	V_m	22.414 L/mol	22.414 × 10⁻³ m³/mol 22.414 dm³/mol
neutron rest mass	m_n	1.67495 × 10⁻²⁴ g 1.008665 amu	1.67495 × 10⁻²⁷ kg
Planck's constant	h	6.6262 × 10⁻²⁷ erg s	6.6262 × 10⁻³⁴ J s
proton rest mass	m_p	1.6726 × 10⁻²⁴ g 1.007277 amu	1.6726 × 10⁻²⁷ kg
Rydberg constant	R_∞	3.289 × 10¹⁵ cycles/s 2.1799 × 10⁻¹¹ erg	1.0974 × 10⁷ m⁻¹ 2.1799 × 10⁻¹⁸ J
velocity of light (in a vacuum)	c	2.9979 × 10¹⁰ cm/s (186,281 miles/second)	2.9979 × 10⁸ m/s

π = 3.1416   2.303 R = 4.576 cal/mol K = 19.15 J/mol K
e = 2.7183   2.303 RT (at 25°C) = 1364 cal/mol = 5709 J/mol
ln X = 2.303 log X

Appendix E   Some Physical Constants for Water and a Few Common Substances

Vapor Pressure of Water at Various Temperatures

Temperature °C	Vapor Pressure torr	Temperature °C	Vapor Pressure torr	Temperature °C	Vapor Pressure torr	Temperature °C	Vapor Pressure torr
−10	2.1	21	18.7	51	97.2	81	369.7
−9	2.3	22	19.8	52	102.1	82	384.9
−8	2.5	23	21.1	53	107.2	83	400.6
−7	2.7	24	22.4	54	112.5	84	416.8
−6	2.9	25	23.8	55	118.0	85	433.6
−5	3.2	26	25.2	56	123.8	86	450.9
−4	3.4	27	26.7	57	129.8	87	468.7
−3	3.7	28	28.3	58	136.1	88	487.1
−2	4.0	29	30.0	59	142.6	89	506.1
−1	4.3	30	31.8	60	149.4	90	525.8
0	4.6	31	33.7	61	156.4	91	546.1
1	4.9	32	35.7	62	163.8	92	567.0
2	5.3	33	37.7	63	171.4	93	588.6
3	5.7	34	39.9	64	179.3	94	610.9
4	6.1	35	42.2	65	187.5	95	633.9
5	6.5	36	44.6	66	196.1	96	657.6
6	7.0	37	47.1	67	205.0	97	682.1
7	7.5	38	49.7	68	214.2	98	707.3
8	8.0	39	52.4	69	223.7	99	733.2
9	8.6	40	55.3	70	233.7	100	760.0
10	9.2	41	58.3	71	243.9	101	787.6
11	9.8	42	61.5	72	254.6	102	815.9
12	10.5	43	64.8	73	265.7	103	845.1
13	11.2	44	68.3	74	277.2	104	875.1
14	12.0	45	71.9	75	289.1	105	906.1
15	12.8	46	75.7	76	301.4	106	937.9
16	13.6	47	79.6	77	314.1	107	970.6
17	14.5	48	83.7	78	327.3	108	1004.4
18	15.5	49	88.0	79	341.0	109	1038.9
19	16.5	50	92.5	80	355.1	110	1074.6
20	17.5						

Specific Heats and Heat Capacities for Some Common Substances

Substance	Specific Heat		Heat Capacity	
	cal/g °C	J/g °C	cal/mol °C	J/mol °C
Al (s)	0.215	0.900	5.81	24.3
Ca (s)	0.156	0.653	6.25	26.2
Cu (s)	0.092	0.385	5.85	24.5
Fe (s)	0.106	0.444	5.92	24.8
Hg (ℓ)	0.0331	0.138	6.62	27.7
H_2O (s), ice	0.500	2.09	9.00	37.7
H_2O (ℓ), water	1.00	4.18	18.0	75.3
H_2O (g), steam	0.484	2.03	8.71	36.4
C_6H_6 (ℓ), benzene	0.415	1.74	32.4	136
C_6H_6 (g), benzene	0.249	1.04	19.5	81.6
C_2H_5OH (ℓ), ethanol	0.587	2.46	27.0	113
C_2H_5OH (g), ethanol	0.228	0.954	10.5	420
$(C_2H_5)_2O$ (ℓ), diethyl ether	0.893	3.74	41.1	172
$(C_2H_5)_2O$ (g), diethyl ether	0.561	2.35	25.8	108

Heats of Transformation and Transformation Temperatures of Several Substances

Substance	M.P. °C	Heat of Fusion		ΔH_{fus}		B.P. °C	Heat of Vaporization		ΔH_{vap}	
		cal/g	J/g	kcal/mol	kJ/mol		cal/g	J/g	kcal/mol	kJ/mol
Al	658	94.5	395	2.54	10.6	2467	2515	10520	67.9	284
Ca	851	55.7	233	2.23	9.33	1487	963	4030	38.6	162
Cu	1083	49.0	205	3.11	13.0	2595	1146	4790	72.8	305
H_2O	0.0	79.8	333	1.44	6.02	100	540	2260	9.73	40.7
Fe	1530	63.7	267	3.56	14.9	2735	1515	6340	84.6	354
Hg	−39	2.7	11	5.57	23.3	357	69.8	292	14.0	58.6
CH_4	−182	14.0	58.6	0.22	0.92	−164	—	—	—	—
C_2H_5OH	−117	26.1	109	1.20	5.02	78.0	204	855	9.39	39.3
C_6H_6	5.48	30.4	127	2.37	9.92	80.1	94.3	395	7.36	30.8
$(C_2H_5)_2O$	−116	23.4	97.9	1.83	7.66	35	83.9	351	6.21	26.0

Appendix F   Ionization Constants for Weak Acids at 25°C

Acid	Formula and Ionization Equation	K_a
acetic	$CH_3COOH \rightleftharpoons H^+ + CH_3COO^-$	1.8×10^{-5}
arsenic	$H_3AsO_4 \rightleftharpoons H^+ + H_2AsO_4^-$	$K_1 = 2.5 \times 10^{-4}$
	$H_2AsO_4^- \rightleftharpoons H^+ + HAsO_4^{2-}$	$K_2 = 5.6 \times 10^{-8}$
	$HAsO_4^{2-} \rightleftharpoons H^+ + AsO_4^{3-}$	$K_3 = 3.0 \times 10^{-13}$
arsenous	$H_3AsO_3 \rightleftharpoons H^+ + H_2AsO_3^-$	$K_1 = 6.0 \times 10^{-10}$
	$H_2AsO_3^- \rightleftharpoons H^+ + HAsO_3^{2-}$	$K_2 = 3.0 \times 10^{-14}$
benzoic	$C_6H_5COOH \rightleftharpoons C_6H_5COO^-$	6.3×10^{-5}
boric	$H_3BO_3 \rightleftharpoons H^+ + H_2BO_3^-$	$K_1 = 7.3 \times 10^{-10}$
	$H_2BO_3^- \rightleftharpoons H^+ + HBO_3^{2-}$	$K_2 = 1.8 \times 10^{-13}$
	$HBO_2^{2-} \rightleftharpoons H^+ + BO_3^{3-}$	$K_3 = 1.6 \times 10^{-14}$
carbonic	$H_2CO_3 \rightleftharpoons H^+ + HCO_3^-$	$K_1 = 4.2 \times 10^{-7}$
	$HCO_3^- \rightleftharpoons H^+ + CO_3^{2-}$	$K_2 = 4.8 \times 10^{-11}$
citric	$H_3C_6H_5O_7 \rightleftharpoons H^+ + H_2C_6H_5O_7^-$	$K_1 = 7.4 \times 10^{-3}$
	$H_2C_6H_5O_7^- \rightleftharpoons H^+ + HC_6H_5O_7^{2-}$	$K_2 = 1.7 \times 10^{-5}$
	$HC_6H_5O_7^{2-} \rightleftharpoons H^+ + C_6H_5O_7^{3-}$	$K_3 = 4.0 \times 10^{-7}$
cyanic	$HOCN \rightleftharpoons H^+ + OCN^-$	3.5×10^{-4}
formic	$HCOOH \rightleftharpoons H^+ + HCOO^-$	1.8×10^{-4}
hydrazoic	$HN_3 \rightleftharpoons H^+ + N_3^-$	1.9×10^{-5}
hydrocyanic	$HCN \rightleftharpoons H^+ + CN^-$	4.0×10^{-10}
hydrofluoric	$HF \rightleftharpoons H^+ + F^-$	7.2×10^{-4}
hydrogen peroxide	$H_2O_2 \rightleftharpoons H^+ + HO_2^-$	2.4×10^{-12}
hydrosulfuric	$H_2S \rightleftharpoons H^+ + HS^-$	$K_1 = 1.0 \times 10^{-7}$
	$HS^- \rightleftharpoons H^+ + S^{2-}$	$K_2 = 1.3 \times 10^{-13}$
hypobromous	$HOBr \rightleftharpoons H^+ + OBr^-$	2.5×10^{-9}
hypochlorous	$HOCl \rightleftharpoons H^+ + OCl^-$	3.5×10^{-8}
nitrous	$HNO_2 \rightleftharpoons H^+ + NO_2^-$	4.5×10^{-4}
oxalic	$(COOH)_2 \rightleftharpoons H^+ + (COOH)COO^-$	$K_1 = 5.9 \times 10^{-2}$
	$(COOH)COO^- \rightleftharpoons H^+ + (COO)_2^{2-}$	$K_2 = 6.4 \times 10^{-5}$
phenol	$HC_6H_5O \rightleftharpoons H^+ + C_6H_5O^-$	1.3×10^{-10}
phosphoric	$H_3PO_4 \rightleftharpoons H^+ + H_2PO_4^-$	$K_1 = 7.5 \times 10^{-3}$
	$H_2PO_4^- \rightleftharpoons H^+ + HPO_4^{2-}$	$K_2 = 6.2 \times 10^{-8}$
	$HPO_4^{2-} \rightleftharpoons H^+ + PO_4^{3-}$	$K_3 = 3.6 \times 10^{-13}$
phosphorous	$H_3PO_3 \rightleftharpoons H^+ + H_2PO_3^-$	$K_1 = 1.6 \times 10^{-2}$
	$H_2PO_3^- \rightleftharpoons H^+ + HPO_3^{2-}$	$K_2 = 7.0 \times 10^{-7}$
selenic	$H_2SeO_4 \rightleftharpoons H^+ + HSeO_4^-$	$K_1 =$ very large
	$HSeO_4^- \rightleftharpoons H^+ + SeO_4^{2-}$	$K_2 = 1.2 \times 10^{-2}$
selenous	$H_2SeO_3 \rightleftharpoons H^+ + HSeO_3^-$	$K_1 = 2.7 \times 10^{-3}$
	$HSeO_3^- \rightleftharpoons H^+ + SeO_3^{2-}$	$K_2 = 2.5 \times 10^{-7}$
sulfuric	$H_2SO_4 \rightleftharpoons H^+ + HSO_4^-$	$K_1 =$ very large
	$HSO_4^- \rightleftharpoons H^+ + SO_4^{2-}$	$K_2 = 1.2 \times 10^{-2}$
sulfurous	$H_2SO_3 \rightleftharpoons H^+ + HSO_3^-$	$K_1 = 1.2 \times 10^{-2}$
	$HSO_3^- \rightleftharpoons H^+ + SO_3^{2-}$	$K_2 = 6.2 \times 10^{-8}$
tellurous	$H_2TeO_3 \rightleftharpoons H^+ + HTeO_3^-$	$K_1 = 2 \times 10^{-3}$
	$HTeO_3^- \rightleftharpoons H^+ + TeO_3^{2-}$	$K_2 = 1 \times 10^{-8}$

Appendix G   Ionization Constants for Weak Bases at 25°C

Base	Formula and Ionization Equation	K_b
ammonia	$NH_3 + H_2O \rightleftharpoons NH_4^+ + OH^-$	1.8×10^{-5}
aniline	$C_6H_5NH_2 + H_2O \rightleftharpoons C_6H_5NH_3^+ + OH^-$	4.2×10^{-10}
dimethylamine	$(CH_3)_2NH + H_2O \rightleftharpoons (CH_3)_2NH_2^+ + OH^-$	7.4×10^{-4}
ethylenediamine	$(CH_2)_2(NH_2)_2 + H_2O \rightleftharpoons (CH_2)_2(NH_2)_2H^+ + OH^-$	$K_1 = 8.5 \times 10^{-5}$
	$(CH_2)_2(NH_2)_2H^+ + H_2O \rightleftharpoons (CH_2)_2(NH_2)_2H_2^{2+} + OH^-$	$K_2 = 2.7 \times 10^{-8}$
hydrazine	$N_2H_4 + H_2O \rightleftharpoons N_2H_5^+ + OH^-$	$K_1 = 8.5 \times 10^{-7}$
	$N_2H_5^+ + H_2O \rightleftharpoons N_2H_6^{2+} + OH^-$	$K_2 = 8.9 \times 10^{-16}$
hydroxylamine	$NH_2OH + H_2O \rightleftharpoons NH_3OH^+ + OH^-$	6.6×10^{-9}
methylamine	$CH_3NH_2 + H_2O \rightleftharpoons CH_3NH_3^+ + OH^-$	5.0×10^{-4}
pyridine	$C_5H_5N + H_2O \rightleftharpoons C_5H_5NH^+ + OH^-$	1.5×10^{-9}
trimethylamine	$(CH_3)_3N + H_2O \rightleftharpoons (CH_3)_3NH^+ + OH^-$	7.4×10^{-5}

Appendix H   Solubility Product Constants for Some Inorganic Compounds at 25°C

Substance	K_{sp}	Substance	K_{sp}
Aluminum compounds		**Calcium compounds**	
$AlAsO_4$	1.6×10^{-16}	$Ca_3(AsO_4)_2$	6.8×10^{-19}
$Al(OH)_3$	1.9×10^{-33}	$CaCO_3$	4.8×10^{-9}
$AlPO_4$	1.3×10^{-20}	$CaCrO_4$	7.1×10^{-4}
		$CaC_2O_4 \cdot H_2O^*$	2.3×10^{-9}
Antimony compounds		CaF_2	3.9×10^{-11}
Sb_2S_3	1.6×10^{-93}	$Ca(OH)_2$	7.9×10^{-6}
		$CaHPO_4$	2.7×10^{-7}
Barium compounds		$Ca(H_2PO_4)_2$	1.0×10^{-3}
$Ba_3(AsO_4)_2$	1.1×10^{-13}	$Ca_3(PO_4)_2$	1.0×10^{-25}
$BaCO_3$	8.1×10^{-9}	$CaSO_3 \cdot 2H_2O^*$	1.3×10^{-8}
$BaC_2O_4 \cdot 2H_2O^*$	1.1×10^{-7}	$CaSO_4 \cdot 2H_2O^*$	2.4×10^{-5}
$BaCrO_4$	2.0×10^{-10}		
BaF_2	1.7×10^{-6}	**Chromium compounds**	
$Ba(OH)_2 \cdot 8H_2O^*$	5.0×10^{-3}	$CrAsO_4$	7.8×10^{-21}
$Ba_3(PO_4)_2$	1.3×10^{-29}	$Cr(OH)_3$	6.7×10^{-31}
$BaSeO_4$	2.8×10^{-11}	$CrPO_4$	2.4×10^{-23}
$BaSO_3$	8.0×10^{-7}		
$BaSO_4$	1.1×10^{-10}	**Cobalt compounds**	
		$Co_3(AsO_4)_2$	7.6×10^{-29}
Bismuth compounds		$CoCO_3$	8.0×10^{-13}
$BiOCl$	7.0×10^{-9}	$Co(OH)_2$	2.5×10^{-16}
$BiO(OH)$	1.0×10^{-12}	$CoS (\alpha)$	5.9×10^{-21}
$Bi(OH)_3$	3.2×10^{-40}	$CoS (\beta)$	8.7×10^{-23}
BiI_3	8.1×10^{-19}	$Co(OH)_3$	4.0×10^{-45}
$BiPO_4$	1.3×10^{-23}	Co_2S_3	2.6×10^{-124}
Bi_2S_3	1.6×10^{-72}		
		Copper compounds	
Cadmium compounds		$CuBr$	5.3×10^{-9}
$Cd_3(AsO_4)_2$	2.2×10^{-32}	$CuCl$	1.9×10^{-7}
$CdCO_3$	2.5×10^{-14}	$CuCN$	3.2×10^{-20}
$Cd(CN)_2$	1.0×10^{-8}	$Cu_2O (Cu^+ + OH^-)†$	1.0×10^{-14}
$Cd_2[Fe(CN)_6]$	3.2×10^{-17}	CuI	5.1×10^{-12}
$Cd(OH)_2$	1.2×10^{-14}	Cu_2S	1.6×10^{-48}
CdS	3.6×10^{-29}	$CuSCN$	1.6×10^{-11}

Table continued on next page

Appendix H Continued

Substance	K_{sp}	Substance	K_{sp}
$Cu_3(AsO_4)_2$	7.6×10^{-36}	Hg_2Cl_2	1.1×10^{-18}
$CuCO_3$	2.5×10^{-10}	Hg_2CrO_4	5.0×10^{-9}
$Cu_2[Fe(CN)_6]$	1.3×10^{-16}	Hg_2I_2	4.5×10^{-29}
$Cu(OH)_2$	1.6×10^{-19}	$Hg_2O \cdot H_2O$ (Hg_2^{2+}	
CuS	8.7×10^{-36}	$+ 2OH^-$)†	1.6×10^{-23}
		Hg_2SO_4	6.8×10^{-7}
Gold compounds		Hg_2S	5.8×10^{-44}
$AuBr$	5.0×10^{-17}	$Hg(CN)_2$	3.0×10^{-23}
$AuCl$	2.0×10^{-13}	$Hg(OH)_2$	2.5×10^{-26}
AuI	1.6×10^{-23}	HgI_2	4.0×10^{-29}
$AuBr_3$	4.0×10^{-36}	HgS	3.0×10^{-53}
$AuCl_3$	3.2×10^{-25}		
$Au(OH)_3$	1×10^{-53}	**Nickel compounds**	
AuI_3	1.0×10^{-46}	$Ni_3(AsO_4)_2$	1.9×10^{-26}
		$NiCO_3$	6.6×10^{-9}
Iron compounds		$Ni(CN)_2$	3.0×10^{-23}
$FeCO_3$	3.5×10^{-11}	$Ni(OH)_2$	2.8×10^{-16}
$Fe(OH)_2$	7.9×10^{-15}	$NiS (\alpha)$	3.0×10^{-21}
FeS	4.9×10^{-18}	$NiS (\beta)$	1.0×10^{-26}
$Fe_4[Fe(CN)_6]_3$	3.0×10^{-41}	$NiS (\gamma)$	2.0×10^{-28}
$Fe(OH)_3$	6.3×10^{-38}		
Fe_2S_3	1.4×10^{-88}	**Silver compounds**	
		Ag_3AsO_4	1.1×10^{-20}
Lead compounds		$AgBr$	3.3×10^{-13}
$Pb_3(AsO_4)_2$	4.1×10^{-36}	Ag_2CO_3	8.1×10^{-12}
$PbBr_2$	6.3×10^{-6}	$AgCl$	1.8×10^{-10}
$PbCO_3$	1.5×10^{-13}	Ag_2CrO_4	9.0×10^{-12}
$PbCl_2$	1.7×10^{-5}	$AgCN$	1.2×10^{-16}
$PbCrO_4$	1.8×10^{-14}	$Ag_4[Fe(CN)_6]$	1.6×10^{-41}
PbF_2	3.7×10^{-8}	Ag_2O ($Ag^+ + OH^-$)†	2.0×10^{-8}
$Pb(OH)_2$	2.8×10^{-16}	AgI	1.5×10^{-16}
PbI_2	8.7×10^{-9}	Ag_3PO_4	1.3×10^{-20}
$Pb_3(PO_4)_2$	3.0×10^{-44}	Ag_2SO_3	1.5×10^{-14}
$PbSeO_4$	1.5×10^{-7}	Ag_2SO_4	1.7×10^{-5}
$PbSO_4$	1.8×10^{-8}	Ag_2S	1.0×10^{-49}
PbS	8.4×10^{-28}	$AgSCN$	1.0×10^{-12}
Magnesium compounds		**Strontium compounds**	
$Mg_3(AsO_4)_2$	2.1×10^{-20}	$Sr_3(AsO_4)_2$	1.3×10^{-18}
$MgCO_3 \cdot 3H_2O$*	4.0×10^{-5}	$SrCO_3$	9.4×10^{-10}
MgC_2O_4	8.6×10^{-5}	$SrC_2O_4 \cdot 2H_2O$*	5.6×10^{-8}
MgF_2	6.4×10^{-9}	$SrCrO_4$	3.6×10^{-5}
$Mg(OH)_2$	1.5×10^{-11}	$Sr(OH)_2 \cdot 8H_2O$*	3.2×10^{-4}
$MgNH_4PO_4$	2.5×10^{-12}	$Sr_3(PO_4)_2$	1.0×10^{-31}
		$SrSO_3$	4.0×10^{-8}
Manganese compounds		$SrSO_4$	2.8×10^{-7}
$Mn_3(AsO_4)_2$	1.9×10^{-11}		
$MnCO_3$	1.8×10^{-11}	**Tin compounds**	
$Mn(OH)_2$	4.6×10^{-14}	$Sn(OH)_2$	2.0×10^{-26}
MnS	5.1×10^{-15}	SnI_2	1.0×10^{-4}
$Mn(OH)_3$	$\sim 1 \times 10^{-36}$	SnS	1.0×10^{-28}
		$Sn(OH)_4$	1×10^{-57}
Mercury compounds		SnS_2	1×10^{-70}
Hg_2Br_2	1.3×10^{-22}		
Hg_2CO_3	8.9×10^{-17}		

Appendix H Continued

Substance	K_{sp}	Substance	K_{sp}
Zinc compounds		$Zn_2[Fe(CN)_6]$	4.1×10^{-16}
$Zn_3(AsO_4)_2$	1.1×10^{-27}	$Zn(OH)_2$	4.5×10^{-17}
$ZnCO_3$	1.5×10^{-11}	$Zn_3(PO_4)_2$	9.1×10^{-33}
$Zn(CN)_2$	8.0×10^{-12}	ZnS	1.1×10^{-21}

* Since $[H_2O]$ does not appear in equilibrium constants for equilibria in aqueous solution in general, it does *not* appear in the K_{sp} expressions for hydrated solids.

† Very small amounts of oxides dissolve in water to give the ions indicated in parentheses. Solid hydroxides are unstable and decompose to oxides as rapidly as they are formed.

Appendix I Dissociation Constants for Complex Ions

Dissociation Equilibrium	K_d
$[AgBr_2]^- \rightleftharpoons Ag^+ + 2\,Br^-$	7.8×10^{-8}
$[AgCl_2]^- \rightleftharpoons Ag^+ + 2\,Cl^-$	4.0×10^{-6}
$[Ag(CN)_2]^- \rightleftharpoons Ag^+ + 2\,CN^-$	1.8×10^{-19}
$[Ag(S_2O_3)_2]^{3-} \rightleftharpoons Ag^+ + 2\,S_2O_3^{2-}$	5.0×10^{-14}
$[Ag(NH_3)_2]^+ \rightleftharpoons Ag^+ + 2\,NH_3$	6.3×10^{-8}
$[Ag(en)]^+ \rightleftharpoons Ag^+ + en^*$	1.0×10^{-5}
$[AlF_6]^{3-} \rightleftharpoons Al^{3+} + 6\,F^-$	2.0×10^{-24}
$[Al(OH)_4]^- \rightleftharpoons Al^{3+} + 4\,OH^-$	1.3×10^{-34}
$[Au(CN)_2]^- \rightleftharpoons Au^+ + 2\,CN^-$	5.0×10^{-39}
$[Cd(CN)_4]^{2-} \rightleftharpoons Cd^{2+} + 4\,CN^-$	7.8×10^{-18}
$[CdCl_4]^{2-} \rightleftharpoons Cd^{2+} + 4\,Cl^-$	1.0×10^{-4}
$[Cd(NH_3)_4]^{2+} \rightleftharpoons Cd^{2+} + 4\,NH_3$	1.0×10^{-7}
$[Co(NH_3)_6]^{2+} \rightleftharpoons Co^{2+} + 6\,NH_3$	1.3×10^{-5}
$[Co(NH_3)_6]^{3+} \rightleftharpoons Co^{3+} + 6\,NH_3$	2.2×10^{-34}
$[Co(en)_3]^{2+} \rightleftharpoons Co^{2+} + 3\,en^*$	1.5×10^{-14}
$[Co(en)_3]^{3+} \rightleftharpoons Co^{3+} + 3\,en^*$	2.0×10^{-49}
$[Cu(CN)_2]^- \rightleftharpoons Cu^+ + 2\,CN^-$	1.0×10^{-16}
$[CuCl_2]^- \rightleftharpoons Cu^+ + 2\,Cl^-$	1.0×10^{-5}
$[Cu(NH_3)_2]^+ \rightleftharpoons Cu^+ + 2\,NH_3$	1.4×10^{-11}
$[Cu(NH_3)_4]^{2+} \rightleftharpoons Cu^{2+} + 4\,NH_3$	8.5×10^{-13}
$[Fe(CN)_6]^{4-} \rightleftharpoons Fe^{2+} + 6\,CN^-$	1.3×10^{-37}
$[Fe(CN)_6]^{3-} \rightleftharpoons Fe^{3+} + 6\,CN^-$	1.3×10^{-44}
$[HgCl_4]^{2-} \rightleftharpoons Hg^{2+} + 4\,Cl^-$	8.3×10^{-16}
$[Ni(CN)_4]^{2-} \rightleftharpoons Ni^{2+} + 4\,CN^-$	1.0×10^{-31}
$[Ni(NH_3)_6]^{2+} \rightleftharpoons Ni^{2+} + 6\,NH_3$	1.8×10^{-9}
$[Zn(OH)_4]^{2-} \rightleftharpoons Zn^{2+} + 4\,OH^-$	3.5×10^{-16}
$[Zn(NH_3)_4]^{2+} \rightleftharpoons Zn^{2+} + 4\,NH_3$	3.4×10^{-10}

* en represents ethylenediamine, $H_2NCH_2CH_2NH_2$

Appendix J Standard Reduction Potentials in Aqueous Solution at 25°C

Acidic Solution	Standard Reduction Potential, $E°$ (volts)
$Li^+ (aq) + e^- \rightarrow Li (s)$	-3.045
$K^+ (aq) + e^- \rightarrow K (s)$	-2.925
$Rb^+ (aq) + e^- \rightarrow Rb (s)$	-2.925

Table continued on next page

Appendix J Continued

Acidic Solution	Standard Reduction Potential, $E°$ (volts)
Ba^{2+} (aq) $+ 2e^- \rightarrow$ Ba (s)	-2.90
Sr^{2+} (aq) $+ 2e^- \rightarrow$ Sr (s)	-2.89
Ca^{2+} (aq) $+ 2e^- \rightarrow$ Ca (s)	-2.87
Na^+ (aq) $+ e^- \rightarrow$ Na (s)	-2.714
Mg^{2+} (aq) $+ 2e^- \rightarrow$ Mg (s)	-2.37
H_2 (g) $+ 2e^- \rightarrow 2H^-$ (aq)	-2.25
Al^{3+} (aq) $+ 3e^- \rightarrow$ Al (s)	-1.66
Zr^{4+} (aq) $+ 4e^- \rightarrow$ Zr (s)	-1.53
ZnS (s) $+ 2e^- \rightarrow$ Zn (s) $+ S^{2-}$ (aq)	-1.44
CdS (s) $+ 2e^- \rightarrow$ Cd (s) $+ S^{2-}$ (aq)	-1.21
V^{2+} (aq) $+ 2e^- \rightarrow$ V (s)	-1.18
Mn^{2+} (aq) $+ 2e^- \rightarrow$ Mn (s)	-1.18
FeS (s) $+ 2e^- \rightarrow$ Fe (s) $+ S^{2-}$ (aq)	-1.01
Cr^{2+} (aq) $+ 2e^- \rightarrow$ Cr (s)	-0.91
Zn^{2+} (aq) $+ 2e^- \rightarrow$ Zn (s)	-0.763
Cr^{3+} (aq) $+ 3e^- \rightarrow$ Cr (s)	-0.74
HgS (s) $+ 2H^+$ (aq) $+ 2e^- \rightarrow$ Hg (ℓ) $+ H_2S$ (g)	-0.72
Ga^{3+} (aq) $+ 3e^- \rightarrow$ Ga (s)	-0.53
$2CO_2$ (g) $+ 2H^+$ (aq) $+ 2e^- \rightarrow (COOH)_2$ (aq)	-0.49
Fe^{2+} (aq) $+ 2e^- \rightarrow$ Fe (s)	-0.44
Cr^{3+} (aq) $+ e^- \rightarrow Cr^{2+}$ (aq)	-0.41
Cd^{2+} (aq) $+ 2e^- \rightarrow$ Cd (s)	-0.403
Se (s) $+ 2H^+$ (aq) $+ 2e^- \rightarrow H_2Se$ (aq)	-0.40
$PbSO_4$ (s) $+ 2e^- \rightarrow$ Pb (s) $+ SO_4^{2-}$ (aq)	-0.356
Tl^+ (aq) $+ e^- \rightarrow$ Tl (s)	-0.34
Co^{2+} (aq) $+ 2e^- \rightarrow$ Co (s)	-0.28
Ni^{2+} (aq) $+ 2e^- \rightarrow$ Ni (s)	-0.25
$[SnF_6]^{2-}$ (aq) $+ 4e^- \rightarrow$ Sn (s) $+ 6F^-$ (aq)	-0.25
AgI (s) $+ e^- \rightarrow$ Ag (s) $+ I^-$ (aq)	-0.15
Sn^{2+} (aq) $+ 2e^- \rightarrow$ Sn (s)	-0.14
Pb^{2+} (aq) $+ 2e^- \rightarrow$ Pb (s)	-0.126
N_2O (g) $+ 6H^+$ (aq) $+ H_2O + 4e^- \rightarrow 2NH_3OH^+$ (aq)	-0.05
$2H^+$ (aq) $+ 2e^- \rightarrow H_2$ (g) **(reference electrode)**	0
AgBr (s) $+ e^- \rightarrow$ Ag (s) $+ Br^-$ (aq)	0.10
S (s) $+ 2H^+$ (aq) $+ 2e^- \rightarrow H_2S$ (aq)	0.14
Sn^{4+} (aq) $+ 2e^- \rightarrow Sn^{2+}$ (aq)	0.15
Cu^{2+} (aq) $+ e^- \rightarrow Cu^+$ (aq)	0.153
SO_4^{2-} (aq) $+ 4H^+$ (aq) $+ 2e^- \rightarrow H_2SO_3$ (aq) $+ H_2O$	0.17
SO_4^{2-} (aq) $+ 4H^+$ (aq) $+ 2e^- \rightarrow SO_2$ (g) $+ 2H_2O$	0.20
AgCl (s) $+ e^- \rightarrow$ Ag (s) $+ Cl^-$ (aq)	0.222
Hg_2Cl_2 (s) $+ 2e^- \rightarrow$ 2Hg (ℓ) $+ 2Cl^-$ (aq)	0.27
Cu^{2+} (aq) $+ 2e^- \rightarrow$ Cu (s)	0.337
$[RhCl_6]^{3-}$ (aq) $+ 3e^- \rightarrow$ Rh (s) $+ 6Cl^-$ (aq)	0.44
Cu^+ (aq) $+ e^- \rightarrow$ Cu (s)	0.521
TeO_2 (s) $+ 4H^+$ (aq) $+ 4e^- \rightarrow$ Te (s) $+ 2H_2O$	0.529
I_2 (s) $+ 2e^- \rightarrow 2I^-$ (aq)	0.535
H_3AsO_4 (aq) $+ 2H^+$ (aq) $+ 2e^- \rightarrow H_3AsO_3$ (aq) $+ H_2O$	0.58
$[PtCl_6]^{2-}$ (aq) $+ 2e^- \rightarrow [PtCl_4]^{2-}$ (aq) $+ 2Cl^-$ (aq)	0.68
O_2 (g) $+ 2H^+$ (aq) $+ 2e^- \rightarrow H_2O_2$ (aq)	0.682
$[PtCl_4]^{2-}$ (aq) $+ 2e^- \rightarrow$ Pt (s) $+ 4Cl^-$ (aq)	0.73
$SbCl_6^-$ (aq) $+ 2e^- \rightarrow SbCl_4^-$ (aq) $+ 2Cl^-$ (aq)	0.75
Fe^{3+} (aq) $+ e^- \rightarrow Fe^{2+}$ (aq)	0.771
Hg_2^{2+} (aq) $+ 2e^- \rightarrow$ 2Hg (ℓ)	0.789
Ag^+ (aq) $+ e^- \rightarrow$ Ag (s)	0.7994

Appendix J   Continued

Acidic Solution	Standard Reduction Potential, $E°$ (volts)
$Hg^{2+} (aq) + 2e^- \rightarrow Hg (\ell)$	0.855
$2Hg^{2+} (aq) + 2e^- \rightarrow Hg_2^{2+} (aq)$	0.920
$NO_3^- (aq) + 3H^+ (aq) + 2e^- \rightarrow HNO_2 (aq) + H_2O$	0.94
$NO_3^- (aq) + 4H^+ (aq) + 3e^- \rightarrow NO (g) + 2H_2O$	0.96
$Pd^{2+} (aq) + 2e^- \rightarrow Pd (s)$	0.987
$AuCl_4^- (aq) + 3e^- \rightarrow Au (s) + 4Cl^- (aq)$	1.00
$Br_2 (\ell) + 2e^- \rightarrow 2Br^- (aq)$	1.08
$ClO_4^- (aq) + 2H^+ (aq) + 2e^- \rightarrow ClO_3^- (aq) + H_2O$	1.19
$IO_3^- (aq) + 6H^+ (aq) + 5e^- \rightarrow \frac{1}{2}I_2 (aq) + 3H_2O$	1.195
$Pt^{2+} (aq) + 2e^- \rightarrow Pt (s)$	1.2
$O_2 (g) + 4H^+ (aq) + 4e^- \rightarrow 2H_2O$	1.229
$MnO_2 (s) + 4H^+ (aq) + 2e^- \rightarrow Mn^{2+} (aq) + 2H_2O$	1.23
$N_2H_5^+ (aq) + 3H^+ (aq) + 2e^- \rightarrow 2NH_4^+ (aq)$	1.24
$Cr_2O_7^{2-} (aq) + 14H^+ (aq) + 6e^- \rightarrow 2Cr^{3+} (aq) + 7H_2O$	1.33
$Cl_2 (g) + 2e^- \rightarrow 2Cl^- (aq)$	1.360
$BrO_3^- (aq) + 6H^+ (aq) + 6e^- \rightarrow Br^- (aq) + 3H_2O$	1.44
$ClO_3^- (aq) + 6H^+ (aq) + 5e^- \rightarrow \frac{1}{2}Cl_2 (g) + 3H_2O$	1.47
$Au^{3+} (aq) + 3e^- \rightarrow Au (s)$	1.50
$MnO_4^- (aq) + 8H^+ (aq) + 5e^- \rightarrow Mn^{2+} (aq) + 4H_2O$	1.51
$NaBiO_3 (s) + 6H^+ (aq) + 2e^- \rightarrow Bi^{3+} (aq) + Na^+ (aq) + 3H_2O$	~1.6
$Ce^{4+} (aq) + e^- \rightarrow Ce^{3+} (aq)$	1.61
$2HClO (aq) + 2H^+ (aq) + 2e^- \rightarrow Cl_2 (g) + 2H_2O$	1.63
$Au^+ (aq) + e^- \rightarrow Au (s)$	1.68
$PbO_2 (s) + SO_4^{2-} (aq) + 4H^+ (aq) + 2e^- \rightarrow PbSO_4 (s) + 2H_2O$	1.685
$NiO_2 (s) + 4H^+ (aq) + 2e^- \rightarrow Ni^{2+} (aq) + 2H_2O$	1.7
$H_2O_2 (aq) + 2H^+ (aq) + 2e^- \rightarrow 2H_2O$	1.77
$Pb^{4+} (aq) + 2e^- \rightarrow Pb^{2+} (aq)$	1.8
$Co^{3+} (aq) + e^- \rightarrow Co^{2+} (aq)$	1.82
$F_2 (g) + 2e^- \rightarrow 2F^- (aq)$	2.87

Basic Solution	
$SiO_3^{2-} (aq) + 3H_2O + 4e^- \rightarrow Si (s) + 6OH^- (aq)$	−1.70
$Cr(OH)_3 (s) + 3e^- \rightarrow Cr (s) + 3OH^- (aq)$	−1.30
$[Zn(CN)_4]^{2-} (aq) + 2e^- \rightarrow Zn (s) + 4CN^- (aq)$	−1.26
$Zn(OH)_2 (s) + 2e^- \rightarrow Zn (s) + 2OH^- (aq)$	−1.245
$[Zn(OH)_4]^{2-} (aq) + 2e^- \rightarrow Zn (s) + 4OH^- (aq)$	−1.22
$N_2 (g) + 4H_2O + 4e^- \rightarrow N_2H_4 (aq) + 4OH^- (aq)$	−1.15
$SO_4^{2-} (aq) + H_2O + 2e^- \rightarrow SO_3^{2-} (aq) + 2OH^- (aq)$	−0.93
$Fe(OH)_2 (s) + 2e^- \rightarrow Fe (s) + 2OH^- (aq)$	−0.877
$2NO_3^- (aq) + 2H_2O + 2e^- \rightarrow N_2O_4 (g) + 4OH^- (aq)$	−0.85
$2H_2O + 2e^- \rightarrow H_2 (g) + 2OH^- (aq)$	−0.8277
$Fe(OH)_3 (s) + e^- \rightarrow Fe(OH)_2 (s) + OH^- (aq)$	−0.56
$S (s) + 2e^- \rightarrow S^{2-} (aq)$	−0.48
$Cu(OH)_2 (s) + 2e^- \rightarrow Cu (s) + 2OH^- (aq)$	−0.36
$CrO_4^{2-} (aq) + 4H_2O + 3e^- \rightarrow Cr(OH)_3 (s) + 5OH^- (aq)$	−0.12
$MnO_2 (s) + 2H_2O + 2e^- \rightarrow Mn(OH)_2 (s) + 2OH^- (aq)$	−0.05
$NO_3^- (aq) + H_2O + 2e^- \rightarrow NO_2^- (aq) + 2OH^- (aq)$	0.01
$O_2 (g) + H_2O + 2e^- \rightarrow OOH^- (aq) + OH^- (aq)$	0.076
$HgO (s) + H_2O + 2e^- \rightarrow Hg (\ell) + 2OH^- (aq)$	0.0984
$[Co(NH_3)_6]^{3+} (aq) + e^- \rightarrow [Co(NH_3)_6]^{2+} (aq)$	0.10
$N_2H_4 (aq) + 2H_2O + 2e^- \rightarrow 2NH_3 (aq) + 2OH^- (aq)$	0.10
$2NO_2^- (aq) + 3H_2O + 4e^- \rightarrow N_2O (g) + 6OH^- (aq)$	0.15
$Ag_2O (s) + H_2O + 2e^- \rightarrow 2Ag (s) + 2OH^- (aq)$	0.34
$ClO_4^- (aq) + H_2O + 2e^- \rightarrow ClO_3^- (aq) + 2OH^- (aq)$	0.36

Table continued on next page

Appendix J   Continued

Basic Solution	Standard Reduction Potential, $E°$ (volts)
O_2 (g) $+ 2H_2O + 4e^- \rightarrow 4OH^-$ (aq)	0.40
Ag_2CrO_4 (s) $+ 2e^- \rightarrow 2Ag$ (s) $+ CrO_4^{2-}$ (aq)	0.446
NiO_2 (s) $+ 2H_2O + 2e^- \rightarrow Ni(OH)_2$ (s) $+ 2OH^-$ (aq)	0.49
MnO_4^- (aq) $+ e^- \rightarrow MnO_4^{2-}$ (aq)	0.564
MnO_4^- (aq) $+ 2H_2O + 3e^- \rightarrow MnO_2$ (s) $+ 4OH^-$ (aq)	0.588
ClO_3^- (aq) $+ 3H_2O + 6e^- \rightarrow Cl^-$ (aq) $+ 6OH^-$ (aq)	0.62
$2NH_2OH$ (aq) $+ 2e^- \rightarrow N_2H_4$ (aq) $+ 2OH^-$ (aq)	0.74
OOH^- (aq) $+ H_2O + 2e^- \rightarrow 3OH^-$ (aq)	0.88
ClO^- (aq) $+ H_2O + 2e^- \rightarrow Cl^-$ (aq) $+ 2OH^-$ (aq)	0.89

Appendix K   Selected Thermodynamic Values

Species	$\Delta H^0_{f298.15}$ kJ/mol	$S^0_{298.15}$ J/mol K	$\Delta G^0_{f298.15}$ kJ/mol
ALUMINUM			
Al (s)	0	28.3	0
$AlCl_3$ (s)	−704.2	110.7	−628.9
Al_2O_3 (s)	−1676	50.92	−1582
BARIUM			
$BaCl_2$ (s)	−860.1	126	−810.9
$BaSO_4$ (s)	−1465	132	−1353
BERYLLIUM			
Be (s)	0	9.54	0
$Be(OH)_2$ (s)	−907.1	—	—
BROMINE			
Br (g)	111.8	174.9	82.4
Br_2 (ℓ)	0	152.23	0
Br_2 (g)	30.91	245.4	3.14
BrF_3 (g)	−255.6	292.4	−229.5
HBr (g)	−36.4	198.59	−53.43
CALCIUM			
Ca (s)	0	41.6	0
Ca (g)	192.6	154.8	158.9
Ca^{2+} (g)	1920	—	—
CaC_2 (s)	−62.8	70.3	−67.8
$CaCO_3$ (s)	−1207	92.9	−1129
$CaCl_2$ (s)	−795.0	114	−750.2
CaF_2 (s)	−1215	68.87	−1162
CaH_2 (s)	−189	42	−150
CaO (s)	−635.5	40	−604.2
CaS (s)	−482.4	56.5	−477.4
$Ca(OH)_2$ (s)	−986.6	76.1	−896.8
$Ca(OH)_2$ (aq)	−1002.8	76.15	−867.6
$CaSO_4$ (s)	−1433	107	−1320

Appendix K Continued

Species	$\Delta H^0_{f298.15}$ kJ/mol	$S^0_{298.15}$ J/mol K	$\Delta G^0_{f298.15}$ kJ/mol
CARBON			
C (s, graphite)	0	5.740	0
C (s, diamond)	1.897	2.38	2.900
C (g)	716.7	158.0	671.3
CCl$_4$ (ℓ)	−135.4	216.4	−65.27
CCl$_4$ (g)	103	309.7	−60.63
CHCl$_3$ (ℓ)	−134.5	202	−73.72
CHCl$_3$ (g)	−103.1	295.6	−70.37
CH$_4$ (g)	−74.81	186.2	−50.75
C$_2$H$_2$ (g)	226.7	200.8	209.2
C$_2$H$_4$ (g)	52.26	209.5	68.12
C$_2$H$_6$ (g)	−84.86	229.5	−32.9
C$_3$H$_8$ (g)	−103.8	269.9	−23.49
C$_6$H$_6$ (ℓ)	49.03	172.8	124.5
C$_8$H$_{18}$ (ℓ)	−268.8	—	—
C$_2$H$_5$OH (ℓ)	−277.7	161	−174.9
C$_2$H$_5$OH (g)	−235.1	282.6	−168.6
CO (g)	−110.5	197.6	−137.2
CO$_2$ (g)	−393.5	213.6	−394.4
CS$_2$ (g)	117.4	237.7	67.15
COCl$_2$ (g)	−223.0	289.2	−210.5
CESIUM			
Cs$^+$ (aq)	−248	133	−282.0
CsF (aq)	−568.6	123	−558.5
CHLORINE			
Cl (g)	121.7	165.1	105.7
Cl$^-$ (g)	−226	—	—
Cl$_2$ (g)	0	223.0	0
HCl (g)	−92.31	186.8	−95.30
HCl (aq)	−167.4	55.10	−131.2
CHROMIUM			
Cr (s)	0	23.8	0
(NH$_4$)$_2$Cr$_2$O$_7$ (s)	−1807	—	—
COPPER			
Cu (s)	0	33.15	0
CuO (s)	−157	42.63	−130
FLUORINE			
F$^-$ (g)	−322	—	—
F$^-$ (aq)	−329.1	—	276.5
F (g)	78.99	158.6	61.92
F$_2$ (g)	0	202.7	0
HF (g)	−271	173.7	−273
HF (aq)	−329.1	—	−276.5
HYDROGEN			
H (g)	218.0	114.6	203.3
H$_2$ (g)	0	130.6	0
H$_2$O (ℓ)	−285.8	69.91	−237.3
H$_2$O (g)	−241.8	188.7	−228.6
H$_2$O$_2$ (ℓ)	−187.8	109.6	−120.4

Table continued on next page

Appendix K   Continued

Species	$\Delta H^0_{f298.15}$ kJ/mol	$S^0_{298.15}$ J/mol K	$\Delta G^0_{f298.15}$ kJ/mol
IODINE			
I (g)	106.6	180.66	70.16
I$_2$ (s)	0	116.1	0
I$_2$ (g)	62.44	260.6	19.36
ICl (g)	17.78	247.4	−5.52
IRON			
Fe (s)	0	27.3	0
FeO (s)	−272	—	—
Fe$_2$O$_3$ (s)	−824.2	87.40	−742.2
Fe$_3$O$_4$ (s)	−1118	146	−1015
FeS$_2$ (s)	−177.5	122.2	−166.7
Fe(CO)$_5$ (ℓ)	−774.0	338	−705.4
Fe(CO)$_5$ (g)	−733.8	445.2	−697.3
LEAD			
Pb (s)	0	64.81	0
PbCl$_2$ (s)	−359.4	136	−314.1
PbO (s, yellow)	−217.3	68.70	−187.9
Pb(OH)$_2$ (s)	−515.9	88	−420.9
PbS (s)	−100.4	91.2	−98.7
LITHIUM			
Li (s)	0	28.0	0
LiOH (s)	−487.23	50	−443.9
LiOH (aq)	−508.4	4	−451.1
MAGNESIUM			
Mg (s)	0	32.5	0
MgCl$_2$ (s)	−641.8	89.5	−592.3
MgO (s)	−601.8	27	−569.6
Mg(OH)$_2$ (s)	−924.7	63.14	−833.7
MgS (s)	−347	—	—
MERCURY			
Hg (ℓ)	0	76.02	0
HgCl$_2$ (s)	−224	146	−179
HgO (s, red)	−90.83	70.29	−58.56
HgS (s, red)	−58.2	82.4	−50.6
NICKEL			
Ni (s)	0	30.1	0
NiO (s)	−244	38.6	−216
NITROGEN			
N$_2$ (g)	0	191.5	0
N (g)	472.704	153.19	455.579
NH$_3$ (g)	−46.11	192.3	−16.5
N$_2$H$_4$ (ℓ)	50.63	121.2	149.2
(NH$_4$)$_3$AsO$_4$ (aq)	−1268	—	—
NH$_4$Cl (s)	−314.4	94.6	−201.5
NH$_4$Cl (aq)	−300.2	—	—
NH$_4$I(s)	−201.4	117	−113
NH$_4$NO$_3$ (s)	−365.6	151.1	−184.0
NO (g)	90.25	210.7	86.57
NO$_2$ (g)	33.2	240.0	51.30
N$_2$O (g)	82.05	219.7	104.2
N$_2$O$_4$ (g)	9.16	304.2	97.82

Appendix K   Continued

Species	$\Delta H^0_{f298.15}$ kJ/mol	$S^0_{298.15}$ J/mol K	$\Delta G^0_{f298.15}$ kJ/mol
N_2O_5 (g)	11	356	115
N_2O_5 (s)	−43.1	178	114
NOCl (g)	52.59	264	66.36
HNO_3 (ℓ)	−174.1	155.6	−80.79
HNO_3 (g)	−135.1	266.2	−74.77
HNO_3 (aq)	−206.6	146	−110.5
OXYGEN			
O (g)	249.2	161.0	231.8
O_2 (g)	0	205.0	0
O_3 (g)	143	238.8	163
OF_2 (g)	23	246.6	41
PHOSPHORUS			
P (g)	58.91	279.9	24.5
P_4 (s, white)	0	177	0
P_4 (s, red)	−73.6	91.2	−48.5
PCl_3 (g)	−306.4	311.7	−286.3
PCl_5 (g)	−398.9	353	−324.6
PH_3 (g)	5.4	210.1	13
P_4O_{10} (s)	−2984	228.9	−2698
H_3PO_4 (s)	−1281	110.5	−1119
POTASSIUM			
K (s)	0	63.6	0
KCl (s)	−436.5	82.6	−408.8
$KClO_3$ (s)	−391.2	143.1	−289.9
KI (s)	−327.9	106.4	−323.0
KOH (s)	−424.7	78.91	−378.9
KOH (aq)	−481.2	92.0	−439.6
SILICON			
Si (s)	0	18.8	0
$SiBr_4$ (ℓ)	−398	—	—
SiC (s)	−65.3	16.6	−62.8
$SiCl_4$ (g)	−657.0	330.6	−617.0
SiH_4 (g)	34	204.5	56.9
SiF_4 (g)	−1615	282.4	−1573
SiI_4 (g)	−132	—	—
SiO_2 (s)	−910.9	41.84	−856.7
H_2SiO_3 (s)	−1189	134	−1092
Na_2SiO_3 (s)	−1079	—	—
H_2SiF_6 (aq)	−2331	—	—
SILVER			
Ag (s)	0	42.55	0
SODIUM			
Na (s)	0	51.0	0
Na (g)	108.7	153.6	78.11
Na^+ (g)	601	—	—
NaBr (s)	−359.9	—	—
NaCl (s)	−411.0	72.38	−384
NaCl (aq)	−407.1	115.5	−393.0
Na_2CO_3 (s)	−1131	136	−1048
NaOH (s)	−426.7	—	—
NaOH (aq)	−469.6	49.8	−419.2

Table continued on next page

Appendix K Continued

Species	$\Delta H^0_{f\,298.15}$ kJ/mol	$S^0_{298.15}$ J/mol K	$\Delta G^0_{f\,298.15}$ kJ/mol
SULFUR			
S (s, rhombic)	0	31.8	0
S (g)	278.8	167.8	238.3
S_2Cl_2 (g)	−18	331	−31.8
SF_6 (g)	−1209	291.7	−1105
H_2S (g)	−20.6	205.7	−33.6
SO_2 (g)	−296.8	248.1	−300.2
SO_3 (g)	−395.6	256.6	−371.1
$SOCl_2$ (ℓ)	−206	—	—
SO_2Cl_2 (ℓ)	−389	—	—
H_2SO_4 (ℓ)	−814.0	156.9	−690.1
H_2SO_4 (aq)	−907.5	17	−742.0
TIN			
Sn (s)	0	51.55	0
$SnCl_2$ (s)	−350	—	—
$SnCl_4$ (ℓ)	−511.3	258.6	−440.2
$SnCl_4$ (g)	−471.5	366	−432.2
SnO_2 (s)	−580.7	52.3	−519.7
TITANIUM			
$TiCl_4$ (ℓ)	−804.2	252.3	−737.2
$TiCl_4$ (g)	−763.2	354.8	−726.8
TUNGSTEN			
W (s)	0	32.6	0
WO_3 (s)	−842.9	75.90	−764.1
ZINC			
ZnO (s)	−348.3	43.64	−318.3
ZnS (s)	−205.6	57.7	−201.3

Answers for Even-Numbered Numerical Exercises

Chapter 1

1–18. (a) 4.32×10^3 (b) 6.87×10^3 (c) 0.174 (d) 7.89 (e) 9.24×10^{-3} (f) 3.00×10^{-2}

1–20. (a) 3 sig fig (b) 2 sig fig (c) 3 sig fig (d) 4 sig fig

1–22. (a) 1.0×10^3 (b) 4.39×10^4 (c) 2.86×10^{-4} (d) 9.8765×10^{-5} (e) 1.0000×10^4 (5 sig fig), 1.000×10^4 (4 sig fig), 1.00×10^4 (3 sig fig), 1.0×10^4 (2 sig fig), 1×10^4 (1 sig fig)

1–24. (a) 19.79 (b) 0.305 (c) 0.49

1–26. (a) 3.65×10^3 (b) 1.75×10^3 (c) 5.28 (d) 1.08×10^3

1–28. (a) kilo (b) milli (c) deca (d) deci (e) centi (f) deci (g) milli (h) micro

1–32. (a) 7.58×10^3 m (b) 7.58×10^4 cm (c) 0.478 kg (d) 9.87×10^3 g (e) 1.386 L (f) 3.692×10^3 mL (g) 1.126 L (h) 786 cm^3

1–34. 91.4 m

1–36. 88 km/hr

1–38. (a) 3.78×10^3 mL (b) 131 mL, 2.00 in (c) 3.06×10^6 cm^3 (d) 61.0 in^3

1–40. $1.51/gal

1–42. (a) 1.00×10^3 g, 1.00 kg (b) 4.25×10^3 g, 4.25 kg (c) 1.4×10^3 g, 1.4 kg (d) 113 g, 0.113 kg

1–44. (a) 45.4 kg (b) 220 lb (c) 2.27×10^4 cg (d) 2.02×10^9 mg

1–48. 8.88×10^7 Al atoms

1–54. 7.20 g/cm^3

1–56. 13.6 g Hg, 7.20 g Cr, 1.89 g Hg/g Cr

1–60. 29 mL

1–62. 66.3 g

1–64. 0.48 kg for 25 mL

1–66. (a) 7.8 cm^3 (b) 7.9 g/cm^3

1–68. (a) 2.35×10^3 cm^3 (b) 13.3 cm, 5.24 in

1–70. (a) 37.778°C (b) 93.333°C (c) −173.2°C, −279.8°F (d) 37.8°C, 311 K

1–72. Answers for vertical columns: 4.92×10^3 ft, 1.48×10^4 ft, 3.00×10^4 ft; 305 m, 3.05×10^3 m, 6.00×10^3 m, 1.08×10^4 m; 41°F, 23°F, −15°F, −69°F; 13°C, −15°C, −44°C

1–78. 13.33 g, 17.33 g, 24.67 g

1–80. (a) 82.9 g CaCl$_2$ (b) 52.9 g Cl$_2$

1–82. (a) All P is consumed. (b) 3.5 g P (c) 11.5 g compound

1–84. 10.0 g + 5.0 g =15.0 g = 11.5 g + 3.5 g

1–86. 0.876 g N/1.00 g O, 0.437 g N/1.00 g O, and 0.350 g N/1.00 g O yield nitrogen mass ratios of 10 : 5 : 4 per 1.00 g O

Chapter 2

2–4. Answers for vertical columns: S, Ag; 10.81 amu, 55.847 amu, 107.868 amu; 10.81 g, 32.06 g, 55.847 g

2–6. (a) 5.50 mol (b) 3.0×10^{23} atoms (c) 9.03×10^{23} atoms (d) 8.67×10^{22} atoms (e) 2.53×10^{21} atoms (f) 34.3 : 1 or 103 : 3

2–8. (a) 0.0833 mol (b) 1.66×10^{-24} mol (c) 9.40×10^{-4} mol

2–12. (a) 60.3% Mg, 39.7% O (b) 69.9% Fe, 30.1% O (c) 32.4% Na, 22.6% S, 45.0% O (d) 29.2% N, 8.3% H, 12.5% C, 50.0% O (e) 8.1% Al, 14.5% S, 72.0% O, 5.4% H

2–14. TiO$_2$

2–16. Fe$_2$O$_3$

2-18. Illustrates Law of Multiple Proportions. 0.669 g O/1.00 g Ti and 0.502 g O/1.00 g Ti yield an oxygen mass ratio of 4 : 3 per 1.00 g Ti, 0.430 g O/1.00 g Fe and 0.382 g O/1.00 g Fe yield an oxygen mass ratio of 9 : 8 per 1.00 g Fe

2-20. $MgSO_3$

2-22. MgS_2O_3

2-24. (a) C_2H_2O (b) $C_4H_8O_2$

2-26. 2.5 mol

2-28. 3.00×10^{24} atoms

2-30. 1.93×10^{24} atoms

2-32. $C_6H_{14}O_2N_2$

2-40. 112 g

2-42. 125 g

2-44. 12.8 g

2-46. 45.3 g, 27.2 g

2-48. 14.4%

2-50. 292 g

2-52. 10 g

2-54. (a) 11.0 g CO_2, 32.0 g SO_2, 43.0 g total (b) 15.2 g

2-56. C_6H_7N

2-58. 0.59 g $CaCO_3$, 0.43 g $MgCO_3$

2-60. 1.950 g

2-62. (a) 5.60 g (b) 5.60 g (c) 6.6 g (d) 19.2 g

2-64. 27.4 g $K_2Cr_2O_7$, 373 g H_2O

2-66. 109 g soln, 9.05 g NH_4Cl

2-68. 39.0 g

2-70. 641 mL

2-72. 2.50 M

2-74. 0.0410 M

2-76. 250 g

2-78. 50.9 mL

2-80. 39.9 mL

2-82. 863 mL

2-84. 40.0 mL

2-86. 150 mL

2-88. 192 mL

2-90. 4.06 M

Chapter 3

3-2. -3.14 kJ, -750 cal

3-4. 43.8°C

3-6. (a) 897 J (b) 2.72×10^3 J (c) 30.2

3-8. 300 g

3-10. 0.487 J/g·°C

3-22. 6.94 g

3-24. -4.58×10^5 J/mol Mg

3-30. (a) 459.4 kJ (b) 1707 kJ (c) -1103 kJ

3-32. -90.6 kJ, exothermic

3-34. -112 kJ

3-36. -1042 kJ

3-38. -375 kJ

3-40. (a) -9.32 kJ (b) -8.46×10^6 kJ

3-42. -64.9 kJ

3-44. -114 kJ

3-46. -24.8 kJ

3-48. -509 kJ

3-50. -128 kJ

3-52. -399 kJ

3-54. -65.3 kJ

Chapter 4

4-18. nuclear density = 1.45×10^{13} g/cm³, atomic density = 14.5 g/cm³

4-28. 79.907 amu

4-30. 28.09 amu

4-32. 7.49% ^6Li

4-34. 10.8 amu, 35.5 amu

4-40. 0.751 amu/atom, 0.751 g/mol

4-42. 0.439 amu/atom, 0.439 g/mol, 3.95×10^{10} kJ/mol

Chapter 5

5-4. 23 ft

5-6. (a) 3.00×10^{14} Hz (b) 5.77×10^{14} Hz (c) 5.66×10^{14} Hz (d) 3.48×10^8 Hz (e) 1.50×10^{17} Hz

5-10. 4.3 light yr

5-18. 2.179×10^{-18} J

5-20. 2.47×10^{15} s⁻¹, 2.92×10^{15} s⁻¹, 3.08×10^{15} s⁻¹; 1.63×10^{-18} J, 1.94×10^{-18} J, 2.04×10^{-18} J

5-26. 239 kJ/mol, 57.1 kcal/mol

5-32. 1.3×10^{-31} nm

Chapter 6

6-102. (a) $+3$ (b) $+7$ (c) $+7$ (d) $+6$ (e) $+3$ (f) $+4$ (g) $+2$

Chapter 7

7-46. 391 kJ/mol bonds

7-48. -93 kJ

7-50. 294 kJ/mol bonds

Chapters 8, 9

None

Chapter 10

10-10. (a) $0.442\,M$ (b) $0.122\,M$ (c) $0.0666\,M$
10-12. 2.0×10^3 g
10-14. $2.32\,M$
10-16. $0.667\,M$
10-18. $0.500\,M$
10-20. 3.00 L NaOH; 2.00 L H_3PO_4
10-22. 618 mL
10-24. 100 mL
10-26. 160 mL
10-28. 30 mL
10-30. 25.2%
10-32. $0.09227\,M$
10-34. $0.03773\,M$
10-36. 0.393 g
10-38. 0.65 tablet, 65%
10-40. $1.50\,N$
10-42. $0.333\,M$; $1.00\,N$
10-44. $0.050\,M$
10-46. $0.100\,N$
10-48. $0.200\,M$; $0.200\,N$
10-50. 61.3%
10-52. $0.200\,M$
10-54. (a) 2.96×10^{-3} mol (b) 0.361 g (c) 60.2%
10-56. $0.08482\,M$
10-58. (a) $2.420\,M$ (b) $1.612\,M$
10-60. (a) $+2, +3, +4, -3, -2, -1, +5$
(b) $+2, +4, 0, -2, -2, +3, +4$
(c) $0, -2, +4, +6, +4, +6, +6$
10-62. (a) $-2, +4, +6, +2, +2.5$
(b) $+3, +3, +6, +6$
(c) $+3, +3, +3$
10-76. 4.0 mL
10-78. 10 mL
10-80. 0.548 g Fe; 10.9%
10-82. 59.3%
10-84. $0.200\,M$; $1.00\,N$
10-86. $0.533\,M$; $0.533\,N$
10-88. Answers for vertical columns, formula weights: 158 g, 294 g, 248 g, 64.1 g, 37.0 g; equiv per mole: 5,6,1,2,2; equiv wt.: 31.6 g, 49.0 g, 248 g, 32.0 g, 18.5 g; mass of 0.1500 equiv: 4.74 g, 7.35 g, 37.2 g, 4.80 g, 2.78 g
10-90. 90.9 g/eq

Chapter 11

11-14. 7.73 L
11-16. 345 torr, 0.454 atm, 4.60×10^4 Pa, 46.0 kPa
11-18. 2, 100%
11-24. 462 mL (a) Celsius scale: 2.0, 100%, Kelvin scale: 1.15, 15.5%
11-26. 212 K
11-28. (b) 25°C, 298 K, 77°F (c) 760 torr, 760 mm Hg, 1 atm, 1.013×10^5 Pa, 101.3 kPa
11-32. 1.86×10^3 K, 1.59×10^3°C
11-34. 3 atm
11-38. 1.43 g/L, 1.43×10^{-3} g/mL
11-40. (a) 3.74 mol (b) 120 g
11-42. 20.0 g/mol
11-44. 320 g
11-48. 65.7 L
11-50. 0.0200 mol N_2; 2.41×10^{22} atoms
11-52. 148 L
11-54. 29.6 g/mol, 1.33% error
11-58. C_4H_{10}
11-62. 1.75 atm
11-64. 84 g
11-66. 0.0594 g
11-74. $R_{HI}/R_{CH_3OH} = 0.500$
11-76. 0.250 mi/s
11-82. 15.4 atm, 15.3 atm, 0.7% difference
11-86. (a) 100 L (b) 150 L
11-88. 6.72 L
11-90. 82.0%
11-92. 5.16 L
11-94. 4.48 L
11-96. 17.9 L H_2, 31.0 L C_2H_2

Chapter 12

12-32. $4Na^+ : 4Cl^- = NaCl$
12-34. (a) $1Cs^+ : 1Cl^- = CsCl$ (b) $4Na^+ : 4Cl^- = NaCl$
(c) $4Zn^{2+} : 4S^{2-} = ZnS$
12-42. 8.28 cm³/mol
12-44. 12.0 g/cm³
12-46. 24.3 g/mol, Mg
12-48. 68.0% occupied, 32.0% empty space
12-50. 0.1246 nm, 8.101×10^{-3} nm³
12-52. 0.23490 nm
12-54. -644 kJ
12-56. -2483 kJ
12-62. 46.7°C
12-64. 7.51 kJ
12-66. -2.59×10^4 J
12-68. 58.7°C
12-70. 24.6°C

Chapter 13

13–18.   0.222
13–20.   111 g
13–24.   (a) 0.306 torr (b) 23.45 torr
13–28.   (a) $\approx$ 36 torr (b) $\approx$ 264 torr (c) $\approx$ 300 torr
13–30.   0.103 atm
13–32.   0.328 m
13–34.   7.41 m
13–36.   25.6 g
13–42. $-3.72°C$, $101.0°C$
13–44. $0.36°C$, $82.63°C$
13–46.   40 amu
13–48. $106.4°C$, $223.5°F$
13–50.   (a) $i = 3$ (b) $i = 2$ (c) $i = 5$ (d) $i = 3$
13–54.   1.3%
13–56. $i = 1.87$; 87%
13–58. $i = 2.85$; 61.7%
13–64. 3.78×10^{-3} atm, 2.87 torr
13–66. 1.46×10^3 g/mol

Chapter 14

14–12. $q = 854$ J, $w = 63.3$ J, $\Delta E = 791$ J
14–16. $w = 608$ J, $q = 608$ J
14–20.   (b) -41.85 kJ/g, -3264 kJ/mol
14–22. $\Delta H = 6.02$ kJ, $\Delta E = 6.18$ kJ
14–24.   (a) -3267 kJ (b) -3267 kJ vs. -3264 kJ
14–44.   (a) $\Delta H^0 = -277.8$ kJ, $\Delta S^0 = 0.0214$ kJ/K, $\Delta G^0 = -284.2$ kJ
        (b) $\Delta H^0 = 25.2$ kJ, $\Delta S^0 = 0.3069$ kJ/K, $\Delta G^0 = -66.3$ kJ
14–48.   (a) below 3.27×10^3 K
14–50.   (a) 370 K or $97°C$
14–52.   432 K or $159°C$
14–58.   (a) 446 J/K (b) 1040.6 J/K
14–60.   (a) -959.3 kJ, -70.54 kJ, 8.3 kJ
        (b) -2.05×10^3 kJ
14–62.   (a) -268 J/K (b) 1.2×10^2 J/K (c) 67.1 J/K
        (d) -167 J/K (e) 32 J/K
14–64.   (a) -86.7 kJ (b) -75.1 kJ (c) -64.1 kJ
        (d) $500°C$ (Note: Reaction is gas phase reaction at $500°C$ and $1000°C$ but not $25°C$.)

Chapter 15

15–12.   (a) $time^{-1}$ (b) $M^{-1} \cdot time^{-1}$ or L/mol·time
        (c) $M^{-2} \cdot time^{-1}$ or $L^2/mol^2 \cdot time$
        (d) $M^{-0.5} \cdot time^{-1}$ or $L^{0.5}/mol^{0.5} \cdot time$
15–14.   Rate $= k[A][B][C]° = (2.0 \times 10^{-2} \, M^{-1} \cdot min^{-1})[A][B]$
15–16.   first order

15–18.   increase by factor of $3 \times 3^2 = 27$
15–20.   Rate $= k[A][B]^2 = (2.5 \times 10^{-3} \, M^{-1} \cdot s^{-1})[A][B]^2$
15–22.   Rate $= k[A]^2[B]^3 = (20 \, M^{-4} \cdot s^{-1})[A]^2[B]^3$
15–24. $k = 0.5 \, M^{-1} \cdot s^{-1}$, $Rate_2 = 0.20 \, M \cdot s^{-1}$, $Rate_3 = 0.80 \, M \cdot s^{-1}$
15–28.   (a) Graph of $1/[CH_3CHO]$ vs. time is linear. So, reaction is second order, and Rate $= k[CH_3CHO]^2$.
        (b) 6.8 L/mol·min or $6.8 \, M^{-1} \cdot min^{-1}$
        (c) 1.5 min (d) 1.1 min
15–30.   1680 s or 28 min
15–32.   (a) 2.5×10^6 s (b) 1.8×10^6 s or 21 days
        (c) 0.23 g (d) 0.43 g
15–36.   (a) 3.62×10^{-15} s^{-1} (b) 1.91×10^{14} s
        (c) 0.471 s^{-1} (d) 1.47 s
15–38.   268 kJ/mol
15–40. $k_2 = 3.65(k_1)$

Chapter 16

16–6. $b < e < a < d < c$
16–8. $Q = K$, equilibrium; $Q < K$, net forward reaction; $Q > K$, net reverse reaction
16–10.   (a) false (b) false (c) false (d) false ($Q = 0.75$)
16–12. $K_c = 2.5 \times 10^{-3}$
16–14. $[CO] = [H_2O] = 0.424 \, M$, $[CO_2] = [H_2] = 0.576 \, M$
16–16.   (a) $2 \, SO_3(g) \rightleftharpoons 2 \, SO_2(g) + O_2(g)$ (b) 0.050
        (c) $[PCl_5] = 0.08 \, M$, $[PCl_3] = [Cl_2] = 0.065 \, M$
16–18. $K_c = 2.51 \times 10^{-2}$
16–20. $[SbCl_5] = 7.40 \times 10^{-4} \, M$,
        $[SbCl_3] = 6.99 \times 10^{-3} \, M$,
        $[Cl_2] = 2.60 \times 10^{-3} \, M$
16–22.   68.5% dissociated
16–28.   (a) $K_c = 2.48 \times 10^{-4}$ (b) $K_p = 1.83 \times 10^{-2}$
16–30. $K_p = 4.0 \times 10^{-3}$
16–32.   (a) $P_{Cl_2} = 0.24$ atm (b) $[Cl_2] = 9.8 \times 10^{-3} \, M$
        (c) 159 g ICl
16–34.   (a) $K_p = K_c = 50$
        (b) $P_{H_2} = P_{I_2} = 0.7$ atm, $P_{HI} = 4.6$ atm,
        $P_{total} = 6.0$ atm (c) 0.12 mol I_2, 30 g I_2
16–42.   (a) $K_c = 64$ (b) $[A] = 0.22 \, M$
16–44.   (a) $K_c = 0.45$ (b) $[A] = 0.07 \, M$,
        $[B] = [C] = 0.18 \, M$ (c) $[A] = 0.52 \, M$,
        $[B] = [C] = 0.48 \, M$
16–46.   (a) $K_p = 1.7$
        (b) $P_{Cl_2} = P_{PCl_3} = 6.4$ atm, $P_{PCl_5} = 24$ atm
        (c) $P_{total} = 37$ atm
        (d) $P_{PCl_5} = 30$ atm
16–48.   (a) $[N_2O_4] = 0.0102 \, M$, $[NO_2] = 0.00688 \, M$
        (b) $[N_2O_4] = 4.51 \times 10^{-3} \, M$,
        $[NO_2] = 4.59 \times 10^{-3} \, M$
        (c) $[N_2O_4] = 0.0222 \, M$, $[NO_2] = 0.0102 \, M$
16–50.   (a) $K \gg 1$, ΔG^0 is negative, net forward

reaction favors equilibrium (b) $K \ll 1$, ΔG^0 is positive, net reverse reaction favors equilibrium

16–52. $K_p = 7.11 \times 10^{24}$

16–54. (a) At 600 K, $K_p = 9.26 \times 10^6$. At 800 K, $K_p = 3.06 \times 10^4$. (b) 2.01×10^3 K

16–56. At 500°C, $K_p = 0.11$ and $K_c = 1.7 \times 10^{-3}$. At 1000°C, $K_p = 1.3 \times 10^2$ and $K_c = 1.2$.

16–58. (a) $\Delta G^0 = -38$ kJ (b) $K = 1.2 \times 10^{-3}$
(c) $\Delta G^0 = 54$ kJ

Chapter 17

17–20. (a) pH = 1.00 (b) pH = 0.43 (c) pH = 2.85
(d) pH = 8.08

17–24. (a) $[H_3O^+] = 1.0 \times 10^{-4}$ M,
$[OH^-] = 1.0 \times 10^{-10}$ M, pH = 4.00,
pOH = 10.00
(b) $[H_3O^+] = 1.7 \times 10^{-9}$ M,
$[OH^-] = 5.8 \times 10^{-6}$ M, pH = 8.76,
pOH = 5.24
(c) $[H_3O^+] = 4.8 \times 10^{-9}$ M,
$[OH^-] = 2.1 \times 10^{-6}$ M, pH = 8.32,
pOH = 5.68
(d) $[H_3O^+] = 5.1 \times 10^{-11}$ M,
$[OH^-] = 2.0 \times 10^{-4}$ M, pH = 10.29,
pOH = 3.71

17–26. (a) $[H^+] = [I^-] = 0.10$ M
(b) $[K^+] = [OH^-] = 0.050$ M
(c) $[Sr^{2+}] = 0.010$ M $[OH^-] = 0.020$ M
(d) $[Ba^{2+}] = 0.0020$ M, $[NO_3^-] = 0.0040$ M
(e) $[H^+] = 0.00060$ M, $[SO_4^{2-}] = 0.00030$ M
(f) $[Fe^{3+}] = 0.0070$ M, $[SO_4^{2-}] = 0.010$ M

17–28. $K_a = 4.1 \times 10^{-5}$

17–30. $K_a = 2.6 \times 10^{-5}$

17–32. $[H_3O^+] = [OCl^-] = 8.4 \times 10^{-5}$ M,
$[HOCl] = 0.20$ M

17–34. $[H_3O^+] = [CN^-] = 4.5 \times 10^{-6}$ M,
$[HCN] = 0.050$ M, $[OH^-] = 2.2 \times 10^{-9}$ M,
pH = 5.35, pOH = 8.65, 9.0×10^{-3} % ionization

17–36. (a) $[H_3O^+] = 6.3 \times 10^{-6}$ M, pH = 5.20,
6.3×10^{-3} % ionization
(b) $[H_3O^+] = 8.1 \times 10^{-3}$ M, pH = 2.09, 8.1% ionization
(c) $[H_3O^+] = 1.6 \times 10^{-5}$ M, pH = 4.80,
0.016% ionization
(d) $[H_3O^+] = 1.4 \times 10^{-3}$ M, pH = 2.85, 1.4% ionization

17–38. $K_a = 3.6 \times 10^{-6}$

17–40. $[NH_4^+] = [OH^-] = 9.5 \times 10^{-4}$ M,
$[NH_3] = 0.050$ M, $[H_3O^+] = 1.0 \times 10^{-11}$ M,
pH = 10.98, pOH = 3.02, 1.9% ionization

17–44. $[H_3O^+] = [HCO_3^-] = 1.4 \times 10^{-4}$ M,
$[H_2CO_3] = 0.050$ M, $[CO_3^{2-}] = 4.8 \times 10^{-11}$ M,
$[OH^-] = 7.1 \times 10^{-11}$ M

17–46. **Concentrations of Species in 0.100 M H_3XO_4**

Species	Concentration (mol/L)
H_3PO_4	0.076
H_3O^+	0.024
$H_2PO_4^-$	0.024
HPO_4^{2-}	6.2×10^{-8}
OH^-	4.2×10^{-13}
PO_4^{3-}	9.3×10^{-19}
H_3AsO_4	0.095
H_3O^+	0.0049
$H_2AsO_4^-$	0.0049
$HAsO_4^{2-}$	5.6×10^{-8}
OH^-	2.0×10^{-12}
AsO_4^{3-}	3.4×10^{-18}

17–52. (a) $K_b = 1.4 \times 10^{-11}$ (b) $K_b = 2.9 \times 10^{-7}$
(c) $K_b = 2.5 \times 10^{-5}$

17–54. (a) 9.3×10^{-4} % hydrolysis
(b) 0.14% hydrolysis (c) 1.3% hydrolysis

17–58. (a) $[H_3O^+] = 9.2 \times 10^{-6}$ M, pH = 5.04
(b) $[H_3O^+] = 1.7 \times 10^{-6}$ M, pH = 5.77
(c) $[H_3O^+] = 1.0 \times 10^{-3}$ M, pH = 3.00

17–62. (a) pH = 3.11, 1.5% hydrolysis
(b) pH = 5.00, 5.0×10^{-3} % hydrolysis
(c) pH = 6.17, 4.5×10^{-4} % hydrolysis

Chapter 18

18–4. (a) $[H_3O^+] = 4.0 \times 10^{-10}$ M, and pH = 9.40 compared to $[H_3O^+] = 6.3 \times 10^{-6}$ M and pH = 5.20 for 0.10 M HCN.
(b) $[H_3O^+] = 7.2 \times 10^{-4}$ M and pH = 3.14 compared to $[H_3O^+] = 8.1 \times 10^{-3}$ M and pH = 2.09 for 0.10 M HF.

18–6. (a) pH = 3.76 (b) pH = 9.40

18–8. (a) pOH = 5.04, pH = 8.96,
$[OH^-] = 9.1 \times 10^{-6}$ M
(b) pOH = 5.04, pH = 8.96,
$[OH^-] = 9.1 \times 10^{-6}$ M

18–10. (a) $[H_3O^+]$ goes from 1.8×10^{-5} M to 2.2×10^{-5} M and pH from 4.74 to 4.66
(b) $[OH^-]$ goes from 9.0×10^{-6} M to 6.5×10^{-6} M and pH from 8.95 to 8.81
(c) $[OH^-]$ goes from 9.0×10^{-6} M to 1.2×10^{-5} M and pH from 8.95 to 9.08

18–12. $[H_3O^+] = 1.0 \times 10^{-10}$ M, pH = 10.00

18–14. (a) $[H_3O^+] = 1.7 \times 10^{-12}$ M,
$[OH^-] = 6.0 \times 10^{-3}$ M, pH = 11.78,
pOH = 2.22
(b) $[H_3O^+] = 2.8 \times 10^{-10}$ M,
$[OH^-] = 3.6 \times 10^{-5}$ M, pH = 9.56,

pOH = 4.44
(c) $[H_3O^+] = 5.6 \times 10^{-10}\ M$,
$[OH^-] = 1.8 \times 10^{-5}\ M$, pH = 9.26,
pOH = 4.74
(d) $[H_3O^+] = 6.2 \times 10^{-11}\ M$,
$[OH^-] = 1.6 \times 10^{-4}\ M$, pH = 10.20,
pOH = 3.80
(e) $[H_3O^+] = 3.7 \times 10^{-10}\ M$,
$[OH^-] = 2.7 \times 10^{-5}\ M$, pH = 9.43,
pOH = 4.57
(f) $[H_3O^+] = 4.5 \times 10^{-10}\ M$,
$[OH^-] = 2.2 \times 10^{-5}\ M$, pH = 9.34,
pOH = 4.66

18-16. 7.4 g $NaCH_3COO$

18-18. 3.5×10^2 mL 0.10 M NaCNO

18-22.

indicator	pH = 2.00	pH = 11.00
methyl red	red	yellow
neutral red	red	yellow
phenolphthalein	colorless	red

18-26.

mL of 0.100 M $HClO_4$ added	mmol acid	mmol excess base or acid	pH
0.0	0	2.50 OH^-	13.00
5.0	0.50	2.00	12.82
10.0	1.00	1.50	12.63
12.5	1.25	1.25	12.52
20.0	2.00	0.50	12.05
24.0	2.40	0.10	11.31
24.9	2.49	0.01	10.30
25.0	2.50	0.00	7.00
25.1	2.51	0.01 H^+	3.70
27.0	2.70	0.20	2.41
30.0	3.00	0.50	2.04

18-28. Some extra points have been added to give a better titration curve.

mol NaOH	$[H^+]$	$[OH^-]$	pH	pOH
none	$4.2 \times 10^{-4}\ M$	$2.4 \times 10^{-11}\ M$	3.37	10.62
0.00200	7.2×10^{-5}	1.4×10^{-10}	4.14	9.86
0.00300	4.2×10^{-5}	2.4×10^{-10}	4.38	9.62
0.00400	2.8×10^{-5}	3.6×10^{-10}	4.56	9.44
0.00500	1.8×10^{-5}	5.5×10^{-10}	4.74	9.26
0.00600	1.2×10^{-5}	8.3×10^{-10}	4.92	9.08
0.00700	7.7×10^{-6}	1.3×10^{-9}	5.11	8.89
0.00800	4.6×10^{-6}	2.2×10^{-9}	5.34	8.66
0.00900	2.0×10^{-6}	4.9×10^{-9}	5.69	8.31
0.00950	9.5×10^{-7}	1.1×10^{-8}	6.02	7.98
0.0100	4.2×10^{-9}	2.4×10^{-6}	8.37	5.63
0.0105	2.0×10^{-11}	5.0×10^{-4}	10.70	3.30
0.0110	1.0×10^{-11}	1.0×10^{-3}	11.00	3.00
0.0120	5.0×10^{-12}	2.0×10^{-3}	11.30	2.70
0.0130	3.3×10^{-12}	3.0×10^{-3}	11.48	2.52
0.0150	2.0×10^{-12}	5.0×10^{-3}	11.70	2.30
0.0300	5.0×10^{-13}	2.0×10^{-2}	12.30	1.70

18-30. Some extra points have been added to give a better titration curve.

mol NH_3	$[H^+]$	$[OH^-]$	pH	pOH
none	$4.2 \times 10^{-4}\ M$	$2.4 \times 10^{-11}\ M$	3.37	10.62
0.00100	1.6×10^{-4}	6.2×10^{-11}	3.79	10.21
0.00200	7.2×10^{-5}	1.4×10^{-10}	4.14	9.86
0.00300	4.2×10^{-5}	2.4×10^{-10}	4.38	9.62
0.00400	2.7×10^{-5}	3.7×10^{-10}	4.57	9.43
0.00500	1.8×10^{-5}	5.6×10^{-10}	4.74	9.26
0.00700	7.7×10^{-6}	1.3×10^{-9}	5.11	8.89
0.00800	4.5×10^{-6}	2.2×10^{-9}	5.35	8.65
0.00900	2.0×10^{-6}	5.0×10^{-9}	5.70	8.30
0.00950	9.5×10^{-7}	1.1×10^{-8}	6.02	7.98
0.0100	1.0×10^{-7}	1.0×10^{-7}	7.00	7.00
0.0105	1.1×10^{-8}	9.0×10^{-7}	7.95	6.05
0.0110	5.6×10^{-9}	1.8×10^{-6}	8.26	5.74
0.00115	3.7×10^{-9}	2.7×10^{-6}	8.43	5.57
0.00130	1.8×10^{-9}	5.4×10^{-6}	8.74	5.26

18-36. (a) $K_{sp} = 2.0 \times 10^{-16}$ (b) $K_{sp} = 2.6 \times 10^{-10}$
(c) $K_{sp} = 2.0 \times 10^{-6}$ (d) $K_{sp} = 3.7 \times 10^{-11}$

18-40. (b) $Q_{sp} = 2.5 \times 10^{-7}$ is $> (1000)(1.8 \times 10^{-10})$, precipitate will form that can be seen
(c) $Q_{sp} = 1.0 \times 10^{-6}$ is $> 8.7 \times 10^{-9}$ but $< (1000)(8.7 \times 10^{-9})$, precipitate will be formed but probably not be seen
(d) $Q_{sp} = 1.0 \times 10^{-12}$ is $> (1000)(1.3 \times 10^{-20})$, precipitate will form that can be seen

18-42. (a) $[Cl^-] = 1.8 \times 10^{-4}\ M$
(b) 1.1×10^{-4} g Ag^+/L (c) 9.1×10^3 L

18-44. $[CO_3^{2-}] = 4.8 \times 10^{-4}\ M$, 2.5×10^3 L

18-48. (a) AuCl
(b) $[Au^+] = 1.1 \times 10^{-5}\ M$, 99.89% Au^+ pptd
(c) $[Au^+] = 1.1 \times 10^{-8}\ M$, $[Ag^+] = 9.5 \times 10^{-6}\ M$

18-50. $Q_{sp} = 3.6 \times 10^{-12}$ is $< 1.5 \times 10^{-11}$, $Mg(OH)_2$ will not precipitate, pH = 8.78

18-52. (a) 1.5 M NH_4NO_3 (b) 120 g NH_4NO_3
(c) pH = 9.08

18-54. $Q_{sp} = 1.7 \times 10^{-10}$ is $> (1000)(4.6 \times 10^{-14})$, precipitate will form that can be seen

Chapter 19

19-18. $6.241 \times 10^{18}\ e^-$

19-20. (a) 5.19×10^3 coul (b) 3.04×10^3 coul
(c) 1.52×10^3 coul (d) 893 coul
(e) 1.07×10^4 coul

19-22. 2.33 g, 1.63×10^3 mL

19-24. 5.74×10^5 s or 6.65 days

19-26. (a) 0.0961 amp (b) 0.155 amp (c) 0.0248 amp
(d) 0.0335 amp

19-28. 207 g/mol, Pb^{2+}

19-30. 0.242 g H_2, 1.94 g O_2; 2.72 L H_2, 1.36 L O_2

19-32. 0.672 g Ag

19-34. 3.17×10^4 coul

19-54. $E^0 = +0.621$ V, spontaneous

19-56. $E^0 = +0.93$ V, spontaneous

19-58. $E^0 = -0.18$ V, nonspontaneous

19-60. (a) $E^0 = -1.18$ V, nonspontaneous
(b) $E^0 = +1.18$ V, spontaneous
(c) $E^0 = -1.66$ V, nonspontaneous
(d) $E^0 = -2.25$ V, nonspontaneous
(e) $E^0 = +0.28$ V, spontaneous

19-62. (a) Cd^{2+} (b) Sn^{4+} (c) H^+ (d) Cl_2
(e) MnO_4^-, acidic (f) F_2

19-64. (a) H_2 (b) Pt (c) Hg (d) Cl^-, basic (e) Cu
(f) Rb

19-68. (a) $E = -0.43$ V (b) $E = +0.10$ V
(c) $E = -0.65$ V (d) $E = -1.11$ V
(e) $E = +0.384$ V (f) $E = +1.21$ V

19-70. (a) $SnF_6^{2-} + 2Fe\ (s) \rightarrow Sn\ (s) + 6F^- + 2Fe^{2+}$,
$E = +0.53$ V
(b) $2MnO_4^- + 16H^+ + 5Cd\ (s) + 5S^{2-} \rightarrow$
$2Mn^{2+} + 8H_2O + 5CdS\ (s)$, $E = +1.86$ V
(c) $2O_2\ (g) + Te\ (s) + 2H_2O\ (\ell) \rightarrow$
$2H_2O_2 + TeO_2\ (s)$, $E = +0.227$ V
(d) $2NO_3^- + 2H^+ + 3H_2\ (g) \rightarrow$
$2NO\ (g) + 4H_2O\ (\ell)$, $E = +0.37$ V

19-72. (a) $E = 2.418$ V, spontaneous
(b) $E = 0.24$ V, spontaneous

19-76. (a) $E^0 = -0.62$ V, $\Delta G^0 = 1.2 \times 10^2$ kJ,
$K = 1.1 \times 10^{-21}$
(b) $E^0 = +0.35$ V, $\Delta G^0 = -34$ kJ,
$K = 8.3 \times 10^5$
(c) $E^0 = +1.833$ V, $\Delta G^0 = -1.061 \times 10^3$ kJ,
$K = 1.008 \times 10^{186}$

19-78. (a) $E^0 = +0.80$ V, $\Delta G^0 = -4.6 \times 10^2$ kJ,
$\Delta G^0 = -4.6 \times 10^2$ kJ/mol $K_2Cr_2O_7$,
$\Delta G^0 = -77$ kJ/mol NaI
(b) $E^0 = +1.34$ V, $\Delta G^0 = -1.29 \times 10^3$ kJ,
$\Delta G^0 = -6.45 \times 10^2$ kJ/mol $KMnO_4$,
$\Delta G^0 = -2.58 \times 10^2$ kJ/mol H_2SO_3
(c) $E^0 = +2.00$ V, $\Delta G^0 = -1.93 \times 10^3$ kJ,
$\Delta G^0 = -965$ kJ/mol $KMnO_4$,
$\Delta G^0 = -386$ kJ/mol $(COOH)_2$

19-80. (a) $K = 6.43 \times 10^{25}$ (b) $K = 4.24 \times 10^{15}$
(c) $K = 5.2 \times 10^3$

19-82. (a) $E^0 = +0.740$ V, $K = 3.77 \times 10^{62}$
(b) $E^0 = +0.794$ V, $K = 3.72 \times 10^{80}$
(c) $E^0 = +1.72$ V, $K = 3.44 \times 10^{174}$

19-84. $\Delta G^0 = 15.6 \times 10^4$ J, $K_{sp} = 4.57 \times 10^{-28}$

Chapter 20

20-12. (a) $\Delta H^0 = -446.2$ kJ (b) $\Delta H^0 = -390.8$ kJ
(c) $\Delta H^0 = -431.2$ kJ

20-30. 569 g Co_3O_4

20-32. $[H^+] = 1.0 \times 10^{-11}$ M,
$[Cr_2O_7^{2-}]/[CrO_4^{2-}] = 4.2 \times 10^{-8}$

20-36. $\Delta G^0 = -1.8 \times 10^2$ kJ

20-54. time $= 1.8 \times 10^8$ s or 2.1×10^3 days, volume
$O_2 = 7.2 \times 10^3$ L

Chapter 21

21-10. 2.33 g $XeOF_4$, 1.88 L HF

21-38. $R_{Cl_2}/R_{F_2} = 0.732$, $R_{Br_2}/R_{F_2} = 0.488$,
$R_{I_2}/R_{F_2} = 0.387$, $R_{At_2}/R_{F_2} = 0.301$

21-40. 1.36×10^3 mol Cl_2, 2.72×10^3 mol NaCl,
1.59×10^5 g NaCl, 350 lb NaCl, 138 lb Na,
2.92×10^5 hr

21-42. 131 L $NaHSO_3$ soln, 870 g $NaIO_3$

21-44. 5.20 g SiF_4, 1.12 L SiF_4

21-46. 4.469 g $KClO_4$

21-66. 0.09 g, 0.0774% Ag tarnished

Chapter 22

22-10. (a) 0 (b) +2 (c) +4 (d) +5 (e) +3

22-12. 946 kJ/mol bonds

22-28. (a) 16.5% N in $NaNO_3$, 35.0% N in NH_4NO_3
(b) $[H_3O^+] = 7.5 \times 10^{-6}$ M, pH $= 5.12$

22-30. 946 kJ/mol $N\equiv N$ bonds, 498.4 kJ/mol $O=O$
bonds, 631.7 kJ/mol $N=O$ bonds

22-32. 23.9 g $NaNH_2$

22-34. 26.3 g superphosphate

22-36. 43.7% P in P_4O_{10}, 31.6% P in H_3PO_4

22-52. 0.0563 mol CO_2, 4.16 g Li_2CO_3,
73.9% Li_2CO_3, 26.1% LiCl

22-58. For X = F, $\Delta G^0 = \Delta G_f^0$ SiF_4 (g) $= -1615$ kJ
and X = Cl, $\Delta G^0 = \Delta G_f^0$ $SiCl_4$ (g) $= -657$ kJ.
Formation of SiF_4 (g) is the more spontaneous
process.

22-66. $[H_3BO_3] = 0.10$ M, $[H_3O^+] = [H_2BO_3^-] = 8.5 \times 10^{-6}$ M, $[HBO_3^{2-}] = 1.8 \times 10^{-13}$ M,
$[BO_3^{3-}] = 3.4 \times 10^{-22}$ M, pH $= 5.07$

22-68. BH_3 simplest formula, 41.5 g/mol, B_3H_9
molecular formula

Chapter 23

23-38. (a) -657 kJ/mol + 2P (b) -194 kJ/mol
(c) -253 kJ/mol (d) -900 kJ/mol + 2P
(e) -95.7 kJ/mol (f) -314 kJ/mol
(g) -93.3 kJ/mol (h) -172 kJ/mol

23-40. $[NH_3] = 6.7 \times 10^{-4}$ M, $[NH_4^+] = [OH^-] = 1.3 \times 10^{-4}$ M, pOH $= 3.9$, pH $= 10.1$

Chapter 24

24–24. $k = 0.032$ min^{-1}, 0.56 μg Fr remain
24–26. $k = 1.21 \times 10^{-4}$ yr^{-1}, 1.8×10^4 yr
24–28. $k = 5.60 \times 10^{-2}$ hr^{-1}, $t_{1/2} = 12.4$ hr

Chapter 25

None

Illustration and Table Credits

Chapter 1 Figure 1–1, photo by Jim Mergenthaler. Figure 1–7, from the CHEM Study film, "Chemical Families." Figure 1–10, courtesy of U.S. National Bureau of Standards. Figure 1–11, courtesy of Arthur H. Thomas Company. Figure 1–14, from F. Brescia, S. Mehlman, F. C. Pellegrini, and S. Stambler, *Chemistry: A Modern Introduction*, 2nd ed., Saunders College Publishing, 1978. Figure 1–15, photo of Nalgene® graduated cylinder and buret supplied by Nalge Company, Division of Sybron Corporation.

Chapter 2 Figure 2–4, from W. L. Masterton, E. J. Slowinski, and C. L. Stanitski, *Chemical Principles*, 5th ed., Saunders College Publishing, 1981.

Chapter 4 Figure 4–3, from Masterton, Slowinski, and Stanitski. Figures 4–6, 4–7, from E. I. Peters, *Introduction to Chemical Principles*, 3rd ed., Saunders College Publishing, 1982. Figure 4–9, adapted from D. B. Murphy, V. Rousseau, and W. F. Kieffer, *Foundations of College Chemistry*, 1st ed., John Wiley & Sons, Inc., 1969. Figures 4–10, 4–12, 4–13, from Masterton, Slowinski, and Stanitski.

Chapter 5 Figure 5–4, courtesy of American Iron and Steel Institute. Figure 5–10, courtesy of J. K. Bates and J. Thallemer, Chemical Technology Division, Argonne National Laboratory; reprinted with permission from the cover of *Science*, 1 October 1982 (Vol. 218, No. 4567, p. 51). Figures 5–14, 5–19, from Masterton, Slowinski, and Stanitski.

Chapter 6 Figure 6–1, from *Annalen der Chemie und Pharmacie*, VIII, Supplementary Volume for 1872, p. 151. Figures 6–2, 6–5, 6–6 from Masterton, Slowinski, and Stanitski. Table 6–2, adapted from Brescia, Mehlman, Pellegrini, and Stambler. Tables 6–3, 6–7, adapted from K. F. Purcell and J. C. Kotz, *Inorganic Chemistry*, Saunders College Publishing, 1977.

Chapter 7 Figure 7–2, after Masterton, Slowinski, and Stanitski.

Chapter 9 Figures on pp. 279, 282, 283, 284, 297, 301, from H. C. Metcalfe, J. E. Williams, and J. F. Castka, *Modern Chemistry*, Holt, Rinehart and Winston, 1982. Figures on pp. 280, 283, from M. M. Jones, D. O. Johnston, J. T. Netterville, and J. L. Wood, *Chemistry, Man and Society*, 4th ed., Saunders College Publishing, 1983. Figure on p. 302, from P. P. Berlow, D. J. Burton, and J. I. Routh, *Introduction to the Chemistry of Life*, Saunders College Publishing, 1982. Figure on p. 305, courtesy of Los Angeles County Air Pollution Control District. Figure 9–2, from Brescia, Mehlman, Pellegrini, and Stambler. Figure 9–4, after Masterton, Slowinski, and Stanitski. Table 9–1, from W. L. Masterton and E. J. Slowinski, *Chemical Principles*, 2nd ed., Saunders College Publishing, 1969.

Chapter 11   Figures 11–2, 11–7, 11–11, from Masterton, Slowinski, and Stanitski. Figures 11–3, 11–6, 11–8, 11–12, 11–13 from Brescia, Mehlman, Pellegrini, and Stambler. Figure 11–14 from W. L. Masterton, E. J. Slowinski, and E. T. Walford, *Chemistry*, Holt, Rinehart, and Winston, 1980. Table 11–4, from J. A. Campbell, *Chemical Systems — Energetics, Dynamics and Structure*, W. H. Freeman and Co.*, 1970, courtesy of the author.

Chapter 12   Figure 12–14(a), courtesy of Eastman Kodak Company.

Chapter 13   Figure 13–3, after Peters. Figure 13–17, from Brescia, Mehlman, Pellegrini, and Stambler.

Chapter 14   Figure 14–5, from Masterton, Slowinski, and Stanitski.

Chapter 15   Figure 15–19, from Masterton, Slowinski, and Stanitski.

Chapter 17   Figure 17–1, courtesy of Arthur H. Thomas Company.

Chapter 18   Table 18–4, from Brescia, Mehlman, Pellegrini, and Stambler.

Chapter 19   Figures 19–1, 19–2, from Brescia, Mehlman, Pellegrini, and Stambler. Figures 19–3, 19–14, 19–17, from Masterton, Slowinski, and Stanitski. Figures 19–4, 19–15, from Masterton, Slowinski, and Walford. Table 19–2, from G. Charlot, *IUPAC Supplement,* ''Selected Constants, Oxidation-Reduction Potentials in Aqueous Solution,'' 1971.

Chapter 20   Middle figure on p. 668, from Berlow, Burton, and Routh. Figure on p. 670, courtesy of Aluminum Company of America. Figure on p. 671, courtesy ORGOTHERM INC. Figure on p. 688, from Metcalfe, Williams, and Castka. Figure on p. 679, courtesy of Smithsonian Institution Oceanographic Sorting Center. Figure on p. 680, from R. A. Serway, *Physics for Scientists and Engineers,* Saunders College Publishing, 1982. Figures on p. 690, courtesy of American Iron and Steel Institute.

Figure 20–6, from Masterton, Slowinski, and Stanitski. Figures 20–8, 20–11, from Masterton, Slowinski, and Walford. Figures 20–9, 20–13, 20–15, from Jones, Johnston, Netterville, and Wood. Figures 20–12, 20–14, from E. G. Rochow, *Modern Descriptive Chemistry,* Saunders College Publishing, 1977. Table 20–1, reprinted with permission from *Chemical and Engineering News,* 26 March 1979, copyright 1979 American Chemical Society. Table 20–13, from *Chemistry: A Contemporary Approach* by G. Tyler Miller, Jr. © 1976 by Wadsworth Publishing Company, Inc., Belmont, CA 94002; reprinted by permission. Table 20–14, from R. G. Gymer, *Chemistry in the Natural World,* D. C. Heath and Company, 1976.

Chapter 21   Figure 21–8, from E. M. Winkler, *Stone,* Springer-Verlag, 1973. Table 21–1, from T. Moeller, *Advanced Inorganic Chemistry,* John Wiley & Sons, Inc., 1952. Tables 21–3, 21–4, adapted from Moeller and from Murphy, Rousseau, and Kieffer. Tables 21–5, 21–7, adapted from Moeller and by permission of the publisher from *Inorganic Chemistry* by Jacob Kleinberg, William J. Argersinger, Jr., and Ernest Griswold (Lexington, MA: D. C. Heath and Company, 1960). Table 21–6, adapted from Moeller. Table 21–8, adapted from Rochow. Table 21–9, adapted from Moeller. Table 21–10, adapted and reprinted by permission of the publisher from *Inorganic Chemistry* by Jacob Kleinberg, William J. Argersinger, Jr., and Ernest Griswold (Lexington, MA: D. C. Heath and Company, 1960).

Chapter 22   Figure 22–2, from Masterton, Slowinski, and Walford. Figure 22–5, from Purcell and Kotz. Table 22–1, adapted from Rochow. Table 22–4, adapted from B. Mason, *Principles of Geochemistry*, 3rd ed., John Wiley & Sons, Inc., 1966.

Chapter 23   Figure 23–1, from M. F. Perutz et al., reprinted by permission from *Nature,* vol. 185, p. 416, copyright © 1960 Macmillan Journals, Limited. Figure 23–4, from Brescia, Mehlman, Pellegrini, and Stambler. Figure 23–8, adapted from P. George and D. S. McClure, *Progress in Inorganic Chemistry,* vol.

1, p. 381, 1959, F. A. Cotton, ed., Wiley (Interscience).

Chapter 24   Figure 24–1, from Masterton, Slowinski, and Walford. Figures 24–3, 24–10, 24–12, from Brescia, Mehlman, Pellegrini, and Stambler. Figures 24–6, 24–11, from M. Merken, *Physical Science with Modern Applications,* 2nd ed., Saunders College Publishing, 1980. Figure 24–7, courtesy of the Lawrence Berkeley Laboratory, University of California, Berkeley. Table 24–3, adapted with permission from R. C. Weast, Ed., *CRC Handbook of Chemistry,* 53rd ed., copyright CRC Press, Inc., Boca Raton, FL, 1973; and from J. A. Dean, Ed., *Lange's Handbook of Chemistry,* 11th ed., copyright McGraw-Hill Book Company, 1973.

Chapter 25   Figure 25–10, courtesy of Procter & Gamble Company. Figure 25–12, from Masterton, Slowinski, and Stanitski. Figure 25–15, reprinted with permission of Macmillan Publishing Company from Conant and Blatt, *The Chemistry of Organic Compounds,* 5th ed., copyright © 1959 by Macmillan Publishing Company. Figure 25–23, from Masterton, Slowinski, and Walford. Table 25–6, reprinted with permission of Macmillan Publishing Company from Conant and Blatt, *The Chemistry of Organic Compounds,* 5th ed., copyright © 1959 by Macmillan Publishing Company. Table 25–7, from H. Hart and R. D. Schuetz, *Organic Chemistry,* 3rd ed., p. 101, copyright © 1966 by Houghton Mifflin Company; used by permission.

Index

Note: Page numbers in *italics* indicate illustrations; page numbers followed by t indicate tables.

The Electronic Configurations of the Atoms of the Elements

Element	Atomic Number	1s	2s	2p	3s	3p	3d	4s	4p	4d	4f	5s
						Populations of Subshells						
H	1	1										
He	2	2										
Li	3	2	1									
Be	4	2	2									
B	5	2	2	1								
C	6	2	2	2								
N	7	2	2	3								
O	8	2	2	4								
F	9	2	2	5								
Ne	10	2	2	6								
Na	11	Neon core			1							
Mg	12				2							
Al	13				2	1						
Si	14				2	2						
P	15				2	3						
S	16				2	4						
Cl	17				2	5						
Ar	18	2	2	6	2	6						
K	19	Argon core						1				
Ca	20							2				
Sc	21						1	2				
Ti	22						2	2				
V	23						3	2				
Cr	24						5	1				
Mn	25						5	2				
Fe	26						6	2				
Co	27						7	2				
Ni	28						8	2				
Cu	29						10	1				
Zn	30						10	2				
Ga	31						10	2	1			
Ge	32						10	2	2			
As	33						10	2	3			
Se	34						10	2	4			
Br	35						10	2	5			
Kr	36	2	2	6	2	6	10	2	6			
Rb	37	Krypton core										1
Sr	38											2
Y	39									1		2
Zr	40									2		2
Nb	41									4		1
Mo	42									5		1
Tc	43									5	2	1
Ru	44									7	1	1
Rh	45									8		
Pd	46									10		
Ag	47									10		1
Cd	48									10		2